SMART STRUCTURES THEORY

The twenty-first century might be called the "Multifunctional Materials Age." The inspiration for multifunctional materials comes from nature and therefore these are often referred to as "bio-inspired materials." Bio-inspired materials encompass smart materials and structures, multifunctional materials, and nano-structured materials. This is a dawn of revolutionary materials that may provide a "quantum jump" in performance and multi-capability. This book focuses on smart materials, structures, and systems, which are also referred to as intelligent, adaptive, active, sensory, and metamorphic. The purpose of these materials from the perspective of smart systems is their ability to minimize life-cycle cost and/or expand the performance envelope. The ultimate goal is to develop biologically inspired multifunctional materials with the capability to adapt their structural characteristics (e.g., stiffness, damping, and viscosity) as required, monitor their health condition, perform self-diagnosis and self-repair, morph their shape, and undergo significant controlled motion over a wide range of operating conditions.

Dr. Inderjit Chopra is an Alfred Gessow Professor in Aerospace Engineering and Director of the Alfred Gessow Rotorcraft Center at the University of Maryland. Dr. Chopra works on fundamental problems related to aeromechanics of helicopters including aeromechanical stability; active vibration control; modeling of composite blades; rotor-head health monitoring; aeroelastic optimization; smart structures; micro air vehicles; and comprehensive aeromechanics analyses of bearingless, tilt-rotor, servo-flap, coaxial, compound, teetering, and circulation control rotors. Dr. Chopra has authored more than 160 archival journal papers and 260 conference-proceedings papers. He has been an associate editor of the *Journal of the American Helicopter Society* (1987–91), the *AIAA Journal of Aircraft* (1987–present), and the *Journal of Intelligent Materials and Systems* (1997–present). He has been elected Fellow in the following societies: American Institute of Aeronautics and Astronautics, American Helicopter Society, American Society of Mechanical Engineers, Aeronautical Society of India, and National Institute of Aerospace.

Dr. Jayant Sirohi is an expert in high-power density–solid state actuators, multifunctional structures, multi-mission–capable micro/unmanned aerial vehicles, and rotorcraft design. He is a professor in the Aerospace Engineering and Engineering Mechanics Department at the University of Texas at Austin. Dr. Sirohi previously worked at Sikorsky Aircraft Corporation, where he was a Staff Engineer in the Advanced Concept group.

Cambridge Aerospace Series

Editors: Wei Shyy and Vigor Yang

1. J. M. Rolfe and K. J. Staples (eds.): *Flight Simulation*
2. P. Berlin: *The Geostationary Applications Satellite*
3. M. J. T. Smith: *Aircraft Noise*
4. N. X. Vinh: *Flight Mechanics of High-Performance Aircraft*
5. W. A. Mair and D. L. Birdsall: *Aircraft Performance*
6. M. J. Abzug and E. E. Larrabee: *Airplane Stability and Control*
7. M. J. Sidi: *Spacecraft Dynamics and Control*
8. J. D. Anderson: *A History of Aerodynamics*
9. A. M. Cruise, J. A. Bowles, C. V. Goodall, and T. J. Patrick: *Principles of Space Instrument Design*
10. G. A. Khoury (ed.): *Airship Technology*, Second Edition
11. J. P. Fielding: *Introduction to Aircraft Design*
12. J. G. Leishman: *Principles of Helicopter Aerodynamics*, Second Edition
13. J. Katz and A. Plotkin: *Low-Speed Aerodynamics*, Second Edition
14. M. J. Abzug and E. E. Larrabee: *Airplane Stability and Control: A History of the Technologies that Made Aviation Possible*, Second Edition
15. D. H. Hodges and G. A. Pierce: *Introduction to Structural Dynamics and Aeroelasticity*, Second Edition
16. W. Fehse: *Automatic Rendezvous and Docking of Spacecraft*
17. R. D. Flack: *Fundamentals of Jet Propulsion with Applications*
18. E. A. Baskharone: *Principles of Turbomachinery in Air-Breathing Engines*
19. D. D. Knight: *Numerical Methods for High-Speed Flows*
20. C. A. Wagner, T. Hüttl, and P. Sagaut (eds.): *Large-Eddy Simulation for Acoustics*
21. D. D. Joseph, T. Funada, and J. Wang: *Potential Flows of Viscous and Viscoelastic Fluids*
22. W. Shyy, Y. Lian, H. Liu, J. Tang, and D. Viieru: *Aerodynamics of Low Reynolds Number Flyers*
23. J. H. Saleh: *Analyses for Durability and System Design Lifetime*
24. B. K. Donaldson: *Analysis of Aircraft Structures*, Second Edition
25. C. Segal: *The Scramjet Engine: Processes and Characteristics*
26. J. F. Doyle: *Guided Explorations of the Mechanics of Solids and Structures*
27. A. K. Kundu: *Aircraft Design*
28. M. I. Friswell, J. E. T. Penny, S. D. Garvey, and A. W. Lees: *Dynamics of Rotating Machines*
29. B. A. Conway (ed.): *Spacecraft Trajectory Optimization*
30. R. J. Adrian and J. Westerweel: *Particle Image Velocimetry*
31. G. A. Flandro, H. M. McMahon, and R. L. Roach: *Basic Aerodynamics*
32. H. Babinsky and J. K. Harvey: *Shock Wave–Boundary-Layer Interactions*
33. C. K. W. Tam: *Computational Aeroacoustics: A Wave Number Approach*
34. A. Filippone: *Advanced Aircraft Flight Performance*
35. I. Chopra and J. Sirohi: *Smart Structures Theory*
36. W. Johnson: *Rotorcraft Aeromechanics*
37. W. Shyy, H. Aono, C. K. Kang, and H. Liu: *An Introduction to Flapping Wing Aerodynamics*
38. T. C. Lieuwen and V. Yang: *Gas Turbine Engines*

Smart Structures Theory

Inderjit Chopra
University of Maryland

Jayant Sirohi
The University of Texas at Austin

CAMBRIDGE
UNIVERSITY PRESS

CAMBRIDGE
UNIVERSITY PRESS

32 Avenue of the Americas, New York, NY 10013-2473, USA

Cambridge University Press is part of the University of Cambridge.

It furthers the University's mission by disseminating knowledge in the pursuit of education, learning, and research at the highest international levels of excellence.

www.cambridge.org
Information on this title: www.cambridge.org/9780521866576

First published 2014

Printed in the United States of America

A catalog record for this publication is available from the British Library.

Library of Congress Cataloging in Publication Data
Chopra, Inderjit.
Smart structures theory / Inderjit Chopra, Jayant Sirohi. – First edition.
pages cm. – (Cambridge aerospace series ; 35)
Includes bibliographical references and index.
ISBN 978-0-521-86657-6 (hardback)
1. Smart materials – Industrial applications. 2. Smart structures – Industrial applications. I. Sirohi, Jayant. II. Title.
TA418.9.S62C47 2013
620.1′12–dc23 2013018869

ISBN 978-0-521-86657-6 Hardback

Contents

Preface

In 1990, a pilot project was started at the Alfred Gessow Rotorcraft Center (University of Maryland) to build a smart rotor with embedded piezoelectric strips. Soon, it attracted the attention of Dr. Gary Anderson of the Army Research Office (ARO). He encouraged us to put together outlines for a major initiative in the smart structures area, which subsequently resulted in the award of a multi-year (1992–1997) University Research Initiative (URI). This provided us an opportunity to develop an effective team of interdisciplinary faculty from Aerospace, Mechanical, Electrical, and Material Engineering. As a result, there was an enormous growth of smart structures research activities on our campus. Following the success of this URI, we were awarded another multi-year (1996–2001) Multi University Research Initiative (MURI) in smart structures by ARO. For this major program, we collaborated with Penn State and Cornell University. This further nurtured the ongoing smart structures activities at Maryland. We deeply acknowledge the support and friendship of many faculty colleagues at Maryland: Appa Anjannappa, Bala Balachandran, James Baeder, Amr Baz, Roberto Celi, Ramesh Chandra, Abhijit Dasgupta, Allison Flatau, James Hubbard, P. S. Krishnaprasad, Gordon Leishman, V. T. Nagaraj, Darryll Pines, Don Robbins, Jim Sirkis, Fred Tasker, Norman Wereley, and Manfred Wuttig.

While the research frontier in smart structures was expanding at the Alfred Gessow Rotorcraft Center, we also initiated classroom teaching at the graduate level in the smart structures area. This textbook was developed from material covered in early versions of these class notes, and it aims to give a broad overview of smart materials and their applications in smart structures and integrated systems. The focus is on the fundamental physical phenomena observed in active materials and on the mathematical modeling of the coupled behavior of a smart structure with active material actuators and sensors. Simplistic descriptions of the physical mechanisms are given so that the reader can obtain an intuitive grasp of the fundamentals without having to delve deeply into rigorous solid mechanics concepts.

The research activities generated a large cadre of dissertations; many of these were pioneering foundational efforts in smart structures. We fondly acknowledge the contributions of our graduates: Jayasimha Atulsimha (VCU), Ron Barrett (Kansas), Oren Ben-Zeev (NAVAIR), Andy Bernhard (Sikorsky), Mike Bothwell (Bell), Peter Chen (IAI), Peter Copp (UMD), Ron Couch (APL), Anubhav Datta (NASA-Ames), Jeanette Epps (NASA-Astronaut), Farhan Gandhi (RPI),

Ranjan Ganguli (IISc), Gopal Kamath (Bombardier), Nikhil Koratkar (RPI), Taeoh Lee (Bell), Judah Milgram (NSWC-Carderock), Harsha Prahlad (SRI), Beatrice Roget (Wyoming), Jinwei Shen (NIA), Kiran Singh (Cambridge), Ed Smith (Penn State), Burtis Spencer (Air Force), Mike Spencer (Orbital Science), Curtis Walz (Boeing-Philadelphia), and Gang Wang (U. Huntsville).

During the 1990s, there was tremendous growth of smart structures activities in the United States and abroad. Many new conferences and workshops were initiated during this period, including: ARO Workshop in Smart Structures, SPIE Symposium in Smart Structures and Materials, AIAA Adaptive Structures Forum, ASME Adaptive Structures and Materials Systems (now called SMASIS), and ICAST (International Conference on Adaptive Structures and Technologies). These conferences and workshops not only helped to communicate our activities in smart structures but also provided avenues for meeting many great friends in this discipline. Over the years, we enjoyed the warmth of many friends in the United States and abroad, including V. K. Aatre (IISc), H. Abramovich (Technion), Diann Brei (Michigan), Flavio Campanile (EMPA), Greg Carman (UCLA), Carlos Cesnik (Michigan), Aditi Chattopadhyay (ASU), Eric Cross (Penn State), Marcello Dapino (OSU), Paolo Ermanni (ETH), Mary Frecker (Michigan), Mike Friswell (Swansea), Ephrahim Garcia (Cornell), Paolo Gaudenzi (U. Rome), Victor Giurgiutiu (South Carolina), S. Gopalakrishnan (IISc), Z. Gurdal (Delft), Dan Inman (Michigan), Seung Jo Kim (KARI), A. V. Krishnamurthy (IISc), Dimitris Lagoudas (Texas A&M), C. K. Lee (National Taiwan), In Lee (KAIST), Jinsong Leng (Harbin), Don Leo (VPI), George Lesieutre (Penn State), Wei-Hsin Liao (Chinese University of Hong Kong), Chris Lynch (UCLA), John Main (VPI), Dave Martinez (Sandia), Yuji Matsuzaki (Nagoya), Peter Monner (DLR), M. C. Natori (Waseda), Fred Nitzsche (Carleton), Roger Ohayon (CNAM), Zoubeida Ounaies (Penn State), K. C. Park (Colorado), Jinhao Qui (Nanjing), Dimitris Saravanos (U. Patras), Janet Sater (IDA), Jonghwan Suhr (Delaware), J. Tani (Tohoku), Horn-Sen Tzou (Zhejiang), A. R. Upadhya (NAL), Ben Wada (JPL), Kon-Well Wang (Michigan), and Wenbin Yu (Utah).

We also collaborated with rotorcraft and other aerospace industries to transition this technology to full-scale systems. Under the DARPA Smart Rotor Program, Friedrich Straub and Hieu Ngo actively collaborated with the Alfred Gessow Rotorcraft Center and injected enthusiasm among our students. We again fondly acknowledge industrial friends in the United States and abroad, including Eric Anderson (CSA), Dan Clingman (Boeing), L. Porter Davis (Honeywell), Peter Jaenker (EADS), Shiv Joshi (NextGen), and Jay Kudva (NextGen). We would also like to thank the University of Maryland and the University of Texas at Austin, where we worked on material for this textbook.

Finally, we acknowledge our deep appreciation for the support and encouragement that we received from Dr. Gary Anderson, a true gentleman, who spearheaded the growth of smart structures activities in the United States. This book is dedicated to him.

Inderjit Chopra (University of Maryland)
Jayant Sirohi (University of Texas at Austin)

1 Historical Developments and Potential Applications: Smart Materials and Structures

The quest for superior capability in both civil and military products has been a key impetus for the discovery of high performance new materials. In fact, the standard of living has been impacted by the emergence of high performance materials. There is no doubt that the early history of civilization is intertwined with the evolution of new materials. For example, different eras of civilization are branded with their material capabilities, and these periods are referred to as the Stone Age, the Bronze Age, the Iron Age, and the Synthetic Material Age. The Stone Age represents the earliest known period of human civilization that stretches back to one million years BC, when tools and weapons were made out of stone. The Bronze Age (sometimes called the Copper Age) spans 3500–1000 BC. Weapons and implements were made of bronze (an alloy of copper and tin) during this period. The alloy is stronger than either of its constituents. Bronze was used to build weapons such as swords, axes, and arrowheads; implements such as utensils and sculptures; and other industrial products. The Iron Age followed the Bronze Age around 1000 BC and was characterized by the introduction of iron metallurgy. Iron ores were plentiful (cheap) but required high-temperature (2800°F) furnaces as compared to copper, which required lower-temperature (1900°F) furnaces. The Iron Age was the age of the industrial revolution, and many of the initial design tools, mechanics-based analyses, and material characterizations were formulated during this period. The Synthetic Material Age started in the early part of the twentieth century with the development of a wide range of man-made synthetic materials. This era saw an explosion of technological developments that touched every phase of human endeavor. Most of the high-performance engineering products, such as aerospace, computers, telecommunication, and medical and power systems, were the result of the development of advanced materials. This was an era of consolidation in terms of the development of comprehensive design tools, material characteristics, and mechanics-based analyses. During this period, the aerospace industry pioneered the development of composite materials and structures that had direct impact on structural capability (e.g., specific strength and specific stiffness) as well as manufacturing and maintenance costs. This translated into an increase in performance, payload, speed, range and a reduction in life-cycle cost.

The twenty-first century may be visualized as the Multifunctional Materials Age. The inspiration for multifunctional materials comes from nature; hence, these are often referred to as "bio-inspired materials." This category encompasses smart materials and structures, multifunctional materials, and nano-structured materials. This is a dawn of revolutionary materials that may provide a "quantum jump" in performance and multi-capability. This book focuses only on smart materials and structures. These are also referred to as intelligent, adaptive, active, sensory, and metamorphic structures and materials and/or systems. The purpose of these materials from the perspective of smart systems is their ability to minimize life-cycle cost and/or expand the performance envelope. The ultimate goal is to develop biologically inspired multifunctional materials with the capability to adapt their structural characteristics (e.g., stiffness, damping, and viscosity) as required, monitor their health condition, perform self-diagnosis and self-repair, morph their shape, and undergo significant controlled motion over a wide range of operating conditions.

Since the 1990s, there has been a major growth in smart structures technology, in both individual technological constituents and their applications in various disciplines. Applications include vibration and noise suppression, stability and damping augmentation, shape control, structural integrity monitoring, and condition-based maintenance. Relevant disciplines include space vehicles, fixed-wing aircraft, rotary-wing aircraft, civil structures, marine systems, automotive systems, robotic systems, machine tools, and medical systems. Major goals have been to enhance system performance (beyond current levels) at a low cost, increase comfort level (minimize noise and vibration) with minimum weight penalty, reduce life-cycle cost (decrease vibratory loads and perform condition-based maintenance), improve precision pointing (space telescope), improve low observable characteristics, and increase product reliability (damage detection, mitigation, and repair).

Development of smart materials and structures is possible through one of three approaches. In the first approach, the new materials with smart functionality can be synthesized at the atomic and molecular levels. Sometimes this is referred to as a nano-structured material. A lot of the relevant methodology is hypothesized and is in an embryonic state at this time. In the second approach, actuators and sensors are attached to a conventional structure that adaptively responds to external disturbances. The actuators and sensors normally do not constitute the load-carrying structure. Even though this is a relatively mature methodology, it is not expected to be a structurally efficient scheme. In the third approach, active plies representing actuators and sensors are synthesized with non-active plies to form a laminated structure. A major drawback is that once the structure is cured, it is not possible to replace nonfunctional plies. Even though this approach appears attractive in terms of structural efficiency, there are issues related to the integrity of the system.

The key elements of smart structures are actuators, sensors, power conditioning, control logics, and computers. Conventional displacement actuators are electromagnetic (including voice coils), hydraulic, and servo- or stepper motors. The principal disadvantages of conventional actuators are their weight, size, and slow response time. Their advantages are their large stroke, reliability, familiarity, and low cost. Smart material actuators are normally compact and change their characteristics under external fields such as electric, magnetic, and thermal. Typical smart material actuators are piezoelectric, electrostrictive, magnetostrictive, shape memory alloys, and electrorheological/magnetorheological (ER/MR) fluids. Conventional sensors are strain gauges, accelerometers, and potentiometers, whereas smart material sensors can be fiber optics, piezoelectrics (ceramics and polymers), and magnetostrictives. There is a wide variation of power requirements for different actuators. Key factors for a power conditioning system are compactness, efficiency, and cost. For an efficient adaptive system, the modeling and implementation of robust feedback control strategies are important. A centralized, compact, and lightweight computer is vital to generate input signals for actuators, perform system identification techniques with output data from sensors, and implement control-feedback strategies.

The basic idea of the synthesis of smart structures appears to have been first conceptualized by Clauser in 1968 [1]. Seven years later, Clauser himself demonstrated the concept [2]. After this work, activity in this area started increasing and grew rapidly in the 1990s.

The historical development of key smart materials is discussed first, followed by their applications in various industrial disciplines. Even though the discovery of many of the smart materials took place during the past century, the commercial availability, cost, and understanding of their behavior have been major impediments to their widespread use in commercial products. Today, one of the most popular smart materials is polycrystalline piezoceramic, which exhibits strong piezoelectric properties. Other popular smart materials include electrostrictives, magnetostrictives, shape memory alloys, and ER/MR fluids.

1.1 Smart Structures

A smart structure involves distributed actuators and sensors as well as one or more microprocessors that analyze the responses from the sensors and use integrated control theory to command the actuators to apply localized strains or displacements to alter system response. A smart structure has the capability to respond to a changing external environment (e.g., load or shape change) as well as to a changing internal environment (e.g., damage or failure). It incorporates smart material actuators that allow the alteration of system characteristics (e.g., stiffness or damping) as well as of system response (e.g., strain or shape) in a controlled manner. Thus, a smart structure involves five key elements: actuators, sensors, control strategies, power- and signal-conditioning electronics, and a computer. Many types of actuators and sensors, such as piezoelectric materials, shape memory alloys, electrostrictive materials, magnetostrictive materials, ER/MR fluids, and fiber optics, are being considered for various applications. These can be integrated with main load-carrying structures by surface bonding or embedding without causing any significant changes in the mass or structural stiffness of the system.

Numerous applications of smart structures technology to various physical systems are evolving to actively control vibration, noise, aeroelastic stability, damping, shape change, and stress distribution. Applications range from space systems to fixed-wing and rotary-wing aircraft, automotive, civil structures, machine tools, and medical systems. At this time, servovalve hydraulic actuators are widely used in aerospace and other applications because of their reliable performance across a large range of force, stroke, and bandwidth. Their drawbacks, such as mechanical complexity, need for hydraulic tubing and reservoir, and size and weight, present an opportunity to search for lightweight compact actuators such as smart material actuators. A "smart material" is defined as a material that transforms its characteristics, such as mechanical states (strain, position, or velocity) or material characteristics, (stiffness, damping, or viscosity) under external field (electric, magnetic, or thermal). Much of the early development of smart structures methodology was driven by space applications such as vibration and shape control of large flexible space structures, but now wider applications are envisaged for aeronautical and other systems.

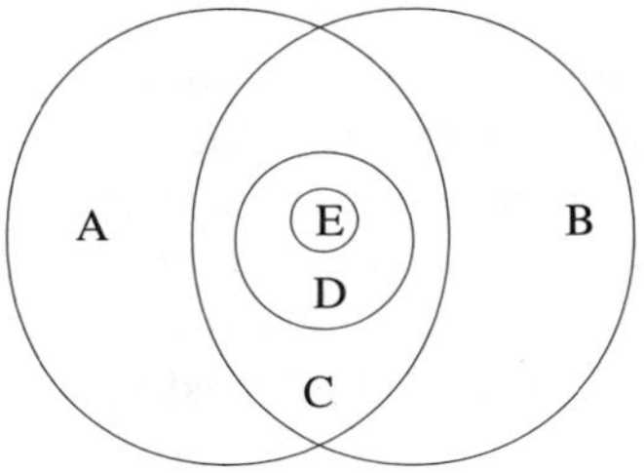

Figure 1.1. Classification of smart structures.

Embedded or surface-bonded smart actuators on an airplane wing or a helicopter blade, for example, can induce airfoil twist/camber change that in turn can cause a variation of lift distribution and may help control static and dynamic aeroelastic problems.

Applications of smart structures technology to aerospace and other systems are expanding rapidly. Major barriers include low actuator stroke, the lack of a reliable smart material characteristics database, nonavailability of robust distributed adaptive control strategies, and inadequate mathematical modeling and analysis of smart systems.

A smart or intelligent structure incorporates distributed actuators and sensors as well as control logic, processors, and power electronics. Figure 1.1 defines various types of structures, as follows:

Adaptive Structures (A): have distributed actuators to alter characteristics in a prescribed manner. They may not have sensors. Examples are conventional aircraft wings with flaps and ailerons, and rotor blades with servoflaps.

Sensory Structures (B): have distributed sensors to monitor the characteristics of the structure (i.e., health monitoring). Sensors may detect strain, displacement, acceleration, temperature, electromagnetic properties, and extent of damage.

Controlled Structures (C): overlap both adaptive and sensory structures. These constitute actuators, sensors, and a feedback control system to actively control the characteristics of the structure.

Active Structures (D): are a subset of controlled structures. Integrated actuators and sensors have load-carrying capability (i.e., structural functionality).

Intelligent or Smart Structures (E): are a subset of active structures. Additionally, they have highly integrated control logic and power electronics.

1.1.1 Smart Material Actuators and Sensors

Piezoelectrics are the most popular smart materials. They undergo deformation (strain) when an electric field is applied across them and, conversely, produce voltage when strain is applied; thus, they can be used as both actuators and sensors. Under an applied field, these materials generate a very low strain but cover a wide range of actuation frequency. Piezoelectric materials are relatively linear (at low fields) and bipolar (positive and negative strain) but exhibit hysteresis. To achieve high actuation force, piezoceramics (ferroelectric ceramic materials) are used. The most widely used piezoceramics (e.g., lead zirconate titanate, or PZT) are mostly available in the form of thin sheets that can be readily attached or embedded in

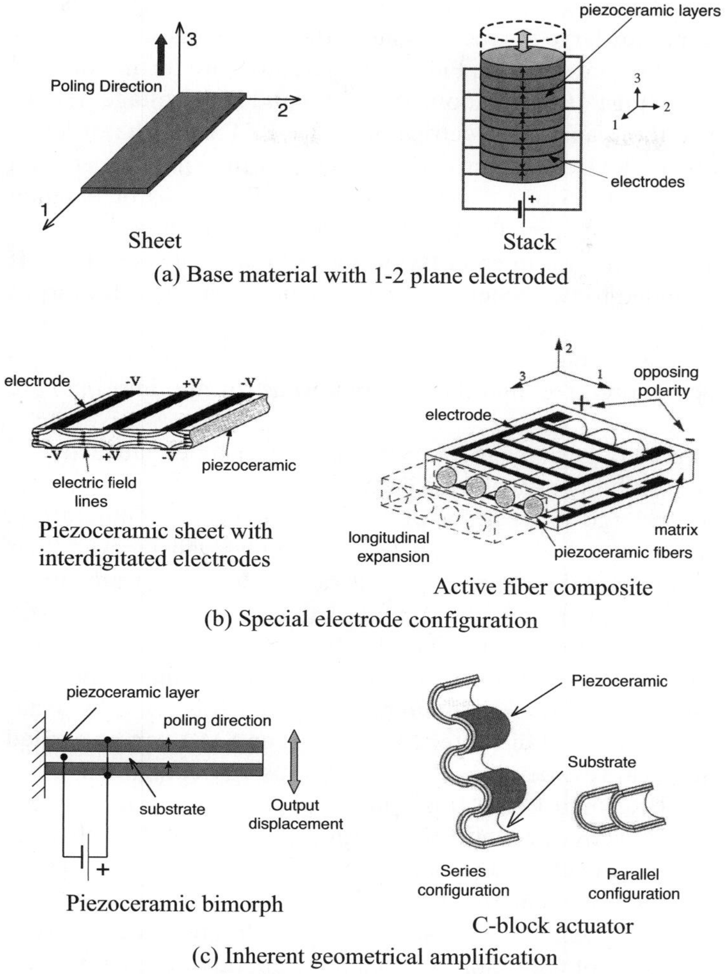

Figure 1.2. Typical piezoceramic actuators.

composite structures or stacked to form discrete piezostack actuators (Fig. 1.2). These sheets generate isotropic strains on the surface and a non-Poisson strain across the thickness. It is possible, however, to generate directional in-plane induced strains with piezoceramics using electrode arrangement, specially shaped piezos, bonding arrangement, and embedded fibers (see Fig. 1.2). Electrostrictives such as lead magnesium niobate (PMN) also require electric field to cause induced strain and have about the same induced-strain capability as piezoelectric materials. However, they are a nonlinear function of field (typically varying quadratically with field) and monopolar. Also, electrostrictive materials are very sensitive to temperature but exhibit negligible hysterisis.

Piezoelectric and electrostrictive materials are also available in the form of "stacks," in which many layers of materials and electrodes are assembled together. Typically, stacks are built using one of two methods. In the first method, the sheets of active material and electrodes are bonded together using an adhesive (normally of lower stiffness than the active material). In the second method, the layers of active material and the electodes are co-fired in the presence of high isostatic pressure. The stacks generate large forces but small displacements in the direction normal to the top and bottom surfaces. Piezostack actuators are further divided into two categories: low-voltage devices (about 100 volts) and high-voltage devices (about 1000 volts). Since the maximum electrical field for PZT is on the order of 1 to 2 kV/mm, low-voltage devices consist of 20 to 100 μm thickness sheets and high-voltage devices consist of 0.5 of 1.0 mm thick sheets. Bimorphs or bending actuators are also available commercially, in which two layers of these materials (piezoceramic) are stacked with a thin shim (typically of brass) between them. If an opposite polarity is applied to two sheets, a bending action is created. Bimorphs cause larger displacement and smaller force as compared to single piezo elements. The bending displacement is the highest at the tip of the cantilevered bimorph actuator. To increase the actuation force, multilayered bimorphs (or multimorphs) are used.

Among other smart materials, shape memory alloys (SMA) appear attractive as actuators because of the possibility of achieving large excitation forces and displacements. These materials undergo phase transformation at a specific temperature. When plastically deformed at a low temperature, these alloys will recover their original undeformed condition if their temperature is raised above the transformation temperature. This process is reversible. A remarkable characteristic of SMA is its large change of modulus of elasticity when heated above phase transformation temperature (typically two to four times the low temperature value). The most common SMA material is Nitinol (nickel titanium alloy), which is typically available in the form of wires of different diameters. Heating of an SMA can be carried out both internally (electrical resistance) and externally (using heating coils), but the response is very slow (less than 1 Hz). It is sometimes possible to speed up the response through forced convective or conductive cooling of material. Magnetostrictive materials such as Terfenol-D elongate when exposed to a magnetic field. These materials are monopolar and nonlinear, and they exhibit some hysteresis (less than piezoelectric). These materials generate low strains and moderate forces across a wide frequency range. Because of the required coil and magnetic return path, these actuators are often bulky. Electrorheological (ER) fluid consists of suspensions of fine dielectric particles in an insulating fluid that exhibits controlled rheological behavior in the presence of large applied electric fields (up to 1–4 kv/mm). Application of an electric field results in a significant change of shear-loss factor (fluid viscosity) that helps alter damping of the system. Magnetorheological (MR) fluid consists of suspensions of ferrous particles in fluid and exhibits change in shear-loss factor due to magnetic fields (low fields but moderately large currents). MR fluids, like ER fluids, are primarily envisaged as augmenting damping in a system. Fiber optics are becoming popular as sensors because they can be easily embedded in composite structures with little effect on structural integrity, and they also have the potential of multiplexing.

Smart structures are becoming feasible because of the (1) availability of smart materials commercially, (2) ease of embedding devices in laminated structures, (3) exploitation of material couplings such as between mechanical and electrical properties, (4) potential of a substantial increase in performance at a small price (e.g., weight penalty), and (5) advances in microelectronics, information processing, and sensor technology. Key elements in the application of smart structures technology to a system are actuators, sensors, control methodology, and hardware (computer and power electronics).

Table 1.1. *Comparison of actuators*

Actuators	Piezoceramic PZT	Piezofilm PVDF	Electrostrictive PMN	Magnetostrictive Terfenol-D	Shape Memory Nitinol
Ferroic class	Ferroelectric	Ferroelectric	Ferroelectric	Ferromagnetic	Ferroelastic
Field	Electric	Electric	Electric	Magnetic	Thermal
Maximum Free Strain %	0.1	0.07	0.1	0.2	8
Response time	μs	μs	μs	μs	s
Young's Modulus E (GPa)	68.9	2.1	117.2	48.3	27.6 for martensite 89.6 for austenite
Strain-voltage characteristic	First-order linear	First-order linear	Nonlinear	Nonlinear	Nonlinear

1.1.2 Smart Actuators

Typical actuators consist of piezoceramics, magnetostrictives, electrostrictives, and shape memory alloys. These normally convert electric/magnetic/thermal inputs into actuation strain/displacement that is transmitted to the host structure, affecting its mechanical state. Piezoelectrics and electrostrictors are available as ceramics, whereas magnetostrictors and shape memory alloys are available as metal alloys. Piezoelectrics are also available in polymer form as thin soft film. Important performance parameters of actuators include maximum stroke or strain (free condition), maximum blocked force (restrained condition), stiffness, and bandwidth. Somewhat less important parameters include linearity, sensitivity to temperature, brittleness and fracture toughness (fatigue life), repeatability and reliability, power density, compactness, heat generation, field requirement, and efficiency. The induced strain is often treated like thermal strain. The total strain in the actuator is assumed to be the sum of the mechanical strain caused by the stress plus the induced strain caused by the electric field. The strain in the host structure is obtained by establishing the displacement compatibility between the host material and the actuator. In a piezoelectric material, when an electric field is applied, the dipoles of the material try to orient themselves along the field, causing strain in the material. This relation of strain versus voltage is linear to the first order. In an electrostrictive material, there is an interaction between the electric field and electric dipoles, which is inherently nonlinear. The magnetostrictive response is based on the coupling of magnetic field and magnetic dipoles in the material, again a nonlinear effect. Shape memory is a result of phase transformation due to temperature change of the material (caused by a thermal field). Phase transformation is very much a nonlinear phenomenon.

A common piezoceramic material is lead zirconate titanate (PZT), and its maximum actuation strain is about one-thousand microstrain. Polyvinylidenefluoride (PVDF) is a polymer piezoelectric film and its maximum actuation strain is about seven-hundred microstrain. A common ceramic electrostrictive material is lead magnesium niobate (PMN) and its maximum actuation strain is about one-thousand microstrain. PZT and PMN are available in the form of thin sheets, which can be either bonded or embedded in a structure.

The PZTs require initial polarization (with high electric field), whereas no such polarization is needed for PMNs. Terfenol, a rare earth material, can create a maximum actuation strain of about two-thousand microstrain. It needs a large magnetic field in the axial direction to cause this actuation strain. Nitinol (nickel titanium alloy), normally available in the form of wires, can create free strain from 20,000 to 60,000 microstrain (2–6%). Table 1.1 shows a comparison of characteristics for

Table 1.2. *Comparison of sensors*

Sensor	Resistance gauge 10 V excitation	Semiconductor gauge 10 V excitation	Fiber Optics 0.04″ interferometer gauge length	Piezofilm 0.001″ thickness	Piezoceramics 0.001″ thickness
Sensitivity	30 V/ε	1000 V/ε	10^6 deg/ε	10^4 V/ε	2×10^4 V/ε
Localization (inches)	0.008	0.03	0.04	<0.04	<0.04
Bandwidth	0 Hz-acoustic	0 Hz-acoustic	0 Hz-acoustic	0.1 Hz-GHz	0.1 Hz-GHz

different smart actuators. Giurgiutiu et al. [3] compared the characteristics of various commercially available piezoelectric, electrostrictive, and magnetostrictive actuators. The comparison was carried out in terms of output energy density. Typically, the energy density per unit mass was found to be in the range of 0.233 to 0.900 J/kg. There is a wide variation in the performance of actuators among manufacturers. Near [4] provided an overview on piezoelectric actuator technology.

1.1.3 Sensors

Typical sensors consist of strain gauges, accelerometers, fiber optics, piezoelectric films, and piezoceramics. Sensors convert strain or displacement (or their time derivatives) into an electric field. Resistance (foil) and semiconductor strain gauges depend on a change of resistance due to the strain, and these require a DC excitation field for measurement. Piezoceramics and piezofilms are based on the variation of piezoelectric charge generated as a result of change in strain, and these do not require any external field. Fiberoptic gauges rely on a mechanical/optical coupling effect where output is expressed in terms of the phase lag of a monochromatic wave passed through the fiber as a result of the strain.

Piezoelectric strain sensors are generally made of polymers such as PVDF, and are very flexible (low stiffness). They can be easily formed into very thin sheets (films) and adhered to any surface. Key factors for sensors are their sensitivity to strain or displacement, bandwidth, and size. Other less important factors include temperature sensitivity, linearity, hysteresis, repeatability, electromagnetic compatibility, embeddability, and associated electronics (size and power requirement). Typically, the sensitivity for a resistor gauge is approximately 30 volts per strain; for a semiconductor gauge, it is 10^3 volts per strain; and for piezoceramic gauges, it is 10^4 volts per strain. The sensitivity of fiber-optic sensors is defined differently and is about 10^6 degrees per strain. Associated electronics may weigh against fiber-optic sensors. Discrete shaped sensors that apply weighting to the sensors' output can help increase sensitivity for a specific application. For example, a modal sensor can magnify the strain of a particular mode. Table 1.2 shows a comparison of characteristics of different sensors for typical excitation voltages, gauge lengths, and sensor thicknesses.

1.1.4 Actuator-Sensor Synthesis

In some cases, the same device can be used simultaneously as both an actuator and a sensor. This is referred to as self-sensing actuation and can be quite advantageous for active control applications because actuation and sensing actions are perfectly collocated [5]. For example, the piezoelectric material can be considered as a transformer between the structural states (stress and strain) and the electric states (voltage and charge). A piezoelectric self-sensing actuator can be created by incorporating two identical piezoelectric elements in a bridge circuit. The objective is to identify the difference in the charge components created by the applied electric field and the mechanical strain. Actuation force can be in the form of force, moment, or distributed strain, and sensing can be in the form of displacement, slope or strain, and their derivatives. For example, displacement, velocity, and acceleration are three separate output components. Hence, there can be a total of nine sensor output components and three force input components. Gupta et al. [6] outlined six criteria for optimal placement of piezoelectric actuators and sensors. These included (1) maximizing modal forces/moments, (2) maximizing deflection of the host structure, (3) minimizing control effort, (4) maximizing degree of controllability, (5) maximizing observability, and (6) minimizing spillover effects. It is important to place piezoelectric actuators in the region where the average modal strains are highest, which would result in maximum modal forces/moments. Placing actuators at the antinodes results in maximum deflection. It is advisable to place sensors at locations where the observability can be maximized. Boundary conditions also play an important role in the optimal placement of actuators and sensors.

1.1.5 Control Methodologies

For smart material applications, distributed control functionality is a key ingredient. There are three levels of control strategies: local control, global control, and higher cognitive functions. In local control, the objectives can be to augment damping, absorb energy, and minimize residual displacements. The objectives of global control can be to stabilize structural response, control shape, and minimize disturbances. The objectives of cognitive functions could be the ability to diagnose component failure and reconfigure and adapt after failures.

In the case of a system with single input and single output, local control can be established through a transfer function. The phase and amplitude of input actuation are adjusted to minimize the single output. Local control is used for adding damping and for low authority control. For the global control, there are several limiting cases of distributed control. The first one is a centralized controller in which the output from all sensors are processed by a centralized processor that provides control outputs to the distributed actuators. The second one is a decentralized controller in which the local control is carried out in an independent manner. However, it is computationally inefficient. Conversely, in the centralized controller, the computer has to process signals at rates corresponding to the highest mode of interest. To avoid these issues, one can arrive at a compromise controller straddling the two approaches of completely centralized and completely decentralized controllers – this is referred to as hierarchical or multilevel control architecture. This control strategy features a centralized controller for overall performance and a distributed processor for localized control. An average response within each element is then passed on to the global processor. This approach appears quite practical for many applications [7].

1.2 Manufacturing Issues

There are several issues concerning building of smart structures. These are as follows:

1. Electrical contact on both sides of the piezo is required. One way to overcome this problem can be to drill a hole in the substructure and use conducting epoxy. The second way is to introduce a thin layer of conductor between the piezo and the substructure and use a conducting bond layer.
2. The piezo has to be insulated from the structure. By anodizing or coating the structure, this problem can be solved.
3. For proper transfer of induced strain to main structure, bond-layer thickness needs to be thin and uniform. For this, pressure is applied during curing.

Embedding versus Surface-Mounting: With surface-mounted actuators, there is an ease of manufacturing, access for inspection, and less maintenance cost. Because of exposure, the actuators are more susceptible to damage. Also, the functioning of the actuators is dependent on the structural surface. With embedded actuators, the piezo becomes inaccessible for inspection. The devices, however, are better protected and interconnections with other devices become easy.

Embedding Electronics: For embedding integrated circuits, it is essential to ensure their electrical insulation and mechanical isolation. For minimal degradation of structure, it is important to have minimum ply interruption.

1.3 Piezoelectricity

Pierre and Paul-Jacques Curie (Fig. 1.3) discovered in 1880 (at the Sorbonne, France) that some crystals (e.g., Rochelle salt, topaz, tourmaline, cane sugar, quartz, sodium chlorate, and zinc blende), when compressed in certain directions, produce electric charges (positive and negative) on specific parts of their surfaces. The electric charges were found to be proportional to the applied pressure and vanished when pressure was removed. Furthermore, if the sign of pressure or strain was changed (e.g., from compression to tension), the developed charges also changed sign. This phenomenon was subsequently named, piezoelectricity, (pressure electricity, as piezo is a Greek word meaning "to press"). Piezoelectricity is different from contact and friction electricity. This effect of generation of charges due to applied pressure or stress is referred to as the "direct effect." In piezoelectric materials, there is also a "converse effect" (sometimes referred to as reciprocal or inverse effect) wherein a strain (or deformation) is caused in the material when it is exposed to an electric field. Again, induced strain is proportional to applied electric charge (polarizing field). The converse effect in piezoelectric crystals was first mathematically predicted by Lippmann in 1881 using fundamental laws of thermodynamics, and the Curie brothers experimentally demonstrated it in the same year. To demonstrate this, flat plates were cut according to a specific crystal orientation and surface bonded with tin foils as electrodes. For thirty years following its invention, until the First World War, piezoelectricity remained a scientific curiosity. Then there was a spurt of research activities in piezoelectricity, especially for applications in underwater ultrasonic detection. Using the converse effect, quartz and Rochelle-salt plates were excited at high frequencies (in the range of one megaHertz) to produce high-frequency sound waves for underwater detection. Paul Langevin and his co-workers in France developed ultrasonic

(a) *Pierre Curie (1859–1906) was born in Paris and became Professor of Physics in the Sorbonne. He was a pioneer in crystallography, magnetism, and radioactivity. He, along with his older brother Jacques, discovered piezoelectricity; direct effect in 1880 and converse effect in 1881. He married Marie Sklodowska in 1895 and they together shared a Nobel prize in Physics in 1903 for their work on radioactive elements.*

(b) *Paul-Jacques Curie (1856–1941) was born in Paris and became Professor of Physics at the University of Montpellier. He, along with his brother Pierre, discovered piezoelectricity in 1880.*

Figure 1.3. The Curie brothers.

submarine detectors using piezoelectricity. These transducers were built from quartz crystals and were excited near their resonance frequency (about 50 kHz) to transmit high-frequency chirp signals into the water and the location of submarines was measured from the timing of return echos. However, they could not perfect this device until the end of the First World War. This echo method became a valuable tool to locate submerged objects as well as to explore the bottom of the ocean. Between the two World Wars, there were other applications using piezoelectric resonators and oscillators in chemistry, biology, and industry. Applications ranged from radio transmitter stations and explosive pressure measurement to many kinds of electrical measurements, microphones, and accelerometers. Around the Second World War, the discovery of polycrystalline piezoceramic materials provided tremendous momentum to this field. These materials with high dielectric constants could be manufactured in high volumes. However, the raw piezoceramics are isotropic and do not possess piezoelectric properties. These ceramics need to be polarized with the application of a strong electric field for a short period of time and then these materials become anisotropic. The advent of piezoceramics expanded the domain of their applications to powerful sonar, ignition systems, hydrophones, and phono cartridges.

Beginning about 1925, Bell Telephone Laboratories used piezoelectric crystals to develop wave filters for multichannel telephony. The properties of exceptionally high dielectric constant and dielectric hysteresis in Rochelle salt were discovered by P. Seignette in 1917. In the early years, the substance was referred to as Seignette salt, and it was widely used in microphones and phonograph pickups. Brush Development Company (Cleveland, Ohio) played a major role in the growth of these applications in the 1930s. During the Second World War, research groups in the United States, Russia, and Japan independently discovered new man-made materials, often referred to as ferroelectrics, which exhibited piezoelectric effects many times higher than those found in natural materials. In the 1940s, Arthur von Hippel and coworkers at MIT discovered the ferroelectric characteristics of a refractory material, barium titanate ($BaTiO_3$), and a relative permittivity in excess of 1000 was determined in this material. Also, it was found that this ceramic material could be depolarized and subsequently repolarized in the opposite direction by applying a high electric field. This discovery was the beginning of the commercial development of piezoelectric crystals in a range of shapes and sizes. This material loses its piezoelectric characteristics at a temperature above 120°C (called the Curie temperature). This limitation was overcome in the late 1950s with the discovery of piezoelectric effects in lead metaniobate ($PbNb_2O_6$) and lead zirconate titanate [$Pb(Ti,Zr)O_6$], which had a Curie temperature of about 250°C. By the late 1950s, ceramic materials with piezoelectric characteristics started becoming available commercially, spurring the growth of their applications.

Polymer polyvinylidenefluoride (PVDF) $(CH_2CF_2)_n$ was discovered in 1969 by Kawai [8]. It is elastically a very soft material with strong piezoelectric effects and is often referred to as piezoelectric film or ferroelectric fluoride. PVDF is a semicrystalline material and is available in a broad range of thin sheets with thicknesses ranging from sub-micron to 1 mm. It can be easily cast into different geometric shapes. Because of their low stiffness, PVDFs are normally used as sensors.

Initially, the basic thermodynamics-based phenomenological theory of piezoelectricity was enunciated by Lord Kelvin (born in Scotland in 1824). Woldemar Voigt (born in Germany in 1850) formulated a comprehensive set of phenomenonologically based constitutive relations for piezoelectric crystals in 1894 using electric (field and polarization), elastic (stress and strain), and piezoelectric coefficients. Of the thirty-two crystal classes, only twenty possess piezoelectric characteristics, and Voigt identified their nonzero piezoelectric coefficients among a maximum number of eighteen coefficients, relating six mechanical stress components to three electric polarization components. In these crystals, the unit cells are nonsymmetric about at least one axis. During the first thirty years after the invention of piezoelectricity, there were major developments in thermodynamic-based tools to describe the behavior of piezoelectric crystals.

In the early years, there was confusion between piezoelectricity and pyroelectricity. Soon it became clear that piezoelectricity is different from pyroelectricity, contact electricity, and electrostriction. Pyroelectricity is a state of electric polarity produced in certain crystals due to a change in temperature. Pyroelectricity has been observed since medieval times (several hundred years BC). For example, tourmaline powder, when placed in hot ashes, produced sparks. This effect was recorded in Europe in 1703 when Dutch merchants brought tourmaline powder from Ceylon and India. The contact electricity is static electricity generated by friction. Again, this effect has been observed since the Stone Age and was referred to as amber (electron)

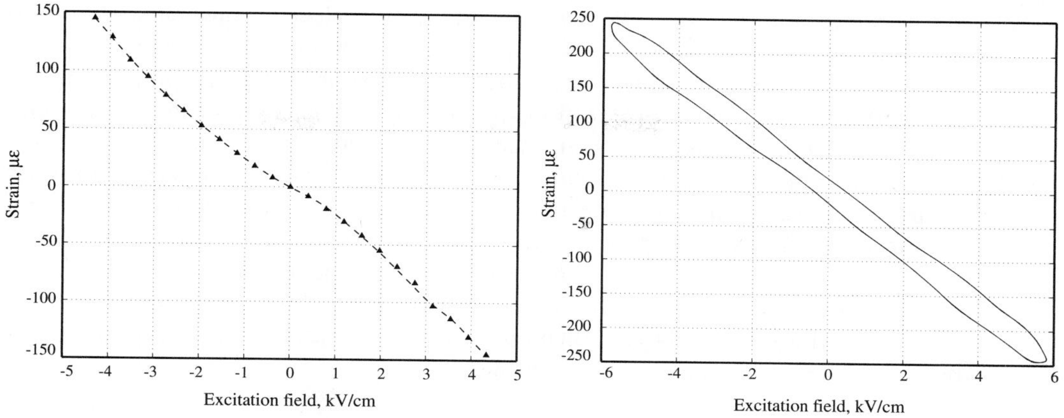

(a) Static strain (b) Dynamic strain

Figure 1.4. Induced free strain in piezoelectric materials.

by the ancient Greeks. Electrostriction is the induced deformation in dielectrics due to electric field, and the deformation is a quadratic function of electric field (same for positive and negative charges). It is especially important in materials with a large dielectric constant. Ferroelectricity has a close analogy to ferromagnetism: the spontaneous creation of electric moment in a crystal by the application of external electric field. Even though electrostriction is evident in piezoelectric materials, its magnitude is negligible, especially at low electric fields.

Piezoelectric materials belong to a major class of materials called ferroelectrics, which consist of randomly oriented dipoles (local charge separation). When the material is heated above a certain temperature (Curie temperature) and a very strong electric field is applied, the electric dipoles reorient themselves relative to the electric field. This is called "poling." Once the material is poled, an applied electric field on any one of the surfaces induces an expansion or contraction of the material.

A piezoelectric crystal has a certain "one-wayness," bias, or polarity in its internal crystal structure, which determines whether a specific region on the surface shows a positive or a negative charge on compression; or, alternatively, which determines the sign of deflection when an electric field is applied to the crystal. Although this polarity is inherent in piezoelectric crystals, it is absent in polycrystalline piezoceramics in their virgin form. Hence, these materials need to be initially polarized, typically by applying a large DC field for a brief period of time. Piezoelectric materials can be broadly classified into three categories: natural crystals, ceramic-based materials, and electro-polymers. Piezoelectric materials are relatively linear and bipolar but exhibit hysteresis (Fig. 1.4). Due to the converse effect, the induced strain is extensional for a positive field in the polarization direction and compressional in the direction normal to the polarization direction. They exhibit small strains and large bandwidth. The piezoelectric coupling coefficients depend on the level of impurities present in the material, preparation procedure, the size and frequency of applied electric/mechanical field, temperature, and aging time.

Piezoelectric materials have been widely used in sensors that include strain gauges, pressure transducers, and accelerometers. Piezoceramics (e.g., PZT) are stiffer than piezoelectrics and are extensively used as compact actuators in a wide range of applications. Piezoelectric films (e.g., PVDF), because of their very low stiffness, are normally used as sensors. Today, piezoelectric transducers are used in a wide range of applications that include structural vibration control; precision positioning; active control of noise; shape control to enhance performance; sensors to determine local strain, acceleration, or velocity; and many other industrial applications such as crystal oscillators, Surface Acoustic Wave (SAW) devices, piezoelectric inkjet printer heads, and piezoelectric accelerometers. In comparison to other smart materials, piezoelectrics have a higher bandwidth than shape memory alloys, are more compact than magnetostrictives, and are bidirectional, unlike electrostrictives. Piezoelectric materials are insensitive to electromagnetic fields and radiation, enabling applications in harsh environments. One major disadvantage of these materials is that they normally cannot be used for static response and measurements because of the leakage of charge with time.

1.4 Shape Memory Alloys

A shape memory alloy (SMA) refers to a specific category of material that has the ability to remember a specific shape even after severe deformation at low temperatures. These materials stay deformed until heated to a moderate increase in temperature, whereupon they recover their original predeformation shape. The parent state of the material is at high temperatures and it is referred to as the austenite state, whereas at low temperatures, the material is considered to be in the martensite state. The austenite phase exhibits a cubic crystalline structure and the martensite phase exhibits a tetragonal or monoclinic crystalline structure. From the austenite phase, the transformation to martensite is a shear-dominated diffusionless transformation and may lead to twinned martensite in the absence of stress and detwinned martensite in the presence of a sufficient level of stress. The material shows a marked difference in mechanical behavior in the two states. The shape memory effect occurs as a consequence of a crystallographically reversible phase transformation in solid state.

The discovery of the shape memory effect was made in 1932 by Swedish researcher Arne Olander in a gold-cadmium alloy. Greninger and Moordian observed the formation of and disappearance of the martensite phase in a copper-zinc alloy through the variation of temperature in 1938. The next discovery of SMA appears to have been reported in 1951 by Chang and Read [9], who found the shape memory effect in a gold (Au) and cadmium (Cd) alloy. Buehler et al. [10] (Fig. 1.5) and Buehler and Wiley [11] at the Naval Ordnance Laboratory showed the shape memory effect in a nickel (Ni) and titanium (Ti) alloy in 1962. It is reported that the discovery of Nitinol occurred by accident, when a strip of nickel-titanium alloy was bent out of shape. When subsequently heated (by David Muzzey with his pipe lighter at a laboratory meeting), the strip stretched back to its original form.

Since that time, the shape memory effect has been observed in many other alloys, which include copper (Cu) and zinc (Zn) alloys [12]; copper (Cu), aluminum (Al), and nickel (Ni) alloys [13]; and indium (In) and thallium (Tl) alloys [14, 15]. Buehler and Wiley [11] received a patent on a nickel-titanium alloy called Nitinol (nickel-titanium alloy developed at Naval Ordinance Laboratory) in 1965. Among the many different SMAs, Nitinol attracted the most attention because of its superior mechanical characteristics (e.g., strength and electric resistivity) in comparison with other SMA materials. Nitinol has a very high recovery strain up to 8% or a very high recovery stress up to 800 MPa, a very high ultimate tensile stress of up to 1000 MPa,

Figure 1.5. William J. Buehler was born in 1923 in Detroit, Michigan. He, along with his co-workers at Naval Ordinance Lab (NOL, later called NSWC) in White Oak, Maryland, discovered nickel-titanium alloy (later named Nitinol) in 1962.

a large elongation prior to failure reaching up to 50%, higher corrosion resistance, easy workability, and great damping capacity. In addition to Nitinol, copper-based shape memory alloys, such as Cu-Zn-Al and Cu-Al-Ni, are available commercially.

Including copper as a ternary element in binary Nitinol results in a reduction of the hysteresis effect and a significant reduction in yield stress. A small addition (1–3%) of ternary elements that are chemically similar to Ti or Ni such as Co, Fe, and Cr, are shown to lower the martensite transformation temperature such that the shape memory effect can occur below ambient condition.

Two key characteristics of SMAs are "shape memory effect" (SME) and "pseudoelasticity" (or superelasticity). Both of these diffusionless (displacive) phase transformation effects are induced by temperature and stress. After deformation at low temperature (formation of residual plastic strain after loading and unloading), the SME allows the material to regain its original shape when heated above the phase transition temperature. The basis of the SME is the crystalline phase change on heating and cooling. In Nitinol, this phase change is from an ordered body-centered cubic structure in the austenite phase (at high temperature), to a face-centered cubic structure (monoclinic structure) in the martensite phase (at low temperature). Nitinol exhibits not only high SME, but also high strength, ductility, and resistance to corrosion. The formation of stress-induced martensite from the austenite phase is referred to as pseudoelasticity or superelasticity. This means that at a high temperature, a specimen exhibits a large apparent plastic strain on loading, which is fully recovered in a hysteresis loop on unloading. The behavior is not triggered by temperature and is only stress dependent. The hysteresis behavior is due to the forward phase transformation from austenite to stress-induced martensite taking place at high stress level and the reverse phase transformation to austenite phase taking place at low stress level. One of the widely exploited concepts in SMAs is constrained recovery force. If an initially deformed SMA specimen is constrained to return to its original shape on heating, it will generate high recovery force. As a consequence of this, the first successful commercial applications were pipe couplings and mechanical fasteners.

Phase transformation in SMAs can be induced by the application of change of temperature or stress, or by a combination of both. In the absence of an applied stress, transformation through temperature is characterized by four characteristic temperatures martensite start temperature M_s, martensite finish temperature M_f, austenite start temperature A_s, and austenite finish temperature A_f. These transformation temperatures change with the presence of applied stress, normally increasing with tensile stress. The temperature affects the chemical-free energy, the applied stress affects mechanical potential energy, and the sum total of two types of energy determines the state of phase transformation of SMAs. The parent phase of the material is austenite (A). When the temperature is reduced, the material transforms into martensite (M); this isothermal process is called "forward phase transformation" ($A \rightarrow M$). The reverse phase transformation ($M \rightarrow A$) is endothermal. During the phase transformation, there is an evolution or absorption of a significant amount of latent heat. The martensite phase consists of a total of twenty-four different variants. An untrained Nitinol wire, when cooled from the parent phase, will revert to a combination of several variants of martensite and form twin bands in the material, and there is no net change of overall shape of the specimen. A key feature of an SMA is that the thermoelastic properties such as Young's modulus, electric resistivity, thermal conductivity, and heat capacity are different in both transformed states. These properties, including transformation temperatures, can be a function of chemical composition, cold work, heat treatment, and thermo-mechanical cycling. Because the variation of these characteristics takes place in a very narrow change of temperatures (based on stress level), SMA actuators are often called "bang-bang" and "off-on" actuators. In some cases, an intermediate R-phase (rhombohedral structure) is also present but normally this is not a major phase to describe the behavior of SMAs.

Since the 1990s, applications of SMAs in the mechanical, medical, and aerospace systems have proliferated. Specific applications include appliance controllers, eyeglass frames, medical wires, electrical switches, pipe couplings, and electronic connectors. Most of these applications have been 1-D in nature in which wires, rods/tubes, and strips are used as active actuators, primarily to cause static-induced deflections. One of the major inhibitions of a widespread use of SMAs in commercial applications has been the inadequate understanding and repeatability of their thermo-mechanical characteristics, especially under a range of loading conditions. Most of the shape memory phenomena are related to one-way SME. The material initially deformed at low temperature (martensite state) recovers its original shape on heating (austenite state). On cooling from the high temperature state, there is no apparent change of shape of the material. This is the one-way SME. The undeformed shape remains constant when the sample is subjected to thermal cycling. It appears possible to condition a two-way SME in SMAs, in which the material remembers both a high-temperature shape and a low-temperature shape. This means a specimen is deformed one way on heating and deformed the other way on cooling.

Two-way SME arises as a result of cyclic thermomechanical transformation (i.e., training), which induces a favorable residual stress field within the material. Hebda and White [16] showed that it requires about 2000 thermal cycles to achieve stable two-way effect, which can last up to 10,000 actuation cycles. Furthermore, it was shown that a very small bias stress during cycling can enhance the amount of retention of transformation strain in the wire. Another potential application of an SMA is for passive damping augmentation in a structure. SMA can provide damping capacities on the order of 10% [17]. In metallic materials, dissipation mechanisms include viscoelastic effects, Coulomb friction, and plastic deformation. For an SMA, the time delay of strain with respect to stress results in a dissipation of energy. Oberaigner et al. [18] showed that dissipation rate and dissipation energy become maximum at a certain fixed temperature between the martensite start and the martensite finish temperatures.

For various applications, it is important to have comprehensive constitutive models of SMAs, which can accurately represent the thermomechanical behavior of the materials in a mathematical form that is readily amenable for inclusion in engineering analyses. Most of these models describe quasi-static (thermodynamic equilibrium), one-way shape memory behavior under uniaxial loading, and these are broadly classified into three categories: phenomenology based macro-mechanics models, thermodynamic-based micro-mechanics models, and micromechanics-based macroscopic models. First-category models are built on phenomenological thermodynamics and are expressed in terms of engineering material constants. These are mostly defined using experimental test data (curve fitting to test data) and are quite amenable for inclusion in engineering analyses. In general, the behavior of the material is primarily a nonlinear function of three variables (assumed independent) and their associated rates of change: stress, strain, and temperature. The properties of a particular alloy depend on the composition of constituent elements, the processing technique, and factors involving manufacturing and heat treatment. Typically, the volume fraction of the martensite phase is used as the internal variable, and most of these models are perfected for uniaxial loading. Under the first category, some of the models are due to Tanaka [19], Liang and Rogers [20], Brinson [21], Boyd and Lagoudas [22], and Ivshin and Pence [23]. In these models, it is assumed that strain, temperature, and the martensite volume fraction are the only state variables. One of the pioneering models is due to Tanaka, which was derived from the second law of thermodynamics expressed in Helmholtz free-energy format, and in which the variation of martensite volume fraction with stress and temperature is expressed in exponential form. It is based on the Clausius–Duhem inequality. Liang and Rogers made a change to the development of martensite volume fraction from exponential form (Tanaka) to cosine form. Neither of these models captures the stress-induced detwinning of the martensite phase. The Tanaka model was modified by Tobushi et al. [24] to include R-phase transformations that are often seen in SMAs. In this model, there are two distinct variables for R-phase and detwinning martensite, which makes it possible to predict the R-phase and SME simultaneously.

Brinson divided martensite volume fraction into two parts: stress-induced and temperature-induced, and modified the Tanaka model accordingly. This model captured the detwinning effect. Epps and Chopra [25], Prahlad and Chopra [26], and Zak et al. [27] made a comparison of these three models with test data obtained from Nitinol wires. The deficiency of prediction of SME using the Tanaka model and the Liang and Rogers model can be overcome if the variation of transformation temperature at low stress (introduced by Brinson) is included. Under this category, Malovrh and Gandhi [28] developed a hierarchy of mechanism-based phenomenological models, comprising linear, piece-wise linear, and nonlinear springs and friction elements to represent the pseudoelastic behavior of SMAs. This approach is similar to the approach followed for elastomeric materials and ER/MR fluids, and the model parameters are identified using experimental test data (hysteresis cycles). The three-element model (comprising a lead spring in series with a unit consisting of a spring in parallel with a friction element) was the most basic model that could reproduce the generic hysteresis behavior. Chang et al. [29] developed a comprehensive coupled thermodynamic model for an SMA wire under uniaxial loading in a finite element framework and validated it systematically with mechanical and infrared experimental test data obtained from a typical polycrystalline NiTi wire. This one-dimensional strain-gradient continuum model was used to satisfactorily validate the SME and pseudoelastic behavior as a function of applied displacement rate and environmental parameters.

Second-category models are detailed, often quite complex, and are constructed using thermodynamic principles. They are less amenable for inclusion in engineering analyses. These are focused on microscale behavior such as nucleation, interface motion, and growth of martensite state. Under the second category, some of the models are due to Falk [30], Ball and James [31], Abeyaratne and Knowles [32], Barsch and Krumhansl [33], and Sadjadpour and Bhattacharya [34].

Third-category models are hybrids of the first two categories, which use thermodynamic phenomena to describe transformation and incorporate several assumptions to simplify micromechanics. Typical models in this category are due to Patoor et al. [35], Sun and Hwang [36, 37], and Huang and Brinson [38].

Most of the constitutive models of SMA are developed for uniaxial loading condition. It is important that the models should be simple and capable of being implemented in standard structural-mechanics analyses; they should incorporate realistic physics and be applicable in a wide range of temperatures and stresses to capture both the SME and pseudoelasticity. They should be adaptable to a wide range of materials and textures in both single crystals and polycrystals. However, in some applications, material may be subjected to a three-dimensional (3-D) stress condition and, as such, a one-dimensional (1-D) model may not be able to accurately estimate behavior. There are some 3-D models available such as those developed by Sun and Hwang [36, 37], Boyd and Lagoudas [39], Graesser and Cozzarelli [40], and Patoor et al. [41]. For example, Boyd and Lagoudas derived the model from free energy and a dissipation potential; they utilized the Gibbs free energy instead of the Helmholtz free energy (utilized by Tanaka). Zhou et al. [42] developed a three-dimensional constitutive model for SMAs based on the results of Differential Scanning Calorinetry (DSC) tests and Brinson's phase transformation relations. It appears attractive to utilize the vast methodology of plasticity available in the literature; however, there are significant differences in the underlying mechanisms affecting material behavior.

By embedding SMA wires in composite laminates, it may be possible to control the structural properties of such SMA hybrid composites (SMAHC) [43]. There are two issues: the bonding of SMA with the composite resin and the curing temperature far above austenite temperature. To overcome the first problem, one needs to incorporate an effective surface-bonding treatment. Jonnalagadda et al. [44] tried four surface treatments: untreated, acid-etched, hand-sanded, and sandblasted. Using standard axial tensile tests, the average interfacial bond strength of SMA wires embedded in an epoxy matrix was measured. Sandblasting significantly increased the bond shear strength, whereas hand-sanding and acid-etching reduced the interface strength. Using photoelasticity and heterodyne interferometry, the resulting stresses induced in the polymer matrix were measured. Increased wire adhesion resulted in lower axial wire displacement and higher interfacial stresses. If the prestrained SMA needs to be embedded in a laminated structure, there are at least three possible ways: (1) hold each SMA wire at two ends during curing, (2) cure at room temperature using special resins, and (3) use silica or teflon tubes with inserted steel wires during curing. For the third method, once the composite structure is cured, replace the steel wires with prestrained SMA wires [45]. Ogisu et al. [46] investigated carbon fiber–reinforced plastic (CFRP) laminates with embedded prestrained SMA foils for their fatigue characteristics. Using a prestrain of 2%, there was a remarkable delay in transverse crack onset strain (more than 30%) and the delamination onset fatigue cycles.

It is now well established that unstable mechanical behavior in SMAs can take place during stress-induced transformations in uniaxial loading [47]. Hence, the material behavior is extremely sensitive to the ambient environment and loading rate. The transformation processes often lead to distinctly nonuniform deformation and temperature fields, which in turn can lead to mechanical instabilities and phase-transformation fronts. Iadicola and Shaw [48] used optical and infrared imaging techniques to determine specimen deformation and temperature fields. It was shown that the grips of the testing machine had a major influence on the induced temperature field. It was also shown that nucleation events for the forward and reverse pseudoelastic transformation can be measured in a single experiment as long as the measurements are made on a part of the specimen that is free of residual strain.

Some researchers examined the concept of a solid engine using SMAs to convert low-grade thermal energy into mechanical energy [49]. There were numerous patents on this topic in 1980s. The underlying principle is a solid-state phase transformation that converts heat into motion. The source of thermal energy can be solar, geothermal or industrial exhaust. However, this concept has limitations that include the low-energy conversion efficiency and fatigue life of the material.

1.5 Electrostrictives

The electrostriction effect is an induced deformation in a dielectric material under the influence of an applied electric field. This effect is present in almost all materials, although it is normally very small for any practical application. Unlike the piezoelectric effect, which is linear with electric field, the electrostrictive effect is quadratic with electric field. This is a property found primarily in centrosymmetric dielectric materials. All ceramic piezoelectrics are in fact polarization-biased electrostrictors. Pioneering work towards the direct use of electrostriction in transducers was carried out by Cross and his team [50, 51] (Fig. 1.6). Lead magnesium niobate (PMN) and its doped derivatives are normally referred to as electrostrictive materials. The relaxor ferroelectric, lead magnesium niobate in solid solution with lead titanate (PMN-PT), was the key breakthrough in developing large electrostriction. It was a very difficult material to make without pyrochlore contamination until the breakthrough by Swartz and Shrout [52], introducing the Columbite method. A second key advance occurred due to Pan [53], who showed that unlike all piezoelectric ceramics that are poled into a metastable domain state, relaxor PMN is a compound that is used in its ground state and can be fabricated to be free from aging.

Among the derivatives, the solid solution of PMN and lead titanate is the most popular one (PMN-PT). These are categorized as relaxor ferroelectrics and not only

Figure 1.6. L. E. Cross is a recognized authority in ferroelectricity and dielectric materials. He, along with his co-workers at Penn State, pioneered numerous developments toward the development of PMN-PT (lead magnesium niobate-lead titanate).

have very high electrostrictive coefficients but also possess high relative permittivities (20,000 to 35,000). These materials show induced strains due to electric field quite comparable to piezoelectrics (0.1%). Unlike piezoelectrics, however, they do not show spontaneous polarization and, as a result, they display a very low hysteresis effect even at high operating frequencies. Electrostrictors normally elongate in the direction of field and contract normal to the field, irrespective of whether the field is positive or negative (Fig. 1.7). A major limitation of electrostrictive materials is their temperature sensitivity. For most applications, the temperature needs to be maintained within ±10°. As a result, electrostrictives are often used in underwater and in vivo applications. As the temperature increases, the induced strain decreases.

Even though electrostrictive materials are nonlinear and monopolar, they exhibit negligible hysteresis and creep. For motion-control applications such as micropositioning systems, one can expect repeatable performance (contrary to piezoelectrics). The electrostrictive coupling coefficients depend on the level of impurities present in the material, preparation procedure, size and frequency of applied electric/mechanical field, temperature, and aging time.

Scortesse et al. [54] found that electrostrictive ceramics such as 0.9PMN-0.1PT undergo a large reduction of the apparent Young's modulus (more than 50%) as a function of the static electric field, but there is an increase of modulus (more

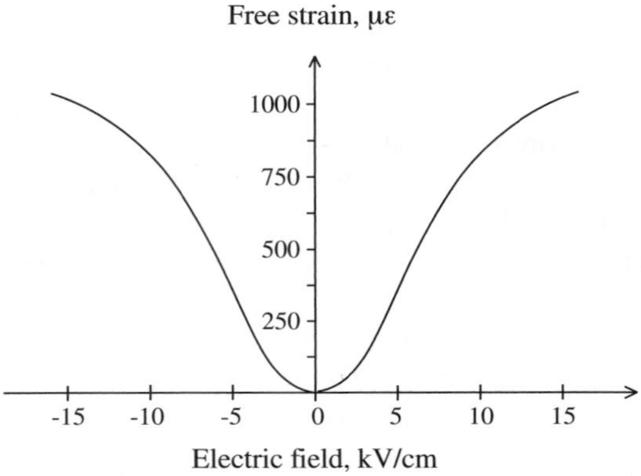

Figure 1.7. Induced free strain in electrostrictive materials.

Figure 1.8. James Prescott Joule (1818–1889) was born in Salford, United Kingdom, and formulated the theory of conservation of energy (i.e., first law of thermodynamics). He discovered ferromagnetism in an iron bar in 1842. Also, in 1840, he formulated Joule's Law of electric heating (i.e., heat generated in a wire is the product of the square of current and resistance).

than 20%) with an application of axial compressive stress (e.g., 30 MPa). The change of elastic modulus in the direction perpendicular to the electric field was found to be small (less than 6%).

Electrostictive materials are described by an even rank tensor; the electrostrictive effect is limited by symmetry and as such is nonlinear (quadratic). The electrostrictive actuator may consist of a multilayered configuration in the form of a stack consisting of many thin layers (125 to 250) that are diffusion bonded during the manufacturing process. Unlike piezoelectrics, PMN is not initially poled. Hence, for both positive and negative voltages, elongation takes place along the applied field direction. Also, PMN actuators show an electric capacitance four to five times as high as that of piezoelectric actuators. Applications include sensors, transducers, actuators, robotics, and artificial muscles.

A good example of the application of electrostrictives is the tilt mirrors built into the wide field and planetary camera II in the Hubble Space Telescope. Initially, Hubble was launched into space with an incorrectly configured primary mirror. Subsequently, Hubble was repaired with six PMN-PT actuators that provided full ground control of the two tilt mirrors in the camera II replacement unit. This feat saved an investment of $7 billion. Stability, no aging, and very low thermal expansion more than compensate for the nonlinear and hysteretic response in applications to optical systems in which PMN-based compositions are still widely used. However, these actuators have not yet penetrated widely into other application areas.

1.6 Magnetostrictives

Magnetostriction is the phenomenon associated with ferromagnetic materials that undergo deformation (or strain) when magnetized (i.e., in response to a change in its magnetic state). The magnetostriction strain arises from a reorientation of the atomic magnetic moments. James Joule (Fig. 1.8), who first reported this phenomenon in 1842, found that an iron bar underwent a change of length when magnetized. This effect can be used in actuator applications. Villari discovered a reciprocal effect in the 1900s, in which the stress-induced dimensional change (or strain) in a ferromagnetic material results in a change in its magnetization. This behavior is called the

"Villari effect" and it can be used in sensor applications. Thus, magnetostrictive materials can convert magnetic energy into mechanical energy and vice versa, which provides capability for both actuation and sensing. Early examples of magnetostrictive materials were iron (Fe), nickel (Ni), cobalt (Co), and their alloys. These materials have very low magnetostriction, defined in terms of maximum strain (i.e., ppm: parts per million). Despite their low magnetostriction, these materials were used in many applications in the first half of the twentieth century including telephone receivers, hydrophones, sonar, torque meters, oscillators, and foghorns. In fact, Philipp Reis tested the first telephone receiver in the 1860s based on magnetostriction. In 1888, Ewing used a magnetostrictive device made of iron and nickel as a force sensor. During the Second World War, sonar transducers were built using nickel with a magnetostriction of about −40 ppm.

1.6.1 Terfenol-D

Around 1963–1964, it was discovered that rare earth metals such as dysprosium (Dy) and terbium (Tb) exhibit giant magnetostriction (>10,000 ppm) at cryogenic temperatures. However, this limitation of very low temperatures hindered their widespread application. During this period, a major effort was undertaken by the U.S. Navy to enhance sonar technology through the development of new magnetostrictive materials that have large magnetostriction at room temperature. In 1971, Clark and Belson at the Naval Ordinance Laboratory (NOL, later called the Naval Surface Warfare Center) and Koon, Schindler, and Carter at the Naval Research Laboratory (NRL) discovered an alloy of rare earth metals that had a giant magnetostriction at room temperature. This magnetostrictive material is now referred to as Terfenol-D (i.e., Te for Terbium, Fe for iron, NOL for Naval Ordinance Laboratory, and D for Dysprosium). This alloy exhibited a maximum strain of 2000 ppm (0.2%) at room temperature. The stoichiometry of Terfenol-D is $Tb_xDy_{1-x}Fe_y$, where x varies from 0.27 to 0.3 and y varies from 1.9 to 2.0. With a change in stoichiometry, a wide range of properties could be achieved. Terfenol became commercially available through ETREMA (a company in Iowa) in the 1980s. It is now used in a wide range of applications that include sonar (low-frequency underwater communication); hearing aids; load sensors; accelerometers; torque sensors; proximity sensors; active vibration and noise control; ultrasonic cleaning; machining and welding; micropositioning and linear and rotational motors and sensors to detect motion, force, and magnetic field. Furthermore, magnetostrictive amorphous wire and thin film are being used in a wide variety of sensing applications. Calkins et al. [55] provided an overview of commercial magnetostrictive applications that includes noncontact torque sensors, motion and position sensors, magneto-elastic strain gauge, force and stress sensors, material characterizing sensors, and magnetic field sensors.

Magnetostrictive materials elongate in the direction of the applied field, whether positive or negative, and contract along the direction normal to the applied field such that the net change in volume is nearly invariant (Fig. 1.9). Magnetostrictive materials exhibit a change in magnetic permeability, magnetomechanical coupling, piezomagnetic coefficients, and mechanical damping with variation of applied DC and AC magnetic fields, static and dynamic structural loads, and temperature. The magnetization is expressed in terms of the volume density of atomic magnetic moment, and changes as a result of the reorientation of magnetic moments in the material through the application of magnetic fields, thermal energy, or stresses. There is a significant

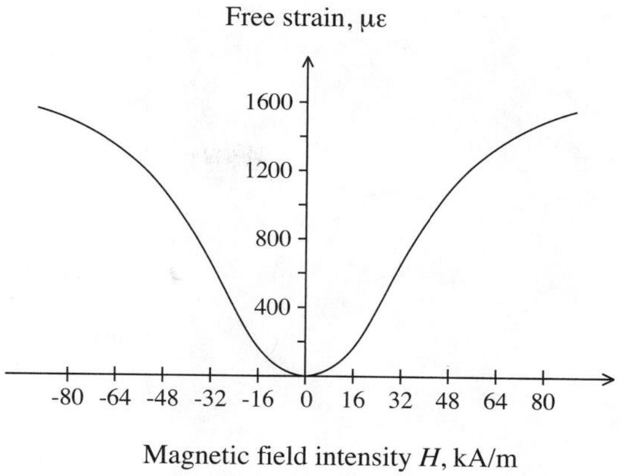

Figure 1.9. Induced strain in magnetostrictive materials.

change in the stiffness characteristics of magnetostrictive materials when the alloy is magnetized, called the Δ-E effect. For example, the Young's modulus is higher under the application of DC magnetic field than under no field. The stiffness of the magnetostrictive material also depends on the state of material; it appears stiffer in a mechanically clamped condition (zero strain) than in a mechanically free condition (zero external force). The behavior of magnetostrictive material depends on the type of field applied (i.e., electromagnetic, elastic, or thermal) and how it is applied. To fully utilize the desirable characteristics of magnetostrictive materials, it is important to characterize their electric, magnetic, thermal, and elastic behavior. A simple constitutive model for magnetostrictive material is the one most commonly used, the linear piezomagnetic model. Even though the actual behavior is intrinsically nonlinear and hysteretic, this quasi-linear model is quite insightful, especially at low signal regimes. Accurate comprehensive magnetostrictive models covering precise coupling among the electric, magnetic, thermal, and elastic regimes at all levels are not readily available. Carman and Mitrovic [56] and Kannan and Dasgupta [57] extended the linear constitutive modeling by including specific nonlinear effects. Another approach to modeling nonlinear dynamic behavior is to use a phenomenological approach following a generalized Preisach operator [58, 59, 60]. These models are normally restrictive and cumbersome.

In 1978, Clark (Fig. 1.10) and co-workers developed another magnetostrictive material, an alloy of amorphous metal (produced by rapid cooling) of iron, nickel, and cobalt together with one or more of the elements silicon, boron, and phosphorus. This alloy is commercially known as metglas (metallic glass) and is normally produced in thin ribbons. This material has an extremely high coupling coefficient (greater than 0.92 for metglas versus 0.7 for Terfenol-D). As a result, metglas is the preferred material for sensor applications.

Because of the bidirectional exchange of energy between elastic and magnetic states, magnetostrictive materials can be used for both actuation and sensing applications. Due to the requirement of magnetic field generation components, magnetostrictive transducers are usually heavy and bulky in comparison to piezoelectric and electrostrictive counterparts. Hence, these materials are mostly used in applications in which weight is not a primary issue and high forces and strains are required. Furthermore, the presence of a magnetic field–generating coil induces noise into adjacent electronic circuits and devices.

Figure 1.10. Arthur E. Clark, along with his co-workers at the Naval Ordinance Lab (later called NSWC), discovered Terfenol-D in the 1970s. Their group also discovered metglas in 1978 and Galfenol in 1998.

Advanced crystalline materials are often manufactured using crystal growth techniques to achieve directional solidification along the drive axis, and these processes plus the requirement of precision machining increase the cost of transducers. These technological and cost issues have migrated toward the development of alternate manufacturing techniques and materials, including crystalline thin films, magnetostrictive sintered powder, and particle-aligned polymer composite structures. Recent advances offer the prospect of new compounds to minimize magnetic anisotropy and hysteresis, and new fabrication processes to produce Terfenol-D thin films efficiently. As a result, quaternary compounds like Terfenol-DH are being developed in which Terbium and Dysprosium are substituted with Holmium. Also, manufacturing processes are being refined to build multilayered driver rods that lead to reduced dynamic losses, especially for operation in the high-frequency spectrum (in MHz range).

1.6.2 Galfenol

A new class of magnetostrictive alloys called iron-gallium alloys (known as FeGa alloys, or Galfenol) has recently been developed by researchers at the Naval Surface Warfare Center [61]. These alloys exhibit moderate magnetostriction ($\sim 350 \times 10^{-6}$) under very low magnetic fields ($\sim$100 Oe), have very low hysteresis, demonstrate high tensile strength ($\sim$500 MPa), and exhibit limited variation in magnetomechanical properties for temperatures between -20°C and 80°C [62, 63, 64].

Atulasimha and Flatau [65] reviewed developments in iron-gallium alloys and described challenges in their processing, methods of characterizing and modeling these materials, as well as actuation and sensing applications. In contrast to conventional magnetostrictive materials like Terfenol, Galfenol is highly ductile, machinable, and weldable. In addition, Galfenol can withstand shock loads, has a high Curie temperature, and is resistant to corrosion. As a result, there have been increasing applications of magnetostrictive materials in a wide range of fields, including areas where they need to be attached to other components or used as load-bearing structures.

Sensing applications of magnetostrictive FeGa alloys include torque sensors in rotor and automobile transmission shafts and sonar devices for detection of underwater explosions. The low bias field required for Galfenol can be achieved with a small permanent magnet, enabling it to be used in compact devices. Galfenol can be deposited epitaxially on a silicon substrate, which makes it well suited for microscale sensing applications. Electrodeposited nanowires made of Galfenol or FeGa/NiFe and FeGa/CoFeB can be used for miniature acoustic and tactile sensors.

The magnetostriction of Galfenol depends strongly on the content of Ga in the alloy, as well as the heat treatment and the applied compressive stress. This complicated dependence on different parameters makes it challenging to characterize the actuation and sensing properties of Galfenol for use in engineering analyses. For sensing applications, characterization of the interaction between the transducer and the sample is very important. Typically, the high permeability of Galfenol makes its reluctance comparable to that of the magnetic circuit.

Most of the material characterization is performed on single-crystal samples of Galfenol; however, polycrystalline material will be necessary for use in real-world applications. There are several ongoing research efforts for processing and fabricating polycrystalline Galfenol sheets, ribbons, and rods. Early efforts focused on directional solidification, followed by investigations into extrusion, forging, rolling, and sintering. Currently, production-grade polycrystal rods are produced using free-standing zone melting or directional solidification.

1.7 ER and MR Fluids

The basics of ER and MR fluids were discovered in the late 1940s and early 1950s [66, 67, 68, 69]; however, the early focus was primarily on ER fluids due to their ready availability in the laboratories. A key characteristic of these fluids is a dramatic change in fluid viscosity with the application of electric and magnetic field, respectively, for ER and MR fluids. When there is no field (electric/magnetic), the suspended particles are randomly distributed in the fluid, and in the presence of field, they form chains. As a result, the rheological properties change with applied field; these fluids can change from liquid to gel and back with response times on the order of milliseconds. They are also called "smart fluids." An ER fluid consists of a low-viscosity insulating base fluid, mixed with nonconducting particles typically in the range of 1–10 μm diameter. On application of electric field, these particles become polarized and increase the yield stress in shear (typically about 10 kPa for static loading and 5 kPa for dynamic loading). Since the initial patent by Willis Winslow in 1947 (i.e., mixing of starch with mineral oil), there have been numerous patents over the years on ER fluids. In early investigations, Winslow (Fig. 1.11) used a range of solid particulates (e.g., starch, lime, gypsum, silica, and carbon) dispersed in a variety of insulating oils (e.g., mineral oil, paraffin, and kerosene) to show significant ER characteristics. Water was also added to ER fluid to modify its electrical resistivity as well as to bond together the constituents. These ER fluids suffered from abrasiveness, chemical instability, and rapid deterioration in properties with time. As a result, there were few commercial applications of ER fluids early on. However, in the 1980s, there were significant improvements in both solid particulates and insulating oil. For example, Stangroom [70] demonstrated the use of nonabrasive polymer particles dispersed in silicon oil to achieve significant ER characteristics. Brooks [71] reported the application of new-generation ER fluids in various devices. Particles ranging from 5 μm to 50 μm dispersed in oil constituted the ER fluid. Larger size

Figure 1.11. Willis M. Winslow was born in Wheat Ridge, Colorado, in 1904. While working at the Public Service Company at Denver, he discovered the ER effect in 1942 and received his first patent in 1947.

particles are more liable to sedimentation, whereas smaller particles are liable to execute Brownian motion.

Because the addition of water can have detrimental effects due to the varying thermal environment (i.e., below freezing point and above boiling point), modern ER fluids do not include water. It has been suggested that acenequinone radical polymers (PARQRs), when dispersed in silicone or partially chlorinated petroleum, result in a good ER fluid [72]. It has been shown that an increase in temperature dramatically increases the current drawn and may be detrimental to the operational integrity of ER fluid. The recommended concentration of particles in carrier fluid is about 40% by volume in order to achieve large shear stress. Increasing the volume fraction increases the zero-field viscous characteristics and also affects current drawn and heat generation. There is a decrease in ER effect at high frequencies because of insufficient time for particles to polarize. Powell [73] showed that in activated ER fluids, the sustainable yield stress increased linearly as a function of the square of the electric-field strength. For very large-strain amplitudes, the magnitude of the yield stress decreased somewhat. An elastic force was generated on the application of an electric field, and this behavior was quite nonlinear in nature; the equivalent modulus decreased with increasing amplitude. In 2003, the giant electrorheological effect (GER) was discovered, capable of sustaining higher yield strengths at lower fields (lower current densities) than widely used ER fluids [74]. The GER fluid consists of urea-coated nanoparticles of barium titanate oxalate suspended in silicone oil. The urea-coated small-size particles result in a high-yield stress due to an increased dielectric constant.

ER fluids are being used in a wide range of applications that include valves with no moving parts, clutches and brakes [67], tunable engine mounts, shock absorbers [75, 76], robotic devices [77], machine tools, and aerospace structures [78].

MR fluids consist of noncolloidal suspensions of micron-sized, paramagnetic particles dispersed in a carrier fluid such as silicone or mineral oil. Since the 1990s, the focus has shifted to MR fluids because their maximum yield stress is twenty to fifty times larger than that of ER fluids and they can operate in a wide range of temperatures (−40°C to 150°C). Also, the ER fluids require a very high voltage (about 4 KV/mm), whereas the MR fluids can be controlled with a low field (12–24 V

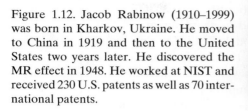

Figure 1.12. Jacob Rabinow (1910–1999) was born in Kharkov, Ukraine. He moved to China in 1919 and then to the United States two years later. He discovered the MR effect in 1948. He worked at NIST and received 230 U.S. patents as well as 70 international patents.

with current 1–2 amp). Furthermore, MR fluids are less sensitive to impurities or additives needed to enhance some characteristics. MR materials show yield stress of up to 100 kPa for an applied magnetic field of 0.5–1.0 T and thereby impact the viscosity of the fluid [79]. The credit for the discovery of MR fluid goes to Jacob Rabinow in 1948 [80]. A typical MR fluid developed and used by Rabinow (Fig. 1.12) consisted of nine to ten parts by weight of carbonyl iron to one part of silicone oil, mineral oil, or kerosene, with a small addition of grease or other thixotropic additive to improve settling stability. This resulted in a yield strength of about 100 kPa. If the applied magnetic field is reduced, the yield stress also diminishes. Without the magnetic field, yield stress is about 2–3 kPa. With the change of magnetic field, the viscosity changes too; it varies from 0.1 to 1.0 Pa-s for field from 0 to 1.0 T. Because the commercial availability of MR fluid is quite recent (under the trademark of Lord Corporation), there has been a growing number of applications. There is no doubt that MR is now preferred over ER fluid in most applications that include rotary brakes and linear dampers.

MR fluids have many attractive features that include high-yield stress, low off-state viscosity, and stable hysteretic behavior over a broad temperature range. However, they are more expensive than ER fluids. Both ER and MR fluids have quick response and reversible behavioral changes when subjected to electric or magnetic fields. Another desirable characteristic of ER/MR fluids is nonsettling of suspended particles. However, except for very special cases such as seismic dampers (i.e., the devices sit quiescent for long periods), suspension stability is not an overriding issue. Durability and longevity of the fluid are other important considerations. In the past decade, many different devices have been built using ER/MR fluids for industrial applications. These devices work according to one of the three flow modes: the shear mode (Couette flow), the flow mode (Poiseuille flow), or the squeeze mode. Jordan and Shaw [81] innovated ER technology toward the development of a flow-control valve by controlling the resistance to flow by changing the applied electric field. The conductors are stationary and the flow passes between them; this is referred to as "flow mode." If the electrodes are moving or rotating with respect to each other (with constant gap); this results in shearing of the fluid and is referred to as "shear mode." Most of the devices use one of these two modes of operation. A third mode of operation is squeeze mode, in which the conductors move with respect to each other in the direction of the field (varying the gap). Some experimental validation of the three modes emerged in the 1990s: flow mode [82], shear mode [83], and squeeze mode [84].

The general force-velocity characteristics of both ER and MR fluids are quite nonlinear. Two classes of models are used to characterize ER/MR devices: first-principle models and phenomenological models. The first-principle models are based on fundamental fluid-mechanics principles (i.e., conservation of mass, momentum, and energy) and these are quite complex as well as less tractable for specific devices. One has to incorporate many heuristic assumptions to make these practicable. These models require systematic validation studies to make them robust as design tools. Conversely, phenomenological-based models are widely used for their simplicity and adaptability to a specific device. These models consist of building blocks such as masses, springs, and dashpots arranged in series and parallel configurations, and their characteristics are normally identified from test data. A simple model to characterize the behavior of ER/MR fluids is the Bingham plastic model, a combination of both viscous and Coulomb damping effects. An alternate scheme to represent phenomenological-based models can be an electrical paradigm involving resistors, capacitors, and inductors.

Stanway et al. [85] attempted to model the response of an ER shear-type damper by modeling it as a viscous damper and a Coulomb damper in parallel, and a nonlinear filtering technique was used to estimate friction force and damping force, which were functions of electric field. Ehrgott and Masri [86] used three approaches to model the oscillatory dynamic behavior of the ER damper: first, a global equivalent linear system approach; second, a parametric identification model; and third, a non-parametric method that approximates the experimentally measured nonlinear response force.

In his very first patent, Winslow [66] described the ER phenomenon with reference to a brake/clutch mechanism. In his later patent in 1953, Winslow described a field-controlled hydraulic device that can act as a vibration damper [87]. There were numerous other attempts in the 1960s and 1970s to apply ER technology in commercial devices but only limited success was achieved. Since the 1990s, there has been a growing application of ER technology in commercial devices.

The MR fluid technology is scalable and, for example, a 20-ton MR fluid damper has been designed and successfully built for civil-engineering applications. This damper is of simple geometry in which the outer cylindrical housing is the magnetic circuit and the effective fluid orifice is the entire annular space between the outside diameter of the piston and the cylindrical housing. Controllable shock absorbers are being examined for potential applications in automotive systems, sports equipment, wind turbines, armament, steering wheels, and washing machines. For example, MR dampers are being examined to control gun recoil on naval gun turrets and field artillery.

The MR sponge consists of MR fluid constrained by capillary action in an absorbent matrix such as sponge (open-celled foam), which allows a minimum volume of MR fluid to be operated in a direct shear mode without seals or precision mechanical tolerances and is less susceptible to sedimentation of suspension particles [88]. Shen et al. [89] introduced fabrication techniques to develop two different MR elastomers. One elastomer was made of polyurethane and the other was made of natural rubber. There was a significant change in the Young's modulus of the polyurethane elastomer (about 30%) under a strong magnetic field, whereas there was a minimal chage of modulus of the rubber elastomer.

In parallel, there have been foundational developments in control theory. The developments in linear system theory and its application to vibration control and structural dynamics took place in the first half of the twentieth century. A major impetus for these developments was to improve ride quality in airplanes and automobiles. In fact, it was during the Second World War that concepts such as vibration isolators, vibration absorbers and vibration dampers were effectively applied in aeronautical systems. Design requirements for strength and safety may often conflict with demands for low vibration and extended fatigue life.

The shear mode is probably most widely investigated and the squeeze mode is less understood. The forces experienced through the ER fluid with AC field excitation are less than those experienced with DC field; however, the forces are functions of voltage amplitude, excitation frequency, and shape of input waveform. The ER fluids are normally more resistant to compressive forces than tensile forces. ER-based devices require a large applied electric field of up to 8 kV/mm of inter-electrode gap. To produce this level of field strength requires a very high voltage (in kV), which deters many potential users because of safety issues. Conversely, MR-based devices require low electric voltage (on the order of tens of volts) to generate a magnetic field of the required field strength. Also, MR fluids generate significantly larger dynamic force level than ER fluids. As a result of these two factors, today we have a large number of commercial applications of MR fluids. After repeated use, the MR fluid progressively thickens until it eventually becomes an unworkable paste. This problem is called "in-use thickening," and it was a major barrier in many early applications. Eventually, Lord Corporation solved this problem, and now MR fluids can operate for a long time [90].

1.8 Capability of Currently Available Smart Materials

Displacement transducers are typically classified into two categories: conventional displacement transducers and solid-state transducers. Smart material actuators fall under the category of solid-state actuators. The capabilities of currently available smart materials are limited. Ferromagnetics such as Terfenol-D have a fast response (60 kHz), but their maximum actuation strain is about 0.2%. Ferroelectrics such as PZT can achieve very high frequency (MHz), but their maximum induced strain is less than 0.1%. SMAs such as Nitinol can achieve large recoverable strains of about 6–8% but at a very low frequency (less than 1 Hz) due to slow heating and cooling processes.

Smart material actuators are superior to traditional electrodynamic and hydraulic actuators in terms of compactness and adaptability to laminated structures, but they lack the wide knowledge base of their basic characteristics. Most of the smart materials have a low strain output and, as such, internal or external amplifying mechanisms are needed to increase the stroke, or output displacement, for most applications. Energy is transferred from the active element to the load through a number of stroke-amplification stages. A wide variety of structural amplification mechanisms are used, which include flextensional shells, two-layered or multilayered bimorphs, Rainbow actuators, and C-block actuators [91, 92, 93].

Flextensional actuators are kinematic amplifiers that couple the longitudinal displacement of active ceramic material in the form of a disk or bar to the

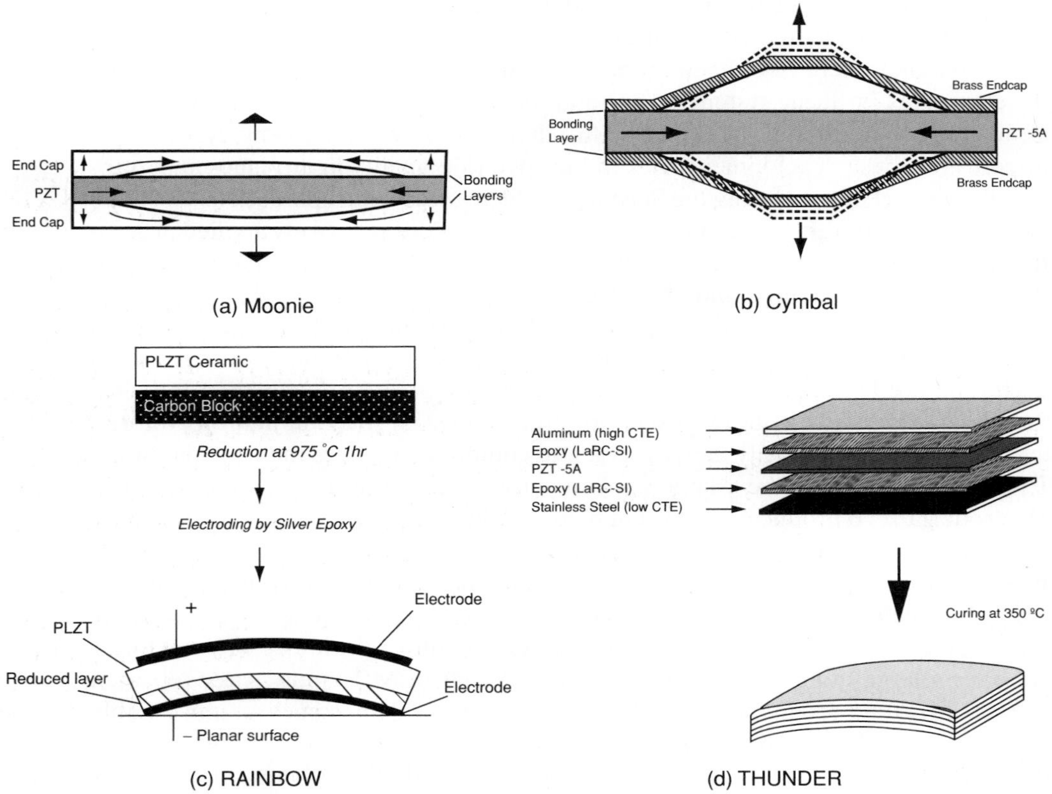

Figure 1.13. Flextensional transducers, from Ref. [91].

radial flexure of a metallic shell (Fig. 1.13). The concept of a flextensional transducer was originated in the early 1920s and applied to an electroacoustic foghorn for ship navigation [94]. The basic principles of the flextensional transducer, especially using magnetostrictive materials, were explained by Hayes in 1936 [95]. Toulis expanded the application of flextensional transducers to underwater acoustic detection in 1966 [96]. Flextensional transducers are classified into five categories based on their shape and mode of operation [94]. Widely used and simplified versions of the flextensional transducer emerged in 1990 in the form of RAINBOW, THUNDER and Moonie actuators. Newnham et al. [97] devised a compact Moonie actuator. A Moonie actuator consists of a piezoceramic or electrostrive ceramic disk sandwiched between two metal endcaps, each having a moon-shaped cavity on its inner surface. The two endcaps serve as displacement amplifiers to transform the lateral displacement of active disk piezoceramic (d_{31} effect) into a large axial displacement normal to the endcaps. On top of this, the "d_{33} effect" of the disk is also superposed. Stroke increases exponentially with an increase in cavity diameter, increases linearly with an increase in cavity depth, and is inversely proportional to the endcap thickness. An improved version of the Moonie actuator was devised as a Cymbal actuator with higher efficiency, more displacement and larger generative force [98]. The Moonie and Cymbal have been used as hydrophones, transceivers, and actuators. Another flextensional design, referred to as the RAINBOW actuator, is constructed by bonding a piezoceramic layer and a chemically reduced layer [99]. The chemically reduced layer is formed using an oxidizing atmosphere at an elevated temperature; it loses its piezoelectric property, and it acts like the shim of a unimorph. They are also categorized as monomorph actuators. It is a pre-stressed, monolithic, axial-mode bender and, because of its dome or saddle-shaped configuration, it is able to produce more displacement and a moderate block force. The RAINBOW actuator is dome-shaped (circular) with the piezoelectric layer on the convex side. Although this actuator can produce large axial displacement (normal to dome surface), it has structural integrity problems under cyclic loading. Another flextensional actuator is the unimorph-type THUNDER actuator, initially developed by the National Aeronautics and Space Administration (NASA) (Langley) [100, 101]. THUNDER is a curved device composed of three layers: a metallic layer (typically aluminum) at the top, bonded to a pre-stressed piezoceramic layer using high performance epoxy (LaRC-SI), and a metallic layer (typically steel) at the bottom surface. Because of the difference in the coefficients of thermal expansion and Young's modulus between materials, the composite actuator deforms to a shallow dome shape during the cooling process. THUNDER is expected to be an improved version of RAINBOW with 10–25% improved performance. To achieve a positive longitudinal displacement, the applied field is in the opposite direction to the polarization in the RAINBOWs and THUNDERs but in the same direction as polarization in the Moonie and Cymbal designs. All of these flextensional actuators provide moderate generative force and displacement values, and their actuation capabilities lie between those of multilayer stacks and bimorph actuators.

1.9 Smart Structures Programs

Applications of smart structures cover a wide range of areas that include aerospace systems (spacecraft, airplanes, helicopters, and jet engines), civil structures (buildings and bridges), machine tools, pipelines, automotives, marine systems (ships and submarines), and medical devices. During the 1990s, there were focused sponsored activities in the United States, Europe, and Asia to foster smart structures activities in the respective regions. In the United States, the basic research activities were carried out through Department of Defense (DoD) funding agencies such as the Army Research Office (ARO), the Office of Naval Research (ONR), and the Air Force Office of Scientific Research (AFOSR), whereas applications-oriented research activities were carried out by the Defense Advanced Research Project Agency (DARPA). Most of the early research programs in smart structures were initiated by ARO and supplemented by DARPA. The following were the early major programs focused on smart structures.

URI in Smart Structures: ARO initiated multidisciplinary research programs in smart structures under URI (the University Research Initiative) in 1992. These were five-year (1992–1997) programs, and three teams, headed respectively by the University of Maryland, the Virginia Polytechnic Institute and State University, and Rensselaer Polytechnic Institute, were selected to foster basic smart structures technology. This truly seeded smart structures activities in the United States.

SPICES: A two-year program (1993–1995) called the Synthesis and Processing of Intelligent Cost Effective Structures (SPICES) was sponsored by the Advanced Research Project Agency (ARPA) and was led by McDonnell Douglas (East) [102]. The objective of the consortium, consisting of ten different organizations, was to establish cost-effective design processes using this multidisciplinary technology for each member's respective product lines, involving manufacturing, modeling, actuation, sensing, signal processing, and control. To demonstrate the technology transition, two tasks were carried out. The first task consisted of an active panel to reduce transmission of broadband high-frequency vibration in the range 1–4 kHz by 30 dB. The second task consisted of a pair of active rails designed to isolate low-frequency modal excitation in the range 5–100 Hz. The first task covered three different composite plates containing a combination of piezoelectric actuators, fiber-optic sensors, SMAs, and piezoelectric shunts, and tests were carried out for damping augmentation, frequency shifting, and active vibration control. The second task covered two composite trapezoidal rails containing a combination of piezoelectric vibration control, piezoelectric shunting, SMA positioning, and frequency shifting. The program successfully demonstrated several test configurations.

ASSET: Applications for Smart Structures in Engineering and Technology was set up to exploit the smart structures technologies within the European Union under the IMT (Industrial Materials and Technologies) research program [103]. About fifty organizations (i.e., academia, government research institutes, and industry) participated in this program with the principal objectives of providing a forum and funds for communication, infrastructure, and exchange of information among partners. There was strong representation from the United Kingdom, France, Germany, and Italy.

CHAP: Compact Hybrid Actuator Program (2000–2003): Smart material actuators such as piezoelectric, electrostrictive, and magnetostrictive have a high energy density but suffer from low stroke. They have been successfully integrated into systems that require low stroke and low force such as the fine positioning of optics and sonar array. The goal of this DARPA-sponsored program was to exploit them in devices that require transducers with high power density or high specific power (superior to traditional electromagnetic- and hydraulic-based actuation).

ADAPTRONIK (1998–2002): This program involved twenty-four partners from industry and research institutions and was conducted under the leadership of the German Aerospace Center (DLR) (Fig. 1.14) [104]. The objective of the program was to develop new self-adapting smart structures with integrated piezoelectric fibers and patches and control logic for active vibration and noise control shape deformation and stabilization micropositioning, and ultrasonic sensing for various industrial applications including aerospace, automotive, rail vehicles, and medical and machine tools.

CLAS: Conformal Load-Bearing Antenna Structures: These involve concurrent consideration of structural and antenna issues, such that a load-carrying structural panel also carries antenna elements and is placed at an appropriate location for superior performance. For this study, a fuselage panel of F/A-18 was selected. It was shown that a large complex RF antenna panel could sustain severe structural loads without loss of avionics performance. This study was carried out by Northrop Grumman under the sponsorship of the Air Force Research Laboratory's (AFRL) Smart Skins Structures Technology Demonstrator (S^3TD) program [105]. Several issues were identified, including airframe panel location, airframe configuration, EMI/lightning, repairability, and risk.

Smart structures research activities in Japan, which started at the same time as in the United States in the early 1980s, initially focused mostly on deployable space structures such as variable-geometry truss (VGT) structures and hingeless masts. A large number of space-related adaptive structures activities were carried out at the Institute of Space and Astronautical Science (ISAS), and these were reported

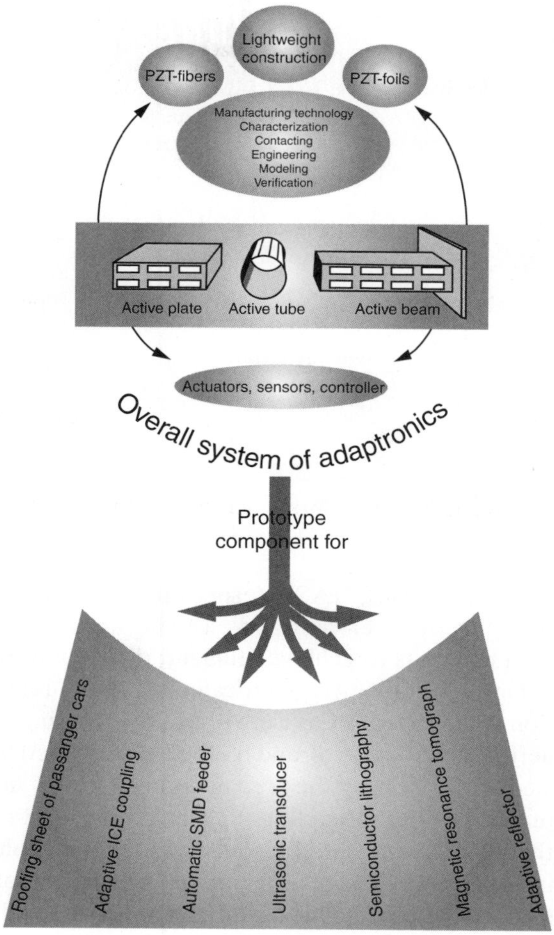

Figure 1.14. Scientific and industrial implementation of ADAPTRONIK program, from Ref. [104].

by Miura [106] for a period from 1984 to 1990. Utku and Wada [107] provided an overview of early smart structures activities in Japan stretching to 1991–1992. Matsuzaki [108] presented smart structures research and development activities in Japan for the following period between 1992 and 1996. During this period, the breadth and depth of research activities dramatically increased. In 1996, MITI (the Ministry of International Trade and Industry) funded a major eight-year national project of Smart Structural Systems (SSS) involving fifty members from national research institutes, universities, and industry (i.e., aerospace, automobile, machine tools, construction, steel, and materials companies). MITI adopted a five-year university-based international R&D program on Smart Materials and Structural Systems (SMSS) in 1998 [109]. Four teams were selected to carry out this program: (1) Health Monitoring Group centered at University of Tokyo, (2) Smart Manufacturing Group at Osaka City University, (3) Active/Adaptive Structures Group at Nagoya University, and (4) Actuator Materials Group at Tohoku University. As an example, the Active/Adaptive Structures program was actively coordinated between Nagoya University and Daimler-Chrysler in Munich (Germany). The goal of this research was to examine passive and active vibration control of beams using surface-bonded/embedded SMA wires/films.

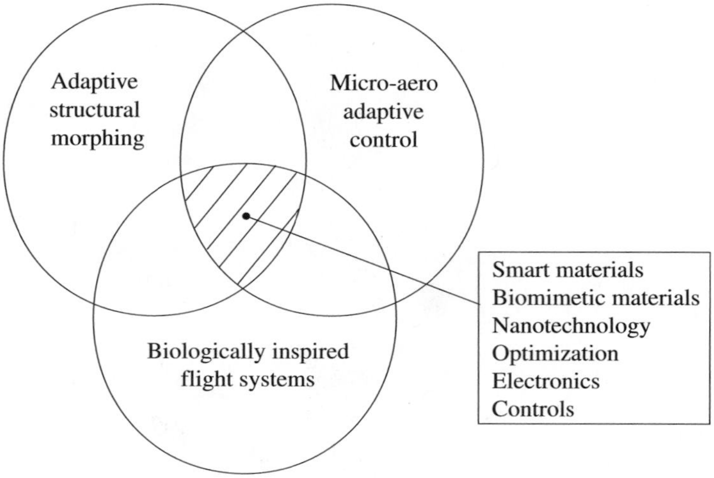

Figure 1.15. NASA Morphing Project, adapted from McGowan et al. [110].

In the 1990s, NASA (especially Langley) initiated a morphing project with the objective of developing and assessing advanced technologies and integrated component concepts to achieve efficient, multipoint mission adaptability in air and space vehicles. Morphing is generically defined as a significant shape change or transfiguration. Three focus areas were pursued, including adaptive structural morphing, micro-aero adaptive control, and biologically inspired flight systems (Fig. 1.15). These areas were supported by the core enabling areas of smart nano- and biologically-inspired materials, multidisciplinary optimization, controls, and electronics [110]. Some of the major barriers to the advancement of morphing were identified. These included insufficient authority and compactness of actuators, inadequate robustness of sensors, and insufficient understanding of associated phenomena. Technology roadmaps were prepared to address technical challenges such as actuators, design tools, control approaches, electronics, and integrated hardware products. Smart materials were viewed as the foundation of the morphing project; focused research activities were undertaken toward the application of smart adaptive materials and structures in aerospace systems.

For example, vibration control of a flexible structure can be carried out using one of three approaches: namely, passive, active, and semi-active control. Applications include automobiles (in the chassis from the engine and tires), helicopters (in the airframe, rotor-induced), aircraft (in the airframe, engine-induced and due to gust), and ships (in the cabin, induced by marine engine and waves). In a passive approach, dissipation mechanisms such as viscous dampers, frictional dampers, and composite damping are introduced through off-line design techniques and remain invariant with the operating environment. Because this is a fixed-design approach, the damping will not be optimal when the operating conditions change. In an active vibration control system, force inputs from actuators are used to suppress vibration based on online measurements from sensors. The controllers provide input signals to actuators to minimize a performance function, such as a weighted sum of vibration amplitudes, at selected stations. The advantage of an active system over a passive system is that it can adapt to system changes and is expected to be more effective in controlling vibration. However, the controller is modeled on the basis of an approximate model

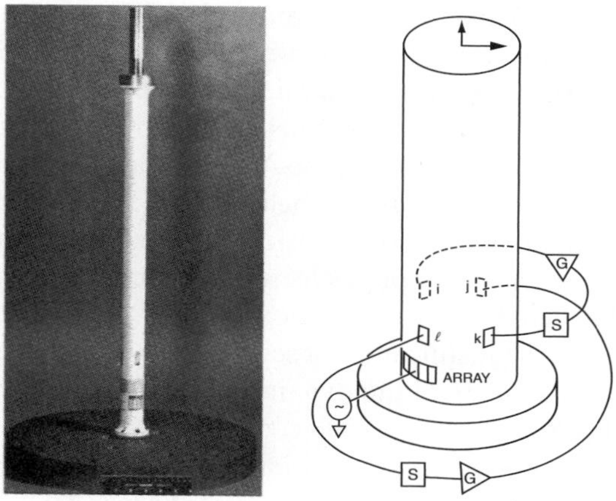

(a) (b)

Figure 1.16. (a) End-loaded cylindrical fiberglass mast on shake plate with piezoelectric ceramic strain transducers attached. (b) Schematic of actuators, sensors, and feedback loop, from Ref. [111].

as a discretized representation consisting of a few key degrees of freedom. Hence, it may sometimes result in undesirable spillover problems in which the structure could get excited due to the interaction between the controlled and uncontrolled modes (i.e., residual or unmodeled modes). Additionally, there can be uncertainties in structural parameters and external disturbances, which can further complicate vibration control. In some problems, there is a merit of combining the two approaches of active feedback system and adaptable energy dissipation. In this approach, damping and stiffness characteristics are varied according to the control commands. Such an approach is expected to have fewer spillover problems.

1.9.1 Space Systems

Many of the early developments in smart structures technology were driven by space applications. One of the key applications in aerospace systems has been active vibration control, especially of large, lightweight truss structures. Because of their high specific actuation energy, compactness, and moderate field requirements, smart material actuators appeared appropriate for space applications. Major drawbacks were the lack of a database of material characteristics, inadequate understanding of material behavior and modeling, structural integrity and reliability issues, and insufficient stroke of actuators. Pioneering work on the modeling of smart structures was carried out by Forward [111], Bailey and Hubbard [112], and Crawley and de Luis [113]. The primary goal of these studies was to sense and control dynamic strains caused by structural vibrations using piezoelectric devices. This approach was sometimes referred to as “electronic damping control.”

Forward [111] carried out an experimental investigation of actively controlling the damping of two closely spaced bending modes in an end-supported cylindrical mast using four pairs of co-located piezoceramic plate elements (Fig. 1.16). Four piezoceramics were used as sensors while an other four acted as actuators. Despite the proximity of modal frequencies (33.85 and 34.12 Hz), a decrease of more than 30 dB in the peak vibration amplitude was demonstrated. Bailey and Hubbard [112] implemented an electronic damping control in the first bending mode of a cantilevered beam using distributed piezoelectric polymer film (PVF_2). Through an application of the distributed-parameter control theory, a significant augmentation of damping from the baseline value (inherent material damping) was demonstrated experimentally (double for large vibration amplitude and forty times for small amplitude). Also, a consistent beam model with induced-strain actuation was formulated.

Crawley and de Luis [113] formulated a systematic model of a beam with induced-strain actuation with both surface-bonded and embedded segmented-piezoelectric actuators including the shear lag effect of the finite thickness bond layer. Both bending and extension of the substructure was considered. The optimal span-wise location of piezoelectric actuators to minimize response of a selected mode was predicted to occur at regions of high average strains, away from areas of zero strains. Also, the justification of segmented actuation over continuous–over-the-length actuation to control the dynamic response of the flexible structure was pointed out. For the selection of piezoelectric material to achieve a high effectiveness in actuation, the important factors were identified as maximum free strain, high modulus of elasticity, and large piezoelectric coefficient (d_{31}). To validate the analytical model, three cantilevered beams were built: an aluminum beam with four pairs (i.e., eight elements) of surface bonded actuators (G-1195), a glass/epoxy beam (nonconducting) with two pairs of embedded actuators, and a graphite/epoxy beam with one pair of embedded actuators (insulated with Kapton film). The predicted resonance response of beams for the first two modes compared well with test data. Furthermore, even though the embedded actuators reduced the ultimate strength of the laminate by 20%, there was very little effect on the global stiffness of the beam. Hanagud et al. [114] and Baz and Poh [115] developed numerical simulations for a cantilevered elastic beam with surface-bonded piezoelectric actuators and demonstrated the effectiveness of closed-loop adaptive systems to actively control structural vibrations (vibration amplitudes).

The shunted piezoelectric damping concept was initially introduced by Forward [116] and later expanded by Hagood and Crawley [117] to add damping to a specific mode of vibration. This concept is also referred to as "passive electronic damping." It is implemented by suitably matching the mechanical and electrical impedances and tuning the electrical circuit to the desired frequency. If damping is to be introduced in more than one mode, then several shunted piezoelectric damping circuits are needed, respectively, one for each mode.

In addition to active control of the vibration of large space structures, applications for space systems include adaptive geometric truss configuration, precision pointing of telescopes and mirrors, structural-integrity monitoring and condition-based maintenance. Large space structures consist of multi-member lightweight flexible trusses as substructures for the support of precision equipment to carry out various space-related missions. The dimensions of many of these space structures, such as the space station, may range up to 100 meters, as shown in Fig. 1.17 [118]. These structures consist of a large number of closely spaced low-frequency natural modes that continually change with changing payload. To achieve a high degree of performance due to changing external environments (e.g., thermal gradients) and internal disturbances (e.g., loose joints), it appeared attractive to incorporate compact smart material actuators at discrete locations to adaptively control the geometry, stiffness, and damping of truss members. The goal was to control both the rigid body and the elastic deformations of large precision space structures using an array of distributed compact actuators, sensors, and processor networks, in conjunction with feedback control strategies. In fact, the application of compact lightweight actuators in large space structures pioneered the area of smart structures in the late 1980s and early 1990s [119, 120, 121].

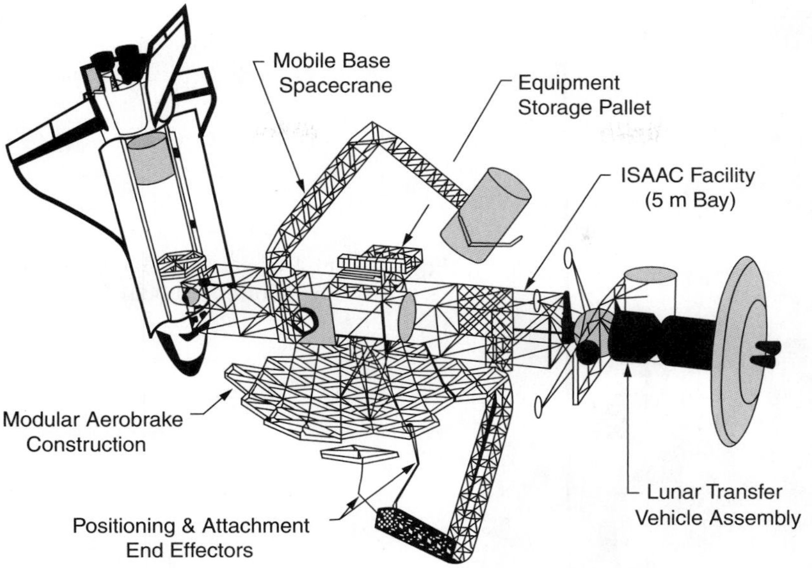

Figure 1.17. Example of a system assembled in space, from Ref. [118].

Using a distributed network of lightweight, compact SMAs and sensors, the structural efficiency and performance, such as the pointing accuracy, of a large space antenna could be enhanced. The application of a large number of piezoelectric actuators to actively control a space structure to meet the precision requirements of mirror pointing was first reported in 1974 [122]. Sato et al. [123] demonstrated a 1-D compact deformable mirror control using multilayered polyvinylidene fluoride (PVDF) films. Early efforts were focused on the development of basic technology to actively control the damping and stiffness of space structures using piezoelectric-type materials and other actuators [124, 125, 126, 127, 128]. At the Jet Propulsion Laboratory, two active space-truss model structures were built, incorporating active load-carrying truss members using piezoceramic actuators.

Fanson et al. [129] successfully carried out precision control of a truss in the laboratory. The truss contains two active piezoelectric struts, and each strut has a collocated displacement and force feedback. The objective of the control strategy was rejection of disturbances in a precision interferometer due to onboard machinery in a spacecraft. Using the two strut-closed loop responses, both the first and the second structural-modes response was reduced by 40 dB.

Deployable space structures such as space antennas often undergo large controlled kinematic changes from an initially compact configuration to a final geometrically expanded configuration [130]. These structures normally have the capability to adjust the length of individual truss members, to achieve the final compatible configuration, and/or to adjust axial preloads in individual truss members to alleviate undesirable vibratory motions and stresses. The variable geometry (VG) adaptive truss consists of repetition of an octahedral truss module in which the lengths of some of the truss members can be continuously adjusted using actuators. For the structure to be adaptive, the necessary condition for a statically determinate truss is $M - 3J + 6 = 0$, where M is the number of truss members and J is the number of joints. This provides a tool for deducing the topological construction of an adaptive truss.

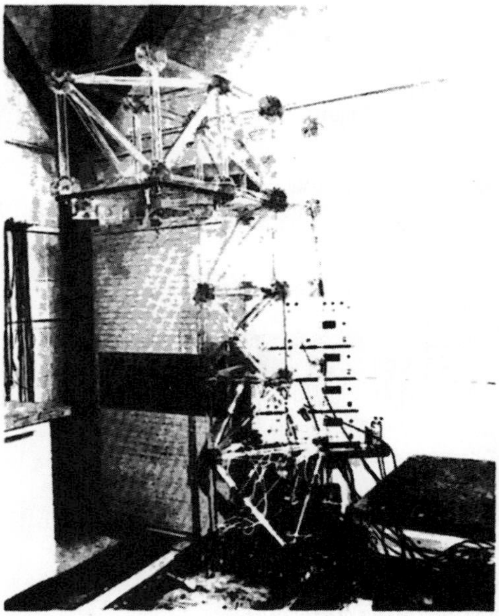

Figure 1.18. The Sandia Gamma Truss Controlled Structure Testbed, from Ref. [135].

Pyrotechnic shock-release mechanisms are quite prevalent in the design of spacecraft. Until 1984, about 15% of space missions experienced some type of shock failures that resulted in the abort of half of the missions. SMA appears to be a natural replacement for pyrotechnics, which can be used to develop compact, gradual release mechanisms for satellites [131]. SMA-based release devices such as Micro Sep-Nut and QWKNUT were developed for microsatellites. In both devices, the active members were initially deformed (detwinned) and, upon heat activation, their shapes were recovered due to SME. Another application of the SMA wire-actuated stepper motor for the orientation of solar flaps in a spacecraft was successfully demonstrated [131]. An identical effort was carried out by NASA Goddard and Lockheed Martin, called the shape memory alloy thermal tailoring experiment (SMATTE) [132]. It demonstrated that a panel can achieve bistable shape via actuation of an SMA foil attached to only one surface of the panel. This could facilitate tailoring of the shape of spacecraft antennas. Hartl and Lagoudas [132] also identified many more applications of SMAs in spacecraft.

To minimize the possibility of tensile stress in piezoceramic elements, the active members were preloaded with compressive stress. Fanson et al. [133] and Chen [134] successfully demonstrated active shape control and active damping control, respectively, using digital control and analogue control feedback schemes. At Sandia National Laboratory, a space-truss model was built that incorporated surface-bonded piezoceramic actuators in outer truss elements in conjunction with collocated piezoelectric polymer-film sensors. Peterson et al. [135] successfully demonstrated active shape control of a truss (Fig. 1.18). Using a realistic space-erectable truss structure, Salama et al. [136] demonstrated the ability of a limited number of actuator/sensor pairs to achieve the desired shape correction with good accuracy. In a few cases, micron-level nonlinearities were observed in the truss behavior, which could be corrected adaptively by the active members to the desired degree of accuracy. Hom et al. [137] examined an adaptive deformable mirror using distributed

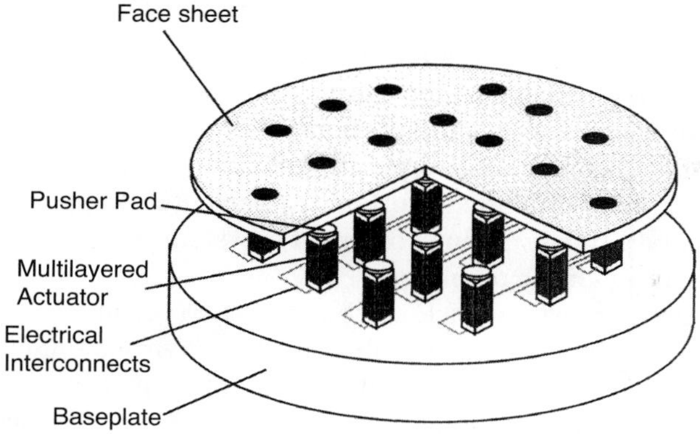

Figure 1.19. Deformable mirror actuated by array of electrostrictive stacks, from Ref. [137].

multilayered electrostrictive actuators. A fully coupled analysis simulating the nonlinear, electromechanical behavior of the actuators and a finite element model of the continuous mirrored face sheet were developed to estimate the system-level performance of the deformable mirror and the commanded voltages required. Each electrostrictive stack consisted of 425 layers of PMN sandwiched between alternating positive and negative electrodes using a co-fired process, and 21 actuators were used to adaptively control the mirrored face sheet shown in Fig. 1.19.

Inflatable space structures such as solar antennas and optical mirrors are subjected to high vibrations caused by changing external environment (e.g., direct sunlight), impacts with space debris, time-varying guiding force, and transient states during inflation. Control of the vibration and shape of space structures is critical to their performance. Because of their extreme flexibility, light weight, and high damping properties, these inflatable structures pose problems to control their vibration. Park et al. [138] used PVDF films as both actuators and sensors to control vibration of an inflated structure and showed promising results. If the actuators and sensors are not placed judiciously, it can lead to a loss of observability and controllability. Jha and Inman [139] carried out a study to find the optimal sizes and placements of piezoelectric actuators and sensors for an inflated torus using a genetic algorithm. Using a cumulative performance index for all of the controlled and observed modes, optimal design solutions to suppress vibration were obtained. Attenuation of vibration was successfully demonstrated using five optimal actuators and five optimal sensors.

Martin and Main [140] used a noncontact electric-gun actuation of a bimorph mirror structure composed of two PVDF layers to induce controlled bending strain. The charge was applied to the PVDF by controlling the potential of a nickel-copper electrode (reflecting surface) on one side and subjecting the opposite side of the thin film mirror to an electron flux. Sufficient actuation authority of the thin film mirror was demonstrated (i.e., deflection of 1/2 cm in a 10 cm long mirror). An electron gun provided actuation capability over a discrete area in a mirror. However, PVDF does not have the capability to sustain a changing temperature environment in space. Andoh et al. [141] examined active shape control of distributed-reflector antennas using a limited number of discrete actuators. To optimize the actuator locations, reference input was represented in terms of eigenfunctions and maximization of controllability was performed through the maximization of singular values of the controllability matrix. A stereo-photogrammetry system was successfully used to measure the steady-state deflection field of a doubly curved reflector antenna prototype. Gaudenzi and Giarda [142] carried out a feasibility study for the vibration control of the SILEX optical payload installed on the European ARTEMIS telecommunication satellite using piezoceramic strain actuators bonded on the surface of the support structure. The goal was to achieve desired pointing accuracy (within a few microradians) in the presence of microvibration disturbances. The authors applied a finite element analysis and showed a significant improvement in system performance using a full-state feedback–linear quadratic regulator.

Acoustic levels inside fairings of a space vehicle can be excessive (i.e., more than 140 dB) during the initial seconds of launch. Such a vibroacoustic environment can damage the payload and, degrade the performance of instrumentation. Typically, acoustic blankets are attached to fairings to minimize the impact of these acoustics. Not only does this result in a weight penalty, but such passive techniques are also less effective at low frequency. The Air Force Research Laboratory (AFRL) initiated a study in the late 1990s to apply active structural-acoustic control to mitigate interior noise more than 0-300 Hz bandwidth in fairings. Lane et al. [143, 144] developed fully coupled structural-acoustic models of a composite fairing using conventional piezoelectric actuators and single-crystal piezoelectric actuators. Simulation results using full-state feedback control showed a reduction of about 10 dB in the internal-acoustics response for the complete frequency range.

Niezrecki and Cudney [145] carried out a feasibility study to control internal acoustics in a launch vehicle using piezoelectric actuators. To demonstrate the concept, the internal-acoustic response of a closed simply-supported cylinder was investigated with PZT actuation at frequencies between 35 and 400 Hz. The sound-pressure levels at the acoustic resonant frequencies were only mildly reduced. This study showed that PZT actuators do not have the ability to control the payload-fairing internal acoustics below about 400 Hz.

A health-monitoring system to monitor the thermal and structural condition of a satellite with a view to reducing the life-cycle cost and increasing the reliability of the system was investigated using fiber Bragg grating (FBG) sensors. Damage due to thermal stress was successfully detected in a typical satellite structure. For this, three critical technologies were developed that included an embeddable optical fiber connector for composite laminated structure, a FBG sensor system to measure strain and temperature, and a damage detection algorithm.

The AFRL in collaboration with other federal agencies (NASA, DARPA, and Balistic Missile Defense Organization [BMDO]), industry, and academia demonstrated the potential of smart structures technology in three space applications: (1) vibration isolation, suppression, and steering (VISS); (2) space experiment and mid-deck active control experiment (MACE); and (3) satellite ultra-quiet isolation technology experiment (SUITE) [146]. The goal of VISS was to demonstrate the vibration isolation of an optical system from broad-base disturbances by a minimum of 20 dB over 1–200 Hz for a space telescope. It was the first successful space-related demonstration of active-vibration isolation using a hexapod Stewart platform. The SUITE consisted of a hexapod assembly of six hybrid active/passive struts involving piezostacks to provide vibration isolation as well as six degrees of controlled motion of the platform.

The objective of the MACE program was to demonstrate adaptive structural control in a microgravity space environment. The follow-up to the VISS program was the miniature vibration isolation system (MVIS) for space applications incorporating piezoelectric actuators, hexapod mounts, microelectronics, and micro electro mechanical system (MEMS) sensors.

Adaptive Reflector: Orbital structures are subject to harsh temperature cycles in the range of ±150°C. For a lightweight reflector, holding the fixed geometry contour is an enormously challenging task. Increasing stiffness results in weight penalty. Monner and Breitbach [147] described the development of an active satellite structure for shape control with piezoceramic actuators and sensors.

Adaptive Satellite Mirror: High-resolution interferometric optical and infrared astronomical instruments for space missions require optical path length accuracy on the order of a few nanometers in mirrors of structural dimension of several meters. This puts extreme requirements on structural deformations under static and dynamic loads. Durr et al. [148, 149] evaluated two different structural designs in an adaptive mirror, one using carbon-fiber-reinforced plastics and another using carbon-fiber-reinforced ceramics, in conjunction with piezoceramic actuators for shape control. The second design was unable to meet the requirements satisfactorily. This study was a part of the German ADAPTRONIK effort.

These are some sample applications to space systems. Many of these studies have demonstrated the potential of smart structures technology in space systems to enhance performance, improve payload, increase structural integrity, and increase mission adaptability. Before this technology gains wide acceptance in space systems, it is important to demonstrate system reliability and robustness.

1.9.2 Fixed-Wing Aircraft

For fixed-wing aircraft applications include active vibration control, gust alleviation, wing-flutter stability augmentation, increasing static divergence, increasing panel flutter stability, stabilizing tail buffeting, interior noise control, shape control for performance enhancement, and structural-integrity monitoring. Applications of smart structures technology in fixed-wing aircraft are envisaged to help increase the payload or, alternately, enhance range and endurance; allow condition-based maintenance encompassing damage detection; mitigation and repair thereby increasing system reliability; minimize downtime and improve operating cost; enhance passenger and crew comfort by reducing cabin vibration and noise; increase the structural life of components by reducing vibratory loads and response due to buffet, panel flutter, and gust response; increase precision pointing and accuracy of weapons (i.e., airborne missiles) in military aircraft; increase performance, maneuverability, and flight envelope by delaying stall and compressibility effects through active shape and twist control; and increase speed by stabilizing wing flutter. Before these smart concepts can gain wide acceptance in production aircraft, there are many issues that need to be addressed, which include inadequate materials and devices characterization and documentation; lack of designer familiarity with this technology; inadequate understanding of materials under combined electrical, thermal, mechanical, and aerodynamic loading; and insufficient information on the system-weight and cost penalties, reliability, and serviceability of such devices. Other concerns are power requirements and conditioning, the robustness of adaptive control strategies, and data acquisition and processing. There is no doubt that airplanes with fixed-geometry wings result in suboptimal response for a wide range of flight conditions. To overcome this problem, a variable-geometry adaptive wing that includes wing warping, camber shaping, leading- and trailing-edge shape control, variable sweep, and spanwise twist distribution may be needed for different flight modes. For example, to obtain a high lift coefficient in low-speed flight, an airfoil shape with large camber, leading-edge radius and thickness are needed. Conversely, in high-speed flight, low camber, leading-edge radius, and thickness are needed to reduce drag. At flow conditions with high angle-of-attack, it may be desirable to stall the inboard sections of the wing as compared to the outboard sections, where the ailerons are located. Induced washout twist (i.e., lower angles of attack at the tip) using smart actuators may be beneficial. Spanwise redistribution of lift will also be beneficial from a structural point of view (i.e., lower root-bending moments). Wing-shape control can also be used to reduce drag. Barbarino et al. [150] used SMA actuators to induce a bump on an airfoil profile to reduce transonic drag. This concept can help to maximize the aerodynamic efficiency in different flight conditions. Note that the first application of smart materials in fixed-wing aircraft took place in 1971, when the hydraulic tubing coupling used in the F-14 was replaced with SMA coupling [151].

Early applications in fixed-wing aircraft include active control of wing flutter, increasing static divergence, panel flutter control, and interior structure-borne noise control. Panel flutter is the dynamic aeroelastic instability of a thin-skin panel of a flight vehicle exposed to the supersonic flow on one of its surfaces, and it results in limit-cycle oscillations. A small amount of damping can often delay the onset of this instability. Hajela and Glowasky [152] conducted a parametric study to control panel flutter using piezoelectric sheet actuators in conjunction with an optimization technique. They determined the best panel configuration and actuator thickness for both structural weight reduction and maximum flutter speed. Using numerical simulation, Frampton et al. [153] examined active control of panel flutter with surface-bonded piezoelectric sheets that are used both as actuators and sensing elements. It was shown that with closed-loop control, a significant increase of flutter-dynamic pressure with piezoelectric actuation is possible. Through analysis, Scott and Weisshaar [154] demonstrated the control of panel flutter actively, using embedded piezoelectric sheet actuators (i.e., PZT and PVDF), as well as passively using shape memory alloy actuators (i.e., Nitinol) by stiffness variation and recovery forces. Suleman and Venkayya [155] carried out flutter analysis of composite panels with piezoelectric actuators/sensors. An active control of panel flutter using a smart material patch can be a "retrofit solution" on an operational aircraft.

Two major aeroelastic instabilities of an airplane wing are static divergence and bending-torsion flutter. Using a numerical study, Ehlers and Weisshaar [156] examined static aeroelastic (e.g., lift effectiveness, divergence, and roll effectiveness) control of an airplane wing using embedded sheets of piezoelectric actuators in an idealized laminated-composite wing structure (box beam). The amount of lift change due to induced-strain actuation was found to be small because of constraints due to limited piezoelectric stiffness (low), low electromechanical coupling, and limited maximum applied field. It was pointed out that available active materials fall short of the actuation authority expected from them for active divergence control of a full-scale wing. Lazarus and Crawley [157] examined aeroelastic stability via active strain actuation using PZT patches, both analytically as well as through low-speed

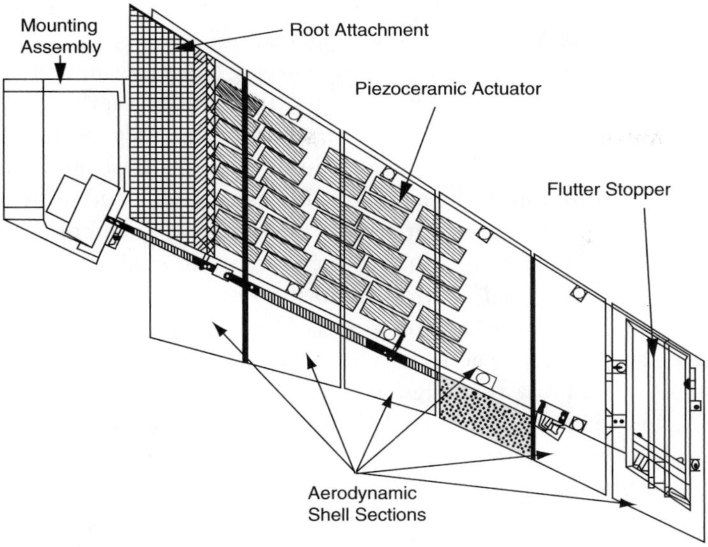

Figure 1.20. Schematic of strain-actuated active aeroelastic wing, from Ref. [159].

wind-tunnel tests on a uniform cantilevered wing. A typical high-performance wing was built out of a graphite epoxy laminate with three banks of piezoceramic actuators distributed over about 70% of its surface. The three tip-displacement measurements were used in a feedback controller implemented in a reduced-order fourteen-state Linear Quadratic Guassian (LQG) controller. The control objective was gust-disturbance alleviation and flutter suppression. Using induced-strain actuation, the flutter speed was increased by 11%. The root mean square (RMS) response, covering a bandwidth of 100 Hz, was reduced by 8 dB.

Nam et al. [158] investigated active flutter suppression of a composite plate wing with segmented piezoelectric sheet actuators bonded on its top and bottom surfaces. The optimization methodology was applied to determine the best size (length, width, and thickness) and placement of piezo actuators for flutter suppression. This numerical study demonstrated a substantial increase in flutter speed (more than 50%) as well as savings in control effort with optimal actuation. Lin, Crawley, and Heeg [159] demonstrated the use of piezoelectric actuation (induced-strain) technology for flutter suppression and gust alleviation on a model wing, shown in Fig. 1.20. Open- and closed-loop tests were carried out in NASA's Transonic Dynamics Tunnel. Significant vibration suppression and load alleviation were demonstrated, reducing the power-spectral density of response of the first mode by an order of magnitude. The flutter dynamic pressure was increased by 12%. The actuation authority of piezoelectric actuators was identified as one of the key barriers to implementation of this technology in full-scale systems.

Suleman et al. [160] carried out wind-tunnel testing on a wing model with adaptive-stressed skin using embedded PZT sheet actuators. There were two ailerons, pivoted about their 30% chord point, which were actuated in phase with each other by two servos located outside of the wing. Thus, using an adaptive skin, the control authority of the ailerons was supplemented. Another problem that has been examined using smart actuation is tail buffeting. It is an aerodynamically forced vibration of the vertical or horizontal tail surfaces caused by impinging of the unsteady shedding of wake from the wings and fuselage components. This problem can become a serious issue at high angle-of-attack, roll, and/or yaw flights.

To achieve optimum performance from a wing under varying flow environments, it is imperative to morph its cross-sectional shape (airfoil profile) based on the flow condition. Airfoils are normally designed for cruise condition and they perform suboptimally at other flight conditions such as takeoff, landing, climb, descent, and other flight maneuvers. The adaptation of the airfoil profile can be achieved in two ways: structural and aerodynamic. In the first approach, the physical airfoil profile is altered, resulting in a reconfigurable wing. Such a technology can increase aerodynamic efficiency, maneuverability, and control authority, but it may result in a weight penalty, additional cost, and structural restraints. A major barrier has been the unavailability of compact large-stroke actuators. Strelec et al. [161] used the two-way effect in SMA wires to develop a reconfigurable wing. The SMA wires were attached to points on the inside of the airfoil (NACA 0012) and the airfoil profile was altered upon heating and cooling, thereby achieving an increase in lift-to-drag ratio at subsonic flow conditions. A wind-tunnel model was built and test results demonstrated the potential of this concept. A 9% increase in lift at a constant angle of attack of 5° was measured in the tunnel. Rossi et al. [162] reduced the drag of a fighter aircraft wing in the transonic regime by altering the airfoil profile using a magnetostrictive adaptive truss for a wing rib. In the second approach, aerodynamic shape control is achieved by a virtual change of shape. The flow is affected using either synthetic jets or circulation control. For morphing of the wing, one of the key challenges is a flexible skin [163]. For an application where camber change is desired, strain requirements are modest, and stretched-elastomers with fiber reinforcement may be adequate. Conversely, large-area morphing applications such as variable span, chord, and sweep require a flexible skin that undergoes large in-plane strain with low actuation force. Simultaneously, the skin should be capable of carrying large out-of-plane aerodynamic loads. One possible solution includes a stretchable elastomeric face sheet with fiber reinforcement supported by a deformable understructure. A small panel size is preferred to limit the unsupported area and maximum out-of-plane deflection. High face-sheet pretension is needed to alleviate wrinkling during morphing, but this also increases the actuation force requirement. The behavior and reliability of the elastomeric face sheet in an operational environment is a major issue. Shape memory polymer flexible skins that can be soft during morphing and stiff while bearing loads are attractive [164]. Other material considerations such as toughness; resistance to erosion, fatigue, and weather; and repeatable recovery of high strain are also important for practical applications.

Integrated vehicle health management (IVHM) is now recognized by commercial and military aircraft users of both new and aging fleets as a way of reducing a vehicle's total life-cycle cost despite IVHM's higher initial acquisition cost. IVHM not only increases flight safety, system reliability, and efficiency, but also results in savings in operational and support (O&S) cost, decreasing the cost and time of inspection and maintenance, and extending the life of an aging aircraft. A key element of IVHM is structural health monitoring (SHM) of the system involving numerous sensors at appropriate locations, data processing and interpretation techniques, and automated filteration of false signals. The prognostics methodology monitors the usage and damage, enabling condition-based inspection and maintenance. Even though IVHM originated with military aircraft, it was subsequently implemented in commercial aircraft. Smart material sensors can play a major role in

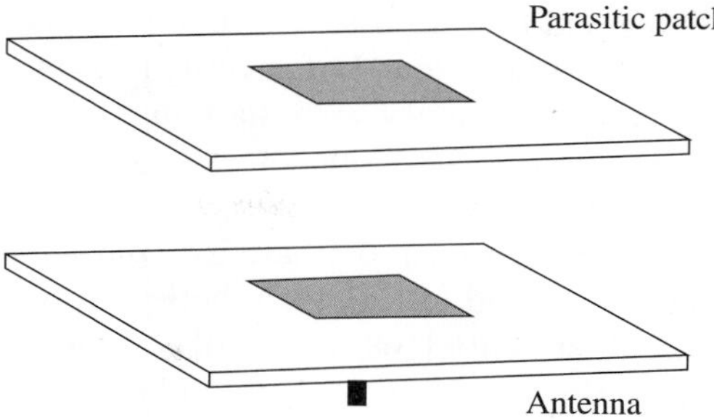

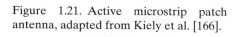

Figure 1.21. Active microstrip patch antenna, adapted from Kiely et al. [166].

IVHM implementation. For example, Boeing used smart-patch technology consisting of an array of piezoelectric sensors and actuators to assess bond-line integrity, especially in areas that are inaccessible to conventional non-destructive inspection (NDI) methods [165]. Using piezoelectric actuators, controlled diagnostic signals are generated and collected using built-in sensors and analyzed using signal-processing techniques to detect faults.

Multifunctional structures can undertake multiple roles in addition to load-carrying capability, such as radio frequency antennas, signal processors, and sensors. A military aircraft carries a large number of antennas: the F-18 has sixty-six antenna apertures located at thirty-seven sites covering a frequency band from 200 MHz to 18 GHz. These antenna apertures, located at myriad sites, can degrade structural integrity. Also, these antennas require local structural reinforcements, resulting in weight and drag penalties as well as increased maintenance cost and vehicle signature. Conformal load-carrying antennas can provide mission flexibility, reduced drag, lower weight penalty, and low observability. Candidate sites for smart antennas (skin panels) in aircraft are the dorsal deck, centerline, weapon bay door, front landing-gear bay door, outer wing, radome, forward wing, wing root lower surface, trailing-edge flaps, and vertical tail. Typically, the mechanical vibration spectrum has a bandwidth of 10–2000 Hz and amplitude could be as large as a few millimeters. Conversely, the transmission frequencies of antennas are normally above 3 GHz and deformations less than 1 millimeter. High reliability, structural integrity, and survivability of embedded devices are key factors for successful operation of these antennas. Kiely et al. [166] carried out the design, modeling, and testing of adaptive materials based smart electromagnetic antennas. Four proof-of-concept designs were considered, which included an active aperture PVDF antenna, an active aperture PZT antenna, an active micro-strip patch antenna, and an electrically active conformal patch antenna (Fig. 1.21). The design objective of these antennas was to achieve the multifunctional capability of variable scanning, variable focusing, and variable resonance frequency. High-precision reflector surfaces for radio frequency antennas and the resulting radiation patterns were systematically evaluated.

To exploit unsteady aerodynamics, the key is the development and integration of innovative, compact, and lightweight actuators and sensors, as well as novel control strategies. Many dynamic actuation concepts for application to active flow control have been investigated, which include synthetic jets, pulsed jets, active control surfaces, and plasma jets. The synthetic jet consists of vibrating a thin member in a cavity in an airfoil, pumping fluid in and out of the orifice (jet-like flow), resulting in a zero–net-mass flux. Synthetic jets are found to be effective to control boundary-layer mixing and flow separation [167]. Compact SMAs are key to the successful operation of synthetic jets. Major design factors are diaphragm motion, tuning of cavity resonance frequency with diaphragm frequency, cavity size and shape, and membrane stiffness. Synthetic jets have the potential of active control of separation and turbulence in boundary layers, mixing enhancement, and thrust vectoring. Piezoelectric disks used as oscillating diaphragms offer the potential for lightweight, high-bandwidth, and efficient synthetic jets to cause volumetric displacement within a fluid-filled cavity [168, 169, 170]. As the piezoelectric actuates, it alternately draws in and blows out the ambient fluid in the cavity. Mane et al. [171] used piezoelectric composite diaphragms (i.e., bimorph and THUNDER) to develop synthetic jets. This provides structural stiffness and durability to the system. They systematically examined the effect of cavity size (height and orifice diameter) on synthetic jet peak velocities, numerically and experimentally. With bimorph actuation, only the orifice diameter was found to be important. Schaeffler et al. [172] developed several synthetic jet designs with the goal of integrating them into an airfoil for wind-tunnel testing. Tests have shown significant potential for these devices.

"Active flow control (AFC)," also known as "adaptive flow control" or "micro-adaptive flow control," in an aircraft appears to show promise of enhancing performance and capability and reducing life-cycle cost through the application of smart structures technology [173, 174]. Active flow control introduces small amounts of energy locally in an adaptive manner to attempt nonlocal changes in the flow field with possible performance gains. It is an interdisciplinary field involving fluid dynamics, material science, structural mechanics and control theory. Targeted goals could include noise control, flutter and gust alleviation, performance enhancement, increased maneuverability, damage tolerance, increased safety, and improved reliability. However, active-flow control devices add complexity to the design and increase life-cycle cost (acquisition and operational cost). The goal should be to develop active flow control devices that are easy to build, compact, inexpensive, and require low power to operate. The basic idea is to cause local flow control and thereby result in delay in flow separation, viscous drag reduction, and control of shock/boundary-layer interaction, as well as control of transition to turbulent state and noise-generating shear layer.

Compact smart actuation is being investigated in the development of microsystems, especially micro-aerial vehicles. Toward the development of an insect mimicking flapping-wing system, Nguyen et al. [175] used piezoceramic unimorph actuators.

Since the 1990s, DARPA has played a leading role in the development of smart structures technology, especially toward the application of smart structures technology in fixed-wing and rotary-wing systems [176]. DARPA categorized these programs as Smart Materials Demonstration Programs. Other groups such as DLR (Germany), the AFRL, and NASA also had their own major programs in aeronautical systems.

Smart Wing Program: This program, consisting of two phases, was initiated by DARPA in 1995 to develop smart material–based control surfaces that could provide improved aerodynamic and aeroelastic performance [177] (Fig. 1.22). The work was carried out by a team led by Northrop Grumman. In the first phase (1995–99), two issues were investigated: active wing twist control using SMA torque tubes and control of hingeless smooth contoured trailing-edge surfaces using SMA wires. To demonstrate the concept, two 16% scale models

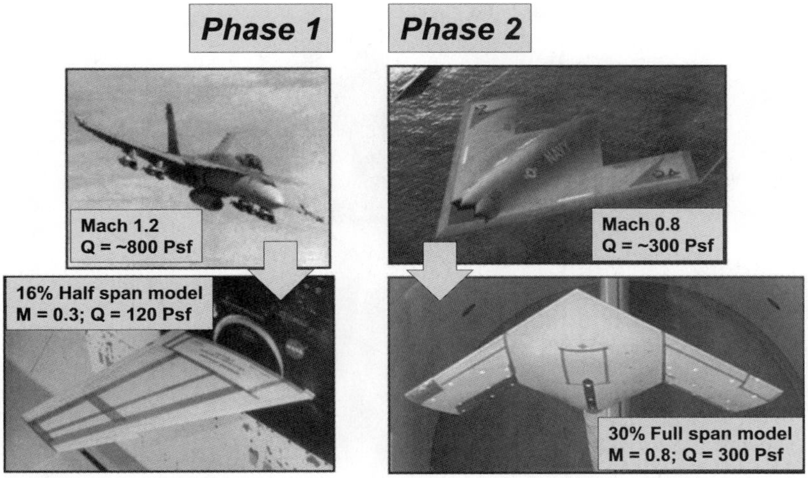

Figure 1.22. Wind-tunnel model scales and test parameters for Phases 1 and 2, from Ref. [177].

of a typical fighter wing (wing semi-span 3.1 ft, wing area 5.92 ft^2) were fabricated and tested in the NASA Langley Transonic Dynamics Tunnel (TDT) (16 ft diameter test section). A tip twist from 1.4° to 3.6° was achieved in the wind-tunnel tests, with a maximum increase of lift of 11.5%. Tests demonstrated the effectiveness of the contoured, hingeless control surfaces with embedded actuation in comparison to the conventional hinged design. A major limitation of this methodology was a low bandwidth (less than 1 Hz) with SMA actuation.

In the second phase (1997–2001), a 30% scale full-span model representative of an unmanned combat air vehicle (UCAV) (wing span 9.2 ft, wing area 34.3 ft^2, and maximum wing thickness 3.4 in) was built for an active smart control surface on the right wing and a conventional control surface on the left wing (Fig. 1.23). The hingeless control system concept and the distributed deflections and vertical forces were created by a transmission technique called "eccentuation." To reduce the actuation requirement, the flexible skin-flexcore trailing-edge surface was designed, which consisted of three parts: an elastomeric outer skin, a flexible honeycomb, and a central fiberglass-leaf spring. In the first test in the NASA Langley Transonic Dynamics Tunnel, SMA-actuated hingeless, smoothly contoured, flexible leading, and trailing-edge control surfaces were evaluated for Mach numbers varying from 0.3 to 0.8. This test demonstrated the effectiveness of a leading-edge control surface to compensate for the loss of aileron effectiveness at high dynamic pressures.

In the second test, the desired combination of bending and twisting deformation of the trailing-edge control surface was achieved by two eccentuators (Fig. 1.24) actuated, respectively, with piezoelectric-based ultrasonic motors (SPL-801 by Sensei Corp.). This test demonstrated spanwise and chordwise shape control with smart trailing-edge control surface at deflection rates as high as 80° per second and a maximum deflection of 20°. This Mach-scale aeroelastic model was quite heavily instrumented, consisting of 160 pressure transducers, 16 fiber-optic strain gages, 10 accelerometers, and 6 inclinometers. Also, this phase demonstrated performance enhancements

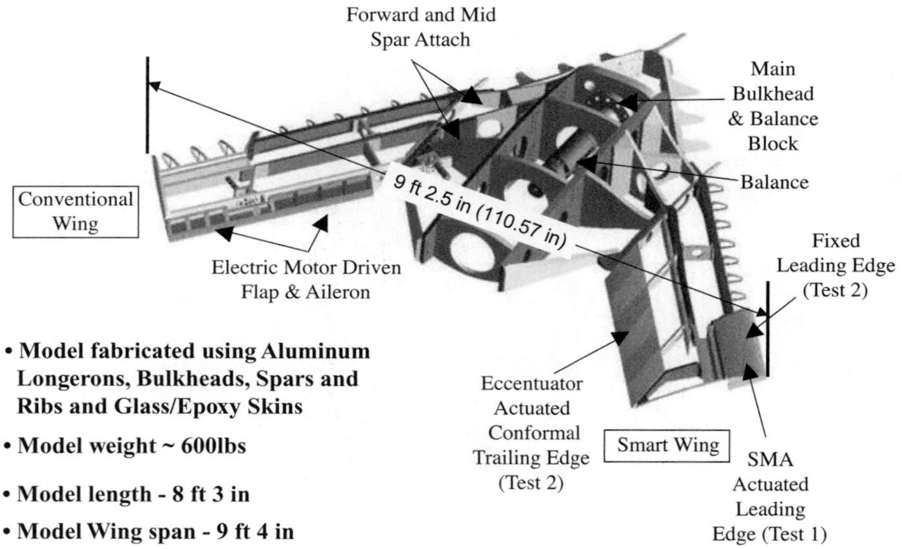

Figure 1.23. Phase 2 wind-tunnel model structural layout, from Ref. [177].

in terms of increased rolling and pitching moments and lower control surface deflections. Several key issues were identified that needed to be addressed before the smart-wing concept could be implemented in an operational vehicle. These included power supplies, cost, the fatigue life of the piezoelectric motor (friction-based), system reliability, and overall system integration. More details about this program including design, fabrication, wind-tunnel testing, performance evaluation, and power requirements can be found in the literature [178, 179, 180, 181].

AAW: The Active Aeroelastic Wing program was envisioned to be a multidisciplinary approach that could integrate wing flexibility, a distributed actuation system, and aeroelasic couplings in order to simultaneously control the camber and twist of a thin high–aspect ratio wing. The program was led by Boeing Phantom Works and was supported by AFRL and NASA Dryden (1984–1993). The goal was to enhance maneuver performance and minimize induced drag by controlling the lift distribution of a lightweight flexible wing. Also, it could lead to a more efficient structural design. The technology was flight demonstrated on an F/A-18 wing using integrated actuators and sensors.

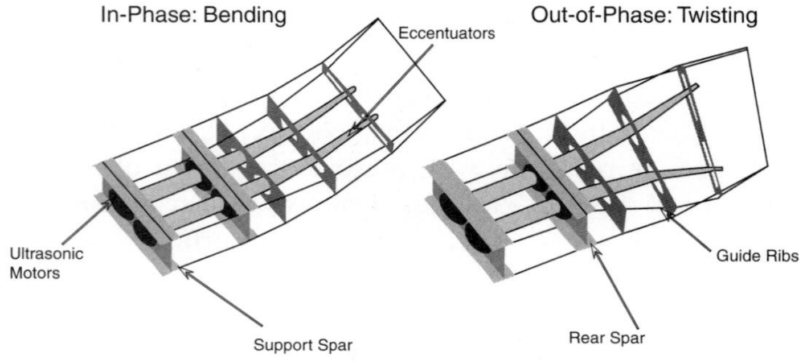

Figure 1.24. Trailing-edge segment design incorporating two eccentuators, from Ref. [177].

Figure 1.25. 1/6-scale F/A-18 model mounted in the transonic dynamics tunnel, from Ref. [185].

The wings were modified with additional actuators to differentially control the split leading-edge flaps, and they had thinner skins at the outer wing panels to twist the wing up to five degrees. In this case, in addition to the traditional trailing-edge ailerons, the leading- and trailing-edge flaps are used to provide aerodynamic forces to twist or warp the wing. In two phases of the project, eighty flights were successfully flown, covering a number of subsonic and supersonic flight conditions.

ENABLE (Evaluation of New Actuators in Buffet Loads Environment): This program was led by the U.S. Air Force (USAF) with participation from Boeing, NASA, Australia, and Canada. High-performance aircraft at high angle of attack emanate strong vortices from wing leading-edge extensions, which often burst and immerse the vertical tail and fin into their wakes. These leading-edge vortices (LEX) can cause enormous buffet loads (unsteady differential pressures) on the tail, which can cause premature fatigue to the airframe structure and also can increase maintenance costs (inspection and repair costs). Extensive flight and wind-tunnel tests were carried out to quantify buffet loads on the vertical tail of the high-performance aircraft, the F-15, F/A-18, and F-22 [182, 183, 184]. Sometimes these vortices, in combination with other vortices emanating from the engine inlets and airframe, can cause even more severe dynamic loads on tail surfaces. To alleviate buffet loads or the structural response (i.e., buffeting) of the fins, a variety of techniques were attempted. One technique is to actively control the fins' structural response using piezoelectric actuators. In early studies, piezoelectric sheet elements were surface-mounted on the full-scale fin. Ground tests showed that these actuators did not have enough authority to suppress buffeting. The objective of the ENABLE program was to examine the performance of two embedded sheet actuators: macro-fiber composite (MFC) actuator and active fiber composite (AFC) actuators. The ground tests using simulated aerodynamic loads on a full-scale fin showed that both type of actuators satisfactorily reduced the structural response due to buffet loads. During the wind-tunnel tests (Fig. 1.25), peak values of power spectral–density fluctuations for tip acceleration due to wing buffeting were reduced by 85% [185]. Two drawbacks with embedded sheet actuators were pointed out as the inaccessibility of actuators in case of any repair and their high-voltage (in kV) requirement.

ACROBAT (Actively Controlled Response of Buffet Affected Tail): This was sponsored by NASA and was carried out jointly by the Wright Laboratory and

Daimler-Chrysler using a 1/6th-scale model of F/A-18 wing model, which was built with piezoelectric sheet actuators placed at an angle at its root to cause bending and torsion deflection. It was tested in the NASA Langley Transonic Dynamics Tunnel in 1995–1996. A single-input/single-output control resulted in a significant alleviation of buffet loads over the entire range of angle-of-attack [186]. To supplement this program, there was an Active Vertical Tail (AVT) program led by McDonnell Douglas Aerospace to actively reduce buffet loads in a 5% scale, aeroelastically scaled tail representative of a typical fighter aircraft. The piezoelectric actuators were attached to the spar to control the first two bending and first torsion modes. A twin-tail aircraft model was tested in a low-speed wind tunnel and successful (up to 65%) alleviation of vibratory peak strain was achieved for a range of flight conditions [187].

German Buffet Suppression Program: Using piezoelectric patch actuators distributed across the surface on a full-scale vertical tail, Suleman et al. [188] carried out an experimental investigation of active suppression of wing flutter and vertical tail buffet. Wind-tunnel tests on simplified wing models demonstrated a 30% reduction in buffeting attenuation and a 6% increase in critical flutter speed.

NASA's Morphing Program: This program was initiated in the early 1990s by Langley Research Center and was focused on the development and assessment of advanced technologies to enable efficient, multipoint adaptability in aerospace vehicles [110]. This encompassed smart materials, nanotechnology, adaptive structures, microflow control, biomimetic devices, structural optimization, controls, and electronics. The goal was for vehicles to efficiently adapt to diverse mission scenarios. Major issues were identified: insufficient understanding of unsteady and nonlinear aerodynamics and their interaction with actuation; adequate energy-efficient flow-control actuators with sufficient authority; a lack of robust sensors and nonintrusive electronics; a weak knowledge base on reliability and maintainability issues; and unavailability of nontraditional design practices to exploit the tailoring of composite couplings, rapid prototyping, and new actuation approaches. It was concluded that interdisciplinary interaction among adaptive structures, smart materials, flow control, and biological systems may provide a fascinating palette for future innovations in aerospace systems.

Active Interior Noise Control: Major sources of noise in the interior of an aircraft are the engine, turbulent boundary layer, and avionics/air conditioning. Turboprop engines produce low-frequency noise (less than 100 to 500 Hz), whereas turbofan noise is quite different. The turbulent boundary–layer noise is generated by unsteady pressure that induces high-frequency vibration in the fuselage structure, which in turn produces a sound field. This noise is random in nature, with a bandwidth from medium to high frequency (less than 1000 Hz). The third source is the forced-air convection system, especially as used to cool down avionics and the air conditioning of the cabin. Although random in nature, this noise has a narrow frequency band. The level of noise is different at different stations in the aircraft. For a propeller-driven aircraft, the excitation frequencies are the blade-passage frequency and its higher harmonics. The high tonal-noise levels occur at these frequencies. To obtain an acceptable noise level inside the cabin for the comfort of passengers, and to ensure work safety in the cabin, passive noise-suppression methods are routinely used. These include cabin linings with high-damping or vibration-absorption materials. This helps to reduce low-frequency noise but results in a significant weight penalty. Smart structures have emerged as a promising active technique to minimize radiated noise. In active noise control, one deploys actuators and sensors in conjunction with a controller. There have been only a few commercial applications of active noise control. Key issues are system integration, limitation of control algorithms, limited actuator authority, and bulkiness of power electronics (i.e., amplifiers, controllers, and real-time computing platform), and acquisition cost. An active noise control in the form of noise cancellation using loudspeakers (located behind the trim panels), called Ultraquiet^{TM}, was used in the Dash 8 and Saab 2000. This anti-sound system works in the low-frequency domain. Piezoelectric materials are widely used as actuators and sensors in such a structure [189]. Such an approach becomes less effective at high frequencies due to the complexity of the controller and the low actuation authority of the actuators. It appears attractive to combine both passive and active techniques, where the passive technique is effective at high frequencies and the active technique becomes effective at low frequencies. Gentry et al. [190] used this approach in the development of a smart foam that uses polyurethane foam and PVDF actuators with a controller. Kim and Lee [191, 192] developed piezoelectric smart panels featuring piezoelectric shunt damping and passive sound-absorbing material. When the sound impinges on a panel, it starts vibrating and the attached piezoelectric patch produces an electric charge, which is effectively dissipated as heat via an electric shunt circuit.

Petitjean and Greffe [193] used active trim panels to actively control noise in an aircraft cabin. An active trim panel consists of a sandwich structure with a lightweight honeycomb core and outer fiberglass skins with embedded piezoelectric patches. Exciting panels at high frequency generates acoustics signals that help to cancel noise at a station. Single-frequency sound levels were reduced by 20 to 49 dB. Because of the large number of actuators and sensors involved, their positioning in the airframe plays a major role in the overall performance of the active noise-reduction system. Bohme, Sachau, and Breitbach [194] used a cooperative simulated annealing (COSA) algorithm to minimize noise in a cabin.

1.9.3 Jet Engines

A fixed-geometry engine inlet results in suboptimal response for a wide range of operational flight conditions (i.e., Mach number, altitude, angle-of-attack, angle-of-slip, and engine airflow condition). To overcome this problem, variable shape control of the engine inlet, such as inlet lip blunting and inlet wall shaping, is used. At low speeds, large inlets with very blunt lips are needed to obtain high inflows without flow separation, especially during a takeoff condition. Conversely, at subsonic cruise condition, sharp inlet lips are desired to reduce drag. At supersonic flight conditions, the flow needs to be decelerated to subsonic conditions in the inlet because a rapid deceleration can result in a substantial loss in pressure and thrust. Mechanical complexity; weight penalty; acquisition cost; and actuator authority, and reliability are some of the critical issues for the implementation of smart

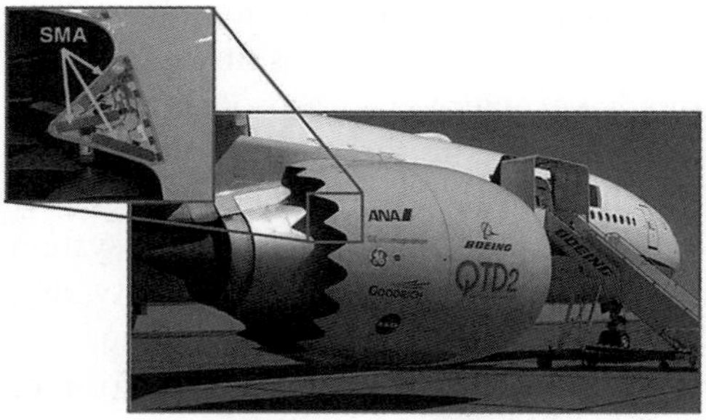

Figure 1.26. Variable geometry chevron using SMA actuators by Boeing, from Ref. [132].

structures to build VG nozzles. Turner et al. [195] carried out fabrication, benchtop testing, and numerical validation of an adaptive jet-engine chevron concept with embedded prestrained shape-memory alloy actuators in a composite laminate. Nitinol ribbons were embedded on one side of the mid-plane of the composite laminate such that thermal excitation induced bending deflection. During the fabrication of the laminate using a hot press, an integrated end constraint was included to restrain the Nitinol at elevated temperatures. Satisfactory agreement was achieved between the predicted and the measured chevron deflection.

Variable Geometry Chevrons (VGC): The objective is to autonomously morph the shape of chevrons using compact SMA actuators to optimize the acoustics and performance of a jet engine for multiple flight conditions. One of the goals is to reduce operational noise during aircraft takeoff and landing conditions. For a commercial jetliner involving high-bypass-ratio turbofan engines, one of the major sources of noise is the turbulent mixing of the hot jet exhaust, fan stream, and ambient air. Serrated aerodynamic surfaces, found to reduce shock-cell noise and located along the trailing-edge of the jet engine primary and secondary nozzles, are referred to as "chevrons." However, chevrons also result in drag or thrust losses because they are normally immersed into the fan flow. Crucial challenges involve the harsh environment of elevated temperatures, loads via vibration levels, and high integrity, lightweight, aerodynamic and structurally efficient design. With VGC, the surfaces are morphed using SMA actuators autonomously by heating with exhaust temperature (Fig. 1.26). They are immersed into the flow during takeoff to lower noise and retracted at cruise to reduce thrust losses. This concept was successfully flight tested on a Boeing 777-300ER with GE-115B engines using 60 Nitinol strip actuators [196]. Hartl et al. [197, 198] carried out the training and thermomechanical characterization of nickel-rich Nitinol (Ni60Ti) for application in the Boeing VGC. After fifty thermomechanical cycles, the response was found to be quite stable with repeatable strain up to 1.6% over a wide range of applied stresses.

Webster [199] examined a similar type of approach for Rolls-Royce engines. Schiller et al. [200] developed a piezoelectric actuated–liquid-fuel modulation system for active combustion-control applications. Through a systematic performance evaluation, a piezoelectric stack actuator was found to be superior for combustion control due to its compact size, high bandwidth, and relatively low cost. It was possible to achieve a fuel modulation of more than 75% of the mean flow rate and successfully stabilize a single-nozzle kerosene combustor.

The SAMPSON Program (Smart Aircraft and Marine Propulsion System demONstration project) (1997–2000) was carried out to explore concepts for shape control using smart structures technology for gas turbine engine inlets for a typical tactical aircraft (F-15 Eagle) and applications for a large-scale marine propulsion

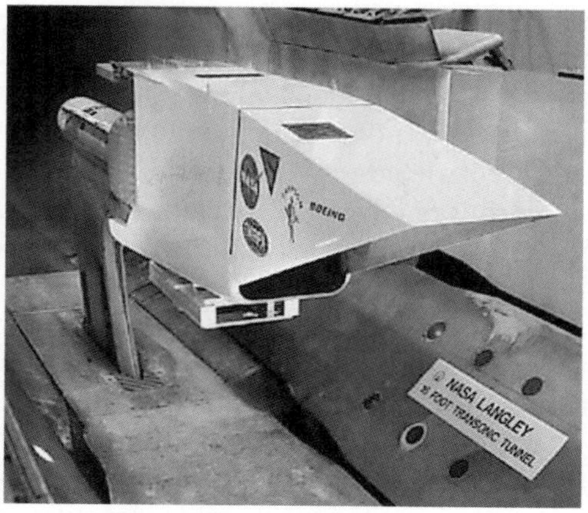

Figure 1.27. F-15 engine inlet used in the SAMPSON project, from Ref. [176].

system and a hydrodynamic maneuvering system. A team led by Boeing carried out this three-year program. Potential envisaged benefits included a 20% increase in range, enhanced maneuverability, and reduced noise signature. The engine inlet is expected to have a direct impact on the vehicle's flight performance, mission effectiveness, and life-cycle cost. Fixed-inlet geometry cannot provide the best performance under all flight conditions. For example, at low-speed takeoff conditions, large inlets with blunt lips appear suited to cause high inflow without flow separation. For subsonic cruise flight conditions, sharp inlet lips are needed to reduce drag as well as radar cross-section (RCS). At a supersonic flight condition, variable geometry inlets are again needed to reduce losses in pressure recovery and to minimize induced drag from rapidly decelerating inlet flow. A full-scale adaptive inlet (16 ft long, 3.5 ft high, and 3 ft wide) of an F-15 Eagle aircraft was built using SMA wire actuators in an antagonistic manner (one SMA cable is set in opposition to another using SME), and actuated using internal heating to control the inlet area, leading-edge blunting, and inlet contour (Fig. 1.27). The SMA actuator concept (consisting of thirty-four wires/rod actuators) had the capability of up to 20,000 lb of force and 6 inches of displacement. The concept was flight tested successfully over a range of Mach numbers. Also, two wind-tunnel tests, which were carried out in the NASA Langley 16-ft TDT, demonstrated the desired two-way control of the inlet cowl and lower lip (extension and retraction). It took 30 seconds to move the cowl by 9 degrees; however, uniformity of temperatures among wires and cooling time were major concerns [201, 202]. Additionally, as part of the SAMPSON project, the use of SMA cables wrapped circumferentially around the aft portion of the fan cowling of a high-bypass turbine engine to change the fan nozzle area was examined [203]. The high exhaust temperature during takeoff and landing was used to transform the SMAs into the austenite phase, thus providing recovery forces to open the nozzle to its maximum cross-sectional area. In high-altitude cruise flight conditions, the exhaust temperature becomes lower, transforming the SMAs into the martensite phase, which allows the nozzle to close. Optimum performance was obtained over a wide range of flight conditions.

1.9.4 Rotary-Wing Aircraft

The structural, mechanical, and aerodynamic complexity and the interdisciplinary nature of rotorcraft offer numerous opportunities for the application of smart structures technologies with the potential for substantial payoffs in system effectiveness [204, 205]. Compared to fixed-wing aircraft, helicopters suffer from severe vibration and fatigue loads, more susceptibility to aeromechanical instability, excessive noise levels, poor flight-stability characteristics, weak aerodynamic performance, and a restricted flight envelope. To reduce these problems to an acceptable level, numerous passive and active devices, and many ad hoc design fixes, are resorted to with resultant weight penalties and reduced payloads. The primary source for all of these problems is the main rotor, which operates in an unsteady and complex aerodynamic environment leading to stalled and reversed flow on the retreating side of the disk, transonic flow on the advancing blade tips, highly yawed flow on the front and rear part of the disk, and blade-vortex interactions under certain flight conditions. Hence, most of the research activities are focused on the application of smart structures technology to rotor systems to improve their performance and effectiveness. Because the rotor is a flexible structure, changes in shape, mechanical properties, and stress/strain fields can be imposed on it. These in turn will alter the vibratory modes, aeroelastic interactions, aerodynamic properties, and dynamic stresses of the rotor. Smart structures technologies may enable these imposed changes to be tailored to conditions sensed in the rotor itself. Furthermore, because the smart actuators and sensors can be distributed over each individual rotor blade, control can be imposed over a much larger bandwidth than with current swashplate-based controls, which are limited to N/rev for an N-bladed rotor. This opens up a hitherto unavailable domain for vibration control, aeromechanical stability augmentation, handling qualities enhancement, stall alleviation, and acoustic suppression. The use of smart structures also offers the prospect of in-flight tracking of main rotor blades and detection of structural damage in the rotor, drivetrain, and other critical components. The pilot can then be alerted to take suitable action. A further very promising application of smart structures is to actively control the interior noise of a rotorcraft. Structure-borne noise can be minimized by actively controlling the response of airframe panels [206]. A source of high-frequency interior noise can be the drivetrain system (i.e., gearbox-meshing tonal frequencies). Actively tuning transmission strut systems with smart actuators may minimize noise in the cabin (i.e., high-frequency noise cancellation) [207].

Three types of smart-rotor concepts have been developed: leading- and trailing-edge flaps actuated with smart material actuators, controllable camber/twist blades with embedded piezoelectric elements/fibers, and active blade tips actuated with tailored smart actuators. The performance of these actuation systems degrades rapidly at high rotational speeds because of increased centrifugal force, dynamic pressure, and frictional moments. For flap actuation, actuators range from piezo-bimorphs to piezostacks and piezoelectric-/magnetostrictive-induced composite coupled systems. Most of these concepts were demonstrated on scaled rotor models (e.g., Froude- and Mach-scaled) and a few were also attempted in full-scale rotor systems. Most smart-material actuators are moderate-force and extremely small-stroke devices; hence, some form of mechanical/fluidic/hybrid amplification of stroke is needed to achieve practicable flap deflections. Because of compactness and weight considerations, the stroke-amplification mechanism and high-energy density–actuators have been key barriers for application to rotor blades.

Koratkar et al. [208, 209, 210] and Roget and Chopra [211] built 6-ft diameter, dynamically scaled rotor models with trailing-edge flaps actuated by multi-layered

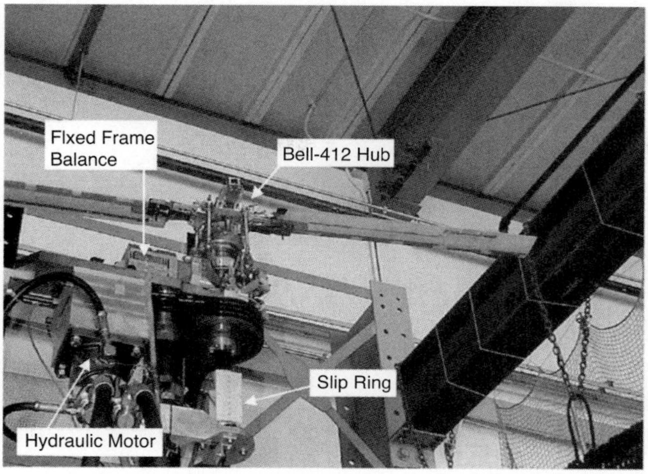

Figure 1.28. Froude-scaled rotor model (6-ft diameter) on hover tower with piezobimorph actuated flaps, from Ref. [205].

piezobimorphs. Initially, Froude-scaled rotor models were built and successfully tested in a vacuum chamber and on a hover tower (Fig. 1.28). Finally, Mach-scaled rotor models were demonstrated in closed-loop testing in the wind tunnel (Fig. 1.29). The flaps spanned about 10% of the rotor radius, were centered at 75% of blade length, and showed more than $\pm 3^\circ$ deflection at 4/rev excitation using 3:1 AC bias (3 to 1 field in the polarized direction) at an RPM of 2150. Using a neural-network-based–adaptive feedback controller, individual blade control resulted in more than 80% reduction in vibratory hub loads in the Glenn L. Martin wind tunnel. A Froude-scaled rotor model was also tested successfully in an open-loop investigation by Fulton and Ormiston [212].

Lee et al. [213, 214, 215] built a model of blade section of length 12 in and chord 12 in with trailing-edge flap (span 4 in and chord 3 in) actuated by piezostacks in conjunction with a double-lever (L-L) amplification mechanism. The model was

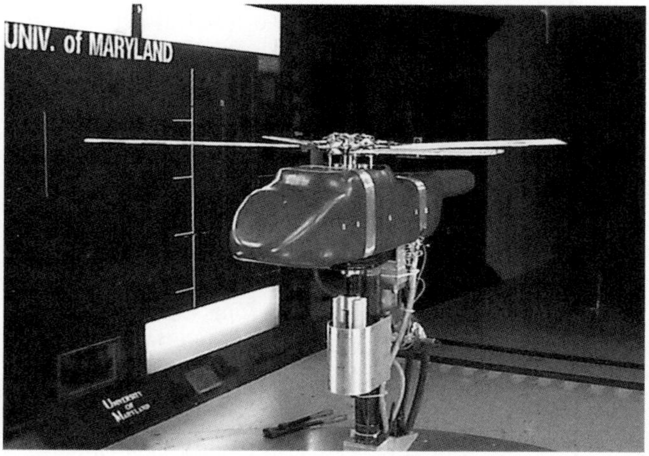

Figure 1.29. Piezobimorph-actuated flap: 6-ft diameter rotor model test in Glenn L. Martin wind tunnel; successfully tested in both open- and closed-loop studies, from Ref. [205].

Figure 1.30. Piezostack-actuated flap: full-scale wing-section model tested in open-jet wind tunnel; produced ±6 deg flap deflection, from Ref. [205].

tested in a vacuum chamber to simulate the full-scale centrifugal field (600 g) and showed the desired stroke-amplification factor of about 20 at all rotor harmonics (i.e., up to 6). The model was tested in an open-jet wind tunnel and successfully demonstrated flap performance of about ±10° at 120 ft/sec. To improve bi-directional performance of this actuation device, a dual L-L amplification system was built and successfully tested in a vacuum chamber and a wind tunnel (Fig. 1.30). This new actuation system showed a significant improvement in flap performance at different operating conditions [215]. Straub et al. [216] built a full-scale smart rotor system for the MD-900 Explorer (five-bladed, 34-ft diameter) with piezostack-actuated flaps to actively control its vibration and noise. To amplify the stroke of piezostacks, a biaxial X-frame mechanism was incorporated. The system was successfully tested on a hover stand to check its performance in rotating environment. Hall and Prechtl [217] built a 1/6th Mach scale rotor model with trailing-edge flaps actuated with X-frame actuators and successfully tested it on a hover stand. Flap deflections of ±2.4° were achieved. Also, Janker et al. [218] developed a novel piezostack-based flexural actuator for the actuation of trailing-edge flaps.

Bernhard et al. [219, 220, 221] built a 6-ft diameter Mach-scaled smart active blade tip (10%) rotor actuated with piezo-induced bending-torsion-coupled composite beam (Fig. 1.31). A novel spanwise variation in the ply lay-up of the composite beam was used with appropriate phasing of surface-mounted piezoceramic actuators to convert the bending-torsion coupled beam into a pure twist actuator. At 2000 RPM in hover, blade-tip deflections of 1.7° to 2.9° were achieved at the first four harmonics (for an excitation of 125 V_{rms}). The associated changes in blade lift corresponded to an aerodynamic thrust authority of up to 30%. This concept appears promising as an auxiliary device for the partial control of noise and vibration.

In the integral-actuation concept, an array of actuators is embedded into the skin or bonded to the spar to achieve a smooth-distributed induced twist, which in turn changes the aerodynamic loads necessary to suppress rotor vibration. Distributed actuators should have enough authority to overcome the inherent stiffness of the blades. Chen et al. [222, 223] built a 6-ft diameter Froude-scaled rotor model with controllable twist blades (Fig. 1.32). For this concept, banks of specially shaped

Figure 1.31. Smart tip rotor model (dia. 6 ft) on hover stand, actuated with composite bending-torsion coupled beam and piezos; produced ±2 deg tip deflections up to 5/rev excitation, from Ref. [205].

(large aspect ratio) multilayered piezoceramic elements were embedded at ±45° relative to blade axis, respectively, over the top and bottom surfaces; an in-phase activation resulted in pure twist in the blade. The model was successfully tested on a hover stand and in the Glenn L. Martin wind tunnel. A tip twist on the order of ±0.4° at 4/rev was obtained in both hover and forward flight ($\mu = 0.33$) that amounted to more than 10% rotor vibratory thrust authority. Although the oscillatory twist amplitudes attained in the forward-flight experiments were less than the target value (i.e., 1° of tip twist for complete vibration suppression), these tests showed the potential for partial vibration suppression. Hagood et al. [224, 225, 226] built a controllable-twist Mach-scaled model rotor (1/6th scale of CH-47D) by embedding AFC plies (i.e., four active plies, each consisting of six 45° AFC actuators) in the top and bottom of the spar laminate and tested on a hover stand. Even though it did not achieve the projected tip twist of ±2°, it showed enormous potential for full-scale rotor applications. Cesnik et al. [227, 228] further improved this technology and successfully tested a Mach-scaled rotor model with embedded active fibers in the TDT wind tunnel in both open- and closed-loop investigations. They also refined analytical tools related to this rotor system.

SMAs show enormous potential in providing large induced strains (up to 6%) but are limited to low-frequency (less than 1 Hz) applications such as tab adjustment for rotor tracking. Epps et al. [229, 230] systematically investigated

Figure 1.32. Active twist blade with embedded piezoactuators (6-ft diameter) rotor test in the Glenn L. Martin wind tunnel; produced ±0.5 deg blade twist up to 5/rev excitation, from Ref. [205].

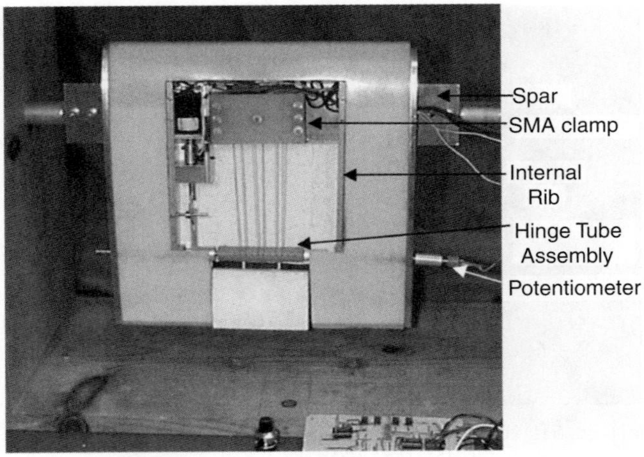

Figure 1.33. Blade tab actuated with SMA actuator, wing section tested in open-jet wind tunnel; produced tab deflections of more than 20 deg, from Ref. [205].

the development of an SMA-actuated trailing-edge tab for in-flight blade tracking. They built a model of the blade section of span and chord of 12 in with a tab of span 4 in and chord 2.4 in, actuated with two to five Nitinol wires of diameter 0.015 on both the top and bottom surfaces (Fig. 1.33). To lock the tab at a desired angle (in power-off condition), a gear-locking mechanism consisting of spur gears, pulling solenoid, and pawl was built. A displacement feedback controller was developed to fine tune the tab deflection in about 10 seconds. This wing section was tested in the open-jet wind tunnel and tab deflections on the order of 20° were obtained at a speed of 120 ft/sec. Singh et al. [231] improved this design and successfully tested it in the wind tunnel for a repeatable open- and closed-loop performance.

Liang et al. [232] investigated the use of a pre-twisted SMA rod to torsionally deflect the blade tab. Two concepts were examined using analysis, respectively, incorporating one- and two-way memory actuators. In the first concept, one SMA rod was used as an actuator while the second one served as a restoring spring (differential bias). In the second concept, one SMA rod in conjunction with a locking arrangement was used. Another study [233] was undertaken to develop an SMA-actuated trailing-edge tab (length 4% radius) for the MD-900 rotor system (five-bladed bearingless rotor) for in-flight blade tracking (Fig. 1.34). This tab was located at 72% radial position and was driven by a SMA torsional actuator (developed by Memry). A locking mechanism was developed to keep the tab in position without power to the actuator. It was designed to undergo $\pm$ 7.5° of twist in steps of 0.25°. Maximum torsional actuation moments expected were 5 in lb during forward flight and 9 in lb during maneuvers. Improved fabrication of the SMA tubes, end attachments, and loading system were developed. To overcome bias forces, two biaxial SMA tubes were used. The actuator had its own integrated microprocessor control. Based on bench-top tests using a spring to simulate aerodynamic forces, the concept appeared feasible.

Prahlad and Chopra [234, 235] examined an SMA torsional-tube actuator integrated into the blade to actively control twist distribution in a tiltrotor between hover and forward flight modes, providing improved aerodynamic performance in both modes of flight. Benchtop testing of the actuator showed the feasibility of the

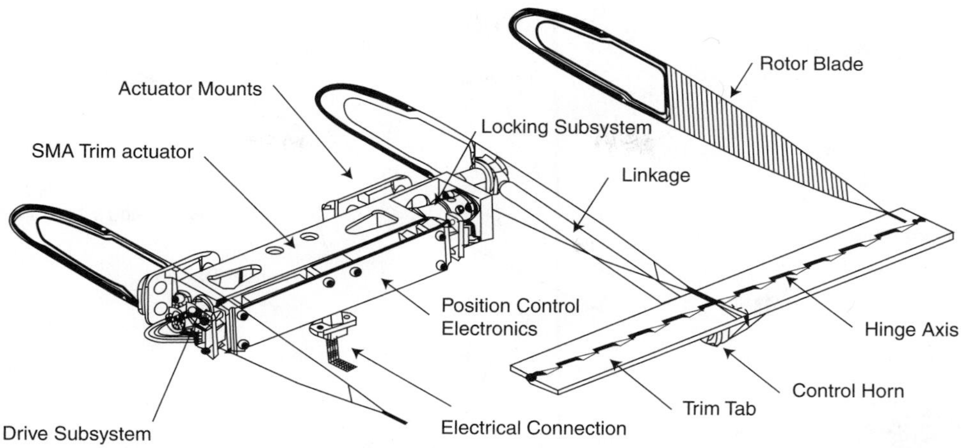

Figure 1.34. SMART trim-actuation system, from Ref. [233].

large recovery torques needed for this application. Torsional behavior of an SMA tube actuator was influenced by heat treatment, twist rate, loading pattern, and two-way SME. It was shown that two tube SMA actuators with an outer diameter of 1.5 inch could produce a twist of 10° in the XV-15 blade.

There are other potential applications of smart structures technology to rotary-wing systems that may yield enormous payoffs in performance improvement and cost savings. These include external noise suppression [236], internal noise suppression [206, 237], primary rotor controls [238, 239, 240, 241], performance enhancement including dynamic stall delay [242], active transmission mounts [243, 244], and active/passive damping augmentation [245, 246]. Kim et al. [247] examined the stabilization of ground resonance instability using a piezoelectric lag damper (based on piezoshunting) and compared its performance with elastomeric lag dampers. A two-bladed rotor was built and piezoceramic elements were bonded to the rotor flexure. The piezoelectric lag damper showed superior performance in stabilizing the weakly damped lag mode, compared to an elastomeric damper.

The interior noise in a helicopter can be divided into two parts: a frequency range of 50 to 500 Hz, caused primarily by the main rotor, tail rotor, and engine; and a frequency range above 500 Hz, generated primarily by the geartrains and transmission system. The sound and vibration energy is propagated to cabin panels either through structure-borne transmission or direct radiation. A likely approach to minimize noise transmission in a cabin may use an active control approach in the low-frequency range (below 500 Hz) and a hybrid active/passive approach for the high-frequency range (above 500 Hz). Passive noise control is now widely used to suppress vibration and noise in the cabin. The approaches include stiffening and isolation of structures, damping augmentation, and sound-proofing treatments. These methods result in a weight penalty and also a restriction of available space for insulation and surface treatments. These passive control approaches normally become less effective in the low-frequency range. Sampath and Balachandran [248] examined an active control of interior noise below Schroeder frequency (i.e., below about 100 Hz) in a three-dimensional enclosure with surface-bonded PZT patches on flexible panels. Three global and one local performance functions were examined. Based on this numerical study with oscillatory excitation, it was concluded that

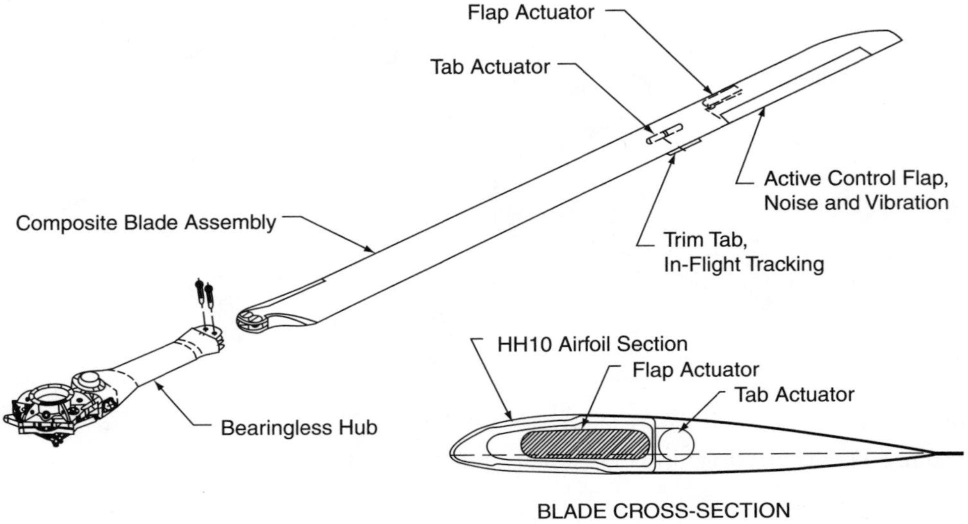

Figure 1.35. MD900 blade with trailing-edge flap and trim tab, from Ref. [249].

energy-based performance functions resulted in superior sound-pressure levels and zones of quietness. The number and location of actuators and sensor play a major role in noise reduction [248].

Smart Material Actuated Rotor Technology (SMART)

The objective was to demonstrate the feasibility of smart material actuated tabs and flaps in a full-scale helicopter rotor system in order to actively control vibration and Blade Vortex Interaction (BVI) noise and to perform in-flight blade tracking. The program was sponsored by DARPA and the team was led by Boeing. A full-scale MD900 light utility helicopter was chosen as the demonstration platform (Fig. 1.35). In Phase 1, two actuation concepts were examined: trailing-edge flaps actuated with piezostacks and the active twisting of blades with embedded piezocomposite fibers. In Phase 2, active flaps with X-frame actuators were used to control vibration and noise. Several two-dimensional airfoil and flap/tab models were tested in the wind tunnel, and test data were used to develop active flap and trim tab systems [249]. Initially, the flap actuation used two biaxial piezoelectric stack columns operating in a push-pull arrangement with a stroke-amplification mechanism (Fig. 1.36).

The flap system consisted of a span of 18% rotor radius and a chord length of 35% of blade chord; it was located at 83% rotor radius. Each blade contained an embedded 2X-Frame actuator with four piezoelectric stack columns. Each actuator was located at 74% radius and connected to the flap by a mechanical linkage. The actuators were powered by a two-channel switching-power amplifier. The closed-loop controller was implemented using a dSPACE 20-channel, single-board controller. The smart rotor was tested successfully in the 40×80-ft wind tunnel at NASA Ames in 2009, demonstrating the effectiveness of active flap control to suppress vibration and noise. Results showed reductions of vibratory loads of about 80% as well as reductions up to 6 dB in blade-vortex interaction and in-plane noise. The impact of the active flap on rotor performance, rotor smoothing, and control power was also demonstrated.

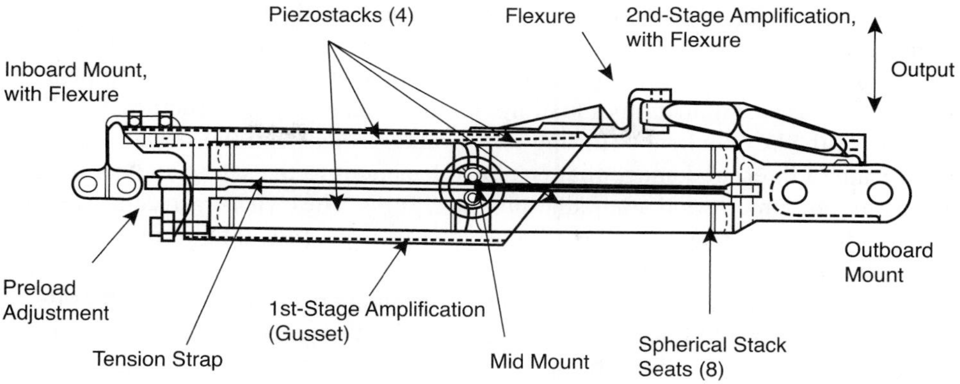

Figure 1.36. Biaxial piezoelectric flap actuator schematic, from Ref. [249].

Under the sponsorship of DARPA, Boeing showed the potential of active flow control in a tiltrotor (V-22) to minimize hover download-lift on wings during take-off conditions and thereby significantly improve payload capability [250]. Using the principle of the synthetic jet, the flow separation over the airfoil was controlled with low power, highly distributed, redundant actuation systems. Two types of actuation schemes were considered; a synthetic jet and a flaperon, mounted on the upper surface near the trailing edge of the wing. The synthetic jet is located at the leading edge of the flaperon that is deflected up to 70° during hover. Using a vibrating membrane (≈50 Hz) located in a cavity in the airfoil surface between the wing and the flaperon, air is expelled and retracted periodically into the flow field, generating a zero–net-mass flux. Two types of smart material actuators were used; a multilayered PVDF cantilevered bender and a single-crystal (PMN-PT and PZN-PT) piezoelectric-poled wafer on a cantilevered-spring steel substrate. The wind-tunnel tests on a 1/10th-scale model with a single-crystal actuated flaperon demonstrated a 10% increase in lift and a 20% increase in angle-of-attack capability [251].

Before this technology is implemented in a full-scale system, there are many important issues that need to be resolved, which include structural integrity and fatigue life of the actuators and built-up systems, life-cycle cost; long-term product reliability; maintainability; and robust, reliable design tools.

1.9.5 Civil Structures

Civil structures include bridges, buildings, dams, industrial chimneys, and nuclear power plants. Applications of smart materials and structures in civil structures include structural-health monitoring, vibration monitoring and suppression, minimization of vibratory loads, and earthquake mitigation. Housner et al. [252] provided an extensive review of the structural control and monitoring of civil-engineering structures as a part of new structures or retrofits of existing structures. Enormously destructive seismic occurrences in Northridge, California, in 1994 and Kobe (Japan) in 1995 have demonstrated the importance and value of structural control in new and existing civil structures. Through an application of adaptive structures technology, the goal is to maximize civil structures' performance, control their motion, and monitor their health, thereby minimizing their life-cycle cost and increase their overall safety. Annamdas and Soh [253] provided a review of the advances in PZT-based structural health monitoring of engineering structures, including civil structures. Using the self-actuating and sensing capabilities of PZT transducers, the electromagnetic impedance, and guided ultrasonic–wave propagation techniques help provide real-time, in-service detection of loadings on and damage in the structure.

One of the key challenges for big civil structures has been the structural health monitoring of internal damage that can be a detriment to the safety and comfort of occupants, equipment, and other adjoining structures. The civil infrastructure deteriorates with time as a result of aging of materials, overstress and fatigue, excessive use, inadequate inspection and maintenance, and unexpected weather-related changes. Health monitoring, repair, retrofitting, and replacement become necessary for safety. It is envisaged to build intelligent civil structures with embedded distributed smart material sensors such as fiber-optic sensors to monitor the structural health of these systems. Structural health monitoring of civil structures poses numerous challenges due to their large size, the diversity and heterogeneity of their material components, and their difficult construction environment. Monitoring of strains, deformations, deflections, and frequencies provides clues about the health of structures. Huston et al. [254] showed the application of fiber-optic sensors to monitor internal structural damage, stresses and external applied loads for a wide range of civil structures.

Fiber-optic sensors can be embedded in or attached to a structure, and they offer the flexibility of size, the potential of multiplexing of sensing, power supply and communication signals, low weight, high bandwidth, resistance to corrosion, and immunity from electromagnetic interference (EMI). The enormity of data processing, limited knowledge base and unproven track record, wiring, and integrity of sensors during and after the construction process has been pointed out. Structural damage may result from gravity loading, the construction process, earthquakes, weather, traffic, floods, fires, waves, and chemical attack (i.e., corrosion). The early application of fiber optics to civil structures (e.g., embedding them in concrete) was reported by Mendez et al. [255] in 1989 and Houston et al. in 1992 [256]. In the 1990s, there was a spurt in the deployment of a wide range of optical-fiber sensors for civil-structure health-monitoring applications [257, 258]. Fiber optics were also embedded in full-scale structures, such as dams, to measure shifting between segments [259], and to measure pressure and invibration [260] (Fig. 1.37), as well as into buildings (e.g., the Stafford Building, University of Vermont) [261] (Fig. 1.38) to determine their in-service loading, vibratory response, wind pressures, and building health. deVries et al. [262] tabulated a comparison of the characteristics of commercially available fiber-optic sensors. Depending on the application and prevailing environment, there are many other smart material sensors that are used in civil structures [263, 264]. Zhou et al. [265] used the particle-tagging approach to monitor the health of structural systems. It is based on embedding micron-sized smart particles in the host structure, for example, one made of concrete or composite, which can be subsequently interrogated to assess its condition (e.g., voids, internal stress, and delaminations). To increase the reliability and safety as well as to expedite the structural health monitoring of civil structures, smart sensors with embedded microprocessors and wireless communication links were proposed [266].

Chong et al. [267] provided an overview of research activities in structural health monitoring of civil infrastructures in the United States. First, the National Science Foundation (NSF) sponsored research activities in non-destructive evaluation (NDE) civil structures were described. Second, the research efforts of the National

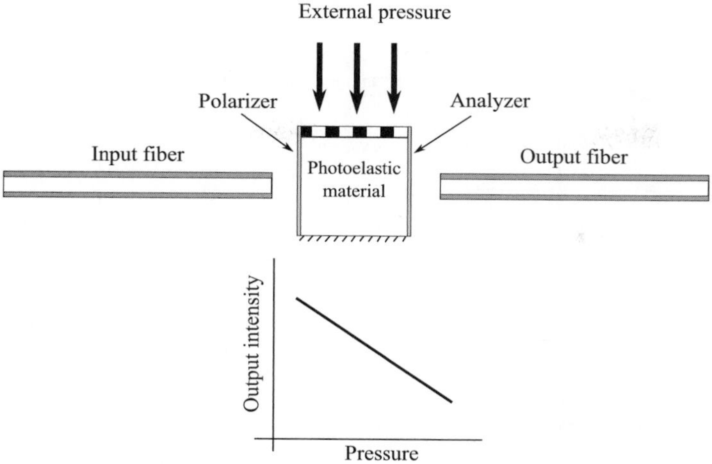

Figure 1.37. Typical fiber-optic photoelastic-based pressure sensor, adapted from Fuhr and Huston [260].

Institute of Standards and Technology (NIST) to develop advanced NDE techniques for the evaluation of concrete structures as well as advances in stress-wave methods were presented. Third, research efforts of the Federal Highway Administration toward condition-based assessment technologies for highway bridges were described. Since the collapse of Silver Bridge in Point Pleasant (West Virginia) in 1967, which resulted in a loss of 46 lives, a database on size, construction, and general condition of about 590,000 bridges and culverts in the United States has been maintained in the National Bridge Inventory (NBI). At least 104,000 have been found to be structurally deficient. These data provide further evidence of the value and urgency of health-monitoring technologies. Tennyson et al. [268] provided an

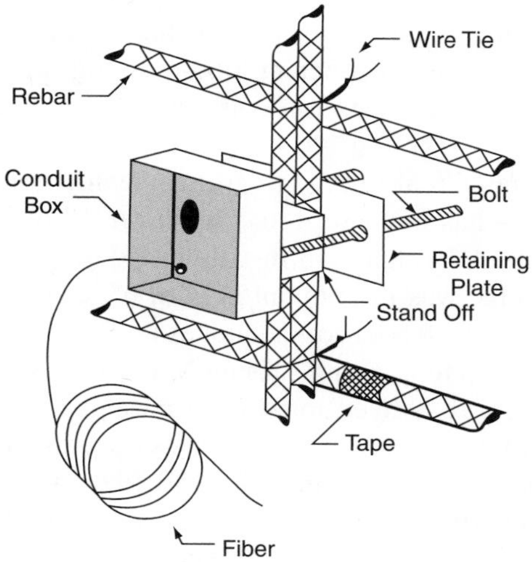

(a) Schematic of the installation of fiber optics into the rebar grid.

(b) Photograph of fiber installed in the rebar grid.

Figure 1.38. Fiber-optic stress sensors embedded in concrete building wall to monitor the structural health of the Stafford Building at the University of Vermont, from Ref. [261].

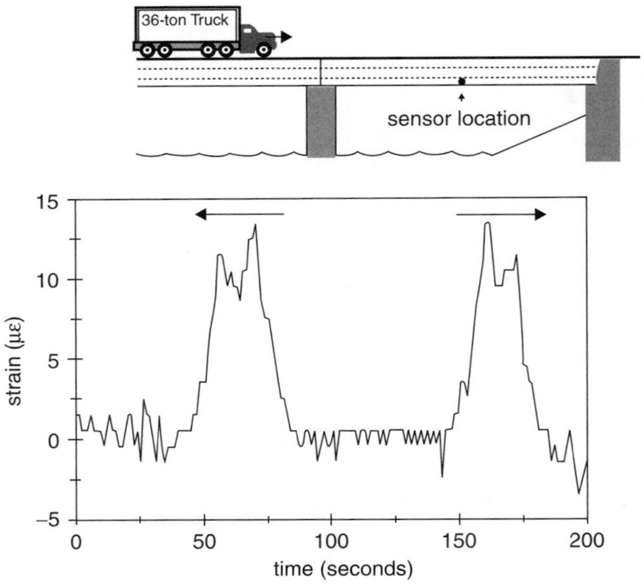

Figure 1.39. Dynamic truck-passage data measured by embedded fiber-optic sensor, from Ref. [268].

overview of structural health monitoring of bridges in Canada using fiber-optic sensors. Fiber Bragg gratings with gage lengths varying from 1 to 20 m were used to measure static and dynamic loads on bridge decks and columns (Fig. 1.39). Sixteen bridges were instrumented across Canada with these sensors. Overall, performance of these sensors was quite satisfactory during a period of six years. Since the construction of the first modern cable-stayed bridge, Stromsund Bridge (SB) in Sweden in 1958, more than 300 cable-stayed bridges have been built all over the world. In the aftermath of the collapse of the Rainbow Bridge in China (QuJiang County) in January 1999, the monitoring of the safety and durability of bridges has become a major issue. This failure was partly attributed to the insufficient load-carrying capability of the cables and, as such, monitoring of cable tension could have provided sufficient warning before its collapse.

The structural behavior, including automated real-time – integrating health monitoring of cable-stayed suspension bridges – has been reported by several investigators. For example, Norway's Skarnsundet Bridge was instrumented with 37 conventional sensors, Korea's Haengju Bridge was instrumented with 65 sensors, Korea's Namhae Bridge was instrumented with 82 sensors, Thailand's Rama IX Bridge was instrumented with 16 sensors, and Hong Kong's Tsing Ma Bridge was instrumented with 265 sensors. All of these were conventional sensors (e.g., strain gages, accelerometers, and inclinometers). Only recently has there been growing use of smart material sensors. Examples include Switzerland's Stork Bridge, which used fourteen optical sensors in conjunction with a remote wireless system. Wang et al. [269] carried out structural health monitoring of a cable-stayed bridge via in situ measurement of cable tension using PVDF piezoelectric film sensors. PVDF film is not only flexible, it is also tough, corrosion resistant, and shock tolerant. The frequency analysis of measured cable tension with PVDF films and comparison with accelerometer data demonstrated the robustness of these sensors, especially in the case of cable sagging.

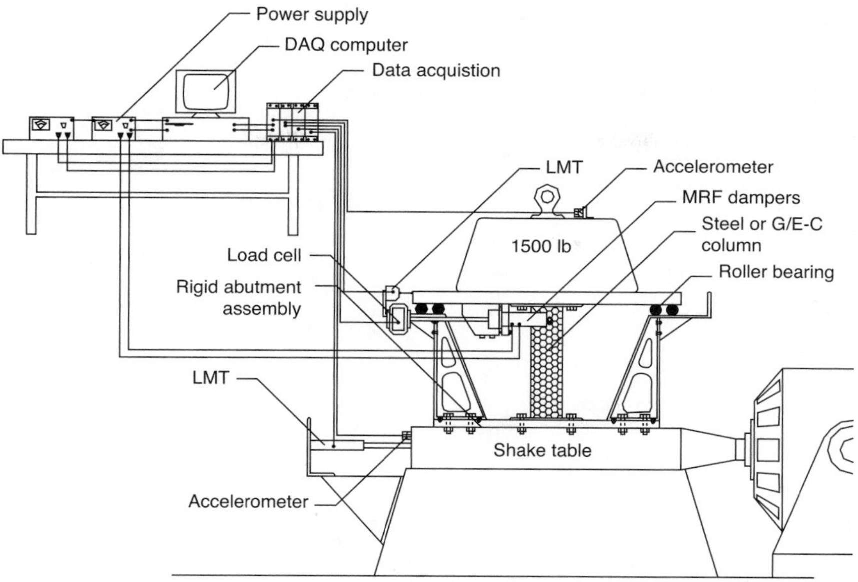

Figure 1.40. Schematic of experimental setup of the two-span scaled bridge, from Ref. [277].

Liao et al. [270] used PVDF film sensors in cable-stayed bridges to wirelessly monitor the dynamic response (i.e., tensile load distribution) of the stayed cables. Betti and Testa [271] used both passive and active methods for the vibration and damage control of long-span bridges. Ambrose et al. [272] used smart sensing networks and techniques to monitor the construction-site shoring systems. This can help to identify dangerous situations (e.g., structures ready to collapse due to overload and weak shoring structure), so corrective action can then be taken, thereby minimizing the risk of injury or loss of life at construction sites. Soh et al. [273] carried out health monitoring and damage detection of a reinforced-concrete (RC) bridge instrumented with PZT patches and built a prototype bridge to validate their predictions. The patches were excited at high frequencies, on the order of kHz. The admittance (conductance) response of patches located near the vicinity of the damage showed drastic changes from the baseline response of a healthy bridge. Pines and Lovell [274] examined a remote wireless damage detection approach to assess the structural integrity of large civil structures using spread spectrum wireless modems, a PC-based data acquisition system, communication software, and sensors. They successfully demonstrated the remote monitoring system over a distance of 1 mile.

An application of the semi-active MR damper is in cable-stayed bridges (i.e., modern bridge construction with spans up to 1000 m) to stabilize large amplitude motions due to high, gusty winds and traffic, thereby helping to decrease the fatigue and corrosion of the strands and increase the safety and durability of bridges [275]. The first application of the MR damper in a cable-stay bridge occurred on the Dongting Lake Bridge in Hunan, China [276]. A total of 312 MR dampers (Lord SD-1005 MR) were installed on 156 stayed cables.

Gordaninejad et al. [277] used MR dampers to control the vibration of a two-span, 1/12-scale bridge model using a combination of passive and semi-active damping capabilities (Fig. 1.40). The two MR dampers provided a controllable damping capability and the steel and graphite/epoxy-concrete (G/E-C) columns provided

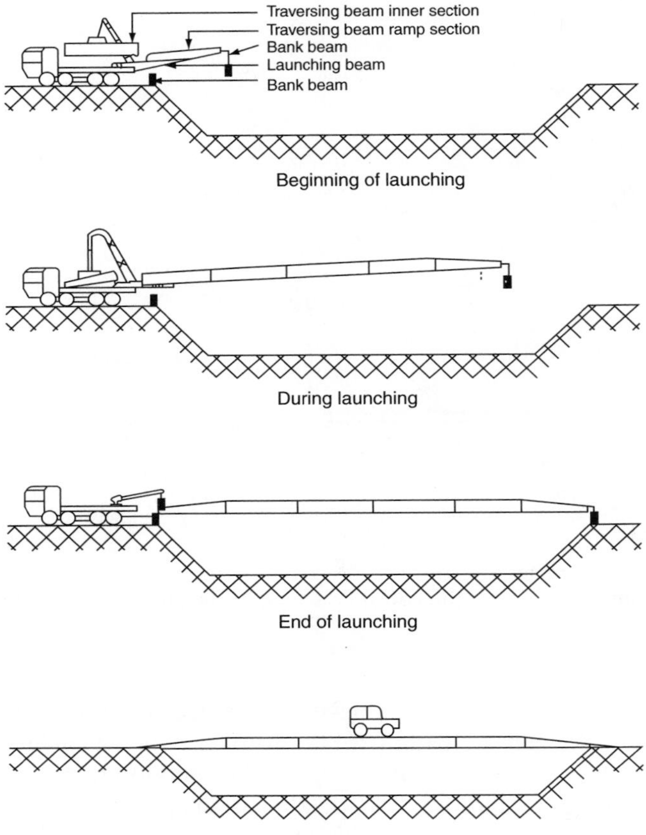

Figure 1.41. Conventional traversing beam, from Ref. [278].

passive structural damping. The MR dampers proved very effective in reducing the relative displacement between the deck and the abutments.

To model the launching and crossing over operations over long support bridges, Baz [278] formulated a finite-element analysis for multisegmented smart traversing beams with built-in wires to monitor and control (actuators and sensors) the beam deflection (Fig. 1.41). Optimal design strategies were applied to minimize control actions at each segment to lessen stresses and deflections over the entire span of the beam. The numerical results show the potential of smart beam concept (bridge) in providing low weight, high strength, and low deflections.

Mitigation of structural disturbances in terms of the extreme loads and vibrations produced by earthquakes or wind in civil structures can be carried out by various methods, which include modification of stiffness, inertia, damping, and shape distributions and by passive and active actuation forces [279]. Many of these methods have been used successfully, and many new methods offer the potential to increase the efficiency and life of large civil structures. Housner et al. [252] carried out a comprehensive review on the assessment of the control and monitoring of civil-engineering structures and identified key points of comparison between civil structural control and other fields of control theory. Among new methods, the application of smart structures technology to civil structures is discussed. Since the 1995 Hyogoken-nanbu (Kobe) earthquake, seismic isolation and passive response-control of buildings have become more prevalent in Japan. Passive response-control systems include seismic isolators, tuned mass dampers, and energy dissipaters.

Since the 1960s, the base-isolation techniques have been routinely adopted in low-rise and medium-rise buildings, and bridges, to minimize the impact of high frequency components of ground motion. By building a sufficiently flexible base, the natural period of vibration is increased to about 4 seconds. However, this may not be adequate for low-frequency disturbances, especially those associated with earthquakes. Further softening the base stiffness to increase the natural period to, say, more than 4 seconds may result in unacceptable large amplitude motion, especially with a large velocity impulse. Since 1970s, for large flexible tall structures, auxiliary dampers, and tuned mass dampers (dynamic absorbers) have been successfully deployed at strategic locations to increase energy dissipation during strong winds. This second method suffers from the drawback of a narrow stationary frequency-band motion. Recent focus has been to incorporate structural control technology to increase the life and safety of civil structures. This is achieved by applying counterforces or through modification of vibration characteristics. The stringent static and dynamic requirements in the design of civil structures subjected to internal and external loads, including high winds and earthquakes, show the potential of active control technology to ensure occupants' safety and/or comfort and the structural integrity and survivability of buildings as well as the equipment within them. Specifically, the objective is to maintain the stresses, strains, accelerations, and displacements within the specified bounds (e.g., peak and RMS) at a specified set of locations due to internal and external excitations. An active control system incorporates actuators, sensors, A/D (analog to digital), and D/A converters, a computer, and a power source. With the availability of high-performance smart-material actuators and sensors and ER/MR dampers, their potential applications to civil structures are expanding. Note that full-scale implementation of active control systems has been accomplished in a number of buildings and bridges; however, cost-effectiveness and reliability considerations have restricted their widespread application. Because of their mechanical simplicity, low power requirement, and system robustness, semi-active controls are preferred at this time for the alleviation of wind and seismic response of buildings and bridges. Agrawal et al. [280] investigated the optimal placement of passive energy-dissipation systems (dampers) to minimize the response of wind-excited buildings using combinatorial optimization techniques. Passive dampers are represented by equivalent damping and stiffness coefficients.

In the ER and MR fluids, the viscosity can be controlled by altering their yield stress through the application of electric and magnetic fields, respectively. This property suits them to use in controllable dampers. A number of ER dampers have been built for applications in civil structures [86, 281, 282]. Today, MR dampers are becoming popular for applications in civil structures because of their superior damping characteristics, especially for active seismic alleviation [283] (Fig. 1.42). Carlson and Spencer [284] reported the design of a full-scale 20-ton MR damper (inside diameter of 20.3 cm and stroke of $\pm$8 cm) emphasizing the scalability of these dampers for large civil structures. The first full-scale application of an MR damper for civil structures was carried out in 2001 by installing two 30-ton MR dampers (built by Sanwa Tekki using Lord MR fluid) between the third and fifth floors of the Tokyo National Museum of Emerging Science and Innovation [276].

The seismic control of a large-scale building was performed on a model building by Nishimura et al. [285] using a simulated earthquake disturbance. Five accelerometers were used to monitor the response of the structure, of which two were used

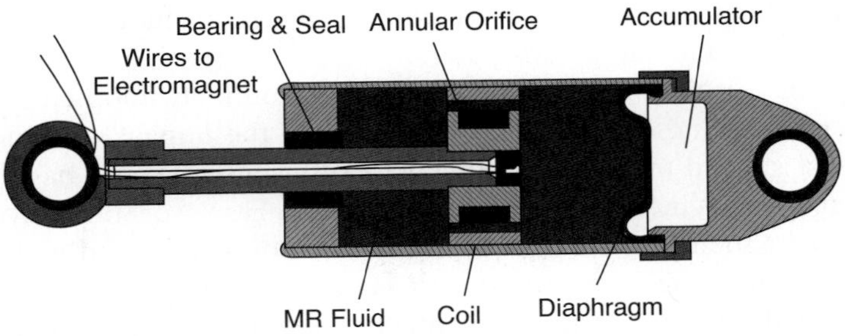

Figure 1.42. Schematic of 20 Ton MR damper with internal diameter of 20.3 cm and stroke of 8 cm, from Ref. [283].

for feedback control. The control objective was to minimize the building response (acceleration) due to a disturbance. As a result of the closed-loop feedback control, the damping factor was increased from nearly zero to 20% in the first three modes. Overall, the low-frequency response was significantly reduced. The hysteretic characteristics of SMA are often exploited for both passive and active structural damping applications. Aiken et al. [286] and Witting and Cozzarelli [287, 288] investigated the application of SMAs for seismic-resistant design of civil structures.

High-performance smart materials are being developed as construction materials to improve the structural integrity of civil structures, especially those subjected to severe dynamic loads. Examples include honeycomb sandwich laminates for bridges; composite column wraps; and fiber-reinforced concrete and plastics for gas storage tanks and pipes, which monitor leakage and damage of underground pipes for water, oil, and gas.

For civil structures, robustness, cost, and maintenance flexibility are key issues for implementation of smart structures technology.

1.9.6 Machine Tools

There have been growing applications of smart structures technology in machine tools to improve their performance. During precise machining of structural components, chattering and ringing of the tool needs to be suppressed. For example, during the dicing of semiconductor wafers, it is necessary to incorporate stiff machine components, which produce ringing during the motion of the machine-bed components from one cut to the next. Controlling of the vibratory motion is extremely important. Also, machine tools are subject to vibratory loads and temperature variations, which result in wear, drift, and settling due to changes in the mounting conditions. Again, for precise machining, it will be important to adjust the tool position using smart actuators.

The quality and speed of glass cutting can be improved by using an adaptive cutter head and an active bridge. Using a piezostack in the z-direction and an electromechanical shaker in the x and y directions along with eleven accelerometers, the cutting speed was improved [147]. Zhang and Sims [289] carried out an experimental investigation to actively control vibration using piezoelectric actuators to mitigate workplace chatter in high-speed machining (Fig. 1.43). Using a positive position feedback controller, a series of milling tests was performed, which demonstrated a sevenfold improvement in the limiting stable depth of cut.

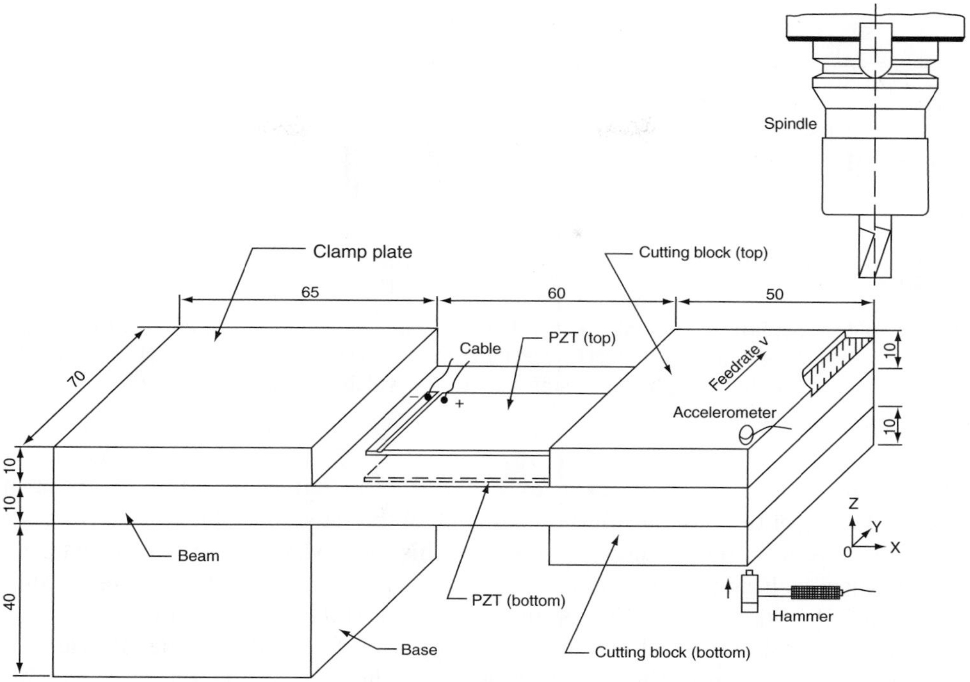

Figure 1.43. Workpiece with a pair of collocated piezoelectric elements, from Ref. [289].

Boring bars are metal cutting tools used to machine precision holes, where one end of the tool is normally fixed to a rotating spindle and the other free end is attached with a cutting insert. Boring bars with large length-to-diameter (L/D) ratios typically have low dynamic stiffness, making them susceptible to vibration, which in turn can have a detrimental effect on the quality of machining and tool life. O'Neal et al. [290] developed an intelligent boring bar, utilizing a micropositioner (consisting of a piezoelectric actuator and photosensitive detectors) to actively control a cutting insert. Subsequently, an integrated structural/control optimization scheme was used to design the micropositioner. This helped to extend the range of the boring tool (40% longer length) while maintaining allowable error within 95%.

In many rotary or linear drive systems in machine tools, it becomes imperative to control and stabilize the velocity due to rapid load changes during machining. A rapid change of load results in a jump of the movable elements of the machine. This problem is quite visible in electro-hydraulic servo drives because the stiffness of the fluid is relatively low. Additional sensors or alternate control methodologies may not be satisfactory for highly dynamic systems. To solve this problem of velocity jump, Milecki and Sedziak [291] used an MR damper, whose characteristics can be adapted in real time. They showed the successful demonstration of MR dampers in stabilization of servo drive velocity in different machines, especially during rapid load reductions.

The wire cutter discharge machine (WEDM) is used to carry out high-performance machining by electric discharge between a thin electrode wire (i.e., 0.05-0.3 mm diameter) and the workpiece. The quality of machining of especially small, delicate, and complex parts depends on an appropriate control of tension in the electrode wire. Due to repulsive force and fluid injection during machining

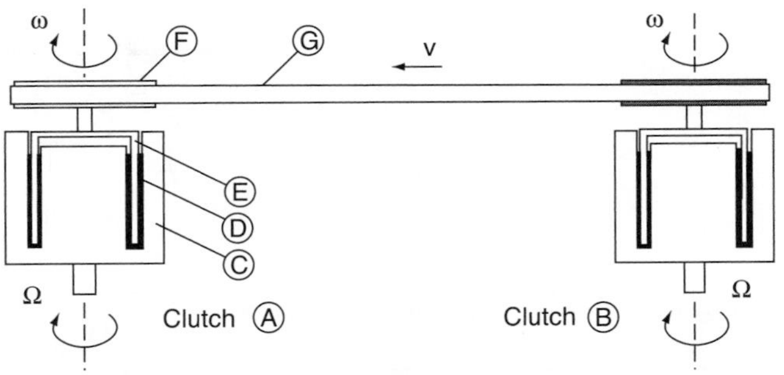

Figure 1.44. Schematic view of high-speed ER traversing mechanism: (C) input rotor – rotating at a constant angular velocity Ω; (D) ER fluid; (E) lightweight output rotor; (F) lightweight output pulley; (G) lightweight traversing belt (onto which a guide can be attached), from Ref. [293].

(for cooling), there is a significant level of vibration that impacts tension force in the wire. One of the popular methods to control the tension in the wire is to utilize electromagnetic brake actuators. However, this method is vulnerable to external disturbance such as repulsive force. Kim et al. [292] controlled the wire tension using an ER brake actuator in conjunction with a sliding mode controller. It was demonstrated using a cylindrical shear-mode-type ER brake that as the electric field increases, the tension in the wire increases and the discharging gap and straightness of the workpiece decreases. The machining performance of the WEDM is far improved with an appropriate control of field.

Flexibility and controlability in operation are important in the design of high-speed mechanisms and devices. A high-speed traversing mechanism is used for winding textile filaments onto spinning or weaving bobbins. Typically, the traverse speed is about 5 m/s, a turnaround period is 10–20 ms, and a traversing length is 250 mm. Johnson et al. [293] successfully used ER fluids to achieve a high-speed traversing mechanism (Fig. 1.44). Simplification of the control strategy for such a mechanism is considered to cover both startup and continous running conditions.

Fielder et al. [294] embedded fiber-optic pressure sensors into the grinding plates of an operational paper mill to monitor the pulp-grinding process in real time, thereby improving the quality and consistency of the pulp online in an active manner. To bear a harsh environment (i.e., pressure up to 175 psi), the sensors were 1.65 mm in diameter with titanium housing, which were installed into the grooves of the grinding plates. Pressure pulses due to the relative motion between the grooves and channels on two grinding plates were measured (i.e., spikes up to 175 psi), which helped to determine the consistency, size, distribution, and quality of paper pulp. Hence, by monitoring the pressure fluctuations, grinding plates can be dynamically controlled, producing a "smart paper mill."

Neugebauer and Hoffmann [295] actively manipulated sheet metal-forming processes using high-performance piezoactuators, thereby minimizing the number of product rejects. Replacing selected drawpins with piezoactuators, controlled force progression at critical forcing points was achieved during sheet draw-in. This refined the deforming process, especially during deep drawing operations in automobile-body production. This results in an "intelligent press" in which the process of sheet draw-in is accurately regulated. It is important to develop tool/die systems to obtain flexible and accurate metal forming. Yang et al. [296] proposed an intelligent tool system for flexible L-bending processing of metal sheets using several sensors incorporated in the tools. The system is autonomous and is capable of changing the shape and pressure of the tools to optimize the forming process and achieve the desired forming accuracy.

Nitinol exhibits far superior resistance to wear than conventional engineering materials such as steel, nickel-based, and cobalt-based tribo-alloys. This characteristic is due to its special pseudoelastic behavior. Additionally, Nitinol demonstrates excellent corrosion-resistance characteristics. Li [297] advanced the use of Nitinol for tribo-logical engineering that included corrosion, erosion, and wear phenomena. Investigations were made to develop tribo-composites using NiTi alloy as the matrix, reinforced by hard ceramic particles, including nano-structured particles. These composites possessed enhanced wear resistance.

Kordonski and Golini [298] developed a precision polishing method called magnetorheological finishing (MRF) to produce surface finish on the order of 10 nm peak-to-valley with surface micro-roughness less than 10 Å on optical glasses, single crystals, and ceramics. In this technique, MR fluid performs the primary function of material removal. In certain conditions, material removal occurs by capture of molecularly small fragments of a hydrated silicon layer by polishing particles instead of by the mechanical scratching used in classical polishing methods. As a result, the surface normal stress and the surface indentation are not important in the process of material removal. Shimada et al. [299] proposed a magnetic compound fluid (MCF) for uniform microscopic polishing of a rotating disk surface with fluctuating magnetic field. The MCF consists of nm-size magnetic and μm-size iron particles in solvent, and its characteristics lie between those of magnetic fluid and MR fluid.

Applications of smart structures technology in machine tools are growing with potential payoffs in the quality, speed, and cost of finished products. Again, a major drawback is the limited stroke of the smart actuators. Other factors can be robustness, cost, and aging of smart materials.

1.9.7 Automotive Systems

In an automobile, there are numerous actuators and sensors to enhance occupants' comfort; to improve performance; to increase safety and reliability; and to control the engine, transmission, suspension, washers, and windows. Among the actuators are electric motors, solenoids, thermobimetals, wax motors, and pressure or vacuum actuators. Among the sensors are thermocouples and analog and digital sensors. Smart material actuators and sensors have not widely penetrated the automobile industry at this time; this is partly due to low awareness of this technology, lack of reliable materials data and constitutive models, and cost and safety concerns. However, it is envisaged that as the scientific community becomes more knowledgeable about this technology, it will find increased application in automobiles.

Specific smart structures applications to automotive systems include active control of vibration and noise, active suspension and engine mount, and fuel injectors for diesel and gasoline engines. Vibration in an automobile is caused by road roughness, wind excitation, the engine exhaust system, and an imbalance of the engine rotor and tires. The purpose of vehicle suspensions is to attenuate vibrations due to various road conditions. Three types of suspension have been attempted: passive, active, and semi-active. The passive type involves hydraulic dampers. It is simple

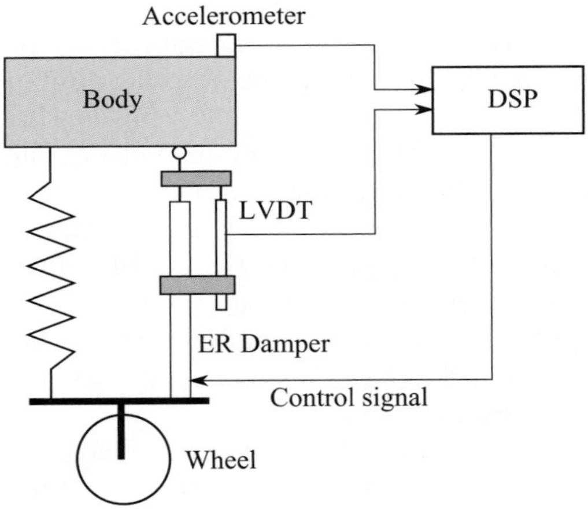

Figure 1.45. Schematic diagram of semi-active suspension system, adapted from Suh and Yeo [302].

in design but limited in performance, especially at a high frequency. The active system is more complex due to requirements of power source and sophisticated control systems (expensive hardware), but it provides superior performance over a wide operating range. The semi-active system lies somewhere in the middle in terms of performance gains and hardware expenses. Nakano et al. [300] built a quarter-car suspension system model using ER dampers and examined its performance characteristics using two different control strategies. Petek et al. [301] and Suh and Yeo [302] constructed a semi-active full suspension system consisting of four ER dampers, showing that vehicle vibration can be satisfactorily suppressed using the skyhook control algorithm (Fig. 1.45). Carlson et al. [303] and Lee and Choi [304] included MR dampers for a vehicle suspension system and showed a sufficient level of damping for a passenger vehicle (Fig. 1.46). Nguyen and Choi [305] optimized the design of MR shock absorbers for application to vehicle suspension considering the damping force, dynamic range, and inductive time constant. There have been several other attempts to develop improved semi-active suspensions based on ER or MR dampers (Fig. 1.47) [306, 307, 308].

Zhu et al. [309] carried out active control of a steering wheel using multilayered piezoelectric actuators. The actuators were bonded to spokes to successfully suppress the vibration of the wheel. An MR clutch has the capability of changing its torque transmissibility continuously within a certain range through control of shear stress of the MR fluid. Lee et al. [310] developed an MR clutch and demonstrated its adaptive torque transmissibility through control of the intensity of the applied magnetic field.

An active isolator incorporates an external energy source in conjunction with an actuator to generate forces on the system subjected to unwanted vibration. Conversely, a semi-active mount does not inject mechanical energy into the system. Ahn et al. [311] conceptualized a small-sized, variable-damping MR fluid mount for the precision equipment of an automobile. On application of high current ($\approx$2 A) in the coil, the peak amplitude of transmissibility decreased by about 70% and the resonance frequency increased by 40%.

Stelzer et al. [312] designed a semi-active MR (flow mode) isolator in conjunction with soft rubber to mount a compressor on an automobile body to isolate the body from high-frequency vibration. Other applications of such an isolator include engine mounts, pumps, and fans in automobiles. Sassi et al. [313] developed a semi-active

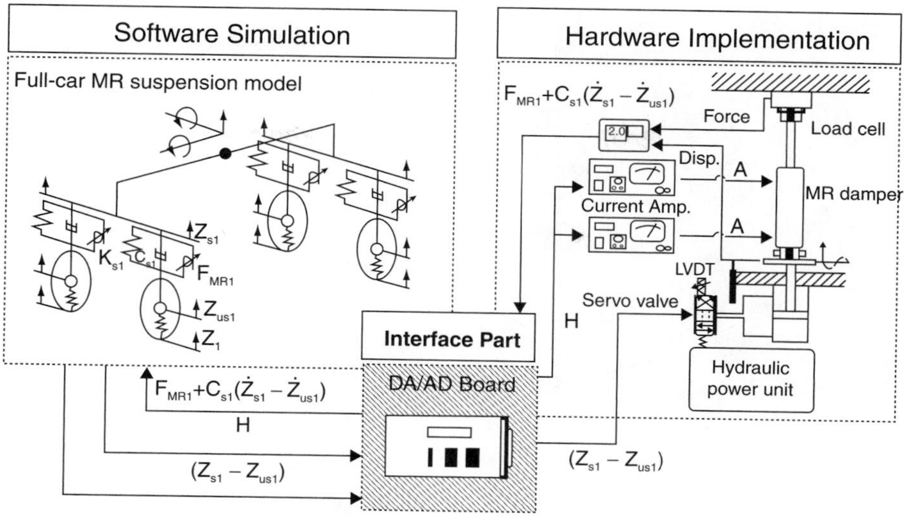

Figure 1.46. Schematic diagram of hardware-in-the-loop-simulation (HILS) for the full-car MR suspension system, from Ref. [304].

MR suspension system for automotive systems and carried out systematic testing to optimize dynamic response (Fig. 1.48).

MR fluid-based adaptive shock absorbers became available commercially in 1998, in the seats of large Class 8, 18-wheeler trucks [314]. In 1999, such MR shock absorbers were introduced in stock-car and drag-race vehicles [315]. Han et al. [316] examined an MR seat damper in conjunction with the primary ER suspension to isolate vibration in a commercial vehicle such as a large truck. A skyhook controller was designed for each damper. It was demonstrated that both vertical displacement and acceleration at the driver's seat were considerably reduced during testing.

Ushijima and Kumakawa [317] used piezoelectric technology for an active engine mount to minimize vibration in the chassis due to the engine. Because of the high-force and low-displacement characteristics of piezoceramic stacks, a hydraulic gearing mechanism was incorporated into the mount. Sproston et al. [318] used ER fluid to develop a prototype engine mount. The fluid is sandwiched between two

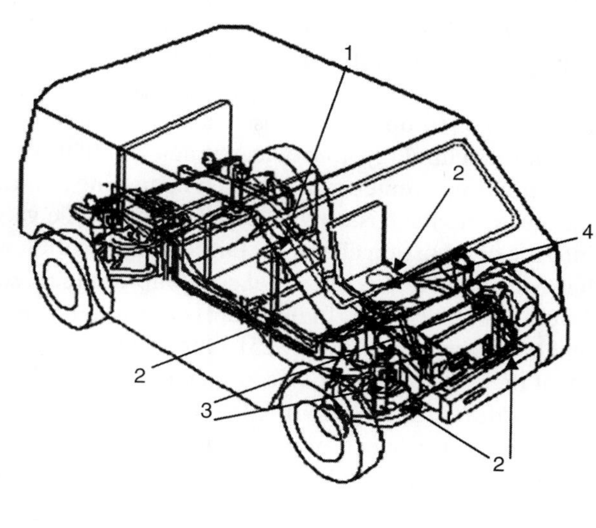

Figure 1.47. Installation of semi-active suspension: 1 – controller; 2 – accelerometer; 3 – MR damper; 4 – sensor for ride comfort evaluation, from Ref. [308].

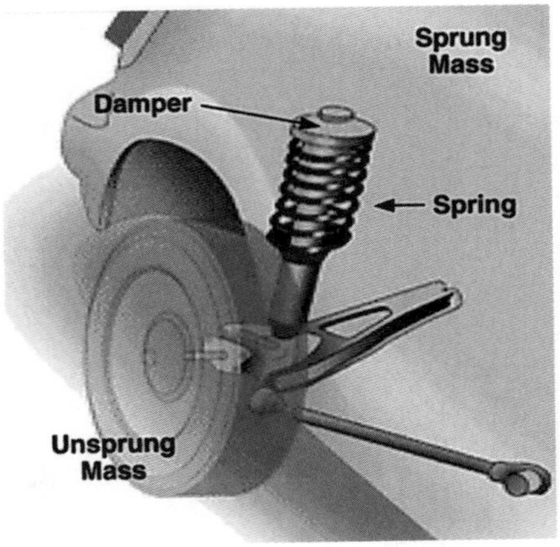

Figure 1.48. Vehicle suspension, from Ref. [313].

electrodes. When actuated by an electric field, the transmissibility of the engine mount is minimized.

Active control of noise may be achieved through active roof panels. Manz and Breitbach [319] and Giovanardi et al. [320] examined interior noise control (30–250 Hz) in an automobile using surface-mounted piezoceramic patches on roof panels. Using multiple input–multiple output (MIMO) control algorithms, in conjunction with a digital signal processor (DSP), the overall sound level radiated from the panel was reduced by 3 dB.

Currently, automotive systems rely primarily on controllable actuation mechanisms such as hydraulic systems, to achieve good braking and clutching characteristics. Hydraulic actuated control systems, however, have high power density, but have poor efficiency and low robustness (varying bulk modulus of pressurized fluid), and they require significant pumping hardware, valves and tubes. Recently, there have been selected investigations to achieve clutching and braking using MR fluids [321]. However, MR fluids have uncertain operation characteristics (loss of homogeneity) in a high centrifugal force (CF) environment due to micron-sized ferrous particles suspended in the carrier fluid plus associated sealing problems. Neelakantan and Washington [323] tried to solve this problem by developing an MR clutch design in which the fluid is encapsulated in a highly absorbent polyurethane foam. Piezoelectric actuators with high block force and quick response were investigated to control clutching and braking requirements [324].

Fuel injectors based on piezoelectric actuators are used in internal combustion engines to open and close fuel-injection valves. As compared to conventional solenoid technology, piezoelectric stack actuators can provide adequate force, and potential impovements in precise positioning and rapid response. The stroke of the piezoelectric stack is used to activate the needle valve that controls the fuel injected into the combustion chamber. Direct fuel injection with piezoelectric stack actuators has been shown to reduce fuel consumption in diesel and gas engines by up to 15%. Overall, piezoelectric fuel injectors are quieter, more economical, more powerful, and have fewer emissions [325]. One concern has been the performance and reliability of these piezoelectric-based fuel injectors under actual operating conditions with temperatures ranging from $-30°$C to $125°$C. Senousy et al. [326] systematically examined the performance of piezostack actuators as pertaining to fuel injection under sinusoidal and trapezoidal driving fields over a temperature range from $-30°$C to

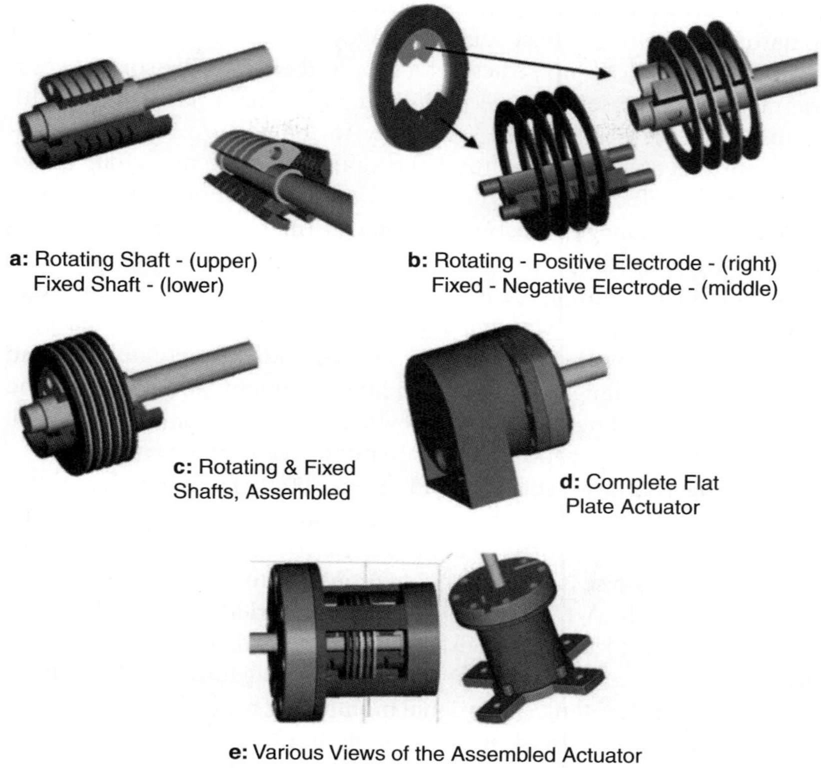

Figure 1.49. Components and assembly of multiple flat-plate ERF actuator, from Ref. [327].

80°C for various frequencies, rise times, and duty cycles. Reducing the duty cycle was shown to significantly decrease the heat generated in the actuators.

In the automotive industry, haptic feedback devices can be used to enhance the human-vehicle interface. Weinberg et al. [327] developed ER fluid-based–rotary resistive actuators for haptic interfaces for vehicle instrument control (Fig. 1.49). They developed the prototype of a haptic joystick mechanism with two degrees of freedom; each degree required a separate resistor actuator. Ahmadkhanlou et al. [328] designed a haptic system based on the MR fluid sponge damper for an automobile steer-by-wire application. The goal was to replace the bulky hydraulic system with a centralized computer that receives electronic inputs from the driver. Results showed good force feedback control and stability of the system.

The growth of applications of smart structures in automotive systems can provide a tremendous impetus to this discipline but, more importantly, can be a major driver to lower the cost of smart materials. However, product reliability, robustness, maintenance flexibility, and affordability can be key factors for widespread application in automotive systems.

1.9.8 Marine Systems

The applications of smart structures in marine systems cover structural acoustic control, the control of machinery vibration, radiated noise reduction, shape control/flow control to increase maneuverability, and health monitoring and condition-based maintenance. Affordability, simplicity of design, and robustness are key factors for these applications. Many of these applications require high-strain/displacement and large force actuators, as well as robust sensors. Kageyama et al. [329] carried out structural health monitoring of ship structures using a fiber-optic sensor network. Potential advantages of these sensors are unlimited gage length, large bandwidth (measure dynamic strains), and reduced sensitivity to temprature. To accommodate the huge dimensions of marine structures (i.e., more than 200 m), a long gauge fiber-optic laser-Doppler velocimeter was developed. An optical time-domain reflectometer was applied to the damage detection of composite material, thereby allowing breakage of fibers to be monitored.

Under the SAMPSON Program, the underwater performance of a trailing-edge tab assisted control (TAC) surface was evaluated. The tab, which constituted 10% chord, spanned the entire length of the control surface and was actuated with an SMA actuator, significantly changed lift, reduced hinge moment, and increased the maneuver capability of the trailing-edge flap. A new tab design as a contoured control surface (without the hinge) was proposed to further enhance its performance.

S^2DS: Smart Sleeve Demonstration System (1998–2000): The objective of this program was to develop a significant quieting improvement (about 10 dB reduction) in self-generated underwater torpedo noise using a compact and less expensive system. A primary source of noise is pressure fluctuations due to the turbulent boundary layer on the torpedo hull. A team led by Lockheed carried out this DARPA-sponsored program. The closed-loop demonstration was carried out underwater on a torpedo hull using 60 reference sensors, 12 error sensors, 2 staggered rings involving 36 actuators, a digital control system, and a signal monitoring system. Single-channel tests showed a noise reduction of 18 dB in a selectable frequency band. Test results showed significant quieting capability across all frequencies of interest.

Quackenbush et al. [330] investigated the SMA-actuated vortex-wake deformable hydrofoil as a control scheme for a lifting surface for submarines. This active forcing is referred to as a smart vortex leveraging tab (SVLT). A prototype was built and tested in hydrodynamic conditions.

To develop a highly maneuverable underwater vehicle, it is important to design the vehicle based on the undulatory swimming techniques and anatomic structure of fish. Rediniotis et al. [331] built a biomimetic active hydrofoil using SMA wire actuators. The vehicle consists of a skeletal structure similar to that of an aquatic animal and SMA actuators for muscles. Controlled heating and cooling of SMA wires generate bi-directional rotation of the vertebrae, which in turn changes the shape of the hydrofoil. This work demonstrated the potential of SMAs as artificial muscles in underwater applications.

Balakrishnan and Niezrecki [332] examined the application of THUNDER actuators as underwater propulsors. Two THUNDER actuators placed in a clamshell configuration were used to propel water. It was found that the actuators had a peak flow rate of about 1500 cm^3/s and a peak thrust of about 4.5 N. The average electric power consumed by two THUNDER actuators (operating at 14 Hz) is far less than that consumed by other propulsion systems. The displacement response and the current drawn were quite nonlinear. Overall, the results show the potential of THUNDER actuators as underwater propulsors.

Kim et al. [333] examined MR inserts to minimize shock-wave propagation in warship structures. An MR insert was made from an aluminum plate, and two piezoelectric disks were used as the transmitter and receiver. The MR insert showed the capability of shock-wave reduction.

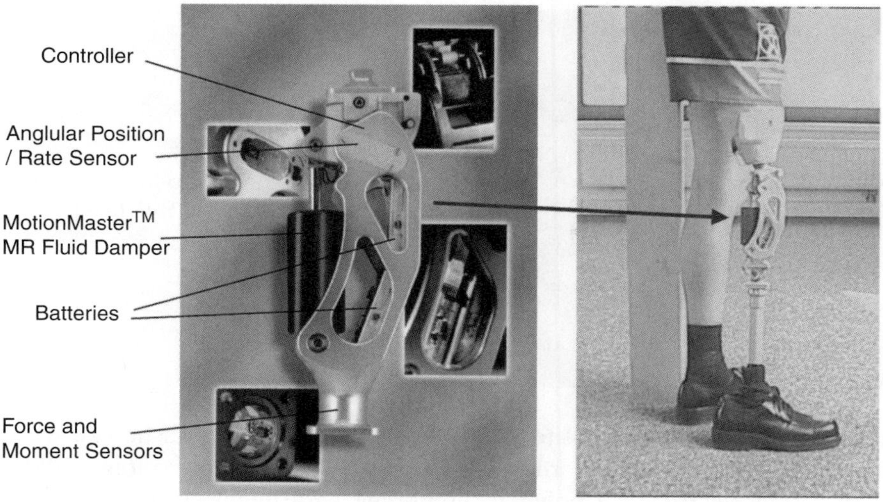

Figure 1.50. Commercial Smart Magnetix™ above-knee prosthesis with real-time control provided by a MotionMaster™ MR fluid damper, from Ref. [337].

1.9.9 Medical Systems

There is a wide variety of applications of smart structures technology in the medical field. Key factors for applications are compactness, low weight, precise control, durability, repeatable operation, and minimum invasiveness. Often, adaptive materials need to be soft with large-strain capability. Applications include prosthetic devices such as artificial hands, knees, and fingers; robotic eyes; the artificial anal sphincter and urethral valve; rehabilitation therapy microrobots; telerobotic surgery; cancer therapy; microrobots swimming in blood vessels; eyeglass frames; orthopedic implants; orthodontic treatments; and tissue fixators.

Many of the externally powered prosthetic devices are actuated by electric servomotors to achieve precise kinematic performance. However, their major drawback is that they are heavy and bulky. To minimize this problem, several alternate lightweight actuators have been investigated for use in artificial muscles. These include electroactive polymers [334], pneumatic actuators [335], and SMA actuators [336]. One possible application of MR fluid technology aims for a compact adaptive damper application to develop an artificial knee that automatically adapts to changing gait conditions [337]. The prosthesis consists of a thigh, a knee joint, a lower leg assembly, and a foot (Fig. 1.50). An array of sensors are used to determine the instantaneous state of the knee, which includes knee angle, swing velocity, axial force, and bending moment. Using a microprocessor-based controller, the MR damper is adapted in real time based on walking speed, weight on the leg, stairs, and slope of terrain. Zite et al. [338] also examined rotary MR fluid–based shear dampers for an orthopedic-active knee brace. The device was designed based on maximum yield stress, corresponding magnetic field, torque, and fluid viscosity. This device could generate variable resistive torque and can fit the requirements of any type of individual. Price et al. [339] presented an SMA ribbon-woven artificial muscle braid, which is capable of achieving strains of more than 30%. Also, they showed that this actuator achieved a 270° angular displacement when applied in a prosthetic elbow joint.

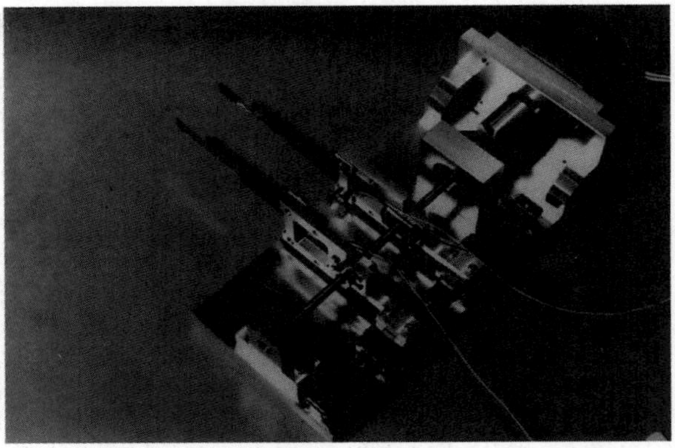

Figure 1.51. Two-fingered parallel gripper, from Ref. [344].

Ionic polymeric–metal composite (IPMC) is a soft, lightweight, plastic-like material, which can undergo large deformations with application of low voltage. IPMC consists of a thin electroactive polymer membrane with metal electrodes plated chemically on both faces. On the application of field across the thickness of IPMC, it bends to the anode side as a result of cation migration toward the cathode in the polymeric network (the composite swells on the cathode side and shrinks on the anode side). Conversely, IPMC can also produce charge when it is deformed. However, they generate low actuation force. Kottke et al. [340] and Lee et al. [341] have examined IPMC for application in artificial muscles. Dielectric gels (non-ionic) are shown to be electrically active actuators that undergo large strains (several hundred percent) and fast response (less than 100 ms) under low current (μA to nA). Hirai [342] studied these soft polymeric materials for application in artificial muscles.

Artificial hands and fingers that are capable of grasping objects are mostly built using servomotors, stepping motors, or pneumatic cylinders in conjunction with tendons, gears, and linear ball bearings. For these devices, the effects of compliance, backlash, and hysteresis can be quite critical. Okamoto et al. [343] used SMA wires to build a two-fingered gripper, primarily for low-speed motion. Chonan et al. [344] built a two-fingered miniature gripper driven by piezoceramic bimorph strips to achieve hybrid position/force control (Fig. 1.51). The fingers are flexible cantilevers actuated by bimorphs at the base and are supported by linear ball bushings that ride on a steel shaft. They demonstrated a grasping force on the order of 0.01 N at the fingertip and a Proportional–Integral–Derivative feedback (PID) controller performed satisfactorily to drive the gripper to achieve stable grasp of an object. Price et al. [345] carried out the design, instrumentation, and control related to the application of SMA toward the development of artificial muscles in a three-fingered robot hand for prosthetic applications (Fig. 1.52). Lee et al. [346] fabricated thick IPMC by stacking several Nafion thin films using hot-pressing system. The application of the IPMC actuator was successfully demonstrated by building artificial fingers with three joints using five-film stacked IPMC actuators (Fig. 1.53). Thayer and Priya [347] developed a biomimetic humanoid hand with the prime objective of typing on a computer keyboard. Each finger has four joints with three degrees of freedom, while the thumb has an additional degree of freedom. The hand consisted of sixteen servomotors dedicated to finger motion and three motors for wrist motion.

Wolfe et al. [348] investigated the deployment of an SMA actuator for a robotic-eye orbital prosthesis. The goal is to create an intelligent prosthesis that can execute

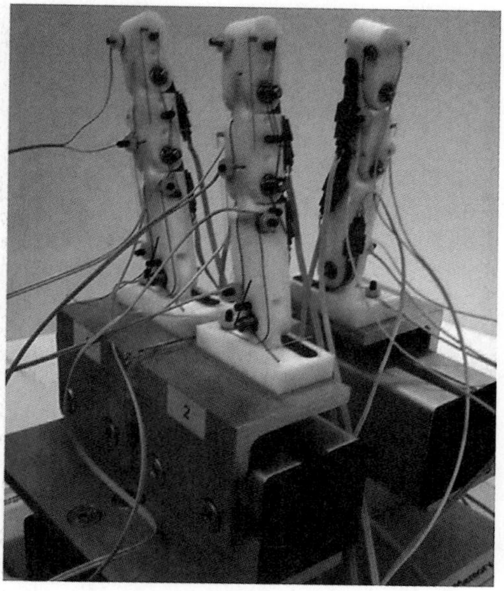

Figure 1.52. SMA actuated artificial hand, from Ref. [345].

vertical and horizontal motion to accommodate 300° of eye rotation. It follows the design philosophy of biomimicry, in which the SMA wires contract on heating and return to their original position when cooled. The precise control of the SMA wire actuator was carried out using pulse-width modulation. A large-scale prototype was built and was successfully tested. To fit an actual model into the orbital cavity, it was proposed to use 100μm-diameter high-temperature SMA wires.

Luo et al. [349] presented the thermal control of an artificial anal sphincter using SMAs to resolve problems of severe fecal incontinence. The artificial sphincters could be fitted around intestines, performing an occlusion function at body temperature and a release function on SMA heating (20°C above body temperature). The device consisted of two SMA plates in conjunction with attached heaters and a reed switch to resolve overheating problems. A successful thermal control was demonstrated in both in-vitro and in-vivo experiments. These artificial sphincters were implanted and successfully tested in animals. Chonan et al. [350] developed an

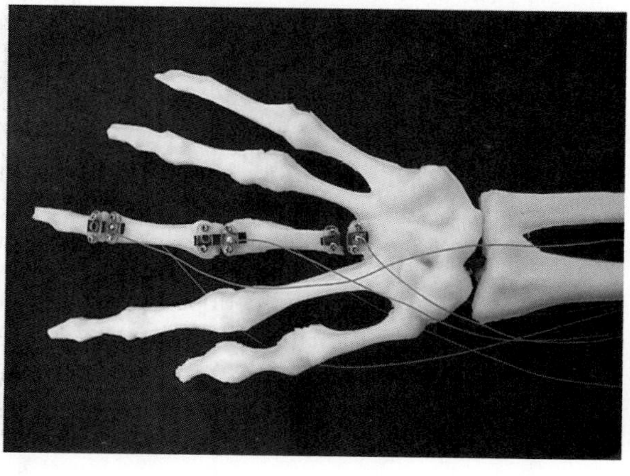

Figure 1.53. Model of artificial finger, from Ref. [346].

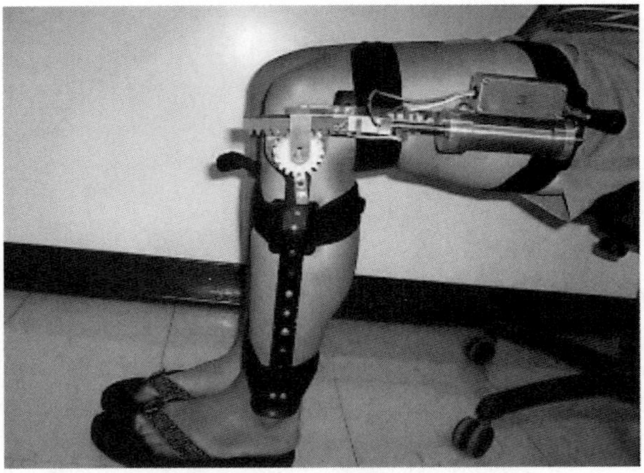

Figure 1.54. A variable resistance knee brace with MR damper, from Ref. [353].

artificial urethral valve using thin Nitinol plates (thickness 0.3 mm), which closes to block the discharge of urine at body temperature (martensite phase) and opens to release urine when the SMA plate is heated (austenite phase) using external heating through an attached Nichrome wire. The functioning of the valve was successfully demonstrated by animal experiments, both in-vitro and in-vivo. Tanaka et al. [351] developed an implantable artificial urethral valve via a transcutaneous power transmission system with closed-loop thermal control. Using the transcutaneous system, electric energy is supplied to the valve without wires penetrating the patient's body. Using an implanted temperature-monitoring circuit, the overheating of the SMA actuator during prolonged urination is prevented. The valve and the closed-loop power transmission were successfully demonstrated through animal experiments, both in-vitro and in-vivo.

It is expected that in the future, robots will be of great use in medical fields such as rehabilitation therapy, nursing, and day-to-day support of elderly people. For example, soft robot hands that have dexterity similar to human hands and are able to perform complex movements will be of immense value in nursing and welfare. Such a robot has to be lightweight, mobile, and soft. Saga [352] developed a tendon-driven robot hand using a pneumatic balloon as a directly operated actuator. It is a simple compact system using flexible silicon rubber material and is quite comparable to the biological human muscle. Dong et al. [353] developed a smart variable-resistance exercise machine using MR-fluid dampers for rehabilitation of patients with neuromuscular and orthopedic conditions (Fig. 1.54). An intelligent controller is incorporated to regulate the resistive force or torque of the device such that it provides both isometric and isokinetic strength training for the human joints, including the elbow, knee, hip, and ankle. Bose and Berkemeier [354] designed and built an ER fluid–based haptic device. The device is similar to a joystick, in which the user feels resistance forces against the motion of the stick through the change of rheological properties of the ER fluid due to an electric field. Large forces can be quickly realized by applying electric fields of different strengths. Such haptic devices can be used in various applications such as supporting tools for the operation of machines, in virtual reality, and computer games, as well as for assistive interfaces for blind persons working with a computer.

To carry out telerobotic surgical procedures efficiently, a force feedback, in addition to visual feedback, is essential. Thus, the user not only sees the movement of the end effector in a video interface monitor but also can feel the forces encountered by the end effector. This way, surgeons can feel whether their tools have hit a hard bone or a soft tissue, and they can avoid any unnecessary complications in surgery. Rov et al. [322] demonstrated the application of MR fluid devices as a force feedback system for telerobotic surgery. For this test, they built an MR sponge damper consisting of polyurethane foam soaked and saturated in MR fluid and wound around an electromagnetic piston, and they demonstrated its effectiveness as a force feedback system.

Minimally invasive surgery (MIS) is routinely used in abdominal procedures such as gall bladder removal, in which surgical procedures are carried out using small surgical tools and viewing equipment in conjunction with long slender tubes. These tubes are inserted into the body through a few small incisions (5–10 mm). There are enormous benefits, including reduced tissue trauma and recovery time, of the MIS technique over conventional open surgery. However, there are limitations with MIS, which include a lack of dexterity and localized actuation of the surgical end-effector and a lack of haptic feedback to the surgeon. Hence, it is quite difficult to expand current MIS technique to more complex surgical procedures such as coronary artery bypass operations. Edinger et al. [355] developed new MIS tool designs incorporating compliant and smart structures technologies. An active grasping tool consisting of a single-piece compliant end-effector and localized actuation and force sensing using a small PZT inchworm actuator was built as a telerobotic system. Rubio et al. [356] developed an electro-thermally driven microgripper using topology-optimized design and laser microfabrication. The design is a symmetric monolithic 2D structure that consists of a complex combination of rigid links integrating both the actuating and gripping mechanisms. The microgripper had overall dimensions of 2.5 mm (width) and 1 mm (length) and was able to deliver the maximum tweezing and actuating displacements of 25.5 μm and 33.2 μm with a power draw of 2.3 W (Fig. 1.55) [356].

Flores et al. [357, 358] suggested a novel cancer-therapeutic approach by injecting MR fluids into the blood vessels supplying the tumor and, through the application of magnetic field at the tumor site, blocking the blood flow within the vessels. The biocompatible MR fluid is made of magnetic particles (0.25–1.0 μm) coated with starch and suspended in water or sheep's blood. The objective is to starve the tumor from the flow of blood. The sealing effect to the fluid flow is achieved at low particle concentrations with strong pressure resistance. The characteristics of the magnetic seal depend on the properties of the fluid (i.e., particle size, volume fraction, viscosity, and susceptibility), the strength of the magnetic field, and flow rates. As flow rate increases, the seal becomes unstable.

Tanaka et al. [359] developed an active palpation sensor for the detection of prostatic cancer and hypertrophy. The receptor of the sensor is a PVDF film placed on the surface of a sponge-rubber layer (Fig. 1.56). The sensor is inserted into the rectum and is driven by the motor at about 50 Hz with a constant peak-to-peak amplitude of 2 mm in order to measure the output voltage that determines the stiffness of the gland. Clinical tests demonstrated the effectiveness of the sensor. Wang et al. [360] developed a PVDF piezopolymer sensor for unconstrained in-sleep cardiorespiratory monitoring such as respiration and heart rate. The objective is to use the sensor on an ordinary bed, under the sheets, at the location of the thorax to pick up the fluctuation of pressure. Wavelet multiresolution–decomposition analysis is used to detect respiration and heartbeat from the sensor output.

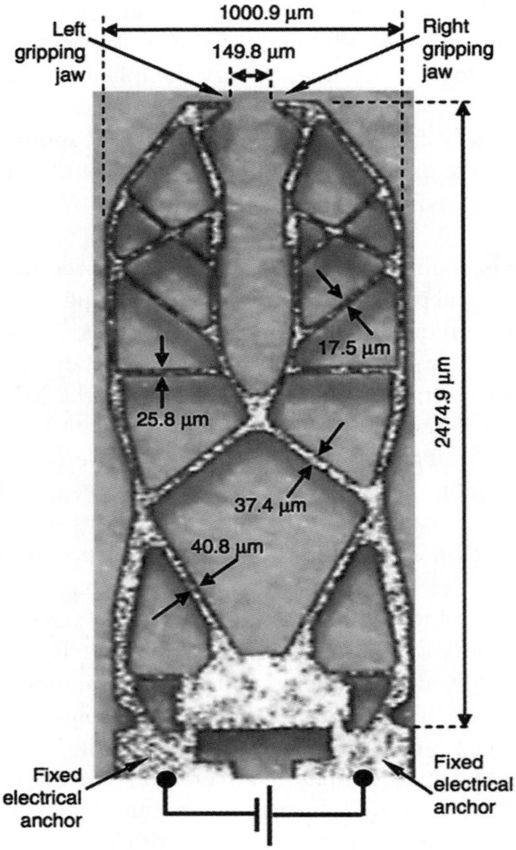

Figure 1.55. Fabricated microgripper prototype, from Ref. [356].

Sudo et al. [361] examined the development of a microrobot working in human blood vessels. It is a wireless swimming mechanism in which the locomotion characteristics are provided by a permanent neodynium magnet in conjunction with an alternating magnetic field (Fig. 1.57). It was found that the swimming velocity of the microrobot depends on the tail width, the tail length, and the amplitude of bending oscillation. The test results showed that the magnetic robot can move through the aorta, arteries, veins, and vena cavae of the human body. However, a further miniaturization of the magnet is needed to freely move the robot through small arteries, arterioles, capillaries, and venules.

An SMA microcoil actuator can induce bending, extension/contraction, torsion, and stiffness variation. Using this concept, a hollow flexible small tube, called a catheter, was built by Haga et al. [362] for minimally invasive diagnosis and treatment of the diseased site. The catheters were 0.3–3.0 mm in diameter and 1.5 m in length and moved like a snake in blood vessels (Fig. 1.58). The tip of the tube could be controlled from outside the body. One potential application of this tube, due to its easy passage at the lower end of the stomach (pylorus), is the treatment of intestinal obstruction. At this time, endoluminal devices are being used for drug delivery, diagnosis, and surgical applications, especially for the gastrointestinal tract. These are in the form of miniaturized and swallowable capsules that can move easily inside the human body. These are normally composed of a cylindrical shell including a camera, an illumination system, a wireless transmitter, and a battery.

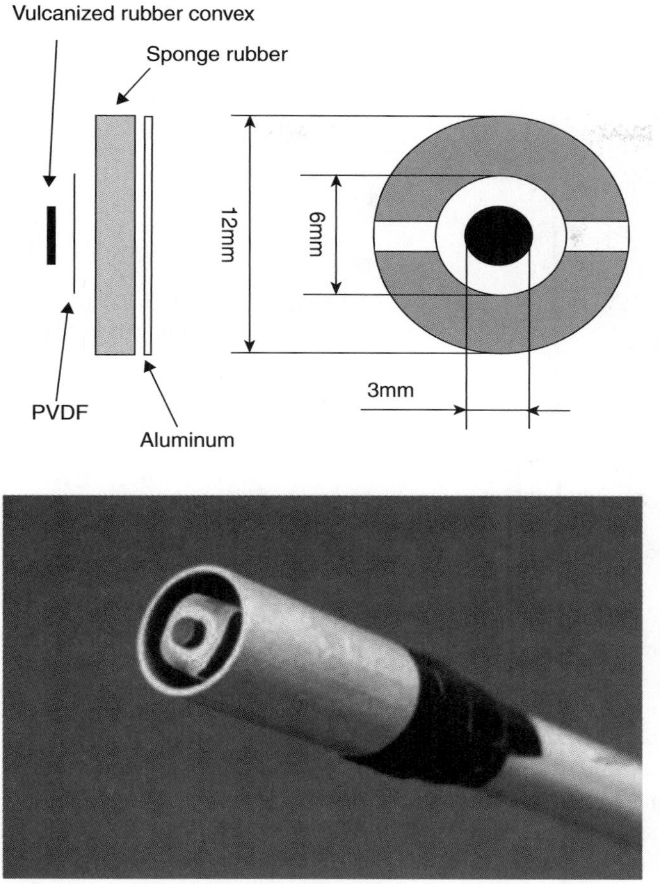

Figure 1.56. Schematic diagram of the geometry of receptor of sensor 1 and a close-up photograph of the sensor head, from Ref. [359].

Buselli et al. [363] developed an alternate approach of a self-propelled, legged endoscopic capsule using superelastic Nitinol in conjunction with microfabrication (Fig. 1.59). Nitinol appears to show good biocompatibility due to its large obtainable strains and adjustable superelastic properties. Sputtered tubes have a high potential for application as vascular implants, such as stents. Miranda et al. [364] fabricated thin Nitinol film stents of thicknesses varying from 5 to 15 μm and a diameter from 1 to 5 mm using magnetron sputtering, 3-D lithography, and wet etching.

The use of SMA in eyeglass frames dates back to 1975 when the first patent was filed [365]. Normally, the frame is built using the superelastic property of SMA, whereby a frame that is accidently bent recovers its original shape. Also, because Nitinol is corrosion resistant, the frame does not require any additional electroplating or coating. It is lightweight as compared to other metals. However, a drawback can be its performance in cold weather because the material becomes very soft in its martensite phase.

Haasters et al. [366] carried out the use of Nitinol as an implant material in orthopedics. Specific applications include osteosynthesis plates (surgical treatment of bone fracture), jaw plates (fixation of lower-jaw fracture), staples (simple clamps of the lower extremities), medullary nails, and spacers. They also carried out several

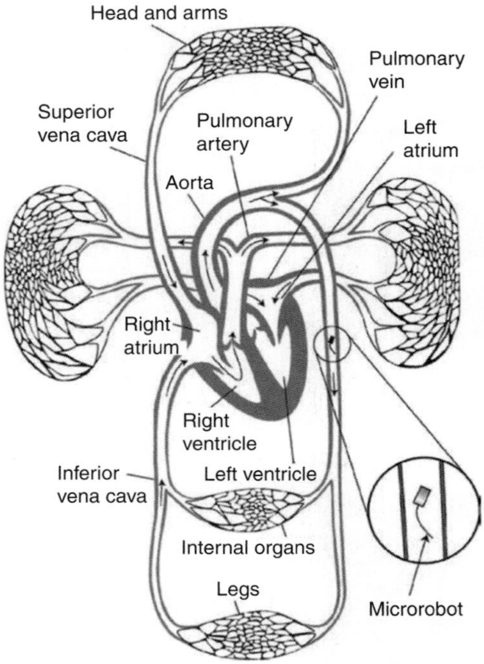

Figure 1.57. Schematic of circulatory system and magnetic swimming robot, from Ref. [361].

applications of Nitinol in animals involving the implantation of jaw plates, staples, and spacers. Other clinical applications of Nitinol as implant material include operative ankylosis of the foot and corrective osteotomy at the knee joint in genu valgum. Sachdeva et al. [367] used superelastic NiTi alloys in dentistry for orthodontic treatment. An effective controlled alignment of teeth requires the application of specific low applied forces (on the order of several Newtons) that are within physiological limits and act over longer periods of time. Too large an applied force can cause damage to the supporting tissues, and too low a force can slow down the alignment process. Traditionally, gold, stainless steel, elastomers, titanium-molybdenum alloys, and chrome cobalt nickel were used for orthodontic therapy. Because of its excellent superelastic material characteristics, such as springback, low stiffness, and constant maximum stress over a wide range of deformations, Nitinol appeared suited for orthodontic mechanotherapy. Superelastic alloys (austenite temperature below

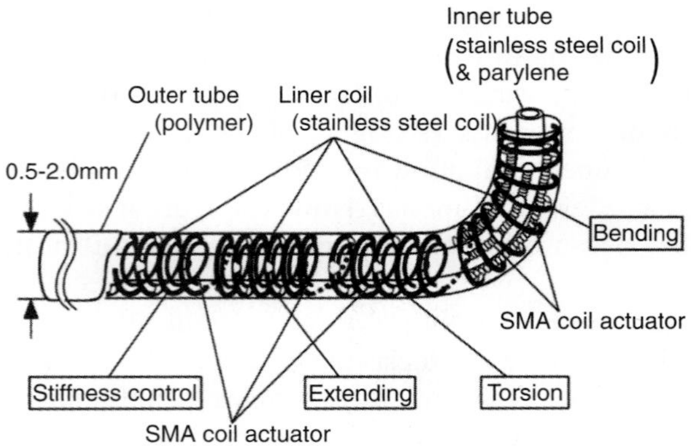

Figure 1.58. Bending, torsional, and extending active catheter using SMA coil actuators, from Ref. [362].

Figure 1.59. Three endoscopic capsules that range from 11 to 12 mm in diameter and 25 to 40 mm in length, from Ref. [363].

room temperature) are used in the fabrication of orthodontic archwires. The efficiency of orthodontic tooth movement is greatly enhanced by employing this alloy through the application of low and continuous forces with large recoverable strains (6–8%). Raboud et al. [368] carried out the simulation of superelastic Nitinol wires used as orthodontic springs to apply the necessary force systems to effectively move teeth. They included both bending and twisting deformations in the model. Fokuyo et al. [369] used Nitinol for a dental endosseous implant. The shape-memory implants appeared to show strong and continuous forces of mastification; they can be easily installed and have a good stress dispersion. There are other medical-related applications of superelastic Nitinol wires, such as the Homer Mammalok needle (used for breast cancer treatment) [370], guidewires, and arthroscopic instrumentation [371].

Song et al. [372] developed a tissue-fixator in minimal access surgery using a SMA. It provides an alternative to conventional thread-based suturing of human tissue. The fixator is made of round NiTi wire and its open and closed forms are shown in Fig. 1.60. Its deployment may be faster than the conventional approach of sutures.

Shahinpoor and Kim [373] presented a review of potential applications of ionic polymer-metal composites in the medical field, including mechanical grippers (at micro- and macro-levels), robotic swimming structures (robotic fish), artificial ventricular or cardiac-assist muscles, surgical tools, peristaltic pumps, and bionic eyes.

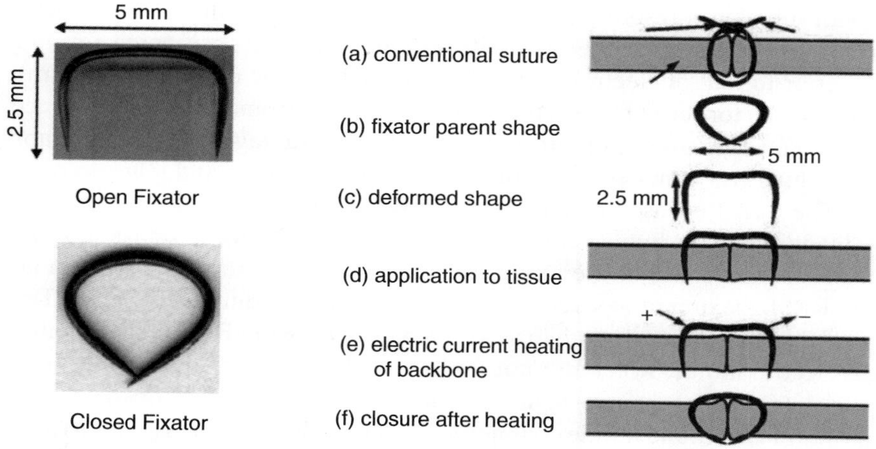

Figure 1.60. The SMA surgical fixator prototype and working principle, from Ref. [372].

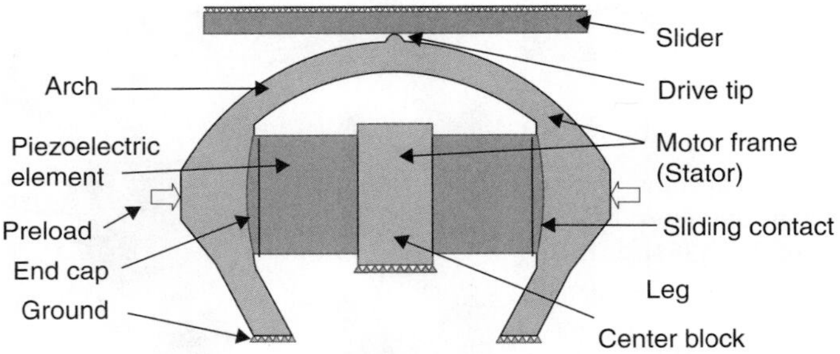

Figure 1.61. Basic ultrasonic motor design components, from Ref. [378].

In each of these devices, electrical energy is converted into mechanical energy to perform useful work. The strength of ionic polymers is large induced strains and the major weaknesses are the low stiffness and high field requirements.

1.9.10 Electronics Equipment

Much commercial electronics equipment is being built using smart structures technology. Key factors for applications are cost, design complexity, power requirements, expanded capability, durability, and precision control. Smart material–based electronic devices include ultrasonic motors, exercise bicycles, CD-ROM drives, the backlight inverter for large LCD-TVs, and active antennas.

Optical systems require extremely precise movements, whether under static or dynamic loads. The incorporation of adaptive structural systems can improve the performance of optical systems. Conventional electromagnetic motors cannot be easily miniaturized (smaller than 1 cubic centimeter) and are too inefficient, especially for more precise positioning and quiet operation. Piezoelectric motors are a possible solution to achieve efficient compact motors at small size. They can be divided into three categories: impact-drive mechanisms, inchworm mechanisms, and ultrasonic motors. Among these, the most popular is the ultrasonic motor. Ultrasonic motors consist of straight metal bars or plates bonded with piezoceramic elements used as stator. The induced displacement is amplified by two teeth and transmitted by the frictional force between the motor and the rail in a linear motion, at a velocity of up to 1 m/s with submicron resolution. In the case of an ultrasonic motor, efficiency is insensitive to size and specific power, and its response time and positioning accuracy are far superior to that of electromagnetic motors. Ultrasonic motors can generate low speed and high torque with no additional gears, no electromagnetic interference, compact size, and short start-stop times. The major disadvantages are that they normally need a high-frequency power supply, and significant wear and tear can occur over time. The term "ultrasonic" means high driving frequency in the range of 20–200 kHz (inaudible to the human ear). The basic ultrasonic motor design is shown in Fig. 1.61. The performance of the motor is characterized in terms of maximum no-load (free) RPM, maximum blocked force or torque, and maximum efficiency. Typically, speed decreases linearly with load. In 1948, Williams and Brown filed a patent on a piezoelectric motor, which did not find an application for a long time. After several attempts by different investigators, Kanazawa et al. [374] developed a refined ultrasonic motor that found a wide commercial application in cameras (Canon) as an autofocus drive. The ultrasonic motors are now finding a wide range of applications in robots, medical instruments, cameras, and aerospace systems. Uchino [375, 376] provided a comprehensive overview of piezoelectric ultrasonic motors that includes historical developments; low-speed/high-torque and high-speed/low-torque motors; rotary- and linear-type motors; standing wave and traveling wave-based devices; and rod type, π-shaped, ring, and cylindrical geometry motors.

Overall, piezoelectric ultrasonic motors have high specific thrust, high displacement resolution, no parasitic magnetic field, and an absence of frictional locking in power-off condition. As such, they find applications in precision micromechanical systems. Many miniature ultrasonic motors are built based on bending modes of piezoelectrically excited beams or plates. Dong et al. [377] used a wobbling motion to develop an ultrasonic rotary motor. It is based on the excitation of a bending-bending mode (two orthogonal modes) of a hollow PZT cylinder, which combine to produce a wobbling motion that drives the motor. The prototype was successfully tested over a range of operating conditions. Sharp et al. [378] provided an overview of the design of ultrasonic piezoelectric motors, including the selection of materials for different motor components. They built a simple motor that consisted of an arched frame, a center ground, and two piezoelectric elements connected to the center ground (see Fig. 1.61). They also carried out Finite Element Method (FEM) analysis to predict the performance of the motor.

Tian et al. [379] developed an exercise bicycle using ER fluids based on zeolite and silicone oil. Changing the strength of the applied electric field results in an active control of the resistance.

Pires et al. [380] designed a miniature piezoelectric bimorph–actuated precision-flow pump to cool down the Light Emitting Diode (LED) set inside a headlight system for medical applications. This is a compact, low-noise, and low-power-consumption system. The flow measurement in a prototype system showed satisfactory performance.

In a high-capacity CD-ROM drive (optical storage device), the elimination or suppression of vibration in the feeding system is a key to achieving desirable performance. The vibration, which is affected by the unbalanced flexible disk rotating at high speed and external excitation to the case frame, restricts the tracking and focusing of the servo. It is quite difficult to achieve satisfactory performance at the resonance frequency of the feeding system using the conventional passive rubber system. Lim et al. [381] used an ER fluid mount for vibration control of a CD-ROM feeding system. Its effectiveness for vibration suppression was demonstrated through hardware-in-the-loop simulation associated with a skyhook controller. Yang et al. [382] developed an optical disk drive using multilayered PZT bimorphs. A novel flexure-hinge mechanism was used to amplify the stroke for both tracking and focusing motions.

Huang et al. [383] examined the development of a high-voltage, high-powered, low-cost backlight inverter for lighting long, cold, cathode fluorescent lamps on large size LCD-TVs using piezoelectric transformer (PT) technology. Benefits of PT technology include high energy-transfer efficiency, very low temperature rise, compact size, and superior safety. This efficient design employs a single-layer PT to drive long, cold cathode fluorescent lamps under high voltage and high power while keeping the material and manufacturing costs competitive as compared to conventional coil-based designs.

Aperture-type antennas are normally rigid, consisting of parabolic, paraboloidal, cylindrical, or spherical shapes. A major flaw with this type of antenna is that the

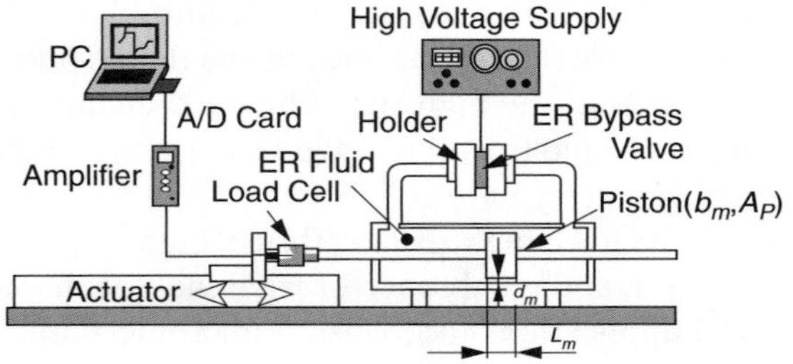

(a) Setup for measuring fundamental characteristics of prototype

(b) Configuration of prototype

Figure 1.62. Prototype ER buffer for railway vehicles, from Ref. [386].

whole structure has to be moved to scan a signal from an alternate point. Conversely, an active antenna has the ability not only to scan but also to vary focus. Such an antenna can be adjusted to compensate for varying atmospheric conditions. Reflector surface adaptation with compact smart material actuators can achieve performance gains, compactness, simplicity, and cost advantage over phase-array antennas. A class of antennas capable of variable directivity (beam steering) and power density (beam shaping) has been developed with adaptive materials such as PVDF film in conjunction with metalized Mylar substrate, piezoelectric stacks, and electrostrictives [384].

1.9.11 Rail

There have been some applications of smart structures technology in rail systems. Key factors for applications are robustness, maintenance cost, and durability.

Peel et al. [385] described the development of a dynamic model of an ER-based controllable-vibration damper for ground-vehicle suspension systems. The phenomenonological-based model is developed for characterizing the behavior of ER fluid in a flow-control valve by taking into account ER fluid inertia and compressibility; an iterative procedure is adopted to solve nonlinear equations. The application of this approach is carried out to control the lateral vibrations of a rail vehicle.

Chonan et al. [386] developed an active "relief buffer" for railway vehicles using ER fluid to control the coupler force. The goal is to lower the coupler force acting between the failed and the relief train set. The buffer system is expected to support a maximum coupler force of 500 kN. To evaluate the feasibility of a buffer system, a prototype consisting of a hydraulic cylinder, an ER bypass slit valve, and a PID controller was built (Fig. 1.62).

Performance of an active buffer is investigated both theoretically and experimentally. A Bingham plastic model was used to model ER fluid flow through the bypass slit and the Newtonian mixed-flow model was used to model flow through the piston-cylinder gap. Results demonstrated that the coupler force of the railway vehicle could be controlled effectively by using the ER bypass damper.

Fotoohi et al. [387] designed an MR damper in conjunction with a skyhook controller for a rail-suspension system.

Peiffer et al. [388] carried out technology development and evaluation of an active vibration control system for high-speed train-bogies. A transfer path analysis was initially carried out to identify the main paths of noise and vibration transmission, and then a detailed finite element analysis was performed for the integrated system to evaluate the performance of several actuators. The objective was to mitigate the structure-borne noise, which is generated at the wheels and transmitted via the primary and secondary suspension system of the bogies to the car body. The system was integrated into one axle of a train bogie and was successfully tested. The structure-borne noise at the wheels, which was found to be the key source for the bogies' interior noise, was a function of wheel threads, vehicle design, type of rail corrugation, and track construction.

Vibration is one of the major issues in high-speed trains, which affects not only the ride quality but also has a significant impact on the ride stability and maintenance cost of the tracks. Various types of suspension linking the bogies and the car bodies have been designed to increase passenger comfort. The most routinely used suspensions are passive in nature and involve springs and pneumatic or oil dampers. These are cost effective and simple in design. However, their performance over a wide frequency range is limited. Conversely, active suspensions for railway vehicles could provide superior performance over a wide frequency range, but these require actuation power and robust control strategies. Liao and Wang [389] examined semi-active suspension systems based on MR fluid dampers. An LQG control law using an acceleration feedback controller was adopted. Through a numerical simulation, they demonstrated the effectiveness of controlled MR dampers under random and periodic track irregularities.

1.9.12 Robots

Applications of smart structures in robotic systems, especially at mini- and microscales, are growing rapidly. These robots are being built for the medical field, computers, surveillance vehicles, automotive systems, and machine tools. Key factors for smart structures applications are stroke and actuation authority, robustness, maintenance cost, durability, precision control, and power requirements.

The robotic gripper is the end-effector of a robotic arm and needs to be high in energy density (power-to-weight ratio), flexible, and complex in kinematic motion. Yan et al. [390] developed a miniature-step mobile robot for micropositioning with three degrees of freedom using a piezostack actuator. The device deploys a rhombic flexure-hinge frame and four electromagnetic legs to achieve large-stroke translation and rotation. An electrical circuit was developed to control the electromagnets to achieve the inchworm principle (i.e., clamp and release from the platform). Ashrafiuon et al. [391] built a small-scale, three-degrees-of-freedom robot with two SMA bias actuators and a servomotor. The nonlinear behavior of the SMA, including hysteresis, requires a controller even for one-way actuation. Several tests were carried

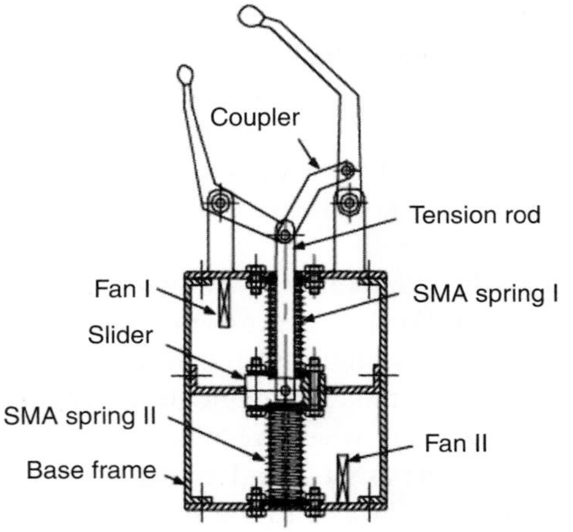

Figure 1.63. Schematic of SMA gripper, from Ref. [393].

out with a desktop prototype robot, and robust performance was obtained despite significant modeling inaccuracies.

Rastegar and Lifang [392] presented a systematic approach for optimal integration of active materials into the structure of a robotic manipulator to minimize higher harmonic components of the required actuating torques. This method may allow robotic manipulators to operate at higher speeds, with greater tracking precision and with minimal vibration.

Yan et al. [393] developed a gripper actuated by a pair of differential SMA springs that are heated by electric current (internal heating) and cooled by fans. The gripper consists of a pair of fingers, a coupler, a tension rod, a slider, and a frame (Fig. 1.63). A PI controller was used to control the output position of the gripper. The opening and closing motions of the two fingers are driven by an SMA-differential actuator in conjunction with a six-bar linkage. A prototype was built and performance was evaluated. Good control of the gripper position was obtained using a feedback system.

1.9.13 Energy Harvesting

Harvesting energy from the environment, especially from wind using windmills to grind grains or pump water, dates back to 500–900 AD in Persia. Because of recent developments in low-power and efficient microelectronics, there has been a renewed interest in energy harvesting using smart materials. One approach is to use piezoelectric materials to generate electric energy from the mechanical vibrations of the host structure (direct effect) [394]. This results in straining of material, which in turn is transformed into electric energy. Such a power generator will perform poorly at low frequencies and low amplitudes. Typically, the energy generated through the piezoelectric effect is not adequate for most applications. Thus, it is important to accumulate and store the harvested energy until a sufficient amount of energy becomes available to power the portable electronics. Sodano et al. [395] investigated two methods of accumulating the energy produced by a piezoelectric plate. The first method uses a capacitor to store energy, and the second method uses rechargeable nickel metal-hydride batteries. Through the excitation of an aluminum plate (0.98 mm thickness) attached to a piezoceramic plate (62 mm $\times$ 40 mm $\times$ 0.257 mm), it was demonstrated that a 40 mAh battery could be charged within a couple of hours.

A fundamental problem of generating electric power with piezoelectric material is that it stores most of the generated electric energy, which causes supplemental excitation opposite to straining direction and thereby reduces the effectiveness of the energy generator. Therefore, to increase the efficiency of a power generator, it is important to minimize the energy storage by the material. This task requires circuitry to remove and store energy generated by piezoelectric material. For example, Umeda et al. [396] developed a bridge rectifier and capacitor to store energy. Elvin et al. [397] used PVDF piezo film attached to a simply supported Plexiglas beam to generate electric energy from bending and accumulate it using a capacitor in conjunction with a cut-off switch. Normally, the storage of energy using a capacitor is not an efficient approach. It has poor power-storage characteristics because of its quick discharge time. Sodano et al. [398] demonstrated that direct storage of energy generated by a piezoelectric sheet using a rechargeable battery (nickel-hydride) is superior to a capacitor approach in terms of the extent of stored power and the fast discharge rate. They also compared power generation using macro-fiber composite (MFC) and monolithic piezoceramic material (PZT) and showed that MFC, despite its high piezoelectric coefficient, is less efficient because of lower current levels [394]. In a later paper, they [399] showed that because of the low capacitance characteristics of MFC, interdigitated electrodes (IDE) design dominates the power-harvesting properties. A careful design can show improved performance of MFC. Amirtharajah and Chandrakasan [400] developed another circuit to store energy by integrating a finite impulse response (FIR) filter, power field-effect transistors (FETs), and pulse-width modulation (PWM) control. Ottman et al. [401, 402] and Lesieutre et al. [403] developed alternate circuitries to store energy. Some researchers have also examined implantable and wearable power-harvesting devices, embedding them into clothing, implanting them inside biological systems, and embedding them in shoes [394]. For example, Kymissis et al. [404] examined the use of piezoelectric actuators embedded in the sole of a shoe to harvest energy during walking. The peak powers were measured to be 20 mW and 80 mW, respectively, with a PVDF stack and a PZT unimorph. Note that when a power-harvesting system is integrated into a structure, it results in an increase in the net damping of the system, comparable to resistive shunting [403]. Doubtless, with the advances in low-power electronics and wireless technology, power harvesting is a key link to developing a completely self-powered system.

To assess the performance of power-harvesting systems generating power from mechanical vibration, a two-port network (coupled electromechanical) model of the transducer, which can interface between the structure and the electrical load in a unified manner, is developed [405]. The power harvested by two different transducers, respectively representing piezoelectric and magnetostrictive materials, on a simply supported beam is calculated for optimal matching conditions. Based on this numerical study, the piezoelectric transducer was found to be superior to the magnetostrictive transducer of equal mass for energy harvesting, provided the transducer parameters are optimally tuned.

A new class of piezoelectric-based energy-harvesting power sources for mounting on platforms that vibrate at very low frequencies (less than 1 Hz) is being developed [394, 406]. The platforms can be ships, trains, or trucks, which rock at low frequency. The mechanical energy available for harvesting is a function of the amplitude and frequency of the platform and the size and mass of the power transducer.

Recently, there has been increased research activity toward ambient energy harvesting using smart material transducers as power supplies for low-power electronics. However, there are numerous barriers that need to be overcome before such energy harvesting becomes a viable option.

BIBLIOGRAPHY

[1] H. R. Clauser. Modern materials concepts make structure key to progress. *Materials Engineering*, 68(6):38–42, 1968.

[2] H. R. Clauser. From static to dynamic materials in design. *Mechanical Engineering*, 20–26, 1975.

[3] V. Giurgiutiu, C. A. Rogers, and Z. Chaudhry. Energy-based comparison of solid-state induced-strain actuators. *Journal of Intelligent Material Systems and Structures*, 7(1): 4–14, 1996.

[4] C. D. Near. Piezoelectric actuator technology. *Proceedings of the SPIE Smart Structures and Materials Symposium*, 2717:246–258, San Diego, CA, 1996.

[5] J. J. Dosch, D. J. Inman, and E. Garcia. A self-sensing piezoelectric actuator for collocated control. *Journal of Intelligent Material Systems and Smart Structures*, 3: 166–185, January 1992.

[6] V. Gupta, M. Sharma, and N. Thakur. Optimization criteria for optimal placement of piezoelectric sensors and actuators on a smart structure: A technical review. *Journal of Intelligent Material Systems and Structures*, 21(12):1227–1243, August 2010.

[7] S. R. Hall, E. F. Crawley, and J. P. How. Hierarchic control architecture for intelligent structures. *Journal of Guidance, Control and Dynamics*, 14(3):503–512, 1991.

[8] H. Kawai. The piezoelectricity of poly(vinylidene) fluoride. *Japanese Journal of Applied Physics*, 8:975–976, 1969.

[9] L. C. Chang and T. A. Read. Plastic deformation and diffusionless phase change in metals – the gold-cadmium beat phase. *Transactions of the AIME, Journal of Metals*, 191:47–52, 1951.

[10] W. J. Buehler, J. V. Gilfrich, and R. C. Wiley. Effect of low-temperature phase change on the mechanical properties of alloys near composition TiNi. *Journal of Applied Physics*, 34:1475–1477, 1963.

[11] W. J. Buehler and R. C. Wiley. Nickel-base alloys. U.S. Patent 3,174,851, 1965.

[12] E. Hornbogen and G. Wassermann. *Uber den einflus von spannungen und das auftretten von umwandlungsplastizitat bei beta 1-beta-umwandlung des messings. Zeitung für Metallkunde*, 47:427–433, 1956.

[13] C. W. Chen. Some characteristics of the martensite transformation of Cu-Al-Ni alloys. *Journal of Metallurgy*, 9:1201–1203, 1957.

[14] M. W. Burkart and T. A. Read. Diffusionless phase change in the indium-thallium alloys. *AIME J. Met.*, 197:1516–1524, 1953.

[15] Z. S. Basinski and J. W. Christian. Experiments on the martensitic transformation in single crystals of indium-thallium alloys. *Acta Metallurgica*, 2(1):148–166, 1954.

[16] D. A. Hebda and S. R. White. Effect of training conditions and extended thermal cycling on Nitinol two-way shape memory behavior. *Smart Materials and Structures*, 4(4):298–304, 1995.

[17] M. Wuttig. Some notes on mechanical damping of shape memory alloys. In *Engineering Aspects of Shape Memory Alloys*, 488–489, 1990.

[18] E. R. Oberaigner, K. Tanaka, and F. D. Fischer. Investigation of the damping behavior of a vibrating shape memory alloy rod using a micromechanical model. *Smart Materials and Structures*, 5(4):456–463, 1996.

[19] K. Tanaka. A thermomechanical sketch of shape memory effect: One-dimensional tensile behavior. *Res Mechanica*, 18:251–263, 1986.

[20] C. Liang and C. A. Rogers. One-dimensional thermomechanical constitutive relations for shape memory materials. *Journal of Intelligent Material Systems and Structures*, 8(4):285–302, 1997.

[21] L. C. Brinson. One-dimensional constitutive behavior of shape memory alloys: Thermomechanical derivation with non-constant material functions. *Journal of Intelligent Materials and Structures*, 1:207–234, 1990.

[22] J. G. Boyd and D. C. Lagoudas. A thermodynamic constitutive model for the shape memory materials, Part I: the monolithic shape memory alloys. *International Journal of Plasticity*, 12(6):805–842, 1998.

[23] Y. Ivshin and T. J. Pence. Constitutive model for hysteretic phase transition behavior. *International Journal of Engineering Science*, 32(4):681–704, 1994.

[24] H. Tobushi, S. Yamada, T. Hachisuka, A. Ikai, and K. Tanaka. Thermomechanical properties due to martensitic and R-phase transformations of TiNi shape memory alloy subjected to cyclic loadings. *Smart Materials and Structures*, 5(6):788–795, 1996.

[25] J. J. Epps and I. Chopra. Comparative evaluation of shape memory alloy constitutive models with test data. *Paper # AIAA-1997-1194, Proceedings of the 38th AIAA/ASME/ASCE/AHS/ASC Structures, Structural Dynamics and Materials Conference*, Kissimmee, FL, April 1997.

[26] H. Prahlad and I. Chopra. Experimental characterization of Ni-Ti shape memory alloy wires under uniaxial loading conditions. *Journal of Intelligent Material Systems and Structures*, 11(4):272–282, 2000.

[27] A. J. Zak, M. P. Cartmell, W. M. Ostachowicz, and M. Wiercigroch. One-dimensional shape memory alloy models for use with reinforced composite structures. *Smart Materials and Structures*, 12(3):338–346, 2003.

[28] B. Malovrh and F. Gandhi. Mechanism-based phenomenological models for the pseudoelastic hysteresis behavior of shape memory alloys. *Journal of Intelligent Material Systems and Structures*, 12(1):21–30, 2001.

[29] B-C. Chang, J. A. Shaw, and M. A. Iadicola. Thermodynamics of shape memory alloy wire: Modeling, experiments and application. *Continuum Mechanics & Thermodynamics*, 18(1/2):83–118, 2006.

[30] F. Falk. Ginzburg-Landau theory of static domain walls in shape-memory alloys. *Zeitschrift für Physik B Condensed Matter*, 51(2):177–185, 1983.

[31] J. M. Ball and R. D. James. Fine phase mixtures as minimizers of energy. *Archive for Rational Mechanics and Analysis*, 100(1):13–52, 1987.

[32] R. Abeyaratne and J. K. Knowles. On the driving traction acting on a surface of strain discontinuity in a continuum. *Journal of the Mechanics and Physics of Solids*, 38(3): 345–360, 1990.

[33] G. R. Barsch and J. Krumhansl. Martensite Chapter, *Nonlinear Physics in Martensitic Transformations*. In *ASM International, Martensite (USA)*, 125–147, 1992.

[34] A. Sadjadpour and K. Bhattacharya. A micromechanics-inspired constitutive model for shape-memory alloys: The one-dimensional case. *Smart Materials and Structures*, 16(1):S51–S62, February 2007.

[35] E. Patoor, A. Eberhardt, and M. Berveiller. Thermomechanical behaviour of shape memory alloys. *Archiwum Mechaniki Stosowanej*, 40(5):775–794, 1988.

[36] Q. P. Sun and K. C. Hwang. Micromechanics modelling for the constitutive behavior of polycrystalline shape memory alloys. I: Derivation of general relations. *Journal of the Mechanics and Physics of Solids*, 41(1):1–17, 1993.

[37] Q. P. Sun and K. C. Hwang. Micromechanics modelling for the constitutive behavior of polycrystalline shape memory alloys. II. Study of the individual phenomena. *Journal of the Mechanics and Physics of Solids*, 41(1):19–33, 1993.

[38] M. Huang and L. C. Brinson. A multivariant model for single crystal shape memory alloy behavior. *Journal of the Mechanics and Physics of Solids*, 46(8):1379–1409, 1998.

[39] J. G. Boyd and D. C. Lagoudas. Thermomechanical response of shape memory composites. *Journal of Intelligent Material Systems and Structures*, 5(3):333–346, 1994.

[40] E. J. Graesser and F. A. Cozzarelli. A proposed three-dimensional constitutive model for shape memory alloys. *Journal of Intelligent Material Systems and Structures*, 5(1): 78–89, 1994.

[41] E. Patoor, A. Eberhardt, and M. Berveiller. Micromechanical modelling of the shape memory behavior. *International Mechanical Engineering Congress and Exposition, Chicago, IL, ASME Applied Mechanics Division*, 189:23–37, 1994.

[42] B. Zhou, S. H. Yoon, and J. S. Leng. A three-dimensional constitutive model for shape memory alloys. *Smart Materials and Structures*, 18(9):095016, 2009.

[43] J. S. N. Paine and C. A. Rogers. Review of multi-functional SMA hybrid composite materials and their applications. *Adaptive Structures and Composite Materials: Analysis and Application, AD-Vol. 45/MD-Vol. 54, ASME*, pages 37–45, 1994.

[44] K. Jonnalagadda, G. E. Kline, and N. R. Sottos. Local displacements and load transfer in shape memory alloy composites. *Experimental Mechanics*, 37(1):78–86, 1997.

[45] J. Epps and R. Chandra. Shape memory alloy actuation for active tuning of composite beams. *Smart Materials and Structures*, 6(3):251–264, 1997.

[46] T. Ogisu, M. Shimanuki, S. Kiyoshima, and N. Takeda. A basic study of CFRP laminates with embedded prestrained SMA foils for aircraft structures. *Journal of Intelligent Material Systems and Structures*, 16(2):175–185, 2005.

[47] J. A. Shaw and S. Kyriakides. Thermomechanical aspects of NiTi. *Journal of the Mechanics and Physics of Solids*, 43(8):1243–1281, 1995.

[48] M. A. Iadicola and J. A. Shaw. An experimental setup for measuring unstable thermomechanical behavior of shape memory alloy wire. *Journal of Intelligent Material Systems and Structures*, 13(2):157–166, 2002.

[49] Y. Liu. The work production of shape memory alloy. *Smart Materials and Structures*, 13(3):552–561, June 2004.

[50] S. Nomura, K. Tonooka, J. Kuwata, L. E. Cross, and R. E. Newnham. Electrostriction in $Pb(Mg_{1/3}Nb_{2/3})O_3$ ceramics. *Proceedings of the 2nd Meeting on Ferroelectric Materials and Their Applications*, pages 133–138, 1979.

[51] S. J. Jang, L. E. Cross, K. Uchino, and S. Nomura. Dielectric and electrostrictive properties of ferroelectric relaxors in the system Lead Magnesium Niobate ($Pb(Mg_{1/3}Nb_{2/3})O_3$) – Lead Titanate ($PbTiO_3$) – Barium Zinc Niobate ($Ba(Zn_{1/3}Nb_{2/3})O_3$). *Journal of the American Ceramic Society*, 64(4):209–212, 1981.

[52] S. L. Swartz and T. R. Shrout. Fabrication of perovskite Lead Magnesium Niobate. *Materials Research Bulletin*, 17(10):1245–1250, October 1982.

[53] W. Pan, E. Furman, G. O. Dayton, and L. E. Cross. Dielectric aging effects in doped Lead Magnesium Niobate: Lead Titanate relaxor ferroelectric ceramics. *Journal of Materials Science Letters*, 5(6):647–649, 1986.

[54] J. Scortesse, J. F. Manceau, F. Bastien, and M. Lejeune. Apparent Young's modulus in PMN-PT electrostrictive ceramics. *The European Physical Journal of Applied Physics*, 14:155–158, 2001.

[55] F. T. Calkins, A. B. Flatau, and M. J. Dapino. Overview of magnetostrictive sensor technology. *Journal of Intelligent Materials and Smart Structures*, 18(10):1057–1066, October 2007.

[56] G. P. Carman and M. Mitrovic. Nonlinear constitutive relations for magnetostrictive materials with application to 1-D problems. *Journal of Intelligent Material Systems and Structures*, 6(5):673–683, 1995.

[57] K. S. Kannan and A. Dasgupta. Continuum magnetoelastic properties of Terfenol-D; what is available and what is needed. *Proceedings of the ASME Adaptive Materials Symposium, Summer Meeting*, ASME-AMD-MD (UCLA), 1995.

[58] J. B. Restoff, H. T. Savage, A. E. Clark, and M. Wun-Fogle. Preisach modeling of hysteresis in Terfenol-D. *Journal of Applied Physics*, 67(9):5016–5018, 1990.

[59] A. A. Adly and I. D. Mayergoyz. Magnetostriction simulation using anisotropic vector Preisach-type models. *IEEE Transactions on Magnetics*, 32(5):3773–3775, 1996.

[60] R. C. Smith. Modeling techniques for magnetostrictive actuators. *Proceedings of SPIE Symposium on Smart Structures and Materials*, 3041:243–255, 1997.

[61] A. E. Clark, M. Wun-Fogle, J. B. Restorff, and T. A. Lograsso. Magnetic and magnetostrictive properties of Galfenol alloys under large compressive stresses. *Proceedings of the International Symposium on Smart Materials-Fundamentals and System*

Applications, Pacific Rim Conference on Advanced Materials and Processing (PRICM-4), Honolulu, Hawaii, December 2001.

[62] R. A. Kellogg, A. B. Flatau, A. E. Clark, M. Wun-Fogle, and T. A. Lograsso. Quasi-static transduction characterization of Galfenol. *Journal of Intelligent Material Systems and Structures*, 16(6):471–479, June 2005.

[63] R. A. Kellogg, A. M. Russell, T. A. Lograsso, A. B. Flatau, A. E. Clark, and M. Wun-Fogle. Tensile properties of magnetostrictive iron-gallium alloys. *Acta Materialia*, 52(17):5043–5050, October 2004.

[64] A. E. Clark, J. B. Restorff, M. Wun-Fogle, T. A. Lograsso, and D. L. Schlagel. Magnetostrictive properties of body-centered cubic Fe-Ga and Fe-Ga-Al alloys. *IEEE Transactions on Magnetics*, 36(5):3238–3240, September 2000.

[65] J. Atulasimha and A. B. Flatau. A review of magnetostrictive Iron-Gallium alloys. *Smart Materials and Structures*, 20(4):043001, 2011.

[66] W. M. Winslow. Method and means for translating electrical impulses into mechanical forces. U.S. Patent 2,417,850, 1947.

[67] W. M. Winslow. Induced fibration of suspensions. *Journal of Applied Physics*, 20: 1137–1140, 1949.

[68] W. Winslow. Field responsive fluid couplings. U.S. Patent 2,886,151, 1959.

[69] J. Rabinow. Magnetic fluid torque and force transmitting device. U.S. Patent 2,575,360, 1951.

[70] J. E. Stangroom. Electrorheological fluids. *Physics in Technology*, 14(6):290–296, 1983.

[71] D. Brooks. Electro-rheological devices. *Chartered Mechanical Engineer*, 29:91–93, 1982.

[72] R. Stanway, J. L. Sproston, and A. K. El-Wahed. Applications of electro-rheological fluids in vibration control: A survey. *Smart Materials and Structures*, 5(4):464–482, 1996.

[73] J. A. Powell. The mechanical properties of an electrorheological fluid under oscillatory dynamic loading. *Smart Materials and Structures*, 2(4):217–231, 1993.

[74] W. Wen, X. Huang, S. Yang, K. Lu, and P. Sheng. The giant electrorheological effect in suspensions of nanoparticles. *Nature Materials*, 2(11):727–730, November 2003.

[75] T. G. Duclos. Design of devices using electrorheological fluids. *Society of Automotive Engineering Transactions, Sec. 2, SAE Paper No. 881134*, 97:2532–2536, 1988.

[76] Z. P. Shulman, R. G. Gorodkin, and V. K. Gleb. The electro-rheological effect and its possible uses. *Journal of Non-Newtonian Fluid Mechanics*, 8:29–41, 1981.

[77] M. V. Gandhi, M. V. Thompson, S. B. Choi, and S. Shakir. Electro-rheological fluid-based articulating robotic systems. *ASME Journal of Mechanisms, Transmissions and Automation in Design*, 111:328–336, 1989.

[78] M. V. Gandhi and M. V. Thompson. Ultra-advanced composite materials incorporating electro-rheological fluids. *Proceedings of the 4th US-Japan Conference on Composite Materials*, Washington, DC, June 1988.

[79] M. R. Jolly, J. W. Bender, and J. D. Carlson. Properties and applications of commercial magnetorheological fluids. *Journal of Intelligent Material Systems and Structures*, 10(1): 5–13, 1999.

[80] J. Rabinow. Magnetic fluid clutch. *National Bureau of Standards Technical News Bulletin*, 32(4):54–60, 1948.

[81] T. C. Jordan and M. T. Shaw. Electrorheology. *IEEE Transactions on Electrical Insulation*, 24(5):849–878, 1989.

[82] M. Whittle, R. Firoozian, D. J. Peel, and W. A. Bullough. A model for the electrical characteristics of an ER valve. *International Journal of Modern Physics B*, 6(15–16): 2683–2704, August 1992.

[83] A. Hosseini-Sianaki, W. A. Bullough, R. C. Tozer, M. Whittle, and J. Makin. Experimental investigation into the electrical modelling of electrorheological fluids in the shear mode. *IEE Proceedings: Science, measurement and technology*, 141(6):531–537, 1994.

[84] J. Clack, R. Stanway, and J. L. Stroston. The electrical transfer characteristics of electro-rheological fluids in the squeeze-flow mode. *Journal of Intelligent Material Systems and Structures*, 5(5):713–722, 1994.
[85] R. Stanway, J. Sproston, and R. Firoozian. Identification of the damping law of an electro-rheological fluid: A sequential filtering approach. *Journal of Dynamic Systems, Measurement and Control*, 111(1):91–96, 1989.
[86] R. C. Ehrgott and S. F. Masri. Modeling the oscillatory dynamic behavior of electrorheological materials in shear. *Smart Materials and Structures*, 1(4):275–285, 1992.
[87] W. M. Winslow. Field-controlled hydraulic device. U.S. Patent 2,661,596, 1953.
[88] J. D. Carlson. Low-cost MR fluid sponge devices. H. Borgmann, editor, *6th International Conference on New Actuators*, pages 417–421, 1998.
[89] Y. Shen, M. F. Golnaraghi, and G. R. Heppler. Experimental research and modeling of magnetorheological elastomers. *Journal of Intelligent Material Systems and Structures*, 15(1):27–35, 2004.
[90] J. D. Carlson. What makes a good MR fluid? *Journal of Intelligent Material Systems and Structures*, 13(7–8):431–435, 2002.
[91] A. Dogan, J. Tressler, and R. E. Newnham. Solid-state ceramic actuator designs. *AIAA Journal*, 39(7):1354–1362, 2001.
[92] A. J. Moskalik and D. Brei. Quasi-static behavior of individual C-block piezoelectric actuators. *Journal of Intelligent Material Systems and Structures*, 8(7):571, 1997.
[93] A. J. Moskalik and D. Brei. Dynamic performance of C-block array architectures. *Journal of Sound and Vibration*, 243(2):317–346, 2001.
[94] K. D. Rolt. History of flextensional electro-acoustic transducers. *Journal of the Acoustical Society of America*, 87(3):1340–1345, 1990.
[95] H. C. Hayes. Sound generating and directing apparatus. U.S. Patent 2,064,911, 1936.
[96] W. J. Toulis. Flexural-extensional electromechanical transducer. U.S. Patent 3,277,433, 1966.
[97] R. E. Newnham, Q. C. Xu, and S. Yoshikawa. Transformed stress direction acoustic transducer. U.S. Patent 4,999,819, 1991.
[98] A. Dogan, K. Uchino, and R. E. Newnham. Composite piezoelectric transducer with truncated conical endcaps cymbal. *IEEE Transactions on Ultrasonics, Ferroelectrics and Frequency Control*, 44(3):597–605, 1997.
[99] G. H. Haertling. RAINBOW ceramics – A new type of ultra-high-displacement actuator. *American Ceramic Society Bulletin*, 73(1):93–96, 1994.
[100] R. F. Hellbaum, R. G. Bryant, and R. L. Fox. Thin layer composite unimorph ferroelectric driver and sensor. US Patent 5,632,841, 1997.
[101] K. M. Mossi, G. V. Selby, and R. G. Bryant. Thin-layer composite unimorph ferroelectric driver and sensor properties. *Materials Letters*, 35(1):39–49, 1998.
[102] J. P. Dunne, S. W. Jacobs, and E. W. Baumann. Synthesis and processing of efficient cost-effective structures Phase II (SPICES II): Smart Materials Aircraft Applications Evaluation. *Proceedings of the SPIE Smart Structures and Materials Conference, Industrial and Commercial Applications*, San Diego, CA, 3326:2–12, 1998.
[103] B. Culshaw. ASSET: Collaboration in Europe on smart structures. *Proceedings of the SPIE Smart Structures and Materials Symposium*, 4332:22–28, 2001.
[104] H. Hanselka and D. Sachau. German industrial research project ADAPTRONIK: Content, results, and outlook. *Proceedings of the SPIE Smart Structures and Materials Symposium*, 4332:29–36, 2001.
[105] A. J. Lockyer, K. H. Alt, J. N. Kudva, and J. Tuss. Air vehicle integration issues and considerations for CLAS successful implementation. *Proceedings of the SPIE Smart Structures and Materials Symposium*, 4332:48–59, 2001.
[106] K. Miura. Adaptive structures research at ISAS, 1984–1990. *Journal of Intelligent Material Systems and Structures*, 3(1):54–74, 1992.
[107] S. Utku and B. K. Wada. Adaptive structures in Japan. *Journal of Intelligent Material Systems and Structures*, 4(4):437–451, 1993.

[108] Y. Matsuzaki. Smart structures research in Japan. *Smart Materials and Structures*, 6(4): 1–10, 1997.

[109] Y. Matsuzaki, T. Ikeda, and C. Boller. New technological development of passive and active vibration control: Analysis and test. *Smart Materials and Structures*, 14(2): 343–348, April 2005.

[110] A. M. R. McGowan, A. E. Washburn, L. G. Horta, R. G. Bryant, D. E. Cox, E. J. Siochi, S. L. Padula, and N. M. Holloway. Recent results from NASA's Morphing Project. *Proceedings of the SPIE Smart Structures and Materials Symposium*, 4698:97–111, 2002.

[111] R. L. Forward. Electronic damping of orthogonal bending modes in a cylindrical mast-experiment. *Journal of Spacecraft and Rockets*, 18(1):11–17, 1981.

[112] T. Bailey and J. E. Hubbard Jr. Distributed piezoelectric-polymer active vibration control of a cantilevered beam. *Journal of Guidance and Control*, 8(5):605–611, 1985.

[113] E. Crawley and J. de Luis. Use of piezoceramic actuators as elements of intelligent structures. *AIAA Journal*, 25(10):1373–1385, October 1987.

[114] S. Hanagud, M. W. Obal, and A. J. Calise. Optimal vibration control by the use of piezoceramic sensors and actuators. *Journal of Guidance, Control and Dynamics*, 15(5):1199–1206, 1992.

[115] A. Baz and S. Poh. Performance of an active control system with piezoelectric actuators. *Journal of Sound and Vibration*, 126(2):327–343, 1988.

[116] R. L. Forward. Electronic damping of vibrations in optical structures. *Applied Optics*, 18(5):690–697, 1979.

[117] N. Hagood and E. Crawley. Experimental investigation into passive damping enhancement for space structures. *Paper # AIAA-1989-3436, Proceedings of the AIAA Guidance, Navigation and Control Conference*, Boston, MA, 1989.

[118] B. K. Wada. Application of adaptive structures in the design of space structures. *Paper # AIAA-1992-1225, Proceedings of the AIAA Aerospace Design Conference*, Irvine, CA, February 1992.

[119] B. K. Wada, J. L. Fanson, and E. F. Crawley. Adaptive structures. *Journal of Intelligent Material Systems and Structures*, 1(2):157–174, April 1990.

[120] B. K. Wada. Adaptive structures: An overview. *Journal of Spacecraft*, 27(3):330–337, May-June 1990.

[121] L. Y. Lu, S. Utku, and B. K. Wada. On the placement of active members in adaptive truss structures for vibration control. *Smart Materials and Structures*, 1(1):8–23, 1992.

[122] D. R. Dean and L. T. James. Adaptive Laser Optical Technique (ALOT). *1st DOD Conference on High Energy Laser Technology, Naval Training Center*, San Diego, CA, USA, October 1974.

[123] T. Sato, H. Ishikawa, and O. Ikeda. Multilayered deformable mirror using PVDF films. *Applied Optics*, 21(20):3664–3668, October 1982.

[124] J. C. Chen. Response of large space structures with stiffness control. *Journal of Spacecraft and Rockets*, 21(5):463–467, September–October 1984.

[125] C. S. Major and S. S. Simonian. An experiment to demonstrate active and passive control of a flexible structure. *American Control Conference, IEEE*, New York, 1984.

[126] M. D. Rhodes and M. M. Jr. Mikulas. Deployable controllable geometry truss beam. NASA Langley Research Center Report, NASA TM-86366, June 1985.

[127] K. Miura, H. Furuya, and H. Suzuki. Variable geometry truss and its applications to deployable truss and space crane arm. *Acta Astronautics*, 12(7/8):599–607, 1985.

[128] M. Natori, K. Iwasaki, and F. Kuwao. Adaptive planar truss structures and their vibration characteristics. *Paper # AIAA-1989-0743, Proceedings of the 28th AIAA/ASME/ASCE/AHS/ASC Structures, Structural Dynamics and Materials Conference*, New York, April 1987.

[129] J. L. Fanson, C. Cheng-Chih, B. J. Lurie, and R. S. Smith. Damping and structural control of the JPL Phase 0 testbed structure. *Journal of Intelligent Material Systems and Structures*, 2(3):281–300, 1991.

[130] K. Miura and H. Furuya. Adaptive structure concept for future space applications. *AIAA Journal*, 26(8):995–1002, August 1988.

[131] O. J. Godard, M. Z. Lagoudas, and D. C. Lagoudas. Design of space systems using shape memory alloys. *Proceedings of the SPIE Smart Structures and Materials Symposium*, 5056:545–558, 2003.

[132] D. J. Hartl and D. C. Lagoudas. Aerospace applications of shape memory alloys. *Proceedings of the Institution of Mechanical Engineers, Part G: Journal of Aerospace Engineering*, 221(4):535–552, 2007.

[133] J. L. Fanson, G. H. Blackwood, and C. C. Chu. Active-member control of precision structures. *Paper # AIAA-1989-1329, Proceedings of the 30th AIAA/ASME/ASCE/AHS/ASC Structures, Structural Dynamics and Materials Conference*, Mobile, AL, April 1989.

[134] G-S. Chen, B. J. Lurie, and B. K. Wada. Experimental studies of adaptive structures for precision performance. *Paper # AIAA-1989-1327, Proceedings of the 30th AIAA/ASME/ASCE/AHS/ASC Structures, Structural Dynamics and Materials Conference*, Mobile, AL, April 1989.

[135] L. D. Peterson, J. J. Allen, J. P. Lauffer, and A. K. Miller. An experimental and analytical synthesis of controlled structure design. In *AIAA-1989-1170, Proceedings of the 30th AIAA/ASME/ASCE/AHS/ASC Structures, Structural Dynamics and Materials Conference*, Mobile, AL, April 1989.

[136] M. Salama, J. Umland, R. Bruno, and J. Garba. Shape adjustment of precision truss structures: Analytical and experimental validation. *Smart Materials and Structures*, 2(4):240–248, 1993.

[137] C. L. Hom, P. D. Dean, and S. R. Winzer. Simulating electrostrictive deformable mirrors: I. Nonlinear static analysis. *Smart Materials and Structures*, 8(5):691–699, 1999.

[138] G. Park, M. H. Kim, and D. J. Inman. Integration of smart materials into dynamics and control of inflatable space structures. *Journal of Intelligent Material Systems and Structures*, 12(6):423–433, 2001.

[139] A. K. Jha and D. J. Inman. Optimal sizes and placements of piezoelectric actuators and sensors for an inflated torus. *Journal of Intelligent Material Systems and Structures*, 14(9):563–576, 2003.

[140] J. W. Martin and J. Main. Noncontact electron gun actuation of a piezoelectric polymer thin-film bimorph structure. *Journal of Intelligent Material Systems and Structures*, 13(6):329–337, 2002.

[141] F. Andoh, G. Washington, H. S. Yoon, and V. Utkin. Efficient shape control of distributed reflectors with discrete piezoelectric actuators. *Journal of Intelligent Material Systems and Structures*, 15(1):3–15, 2004.

[142] P. Gaudenzi, D. Giarda, and F. Morganti. Active microvibration control of an optical payload installed on the ARTEMIS spacecraft. *Journal of Intelligent Material Systems and Structures*, 9(9):740–748, 1999.

[143] S. A. Lane, S. Griffin, and D. Leo. Active structural acoustic control of a launch vehicle fairing using monolithic piezoceramic actuators. *Journal of Intelligent Material Systems and Structures*, 12(12):795–806, 2001.

[144] S. A. Lane, S. Griffin, and D. J. Leo. Active structural-acoustic control of composite fairings using single-crystal piezoelectric actuators. *Smart Materials and Structures*, 12(1):96–104, 2003.

[145] C. Niezrecki and H. H. Cudney. Feasibility to control launch vehicle internal acoustics using piezoelectric actuators. *Journal of Intelligent Material Systems and Structures*, 12(9):647–660, 2001.

[146] B. K. Henderson and K. K. Denoyer. Recent transitions of smart structures technologies through flight experiments. *Proceedings of the SPIE Smart Structures and Materials Symposium*, 4332:153–158, 2001.

[147] H. P. Monner, H. Hanselka, and E. J. Breitbach. New results and future plans of the German major project ADAPTRONICS. *Proceedings of the SPIE Smart Structures and Materials Symposium*, 4698:313–324, 2002.

[148] J. K. Dürr, R. Honke, M. V. Alberti, and R. Sippel. Analysis and design of an adaptive lightweight satellite mirror. *Proceedings of the SPIE Smart Structures and Materials Symposium*, 4698:351–363, 2002.

[149] J. K. Dürr, R. Honke, M. Von Alberti, and R. Sippel. Development and manufacture of an adaptive lightweight mirror for space application. *Smart Materials and Structures*, 12(6):1005–1016, 2003.

[150] S. Barbarino, S. Ameduri, L. Lecca, and A. Concillo. Wing shape control through and SMA-based device. *Journal of Intelligent Materials and Smart Structures*, 20(3): 283–296, 2009.

[151] K. N. Melton. Ni-Ti Based Shape Memory Alloys. *Engineering Aspects of Shape Memory Alloys*, Edited by T. W. Duerig. Butterworth-Heinemann Limited, 1990.

[152] P. Hajela and R. Glowasky. Application of piezoelectric elements in supersonic panel flutter suppression. *Paper # AIAA 91-3191, Proceedings of the AIAA, AHS and ASEE, Aircraft Design Systems and Operations Meeting*, Baltimore, MD, 1991.

[153] K. D. Frampton, R. L. Clark, and E. H. Dowell. Active control of panel flutter with piezoelectric transducers. *Journal of Aircraft*, 33(4):768–774, July–August 1996.

[154] R. C. Scott and T. A. Weisshaar. Panel flutter suppression using adaptive material actuators. *Journal of Aircraft*, 31(1):213–222, January–February 1994.

[155] A. Suleman and V. B. Venkayya. Flutter control of an adaptive laminated composite panel with piezoelectric layers. *Paper # AIAA 94-1744 Proceedings of the 6th AIAA/USAF/NASA/ISSMO Symposium on Multidisciplinary Analysis and Optimization*, pages 4–6, 1996.

[156] S. M. Ehlers and T. A. Weisshaar. Static aeroelastic control of an adaptive lifting surface. *Journal of Aircraft*, 30(4):534–540, July–August 1993.

[157] K. B. Lazarus, E. F. Crawley, and C. Y. Lin. Multivariable active lifting surface control using strain actuation: Analytical and experimental results. *Journal of Aircraft*, 34(3): 313–321, 1997.

[158] C. Nam, Y. Kim, and T. A. Weisshaar. Optimal sizing and placement of piezo-actuators for active flutter suppression. *Smart Materials and Structures*, 5(2):216–224, 1996.

[159] C. Y. Lin, E. F. Crawley, and J. Heeg. Open- and closed-loop results of a strain-actuated active aeroelastic wing. *Journal of Aircraft*, 33(5):987–994, September–October 1996.

[160] A. Suleman, C. Crawford, and A. P. Costa. Experimental aeroelastic response of piezoelectric and aileron controlled 3-D wing. *Journal of Intelligent Material Systems and Structures*, 13(2):75–83, 2002.

[161] J. K. Strelec, D. C. Lagoudas, M. A. Khan, and J. Yen. Design and implementation of a shape memory alloy actuated reconfigurable airfoil. *Journal of Intelligent Material Systems and Structures*, 14(4):257–273, 2003.

[162] M. J. Rossi, F. Austin, and W. van Nostrand. Active rib experiment for shape control of an adaptive wing. *Paper # AIAA-1993-1700, Proceedings of the 34th AIAA/ASME/ASCE/AHS/ASC Structures, Structural Dynamics and Materials Conference*, La Jolla, CA, April 1993.

[163] C. Thill, I. Bond, E. Potter, and P. Weaver. Morphing skins. *The Aeronautical Journal*, 112:117–139, 2008.

[164] R. Beblo, K. Gross, and W. L. Mauck. Mechanical and curing properties of a styrene-based shape memory polymer. *Journal of Intelligent Material Systems and Structures*, 21(7):677–683, 2010.

[165] R. Ikegami and E. D. Haugse. Structural health management for aging aircraft. *Proceedings of the SPIE Smart Structures and Materials Symposium*, 4332:60–67, 2001.

[166] E. Kiely, G. Washington, and J. Bernhard. Design and development of smart microstrip patch antennas. *Smart Materials and Structures*, 7(6):792–800, 1998.

[167] F. J. Chen, C. Yao, G. B. Beeler, R. G. Bryant, and R. L. Fox. Development of synthetic jet actuators for active flow control at NASA Langley. *Paper# AIAA-2000-2405, Proceedings of Fluids 2000*, Denver, CO, June 2000.

[168] R. Rathnasingham and K. Breuer. Coupled fluid-structural characteristics of actuators for flow control. *AIAA Journal*, 35(5):832–837, May 1997.

[169] B. L. Smith and A. Glezer. The formation and evolution of synthetic jets. *Physics of Fluids*, 10(9):2281–2297, September 1998.

[170] A. Crook and A. M. Sadri. The development and implementation of synthetic jets for control of separated flow, *AIAA Paper # 1999-3176 17th Applied Aerodynamics Conference*, Norfolk, VA, July, 1999.

[171] P. Mane, K. Mossi, A. Rostami, R. Bryant, and N. Castro. Piezoelectric actuators as synthetic jets: Cavity dimension effects. *Journal of Intelligent Material Systems and Structures*, 18(11):1175–1190, 2007.

[172] N. Schaeffler, M. Kegerise, T. Hepner, and G. Jones. Overview of active flow control actuator development at NASA Langley Research Center. *Paper # AIAA-2002-3159, Proceedings of the 1st AIAA Flow Control Conference*, Saint Louis, MO, 2002.

[173] A. Kumar and J. N. Hefner. Future challenges and opportunities in aerodynamics. In *22nd International Congress of Aeronautical Sciences, ICAS Paper 2000-0.2*, August 2000.

[174] A. E. Washburn. NASA micro-aero-adaptive control. *Proceedings of the SPIE Smart Structures and Materials Symposium*, 4332:326–344, 2001.

[175] V. Q. Nguyen, M. Syaifuddin, H. C. Park, D. Y. Byun, N. S. Goo, and K. J. Yoon. Characteristics of an insect-mimicking flapping system actuated by a unimorph piezoceramic actuator. *Journal of Intelligent Material Systems and Structures*, 19(10):1185–1193, 2008.

[176] B. Sanders, R. Crowe, and E. Garcia. Defense Advanced Research Projects Agency – smart materials and structures demonstration program overview. *Journal of Intelligent Material Systems and Structures*, 15(4):227–233, April 2004.

[177] J. N. Kudva. Overview of the DARPA smart wing project. *Journal of Intelligent Material Systems and Structures*, 15(4):261–267, April 2004.

[178] C. A. Martin, B. J. Hallam, J. S. Flanagan, and J. D. Bartley-Cho. Design, fabrication and testing of a scaled wind-tunnel model for the smart wing project. *Journal of Intelligent Material Systems and Structures*, 15(4):269–278, April 2004.

[179] J. D. Bartley-Cho, D. P. Wang, C. A. Martin, J. N. Kudva, and M. N. West. Development of high-rate, adaptive trailing-edge control surface for the smart wing Phase 2 wind-tunnel model. *Journal of Intelligent Material Systems and Structures*, 15(4): 279–291, April 2004.

[180] B. Sanders, D. Cowan, and L. Scherer. Aerodynamic performance of the smart wing control effectors. *Journal of Intelligent Material Systems and Structures*, 15(4):293–303, April 2004.

[181] A. J. Lockyer, C. A. Martin, D. K. Linder, P. S. Walia, and B. F. Carpenter. Power system requirement for integration of smart structures into aircraft. *Journal of Intelligent Material Systems and Structures*, 15(4):305–315, April 2004.

[182] W. E. Triplett. Pressure measurements of twin vertical tails in buffeting flow. *Journal of Aircraft*, 20(11):920–925, November 1983.

[183] C. L. Pettit, M. Bandford, D. Brown, and E. Pendleton. Full-scale wind-tunnel pressure measurements on an F/A-18 tail during buffet. *Journal of Aircraft*, 33(6):1148–1156, November–December 1996.

[184] R. W. Moses and L. Huttsell. Fin-buffeting features of an early F-22 model. *Paper # AIAA-2000-1695, Proceedings of the 41st AIAA/ASME/ASCE/AHS/ASC Strucurres, Structural Dynamics and Materials Conference*, Atlanta, GA, April 2000.

[185] R. W. Moses, C. D. Wieseman, A. A. Bent, and A. E. Pizzochero. Evaluation of new actuators in a buffet loads environment. *Proceedings of the SPIE Smart Structures and Materials Symposium*, 4332:10–21, 2001.

[186] R. W. Moses. Vertical tail buffeting alleviation using piezoelectric actuators – some results of the Actively Controlled Response of Buffet-Affected Tails (ACROBAT) program, NASA TM 110336, April 1997.

[187] R. M. Hauch, J. H. Jacobs, K. Ravindra, and C. Dima. Reduction of vertical tail buffet response using active control. *Paper # AIAA-1995-1080, Proceedings of the 36th AIAA/ASME/ASCE/AHS/ASC Structures, Structural Dynamics and Materials Conference*, April 1995.

[188] A. Suleman, A. P. Costa, and P. A. Moniz. Experimental flutter and buffeting suppression using piezoelectric actuators and sensors. *Proceedings of the SPIE Smart Structures and Materials Symposium*, 3674:72–83, 1999.

[189] C. R. Fuller. Active control of sound transmission/radiation from elastic plates by vibration inputs: I. Analysis. *Journal of Sound and Vibration*, 136(1):1–15, 1990.

[190] C. A. Gentry, C. Guigou, and C. R. Fuller. Smart foam for applications in passive-active noise radiation control. *The Journal of the Acoustical Society of America*, 101(4):1771–1778, April 1997.

[191] J. Kim and J. K. Lee. Broadband transmission noise reduction of smart panels featuring piezoelectric shunt circuits and sound-absorbing material. *The Journal of the Acoustical Society of America*, 112(3):990–998, 2002.

[192] J. Kim and Y. C. Jung. Piezoelectric smart panels for broadband noise reduction. *Journal of Intelligent Material Systems and Structures*, 17(8–9):685–690, 2006.

[193] B. Petitjean and C. Greffe. Active interior noise control: An industrial perspective. *Proceedings of the SPIE Smart Structures and Materials Symposium*, 4698:133–142, 2002.

[194] S. Bohme, D. Sachau, and H. Breitbach. Optimization of actuator and sensor positions for an active noise reduction system. *Proceedings of the SPIE Smart Structures and Materials Symposium*, 6171:61710N-1 to 61710N-11, 2006.

[195] T. L. Turner, R. D. Buehrle, R. J. Cano, and G. A. Fleming. Modeling, fabrication and testing of a SMA hybrid composite jet engine chevron concept. *Journal of Intelligent Material Systems and Structures*, 17(6):483–497, 2006.

[196] F. T. Calkins, J. H. Mabe, and G. W. Butler. Boeing's variable geometry chevron: Morphing aerospace structures for jet noise reduction. *Proceedings of the SPIE Smart Structures and Materials Symposium*, 6171:61710O-1 to 61710O-12, 2006.

[197] D. J. Hartl, D. C. Lagoudas, F. T. Calkins, and J. H. Mabe. Use of a Ni60Ti shape memory alloy for active jet engine chevron application: I. Thermomechanical characterization. *Smart Materials and Structures*, 19(1):015020, 2010.

[198] D. J. Hartl, D. C. Lagoudas, F. T. Calkins, and J. H. Mabe. Use of a Ni60Ti shape memory alloy for active jet engine chevron application: II. Experimentally validated numerical analysis. *Smart Materials and Structures*, 19(1):015021, 2010.

[199] J. Webster. High integrity adaptive SMA components for gas turbine applications. *Proceedings of the SPIE Smart Structures and Materials Symposium*, 6171:61710F-1 to 61710F-8, 2006.

[200] N. H. Schiller, W. R. Saunders, and W. Chishty. Development of a piezoelectric-actuated fuel modulation system for active combustion control. *Journal of Intelligent Material Systems and Structures*, 17(5):403–410, 2006.

[201] D. M. Pitt, J. P. Dunne, E. V. White, and E. Garcia. Wind-tunnel demonstration of the SAMPSON smart inlet. *Proceedings of the SPIE Smart Structures and Materials Symposium*, 4332:345–356, 2001.

[202] D. M. Pitt, J. P. Dunne, E. V. White, and E. Garcia. SAMPSON smart inlet SMA powered adaptive lip design and static test. *Paper # AIAA-2001-1359, Proceedings of the 42nd AIAA/ASME/ASCE/AHS/ASC Structures, Structural Dynamics and Materials Conference and Exhibit*, Seattle, WA, April 16–19, 2001.

[203] N. Rey, G. Tillman, R. M. Miller, T. Wynosky, M. J. Larkin, J. D. Flamm, and L. S. Bangert. Shape memory alloy actuation for a variable area fan nozzle. *Proceedings of the SPIE Smart Structures and Materials Symposium*, 4332:371–382, 2001.

[204] I. Chopra. Status of application of smart structures technology to rotorcraft systems. *Journal of the American Helicopter Society*, 45(4):228–252, October 2000.

[205] I. Chopra. Review of state of art of smart structures and integrated systems. *AIAA Journal*, 40(11):2145–2187, November 2002.

[206] B. Balachandran, A. Sampath, and J. Park. Active control of interior noise in a three-dimensional enclosure. *Smart Materials and Structures*, 5(1):89–97, 1996.

[207] W. Gembler, H. Schweitzer, R. Maier, M. Pucher, P. Jaenker, and F. Hermle. Smart struts: The solution for helicopter interior noise problems. In *Proceedings of the 25th European Rotorcraft Forum*, Rome, Italy, 1, September 1999.

[208] N. A. Koratkar and I. Chopra. Development of a Mach-scaled model with piezoelectric bender actuated trailing-edge flaps for helicopter individual blade control (IBC). *AIAA Journal*, 38(7):1113–1124, July 2000.

[209] N. A. Koratkar and I. Chopra. Wind-tunnel testing of a Mach-scaled rotor model with trailing-edge flaps. *Smart Materials and Structures*, 10(1):1–14, Febuary 2001.

[210] N. A. Koratkar and I. Chopra. Wind-tunnel testing of a Mach-scaled rotor model with trailing-edge flaps. In *Proceedings of the 57th Annual Forum of the American Helicopter Society*, Washington, DC, May 2001.

[211] B. Roget and I. Chopra. "Individual blade-control methodology of a rotor with dissimilar blades," *Journal of the American Helicopter Society*, 48(3): 176–185, July 2003.

[212] M. Fulton and R. A. Ormiston. Small-scale rotor experiments with on-blade elevons to reduce blade vibratory loads in forward flight. *Journal of the American Helicopter Society*, 46(2):96–106, April 2001.

[213] T. Lee and I. Chopra. Design issues of a high-stroke, on-blade piezostack actuator for helicopter rotor with trailing-edge flaps. *Journal of Intelligent Material Systems and Structures*, 11(5):328–342, May 2000.

[214] T. Lee and I. Chopra. Design of piezostack-driven trailing-edge flap actuator for helicopter rotors. *Smart Materials and Structures*, 10(1):15–24, February 2001.

[215] T. Lee and I. Chopra. Wind-tunnel test of blade sections with piezoelectric trailing-edge flap mechanism. In *Proceedings of the 57th Annual Forum of the American Helicopter Society*, Washington, DC, May 2001.

[216] F. K. Straub, H. T. Ngo, V. Anand, and D. B. Domzalski. Development of a piezoelectric actuator for trailing-edge flap control for full-scale rotor system. *Smart Materials and Structures*, 10(1):25–34, February 2001.

[217] S. R. Hall and E. F. Prechtl. Preliminary testing of a Mach-scaled active rotor blade with a trailing edge servo-flap. *Proceedings of the SPIE Smart Structures and Materials Symposium*, 3668:14–21, 1999.

[218] P. Janker, V. Kloppel, F. Hermle, T. Lorkowski, S. Storm, M. Christmann, and M. Wettemann. Development and evaluation of a hybrid piezoelectric actuator for advanced flap control technology. In *Proceedings of the 25th European Rotorcraft Forum*, Rome, Italy, September 1999.

[219] A. P. F. Bernhard and I. Chopra. Trailing-edge flap activated by a piezo-induced bending-torsion coupled beam. *Journal of the American Helicopter Society*, 44(1):3–15, January 1999.

[220] A. P. F. Bernhard and I. Chopra. Hover test of an active-twist, Mach-scale rotor using a piezo-induced bending-torsion actuator beam. In *Proceedings of the AHS Northeast Region Technical Specialists' Meeting on Improving Rotorcraft Acceptance Through Active Control Technology*, Bridgeport, CT, October 2000.

[221] A. P. F. Bernhard and I. Chopra. Analysis of a bending-torsion coupled actuator for a smart rotor with active blade tips. *Smart Materials and Structures*, 10(1):35–52, February 2001.

[222] P. C. Chen and I. Chopra. Wind-tunnel test of a smart rotor with individual blade twist control. *Journal of Intelligent Material Systems and Structures*, 8(5):414–425, May 1997.

[223] P. C. Chen and I. Chopra. Hover testing of smart rotor with induced-strain actuation of blade twist. *AIAA Journal*, 35(1):6–16, January 1997.

[224] R. C. Derham and N. W. Hagood. Rotor design using smart materials to actively twist blades. In *Proceedings of the 52nd Annual Forum of the American Helicopter Society*, Washington, DC, June 1996.

[225] J. P. Rodgers and N. W. Hagood. Preliminary Mach-scale hover testing of an integral twist-actuated rotor blade. *Proceedings of the SPIE Smart Structures and Materials Symposium*, 3329:291–308, 1998.

[226] J. P. Rodgers, N. W. Hagood and D. Weems. Design and manufacture of an integral twist-actuated rotor blade. *Paper # AIAA-1997-1264, Proceedings of the 38th AIAA/ASME/ASCE/AHS/ASC Structures, Structural Dynamics and Materials Conference*, Kissimmee, FL, April 1997.

[227] C. E. S. Cesnik, S. J. Shin, and M. L. Wilbur. Dynamic response of active twist rotor blades. *Smart Materials and Structures*, 10(1):62–76, February 2001.

[228] C. E. S. Cesnik and S. J. Shin. Control of integral twist-actuated helicopter blades for vibration reduction. In *Proceedings of the 58th American Helicopter Society Annual Forum*, Montreal, Canada, May 2002.

[229] J. J. Epps and I. Chopra. In-flight tracking of helicopter rotor blades using shape memory alloy actuators. In *Proceedings of the 56th American Helicopter Society Annual Forum*, Virginia Beach, VA, May 2000.

[230] J. J. Epps and I. Chopra. In-flight tracking of helicopter rotor blades using shape memory alloy actuators. *Smart Materials and Structures*, 10(1):104–111, February 2001.

[231] K. Singh, J. Sirohi, and I. Chopra. An improved shape memory alloy actuator for rotor blade tracking. *Journal of Intelligent Material Systems and Structures*, 14(12):767–786, December 2003.

[232] C. Liang, F. Davidson, L. M. Schetky, and F. K. Straub. Applications of torsional shape memory alloy actuators for active rotor blade control – opportunities and limitations. *Proceedings of the SPIE Smart Structures and Materials Symposium*, 2717:91–100, 1996.

[233] D. K. Kennedy, F. K. Straub, L. M. Schetky, Z. Chaudhry, and R. Roznoy. Development of an SMA actuator for in-flight rotor blade tracking. *Journal of Intelligent Materials and Smart Structures*, 15:235–260, April 2004.

[234] H. Prahlad and I. Chopra. Characterization of SMA torsional actuators for variable twist tilt rotor (VTTR) blades. *Paper # AIAA-2002-1445, Proceedings of the 43rd AIAA/ASME/ASCE/AHS/ASC Structures, Structural Dynamics and Materials Conference*, Denver, CO, April 2002.

[235] H. Prahlad and I. Chopra. Modeling and experimental characterization of SMA torsional actuators. *Journal of Intelligent Material Systems and Structures*, 18(1):29–38, January 2007.

[236] P. C. Chen, R. A. D. Evans, J. Niemczuk, and J. D. Baeder. Blade-vortex interaction noise reduction with active twist smart rotor technology. *Smart Materials and Structures*, 10(1):77–85, 2001.

[237] A. Sampath and B. Balachandran. Active control of multiple tones in an enclosure. *The Journal of the Acoustical Society of America*, 106(1):211–225, 1999.

[238] J. Shen and I. Chopra. Actuation requirements for a swashplateless helicopter control system with trailing-edge flaps. *Paper # AIAA-2002-1444, Proceedings of the 43rd AIAA/ASME/ASCE/AHS/ASC Structures, Structural Dynamics and Materials Conference*, Denver, CO, April 2002.

[239] J. Shen and I. Chopra. Aeroelastic stability of trailing-edge flap helicopter rotors. *Journal of the American Helicopter Society*, 48(4):236–243, 2003.

[240] J. Shen and I. Chopra. Swashplateless helicopter rotor with trailing-edge flaps. *Journal of Aircraft*, 41(2):208–214, 2004.

[241] J. Shen and I. Chopra. A parametric design study for a swashplateless helicopter rotor with trailing-edge flaps. *Journal of the American Helicopter Society*, 49(1):43–53, 2004.

[242] R. Kube and V. Kloeppel. On the role of prediction tools for adaptive rotor system developments. *Smart Materials and Structures*, 10(1):137–144, 2001.

[243] I. Pelinescu and B. Balachandran. Analytical study of active control of wave transmission through cylindrical struts. *Smart Materials and Structures*, 10(1):121–136, 2001.

[244] D. R. Mahapatra, S. Gopalakrishnan, and B. Balachandran. Active feedback control of multiple waves in helicopter gearbox support struts. *Smart Materials and Structures*, 10(5):1046–1058, 2001.

[245] G. M. Kamath, N. M. Wereley, and M. R. Jolly. Characterization of magnetorheological helicopter lag dampers. *Journal of the American Helicopter Society*, 44(3):234–248, 1999.

[246] P. Konstanzer, B. Grohmann, and B. Kroplin. Decentralized vibration control and coupled aeroservoelastic simulation of helicopter rotor blades with adaptive airfoils. *Journal of Intelligent Material Systems and Structures*, 12(4):209, 2001.

[247] S. J. Kim and C. Y. Yun. Performance comparison between piezoelectric and elastomeric lag dampers on ground resonance stability of helicopter. *Journal of Intelligent Material Systems and Structures*, 12(4):215, 2001.

[248] A. Sampath and B. Balachandran. Studies on performance functions for interior noise control. *Smart Materials and Structures*, 6(3):315–332, 1997.

[249] F. K. Straub, D. K. Kennedy, D. B. Domzalski, A. A. Hasan, H. Ngo, V. Anand, and T. Birchette. Smart material actuated rotor technology – SMART. *Journal of Intelligent Materials and Smart Structures*, 15(4):249–260, 2004.

[250] D. Jacot, T. Calkins, and J. Mabe. Boeing active flow control system BAFC-II. *Proceedings of the SPIE Smart Structures and Materials Symposium*, 4332:317–325, 2001.

[251] F. T. Calkins and D. J. Clingman. Vibrating surface actuators for active flow control. *Proceedings of the SPIE Smart Structures and Materials Symposium*, 4698:85–96, 2002.

[252] G. W. Housner, L. A. Bergman, T. K. Caughey, A. G. Chassiakos, R. O. Claus, S. F. Masri, R. E. Skelton, T. T. Soong, B. F. Spencer, and J. T. P. Yao. Structural control: Past, present, and future. *Journal of Engineering Mechanics*, 123(9):897–971, 1997.

[253] V. G. M. Annamdas and C. K. Soh. Application of electromechanical impedance technique for engineering structures: Review and future issues. *Journal of Intelligent Material Systems and Structures*, 21(1):41–59, 2010.

[254] D. R. Huston, P. L. Fuhr, T. P. Ambrose and D. A. Barker. Intelligent civil structures – activities in Vermont. *Smart Materials and Structures*, 3(2):129–139, 1994.

[255] A. Mendez, T. F. Morse, and F. Mendez. Applications of embedded optical fiber sensors in reinforced concrete buildings and structures. *Proceedings of the SPIE Fiber Optic Smart Structures and Skins II Symposium*, 1170:60–69, 1989.

[256] D. R. Houston, P. L. Fuhr, P. J. Kajensky, and D. Snyder. Concrete beam testing with optical-fiber sensors. In *Proceedings of the ASCE Minisymposium on the Nondestructive Testing of Concrete*, San Antonio, TX, April 1992.

[257] X. Chen, F. Ansari, and H. Ding. Embedded fiber-optic displacement sensor for concrete elements. *Proceedings of the 11th Engineering Mechanics Conference, ASCE*, pages 359–364, 1996.

[258] R. M. Measures. Structural sensing with fiber-optic systems. *Proceedings of the 11th Engineering Mechanics Conference, ASCE*, pages 224–227, 1996.

[259] A. Holst and R. Lessing. Fiber-optic intensity modulated sensors for continuous observation of concrete and rock-fill dams. In *Proceedings of the 1st European Conference on Smart Structures and Materials*, Glasgow, Scotland, 1992.

[260] P. L. Fuhr and D. R. Huston. Multiplexed fiber optic pressure and vibration sensors for hydroelectric dam monitoring. *Smart Materials and Structures*, 2(4):260–263, 1993.

[261] P. L. Fuhr, D. R. Huston, P. J. Kajenski, and T. P. Ambrose. Performance and health monitoring of the Stafford Medical Building using embedded sensors. *Smart Materials and Structures*, 1(1):63–68, 1992.

[262] M. deVries, M. Nasta, V. Bhatia, T. Tran, J. Greene, R. O. Claus and S. Masri. Performance of embedded short-gage-length optical-fiber sensors in a fatigue-loaded reinforced concrete specimen. *Smart Materials and Structures*, 4(1A):107–113, 1995.

[263] G. Song, H. Gu, and Y-L. Mo. Smart aggregates: Multifunctional sensors for concrete structures – a tutorial and a review. *Smart Materials and Structures*, 17(3):033001, 2008.

[264] S. Aizawa, T. Kakizawa, and M. Higasino. Case studies of smart materials for civil structures. *Smart Materials and Structures*, 7(5):617–626, 1998.

[265] S. Zhou, Z. Chaudhry, C. A. Rogers, and R. Quattrone. Review of embedded particle tagging methods for nondestructive evaluation (NDE) of composite materials and structures. *Proceedings of the SPIE Smart Structures and Materials Symposium*, 2444:39–52, 1995.

[266] S. J. Dyke. Recent R&D activities on smart sensors and monitoring in the US. In *Proceedings of US–Korea Joint Seminar/Workshop on Smart Structures Technologies, Coordinators: C. B. Yun and L. A. Bergman*, Seoul, Korea, 2–4 September 2004.

[267] K. P. Chong, N. J. Carino, and G. Washer. Health monitoring of civil infrastructures. *Smart Materials and Structures*, 12(3):483–493, 2003.

[268] R. C. Tennyson, A. A. Mufti, S. Rizkalla, G. Tadros, and B. Benmokrane. Structural health monitoring of innovative bridges in Canada with fiber-optic sensors. *Smart Materials and Structures*, 10(3):560–573, 2001.

[269] D. Wang, J. Liu, D. Zhou, and S. Huang. Using PVDF piezoelectric film sensors for in-situ measurement of stayed-cable tension of cable-stayed bridges. *Smart Materials and Structures*, 8(5):554–559, 1999.

[270] W. H. Liao, D. H. Wang, and S. L. Huang. Wireless monitoring of cable tension of cable-stayed bridges using PVDF piezoelectric films. *Journal of Intelligent Material Systems and Structures*, 12(5):331–339, May 2001.

[271] R. Betti and R. B. Testa. Vibration and damage control for long-span bridges. *Smart Materials and Structures*, 4(1A):91–100, 1995.

[272] T. P. Ambrose, D. R. Houston, P. L. Fuhr, E. A. Devino, and M. P. Werner. Shoring systems for construction-load monitoring. *Smart Materials and Structures*, 3(1):26–34, 1994.

[273] C. K. Soh, K. K. H. Tseng, S. Bhalla, and A. Gupta. Performance of smart piezoceramic patches in health monitoring of a RC bridge. *Smart Materials and Structures*, 9(4):533–542, 2000.

[274] D. J. Pines and P. A. Lovell. Conceptual framework of a remote wireless health-monitoring system for large civil structures. *Smart Materials and Structures*, 7(5):627–636, 1998.

[275] Y. Fujino. Vibration and control of long-span bridges. *Advances in Structural Dynamics*, Edited by J. M. Ko and Y. L. Xu, volume 1, pages 55–66. Elsevier Science Ltd., Oxford (UK), 2000.

[276] S. Nagarajaiah, B. F. Spencer, and J. Yang. Recent advances in control of civil infrastructures in USA. In C. B. Yun and L. A. Bergman, editors, *Proceedings of US–Korea Joint Seminar/Workshop on Smart Structures Technologies*, Seoul, South Korea, 2–4 September 2004.

[277] F. Gordaninejad, M. Saiidi, B. C. Hansen, E. O. Ericksen, and F. K. Chang. Magneto-rheological fluid dampers for control of bridges. *Journal of Intelligent Material Systems and Structures*, 13(2):167–180, 2002.

[278] A. Baz. Optimal deflection control of multi-segment traversing beams. *Smart Materials and Structures*, 4(2):75–82, 1995.

[279] M. Sakamoto and T. Kobori. Research, development and practical applications on structural response control of buildings. *Smart Materials and Structures*, 4(1A):58–74, 1995.

[280] A. K. Agrawal and J. N. Yang. Optimal placement of passive dampers on seismic and wind-excited buildings using combinatorial optimization. *Journal of Intelligent Material Systems and Structures*, 10(12):997–1014, 1999.

[281] G. Leitmann and E. Reithmeier. Semi-active control of a vibrating system by means of electrorheological fluids. *Dynamics and Control*, 3(1):7–33, 1993.

[282] H. P. Gavin. Design method for high-force electrorheological dampers. *Smart Material and Structures*, 7(5):664–673, 1998.

[283] S. J. Dyke, B. F. Spencer Jr., M. K. Sain, and J. D. Carlson. An experimental study of MR dampers for seismic protection. *Smart Materials and Structures*, 7(5):693–703, 1998.

[284] J. D. Carlson and B. F. Spencer, Jr. Magneto-rheological fluid dampers for semi-active seismic control. *Proceedings of the 3rd International Conference on Motion and Vibration Control*, pages 35–40, 1996.

[285] I. Nishimura, A. M. Abdel-Ghaffar, S. F. Masri, R. K. Miller, J. L. Beck, T. K. Caughey, and W. D. Iwan. An experimental study of the active control of a building model. *Journal of Intelligent Material Systems and Structures*, 3(1):134–165, 1992.

[286] I. D. Aiken, D. K. Nims, A. S. Whittaker, and J. M. Kelly. Testing of passive energy dissipation systems. *Earthquake Spectra*, 9(3):335–370, 1993.

[287] P. R. Witting and F. A. Cozzarelli. Technical Report NCEER-92-0013. National Center for Earthquake Engineering Research, State University of New York, Buffalo, 1992.

[288] P. R. Witting and F. A. Cozzarelli. Design and seismic testing of shape memory structural dampers. *Proceedings of Damping-93*, San Francisco, CA, 1993.

[289] Y. Zhang and N. D. Sims. Milling workpiece chatter avoidance using piezoelectric active damping: A feasibility study. *Smart Materials and Structures*, 14(6):N65–N70, December 2005.

[290] G. P. O'Neal, B. K. Min, Z. J. Pasek, and Y. Koren. Integrated structural/control design of micro-positioner for boring bar tool insert. *Journal of Intelligent Material Systems and Structures*, 12(9):617–627, September 2001.

[291] A. Milecki and D. Sedziak. The use of magnetorheological fluid dampers to reduce servo drive velocity jumps due to load changes. *Journal of Intelligent Material Systems and Structures*, 16(6):501–510, June 2005.

[292] K. S. Kim, S. B. Choi, and M. S. Cho. Vibration control of a wire cut discharge machine using ER brake actuator. *Journal of Intelligent Material Systems and Structures*, 13(10): 621–624, October 2002.

[293] A. R. Johnson and W. Bullough. Design and control considerations for a high-speed electrorheological traversing mechanism. *Journal of Intelligent Material Systems and Structures*, 13(11):725–735, 2002.

[294] R. S. Fielder, C. Boyd, M. Palmer, and O. Eriksen. Real time pressure monitoring of dynamic control during paper mill operation using fiber optic pressure sensors. *Proceedings of the SPIE Smart Structures and Materials Symposium*, 6171:61710B-1 to 61710B-11, 2006.

[295] R. Neugebauer, M. Hoffmann, H.-J. Roscher, S. Scheffler and K. Wolf. Control of sheet-metal forming processes with piezoactuators in smart structures. *Proceedings of the SPIE Smart Structures and Materials Symposium*, 6171:61710E-1 to 61710E-9, 2006.

[296] M. Yang, K. Manabe, and H. Nishimura. Development of an intelligent tool system for flexible L-bending process of metal sheets. *Smart Materials and Structures*, 7(4): 530–536, 1998.

[297] D. Y. Li. Exploration of Ti-Ni shape memory alloy for potential application in a new area: Tribological engineering. *Smart Materials and Structures*, 9(5):717–726, 2000.

[298] W. I. Kordonski and D. Golini. Fundamentals of magnetorheological fluid utilization in high precision finishing. *Journal of Intelligent Material Systems and Structures*, 10(9): 683–689, 2000.

[299] K. Shimada, H. Nishida, and Y. Akagami. Electrorheological rotating disk clutch mechanism under AC electric field: II. Efficiency metrics. *Journal of Intelligent Material Systems and Structures*, 13(7–8):497–502, July 2002.

[300] M. Nakano. A novel semi-active control of automotive suspension using an electrorheological shock absorber. *Proceedings of the 5th International Conference on ER Fluids, MR Suspensions and Associated Technologies, SAE Technical Papers Series 920275*, 1995.

[301] N. K. Petek, D. J. Romstadt, M. B. Lizell, and T. R. Weyenberg. Demonstration of an automotive semi-active suspension using electrorheological fluid. *SAE Transactions*, 104(6):987–992, 1995.

[302] M. S. Suh and M. S. Yeo. Development of semi-active suspension systems using ER fluids for the wheeled vehicle. *Journal of Intelligent Material Systems and Structures*, 10(9):743–747, 1999.

[303] J. D. Carlson, D. M. Catanzarite, and K. A. St. Clair. Commercial magneto-rheological fluid devices. *Proceedings of the 5th International Conference on ER Fluids, MR Fluids and Associated Technologies, World Scientific*, Singapore, pages 20–28, 1995.

[304] H. S. Lee and S. B. Choi. Control and response characteristics of a magneto-rheological fluid damper for passenger vehicles. *Journal of Intelligent Material Systems and Structures*, 11(1):80–87, 2000.

[305] Q.-H. Nguyen and S.-B. Choi. Optimal design of MR shock absorber and application to vehicle suspension. *Smart Materials and Structures*, 18(3):035012, 2009.

[306] M. S. Yeo, H. G. Lee, and M. C. Kim. A study on the performance estimation of semi-active suspension system considering the response time of elelctro-rheological fluid. *Journal of Intelligent Material Systems and Structures*, 13(7):485–489, July 2002.

[307] Y. Shen, M. F. Golnaraghi, and G. R. Heppler. Analytical and experimental study of the response of a suspension system with a magnetorheological damper. *Journal of Intelligent Material Systems and Structures*, 16(2):135–147, 2005.

[308] M. Yu, C. R. Liao, W. M. Chen, and S. L. Huang. Study on MR semi-active suspension system and its road testing. *Journal of Intelligent Material Systems and Structures*, 17 (8-9):801–806, 2006.

[309] Y. Zhu, J. Qiu, J. Tani, S. Suzuki, Y. Urushiyama, and Y. Hontani. Vibration control of a steering wheel using piezoelectric actuators. *Journal of Intelligent Material Systems and Structures*, 10(2):92–99, 1999.

[310] U. Lee, D. Kim, N. Hur, and D. Jeon. Design analysis and experimental evaluation of an MR fluid clutch. *Journal of Intelligent Material Systems and Structures*, 10(9): 701–707, 1999.

[311] Y. K. Ahn, B. S. Yang, M. Ahmadian, and S. Morishita. A small-sized variable-damping mount using magnetorheological fluid. *Journal of Intelligent Material Systems and Structures*, 16(2):127–133, 2005.

[312] G. J. Stelzer, M. J. Schulz, J. Kim, and R. J. Allemang. A magnetorheological semi-active isolator to reduce noise and vibration transmissibility in automobiles. *Journal of Intelligent Material Systems and Structures*, 14(12):743–765, December 2003.

[313] S. Sassi, K. Cherif, L. Mezghani, M. Thomas, and A. Kotrane. An innovative magneto-rheological damper for auotmotive suspension: From design to experimental characterization. *Smart Materials and Structures*, 14(4):811–822, 2005.

[314] D. F. Leroy, R. Marjoram, and K. St Clair. Giving truck drivers a smooth ride. *Machine Design*, 71(20):7–11, 1999.

[315] Anon. Magnetic fluid shocks. In *Mechanical Engineering Magazine, 32-3*. ASME, 1999.

[316] Y. M. Han, M. H. Nam, S. S. Han, H. G. Lee, and S. B. Choi. Vibration control evaluation of a commercial vehicle featuring MR seat damper. *Journal of Intelligent Material Systems and Structures*, 13(9):575–579, 2002.

[317] T. Ushijima and S. Kumakawa. Active engine mount with piezo-actuator for vibration control, SAE Paper No. 93-0201, 1993.

[318] J. L. Sproston, M. J. Stanway, M. J. Prendergast, J. R. Case, and C. E. Wilne. A prototype engine mount using electrorheological fluids. *Journal of Intelligent Materials and Smart Structures*, 4(3):418–419, July 1993.

[319] H. Manz and E. Breitbach. Application of smart materials in automotive structures. *Proceedings of the SPIE Smart Structures and Materials Symposium*, 4332:197–204, 2001.

[320] M. Giovanardi, K. Schmidt, and H. Kunze. Narrow and broad band sound reduction in automotive panels. *Proceedings of the SPIE Smart Structures and Materials Symposium*, 4332: 205–216, 2001.

[321] V. A. Neelakantan and G. N. Washington. Effect of centrifugal force on magneto-rheological fluid clutches. In *Proceedings of the ASME International Mechanical Engineering Congress and Exposition (IMECE 02)*, New Orleans, LA, November 2002.

[322] A. M. Rov, V. A. Neelakantan, and G. N. Washington, Design and Development of Passive and Active Force Feedback Systems Using Magnetorheological Fluids, Paper No. IMECE2003-43210, *Proceedings of ASME 2003 International Mechanical Engineering Congress and Exposition Aerospace*, Washington, DC, USA, November 15–21, 2003, pp. 333–339.

[323] V. A. Neelakantan and G. N. Washington. Modeling and reduction of centrifuging in magnetorheological (MR) transmission clutches for automotive applications. *Journal of Intelligent Material Systems and Structures*, 16(9):703–711, September 2005.

[324] V. A. Neelakantan, G. N. Washington, and N. K. Bucknor. Two-stage actuation system using DC motors and piezoelectric actuators for controllable and automotive brakes

and clutches. *Proceedings of the SPIE Smart Structures and Materials Symposium*, 5762:275–286, 2005.

[325] F. Boecking and B. Sagg. Piezo actuators: A technology prevails with injection valves for combustion engines. *Proceedings of the 10th International Conference on New Actuators*, pages 171–176, 2006.

[326] M. S. Senousy, F. X. Li, D. Mumford, M. Gadala, and R. Rajapakse. Thermo-electro-mechanical performance of piezoelectric stack actuators for fuel-injector applications. *Journal of Intelligent Material Systems and Structures*, 20(4):387–399, 2009.

[327] B. Weinberg, J. Nikitczuk, A. Fisch, and C. Mavroidis. Development of electro-rheological fluidic resistive actuators for haptic vehicular instrument controls. *Smart Materials and Structures*, 14(6):1107–1119, December 2005.

[328] F. Ahmadkhanlou, G. N. Washington, S. E. Bechtel, and Y. Wang. Magnetorheological fluid based automotive steer-by-wire systems. *Proceedings of the SPIE Smart Structures and Materials Symposium*, 6171:61710I-1 to 61710I-12, 2006.

[329] K. Kageyama, I. Kimpara, T. Suzuki, I. Ohsawa, H. Murayama, and K. Ito. Smart marine structures: An approach to the monitoring of ship structures with fiber-optic sensors. *Smart Materials and Structures*, 7(4):472–478, August 1998.

[330] T. R. Quackenbush, A. J. Bilanin, and B. F. Carpenter. Test results for an SMA-actuated vortex wake control system. *Proceedings of the SPIE Smart Structures and Materials Symposium*, 3674:84–94, 1999.

[331] O. K. Rediniotis, L. N. Wilson, D. C. Lagoudas, and M. M. Khan. Development of a shape-memory-alloy actuated biomimetic hydrofoil. *Journal of Intelligent Material Systems and Structures*, 13(1):35–49, 2002.

[332] S. Balakrishnan and C. Niezrecki. Investigation of THUNDER actuators as underwater propulsors. *Journal of Intelligent Material Systems and Structures*, 13(4):193–207, 2002.

[333] J. H. Kim, B. W. Kang, K. M. Park, S. B. Choi, and K. S. Kim. MR inserts for shock wave reduction in warship structures. *Journal of Intelligent Material Systems and Structures*, 13(10):661–665, October 2002.

[334] Y. Bar-Cohen, T. Xue, M. Shahinpoor, J. Simpson, and J. Smith. Flexible, low-mass robotic arm actuated by electroactive polymers and operated equivalently to human arm and hand. In *Proceedings of the 3rd Conference and Exposition, Demonstration on Robotics for Challenging Environments*, Albuquerque, NM, pages 15–21, April 1998.

[335] C. P. Chou and B. Hannaford. Measurement and modeling of Mckibben pneumatic artificial muscles. *IEEE Transactions on Robotics and Automation*, 12(1):90–102, 1996.

[336] K. DeLaurentis and C. Mavroidis. Mechanical design of a shape memory alloy actuated prosthetic hand. *Technol. Health Care*, 10(2):91–106, 2002.

[337] J. D. Carlson, W. Matthis, and J. R. Toscano. Smart prosthetics based on magnetorheological fluids. *Proceedings of the SPIE Smart Structures and Materials Symposium*, 4332:308–316, 2001.

[338] J. L. Zite, F. Ahmadkhanlou, V. A. Neelakantan, and G. N. Washington. A magnetorheological fluid-based orthopedic active knee brace. *Proceedings of the SPIE Smart Structures and Materials Symposium*, 6171:61710H-1 to 61710H-9, 2006.

[339] A. Price, A. Edgerton, C. Cocaud, H. Naguib, and A. Jnifene. A study on the thermo-mechanical properties of shape memory alloy-based actuators used in artificial limbs. *Journal of Intelligent Material Systems and Structures*, 18(1):11–18, January 2007.

[340] E. Kottke, L. D. Partridge, and M. Shahinpoor. Bio-potential neural activation of artificial muscles. *Journal of Intelligent Material Systems and Structures*, 18(2):103–109, February 2007.

[341] M. J. Lee, S. H. Jung, G. S. Kim, I. Moon, S. Lee, and M. S. Mun. Actuation of the artificial muscle based on ionic polymer metal composite by electromyography (EMG) signals. *Journal of Intelligent Material Systems and Structures*, 18(2):165–170, February 2007.

[342] T. Hirai. Electrically active non-ionic artificial muscle. *Journal of Intelligent Material Systems and Structures*, 18(2):117–122, February 2007.

[343] T. Okamoto, O. Kitani, and T. Totii. Robotic transplanting of orchid in mericlon culture. *Journal of the Society of Agricultural Machinery, Japan*, 55:103–110, 1993.

[344] S. Chonan, Z. W. Jiang, and M. Koseki. Soft-handling gripper driven by piezoceramic bimorph strips. *Smart Materials and Structures*, 5(4):407–414, 1996.

[345] A. D. Price, A. Jnifene, and H. E. Naguib. Design and control of a shape memory alloy based dexterous robot hand. *Smart Materials and Structures*, 16(4):1401–1414, August 2007.

[346] S. J. Lee, M. J. Han, S. J. Kim, J. Y. Jho, H. Y. Lee, and Y. H. Kim. A new fabrication method for IPMC actuators and application to artificial fingers. *Smart Materials and Structures*, 15(5):1217–1224, 2006.

[347] N. Thayer and S. Priya. Design and implementation of a dextrous anthropomorphic robotic typing (DART) hand. *Smart Materials and Structures*, 20(3):035010, 2011.

[348] T. B. Wolfe, M. G. Faulkner, and J. Wolfaardt. Development of a shape memory alloy actuator for a robotic eye prosthesis. *Smart Materials and Structures*, 14(4):759–768, August 2005.

[349] Y. Luo, T. Okuyama, T. Takagi, T. Kamiyama, K. Nishi, and T. Yambe. Thermal control of shape memory alloy artificial anal sphincters for complete implantation. *Smart Materials and Structures*, 14(1):29–35, February 2005.

[350] S. Chonan, Z. W. Jiang, J. Tani, S. Orikasa, Y. Tanahashi, T. Takagi, M. Tanaka, and J. Tanikawa. Development of an artificial urethral valve using SMA actuators. *Smart Materials and Structures*, 6(4):410–414, 1997.

[351] M. Tanaka, F. Wang, K. Abe, Y. Arai, H. Nakagawa, and S. Chonan. A closed-loop transcutaneous power transmission system with thermal control for artificial urethral valve driven by SMA actuator. *Journal of Intelligent Material Systems and Structures*, 17(8–9):779–786, 2006.

[352] N. Saga. Development of tendon-driven system using a pneumatic balloon. *Journal of Intelligent Material Systems and Structures*, 18(2):171–174, February 2007.

[353] S. Dong, K. Lu, J. Q. Sun, and K. Rudolph. Smart rehabilitation devices: Part I. Force tracking control. *Journal of Intelligent Material Systems and Structures*, 17(6):543–552, 2006.

[354] H. Bose and H. J. Berkemeier. Haptic device working with an electrorheological fluid. *Journal of Intelligent Material Systems and Structures*, 10(9):714–717, 1999.

[355] B. Edinger, M. Frecker, and J. Gardner. Dynamic modeling of an innovative piezoelectric actuator for minimally invasive surgery. *Journal of Intelligent Material Systems and Structures*, 11(10):765–770, October 2000.

[356] W. M. Rubio, E. C. N. Silva, E. V. Bordatchev, and M. J. F. Zeman. Topology optimized design, microfabrication and characterization of electro-thermally driven microgripper. *Journal of Intelligent Material Systems and Structures*, 20(6):669–681, 2009.

[357] G. A. Flores, R. Sheng, and J. Liu. Medical applications of magnetorheological fluids: A possible new cancer therapy. *Journal of Intelligent Material Systems and Structures*, 10(10):708–713, September 1999.

[358] G. A. Flores and J. Liu. Embolization of blood vessels as a cancer therapy using magnetorheological fluids. *Journal of Intelligent Material Systems and Structures*, 13(10):641–646, October 2002.

[359] M. Tanaka, M. Furubayashi, Y. Tanahashi, and S. Chonan. Development of an active palpation sensor for detecting prostatic cancer and hypertrophy. *Smart Materials and Structures*, 9(6):878–884, 2000.

[360] F. Wang, M. Tanaka, and S. Chonan. Development of a PVDF piezopolymer sensor for unconstrained in-sleep cardiorespiratory monitoring. *Journal of Intelligent Material Systems and Structures*, 14(3):185–190, 2003.

[361] S. Sudo, S. Segawa, and T. Honda. Magnetic swimming mechanism in a viscous liquid. *Journal of Intelligent Material Systems and Structures*, 17(8):729–736, August 2006.

[362] Y. Haga, M. Mizushima, T. Matsunaga, and M. Esashi. Medical and welfare applications of shape memory alloy microcoil actuators. *Smart Materials and Structures*, 14(5):S266–S272, 2005.

[363] E. Buselli, P. Valdastri, and M. Quirini. Superelastic leg design optimization for an endoscopic capsule with active locomotion. *Smart Materials and Structures*, 18(1):015001, 2009.

[364] R. L. D. Miranda, C. Zamponi, and E. Quandt. Fabrication of TiNi thin film stents. *Smart Materials and Structures*, 18(10):104010, 2009.

[365] Nihon Kinzoku Co., Ltd. Japanese Patent Application No. U50-133425, 1975.

[366] J. Haasters, G. V. Salis-Solio, and G. Bensmann. The use of Ni–Ti as an implant material in orthopedics. In *Engineering Aspects of Shape Memory Alloys*, ed. T. W. Duerig, K. N. Melton, D. Stöckel, and C. M. Wayman, Butterworth-Heinemann, pages 426–444, 1990.

[367] R. C. L. Sachdeva and S. Miyazaki. Superelastic Ni–Ti alloys in orthodontics. In *Engineering Aspects of Shape Memory Alloys*, ed. T. W. Duerig, K. N. Melton, D. Stöckel, and C. M. Wayman, Butterworth-Heinemann, pages 452–469, 1990.

[368] D. W. Raboud, M. G. Faulkner, and A. W. Lipsett. Superelastic response of NiTi shape memory alloy wires for orthodontic applications. *Smart Materials and Structures*, 9(5): 684–692, 2000.

[369] S. Fukuyo, E. Sairenji, Y. Suzuki, and K. Suzuki. Shape memory implants. In *Engineering Aspects of Shape Memory Alloys*, ed. T. W. Duerig, K. N. Melton, D. Stöckel, and C. M. Wayman, Butterworth-Heinemann, 1990, pages 470–476, 1990.

[370] J. P. O'Leary, J. E. Nicholson, and R. F. Gatturna. The use of Ni–Ti in the Homer Mammalok. In *Engineering Aspects of Shape Memory Alloys*, ed. T. W. Duerig, K. N. Melton, D. Stöckel, and C. M. Wayman, Butterworth-Heinemann, 1990, pages 477–482, 1990.

[371] J. Stice. The use of superelasticity in guidewires and arthroscopic instrumentation. In *Engineering Aspects of Shape Memory Alloys*, ed. T. W. Duerig, K. N. Melton, D. Stöckel, and C. M. Wayman, Butterworth-Heinemann, 1990, pages 483–487, 1990.

[372] C. Song, P. A. Campbell, T. G. Frank, and A. Cuschieri. Thermal modelling of shape memory alloy fixator for medical application. *Smart Materials and Structures*, 11(2): 312–316, 2002.

[373] M. Shahinpoor and K. J. Kim. Ionic polymer-metal composites: IV. Industrial and medical applications. *Smart Materials and Structures*, 14(1):197–214, 2005.

[374] H. Kanazawa, T. Tsukimoto, T. Maeno, and A. Miyake. Tribology of ultrasonic motors. *Japanese Journal of Tribology*, 38:315–324, 1993.

[375] K. Uchino. Piezoelectric ultrasonic motors: Overview. *Smart Materials and Structures*, 7(3):273–285, June 1998.

[376] K. Uchino. *Piezoelectric Actuators and Ultrasonic Motors*. Kluwer Academic Publishers, 1997.

[377] S. Dong, S. Cagatay, K. Uchino, and D. Viehland. A 'center-wobbling' ultrasonic rotary motor using a metal tube-piezoelectric plate composite stator. *Journal of Intelligent Material Systems and Structures*, 13(11):749–755, 2002.

[378] S. L. Sharp, J. S. N. Paine, and J. D. Blotter. Design of a linear ultrasonic piezoelectric motor. *Journal of Intelligent Material Systems and Structures*, 21(10):961–973, 2010.

[379] Y. Tian, H. Yu, Y. Meng, and S. Wen. A prototype of an exercising bicycle based on electrorheological fluids. *Journal of Intelligent Material Systems and Structures*, 17 (8–9):807–811, 2006.

[380] R. F. Pires, P. H. Nakasone, C. R. de Lima, and C. N. Silva. A miniature bimorph piezoelectrically actuated flow pump. *Proceedings of the SPIE Smart Structures and Materials Symposium*, 6171:617109-1 to 617109-12, 2006.

[381] S. C. Lim, J. S. Park, S. B. Choi, and Y. P. Park. Vibration control of a CD-ROM feeding system using electro-rheological mounts. *Journal of Intelligent Material Systems and Structures*, 12(9):629–637, 2001.

[382] W. Yang, S. Y. Lee, and B. J. You. A piezoelectric actuator with motion-decoupling amplifier for optical disk drives. *Smart Materials and Structures*, 19(6):065027, 2010.

[383] Y. T. Huang, C. K. Lee, and W. J. Wu. High-powered backlight inverter for LCD-TVs using piezoelectric transformers. *Journal of Intelligent Material Systems and Structures*, 18(6):601–609, 2007.

[384] G. Washington. Smart aperture antennas. *Smart Materials and Structures*, 5(6):801–805, 1996.

[385] D. J. Peel, R. Stanway, and W. A. Bullough. Dynamic modelling of an ER vibration damper for vehicle suspension applications. *Smart Materials and Structures*, 5(5):591–606, 1996.

[386] S. Chonan, M. Tanaka, T. Naruse, and T. Hayase. Development of an electrorheological active buffer for railway vehicles: Estimation of the capacity from prototype performance. *Smart Materials and Structures*, 13(5):1195–1202, October 2004.

[387] A. Fotoohi, A. Yousefi-Koma, and N. Yasrebi. Active control of train bogies with MR dampers. *Proceedings of the SPIE Smart Structures and Materials Symposium*, 6171:61710J-1 to 61710J-9, 2006.

[388] A. Peiffer, S. Storm, A. Röder, R. Maier, and P. G. Frank. Active vibration control for high-speed train bogies. *Smart Materials and Structures*, 14(1):1–18, February 2005.

[389] W. H. Liao and D. H. Wang. Semiactive vibration control of train suspension systems via magnetorheological dampers. *Journal of Intelligent Material Systems and Structures*, 14(3):161–172, 2003.

[390] S. Yan, F. Zhang, Z. Qin, and S. Wen. A 3-DOFs mobile robot driven by a piezoelectric actuator. *Smart Materials and Structures*, 15(1):N7–N13, February 2006.

[391] H. Ashrafiuon, M. Eshraghi, and M. H. Elahinia. Position control of a three-link shape memory alloy actuated robot. *Journal of Intelligent Material Systems and Structures*, 17(5):381–392, 2006.

[392] J.S. Rastegar and Y. Lifang. A systematic method for the design of piezostack actuator integrated robots for high-speed and precision operation. *Journal of Intelligent Material Systems and Structures*, 12(12):835–846, 2001.

[393] S. Yan, X. Liu, F. Xu, and J. Wang. A gripper actuated by a pair of differential SMA springs. *Journal of Intelligent Material Systems and Structures*, 18(5):459–466, 2007.

[394] H. A. Sodano, D. J. Inman, and G. Park. A review of power harvesting from vibration using piezoelectric materials. *The Shock and Vibration Digest*, 36(3):197–205, 2004.

[395] H. A. Sodano, D. J. Inman, and G. Park. Generation and storage of electricity from power-harvesting devices. *Journal of Intelligent Material Systems and Structures*, 16(1): 67–75, 2005.

[396] M. Umeda, K. Nakamura, and S. Ueha. Energy-storage characteristics of a piezo-generator using impact-induced vibration. *Japanese Journal of Applied Physics*, 36(5B):3146–3151, 1997.

[397] N. G. Elvin, A. A. Elvin, and M. Spector. A self-powered mechanical strain energy sensor. *Smart Materials and Structures*, 10(2):293–299, 2001.

[398] H. A. Sodano, E. A. Magliula, G. Park, and D. J. Inman. Electric power generation from piezoelectric materials. In *Proceedings of the 13th International Conference on Adaptive Structures and Technologies*, Potsdam, Germany, 7–9 October 2002.

[399] H. A. Sodano, J. Lloyd, and D. J. Inman. An experimental comparison between several active composite actuators for power generation. *Smart Materials and Structures*, 15(5):1211–1216, October 2006.

[400] R. Amirtharajah and A. P. Chandrakasan. Self-powered signal processing using vibration-based power generation. *IEEE Journal of Solid-State Circuits*, 33(5):687–695, 1998.

[401] G. K. Ottman, H. F. Hofmann, A. C. Bhatt, and G. A. Lesieutre. Adaptive piezoelectric energy-harvesting circuit for wireless remote power supply. *IEEE Transactions on Power Electronics*, 17(5):669–676, 2002.

[402] G. K. Ottman, H. F. Hofmann, and G. A. Lesieutre. Optimized piezoelectric energy-harvesting circuit using step-down converter in discontinuous conduction mode. *IEEE Transactions on Power Electronics*, 18(2):696–703, 2003.

[403] G. A. Lesieutre, H. F. Hofmann, and G. K. Ottman. Electric power generation from piezoelectric materials. In *Proceedings of the 13th International Conference on Adaptive Structures and Technologies*, Potsdam, Germany, 7–9 October 2002.

[404] J. Kymissis, C. Kendall, J. Paradiso, and N. Gershenfeld. Parasitic power harvesting in shoes. In *Proceedings of the 2nd IEEE International Symposium on Wearable Computers*, Pittsburgh, PA, pages 132–139, 19–20 October 1997.

[405] K. Nakano, S. J. Elliott, and E. Rustighi. A unified approach to optimal conditions of power harvesting using electromagnetic and piezoelectric transducers. *Smart Materials and Structures*, 16(4):948–958, August 2007.

[406] J. Rastegar, C. Pereira, and H.-L. Nguyen. Piezoelectric-based power sources for harvesting energy from platforms with low frequency vibration. *Proceedings of the SPIE Smart Structures and Materials Symposium*, 6171: 617101-1 to 617101-7, 2006.

2 Piezoelectric Actuators and Sensors

2.1 Fundamentals of Piezoelectricity

The term "piezoelectricity" translates roughly to "pressure electricity" and refers to an effect observed in many naturally occurring crystals; that is, the generation of electricity under mechanical pressure. The effect was first predicted and then experimentally measured by the brothers Pierre and Jacques Curie in 1880. The research was prompted by investigations into a closely related effect, the pyroelectric effect, which is the generation of electricity as a result of a change in temperature. The effect observed by the Curie brothers is also known as the direct piezoelectric effect. A strict definition of the direct effect is "electric polarization produced by mechanical strain, being directly proportional to the applied strain." A converse piezoelectric effect also exists and is the appearance of mechanical strain as a result of an applied electric field.

The origin of the piezoelectric effect can be traced to fundamental geometric properties of certain crystals. Based on their geometry, crystals are normally classified into seven categories: triclinic, monoclinic, orthorhombic, tetragonal, trigonal, hexagonal, and cubic. A structure is called centrosymmetric if it has symmetry with respect to a single point. Based on their symmetry with respect to a point, the crystals are further classified into 32 classes, of which only 20 classes can exhibit piezoelectricity. The unit cell of these crystals possesses a certain degree of asymmetry, leading to a separation of positive and negative charges that results in a permanent polarization. A crystal that is centrosymmetric by definition cannot have any asymmetry and therefore cannot be piezoelectric. Hence, no piezoelectricity is exhibited in any of the crystal classes that are centrosymmetric.

The terms "direct" and "converse" bear only an historical importance. The converse piezoelectric effect was predicted by Lippmann and experimentally verified in 1881 by the Curie brothers. Measurements on quartz showed that the piezoelectric coefficients for the direct and converse effects were equal. The first quantitative measurements were made on quartz and tourmaline, but a large number of naturally occurring crystals exhibiting the piezoelectric effect were subsequently identified, such as Rochelle's salt, tourmaline, quartz, cane sugar, and tartaric acid. Based on thermodynamic principles, a phenomenological theory of piezoelectricity was developed by several researchers such as Lord Kelvin, P. Duhem, and F. Pockels, and a comprehensive treatment of the subject was given by W. Voigt in 1894.

The piezoelectric effect remained a subject of purely academic interest untill the First World War. Increased interest in locating underwater objects and exploring the ocean floor led to the development of piezoelectric devices for emitting and receiving ultrasonic waves underwater, the precursor to modern sonar equipment. These applications were pioneered by Langevin, who developed ultrasonic emitters and detectors driven by quartz plates. Several designs of piezoelectric resonators, oscillators, and transformers were subsequently developed during the next few decades and are now used in a wide range of applications.

Most piezoelectric materials are crystalline in nature; they can be either single crystals or polycrystalline. They can be formed in nature or formed by synthetic processes. One of the types of piezoelectric materials widely used in technological applications is piezoceramics, also known as ferroelectric ceramics. These were developed in the second half of the twentieth century and have much larger piezoelectric coefficients than natural crystals. In their original unprocessed form, these materials do not possess piezoelectric characteristics, and they are isotropic. They need to be polarized through the application of a strong electric field. When the field is removed, the ceramic material becomes piezoelectric, a permanent deformation takes place, and the material becomes anisotropic. Once polarized, they can be associated with a well-defined crystal-axis system and their behavior can be expressed in terms of this axis system. The material requires metal electrodes deposited on appropriate surfaces for application of an electric field.

The piezoelectric effect can be expressed in terms of constitutive relations that can be derived from basic thermodynamic relations. It is convenient to express the mechanical strain and the electric displacement as independent variables, resulting in forms of the constitutive relations that bear a one-to-one correspondence with the converse and direct piezoelectric effects. A standard way of writing these equations is [1]

$$S_{ij} = s^{\mathbb{E}}_{ijkl} T_{kl} + d_{kij} \mathbb{E}_k \tag{2.1}$$

$$D_i = d_{ikl} T_{kl} + \epsilon^{T}_{ik} \mathbb{E}_k \tag{2.2}$$

where S_{ij} is the mechanical strain tensor, T_{kl} is the mechanical stress tensor, ϵ^T_{ik} is the permittivity tensor, $s^{\mathbb{E}}_{ijkl}$ is the compliance tensor, and d_{kij} or d_{ikl} is the piezoelectric coefficients tensor. Equation (2.1) is the actuator equation and Eq. (2.2) is the sensor equation. Actuator applications are based on the converse piezoelectric effect. The actuator is bonded to a structure and an external electric field is applied to it, which results in an induced-strain field. Sensor applications are based on the direct effect. The sensor is exposed to a stress field and generates a charge in response, which is measured. Note that the superscripts $\mathbb{E}$ and T imply that the corresponding quantities are measured at constant electric field and constant stress, respectively.

The actuator equation (Eq. 2.1) is based on the assumption that the total strain is the sum of the strain caused by the mechanical stress and the controllable actuation strain induced by the applied electric field. Similarly, the sensor equation is based on the assumption that the total electric charge (or displacement) is the sum of the charge induced due to the mechanical stress and the charge generated due to the external electric field.

The basic piezoelectric constitutive relations are assumed linear, in which elastic, piezoelectric, and dielectric coefficients are assumed constant and independent of applied mechanical stress as well as the electric field. In reality, they are often nonlinear, especially under high electrical or mechanical fields. In addition, piezoelectric materials exhibit hysteresis effects, electrical aging, and magneto-mechano-electric interactions. The nonlinearities can be incorporated into analytical models in several different ways, the simplest of which is by including field-dependent coefficients in the equations. The linear constitutive relations are valid only for low electric field and low mechanical stress levels. Furthermore, the constitutive relations are quasi-static and do not represent any dynamic effects.

The remainder of this book uses engineering notation and expresses tensors in Voigt notation to simplify the subscripts. The piezoelectric constitutive equations are rewritten as

$$\epsilon_i = s_{ij}^{\mathbb{E}}\sigma_j + d_{ik}\mathbb{E}_k \qquad \text{(actuator equation)} \tag{2.3}$$

$$D_k = d_{ki}\sigma_i + e_{kl}^{\sigma}\mathbb{E}_l \qquad \text{(sensor equation)} \tag{2.4}$$

where the indices $i, j = 1, 2 \ldots 6$ and $k, l = 1, 2, 3$. In matrix notation, this can be written as

$$\boldsymbol{\epsilon} = \boldsymbol{s}^{\mathbb{E}}\boldsymbol{\sigma} + \boldsymbol{d}^c\,\mathbb{E} \qquad \text{(actuator equation)} \tag{2.5}$$

$$\boldsymbol{D} = \boldsymbol{d}^d\,\boldsymbol{\sigma} + \boldsymbol{e}^{\sigma}\mathbb{E} \qquad \text{(sensor equation)} \tag{2.6}$$

where the strain vector $\boldsymbol{\epsilon}$ (dimensionless), in terms of standard engineering notation, is given by

$$\boldsymbol{\epsilon} = \begin{Bmatrix} \epsilon_1 \\ \epsilon_2 \\ \epsilon_3 \\ \epsilon_4 \\ \epsilon_5 \\ \epsilon_6 \end{Bmatrix} = \begin{Bmatrix} \epsilon_1 \\ \epsilon_2 \\ \epsilon_3 \\ \gamma_{23} \\ \gamma_{31} \\ \gamma_{12} \end{Bmatrix} \tag{2.7}$$

where ϵ_1, ϵ_2, and ϵ_3 are direct strains along the mutually orthogonal right-handed axes 1, 2, and 3, respectively, and γ_{23}, γ_{31}, and γ_{12} are shear strains. Note that axis "3" is oriented along the polarization direction of the piezoelectric element. Similarly, the stress vector $\boldsymbol{\sigma}$ (N/m^2) is given by

$$\boldsymbol{\sigma} = \begin{Bmatrix} \sigma_1 \\ \sigma_2 \\ \sigma_3 \\ \sigma_4 \\ \sigma_5 \\ \sigma_6 \end{Bmatrix} = \begin{Bmatrix} \sigma_1 \\ \sigma_2 \\ \sigma_3 \\ \tau_{23} \\ \tau_{31} \\ \tau_{12} \end{Bmatrix} \tag{2.8}$$

where σ_1, σ_2, and σ_3, are direct stresses and τ_{23}, τ_{31}, and τ_{12} are shear stresses. For the most general case, the compliance matrix $\boldsymbol{s}^{\mathbb{E}}$ (m^2/N) is given by

$$\boldsymbol{s}^{\mathbb{E}} = \begin{bmatrix} s_{11}^{\mathbb{E}} & s_{12}^{\mathbb{E}} & s_{13}^{\mathbb{E}} & s_{14}^{\mathbb{E}} & s_{15}^{\mathbb{E}} & s_{16}^{\mathbb{E}} \\ s_{21}^{\mathbb{E}} & s_{22}^{\mathbb{E}} & s_{23}^{\mathbb{E}} & s_{24}^{\mathbb{E}} & s_{25}^{\mathbb{E}} & s_{26}^{\mathbb{E}} \\ s_{31}^{\mathbb{E}} & s_{32}^{\mathbb{E}} & s_{33}^{\mathbb{E}} & s_{34}^{\mathbb{E}} & s_{35}^{\mathbb{E}} & s_{36}^{\mathbb{E}} \\ s_{41}^{\mathbb{E}} & s_{42}^{\mathbb{E}} & s_{43}^{\mathbb{E}} & s_{44}^{\mathbb{E}} & s_{45}^{\mathbb{E}} & s_{46}^{\mathbb{E}} \\ s_{51}^{\mathbb{E}} & s_{52}^{\mathbb{E}} & s_{53}^{\mathbb{E}} & s_{54}^{\mathbb{E}} & s_{55}^{\mathbb{E}} & s_{56}^{\mathbb{E}} \\ s_{61}^{\mathbb{E}} & s_{62}^{\mathbb{E}} & s_{63}^{\mathbb{E}} & s_{64}^{\mathbb{E}} & s_{65}^{\mathbb{E}} & s_{66}^{\mathbb{E}} \end{bmatrix} \tag{2.9}$$

Because $s_{ij}^{\mathbb{E}} = s_{ji}^{\mathbb{E}}$, the 36 constants in the compliance matrix reduce to 21. The electric displacement, $\boldsymbol{D}$ (C/m^2 or Coulombs per square meter), and electric field, $\mathbb{E}$ (N/C or V/m), are vectors given by

$$\boldsymbol{D} = \begin{Bmatrix} D_1 \\ D_2 \\ D_3 \end{Bmatrix} \tag{2.10}$$

$$\mathbb{E} = \begin{Bmatrix} \mathbb{E}_1 \\ \mathbb{E}_2 \\ \mathbb{E}_3 \end{Bmatrix} \tag{2.11}$$

The electric permittivity matrix, $\boldsymbol{e}$ (F/m or C^2/N-m^2), is

$$\boldsymbol{e} = \begin{bmatrix} e_{11} & e_{12} & e_{13} \\ e_{21} & e_{22} & e_{23} \\ e_{31} & e_{32} & e_{33} \end{bmatrix} \tag{2.12}$$

The terms $e_{ij}, i \neq j$ are called cross-permittivities, and $e_{ij} = e_{ji}$. The electric permittivity is a measure of the charge density due to an electric field. The permittivity e_{ij} defines the charge generated or electric displacement on electrodes normal to the i-axis due to an electric field in the j-direction. For most of the piezoelectric materials, a field along one axis results in electric displacement only along the same axis. This means that $e_{ij} = 0$ for $i \neq j$. In addition, the mechanical-boundary conditions play an important role in the interchange of electrical and mechanical energy. When the piezoelectric material is mechanically unrestrained, the electric permittivity is higher than when the material is mechanically restrained. This can be written as

$$e_{ii}^{\sigma} > e_{ii}^{\epsilon} \tag{2.13}$$

where the superscript σ indicates a condition of constant stress (no mechanical restraint) and the superscript ϵ indicates a condition of constant strain (completely restrained).

The piezoelectric coefficient matrices, $\boldsymbol{d}$, represent the electromechanical coupling inherent in the material. The matrix $\boldsymbol{d}^c$ (m/V) is called the converse piezoelectric coupling matrix, and the matrix $\boldsymbol{d}^d$ (C/N) is called the direct piezoelectric coupling matrix. In the converse piezoelectric effect, the piezoelectric constant d_{ik}^c

represents the mechanical strain produced along the i-axis by an applied electric field along the k-axis

$$d^c_{ik} = \frac{\text{strain induced in } i\text{-direction}}{\text{electric field applied in } k\text{-direction}}, \quad \frac{1}{\text{volts/meter}} \quad \text{or} \quad \frac{\text{m}}{\text{V}} \tag{2.14}$$

In the direct piezoelectric effect, d^d_{ki} represents the charge generated on the electrodes normal to the k-axis due to an applied mechanical stress σ_i

$$d^d_{ki} = \frac{\text{charge generated in } k\text{-direction}}{\text{mechanical stress applied in } i\text{-direction}}, \quad \frac{\text{Coulomb/square meter}}{\text{Newton/square meter}} \quad \text{or} \quad \frac{\text{C}}{\text{N}} \tag{2.15}$$

It has been experimentally verified that for most practical purposes, if $\boldsymbol{d}^d = \boldsymbol{d}$, then $\boldsymbol{d}^c = \boldsymbol{d}^T$; that is, the direct piezoelectric coupling matrix is the transpose of the converse piezoelectric coupling matrix. Note that it is theoretically possible for a material to have 18 independent piezoelectric constants. Thus,

$$\boldsymbol{d}^d = \begin{bmatrix} d_{11} & d_{12} & d_{13} & d_{14} & d_{15} & d_{16} \\ d_{21} & d_{22} & d_{23} & d_{24} & d_{25} & d_{26} \\ d_{31} & d_{32} & d_{33} & d_{34} & d_{35} & d_{36} \end{bmatrix} = d \tag{2.16}$$

$$\boldsymbol{d}^c = \boldsymbol{d}^T = \begin{bmatrix} d_{11} & d_{21} & d_{31} \\ d_{12} & d_{22} & d_{32} \\ d_{13} & d_{23} & d_{33} \\ d_{14} & d_{24} & d_{34} \\ d_{15} & d_{25} & d_{35} \\ d_{16} & d_{26} & d_{36} \end{bmatrix} \tag{2.17}$$

From these equations, it is clear that the units of the piezoelectric coefficient d_{ij} can be expressed in terms of $m/V \equiv C/N$. It should be noted that these forms of $\boldsymbol{s}$, $\boldsymbol{e}$, and $\boldsymbol{d}$ represent the most general case. Depending on the symmetry present in specific crystals, many of these coefficients may be equal to each other or even zero. These coefficients are defined in Table 2.1.

2.2 Piezoceramics

With the discovery of piezoceramics exhibiting a much larger piezo effect than natural materials, the domain of applications expanded considerably. Piezoceramic elements can also be manufactured easily in large quantities and in specific shapes, which makes them ideally suited for adaptive-structures applications, in the form of actuators as well as sensors. Typical piezoceramics include barium titanate ($BaTiO_3$), which was one of the first piezoceramics to be extensively investigated, and lead zirconate titanates, or PZTs ($PbZr_{1-x}Ti_xO_3$). These compositions fall in a broad category of compounds called perovskites, which consist of a combination of tetravalent metals (e.g., titanium or zirconium), divalent metals (e.g., lead or barium), and oxygen. Recently, relaxor materials such as lead magnesium niobate (PMN), exhibiting superior performance compared to PZT-based ceramics, are being developed.

Because dopants and defect structure have an enormous influence on domain wall motion, they markedly affect the magnitude of the piezoelectric coefficients.

Table 2.1. *Definition of symbols (indices: $i, j = 1, 2, \ldots 6$ and $k, l = 1, 2, 3$)*

Symbol	Name	Piezoelectric Effect	Definition	Size	Units
d^d_{ki}	Piezoelectric coefficient	Direct	Charge accumulated on surface electrodes normal to k-axis due to stress component i	3×6	C/N
d^c_{ik}	Piezoelectric coefficient	Converse	Induced-strain component i due to electric field along k-axis (applied on electrodes normal to k-axis)	6×3	m/V
D_k	Electric displacement	Direct	Charge accumulated on surface electrodes normal to the k-axis	3×1	C/m^2
e^{σ}_{kl}	Electric permittivity	Direct	Ratio of charge accumulated on surface electrodes normal to k-axis to electric field along l-axis at constant stress	3×3	F/m
$\mathbb{E}_k$	Electric field	Direct/Converse	Electric field applied on surface electrodes normal to the k-axis	3×1	V/m or N/c
$s^{\mathbb{E}}_{ij}$	Elastic compliance	Converse	Ratio of mechanical strain component i to stress component j at constant electric field	6×6	m^2/N
ϵ_i	Strain	Converse	$\epsilon_1, \epsilon_2, \epsilon_3$ (direct strains), $\epsilon_4, \epsilon_5, \epsilon_6$ (shear strains)	6×1	–
σ_i	Stress	Direct/Converse	$\sigma_1, \sigma_2, \sigma_3$ (direct stresses), $\sigma_4, \sigma_5, \sigma_6$ (shear stresses)	6×1	N/m^2

This, in turn, also has a major influence on the nature of hysteresis loops in the material. Such an interaction between the domain walls and defects leads to "soft" and "hard" piezoelectric compositions. In a "soft" piezoelectric material, piezoelectric coefficients are large and the material exhibits high hysteresis. In a "hard" piezoelectric material, the piezoelectric coefficients are small and the material exhibits low hysteresis. In such materials, the domain wall motion is inhibited. Soft piezoelectrics are preferred for most of the actuator applications because of their larger induced strain. Hard piezoelectrics are preferred where low hysteretic response is desired.

Piezoceramics are polycrystalline in nature and do not have piezoelectric characteristics in their original state. Piezoelectric effects are induced in these materials by electrical poling (the application of high electric field results in polarization). The most commonly used piezoceramics are based on PZT compounds. These materials have been widely used as actuators in adaptive-structures applications. Piezoceramics are available commercially in a variety of basic shapes such as sheets, discs, and cylinders, as well as in the form of assembled actuators such as piezoceramic stacks, benders, unimorphs, and torque tubes. The remainder of

Table 2.2. *Comparison of the characteristics of soft and hard piezoelectric ceramics*

Characteristic	Soft Ceramics	Hard Ceramics
Piezoelectric coefficient (d_{ij})	Large	Small
Curie temperature	Low	High
Electric permittivity (e_{ij})	Large	Small
Resistivity	Large	Small
Coercive field	Low	High
Linearity	Poor	Good
Polarization/depolarization	Easy	Difficult
Electromechanical coupling factor	Large	Small

this chapter focuses on the properties and behavior of a typical PZT composition, PZT-5H.

2.3 Soft and Hard Piezoelectric Ceramics

The piezoelectric properties of ceramics are a function of their constituents. A small amount of a dopant material added to a piezoceramic can make it either a soft or a hard piezoceramic. In general, soft piezoceramics are characterized by a large electromechanical coupling factor, large piezoelectric constant, high electric permittivity, low modulus of elasticity, low Curie temperature, and poor linearity. Soft ceramics produce larger maximum strains, exhibit greater hysteresis, and are more susceptible to depolarization than hard ceramics. Generally, large values of permittivity and dielectric dissipation may exclude these ceramics from applications requiring high-frequency input in combination with high electric fields. Typically, hard ceramics are suited for high-force actuation and soft ceramics are suited for sensing applications. Table 2.2 summarizes the differences between soft and hard piezoceramics.

2.4 Basic Piezoceramic Characteristics

Piezoceramics based on PZT are solid solutions of lead zirconate and lead titanate, often doped with other elements to obtain specific properties. The material is manufactured by mixing a powder of lead, zirconium, and titanium oxides and then heating the mixture to around 800–1000°C. It then transforms to perovskite PZT powder, which is mixed with a binder, sintered into the desired shapes, and cooled. As the temperature of the material drops below the Curie temperature (which is specific to the material composition), it undergoes a phase transformation in which the cubic unit cells become tetragonal. A typical PZT unit cell at room temperature is shown in Fig. 2.1(a), with the three reference axes (*a*, *b*, and *c* axes). The sides of the unit cell along the *a* and *b* axes are equal in length, whereas the side along the *c*-axis is slightly longer. For this reason, the *c*-axis is also sometimes called the "long' axis.

As the titanium ion is slightly displaced from the center of the unit cell, a seperation of charge occurs between the positively charged titanium ions and the negatively charged oxygen ions. As a result, the unit cell has a permanent dipole moment oriented along the *c*-axis, as shown in Fig. 2.1(b). The dipole moment $\boldsymbol{p}$ is

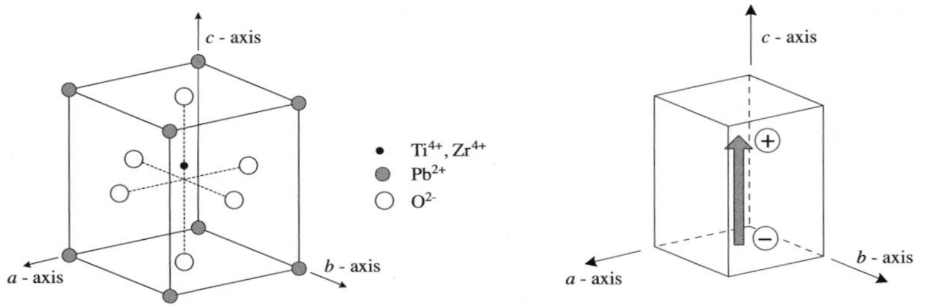

(a) PZT unit cell (b) Equivalent dipole moment

Figure 2.1. Spontaneous polarization in PZT.

a vector quantity defined as

$$\boldsymbol{p} = q \times \boldsymbol{r} \tag{2.18}$$

where q is the magnitude of each charge and $\boldsymbol{r}$ is the displacement vector of the positive charge with respect to the negative charge. Therefore, the dipole moment of the PZT unit cell is denoted as an arrow pointing along the positive c-axis, from the negative charge to the positive charge.

A volume of unit cells with their dipole moments oriented in the same direction is called a "domain." Due to the dipole moment inherent in the unit cells, each domain in the material has a spontaneous polarization, which is defined as the dipole moment per unit volume. Note that the polarization is also a vector quantity and has the same direction as the dipole moment. It is this polarization that is responsible for the piezoelectric characteristics of a single domain. In any bulk material, imperfections always exist in the form of breaks in the lattice structure. Although such imperfections can exist even in naturally occurring crystals, they are even more prevalent in polycrystalline materials such as the piezoceramic, resulting in a large number of domains. The direction of the unit cells can change from one region (domain) to the next. Hence, a bulk sample of unpoled piezoceramic will contain a large number of randomly oriented domains, each with a dipole moment (Fig. 2.2(a)). As a result, the net spontaneous polarization of the sample, P_s, is zero.

The next step in the manufacturing process is the application of electrodes on the surface by using an electroplating or sputtering process. Application of a high electric field (typically more than 2000 V/mm) results in the realignment of most of the domains in such a way that their dipole moments are oriented mostly parallel to the applied field, as shown in Fig. 2.2(b). This process is called "poling," and it imparts a permanent polarization to the ceramic (analogous to magnetization of a

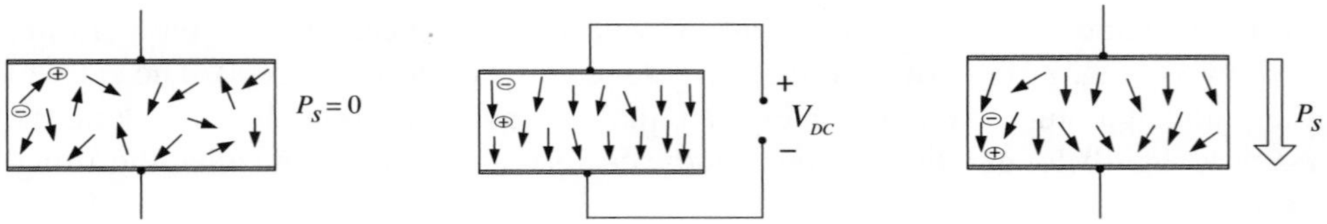

(a) Randomly oriented domains before poling, $P_s = 0$ (b) Domains are aligned along positive poling field (c) Removal of poling field results in remnant polarization, $P_s \neq 0$

Figure 2.2. Effect of poling on domains.

ferrous material with a permanent magnet). Because the c-axis of the unit cell is longer than the a and b axes, the reorientation of the domains creates a mechanical distortion. A large electric field applied opposite to the direction of poling can result in a complete reorientation of the domains, destroying the net polarization of the material. This is called "depoling of the piezoceramic." Similarly, depoling can also be achieved by applying a large mechanical stress to the material.

Poled piezoceramics exhibit both direct and converse piezoelectric effects. The mechanisms that provide both sensing and actuation capabilities are due to the non-centrosymmetric nature of the material. Polar changes due to the applied field cause displacement of ions as a result of their alignment with the field. Conversely, the application of stresses causes deformations in the material, which in turn alter the polarization, resulting in separation of charge. The actuation phase (converse effect) consists of three parts. The first is called intrinsic effect and involves the deformation of dipoles in the unit cells. The second is called extrinsic effect and involves the motion of domain walls, caused by reorientation of domains. It is a major source of nonlinearity and losses in piezoceramics. The third effect is due to the electrostriction of materials and, as a result of this effect, the deformation is generally proportional to the square of the electric field. Electrostriction effects are much smaller than the other two effects and are discussed in detail in a subsequent chapter. A simplistic interpretation of the mechanism of the direct and converse piezoelectric effects is shown in Figs. 2.3 and 2.4. It is important to note that the main goals of these simplistic interpretations are to facilitate a physical understanding of the phenomena and to serve as a simple, intuitive memory aid. The actual mechanisms involved are complex and require a comprehensive understanding of material science at the micromechanical scale.

Consider a sample of piezoceramic material with its poling direction marked with either an arrow or a dot on the positive electrode (Fig. 2.3(a)). A compressive stress applied along the poling direction has the effect of "flattening" the domains; that is, the domains tend to orient themselves so that their long axes are perpendicular to the direction of applied stress. Complete reorientation of the domains is prevented by internal elastic stresses, and the domains attain a final orientation in which the internal and external stresses are in equilibrium. From Figs. 2.2(b) and 2.2(c), it can be seen that the reorientation of the domains in this manner results in the negatively charged ends moving away from the top electrode. Consequently, an effective positive charge is built up on the top electrode, producing a positive voltage, which is in the same direction as the poling voltage. Note that a compressive stress along the poling direction produces similar behavior as a tensile stress perpendicular to the poling direction.

In a similar way, the mechanism of the converse piezoelectric effect can be examined. Application of a voltage of the same polarity as the poling voltage tends to align the domains along the poling direction. Consequently, the dimension of the sample increases by Δl along the poling direction and decreases perpendicular to the poling direction (Fig. 2.4(a)). Application of a voltage with a polarity opposite to that of the poling direction tends to align the domains perpendicular to the poling direction, resulting in a decrease in dimension along the poling direction and an increase in dimension perpendicular to the poling direction (Fig. 2.4(b)). Vibration of the piezoceramic sheet along a polar direction is referred to as "longitudinal mode," whereas vibration in a direction normal to the polar axis is referred to as "transverse" or "lateral mode."

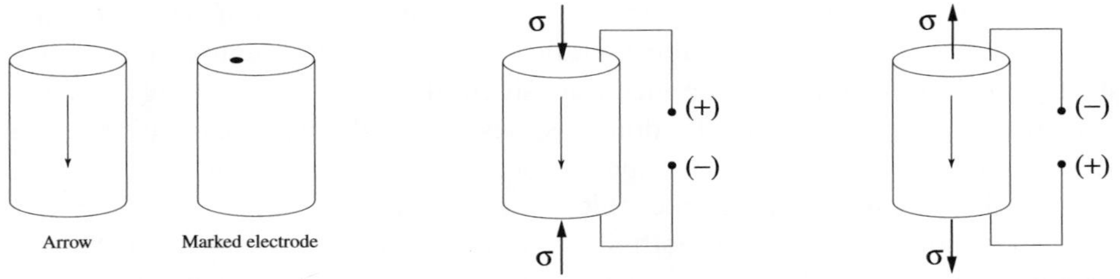

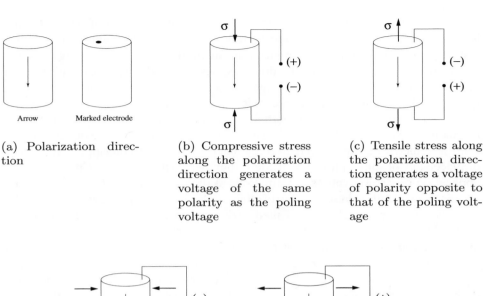

(a) Polarization direction

(b) Compressive stress along the polarization direction generates a voltage of the same polarity as the poling voltage

(c) Tensile stress along the polarization direction generates a voltage of polarity opposite to that of the poling voltage

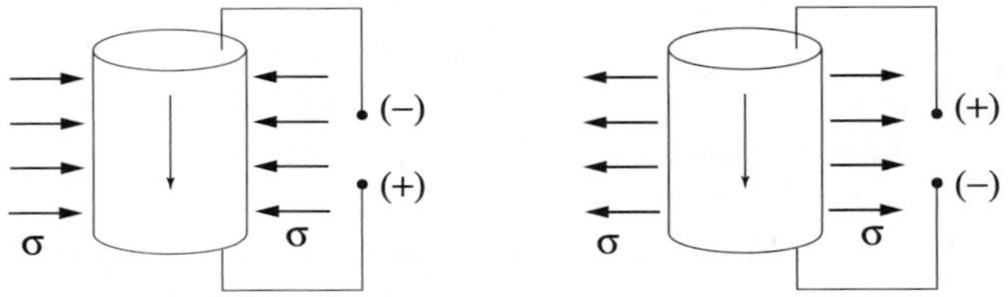

(d) Compressive stress perpendicular to polarization direction generates a voltage of opposite polarity to the poling voltage

(e) Tensile stress perpendicular to the polarization direction generates a voltage of the same polarity as the poling voltage

Figure 2.3. Effect of applied stress on voltage generated (direct piezoelectric effect).

2.5 Electromechanical Constitutive Equations

Piezoceramic materials are relatively well behaved and linear at low electric fields and low mechanical stress levels; they show considerable nonlinearity at high values of electric field and mechanical stress. The linear piezoelectric constitutive relations, Eqs. 2.3 and 2.4, can be used to model the behavior of the piezoceramics at low excitation levels. A convenient way of modeling the nonlinearities at high electric field is to use the linear constitutive equations (Eqs. 2.3 and 2.4) with electric field-dependent constants. The actuation strain can also be modeled like an equivalent thermal strain, and this representation is often used in commercial FEM software.

A typical piezoceramic sheet is shown in Fig. 2.5. The initial polarization direction is expressed as the z-axis (or 3-axis). The axes x and y (or 1-axis and 2-axis) are defined in a plane normal to the z-axis, in a conventional right-handed system. To polarize the material, a high DC field is applied between the electroded faces. As the sheet is polarized along the z-axis, the electroded faces are in the $x-y$ plane (normal to the direction of polarization). Note that for shear actuation, these poling electrodes must be removed and subsequently replaced with a pair of electrodes deposited on faces normal to the x-axis or y-axis. The direction of polarization is

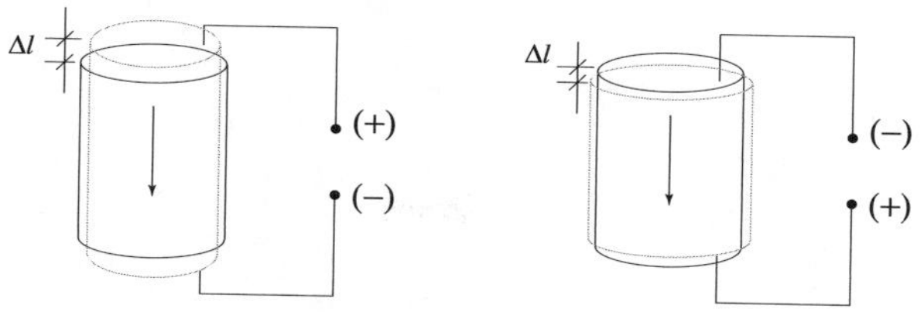

Figure 2.4. Effect of applied voltage on change in dimensions (converse piezoelectric effect).

(a) Voltage of the same polarity as the poling voltage causes an extension along the poling direction and contraction perpendicular to the poling direction

(b) Voltage of the opposite polarity as the poling voltage causes a contraction along the poling direction and extension perpendicular to the poling direction

identified by an arrow in the negative z-direction and is indicated on the piezoceramic sheet by a dot on one of the electrodes, as shown in Fig. 2.5. Note that a positive electric field $\mathbb{E}_3$ results if the electrode marked by the dot is at a higher voltage than the electrode on the opposite face of the sheet. Axes 4, 5, and 6 represent right-hand rotations, respectively, about axes 1, 2, and 3. The application of the electric fields $\mathbb{E}_1$, $\mathbb{E}_2$, and $\mathbb{E}_3$ is depicted in Fig. 2.6. However, these fields can only be applied if the material has electrodes in a plane normal to the desired field direction.

The piezoceramic sheet is polycrystalline and needs to be poled to induce piezoelectric effect. The spontaneous polarization P_s is imparted to the material by applying a high DC voltage between a pair of electrode faces on the 1-2 plane. During the poling process, the piezo ceramic sheet undergoes a permanent change in dimensions. Prior to polarization, the piezoceramic material is isotropic and becomes anisotropic after polarization. A piezoceramic sheet poled along the 3-axis can be idealized as a transversely isotropic (in the 1-2 plane) material. Because the majority of the c-axes of the unit cells are oriented along the 3-axis of the element, properties along this axis are different from properties along the 1-axis and the 2-axis. In addition, because the a-axis and b-axis of the unit cell are equal, properties of the bulk material along the 1-axis and 2-axis are equal, resulting in the idealization of the sheet element as transversely isotropic.

Once polarized, voltage of the same polarity as the poling voltage causes a temporary extension in the poling direction and contraction in the plane parallel to electrodes (Fig. 2.4(a)). The piezoceramic element returns to its original poled

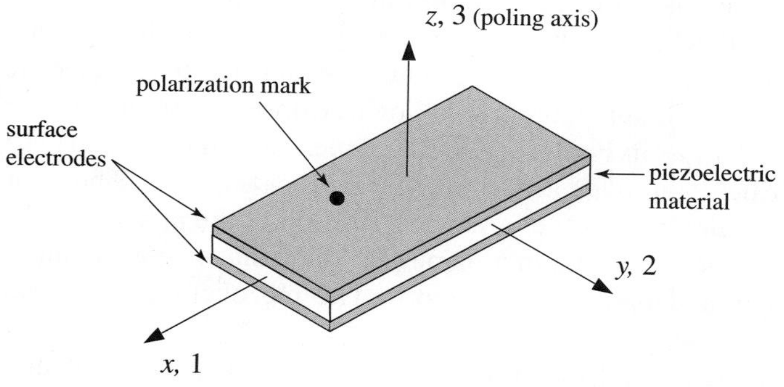

Figure 2.5. Definition of co-ordinate axes and poling direction for a piezoceramic sheet.

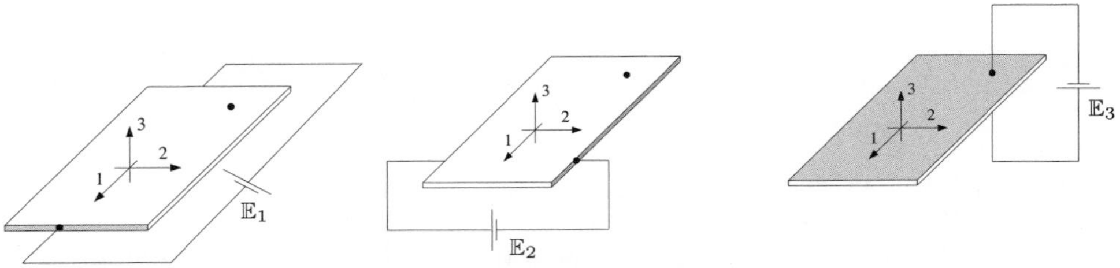

(a) Electric field $\mathbb{E}_1$ (b) Electric field $\mathbb{E}_2$ (c) Electric field $\mathbb{E}_3$

Figure 2.6. Possible electric field directions for a piezoceramic sheet.

dimensions after the removal of voltage. Strains in two directions are unequal, resulting in a small change of volume with the application of voltage. If a compressive force is applied in the poling direction or a tensile force is applied in the plane normal to the poling direction (parallel to the electrodes), a positive voltage is generated (same polarity as the original poling voltage). Shear strain is produced when the field is applied normal to the polarized direction (axis 1 or 2). However, it requires deposition of electrodes normal to axis 1 or 2. Often, one requires a very high voltage (several kV), depending on the width of the plate, to induce significant shear strains.

The properties of the piezoceramic gradually change with time (logarithmic function of time) after the original polarization of material. After some time from initial poling, the material becomes quite stable. Unless the stress level is very high, the properties of piezoceramic material are independent of stress. Each piezoceramic material has a specific temperature called the Curie temperature, beyond which the material loses its piezoelectric characteristics due to a change in lattice structure that destroys the inherent asymmetry of the unit cell. However, it is suggested to operate at a temperature far below the Curie temperature to avoid accelerated aging of material, increased electrical losses, and reduced safe stress that occurs at an elevated temperature.

The piezoceramic sheet element shown in Fig. 2.5 can be used both as an actuator and as a sensor. The constitutive equations for both cases are discussed in the next subsection.

2.5.1 Piezoceramic Actuator Equations

The constitutive relations are based on the assumption that the total strain in the actuator is the sum of the mechanical strain induced by the stress, the thermal strain due to temperature, and the controllable actuation strain due to the electric voltage. The piezoceramic material is assumed to be linear. Consider a piezoceramic sheet poled along its thickness, as in Fig. 2.5. The 1-2 planes (top and bottom) are electroded, and an electric field is applied across the thickness of the sheet. The 3 (or z or c) axis is assigned to the direction of the initial polarization of the piezoceramic, and 1 (or x or a) and 2 (or y or b) axes lie in a plane perpendicular to z axis. Shear about these three axes is represented by 4, 5, and 6. The piezoceramic material constants are defined using these axes.

When the material is used as an actuator, the electric field is an input and mechanical strain is the output. The corresponding constitutive relation for the material can be written from the general case (Eq. 2.5) as

$$\boldsymbol{\epsilon} = \boldsymbol{s}^{\mathbb{E}}\boldsymbol{\sigma} + \boldsymbol{d}^{c}\mathbb{E} \tag{2.19}$$

Including the effects of thermal expansion, the constitutive relations for a piezoceramic actuator are

$$\boldsymbol{\epsilon} = \boldsymbol{s}^{\mathbb{E}}\boldsymbol{\sigma} + \boldsymbol{d}^{c}\mathbb{E} + \boldsymbol{\alpha}\Delta T \tag{2.20}$$

Due to the specific crystal structure of PZT, several elements of the coefficient matrices become equal to each other or to zero. The compliance matrix $\boldsymbol{s}^{\mathbb{E}}$ (m^2/N) defines the mechanical compliance of the material, with the superscript $\mathbb{E}$ indicating that the quantity is measured at constant electric field. The compliance $s^{\mathbb{E}}_{km}$ is defined as elastic strain in direction-k due to unit stress in direction-m with electrodes short-circuited. These coefficients are a function of Young's modulus ($E^{\mathbb{E}}_k$) and Poisson's ratio (ν_{km}), which can be different in different directions. In some cases, the Young's modulii $E^{\mathbb{E}}_k$ are specified instead of the compliances. For the remainder of this discussion, to simplify the notation, the symbol E_k with no superscript denotes the Young's modulus at constant electric field.

In the general case, $\boldsymbol{s}^{\mathbb{E}}$ is given by Eq. 2.9; however, because the piezoceramic is isotropic in a plane perpendicular to the poling direction, the compliance matrix becomes

$$\boldsymbol{s}^{\mathbb{E}} = \begin{bmatrix} s^{\mathbb{E}}_{11} & s^{\mathbb{E}}_{12} & s^{\mathbb{E}}_{13} & 0 & 0 & 0 \\ s^{\mathbb{E}}_{12} & s^{\mathbb{E}}_{11} & s^{\mathbb{E}}_{13} & 0 & 0 & 0 \\ s^{\mathbb{E}}_{13} & s^{\mathbb{E}}_{13} & s^{\mathbb{E}}_{33} & 0 & 0 & 0 \\ 0 & 0 & 0 & s^{\mathbb{E}}_{44} & 0 & 0 \\ 0 & 0 & 0 & 0 & s^{\mathbb{E}}_{44} & 0 \\ 0 & 0 & 0 & 0 & 0 & s^{\mathbb{E}}_{66} \end{bmatrix}$$

$$= \begin{bmatrix} \frac{1}{E_1} & -\frac{\nu_{12}}{E_1} & -\frac{\nu_{31}}{E_3} & 0 & 0 & 0 \\ -\frac{\nu_{12}}{E_1} & \frac{1}{E_1} & -\frac{\nu_{31}}{E_3} & 0 & 0 & 0 \\ -\frac{\nu_{31}}{E_3} & -\frac{\nu_{31}}{E_3} & \frac{1}{E_3} & 0 & 0 & 0 \\ 0 & 0 & 0 & \frac{2(1+\nu_{31})}{E_3} & 0 & 0 \\ 0 & 0 & 0 & 0 & \frac{2(1+\nu_{31})}{E_3} & 0 \\ 0 & 0 & 0 & 0 & 0 & \frac{2(1+\nu_{12})}{E_1} \end{bmatrix} \tag{2.21}$$

where E_1 is the Young's modulus in a plane normal to polarized direction (note $E_1 = E_2$) and E_3 is in the polarized direction. The piezoelectric coefficient matrix, $\boldsymbol{d}^{c}$ (m/Volt) defines strain per unit field at constant stress. The superscript c has been added to identify it as the converse piezoelectric effect. The piezoelectric coefficient matrix, $\boldsymbol{d}^{c}$, is given by

$$\boldsymbol{d}^{c} = \begin{bmatrix} 0 & 0 & d_{31} \\ 0 & 0 & d_{32} \\ 0 & 0 & d_{33} \\ 0 & d_{24} & 0 \\ d_{15} & 0 & 0 \\ 0 & 0 & 0 \end{bmatrix} \tag{2.22}$$

The coefficient d_{31} characterizes strain in the 1-axis due to an electric field $\mathbb{E}_3$ along the 3-axis, the coefficient d_{32} characterizes strain in the 2-axis due to an electric field $\mathbb{E}_3$ along the 3-axis, and the coefficient d_{33} relates strain in the 3-axis due to an electric field along the 3-axis. The coefficients d_{24} and d_{15} characterize shear strains in the planes 2-3 and 3-1 due to field $\mathbb{E}_2$ and $\mathbb{E}_1$, respectively. In the case of a piezoceramic material, transverse isotropy results in $d_{31} = d_{32}$ and $d_{24} = d_{15}$. It is important to note that there can be no induced shear in the 1-2 plane. In the present actuator configuration, with electrodes only on the 1-2 planes, it is only possible to apply an electric field in the 3-direction, $\mathbb{E}_3$. Therefore, it is not possible to obtain any shear in the actuator configuration under consideration.

The vector $\boldsymbol{\alpha}$ (1/°K) represents thermal coefficients of expansion and ΔT is the temperature change (Kelvin or °K). It is also possible to introduce modified coefficients to combine thermal and induced strain. Because of the transverse isotropy

$$\boldsymbol{\alpha} = \begin{Bmatrix} \alpha_1 \\ \alpha_1 \\ \alpha_3 \\ 0 \\ 0 \\ 0 \end{Bmatrix} \tag{2.23}$$

The actuator constitutive equations are

$$\epsilon_i = s_{ij}^{\mathbb{E}} \sigma_j + d_{ik} \mathbb{E}_k + \alpha_i \tag{2.24}$$

where the indices $i, j = 1, 2 \ldots 6$ and $k = 1, 2, 3$. Expanding these equations

$$\begin{Bmatrix} \epsilon_1 \\ \epsilon_2 \\ \epsilon_3 \\ \epsilon_4 \\ \epsilon_5 \\ \epsilon_6 \end{Bmatrix} = \begin{bmatrix} s_{11}^{\mathbb{E}} & s_{12}^{\mathbb{E}} & s_{13}^{\mathbb{E}} & 0 & 0 & 0 \\ s_{12}^{\mathbb{E}} & s_{11}^{\mathbb{E}} & s_{13}^{\mathbb{E}} & 0 & 0 & 0 \\ s_{13}^{\mathbb{E}} & s_{13}^{\mathbb{E}} & s_{33}^{\mathbb{E}} & 0 & 0 & 0 \\ 0 & 0 & 0 & s_{44}^{\mathbb{E}} & 0 & 0 \\ 0 & 0 & 0 & 0 & s_{44}^{\mathbb{E}} & 0 \\ 0 & 0 & 0 & 0 & 0 & 2(s_{11}^{\mathbb{E}} - s_{12}^{\mathbb{E}}) \end{bmatrix} \begin{Bmatrix} \sigma_1 \\ \sigma_2 \\ \sigma_3 \\ \sigma_4 \\ \sigma_5 \\ \sigma_6 \end{Bmatrix}$$
$$+ \begin{bmatrix} 0 & 0 & d_{31} \\ 0 & 0 & d_{31} \\ 0 & 0 & d_{33} \\ 0 & d_{15} & 0 \\ d_{15} & 0 & 0 \\ 0 & 0 & 0 \end{bmatrix} \begin{Bmatrix} \mathbb{E}_1 \\ \mathbb{E}_2 \\ \mathbb{E}_3 \end{Bmatrix} + \begin{Bmatrix} \alpha_1 \\ \alpha_2 \\ \alpha_3 \\ 0 \\ 0 \\ 0 \end{Bmatrix} \Delta T \tag{2.25}$$

It can be seen that for a piezoceramic sheet at constant temperature, with no external mechanical stress, an electric field $\mathbb{E}_3$ causes direct strains ϵ_1, ϵ_2, and ϵ_3. This is very similar to thermal strain. If an electric field $\mathbb{E}_1$ or $\mathbb{E}_2$ is applied, the material reacts with shear strain ϵ_4 and ϵ_5, respectively. Again, it is not possible to obtain shear strain in the 1-2 plane by the application of an electric field. For orthotropic materials, there are no corresponding thermal strains. To overcome this problem, it is better to assume piezoelectric materials as anisotropic. Note that unless the level of

Table 2.3. *Small signal PZT-5H characteristics*

d_{31} ($\times 10^{-12}$ m/V, pC/N)	−274
d_{33} ($\times 10^{-12}$ m/V, pC/N)	593
d_{15} ($\times 10^{-12}$ m/V, pC/N)	741
$s_{11}^{\mathbb{E}}$ ($\times 10^{-12}$ m^2/N)	16.5
$s_{33}^{\mathbb{E}}$ ($\times 10^{-12}$ m^2/N)	20.7
$s_{44}^{\mathbb{E}}$ ($\times 10^{-12}$ m^2/N)	58.3
$s_{66}^{\mathbb{E}}$ ($\times 10^{-12}$ m^2/N)	42.5
$s_{12}^{\mathbb{E}}$ ($\times 10^{-12}$ m^2/N)	−4.78
$s_{13}^{\mathbb{E}}$ ($\times 10^{-12}$ m^2/N)	−8.45
E_1 (GPa)	60.6
E_3 (GPa)	48.3
ν_{12}	0.2896
ν_{31}	0.4082
ρ (kg/m^3)	7500
Curie point (°C)	193
K_{11}^{σ}	3130
K_{33}^{σ}	3400
Compressive strength (MPa)	>517
Static tensile strength (MPa)	~75
Poling field (kV/cm)	~12
Dielectric breakdown (kV/cm)	~20
Depoling field (DC) (kV/cm)	~5.5

mechanical force is high, there is no effect of mechanical bias strain on piezoceramic properties, and the piezoelectric coefficients are assumed to be constant.

Piezoceramics are available commercially in the form of thin sheets (e.g., thickness 0.254 mm) such as PZT-5H, and the manufacturer-supplied characteristics of a typical sample are shown in Table 2.3. Among piezoceramics, PZT-5H is most widely used because of its lower electric-field requirement than other actuators for the same strain. PZT-5A has a high sensitivity, high time stability, and high resistivity at elevated temperatures. PZT-4 has a high resistance to depolarization under mechanical stress and exhibits low dielectric losses under high electric field. Frequently, it is used in deep-sea acoustic transducers and as an active element in electrical-power-generation systems. Table 2.4 shows the properties of several commercially available piezoceramic compositions. PZT-8 requires a higher field than PZT-5H but will need less power because of its lower dielectric constant. One major disadvantage of PZT-5H is that its dissipation factor is relatively large and increases with the applied electric field. This can lead to self-heating problems in the actuator. The choice of an appropriate material for any application must therefore be based on multiple factors.

WORKED EXAMPLE: A piezoelectric sheet of length 2 in (0.0508 m), width 1 in (0.0254 m), and thickness 0.01 in (0.000254 m), as shown in Fig. 2.6(c), is subjected to a force F along the "1" direction, and a voltage $V = 100$ volts is applied to the electrodes. Calculate the free strain and blocked force of the sheet and plot the variation of strain along the "1" direction with applied force. Use the material properties for PZT-5H as given in Table 2.3.

Table 2.4. *Typical commercially available piezoceramics*

	PZT 4	PZT 8	Navy Type I	Navy Type II (PZT 5A)	Navy Type III	Navy Type V	Navy Type VI (PZT 5H)	PKI 100	PKI 700	PKI 906	BM500	BM740	BM800
Density ($\times 10^3$ Kg/m^3)	7.5	7.6	7.6	7.7	7.6	7.6	7.6	6	7.6	7.8	7.65	7.65	7.6
Curie Temperature (°C)	328	300	350	350	350	220	200	450	350	150	360	340	325
Permittivity Factor K_{33}^{σ}, at 1 KHz (–)	1300	1000	1250	1800	1000	2700	3400	300	500	5500	1750	425	1000
Dissipation Factor $\tan\delta$, at 1 KHz (%)	0.004	0.004	0.5	1.5	0.4	2	2.2	<1.5	1.5	2.3	1.6	0.5	0.3
Transverse Coupling Factor k_{31} (–)	–0.334	0.3	0.33	0.34	0.3	0.36	0.36	<0.10	0.3	0.35	0.37	0.25	0.29
Transverse Charge Coefficient d_{31} ($\times 10^{-12}$ m/V)	–123	–0.97	–120	–175	–100	–230	–270	0.35	–60	0.7	–160	–55	–80
Longitudinal Charge Coefficient d_{33} ($\times 10^{-12}$ m/V)	289	225	275	400	220	490	550	85	150	660	365	160	220
Shear Charge Coefficient d_{15} ($\times 10^{-12}$ m/V)	496	330	480	580	320	670	720	105	362	700	–	–	–
Young's Modulus E ($\times 10^{10}$ N/m^2)			7.6	7.1	7.2	6.3	6	5.6	8.6	5.5	–	–	–
Poisson's Ratio (–)			0.31	0.31	0.31	0.31	0.31	–	0.25	0.22	–	–	–
Elastic Compliance s_{11}^{E} ($\times 10^{-12}$ m^2/N)	12.3	11.5	11.5	15.4	10.4	15.9	15.9	–	10.8	15.4	15.5	10.5	11
Elastic Compliance s_{33}^{E} ($\times 10^{-12}$ m^2/N)	15.5	13.5	15	18.4	13.5	18	20.2	–	13.9	18.2	19	14	13.5
Maximum AC Field (KV/m [V/mil])	–	–	350 [9]	300 [8]	400 [10]	200 [5]	200 [5]	–	350 [9]	160 [4]	–	–	–
Maximum DC Field – forward (KV/m [V/mil])	–	–	700 [18]	600 [15]	800 [20]	400 [10]	400 [10]	–	700 [18]	320 [8]	–	–	–
Maximum DC Field – reverse (KV/m [V/mil])	–	–	350 [9]	300 [8]	400 [10]	200 [5]	200 [5]	–	350 [9]	160 [4]	–	–	–

Navy types and PKI: www.piezo-kinetics.com/materials.htm
BM types: www.sensortech.ca/index.html
Others: www.piezo.com, www.matsysinc.com, www.trstechnologies.com

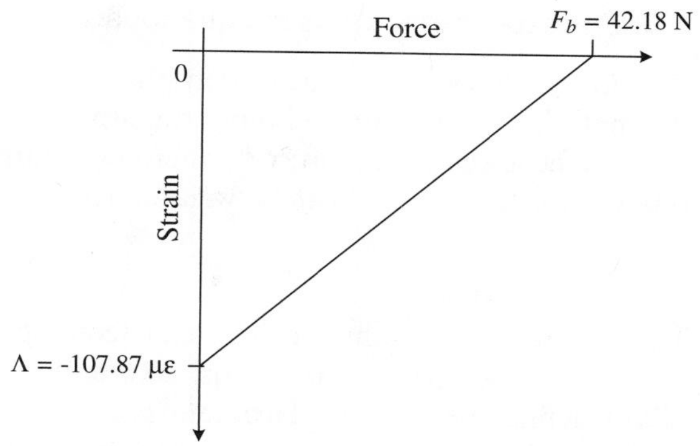

Figure 2.7. Variation of strain with force (in 1 direction) of a piezoceramic sheet actuator, $V = 100$ volts.

SOLUTION: The actuator equation (Eq. 2.5) along the 1 direction, with an electric field along the 3 direction, can be written as

$$\epsilon_1 = s_{11}^{\mathbb{E}}\sigma_1 + d_{31}\mathbb{E}_3 \tag{2.26}$$

For a sheet of length l_c, width b_c, and thickness t_c, the previous equation becomes

$$\epsilon_1 = s_{11}^{\mathbb{E}}\frac{F}{b_c t_c} + d_{31}\frac{V}{t_c} \tag{2.27}$$

The free strain Λ is obtained when there is no force acting on the sheet. Setting $F = 0$ yields

$$\Lambda = d_{31}\frac{V}{t_c} = -274 \times 10^{-12}\frac{100}{0.01 \times 0.0254} = -107.87\mu\epsilon \tag{2.28}$$

The blocked force F_b is obtained by setting the strain $\epsilon_1 = 0$.

$$s_{11}^{\mathbb{E}}\frac{F_b}{b_c t_c} + d_{31}\frac{V}{t_c} = 0 \rightarrow F_b = -\frac{d_{31}Vb_c}{s_{11}^{\mathbb{E}}} \tag{2.29}$$

Substituting the material properties yields the blocked force as

$$F_b = -\frac{(-274 \times 10^{-12}).100.(1 \times 0.0254)}{16.5 \times 10^{-12}} = 42.18 \text{ N} \tag{2.30}$$

Note that due to the negative sign of d_{31}, the free strain is compressive and the blocked force is tensile. This means that the sheet tends to contract in the 1 direction when a positive field is applied on the electrodes, and a tensile force in the 1 direction is required to restrain the sheet. The variation of strain with applied force is plotted in Fig. 2.7 and is given by

$$\epsilon_1 = 2.557 \times F - 107.87 \quad (\mu\epsilon) \tag{2.31}$$

2.5.2 Piezoceramic Sensor Equations

Consider a piezoceramic sheet as in Fig. 2.5 that is poled across its thickness. To use the material as a sensor, the input quantity is the mechanical stress and the output is an electric displacement, or generated charge. The corresponding constitutive relation for the material can be written from the general case (Eq. 2.6) as

$$\boldsymbol{D} = \boldsymbol{d}^d \boldsymbol{\sigma} + \boldsymbol{e}^\sigma \mathbb{E} \tag{2.32}$$

The general expressions for the piezoelectric coefficient matrix $\boldsymbol{d}^d$ and the permittivity matrix $\boldsymbol{e}^\sigma$ are given by Eqs. 2.16 and 2.12. The piezoelectric coefficient $\boldsymbol{d}^d$ (C/N) defines electric displacement per unit stress at constant electric field. The superscript d has been added to identify the coefficient as that corresponding to the direct piezoelectric effect. It has been found experimentally that the matrix $\boldsymbol{d}^d$ is the transpose of $\boldsymbol{d}^c$. Note that measurement of piezoelectric coefficients by the direct effect is usually more difficult and less accurate than measurement based on the converse effect. This is because application of a pure uniaxial stress is difficult and some of the charge generated by the application of stress can leak off before it is measured. However, while using the converse effect, it is much easier to apply a uniform electric field and assure a zero stress state within the sample. It is also difficult to control the electrical-boundary conditions during static testing. However, static measurements can be made more accurate by superimposing a low-frequency alternating electric field or mechanical stress.

For PZT, the piezoelectric coefficient matrix $\boldsymbol{d}^d$ is given by

$$\boldsymbol{d}^d = \begin{bmatrix} 0 & 0 & 0 & 0 & d_{15} & 0 \\ 0 & 0 & 0 & d_{24} & 0 & 0 \\ d_{31} & d_{32} & d_{33} & 0 & 0 & 0 \end{bmatrix} \tag{2.33}$$

where the elements d_{ij} have the same values as in $\boldsymbol{d}^c$. Note that shear stress in the 1-2 plane, σ_6 (or σ_{12}), is not capable of generating any electric response.

Piezoelectric coefficients g_{ij} are sometimes used to quantify the sensitivity of a piezoceramic sensor material. These coefficients denote the electric field developed along the i-axis (electrodes perpendicular to the i-axis) due to an applied stress along the j-axis, provided that all other external stresses are constant. It also expresses the strain developed along the j-axis due to a unit electric charge per unit area of electrodes applied along the i-axis (electrodes perpendicular to the i-axis). For example, g_{33} denotes a field developed in direction 3 due to an applied stress in direction 3 when all other stresses are zero. In the context of the converse effect, it also denotes the strain developed in direction 3 due to a unit charge per unit area of electrodes normal to direction 3.

The electric permittivity matrix for a piezoceramic is given by

$$\boldsymbol{e}^\sigma = \begin{bmatrix} e^\sigma_{11} & 0 & 0 \\ 0 & e^\sigma_{22} & 0 \\ 0 & 0 & e^\sigma_{33} \end{bmatrix} \tag{2.34}$$

For a PZT, transverse isotropy results in $e^\sigma_{11} = e^\sigma_{22}$. Usually, the permittivities are specified in terms of a relative permittivity, K^σ_{ij}. This is the ratio of the corresponding electric permittivity and the permittivity of free space, e_0.

$$K^\sigma_{ij} = e^\sigma_{ij}/e_0 \tag{2.35}$$

Note that the units of permittivity are Farad/m (F/m) or C.V/m.

The sensor constitutive relations for a PZT are

$$D_k = d_{ki}\sigma_i + e^{\sigma}_{kl}\mathbb{E}_l \tag{2.36}$$

where the indices $i = 1, 2, \ldots 6$ and $k, l = 1, 2, 3$. Expanding these equations

$$\begin{Bmatrix} D_1 \\ D_2 \\ D_3 \end{Bmatrix} = \begin{bmatrix} 0 & 0 & 0 & 0 & d_{15} & 0 \\ 0 & 0 & 0 & d_{15} & 0 & 0 \\ d_{31} & d_{31} & d_{33} & 0 & 0 & 0 \end{bmatrix} \begin{Bmatrix} \sigma_1 \\ \sigma_2 \\ \sigma_3 \\ \sigma_4 \\ \sigma_5 \\ \sigma_6 \end{Bmatrix} + \begin{bmatrix} e^{\sigma}_{11} & 0 & 0 \\ 0 & e^{\sigma}_{11} & 0 \\ 0 & 0 & e^{\sigma}_{33} \end{bmatrix} \begin{Bmatrix} \mathbb{E}_1 \\ \mathbb{E}_2 \\ \mathbb{E}_3 \end{Bmatrix} \tag{2.37}$$

This equation summarizes the principle of operation of piezoceramic sensors. Typically, no external electric field is applied to the sensor, and a stress field causes an electric displacement to be generated as a result of the direct piezoelectric effect. In the general case, the charge generated q, is related to the displacement D_3 by the relation

$$q = \iint \{D_1\ D_2\ D_3\} \cdot \begin{Bmatrix} dA_1 \\ dA_2 \\ dA_3 \end{Bmatrix} \tag{2.38}$$

where dA_1, dA_2, and dA_3 are the components of the electrode area in the 2-3 plane, 1-3 plane, and 1-2 plane, respectively. It can be seen that the charge generated depends on only the component of the electrode area normal to the displacement. In the case of the piezoceramic sheet (see Fig. 2.5), only D_3 appears. The charge q and the voltage V_c generated across the sensor electrodes are related by the capacitance of the sensor, C_p as

$$V_c = q/C_p \tag{2.39}$$

Therefore, by measuring the charge generated by the piezoceramic material, from Eqs. 2.37 and 2.38, it is possible to calculate the stress in the material. From these values, knowing the compliance of the material, the strain in the material is calculated.

The sensors described in this book are all in the form of sheets as shown in Fig. 2.5, with their two faces coated with thin electrode layers. The 1- and 2-axes of the piezoelectric material are in the plane of the sheet. The capacitance of a sheet element is found by treating it as a parallel plate capacitor and is given by

$$C_p = \frac{e^{\sigma}_{33} l_c b_c}{t_c} \text{ (Farad)} \tag{2.40}$$

In the case of a uniaxial stress field, the correlation between strain and charge developed is simple. However, for the case of a general plane stress distribution in the 1-2 plane, this correlation is complicated by the presence of the d_{32} term in the

$\boldsymbol{d}^d$ matrix. The voltage generated by different stress fields on a piezoelectric sheet actuator with electrodes on the 1-2 planes is shown here.

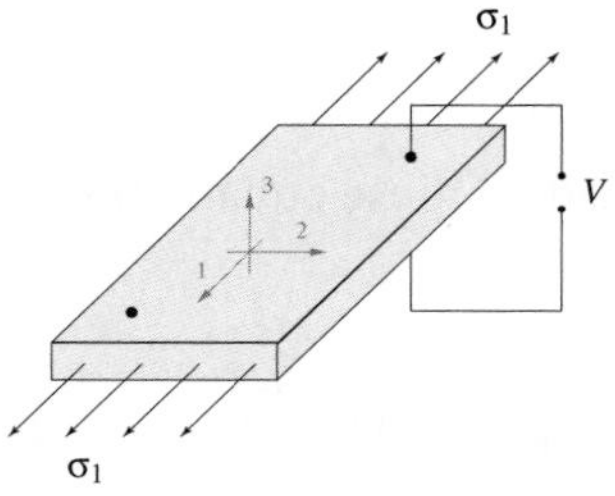

voltage due to σ_1 alone $\quad V = \dfrac{\sigma_1}{e_{33}^{\sigma}} d_{31} t_c = \sigma_1 g_{31} t_c$

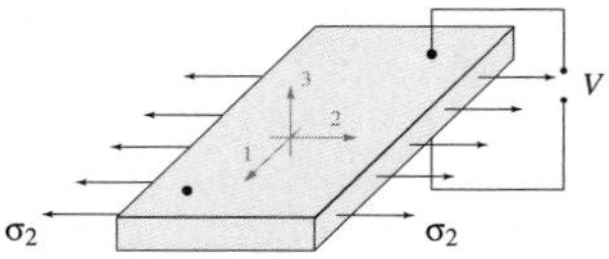

voltage due to σ_2 alone $\quad V = \dfrac{\sigma_2}{e_{33}^{\sigma}} d_{31} t_c = \sigma_2 g_{31} t_c$

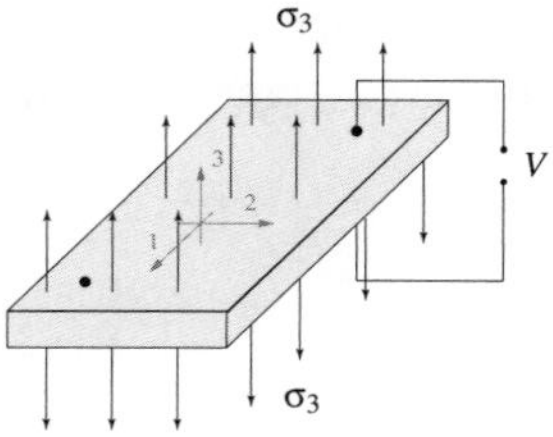

voltage due to σ_3 alone $\quad V = \dfrac{\sigma_3}{e_{33}^{\sigma}} d_{33} t_c = \sigma_3 g_{33} t_c$

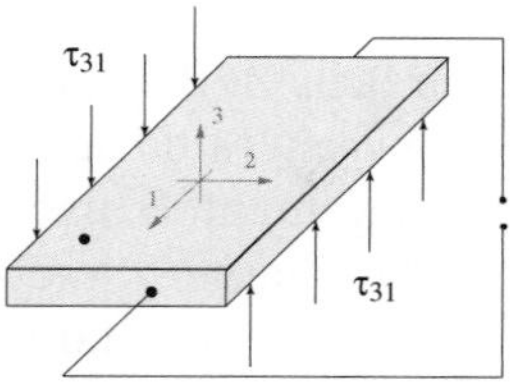

voltage due to ζ_{23} (or σ_4) alone $\quad V = \dfrac{\tau_{31}}{e_{11}^{\sigma}} d_{15} l_c = \tau_{31} g_{15} l_c$

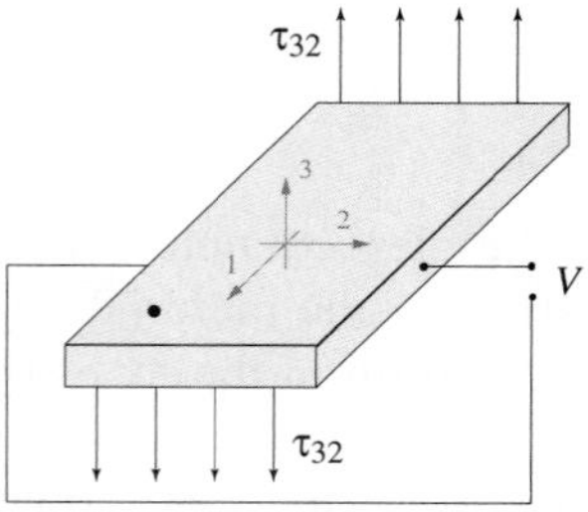

voltage due to ζ_{31} (or σ_5) alone $\quad V = \dfrac{\tau_{32}}{e_{22}^{\sigma}} d_{15} b_c = \tau_{32} g_{15} b_c$

WORKED EXAMPLE: A piezoelectric sheet of length 2 in (0.0508 m), width 1 in (0.0254 m), and thickness 0.01 in (0.000254 m), as shown in Fig. 2.8, is subjected to a force F along the 1 direction. Assume that an electronic circuit moves all the charge generated by the piezoelectric sheet to a capacitance C. As a result, we can assume that $\mathbb{E} = 0$ for the sheet. Calculate the voltage V developed due to a force $F = 25$N for a capacitance $C = 100$nF. Use the material properties for PZT-5H as given in Table 2.3.

SOLUTION: The sensor equation (Eq. 2.6) along the 1 direction, for a sheet with electrodes normal to the 3 direction, can be written as

$$D_3 = d_{31}\sigma_1 + e_{33}^{\sigma}\mathbb{E} \tag{2.41}$$

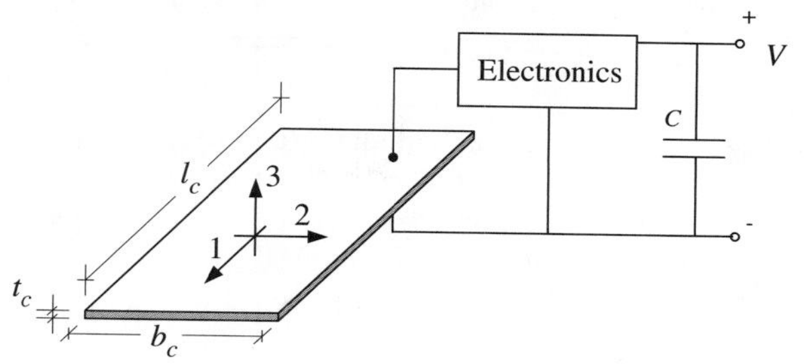

Figure 2.8. Piezoceramic sheet element as a sensor.

For a sheet of length l_c, width b_c, and thickness t_c, assuming all the charge q generated by the applied stress is transferred to the capacitance C

$$D_3 = \frac{q}{b_c l_c} = d_{31}\frac{F}{b_c t_c} \tag{2.42}$$

In addition, the charge on the capacitance is related to the voltage generated by

$$q = D_3 l_c b_c = CV \tag{2.43}$$

Therefore, the voltage developed on the capacitance is

$$V = \frac{d_{31} F l_c}{C t_c} = \frac{(-274 \times 10^{-12}).25.(2 \times 0.0254)}{(100 \times 10^{-9}).(0.01 \times 0.0254)} = -13.7 \text{ volts} \tag{2.44}$$

Although any piezoelectric material can be used as a sensor, two types of piezoelectric materials are described herein. These two materials are a typical piezoceramic, PZT-5H, and PVDF, a piezoelectric polymer film.

1. Piezoceramic sensors: Piezoceramic sensors exhibit most of the characteristics of ceramics, namely a high elastic modulus, brittleness, and low tensile strength. The material itself, by virtue of the poling process, is assumed transversely isotropic in the plane normal to the poling direction. For PZT sensors, the values of $\boldsymbol{d}^d$ and $\boldsymbol{e}^\sigma$ are as described in Eqs. 2.33 and 2.34. Typical values for PZT-5H are in Table 2.3.
2. PVDF sensors: PVDF is a polymer (polyvinylidene fluoride) consisting of long chains of the repeating monomer $(-CH_2 - CF_2-)$. The hydrogen atoms are positively charged and the fluorine atoms are negatively charged with respect to the carbon atoms, which leaves each monomer unit with an inherent dipole moment. PVDF film is manufactured by solidification of the film from a molten phase, which is then stretched in a particular direction and finally poled. In the liquid phase, the individual polymer chains are free to take up any orientation; therefore, a given volume of liquid has no net dipole moment. After solidification and stretching the film in one direction, the polymer chains are mostly aligned along the direction of stretching. Sometimes, stretching is carried out in two axes to achieve desired characteristics. This, combined with the poling, imparts a permanent dipole moment to the film, which then behaves like a piezoelectric material.

   The process of stretching the film, which orients the polymer chains in a specific direction, renders the material piezoelectrically orthotropic, which means $d_{31} \neq d_{32}$. The stretching direction is taken as the 1-direction. For small strains, however, the material is considered mechanically isotropic.

Table 2.5. *Typical properties at 25°C*

	PZT-5H	PVDF
Young's Modulus (GPa) $E_{11}^{\mathbb{E}}$	60.6	4–6
d_{31} (pC/N)	−274	18–24
d_{32} (pC/N)	−274	2.5–3
d_{33} (pC/N)	593	−33
e_{33}^{σ} (nF/m)	30.1	0.106

The typical characteristics of PZT and PVDF are compared in Table 2.5. The Young's modulus of the PZT material is comparable to that of aluminum, whereas that of PVDF is approximately 1/12th that of aluminum. It is therefore much more suited to sensing applications because it is less likely to influence the dynamics of the host structure as a result of its own stiffness. It is also very easy to shape PVDF film for any desired application. These characteristics make PVDF films more attractive for sensor applications compared to PZT, despite their lower piezoelectric coefficients (approximately 1/10th of PZT). Also, PVDF is pyroelectric, which translates into a highly temperature-dependent performance compared to PZT sensors. PVDF is not only soft, it also tough and chemically resistant. PVDF is widely used as a sensor material in hydrophones. The electromechanical characteristics of PVDF are quite nonlinear and time-dependent. Both biaxially stretched and uniaxially stretched PVDF sheets exhibit strong dependence of their electromechanical coefficients on pre-stress and the compliance changes with stress level. The dynamic moduli are highly sensitive to frequencies but are insensitive to pre-stressing [2].

2.5.3 Alternate Forms of the Constitutive Equations

The linear constitutive equations for a piezoelectric material are

$$\epsilon_i = s_{ij}^{\mathbb{E}}\sigma_j + d_{ik}\mathbb{E}_k \tag{2.45}$$

$$D_k = d_{ki}\sigma_i + e_{kl}^{\sigma}\mathbb{E}_l \tag{2.46}$$

These can be rewritten as

$$\epsilon_i = s_{ij}^{D}\sigma_j + g_{ik}D_k \tag{2.47}$$

$$\mathbb{E}_k = g_{ki}\sigma_i + \beta_{kl}^{\sigma}D_l \tag{2.48}$$

where the indices $i, j = 1, 2, \ldots 6$ and $k, l = 1, 2, 3$. The g coefficient in the converse piezoelectric relation is defined as the ratio of strain developed to the applied charge density. The piezoelectric voltage coefficient g_{ik} is defined as the induced strain per unit of electric displacement applied or, alternatively, the electric field generated per unit of mechanical stress applied.

$$g_{ik} = \frac{\text{strain component } i}{\text{applied charge density in direction } k}, \quad \frac{\text{m}^2}{\text{C}} \tag{2.49}$$

The g coefficient in the direct piezoelectric effect is defined as the ratio of the open-circuit electric field produced to the applied mechanical stress.

$$g_{ki} = \frac{\text{open-circuit electric field in direction } k}{\text{applied mechanical stress component } i}, \quad \frac{\text{Vm}}{\text{N}} \tag{2.50}$$

Note that both of these coefficients are identical, similar to the direct and converse piezoelectric coefficients. For example, g_{51} represents the shear strain (γ_{31})

Table 2.6. *Definition of symbols (indices: $i, j, r = 1, 2, \ldots 6$ and $k, l, m = 1, 2, 3$)*

Symbol	Meaning	Size	Units
c_{ij}	Elastic stiffness	6×6	N/m^2
d_{ki}	Piezoelectric coefficient	3×6	m/V or C/N
D_k	Electric displacement	3×1	C/m^2
e_{kl}	Electric permittivity	3×3	F/m
$\bar{e}_{ki}$	Piezoelectric constant	3×6	C/m^2
$\mathbb{E}_k$	Electric field	3×1	V/m
g_{ki}	Piezoelectric constant	3×6	V/m or m^2/C
h_{ki}	Piezoelectric constant	3×6	V/m or N/C
s_{ij}	Elastic compliance	6×6	m^2/N
ϵ_i	Strain	6×1	–
σ_i	Stress	6×1	N/m^2
β_{kl}	Impermittivity	3×3	m/F

produced due to the charge $\mathbb{D}_1$ (on a surface normal to 1-axis). Also, g_{15} defines the open-circuit electric field produced along the 1 direction (on electrodes normal to the 1 direction) due to an applied shear stress τ_{31}. Another form of the equations is

$$\sigma_i = c_{ij}^D \epsilon_j - h_{ik} D_k \tag{2.51}$$

$$\mathbb{E}_k = -h_{ki}\epsilon_i + \beta_{kl}^{\epsilon} D_l \tag{2.52}$$

where the indices $i, j = 1, 2, \ldots 6$ and $k, l = 1, 2, 3$. The superscripts $\mathbb{E}$, D, σ, and ϵ, respectively, indicate states or measurements taken at constant-electric-field (short-circuit), constant-electric-displacement (open-circuit), constant-stress, and constant-strain condition. Many of the coefficients (listed in Table 2.6) are interrelated, as follows:

$$s_{ij}^D = s_{ij}^{\mathbb{E}} - d_{ir} g_{rj} \tag{2.53}$$

$$c_{ir}^{\mathbb{E}} s_{rj}^{\mathbb{E}} = \delta_{ij} \tag{2.54}$$

$$c_{ir}^D s_{rj}^D = \delta_{ij} \tag{2.55}$$

$$\beta_{km}^{\sigma} e_{ml}^{\sigma} = \delta_{kl} \tag{2.56}$$

$$\beta_{km}^{\epsilon} e_{ml}^{\epsilon} = \delta_{kl} \tag{2.57}$$

$$c_{ij}^D = c_{ij}^{\mathbb{E}} + \bar{e}_{ir} h_{rj} \tag{2.58}$$

$$g_{ki} = \beta_{km} d_{mi} \tag{2.59}$$

$$e_{kl}^{\sigma} = e_{kl}^{\epsilon} + D_{kj} \bar{e}_{jl} \tag{2.60}$$

$$\beta_{kl}^{\sigma} = \beta_{kl}^{\epsilon} - g_{kj} h_{jl} \tag{2.61}$$

$$\bar{e}_{ki} = d_{kj} c_{ji}^{\mathbb{E}} \tag{2.62}$$

$$d_{ki} = e_{km}^{\sigma} g_{mi} \tag{2.63}$$

$$h_{ki} = g_{kj} c_{ji}^D \tag{2.64}$$

where the indices $i, j, r = 1, 2, \ldots 6$ and $k, l, m = 1, 2, 3$. δ_{ij} is a 6×6 unit matrix and δ_{kl} is a 3×3 unit matrix. In these equations, c_{ij} is the elastic stiffness matrix (N/m^2), s_{ij} is the elastic compliance matrix (m^2/N), g_{ki} is a piezoelectric constant matrix (Vm/N or m^2/C), $\mathbb{E}$ is the electric field (V/m), d_{ki} is the piezoelectric coefficient matrix (m/V or C/N), D_k is the electric displacement (C/m^2), e_{kl} is the electric permittivity (F/m), h_{ki} is a piezoelectric constant (V/m or N/C), and $\bar{e}_{ki}$ is another piezoelectric constant (C/m^2).

For example, the alternate form of the constitutive equations (Eqs. 2.47 and 2.48) can be expanded as

$$\begin{Bmatrix}\epsilon_1\\\epsilon_2\\\epsilon_3\\\epsilon_4\\\epsilon_5\\\epsilon_6\end{Bmatrix}=\begin{bmatrix}s_{11}^D & s_{12}^D & s_{13}^D & 0 & 0 & 0\\ s_{12}^D & s_{11}^D & s_{13}^D & 0 & 0 & 0\\ s_{13}^D & s_{13}^D & s_{33}^D & 0 & 0 & 0\\ 0 & 0 & 0 & s_{44}^D & 0 & 0\\ 0 & 0 & 0 & 0 & s_{44}^D & 0\\ 0 & 0 & 0 & 0 & 0 & 2(s_{11}^D-s_{12}^D)\end{bmatrix}\begin{Bmatrix}\sigma_1\\\sigma_2\\\sigma_3\\\sigma_4\\\sigma_5\\\sigma_6\end{Bmatrix}+\begin{bmatrix}0 & 0 & g_{31}\\ 0 & 0 & g_{31}\\ 0 & 0 & g_{33}\\ 0 & g_{15} & 0\\ g_{15} & 0 & 0\\ 0 & 0 & 0\end{bmatrix}\begin{Bmatrix}D_1\\D_2\\D_3\end{Bmatrix} \tag{2.65}$$

$$\begin{Bmatrix}\mathbb{E}_1\\\mathbb{E}_2\\\mathbb{E}_3\end{Bmatrix}=\begin{bmatrix}0 & 0 & 0 & 0 & g_{15} & 0\\ 0 & 0 & 0 & g_{15} & 0 & 0\\ g_{31} & g_{31} & g_{33} & 0 & 0 & 0\end{bmatrix}\begin{Bmatrix}\sigma_1\\\sigma_2\\\sigma_3\\\sigma_4\\\sigma_5\\\sigma_6\end{Bmatrix}+\begin{bmatrix}\beta_{11}^\sigma & 0 & 0\\ 0 & \beta_{11}^\sigma & 0\\ 0 & 0 & \beta_{33}^\sigma\end{bmatrix}\begin{Bmatrix}D_1\\D_2\\D_3\end{Bmatrix} \tag{2.66}$$

Note that $c_{ij}^D s_{jr}^D = \delta_{ir}$:

$$\begin{bmatrix}c_{11}^D & c_{12}^D & c_{13}^D & c_{14}^D & c_{15}^D & c_{16}^D\\ c_{21}^D & c_{22}^D & c_{23}^D & c_{14}^D & c_{15}^D & c_{16}^D\\ c_{31}^D & c_{32}^D & c_{33}^D & c_{14}^D & c_{15}^D & c_{16}^D\\ c_{41}^D & c_{42}^D & c_{43}^D & c_{14}^D & c_{15}^D & c_{16}^D\\ c_{51}^D & c_{52}^D & c_{53}^D & c_{14}^D & c_{15}^D & c_{16}^D\\ c_{61}^D & c_{62}^D & c_{63}^D & c_{64}^D & c_{65}^D & c_{66}^D\end{bmatrix}\begin{bmatrix}s_{11}^D & s_{12}^D & s_{13}^D & s_{14}^D & s_{15}^D & s_{16}^D\\ s_{21}^D & s_{22}^D & s_{23}^D & s_{14}^D & s_{15}^D & s_{16}^D\\ s_{31}^D & s_{32}^D & s_{33}^D & s_{14}^D & s_{15}^D & s_{16}^D\\ s_{41}^D & s_{42}^D & s_{43}^D & s_{14}^D & s_{15}^D & s_{16}^D\\ s_{51}^D & s_{52}^D & s_{53}^D & s_{14}^D & s_{15}^D & s_{16}^D\\ s_{61}^D & s_{62}^D & s_{63}^D & s_{64}^D & s_{65}^D & s_{66}^D\end{bmatrix}=\begin{bmatrix}1 & 0 & 0 & 0 & 0 & 0\\ 0 & 1 & 0 & 0 & 0 & 0\\ 0 & 0 & 1 & 0 & 0 & 0\\ 0 & 0 & 0 & 1 & 0 & 0\\ 0 & 0 & 0 & 0 & 1 & 0\\ 0 & 0 & 0 & 0 & 0 & 1\end{bmatrix} \tag{2.67}$$

For piezoceramics:

$$\begin{bmatrix}s_{11}^D & s_{12}^D & s_{13}^D & 0 & 0 & 0\\ s_{12}^D & s_{11}^D & s_{13}^D & 0 & 0 & 0\\ s_{13}^D & s_{13}^D & s_{33}^D & 0 & 0 & 0\\ 0 & 0 & 0 & s_{44}^D & 0 & 0\\ 0 & 0 & 0 & 0 & s_{44}^D & 0\\ 0 & 0 & 0 & 0 & 0 & 2(s_{11}^D-s_{12}^D)\end{bmatrix}=\begin{bmatrix}s_{11}^{\mathbb{E}} & s_{12}^{\mathbb{E}} & s_{13}^{\mathbb{E}} & 0 & 0 & 0\\ s_{12}^{\mathbb{E}} & s_{11}^{\mathbb{E}} & s_{13}^{\mathbb{E}} & 0 & 0 & 0\\ s_{13}^{\mathbb{E}} & s_{13}^{\mathbb{E}} & s_{33}^{\mathbb{E}} & 0 & 0 & 0\\ 0 & 0 & 0 & s_{44}^{\mathbb{E}} & 0 & 0\\ 0 & 0 & 0 & 0 & s_{44}^{\mathbb{E}} & 0\\ 0 & 0 & 0 & 0 & 0 & 2(s_{11}^{\mathbb{E}}-s_{12}^{\mathbb{E}})\end{bmatrix}$$
$$-\begin{bmatrix}0 & 0 & d_{31}\\ 0 & 0 & d_{31}\\ 0 & 0 & d_{33}\\ 0 & d_{15} & 0\\ d_{15} & 0 & 0\\ 0 & 0 & 0\end{bmatrix}\begin{bmatrix}0 & 0 & 0 & 0 & g_{15} & 0\\ 0 & 0 & 0 & g_{15} & 0 & 0\\ g_{31} & g_{31} & g_{33} & 0 & 0 & 0\end{bmatrix} \tag{2.68}$$

where

$$\begin{bmatrix} 0 & 0 & 0 & 0 & g_{15} & 0 \\ 0 & 0 & 0 & g_{15} & 0 & 0 \\ g_{31} & g_{31} & g_{33} & 0 & 0 & 0 \end{bmatrix} = \begin{bmatrix} \beta^{\sigma}_{11} & 0 & 0 \\ 0 & \beta^{\sigma}_{11} & 0 \\ 0 & 0 & \beta^{\sigma}_{33} \end{bmatrix} \begin{bmatrix} 0 & 0 & 0 & 0 & d_{15} & 0 \\ 0 & 0 & 0 & d_{15} & 0 & 0 \\ d_{31} & d_{31} & d_{33} & 0 & 0 & 0 \end{bmatrix}$$

$$= \begin{bmatrix} e^{\sigma}_{11} & 0 & 0 \\ 0 & e^{\sigma}_{11} & 0 \\ 0 & 0 & e^{\sigma}_{33} \end{bmatrix}^{-1} \begin{bmatrix} 0 & 0 & 0 & 0 & d_{15} & 0 \\ 0 & 0 & 0 & d_{15} & 0 & 0 \\ d_{31} & d_{31} & d_{33} & 0 & 0 & 0 \end{bmatrix}$$

$$= \begin{bmatrix} 0 & 0 & 0 & 0 & d_{15}/e^{\sigma}_{11} & 0 \\ 0 & 0 & 0 & d_{15}/e^{\sigma}_{11} & 0 & 0 \\ d_{31}/e^{\sigma}_{33} & d_{31}/e^{\sigma}_{33} & d_{33}/e^{\sigma}_{33} & 0 & 0 & 0 \end{bmatrix} \tag{2.69}$$

2.5.4 Piezoelectric Coupling Coefficients

A piezoelectric transducer is basically an energy-conversion device. The direct piezoelectric effect results in the conversion of mechanical energy into electrical energy, whereas the converse piezoelectric effect results in the conversion of electrical energy to mechanical energy. The coupling coefficients k_{ij} are a measure of the efficiency of this energy conversion.

A simple one-dimensional analysis of the piezoceramic sheet element described in Section 2.5.1 illustrates the significance of the coupling coefficient. Consider a uniform uni-directional stress σ_1 applied to the piezoceramic sheet (Fig. 2.9). The one-dimensional constitutive relations along the 1-axis for the piezoceramic sheet can be written as

$$\epsilon_1 = s^{\mathbb{E}}_{11}\sigma_1 + d_{31}\mathbb{E}_3 \tag{2.70}$$

$$D_3 = d_{31}\sigma_1 + e^{\sigma}_{33}\mathbb{E}_3 \tag{2.71}$$

$\mathbb{E}_3$ can be eliminated from these equations. From Eq. 2.71

$$\mathbb{E}_3 = \frac{D_3}{e^{\sigma}_{33}} - \frac{d_{31}}{e^{\sigma}_{33}}\sigma_1 \tag{2.72}$$

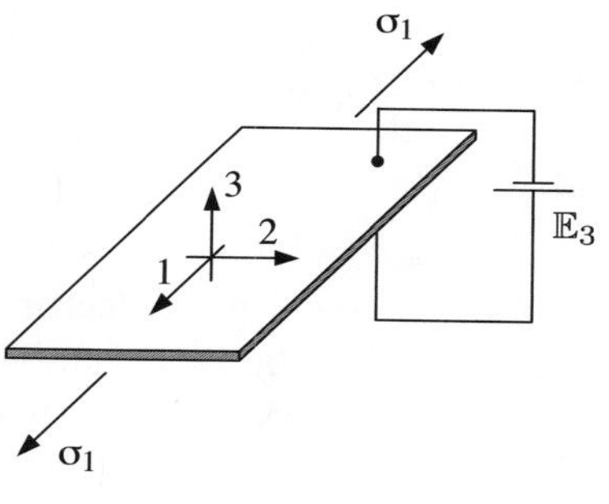

Figure 2.9. Piezoceramic sheet element under a uniaxial stress.

which can be substituted into Eq. 2.70, giving

$$\begin{aligned}\epsilon_1 &= s_{11}^{\mathbb{E}}\left[1 - \frac{d_{31}^2}{s_{11}^{\mathbb{E}} e_{33}^{\sigma}}\right]\sigma_1 + \frac{d_{31}}{e_{33}^{\sigma}} D_3 \\ &= s_{11}^{D}\sigma_1 + \frac{d_{31}}{e_{33}^{\sigma}} D_3\end{aligned} \tag{2.73}$$

The quantity s_{11}^{D} is the compliance of the material in the 1-axis at a constant electric displacement and is given by

$$s_{11}^{D} = s_{11}^{\mathbb{E}}\left[1 - \frac{d_{31}^2}{s_{11}^{\mathbb{E}} e_{33}^{\sigma}}\right] \tag{2.74}$$

This equation illustrates the importance of electrical-boundary conditions for the behavior of the piezoelectric material. The condition of constant electric displacement is also referred to as the open-circuit condition. This arises because if the electrodes are open, charge developed on the electrodes due to mechanical deformation of the material remains on the electrodes, resulting in a constant electric displacement in the material. In this condition, the separation of charge across electrodes results in an electric field across the thickness of the material.

The condition of constant electric field is also referred to as the short-circuit condition. When the electrodes are shorted together, no charge separation can occur between the electrodes; therefore, the electric field across the material remains constant at zero.

When a piezoelectric element is used as an actuator, the voltage across the electrodes is controlled by a power supply. Charge is either supplied or removed from the electrodes by the power supply in order to maintain the specified voltage. Therefore, an actuator connected in this manner is subject to electrical-boundary conditions equivalent to a short-circuit or constant electric field.

From Eq. 2.74, the relationship between open-circuit and short-circuit compliance can be written as

$$s_{11}^{D} = s_{11}^{\mathbb{E}}\left[1 - k_{31}^2\right] \tag{2.75}$$

This can be rewritten as

$$\frac{E^{\mathbb{E}}}{E^{D}} = \left[1 - k_{31}^2\right] \tag{2.76}$$

where k_{31} is the electro-mechanical coupling coefficient, given by

$$k_{31}^2 = \frac{d_{31}^2}{s_{11}^{\mathbb{E}} e_{33}^{\sigma}} \tag{2.77}$$

Alternatively, k_{31} can be expressed in terms of the open-circuit and short-circuit compliances

$$k_{31}^2 = 1 - \frac{s_{11}^{D}}{s_{11}^{\mathbb{E}}} \tag{2.78}$$

For a typical piezoelectric ceramic, PZT-5H, $k_{31}^2 = 0.15$. From the previous equation, it can be seen that for this material, the ratio of open-circuit and short-circuit compliances is 85%, which is a significant change in a material property that in passive materials is normally assumed to be invariant. This means that the short-circuit Young's modulus is 85% of the open-circuit Young's modulus. This characteristic of the piezoelectric is exploited in several applications, such as passive damping.

The previous one-dimensional analysis can be extended to calculate the energy stored in the material. Consider a piezoelectric element poled as described above, that is rigidly clamped so it cannot deform along the 1-axis. An electric field $\mathbb{E}$ is applied across the electrodes. It is convenient to calculate the strain energy stored in the element by first allowing it to deform under the applied electric field without any mechanical constraint. The constraint force is then applied to the element in order to return it to its initial dimensions. The work done in this process by the constraint force will be equal to the strain energy stored in the element.

As a result of this approach, the stress in the element is given by the free strain $\Lambda_1 = d_{31}\mathbb{E}_3$ divided by the short-circuit compliance $s_{11}^{\mathbb{E}}$ of the material.

$$\sigma_1 = \frac{d_{31}\mathbb{E}_3}{s_{11}^{\mathbb{E}}} \tag{2.79}$$

The strain energy of the element, U_{mech}, can be calculated as

$$U_{\text{mech}} = \frac{1}{2}\int_{\mathbb{V}} \sigma_1 \epsilon_1 d\mathbb{V} \tag{2.80}$$

where $\mathbb{V}$ represents the volume of the element. Substituting from Eq. 2.79

$$U_{\text{mech}} = \frac{1}{2}\frac{d_{31}^2}{s_{11}^{\mathbb{E}}}\mathbb{E}_3^2\mathbb{V} \tag{2.81}$$

The electrical energy stored in the element, U_{elect}, can be calculated as

$$\begin{aligned} U_{\text{elect}} &= \frac{1}{2}\int_{\mathbb{V}} e_{33}^{\sigma}\mathbb{E}_3^2 d\mathbb{V} \\ &= \frac{1}{2}e_{33}^{\sigma}\mathbb{E}_3^2\mathbb{V} \end{aligned} \tag{2.82}$$

It is important to note that the electrical energy stored in the element can be calculated in this manner because the element is mechanically clamped. As a result of this constraint, there is no work done in reorienting the dipoles in the material. The only work done is in placing the appropriate charge on the electrodes of the element. Consequently, the spontaneous polarization of the material does not appear in the calculation of electrical energy.

The efficiency of energy conversion can be estimated by the ratio of strain energy and electrical energy. From Eqs. 2.81 and 2.82

$$\frac{U_{\text{mech}}}{U_{\text{elect}}} = \frac{d_{31}^2}{s_{11}^{\mathbb{E}}e_{33}^{\sigma}} = k_{31}^2 \tag{2.83}$$

Therefore, the electromechanical coupling coefficient gives a measure of the energy-conversion efficiency of the material. For an actuator, this quantity sets an upper limit on the mechanical-power output for a given electrical input. However, it is shown later that, in general, the maximum energy that can be extracted from a piezoelectric actuator is equal to only half of the value in Eq. 2.81. In the previous example, the energy conversion efficiency is 15%.

The coupling coefficients are expressed in terms of piezoelectric, dielectric, and elastic constants. They are nondimensional coefficients that define conversion between mechanical and electrical energy, or vice versa, for a particular stress and electric-field configuration. Consequently, different coupling factors can be defined in terms of the components of stress and electric field. In general, the electromechanical coupling coefficient can be written as

$$k_{ij} = \sqrt{\frac{\text{mechanical energy stored in direction } j}{\text{electrical energy applied in direction } i}} \quad \text{(or)} \tag{2.84}$$

$$= \sqrt{\frac{\text{electrical energy stored in direction } j}{\text{mechanical energy applied in direction } i}} \tag{2.85}$$

For example, the extensional coupling coefficient with stress applied along the 1 direction and electrodes normal to the 3 axis is

$$k_{31} = \frac{d_{31}}{\sqrt{e^{\sigma}_{33} s^{\mathbb{E}}_{11}}} \tag{2.86}$$

Similarly, the extensional coupling coefficient with stress applied along the 3 direction and electrodes normal to the 3 axis is

$$k_{33} = \frac{d_{33}}{\sqrt{e^{\sigma}_{33} s^{\mathbb{E}}_{33}}} \tag{2.87}$$

In the case of a shear stress in the 1-3 plane and with electrodes normal to the 1 axis

$$k_{51} = \frac{d_{51}}{\sqrt{e^{\sigma}_{11} s^{\mathbb{E}}_{44}}} \tag{2.88}$$

Table 2.7 shows a comparison of the piezoelectric coefficients, stiffnesses, and coupling coefficients of different PZTs. k_p represents the coupling between the electric field along the poling axis (3-axis) and mechanical action simultaneously in the 1- and 2-axes.

2.5.5 Actuator Performance and Load Line Analysis

This section describes several important concepts regarding the application of actuators to external loads. Although the discussion is focused on piezoelectric actuators, the basic principles are valid for any kind of actuator.

Blocked Force and Free Displacement of an Actuator

A piezoelectric actuator is normally specified in terms of two key parameters, blocked force, F_{bl}, and free displacement, δ_f. Blocked force is the force required to fully constrain the piezoelectric actuator and prevent it from deforming under the application of an electric field. Free displacement is the maximum induced displacement due to piezoelectric effect at a specified field with no external load.

Substituting $\sigma = 0$ in the actuator constitutive relation, Eq. 2.25, and ignoring thermal effects gives an expression for the free strain, $\mathbf{\Lambda}$, that corresponds to the free

Table 2.7. *Coupling coefficients of different PZTs*

	PZT-4	PZT-5A	PZT-5H	PZT-8
d_{31} (pC/N)	−123	−171	−274	−97
d_{33} (pC/N)	289	374	593	225
d_{15} (pC/N)	496	584	715	330
$s_{11}^{\mathbb{E}}$ ($\times 10^{-12}$ m^2/N)	12.3	16.4	16.5	11.5
$s_{33}^{\mathbb{E}}$ ($\times 10^{-12}$ m^2/N)	15.5	18.8	20.7	13.5
$s_{44}^{\mathbb{E}}$ ($\times 10^{-12}$ m^2/N)	39.0	47.5	43.5	31.9
$s_{12}^{\mathbb{E}}$ ($\times 10^{-12}$ m^2/N)	−4.05	−5.74	−4.78	−3.7
$s_{13}^{\mathbb{E}}$ ($\times 10^{-12}$ m^2/N)	−5.31	−7.22	−8.45	−4.8
k_{33}	0.70	0.71	0.75	0.64
k_{33}^2	0.49	0.50	0.56	0.41
k_{31}	−0.33	−0.34	−0.39	−0.30
k_{31}^2	0.11	0.11	0.15	0.09
k_{15}	0.71	0.69	0.68	0.55
k_{15}^2	0.50	0.48	0.46	0.30
k_p	0.58	0.60	0.65	0.51
k_p^2	0.34	0.36	0.42	0.26

displacement.

$$\mathbf{\Lambda} = \begin{Bmatrix} \Lambda_1 \\ \Lambda_2 \\ \Lambda_3 \\ \Lambda_4 \\ \Lambda_5 \\ \Lambda_6 \end{Bmatrix} = \begin{Bmatrix} d_{31}\mathbb{E}_3 \\ d_{31}\mathbb{E}_3 \\ d_{33}\mathbb{E}_3 \\ d_{15}\mathbb{E}_2 \\ d_{15}\mathbb{E}_1 \\ 0 \end{Bmatrix} \tag{2.89}$$

For piezoelectric film, PVDF, the free displacement strain is non-isotropic on the surface of the sheet because d_{31} is not equal to d_{32} and d_{24} is not equal to d_{15}. The free strain is expressed as

$$\mathbf{\Lambda} = \begin{Bmatrix} d_{31}\mathbb{E}_3 \\ d_{32}\mathbb{E}_3 \\ d_{33}\mathbb{E}_3 \\ d_{24}\mathbb{E}_2 \\ d_{15}\mathbb{E}_1 \\ 0 \end{Bmatrix} \tag{2.90}$$

The blocked force is given by the product of the free displacement and the stiffness of the actuator. The stiffness of the actuator and its relation to electrical-boundary conditions is discussed in more detail in the following subsections.

Actuator Load Line

To evaluate the performance of an actuator and to assess its suitability to a particular application, it is important to understand the concept of an actuator load line. At

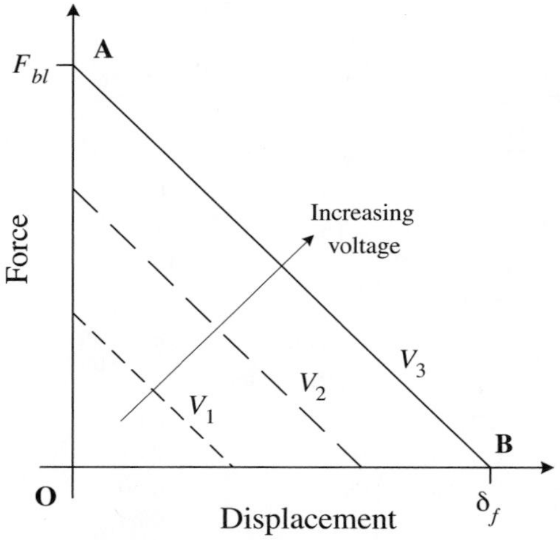

Figure 2.10. Actuator load line.

a given actuation voltage, the force and displacement of the actuator under any loading condition will lie on the load line. By plotting the load line of the actuator along with the external load, it is possible to visualize the mechanism of operation of the actuator.

The force and displacement of a typical piezoelectric actuator are shown in Fig. 2.10. At a given actuation voltage V_3 , the load line of the actuator is given by the straight-line segment **AB**. Note that, in general, the actuator load line can be a curve of any shape. For the specific case of a piezoelectric actuator, the load line is a straight line. The intercepts of the load line on the force and displacement axes represent the blocked force F_{bl} and free displacement δ_f, respectively. The load lines at actuation voltages V_1 and V_2 are also plotted in Fig. 2.10. As the actuation voltage increases, the load line moves such that the intercepts on the force and displacement axes increase. The origin **O** corresponds to some reference point or the undeformed state of the actuator.

At a constant actuation voltage, the force exerted by the actuator F_o at any point on the load line can be expressed in terms of the actuator displacement δ_o as

$$\begin{aligned} F_o &= F_{bl}\left[1 - \frac{\delta_o}{\delta_f}\right] \\ &= F_{bl} - \delta_o K_{act} \end{aligned} \tag{2.91}$$

where K_{act} is the effective stiffness of the actuator. Because the actuator is connected to a power supply, this stiffness is related to the short-circuit compliance of the material, $s_{11}^{\mathbb{E}}$. This can be seen by expressing the displacement of the actuator in terms of its free displacement, as follows:

$$\begin{aligned} \delta_o &= \delta_f\left[1 - \frac{F_o}{F_{bl}}\right] \\ &= \delta_f - \frac{F_o}{K_{act}} \end{aligned} \tag{2.92}$$

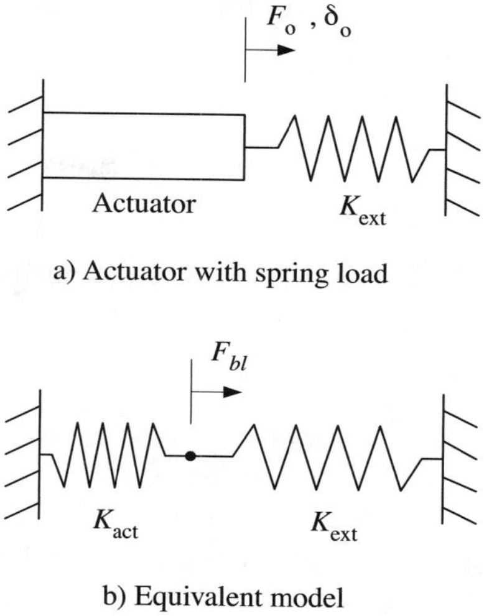

Figure 2.11. Effect of spring load on the actuator.

By comparing the previous equation with the 1-dimensional constitutive relation for an actuator, Eq. 2.70, it can be seen that for an actuator of length l_c and cross-sectional area A_c

$$K_{act} = \frac{A_c}{s_{11}^{\mathbb{E}} l_c} \tag{2.93}$$

External Loads and Impedance Matching

The effect of an external load of stiffness K_{ext} on the actuator can now be analyzed using the actuator load line. Consider the piezoelectric actuator connected to an external spring load as shown in Fig. 2.11(a). It will be shown that the actuator-load system can be modeled as two springs in parallel, under the action of the blocked force of the actuator as in Fig. 2.11(b).

In Fig. 2.12, the force-displacement characteristic of the spring load, which is the line segment **OC**, is plotted on the actuator load line **AB**. The intersection of the actuator load line and the spring characteristic line is the equilibrium point of

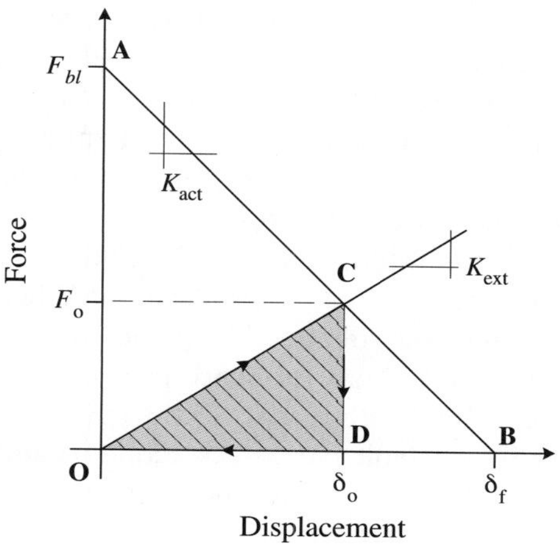

Figure 2.12. Load line analysis of a spring load.

the system. As the applied voltage is varied, the equilibrium point moves along the line **OC**. The coordinates of the point **C** can be found by substituting the spring force in Eq. 2.91

$$\begin{aligned} F_o &= F_{bl} - \delta_o K_{act} \\ &= K_{ext}\delta_o \end{aligned} \tag{2.94}$$

which gives the equilibrium position as

$$\delta_o = \frac{F_{bl}}{K_{act} + K_{ext}} \tag{2.95}$$

Note that this displacement is the same as the displacement of a parallel combination of the springs K_{ext} and K_{act} under the load F_{bl}, as depicted in Fig. 2.11(b).

Considering a complete cycle, the actuator transfers some energy into the load while expanding, and the same energy is returned to the actuator while it contracts. The net work done by the actuator is therefore zero. However, conceptually, if there is a way to retain the energy transferred to the load in each half cycle, the work done by the actuator, ΔW_{act}, will be positive and is given by the shaded area **OCDO** in Fig. 2.12.

$$\Delta W_{act} = \frac{1}{2}\delta_o F_o \tag{2.96}$$

Substituting from Eqs. 2.95 and 2.94, this becomes

$$\Delta W_{act} = \frac{1}{2} F_{bl}^2 \frac{K_{ext}}{(K_{act} + K_{ext})^2} \tag{2.97}$$

To find the load condition at which maximum work is done by the actuator

$$\frac{\partial(\Delta W_{act})}{\partial K_{ext}} = 0 \Rightarrow K_{ext} = K_{act} \tag{2.98}$$

This means that the maximum energy can be extracted from the actuator if the stiffness of the external load equals the stiffness of the actuator. This condition is called "impedance matching." Although the previous analysis was only for static actuation, it can be extended to dynamic actuation.

The load line also provides information about the capability of an actuator to perform mechanical work. Given an impedance-matched working condition, the maximum energy that can be extracted from the actuator is proportional to the product of its blocked force and free displacement. Therefore, the area under the load line or the force-displacement curve of the actuator can be used as a measure of the "available energy" in the actuator and can be quantized by an index. For example, a strain-force index can be defined as the product of the free strain of an actuator, multiplied by its blocked force, normalized by the cross-sectional area of the actuator. This translates to an effective "strain energy" contained in the actuator per unit volume. Several actuators can be compared on this basis, and the one with the best performance can be chosen for the application. A similar index can be defined with respect to the mass of the actuator. Table 2.8 lists several commercially available piezostack actuators and their strain-force indices. It is important to note that the strain-force index is only a relative number for comparison between actuators and does not reflect the actual energy that can be extracted from the actuator.

Table 2.8. *Operating voltage, free strain, blocked force, and strain-force index of commercially available piezostacks*

Piezostack	Maximum Voltage (V)	Strain ($\times 10^{-6}$)	Blocked Force F_{bl} (N)	Normalized F_{bl} (MPa)	Strain-Force Index (kJ/m^3)
MM 8M (70018)	360	254	571	7.26	1.8
MM 5H (70023-1)	200	449	450	5.73	2.6
MM 4S (70023-2)	360	497	637	8.08	4.0
PI P-804.10	100	1035	5052	50.52	52.3
PI PAH-018.102	1000	1358	6711	67.11	91.1
XI RE0410L	100	468	424	35.66	16.7
XI PZ0410L	100	910	312	24.74	22.5
EDO 100P-1 (98)	800	838	687	13.82	11.6
EDO 100P-1 (69)	800	472	223	4.56	2.1
SU 15C (H5D)	150	940	1186	51.69	48.6
SU 15C (5D)	150	1110	1222	53.21	59.0

2.6 Hysteresis and Nonlinearities in Piezoelectric Materials

Ferroelectric materials are noncentrosymmetric in nature, and domain switching takes place in response to an applied field or stress. Polar changes occur when ions displace to align with an applied field, resulting in strain. Conversely, with the application of a stress, the resulting deformation alters the polarization and, as a result, the electric field. These materials exhibit varying levels of hysteresis and nonlinear saturation effects at moderate to high levels of field. The generation of hysteresis is attributed to the impediment of domain-wall movement by inherent material inclusions and stress nonhomogeneities. For higher input fields, irreversible motion of domain walls pinned at inclusions becomes more significant.

Hysteresis models for piezoelectric materials can be divided into three categories: microscopic, macroscopic, and semi-macroscopic. Microscopic models are mostly limited to material stoichiometries (lattice/grain levels) and are not applicable to realistic system-level problems. Macroscopic models are based on phenomenological principles and are applicable to solve system-level problems. Preisach models fall under this category. Semi-macroscopic models use a combination of physics and experimental data. Smith and Ounaies [3] used a semi-macroscopic model of hysteresis behavior of piezoceramic materials. It is clear that piezoceramic materials, when used as actuators, display a significant hysteresis in the transfer function between voltage and displacement. A large number of techniques have been deployed to reduce this hysteresis, including displacement-feedback techniques, Preisach modeling and inversion, phase control, polynomial approximation, and current or charge actuation. The hysteresis in a piezoelectric actuator is reduced if the charge is regulated instead of the voltage [4]. However, the complexity of implementation of this technique has prevented its wide acceptance. It requires additional circuitry to avoid charging of the load capacitor. One possible approach may be to short-circuit the load every 400 ms or so and thereby periodically discharge the load capacitance [5]. This, however, introduces undesirable high-frequency disturbances and may significantly distort the low-frequency charge signals. Fleming and Moheimani [6] adopted a compliance-feedback current driver containing a secondary voltage-feedback loop to prevent DC charging of capacitive loads. Experimental results demonstrated good low-frequency current and charge tracking and a complete rejection of DC offsets.

Ferroelectric ceramics switch their polarization under an applied electric or mechanical field. Classical linear piezoelectricity is not adequate to capture the nonlinear behavior of these materials. As an example, barium titanate ($BaTiO_3$) has a perovskite-type structure with a cubic unit cell above its Curie temperature and is slightly distorted to the tetragonal form below its Curie temperature. An applied field in the direction opposite to the polarization may reverse the direction of polarization (switch through 180°C). However, the application of a compressive stress along the polar axis can switch it through 90°C. A polycrystalline ceramic behaves as a nonpolar material even though its constituents (crystals) are polar. It can be transformed into a polar material through the application of a large electric field – called a process poling. Macroscopic electromechanical behavior is a consequence of this domain structure. At small fields, domain-wall motions are reversible, and the macroscopic strain or electric displacement vanishes after unloading. At higher fields, domain wall-motions are irreversible and macroscopic strain or electric displacement is nonlinear. The hysteretic loops including "butterfly curves" are related to these domain-wall motions. The remnent polarization or remnent strain persists at the macroscopic level. There are several nonlinear constitutive models to describe ferroelectricity and ferroelasticity, and they are categorized into microscopic and macroscopic models. For the microelectromechanical approach, typical models are due to Hwang et al. [7] and Huber et al. [8]. For the macroelectromechanical approach, typical models are due to Bassiouny et al. [9, 10], and Chen and Lynch [11]. Elhadrouz et al. [12] formulated a macroscopic phenomenological approach to describe the nonlinear behavior of ferroelectric and ferroelastic ceramics under high electromechanical loading. To capture the history of dependence and dissipation, two internal variables that are the remnent strain (induced by stress) and the remnent polarization (induced by electric field) are introduced. Dielectric behavior, butterfly curves, ferroelastic hysteresis, and mechanical depolarization are satisfactorily captured with this model.

Li et al. [13] investigated the hysteresis phenomenon of ferroelectric-ferroelastic materials in polarization and developed an experiment-based phenomenological model that includes electrical yielding, mechanical yielding, and isotropic hardening. The nonlinear constitutive relations are expressed in terms of finite-element analysis. Smith et al. [14] developed a homogenized energy framework at mesoscale to model hysteresis and constitutive nonlinearities in ferroelastic materials, by constructing Helmholtz and Gibbs energy relations at the lattice level. The accuracy of the resulting model is demonstrated for both symmetric major loops and biased minor loops using experimental data from PZT-4 and PZT-5H.

2.7 Piezoceramic Actuators

Piezoceramics are potential actuators for a wide range of applications in aerospace, automotive, civil structures, machine tools, and biomedical systems to actively control vibration and noise, improve performance, and augment stability. One of the major barriers for various applications is the small stroke of these actuators [15]. To increase induced strain, these actuators are often driven under high electric fields and sometimes even to extreme limits. In addition, the operating conditions of the system itself may cause high mechanical loads on the actuator. For example, in rotorcraft applications, actuators placed on rotor blades are exposed to high tensile stresses due to centrifugal forces. Although the piezoelectric material is relatively well behaved and linear at low electric fields and low mechanical stress levels, it shows considerable nonlinearity at high values of electric field and mechanical stress. To develop an efficient structural system with piezoceramic actuators, it is necessary to accurately predict the response of the actuators, including magnitude and phase of induced strain, power consumption, and integrity under different excitation and loading levels. Currently, neither the mathematical tools to cover a wide range of operating conditions nor reliable test data to validate these tools are readily available.

This section discusses the behavior of PZT-5H piezoceramic sheet actuators under different types of excitation and mechanical loading. PZT-5H is a typical piezoceramic composition that is widely used in adaptive-structures applications because of its low field requirement. The behavior and characteristics of this actuator are a good representation of any piezoceramic actuator. Therefore, most of the techniques and experiments described herein can be used to obtain preliminary quantitative data about the performance and capabilities of a piezoceramic actuator. These data would be valuable for the initial design of a smart system.

The following discussion is divided into two parts: static behavior and dynamic behavior. The free-strain response of the actuators under DC excitation is experimentally investigated along with the associated drift of the strain over time. The drift phenomenon is especially important in cases of static deflection of control surfaces or blade geometry. The effect of tensile stress on the free-strain response is examined to quantify the effect of the high centrifugal forces experienced by actuators mounted on rotor blades. The magnitude and phase of the free-strain response of the actuator under different excitation fields and frequencies is measured, and a phenomenological model to predict this behavior is developed and validated experimentally. The power consumption of the actuators, which is very important for sizing the electrical slip-ring units in a rotating system, is calculated using an electromechanical impedance method. This is then validated by measuring the power consumption of a free actuator and a pair of actuators surface-bonded to a host structure. The performance of actuators in a practical application is constrained by depoling limits and dielectric breakdown of the actuator material. These aspects are also discussed, along with the feasibility of recovering performance by repoling in the event of accidental depoling.

2.7.1 Behavior under Static Excitation Fields

Piezoceramic actuators are capable of responding to static, or steady, electric fields. However, several phenomena not normally encountered in conventional electromagnetic actuators are observed in a piezoceramic actuator under a static excitation field. Additionally, the presence of significant amounts of hysteresis in the material as a result of its noncentrosymmetric unit cell and associated domain motion requires certain procedures to be followed to obtain meaningful data in static experiments. This section describes important effects related to the static behavior of PZT sheet actuators.

Experimental Sample Preparation

In the following sections, the measured characteristics of piezoceramic (PZT-5H) sheets of dimension 1 in $\times$ 0.5 in $\times$ 0.01 in (25.4 mm $\times$ 12.7 mm $\times$ 0.254 mm)

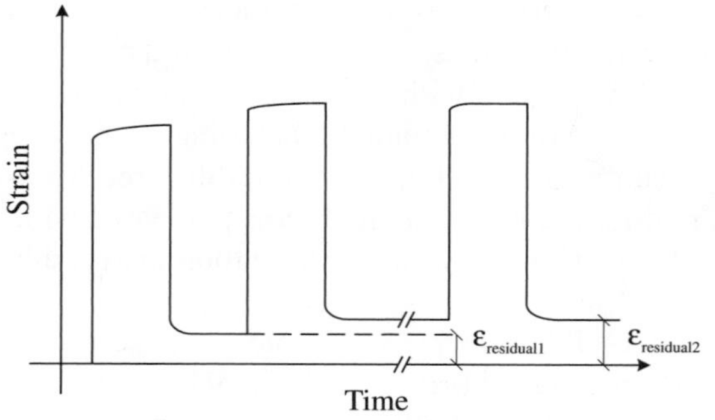

Figure 2.13. Strain during DC cycling.

obtained from Morgan Matroc, Inc., are described. Some of the manufacturer-supplied characteristics of this material are shown in Table 2.3. Most of these data are valid only for small excitation fields. However, these materials frequently encounter high excitation fields when used as actuators. As discussed in subsequent sections, the material properties can vary significantly from the tabulated small-signal values. All of the samples were poled along their thickness. They were excited along the poling direction and strains were measured in a plane perpendicular to the poling direction. Each sample was instrumented with a strain gauge with a gauge length of 0.125 in (3.2 mm) in a quarter-bridge configuration. The excitation leads were soldered to the faces of the sheet and the sample was suspended by means of the excitation leads, so there were no structural-boundary constraints. The experimental data in the following sections represent the average behavior of three randomly selected samples for each test point. A careful averaging process was necessary as a variation of properties of up to 15% was found to be not uncommon in the experimental samples. Before the tests were carried out, the samples were cycled to erase the effect of previous excitations. The cycling can be of two types, depending on whether the properties to be observed are static or dynamic.

Cycling

If the application involves a static excitation, a DC cycling is performed on the actuator. This involves exciting the actuator with its highest operating DC field, switching off the field, and then measuring the residual strain. This process is repeated several times until the residual strain after each cycle has stabilized. A schematic of the actuator strain during the DC cycling process is shown in Fig. 2.13. After each cycle, the difference in residual strain keeps decreasing until it becomes almost zero. The number of cycles needed to stabilize the performance depends on the cycling field and normally increases with increasing cycling field. A cycled actuator has an inherent bias and will show almost zero residual strain on the application of an excitation field less than the cycling field. Note that the polarity of the field is important and that reversing the polarity, either during cycling or during operation of the actuator, will destroy the bias. This treatment is therefore suitable only for unipolar operations. If the actuator is exposed to the DC cycling voltage for a long period of time, it is observed that there is an additional effect of stabilizing the drift. Another type of cycling treatment is AC cycling, which involves exciting the actuator for several cycles under an AC field. The effect of this treatment is that it removes all biases in the actuator and also minimizes residual strain.

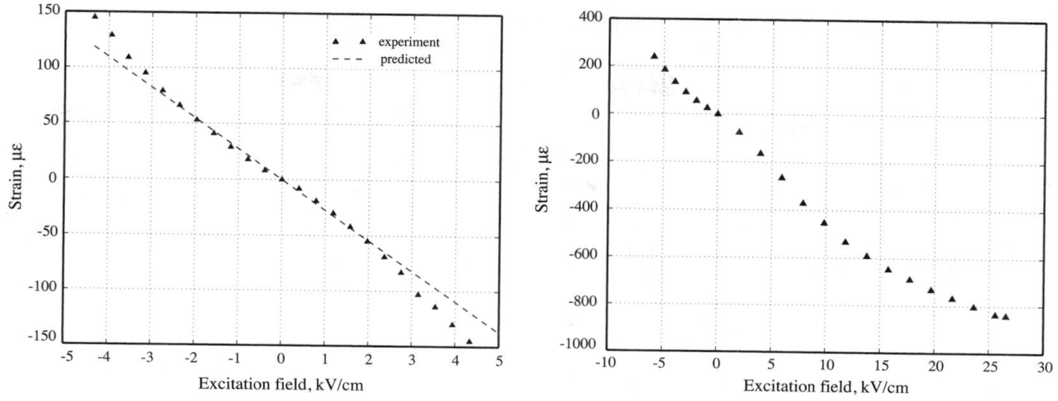

(a) Low excitation levels

(b) High excitation levels

Figure 2.14. Static-free-strain behavior of PZT-5H.

Static Free Strain

A typical DC field (along the polarized direction) versus transverse-free strain (in a plane normal to the polarized direction) plot for a PZT-5H actuator is shown in Fig. 2.14. To obtain the static-free-strain plot, the following steps were taken. DC cycling was initially carried out on the sample to minimize its residual strains. A typical sample required on the order of 10 DC cycles to bring the residual strain to less than $3\mu\epsilon$. Each point on the curve is then obtained by applying a DC field and measuring the resulting strain with a strain gauge. After each reading, the excitation is switched off and the gauge is reset. This effectively ignores the hysteresis and drift and generates a quasi-steady free-strain curve. The curve is almost linear at low applied electric field levels and the linear piezoelectric coefficients can be used to satisfactorily predict this part of the curve. In this region, the slope of the curve is the coefficient d_{31} and the value quoted by the manufacturer [16] is -274×10^{-12} m/V. The strains predicted by this linear relation are also plotted for comparision. The negative value of d_{31} means that a positive electric field in the polarization direction results in a compressive strain on the surface of the PZT sheet. At higher electric fields, nonlinear effects become apparent. These effects are attributed to factors such as reversible domain-wall motion. The reason that such effects are much smaller at lower values of electric field is that non-180° domain-wall motion results in a permanent mechanical distortion of the material and consequently requires a larger energy and, hence, occurs at larger field strengths. At high field strengths, a larger change in induced strain per unit increase in field is expected, both for negative and positive fields. Also, strain values for the same positive and negative fields are not equal, which means the free-strain curve is asymmetric. Such an asymmetry has also been observed in the inverse piezoelectric response of Rochelle salt [17] and is attributed to the permanent electric polarization in the crystal. The asymmetry present in piezoceramics is small for low values of field but becomes larger as the field is increased. In terms of actual voltage applied to the piezoceramic sheet, a voltage of 100 volts corresponds to a field of 3.937 kV/cm.

Figure 2.14(b) shows a free-strain curve spanning a much higher field range. The maximum positive field is limited by the breakdown of the dielectric, which in

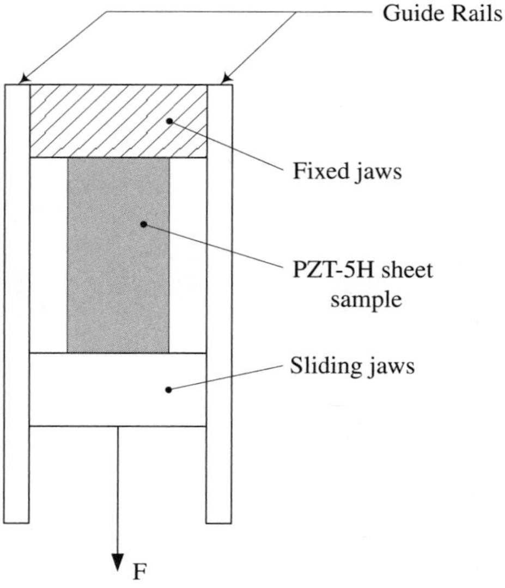

Figure 2.15. Tensile-testing fixture.

this case is the ceramic itself, whereas the maximum negative field is limited by the piezoceramic depoling. Note that the DC depoling field for PZT-5H is approximately −5.5 kV/cm.

Effect of External Stresses

In the literature, investigations have been carried out to examine the effect of external mechanical stresses on the behavior of piezoceramics. However, most of the existing works are focused on the effect of compressive loads because many early applications of piezoceramics were in underwater devices in which the materials are exposed to high hydrostatic stresses [18, 19, 20, 21]. Some results of the effects of compressive stresses are found in available references. Compressive stresses tend to align the *c*-axes of the domains perpendicular to the direction of stress. For example, if the compressive stress acts along the x-axis, say in the plane of the sheet, the *c*-axes of the domains are randomly reoriented parallel to the yz plane, which is across the thickness. This destroys some of the initial polarization and thus reduces the net polarization. Hard ceramics (lower compliance) like PZT-8 and PZT-4 experience large changes in piezoelectric coefficients but show good recovery on removal of the stress [21]. Soft ceramics like PZT-5H show a permanent degradation in properties with stress cycles.

Limited work has been done on the effect of tensile stress on the behavior of piezoceramics, which is pertinent to their application in the development of a smart rotor [22]. Experiments were carried out to observe the effects of tensile loads perpendicular to the poling direction on the free strain. A test fixture was designed to apply tensile loads to a piezoceramic sample while allowing it to strain freely. The ceramic samples tested were 2 in × 1 in × 0.01 in (50.8 mm × 25.4 mm × 0.254 mm) commercially available PZT-5H sheets. Loads were applied to the sample by means of weights suspended from a sliding bracket (Fig. 2.15). The static-free-strain curves were obtained for different values of applied tensile load. The data obtained from 10 samples were averaged out to obtain the final free-strain values. The results are shown in Fig. 2.16. From the plot, it can be seen that there is a slight increase

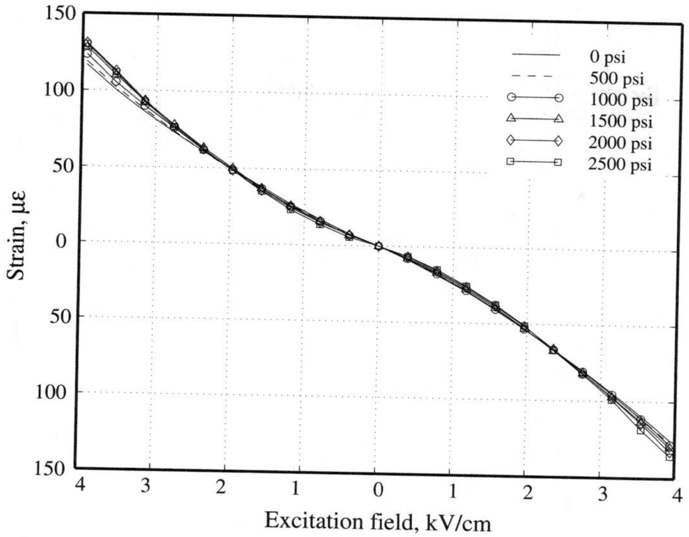

Figure 2.16. Variation of static free strain with transverse tensile stress.

in free strain with an increase in applied tensile load. Although this is a very small increase, the averaging process makes it more likely that this is a phenomenological change rather than experimental scatter. It was not possible to apply stresses higher than 2500 psi (17.2 MPa) because of stress concentration near the end supports and a small amount of bending in the sample due to misalignment. Published data [23, 16] indicate that the static tensile strength of a typical piezoceramic is around 13,000 psi (89.6 MPa), but samples invariably fracture at lower average stresses.

For PZT-5H under a compressive stress parallel to the poling axis (i.e., along the thickness), d_{31} remains constant or shows a slight increase at low values of stress and then drops off at higher values [24]. Such a stress will tend to randomly reorient the *c*-axes of the unit cells in a plane perpendicular to the poling direction (i.e., in the plane of the sheet). A uniaxial tensile stress acting along the length of the sheet will also tend to align the dipoles along the plane of the sheet, with a preference along the direction of stress. Hence, it can be expected that these two stress states will produce similar changes in the properties of the material. Consequently, the behavior seen in Fig. 2.16, with tensile stress in the plane of the piezoceramic sheet, is consistent with previously observed phenomena [24] with compressive stress along the poling direction.

A curve was fit to these data points to empirically predict the variation of free strain with applied external stresses. The curve fit was of the form

$$\epsilon = a + b\mathbb{E} + c\mathbb{E}^2 + d\mathbb{E}^3 \tag{2.99}$$

For simplicity, a linear variation of the coefficients a, b, c, and d with applied tensile stress was fit to the data. This variation is given by

$$a = -1.9637 - 7.039 \times 10^{-4}\sigma \tag{2.100}$$

$$b = -25.82 - 7.54 \times 10^{-4}\sigma \tag{2.101}$$

$$c = -0.1535 + 3.86 \times 10^{-4}\sigma \tag{2.102}$$

$$d = -0.298 - 1.244 \times 10^{-4}\sigma \tag{2.103}$$

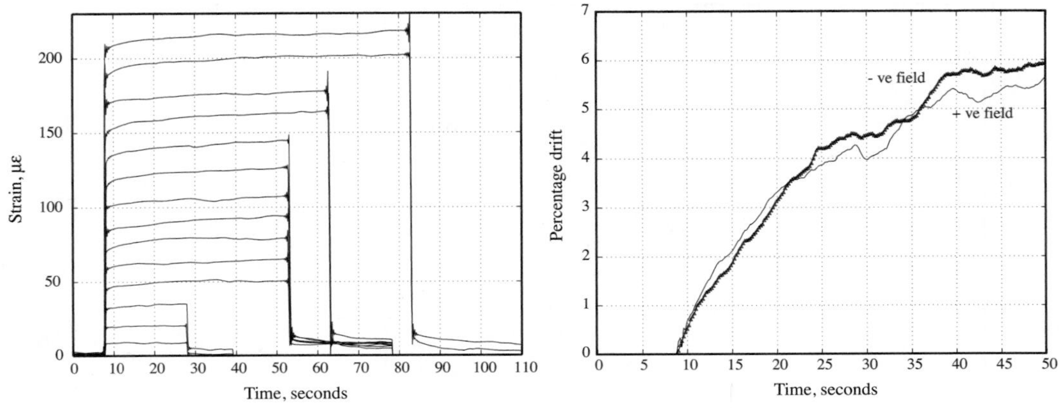

(**a**) Drift in the free strain (**b**) Effect of field direction

Figure 2.17. Static strain drift behavior.

where ϵ is the strain in microstrains, $\mathbb{E}$ is the electric field strength in kV/cm, and σ is the applied tensile stress in psi. As expected, at zero external stress, the free-strain curve reduces to that in Fig. 2.14(a). The increase in free strain is on the order of 10% at 3.937 kV/cm under a tensile stress of 2500 psi. The empirical curve fit as shown here provides a convenient way of incorporating the PZT characteristics into a mathematical model of the entire smart structure.

Drift

An effect often observed experimentally is the drift of the actuator strain in response to a DC excitation. The drift phenomenon is a slow increase of the free strain with time after the application of a DC field. An uncontrolled drift in the actuator position is obviously detrimental to the overall performance of systems wherein the actuator is meant to maintain a certain static deflection, for example, the steady deflection of a trailing-edge flap. The basic drift phenomenon is as follows: After the application of a DC field, the strain jumps to a certain value and then increases slowly with time. When the field is switched off, the strain falls back to some value and then slowly decreases until it stabilizes at some residual strain. The curves describing both the slow increase and the slow decrease are of a similar nature and are roughly logarithmic with time. One of the manufacturers [25] has given the following formula for drift of a piezostack actuator:

$$\Delta\epsilon = \Delta\epsilon_o \left(1 + \gamma \ln \frac{t}{0.1}\right) \tag{2.104}$$

where t is the time in seconds, $\Delta\epsilon_o$ is the strain 0.1 seconds after the application of the field, and γ is a factor that depends on the system's characteristics, typically on the order of a few percentage points. Note that the percentage increase in strain after the application of the field is independent of the field strength. Experimental observations on a PZT sheet (Fig. 2.17(a)) show a family of drift curves for DC excitation fields from 0.4 kV/cm to 5.5 kV/cm. The experimental data show that the percentage drift is roughly the same regardless of excitation field. Figure 2.17(b) shows the percentage drift in response to a positive and a negative field of the same

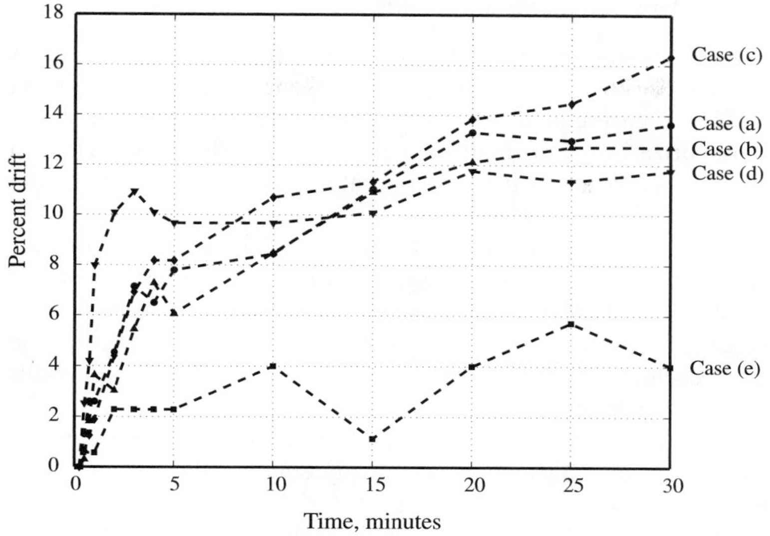

Figure 2.18. Drift under combined excitation.

magnitude, 4 kV/cm. It can be concluded that the direction of applied field has negligible effect on the magnitude or the rate of strain drift.

Similar phenomena have been observed by other researchers during investigations on Rochelle salt [17]. It is believed that the drift phenomenon is largely intrinsic in nature. The drift can be attributed to a gradual change in the permanent polarization of the material. The dependence of such effects on the state of the material as a result of previous mechanical or electrical treatment has been pointed out and referred to as the "fatigue" effect.

In an attempt to stabilize the drift, an AC field of 5% of the steady DC field was superimposed at high as well as low frequencies. The following five different types of excitations were tried out: (a) 3.937 kV/cm DC field, (b) 0.196 kV_{rms}/cm sinusoidal field at 10 Hz riding on a 3.937 kV/cm DC field, (c) 0.196 kV_{rms}/cm sinusoidal field at 500 Hz riding on a 3.937 kV/cm DC field, (d) 1.968 kV_{rms}/cm sinusoidal field at 500 Hz riding on a 3.937 kV/cm DC field, and (e) 3.937 kV/cm DC field on an actuator that was previously exposed to a DC field of the same magnitude for three hours.

The effect of these excitations is shown in Fig. 2.18. The percentage of drift from the instantaneous strain is plotted versus time. After half an hour, the strain signal had increased by approximately 12% to 18% for most cases. It can be seen that there is not much change in the drift due to a pure DC field in comparison to those with different superimposed AC fields. The only exception is case (e), where the percentage of drift is the smallest. The data plotted has been averaged out over three trials. There is some amount of scatter because the repeatability of such drift tests is sensitive to the previous excitation of the sample. The kind of treatment given to the actuator in case (e) is similar to the fatigue effect, and it appears possible to stabilize drift by exposing the piezoceramic continuously to a DC field for several hours before the actual excitation. It was also observed that the drift is similarly reduced for any excitation field of magnitude less than the stabilization field. This kind of DC stabilization introduces a bias in the piezoceramic and zeros out residual strains in response to an excitation in the same direction. The bias, however, is destroyed on reversing the polarity of the field.

2.7.2 Behavior under Dynamic Excitation Fields

One of the biggest advantages in using PZT actuators for adaptive structures applications is their large bandwidth. Because the piezoelectric effect is an electromechanical effect occuring at the unit-cell level, the response of the material is very fast. Characterization of the material behavior under dynamic excitation is therefore very important to fully utilize its operating range. Hysteresis of the material plays a dominant role in its response.

The origin of the hysteresis can be traced to the orientations of the unit cells of the material, which can switch from one orientation to another in response to an electrical field or mechanical stress field of sufficient magnitude. This response to an electric field is also called an extrinsic effect (as opposed to the intrinsic converse piezoelectric effect of each unit cell), resulting in movement of the domain walls and associated nonlinearities in the overall response of the piezoceramic. From the point of view of an actuator, the hysteresis inherent in the material results in energy dissipation in the form of heat. This energy is basically equal to the work done in reorienting the dipoles in the material in the direction of the applied electric field. The energy dissipated is quantified in terms of a dissipation factor, called tan δ. This quantity is related to the non-ideal dielectric nature of the material and is discussed in detail in a later section.

An external dynamic stress will also have the effect of reorienting the domains, and this leads to an effect similar to static friction, which can be observed as a hysteresis in the stress-strain curve of the material (under constant electric field). This mechanical hysteresis results in an effective damping in the material. Hysteresis is a nonlinear phenomenon, in which the induced strain lags behind the applied field.

Piezoelectric materials exhibit varying levels of hysteresis and nonlinear saturation effects at moderate to high levels of applied field. The material hysteresis is often attributed to the impediment of domain-wall movement as a result of inherent material inclusions and stress nonhomogeneities. At low field levels, domain-wall movement is reversible, whereas at high field levels, domain walls move over extended distances. If we restrict the applied field or stress to a sufficiently low level, it minimizes hysteresis. This restricts the range of applications. For certain applications, it becomes necessary to minimize the material hysteresis, a goal that can be achieved indirectly through a feedback mechanism. Smith and Ounaies [3] addressed the modeling of hysteresis and nonlinear constitutive relations in piezoelectric materials based on the quantification of the reversible and irreversible motion of domain walls pinned at inclusions in the material. Basically, the theory characterizes the inherent hysteresis in the relation between the input field and the output polarization.

Strain Hysteresis

To observe the losses and the actual hysteretic behavior of the material, a quasi-steady free-strain test was performed by changing the voltage successively from point to point. In contrast to the earlier static-free-strain curve, the excitation was not switched off and the gauge was not zeroed after each reading. At each point, the strain was allowed to stabilize before taking a measurement. The experiments were carried out on several different samples, and each point is the average of

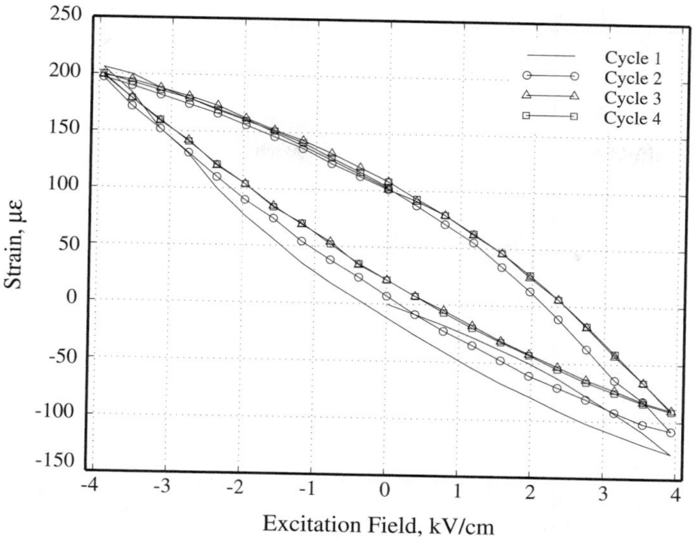

Figure 2.19. Quasi-static hysteresis curve, uncycled PZT.

three measurements. The curve shown in Fig. 2.19 was the response of a brand-new sample (uncycled). The field was slowly changed in steps of 0.3937 kV/cm. The strain response seems to have some bias and takes around three cycles to stabilize. The same experiment was repeated after exciting the actuator with a moderately high $3.15\mathrm{kV_{rms}/cm}$, 5 Hz sinusoidal field. This is referred to as "AC cycling" and is carried out in an attempt to train the actuator to a certain excitation, such that its residual strain and drift are minimized. The quasi-static strain response, shown in Fig. 2.20(a), now seems to have less bias and stabilizes in the second cycle itself. The aspect ratio (lateral width to length ratio) of the hysteresis curves is around 15%. Figure 2.20(b) shows the quasi-steady hysteresis curve at a lower maximum field: the aspect ratio becomes smaller. It can be seen that the normalized area under the curve is larger when the maximum field is larger. Also, the curves are asymmetric with respect to the zero-strain axis, which is similar to the asymmetry observed in the static-free-strain curves and is due to the permanent electric polarization of the ceramic. Another interesting feature is that the curves in Fig. 2.20(b) were obtained by starting with a negative excitation, whereas the curves in Fig. 2.20(a) were obtained by starting with a positive excitation. This difference is seen in the first quarter of the first cycle. The remainder of the curves show no dependence on the sequence of excitation.

The shape of the strain-field hysteresis loop changes with excitation frequency and field. Figure 2.21(a) shows the variation in the experimentally measured strain-field hysteresis loop at 5 Hz for a free actuator. It can be seen that with increasing field, the overall shape of the curve is not affected much, but the mean slope increases with increasing field. Also plotted for the sake of comparison is the DC free-strain curve, which matches closely with the increasing positive excitation and increasing negative excitation segments of the hysteresis curves. This is to be expected, because the DC free-strain curve is obtained from the low-frequency strain-hysteresis response by ignoring the hysteresis. The dynamic-hysteresis curves reduce to the static-free-strain curve as the frequency of excitation is decreased. The effect of frequency is seen in Fig. 2.21(b). The area under the high-frequency curve is less than the area under the low-frequency curve, which means that there are larger energy losses at

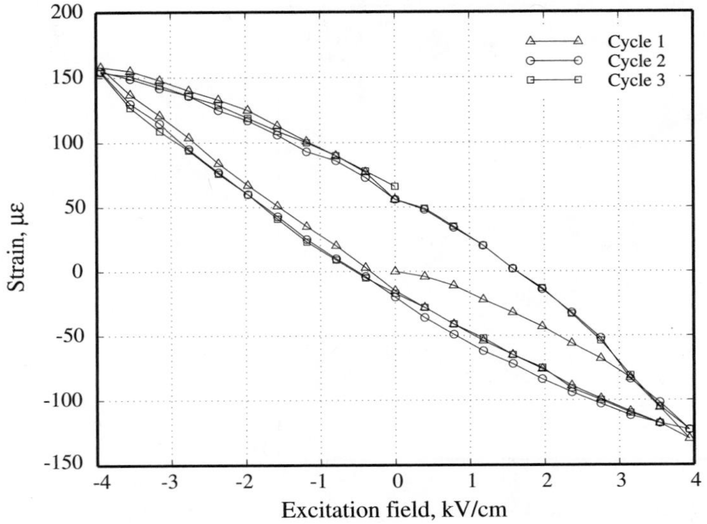

(a) Maximum field=3.937 kV/cm

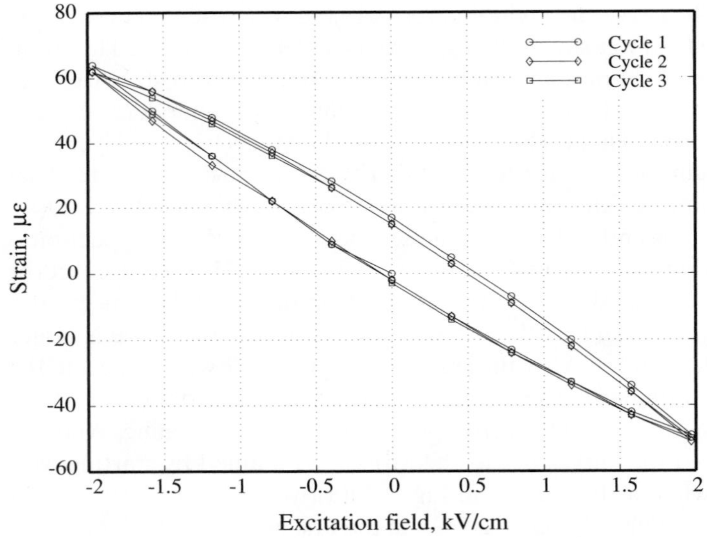

(b) Maximum field=1.968 kV/cm

Figure 2.20. Quasi-static hysteresis curves, after AC cycling.

lower frequencies. These hysteresis curves can be generated using the phenomenological model developed previously, which also traces out the static-free-strain curve when the frequency reaches zero. Vautier and Moheimani [26] showed that using electric charge instead of voltage to drive the actuator can reduce the hysteresis by demonstrating this concept experimentally on a cantilevered beam.

Dynamic Strain Response

To observe the magnitude and phase of the induced strain of a free actuator under AC excitation, the sample was excited by a sinusoidal field stepped from 0 to 200 Hz.

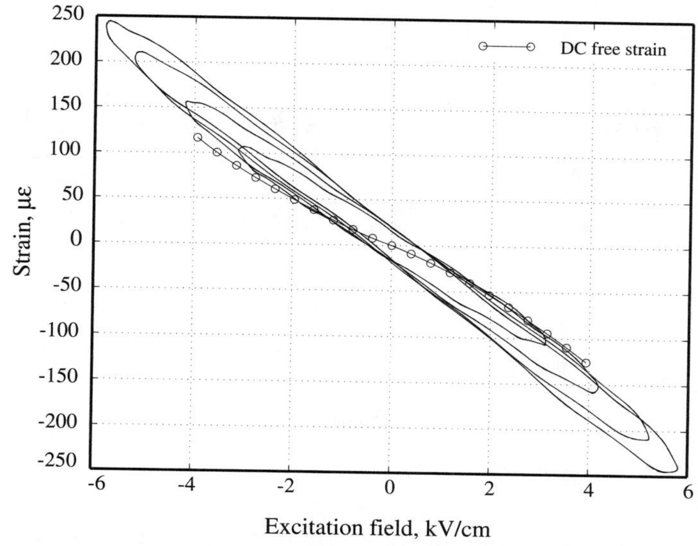

(a) Effect of increasing field, 5 Hz excitation frequency

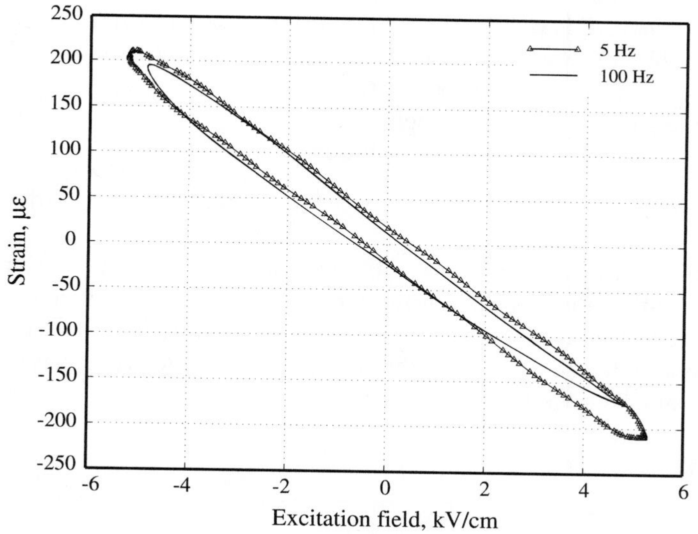

(b) Effect of increasing frequency

Figure 2.21. Variation of strain hysteresis with field and frequency for a free actuator.

This was carried out at different excitation fields to see the effect of excitation field as well as excitation frequency, and these results are summarized in Fig. 2.22. The dependence of crucial material properties such as e^{σ}_{33} and $\tan\delta$ on the magnitude and frequency of the excitation can be calculated from the magnitude and phase of the current drawn by the actuator. These parameters are important for predicting the power consumption of actuators bonded to a structure. From Fig. 2.22(a), it can be seen that the free strain and, hence, d_{31} is relatively independent of the actuation frequency. The slight increase seen at higher frequencies is due to an electro mechanical resonance around 1 kHz. The influence of the resonant dynamics

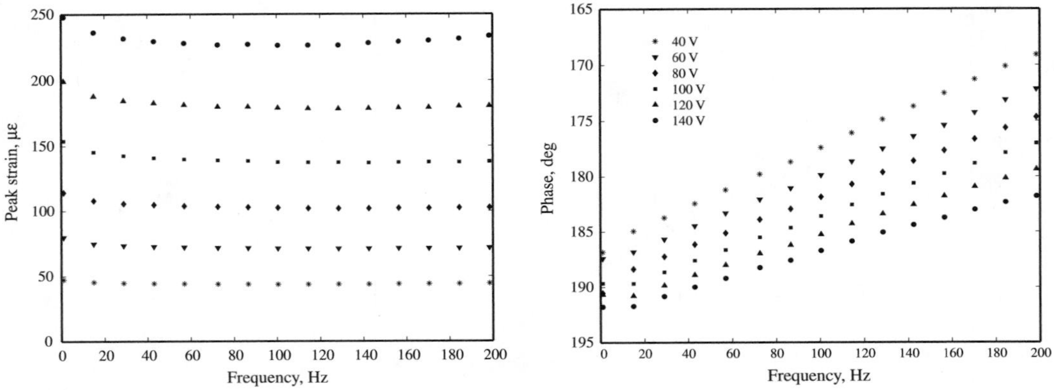

(a) Peak free strain (b) Strain phase

Figure 2.22. Response of a free PZT actuator.

can be seen in Fig. 2.22(b), which is a plot of the phase of the free strain. These dynamic characteristics, although important for a free actuator, are of less interest while investigating the performance of actuators bonded to a structure, where the dynamics of the parent structure is dominant. At low frequencies, the free strain increases by approximately 10% due to poling effects. At higher frequencies and excitation fields, nonlinear effects cause the free strain to increase by as much as 15% compared to the free strain in the linear region.

This variation can be described in terms of the following relation:

$$\mathbb{E} = A\epsilon + B\dot{\epsilon} \tag{2.105}$$

where $\mathbb{E}$ is the excitation field in volts/m and ϵ is the free strain of the actuator in microstrain. By analogy with a mechanical spring-damper system, the coefficients A and B can be considered as an effective stiffness and damping, respectively. The variation of this stiffness and damping with excitation voltage and field is calculated from the experimental data. The values of the stiffness and damping are shown in Fig. 2.23.

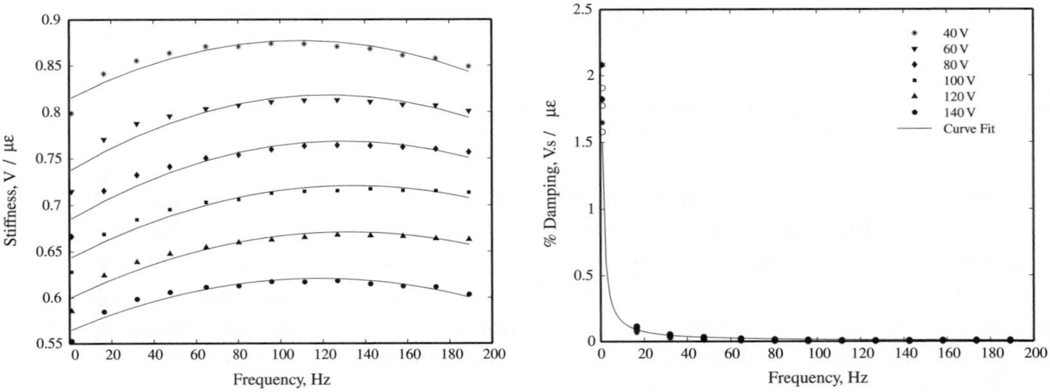

(a) Effective stiffness coefficient, A (b) Effective damping coefficient, B

Figure 2.23. Effective stiffness and damping variation, PZT-5H.

It can be seen that the stiffness depends to a large extent on excitation voltage, as expected, but the frequency dependence is small. A quadratic variation of stiffness with frequency is calculated and is given by

$$A = af^2 + bf + c \tag{2.106}$$

where f is the frequency of excitation in Hertz, and the coefficients a, b, and c are linear functions of the excitation voltage.

$$a = 15.1 \times 10^{-9}\mathbb{E} - 6.09 \times 10^{-6} \tag{2.107}$$

$$b = -2.59 \times 10^{-6}\mathbb{E} + 1.4 \times 10^{-3} \tag{2.108}$$

$$c = -2.44 \times 10^{-3}\mathbb{E} + 0.893 \tag{2.109}$$

The variation of damping with field is not significant, especially at higher values of frequency, so the damping is expressed as

$$B = \frac{1.5}{f} \tag{2.110}$$

This expression for damping is equivalent to stating that the energy lost per cycle per unit displacement amplitude is a constant, a result that is intuitively expected. Using these relations for effective stiffness and damping, the free strain of the actuator can be calculated at any given frequency in the range of 1–200 Hz and excitation voltage in the range of 40–140 V. The comparison of experimentally measured free strain with predictions using this phenomenological model is shown in Fig. 2.24, where the sample is excited at 80 V_{peak} and 120 V_{peak} at frequencies of 5 Hz, 25 Hz, and 100 Hz. It can be seen that at lower frequencies, the damping is slightly underpredicted and, at higher frequencies, the damping is slightly overpredicted. This is due to the rather simple hyperbolic variation of the effective damping parameter assumed in Eq. 2.110. A more complicated variation would yield more accurate results, but the accuracy of the present assumption is considered to be within acceptable limits.

It is worth mentioning here that the static-free-strain values can be obtained from these equations by first setting $\dot{\epsilon} = 0$ and then setting $f = 0$. The resulting equation for static-free strain is

$$\epsilon_{static} = \frac{\mathbb{E}}{-2.44 \times 10^{-3}\mathbb{E} + 0.893} \tag{2.111}$$

The values of static-free strain obtained by this equation are very close to the values obtained from Eq. 2.99, although not precisely the same. This is because Eq. 2.99 represents the static behavior that has an inherent asymmetry between positive and negative excitation voltages due to remnant polarization effects. This asymmetry decreases as the frequency of excitation increases. However, the model represented by Eq. 2.105 assumes a solution that is inherently symmetric. Therefore, for static and low-frequency behavior, Eq. 2.99 should be used to obtain more accurate results whereas Eq. 2.105 should be used for high-frequency behavior.

Viswamurthy et al. [27] modeled the dynamic-hysteresis behavior between the applied electric field and displacement of a piezoceramic stack actuator using a Preisach model. The unknown coefficients of the model were obtained by identification from experimental data. It was demonstrated that ignoring the dynamic hysteresis by using a linear model of the actuator led to an erroneous prediction of the optimal control input in a feedback system.

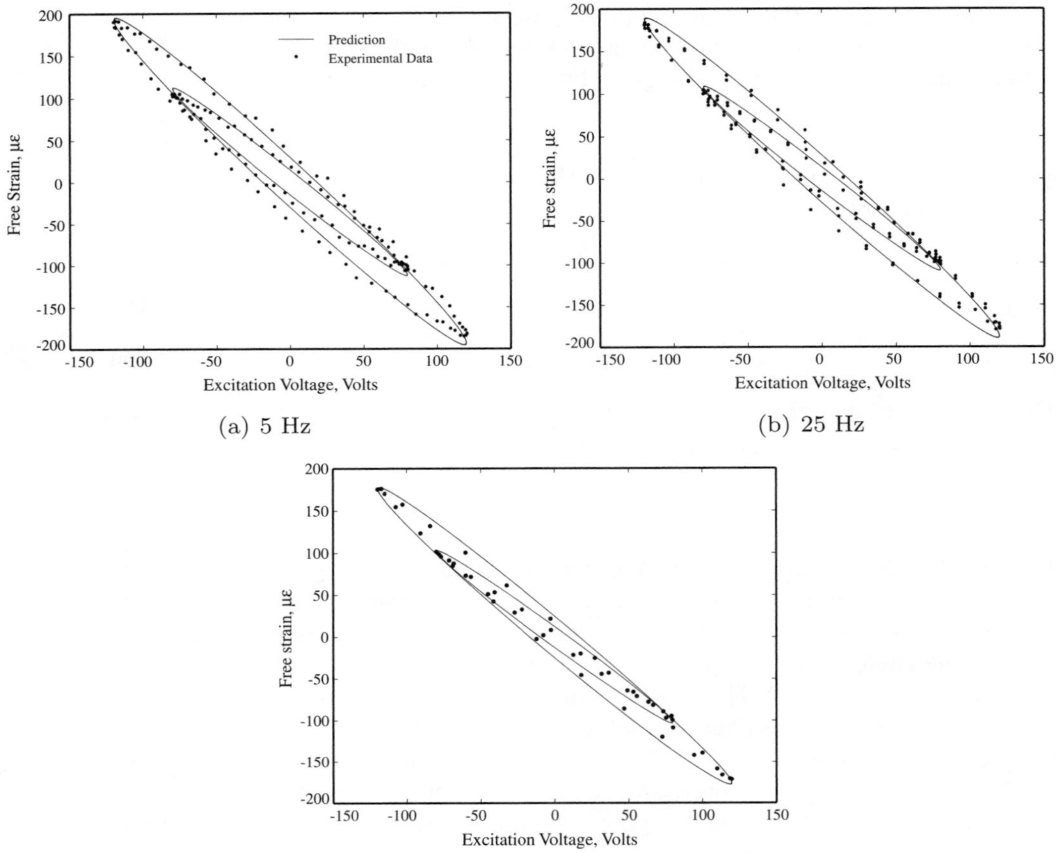

(c) 100 Hz

Figure 2.24. Model validation at three different frequencies and two excitation voltages of 80 V_{peak} and 120 V_{peak}.

There are other methods that are used to model the hysteresis behavior of piezoelectric materials. Maxwell resistor capacitor model [28], phaser approach [29], and describing functions [30]. Instead of using the electric field, the electric charge is controlled for actuation, which minimizes the hysteresis effect. However, because of the increased complexity of implementation, such circuits are not widely used [5]. It has also been shown that by using electric charge to drive piezoelectric actuators for vibration feedback control, negative effects associated with hysteresis can be significantly reduced [26]. By using electric charge, an improved model of the plant is obtained, which in turn increases the robustness of the controller.

Effect of DC Bias

Several studies have reported the beneficial effects of operating piezoceramic actuators with a DC bias field. A DC bias field increases the value of d_{31} under stress [20]. This can be explained by the stabilizing effect that a DC bias field has on the

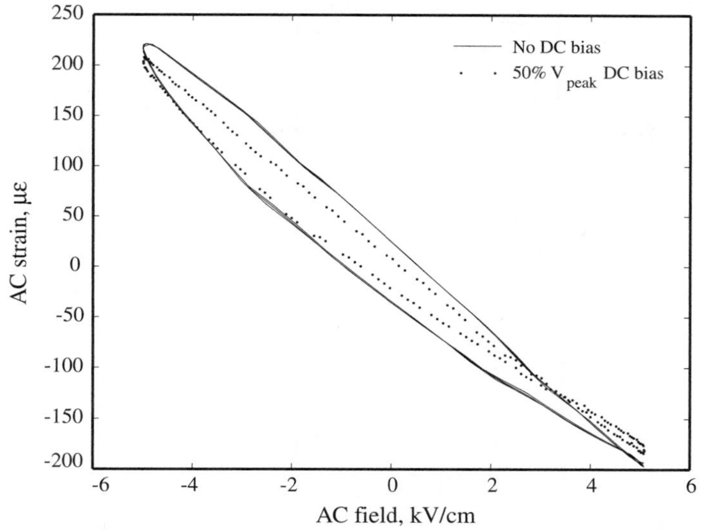

Figure 2.25. Effect of a DC bias field.

pinning of the domains [31]. Under a DC bias, the domains are better aligned and the domain walls become more difficult to move, which effectively reduces the extrinsic contribution to nonlinear effects and losses in actuation. Figure 2.25 shows a strain-field hysteresis loop under a pure AC field and under the same AC field with a superimposed positive DC field of strength $0.5V_{peak}$. Only the AC component of the strain is plotted for comparision. It can be seen that the area under the loop is less for the actuation with a bias than without bias. This shows that the losses have decreased due to the application of a DC bias. There is also a small decrease in the magnitude of peak free strain under bias.

2.7.3 Depoling Behavior and Dielectric Breakdown

Although increasing the applied electric field is necessary to obtain larger deformations from the actuator, the magnitude of the electric field is constrained by two limits: depoling and dielectric breakdown. As the electric field applied along the poling direction increases, dielectric breakdown eventually occurs in the piezoceramic material. However, this field usually corresponds to several hundreds of volts applied to the actuator and is normally not encountered during operation. A more critical constraint is the depoling of the piezoceramic. When exposed to a high electric field opposite to the poling direction, the piezoceramic loses most of its piezoelectric capability. The actuation is accompanied by large dielectric losses and poor efficiency. This is known as depoling of the piezoceramic and is accompanied by a permanent change in dimensions of the sample. This is probably due to large-scale domain switching in the material. The DC depoling field of PZT-5H is approximately 5.5 kV/cm. For an AC excitation, the depoling field depends on the frequency. It is observed that under a dynamic excitation, the depoling field of the actuator becomes lower than the DC value. This trend is shown in Fig. 2.26. As in other experiments, there is a relatively large scatter because of variations in the samples and their previous excitation history. A curve was fit to the experimental

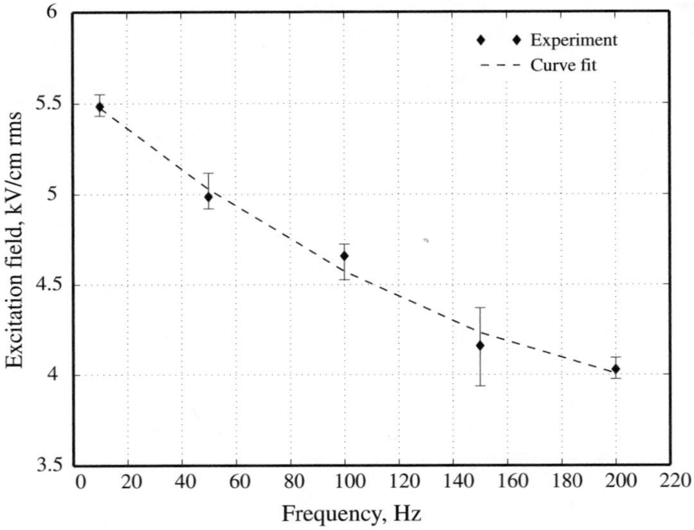

Figure 2.26. Variation of depoling field with excitation frequency.

variation of depoling field with frequency.

$$\mathbb{E}_{dep} = 2.292 \times 10^{-5} f^2 - 1.255 \times 10^{-2} f + 5.6 \tag{2.112}$$

where $\mathbb{E}_{dep}$ is the AC RMS depoling field in kV/cm and f is the excitation frequency in Hertz. Figure 2.27(a) shows the effect of depoling on the actuator strain response. The PZT is excited at 4.7 kV_{rms}/cm at 100 Hz. This is just at the depoling field, and the PZT usually takes a few seconds to depole, during which the strain-field hysteresis loop transforms slowly into the "butterfly loop," which indicates that the sample has depoled. This process is accompanied by a rapid increase in current drawn. It can be seen that the area under the depoled loop is more than the area under the nondepoled loop, indicating that the energy losses are much larger in the case of the depoled actuator.

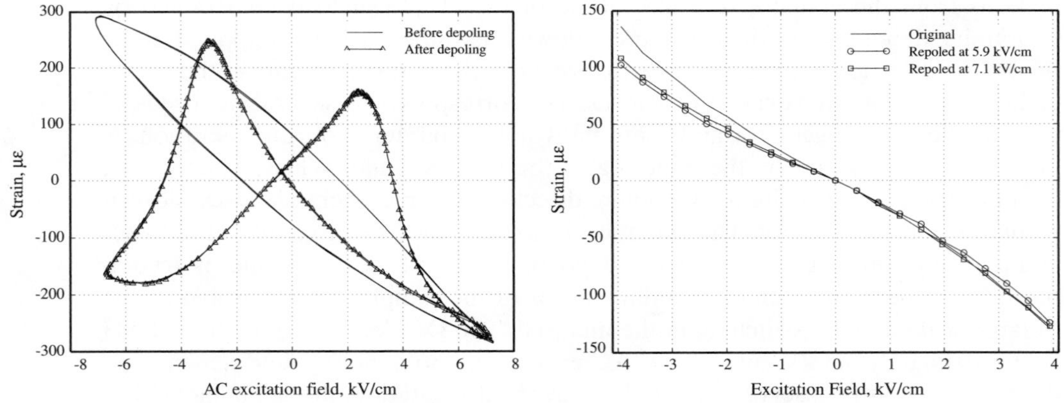

(a) Comparison of depoled and non-depoled strain response

(b) Repoling effectiveness

Figure 2.27. Depoling behavior.

However, the application of an electric field along the initial direction of polarization reorients the domains along the poling direction, thus reversing the depoling action. The sample was depoled and then repoled by exposing it to a DC electric field of 5.9 kV/cm and 7.1 kV/cm for about 5 minutes. The free-strain curves before and after repoling are shown in Fig. 2.27(b). Although these repoling fields are much less than the initial poling field of the ceramic, which is usually on the order of 12 kV/cm, it is possible to recover most of the performance of the actuator. This may be useful in case of accidental depoling of actuators in smart systems. It was also observed that it is possible to repole the samples by means of exciting them with an AC field that is approximately 90% of the AC depoling field. An interesting observation during this kind of repoling is that after the actuator has depoled, it is necessary to shut off the excitation and let the actuator sit for 3–4 minutes before applying the repoling field. If this is not done, the repoling is not effective. Note that it is more difficult to repole actuators bonded onto structures because the tensile stresses created in the actuator by the application of the repoling field tend to impede domain reorientation. Care should be taken not to exceed the tensile-failure stress of the piezoceramic, which will result in cracking of the bonded actuator.

During the poling process, due to significant alignment of domains within the crystallites, there is a permanent change in the dimensions of piezoceramics. For PZT compositions, the ceramic increases in length in the poling direction (z-direction) by 0.47% and decreases in length in all directions perpendicular to this by about 0.20%. Conversely, barium titanate undergoes approximately one-half the distortion of PZT, typically experiencing strains of 0.11% and 0.046%, respectively, in the poling and normal directions. There is more alignment of domains toward applied field in PZT than in barium titanate, resulting in a larger piezoelectric effect in PZT compositions. An applied stress alone cannot polarize a ceramic material; however, stress can depolarize the material. Applied stress in conjunction with field (parallel to field) will either help (tensile stress) or impede (compressive stress) the poling process. In a similar way, applied stress normal to the poling field will either help the poling process (compressive stress) or impede the poling process (tensile stress).

2.7.4 Power Consumption

The prediction of power consumed by the system with bonded piezoactuators is an important part of designing an adaptive structure. Piezoelectric actuators, as demonstrated in subsequent sections, behave as capacitive loads. As a result, special power amplifiers, capable of delivering large currents, are required to drive practical systems incorporating these actuators. The large currents result in excessive heat generation and are a cause of concern in designing rotating actuation systems with slip-rings for power transmission. In this section, the power consumption of piezoelectric actuators is discussed for both bonded and free actuators, and a theoretical model is experimentally validated. Some methods of reducing the current drawn from the power amplifier are described.

Electromechanical Impedance Approach

The net impedance of an actuator bonded to a structure can be divided into two parts: a purely electrical impedance and a purely mechanical impedance. The energy supplied by the power source driving the actuator appears as an increase in electrical energy of the actuator, an increase in strain energy of the actuator and structure, electrical and mechanical losses, and any work output from the structure. It is therefore convenient to treat the impedance of the actuator and structure as a net electromechanical impedance seen by the power source. Once all the mechanical and electrical impedances have been lumped into an effective electromechanical impedance, it becomes easy to calculate the current drawn and, therefore, the power requirements of the actuator-structure combination.

Electrical Impedance of a Free Actuator

A free piezoelectric actuator, by virtue of its physical configuration, behaves primarily as a capacitive load. In the case of a piezoceramic sheet actuator, it can be treated as a parallel plate capacitor.

For an ideal capacitor, the electrical impedance is given by

$$Z = \frac{1}{j\omega C} \tag{2.113}$$

where ω is the angular frequency of the applied field and C is the capacitance in Farads, which for an ideal parallel plate capacitor is given by

$$C = \frac{eA}{t} \tag{2.114}$$

where e is the electric permittivity, A is the area of the plate, and t is the distance between the plates, which in this case is the thickness of the piezoceramic sheet. If a sinusoidal voltage is applied to the capacitor, the current drawn leads the voltage by exactly 90°. Real capacitors with dielectric media, however, have energy losses. These losses are due to conduction currents in the dielectric as well as molecular friction opposing the rotation of dipoles in the material. This causes the current to lead the voltage by a phase angle δ of less than 90°. The non-ideal capacitor is usually modeled by a simplified equivalent circuit incorporating a shunt resistance in parallel with an ideal capacitor. The energy losses appear as Ohmic heating in the shunt resistance. The dissipation factor, given by $\tan\delta$, is therefore a measure of the energy loss in the capacitor and, consequently, the power consumed by the actuator. The impedance of a non-ideal capacitor can be given in terms of a complex electric permittivity [32]

$$\bar{e} = e_o k' - j e_o k'' = \left(e_o - j e \frac{k''}{k'}\right) k' \tag{2.115}$$

where e_o is the permittivity of free space, k' is the relative permittivity of the dielectric, and

$$\tan\delta = \frac{k''}{k'} \tag{2.116}$$

From Eqs. 2.114, 2.115, and 2.116, the electrical impedance of the piezoceramic sheet can be expressed as

$$Z = \frac{t}{j\omega e_{33}^{\sigma}(1 - j\tan\delta)A} \tag{2.117}$$

It is well known that the electric permittivity and the dissipation factor of piezoceramics increase with increasing field. This information is essential for predicting

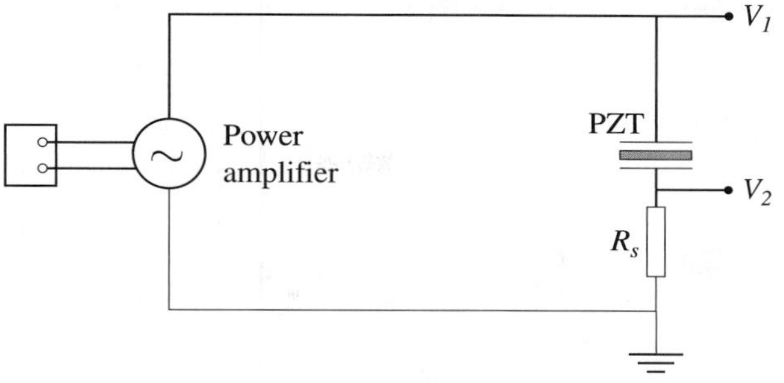

Figure 2.28. Measurement of electromechanical impedance.

the power consumption of systems incorporating piezoceramic actuators. Therefore, it is important to measure the impedance of the piezoeceramic sheets under realistic operating conditions.

Measurement of Actuator Impedance

The impedance of electrical devices is typically measured using an impedance analyzer, or an LCR meter. However, these instruments measure impedance either by measuring the current drawn when a known sinusoidal voltage is applied to the device or by finding the resonant frequency of the circuit using a frequency sweep. In general, the instrument applies a small voltage to the device, in the range of several millivolts to a few volts, either at a specific frequency (1kHz in the case of most LCR meters) or over a range of frequencies. This testing procedure is sufficient for devices that are close to "ideal," for example, a capacitor in which the permittivity of the dielectric is constant with the magnitude and frequency of the applied field. However, in the case of piezoceramics, it is known that the material constants such as electric permittivity are highly dependent on the operating conditions, such as applied electric field and mechanical stress. Therefore, to measure the actual impedance of the actuator, or of the actuator-structure combination, it is essential to maintain the same electrical and mechanical boundary conditions as in the intended application.

A simple way of achieving this is shown in Fig. 2.28. A function generator is connected to a power amplifier that drives the PZT actuator at the intended operating voltage. The PZT actuator could be mechanically unconstrained, in which case the measured impedance would be the electromechanical impedance of the actuator alone; or, it could be bonded to a structure, in which case the measured impedance would be the electromechanical impedance of the actuator-structure combination. A precision-sensing resistance (R_s) is connected between the negative electrode of the PZT and the ground of the circuit. The voltage output from the power amplifier, V_1, and the voltage across the sensing resistance, V_2, are both measured by a data-acquisition system. Typically, most data-acquisition systems have the capability to record both magnitude and phase information. The current passing through the circuit is calculated as

$$i_{circuit} = \frac{V_2}{R_s} \tag{2.118}$$

The voltage across the PZT actuator is

$$V_{PZT} = V_1 - V_2 \tag{2.119}$$

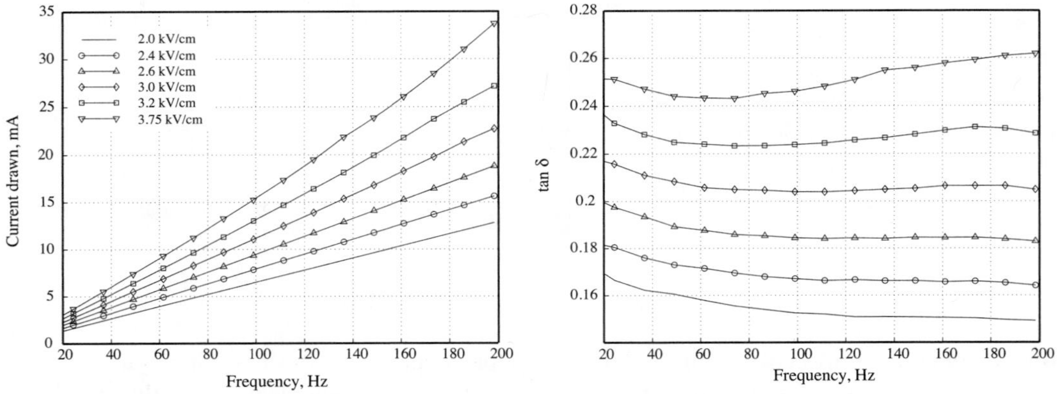

(a) Current drawn (b) Calculated $\tan\delta$

Figure 2.29. Experimentally measured current and $\tan\delta$ for a free piezoelectric sheet.

and the impedance of the actuator, Z_{act}, is given by

$$Z_{act} = \frac{V_{PZT}}{i_{circuit}} \tag{2.120}$$

Alternatively, the magnitude and phase of the impedance can be calculated in a simple way by capturing the two voltage waveforms on an oscilloscope. Using a data-acquisition system, the measurement can be automated, and the impedance over a range of frequencies can be measured using a swept sinusoid from the function generator. Note that a high impedance probe, or a potential divider, is required to measure the voltage V_1, which may typically be in the range of hundreds of volts. However, because the sensing resistance R_s is small the voltage V_2 is small, and can be measured directly.

The measured current drawn and the calculated dissipation factor for a PZT-5H sheet (of dimensions 2 in long, 1 in wide, and 0.01 in thick) is shown in Fig. 2.29. Equating the impedance form Eq. 2.117 to the value of impedance calculated from the experimentally measured voltage and current, the variation of e_{33}^{σ} and $\tan\delta$ with field can be generated and is shown in Fig. 2.30. Curves are fit to the experimental data, and the variation of e_{33}^{σ} and $\tan\delta$ with field can be calculated as:

$$K_e = 5.3187\mathbb{E}^2 - 5.9754\mathbb{E} + 7.32 \tag{2.121}$$

$$\tan\delta = 0.0662\mathbb{E} + 0.0376 \tag{2.122}$$

where K_e is the percent increase in e_{33}^{σ} and $\mathbb{E}$ is the electric field in kV_{rms}/cm.

Electromechanical Impedance of the Actuator

A short description of the derivation of the combined electromechanical impedance is given here [33]. The structural impedance of the active system is derived by considering a PZT actuator deforming along its length only (1-direction), driving a single degree of freedom–spring-mass damper system. For an actuator of length l_c, width b_c and thickness t_c, and with an elastic modulus $E_{11}^{\mathbb{E}}$, the force exerted is given by

$$F = K_A l_c(\epsilon_{mech} - \Lambda) \tag{2.123}$$

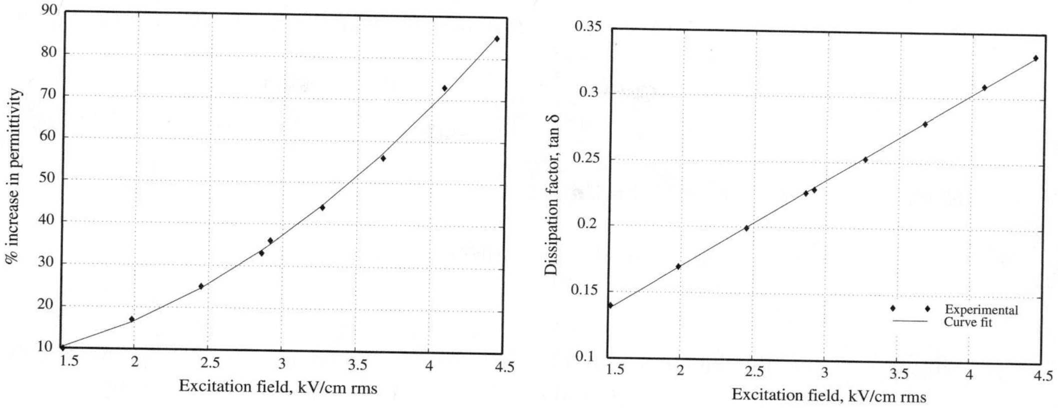

(a) Dielectric Constant (b) Dissipation Factor

Figure 2.30. Variation in piezoelectric material constants with applied electric field.

where K_A is the static stiffness of the PZT, given by $E_{11}^{\mathbb{E}} b_c t_c / l_c$, Λ is the free strain, which is defined as $d_{31} V / t_c$, and ϵ_{mech} is the mechanical strain of the structure at the actuator location. For a PZT sheet being excited along the 3-axis (or z-axis), assuming it deforms only along the 1-axis (or x-axis), the constitutive relations (Eqs. 2.70 and 2.71) can be written as

$$\epsilon_1 = \overline{s}_{11}^{\mathbb{E}} \sigma_1 + d_{31} \mathbb{E}_3 \tag{2.124}$$

$$D_3 = d_{31} \sigma_1 + e_{33}^{\sigma} \mathbb{E}_3 \tag{2.125}$$

In these equations, ϵ_1 is the strain, σ_1 is the stress, and $\overline{s}_{11}^{\mathbb{E}}$ is the complex compliance. The equation of motion for a PZT sheet vibrating in the x-direction is given by

$$\rho \frac{\partial^2 u}{\partial t^2} = \overline{Y}_{11}^{\mathbb{E}} \frac{\partial^2 u}{\partial x^2} \tag{2.126}$$

where ρ is the mass density (kg/m^3) and $\overline{Y}_{11}^{\mathbb{E}}$ is the complex modulus given by

$$\overline{Y}_{11}^{\mathbb{E}} = E_{11}^{\mathbb{E}} (1 + j\eta) \tag{2.127}$$

where η is the mechanical loss factor of the PZT. The complex electric permittivity $\overline{e}_{33}^{\sigma}$ is given by

$$\overline{e}_{33}^{\sigma} = e_{33}^{\sigma} (1 - j\eta) \tag{2.128}$$

Assuming a solution to Eq. 2.126 of the form

$$u = (a_1 \sin kx + a_2 \cos kx) e^{j\omega t} \tag{2.129}$$

where $k^2 = \omega^2 \rho / \overline{Y}_{11}^{\mathbb{E}}$, and applying the appropriate boundary conditions, it is possible to derive expressions for constants a_1 and a_2. A similar derivation can be performed to find the impedance of an actuator (assuming both electrodes shorted) under a constant-force excitation. This gives an actuator mechanical impedance expressed as

$$Z_A = -\frac{K_A(1+\eta j)kl_c}{\omega \tan(kl_c)} j \tag{2.130}$$

Mechanical Impedance of the Structure

The mechanical impedance of the structure is defined as

$$F = Z\dot{x} \tag{2.131}$$

where x is the displacement of the actuator along its length (1-direction). The mechanical impedance of the beam in bending (from Eq. 2.131) is given by

$$Z = \frac{4}{(t_b + t_c)^2} \frac{M}{(\theta_2 - \theta_1) j\omega} \tag{2.132}$$

where M is the actuation moment, θ_2 and θ_1 are the beam slopes at the ends of the actuator, and t_b is the beam thickness. In the theoretical validation, the actuation moment M and the beam slopes θ_1, θ_2 are calculated using the Euler-Bernoulli model.

Electromechanical Impedance of the Actuator-Structure Combination

Using the expressions for the mechanical impedances of the actuator and the structure, and using the derived constants a_1 and a_2, the assumed displacement (Eq. 2.129) can be solved. This is then substituted in the constitutive relations (Eqs. 2.124 and 2.125) to obtain the value of the electric displacement D_3. The current is defined as

$$I = \dot{q} = j\omega \iint D_3 dx dy \tag{2.133}$$

The final expression for the consumed current is

$$I = j\omega \mathbb{E} b_c l_c \left(\frac{d_{31}^2 \overline{Y}_{11}^{\mathbb{E}} Z_A \tan(kl_c)}{(Z + Z_A)kl_c} + \bar{e}_{33}^{\sigma} - d_{31}^2 \overline{Y}_{11}^{\mathbb{E}} \right) \tag{2.134}$$

Note that for a free PZT actuator, although the actuator impedance Z_A is finite, the impedance of the structure, Z, is zero. Also, the factor $\tan(kl_c)/(kl_c)$ is approximately equal to unity. The current drawn by the free PZT (Eq. 2.134) then reduces to

$$I_{\text{freePZT}} = j\omega \mathbb{E} b_c l_c \bar{e}_{33}^{\sigma} \tag{2.135}$$

From this, the impedance of the free PZT can be written as

$$Z_{\text{freePZT}} = \frac{V}{I_{\text{freePZT}}} = \frac{t_c}{j\omega \mathbb{E} b_c l_c \bar{e}_{33}^{\sigma}} \tag{2.136}$$

Note that this expression for the impedance of the free PZT is the same as the expression derived considering the PZT to be a lossy capacitor (Eq. 2.117), considering $\tan\delta \approx \delta$. The experimentally measured current consumed by a free actuator and a pair of actuators bonded to a beam is compared with that predicted using Eq. 2.134. The variation of e_{33}^{σ} and $\tan\delta$ given by Eqs. 2.121 and 2.122 is also incorporated in the theoretical predictions.

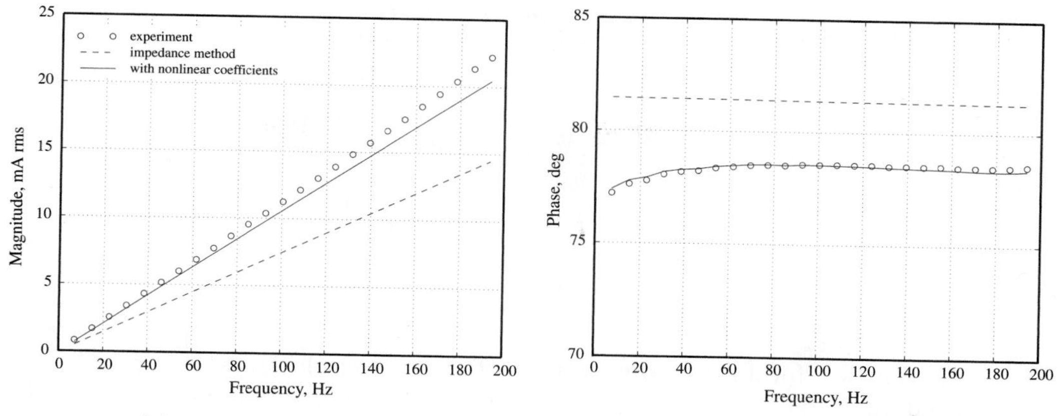

(a) Magnitude of current (b) Phase of current

Figure 2.31. Current consumption predictions for a free actuator, 3.0 kV_{rms}/cm.

Experiments were performed on a free actuator and on a pair of actuators bonded to a cantilevered aluminum beam of dimensions 12 in $\times$ 1 in $\times$ 0.032 in. The comparision of experimental and theoretically predicted current consumption is shown in Figs. 2.31 and 2.32. The predicted values show very good agreement with experiment when the variation of e_{33}^{σ} and $\tan\delta$ is taken into account, and the agreement is poor when these parameters are assumed constant. This emphasizes the importance of incorporating these nonlinearities when attempting to predict the power consumption of such actuator systems.

Reducing the Power Consumption of PZT Actuators

Due to the highly capacitive nature of the actuators, although the actual energy dissipated in the actuator is small, a large current is drawn from the power amplifier driving it. This makes the driving circuit bulky and inefficient, and it poses a challenge to compact smart systems with embedded electronics. The problem becomes even

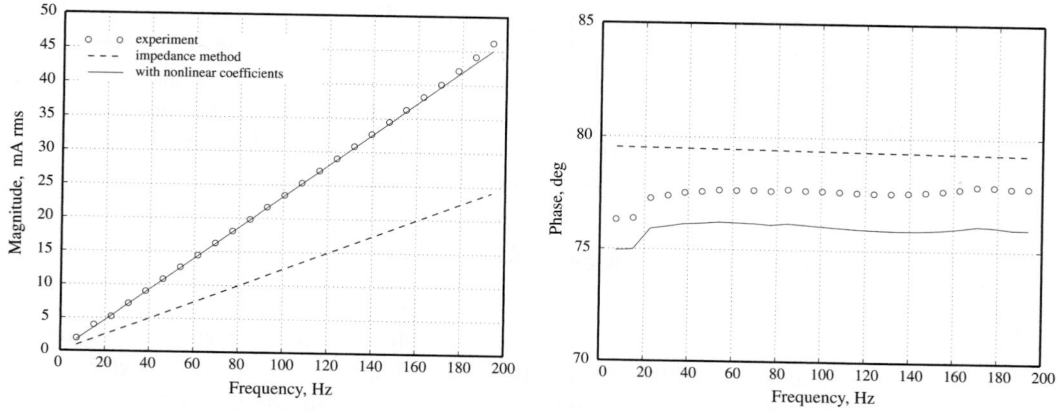

(a) Magnitude of current (b) Phase of current

Figure 2.32. Current consumption predictions for a pair of actuators bonded onto a beam, 3.0 kV_{rms}/cm.

Table 2.9. *Electric and mechanical analogs*

Electrical Quantity	Mechanical Quantity
Charge	Displacement
Voltage	Force
Current	Velocity
Capacitance	Compliance
Inductance	Inertia
Resistance	Damping
Electrical Impedance	Mechanical Impedance

more critical for a rotary-wing smart system in which the transfer of power from fixed frame to rotating frame poses serious restrictions on the slipring unit [34].

Several approaches to address this issue can be found in the literature and can be broadly grouped under two methods: those that involve the design of efficient driving electronics to supply power to the actuator, and those that modify the effective impedance of the actuator by adding components to the actuator circuit. For example, in the first approach, special Pulse Width Modulated (PWM) amplifiers can be designed to decrease the power dissipated in the amplifier and make it more compact than conventional amplifiers. In the past, high power PWM amplifier designs were proposed to drive piezoelectric actuators and electrostrictive actuators [35, 36]. However, these amplifiers do not recover the necessary energy to charge the actuator capacitance, that is wasted on the negative half-cycle of excitation. Hybrid techniques have also been suggested [37], wherein the charge used to displace the actuator is recirculated within the amplifier. In the second approach, the effective impedance of the actuator is changed by adding passive or semi-active components to the actuator driving circuitry, through the modification of the driving circuit using an additional inductor in a series or parallel arrangement [38, 39]. Although the concept is theoretically feasible, the size of the correcting inductance required for practical applications can become prohibitive.

Many of the previous test characteristics were focused on a specific piezoceramic (PZT-5H). One should expect some variation of characteristics among piezoceramics from other manufacturers. Also note that PVDF can have quite different characteristics than PZT. Sathiyanarayan et al. [2] carried out systematic material tests on PVDF sheets and showed nonlinear and time-dependent electromechanical behavior. PVDF sheets exhibited a strong dependence on strain rate in the transverse direction, compared to the longitudinal direction. The biaxially stretched sheets showed transverse isotropy. Dynamic moduli were found to be insensitive to pre-stressing but were sensitive to frequency of oscillation.

2.8 Equivalent Circuits to Model Piezoceramic Actuators

Piezoelectric material characteristics, as well as structural properties, can be represented in terms of an equivalent electric circuit. Mechanical properties are expressed in terms of analogous electrical quantities (see Table 2.9). This provides a convenient way to analyze the effect of the piezoelectric material in conjunction with the rest of the electric circuit. Therefore, this approach finds wide application in systems in which the piezoelectric material is used as a sensor, in which the transduction of some mechanical inputs into electrical quantities, as well as the signal conditioning and

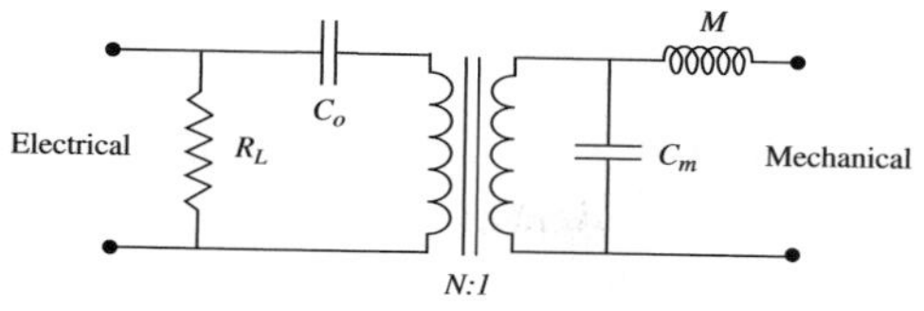

(a) Circuit 1

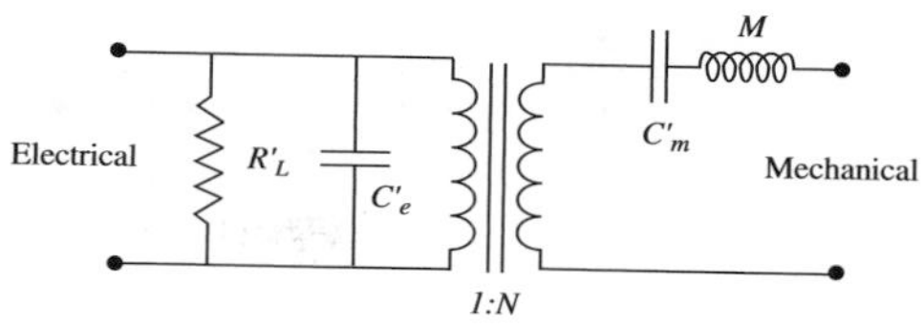

(b) Circuit 2

Figure 2.33. Equivalent circuits to model piezoelectric actuators.

output, is of interest. Numerous equivalent circuit models can be found in the literature, each consisting of a combination of resistors, capacitors, and inductors. The choice of a particular circuit depends on the operating regime of the piezoelectric material; for example, whether the frequency of operation is close to resonance. If low-frequency operation away from resonance is of interest, then the equivalent circuit can be considerably simplified by eliminating the inductive elements. The capacitances represent the dielectric properties of the material; note that a piezoelectric element acts predominantly as a capacitor. The resistive elements are included to model the lossy nature of the dielectric (i.e., its complex electric permittivity). Often, these properties are provided by the manufacturer at low values of electric field, suitable for sensor applications. For actuator applications, the piezoelectric material is subjected to a very high electric field to maximize the strain output. Therefore, the nonlinear variation of electric permittivity with electric field must be incorporated in the elements of the equivalent circuit.

Two possible equivalent electric circuits are shown in Fig. 2.33. The mechanical terminals represent mechanical energy transfer to or from the piezoelectric element. The transformer symbol represents an ideal electromechanical transformer (voltage to force and vice versa). For example, in an equivalent circuit, current is analogous to velocity and vice versa. The transformer ratio N is related to the electromechanical coupling efficiency of the material. Both of the circuits are equivalent and it is a matter of convenience to apply either one to a particular problem. At higher frequencies, one needs to add additional lumped elements to the circuit to represent the dynamic behavior.

Figure 2.34 shows an equivalent circuit for a piezoelectric sensor. The piezoelectric element can be treated as either a charge source or a voltage source, along with a capacitance. The inductance in the circuit is to incorporate the mechanical elements and to simulate high-frequency behavior. The voltage V is the source that is directly proportional to the applied force, pressure, or strain. The output signal is obtained from the source after passing through the equivalent circuit. For example, in a piezoelectric accelerometer, the inductance L_m represents the seismic mass of the sensor, the capacitance C_e is inversely proportional to the mechanical elasticity of the sensor, C_o represents the inherent static capacitance of the transducer, and R_i is the leakage resistance of the sensor element. If the sensor is connected to an output load resistance, then this will form a parallel circuit with the leakage resistance. Figure 2.35 shows the response of this sensor as a function of frequency for a sinusoidal forcing. The flat region of the frequency response is typically the usable region, between the high-pass cutoff (to avoid leakage) and the resonant peak. Note that such a sensor is not capable of yielding a purely static output (no DC response).

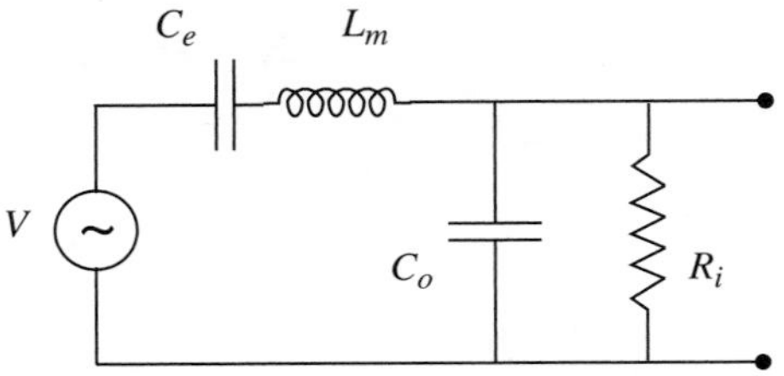

Figure 2.34. Equivalent circuit of a piezoelectric sensor.

2.8.1 Curie Temperature

For each piezoceramic material, there is a specific temperature, called the Curie temperature, above which the material suffers a permanent and complete loss of its piezoelectric characteristics. For practical applications, the operating temperature must be limited to some value substantially below the Curie temperature. At elevated temperatures, the aging process is accelerated, electric losses increase, and the maximum safe stress decreases.

2.8.2 Cement-Based Piezoelectric Composites

To overcome the deficiency in compatibility of traditional piezoelectric materials (piezoceramic, piezo-polymer, and polymer-based composite) with civil engineering materials such as concrete, a PZT/sulfoaluminate cement-based composite has been developed using a compression technique [40]. The piezoelectric properties of the 0–3 cement-based piezoelectric composites are improved by increasing the poling field (>4 kV/mm) and poling time (>45 minutes). The piezoelectric characteristics of the composite are nonlinear functions of the PZT content.

2.8.3 Shape Memory Ceramic Actuators

Field-induced phase transitions in electroceramics can cause large strains. Certain classes of material, due to the metastability of some phases, are capable of retaining a residual strain even after the electric field is completely switched off. For example, lanthanum – and niobium–doped piezoceramics exhibit anti-ferroelectric (AFE) to ferroelectric (FE) phase transitions (Fig. 2.36) and show shape-memory behavior. Examples of these materials are, respectively, lead lanthanum zirconate stannate titanate (PLZST) [41] and lead niobium zirconate stannate titanate (PNZST) [42]. Strains as high as 0.6% have been reported to occur during the phase transition. Unlike ferroelectrics, anti-ferroelectrics do not exhibit any macroscopic polarization. The induced strain depends on many factors that include stress, actuation frequency,

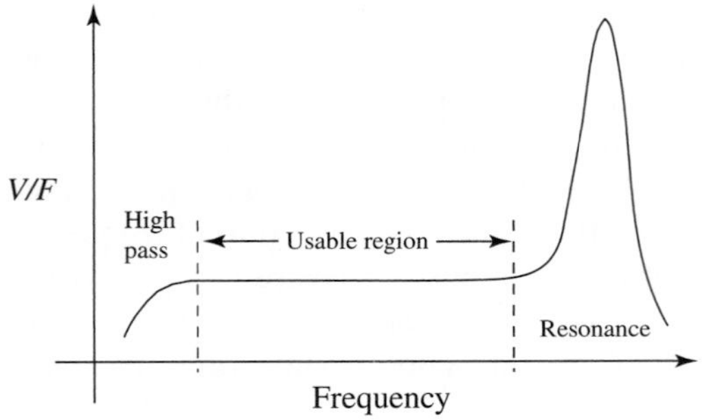

Figure 2.35. Frequency response of a piezoelectric sensor, output voltage V for a sinusoidal forcing of magnitude F.

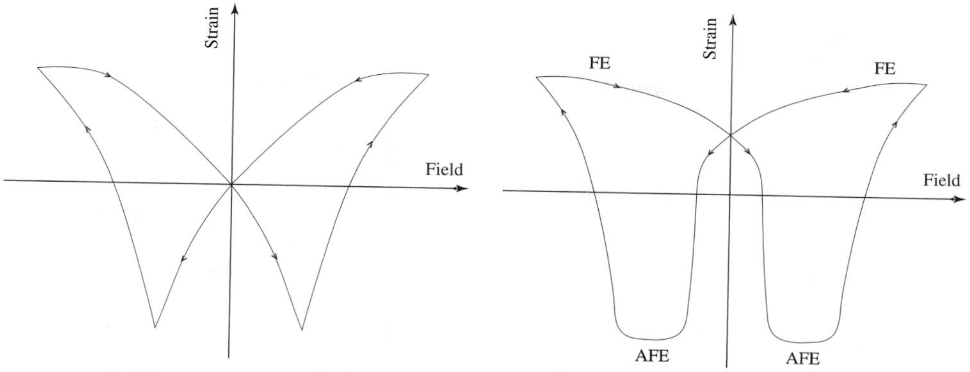

(a) Ferroelectric phase

(b) Anti-ferroelectric to ferroelectric phase transition

Figure 2.36. Field-induced phase transition from anti-ferroelectric (AFE) to ferroelectric (FE).

and temperature [43]. If the ferroelectric phase is stable, the material behaves like a conventional piezoceramic.

2.9 Piezoelectric Sensors

Piezoelectric elements are commonly used in smart structural systems as both sensors and actuators [15]. A key characteristic of piezoelectric elements is the utilization of the converse piezoelectric effect to actuate the structure in addition to the direct effect to sense structural deformation. Typically, piezoceramics are used as actuators and polymer piezo films are used as sensing materials. It is also possible to use piezoceramics for both sensing and actuation, as in the case of self-sensing actuators [44]. Many researchers have used piezoceramic sheet elements as sensors in controllable structural systems [45] and also in health-monitoring applications [46]. Most of these applications rely on the relative magnitudes of either the voltage or rate of change of voltage generated by the sensor, or the frequency spectrum of the signal generated by the sensor. Several investigations have been carried out on discrete piezoelectric sensor systems [45], active control of structures with feedback from piezoelectric sensors [47], and collocated sensors and actuators [44, 48]. It has been shown [49] that piezoceramic strain transducers have a linear response up to the pico-strain level (Fig. 2.37), and their strain sensitivity is several orders of magnitude larger than a conventional resistive strain gauge of similar dimensions.

Piezoelectric strain rate sensors have been investigated in references [50] and [51], wherein their superior noise immunity compared to differentiated signals from conventional foil gauges has been demonstrated. The correlation between the piezoelectric gauge reading and the resistive gauge measurement is quite good; however, the comparison was performed only at one frequency, 25 Hz.

This section discusses the behavior of piezoelectric elements as strain sensors. Strain is measured in terms of the charge generated by the element as a result of the direct piezoelectric effect. Strain measurements from piezoceramic (PZT-5H) and piezofilm (PVDF) sensors are compared with strains from a conventional resistive strain gauge and the advantages of each type of sensor are discussed, along with their limitations. The sensors are surface-bonded to a beam and are calibrated over

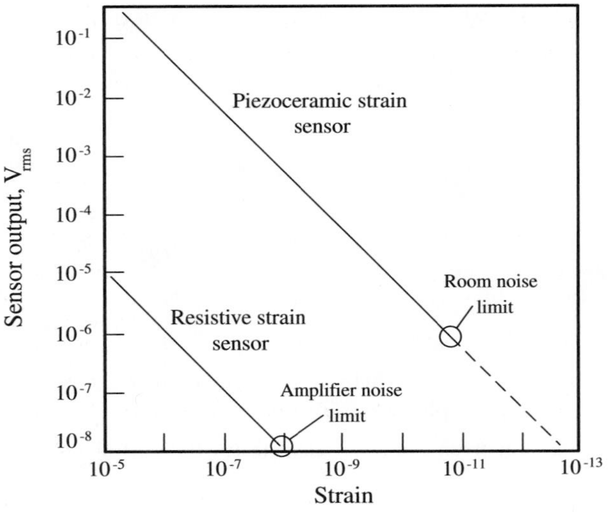

Figure 2.37. Comparison of piezoceramic and resistive strain sensors, adapted from Ref. [49].

a frequency range of 5–500 Hz. Correction factors to account for transverse strain and shear lag effects due to the bond layer are analytically derived and experimentally validated. The effect of temperature on the output of PZT strain sensors is investigated. Additionally, design of signal conditioning electronics to collect the signals from the piezoelectric sensors is described. The superior performance of piezoelectric sensors compared to conventional strain gauges in terms of sensitivity and signal-to-noise ratio is demonstrated.

In addition to the possibility of performing collocated control, such actuators/sensors have other advantages such as compactness, sensitivity over a large strain bandwidth, and ease of embeddability for performing structural health monitoring, as well as distributed active control functions concurrently. These features, combined with the extremely good signal-to-noise ratio of piezoelectric sensors, make them ideally suited for applications involving severe environments and small signals. A resistance-type strain gauge measures an average strain over its gauge length along a specific direction (transverse sensitivity is negligible). Conversely, a piezoceramic sensor measures average strain over its attached surface area and is not directional.

2.9.1 Basic Sensing Mechanism

A description of the basic piezoelectric mechanism is given in Section 2.5.2. The constitutive relation for a piezoelectric sensor (Eq. 2.32) can be written as

$$\boldsymbol{D} = \boldsymbol{d}^d \boldsymbol{\sigma} + \boldsymbol{e}^\sigma \mathbb{E} \tag{2.137}$$

A sheet of piezoelectric material poled across its thickness, as in Fig. 2.38(a), can be used to sense strain or strain rate in the 1-2 plane. The sensor generates a voltage across its electrodes that is measured by appropriate signal-conditioning electronics (Fig. 2.38(b)). In most applications, no electric field is applied to the sensor.

The voltage generated across the electrodes of the piezoelectric sheet is fundamentally due to an electric displacement, or charge, generated in the element as a result of the direct piezoelectric effect. For the case of the piezoelectric sheet, under

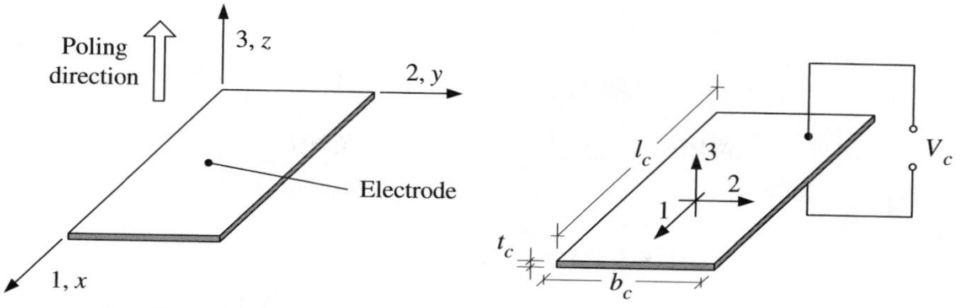

(a) Piezoelectric sheet (b) Measuring output voltage

Figure 2.38. Piezoelectric sheet as a sensor.

no external electric field, the electric displacement across the electrodes, D_3, is given by (Eq. 2.37)

$$D_3 = d_{31}\sigma_1 + d_{32}\sigma_2 + d_{33}\sigma_3 \tag{2.138}$$

Note that a piezoelectric sheet sensor cannot measure shear stresses. Because a sensor of this type is usually bonded onto a structure, $\sigma_3 = 0$, and only direct stresses in the 1-2 plane are measured. It is important to note that the stresses in Eq. 2.138 are the stresses in the piezoelectric sensor itself, not in the structure to which it is bonded. These stresses are caused by strains in the 1-2 plane transferred from the structure to the piezoelectric sheet, multiplied by the appropriate modulus of the piezoelectric, depending on the electrical-boundary conditions imposed by the sensing electronics. Therefore, the piezoelectric sensor is, in reality, a strain sensor, and can be used to measure strains on the surface of a structure.

The electric displacement is related to the generated charge by (Eq. 2.38)

$$q = \iint D_3 dA_3 = \iint D_3 dx dy \tag{2.139}$$

This charge, or an equivalent current, is collected by appropriate sensing electronics. Because the current is the rate of change of charge, measurement of the current will result in a sensor that measures the rate of change of strain. Details of these sensing methods are discussed in subsequent sections.

2.9.2 Bimorph as a Sensor

A bimorph can be used as a sensor to measure bending in response to external stimuli. Let us consider two identical piezoceramic sheets bonded in a parallel arrangement with their polarization axes (z-axes) in the vertical direction. A tip load P in the upward direction will cause bending of the bimorph, which can be measured in terms of the generated voltage.

Assuming Euler-Bernoulli beam bending, the stress across the thickness of the beam with moment of inertia I is:

$$\sigma_1(z) = -\frac{P(l_c - x)}{I} \tag{2.140}$$

Let the two plates of the bimorph be connected electrically in series, as shown in Fig. 2.39(a). The dots on the sides of the piezoelectric sheets indicate the electrode

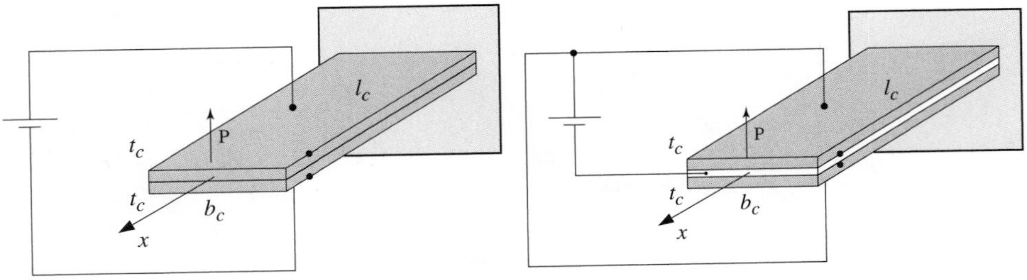

(a) Series connection (b) Parallel connection

Figure 2.39. Electrical connections for piezoelectric bender bimorphs.

of positive polarity. The electric displacement is given by (neglecting lateral effects)

$$D_3 = d_{31}\sigma_1 = -d_{31}\frac{P(l_c - x)}{I} \tag{2.141}$$

The charge generated is obtained by integrating the electric displacement over the electrode area

$$q = \iint_{\text{area}} D_3\, dx\, dy = -\frac{d_{31}t_c l_c^2}{2I}P \tag{2.142}$$

and the voltage is given by

$$V = \frac{q}{C_p} = -\frac{d_{31}t_c l_c^2}{2I}\frac{t_c}{e_{33}^{\sigma} l_c b_c} = -\frac{3}{2}\frac{d_{31}l_c P}{b_c t_c e_{33}^{\sigma}} \tag{2.143}$$

Substituting for the piezoelectric constant

$$g_{31} = \frac{d_{31}}{e_{33}^{\sigma}} \tag{2.144}$$

$$V = -\frac{3}{2}\frac{Pl_c}{b_c t_c}g_{31} \tag{2.145}$$

If the plates are bonded in a parallel arrangement such that their polarized axes face in opposite directions with a common mid-electrode (Fig. 2.39(b)), the voltage generated is given by

$$V = -\frac{3}{4}\frac{Pl_c}{b_c t_c}g_{31} \tag{2.146}$$

2.9.3 Signal-Conditioning Electronics

A piezoelectric sheet behaves like a capacitor with a large internal resistance (on the order of GΩ). When used as a sensor, the sheet generates a charge that appears as a voltage across its electrodes. In the case of a static strain, a DC voltage is generated across the electrodes of the sensor. Due to the large internal resistance of the sensor, this voltage remains on the electrodes for a substantial period of time but eventually leaks off. However, to accurately measure this voltage, the input impedance of the measurement device should be several orders of magnitude larger than the impedance of the piezoelectric sensor. Typically, electrometers with input impedances $>10^{14}\Omega$ can be used to measure these static voltages. Standard multimeters do not have sufficiently high input impedance either to provide an accurate measurement or to prevent the static charge from leaking off. Additionally, measurement of dynamic strains and the need for using standard data-acquisition systems pose further challenges to the use of piezoelectric sensors. Most oscilloscopes and data-acquisition systems have an input impedance of 1 MΩ. These issues necessitate the use of appropriate signal-conditioning electronics between the piezoelectric sensor and the measurement system.

The primary purpose of the signal-conditioning system is to provide a signal with a low output impedance while simultaneously presenting a very high input impedance to the piezoelectric sensor. There are several ways of achieving this. Although many designs of signal-conditioning electronics exist of varying complexity and accuracy, they can be divided into three fundamental groups: measurement of voltage, measurement of charge, and measurement of current. The voltage is measured using a voltage follower and is calibrated to yield the measured quantity, such as force (in a load cell) or acceleration (in an accelerometer). In the second approach, the charge is measured using a charge amplifier, resulting in a sensor capable of measuring strain. In the third approach, the current is measured using a transresistance amplifier, yielding strain-rate measurements. The charge amplifier is the most commonly used type of signal conditioning for commercial piezoelectric sensors. The behavior of the charge amplifier, including its frequency response, is described in detail in the next subsection, along with the operational concept of the voltage follower and transresistance amplifier.

Voltage Follower – Measurement of Voltage

A voltage follower provides a very high input impedance to the piezoelectric sensor. A schematic of the voltage follower circuit is shown in Fig. 2.40 and a detailed analysis of the circuit is presented by Dally et al. [52].

Considering a sensor of length l_c, width b_c, and thickness t_c (Fig. 2.38), the capacitance of the sensor is given by

$$C_p = \frac{e_{33}^{\sigma} l_c b_c}{t_c} \tag{2.147}$$

The relation between charge and voltage generated across the electrodes of the sensor is given by Eq. 2.39

$$V_c = q/C_p \tag{2.148}$$

Assuming only a uniaxial strain along the 1-direction, from Eqs. 2.138, 2.139, 2.147, and 2.39), the voltage generated by the sensor can be expressed as

$$V_c = \frac{d_{31} Y_c b_c}{C_p} \int_{l_c} \epsilon_1 dx \tag{2.149}$$

where Y_c is the Young's modulus of the piezoelectric material, depending on the electrical-boundary conditions of the sensor. In the case of the voltage follower, the sensor is directly connected to the noninverting input of the operational amplifier, which theoretically has an infinite input impedance. Therefore, the sensor exists in an open-circuit condition and $Y_c = E_{11}^{D}$.

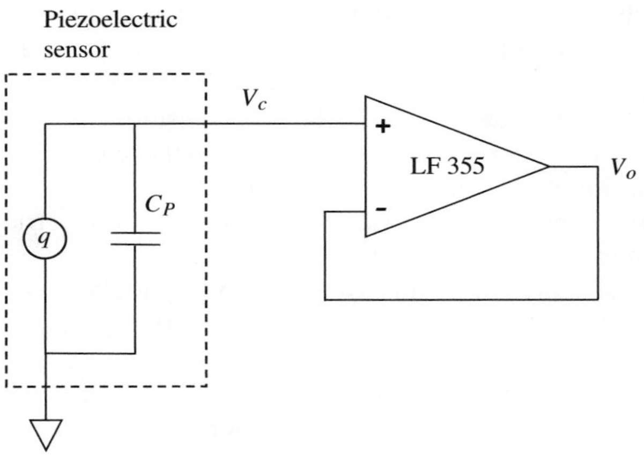

Figure 2.40. Voltage follower circuit.

Assuming the value of ϵ_1 to be averaged over the gauge length and defining a sensitivity parameter

$$S_q = d_{31} E_c^D l_c b_c \tag{2.150}$$

where E_c^D is the Young's modulus of the sensor material in open-circuit condition, the equation relating strain and voltage generated by the sensor is

$$\epsilon_1 = \frac{V_c C_p}{S_q} \tag{2.151}$$

The output voltage of the voltage follower, $V_o = V_c$, which can be calibrated to measure either the strain as described previously, or other quantities of interest such as force in piezoelectric load cells or acceleration in piezoelectric accelerometers. In real applications, the finite input impedance of the amplifier, capacitance of the lead wires, and bias currents of the operational amplifier are important issues to be considered in the operation of the circuit.

Charge Amplifiers – Measurement of Charge

The signal-conditioning circuit used to measure charge is called a, charge amplifier, [52, 53]. The circuit is shown in Fig. 2.41. A piezoelectric sensor can be modeled as a charge generator in parallel with a capacitance, C_p, equal to the capacitance of the sensor. The cables that carry the signal to the charge amplifier act collectively as a capacitance C_c in parallel with the sensor. The charge amplifier has several advantages [52]. First, the charge generated by the sensor is transferred onto the feedback capacitance, C_F. This means that once the value of C_F is known and fixed, the calibration factor is fixed, irrespective of the capacitance of the sensor. Second, the value of the time constant, which is given by $R_F C_F$, can be selected to give the required dynamic frequency range. Note, however, that there is always some finite leakage resistance in the piezoelectric material, which causes the generated charge to leak off. Therefore, although the time constant of the circuit can be made very large to enable operation at very low frequencies, it is not possible to determine a pure static condition. This basic physical limitation exists for all kinds of sensors utilizing the piezoelectric effect. Third, the effect of the lead wire capacitance, C_c, which is always present for any physical measurement system, is eliminated. This has the important

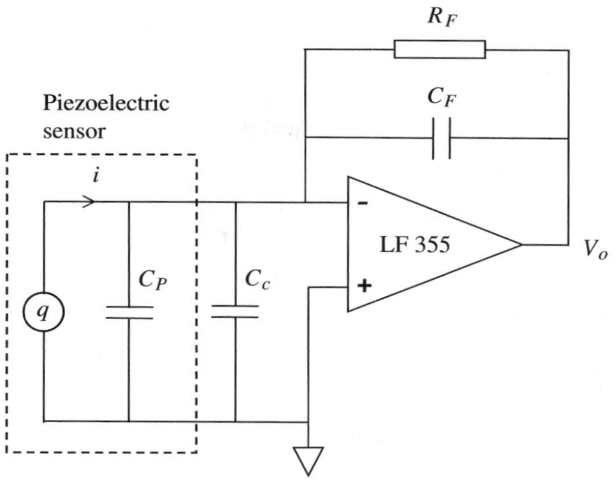

Figure 2.41. Charge amplifier circuit.

consequence that there are no errors introduced in the measurements by the lead wires.

Proceeding in a similar manner as described in the previous section, the current i (Fig. 2.41) can be expressed as

$$i = \dot{q} = d_{31} Y_c l_c b_c \dot{\epsilon}_1 \tag{2.152}$$

$$= S_q \dot{\epsilon}_1 \tag{2.153}$$

In this case, from the circuit diagram (Fig. 2.41), it can be seen that the bottom electrode of the sensor is connected to ground and the top electrode is at a "virtual" ground at the inverting input of the operational amplifier. Because both electrodes of the sensor are effectively grounded, the equivalent Young's modulus of the material is the short-circuit modulus, $Y_c = E_{11}^{\mathbb{E}}$.

Assuming ideal operational-amplifier characteristics, the governing differential equation of the circuit can be derived as

$$\dot{V}_o + \frac{V_o}{R_F C_F} = -\frac{S_q \dot{\epsilon}_1}{C_F} \tag{2.154}$$

which, for harmonic excitation, has the solution

$$\bar{V}_o = -\left(\frac{j\omega R_F C_F}{1 + j\omega R_F C_F}\right) \frac{S_q \bar{\epsilon}_1}{C_F} \tag{2.155}$$

$$= H(\omega)(-S_q^* \bar{\epsilon}_1) \tag{2.156}$$

where the quantities with a bar represent their magnitudes, and ω is the frequency of operation. The quantity S_q^* is called the "circuit sensitivity," representing the output voltage per unit strain input, and is given by

$$S_q^* = \frac{d_{31} E_c^D l_c b_c}{C_F} \tag{2.157}$$

The magnitude and phase of the gain $H(\omega)$ are plotted in Fig. 2.42(a) for different values of time constant, while keeping $R_F = 10M\Omega$. It can be seen that this

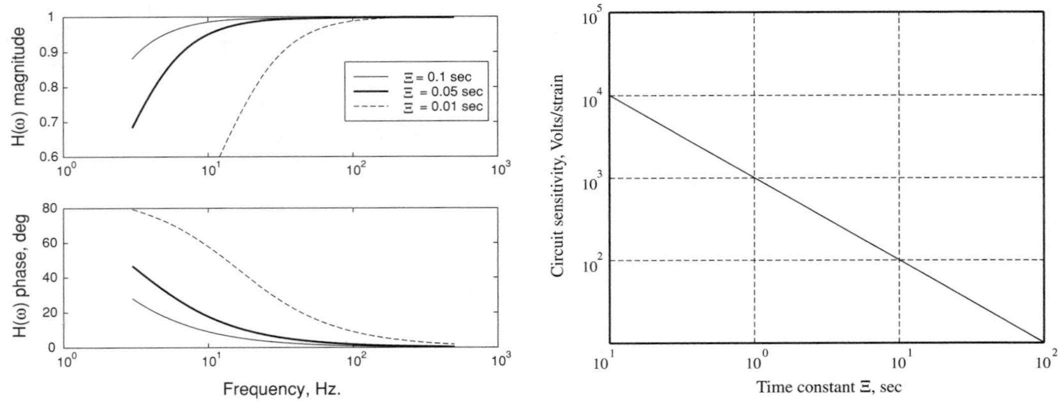

(a) Circuit-transfer function, for different time constants

(b) Trade-off between sensitivity and time constant

Figure 2.42. Circuit characteristics.

represents a high-pass filter characteristic, with a time constant $\Theta = R_F C_F$. As discussed before, the value of this time constant can be made very large for low-frequency measurements. Another point to be noted is that the sensitivity of the circuit depends inversely on the value of the feedback capacitance, C_F. For a given strain, as the value of C_F decreases, the output voltage V_0 will increase. However, this capacitance cannot be decreased indefinitely. From Eq. 2.155, it can be seen that the lower cutoff frequency of the circuit varies directly with C_F. This trade-off is shown in Fig. 2.42(b), assuming a fixed value of R_F of 10MΩ. Although larger time constants are possible with larger values of feedback resistance, it is not practical to increase the value of the feedback resistor R_F beyond the order of tens of megaohms due to various operational constraints. For a time constant on the order of 0.1 seconds, the circuit sensitivity is on the order of 10^4 volts/strain, which translates to an output voltage in the millivolt range in response to a 1-microstrain input. This sensitivity is achievable in a conventional resistive strain gauge only after extensive amplification and signal conditioning is incorporated. It can be seen that for larger time constants, the sensitivity drops, which means that as a pure static condition is approached, the output signal becomes weaker. Hence, it is not possible to measure pure static or quasi-static conditions. The major advantage of the charge amplifier comes from the fact that the circuit sensitivity, and therefore the output voltage, is unaffected by the capacitance of the sensor and stray capacitances such as the input cable capacitance. The output depends only on the feedback capacitor. This makes it easy to use the same circuit with different sensors without changing the calibration factor.

Transresistance Amplifiers – Measurement of Current

A simplified circuit diagram of a transresistance amplifier is shown in Fig. 2.43. The basic concept consists of sensing the current i from the piezoelectric sensor by measuring the voltage drop caused by it across a sensing resistor, R_S.

Assuming an ideal operational amplifier, the output voltage V_o is given by

$$V_o = iR_S = -\dot{q}R_S \tag{2.158}$$

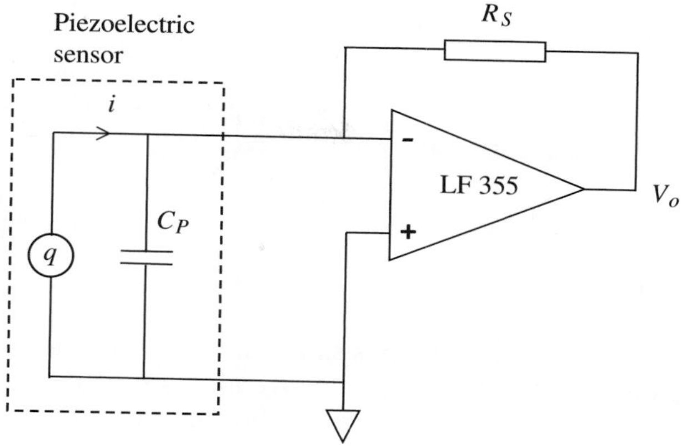

Figure 2.43. Transresistance amplifier circuit.

The current is given by Eq. 2.153. The output voltage becomes

$$V_o = -d_{31} Y_c l_c b_c \dot{\epsilon}_1 R_S \tag{2.159}$$

$$= S_q R_S \dot{\epsilon}_1 \tag{2.160}$$

From the circuit diagram (Fig. 2.43), it can be seen that the "virtual" ground results in a short-circuit condition. Therefore, $Y_c = E_{11}^{\mathbb{E}}$.

It is possible to define a sensitivity parameter as in the previous case. The circuit sensitivity, S_q^*, represents the output voltage per unit strain input and is given by

$$S_q^* = \frac{d_{31} E_{11}^{\mathbb{E}} l_c b_c}{C_F} R_S \tag{2.161}$$

2.9.4 Sensor Calibration

To correlate the measurements from the strain sensor with physical strain values, the sensor must be calibrated in a known strain field. This calibration process ensures that the correct factors are used when converting the measured voltage to physical strain. In addition, the effect of any correction factors required to compensate for phenomena specific to piezoelectric sensors can be quantified. Once the calibration procedure has been carried out and the effect of various parameters quantified, the sensor can be used to measure the strain in any installation under similar mounting conditions. An experimental setup and procedure used to calibrate the piezoelectric sensors, as well as a discussion of correction factors, is presented in the following subsections.

Experimental Setup

A dynamic beam-bending setup is used to calibrate the piezoelectric sensors. A pair of PZT sheets is bonded 20 mm from the root of a cantilevered aluminum beam of dimensions 280 × 11 × 1.52 mm and connected so as to provide a pure bending actuation to the beam. A conventional foil-type strain gauge is bonded on the beam surface at a location approximately 50 mm from the end of the actuators, and a piezoelectric sensor is bonded at the same location on the other face of the beam so that both sensors are exposed to the same strain field. A sketch of

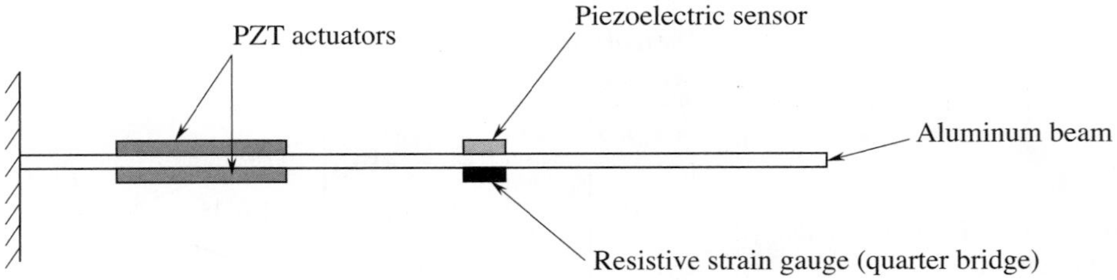

Figure 2.44. Calibration setup.

the experimental setup is shown in Fig. 2.44. The strain reading from the resistive gauge is recorded using a conventional signal-conditioning unit and the strain is calculated using standard calibration formulae. The output of the piezoelectric sensor is measured using conditioning electronics and converted to strain. A sine sweep is performed from 5–500 Hz and the transfer functions of the two sensors are compared.

Correction Factors

Note that the derivation of Eq. 2.151 was based on the assumption that only strain in the 1-direction contributed to the charge generated, the effect of other strain components was negligible, and that there is no loss of strain in the bond layer. In reality, however, a transverse component of strain exists, and there are some losses in the finite thickness bond layer. Hence, the value of strain as calculated by this equation is not the actual strain that is measured by the strain gauge. Correction factors are required to account for transverse strain and shear lag losses in the bond layer. These correction factors are discussed in the following subsections.

Poisson's Ratio Effect

The sensor on the beam is, in reality, exposed to both longitudinal and transverse strains. If the 1-direction is assumed to be aligned with the length of the beam and the 2-direction with the width of the beam, Eq. 2.138 can be rewritten as follows (assuming we are using a charge amplifier for measurement, which means the sensor is in a short-circuit condition)

$$D_3 = d_{31}E_{11}^{\mathbb{E}}\epsilon_1 + d_{32}E_{22}^{\mathbb{E}}\epsilon_2 \tag{2.162}$$

For a longitudinal stress, there will be a lateral strain due to Poisson's effect at the location of the sensor

$$\epsilon_2 = -\nu\epsilon_1 \tag{2.163}$$

where ν is the Poisson's ratio of the host structure material, which, in this case, is aluminum ($\nu = 0.3$). Hence, Eq. 2.151 can be rewritten as

$$\epsilon_1 = \frac{V_o}{K_p S_q^*} \tag{2.164}$$

where K_p is the correction factor due to Poisson's effect. For PZT sensors, it can be seen that

$$K_p = (1 - \nu) \tag{2.165}$$

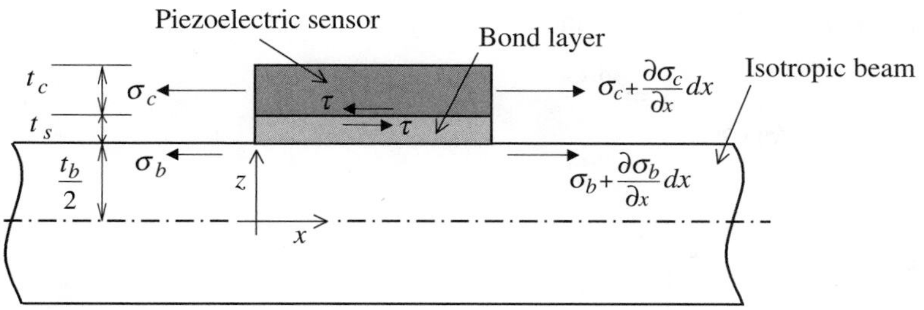

Figure 2.45. Forces and moments acting on the sensor.

for PVDF sensors, K_p is given by

$$K_p = \left(1 - \nu \frac{d_{32}}{d_{31}}\right) \tag{2.166}$$

This is a key distinction between piezoelectric sensors and conventional resistive gauges. The transverse sensitivity of a piezoelectric sensor is of the same order as its longitudinal sensitivity. However, for a conventional strain gauge, the transverse sensitivity is close to zero and is normally neglected. Hence, in a general situation, it is not possible to distinguish the principal strains of a structure using only one piezoelectric sensor. At least two sensors are required, constructed out of a piezoelectrically or mechanically orthotropic material. Therefore, this rules out the use of PZT sensors, in which both longitudinal and transverse strain measurements are required. For calibration, the transverse strain is known a priori, which enables the derivation of a correction factor.

Shear Lag Effect

The shear lag effect accounts for the loss in strain transmitted from the host structure to the sensor as a result of the finite stiffness of the bond layer. Consider a sensor of length l_c, width b_c, thickness t_c, and Young's modulus E_c bonded onto the surface of a beam of length l_b, width b_b, thickness t_b, and Young's modulus E_b. Let the thickness of the bond layer be t_s. Assuming the beam to be actuated in pure bending, the forces and moments acting on the beam can be represented as shown in Fig. 2.45.

Linear-strain distribution across the thickness of the beam is assumed, and the actuator thickness is considered small compared to the beam thickness. The strain is assumed constant across the thickness of the actuator. Force equilibrium in the sensor along the x direction gives

$$\frac{\partial \sigma_c}{\partial x} t_c - \tau = 0 \tag{2.167}$$

and moment equilibrium in the beam gives

$$\frac{\partial \sigma_b}{\partial x} + \tau \frac{3 b_c}{b_b t_b} = 0 \tag{2.168}$$

The strains can be related to the displacements by

$$\epsilon_c = \frac{\partial u_c}{\partial x} \tag{2.169}$$

$$\epsilon_b = \frac{\partial u_b}{\partial x} \tag{2.170}$$

$$\gamma = \frac{1}{t_s}(u_c - u_b) \tag{2.171}$$

where u_c and u_b are the displacements of the sensor and on the beam surface, respectively, and γ is the shear strain in the bond layer.

Substituting Eqs. 2.169–2.171 in Eqs. 2.167 and 2.168 and simplifying leads to the relation

$$\frac{\partial^2 \zeta}{\partial x^2} - \left[\frac{G}{E_c t_c t_s} + \frac{3 b_c G}{E_b b_b t_b t_s}\right]\zeta = 0 \tag{2.172}$$

where G is the shear modulus of the bond-layer material and ζ is defined as the quantity $(\epsilon_c/\epsilon_b - 1)$. Making the substitution

$$\Gamma^2 = \frac{G}{E_c t_c t_s} + \frac{3 b_c G}{E_b b_b t_b t_s} \tag{2.173}$$

leads to the governing equation for shear lag in the bond layer

$$\frac{\partial^2 \zeta}{\partial x^2} - \Gamma^2 \zeta = 0 \tag{2.174}$$

The general solution for this equation is

$$\zeta = A \cosh \Gamma x + B \sinh \Gamma x \tag{2.175}$$

with the boundary conditions

$$\text{at } x = 0 \qquad \zeta = -1 \tag{2.176}$$

$$\text{at } x = l_c \qquad \zeta = -1 \tag{2.177}$$

Solving these gives the complete solution as

$$\zeta = \frac{\cosh \Gamma l_c - 1}{\sinh \Gamma l_c} \sinh \Gamma x - \cosh \Gamma x \tag{2.178}$$

This variation is calculated both along the length and the width of the sensor, and the two effects are assumed to be independent, which means that effects at the corners of the sensor are neglected. The function is plotted in Fig. 2.46(a), along the length, for a PZT sensor of size 6.67 × 3.30 × 0.25 mm, and in Fig. 2.46(b) for a PVDF sensor of the same length and width but of a thickness 56 μm. The variations are plotted for different values of the bond-layer thickness ratio, $\Xi = t_s/t_c$, for both types of sensors. The values of Ξ are calculated by varying the bond-layer thickness for a constant sensor thickness. The PVDF sensor shows a much lower shear lag loss than the PZT sensor for a given bond-layer thickness ratio. This is due to the combined effect of lower sensor thickness and lower E_c in the case of PVDF in Eq. (2.173). As a result, the shear-lag effect is almost negligible for a PVDF sensor.

To quantify the effect of the shear lag, effective dimensions are defined along the length and width of the sensor such that the effective sensor dimensions are

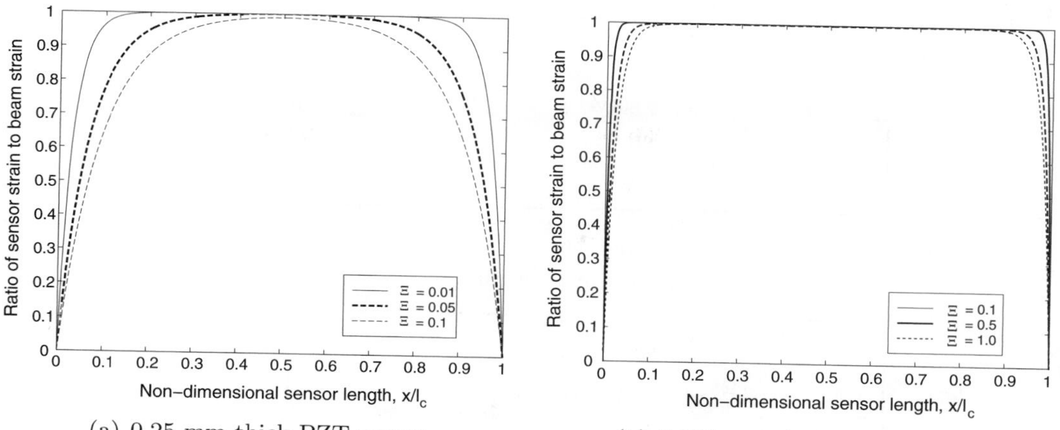

(a) 0.25 mm thick PZT sensor

(b) 0.056 mm thick PVDF sensor

Figure 2.46. Shear-lag effects along sensor length.

subjected to a constant strain, which is the same as the assumed strain on the beam surface. By doing this, the sensor is assumed to be of new dimensions, smaller than the actual geometrical dimensions, over which $\zeta = 0$ identically. The values of the effective length and width fractions, l_{eff} and b_{eff}, respectively, can be obtained by integrating the area under the curves in Fig. 2.46. For the sensor under discussion, which had a bond-layer thickness of 0.028 mm ($\Xi = 0.112$), the effective length fraction is 0.7646 and the effective width fraction is 0.4975. This means that only approximately 76% of the sensor length and 50% of the sensor width contribute to the total sensed strain. Because the whole geometric area of the sensor is no longer effective in sensing the beam-surface strain, these correction factors must be inserted in the calibration equation, Eq. 2.151, which becomes

$$\epsilon_1 = \frac{V_o}{K_b S_q^*} \tag{2.179}$$

where K_b is the correction factor to take care of shear-lag effects in the bond layer. The value of K_b is independent of the material properties of the sensor and is dependent only on its geometry. For both PZT and PVDF sensors, K_b is given by

$$K_b = l_{eff} b_{eff} \tag{2.180}$$

It should be noted here that for a PVDF sensor, the value of K_b is very close to unity (see Fig. 2.46(b)) and the shear-lag effect can be neglected without significant error). The final conversion relation from output voltage to longitudinal strain is

$$\epsilon_1 = \frac{V_o}{K_p K_b S_q^*} \tag{2.181}$$

Signal-to-Noise Ratio

Experiments were performed on the beam-bending setup as described previously. For the sine sweeps, the beam was actuated from 5–500 Hz. A conventional 350Ω resistive strain gauge was used, with a Micro Measurements 2311 signal-conditioning system. For the piezo sensor, a charge amplifier was built using high-input impedance LF355 operational amplifiers, with $R_F = 10\ M\Omega$ and $C_F = 10$nF.

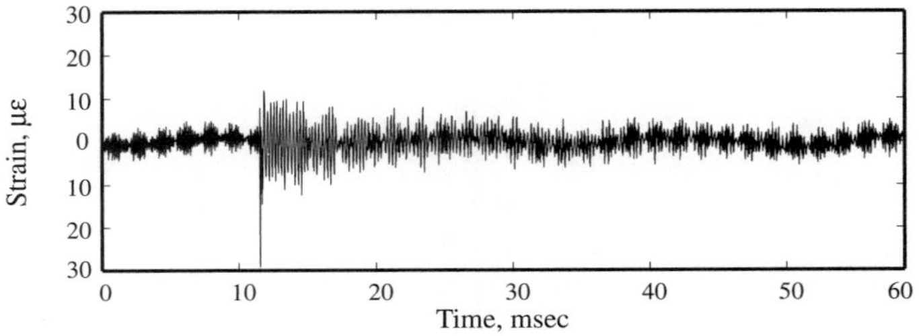

(a) Resistive strain gauge

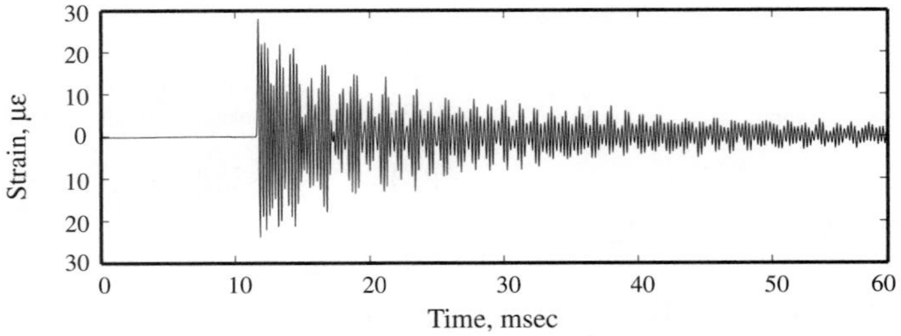

(b) PZT sensor

Figure 2.47. Resistive strain gauge and PZT sensor impulse response in the time domain.

A major advantage of using piezoelectric sensors as opposed to conventional resistive strain gages is their superior signal-to-noise ratio and high-frequency-noise rejection. Shown in Fig. 2.47 is the impulse response from both the conventional resistive strain gauge and a PZT strain sensor. The two readings were taken simultaneously after the beam was impacted at the tip. The responses are unfiltered and show the actual recorded voltages from the signal conditioners. Note the large-amplitude background noise in the resistive gage output and the much higher signal-to-noise ratio of the PZT strain gauge. The foil strain gage operates by sensing an imbalance in a Wheatstone bridge circuit, which is on the order of microvolts. Therefore, at low strain levels, the signal-to-noise ratio of resistive strain gages is quite poor. The superior signal-to-noise ratio of piezoelectric sensors makes them much more attractive in situations where there is a low strain or high noise level. This can be seen more clearly in Fig. 2.48, which shows the frequency response of the beam to a small impulse as recorded by a conventional resistive strain gage and a PZT sensor. The spikes in the frequency response at 60 Hz, 120 Hz, 240 Hz, 360 Hz, and 420 Hz are overtones of the AC powerline frequency. Because the resistive strain gage requires an excitation, its output can become contaminated with a component of the AC powerline signal. PZT sensors are inherently free from this contamination; however, the signal conditioning circuitry introduces some contamination into the PZT sensor output as well. It is worth mentioning here that the signal-conditioning electronics associated with the resistive-strain gage are much more involved and bulky compared to those used in conjunction with the piezoelectric sensor.

The correlations between strain measured by a conventional strain gage and that measured by a PZT sensor are shown in Figs. 2.49–2.51 for a sine sweep ranging from 5 Hz to 500 Hz. Figure 2.49 shows the correlation between the strain measurements from a strain gauge and a PZT sensor after the appropriate correction factors are applied. The dimensions of the PZT sensor are $3.5 \times 6.0 \times 0.23$ mm, and the value of C_F for this experiment is 1.1nF. Also shown for comparison is the strain calculated from the PZT sensor readings without application of the correction factors. It can

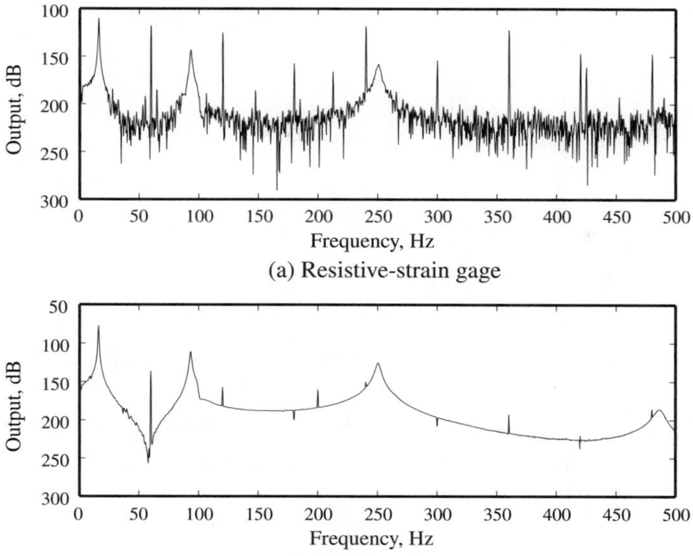

(a) Resistive-strain gage

(b) PZT sensor

Figure 2.48. Resistive-strain gage and PZT sensor impulse response in the frequency domain.

be seen that the correction factors are significant and that after they are applied, there is good agreement between the strains measured by the strain gage and the PZT sensor. Figure 2.50 shows a frequency sweep response at a very low excitation voltage, such that the strain response is only on the order of several microstrain. The sensor in this case is a PZT sensor of size $6.9 \times 3.3 \times 0.23$ mm. Good correlation is observed over the whole frequency range, both in matching resonant frequencies and the magnitude of the measured strain at off-resonant conditions. Hence, it can be concluded that the PZT sensors are capable of accurately measuring both low and high strain levels. Note that the strain gage is not able to accurately pick out the peak amplitudes at low strain levels. This clearly demonstrates the superiority of piezoelectric sensors in such applications. The error between the resistive-strain gauge and the PZT sensor is, at most, between 5% and 10% for off-peak conditions.

Figure 2.51 shows the results of replacing the PZT sensor with a PVDF sensor of size $7.1 \times 3.6 \times 0.056$ mm at higher strain levels. Again, good correlation is observed

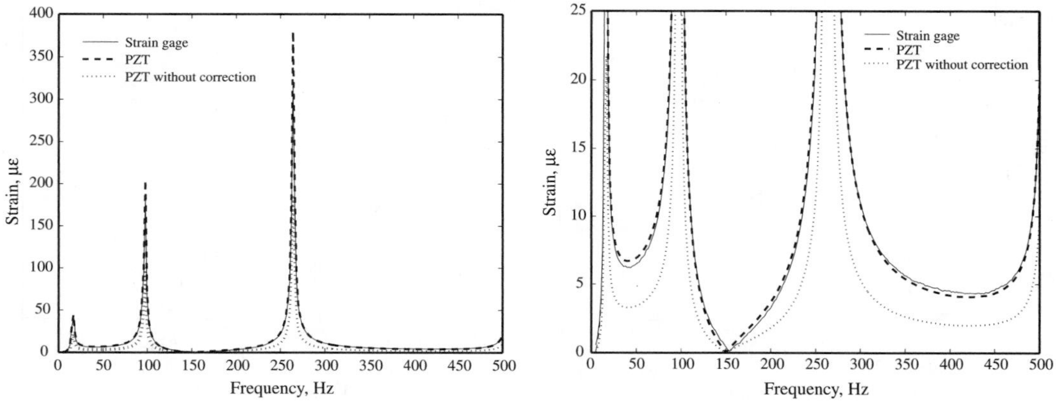

(**a**) Normal vertical scale (**b**) Expanded vertical scale

Figure 2.49. Correlation of PZT sensor and resistive strain gauge with correction factor.

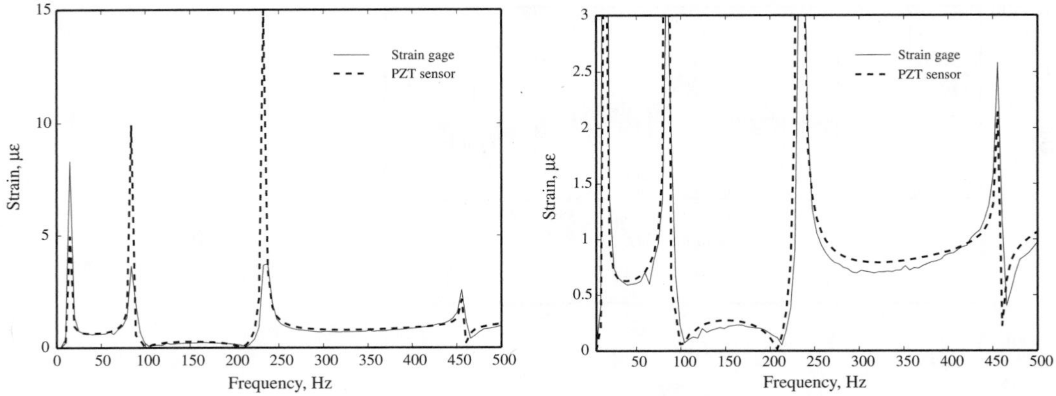

(**a**) Normal vertical scale (**b**) Expanded vertical scale

Figure 2.50. Correlation of PZT sensor and resistive-strain gage for response at low strain levels.

for lower frequencies but, at high frequencies, some discrepancy is apparent. Also plotted on the same figure is a prediction of the strain response at the sensor location calculated by an assumed-modes method. The theoretical model uses a complex modulus with 3% structural damping. It can be seen that both sensors follow the same trend as that of the theoretical prediction. The discrepancy after the third resonant peak can be explained by a slight error in collocation of the resistive-strain gauge and the PVDF sensor. This difference in position gives rise to a shift in the zeros of the transfer function, and the dynamics of these zeros affect the shape of the transfer function in this frequency range.

Effect of Sensor Transverse Length

The size of the sensor is chosen on the basis of the desired gage length. The strain measurements from PZT sensors of three different sizes are shown in Fig. 2.52. All three sensors have the same gage length of 0.125 inch, which is also the same as the

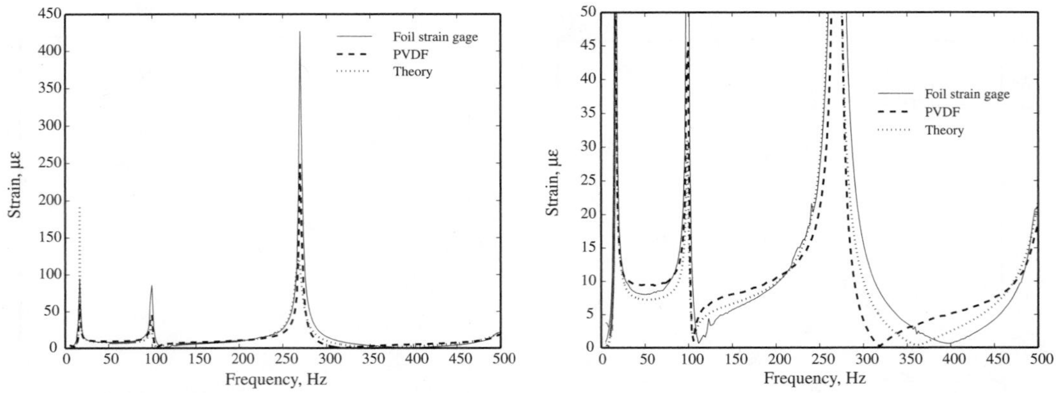

(**a**) Normal vertical scale (**b**) Expanded vertical scale

Figure 2.51. Correlation of PVDF sensor and resistive strain gage response.

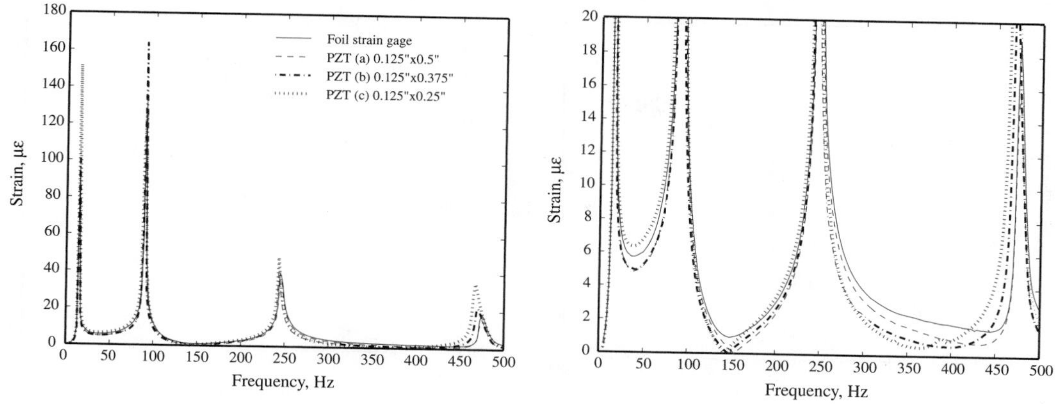

(a) Normal vertical scale (b) Expanded vertical scale

Figure 2.52. Correlation of resistive-strain gage and PZT sensors of different sizes, keeping gauge length constant.

gauge length of the resistive strain gage. The width of the sensors varies from 0.5 inch in case (a) to 0.375 inch in case (b) and 0.25 inch in case (c). It can be seen that there is good correlation between strain gauge and piezoelectric sensor, irrespective of the sensor size. The primary effect of the sensor size can be seen from Eq. 2.157. For a given sensor material, the output depends only on the area of the sensor, $l_c b_c$. A larger sensor would therefore produce a larger sensitivity. Assuming the sensing direction to be along l_c, for a constant gage length, the sensitivity can be increased by increasing the width of the sensor, b_c.

The secondary effect of sensor size can be seen from Eqs. 2.173 and 2.178. For a given sensor thickness t_c and bond thickness t_s, as the sensor dimensions increase, the shear-lag losses decrease and the strain is transferred more efficiently from the surface of the structure to the sensor. The good correlation between PZT sensor strain measurements and conventional resistive strain gage measurements irrespective of sensor size validates the theoretically derived shear lag correction factor.

Hence, it can be concluded that the best strain sensitivity can be achieved by making the sensor area as large as possible, with the constraint of selecting an appropriate gage length for the application. It should also be pointed out that the sensor adds stiffness to the structure, and this additional stiffness increases with sensor size. This can be a significant factor in the case of PZT sensors but will normally be negligible for PVDF sensors.

Effect of Temperature on Sensor Characteristics

The properties of all piezoelectric materials may vary with temperature. In the case of piezoelectric ceramics, variation with temperature is highly dependent on material composition. Both the electric permittivity and the piezoelectric coefficients vary with temperature. Because the charge amplifier effectively transfers the charge from the piezoelectric sensor onto a reference capacitor, the change in electric permittivity, and, hence, the capacitance of the sensor with temperature has no effect on the sensor output. The only dependence of sensor output on temperature is due to the change in piezoelectric coefficients, as seen from Eq. 2.157. As per the datasheets supplied

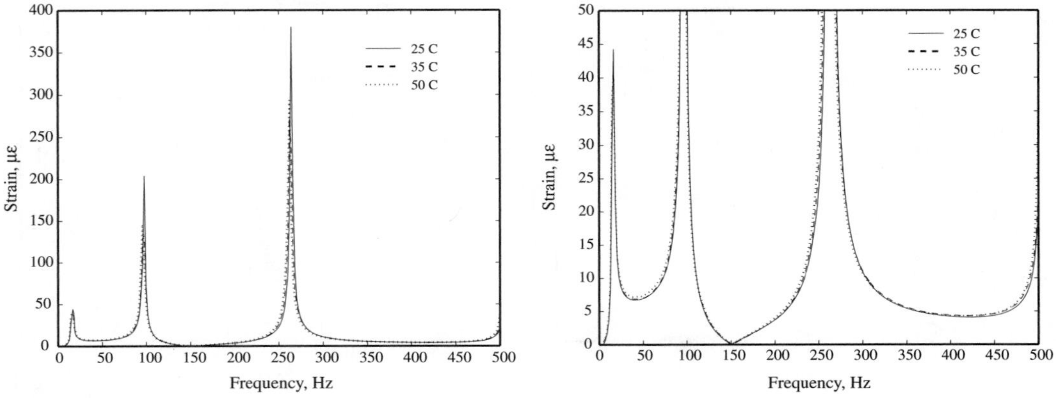

(a) Normal vertical scale (b) Expanded vertical scale

Figure 2.53. PZT sensor output variation with temperature.

by the manufacturer, the magnitude of d_{31} increases by approximately 10% from room temperature (25°C) to 50°C. Tests were carried out in this temperature range by placing the entire experimental setup in an environmental chamber. The results are plotted in Fig. 2.53, which shows a negligible change in sensor output without the use of any temperature correction factors. PVDF film exhibits pyroelectricity in addition to piezoelectricity; hence, it has highly temperature-dependent properties. PVDF film is sometimes used in temperature sensing. Care must be taken, therefore, to take measurements from PVDF sensors at known temperature conditions and to use the appropriate values of the constants for calibration.

PROBLEM

1. A piezoceramic element (PZT-5H) with length $l_c = 2$ in, width $b_c = 1$ in and thickness $t_c = 0.012$ in is applied to an electric field of 150 volts along its polarized direction (3-axis). Manufacturer-supplied material constants are as follows:
   $d_{31} = -274 \times 10^{-12}$ m/V,
   $d_{33} = 593 \times 10^{-12}$ m/V,
   $E_c = 10.5 \times 10^6$ psi
   $e_{33} = 30.1 \times 10^{-9}$ Farad/m
   (a) Calculate its maximum free strains in three directions.
   (b) Calculate blocked force F along the axial direction (1-axis).
   (c) If this piezoceramic element is pulled along the axial direction with a force of 10 lbs, determine the voltage across its surface.

BIBLIOGRAPHY

[1] *IEEE Standard on Piezoelectricity*. ANSI/IEEE, 1987. Std. 176.

[2] S. Sathiyanarayan, S. M. Sivakumar, and C. L. Rao. Nonlinear and time-dependent electromechanical behavior of polyvinylidene fluoride. *Smart Materials and Structures*, 15(3):767–781, June 2006.

[3] R. C. Smith and Z. Ounaies. A domain wall model in piezoelectric material. *Journal of Intelligent Material Systems and Structures*, 11(1):62–79, January 2000.

[4] C. V. Newcomb and I. Flinn. Improving the linearity of piezoelectric actuators using charge feedback. *IEE Electronics Letters*, 18(11):442–444, 1982.

[5] J. A. Main, E. Garcia, and D. V. Newton. Precision position control of piezoelectric actuators using charge feedback. *Journal of Guidance, Control, and Dynamics*, 18(5): 1068–1073, September–October 1995.

[6] A. J. Fleming and S. O. R. Moheimani. Improved current and charge amplifiers for driving piezoelectric loads, and issues in signal processing design for synthesis of shunt damping circuits. *Journal of Intelligent Material Systems and Structures*, 15(2):77–92, 2004.

[7] S. C. Hwang, C. S. Lynch, and R. M. McMeeking. Ferroelectric/ferroelastic interactions and a polarization switching model. *Acta Metallurgica et Materialia*, 43(5):2073–2084, May 1995.

[8] J. E. Huber, N. A. Fleck, C. M. Landis, and R. M. McMeeking. A constitutive model for ferroelectric polycrystals. *Journal of the Mechanics and Physics of Solids*, 47(8): 1663–1697, August 1999.

[9] E. Bassiouny, A. F. Ghaleb, and G. A. Maugin. Thermomechanical formulation for coupled electromechanical hysteresis effects – I. Basic equations, II. Poling of ceramics. *International Journal of Engineering Science*, 26:1279–1306, 1988.

[10] E. Bassiouny and G. A. Maugin. Thermomechanical formulation for coupled electromechanical hysteresis effects – III. Parameter identification, IV. Combined electromechanical loading. *International Journal of Engineering Science*, 27:975–1000, 1989.

[11] W. Chen and C. S. Lynch. A micro–electromechanical model for polarization switching of ferroelectric materials. *Acta Materialia*, 46(15):5303–5311, September 1998.

[12] M. Elhadrouz, T. B. E. N. Zineb, and E. Patoor. Constitutive law for ferroelastic and ferroelectric piezoceramics. *Journal of Intelligent Material Systems and Structures*, 16(3): 221–236, 2005.

[13] Y. C. Li. Nonlinear constitutive law for ferroelectric-ferroelastic materials and its finite element formulation. *Journal of Intelligent Material Systems and Structures*, 16(7–8): 659–671, 2005.

[14] R. C. Smith, A. G. Hatch, B. Mukherjee, and S. Liu. A homogenized energy model for hysteresis in ferroelectric materials: General density formulation. *Journal of Intelligent Material Systems and Structures*, 16(9):713–732, 2005.

[15] I. Chopra. Review of state-of-art of smart structures and integrated systems. *AIAA Journal*, 40(11):2145–2187, November 2002.

[16] Morgan Matroc, Inc. (Electro Ceramics Division). Guide to modern piezoceramic ceramics. Bedford, OH, 1997.

[17] I. Vigness. Dilatations in rochelle salt. *Physical Review*, 48:198–202, 1935.

[18] R. F. Brown. Effect of two-dimensional mechanical stress on the dielectric properties of poled ceramic barium titanate and lead zirconate titanate. *Canadian Journal of Physics*, 39:741–753, 1961.

[19] H. H. A. Kreuger and D. Berlincourt. Effects of high static stress on the piezoelectric properties of transducer materials. *Journal of the Acoustical Society of America*, 33(10): 1339–1344, 1961.

[20] H. H. A. Kreuger. Stress sensitivity of piezoelectric ceramics, Part 1: Sensitivity to compressive stress parallel to the polar axis. *Journal of the Acoustical Society of America*, 42(3):636–645, 1967.

[21] H. H. A. Kreuger. Stress sensitivity of piezoelectric ceramics, Part 3: Sensitivity to compressive stress perpendicular to the polar axis. *Journal of the Acoustical Society of America*, 43(3):583–591, 1968.

[22] P. C. Chen and I. Chopra. Hover testing of smart rotor with induced-strain actuation of blade twist. *AIAA Journal*, 35(1):6–16, January 1997.

[23] R. Gerson, S. R. Burlage, and D. Berlincourt. Dynamic tensile stress of a ferroelectric ceramic. *Journal of the Acoustical Society of America*, 33(11):1483–1485, 1961.

[24] Q. M. Zhang, J. Zhao, K. Uchino, and J. Zheng. Change of the weak-field properties of $Pb(Zr, Ti)O_3$ piezoceramics with compressive uniaxial stresses and its links to the effect of dopants on the stability of the polarizations in the materials. *Journal of Materials Research*, 12(1):226–234, January 1997.

[25] *Products for Micropositioning, US edition*. Physik Instrumente (PI), 1997.

[26] B. J. G. Vautier and S. O. R. Moheimani. Charge-driven piezoelectric actuators for structural vibration control: Issues and implementation. *Smart Materials and Structures*, 14(4):575–586, 2005.

[27] S. R. Viswamurthy, A. K. Rao, and R. Ganguli. Dynamic hysteresis of piezoceramic stack actuators used in helicopter-vibration control: Experiments and simulations. *Smart Materials and Structures*, 16(4):1109–1119, August 2007.

[28] S. H. Lee, T. J. Royston, and G. Friedman. Modeling and compensation of hysteresis in piezoceramic transducers for vibration control. *Journal of Intelligent Material Systems and Structures*, 11(10):781–790, 2000.

[29] J. M. Cruz-Hernández and V. Hayward. Phase control approach to hysteresis reduction. *IEEE Transactions on Control Systems Technology*, 9(1):17–26, 2002.

[30] J. A. Main and E. Garcia. Design impact of piezoelectric actuator nonlinearities. *Journal of Guidance, Control, and Dynamics*, 20(2):327–332, 1997.

[31] S. Li, W. Cao, and L. E. Cross. The extrinsic nature of nonlinear behavior observed in lead zirconate titanate ferroelectric ceramic. *Journal of Applied Physics*, 69(10):7219–7224, 1991.

[32] L. W. Matsch. *Capacitors, Magnetic Circuits and Transformers*. Prentice-Hall Inc., 1964.

[33] C. Liang, F. Sun, and C. A. Rogers. Coupled electromechanical analysis of piezoelectric ceramic actuator-driven systems: Determination of the actuator power consumption and system energy transfer. *Proceedings of the SPIE Smart Structures and Materials Symposium*, 1917:286–298, 1993.

[34] T. Lee and I. Chopra. Design and static testing of a trailing-edge flap actuator with piezostacks for a rotor blade. *Proceedings of the SPIE Smart Structures and Materials Symposium*, 3329:21–332, 1998.

[35] D. J. Clingman and M. Gamble. High-voltage switching piezo drive amplifier. *Proceedings of the SPIE Smart Structures and Materials Symposium*, 3326:472–478, March 1998.

[36] G. A. Zvonar, A. J. Luan, F. C. Lee, D. K. Lindner, S. Kelly, D. Sable, and T. Schelling. High-frequency switching amplifiers for electrostrictive actuators. *Proceedings of the SPIE Smart Structures and Materials Symposium*, 2721:465–475, 1996.

[37] D. Newton, J. Main, E. Garcia, and L. Massengill. Piezoelectric actuation systems: Optimization of driving electronics. *Proceedings of the SPIE Smart Structures and Materials Symposium*, 2717:259–266, February 1996.

[38] C. Niezrecki and H. H. Cudney. Power factor correction methods applied to piezoelectric actuators. Paper# AIAA-1993-1688, *Proceedings of the 34th AIAA/ASME/ASCE/AHS/ASC Structures, Structural Dynamics and Materials Conference* La Jolla, CA, April 1993.

[39] C. Niezrecki and H. H. Cudney. Improving the power consumption characteristics of piezoelectric actuators. *Journal of Intelligent Material Systems and Structures*, 5(4):522–529, 1994.

[40] S. Huang, J. Chang, R. Xu, F. Liu, L. Lu, Z. Ye, and X. Cheng. Piezoelectric properties of 0–3 PZT/Sulfoaluminate cement composites. *Smart Materials and Structures*, 13(2):270–274, 2004.

[41] L. Cross. Polarization-controlled ferroelectric high-strain actuators. *Journal of Intelligent Material Systems and Structures*, 2(3):241–260, 1991.

[42] K.-Y. Oh, A. Furuta, and K. Uchino. Field induced strains in antiferroelectrics. *Proceedings of the IEEE Ultrasonics Symposium, 4–7 December, 1990*, Honolulu, HI, 2:743–746, 1990.

[43] K. Ghandi and N. W. Hagood. Shape memory ceramic actuation of adaptive structures. *AIAA Journal*, 33(11):2165–2172, 1995.

[44] J. J. Dosch, D. J. Inman, and E. Garcia. A self-sensing piezoelectric actuator for collocated control. *Journal of Intelligent Material Systems and Smart Structures*, 3(1): 166–185, January 1992.

[45] J. Qui and J. Tani. Vibration control of a cylindrical shell using distributed piezoelectric sensors and acuators. *Journal of Intelligent Material Systems and Structures*, 6(4):474–481, April 1, 1995.

[46] P. Samuel and D. Pines. Health monitoring/damage detection of a rotorcraft planetary gear train using piezoelectric sensors. *Proceedings of the SPIE Smart Structures and Materials Symposium*, 3041:44–53, 1997.

[47] S. Hanagud, M. W. Obal, and A. J. Calise. Optimal vibration control by the use of piezoelectric sensors and actuators. *Journal of Guidance, Control and Dynamics*, 15(5): 1199–1206, 1992.

[48] E. H. Anderson and N. W. Hagood. Simultaneous piezoelectric sensing/actuation: Analysis and application to controlled structures. *Journal of Sound and Vibration*, 174(5): 617–639, 1994.

[49] R. L. Forward. Picostrain measurements with piezoelectric transducers. *Journal of Applied Physics*, 51(11):5601–5603, November 1980.

[50] C. K. Lee and T. C. O'Sullivan. Piezoelectric strain rate gages. *Journal of the Acoustical Society of America*, 90(2):945–953, August 1991.

[51] C. K. Lee, T. C. O'Sullivan, and W. W. Chiang. Piezoelectric strain rate sensor and actuator designs for active vibration control. Paper# AIAA-1991-1064, *Proceedings of the 32nd AIAA/ASME/ASCE/AHS/ASC Structures, Structural Dynamics and Materials Conference*, Baltimore, MD, April 1991.

[52] J. W. Dally, W. F. Riley, and K. G. McConnell. *Instrumentation for Engineering Measurements*. John Wiley and Sons, 1993.

[53] D. F. Stout. *Handbook of Operational Amplifier Circuit Design*. McGraw Hill, 1976.

3 Shape Memory Alloys (SMAs)

Certain classes of metallic alloys have a special ability to "memorize" their shape at a low temperature and recover large deformations imparted at a low temperature on thermal activation. These alloys are called shape memory alloys (SMAs). The recovery of strains imparted to the material at a lower temperature, as a result of heating, is called the shape memory effect (SME). The SME was first discovered by Chang and Read in 1951 in the Au-Cd (gold-cadmium) alloy system. However, the effect became more well known after the discovery of nickel-titanium alloys.

Buehler and Wiley [1, 2] discovered a nickel-titanium alloy in 1961 called NiTiNOL (i.e., **Ni**ckel **Ti**tanium alloy developed at the **N**aval **O**rdinance **L**ab) that exhibited a much greater SME than previous materials. This material was a binary alloy of nickel and titanium in a ratio of 55% to 45%, respectively. A 100% recovery of strain up to a maximum of about 8% pre-strain was achieved in this alloy. Another interesting feature noticed was a more than 200% increase in Young's modulus in the high-temperature phase compared to the low-temperature phase. Subsequently, it was determined that the percentage of nickel and titanium influences the material properties of Nitinol and can be varied to control the transformation temperatures in the material [3]. Also, the addition of a third or fourth element (most commonly copper) to NiTi can be used to selectively control some properties of SMA wires. For example, the addition of copper as a ternary element not only reduces the temperature hysteresis but also reduces the yield stress. Other alloys exhibiting the SME include Cu-Al-Ni, Cu-Zn-Al, Au-Cd, Mn-Cu, and Ni-Mn-Ga, with recoverable strains of 3–8%. However, NiTi is the most practical material in terms of its superior ductility, higher resistance to corrosion and abrasion, higher tensile strength, and lower susceptibility to grain-boundary fracture.

The first successful application of SMA was carried out by Raychem Corporation in fasteners and tube couplings for the hydraulic system of the F-14 aircraft. Shrinking of the diameter of the Ni-Ti tube at high temperature resulted in sealing of the joint with the couplings. Nowadays, SMA devices are being used in a wide range of applications that include home appliances, automobiles, aerospace systems, railway trains, robotic systems, medical devices, and civil structures. Key advantages of SMA actuators over other conventional actuators are their large force-output/weight ratio, large stroke, large specific-energy density, flexibility in design, compactness, and environmental friendliness (i.e., no dust or noise during operation). SMAs can be

Table 3.1. *Alloying elements commonly used in shape memory materials*

Element	Symbol	Element	Symbol
Aluminum	Al	Manganese	Mn
Cadmium	Cd	Nickel	Ni
Copper	Cu	Titanium	Ti
Gallium	Ga	Zinc	Zn
Gold	Au		

used directly in many applications without additional mechanisms. Furthermore, the power circuitry needed for actuation is comparatively simple. Examples of some applications are orthodontic wires, cardiac stents, eyeglass frames, and antennas for cellular phones.

Two key characteristics of an SMA are the SME and pseudoelasticity. The SME is the material's ability to recover large mechanically induced strains (up to 8%) at low temperatures by moderate increases in temperature (approximately 10–20°C). Pseudoelasticity refers to the material's ability, in a somewhat higher temperature regime, to undergo strains (up to 8%) during loading and then recover on unloading in a hysteresis loop.

There are about 20 alloys that exhibit the properties of SME and superelasticity. These alloys are obtained from elements listed in Table 3.1. NiTi-based SMAs are the most widely used in practical applications. However, NiTi alloys are more expensive than Cu-based alloys. The grain sizes of Cu-based alloys are much larger than those of NiTi alloys, so it becomes easy to see the grain boundaries during testing using optical microscopy. Shape memory alloys are composed of austenite and martensite phases, and the shape-memory characteristics are due to the combination of the individual effects of these two phases. The forward-phase transformation (austenite to martensite) is exothermal (heat-emitting) and the reverse-phase transformation (martensite to austenite) is endothermal (heat-absorbing).

The thermomechanical behavior of SMA material depends on the internal crystalline structure, stress, temperature, and history of the material. Material properties of SMAs can also be a function of chemical composition, cold work, heat treatment, and thermomechanical cycling. The ability of the SMA to recover large strains comes from reversible-phase transformation characteristics. Large recoverable strains offer work densities an order of magnitude larger than conventional approaches. Under the no-stress condition, an SMA exists in the austenite phase (called the parent phase) at high temperatures and the material tranforms to the low-temperature martensite phase on cooling. The high-energy austenite phase is associated with a body-centered cubic crystal structure, whereas the low-energy martensite phase is linked with a face-centered cubic crystal structure (Fig. 3.1). The SME is explained schematically in Fig. 3.2, respectively, for beam bending and beam extension. Note that the original shape is imparted to the material at a high temperature, either as a consequence of the manufacturing process or intentionally by means of a physical deformation. The response of the SMA to thermodynamic states is nonlinear, hysteretic, and path-dependent.

In untrained SMAs, repeated thermal cycling can introduce non-oriented lattice defects in the material as a result of the accumulation of plastic strain. These defects are responsible for the creation of an internal stress field that plays an important

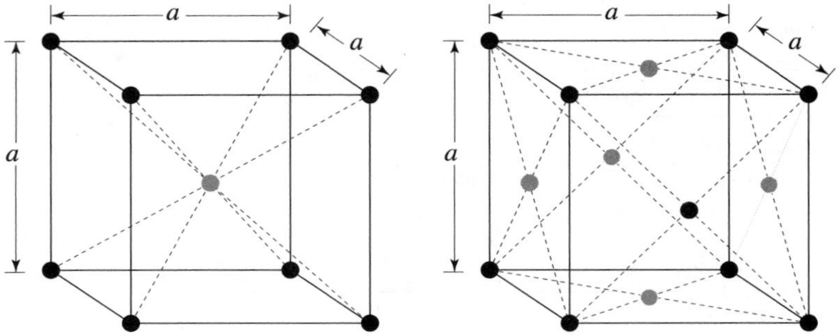

(a) Body centered cubic (austenite phase) (b) Face centered cubic (martensite phase)

Figure 3.1. Crystal structure of SMA phases.

role in the macroscopic behavior of the material. The connection between microscopic and macroscopic behavior of an SMA is very complex and has not been fully understood in terms of analytical modeling. This is primarily because the mechanical response depends on a wide range of parameters, including temperature, loading rate, strain range, specimen geometry, thermomechanical history, and ambient environment. The unique properties of SMAs are directly related to the solid-state displacive (martensite)-phase transformations that can be induced by heating (or cooling) and, in some temperature regimes, by stressing (or unloading). The transformation from the austenite to martensite and back again to the austenite phase during the pseudoelastic state of polycrystalline Nitinol takes place through the nucleation and propagation of phase-transformation fronts, resulting in nonuniform deformation and temperature fields (as a result of the generation or absorption of latent heat at local fronts). This results in self-heating and self-cooling of the alloy.

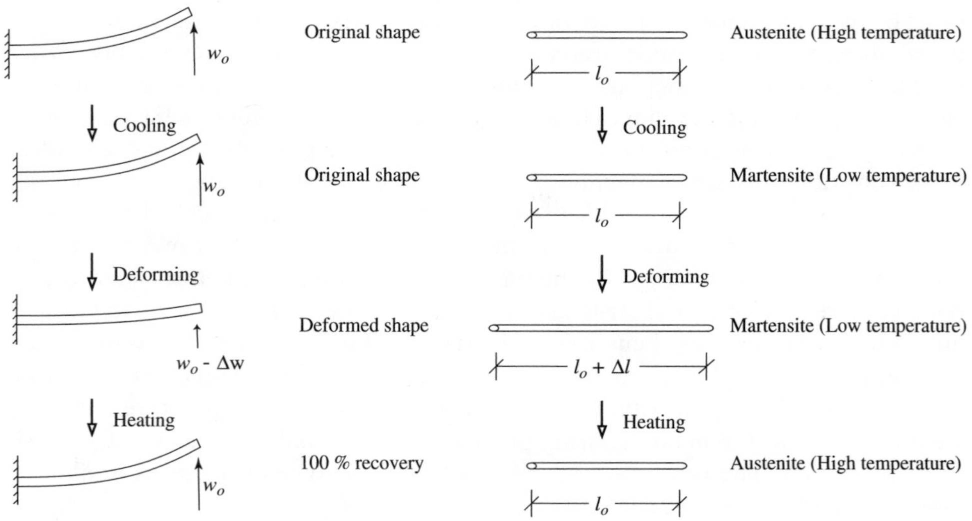

(a) Beam bending (b) Beam extension

Figure 3.2. One-way shape memory effect.

Although SMAs can be manufactured in a single-crystal form, the vast majority of commercially available SMAs are polycrystals. As a result, the macroscopic behavior is the combined effect of all grains.

Shapes memory alloy are capable of providing unique capabilities that can be used in a wide range of applications. Shape memory effect can be used for compact actuation, and pseudoelastic effect can be exploited in vibration isolation, energy dissipation, and large recoverable deformations. SMAs can provide large actuation force over a large stroke (large strain/displacement). One can build actuators that can extend, bend, twist, or perform a combination of these motions. One of the major challenges has been the rapid heat activation of SMAs because these materials have a high heat capacity. As a result, the actuation frequency with SMAs is quite limited, especially when the speed of cooling is slow. Some attempts have been made to expedite the actuation speed, including forced convection with flowing water and forced conduction with thermoelectric cooling modules. Another challenge is the material thermomechanical stability under multiple cycles of transformation.

The martensite phase exists either as a randomly twinned structure (at low temperature and low stress) or a stress-induced detwinned structure that can accommodate relatively large, recoverable strains. The transformation from austenite phase to twinned martensite results in a negligible shape change (at the macroscopic level); it is referred to as self-accommodated martensite. The transformation from twinned martensite to detwinned martensite takes place under the application of a sufficient level of external stress. The martensite phase exists in multiple variants representing twinned and detwinned states.

3.1 Fundamentals of SMA Behavior

The SME occurs as a result of a transformation between two phases in the material. The specific lattice structure of the material results in a deformation behavior that is very different from that of conventional metallic alloys. The material can undergo and recover large deformations, and it exhibits mechanical hysteresis in loading-unloading cycles. Furthermore, the properties of the material depend on temperature. These phenomena can be traced to the lattice structure and associated deformation mechanisms inside the material. To understand the macroscopic behavior of the material in response to external loads and temperature, it is important to look at the underlying phenomena at a microscopic level.

3.1.1 Phase Transformation

The basic phenomenon responsible for the SME is a phase transformation. Although the actual dynamics of the transformation are quite complex, a simplistic description provides physical insight into the fundamental mechanisms responsible for the SME. At high temperatures, the material exists in the austenite phase, usually with a body-centered cubic crystal structure. On cooling, the austenite phase transforms to the martensite phase, which typically has a face-centered cubic crystal structure. The terms austenite and martensite were originally used to refer to phases in steels; however, these terms are now generalized descriptions of material structures with specific properties. In general, the lattice structure of the martensite phase is more disordered and exhibits less symmetry than that of the austenite phase. The transformation to martensite is known as a displacive transformation, wherein the atoms

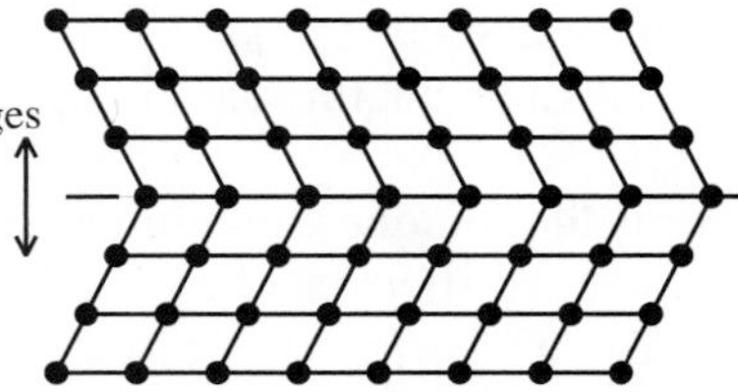

Figure 3.3. Twinning in a two-dimensional lattice.

of the material move by relatively small amounts to form a new stable crystal structure. Because there is no significant migration of atoms, the transformation proceeds at the local speed of sound in the material. However, as the material is thermally activated, the speed of the transformation is limited by the heat-transfer rate in the material, which is typically much slower than the local speed of sound. Consequently, actuation based on the SME is usually very slow, and typical actuators operate at frequencies of less than 1 Hz. The phase transformation from austenite to martensite can also be produced by the application of a mechanical stress. In the unstressed state, the phase change from the high-energy austenite state (parent phase) to the low-energy martensite state results in the formation of multiple martensite variants and twins with no net change of strain. If a tensile stress is applied to this material, when it reaches a certain critical value, the pairs of martensite twins will begin "detwinning" to the stress-preferred twins.

The level of the thermally induced phase transformation depends on the temperature and is not influenced by the length of time for which the temperature is applied. The transformation from the austenite (parent) phase to the martensite phase involves two processes: the Bain strain and the lattice invariant shear [3]. The Bain distortion involves atomic rearrangement (small atomic-scale reshuffling) that produces the new crystal structure of martensite. The lattice invariant shear results in relieving the large amount of strain energy associated with the accommodation of new crystal structures.

3.1.2 Lattice Structure and Deformation Mechanism

A brief description of the concept of twinning is useful in understanding the deformation mechanism of the material. When two unit cells in a lattice are oriented in such a manner that they appear to be mirror images of one an other, they are called a "twin pair." Figure 3.3 is a schematic representation of a two-dimensional lattice that is twinned. The twin boundary is a line of atoms about which the rest of the lattice appears to be mirrored. It is important to note that there is no break in the lattice structure at a twin boundary; rather, there is merely a change in the orientation of the unit cells to another preferred direction. In the example shown in Fig. 3.3, each two-dimensional unit cell can have two preferred orientations. In the case of a real material, the lattice is three-dimensional and several twin variants can exist.

The austenite phase is shown schematically in Fig. 3.4. Let us consider a section of the lattice of initial length l_o, and, for simplicity, assume only deformations in the horizontal direction. The solid circles represent individual atoms and the lines represent bonds between them. Note the positions of the atoms marked by the arrows (A, B, C, D, E, F) and the bonds between these atoms. When the material transforms

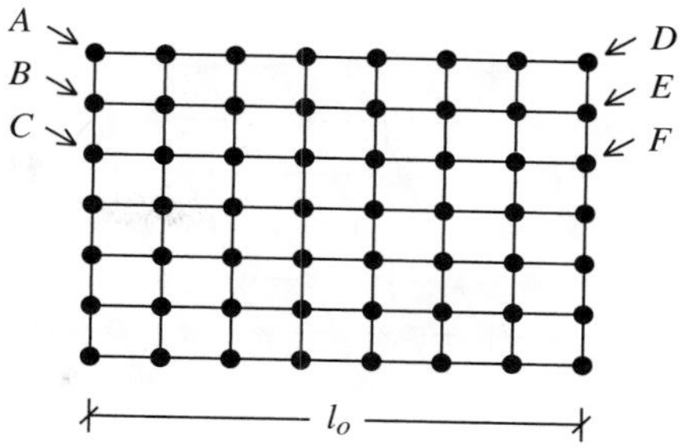

Figure 3.4. Austenite lattice structure.

from austenite to martensite, each atom moves by a small amount, resulting in a twinned martensite phase, as shown in Fig. 3.5(a). Note that the length of the new lattice is almost the same as the original length l_o and that the bonds between the atoms remain the same as in the original austenite phase.

It is also important to note that there are several possible combinations of twins and twin boundaries in the section of the lattice shown in Fig. 3.5(a) that would result in the same length of l_o. The overall shape of the section of the lattice would be different depending on the combination of twins that occur. In contrast, the austenite phase can have only one arrangement of atoms (see Fig. 3.4). Translating this to three dimensions, it follows that for a given volume, the austenite phase can exist in only one shape, whereas the martensite phase can have several different shapes. Thus, the martensite transformation is a diffusionless transformation through a shear-like mechanism from a more symmetric crystal structure (parent phase) to a less symmetric martensitic phase.

Figure 3.5(b) shows the same section of lattice in fully detwinned martensite, which means that all of the twin variants have reoriented into one single direction. It can be seen that the lattice increases in length by an amount Δl, and the atoms must move over larger distances to occupy their position in the new lattice arrangement. Because the atoms must move over smaller distances, the austenite-to-martensite phase transformation results in the formation of twinned martensite. Again, note that the original bonds – for example, those linking the atoms marked by the arrows – remain unchanged.

Assume a horizontal stress is applied to the twinned martensite lattice (see Fig. 3.5(a)). The stress results in some initial deformation of the lattice, after which detwinning starts occurring, and the lattice structure approaches that of fully

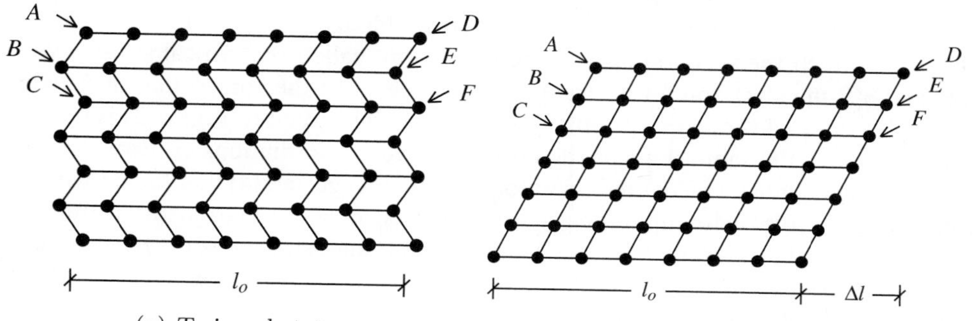

(a) Twinned state (b) Detwinned state

Figure 3.5. Twinning in martensite.

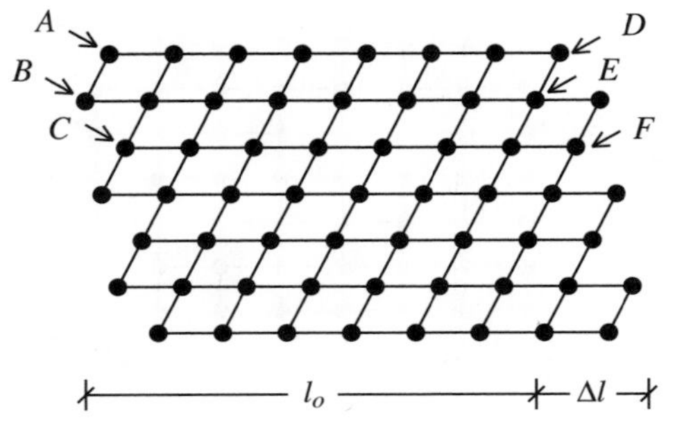

Figure 3.6. Deformation by slip.

detwinned martensite (see Fig. 3.5(b)). A relatively small stress is required to detwin the martensite because bonds between the atoms are not broken in the process. The detwinning process is basically a rearrangement of atoms. At a much higher level of stress, deformation is accompanied by slip, schematically shown in Fig. 3.6. In this case, bonds between atoms are broken and new bonds are formed. As a result, this kind of deformation can be permanent and irreversible. Although many types of martensite form by slip, the SME in SMAs is predominantly caused by twinning.

Almost all of the physical properties of the material, such as Young's modulus, specific heat, and resistivity, differ between the martensite and austenite phases. In structures with embedded SMAs, these changes can be exploited to cause an overall change in properties of the structure for different applications. For example, the Young's modulus of SMA material in the austenite phase is much higher than the modulus in the martensite phase. This is because it is much easier to deform the material by detwinning than by slip. Measurement of the physical properties of the material is a useful way to estimate the amount of each phase present in a given sample of material.

3.1.3 Low-Temperature Stress-Strain Curve

A typical SMA stress-strain curve at low temperatures is shown in Fig. 3.7. At low temperature, the material exists in the martensite phase. As the stress increases, there is a region of elastic deformation (region 1) where the strain typically increases linearly with stress. At a certain stress level, the martensite starts detwinning. Because the twin boundaries can be easily moved in the material, the slope of this region (region 2) of the stress-strain curve is very small, and the material deforms almost plastically. However, this deformation is recoverable by the SME. The stress levels between which the second elastic region exists are called the critical stress levels, σ_s^{cr} and σ_f^{cr}, the start and finish stresses, respectively. At the completion of detwinning, the slope of the stress-strain curve increases. This region (region 3) is usually linear and has the same slope as region 1. However, deformations occurring in this region may be mostly recovered on unloading. It is suggested [4] that the detwinned martensite in this region undergoes two actions: elastic deformation and formation of a new orientation of martensite. After a certain high strain level is reached, slip starts occurring and the material deforms plastically again (region 4). Thus, there are two distinct yield points in the stress-strain curve. Note that the deformation after the second yield point is permanent and cannot be recovered by the SME.

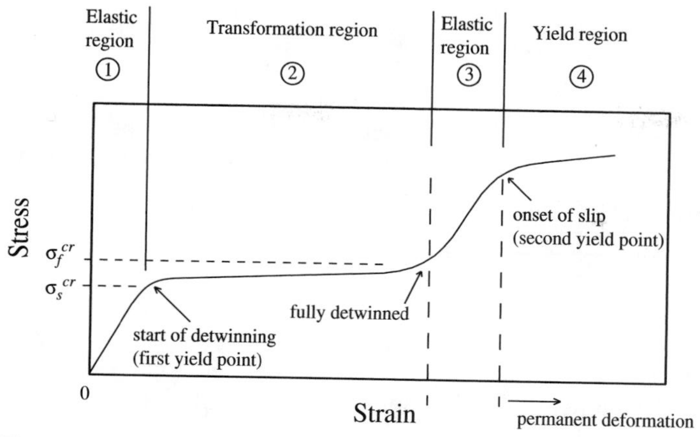

Figure 3.7. Low-temperature SMA stress-strain behavior.

The critical stress at which irreversible plastic strain takes place decreases with an increase of temperature.

3.1.4 Origin of the One-Way SME

A schematic diagram of the changes taking place in the lattice structure of the material during the shape-memory transformation is shown in Fig. 3.8. A deformation imparted to the material in the low-temperature martensite phase is fully recovered on heating as the material completely transforms to the high-temperature austenite phase. On subsequent cooling, the material returns completely to the martensite phase, but there is no further change in the shape of the material. Because the

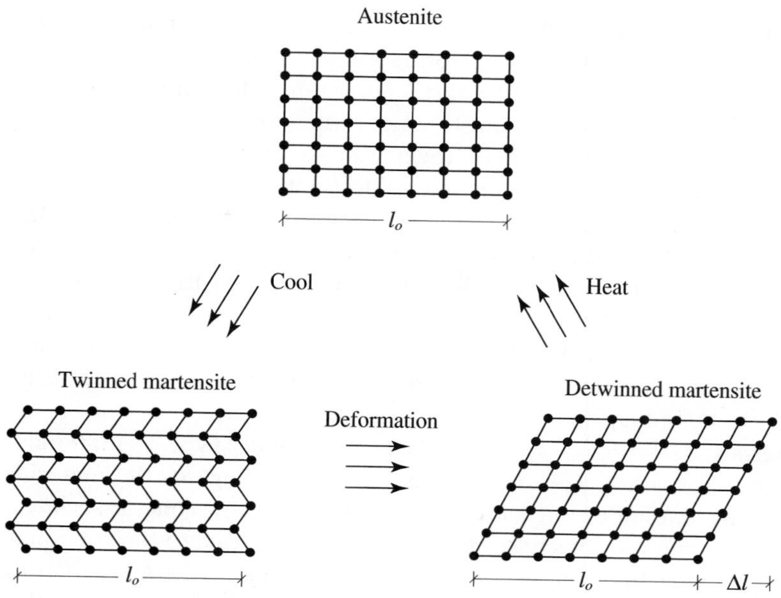

Figure 3.8. Mechanism of one-way SME.

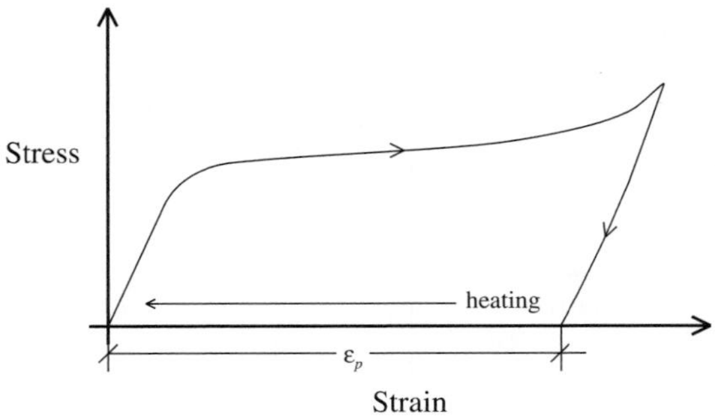

Figure 3.9. Stress-strain behavior in the one-way SME, at low temperature.

shape change occurs only during heating, this transformation is called the one-way SME.

The reason for this behavior can be understood from Section 3.1.2. For a given volume of material, the austenite phase can exist in only one unique shape. The austenite transforms to twinned martensite because the individual atoms must move through the least distances, resulting in the lowest energy state. The undeformed martensite has the same shape as the austenite phase. Therefore, on heating the deformed martensite, the original undeformed shape is recovered and, on cooling, no further shape change takes place.

The stress-strain behavior of a sample of SMA undergoing the one-way SME is shown schematically in Fig. 3.9. The material starts at low temperature in the martensite phase. Under the applied stress, it initially deforms elastically and then starts detwinning, as described in Section 3.1.3. On unloading, the material remains deformed with a strain ϵ_p, also called the pre-strain. Heating the material then causes the pre-strain to be recovered and the sample returns to its original dimensions.

There are four important temperatures related to the phase transformations occurring in the SMA. These transformation temperatures are the martensite start M_s, martensite finish M_f, austenite start A_s, and austenite finish A_f temperatures. The temperatures associated with the transformation from the martensite to the austenite ($M \rightarrow A$) phase are the austenite start temperature, A_s, denoting the start of the phase change, and the austenite finish temperature, A_f, denoting the completion of the phase change. This is called "reverse phase transformation," and it is endothermic. Similarly, the temperatures related to the transformation from the austenite to the martensite ($A \rightarrow M$) phase are the martensite start temperature, M_s, which is indicative of the start of the martensite formation, and the martensite finish temperature, M_f, which marks the completion of the martensite formation. This is called "forward phase transformation" and it is exothermal. These four temperatures are determined through experiment and are also dependent on stress level. Usually, A_s, A_f, M_s, and M_f are defined at zero stress level. The transformation temperatures are shown schematically in Fig. 3.10.

In most materials, $M_f < M_s < A_s < A_f$. These are called "Type I" materials. For a temperature range, $M_s < T < A_s$, there is no phase change of the material (stress-free condition). An important feature is that there is a hysteresis associated with the phase transformation. This hysteresis arises primarily due to the frictional effect involved in moving the twin boundaries in the material. As a result, the phase-transformation temperatures are different for heating and cooling. In the heating cycle (reverse phase), for temperatures below A_s, the material is in the 100%

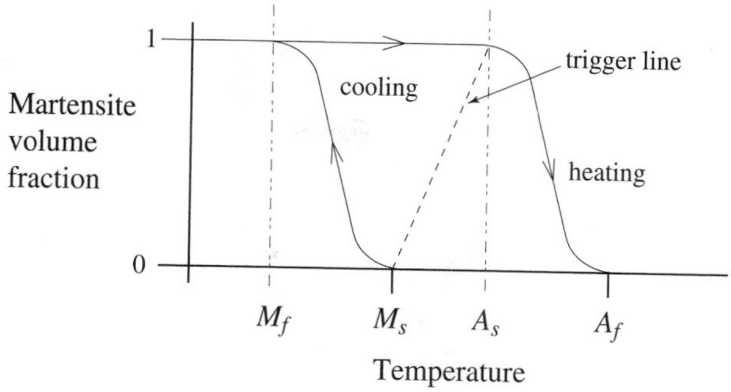

Figure 3.10. Transition temperatures of an SMA, at the no-stress condition.

martensite phase, whereas for temperatures above A_f, the material is in the 100% austenite phase. In the cooling cycle (forward phase), for temperatures above M_s, the material is in 100% austenite phase, whereas for temperatures below M_f, the material is in 100% martensite phase. At any other temperature T ($M_f < T < A_f$), the material can be partly in the martensite phase and partly in the austenite phase. This combination of phases in the material can be characterized by the martensite volume fraction, ξ. The martensite volume fraction refers to the volumetric fraction of the material that is in the martensite phase, and takes on values from 1 (pure martensite) to 0 (pure austenite).

In a typical cycle of the SME, the material is deformed at $T < A_s$ and, on heating, starts recovering the deformation at $T = A_s$. When the temperature reaches A_f, recovery of the deformation is complete, and this shape is retained on cooling to the initial temperature $T < M_f$. The cycle of deformation, heating, recovery, and cooling can now be repeated. In the case of a mixed state of transformation, the trigger line defines the boundary of phase transformation. Although material deformation has been described as an example to illustrate the effect of the transformation temperatures, many other phase-related physical properties (e.g., resistivity, heat conductivity, and Young's modulus) undergo changes as well. In a stress-free state, the martensite phase exists in multiple variants that are crystallographically similar but are oriented in different planes.

3.1.5 Stress-Induced Martensite and Pseudoelasticity

Phase transformations in the SMA are induced by both temperature and mechanical stress. There is an equivalence between temperature and stress. An increase in stress is equivalent to a decrease in temperature and has the effect of stabilizing the martensite phase. Under stress, the phase transformation of the material is changed significantly, and additional heat is needed to deform the SMA specimen against the applied stress. Transformation temperatures M_s, M_f, and A_s are generally a linear function of stress. Also, A_f increases with stress but is a more complex function. The diagonal dashed line connecting M_s to A_s is called the "trigger line," about which forward and reverse transformations occur. The amount of energy required to trigger phase transformation depends on the martensite volume fraction present and the applied stress.

For example, under the no-stress condition, the martensite formation starts at M_s and completes when the temperature M_f is reached. However, if a tensile stress is applied, stress-induced martensite formation starts at a temperature above M_s. At

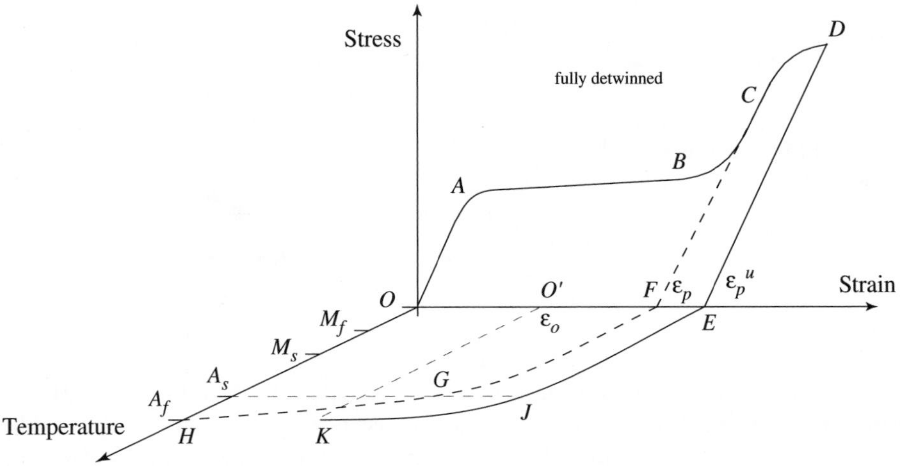

Figure 3.11. Stress-strain-temperature plot, initially 100% twinned martensite.

higher temperatures, larger stresses are required to form stress-induced martensite. The highest temperature at which it is possible to form martensite is called M_d. Beyond this temperature, the critical stress needed to induce martensite is higher than the stress at which permanent dislocations or slips occur. The value of this critical stress decreases with an increase of temperature. This means that if the applied stress exceeds the critical stress, the residual strain is not completely recoverable on unloading or by heating. Figure 3.11 shows the stress-strain-temperature plot up to the critical yield stress, and Table 3.2 lists the key points in the figure. Point C represents the maximum elastic stress condition. On unloading, the elastic portion of the total strain is recovered, and there will be a residual strain ϵ_p as a result of the detwinned martensite (point G). On heating above the austenite finish temperature, the residual strain ϵ_p is completely recovered. For Ni-Ti alloys, this ϵ_p can be on the order of 8%.

Let us consider an alternate path from the detwinned state B. If the temperature is now raised above the austenite finish temperature without any change of stress, there will be some change of strain (decrease in strain) due to the increase of elastic modulus. The material is transformed from stress-induced martensite to the austenite phase. If we lower the temperature without changing the stress, the material will transform into stress-induced martensite, with an increase in strain. Furthermore, if we now lower the stress, there will be no further change of strain, and the material will be in the detwinned state.

The formation of stress-induced martensite from the austenite phase results in a phenomenon called pseudoelasticity, sometimes referred to as superelasticity. This behavior is not triggered by temperature and is only stress-dependent. A schematic diagram of the stress-strain behavior of a SMA undergoing pseudoelasticity is shown in Fig. 3.12. Stress is applied to the material at a temperature above A_s, and it starts deforming elastically. When a critical stress level (σ_1) is reached, the austenite phase becomes unstable and stress-induced martensite starts forming, resulting in a low stiffness region similar to that of plastic deformation. This means that the body-centered cubic lattice transforms into the monoclinic one, which in turn results in a macroscopic elongation. When the stress is removed, the stress-induced martensite becomes unstable and transforms back into austenite. During unloading, the initial

Table 3.2. *Regions on stress-strain-temperature diagram*

$O-A$:	Elastic region.
A:	Initiation of stress-induced martensite (transformation of temperature-induced martensite to stress-induced martensite). Initiation of detwinning process. On the removal of external load, entire strain is recovered.
B:	Completion of stress-induced martensite, completion of detwinning process. On the removal of external load, only a small elastic strain is recovered, leaving the material with a large residual strain ϵ_p.
$B-C$:	Elastic region.
C:	Start of slip and permanent deformation, tangent modulus starts decreasing. On unloading, a small elastic strain is recovered, leaving the material with residual strain ϵ_p that can be recovered on heating.
D:	Yield point with minimum tangent modulus (ultimate stress condition). On unloading, a small elastic strain is recovered, leaving the residual strain ϵ_p^u. On heating, some strain is recovered, leaving behind a permanent strain $\epsilon_p^u - \epsilon_p$.
E:	Residual strain ϵ_p^u from yield point (includes recoverable and nonrecoverable inelastic components).
F:	Residual strain ϵ_p (maximum recoverable strain).
G:	Austenite start condition (initiation of recovery strain).
H:	Austenite finish condition (complete recovery of residual strain).
J:	Austenite start condition.
K:	Austenite finish condition (irrecoverable strain ϵ_o).
$H-O$:	Transformation from austenite phase to temperature-induced martensite, formation of twinned martensite.
$K-O'$:	Transformation from austenite phase to temperature-induced martensite, formation of twinned martensite.

response is elastic, followed by quick recovery of strain (with a small change of stress), and the material transforms back into the austenite phase. If the temperature is above the austenite finish (A_f), the strain in the material can be fully recovered. This is in contrast to conventional metals, in which the strain occurring due to plastic deformation is permanent and cannot be recovered. If the material temperature lies between the ausenite start (A_s) and the austenite finish (A_f), there will be a partial recovery of strain. However, the residual strain ϵ_r is fully recoverable on the application of heat (i.e., raising the temperature above A_f). Due to the large hysteresis in the loading-unloading cycle, the pseudoelastic behavior of SMA has many applications in damping augumentation. The origin of this hysteresis can again be traced to the frictional effect of moving twin boundaries in the martensite phase. Above a critical stress, irreversible plastic slips start taking place. As a result, the

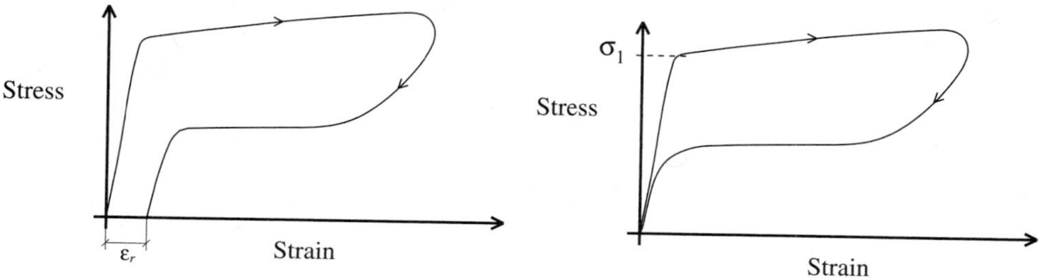

(a) Partial pseudoelasticity, $A_s < T < A_f$ (b) Complete pseudoelasticity, $T > A_f$

Figure 3.12. Pseudoelastic stress-strain behavior.

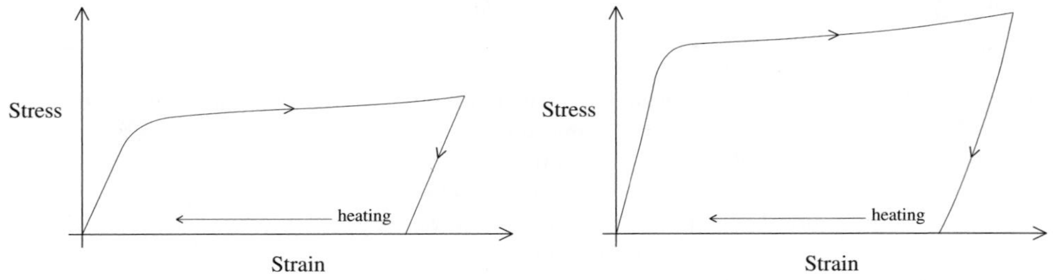

(a) $T < A_s$, initially with 100% martensite (shape memory effect)

(b) $M_s < T < A_s$, initially with 100% austenite

Figure 3.13. Shape memory effect for material starting at 100% martensite and at 100% austenite.

residual strain is not completely recovered on the application of heat. Note that this critical stress decreases with an increase of material temperature. Also, the crystallographic structure may change with the formation of slips. A decrease in temperature has the same effect as an increase in stress.

From the previous discussions, we see that if the material is deformed at a temperature $T < A_s$, there is a residual strain on removal of the stress and the SME can be observed on subsequent heating to $T > A_f$. If the material is deformed at $T > A_s$, partial or complete pseudoelastic behavior can be observed. Another interesting variation of the SME can be observed if the material starts at a 100% austenite phase and at a temperature $M_s < T < A_s$. The difference between the stress-strain behavior of the material in this initial condition and the stress-strain behavior of the material in a 100% martensite phase at $T < A_s$ is shown in Fig. 3.13. At some critical stress, in the case of the material starting from pure martensite, detwinning starts taking place, whereas in the case of the material starting from the 100% austenite phase, stress-induced martensite starts taking place. Therefore, the shape of the stress-strain curve is different in the two cases. On removal of stress, both cases transform into the detwinned martensite phase. Subsequent heating results in recovery of the strain.

The key mechanism of phase transformation is the difference in Gibbs free energies, respectively, of two phases (martensite and austenite), which depend on both temperature and externally applied stress. The austenite phase is stable at high energy levels, whereas the martensite phase is stable at low energy levels. In the forward-phase transformation ($A \rightarrow M$), the driving force is due to the positive Gibbs free energy and is balanced by an increase in elastic strain energy and interfacial energy plus resistance due to any internal motion. Conversely, in the reverse-phase transformation ($M \rightarrow A$), the stored elastic strain energy and interfacial energy are driving forces to increase the Gibbs free energy. Hence, from a thermomechanical point of view, externally applied stress and temperature play equivalent roles in the transformation process. The hysteresis in the SME at low temperature and in the pseudoelastic effect at high temperature has its origins in Gibbs free energy.

Figure 3.14(a) shows temperature-induced transformation for a stress-free condition. The transformation temperatures are M_s, M_f, A_s, and A_f. In this case, M_s and M_f identify the beginning and the end of the forward-phase transformation and A_s and A_f identify the beginning and the end of the reverse-phase transformation.

Figure 3.14. Effect of stress on phase transformation.

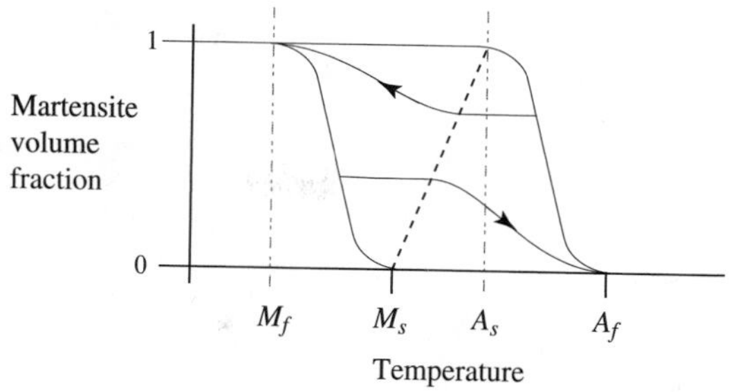

(a) Stress-free temperature-induced phase transformation

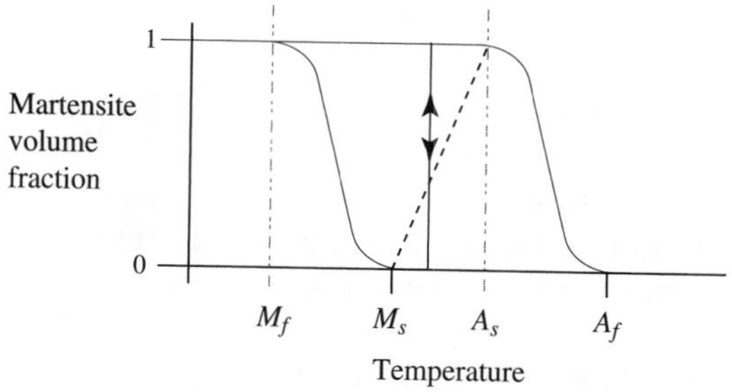

(b) Isothermal stress-induced phase transformation

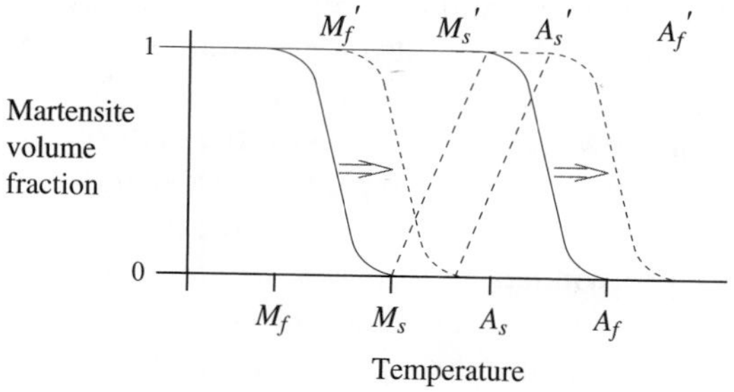

(c) Effect of applied stress on temperature-induced transformation

The diagonal line joining M_s and A_s (the trigger line), is the boundary between two phase transformations. Depending on the initial state of the material, the forward transformation can start at a much higher temperature than M_s and the reverse transformation can start at a much lower temperature than A_s (see Fig. 3.14(a)).

During an isothermal stress-induced transformation, the transformation starts when the mechanical energy due to applied stress becomes equal to the required energy of the opposite phase. As shown in Fig. 3.14(b), the up and down vertical path lines, respectively, represent the forward (loading) and reverse (unloading) transformations. These transformations are accompanied by heat generation. However, it is normally assumed that this heat generation is negligible and does not affect the temperature of the material. Figure 3.14(c) shows the effect of applied stress on transformation temperatures. There is no doubt that the temperature and stress have mutual effects on the transformation mechanisms. As shown, with higher stress, the transformation temperatures increase and the hysteresis loop moves toward the right side.

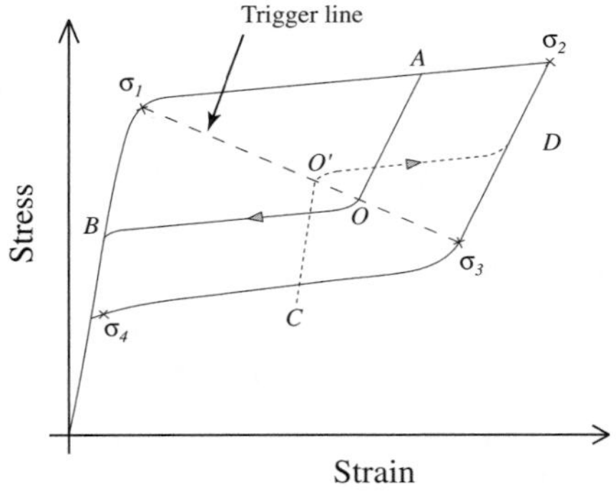

Figure 3.15. Critical stress trigger line in pseudoelastic hysteresis.

As in the temperature-induced transformation, there is also a trigger line for critical stress (Figure 3.15). This is the line joining σ_1 (critical stress for initiation of the martensite phase) in forward-phase transformation and σ_3 (critical stress for initiation of the austenite phase) in reverse-phase transformation. Let us consider a case when the forward-phase transformation stops at A. On reducing the stress, it takes a recovery path AOB. Point O falls on the trigger line. AO is elastic with the Young's modulus of the full recovery part $(\sigma_2 - \sigma_3)$. At the recovery point O, the reverse transformation starts and finishes at point B. In a similar way, if the stress level is increased during unloading, it follows the path $CO'D$. Point O' falls on the trigger line and CO' represents an elastic region. Then, forward transformation takes place. Note that the area under the pseudoelastic hysteresis curve represents the amount of energy dissipation.

Below the critical stress σ_1, the material behaves elastically in the austenite state. Between σ_1 and σ_2, two phases coexist and the deformation is nonhomogeneous. At a certain critical stress σ_2, most of the material is transformed to stress-induced martensite (detwinned martensite). Beyond the critical stress σ_2, the material behavior is again elastic (monoclinic martensite lattice with lower Young's modulus). Elastic distortion continues up to a stress level of σ_5; beyond this stress, the martensite lattices begin to slip and permanent deformation starts taking place (tangent modulus starts decreasing), as shown in Fig. 3.16. At a stress level of σ_6, the tangent modulus reaches a minimum value. A further increase in stress/displacement would lead to failure. On unloading, some pockets of material transform to austenite. Some strain is recovered, but there will be an irrecoverable permanent strain ϵ_p in the specimen. Below the elastic stress limit, σ_5, the material continues to be in the stress-induced martensite state (detwinned) until the stress level σ_3 is reached, and then the transformation from stress-induced martensite to austenite begins. By the time the unloading plateau stress σ_4 is reached, the material is completely transformed to austenite state. On unloading to the zero-stress state, the strain is completely recovered and the material is in the austenite phase.

3.1.6 Two-Way SME

The one-way SME results in a single thermally activated shape change; that is, the material "remembers" only the high-temperature shape. Any deformations introduced in the low-temperature phase are erased on the application of a temperature

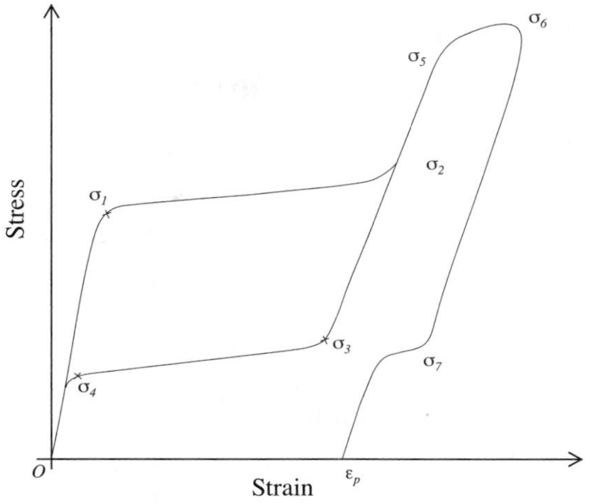

Figure 3.16. Internal recovery in pseudoelastic hysteresis.

high enough to ensure complete transformation to the austenite phase. To provide actuation in a cyclic manner, the material must be mechanically deformed after each strain recovery. Shape memory alloys also exhibit a phenomenon known as the two-way shape memory effect (TWSME) [5, 6, 7, 8], or "reversible shape memory effect" [9, 10]. In the two-way effect, the material "remembers" both a high- and a low-temperature shape. Consequently, the material can continuously cycle between the two shapes as the temperature is raised and lowered, without the need for an external stress. This makes actuators based on the TWSME attractive in a variety of applications. Note that this effect is different from the stress-induced two-way SME, in which the specimen is under applied stress and the temperature is varied, although the specimen recovers to two different shapes on heating and cooling.

A TWSME can be introduced by appropriate "training" of the material, which usually involves some combination of thermal and mechanical cycles. The microstructural changes during TWSME are a matter of continuing research and are discussed in the literature [7, 8]. The training procedure introduces microstresses in the material, which result in the preferential formation of specific martensite twin variants. Stress-induced, or stress-biased, martensite forms the major part of the material at low temperatures, which transforms to austenite at higher temperatures [11]. Figure 3.17 gives a schematic description of the TWSME induced in the material as a result of a specific type of training. The material starts off in the undeformed state (a), at a temperature below M_f. It is then deformed at low temperature to the shape (b). On heating to above A_f, it recovers some of the strain and assumes a shape (c), close to the initial shape. On subsequent cooling to below M_f, it does not retain its shape but instead returns to a shape (d) that is close to the original deformed shape. On further temperature cycles, the material changes between shapes (c) and (d). From this simplistic description, it is obvious that the maximum possible change in strain due to the TWSME must be less than the maximum recoverable strain of the material (one-way shape memory effect).

Sometimes, TWSME can develop in applications that are based on the one-way SME, as a consequence of the thermomechanical environment experienced by the actuator over the course of multiple cycles. As the shape recovery in TWSME occurs

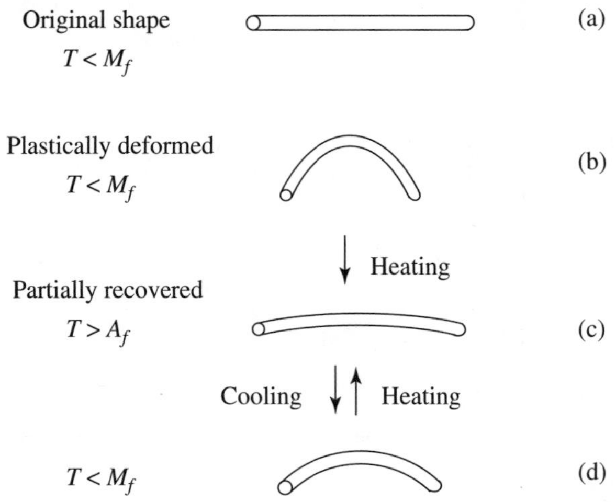

Figure 3.17. Two-way shape memory effect.

in a stress-free condition, such an effect is identified by a progressive decrease in the stress required to cycle the actuator. There are several different training techniques that can be used to intentionally impart TWSME to the material. A detailed description of these training techniques is provided by Perkins and Hodgson [8], and Blonk and Lagoudas [12], and the characteristics of each training method as well as the TWSME properties it imparts have been discussed by several authors. The training techniques can be broadly classified under the following types:

(1) Overdeformation: The material is initially cooled below M_f. It is then deformed plastically, beyond the usual limit for OWSME (one-way shape memory effect). The strains in the material at this point are much higher than the maximum residual strain ϵ_L. As a result, on heating to above A_f, the material only partially recovers its original shape. Subsequent cooling to below M_f will result in the material moving back toward the intial deformed shape. This sequence of events is shown schematically in Fig. 3.17. A similar procedure involves cycling the material through the pseudoelastic region, applying a plastic strain, and finally rapidly heating up the material [13].
(2) Repeated cycling (SME) [14, 15]: The material is initially cooled below M_f and then deformed and unloaded. This is followed by heating to above A_f, which results in complete recovery of the strain (OWSME). The material is then cooled to below M_f and the process is repeated. Around 5–10 cycles are performed, and gradually the material will begin to spontaneously change shape on cooling, moving toward the deformed shape without application of stress. However, the spontaneous deformation will be much less than that introduced for training (typically 1/5 to 1/4). Therefore, because the maximum training strain for SME is 6%, the maximum TWSME is around 1–2%.
(3) Repeated cycling (pseudoelastic range): In this method, the material is heated to above A_f but below M_d, placing it in the pseudoelastic region. The material is repeatedly cycled between the stressed and stress-free condition. After around 5–10 cycles, the material begins to retain memory of both the deformed and the undeformed states. Again, after the training, the maximum strain exhibited by the TWSME is a fraction of ϵ_L.

(4) Combined SME/pseudoelastic cycling [16]: The material is first deformed pseudoelastically, then cooled to below M_f at constant strain. The constraint on the material is then released, and it is heated to above A_f to recover the initial shape via the SME. Several repetitions of this combined SME/pseudoelastic cycle impart the TWSME to the material.
(5) Constant strain, temperature cycling [17]: The material is initially deformed at a temperature below M_f. It is then kept at constant strain while the temperature is cycled between a low temperature (below M_f) and a high temperature (above A_f). Several such cycles will impart the TWSME to the material.

Other techniques have been suggested that are slightly different than those listed here. Mellor [18] proposed a technique involving heating the sample above A_f and imposing a strain (between 0.7% and 2%), followed by a partial anneal accomplished by heating the material to half of the annealing temperature. The constraint on the material is then released and it is allowed to cool. Sun and Hu [19] proposed deforming the material and heating it to $A_f + 30°C$ after unloading, then quenching it in water. Tokuda et al. [20] described the two-way effect obtained by subjecting a thin-walled tube to combined axial and torsional loading. Although these techniques vary in their procedure, the final TWSME behavior that is imparted to the material is the same. As with all SMA phenomena, annealing can erase all of the memory of the material. Therefore, care must be taken not to increase the temperature too high.

It should also be noted that the TWSME is quite easy to introduce inadvertently into the material during the course of experiments and during normal operation in several applications. For example, if the constant-stress-recovery test is performed repeatedly for several different values of stress, it is equivalent to performing repeated SME cycles, which as described previously, can induce the TWSME in the material. However, this can be erased by an appropriate annealing procedure.

A major consequence of the deformation mechanism is that the material can recover more deformation under an external stress when it is heated compared to when it is cooled. Therefore, some applications incorporate a return spring or bias spring to help the material return to its low-temperature deformed state. However, the inclusion of such a spring element partly reduces the advantage of the TWSME over the OWSME. To impart repeatable and consistent TWSME behavior to a number of different samples, their training procedure must be identical to their annealing procedure, which precedes it.

Other important issues regarding the TWSME include the number of cycles that the material takes to stabilize and the degradation of the material response with time. The stabilization period, or the number of cycles required for the training to be complete and the material to exhibit repeatable TWSME behavior, is highly dependent on the training procedure and the alloy composition [17, 21]. The TWSME can also degrade over a number of actuation cycles. In such a case, the material may not be able to deform completely to its low-temperature shape or to completely recover the strain at its high-temperature condition. This degradation in performance is increased by external stress and is larger for a larger change in strain between the low-temperature and high-temperature shapes. A typical number could be 20% degradation in recovered strain during 1000 cycles [22]. The degradation is not constant with time and can accelerate as the number of cycles increases. Scherngell and Kneissl [23] discuss the degradation of TWSME in a binary Ni-50.3% Ti alloy as a function of training parameters. The training was performed by thermal cycling in a constant-stress condition, and the material was subsequently subjected to stress-free thermal cycles to observe the TWSME. It was noted that in the initial stages, there was a large degradation in the TWSME, which stabilized as the number of thermal cycles increased. The actual amount and rate of degradation is highly specific to the material composition, operating environment, and amount of TWSME introduced, and it must be characterized separately for each system.

Kafka [24] used a mesomechanical approach to model the two-way shape memory phenomenon. It is based on a substructure that is continuous and remains elastic throughout the martensite and austenite subvolumes and is applicable to binary alloys (Ni-Ti) for quasi-static processes. A phenomenological model of TWSME was developed by Takagi et al. [25] and was used in a finite element model to predict the thermomechanical response of SMA plates with TWSME. This approach is based on the internal energy of the material and changes in the energy associated with phase transformation. A phenomenological approach is taken by considering an equivalent transformation-induced specific heat, which is a function of the transformation temperatures. The value of the equivalent specific heat can be measured using differential scanning calorimeter (DSC). The R-phase transformation is also modeled using this approach. Blonk et al. [12] adapted an SMA constitutive model to predict the behavior of an elastomeric rod with embedded SMA wires that were trained to exhibit the TWSME. In general, the modeling of the TWSME is very similar to that of the SME, with specific focus on the transformation temperatures and the strain-temperature behavior of the material.

3.1.7 All-Round SME

A phenomenon that is very similar to the TWSME but is exhibited only in specific alloys is called the all-round shape memory effect (ARSME). The basic behavior is the same as in the case of the TWSME: When the material is heated or cooled, it transforms to either a high-temperature or a low-temperature shape, without the need for an external stress to cause the shape change. This phenomenon was first reported by Nishida et al. [26, 27] for a Ti-51%Ni alloy. An important aspect of ARSME is that it is exhibited only by NiTi alloys having more than 50.5% Ni. The training procedure required to impart the ARSME is quite different than that required for TWSME. This procedure [8] is shown schematically in Fig. 3.18.

The material is first deformed at a low temperature (below M_f). It is then constrained in the deformed position while the temperature is raised to about 400°C. It is aged in this condition for about 50–100 hours (similar to annealing). A solutionizing process can be performed before the aging, keeping the material at 800°C for 20 minutes [28]. During this aging process, it is hypothesized that precipitates form in the material, which generate local areas of high internal stress. When the material is cooled down and the constraints are released, it assumes a shape that is opposite to the initial deformed shape. It is believed that the material deforms in this manner to alleviate the internal stresses created during the aging process. The exact micromechanical processes that occur during the ARSME are a subject of ongoing research.

After the material is cooled, it can be cycled between the low-temperature and the high-temperature shape (i.e., the shape the material was constrained in during aging) by appropriately changing the temperature. Typical values of strain obtained

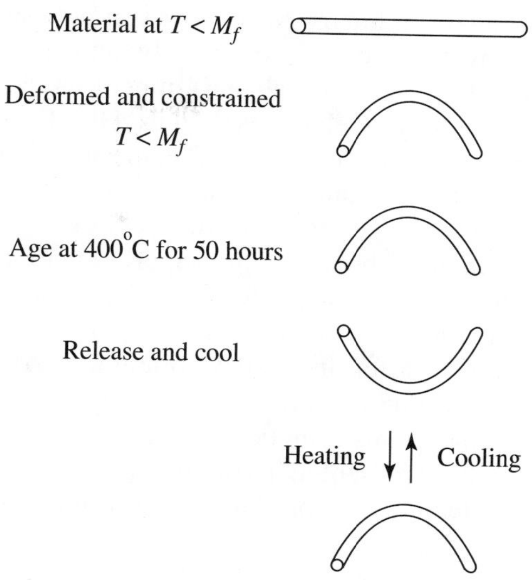

Figure 3.18. Schematic of training procedure to impart ARSME.

during the ARSME are on the order of 0.25%, with a temperature hysterisis of 12°C, although specific values are highly material-dependent.

3.1.8 R-Phase Transformation

The typical austenite-to-martensite transition is characterized by a large temperature hysteresis, which can be as high as 10°–20°C. In some alloys, while cooling down from the austenite state, a transformation can be observed at a temperature $T > A_f$. This transformation is characterized by a much smaller temperature hysteresis, on the order of 1.5°C. During this transformation, the cubic unit cell of the material undergoes a distortion along its diagonal, making it rhombohedral. Therefore, this transformation is called the R-phase transformation. The R-phase transformation occurs between the temperatures T_R and $T'_R (A_f < T_R < T'_R)$, also sometimes called R_s (start of R-phase transformation) and R_f (finish of R-phase transformation).

As the temperature is decreased, the lattice distortion increases, until the martensitic transformation begins at M_s. The material in the R-phase exhibits the SME and pseudoelasticity; however, maximum recoverable strain is on the order of 0.5%–1%. Therefore, although some applications make use of the R-phase due to its inherently low-temperature hysteresis, the maximum achievable strains are much lower than in the case of the martensitic SME. Similar to the case of martensitic SME, changes in other physical properties of the material are observed during the R-phase transition. For example, the electrical resistivity can increase by a factor of almost 300% during the martensite-to-austenite transformation (on heating from below M_f to above A_f) and subsequently can decrease by almost 50% during the R-phase transformation (heating from T_R to T'_R). The R-phase transformation temperatures are typically measured by means of DSC tests.

The appearance of the R-phase can be noticed as a small discontinuity or non-linearity in the stress-strain curve, or in the DSC measurements. The R-phase is also observed primarily at low stress levels. At higher stresses, the material is dominated by the formation of stress-induced martensite. Therefore, the maximum load is also limited in applications utilizing the R-phase transformation. The appearance of the

R-phase in binary NiTi alloys can be encouraged by various procedures [29], such as by cold-working followed by annealing (400°C–500°C), by aging of alloys with greater than 50.5% Ni at a temperature between 400°C–500°C, or by the addition of other alloying elements such as Al or Fe. Only in the case of ternary alloys, such as NiTiFe, NiTiAl, and NiTiCo, does the R-phase occur spontaneously.

The mechanical behavior associated with the R-phase was first investigated by Khachin et al. [30]. Miyazaki et al. [29, 31] made a detailed investigation of mechanical behavior and superelasticity in the R-phase as well as the behavior of single-crystals in the R-phase. The modeling of the R-phase is also similar to that of the martensitic phase. The R-phase is represented by a volume fraction, η, similar to the martensite volume fraction ξ. Assuming the first cycle occurs under zero stress-strain condition ($\eta = 0$, $\xi = 0$), an application of strain below 1% results in R-phase transformation ($\eta = 1, \xi = 0$). The R-phase and martensite transformations can be represented in a unified way [32, 33] using energy functions, with appropriate "switching functions" to handle different transformations.

3.1.9 Porous SMA

Porous SMA offers the potential of higher specific actuation energy and damping capacity. Also, porous SMA can be engineered to match the impedance at the connecting joints to optimize performance. Porous SMAs can be manufactured with open- and closed-pore designs. Fabrication methods include casting, metallic deposition, and powder metallurgy. Three methods are commonly used to manufacture porous Nitinol from powder metallurgy. One method is conventional sintering, which requires a long heating time and produces samples that are limited in terms of shape and pore size. First, a cold mixture of Ni and Ti powder is compacted into pellets, which are then sintered at near melting temperature to produce a binary NiTi phase through diffusion. The porosity of the specimen is varied by the powder compaction pressure and the initial shape and size of the powder. The voids left between powder particles result in the porosity of the specimen. Sintering requires a long heating time of about 48 hours, and the process is limited to small specimens.

The second method is self-propagating high-temperature synthesis (SHS), and it is initiated by a thermal explosion ignited at one end of the specimen, which gradually propagates to the other end in a self-sustained manner. The third method is sintering Ni and Ti powder at elevated pressure using a hot isostatic press (HIP), which compresses and traps argon gas bubbles between neighboring metal powder particles. Lagoudas and Vandygriff [34] refined this technique to develop small and large porous NiTi SMA specimens with varying pore size ranging from 20 μm to 1 mm. The porosity of fabricated specimens varied from 50% to 42%.

3.2 Constrained Behavior of SMA

Shape memory alloy is commercially available in the form of tubes, wires, and bars. The wire form is one of the most widely used in applications. Because there is no need for machining, SMA wires are very easy to incorporate in actuation mechanisms. Thermal activation of the wires can also be accomplished conveniently and compactly by passing an electric current through the wires. The resistance of the wires causes a self-heating that activates the SME. The wires can also be heated externally by setting them up in a thermal chamber. For experimental characterization,

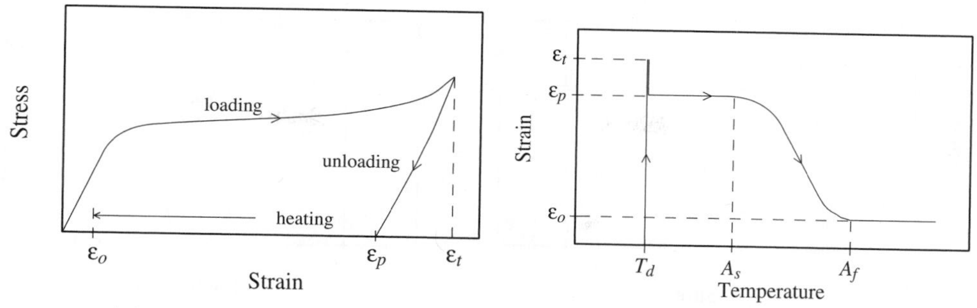

(a) Stress-strain behavior (b) Strain-temperature behavior

Figure 3.19. Schematic of transformation during free recovery of an SMA wire.

testing of SMA wires provides an insight into the fundamental behavior of SMA, while simplifying the loading condition to purely uniaxial quasi-static (low strain rate) external loads.

Before discussing the experimental behavior of SMA wires due to change in temperature, it may be instructive to look qualitatively at the behavior of an SMA wire under unloaded and loaded conditions.

3.2.1 Free Recovery

Let us consider an example in which an extensional load is applied on an SMA wire specimen that is initially in the fully martensitic state, at a temperature $T_d < M_f$. At an applied load, a maximum strain of ϵ_t is generated. On unloading, the wire recovers some amount of elastic strain, and an apparent plastic strain ϵ_p is retained. We refer to this as an "apparent" plastic strain because it is, in reality, recoverable by the SME, except for the plastic strain that occurs beyond the second yield point. Now if the wire is heated to a temperature higher than A_f, it recovers to a final strain ϵ_o. This is shown schematically in Fig. 3.19(a). The variation of strain with temperature during the loading and heating process is shown in Fig. 3.19(b). Thus, the recoverable strain is $(\epsilon_p - \epsilon_o)$. Note that the recoverable strain increases with total strain ϵ_t, reaches a maximum value, and then decreases as the permanent slip starts increasing. For Nitinol, the maximum strain that is fully recoverable is about 8%. For Cu-Al-Ni and Cu-Zn-Al alloys, the maximum recoverable strain is on the order of 3–5%.

3.2.2 Constrained Recovery

While in the free-recovery case, one end of the wire is clamped and the other end is left free. In the constrained-recovery case, both ends of the wire are rigidly clamped. The wire is initially loaded to a strain ϵ_t and unloaded to a pre-strain of ϵ_p. Then, both ends of the wire are clamped and the wire is heated to a temperature above A_f. The wire tries to recover the pre-strain; however, because the ends are clamped, a stress is generated in the wire. This stress is called the "recovery stress" or the blocked stress, σ_{bl}. This recovery stess is much higher than the stess σ_t that generated the pre-strain. A schematic of the stress-strain behavior during constrained recovery is shown in Fig. 3.20(a) and the strain-temperature behavior is shown in Fig. 3.20(b). The constrained-recovery stress is used in a wide range of applications that include fasteners and couplers, as well as very-low-frequency (<1 Hz) actuators.

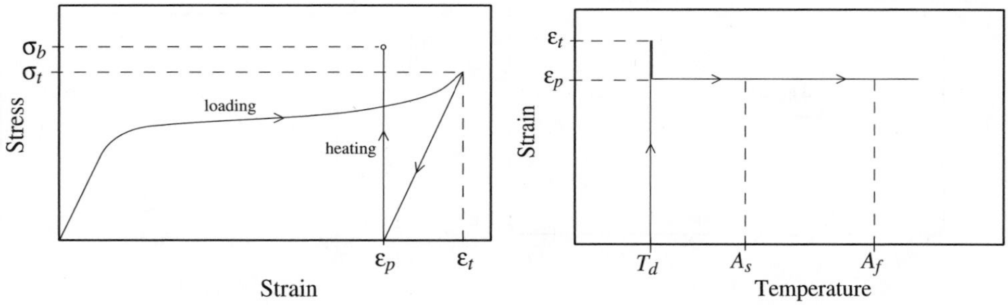

(a) Stress-strain behavior (b) Strain-temperature behavior

Figure 3.20. Schematic of transformation during constrained recovery of an SMA wire.

3.2.3 Effective Load Lines of an SMA Wire Actuator

The free-recovery and constrained-recovery cases represent the two limiting cases of boundary conditions: infinitely soft and infinitely stiff, respectively. When the SMA wire is acting against any other boundary condition, such as a spring of finite stiffness, the behavior will lie between these two limiting cases. This is schematically represented in Fig. 3.21. The path labeled 1 is for the SMA wire acting against an infinitely stiff support, leading to a recovery stress of σ_b. The path labeled 4 is for a free-end condition, where no stress is developed in the wire. For the wire acting against a linear spring of finite stiffness, the stress and strain follow the path 2 on heating, leading to a final stress σ_k in the wire. For the case of a nonlinear spring, such as another SMA wire, the path followed is 3, leading to a final stress value of σ_{SMA}. The stress-strain characteristics of the SMA wire can therefore be treated as effective actuator load lines. Note that these load lines are highly nonlinear, in contrast to the load lines of piezoelectric actuators.

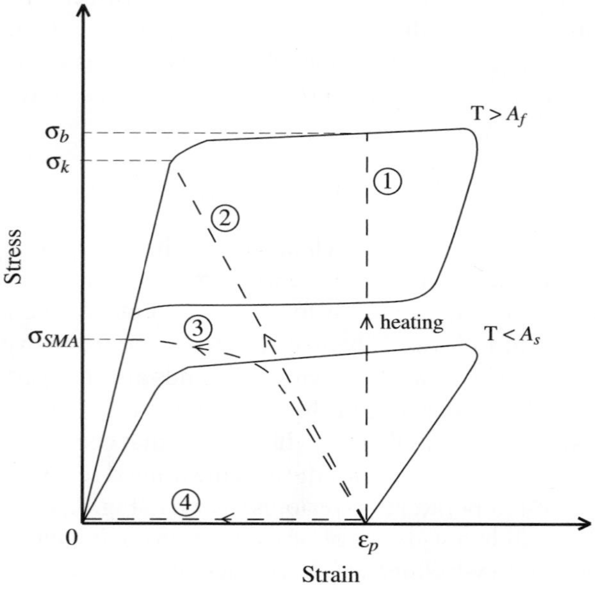

Figure 3.21. Schematic of stress-strain behavior of the SMA wire with different boundary conditions.

3.3 Constitutive Models

Most of the constitutive models proposed to describe the thermomechanical behavior of SMAs are quasi-static and uniaxial for the OWSME. They can be broadly classified into three categories: phenomenological macromechanics models, thermodynamics-based micromechanics models, and micromechanics-based–hybrid macromechanics models. These models are also discussed in Chapter 1.

For structural analysis, the model has to be simple and applicable in standard stress-strain mechanics analyses. It should also incorporate realistic physics and be applicable over a wide range of temperatures and stresses to capture both SME and pseudoelasticity. It should be adaptable to a wide range of materials and textures in both single crystals and polycrystals. The crystallographic symmetry of the austenite phase is higher than that of the martensite phase and, as a result, one can get a number of symmetry-related variants of martensite evolved around the load and temperature history.

Macroscopic Phenomenological Models

These models are based on phenomenological thermodynamics and are mostly defined using experimental data (curve-fitting). These are simple and capture adequate physics. They are quite amenable to inclusion in engineering analyses. In most of these models, the strain, temperature, and martensite volume fraction are the only state variables. They are based on the phase diagram of SMA transformation, which describes the transition from the martensite to austenite (parent phase) and austenite to martensite phases under stressed conditions. Most of these models are developed for uniaxial loading.

Microscopic Thermodynamics Models

These models depend on micro-scale thermodynamics to describe phenomena and as a result are quite involved. These are less amenable to inclusion in engineering analyses. They are beneficial for explaining phenomena at the micro-scale, such as nucleation, interface motion, and growth of a martensite phase.

Micromechanics-Based Hybrid Macroscopic Models

These are hybrid between the first two categories. They capture key details from micro-scale thermodynamics and incorporate several simplifying assumptions to describe phenomena at the macroscopic level. They estimate the interaction energy due to phase transformation of the material at the microstructure level using a group of important variants. They may be amenable for inclusion in engineering analyses.

3.4 Quasi-Static Macroscopic Phenomenological Constitutive Models

This section describes four quasi-static constitutive models that have been proposed to describe the material behavior in the OWSME. These models are chosen due to their common approach and wide applicability to a range of operating conditions. To accommodate the large variations in SMA material properties due to manufacturing, composition, training, heat treatment, and other factors, these models use material parameters that are determined experimentally. For simplicity, only uniaxial loading behavior is considered, and quasi-static deformation is assumed, resulting in an isothermal condition.

In these models, it is assumed that strain, temperature, and the martensite volume fraction are the only state variables. Most of the constitutive models are developed for quasi-static loading and, as such, it is assumed that the material at each instant is in thermodynamic equilibrium. Such an assumption is not applicable if the strain rate is high. Typically, a strain rate below 5×10^{-4}/s for the wire sample represents a quasi-static loading condition. Tanaka [35] developed an exponential expression to describe the stress and temperature rather than determining the free-energy expression. Liang and Rogers [36] presented a model that is based on the rate form of the constitutive equation developed by Tanaka. In their model, Tanaka's equation is integrated with respect to time and it is assumed that the coefficients in the equation are constant. A major drawback of the Tanaka and the Liang and Rogers models in their original form is that they do not capture the stress-induced detwinning of the martensite phase. However, both models describe the phase transformation from martensite to austenite and vice versa. Brinson [37] divided the martensite volume fraction into two parts, the stress-induced and temperature-induced components, and modified the Tanaka model accordingly. Both Liang and Rogers and Brinson used the cosine form of evolution kinetics instead of the exponential form used by Tanaka. Boyd and Lagoudas described the modeling of SMAs using the thermodynamic approach, in which the constitutive relations are derived from free energy and a dissipation potential. Ivshin and Pence [38] used an inverse hyperbolic tangent form of evolution kinetics. Most of the phenomenological models are one-dimensional, but the Boyd-Lagoudas model is applicable to three-dimensional loading.

Because stress is a function of temperature T, martensite volume fraction ξ, and strain ϵ, the material constitutive relation in differential form becomes

$$d\sigma = \frac{\partial \sigma}{\partial \epsilon} d\epsilon + \frac{\partial \sigma}{\partial \xi} d\xi + \frac{\partial \sigma}{\partial T} dT \tag{3.1}$$

This leads to a general expression

$$d\sigma = E(\epsilon, \xi, T) d\epsilon + \Omega(\epsilon, \xi, T) d\xi + \Theta(\epsilon, \xi, T) dT \tag{3.2}$$

where $E(\epsilon, \xi, T)$ represents modulus of the material, $\Omega(\epsilon, \xi, T)$ is the transformation tensor, and $\Theta(\epsilon, \xi, T)$ is thermal coefficient of expansion for the SMA material. Because the strains due to the thermal coefficient of expansion are much lower than the strains due to the phase transformation, this coefficient is normally neglected.

3.4.1 Tanaka Model

One of the popularly used models is Tanaka's model [35], which is derived from thermodynamic considerations. The second law of thermodynamics is written in terms of the Helmholtz free energy and then its rate form is derived. The strain ϵ, temperature T, and martensite volume fraction ξ are assumed to be independent state variables. The stress σ in the material is calculated from these quantities. Because the martensite volume fraction ξ is dependent on the stress, an iterative numerical solution of the equations is necessary.

From Eq. 3.2, the constitutive equation is derived as

$$(\sigma - \sigma_o) = E(\xi)(\epsilon - \epsilon_o) + \Theta(T - T_o) + \Omega(\xi)(\xi - \xi_o) \tag{3.3}$$

where E is the Young's modulus in the elastic regime of the material, Θ is a thermo-elastic constant, and Ω is a phase-transformation constant. The terms associated with subscript o refer to the initial state of the material. Eq. 3.3 shows that the stress consists of three parts: the mechanical stress, the thermo-elastic stress, and the stress due to phase transformation.

Note that the Young's modulus and the phase-transformation coefficient are functions of the martensite volume fraction ξ. The rule of mixtures can be used to calculate the effective modulus of the material containing both austenite and martensite phases, resulting in the expression

$$E(\xi) = E_A + \xi(E_M - E_A) \tag{3.4}$$

where E_A is the Young's modulus in the austenite phase and E_M is the Young's modulus in the martensite phase. The ratio of E_A to E_M is generally greater than 2.

If we consider a material at a temperature below the martensite finish ($T < M_f$) and zero stress/strain condition ($\sigma_o = \epsilon_o = 0$), the material will be completely in martensite phase (detwinned state). This, in turn, helps to determine the transformation constant Ω as

$$\Omega(\xi) = -\epsilon_L E(\xi) \tag{3.5}$$

where ϵ_L is the maximum recoverable strain. Tanaka's model assumes an exponential function for the martensite volume fraction. During the austenite-to-martensite ($A \rightarrow M$) transformation (forward phase), ξ is given by

$$\xi(\sigma, T) = 1 - e^{a_M(M_s - T) + b_M\sigma} \tag{3.6}$$

During the $M \rightarrow A$ (martensite-to-austenite) transformation (inverse phase transformation)

$$\xi(\sigma, T) = e^{a_A(A_s - T) + b_A\sigma} \tag{3.7}$$

where a_M, b_M, a_A, and b_A are empirically determined constants given by

$$a_M = \frac{\ln(0.01)}{(M_s - M_f)} \qquad b_M = \frac{a_M}{C_M} \tag{3.8}$$

$$a_A = \frac{\ln(0.01)}{(A_s - A_f)} \qquad b_A = \frac{a_A}{C_A} \tag{3.9}$$

The C_A and C_M are called the "stress influence coefficients." C_A is the stress influence coefficient (or stress rate) for the austenite phase and is given as

$$C_A = \frac{1}{\frac{dA_s}{d\sigma}} \tag{3.10}$$

C_M is the stress influence coefficient for the martensite phase and is given as

$$C_M = \frac{1}{\frac{dM_s}{d\sigma}} \tag{3.11}$$

These coefficients (C_A and C_M) represent the effect of stress on the transformation temperatures and are the slopes of the austenite and martensite lines,

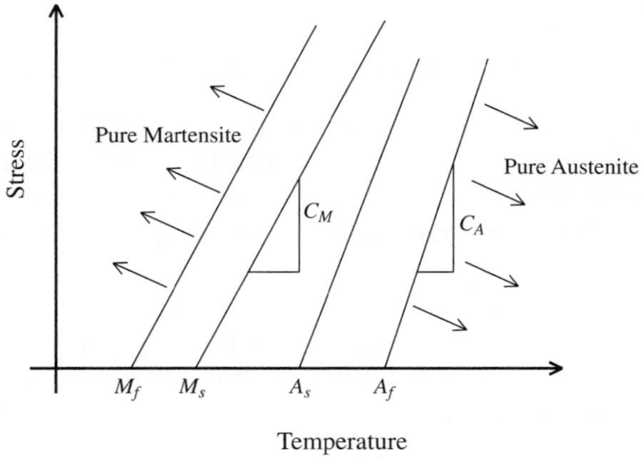

Figure 3.22. Critical stress-temperature phase diagram, Tanaka's model.

respectively, on the critical stress-temperature diagram. A schematic critical stress-temperature phase diagram is shown in Fig. 3.22, and this diagram describes the way in which stress affects the transformation temperatures. On the x-axis, the intercepts of the lines mark the transformation temperatures at zero stress. The regions in which the pure phase exists are marked on the diagram and, in other regions, one or more phases can coexist. It is assumed that there is a linear variation of transformation temperatures with stress. The points on these curves are normally obtained experimentally and are conveniently fit with a straight line, although the actual variation may not be linear. In polycrystalline SMAs, there can be a significant difference in transformation behavior in compression and tension. For most applications, the phase diagram is obtained under tensile stress.

It is frequently assumed that the stress-influence coefficients, C_A and C_M, are equal. Note that a_M, a_A, b_M, and b_A are lumped terms that are defined in terms of other parameters. The Tanaka model has also been modified [39] to include the R-phase transformations that are sometimes seen in SMAs. The R-phase occurs in some SMA compositions under certain specific conditions, during cooling prior to the martensitic transformation. A detailed description of the R-phase is in Ref. [40] as well as in Section 3.1.8.

In the modified Tanaka model, there are two distinct variables for R-phase and detwinned martensite, which make it possible to predict the R-phase and SME simultaneously. However, this model is applicable only to materials exhibiting an R-phase transformation. This was determined by the fact that a typical material with an R-phase transformation exhibits two peaks (one each for the R-phase and martensitic transformations) in each of the heating and cooling cycles of the DSC measurements.

3.4.2 Liang and Rogers Model

The second model was formulated by Liang and Rogers [36]. They utilized the same constitutive relation and form of the evolutionary equation for the martensite volume fraction as was developed by Tanaka. The difference between the two models arises in the modeling of the martensite volume fraction. In the Liang and Rogers model, ξ is modeled as a cosine function.

For the $A \rightarrow M$ transformation, the parameter is defined as

$$\xi = \frac{(1-\xi_A)}{2}\cos\left[a_M(T-M_f)+b_M\sigma\right]+\frac{(1-\xi_A)}{2} \tag{3.12}$$

and for the $M \rightarrow A$ transformation as

$$\xi = \frac{\xi_M}{2}\left\{\cos\left[a_A(T-A_s)+b_A\sigma\right]+1\right\} \tag{3.13}$$

The empirical constants are defined by

$$a_M = \frac{\pi}{(M_s-M_f)} \qquad b_M = -\frac{a_M}{C_M} \tag{3.14}$$

$$a_A = \frac{\pi}{(A_f-A_s)} \qquad b_A = -\frac{a_A}{C_A} \tag{3.15}$$

where ξ_A and ξ_M are the initial martensite volume fraction values for the $A \rightarrow M$ and $M \rightarrow A$ transformation processes, respectively. Usually, these values are obtained by assuming an initial phase. Again, various parameters are determined experimentally by testing SMA specimens.

Note that both the Tanaka model and the Liang and Rogers model have a serious limitation: They do not represent the stress-induced martensite-phase transformation. This means that detwinning of the martensite phase (from initially 100% twinned martensite) is not captured. At a temperature below the martensite finish temperature ($T < M_f$), the stress-strain behavior of material is represented as linear elastic ($\sigma = E\epsilon$) and the material is assumed to be in martensitic-phase ($\xi = 1.0$). In fact, stress-induced martensitic-phase transformation is not covered for temperatures below M_s or even for high temperatures, when any temperature-induced martensite is present.

3.4.3 Brinson Model

The third model was developed by Brinson [37]. This model captures stress-induced martensite at all temperatures. The Brinson model addresses this issue by separating the martensite variable into stress-induced and temperature-induced components.

$$\xi = \xi_s + \xi_T \tag{3.16}$$

The stress-induced martensitic volume fraction ξ_s describes the amount of detwinned or stress-preferred variant of martensite (single variant) present in the sample, and the temperature-induced martensite volume fraction ξ_T describes the amount of martensite (containing all variants) that occurs from the reversible-phase transformation from austenite. The sum of the two martensite volume fraction components is always ≤ 1.0. This model uses the same constitutive equation as the Tanaka model and the Liang and Rogers model, with some modifications. The coefficients of the constitutive equation are assumed to be variable in order to account for both the SME and pseudoelasticity effect. The constitutive equation in differential form becomes

$$d\sigma = \frac{\partial\sigma}{\partial\epsilon}d\epsilon + \frac{\partial\sigma}{\partial\xi_s}d\xi_s + \frac{\partial\sigma}{\partial\xi_T}d\xi_T + \frac{\partial\sigma}{\partial T}dT \tag{3.17}$$

This reduces to

$$d\sigma = E(\xi)d\epsilon + \Omega_s d\xi_s + \Omega_T d\xi_T + \Theta dT \tag{3.18}$$

which can be written as

$$\sigma - \sigma_o = E(\xi)\epsilon - E(\xi_o)\epsilon_o + \Omega_s(\xi_s - \xi_{so}) + \Omega_T(\xi_T - \xi_{T_o}) + \Theta(T - T_o) \tag{3.19}$$

where

$$E(\xi) = \frac{\partial \sigma}{\partial \epsilon} = E_A + \xi(E_M - E_A) \tag{3.20}$$

$$\Omega_S = \frac{\partial \sigma}{\partial \xi_S} \tag{3.21}$$

$$\Omega_T = \frac{\partial \sigma}{\partial \xi_T} \tag{3.22}$$

$$\Theta = \frac{\partial \sigma}{\partial T} \tag{3.23}$$

Assume that the material initially is in the austenite phase, at a temperature $T > A_f$, and in a condition of zero stress and zero strain. Then

$$\xi_{so} = 0, \qquad \xi_{T_o} = 0, \qquad \sigma_o = 0, \qquad \epsilon_o = 0 \tag{3.24}$$

From this initial condition, let the stress on the material be slowly increased. As a result, the material starts transforming from austenite to stress-induced martensite (detwinned). Once the martensite is completely detwinned, let the stress be removed, yielding the residual strain ϵ_L in a stress-free condition. The temperature is maintained constant throughout the stress cycle. The final condition is then given by

$$\xi_s = 1, \qquad \xi_T = 0, \qquad \sigma = 0, \qquad \epsilon = \epsilon_L \tag{3.25}$$

Substituting the initial conditions (Eq. 3.24) and the final conditions (Eq. 3.25) in the constitutive equation (Eq. 3.19), we obtain

$$\Omega_s = -\epsilon_L E \tag{3.26}$$

Now let us consider another case in which the material is at a temperature below the austenite start ($T < A_s$) and in a 100% martensite state, at zero stress and zero strain. The initial conditions are

$$\xi_{so} = 0, \qquad \xi_{T_o} = 1, \qquad \sigma_o = 0, \qquad \epsilon_o = 0 \tag{3.27}$$

A stress is applied and subsequently removed (at a constant temperature), such that the material transforms into completely detwinned martensite, with a residual strain ϵ_L. The final condition is given by

$$\xi_s = 1, \qquad \xi_T = 0, \qquad \sigma = 0, \qquad \epsilon = \epsilon_L \tag{3.28}$$

Substituting the initial conditions (Eq. 3.27) and the final conditions (Eq. 3.28) in the constitutive equation (Eq. 3.19), we obtain

$$\Omega_T = 0 \tag{3.29}$$

The resulting modified constitutive equation becomes

$$\begin{aligned} \sigma - \sigma_o &= E(\xi)\epsilon - E(\xi_o)\epsilon_o + \Omega(\xi)\xi_s - \Omega(\xi_o)\xi_{so} + \Theta(T - T_o) \\ &= E(\xi)\epsilon - E(\xi_o)\epsilon_o - \epsilon_L E(\xi)\xi_s + \epsilon_L E(\xi_o)\xi_{so} + \Theta(T - T_o) \end{aligned} \tag{3.30}$$

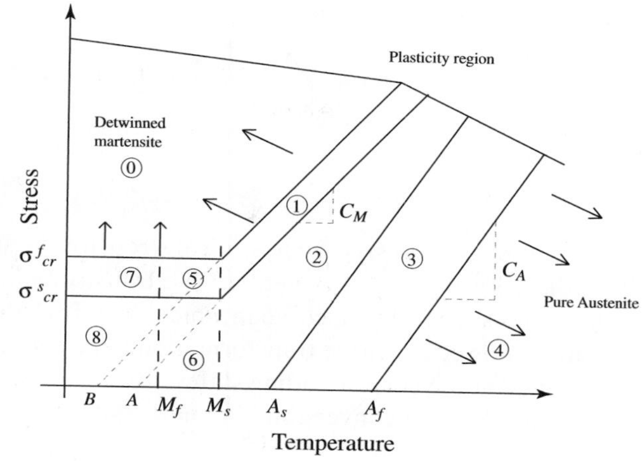

Figure 3.23. Critical stress-temperature phase diagram, Brinson model.

The relations for the Young's modulus and the phase-transformation coefficient are the same as Eqs. 3.4 and 3.5, respectively. Substituting these expressions, the constitutive equation for $\xi_{To} = 1$ and $\xi_{so} = 0$ reduces to

$$\sigma = E(\xi)(\epsilon - \epsilon_L \xi_s) + \Theta(T - T_0) \tag{3.31}$$

The transformation functions can be derived as [41]

$$\begin{aligned} \Omega_s &= (\epsilon - \epsilon_L \xi_s) \frac{\partial E(\xi)}{\partial \xi_s} - \epsilon_L E(\xi) \\ &= (\epsilon - \epsilon_L \xi_s)(E_M - E_A) - \epsilon_L E(\xi) \end{aligned} \tag{3.32}$$

$$\begin{aligned} \Omega_T &= (\epsilon - \epsilon_L \xi_s) \frac{\partial E(\xi)}{\partial \xi_T} \\ &= (\epsilon - \epsilon_L \xi S)(E_M - E_A) \end{aligned} \tag{3.33}$$

The critical stress-temperature diagram used for the Brinson model is shown schematically in Fig. 3.23. A modified cosine model for the martensite volume fraction is used. Note that the stress-induced martensite is single variant, whereas the temperature-induced martensite involves multiple variants.

For the $A \to M$ transformation:

For $T > M_s$ and $\sigma_s^{cr} + C_M(T - M_s) < \sigma < \sigma_f^{cr} + C_M(T - M_s)$,

$$\xi_s = \frac{1 - \xi_{so}}{2} \cos\left[\frac{\pi}{\sigma_s^{cr} - \sigma_f^{cr}}\left[\sigma - \sigma_f^{cr} - C_M(T - M_s)\right]\right] + \frac{1 + \xi_{so}}{2} \tag{3.34}$$

$$\xi_T = \xi_{To} - \frac{\xi_{To}}{1 - \xi_{so}}(\xi_s - \xi_{so}) \tag{3.35}$$

For $T < M_s$ and $\sigma_s^{cr} < \sigma < \sigma_f^{cr}$,

$$\xi_s = \frac{1 - \xi_{so}}{2} \cos\left[\frac{\pi}{\sigma_s^{cr} - \sigma_f^{cr}}(\sigma - \sigma_f^{cr})\right] + \frac{1 + \xi_{so}}{2} \tag{3.36}$$

$$\xi_T = \xi_{To} - \frac{\xi_{To}}{1 - \xi_{so}}(\xi_s - \xi_{so}) + \Delta_{T\xi} \tag{3.37}$$

where, if $M_f < T < M_s$ and $T < T_o$

$$\Delta_{T\xi} = \frac{1 - \xi_{To}}{2}\left\{\cos\left[a_M(T - M_f)\right] + 1\right\} \tag{3.38}$$

else

$$\Delta_{T\xi} = 0 \tag{3.39}$$

In these equations, σ_s^{cr} is the critical stress for the start of the transformation and σ_f^{cr} is the critical stress at the end of the transformation. These values are approximated from a stress-strain curve in which the initial phase was 100% martensite. On the curve, it is clear where transformation begins and ends so that these values are determined at the corresponding states.

For the $M \rightarrow A$ conversion, the martensite volume fraction is determined from the following relations:

For $T > A_s$ and $C_A(T - A_f) < \sigma < C_A(T - A_s)$

$$\xi = \frac{\xi_o}{2}\cos\left[a_A\left(T - A_s - \frac{\sigma}{C_A}\right) + 1\right] \tag{3.40}$$

$$\xi_s = \xi_{so} - \frac{\xi_{so}}{\xi_o}(\xi_o - \xi) \tag{3.41}$$

$$\xi_T = \xi_{To} - \frac{\xi_{To}}{\xi_o}(\xi_o - \xi) \tag{3.42}$$

where all terms with a subscript "o" denote the initial condition and a_A and a_M are equivalent to those defined in the Liang and Rogers model.

$$a_M = \frac{\pi}{(M_s - M_f)} \qquad b_M = -\frac{a_M}{C_M} \tag{3.43}$$

$$a_A = \frac{\pi}{(A_f - A_s)} \qquad b_A = -\frac{a_A}{C_A} \tag{3.44}$$

$$C_M = \frac{1}{dM_s/d\sigma} \qquad C_A = \frac{1}{dA_s/d\sigma} \tag{3.45}$$

For an SMA with material characteristics that are more general functions of the martensite volume fraction, we can derive alternate constitutive relations. Because the Brinson's model differentiates between temperature-induced and stress-induced martensite, there are two parts in the martensite region of the diagram. The start and finish stresses at which the transformation from twinned to detwinned martensite occurs is marked by the lines that are parallel to the x-axis at a constant stress of σ_{cr}^s and σ_{cr}^f, respectively. The region above the line at σ_{cr}^f is the region in which the material exists as pure detwinned martensite (region 0). The transformation from twinned to detwinned martensite on application of stress is reversible only by conversion to austenite and then cooling under low stress; it is not reversible merely on unloading the material from a detwinned state. The region below the stress of σ_{cr}^s (region 8), therefore, can exist in pure twinned martensite (when cooled from austenite below this stress), pure detwinned martensite (when unloaded after loading to a value beyond σ_{cr}^f), or in a mixture of the two (when unloaded after loading to a stress of above σ_{cr}^{s} but below σ_{cr}^{f}). Recall that the material thermal function Θ is comparatively very small (five orders of magnitude smaller than E) and is therefore neglected.

The austenite start and finish regions are the same as those prescribed in the Tanaka model and the Liang and Rogers model. Therefore, the austenite start and finish temperatures, A_s and A_f, are the same as those used for the Tanaka, and the Liang and Rogers models. However, in Brinson's model, the parameters M_s and M_f are defined as the temperature above which the martensite transformation stresses are a linear function of temperature, as shown in Fig. 3.23. The critical stresses σ_{cr}^{s} and σ_{cr}^{f} are assumed constant with decreasing temperature below M_s. Note that some authors have shown a small increase of these two stresses below M_s, which is neglected here. In the Tanaka, and the Liang and Rogers models, these parameters are defined at zero stress and are the temperatures for martensite start and finish obtained by cooling austenite without the application of stress. Therefore, when calculating these constants from the experimental critical points, the numerical values used for the Tanaka, and the Liang and Rogers models for M_s and M_f (i.e., points A and B in Fig. 3.23, respectively) are different from those used for the Brinson model (i.e., points marked M_s and M_f). The values used in the Tanaka, and the Liang and Rogers models should be those obtained by extrapolating the martensite start and finish lines to zero stress. The different values for the models must be used to obtain a fair comparison between the models and to match them to experimental observations. The slopes C_M and C_A are the stress-temperature coefficients for martensite and austenite, respectively. When the temperature is below M_f, the transformation is not due to any phase change; rather, it is due to reorientation of martensite variants in the direction of applied stress. Heating of the SMA above A_f and cooling it below M_f can recover this deformation. Within a temperature range between M_s and A_f, the martensite and austenite can coexist. Above A_f, the material is in the austenite phase, which is a stable state at the zero-stress condition. Regardless of the extent of loading (forward transformation), at the end of unloading (reverse transformation), the material regains the austenite phase (beginning state). There is no physical deformation associated with this transformation process.

Chung et al. [42] pointed out a weakness of the Brinson model in region 5 ($\sigma_s^{cr} < \sigma < \sigma_f^{cr}$ and $M_f < T < M_s$). This is the region where both stress-induced martensite, formed from the austenite phase via the stress increment, and temperature-induced martensite, formed from the austenite phase via the temperature decrement, take place. It was shown that for a fixed temperature, when the stress increases, the Brinson model predicts satisfactory stress-induced results for any set of initial conditions. Conversely, for a fixed stress, the Brinson model does not predict consistent temperature-induced martensite, formed from the austenite phase by a temperature decrease. Results are a function of the initial conditions. Satisfactory results are predicted only when the initial condition of the material is the pure martensite state ($\xi_{So} = 0, \xi_{To} = 1$, or $\xi_{So} = 1, \xi_{To} = 0$). It was pointed out that for some cases, the total value of the martensite volume fraction may exceed 1.0, which is an anomaly.

Note that the stress-induced martensite fraction must be 1.0 at the final critical stress ($\sigma = \sigma_f^{cr}, \xi_S = 1$). The total martensite volume fraction must be 1.0 at the martensite finish temperature ($T = M_f, \xi_T = 0$). As a result, the total martensite volume fraction becomes larger than 1.0 ($\xi = \xi_S + \xi_T = 2.0$) at this condition. The

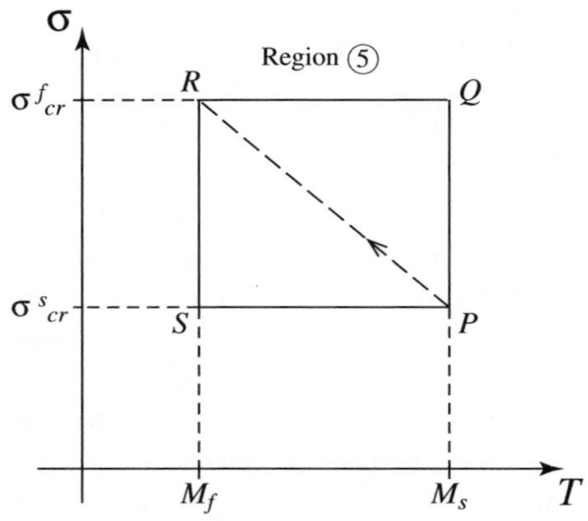

Figure 3.24. Region 5 of the critical stress-temperature diagram, Brinson model.

Brinson model satisfies all conditions when either only the stress increases or only the temperature decreases. In the case of simultaneous change of temperature and stress, Khandelwal et al. [43] and Chung et al. [42] revised the Brinson model.

The critical stress-temperature diagram can be divided into eight regions, as shown in Fig. 3.23. The regions are defined as follows:

Region 0: Stress-induced martensite region (no transformation).
Region 1: Transformation from austenite to martensite, either stress-induced or temperature-induced.
Region 2: Mixture of temperature-induced or stress-induced martensite and austenite phase (no transformation).
Region 3: Transformation from martensite to austenite.
Region 4: Pure austenite region (no transformation).
Region 5: Transformation from austenite to stress-induced martensite, or temperature-induced martensite to stress-induced martensite, and transformation from austenite to temperature-induced martensite.
Region 6: Transformation from austenite to temperature-induced martensite.
Region 7: Transformation from temperature-induced to stress-induced martensite.
Region 8: Mixture of stress-induced martensite and temperature-induced martensite (no transformation).

A closer view of region 5 is shown in Fig. 3.24. The rectangle *PQRS* defines region 5, where simultaneous evolution of the twinned martensite fraction ξ_T and detwinned martensite fraction ξ_S can take place. In this region, a simultaneous increase of stress and a decrease of temperature (as shown by the arrow in Fig. 3.24) can lead to an unrealistic volume fraction ($\xi > 1$). Specifically, at the point *R*, ξ can become 2.0.

The proposed modification to the Brinson model is based on the assumption that there is a coupling between the evolution of ξ_S and ξ_T. The twinned fraction ξ_T can transform into the detwinned fraction ξ_S due to the application of stress, whereas the detwinned fraction ξ_S cannot transform into the twinned fraction ξ_T.

For $T < M_s$ and $\sigma_s^{cr} < \sigma < \sigma_f^{cr}$

$$\xi_S = \frac{1-\xi_{So}}{2}\cos\left[\frac{\pi}{\sigma_s^{cr}-\sigma_f^{cr}}(\sigma-\sigma_f^{cr})\right]+\frac{1+\xi_{So}}{2} \tag{3.46}$$

$$\xi_T = (\xi_{T1}-\xi_{So})\frac{1-\xi_S}{1-\xi_{So}} \tag{3.47}$$

where

$$\xi_{T1} = \frac{1+\xi_o}{2}+\frac{1-\xi_o}{2}\cos\left[a_M(T-M_f)\right]$$

$$\xi_o = \xi_{So}+\xi_{To}$$

Panico and Brinson [44] further modified the 1-D phenomenological model for a 3-D stress environment, which accounts for the evolution of both stress-induced and temperature-induced martensite variants. For the 1-D model, one needs to determine 10 material constants: M_s, M_f, A_s, A_f, σ_{cr}^s, σ_{cr}^f, E_M, E_A, C_M, and C_A.

3.4.4 Boyd and Lagoudas Model

In the Boyd and Lagoudas model [45], the total specific Gibbs free energy is determined by summing the free energy of each phase of shape-memory materials plus the free energy of mixing. The second law of thermodynamics can then be written in terms of the Gibbs free energy, and a constitutive relation can be derived. The total strain ϵ_{ij}^{te} consists of two parts: the mechanical strain, ϵ_{ij}, and the transformation strain, ϵ_{ij}^t, which is a function of the martensite volume fraction.

$$\epsilon_{ij}^{te} = \epsilon_{ij}+\epsilon_{ij}^t \tag{3.48}$$

The constitutive relation can be written as

$$\epsilon_{ij}^{te} = a_{ijkl}^1\sigma_{kl}+a_{ij}^2\Delta T \tag{3.49}$$

where a_{ijkl}^1 is the compliance tensor and a_{ij}^2 is the coefficient of thermal-expansion tensor. The transformation strain rate is assumed to have the following form

$$\epsilon_{ij}^t = \Lambda_{ij}\dot{\epsilon} \tag{3.50}$$

where

$$\Lambda_{ij} = \sqrt{\frac{3}{2}}H \tag{3.51}$$

and H is the maximum uniaxial transformation strain equivalent to the maximum recoverable strain, ϵ_L.

The martensite volume fraction is calculated from

$$\sigma_{ij}^{\text{eff}}\Lambda_{ij}+d^1T-\rho b_1\xi = Y^{**}+d_{ijkl}^3\sigma_{ij}\sigma_{kl}+d_{ij}^4\sigma_{ij}\Delta T \tag{3.52}$$

where d^1T is related to the entropy at a reference state, b_1 is a material constant, ρ is the mass density, Y** is a threshold stress value, d_{ijkl}^3 and d_{ij}^4 are parameters that are related to the changing elastic moduli and the coefficient of thermal expansion during transformation, respectively, and ΔT is the temperature difference.

Although this model is widely applicable to monolithic SMA structures and is a true three-dimensional model, in one-dimensional form, it is quite similar to the Tanaka model and has been shown by Brinson [46] to yield similar results in simulations.

In the model, Y^{**}, d_1, and ρb_1 are defined as follows

For $M \rightarrow A$ transformation

$$Y^{**} = C_A H A_f \tag{3.53}$$

$$d_1 = -HC_A \tag{3.54}$$

$$\rho b_1 = -Y^{**} + d_1 M_f \tag{3.55}$$

For $A \rightarrow M$ transformation

$$Y^{**} = C_M H M_s \tag{3.56}$$

$$d_1 = -HC_M \tag{3.57}$$

$$\rho b_1 = -Y^{**} + d_1 A_s \tag{3.58}$$

3.4.5 Other SMA Models

There are several other macromechanics models that model pseudoelasticity and shape memory effects of SMAs under quasi-static loading. Matsuzaki et al. [47, 48] developed a phase-interaction energy function in terms of the martensite volume fraction covering five crystal phases: austenite, detwinned martensite, twinned martensite, detwinned rhombohedral, and twinned rhombohedral. To examine the effectiveness of this unified thermomechanical model, the predicted results were successfully compared with experimental measurements associated with SME and pseudoelasticity for SMA wires subjected to cyclic loading up to 1 Hz. A key feature of this formulation is the modeling of twinned and detwinned rhombohedral phases.

There are other models that are based on evolutionary plasticity, as suggested by Graesser and Cozzareli [49]. However, only mechanical loading under isothermal conditions could be simulated with constant parameters. A mechanism-based phenomenological model for pseudoelastic behavior was developed by Malorvh and Gandhi [50] comprising linear, piecewise linear, and nonlinear springs and friction elements. The model parameters are identified from experimental hysteresis cycles.

Several studies that compare the relative merits and demerits of each model have also been carried out. Schroeder et al. [51] compared the Landau-Devonshire theory formulation [52] with the Graesser and Cozzareli, Brinson, and Boyd and Lagoudas models in terms of their capabilities and computational effort. Another study [53] comparing the phenomenological model of Grasser and Cozzareli with the Brinson model showed that whereas the phenomenological approach was more suitable for repeated mechanical cycling under isothermal conditions, it could not handle more complex situations involving thermal cycling. A comparison between the thermodynamic model of Boyd and Lagoudas, and Tanaka-based models [54] led to the unification of these approaches under the same broad assumptions, and it highlighted differences in the simulations based on these different approaches. A comparison of these models along with the Ivshin and Pence model [46] led to the observation that most of the constitutive models yielded similar results for most simple simulations, and the main differences were in the formulation of the transformation kinetics.

3.5 Testing of SMA Wires

The behavior of an SMA is a function of its three primary variables: stress, strain, and temperature. Material characterization involves studying the dependence of two of these variables while the third is kept constant. Systematic tests need to be carried out to determine important material parameters that are required for analytical models. For simplicity, quasi-static conditions are maintained in order to eliminate any dynamic effects. In addition to mechanical stress-strain testing, the transformation temperatures of the material are normally determined using DSC.

The thermomechanical properties of the shape memory alloys depend on many variables, such as wire manufacturing process, wire diameter, pre-strain, stress level, temperature, annealing, and whether or not the material has been cycled (thermomechanical history). Prior to any mechanical testing of an SMA wire, it is necessary to cycle the wire to assure repeatable experimental results. The testing described in this section is focused on a binary alloy (Ni-Ti – 50.5%–49.5%) wire.

3.5.1 Sample Preparation, Cycling, and Annealing

As an example, a wire of diameter 0.38 mm (0.015 in) from Dynalloy [55] is used. Due to the manufacturing process of drawing these wires at high temperature and winding them on rollers, the unstrained wire is often slightly curved along its length and does not recover to a completely straight wire even above the austenite temperature. This effect is normally ignored, especially for thin wires. Typically, tensile testing of the wire sample is carried out on a testing length of about 0.127 m (5 in). In addition to this, about 0.0127 m (0.5 in) is required for gripping at either end, making the total wire length 0.1524 m (6 in).

The manufacturing process of the SMA typically results in the formation of an oxide layer. To heat the sample by passing electrical current through it (internal heating), it is important to remove this oxide layer at both ends to ensure a good electrical contact. This can be achieved by lightly sanding the sample at the point of attachment with electrical contacts or by crimping on spade lugs. Alternately, an environmental chamber can be used to heat the wire to a specified constant temperature (also called external heating).

Due to small dislocations and other irregularities inherent in the SMA wire as manufactured, the mechanical characteristics of the wire drift with the increasing number of cycles. Figure 3.25 shows the pseudoelastic characteristics of the wire over repeated cycling. The test was performed using an environmental chamber to keep the wire at a temperature of 90°C, which is well above A_f. Details of the experimental setup used for these data are described in Section 3.5.4. From the figure, we observe that during the first few cycles, the wire does not completely recover its strain on unloading but instead is left with some residual strain (nearly 1%). This residual strain decreases with the increasing number of cycles. After about 10 cycles, no significant deviation in the characteristics of the wire is observed. The wire also shows a complete recovery of its strain when brought down to the zero-stress condition.

To ensure repeatable characteristics, it is important to stabilize the wire. This is accomplished through mechanically cycling the wire before taking data. The cycling procedure consists of extending the wire to a strain of about 4% at a strain rate of

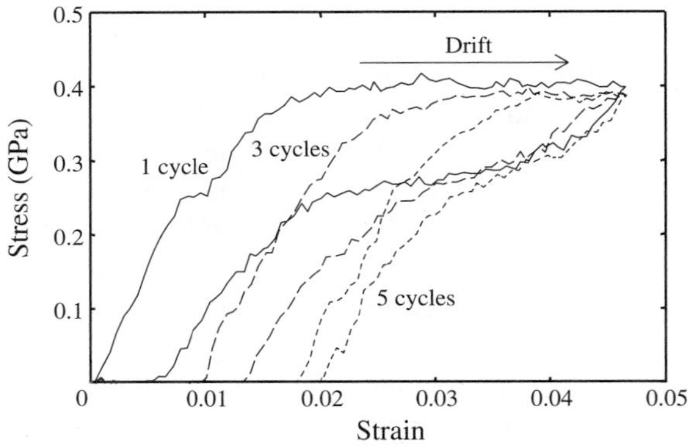

Figure 3.25. Effect of cycling on pseudoelastic characteristics of the wire.

about 0.0005/s and then releasing the wire to zero-stress condition. The cycling is conducted at a constant environmental temperature well above A_f (typically 90°C). The hysteretic pseudoelastic behavior of the wire then stabilizes with the increasing number of cycles, as described previously. This procedure is repeated (typically 20–30 cycles) until no variation in the residual strain characteristics is observed.

An alternate procedure for cycling [54] is to mechanically strain the material at low temperatures (in martensite phase). This is then followed by a thermal cycle under no stress in which the sample is heated to a temperature above A_f and then cooled down to below M_f. The material is then strained again at low temperature. This procedure is repeated about 20–30 times. A simple way of performing this cycling on a wire is by clamping one end and suspending a weight from the other end. The weight is then removed and the wire is heated to recover the deformation. Deforming the wire and recovery of the deformation comprise one cycle. It is important to cycle the wire at a stress level much higher than the stress of interest. Although this procedure is an effective way to cycle the material, it is found that the first method yields similar results and is easier to perform because it does not require thermal cycling. In addition to stabilizing the stress-strain characteristics of the material, cycling also can have the effect of decreasing the area of the hysteresis loop [56].

To ensure repeatability of test data, each experimental point should be repeated two to three times. Although the conditions controlling the tests are expected to be held constant, up to 10% variation in the stress levels may be observed in some cases. The variation could be the result of starting from a slightly different volume fraction of martensite. Because the material state is extremely sensitive to its thermomechanical loading history, it is difficult to obtain exactly the same starting composition of the material at the beginning of each test. This variation could result in the experimental variations observed even after cycling the wire.

Annealing is another procedure that is performed on the raw material to eliminate any thermomechanical history effects and to modify certain properties of the material. A typical annealing process involves heating the raw material to an elevated temperature (e.g., 500°C–800°C), maintaining it at the elevated temperature for a specified period of time (e.g., 40 minutes), and finally quenching it in water or oil (usually to prevent precipitation of Ni). It has been observed that the annealed material exhibits lower transformation stresses and increased transformation temperatures (except, in some cases, M_s). However, the exact effect of annealing is highly dependent on the material composition. Sometimes, the as-manufactured SMA

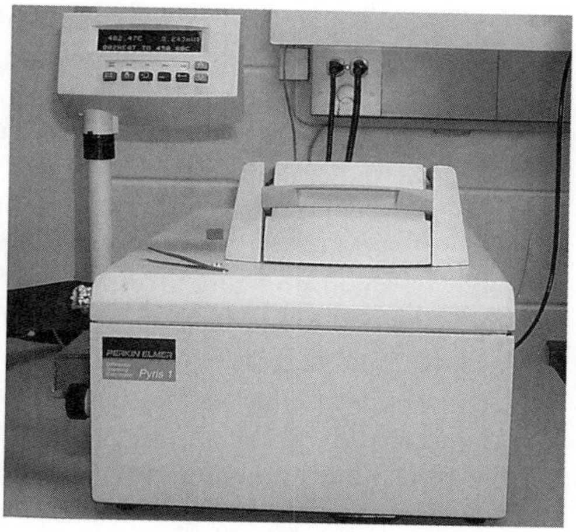

Figure 3.26. Differential scanning calorimeter.

sample has an austenite start temperature A_s that is close to or below room temperature. As a result, the material exhibits partial pseudoelastic behavior at room temperature instead of pure SME. In such cases, an appropriate annealing procedure can be used to increase the M_f above room temperature and obtain pure SME in the sample at room temperature. Due to their sensitivity to mechanical cycling as well as to the annealing procedure, reports of SMAs' experimental characteristics are always accompanied by the cycling and annealing procedures that were performed on the samples prior to the experiments.

3.5.2 Transformation Temperatures under Zero Stress

To characterize the material, it is important to identify the transformation temperatures A_s, A_f, M_s, and M_f, which are the austenite start, austenite finish, martensite start, and martensite finish temperatures, respectively. Note that these transformation temperatures are often defined for the material in a stress-free condition; however, they change as a function of applied stress. These temperatures are important coefficients in the constitutive models and their accurate measurement is necessary for prediction of the material behavior. Several methods have been reported to measure the transformation temperatures, including DSC, the applied loading method, and the electrical resistance method.

To determine the transformation temperatures of the wire under the no-stress condition, the DSC (Perkin-Elmer Pyris 1 shown in Fig. 3.26) is used. This instrument measures the heat flow to and from the material as a function of temperature. A small sample of the material can be heated and cooled between given temperatures at a specific temperature rate that is regulated by a temperature controller. The heating and cooling is performed by controlling a combination of a heating element and a regulated mixture of helium and liquid nitrogen.

Measurement of transformation temperatures using the DSC is based on the latent heat of the material. When a material undergoes a phase transformation, it absorbs (or emits) a specific amount of heat, called the latent heat, at a constant temperature. This is a characteristic of any phase transformation, including transformations among the various phases of water (i.e., liquid, steam, and ice). In the DSC test, the phase transformation appears as a sudden peak in the plot of heat flow versus temperature.

Preparation of the sample for the DSC tests consists of cutting the SMA wire into small pieces of approximate length 0.005 m (0.19 in). These pieces are then placed in a custom pan and sealed from the top with a cover plate using a tabletop press. The weight of the SMA present in the pan is deduced by subtracting the empty weight of the pan from the weight of the pan with the SMA. For best results with the DSC, the weight of the SMA should be between 20 mg and 40 mg. The process of cutting the SMA wire and subsequent pressing of the cover plate onto the SMA may cause a small non-zero stress in the SMA sample. However, testing different weights of the SMA and the application of different amounts of pressure to seal the cover plate could help confirm the repeatability of the DSC results, indicating that the effect of this non-zero stress on the transformation temperatures is negligible.

The procedure for testing consists of first preparing two baseline test pans (without the SMA) and one pan with sample SMA. The pans are weighed in a digital balance. The two baseline pans (without the SMA) are placed inside the thermal chamber of the DSC and sealed from the environment. The temperature of the chamber is then raised from 10°C to 100°C at a rate of 5°C/min, and the baseline heat-flow rate versus temperature is recorded by the data-acquisition computer. After the thermal cycle is completed, one of the baseline pans is removed and replaced with the pan with the SMA. The temperature-cycling procedure is repeated and the resulting heat flow is recorded once again. The quantitative variation of heat flow of the SMA with temperature can then be obtained by subtracting the heat flow of the baseline case from that of the case with the sample at each temperature to yield a subtracted heat flow. This entire procedure is repeated 2–3 times for each sample to ensure consistency. It is important to note that because the baseline heat flow is nearly constant and displays no peaks, the subtracted heat flow is important only for quantitative specific heat and latent heat measurements. The subtracted heat flow rate of the SMA is sufficient to determine the transformation temperatures of the material.

The heating rate (5°C/min) is chosen as representing a quasi-static value because it is observed that slower rates produce no significant change in the material behavior. High heating rates tend to produce inconsistent results. This could be due to the material's inability to attain a constant temperature throughout the sample at very high heating rates and, thus, its inability to attain equilibrium at a particular temperature. When the SMA starts its transformation from austenite (parent phase) to martensite, it emits heat due to the exothermic transition. Conversely on heating from the martensite to the austenite phase, it absorbs heat due to the endothermic transition. Two characteristic spikes in the heat flow are observed.

The unsubtracted and subtracted heat-flow rates are plotted in Figs. 3.27(a) and 3.27(b), respectively. We observe that in the subtracted heat-flow profile, the heat-flow rate during heating is of approximately the same magnitude as that during cooling. The change in sign indicates a change in the heat-flow direction, to or from the sample. The transformation temperatures can be obtained from this by marking the temperatures at which the peak in the heat flow begins and ends, and they are computed directly by the data-acquisition software. The magnitude of the peak can be used to determine the latent heat of transformation of the material. The transformation temperatures are indicated by circles in the plots.

The electrical resistance method is based on a large variation of resistivity over the transformation temperature range. A drop in resistivity takes place as soon as the

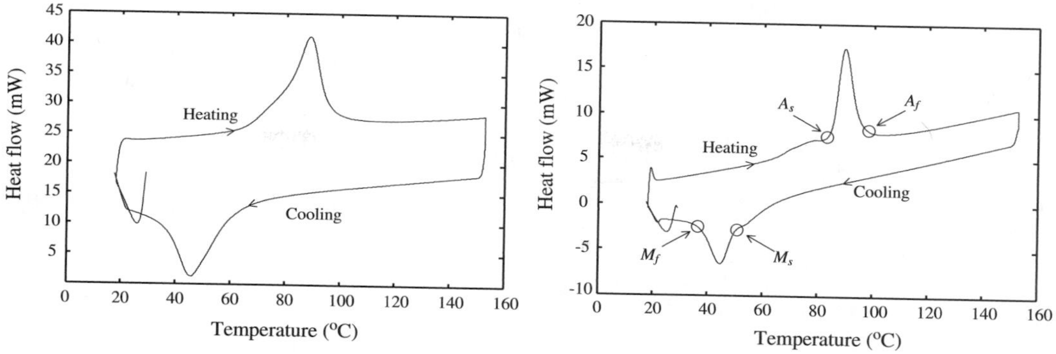

(a) Unsubtracted heat flow (b) Subtracted heat flow

Figure 3.27. Heat flow as a function of temperature measured using DSC.

transformation from austenite to martensite begins (i.e., during the cooling cycle). Conversely, a large increase of resistivity takes place during the transformation from the martensite to the austenite phase. Monitoring the change of resistance with temperature helps to identify transformation temperatures. Although this is a simple approach, it is not easy to precisely identify the transformation temperatures from resistance data. For example, a drastic change in resistivity during the phase transformation also impacts the internal heating of the wire. The thermal conductivity of the material is also affected during the phase transformation; however, the effect on heat capacity is typically quite small. The direction of change in resistivity with phase transformation is a function of the specific alloy. For example, the electrical resistivity of a Nitinol wire from Dynalloy increases when going from martensite to austenite, whereas a wire of another alloy from the same company showed the opposite behavior [57].

Abel et al. [58] examined three methods: DSC, electrical resistance, and applied loading to determine the transformation temperatures of NiTi wire under different heat-treatment conditions. The results showed that the transformation temperatures measured by DSC did not agree with those measured by the other two methods, which were similar. The applied-loading method using a mechanical testing machine was determined to be the most effective method to determine stress-dependent transformation temperatures. The electrical-resistance method provides a better estimate for M_s and M_f than for A_s and A_f.

3.5.3 Variation of Transformation Temperatures with Stress

The transformation temperatures of the SMA are a function of the stress in the material. The previous section described the measurement of the transformation temperatures at zero stress. To obtain a complete picture of the transformation behavior of the SMA, it is also necessary to measure the variation of transformation temperatures with stress. This can be accomplished either by a simple benchtop experiment or by using a tensile testing machine to maintain a constant load. The constant-load, strain-temperature test also provides a direct measurement of the actuation capabilities of the material. Although the actuation load is seldom kept constant in a practical application, the measurement of the actuation stroke

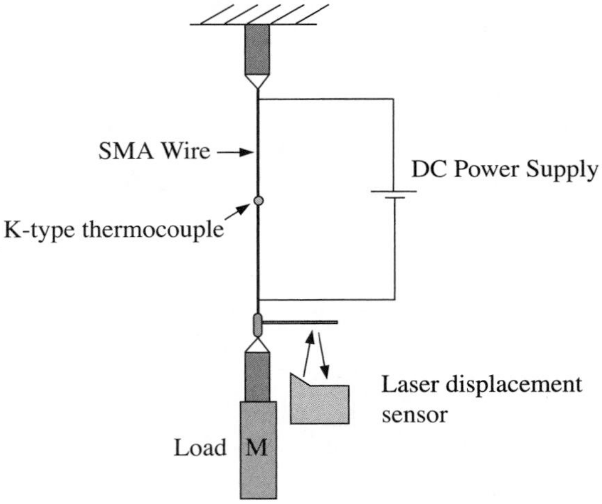

Figure 3.28. Test setup to measure strain as a function of temperature at a constant load.

with temperature under a constant load enables the design of an actuator for any given loading pattern.

A schematic of a simple benchtop test setup to measure the transformation temperatures as a function of stress is shown in Fig. 3.28. The test specimen is clamped to a rigid support by a pin vise. The bottom end of the wire is attached to a pan that supports dead weights. The wire is heated by resistance heating, and a constant voltage is supplied to the wire by a DC power supply. Electrical connections are crimped onto the wire because this assures a good electrical contact with the material. The entire setup is mounted in a plexiglass case to prevent large variations of temperature of the wire due to convective flow in the laboratory.

The wire is initially at room temperature in an unstrained condition. Note that the wire sample is cycled before the experiment, and the length after cycling is taken as the unstrained length. When a weight is added to the pan, the wire strains by a certain amount, which serves as a pre-strain for the OWSME. A voltage is then applied across the ends of the wire, causing the temperature of the wire to increase due to internal heating. As the temperature increases beyond A_s, the wire starts transforming to austenite and starts recovering the initial pre-strain. When the voltage to the wire is reduced, the wire begins to cool down, transforms back into martensite, and returns to the initial strained condition under the influence of the dead weights.

Note that due to significant changes in resistance of the wire during transformation, the current drawn by the wire varies constantly. To precisely control the heat flow rate during this test, a special controller is necessary. However, quasi-static conditions can be ensured by allowing the wire to stabilize at set temperatures before recording the data. The displacement of the weight is measured using a linear potentiometer or a laser displacement sensor, and the average strains in the wire are calculated. The temperature is measured using a K-type thermocouple mounted on the wire through a thermal interface material, such as those frequently used in the heat sinks of electronic devices.

Figure 3.29 shows the recovery strain-temperature curve for an applied load of 5.3 lbs. Note that the recovered strain, or stroke, increases with increasing tensile load. This is because of the larger initial strains, or pre-strains, associated with larger stress values. However, there is a limit to the actuation stresses because very large

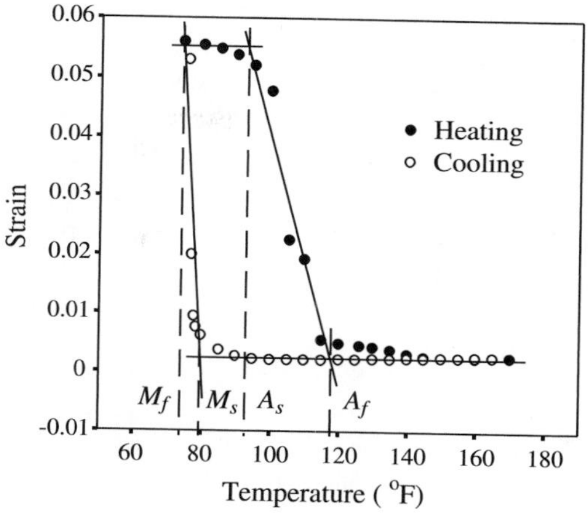

Figure 3.29. Determination of transformation temperatures from the strain-temperature curve, constant load of 5.3 lbs, Nitinol of diameter 0.015 in, and length of 5 in.

stresses produce permanent plastic deformation of martensite at low temperatures, as described in Section 3.1.3.

The critical temperatures are obtained as the temperatures at which the slope of the curve changes dramatically. Note that at high temperatures, when the material is in the pure austenite phase, the strain is non-zero. This strain is the austenite strain associated with the load applied and is numerically equal to the applied stress divided by the austenite's Young's modulus. It is also important to note that during transformation, the strain changes by large amounts at a relatively constant applied load; therefore, it is difficult to precisely control the strain imparted by a constant load in this regime, which is equivalent to the pre-strain imparted to the SMA wire. This has important consequences when using the material as an actuator, where the material must be pre-strained by a precise amount.

Figure 3.30(a) shows the strain-temperature curve at two different stress levels. It can be seen that the transformation temperatures increase with increasing stress. The strain-temperature curve can be obtained at several different stresses, and the transformation temperatures can be extracted at each value of stress. These temperatues are plotted against the stress levels in Fig. 3.30(b), and a straight line can be fitted to the data. Although the test data may not lend themselves to a perfect straight line, thermodynamic relations point to a linear relationship. The slope of this line provides the sensitivity of the transformation temperature with stress, also called the stress-influence coefficient. The slopes C_A and C_M are the austenite and martensite stress-influence coefficients, respectively. Quite often, C_A and C_M are assumed to be equal.

This experimental setup gives good results at low heating rates, which translates to low actuation speeds. At higher rates, additional inertia forces may act on the wire, and the setup may be susceptible to oscillations because the mass and wire behave like a pendulum. These effects make it difficult to maintain a constant stress in the material. The strain-temperature test can be performed more precisely using a tensile testing machine. The temperature of the wire can be controlled either by resistive heating or by enclosing the specimen in a temperature-controlled environmental chamber. The testing machine can be programmed to maintain a constant load on the wire, which will maintain a constant stress in the material irrespective of the heating rate, within the constraints of the bandwidth of the testing machine.

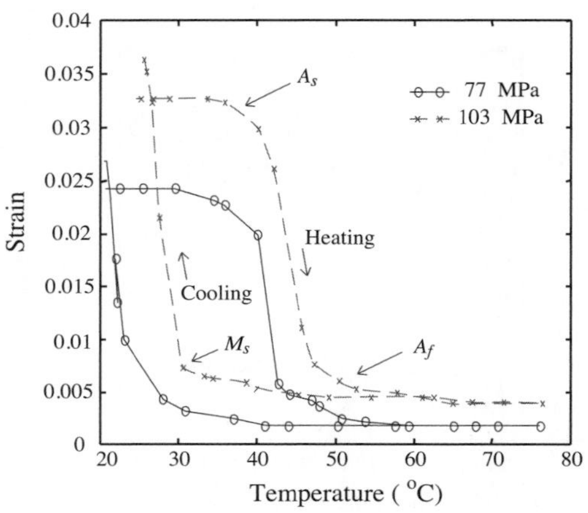

(a) Strain-temperature curves at different stresses

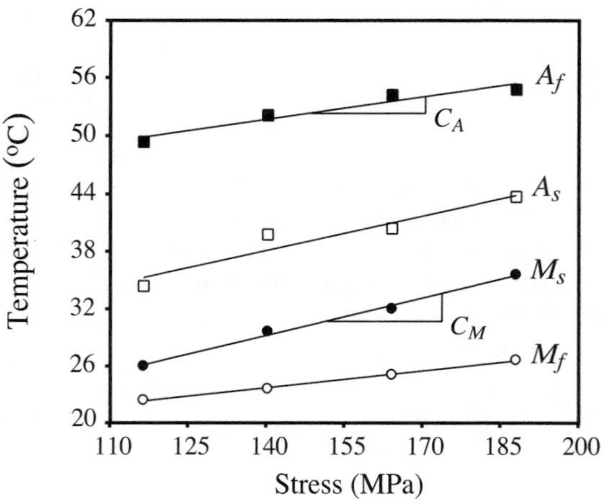

(b) Variation of transformation temperatures with stress

Figure 3.30. Effect of stress on strain-temperature behavior.

3.5.4 Stress-Strain Behavior at Constant Temperature

The transformation temperatures are some of the most important parameters in characterizing the behavior of a given SMA composition. Knowledge of the transformation temperatures is essential to identify the range of temperatures within which the SMA actuator can be operated. In addition to the transformation temperatures, it is also important to determine the stress-strain behavior of the material as a function of temperature. This information is required to size an actuator for a required application.

The stress-strain behavior of the material is qualitatively described in Section 3.1.3 (at a temperature below A_s) and in Section 3.1.5 (at a temperature above A_f). Experimentally, this characterization can be carried out on an SMA wire specimen by straining the specimen in a tensile testing machine, while maintaining it at a constant temperature in a controllable environmental chamber. For example, results of tests carried out using an MTS 810 test machine (Fig. 3.31) with an ATS controllable thermal chamber are described here. The thermal chamber encloses the wire specimen as well as the grips that hold it, enabling a constant temperature to be maintained along the length of the wire. It is difficult to perform this test by resistively heating the wire specimen because the strain imposed on the wire results in a local temperature change. The temperature of the wire can be controlled by varying the current passing through the wire, using temperature feedback from thermocouples mounted on the wire. However, only a finite number of thermocouples can be installed on the wire, and they have an inherently slow response time. As a result, it is difficult to maintain a constant temperature along the length of the wire by resistive heating.

The experimental procedure consists of first cycling the wire specimen by heating the loaded wire to a temperature well above A_f and then cooling it under no stress to below M_f. This establishes an initial condition for all measurements. From this point, the wire is strained to a value that is just below the point of the second yield

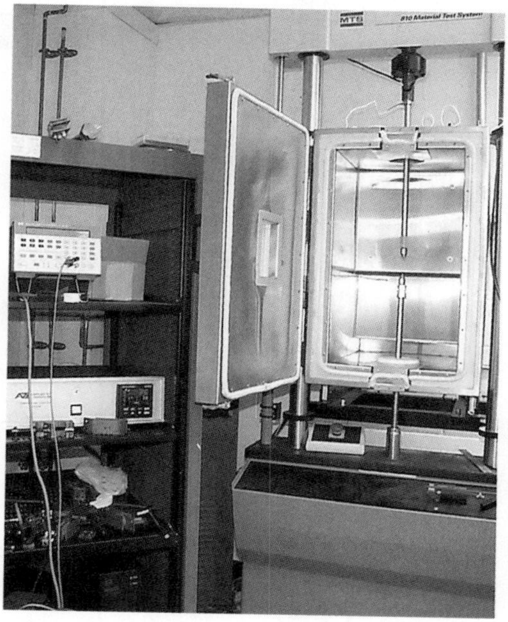

Figure 3.31. MTS testing machine with controllable thermal chamber.

point of the material. This ensures that a maximum of the material detwinning is captured, without the possibility of permanent plastic deformation. For example, for the alloy under consideration, this corresponds to a strain of approximately 5%, whereas permanent plastic deformation begins to occur at approximately 6.5%. After the maximum strain is reached, the specimen is unloaded to zero stress. Each straining and unloading to zero stress represents one test cycle. Stress and strain values are continuously measured during the process while maintaining a constant temperature in the thermal chamber. After each test cycle, the temperature of the material is increased to above A_f and then decreased to below M_f. This ensures that the strain imparted to the specimen is recovered and that the starting condition is consistent for the next test cycle. In a history-dependent material such as an SMA, maintaining consistency of the test procedure is important because the material can be in multiple states at the same stress, temperature, and strain.

Note that these tests are all quasi-static, which implies an isothermal loading and unloading of the wire specimen. For the specimen under discussion, a strain rate of 5×10^{-4}/s is a typical quasi-static value. In general, the strain rate that can be considered quasi-static is a function of the dimensions of the test specimen because the ratio of heat retained to the heat convected is proportional to the ratio of volume to surface area of the specimen. As an example, isothermal conditions can be reached at higher strain rates in wires with lower diameters. Otherwise, higher strain rates cause some differences in material behavior [59] compared to quasi-static measurements, which is discussed in a subsequent section.

The constant-temperature stress-strain plots obtained experimentally are shown in Fig. 3.32. These plots clearly show both the SME at low temperatures and pseudoelasticity at high temperatures. At temperatures below A_s (45°C), the material is purely martensite, and a region of detwinning can clearly be seen occurring at relatively constant stress. At a higher temperature (84°C), the material

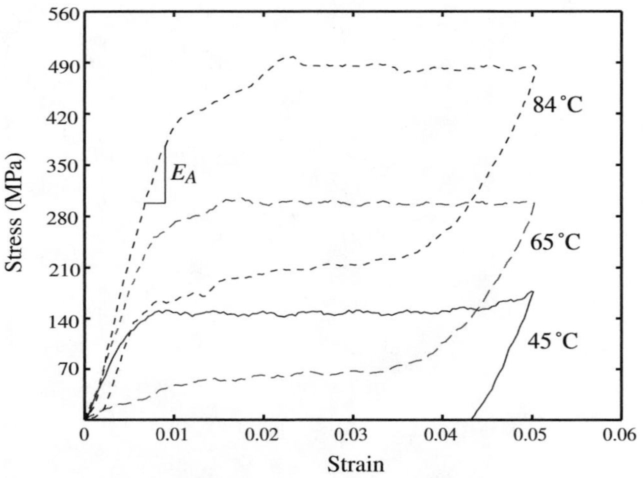

Figure 3.32. Constant temperature stress-strain behavior of Nitinol of diameter 0.015 in and length of 5 in.

is fully austenite to start with and is transformed to stress-preferred martensite. This transformation is reversed on unloading, causing a complete pseudoelastic recovery.

The Young's modulus of the material in the linear elastic region at 84°C is marked as E_A in Fig. 3.32. Similarly, the Young's modulus can be calculated from the data at each temperature. At the lowest temperature, 45°C, the slope of the linear elastic region corresponds to approximately the Young's modulus of pure martensite (E_M). Because the highest temperature plot is close to the austenite finish temperature (A_f), the modulus marked in the figure is approximately the Young's modulus for pure austenite (E_A). At an intermediate temperature, the Young's modulus lies between these two extremes, and its magnitude is determined by the volume fractions of martensite and austenite in the material. The increase in the elastic modulus from martensite to austenite can also be seen in the plot. The values of pure-phase Young's moduli for martensite and austenite calculated from these curves are required for the constitutive models.

From the stress-strain plots at constant temperature, one can obtain various critical stresses ($\sigma_1, \sigma_2, \sigma_3, \sigma_4$). These stresses can be used to plot the critical stress versus temperature diagram. From this diagram, one can determine the transformation temperature at different applied-stress levels (Fig. 3.33).

3.5.5 Stress-Temperature Behavior at Constant Strain

After obtaining the transformation temperatures of the material as described in Sections 3.5.2 and 3.5.3, and measuring the stress-strain behavior as in Section 3.5.4, a designer will have enough information to size an SMA actuator based on given actuation requirements. However, it is also of interest to investigate the constrained-recovery behavior of the material for use in certain applications in which the blocked force of the actuator is important [60]. Constrained recovery means that the wire is kept at constant strain, and the testing involves measurement of stress-temperature characteristics at a given pre-strain.

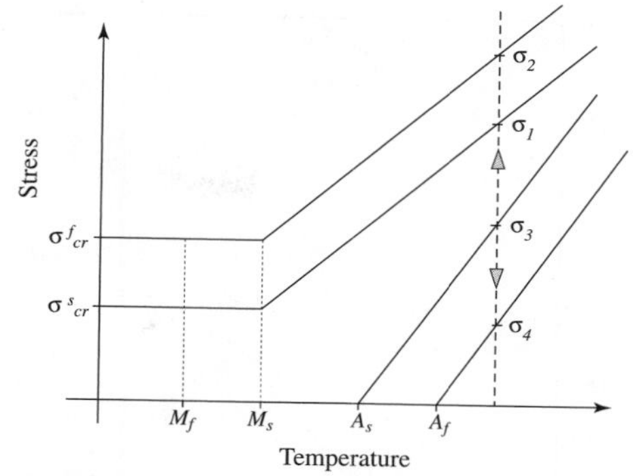

(a) Loading and unloading at a constant temperature

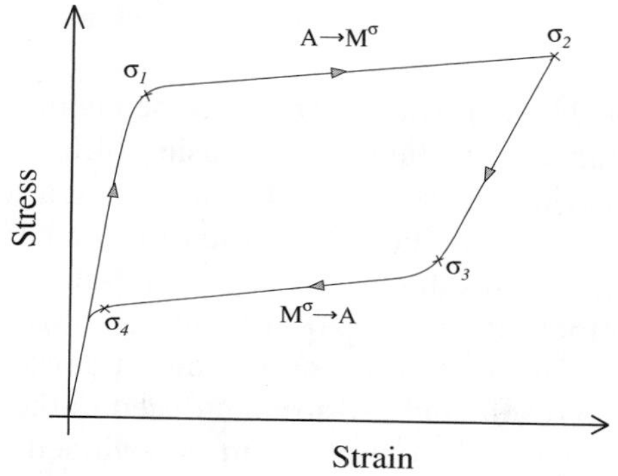

(b) Determination of critical stresses, $T > A_f$

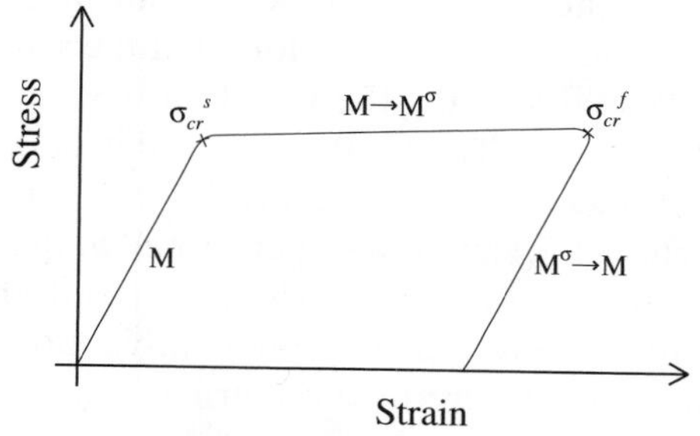

(c) Determination of critical stresses, $T < M_f$

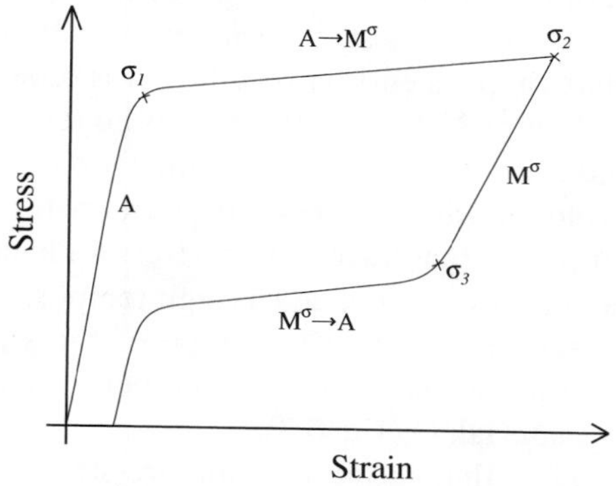

(d) Determination of critical stresses, $A_s < T < A_f$

Figure 3.33. Obtaining critical stresses from constant temperature stress-strain curves (M: martensite, M^{σ}: stress-induced martensite, A: austenite).

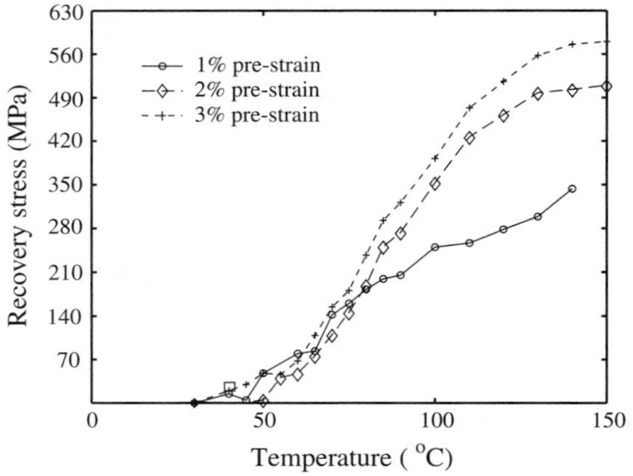

Figure 3.34. Constrained-recovery behavior.

The constrained-recovery behavior involves the study of the blocked force as a function of temperature using different values of pre-strain in the wire. These experiments can be performed using a tensile test machine, such as an MTS 810, in a setup similar to that used for the constant-temperature stress-strain tests. The testing procedure consists of first cycling the temperature of the wire by heating to a temperature above A_f and subsequently cooling it to a temperature below M_f, with the grips of the testing machine unlocked, which ensures a zero-stress condition. As a result, any pre-strain induced in the material based on history disappears, and the sample is completely in the twinned martensite phase before the test. The wire is then pre-strained to a certain length at room temperature ($T < A_s$). To ensure consistency, the pre-strain is conducted at a constant environmental temperature maintained at, say, 30°C for all of the samples. The displacement of the wire is then decreased to bring the wire to a stress-free state and, as a result, the elastic strain is recovered. The total strain at this point is referred to as the pre-strain imparted to the wire. This procedure is similar to the stress-strain procedure described in Section 3.5.4 and is also performed at the same strain rate (5×10^{-4}/s).

After ensuring the wire is just taut, the grips of the testing machine are locked in place so that no further relative movement of the ends of the wire can occur. The temperature of the thermal chamber is then increased and the load applied by the wire on the end constraint measured as a function of the temperature. The heating rate used for this procedure ($\approx$5°C/min) is also the same as that used in the calorimetry tests (see Section 3.5.2) and is therefore consistent with quasi-static behavior. This entire procedure was repeated for different values of pre-strain.

Figure 3.34 shows the measured recovery stress as a function of the temperature for different values of pre-strain. By comparing the values of the final recovery stress developed with the stress-strain characteristics at high temperatures (see Fig. 3.32), it is apparent that the recovery stress for a particular pre-strain is about the same as the maximum stress needed at high temperatures to develop this pre-strain. It is shown in Section 3.4.1 that the slope of the stress-temperature curve before any phase transformation occurs (i.e., temperature below A_s) is the thermoelastic constant of the material, Θ (Eq. 3.3).

Note that above a certain pre-strain level (approximately 2%), the final stresses are relatively independent of pre-strain. However, the path followed for different pre-strains is independent of the pre-strain itself. This behavior implies that there is no significant advantage to increasing pre-strains beyond the threshold level in an application involving constrained recovery. This is a significant observation because lower pre-strains offer the advantage of minimizing permanent plastic deformation and fatigue after repeated cycles [61, 62].

3.5.6 Comparison of Resistive Heating and External Heating

In actuator applications, the most convenient way of heating the wires is by passing a current through them. Due to the internal resistance of the wire, heat is produced, which results in a rise in temperature, leading to the temperature-induced phase transformation. Actuation using this type of resistive heating does not require any additional hardware except the attachment of electrical leads. This also makes resistive heating very attractive for embedded actuator applications. However, the change in resistivity of the SMA during transformation, as well as local changes in resistance of the wire specimen, make it difficult to maintain a constant temperature over the time of the transformation and along the length of the wire. However, it is possible to achieve relatively good control of temperature (about 2–3°C variation) by very slow activation of the wire. This method involves making small changes in electrical input to the wire and waiting for equilibrium to be achieved before acquiring data. The constant-load strain-temperature tests (see Section 3.5.3) describe resistive heating of the wire specimens.

Another way of changing the temperature of the wire is by heating it externally. In the laboratory, this can be accomplished in a controlled manner using a thermal chamber. The temperature of the wire specimen can be precisely controlled and maintained constantly over time as well as along the length of the wire, which makes environmental heating particularly suited for accurate measurements and correlations with predictions. However, it is obvious that this may not be a viable option in many practical actuator applications. External heating, or environmental heating, can also be performed by placing a heating element in proximity to the SMA actuator. Such an arrangement is more involved than that required for resistive heating. The constant-temperature stress-strain tests described in Section 3.5.4 are obtained via environmental heating using a thermal chamber.

It is important to point out that the properties of the SMA, such as transformation temperatures and critical stresses, vary somewhat depending on the type of heating used in the test procedure. Experimental evidence suggests that the difference in the behavior is not an artifact of the measurement technique. The reasons for this difference in behavior are not clear and are a subject of research. The critical temperatures obtained from strain-temperature tests conducted under resistive heating conditions may not directly correlate to those obtained from a test involving environmental heating.

An example of the discrepancy observed with resistive heating and environmental heating is shown in Fig. 3.35. The figure shows two constant-temperature stress-strain curves, measured at a temperature of 45°C. The curve measured using resistive heating shows a pseudoelastic recovery, whereas the curve measured using environmental heating shows the SME. This suggests that the transformation occurs at lower temperatures when using resistive heating.

The critical stress-temperature results obtained using environmental heating are plotted in Fig. 3.36 and those obtained using resistive heating are plotted in Fig. 3.37. By comparing the two heating methods, we see that the entire plot seems

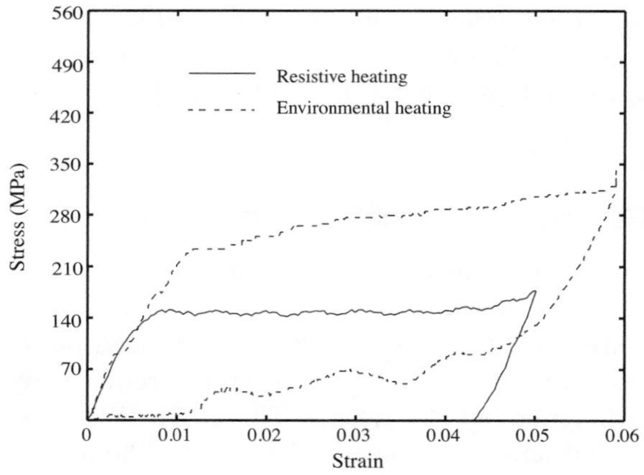

Figure 3.35. Comparison of stress-strain curves obtained using resistive and environmental heating, at 45°C.

shifted toward lower temperatures in the case of resistive heating. At the same stress levels, the material seems to transform at lower temperatures when resistively heated. Indeed, this can also be observed visually. However, the critical stresses for detwinned martensite at low temperatures and the slope of the critical stresses seem unchanged from the environmental-heating method. Although the reason for this discrepancy is not yet sufficiently clear, there are a few plausible explanations. In a polycrystal line structure involving different phases, the resistivity of one phase is quite different from that of the other. Passing a current through the wire will cause most of this total current to flow through low-resistance pathways. Hence, the local temperature profiles at the boundaries of the different phases may be quite different (and higher) than the temperature measured at the surface (because the environmental temperature is room temperature). This could result in an apparent lowering of the transformation temperatures in the resistive-heating case. It is important to note that in both heating methods, temperature was measured using a thermocouple attached through a thermally conductive, electrically resistive material and not directly on the wire. It was verified that in the resistive-heating case, there was no current flow into the thermocouple wire. The data, however, indicate that there are substantial differences between characterizations done using the two

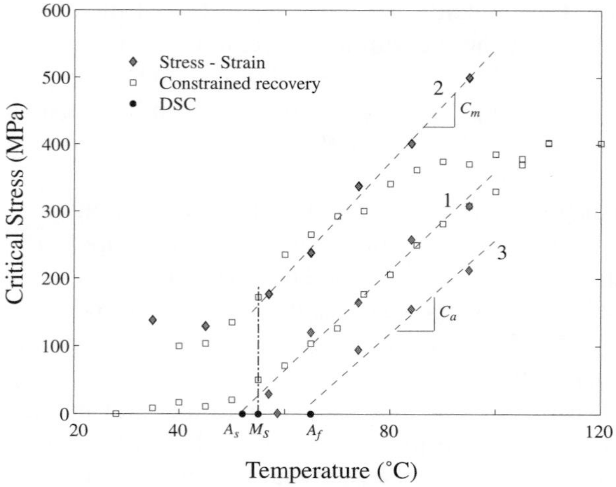

Figure 3.36. Experimental critical stress-temperature diagram, environmental heating.

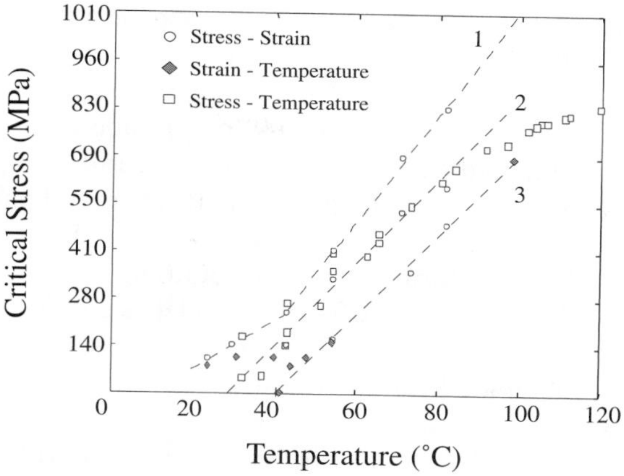

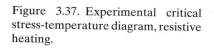

Figure 3.37. Experimental critical stress-temperature diagram, resistive heating.

heating methods, and using the constants derived from environmental heating may yield inaccurate predictions if applied to a test involving resistive heating.

Care must be taken to use the same type of heating technique in all of the experiments to obtain material data for the constitutive models. Although environmental heating appears more appropriate for validation studies, the material characterization using resistive heating [63] may be more convenient to model behavior of the SMA wires.

To effectively design actuators and other systems based on SMAs, it is necessary to understand and model the behavior of the material under mechanical loads and changes in temperature. In uniaxial loading, the SMA behavior is primarily a function of stress, strain, and temperature and their associated time derivatives; the SMA behavior is a nonlinear function of these variables. Many constitutive models have been developed to describe the thermomechanical behavior of SMA materials. In large part, these models are curve fits to experimental data.

Simplified constitutive models assume quasi-static behavior. However, the SMA is a thermomechanically coupled system. A change in temperature causes a mechanical deformation and, conversely, a mechanical deformation causes heat to be generated or absorbed by the material. Depending on the rate of mechanical deformation and the heat-transfer properties of the material and its environment, the rate of change of temperature of the material is affected. This, in turn, can affect its mechanical behavior. Therefore, for more accurate modeling, heat-transfer effects need to be considered. From an energy point of view, the applied stress results in a change of strain energy, whereas temperature affects the chemical-free energy. It is the sum of these two energies that influences the phase transformation.

3.6 Obtaining Critical Points and Model Parameters from Experimental Data

The empirical constants used in the constitutive models described previously are obtained from a series of experiments performed on the material. For example, to fully define the SMA constitutive model using the Brinson approach, we need to determine 11 material coefficients M_s, M_f, A_s, A_f, E_M, E_A, ϵ_L, σ_s^{cr}, σ_f^{cr}, C_M, and C_A. At this point, it is worth summarizing the testing procedure used to obtain these empirical constants. A brief description of the tests that can be performed to obtain these constants, as well as a physical interpretation of the constants, is given as follows:

1. DSC test: The transformation temperatures (M_s, M_f, A_s, A_f) at zero stress are obtained from this test. The test setup and procedure are explained in detail in Section 3.5.2. M_s and M_f are the start and finish temperatures for transformation to martensite at no stress. A_s and A_f are the start and finish temperatures for transformation to austenite at no stress. Note that the constants are used differently in the Tanaka model as compared to the Brinson model.
2. Constant-temperature stress-strain tests: Several important parameters can be obtained from these tests:
   (a) E_M and E_A: These are the Young's moduli of the SMA in pure martensite and austenite phases, respectively. The Young's moduli are measured in the linear elastic region of the stress-strain curves at constant temperature. The low-temperature (i.e., below M_f) curves yield the estimate for E_M, whereas the high-temperature (i.e., above A_f) curves yield the value for E_A.
   (b) Critical stresses of transformation σ_s^{cr} and σ_f^{cr}: These represent the start and completion of detwinned martensite, below the temperature A_s. Below σ_s^{cr}, the material exists in fully twinned martensite state, whereas above σ_f^{cr}, the material is in completely detwinned martensite state. These stresses are modeled as constant values that are invariant with temperature in the Brinson model. The transformation from twinned to detwinned martensite occurs with the application of stress. On removal of the stress, the material stays in a detwinned martensite state. To revert to twinned martensite, the material must be heated to transform into the austenite state and then cooled under the no-stress condition.
   (c) ϵ_L: This is the maximum recoverable strain that can be obtained from the SMA, and it is a material constant. This strain value is obtained from the low-temperature stress-strain curve.
   (d) Stress-influence coefficients C_M and C_A: These are rates of variation of critical stress with temperature. They are given by the slopes of linear curve fits of experimentally determined critical stresses with temperature. From the constant-temperature stress-strain curves, the critical stresses can be found as shown in Figure 3.38. At a low temperature (45°), the critical stresses σ_1 and σ_2 are obtained similar to σ_s^{cr} and σ_f^{cr}. At a high temperature (84°), the critical stresses are obtained from the pseudoelastic curve. These critical stresses are plotted as a function of the temperature, and the values of C_M and C_A are obtained.
   (e) Constant-stress strain-temperature test: The variation of transformation temperatures with applied stress can be measured from this test, also yielding values for C_M and C_A. An alternate definition for C_M and C_A is the inverse of the slopes of linear curve fits of transformation temperatures with applied stress. The value of C_M represents the inverse of the rate of change of the martensite start and finish temperatures with stress, and the value of C_A represents the inverse of the corresponding rate for austenite start and finish temperatures. These quantities are marked in Fig. 3.30. The test procedure is described in Section 3.5.3. An alternate procedure is to determine these constants from test data obtained in a thermally controlled environment (by external heating).

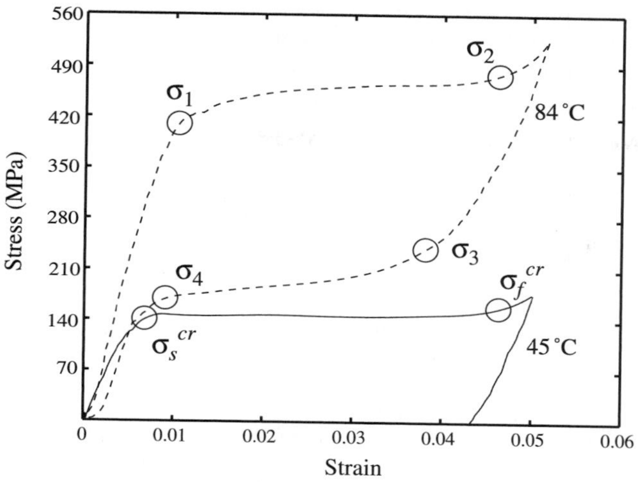

Figure 3.38. Critical stresses from constant-temperature stress-strain curves.

(f) Constrained recovery test: This test, described in Section 3.5.5, does not directly yield any points on the critical diagram, but it is useful as a check for the experimental critical points obtained from the other tests.

A convergence of data from different tests on the same curve indicates that the critical points of the material are unique and show fair agreement with the assumption of linear variation of the critical stresses with temperature that is inherent in all of the models. The recovery stress datapoints must lie in the transformation region between the austenite start and finish curves for heating phase and martensite start and finish curves for the cooling phase. Because all of these tests so far reported were carried out at very low rates (strain rate $= 5 \times 10^{-4}$/s, heating rates = 1°C/min), the conclusions that are drawn from these tests are applicable under quasi-static conditions only. The strain and heating rates were chosen as representing a quasi-static value because it was observed that slower strain rates produced no significant change in the material behavior. Although temperature measurements were not taken during the isothermal tests, this seems to indicate near-isothermal conditions in the wire subjected to these strain rates. Because the ratio of heat retained (proportional to the volume) to heat transferred by convection (proportional to the surface area) is proportional to the diameter of the wire, we expect near-isothermal conditions to be reached at higher strain rates in wires with lower diameters. The choice of a strain rate of 5×10^{-4}/s as quasi-static for this wire is, therefore, consistent with the findings reported by Shaw and Kyriakides [59]. Figure 3.36 shows the experimental critical stress-temperature diagram (environmental heating). Data from four tests – heat flow measurements (i.e., DSC), constant temperature, constrained recovery, and constant stress – are consolidated on this plot. The critical points obtained from the stress-strain plots are plotted in the filled-diamond symbols. The filled-circle symbols denote the points obtained from the heat-flow measurements at no stress, the square symbols are used to plot the constrained-recovery curve for different pre-strains, and the diamond symbols show the critical points obtained from the stress-strain behavior at constant temperature for heating and cooling cycles. The unfilled-circle markers are the points obtained from the strain-temperature characteristics. The

Table 3.3. *Constitutive model parameters used for 0.015 in (0.381 mm) diameter SMA wire, environmental heating*

<table>
<tr><th rowspan="2">Parameter</th><th colspan="2">Value</th></tr>
<tr><th>Tanaka</th><th>Brinson</th></tr>
<tr><td>M_s, °C</td><td>43.5</td><td>55.0</td></tr>
<tr><td>M_f, °C</td><td>40.7</td><td>42.0</td></tr>
<tr><td>A_s, °C</td><td colspan="2">52.0</td></tr>
<tr><td>A_f, °C</td><td colspan="2">65.0</td></tr>
<tr><td>C_A, MPa/°C</td><td colspan="2">8.0</td></tr>
<tr><td>C_M, MPa/°C</td><td colspan="2">12.0</td></tr>
<tr><td>σ_{cr}^s, MPa</td><td>–</td><td>138.0</td></tr>
<tr><td>σ_{cr}^f, MPa</td><td>–</td><td>172.0</td></tr>
<tr><td>E_M, GPa</td><td colspan="2">20.3</td></tr>
<tr><td>E_A, GPa</td><td colspan="2">45.0</td></tr>
<tr><td>ϵ_L</td><td colspan="2">0.067</td></tr>
</table>

lines shown are linear curve fits to data and they define the regions of martensite start and austenite start, respectively. The martensite finish line is not shown here.

3.7 Comparison of Constitutive Models with Experiments

Having obtained the important material parameters from experiments for each of the models, their predictions can be compared with experimental data. As noted in Section 3.5.6, there are significant differences between experiments performed with environmental heating and with resistive heating. To maintain consistency in the comparisons with the models, the parameters used are obtained from tests that use only environmental heating. These constants are listed in Table 3.3.

A comparison of the measured stress-strain curve at 35°C with calculations using Brinson's model is shown in Fig. 3.39(a). The correlation between the measured and calculated values appears satisfactory. Based on the four measured phase-transformation temperatures, it is assumed that the material was initially 100% martensite (before loading). Therefore, the initial temperature-induced martensite volume fraction (ξ_T) equals one and the stress-induced martensite volume fraction (ξ_S) equals zero. During the loading, stress-induced martensite is formed, ξ_T decreases, and ξ_S increases. Figure 3.39(b) shows the calculated variation of the martensite volume fractions for the elastic region. It can be seen that, initially, $\xi_T = 1$ and $\xi_S = 0$. Subsequently, as the critical stress ($\approx$138 MPa) is reached, the stress-induced martensite increases and the temperature-induced martensite decreases. However, their sum equals 1.0 at all times because this temperature is below A_s.

In the Tanaka and the Liang and Rogers models, there is no separate stress-induced martensite and, therefore, the temperature-induced martensite volume fraction stays equal to one for the elastic region and the transformation region. This may lead to incorrect estimations of the behavior. Among the constitutive models, the Brinson model is applicable for predictions at temperatures below A_s when the

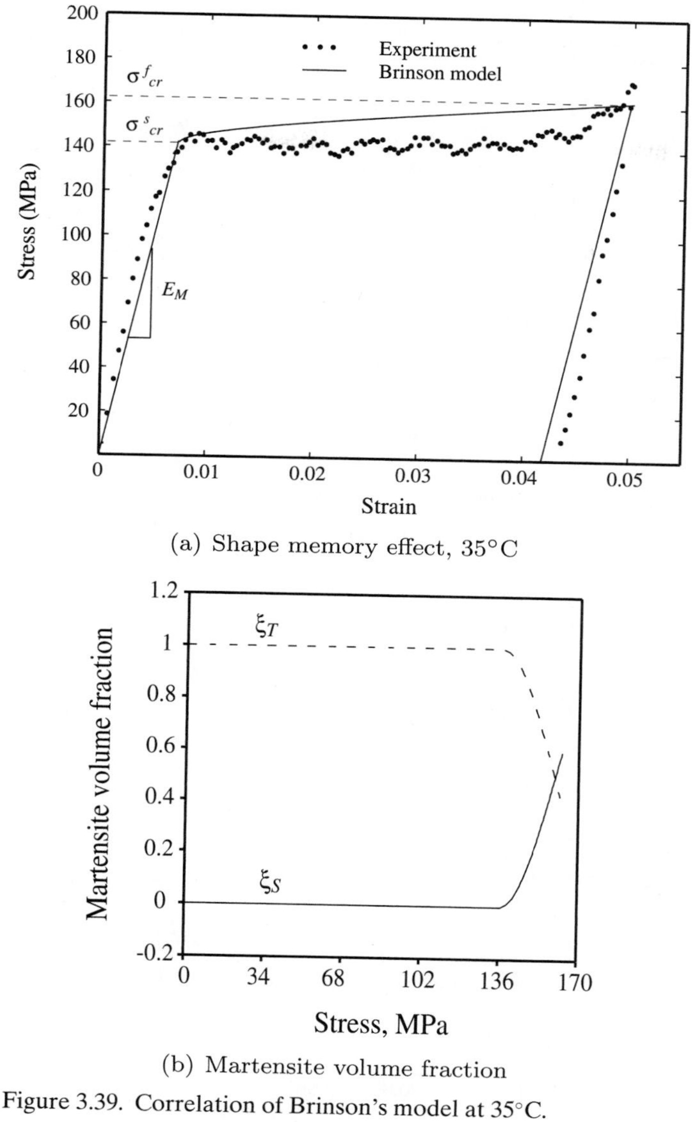

Figure 3.39. Correlation of Brinson's model at 35°C.

material is starting from randomly oriented "twinned" martensite (Fig. 3.39(a)). At temperatures above A_s, however, all of the models are applicable.

The comparison of the experiments with the three constitutive models is shown in Fig. 3.40. The constants M_s and M_f used in the Brinson model are different from those in the Tanaka and the Liang and Rogers models due to the differences in their interpretation of these constants. Among the three constitutive models, only the Brinson model is applicable for predictions at temperatures below A_s starting from randomly oriented "twinned" martensite (Fig. 3.39(b)).

At temperatures above A_s, however, all of the models are applicable and they are compared with experiments at two representative temperatures, one close to A_f (i.e., starting from nearly pure austenite (Fig. 3.40(a)) and the other well above A_f (Fig. 3.40(b)). From the two isothermal comparisons, it can be seen that all of the models match the experimentally measured characteristics of pseudoelasticity

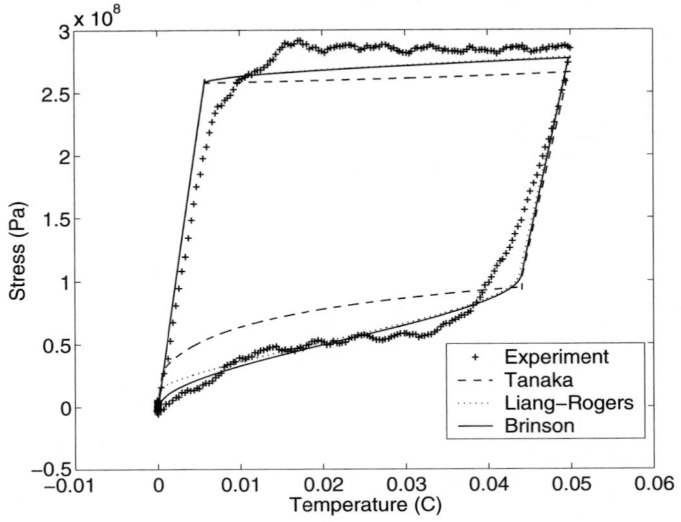

(a) Isothermal test, 65°C

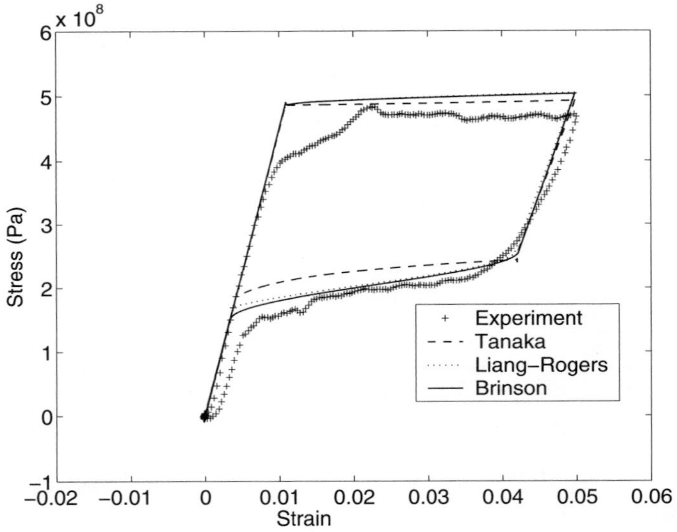

(b) Isothermal test, 84°C

Figure 3.40. Model comparison for isothermal tests, strain rate = 5×10^{-4}/s.

quite closely and that there are only minor differences in the transformation paths that are prescribed. The Brinson and the Liang and Rogers models predict the same path for pseudoelastic transformation above A_f. It may be noted that the only differences between the Brinson model and the Liang and Rogers model is the form of constitutive law – the Liang and Rogers model uses $E(\xi)(\epsilon - \epsilon_o)$, whereas the Brinson model uses $E(\xi)\epsilon - E(\xi_o)\epsilon_o$, and that they both use a cosine model for the transformation kinetics. From the experimental results, it was observed that these models matched more closely with experimentally measured unloading curves. It may also be noted that due to the assumed linear variation of the critical stresses with temperature in all three models, it is difficult to obtain exact matching of the models to experimental data for all temperatures. Therefore, for certain temperatures (see Fig. 3.40(a)), the models slightly underpredict experimental behavior, whereas in other temperature regimes (see Fig. 3.40(b)), they overpredict experimental results.

It should be mentioned here that because the Tanaka model employs an exponential representation for the martensite volume fraction, the results are quite sensitive to small variations in the material parameters, and the numerical-solution scheme can sometimes go unstable and/or unbounded during the calculations for the stress and volume fractions. However, the overall correlation for all of the models is good over the entire temperature range, and the differences are minimal.

3.8 Constrained Recovery Behavior (Stress versus Temperature) at Constant Strain

These experiments were carried out to determine the behavior of an SMA wire when its length is constrained. The wire was first pre-strained to a certain length at room temperature ($T < A_s$). The load used to pre-strain was then removed to bring the wire to a stress-free state. Only a small elastic portion of the total strain was recovered. This was accounted for and the wire was made just taut again. The temperature of the wire was then increased, and the load applied by the wire on the end constraint was measured as a function of the temperature. This gives the actuation-force capability of the wire as a function of temperature when it is not allowed to recover its strain. The load cell attached to the grips was used to measure the stress applied by the wires, and a thermocouple was used to measure the temperature of the wire. A thermal chamber was used to control the temperature of the wire. The tests were carried out at a heating rate of approximately 5°C/min (0.083°C/s).

It was observed that below about 2% pre-strain, although the final stress reached is dependent on the amount of pre-strain, the path followed for different pre-strains is independent of the pre-strain itself. Above a threshold value of about 2%, both the final stress attained and the paths followed are fairly independent of pre-strain. The Tanaka and the Liang and Rogers models predict a maximum recovery stress that is a linear function of the martensitic residual strain or initial strain. In the experiments that were conducted, this was found to be accurate only at low values of pre-strains. One reason for this could be the unavoidable permanent plastic deformations that result when the wire is constrained from recovering its free length. These are likely to yield an overprediction for higher values of strains resulting from these models. A nominal pre-strain of 2% was therefore chosen to compare the models with experimental data. It is useful to note an important point relating to the application of the Tanaka and the Liang and Rogers models to constrained-recovery data. Because these models predict transformation between austenite and detwinned martensite only, their applicability to explain constrained-recovery behavior must be interpreted appropriately. In these simulations, the initial volume fraction of martensite for these two models is not close to 1.0 even though the material is indeed in complete martensite phase.

The volume fraction used for the first simulation is ϵ_o/ϵ_1, which is the proportion of detwinned martensite that would have been used in the Brinson model. This adjustment to the models is necessary to apply them to the first cycle of the recovery stress-temperature behavior. However, in subsequent cycles, because the austenite cools to martensite in the presence of stress, the low-temperature phase is mostly

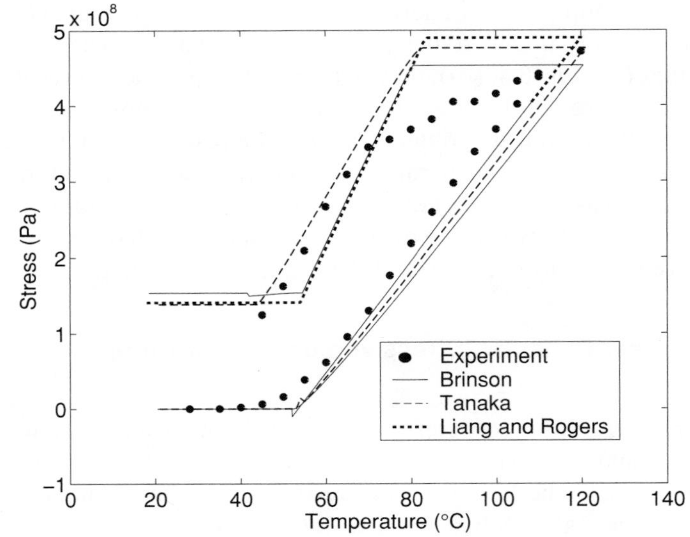

(a) Constrained-recovery curve for pre-strains

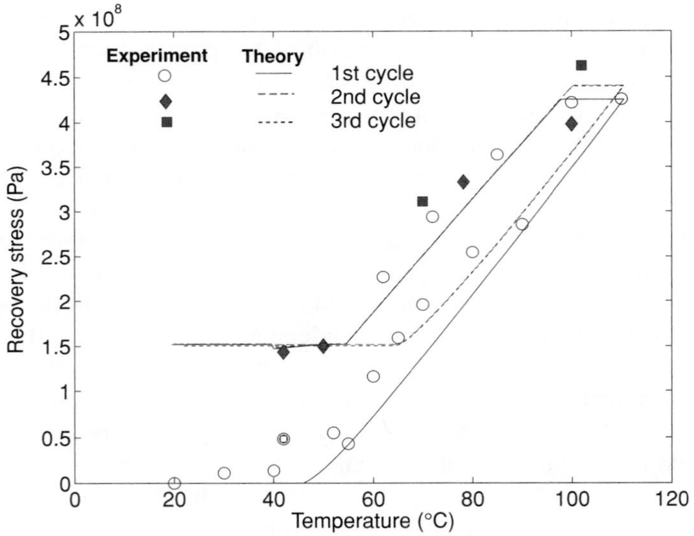

(b) Constrained-recovery behavior over successive loading cycles

Figure 3.41. Model comparisons for constrained-recovery curve of stress versus temperature, 2% pre-strain.

detwinned martensite. This proportion of detwinned martensite is determined by the simulations themselves and can then be used in subsequent simulations with the Tanaka model without loss of generality. However, by applying these models in this manner (which is strictly not correct for the first cycle), one can get the simulations started, which will then approximate the behavior of the SMA correctly in subsequent cycles. Figure 3.41(a) shows a comparison of the recovery-stress predictions with the various models against experimental data. It can be seen that for the temperature range and pre-strain tested, all of the models predict the final recovery stress

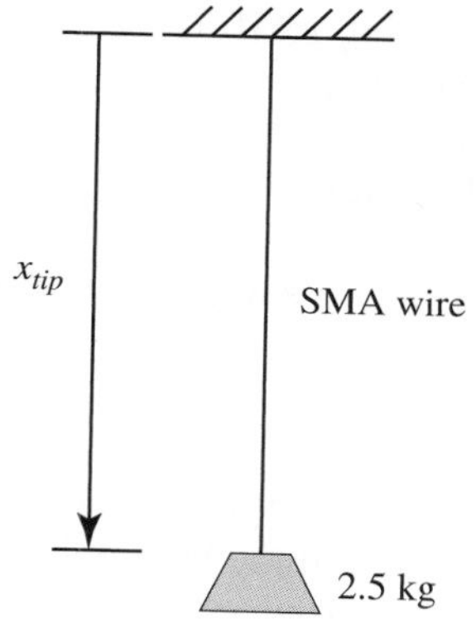

Figure 3.42. SMA wire with a constant load.

fairly well, although there is slight overprediction with the Tanaka and the Liang and Rogers models. The path followed on the thermal-loading cycle is also well predicted by all three models. Note that the predictions of the models can be improved by assigning slightly different values to the experimental constants. However, this often conflicts with better predictions made using the stress-strain characteristics and the critical stress-temperature diagram. The constants here were chosen considering test data from all three sources and therefore reflect a compromise between characteristics over the entire thermomechanical range for the particular material. The unloading cycle, however, is predicted in an idealized manner by the three models. In experiments, the transition between austenite and martensite did not have a unique starting point and occurred over a fairly large range of temperatures. This observation was made for other pre-strain values as well. However, all of the models accurately and fairly predict the final steady slope of the thermal unloading. Figure 3.41(b) also shows the stress history of the wire after repetitive cycling. The wire is pre-strained to 2%, then kept constrained while it is heated, and its constrained-recovery properties are recorded over several cycles. This information is especially relevant in applications involving the use of SMAs as actuators. Because all of the constitutive models were found to yield fairly similar results, the Brinson model is chosen as a representative model in this figure. Theoretical calculations are made with the final state of the previous cycle as the starting input for the current cycle. From the plots, it can be seen that the final stress levels show a slight increase from the first cycle to subsequent cycles. The intermediate path taken is also slightly different between the first and subsequent cycles. However, after three cycles, the material is stabilized, following virtually the same path and attaining the same final stresses. This is also seen in the model; the martensitic volume fractions approach equilibrium values after about the third cycle. This analysis and supporting experimental data demonstrate the feasibility of using SMAs as actuators under repetitive cycling. This test was conducted for a moderately low pre-strain level of 2%.

3.8.1 Worked Example

An SMA wire of diameter 0.5 mm (0.0197 in) and length 0.3 m (11.8 in) is held at one end and loaded vertically with a weight of 2.5 kg (5.625 lb) (Fig. 3.42). Determine the tip position of the wire at room temperature (15°C). Use the Tanaka model to calculate the tip position, x_{tip}, as a function of temperature during quasi-static heating to 70°C followed by cooling to 15°C. Use the material constants given in Table 3.3.

Solution

The stress in the wire is given by

$$\sigma_o = \frac{F}{A} = \frac{2.5 \times 9.81}{\pi(0.5 \times 10^{-3})^2/4} = 124.9 \text{ MPa}$$

Given that $\sigma_s^{cr} = 137.9$ MPa and that $A_s = 34.4$°C, we can conclude that the wire is purely in the twinned martensite phase. Therefore, the strain is given by

$$\epsilon_o = \frac{\sigma_o}{E_M} = \frac{124.9 \times 10^6}{23.7 \times 10^9} = 0.0053$$

and the initial deflection is

$$\Delta l_o = \epsilon_o l = 0.0053 \times 0.3 = 1.6 \text{ mm}$$

Because a weight is suspended from the wire, the stress in the wire is constant at all times. As the wire is heated, it undergoes a martensite-to-austenite transformation. The transformation begins at a temperature T_1 and ends at a temperature T_2. These temperatures can be found from the stress-influence coefficients.

$$T_1 = A_s + \frac{\sigma_o}{C_A} = 34.4 + \frac{124.9}{13.5} = 43.65°\text{C}$$

$$T_2 = A_f + \frac{\sigma_o}{C_A} = 48.3 + \frac{124.9}{13.5} = 57.55°\text{C}$$

Similarly, while cooling, the austenite-to-martensite transformation begins at a temperature T_3 and ends at a temperature T_4 given by

$$T_3 = M_s + \frac{\sigma_o}{C_M} = 26.7 + \frac{124.9}{13.2} = 36.16°\text{C}$$

$$T_4 = M_f + \frac{\sigma_o}{C_M} = 23.3 + \frac{124.9}{13.2} = 32.76°\text{C}$$

Let us first consider the heating of the wire. The initial martensite volume fraction, $\xi_o = 1$, because the wire is completely in the martensite phase. During the $M \to A$ transformation, the martensite volume fraction as a function of temperature T is given by

$$\xi(T) = e^{a_A(A_s - T) + b_A \sigma_o}$$

where the constants a_A and b_A are

$$a_A = \frac{\ln 0.01}{A_s - A_f} = 0.3313 \text{ 1/°C}$$

$$b_A = \frac{a_A}{C_A} = 2.45 \times 10^{-8} \text{ 1/Pa}$$

The strain in the wire is given by the governing equation (Eq. 3.3)

$$(\sigma - \sigma_o) = E(\xi)(\epsilon - \epsilon_o) + \Theta(T - T_o) + \Omega(\xi)(\xi - \xi_o)$$

Because the stress remains constant, $\sigma - \sigma_o = 0$. Neglecting the thermal expansion and substituting for the phase-transformation constant (see Eq. 3.5) yields the strain in the wire between the temperatures T_1 and T_2 as

$$\epsilon(T) = \epsilon_L(\xi - \xi_o) + \epsilon_o$$

Note that there is no change in strain at temperatures less than T_1 and greater than T_2. Similarly, during the cooling of the wire, the initial martensite volume fraction $\xi_o = 0$ because the wire is initially in the pure austenite phase. The $A \to M$ transformation occurs between the temperatures T_3 and T_4. During the phase transformation, the martensite volume fraction is given by

$$\xi(T) = 1 - e^{a_M(M_s - T) + b_M \sigma_o}$$

where the constants a_M and b_M are

$$a_M = \frac{\ln(0.01)}{(M_s - M_f)}$$

$$b_M = \frac{a_M}{C_M}$$

The tip position of the wire and the martensite volume fraction are shown in Fig. 3.43.

3.8.2 Worked Example

An SMA wire of diameter 0.5 mm (0.0197 in) and length 0.3 m (11.8 in) is held between two ends, one end in a vise and the second end using a spring restraint with a linear spring constant $k = 3500$ N/m (20.1 lb/in), as shown in Fig. 3.44. Assume that the initial stress in the wire is zero at room temperature (15°C). Use the Tanaka model to calculate the tip position, x_{tip}, as a function of temperature during quasi-static heating to 70°C followed by cooling to 15°C. Use the material constants given in Table 3.3.

Solution

As the wire is heated above the temperature T_1, it recovers strain and contracts. This causes extension of the linear spring and a corresponding increase in stress in the wire. The stress and strain in the wire are related by the displacement compatibility of the SMA wire and the linear spring.

$$\epsilon = -\frac{\Delta l}{L} = -\frac{F}{kL} = -\frac{\sigma A}{kL}$$

where F is the force in the spring and L is the initial length of the SMA wire. It is important to point out that the transition temperatures T_1, T_2, T_3, and T_4 are functions of the stress in the wire. In the previous example, because the stress was always constant, the transition temperatures were also constant. However, in

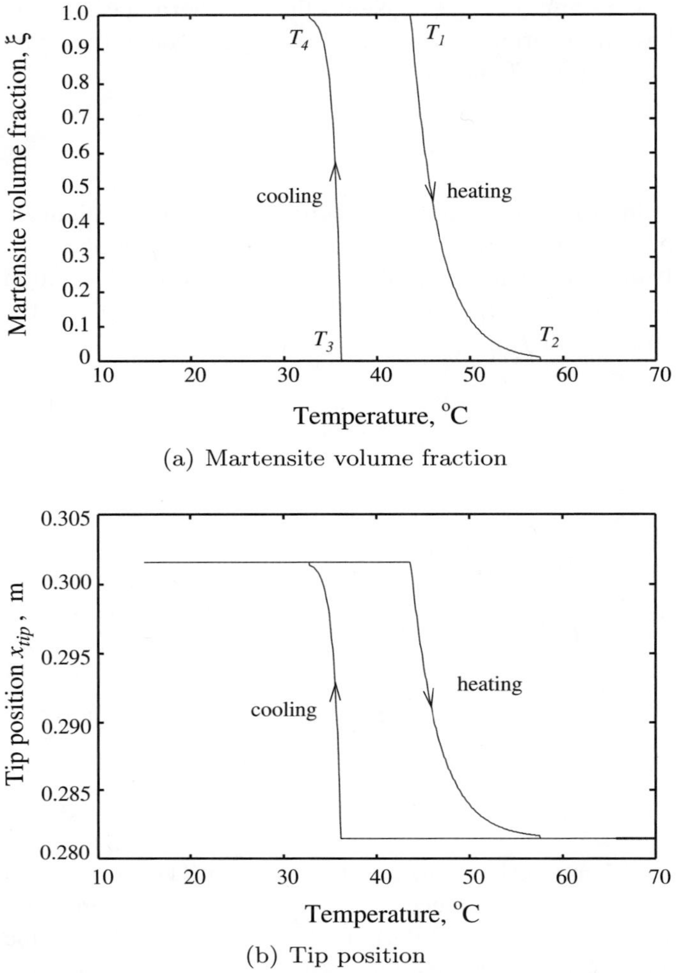

Figure 3.43. Heating and cooling of an SMA wire with a constant load.

the present case, the transition temperatures must be recalculated at each stress value.

For the $M \rightarrow A$ transformation (occurring between the temperatures T_1 and T_2)

$$\sigma_o = 0$$

$$\epsilon_o = 0$$

$$\xi_o = 1$$

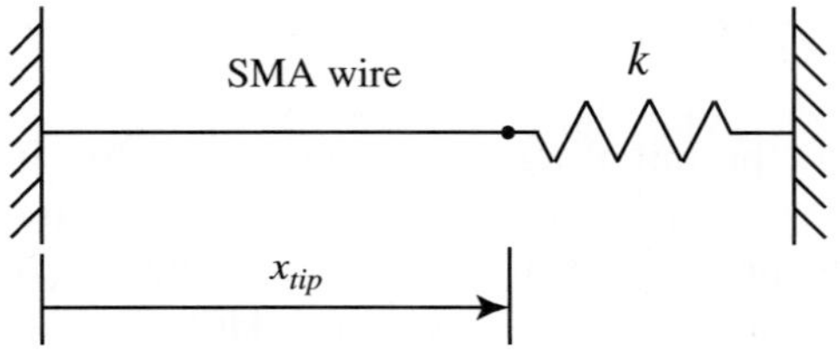

Figure 3.44. SMA wire with a linear spring.

Substituting these in the governing equation (Eq. 3.3) and neglecting thermal expansion, we obtain

$$\sigma = E(\xi)\epsilon + \Omega(\xi)(\xi - 1)$$

Substituting for $\Omega(\xi)$ and ϵ

$$\sigma = -E(\xi)\frac{\sigma A}{kL} - \epsilon_L E(\xi)(\xi - 1)$$

This equation can be solved by iteration using the Newton-Raphson method

$$\sigma_{new} = \sigma_{old} - \frac{f(\sigma_{old})}{f'(\sigma_{old})}$$

where the functions $f(\sigma_{old})$ and $f'(\sigma_{old})$ are obtained as

$$\begin{aligned} f(\sigma) &= \sigma + E(\xi)\frac{\sigma A}{kL} - \epsilon_L E(\xi)(\xi - 1) \\ f'(\sigma) &= \frac{\partial f(\sigma)}{\partial \sigma} \\ &= 1 + E(\xi)\frac{A}{kL} + \frac{\sigma A}{kL}\frac{\partial E(\xi)}{\partial \sigma} + \epsilon_L E(\xi)\frac{\partial \xi}{\partial \sigma} + \epsilon_L \xi \frac{\partial E(\xi)}{\partial \sigma} - \epsilon_L \frac{\partial E(\xi)}{\partial \sigma} \end{aligned}$$

where

$$E(\xi) = E_A - \xi(E_A - E_M)$$

and from the Tanaka model

$$\xi(\sigma, T) = e^{a_A(A_s - T) + b_A \sigma}$$

This gives

$$\begin{aligned} \frac{\partial \xi}{\partial \sigma} &= b_A \xi \\ \frac{\partial E(\xi)}{\partial \sigma} &= (E_M - E_A)\frac{\partial \xi}{\partial \sigma} = (E_M - E_A) b_A \xi \end{aligned}$$

Substituting this in the equation for $f'(\sigma)$, we obtain

$$\begin{aligned} f'(\sigma) = 1 + E(\xi)\frac{A}{kL} + \frac{\sigma A}{kL}(E_M - E_A) b_A \xi + \epsilon_L E(\xi) b_A \xi \\ + \epsilon_L (E_M - E_A) b_A \xi^2 - \epsilon_L (E_M - E_A) b_A \xi \end{aligned}$$

The final stress at the end of the transformation, σ_H^f, is given by substituting $\xi = 0$ in the governing equation

$$\sigma_H^f = -E_A \frac{\sigma_H^f A}{kL} + \epsilon_L E_A$$

which results in

$$\sigma_H^f = \frac{\epsilon_L E_A kL}{kL + E_A A}$$

The $A \rightarrow M$ transformation occurs during the cooling cycle, between the temperatures T_3 and T_4. The initial conditions for this transformation are given by

$$\sigma_o = \sigma_H^f = \frac{\epsilon_L E_A kL}{kL + E_A A}$$

$$\epsilon_o = -\frac{\sigma_H^f A}{kL} = -\frac{\epsilon_L E_A A}{kL + E_A A}$$

$$\xi_o = 0$$

Proceeding as described previously, the governing equation becomes

$$\sigma - \sigma_o = E(\xi)(\epsilon - \epsilon_o) - \epsilon_L E(\xi)\xi$$

The equation is solved using the Newton-Raphson method. The functions $f(\sigma)$ and $f'(\sigma)$ are given by

$$f(\sigma) = \sigma - \sigma_o - E(\xi)\left(-\frac{\sigma A}{kL} - \epsilon_o\right) + \epsilon_L E(\xi)\xi$$

$$f'(\sigma) = 1 - \frac{\partial E(\xi)}{\partial \sigma}\left(-\frac{\sigma A}{kL} - \epsilon_o\right) + E(\xi)\frac{A}{kL} + \epsilon_L E(\xi)\frac{\partial \xi}{\partial \sigma} + \epsilon_L \xi \frac{\partial E(\xi)}{\partial \sigma}$$

However, we know that

$$\frac{\partial E(\xi)}{\partial \sigma} = (E_M - E_A)\frac{\partial \xi}{\partial \sigma}$$

From the Tanaka model, the martensite volume fraction is given by

$$\xi(\sigma, T) = 1 - e^{a_M(M_S - T) + b_M \sigma}$$

This gives

$$\frac{\partial \xi}{\partial \sigma} = b_M(\xi - 1)$$

Substituting in the expression for $f'(\sigma)$

$$f'(\sigma) = 1 - (E_M - E_A)b_M(\xi - 1)\left[-\frac{\sigma A}{kL} - \epsilon_o - \epsilon_L \xi\right] + E(\xi)\frac{A}{kL} + \epsilon_L E(\xi) b_M(\xi - 1)$$

The calculated stress, tip position, and martensite volume fraction are shown in Figs. 3.45–3.47.

3.9 Damping Capacity of SMA

When the material is in the austenite phase, a large tensile stress induces a transformation to the martensite phase (stress-induced martensite). Above the critical stress, a significant increase in strain takes place without much increase in stress. On unloading, the large strain is recovered with a considerable hysteresis. This is the pseudoelastic behavior of the SMA. The area enclosed in the stress-strain hysteresis loop represents the amount of energy dissipated in one cycle, which is transformed into heat. This occurs as a result of the internal friction and is responsible for the damping capacity of the material. This damping capacity depends on a variety of internal and external parameters that include strain amplitude, strain rate

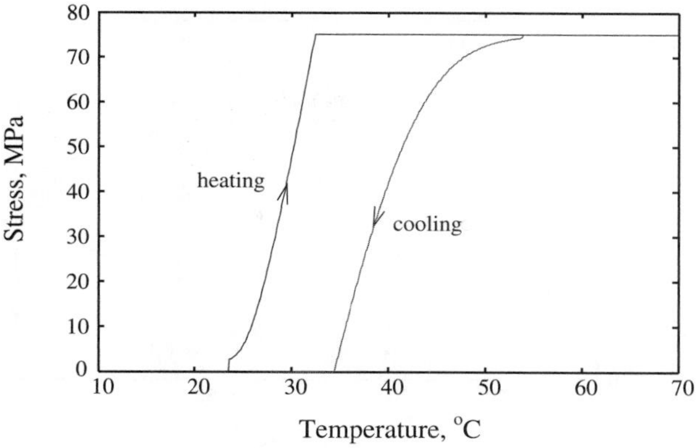

Figure 3.45. Stress in the SMA wire, acting against linear spring.

(or frequency of cyclic loading), alloy composition, grain size, heat treatment, and thermomechanical cycling. It is well known that the forward martensitic transformation is exothermic, whereas the reverse transformation is endothermic. Hence, high strain rates result in temperature changes, which in turn affect the stress-strain behavior. Lammering and Schmidt [64] showed that the area of the hysteresis loop is reduced with increasing strain rate. The damping capacity of the SMA is highly dependent on the vibration amplitude.

In the martensite phase, the damping appears to be caused by the mobility of the twinned-phase interfaces and defects inside the martensite phase. Liu and Humbeeck [65] have shown that the damping level in the martensitic phase is dependent on the strain amplitude and annealing temperature. The damping capacity increases with strain amplitude but decreases with increasing number of cycles until it reaches a stable value. The austenite damping capacity is generally smaller than the martensite value. Pushtshaenko et al. [66] studied the vibration damping of a SMA rod using the Likhachev model [67] that allows description of the one-way effect, pseudoelastic properties, and thermal-loading cycles in a modular manner.

The SMA damping can also be characterized using a complex modulus approach. Gandhi and Wolons [68] and Wolons et al. [69] showed that the hysteretic behavior of Nitinol undergoes a considerable change as the frequency of excitation increased. Hence, using the quasi-static hysteresis would then lead to erroneous predictions of the damping capacity of the material. They characterized the damping behavior of

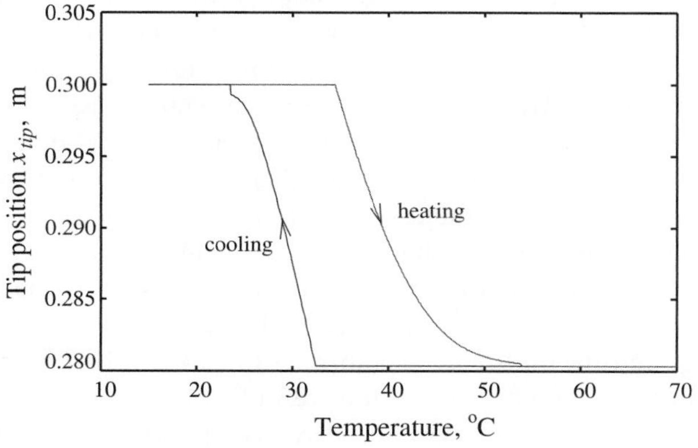

Figure 3.46. Tip position of the SMA wire, acting against linear spring.

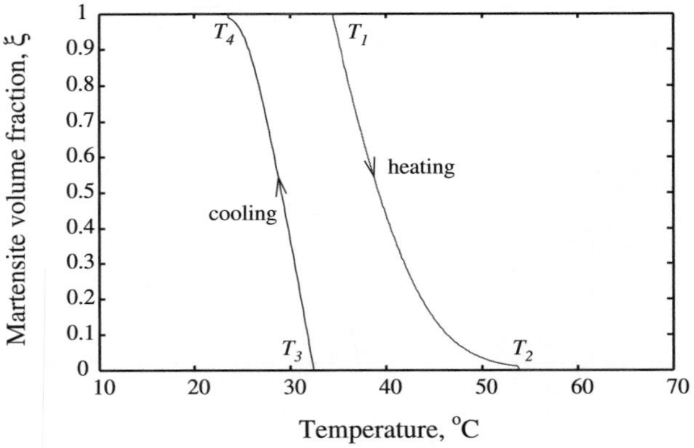

Figure 3.47. Martensite volume fraction, acting against linear spring.

Nitinol above the austenite finish temperature using the complex-modulus approach. The value of loss modulus at 6–10 Hz was found to be about 50% of that at low frequency and did not show any major reduction at higher frequencies. As the strain amplitude increased, the storage modulus (i.e., stiffness) initially decreased rapidly, implying the softening of the material, followed by a much smaller change with amplitude. A change in temperature has a significant effect on both the storage modulus and the loss modulus.

3.10 Differences in Stress-Strain Behavior in Tension and Compression

The discussion in the preceding sections focuses primarily on the behavior of SMA in tension. For experiments and applications involving an SMA wire, it is only possible to apply uniaxial tensile stresses on the material. However, numerous applications make use of SMA in different shapes, such as washers [70, 71], tubes, and bars. Also, for torsional loading, the material is exposed to both compressive and tensile stresses. In these applications, the material is often exposed to a combination of tensile and compressive stresses. Experimental evidence shows that the behavior is extremely sensitive to the type of loading. Several studies have been reported in the literature on the comparison of SMA behavior in tension and compression. Hesse et al. [70] performed tests on SMA (Ni-55.7% Ti) ring and disk samples in both tension and compression at different temperatures and under different cycling and annealing procedures. The shape of the stress-strain curve, transformation stresses, and elastic moduli were shown to depend on whether the applied stress is tensile or compressive. Typical experimental results are shown in Table 3.4 and a schematic of the stress-strain curves in tension and compression is shown in Fig. 3.48. Auricchio and Sacco [72] and Gall et al. [73] reported that the recoverable strain in the case of compression is approximately 2%–3% less than in the case of tension.

The transformation stresses in compression are lower than the corresponding values in tension, whereas the stress at a particular value of strain is higher in the case of compression than in tension. However, the specific variations in behavior are highly material-dependent. In general, the stress-strain behavior in tension is quite different than in compression [74, 65, 10, 75, 76]. The main difference is seen in the shape of the martensite detwinning region. Liu et al. [74] reported that the slope of the transformation region is different in the case of the tension and the

Table 3.4. *Measured properties in compression and tension for Ni-55.7% Ti, from [70]*

	E_M GPa	E_A GPa	E_T GPa	ϵ_L %	σ_s^{cr} MPa
Compression	50	23	7.0	−2.4	−330
Tension	42	25	0.7	4.6	210

compression stress-strain curves. In the case of tension, this detwinning region exists as a nearly horizontal plateau, whereas in the case of compression, no such plateau exists and the detwinning region is characterized by a higher modulus. This is attributed to micromechanical differences in the martensitic detwinning mechanism in the presence of an external stress. In the case of mechanical cycling involving both tensile and compressive stresses, it has been observed that the curve is asymetric about zero.

3.11 Non-Quasi-Static Behavior

The previous sections discuss the experimental characteristics and constitutive modeling of SMA wires under quasi-static conditions, in which the strain rates were relatively low. Most constitutive models describing SMA phenomenology show good agreement with the experimental characteristics of the material under this kind of loading. With high strain rates, the material behavior is significantly different from that observed under quasi-static conditions [77, 78]. It is important to understand these differences in order to design SMA actuators for dynamic applications.

The reasons for this change in the material behavior with the loading rate are not completely understood. It is postulated that high strain rates are accompanied by a significant change in material temperature, which in turn affects the mechanical behavior of the material [59, 62, 61]. This is due to the origination of local nucleation sites with temperature differences along the wire [77]. It has been shown that the material may momentarily reach higher temperatures locally and then settle down to a lower equilibrium condition. It has been demonstrated that the dependence on strain rate disappears when the wire is placed in an effective heat sink [79], further

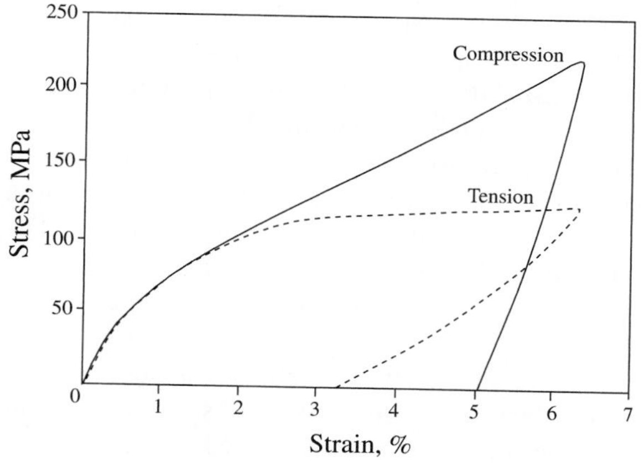

Figure 3.48. Schematic of room-temperature stress-strain curves in tension and compression.

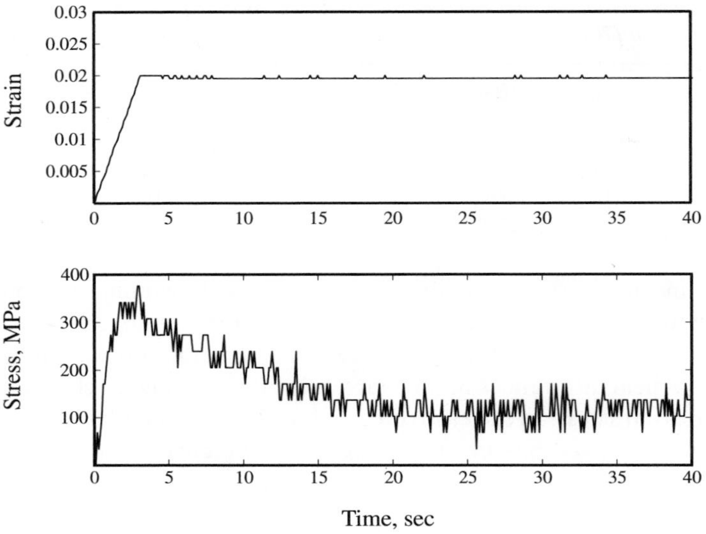

Figure 3.49. Stress-relaxation.

indicating that the cause of the mechanical variation in material characteristics may be due to the nonuniform change of the temperature of the material during non-quasi-static loading. From these observations, it appears that the non-quasi-static mechanical behavior of the SMA is strongly coupled with its temperature response.

It is essential to define the term "quasi-static rate of loading" because this is not an absolute quantity and varies with the test sample size and its thermomechanical condition. A "quasi-static rate" can be defined as a rate below which no significant variation in the stress-strain characteristics of the wire is observed. Note that the quasi-static condition also depends on the size of the specimen, as described in Section 3.5.4. For example, for a Nitinol wire of diameter 0.015 in, this strain rate corresponds to approximately 5×10^{-4}/s.

3.11.1 Stress-Relaxation

One of the effects of a high strain rate can be observed in the stress-strain behavior of the wire. An experimental setup similar to that described in Section 3.5.4 can be used to investigate this effect, while increasing the strain rate significantly higher than the quasi-static value. The strain is increased at a non-quasi-static rate to a certain value and then held constant. The resulting stress-time history is shown in Fig. 3.49. The temperature of the thermal chamber is kept constant at 25°C. It can be seen that the stress increases along with the strain to a maximum value and then slowly decreases with time after the strain rate is turned to zero. The stress stabilizes to a final value that is less than the initial transient peak. The final value of the stress reached has about the same value as obtained in the quasi-static test described in Section 3.5.4. This decrease in stress due to the "stress-relaxation" can be on the order of 70% of the initial stress. Similar behavior is observed at different operating temperatures, with a decrease in the amount of stress-relaxation at higher temperatures.

The relaxation behavior in SMAs has also been reported before in other alloy systems [80, 81] and, more recently, in Ni-Ti alloys [82]. This effect is significant because it implies that the variations of the stress state of the material with non-quasi-static loading are temporary and that the material settles down to its quasi-static value when strain is kept constant.

3.11.2 Effect of Strain Rate

The stress-strain behavior at non-quasi-static strain rates can also be measured with the experimental setup described in Section 3.5.4. Although the temperature of the thermal chamber in which the wire is mounted is kept constant, the temperature of the wire itself may change as a result of the loading. Figure 3.50 shows the stress-strain curves at two different environmental temperatures of 45°C (below M_s) and 70°C (above A_f), respectively. At both temperatures, when the material is loaded at a faster rate, the stress levels are significantly higher than those observed in the quasi-static test (at a strain rate of 0.0005/s). The main difference noticed here is the increased slope of the transformation region. For both temperatures, it can be seen that the transformation stresses remain almost constant.

At high environmental temperatures (Fig. 3.50(b)), the similar trend of increased slope of transformation regions at higher strain rates appears during both loading and unloading cycles. Loading at higher strain rates results in a higher final stress for the same final strain. The change in critical transformation stress with strain rate is relatively small.

The effects of non-quasi-static strain rates can be summarized follows:

1. The transformation stresses in the wire at a constant environmental temperature increase with the rate of strain for non-quasi-static loading.
2. These stresses "relax" to quasi-static values when the strain is kept constant or when the strain rate returns to a quasi-static value.

The differences pointed out in tests carried out at different rates and conditions of loading also serve to emphasize the need to standardize the conditions of testing for an SMA material so that meaningful comparisons and conclusions can be drawn among different test samples.

3.11.3 Modeling Non-Quasi-Static Behavior

The constitutive models described in previous sections are adequate to describe the behavior of the wire under quasi-static conditions. However, they do not include terms dependent on strain rates and hence do not take the non-quasi-static behavior into consideration. For dynamic loading, an accurate prediction of the strain-rate–dependent behavior is important for proper evaluation of the response of any device using SMA.

One approach to modeling non-quasi-static effects is to derive constitutive models with a fundamental dependence on strain rate. This approach has been extensively used in modeling high-rate plastic behavior of metals [83]. The kinetics of the phase transformation can be modeled using strain-energy functions [84] and Phase Interaction Energy Functions (PIEF) [48], among others. Phenomenological models [50] that estimate the effects of frequency and strain rate on the mechanical characteristics of the SMA in a limited temperature range have also been attempted.

A simple way of incorporating the effect of strain rate is to use a heat-transfer model in a thermomechanical approach coupled with the rate form of quasi-static

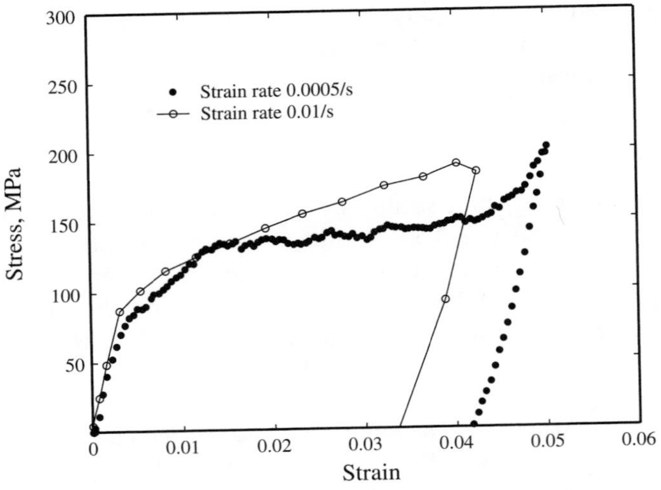

(a) Low Temperature (45°C)

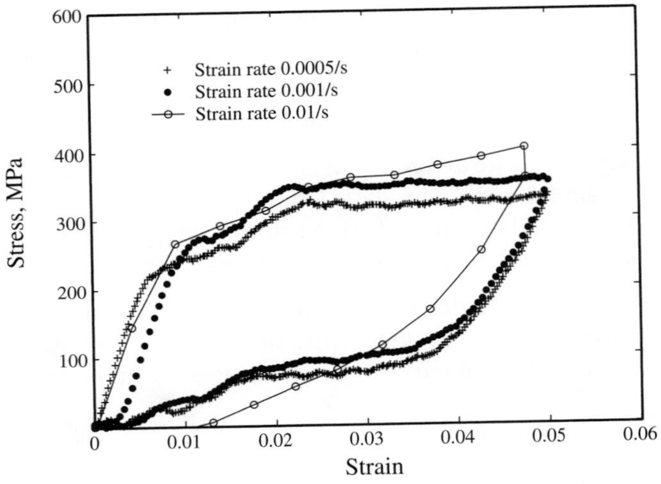

(b) High Temperature (70°C)

Figure 3.50. Variation of stress-strain behavior with strain rates.

models [85, 86]. These models, however, have been applied only to the pseudoelastic regime in most cases [87]. The coupled model has two parts: one describing the rate form of SMA constitutive models to prescribe the stress rate; and the other, an energy analysis to prescribe the temperature rates induced in the wire. Because the differential equations describing stress rates and temperature rates are coupled, a simultaneous solution must be obtained.

The following assumptions are made in the derivation:

1. The temperature of the wire is the same throughout the material.
2. The stress and strain fields inside the material are the same throughout the sample.

3. The critical stresses of transformation of the material are invariant with strain rate.

A brief description of incorporating strain-rate terms in the quasi-static constitutive relations and a derivation of thermomechanical equilibrium is provided in the next two subsections.

3.11.4 Rate Form of Quasi-Static SMA Constitutive Models

The Brinson model [37] is used as a representative model in this discussion, although the formulation can be extended to any other quasi-static model that predicts SMA behavior, such as the Tanaka model [35] and the Liang and Rogers model [36]. In the present formulation, the stress rate is determined as a function of not only the state variables – strain and temperature – but also their associated rates. The constitutive equation can be represented in the following first-order form:

$$\dot{\sigma} = \sigma(\epsilon, T, \dot{\epsilon}, \dot{T}) \tag{3.59}$$

where $\dot{\sigma}$, $\dot{\epsilon}$, and $\dot{T}$ are the rates of stress, strain, and temperature, respectively, and σ, ϵ, and T are their instantaneous values. In the following analysis, the temperature rate and corresponding instantaneous temperature can be either prescribed externally or determined from a coupled heat-transfer analysis described in the following section.

The quasi-static formulation for the Brinson model (see Section 3.4.3) is

$$\sigma - \sigma_o = E(\xi)\epsilon - E(\xi_o)\epsilon_o + \Omega(\xi)\xi_s - \Omega(\xi_o)\xi_{so} + \Theta(T - T_o) \tag{3.60}$$

where the modeling variables and constants are the same as those defined in Section 3.4.3. A rate form of these equations can be derived by taking derivatives with respect to time

$$\dot{\sigma} = E\dot{\epsilon} + \Omega_s\dot{\xi}_s + \Omega_T\dot{\xi}_T + \Theta\dot{T} \tag{3.61}$$

where Ω_T and Ω_S are the transformation stresses associated with the temperature-induced and stress-induced transformations, respectively. Neglecting the contribution of pure-phase thermal expansion ($\Theta\dot{T}$ term) and applying initial conditions to these equations, we can derive the rate-form of the simplified Brinson equation as

$$\dot{\sigma} = E(\xi)(\dot{\epsilon} - \epsilon_L\dot{\xi}_s) \tag{3.62}$$

The equations for the martensite volume fraction rates are obtained by taking time derivatives of the corresponding quasi-static equations. For the conversion to martensite, the martensitic volume fraction rates are given by

For $T > M_s$ and $(\sigma_s^{cr} + C_M(T - M_s)) < \sigma < (\sigma_f^{cr} + C_M(T - M_s))$

$$\begin{aligned} \dot{\xi}_s &= -\left(\frac{1 - \xi_{so}}{2}\right)\left(\frac{\pi}{\sigma_s^{cr} - \sigma_f^{cr}}\right)(\dot{\sigma} - C_M\dot{T}) \\ &\sin\left[\frac{\pi}{\sigma_s^{cr} - \sigma_f^{cr}}(\sigma - \sigma_f^{cr} - C_M(T - M_s))\right] \\ \dot{\xi}_T &= -\left(\frac{\xi_{To}}{1 - \xi_{so}}\right)\dot{\xi}_s \end{aligned} \tag{3.63}$$

For $T < M_s$ and $\sigma_s^{cr} < \sigma < \sigma_f^{cr}$

$$\begin{aligned}\dot{\xi}_s &= -\left(\frac{1-\xi_{so}}{2}\right)\left(\frac{\pi\dot{\sigma}}{\sigma_s^{cr} - \sigma^c r_f}\right) \quad \sin\left[\frac{\pi}{\sigma_s^{cr} - \sigma^c r_f}(\sigma - \sigma_f^{cr})\right] \\ \dot{\xi}_T &= -\left(\frac{\xi_{To}}{1-\xi_{so}}\right)\dot{\xi}_s + \dot{\Delta}_{T\epsilon}\end{aligned} \tag{3.64}$$

where, if $M_f < T < M_s$ and $T < T_0$

$$\dot{\Delta}_{T\epsilon} = -\left(\frac{1-\xi_{To}}{2}\right) a_M \dot{T} \sin(a_M(T - M_f)) \tag{3.65}$$

else

$$\dot{\Delta}_{T\epsilon} = 0 \tag{3.66}$$

For conversion to austenite, these variables then become

For $T > A_s$ and $C_A(T - A_f) < \sigma < C_A(T - A_s)$

$$\begin{aligned}\dot{\xi} &= -\frac{\xi_o}{2} a_A \left(\dot{T} - \frac{\dot{\sigma}}{C_A}\right) \quad \sin\left[a_A\left(T - A_s - \frac{\sigma}{C_A}\right)\right] \\ \dot{\xi}_s &= \frac{\xi_s o}{\xi_o}\dot{\xi} \\ \dot{\xi}_T &= \frac{\xi_T o}{\xi_o}\dot{\xi}\end{aligned} \tag{3.67}$$

where the material constants are the same as defined in the quasi-static Brinson model and are obtained from a comprehensive experimental characterization of the SMA wire [78]. It is important to note that the development of the rate equations yields the same predictions as the quasi-static form of the Brinson model, provided the temperature is held constant (i.e., the temperature rate is zero).

Using the rate formulation and given the instantaneous temperature and rates of temperature and strain, this differential equation can be solved for the instantaneous stresses. However, in reality, the temperature rise of the material is not an independent prescribed function but rather is coupled to the loading, material characteristics, and heat-transfer aspects of the test sample. Describing the instantaneous temperature rates in terms of these states requires an energy analysis of the SMA material [88].

3.11.5 Thermomechanical Energy Equilibrium

The instantaneous temperature and rate of temperature can be calculated by considering an energy equilibrium between the input energy to the material and the energy that is lost or absorbed. Because this is a rate formulation, the equilibrium equation is written in the form of an energy rate, or power equilibrium.

The equilibrium equation can be written as

$$P_{in} = -P_{loss} + P_{absorbed} \tag{3.68}$$

The input power, P_{in}, is composed of two parts: the mechanical power that is causing the deformation (e.g., imposed through the grips of the testing machine), P_{mech}, and the heat supplied by activation of the material, P_{act}. In the case of constant-temperature stress-strain tests, $P_{act} = 0$. This term exists only when the wire is resistively heated, which is discussed in greater detail in Section 3.12.1. The heat lost to the surroundings, P_{loss}, depends on the convective, conductive, and radiative properties of the sample. This depends on the specific application or experimental setup. The heat absorbed by the SMA specimen itself, $P_{absorbed}$, consists of two parts: the specific heat component that causes a rise in temperature of the material, and a latent heat component that is present during a phase transformation and occurs at constant temperature. The rate of strain energy stored in the material, P_{strain}, also appears in terms of the absorbed power. Expanding the terms on each side and assuming that heat losses are dominated by convection, the overall equilibrium equation is

$$P_{mech} + P_{act} = -P_{conv} + P_{strain} + P_{spec} + P_{latent} \tag{3.69}$$

Each of these quantities can be expressed as a function of the material states and validated individually against experimental data. Eq. 3.69 then relates the evolution of the rate of temperature with the absolute stress, strain, and temperature, as well as the rates of strain and stress in the material. By solving this equation and the rate form of the constitutive equation (see Eq. 3.62), the evolution of the temperatures and stresses in the SMA can be computed simultaneously.

Rate of Change of Strain Energy

The rate of change of strain energy stored in the wire is given by

$$P_{strain} = \frac{1}{2}\mathcal{V}(\sigma\dot{\epsilon} + \dot{\sigma}\epsilon) \tag{3.70}$$

where $\mathcal{V}$ is the volume of the material. This quantity is positive during the loading cycle, indicating work done on the wire sample. This quantity becomes negative during unloading, indicating work done by the wire. Note that the stress rates are related by the SMA constitutive behavior described in Eq. 3.62.

Heat Dissipation: Convective Losses

Assuming that convective losses are the dominant form of heat loss, the heat lost from the material is given by

$$P_{loss} = hA\Delta T \tag{3.71}$$

where A is the surface area of the material, ΔT is the difference between sample and environmental temperatures, and h is the convective heat-transfer coefficient. This is a function of the mounting configuration and dimensions of the test sample, as well as the environmental temperature, and can be estimated using empirical formulae for a given configuration.

An SMA wire mounted in a thermal chamber can be treated as a cylinder with free convection [89]. The empirical relationship for heat-transfer coefficient h is given by

$$h = \frac{\bar{Nu}_D k}{D} \tag{3.72}$$

where $\bar{Nu}_D$ is the Nusselt number based on the cylinder diameter and ambient temperature, k is the thermal conductivity of air, and D is the diameter of the cylinder. The Nusselt number and thermal conductivity of air are determined from empirical relationships [90].

$$\bar{Nu}_D = \left[0.57 + \frac{0.377 Ra_D^{\frac{1}{6}}}{\left[1 + \left(\frac{0.539}{Pr}\right)^{\frac{9}{13}}\right]^{\frac{8}{57}}}\right]^2$$
$$Ra_D = \frac{g\beta(T_s - T_{\text{inf}})D^3}{\nu\alpha} \tag{3.73}$$

where Ra_D is the Rayleigh number that represents the degree of turbulence in the thermal boundary layer of the element. The values of Prandtl number Pr, volumetric thermal expansion coefficient β, dynamic viscosity ν, and thermal diffusivity α are determined from the table of thermophysical properties of air [89]. Note that the Nusselt number and thus, the heat-transfer coefficient, is a function of the temperature difference between the sample and the ambient air and therefore needs to be updated constantly as the material temperature varies, reaching a converged solution for each time-step.

From this calculation, an estimate of the heat-transfer coefficient can be obtained for a given configuration. Note that the heat-transfer coefficient varies directly in proportion to the surface area and inversely with the volume and is, therefore, more likely to affect the calculations for a thin wire. The heat-transfer coefficient also varies greatly with surface finish and other properties of the material interface. It is important, therefore, that the heat-transfer coefficient obtained using the empirical formulae be experimentally validated.

Heat Absorbed by the Material

The heat absorbed by the material consists of two parts: the specific-heat and the latent heat. The specific-heat component is the heat that is absorbed or released by the material in order to increase or decrease its temperature. The net specific heat is the sum of the specific heats of the martensite and austenite components of the material. The net specific heat C_p and the heat rates to change the temperature of the material are given by

$$C_p = \xi_M C_{pM} + (1 - \xi_M) C_{pA}$$
$$P_{spec} = m C_p \Delta T \tag{3.74}$$

where m is the total mass and C_{pM}, C_{pA}, and C_p are the specific heats of pure martensite, pure austenite, and mixed phase, respectively.

The latent heat of the material is the heat absorbed or released to change the phase of the material at constant temperature. It appears only during transformation and is proportional to the rate of phase transformation occurring in the material. It

Table 3.5. *Material constants used for 0.015 in (0.381 mm) diameter SMA wire*

Constant	Value		Source	Units
	Martensite	Austenite		
C_p	600	600	DSC tests	J/Kg/K
k	0.086	0.18	SMA manufacturer [92]	W/cm °C
ρ	0.83×10^{-6}	0.77×10^{-6}	Resistance tests	Ω-m
E	20.3	45	Mechanical testing (see Section 3.5.4)	Pa
L	5000		DSC tests	J/Kg
h	10.43[a]		Heat transfer tests	W/m^2K

[a] Heat-transfer coefficient h determined for a wire diameter of 0.015 in, wire temperature = 90°C, room temperature = 25°C

has been shown [59] that the martensite-to-austenite transformation is exothermic (heat-emitting), whereas the austenite-to-martensite transformation is endothermic (heat-absorbing). Because this quantity is related to the transformation process, it is a function of the rate of change of the martensite volume fraction and can be represented as

$$P_{latent} = mL_{M\rightarrow A}\dot{\xi} \tag{3.75}$$

This quantity takes the sign of $\dot{\xi}$ and is positive during transformation from austenite to martensite and negative during the reverse transformation, accurately representing the physical nature of the latent heat. The parameters required to calculate the heat loss and heat absorbed can be measured experimentally. Typical values of these parameters are listed in Table 3.5. Bhattacharyya et al. [91] experimentally determined the convection coefficient for Nichrome and NiTi SMA wires subjected to a constant load, heated by electric current, and cooled by free convection. A simplified phenomenological model of the convection coefficient was developed.

The model can also be applied to any arbitrary loading condition in which the strain rate is prescribed as a function of time. Figure 3.51 shows the predicted stress and temperature profiles for a test involving composite strain rates (in which the strain rate is stepped down from a value of 0.01/s to 0.0005/s during loading). From Figure 3.51(a), we observe good qualitative agreement with the experimental data. However, the rate of stress relaxation is underpredicted, possibly due to the heat-transfer coefficient being underpredicted.

Figure 3.51(b) shows the corresponding theoretical and experimental temperature profiles for this test involving the two different strain rates. As predicted in the model, an instantaneous drop in the temperature was experimentally observed when the loading condition was changed. In addition, the experimental temperature profile is in qualitative agreement with the model predictions, further justifying this modeling approach to predict strain-rate variations. However, the magnitude of the temperature rise and fall is again overpredicted in the model, possibly due to the temperature-measurement issues described previously. However, the good qualitative agreement for this complex temperature profile is a promising result for the current modeling approach.

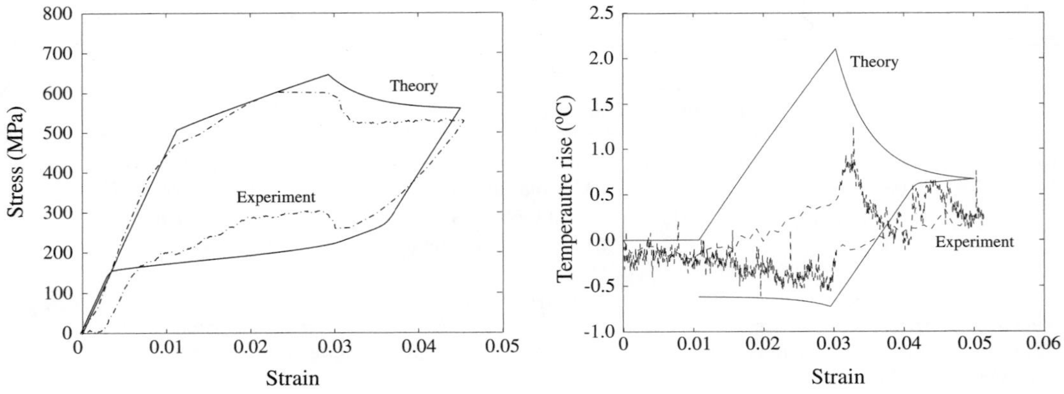

(a) Model stress predictions (b) Model temperature predictions

Figure 3.51. Model predictions for stress and temperature evolution for composite strain rates, 0.01/s and 0.0005/s. Environmental temperature = 84°C.

Viello et al. [93] showed using a Tanaka-based analysis and experimental testing that the strain rate had a great influence on SMA load-displacement response. For prediction with high strain rates, it is important to couple the material-constitutive model with a thermodynamic model that can account for internal heat produced in terms of both a phase-transition latent heat and a mechanical dissipation.

Leo et al. [94] carried out experimental and analytical studies to investigate the effects of temperature and strain rates. In the pseudoelastic range, temperature and strain-rate effects were found to be coupled. They tested two identical 0.0652 mm diameter NiTi wires at 23°C in air and in water, respectively. Three extension rates of 0.5, 5, and 50 mm/min were applied. It was shown that when the temperature of the alloy was held constant, the effect of strain rate on the pseudoelastic behavior was small.

3.11.6 Cyclic Loading

Consider a case of purely tensile stress. The cyclic loading generates a non-quasi-static strain condition, which affects the thermomechanical behavior of the SMA. Miyazaki et al. [95], Perkins and Sponholz [96], and Lim and McDowell [97] have shown that the main reason for a significant change in the thermomechanical behavior of SMA is the generation of defects in its microstructure, resulting in a pile-up of dislocations around defects. This results in an accumulation of residual martensite volume fractions. It is observed that for constant amplitude cyclic strain, there is a reduction in the forward-phase transformation stresses and there is a general work hardening. Conversely, there is an increase of stresses during the reverse-phase transformation. Also, there is an accumulation of residual strain in the direction of loading, which stabilizes with cycles. Furthermore, there is a reduction of the pseudoelastic hysteresis area. However, if the cyclic stresses are in the elastic range (not in pseudoelastic hysteresis), there is a negligible effect on the thermoelastic behavior. Once the pseudoelastic behavior is stabilized at a large cyclic strain amplitude, a low strain amplitude has no effect on it.

3.12 Power Requirements for SMA Activation

The power requirement for operating an SMA actuator depends on the type of heating employed. The basic principle involved is a power balance among the heat supplied, the heat absorbed by the SMA material itself, and the heat loss through the surrounding material. In the case of external heating, the heat-transfer characteristics depend on the position of the heater, the efficiency of the heater, the intervening medium, and other factors. Therefore, calculation of the required power for external heating is highly specific to the application itself.

In the case of resistive heating, the heat supplied by the input electric current as well as the heat absorbed by the SMA material can be calculated. The heat loss from the actuator, however, is dependent on the specific configuration of the actuator. The calculation of required power is complicated by the fact that the resistivity, specific heat capacity, Young's modulus, and other properties of the wire change during the transformation by large amounts.

The thermodynamic equilibrium of the SMA wire can be described by the following equation

$$P_{in} = P_{abs} + P_{loss} \tag{3.76}$$

where P_{in} is the input power in the form of electrical power and P_{abs} is the heat retained by the SMA in the form of specific heat and latent heat. The heat lost to the surroundings, P_{loss}, is in the form of conduction from the wire to the surrounding composite material and convection to the ambient air.

3.12.1 Power Input: Resistance Behavior of SMA Wires

The power input to the SMA wire occurs in the form of Joule heating due to the resistance of the wire itself. The power input is given by

$$P_{in} = i^2 R_{wire} \tag{3.77}$$

where i is the current passing through the wire and R_{wire} is the resistance of the wire. However, previous research [98, 99] has shown that SMAs exhibit a large change in resistance when they undergo transformation. This behavior of electrical resistance is the combined effect of the changing electrical resistivity (ρ) of the material and changes in the length and cross-sectional area of the SMA wire. It can be assumed that the resistivity remains constant in the pure phases and changes only during transformation. This variation can be described by the following piecewise approximation.

For heating cycle

$$\rho(T, \sigma) = \begin{cases} \rho_M & \text{if } T < A_s, \\ \rho_A & \text{if } T > A_f, \\ \rho_M \xi + \rho_A(1 - \xi) & \text{if } A_s < T < A_f \end{cases} \tag{3.78}$$

For cooling cycle

$$\rho(T, \sigma) = \begin{cases} \rho_A & \text{if } T > M_s, \\ \rho_M & \text{if } T < M_f, \\ \rho_M \xi + \rho_A(1 - \xi) & \text{if } M_f < T < M_s \end{cases} \tag{3.79}$$

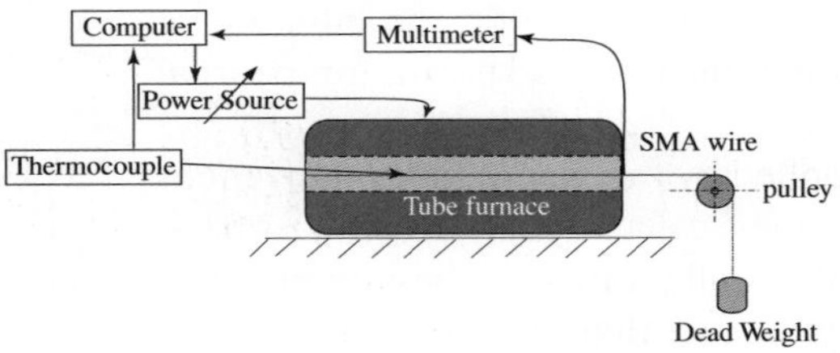

(a) Schematic of the resistance testing equipment

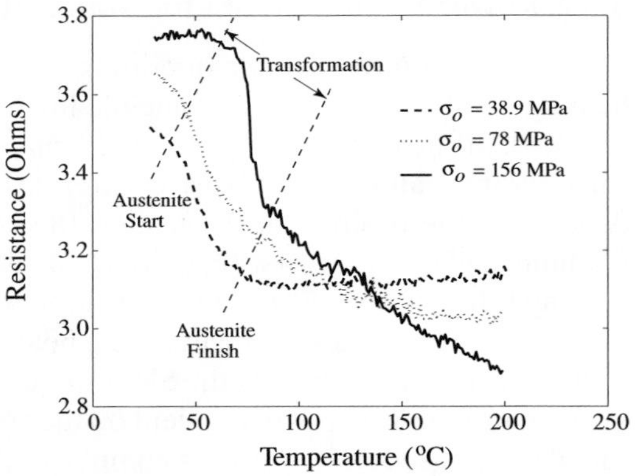

(b) Variation of resistance with temperature and stress

Figure 3.52. Resistance variation in SMAs.

where ρ is the resistivity of the material at a temperature T ρ_M and ρ_A are the resistivities of the wire in martensite and austenite phases, respectively. Typically, ρ_A is approximately 15–20% less than ρ_M. These parameters are listed in Table 3.5. These values depend on the stress in the material and can be found experimentally using a setup similar to the constant-stress tests described in Section 3.5.3. Figure 3.52(a) is a schematic of the setup used to obtain the variation of resistance with temperature. An SMA wire of diameter 0.015 in was heated in a tube furnace (external heating). The furnace was heated using a variable power source and cooled through convection with the outside air (accompanied by a decrease in input heat from the furnace). The heating and cooling was carried out at slow rates ($\approx$0.5°C/min). One end of the wire was fixed to the furnace, and a dead weight was suspended from the other end of the wire through a pulley to maintain a constant stress in the wire. The temperature of the wire was monitored using a K-type thermocouple placed directly on the wire. This was connected to a thermal controller and a data-acquisition computer. Note that it is possible to carry out a similar experiment using resistive heating, but this would require a more complicated experimental setup to extract the resistance of the wire from the voltage supplied and current drawn.

Figure 3.52(b) shows the resulting variation of resistance with temperature during the heating cycle for different stresses applied to the wire. The inflection points in the resistance can be used to detect the transformation temperatures of the wire and their variation with applied stress. Note that, as expected, the transformation temperatures show an increase with increasing stress. The data on the variation of transformation temperatures with stress obtained from this experiment show good agreement with critical stress-temperature data obtained previously from mechanical testing (see Section 3.6).

The overall resistance of the wire is then obtained from its resistivity. It is important to note that the resistance of the wire is a function not only of the resistivity of the wire material but also the length and cross-sectional area of the wire, which change during transformation. Therefore, whereas the resistivity of the martensite phase, for example, is independent of the stress, the resistance of the wire can change depending on the amount of deformation. A strain of ϵ in the wire is accompanied by a corresponding decrease in the cross-sectional area of the material such that the

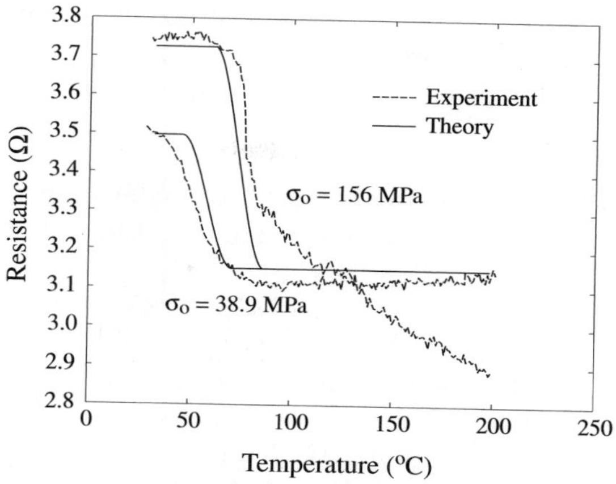

Figure 3.53. Validation of the predicted variation of resistance with temperature.

net volume remains about constant. The resistance of the wire is then represented as

$$\begin{aligned} R(T, \sigma) &= \rho \frac{l'}{A'} \\ &= \rho \frac{(l + \epsilon l)^2}{Al} \end{aligned} \tag{3.80}$$

where l and A represent the original length and area of the unstrained wire, respectively, and l' and A' represent its deformed length and area. Such a formulation can be included with a constitutive model, such as the Brinson model, to calculate the resistance behavior of the SMA wire and, therefore, the input electrical power. Note that the temperature coefficient of resistivity is ignored in the previous discussion but can easily be included if necessary.

Figure 3.53 compares the theoretical predictions of the resistances with experimental data for two different applied loads. From the figure, it is seen that the magnitude of the increase of the resistance with applied load is predicted accurately by the model, especially for lower values of stress. It is also observed that the overall behavior of the resistance during the transformation region shows good agreement with the predictions from the theoretical model. The experimental high-stress curves exhibit a nearly linear variation in the resistance of the material even above A_f, which is not predicted in the models. A possible reason for this discrepancy could arise from the temperature-coefficient of resistivity of the material in pure phase, which is neglected in the current formulation. However, limited available data exist in the literature to estimate this coefficient for a material in pure austenite.

Despite the drawbacks in the current material model, it is seen that the model predicts the behavior of the resistance quite satisfactorily for lower values of stresses. It is therefore a useful tool for making preliminary estimates of the energy requirements for activation of an SMA wire.

3.12.2 Heat Absorbed by the SMA Wire

Another important aspect of determining power requirements for the wire is the variation of heat capacity of the wire with temperature. This issue is also discussed in Section 3.11.5. The heat absorbed by the material has two components: a specific heat and a latent heat.

The specific-heat component is given by

$$P_{spec} = mC_p \Delta T \tag{3.81}$$

where m is the total mass of the material and ΔT is the change in its temperature. The specific heat of the material, C_p, can be assumed to vary linearly with the volume fraction of martensite

$$C_p = \xi C_{pM} + (1 - \xi) C_{pA} \tag{3.82}$$

where C_{pM} and C_{pA} are the specific heats of pure martensite and pure austenite, respectively.

In Section 3.11.5, the latent-heat rate is described in terms of the rate of change of the martensite volume fraction, $\dot{\xi}$, as

$$P_{latent} = mL_{M\rightarrow A}\dot{\xi} \tag{3.83}$$

where $L_{M\rightarrow A}$ is the latent heat of the martensite-to-austenite transformation.

The parameters C_{pA}, C_{pM}, and $L_{M\rightarrow A}$ are all obtained from experiments for a particular sample of SMA. The values of these parameters are listed in Table 3.5.

3.12.3 Heat Dissipation

In the case of an SMA wire in air, such as in the experimental fixture described in Section 3.5.4, it can be assumed that all of the heat dissipation occurs by convection. Because the diameter of the SMA wire is small, heat conducted away from the wire through the end fixtures can be neglected.

Using simple one-dimensional thermal-transfer theory [89], the rate of convective heat loss is

$$P_{loss} = hA(T - T_\infty) \tag{3.84}$$

where h is the effective heat-transfer coefficient of the material, A is the exposed cross-sectional area, and T_∞ is the temperature far away from the SMA. The method of obtaining the heat-transfer coefficient using empirical models for the SMA wire is described in detail in Section 3.11.5. A typical value of the heat-transfer coefficient for the material is given in Table 3.5.

3.13 Torsional Analysis of SMA Rods and Tubes

There are many applications of SMA rods and tubes in torsion. As such, it is important to understand the modeling and analysis of these structures in torsion. A simple torsion model can be developed based on the extension of a one-dimensional formulation, such as the Brinson model, and can incorporate the quasi-steady thermomechanical characteristics of the material.

Although the behavior of a cylindrical structure undergoing pure torsion can be idealized as a one-dimensional problem, there can be some differences between the torsional and extensional characteristics of the material that are accentuated in the case of SMAs. In the extensional case, each element of the structure is strained axially under a constant applied stress. In an axisymmetric structure, such as a rod or a tube undergoing torsional deformation, the extensional stress-strain is not constant

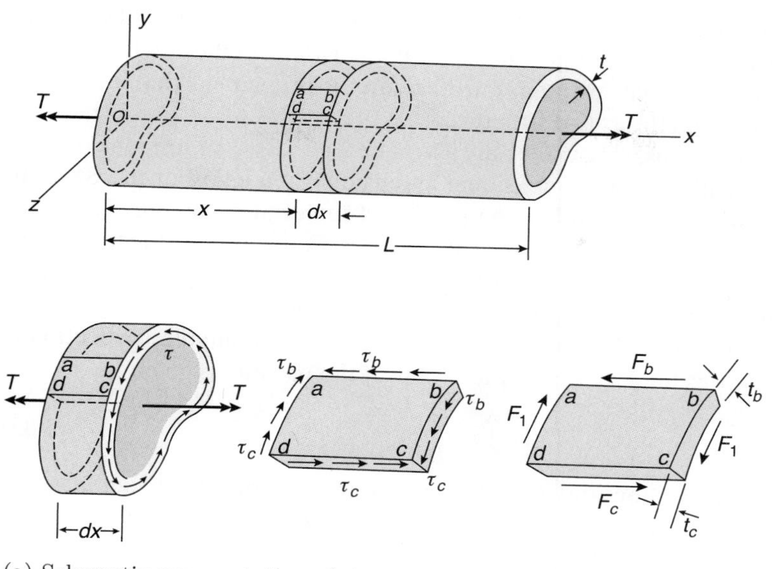

(a) Schematic representation of stresses on a torsional tube, from Ref.[100]

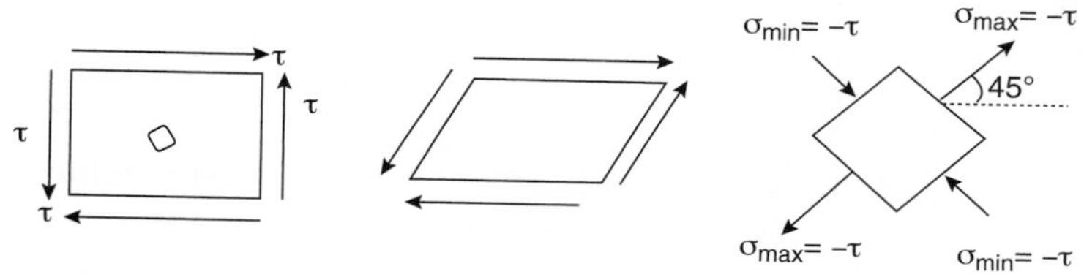

(b) Equivalent extensional and compression stresses for pure shear

Figure 3.54. Representative stresses on a structural element for a rod in torsion.

throughout the structure but rather is a function of the radial location of the material element. Furthermore, the central low-strain region remains elastic while the outer high-strain layers will undergo martensite transformation. The global elastic twist due to the applied torque depends on the state of the material at each local element. As such, the overall response can be viewed as a cumulative behavior of several structural elements.

Consider the behavior of a SMA rod acted on by a pure torque. A condition of pure torsion is assumed at each cross section of the SMA rod, with no axial or radial stress. This means that every structural element is in a state of pure shear loading. This, in turn, can be expressed as the combined effect of pure compression and pure tension, acting at an angle of 45° to the longitudinal axis of the rod (Fig. 3.54). Therefore, to simplify the analysis, it is assumed that the local elemental behavior can be expressed in terms of an extensional constitutive model. Consequently, the stresses, strains, martensite volume fraction, and Young's modulus are all functions of the radial location of the element.

It is also assumed that each radial element acts independently of the radial elements surrounding it. The formulation intrinsically assumes continuity of stress and strain across the radius but does not account for any interaction between the radial elements. Similar to a normal isotropic cylinder in torsion, all radial elements at a particular axial station undergo the same angular deformation, which results in a condition of no sectional warping. Furthermore, it is assumed that a constant temperature exists across the entire structure.

With these assumptions, we expect only a first-order analysis of torsional behavior, and we can extend the one-dimensional extensional modeling of the SMA to the torsional case without any added complexity of material modeling. There are many inherent limitations of such a simplifying analysis. For example, due to thermomechanical coupling as well as thermal-boundary conditions, it is expected that there will be some nonuniformity of temperature within the structure.

For a rod (or tube) of uniform cross section, with a given angular deflection θ, the shear strain γ varies linearly across the radius of the rod and is given by

$$\gamma(r) = \frac{\theta r}{L} \tag{3.85}$$

where r is the radial location and L is the total length of the rod. From classical torsion theory, the shear strain and normal strain ϵ at 45° are related by

$$\epsilon(r) = \frac{\gamma(r)}{2} \tag{3.86}$$

Thus, the normal strains vary linearly across the radius, with the outer surface experiencing the highest strain. The resulting normal stress is a function of material properties and can be expressed as

$$\sigma(r) = \sigma(\epsilon, T, \xi, r) \tag{3.87}$$

Similarly, the shear stress τ at any radial station can be transformed into a combination of normal tensile stress and normal compressive stress

$$\tau(r) = G\gamma(r) \tag{3.88}$$

where G is the shear modulus, which is related to the Young's modulus by

$$G = \frac{E}{2(1+\nu)} \tag{3.89}$$

where ν is the Poisson's ratio of the material. The normal stress can also be written as

$$\sigma(r) = E(r)\epsilon(r) \tag{3.90}$$

This results in

$$\tau(r) = \frac{\sigma(r)}{(1+\nu)} \tag{3.91}$$

The torque T necessary to obtain a desired angular twist θ is obtained as

$$\begin{aligned} T &= \int_{r_i}^{r_o} 2\pi r^2 \tau \, dr \\ &= \int_{r_i}^{r_o} 2\frac{\sigma(r)}{1+\nu}\pi r^2 \, dr \end{aligned} \tag{3.92}$$

where r_i is the inner radius of the tube and r_o is the outer radius of the tube. In the case of a rod, $r_i = 0$.

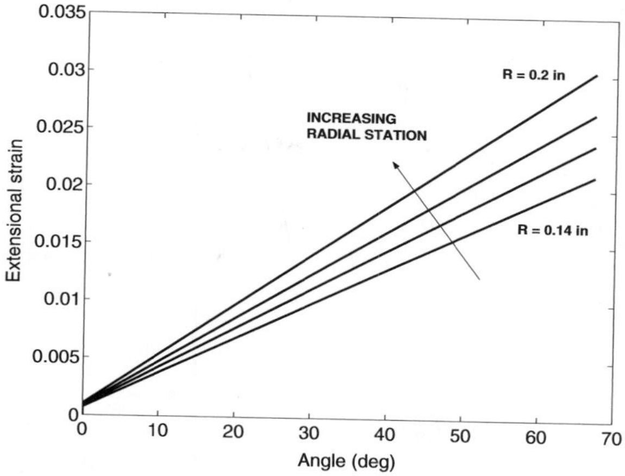

(a) Variation of linear and shear strains with angle

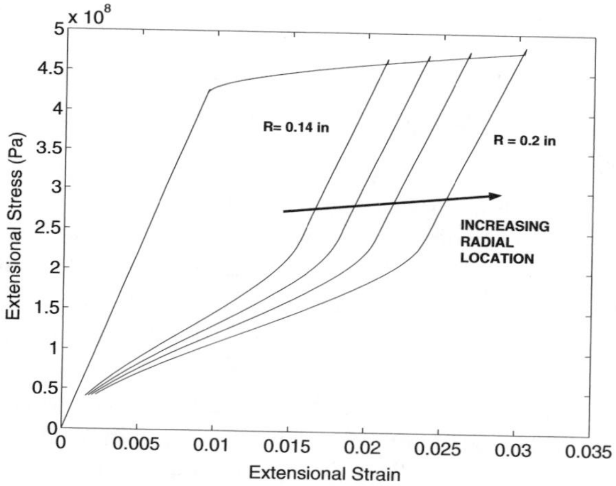

(b) Extensional stress-strain profile for different radial stations

Figure 3.55. Torsional model simulations at different radial locations for a SMA rod undergoing torsional deformation.

Each radial element of the rod (or tube) follows the same stress-strain-temperature profile as those of SMAs under extensional loading. However, the different radial locations execute different stress-strain-temperature loops simultaneously. This is illustrated in Fig. 3.55, which shows the state of stress and strain at four radial stations of a rod of diameter 6.35 mm (0.25 in) undergoing torsional deformations at a constant temperature of 100°C. From Fig. 3.55(a), it can be seen that the strains are larger at greater radial stations and are linearly dependent on the angle of twist. Accordingly, as shown in Fig. 3.55(b), the material at each radial station traces out its own minor loop on the stress-strain diagram. The torque can be calculated numerically by dividing the rod (or tube) into N radial elements.

$$T = \sum_{j=1}^{N} \frac{2\pi}{1+\nu} \frac{r_{o_j}^3 - r_{i_j}^3}{3} \sigma_j \tag{3.93}$$

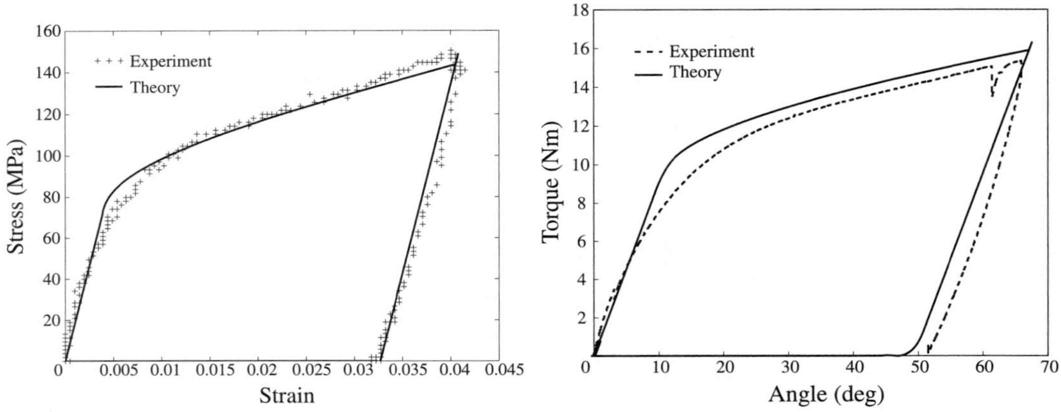

(a) Axial stress-strain behavior for open section tube (sample C)

(b) Torque vs. angle behavior for SMA tube (sample B)

Figure 3.56. Comparison of model prediction with test data.

3.13.1 Validation with Test Data

The SMA constitutive model requires several parameters that are obtained by extensional testing of SMA specimens. The torsion tests were carried out on three specimens. The first one was a solid rod of diameter 6.35 mm (0.25 in), referred to as Sample A. The second one was a thin-walled tube with an outer diameter of 10.2 mm (0.4 in) and inner diameter of 7.1 mm (0.28 in), referred to as Sample B. This tube was constructed out of a solid rod of outer diameter 10.2 mm (0.4 in) and bored on the inside by a wire electron-discharge machining process. As a consequence of the machining process, a tubular open cross-section sample was also obtained, with an outer diameter of 6.35 mm (0.25 in) and inner diameter of 5.3 mm (0.21 in), referred to as Sample C. The extension tests were performed on the same material using different samples with lengths of 10.16 mm (4 in).

The rod and tube samples were gripped using collets in a tensile testing machine and subsequently in a torsion testing machine. Slippage of the sample in the grips, due to a change in the dimensions caused by differential thermal expansion and material-phase transformation, was eliminated by properly tightening the grips at an elevated temperature of 150°C. A K-type thermocouple was used to measure the sample temperature. The strain was measured using an extensometer of gauge length 25.4 mm (1 in) mounted on the sample. A 5000 lb load cell was used to measure the force on the SMA. The tests were carried out at a strain rate of 1.64×10^{-4}/s (1 mm/s). Figure 3.56 shows a comparison of the predictions obtained using the Brinson model with experimental results on Sample C at a test temperature of 35°C. It can be seen that the model yielded good correlation with test data. The constitutive model parameters are presented in Table 3.6.

The torsional testing on the sample was carried out using an Instron torsion testing machine, with an environmental control chamber (Fig. 3.57). By mounting the grip on one end of the rod on a linear slide, any axial constraints were eliminated, thus ensuring a state of pure torsion in the sample. The twist angle was measured using a digital encoder mounted on the actuation head of the testing machine, which

Table 3.6. *Constitutive model parameters for correlation with torsion test data*

(f) Parameter	(f) Rod Specimen (A)	(f) Tube Specimen (B)	(f) Units
M_s	55	58	°C
M_f	35	40	°C
A_s	60	60	°C
A_f	90	120	°C
C_A	20×10^6	4×10^6	Pa
C_M	10×10^6	6×10^6	Pa
σ_{cr}^s	3.7×10^7	5×10^7	Pa
σ_{cr}^f	16.5×10^7	20×10^7	Pa
E_M	35×10^9	18×10^9	Pa
E_A	65×10^9	45.0×10^9	Pa
ϵ_L	0.067		–

[a] Determined from experiments.
[b] Values listed are after cycling; properties immediately after heat treatment may differ.

had an accuracy of 0.1°. The torque was measured using a torque cell of range 203 N.m (2000 in-lb) (measurement resolution 0.25 in-lb) fixed to the grip mounted on the linear slide. To avoid crushing of SMA tube samples, steel plugs were designed to snugly fit inside the SMA tube bore in the gripped portions of the sample. Torsion tests were carried out on Samples A and B with test lengths of 0.127 m (5 in) and 0.1524 m (6 in). A twist angle of 70° corresponds to a maximum axial strain of about 3% and 4% for Sample A and Sample B, respectively.

The test samples were first cycled in torsion by twisting them in both directions at a temperature above A_f (similar to extensional cycling). It was noticed that the material did not show significant deviation in the characteristic response with repeated cycling (unlike extensional tests that showed significant drift in the first few cycles). This shows that the cycling procedure is less important for torsional testing. This may be partly due to an inherently large amount of cold work (about 40%) present in the torsional rods, relative to the previously tested SMA wires (20% cold work).

Figure 3.58 shows predicted and measured torsional responses for the solid rod A at three different temperatures. Results exhibit the SME (at low temperature, 35°C)

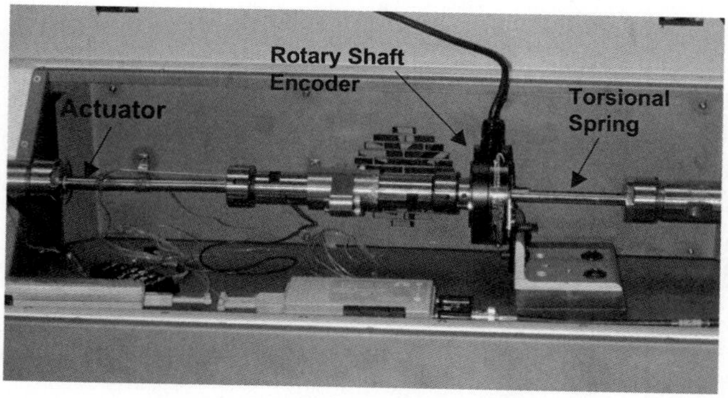

Figure 3.57. Instron torsional test machine with thermal chamber.

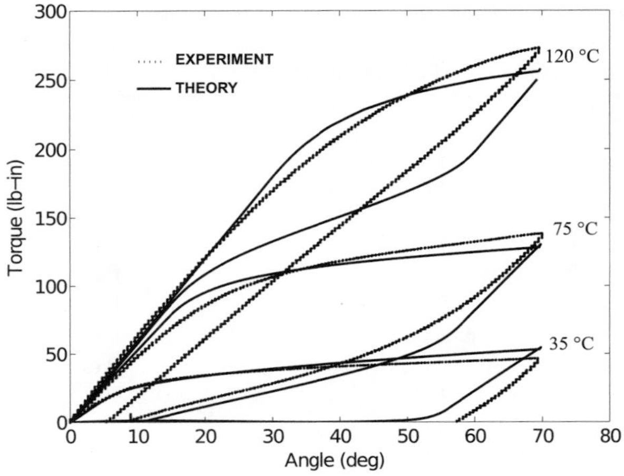

Figure 3.58. Comparison of torsional model predictions with quasi-static data for 0.25 in diameter rod (Sample A). Ramping rate = 0.015 deg/s.

and partial pseudoelastic characteristics at high temperatures (75°C and 120°C). One important difference from the pure extensional results is that complete pseudoelastic recovery was not observed. Again, results showed good agreement with predictions.

An important observation is that after a number of cycles of pre-twisting the sample in one direction, the constant temperature characteristics of the material become asymmetric with respect to the direction of twisting, as shown in Fig. 3.59(a). For example, a torque of about 350 in-lb is needed to obtain a twist angle of 70° (clockwise) after several cycles of twisting and untwisting of Tube B. If this tube is now twisted in the opposite direction, a torque of −225 in-lb is required to achieve a twist of −70°. This observed asymmetry appears to be related to the development of TWSME. The asymmetry in both the mechanical and the recovery properties is not an inherent property of the material but rather is introduced after a number of cycles repeated in the same direction of loading. This behavior can be eliminated by applying an appropriate heat treatment to the sample (Fig. 3.59(b)).

To study the variation of the twist angle-torque behavior with the rate of loading, the quasi-static tests were carried out at different rates. For the rod Sample A, an increase in the measured torque was observed when twisted at high temperatures (Fig. 3.60(a)). The increase in torque primarily occurs during the transformation region, with the linear elastic region being nearly invariant with loading rates. However, this effect appears much smaller than that observed in the extensional behavior of wires. Again, this increase during the transformation region may be attributed to a rise in temperature associated with loading at high strain rates. Figure 3.60(b) shows the corresponding change of surface temperature of the material at two different loading rates. The magnitude of temperature rise is generally larger at the higher twist rate when the temperature is between M_s and A_f. This is due to the higher rates of mechanical-energy input to the material, which manifests itself as higher rates of self-heating in the material. This trend is less when the material is at low and high temperatures because the material is in the pure phase. A key difference from extensional loading is that the effective strain rate is not constant across the material but rather varies linearly with the radial location. Thus, the heat generated is nonuniform across the cross section.

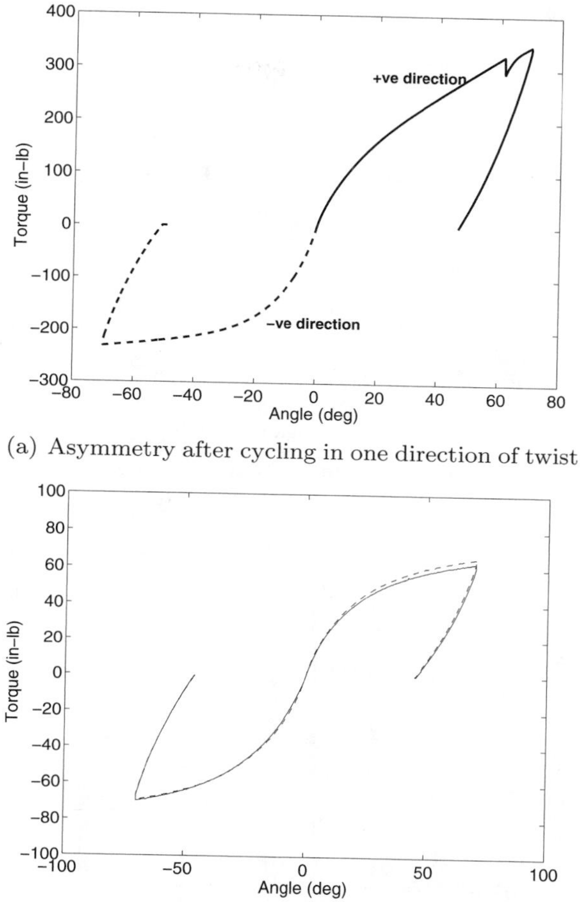

(a) Asymmetry after cycling in one direction of twist

(b) Angle torque characteristics after a heat treatment

Figure 3.59. Asymmetry of behavior with direction of loading for SMA tube with and without heat treatment (Sample B).

3.13.2 Constrained Recovery Behavior

The torsional actuation characteristics of SMA rods (or tubes) are determined by testing their thermal recovery against a constant torque. An understanding of the recovery behavior of a pre-twisted specimen is important for the design of a torsional actuator. Note that for a solid SMA rod (e.g., Sample A), a large portion of the inner radial locations do not undergo sufficient strain to exhibit the recovery characteristics. The test procedure consists of first pre-twisting the sample to a prescribed angle at a constant temperature. The sample is then loaded to the desired test torque. Subsequently, the temperature of the environmental chamber is ramped up while maintaining the torque constant. The recovery angle is therefore obtained as a function of temperature at constant torque. Figure 3.61(a) shows the measured recovery characteristics of the tube (Sample B) for a pre-twist angle of 70° and zero torque (free condition). There appears to be a complete recovery of the pre-twist angle at a high temperature and this recovered angle is almost unchanged on cooling, clearly showing a OWSME. Figure 3.61(b) shows the corresponding recovery characteristics when actuated against a constant torque of 100 in-lb.

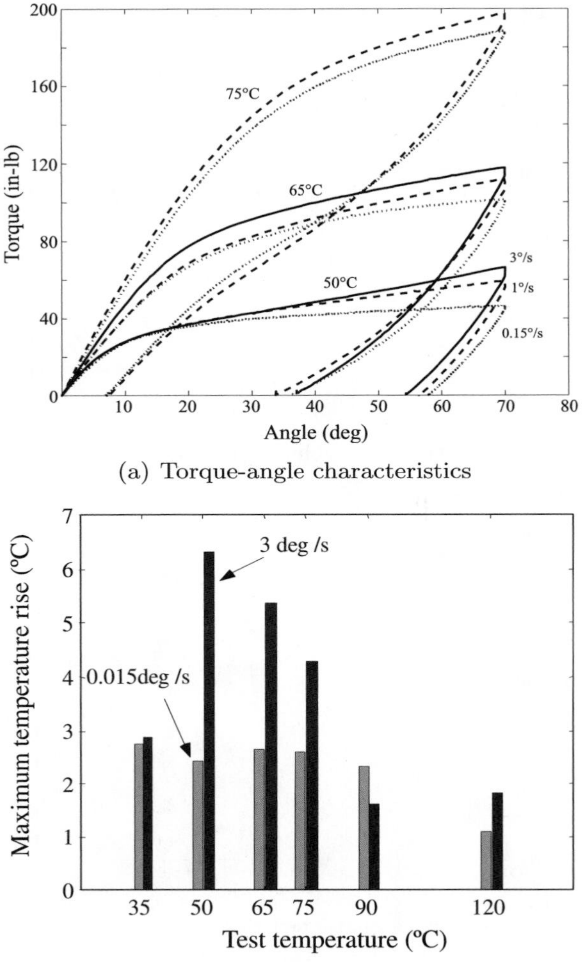

(a) Torque-angle characteristics

(b) Temperature rise

Figure 3.60. Variation of the torque-angle and temperature-rise characteristics of an SMA rod under different twist rates (Sample A).

As shown by Lim and McDowell [97], there is a change of temperature due to latent-heat generation/absorption during phase transformation under mechanical loading/unloading in the pseudoelastic range. Thus, the creep and relaxation phenomena take place during the phase transformation. Lexcellent and Rejzner [101] developed a thermodynamic three-dimensional model of SMA behavior taking into account transformation kinetics laws, asymmetry of stress-strain behavior under tension and compression, and axial-torsion proportional loading of thin tubes. This model requires 13 independent thermodynamic material constants. The thermodynamic coupling between the stress-strain and temperature is determined using the heat equation. Keefe and Carman [102] developed analytical models to evaluate the thermomechanical behavior of SMA torque tubes with varying wall thickness. Tests were conducted in both tension-torsion and compression-torsion to measure recovery torque. The differences in the responses were attributed to the detwinning behavior rather than the loading profile.

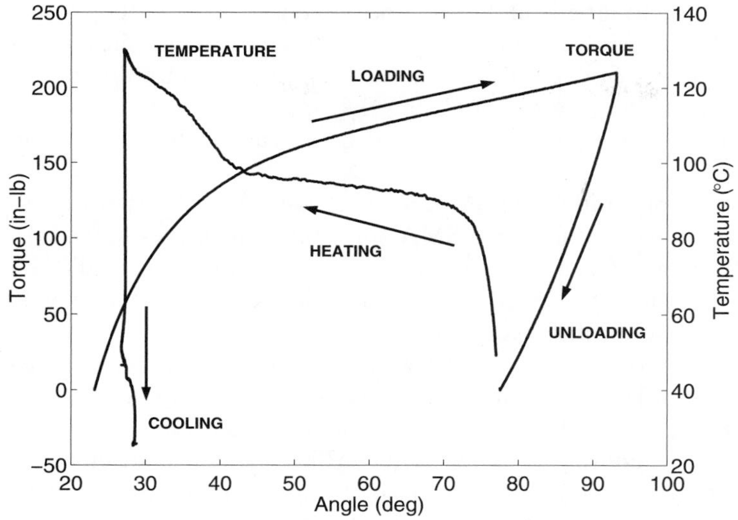

(a)Free recovery (0 in-lb torque), no 2-way effect

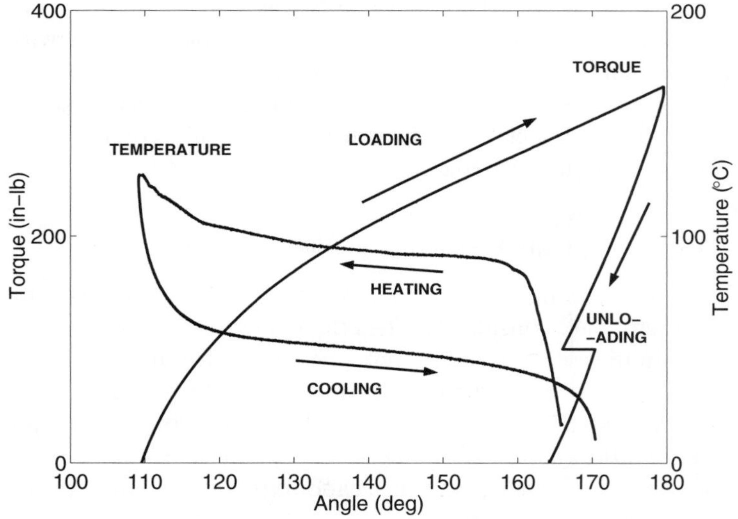

(b)Recovery under constant torque of 100 in-lb

Figure 3.61. Recovery characteristics of the SMA tube under constant torque (Sample B).

3.14 Composite Structures with Embedded SMA Wires

Embedding SMAs into composite structures offers the capability to tune the properties of the structures, such as stiffness and structural damping. This capability has been used in a variety of applications to enhance the functionality of the composite structure. For example, the natural frequencies of a composite structure can be tuned by the activation of embedded SMAs. In addition to being able to tune the dynamic properties of the structure, SMA-embedded composites offer advantages such as structural damping augmentation [103], controlling the buckling in a thin structure [104], structural-acoustic transmission control [105], and delay in the fracture of composites due to fatigue and low-velocity ballistic impact [106]. When combined with the advantages of structural tailoring offered by composites, structures with embedded SMAs can selectively tune their properties in a directional manner. In addition to the advantages of tuning the properties of the structure, embedded SMAs can be used to control the shape of a structure by acting as an actuator for shape control in the material.

Two concepts for incorporating the SMA wires inside the composite are discussed. The first concept involves inserting SMA wires through sleeves in the host structure to take advantage of the variable recovery force in a pre-strained SMA wire. The second concept uses the change in Young's modulus of the SMAs with temperature to alter the overall stiffness of the structure.

Through the activation of embedded SMA wires in a coupled composite structure, one can actively control the shape of the structure. Through a tailored ply layup in a composite beam, one can achieve bending-torsion and extension-torsion coupling [107]. For example, Chandra [108] embedded two SMA bender-plate elements in a bending-torsion coupled solid-section composite beam and induced beam twist with thermal activation. These benders were trained to deform in bending. Internal resistive heating was used to activate the SMAs. Good correlation of the results predicted using the Vlasov theory with the experiments was shown. Ghomshei et al. [109] developed a nonlinear finite element analysis of a composite beam with embedded SMA wires to calculate passive and active response. The model is based on higher order shear-deformation-beam theory together with the von Karman strain model. Satisfactory validation of predictions with experimental results corroborated the nonlinear finite element modeling approach.

3.14.1 Variable Stiffness Composite Beams

To develop a variable stiffness beam, two candidate configurations for interfacing the SMA with the host structure are considered. The effect of both of these concepts is to produce a net change in the stiffness and thereby the natural frequencies of the beam in the bending degree-of-freedom when the SMA wire is activated. However, the degree to which a change in stiffness can be achieved and the requirements for the boundary conditions are different for each of these concepts. These requirements and capabilities must be considered when selecting a candidate configuration for the variable stiffness beam for a particular application.

An SMA wire that is first pre-strained (i.e., strained to plastic deformation at low temperatures) and then heat-activated develops large recovery stresses when its length is constrained. This process of heat activation while maintaining the length of the wire constant is called constrained recovery and can be used as an active force generator [60].

The first concept for developing a variable stiffness beam utilizes this constrained recovery of the wire as an active-force generator [110, 60] to tune the stiffness of the structure. Figure 3.62(a) is a schematic of this concept, which is referred to as the "SMA-in-sleeve concept." This scheme involves SMA wires that are not embedded directly into the structure but instead are inserted into hollow sleeves embedded into the laminated structure. The hollow sleeves are formed by laying up a thin silica tube inside the composite material. During the curing process of the composite structure, steel wires are inserted into the silica tubes; after completion of curing, the steel wires are replaced with pre-strained SMA wires.

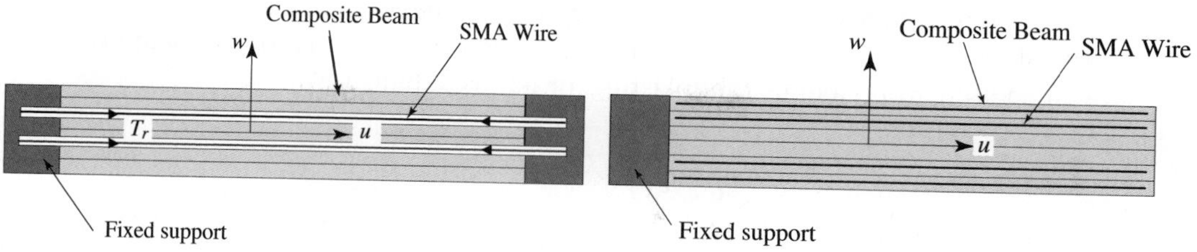

(a) SMA-in-sleeve concept (b) Directly embedded SMA wires

Figure 3.62. Schematic of two different concepts for varying stiffness of a beam using embedded SMAs.

The pre-strained SMA wires are held between fixed supports. When the SMA wires are now heated within the fixed supports, they develop a constrained recovery force T_r due to the SME. The SMA wires, when activated, can therefore be visualized as a string with variable tension. Because the wires are passed through sleeves that are embedded in the structure, they must undergo the same deformation as the host structure (sleeve). Therefore, to deflect the host structure (composite beam) in bending, additional work must be done to deflect the SMA wire in tension. The stiffness therefore has two components – one a fixed stiffness of the composite beam, and the other a variable component due to the SMA wire in tension. The additional work to deflect the SMA wire in tension manifests as an increased stiffness and therefore increased natural frequency for the host structure.

It is important to note that to change the natural frequencies with the SMA-in-sleeve concept, the SMA wires must be held independent of the host structure. The constrained recovery force developed in the wire must react against the fixed supports and not against the host beam itself. If the wire is attached directly to the composite beam, an equilibrating compressive force is developed in the beam, and the natural frequencies of the structure are not affected.

Due to this constraint on the implementation of the SMA-in-sleeve concept, the boundary conditions of the beam to which this concept can be applied are restricted. For the simple case of the beam in bending, this configuration is directly applicable only in the case of a fixed-fixed boundary condition. When the SMA wire is attached to the host structure directly (and thus not held independent of the host structure), negligible change in natural frequency is expected. This configuration, however, can be used to generate control moments on one side of the beam for actuation or active control of the vibrations in a structure [111].

The second concept for a variable stiffness structure involves directly embedding the SMA wires into the host structure [103, 112]. In this concept, the SMA wire is an integral part of the structure (Fig. 3.62(b)) and is co-cured with the composite material. Note that the Young's modulus of the SMA in pure martensite phase is typically two to three times lower than that in pure austenite phase. In the transformation region, the Young's modulus varies in proportion to the martensite volume fractions of the material. When the SMA wires are embedded into a composite structure, this change in the Young's modulus of the SMA results in a corresponding change in the stiffness of the composite structure. Because the SMA wires are now an integral part of the host structure, the mechanism of varying the natural frequencies of the structure is not dependent on the boundary condition. This concept requires the transfer of strain from the wire to the matrix, and maximum interfacial adhesion between the SMA wire and the polymer matrix is needed. The surface treatment of the SMA wire is a major issue. To treat the surface, one of the following approaches can be used: acid etching, sandblasting, or handsanding.

3.14.2 SMA-in-Sleeve Concept

In the first concept (i.e., SMA-in-sleeves), the SMA is not part of the host structure and behaves as a string in tension that is constrained to follow the same displacements as the composite beam. One approach to modeling this concept is to model the beam as a structure on an elastic foundation [110]. The recovery force in the wires can be viewed as an increase in the stiffness of the elastic foundation of the structure. Thus, the change in frequency of the host structure is due to the change in the boundary conditions of the beam (i.e., variation of stiffness of the elastic foundation).

A second method is to model the system using a Hamiltonian approach. Using this approach, the elemental stiffness matrix can be constructed by individually superimposing the contributions from the beam and the SMA wire. Baz et al. [60] derived these equations for the elemental stiffness for a prismatic beam as

$$[K_e] = E_m I_m \int_0^L \{D\}^T \{D\}\, dx \;-\; P_n \int_0^L \{C\}^T \{C\}\, dx \tag{3.94}$$

where E_m and I_m are the Young's modulus and moment of inertia of the baseline beam, respectively; L is the length of the beam; and P_n is the total external axial force acting on the beam. The matrices $\{C\}$ and $\{D\}$ are derived from the matrix of spatial-shape functions (or Hermite cubics), $\{A\}$ as

$$\{C\} = \frac{d}{dx}\{A\} \tag{3.95}$$

$$\{D\} = \frac{d^2}{dx^2}\{A\} \tag{3.96}$$

where A is the matrix of spatial-shape functions for finite element analysis [113] of bending deflections in a uniform beam. The total axial force P_n has a component due to external mechanical forces P_m, thermal expansion of the host structure P_t, and tension in the SMA wire T_r. This is expressed as

$$P_n = (P_m + P_t - T_r) \tag{3.97}$$

Neglecting the contribution of the thermal expansion of the host structure and assuming no external axial mechanical forces, we can find the elemental stiffness matrices $[K_e]$ for a given tension T_r in the SMA wire using Eq. 3.94.

The tension T_r in the SMA wires occurs due to the constrained-recovery stress in the SMA when the material is not allowed to recover its original length. This quantity has been obtained from experimental testing by Baz [60]. The tension T_r can be estimated using constitutive models for the SMA wire behavior.

It is known from experiments that the constrained-recovery force for the SMA wire is proportional to the imparted pre-strain below about 2% and shows no significant increase for higher levels of pre-strain. It was also observed that from the second cycle of activation, the recovery stress consistently oscillates between two non-zero stresses during thermal cycling. This behavior of the SMA is accurately captured by the constitutive models for quasi-static behavior.

From the Brinson model [46], the recovery force and corresponding stresses developed in the wire are given by

$$T_r = \sigma A_{SMA} \tag{3.98}$$

$$\sigma = E(\xi)(\epsilon - \epsilon_L \xi_s) + \Theta(T - T_0) \tag{3.99}$$

where A_{SMA} is the cross-sectional area of the wire. Starting with a pre-strain of ϵ_p and a stress-free condition ($\sigma_0 = 0$) and ignoring the effect of the pure-phase thermal expansion, this equation simplifies to

$$\sigma_r = E(\xi)(\epsilon_p - \epsilon_L \xi_s) \tag{3.100}$$

When the SMA wire is inserted in the sleeve, the displacement of the wire is compatible to the deflections in the host structure. However, because the wire is not completely embedded in the host structure, the stresses in the host structure are not compatible with the recovery stresses in the wire.

For a given temperature of the SMA wires, the elemental stiffness matrices can be obtained from the recovery stress σ_r. The natural frequencies of the entire structure may then be found by the finite element formulation [113]. Note that because the SMA wire is not an integral part of the beam structure, it does not contribute to the mass matrix in the formulation but rather only to the stiffness matrix.

For a beam with composite coupling, the overall bending stiffness EI can be replaced using Classical Laminate Plate Theory (CLPT) [114]. In this case, the formulation varies depending on whether the SMA wires are in the fiber direction in each ply or in the direction parallel to the axis of the entire beam. In the current case, however, the analysis is applied only to an uncoupled beam.

The resulting predictions from the analysis for a representative rectangular beam with SMAs inserted in sleeves are shown in Fig. 3.63. The analysis is carried out for a rectangular beam with a thickness of 0.082 in (0.0021 m) (16 plies in [0] direction), a width of 0.4 in (0.0102 m), and a length of 10 in (0.254 m). The material used is Graphite Epoxy (T300/5208). The SMA wire has a diameter of 0.015 in (3.8×10^{-4} m). For the given dimensions, eight wires correspond to a volume fraction of about 4.19% in the beam. A uniform distribution of SMA wire across the cross section of the beam is assumed in this analysis.

From Fig. 3.63(a), we see that the analysis predicts significant changes in natural frequencies for a rectangular beam when the SMA wires are activated. The analysis predicts a natural frequency increase of 22.5% from the baseline case using eight SMA wires. This change in natural frequency corresponds to an increase of nearly 100% in the effective static stiffness of the material.

Figure 3.63(b) shows the corresponding predicted change in natural frequencies with SMA wire temperature for different numbers of SMA wires. From the figure, a temperature hysteresis for the natural frequencies of the beam is predicted. This hysteresis follows the characteristics of the constrained-recovery behavior for the SMA wire. Recall that when starting from a zero-stress condition, the SMA wire first develops a high recovery stress during the heating cycle. In the first step of thermal cycling, the stress does not come back to zero on cooling but instead stabilizes to a positive value. On subsequent thermal cycling, the stress oscillates between this value of stress at low temperature and the recovery stress at high temperature, thus completing the hysteresis cycle. It was also demonstrated that the model shows good

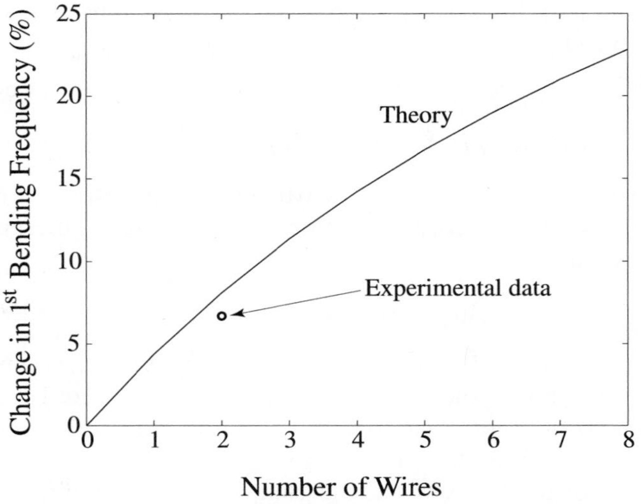

(a) Theoretical predictions for change in natural frequency with number of wires

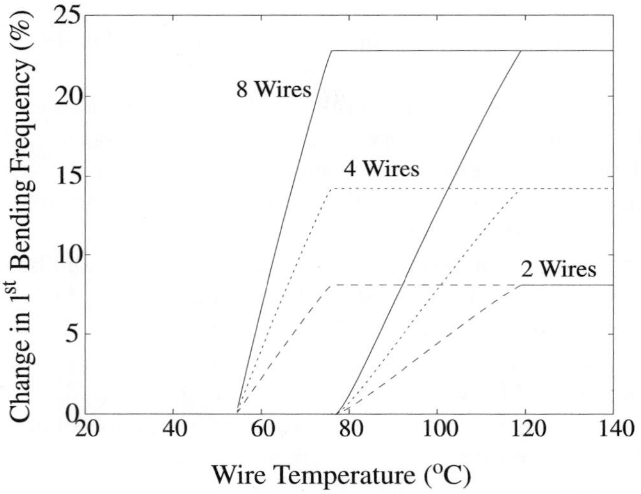

(b) Theoretical variation of natural frequencies with temperature, 2% pre-strain

Figure 3.63. SMA inserted in sleeves, rectangular beam.

prediction with this cycling characteristic of the wire. The model predictions for the natural frequencies of the beam used in this section, therefore, utilize predictions from the second thermal cycle onward to predict the natural frequencies of the beam. The baseline value of natural frequency at low temperature from the second thermal cycle onward is set to zero in the current analysis.

3.14.3 Beams with Embedded SMA Wires

An alternate scheme of changing the natural frequencies of the composite beam occurs when the SMA wires are directly embedded into the structure and co-cured with the host composite structure. In this case, the effect of the SMA in the structure is a change in bending stiffness of the composite structure due to the inherent change in the Young's modulus of the SMA with temperature. Unlike in the first scheme (i.e., SMA-in-sleeves), this scheme does not rely on the recovery stresses developed in the wire but rather on the change in the material properties of the SMA.

We assume here that in order to maintain the integrity of the host structure, the ply strains of the composite structure are low and therefore limited to within the linear elastic region for the SMA (i.e., less than 0.5%). This implies that the Young's modulus of the SMA is only a function of temperature (for zero starting stress) and not a function of strain (or stress) in the host structure. The SMA fibers are also assumed to be oriented in the direction of the fibers in the composite structure for each ply.

The stiffness of the structure can be divided into two parts – a constant stiffness of the passive material (fibers in the composite material) and a variable temperature-dependent stiffness due to the SMA wires. The Young's modulus of each of the plies in the direction of the fibers is then given by a volume-fractions approach (i.e., the mixture rule)

$$E_{ply} = (1 - V_{SMA})E_{fiber} + V_{SMA}E_{SMA} \tag{3.101}$$

where E_{fiber} and E_{SMA} are the constant Young's modulus of the fiber material and the variable Young's modulus of the SMA wire, respectively. V_{SMA} is the volume fraction of SMA wire in each ply. From this equation, it is seen that for a given SMA wire, the change in stiffness of each ply increases with increasing volume fraction of SMA.

The variable component of stiffness in the composite structure is the Young's modulus of the SMA wires. The analytical Young's modulus of an SMA can be obtained as a function of the temperature of the SMA by prescribing a stress-free condition ($\sigma = 0$) to the SMA constitutive models. In most constitutive models for SMAs, the Young's modulus of the SMA wire is related to the volume fraction of martensite in the material by a simple rule of mixtures

$$E(\xi) = E_A + \xi(E_M - E_A) \tag{3.102}$$

where ξ is the volume fraction of martensite in the SMA, and E_M and E_A are the Young's moduli of the material in pure martensite and austenite phase, respectively. The constants E_M and E_A are obtained from material characterization of the SMA wire.

For a given temperature of SMA material, therefore, we can obtain an effective variable Young's modulus for each ply. From this, the stiffness matrix $[K_{ele}]$ can be constructed. Unlike the SMA wires inserted through sleeves in the structure, the SMA also contributes to the mass matrix of the structure. The effective mass density for each ply is calculated similar to the effective stiffness for each ply as

$$\rho_{ply} = (1 - V_{SMA})\rho_{fiber} + V_{SMA}\rho_{SMA} \tag{3.103}$$

where ρ_{fiber} and ρ_{SMA} are the mass densities of the host composite structure and the SMA wires, respectively. The effective mass per unit length of the SMA embedded composite structure can then be calculated from this effective density and used to construct the elemental mass matrices for the structure.

Having obtained the elemental stiffness and mass matrices for the SMA embedded composite structure, an estimate for the natural frequencies of a coupled laminate can be computed using finite element analysis for a beam in bending [113]. In the

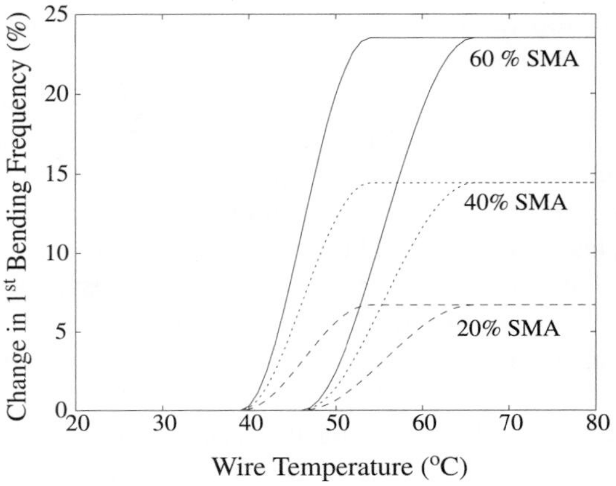

Figure 3.64. Change in natural frequency versus temperature of SMA wires for a rectangular beam with embedded SMAs.

current work, the model is applied only to an uncoupled beam with embedded SMA wires. However, it is also possible to extend this formulation to the case of a composite coupled beam by replacing a single-stiffness matrix with the coupled-stiffness matrices ***A***, ***B***, and ***D*** from CLPT equations.

The maximum change in natural frequencies of the material is obtained when the SMA Young's modulus changes between E_M and E_A. At intermediate temperatures, the SMA Young's modulus increases with temperature between the transformation temperatures. Because the description of the variation of the martensite volume fraction with temperature and stress differs based on the model used, the assumed path of the change in natural frequencies of the structure also differs correspondingly, depending on the SMA constitutive model used.

Figure 3.64 shows the percentage change in first bending natural frequency for a rectangular beam as a function of the temperature of the SMA wires. The dimensions and materials of the rectangular beam are the same as in the SMA-in-sleeve case discussed in Section 3.14.2. From the figure, it is observed that the natural frequencies are also predicted to exhibit hysteresis with the temperature of the SMA wire. By comparing the predictions for the SMA-in-sleeve (see Fig. 3.63(b)) case with the case of the integrally embedded SMAs (Fig. 3.64), it can be observed that for the same change in natural frequencies of the host structure, the volume of SMA wires that are required is much greater in the latter case. This comparison effectively illustrates the advantages of the SMA-in-sleeve concept.

From Fig. 3.64, it is seen that to achieve a greater variation in the natural frequencies of the structure, greater volume fractions of SMA are required. However, as discussed in Section 3.14.5, several manufacturing and strength considerations limit the volume fractions that can be embedded in the SMA. For a given volume fraction of SMA, the maximum benefit of the changing stiffness of the SMA can be derived by increasing the distance between the SMA wires and the neutral axis of the beam. To achieve this, it is possible to optimize the cross section of the beam to maximize the influence of the change in the Young's modulus of the SMA.

An example of a structure designed to take advantage of the change in the Young's modulus of the SMA is a beam with an I-shaped cross section with the SMA wires embedded in the flanges. This arrangement increases the distance of the SMA from the neutral axis, thereby enabling a greater control over the natural frequencies for the same volume fraction of SMA. The number of SMA wires embedded in the flange is assumed to change with the volume fraction of the SMA.

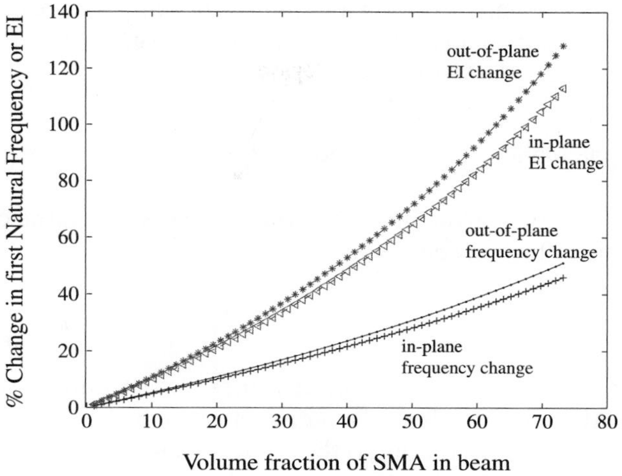

Figure 3.65. Predicted change in natural frequency and bending stiffness, for I-beam with embedded SMA wires.

Figure 3.65 shows the predictions of the model for the change in in-plane and out-of-plane bending stiffness, as well as natural frequencies, for an I-beam with embedded SMA wires. From the figure, it can be seen that the change in natural frequencies of the beam can be different for the two axes of bending. Thus, by varying the geometry of the cross section and placement of the SMA wires, the change in bending natural frequency can be controlled independently around each axis of bending. This indicates that the effect of the SMA may be tuned to produce desired effects in each direction of bending for the composite beam.

The constitutive modeling of the two concepts discussed previously indicates that inserting the SMAs in sleeves produces significantly higher changes in natural frequencies for a given volume fraction of SMA in the structure. However, for this configuration, the SMA must be held independently of the host structure in a fixed-fixed condition. In several applications, such as a rotating environment, this appears infeasible because the supporting structure must be held independent of the beam. Thus, embedding the wire directly into the structure is considered a more feasible solution in these applications. Embedding the SMA wires directly with the composite offers the advantage of structural integrity, and it is not restricted in the geometrical configuration in which it can be used. However, the large volume fractions of SMA required to produce a significant change in natural frequencies of the beam with this concept result in a large weight penalty and manufacturing difficulties (see Section 3.14.5). The constitutive models derived here, however, can be used to predict these effects and to design a structure with variable stiffness using either of the two concepts.

3.14.4 Power Requirements for Activation of SMA in Structures

In applications involving either of the two concepts described previously, it is not possible to monitor the wire temperature inside a composite beam. It is therefore critical to obtain an accurate estimate of the temperature of the wire for a given electrical input to the wire. This is crucial in determining the state of the transformation in the material and, therefore, in determining the stiffness of the entire material. This section outlines an energy analysis similar to the one described in Section 3.12. The energy analysis is used to estimate the temperature of the SMA wire as a function of the input energy, and it is validated with experiments on a composite laminate with embedded SMA wires.

Calculation of Power Required

The thermodynamic equilibrium of the SMA in the composite can be described by the following equation

$$P_{in} = P_{abs} + P_{loss} \tag{3.104}$$

where P_{in} is the input electrical power (heating), P_{abs} is the heat retained by the SMA, and P_{loss} is the heat dissipated to the surroundings.

The power input to the SMA wire is

$$P_{in} = i^2 R_{wire} \tag{3.105}$$

where i is the current passing through the wire and R_{wire} is the resistance of the wire. Note that the resistance of the wire changes during transformation and is a function of the stress and temperature in the wire, as discussed in Section 3.12. The heat retained by the wire is given by the sum of the specific heat (causing a temperature rise) and the latent heat (during transformation)

$$P_{abs} = mC_p \Delta T + mL_{trans}\dot{\xi} \tag{3.106}$$

where m is the total mass of the wire and ΔT is the change in its temperature. Note that the specific heat of the material, C_p, also varies as a function of the martensite volume fraction. The second term in the equation is the latent heat term and is proportional to the rate of change of martensite volume fraction, $\dot{\xi}$. L_{trans} is the latent heat of the appropriate phase transformation. The parameters in the previous equation are all obtained from experiments for a particular sample of SMA. Typical values are listed in Section 3.12.

The heat dissipated from the SMA wires depends on the surrounding material. In the present case, the SMA wires are enclosed in the structure by glass fibers and a thermally insulating matrix. The thermal conductivity of the glass fibers-epoxy system (0.29–0.31 W/m-K) is poor compared to that of the metallic SMAs (8.6–18 W/m-K). It is therefore assumed in the current formulation that all of the heat dissipation of the SMA occurs outside of the beam and that the primary mechanism of heat dissipation is through convective loss of heat to the surrounding air in the exposed parts of the SMA outside the beam. Using simple one-dimensional thermal-transfer theory [89], the rate of convective heat loss is

$$P_{loss} = hA(T - T_{\infty}) \tag{3.107}$$

where h is the effective heat-transfer coefficient of the material, A is the exposed cross-sectional area, T is the temperature of the SMA, and T_{∞} is the temperature of the surrounding air. The value of the heat-transfer coefficient used for the current analysis is the same as that determined in Section 3.12.

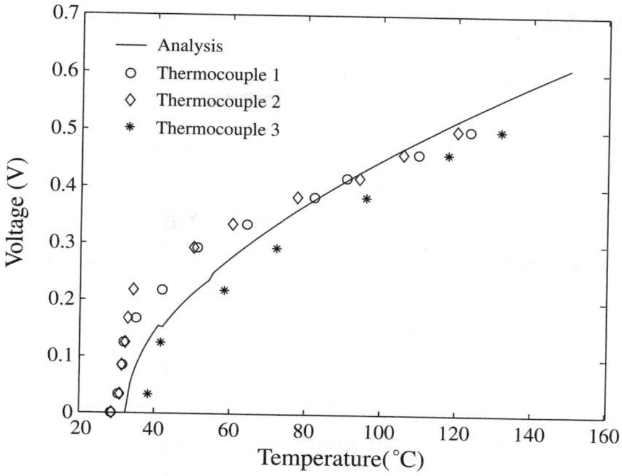

Figure 3.66. Validation of predicted input voltage required for a given temperature.

Experimental Validation

After obtaining expressions to describe the various components in the thermodynamic equation, we can rewrite the equilibrium equation for a given voltage input V_{in} as a function of the temperature T and material stress σ as

$$\frac{V_{in}^2}{R(T,\sigma)} = E_{in}(T,\sigma) + E_{loss}(T) \tag{3.108}$$

This equation can be solved to obtain equilibrium temperatures at any given stress and voltage.

To validate the predictions of the thermal analysis, the temperature profiles as a function of the input voltage were measured in a test coupon. The test coupon was a thin composite laminate fabricated with embedded SMA wires. The coupon laminate consisted of 16 layers of glass fiber prepreg (0/90 weave) of length 6 in (0.1524 m) and width 3 in (0.0762 m). During the layup process, three SMA wires that were longer than the composite coupon specimen were placed at the interface of the innermost two plies (between the fourth and fifth plies). These wires were held taut by end fixtures on either side. To monitor the temperature of the wire inside the sample, thin wire thermocouples (K-type) were attached to the surface of the SMA wires using thermally conductive tape. The test coupon with this combination of embedded SMA wire and thermocouple was then cured for testing.

During the experiment, the three wires were connected in parallel to a DC power supply, and the resulting steady-state temperature was recorded as a function of the applied input voltage. From Fig. 3.66, it is observed that the theoretical predictions show good agreement with experimental values of temperature for all three wires. Thermocouples 1 and 3 were near the two ends of the test coupon, whereas Thermocouple 2 was in the center. The theoretical prediction approximated the measurements from Thermocouple 2 due to the highest insulation provided by the composite coupon at its center.

It is important to note that this test was carried out at zero stress and that comprehensive testing of this under all conditions of stress and temperature was not undertaken. However, the encouraging results obtained here indicate the feasibility of this modeling approach to predict the power requirement of the wire embedded in a composite beam.

3.14.5 Fabrication of Variable Stiffness Composite Beams

SMA-in-Sleeve

For the SMA-in-sleeve case, a rectangular graphite-epoxy beam fabricated in a previous study [110] was used for the experimental testing. This beam had a length of 12 in (0.3048 m) and a rectangular cross section with a width of 0.5 in (0.0127 m) and thickness of 0.0625 in (1.58 mm). The beam was fabricated with 16 layers of unidirectional graphite-epoxy prepreg, which were laid up in the 0° direction (parallel to the axis of the beam). Two hollow silica tubes, referred to as sleeves were embedded along the neutral axis (see Fig. 3.62(a)).

The manufacturing procedure consisted of cutting strips with the required length and width from the graphite-epoxy prepreg. Eight layers of prepreg were layed up on top of each other in one half of a rectangular mold. Two fused silica tubes of 0.02 in (0.508 mm) inner diameter were then placed on top of the stack of prepreg. The silica tubes were inserted with dummy steel wires during the manufacturing process. Eight additional layers were then laid up on top of the silica tubes, and the other half of the mold was pressed down on the stack to create a rectangular space for the beam. The graphite-epoxy prepreg and the silica sleeves were cured together in an oven.

After curing the beam, the dummy steel wires were removed from the sleeves and SMA wires of 0.015 in diameter were inserted in their place. The wires were then pre-strained to about 2% and held in place by attaching them to fixed supports.

Beams with Embedded SMA Wires

To test the second concept of obtaining frequency changes by directly embedding SMA wires into a structure, two beams with different volume fractions of embedded SMA wires were fabricated. As explained in Section 3.14.1, an "I" cross section increases the authority of the SMA wire to change the natural frequencies of the beam. To exploit this advantage, the beams were fabricated with the I-section.

The dimensions of the I-beams fabricated are shown in Fig. 3.67(a), with the same overall dimensions used for both of the beams. As shown in the figure, the SMA wires are concentrated only in the flange region, with the web being made purely from the host composite material. The number of SMA wires embedded in the beam was different in the two beams; the first beam (referred to as Beam A) had four wires embedded in each flange, and the second (referred to as Beam B) had eight in each flange. These configurations correspond to a total SMA volume fraction of 9% and 18% in Beams A and B, respectively.

The beams were fabricated using 0/90 weave of S-glass prepreg. Figure 3.67(b) shows a schematic of the fabrication process for Beam B (eight wires in each flange). The schematic shows the mold used to fabricate the beam. The mold consists of two sets, marked as 1 and 2 in the figure.

The layup process is divided into two parts. First, the beam web is laid up in two "C" configuration halves to cover the web and a part of the flange. Four layers of prepreg were folded into a "C" shape and laid up in each half of the web structure, as shown in the figure. The mold blocks marked as "1" were then brought closer to each other with the application of external pressure, thus compressing the prepreg

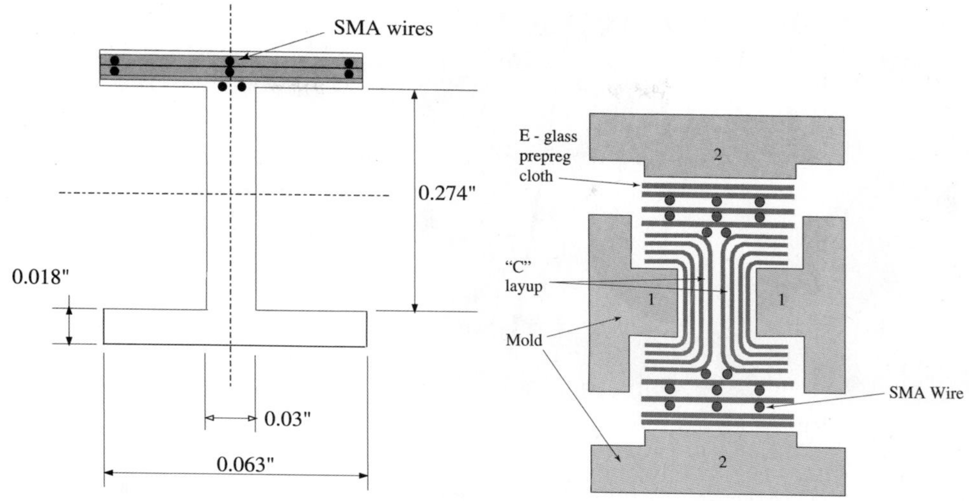

(a) Dimensions of the I-section beam (b) Manufacturing of I-beam with embedded SMA wires

Figure 3.67. Manufacturing of a beam with "I" cross section and embedded SMA wires.

in the central web of the beam. Two SMA wires were laid up in the junction of the flange and the web on either side.

The second step consisted of laying up the prepreg and SMA wires to make up the flange of the beam. This was achieved by successively laying up two sandwiched layers in each flange. These layers each consist of three SMA wires sandwiched between two layers of prepreg material. Two such prepreg-SMA sandwiched layers were laid up successively in each flange, making a total of six additional wires in each flange, as shown in Fig. 3.67(b). One layer of prepreg material was then laid up on top in each flange. The mold blocks marked 2 in the figure were then brought in contact with the other blocks (marked 1) so as to enclose the composite layup with embedded SMAs in the space enclosed by the four blocks. In cases involving lower volume fractions of SMAs (as in Beam A), the number of SMA wires in the flange can be varied appropriately to obtain the desired volume fraction of SMA in the beam.

The molds were cured at a temperature of 250°F for an hour while pressure was applied through the clamps. No apparent delamination was observed at the junctions between the SMA wires and the prepreg glass fiber. After completion of the curing process, the beams were removed from the molds. It is important to note that to actuate the SMA wires, the length of the wires must be greater than the length of the beam. The additional length that projects out of the length of the beam, therefore, must be held in place using grips on either side of the mold during the curing process.

Note that beams with larger volume fractions of SMA wires can exhibit significant problems arising from failure at the interface of the SMA and the glass-fiber composite. At volume fractions greater than about 18%, there was insufficient flow of the epoxy around the SMA fibers, which resulted in a lack of adhesion between the SMA and the host composite material. This resulted in the peeling off of layers above the SMA wires after curing. This problem was also encountered in previous work on SMA composites [103, 115]. Figure 3.68, adapted from a study by

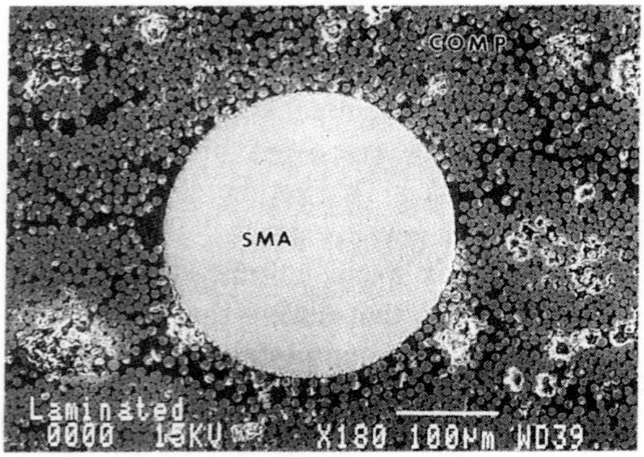

Figure 3.68. Formation of voids around the SMA wire embedded in the composite structure, adapted from [116].

Friend et al. [116], shows a typical SMA-embedded composite structure under a microscope. The formation of voids around the SMA wires can be clearly observed. Void formation decreases the influence of the SMA wires and also compromises the structural integrity of the beam.

The change in natural frequency that can be achieved by the SMA embedded composites, therefore, was limited by the manufacturing constraints of these beams. In addition, significant weight penalties are encountered due to the larger density of SMA (6450 kg/m^3) compared to the host composite S-glass material (1700–2000 kg/m^3).

3.14.6 Experimental Testing of Variable Stiffness Beams

The variable stiffness beams fabricated as described in the previous section were tested to determine the variation in the natural frequencies with activation of SMA wires. The results from the experimental testing were also used to validate the predictions for the change in natural frequencies from the constitutive models. Note that all tests discussed in this section are with reference to the bending mode of the beam. Although the focus of the current study is on the first bending mode, the observations can be extended to higher modes of vibration as well.

The variable stiffness beams were instrumented with strain gauges located near the fixed end of each beam. The SMA wires were heated internally by passing electrical current through them. Because the focus of the current study is to determine the maximum change in natural frequency achievable by activating the SMA wires, the beams were tested only with the SMA in pure martensite phase at room temperature and in pure austenite at high temperature.

SMA-in-Sleeve

The concept of using the constrained-recovery force of the wire to change the natural frequencies of the structure requires specific boundary conditions to be implemented. This boundary condition requirement specifies that the SMA wire should be held independent of the host structure to achieve a change in natural frequency. The boundary condition for the beam in this test, therefore, was a fixed-fixed condition, as shown in Fig. 3.62(a).

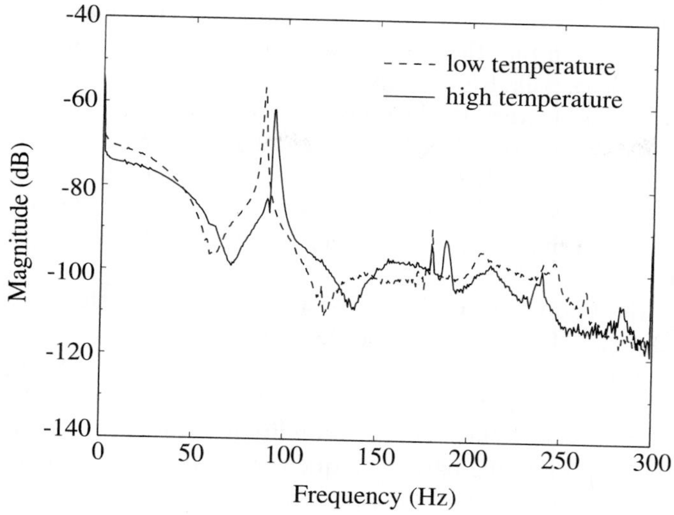

(a) SMA wires independently supported

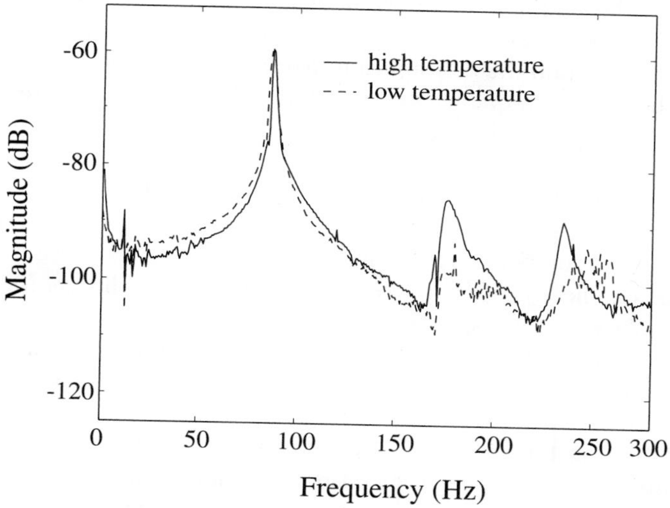

(b) SMA wire attached to one end of the beam

Figure 3.69. Transfer function of rectangular fixed-fixed beam with SMA embedded in sleeves.

The SMA wires were pre-strained to the desired degree by attaching them to a turnbuckle on either end, which was in turn attached to a fixed boundary. By modifying the length of the turnbuckle, the wire length was modified. The length of the turnbuckle was calibrated to estimate the strain imparted to the wire. In the tests reported in this section, the wire was pre-strained by 2%.

The impulse response of the beam was obtained by measuring the signal from the strain gauge after striking the beam with an impulse hammer. Figure 3.69(a) shows the measured transfer function of the beam strain for the wire with SMAs in sleeves before and after activation of the SMA wires. The beam was held in fixed-fixed condition, and the SMAs were held independent of the beam. The impulse response shows a shift of 6.7% in the first natural frequency when the SMAs are heated.

This measured shift in the first natural frequency is compared with the analytical predictions in Fig. 3.63(a). It is seen that the analysis overestimates change in the natural frequency of the beam. This is likely due to the analytical model's assumption of an ideal compatibility of strains between the beam (at the location of the sleeves) and the SMA wire. However, a small gap between the wire and the sleeve is required to manufacture the beam and subsequently insert the SMA wires. The strains in the SMA wire are therefore slightly less than those of the sleeve, thereby reducing the effective stiffness due to the SMA wires. The model therefore overpredicts the effect of the SMA wire on the structure.

To study the effect of the boundary conditions on this concept, an alternate test was performed using the same beam. In this test, the SMA wires were not held between the fixed grips but instead were attached to the beam at the ends. Thus, the recovery force in the wire exerted a compressive stress in the beam. Figure 3.69(b) shows the resulting transfer function for this boundary condition. From the figure, it is seen that in this boundary condition, the natural frequencies of the beam remain almost unchanged when the wire is heated and the constraining force is applied to the beam. This is due to the equilibration of stresses in the wire and the beam. These experimental results validate the assumption of the model in describing the behavior of the beam as the superposed responses of a fixed-fixed composite beam and a string in tension. The results also prove that the concept of utilizing the recovery force of the SMA wires to change the natural frequencies of the beam is feasible only in cases in which the wire is held independent of the host structure.

Beams with Embedded SMA Wires

In the experiments on beams with embedded SMA wires, the beam boundary condition was cantilevered because the wires do not require fixed-fixed conditions for activation. Figure 3.70(a) shows a test specimen of an I-beam with embedded SMA wires. The beam was actuated using PZT-5H sheet actuators, with a sinusoidal frequency sweep from 1 to 200 Hz at a constant input amplitude voltage of 60 V RMS.

Figure 3.70(b) shows the transfer function of Beam A (9% volume fraction of SMA) in the transverse direction. A change in the first-bending natural frequency of 5.6% was observed when the SMA wires were heated. The same test was repeated for Beam B (18% volume fraction of embedded SMA). The change in the first natural frequency was observed to be about 11% in this case. The two experimental data points obtained are plotted against theoretical predictions in Fig. 3.65. The experimental frequency change is in good agreement with the predictions for the volume fractions under consideration. However, beams with larger volume fractions of SMA could not be tested due to the manufacturing constraints discussed in Section 3.14.5.

The experimental testing demonstrates the feasibility of developing variable stiffness structures with embedded SMAs. The two concepts reviewed here are SMA wires inserted in sleeves and held independent of the structure, and SMA wires embedded into the composite structure. Although both concepts appear to show potential to control the natural frequencies of a composite structure, there are significant drawbacks associated with each. In the SMA-in-sleeve case, the SMA wire must be held independent of the host composite structure and therefore requires a fixed-fixed boundary condition. This makes this concept infeasible for many applications. In the case of the integrally embedded composite wires, the volume fractions

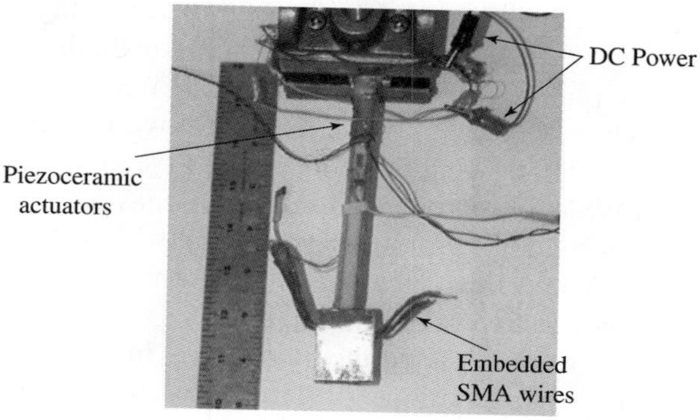

(a) Beam test specimen

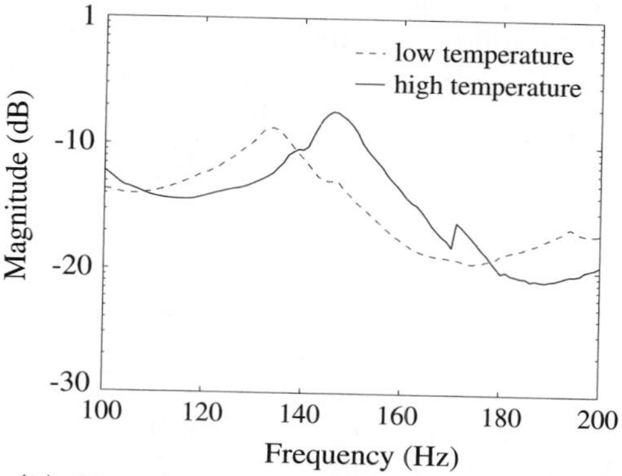

(b) Transfer function of beam bending strain

Figure 3.70. Test specimen and transfer functions for bending strain of an "I"-beam with embedded SMA wires (9% volume fraction of SMA).

and, therefore, natural frequency changes that can be achieved are limited by manufacturing constraints. These factors must be taken into account when designing a variable stiffness structure incorporating SMA wires.

3.15 Concluding Remarks

Shape memory alloys such as Nitinol have large force and stroke and, therefore, have enormous potential for low-frequency (i.e., quasi-static) applications. These materials exhibit highly nonlinear behavior with respect to temperature and stress as well as strain history, and they require fine-tuning (using an adaptive feedback controller) to achieve the desired state. Also, the stiffness varies considerably during the phase transformation. As a consequence, a locking mechanism is required to maintain a desired state. Also, the variation of properties during transformation from martensite to austenite or vice versa is quite abrupt. Hence, it can be quite difficult to achieve refined control in some applications.

Most of the macromechanics constitutive models have been developed for tensile quasi-static loading for the OWSME. Constitutive models for time-varying loading are in an early stage of development. Macromechanics models for TWSME are limited. Although micromechanics models for SME are more detailed and insightful, they have limited practical use in engineering design. These models can help to refine the macromechanics models. Furthermore, validation of both micromechanics and macromechanics models with test data for a range of loadings needs to be expanded to cover tensile, compressive, and shear loading. It may be important to develop simple phenomenological models for TWSME and time-varying loadings.

There have been limited constitutive models for torsional loading. For this, a good understanding of both compressive and tensile loading is required, as well as the TWSME. Micromechanics models may provide guidance to refine the macromechanics models. Systematic validation with test data is necessary to develop robust models for design. This topic has enormous potential toward the development of a mission-adaptive morphing structure.

Embedding shape-memory wires in a laminated structure may expand the domain of applications. So far, there has been limited validation of the response of built-up laminated structures with embedded Nitinol wires. For such structures, local stress-strain distributions using detailed finite element analyses (e.g., three-dimensional solid elements and higher order shear-deformation theory models) may reveal the mechanism of actuation as well as help to establish the integrity of the structure. Again, it may be challenging to develop two-way shape adaptive laminated structures with embedded SMA wires.

BIBLIOGRAPHY

[1] W. J. Buehler, J. V. Gilfrich, and R. C. Wiley. Effect of low-temperature phase changes on the mechanical properties of alloys near composition TiNi. *Journal of Applied Physics*, 34(5):1475–1477, May 1963.

[2] W. J. Buehler, R. C. Wiley, and F. E. Wang. Nickel-based alloys. U.S. Patent No. 3,174,851, March 1965.

[3] C. M. Wayman and T. W. Duerig. An introduction to martensite and shape memory. In *Engineering Aspects of Shape Memory Alloys*, edited by T. W. Duerig. Butterworth-Heinemann, 1990.

[4] K. N. Melton and O. Mercier. Deformation behaviour of NiTi-based alloys. *Metallurgical and Materials Transactions A* 9(10):1487–1488, 1978.

[5] H. Tas, L. Delaey, and A. Deruyttere. Stress-induced transformations and the shape-memory effect. *Journal of Less-Common Metals*, 28:141, 1972.

[6] L. Delaey, R. V. Krishnan, H. Tas, and H. Warlimont. Thermoelasticity, pseudoelasticity and the memory effects associated with martensitic transformations – 1, 2, 3. *Journal of Materials Science*, 9(9):1521–1555, September 1974.

[7] L. Delaey and J. Thienel. Microstructural Changes During SME Behavior. In *Shape memory effects in alloys*, edited by J. Perkins, pages 341–350. Proceedings of the International Symposium on Shape Memory Effects and Applications, Toronto. Plenum Press, New York, 1975.

[8] J. Perkins and D. Hodgson. The two-way shape memory effect. In *Engineering Aspects of Shape Memory Alloys*, edited by T. W. Durig, K. N. Melton, D. Stöckel, and C. M. Wayman, pp. 195–206, Butterworth-Heinemann, 1990.

[9] A. Nagasawa, K. Enami, Y. Ishino, Y. Abe, and S. Nenno. Reversible shape memory effect. *Scripta Metallurgica*, 8(8):1055–1060, September 1974.

[10] R. J. Wasilewski. Effects of applied stress on the martensitic transformation in TiNi. *Metallurgical and Materials Transactions B*, 2(11):2973–2981, November 1971.

[11] C. M. Wayman and H. K. D. H. Bhadeshia. Phase transformations, nondiffusive. In *Physical Metallurgy*, edited by R. W. Cahn and P. Haasen, pp. 1507–1553, Elsevier, 1983.

[12] B. J. de Blonk and D. C. Lagoudas. Actuation of elastomeric rods with embedded two-way shape memory alloy actuators. *Smart Materials and Structures*, 7(6):771–783, 1998.

[13] E. Patoor, P. Barbe, A. Eberhardt, and M. Berveiller. Internal stress effect in the shape memory behavior. *European symposium on martensitic transformation and shape memory properties, J. Physique Coll. IV*, 1(C4):95–100, 1991.

[14] K. Escher and E. Hornbogen. Aspects of two-way shape memory in NiTi-silicone composite materials. *European symposium on martensitic transformation and shape memory properties, J. Physique IV*, 1(C4):427–432, 1991.

[15] K. Escher and E. Hornbogen. Robot grippers – An application of two-way shape memory. In *Progress in Shape Memory*, edited by S. Eucken, pages 301–316. DGM-Informationsgesellschaft, 1991.

[16] R. Stalmans, J. V. Humbeeck, and L. Delaey. Training and the two way memory effect in copper based shape memory alloys. *European symposium on martensitic transformation and shape memory properties, J. Physique Coll. IV*, 1(C4):403–408, 1991.

[17] D. A. Hebda and S. R. White. Effect of training conditions and extended thermal cycling on nitinol two-way shape memory behavior. *Smart Materials and Structures*, 4 (4):298–304, 1995.

[18] B. G. Mellor, J. M. Guilemany, and J. Fernandez. Two way shape memory effect obtained by stabilised stress induced martensite in Cu-Al-Co and Cu-Al-Mn alloys. *European symposium on martensitic transformation and shape memory properties, J. Physique Coll. IV*, 1(C4):457–462, 1991.

[19] L. Sun and K-H Wu. The two-way effect in NiTi-Pd high temperature shape memory alloys. *Proceedings of SPIE Smart Structures and Materials Symposium*, 2189:298–305, 1994.

[20] M. Tokuda, S. Sugino, and T. Inaba. Two-way shape memory behavior obtained by combined loading training. *Journal of Intelligent Materials Systems and Structures*, 12(4): 289–294, April 2001.

[21] X. Meng, F. Chen, W. Cai, L. Wang, and L. Zhao. Two-way shape memory effect and its stability in a Ti-Ni-Nb wide hysteresis shape memory alloy. *Materials Transactions*, 47(3):724–727, 2006.

[22] J. V. Humbeeck, R. Stalmans, M. Chandrasekaran, and L. Delaey. On the stability of shape memory alloys. In T. W. Duerig *Engineering Aspects of Shape Memory Alloys*, pages 96–105, Butterworth-Heinemann, 1990.

[23] H. Scherngell and A. C. Kneissl. Training and stability of the intrinsic two-way shape memory effect in Ni-Ti alloys. *Scripta Materialia*, 39(2):205–212, 1998.

[24] V. Kafka. An overview of applications of mesomechanical approach to shape memory phenomena – Completed by a new application to two-way shape memory. *Journal of Intelligent Materials and Smart Structures*, 19:3–17, January 2008.

[25] T. Takagi, Y. Luo, S. Suzuki, M. Matsumoto, and J. Tani. Modeling and numerical simulation on thermomechanical behavior of SMA plates with two-way shape memory effect. *Journal of Intelligent Material Systems and Structures*, 12(11):721–728, November 2001.

[26] M. Nishida and T. Honma. All-round shape memory effect in Ni-rich TiNi alloys generated by constrained aging. *Scripta Metallurgica*, 18(11):1293–1298, 1984.

[27] M. Nishida and T. Honma. Effect of heat treatment on the all-round shape memory effect in Ti-51 at % Ni. *Scripta Metallurgica*, 18(11):1299–1302, 1984.

[28] Y. Luo, T. Okuyama, T. Takagi, T. Kamiyama, K. Nishi, and T. Yambe. Thermal control of shape memory alloy artificial anal sphincters for complete implantation. *Smart Materials and Structures*, 14(1):29–35, 2005.

[29] S. Miyazaki and K. Otsuka. Deformation and transition behavior associated with the R-phase in Ti-Ni alloys. *Metallurgical Transactions A (Physical Metallurgy and Materials Science)*, 17A(1):53–63, January 1986.

[30] V. N. Khachin, Y. I. Paskal, V. E. Gunter, A. A. Monasevich, and V. P. Sivokha. Structural transformation, physical properties and memory effects in the nickel-titanium and titanium-based alloys. *Physics of Metals and Metallography*, 46(3):49–57, 1978.

[31] S. Miyazaki, S. Kimura, and K. Otsuka. Shape-memory effect and pseudoelasticity associated with the R-phase transition in Ti-50.5 at% Ni single crystals. *Philosophical*

Magazine A: Physics of Condensed Matter, Defects and Mechanical Properties, 57(3): 467–478, March 1988.

[32] H. Naito, Y. Matsuzaki, and T. Ikeda. A unified model of thermomechanical behavior of shape memory alloys. *Proceedings of the SPIE Smart Structures and Materials Symposium*, 4333:291–300, 2001.

[33] P. Sittner, R. Stalmans, and M. Tokuda. An algorithm for prediction of the hysteresis responses of shape memory alloys. *Smart Materials and Structures*, 9(4):452–465, 2000.

[34] D. C. Lagoudas and E. L. Vandygriff. Processing and characterization of NiTi porous SMA by elevated pressure sintering. *Journal of Intelligent Material Systems and Structures*, 13(12):837–850, 2002.

[35] K. Tanaka. A thermomechanical sketch of shape memory effect: One-dimensional tensile behavior. *Res. Mechanica*, 18:251–263, 1986.

[36] C. Liang and C. A. Rogers. One-dimensional thermomechanical constitutive relations for shape memory materials. *Journal of Intelligent Material Systems and Structures*, 8(4):285–302, 1997.

[37] L. C. Brinson. One-dimensional constitutive behavior of shape memory alloys: Thermomechanical derivation with non-constant material functions and redefined martensite internal variable. *Journal of Intelligent Material Systems and Structures*, 4(2):229–242, 1993.

[38] Y. Ivshin and T. J. Pence. Thermomechanical model for a one variant shape memory material. *Journal of Intelligent Material Systems and Structures*, 5(4):455–473, 1994.

[39] H. Tobushi, S. Yamada, T. Hachisuka, A. Ikai, and K. Tanaka. Thermomechanical properties due to martensitic and R-phase transformations of TiNi shape memory alloy subjected to cyclic loadings. *Smart Materials and Structures*, 5(6):788–795, 1996.

[40] K. Otsuka. Introduction to the R-Phase Transition. In T. W. Duerig, *Engineering Aspects of Shape Memory Alloys*, Butterworth-Heinemann Ltd., 1990.

[41] L. Brinson and M. Panico. Comments to the paper "Differential and integrated form consistency in 1-D phenomenological models for shape memory alloy constitutive behavior" by V. R. Buravalla and A. Khandelwal. *International Journal of Solids and Structures*, 46:217–220, 2009.

[42] J.-H. Chung, J.-S. Heo, and J.-J. Lee. Implementation strategy for the dual transformation region in the Brinson SMA constitutive model. *Smart Materials and Structures*, 16(1):N1–N5, February 2007.

[43] A. Khandelwal and V. R. Buravalla. A correction to the Brinson's evolution kinetics for shape memory alloys. *Journal of Intelligent Material Systems and Structures*, 19(1): 43–46, 2008.

[44] M. Panico and L. C. Brinson. A three-dimensional phenomenological model for martensite reorientation in shape memory alloys. *Journal of the Mechanics and Physics of Solids*, 55:2491–2511, 2009.

[45] J. G. Boyd and D. C. Lagoudas. A thermodynamic constitutive model for the shape memory materials, part I: The monolithic shape memory alloys. *International Journal of Plasticity*, 12(6):805–842, 1998.

[46] L. C. Brinson and M. S. Huang. Simplifications and comparisons of shape memory alloy constitutive models. *Journal of Intelligent Material Systems and Structures*, 7(1): 108–114, 1996.

[47] Y. Matsuzaki and H. Naito. Macroscopic and microscopic constitutive models of shape memory alloys based on phase interaction energy function: A review. *Journal of Intelligent Material Systems and Structures*, 15(2):141–155, 2004.

[48] H. Naito, Y. Matsuzaki, and T. Ikeda. A unified constitutive model of phase transformations and rearrangements of shape memory alloy wires subjected to quasi-static load. *Smart Materials and Structures*, 13(3):535–543, 2004.

[49] E. J. Graesser and F. A. Cozzarelli. A proposed three-dimensional constitutive model for shape memory alloys. *Journal of Intelligent Materials Systems and Structures*, 5(1): 78–89, 1994.

[50] B. Malovrh and F. Gandhi. Mechanism-based phenomenological models for the pseudoelastic hysteresis behavior of shape memory alloys. *Journal of Intelligent Material Systems and Structures*, 12(1):21–30, 2001.

[51] B. Schroeder, C. Boller, J. Kramer, and B. Kroplin. Comparative assessment of models for describing the constitutive behavior of shape memory alloys. In *Proceedings, 4th ESSM and MIMR Conference on Shape Memory Alloys*, Harrogate, pages 305–312, 1998.

[52] F. Falk. *Free Boundary Problems – Theory and Applications*, vols. VI–VII. Longman, 1990.

[53] B. J. Hurlbut and M. E. Regelbrugge. Comparison and calibration of models of shape-memory alloy behavior. Seventh International Conference on Adaptive Structures: September 23–25, 1996, Rome, Italy. CRC Press, 1997.

[54] Z. Bo and D. C. Lagoudas. Comparison of different thermomechanical models for shape memory alloys. In *Adaptive Structures and Composite Materials: Analysis and Application: ASME Symposium*, volume 54, pages 9–19, 1994.

[55] Technical Characteristics of Flexinol™ Actuator Wires, Dynalloy, Inc., 2002, http://www.dynalloy.com.

[56] J. Salichs, Z. Hou, and M. Noori. Vibration suppression of structures using passive shape memory alloy energy dissipation devices. *Journal of Intelligent Material Systems and Structures*, 12(10):671–680, 2001.

[57] M. G. Faulkner, J. J. Amalraj, and A. Bhattacharyya. Experimental determination of thermal and electrical properties of Ni-Ti shape memory wires. *Smart Materials and Structures*, 9(5):632–639, October 2000.

[58] E. Abel, H. Luo, M. Pridham, and A. Slade. Issues concerning the measurement of transformation temperatures of NiTi alloys. *Smart Materials and Structures*, 13(5): 1110–1117, October 2004.

[59] J. A. Shaw and S. Kyriakides. Thermomechanical aspects of NiTi. *Journal of the Mechanics and Physics of Solids*, 43(8):1243–1281, 1995.

[60] A. Baz, S. Poh, J. Ro, and J. Gilheany. Control of the natural frequencies of Nitinol-reinforced composite beams. *Journal of Sound and Vibration*, 185(1):171–185, 1995.

[61] H. Tobushi, T. Nakahara, Y. Shimeno, and T. Hashimoto. Low-cycle fatigue of TiNi shape memory alloy and formulation of fatigue life. *Journal of Engineering Materials and Technology*, 122:186–191, 2000.

[62] S. Mikuriya, T. Nakahara, H. Tobushi, and H. Watanabe. The estimation of temperature rise in low-cycle fatigue of TiNi shape-memory alloy. *JSME International Journal, Series A*, 43(2):166–172, 2000.

[63] J. Epps and I. Chopra. Shape memory alloy actuators for in-flight tracking of helicopter rotor blades. *Smart Materials and Structures*, 10(1):104–111, 2001.

[64] R. Lammering and I. Schmidt. Experimental investigations on the damping capacity of niti components. *Smart Materials and Structures*, 10(5):853–859, 2001.

[65] Y. Liu and J. V. Humbeeck. On the damping behaviour of NiTi shape memory alloy. *Journal de Physique IV*, 7(5):519–524, November 1997.

[66] O. Pushtshaenko, E. R. Oberaigner, T. Antretter, F. D. Fischer, and K. Tanaka. Simulation of the damping of a shape memory alloy rod by using the Likhachev model. *Journal of Intelligent Material Systems and Structures*, 13(12):817–823, 2002.

[67] V. A. Likhachev. Structure-analytical theory of martensitic unelasticity. *Journal de physique IV*, 5:137–142, 1995.

[68] F. Gandhi and D. Wolons. Characterization of the pseudoelastic damping behavior of shape memory alloy wires using complex modulus. *Smart Materials and Structures*, 8(1):49–56, 1999.

[69] D. Wolons, F. Gandhi, and B. Malovrh. Experimental investigation of the pseudoelastic hysteresis damping characteristics of shape memory alloy wires. *Journal of Intelligent Material Systems and Structures*, 9(2):116–126, 1998.

[70] T. Hesse, M. Ghorashi, and D. Inman. Shape memory alloy in tension and compression and its application as clamping-force actuator in a bolted joint, Part 1: Experimentation. *Journal of Intelligent Material Systems and Structures*, 15(8):577–587, August 2004.

[71] M. Ghorashi and D. Inman. Shape memory alloy in tension and compression and its application as clamping-force actuator in a bolted joint, Part 2: Modeling. *Journal of Intelligent Material Systems and Structures*, 15(8):589–600, August 2004.

[72] F. Auricchio and E. Sacco. Temperature-dependent beam for shape memory alloys: Constitutive modeling, finite-element implementation and numerical simulations. *Computer Methods in Applied Mechanics and Engineering*, 174:171–190, 1999.

[73] K. Gall, H. Sehitoglu, Y. I. Chumlyakov, and I. V. Kireeva. Tension-compression asymmetry of the stress-strain response in aged single crystal and polycrystalline NiTi. *Acta Materialia*, 47(4):1203–1217, 1999.

[74] Y. Liu, Z. Xie, J. V. Humbeeck, and L. Delaey. Asymmetry of stress-strain curves under tension and compression for NiTi shape memory alloys. *Acta Materialia*, 46(12): 4325–4338, 1998.

[75] K. N. Melton. Ni-Ti based shape memory alloys. In T. W. Duerig, *Engineering Aspects of Shape Memory Alloys*. Butterworth-Heinemann, 1990.

[76] R. Plietsch and K. Ehrlich. Strength differential effect in pseudoelastic NiTi shape memory alloys. *Acta Materialia*, 45(6):2417–2424, June 1997.

[77] J. A. Shaw and S. Kyriakides. Initiation and propagation of localized deformation in elasto-plastic strips under uniaxial tension. *International Journal of Plasticity*, 13(10): 837–871, 1997.

[78] H. Prahlad and I. Chopra. Experimental characteristics of Ni-Ti shape memory alloys under uniaxial loading conditions. *Journal of Intelligent Material Systems and Structures*, 11(4):272–282, 2000.

[79] K. Wu, F. Yang, Z. Pu, and J. Shi. The effect of strain rate on the detwinning and superelastic behavior of NiTi shape memory alloys. *Journal of Intelligent Material Systems and Structures*, 7(2):138–144, 1996.

[80] L. G. Chang. On diffusionless transformation in Au-Cd single crystals containing 47.5 atomic percent cadmium: Characteristics of single-interface transformation. *Journal of Applied Physics*, 23(7):725–728, 1952.

[81] M. W. Burkart and T. A. Read. Diffusionless phase change in the indiium-thallium system. In *Transactions of the AIME*, volume 197, pages 1516–1524, 1953.

[82] K. Bhattacharya, R. D. James, and P. J. Swart. Relaxation in shape memory alloys, Part I: Mechanical model; Part II: Thermomechanical model and proposed experiments. *Acta Materialia*, 45(11):4547–4568, 1997.

[83] J. Hodowany, G. Ravichandran, A. J. Rosakis, and P. Rosakis. Partition of plastic work into heat and stored energy in metals. *Experimental Mechanics*, 40(2):113–123, 2000.

[84] R. Abeyaratne and S. Kim. Cyclic effects in shape-memory alloys: A one-dimensional continuum model. *International Journal of Solids and Structures*, 34(25):3273–3289, 1997.

[85] A. Bekker, L. C. Brinson, and K. Issen. Localized and diffuse thermo-induced phase transformation in 1-D shape memory alloys. *Journal of Intelligent Material Systems and Structures*, 9(5):355–365, 1998.

[86] T. J. Lim and D. L. McDowell. Mechanical behavior of a Ni-Ti shape memory alloy under axial-torsional proportional and nonproportional loading. *Journal of Engineering Materials and Technology, Transactions of the ASME*, 121(1):9–18, 1998.

[87] P. H. Lin, H. Tobushi, K. Tanaka, T. Hattori, and M. Makita. Pseudoelastic behavior of NiTi shape memory alloy subjected to strain variations. *Journal of Intelligent Material Systems and Structures*, 5(5):694–701, 1994.

[88] A. Bhattacharyya and G. J. Weng, An energy criterion for the stress-induced martensitic transformation in a ductile system. *Journal of Mechanics and Physics of Solids*, 42(11): 1699–1724, 1994.

[89] F. P. Incroperra and D. P. DeWitt. *Fundamentals of Heat and Mass Transfer*, 3rd ed. John Wiley and Sons, 1990.

[90] S. W. Churchill and H. H. S. Chu. Correlating equations for laminar and turbulent free convection from a horizontal cylinder. *International Journal of Heat and Mass Transfer*, 18:1049, 1975.

[91] A. Bhattacharyya, L. Sweeney, and M. G. Faulkner. Experimental characterization of free convection during thermal phase transformations in shape memory alloy wires. *Smart Materials and Structures*, 11(3):411–422, 2002.

[92] Technical Characteristics of SMA Wires, Shape Memory Applications, http://www.sma-inc.com/NiTiProperties.html. Inc., 2002.

[93] A. Vitiello, G. Giorleo, and R. E. Morace. Analysis of thermomechanical behaviour of Nitinol wires with high strain rates. *Smart Materials and Structures*, 14(1):215–221, February 2005.

[94] P. H. Leo, T. W. Shield, and O. P. Bruno. Transient heat-transfer effects on the pseudo-elastic behavior of shape-memory wires. *Acta Metallurgica et Materialia*, 41(8):2477–2485, August 1993.

[95] S. Miyazaki, T. Imai, Y. Igo, and K. Otsuka. Effect of cyclic deformation on the pseudoelasticity characteristics of Ti-Ni alloys. *Metallurgical and Materials Transactions A*, 17(1):115–120, January 1986.

[96] J. Perkins and R. O. Sponholz. Stress-induced martensitic transformation cycling and two-way shape memory training in Cu-Zn-Al alloys. *Metallurgical and Materials Transactions A*, 15(2):313–321, February 1984.

[97] T. J. Lim and D. L. McDowell. Path dependence of shape memory alloys during cyclic loading. *Journal of Intelligent Material Systems and Structures*, 6(6):817–830, 1995.

[98] V. Brailovski, F. Trochu, and G. Daigneault. Temporal characteristics of shape memory linear actuators and application to circuit breakers. *Materials and Design*, 17(3):151–158, 1996.

[99] J. Y. Lee, G. C. Mcintosh, A. B. Kaiser, W. Park, M. Kaack, J. Pelzl, C. K. Kim, and K. Nahm. Thermopower behavior for the shape memory alloy Ni-Ti. *Journal of Applied Physics*, 89(11):6223–6227, 2001.

[100] J. M. Gere and B. J. Goodno. *Mechanics of Materials*, 7th ed. Cengage Learning, Toronto, Canada: 2009.

[101] C. Lexcellent and J. Rejzner. Modeling of the strain rate effect, creep and relaxation of a Ni-Ti shape memory alloy under tension (compression) – torsional proportional loading in the pseudoelastic range. *Smart Materials and Structures*, 9(5):613–621, 2000.

[102] A. C. Keefe and G. P. Carman. Thermomechanical characterization of shape memory alloy torque tube actuators. *Smart Materials and Structures*, 9(5):665, 2000.

[103] T. C. Kiesling and C. A. Rogers. Impact failure modes of thin graphite-epoxy composites embedded with superelastic nitinol. *Paper # AIAA-1996-1475, Proceedings of the 37th AIAA/ASME/ASCE/AHS/ASC Structures, Structural Dynamics and Materials Conference*, Salt Lake City, UT, April 1996.

[104] D. C. Lagoudas and I. G. Tadjbakhsh. Deformations of active flexible rods with embedded line actuators. *Smart Materials and Structures*, 2(2):71, 1993.

[105] C. A. Rogers. Active vibration and structural acoustic control of shape memory alloy hybrid composites : Experimental results. *Journal of the Acoustical Society of America*, 88(6):2803–2811, 1990.

[106] H. Jia, F. Lalande, R. L. Ellis, and C. A. Rogers. Impact energy absorption of shape memory alloy hybrid composite beams. In *Proceedings of the 38th AIAA/ASME/AHS/ASC Structures, Structural Dynamics and Materials Conference*, Kissimmee, FL, pages 917–926, 1999.

[107] S. N. Jung, V. T. Nagaraj, and I. Chopra. Assessment of composite rotor blade modeling techniques. *Journal of the American Helicopter Society*, 44(3):188–205, 1999.

[108] R. Chandra. Active shape control of composite blades using shape memory actuation. *Smart Materials and Structures*, 10(5):1018–1024, 2001.

[109] M. M. Ghomshei, A. Khajepour, N. Tabandeh, and K. Behdinan. Finite element modeling of shape memory alloy composite actuators: Theory and experiment. *Journal of Intelligent Material Systems and Structures*, 12(11):761–773, 2001.

[110] J. Epps and R. Chandra. Shape memory alloy actuation for active tuning of composite beams. *Smart Materials and Structures*, 6(3):251, 1997.

[111] A. Baz, K. Imam, and J. McCoy. Active vibration control of beams using shape memory actuators. *Journal of Sound and Vibration*, 140(3):437–456, 1990.

[112] C. Liang, C. A. Rogers, and E. Malafeew. Investigation of shape memory polymers and their hybrid composites. *Journal of Intelligent Material Systems and Structures*, 8 (4):380–386, 1997.

[113] L. Meirovitch. *Elements of Vibration Analysis*, 2nd ed. McGraw-Hill, Singapore, 1986.

[114] P. K. Mallick. *Fiber Reinforced Composite Materials*, 2nd ed. Marcel Dekker, New York, 1993.

[115] K. Kimura, T. Asaoka, and K. Funami. Crack arrest effect of Ti-Ni SMA particle dispersed alloy. In *Proceedings of the International Conference on Thermomechanical Processing of Steels and Other Materials, THERMEC*, volume 2, pages 1675–1680, 1997.

[116] C. M. Friend, A. P. Atkins, and N. B. Morgan. The durability of smart composites containing shape-memory alloy actuators. *Journal de Physique IV*, 5(C8, p2.):1171–1176, 1995.

4 Beam Modeling with Induced-Strain Actuation

A one-dimensional beam with surface-bonded or embedded induced-strain actuators represents a basic and important element of an adaptive structure. Many structural systems, such as helicopter blades, airplane wings, turbo-machine blades, missiles, space structures, and many civil structures, are routinely represented as beams. For example, with induced-strain actuation, it may be possible to actively control aerodynamic shape for vibration suppression, stability augmentation, and noise reduction. Several beam theories have been developed to predict the flexural response of isotropic and anisotropic beams with surface-bonded and embedded induced-strain actuation, which range from simplified models to detailed models involving uniform, linear, and nonlinear displacement distribution through the thickness. First, three simple approaches used to model beams with induced-strain actuators are explained. These are the simple blocked-force model, the uniform-strain model, and the Euler-Bernoulli model. Although these methods are applicable for any kind of induced-strain actuator, the remainder of this chapter illustrates the method of analysis assuming piezoelectric actuation. Then, refined beam models are briefly discussed.

4.1 Material Elastic Constants

For a general anisotropic, linearly elastic material, the stress-strain relations are based on Hooke's law.

$$\begin{Bmatrix} \sigma_x \\ \sigma_y \\ \sigma_z \\ \tau_{yz} \\ \tau_{zx} \\ \tau_{xy} \end{Bmatrix} = \begin{bmatrix} Q_{11} & Q_{12} & Q_{13} & Q_{14} & Q_{15} & Q_{16} \\ Q_{21} & Q_{22} & Q_{23} & Q_{24} & Q_{25} & Q_{26} \\ Q_{31} & Q_{32} & Q_{33} & Q_{34} & Q_{35} & Q_{36} \\ Q_{41} & Q_{42} & Q_{43} & Q_{44} & Q_{45} & Q_{46} \\ Q_{51} & Q_{52} & Q_{53} & Q_{54} & Q_{55} & Q_{56} \\ Q_{61} & Q_{62} & Q_{63} & Q_{64} & Q_{65} & Q_{66} \end{bmatrix} \begin{Bmatrix} \epsilon_x \\ \epsilon_y \\ \epsilon_z \\ \gamma_{yz} \\ \gamma_{zx} \\ \gamma_{xy} \end{Bmatrix} = \boldsymbol{Q} \begin{Bmatrix} \epsilon_x \\ \epsilon_y \\ \epsilon_z \\ \gamma_{yz} \\ \gamma_{zx} \\ \gamma_{xy} \end{Bmatrix} \tag{4.1}$$

where Q_{ij} are the elements of the stiffness matrix $\boldsymbol{Q}$. σ_x, σ_y, and σ_z are direct stresses and τ_{yz}, τ_{zx}, and τ_{xy} are shear stresses. Similarly, ϵ_x, ϵ_y and ϵ_z are direct strains and γ_{yz}, γ_{zx}, and γ_{xy} are shear strains. These constitute 36 material constants that describe the material completely. From energy considerations, the material stiffness matrix must be symmetric. Therefore

$$Q_{ij} = Q_{ji} \tag{4.2}$$

This results in 21 independent material constants. The stress-strain relations are invertible and the components of strain are related to the components of stress through the compliance matrix, $\boldsymbol{s}$, as follows

$$\begin{Bmatrix} \epsilon_x \\ \epsilon_y \\ \epsilon_z \\ \gamma_{yz} \\ \gamma_{zx} \\ \gamma_{xy} \end{Bmatrix} = \begin{bmatrix} s_{11} & s_{12} & s_{13} & s_{14} & s_{15} & s_{16} \\ s_{21} & s_{22} & s_{23} & s_{24} & s_{25} & s_{26} \\ s_{31} & s_{32} & s_{33} & s_{34} & s_{35} & s_{36} \\ s_{41} & s_{42} & s_{43} & s_{44} & s_{45} & s_{46} \\ s_{51} & s_{52} & s_{53} & s_{54} & s_{55} & s_{56} \\ s_{61} & s_{62} & s_{63} & s_{64} & s_{65} & s_{66} \end{bmatrix} \begin{Bmatrix} \sigma_x \\ \sigma_y \\ \sigma_z \\ \tau_{yz} \\ \tau_{zx} \\ \tau_{xy} \end{Bmatrix} \tag{4.3}$$

$$= \boldsymbol{s} \begin{Bmatrix} \epsilon_x \\ \epsilon_y \\ \epsilon_z \\ \gamma_{yz} \\ \gamma_{zx} \\ \gamma_{xy} \end{Bmatrix} \tag{4.4}$$

Note that $\boldsymbol{s} = \boldsymbol{Q}^{-1}$. Again, the compliance matrix is symmetric

$$s_{ij} = s_{ji} \tag{4.5}$$

This results in 21 constants. There can be a further reduction of material constants (stiffness or compliance) due to a specific symmetry in the material, as listed in the following subsections.

Monoclinic Symmetry

This means that the material structure is symmetric with respect to the $x - y$ plane. In this case, the stress-strain relations reduce to

$$\begin{Bmatrix} \sigma_x \\ \sigma_y \\ \sigma_z \\ \tau_{yz} \\ \tau_{zx} \\ \tau_{xy} \end{Bmatrix} = \begin{bmatrix} Q_{11} & Q_{12} & Q_{13} & 0 & 0 & Q_{16} \\ Q_{12} & Q_{22} & Q_{23} & 0 & 0 & Q_{26} \\ Q_{13} & Q_{23} & Q_{33} & 0 & 0 & Q_{36} \\ 0 & 0 & 0 & Q_{44} & Q_{45} & 0 \\ 0 & 0 & 0 & Q_{45} & Q_{55} & 0 \\ Q_{16} & Q_{26} & Q_{36} & 0 & 0 & Q_{66} \end{bmatrix} \begin{Bmatrix} \epsilon_x \\ \epsilon_y \\ \epsilon_z \\ \gamma_{yz} \\ \gamma_{zx} \\ \gamma_{xy} \end{Bmatrix} \tag{4.6}$$

Now there are 13 material constants needed to describe the material. The compliance matrix can also be written in a similar manner. For a monoclinic material, the compliance matrix can be expressed as

$$\begin{Bmatrix} \epsilon_x \\ \epsilon_y \\ \epsilon_z \\ \gamma_{yz} \\ \gamma_{zx} \\ \gamma_{xy} \end{Bmatrix} = \begin{bmatrix} s_{11} & s_{12} & s_{13} & 0 & 0 & s_{16} \\ s_{12} & s_{22} & s_{23} & 0 & 0 & s_{26} \\ s_{13} & s_{23} & s_{33} & 0 & 0 & s_{36} \\ 0 & 0 & 0 & s_{44} & s_{45} & 0 \\ 0 & 0 & 0 & s_{45} & s_{55} & 0 \\ s_{16} & s_{26} & s_{36} & 0 & 0 & s_{66} \end{bmatrix} \begin{Bmatrix} \sigma_x \\ \sigma_y \\ \sigma_z \\ \tau_{yz} \\ \tau_{zx} \\ \tau_{xy} \end{Bmatrix} \tag{4.7}$$

This can also be expressed in terms of moduli of elasticity and Poisson's ratios

$$\begin{Bmatrix} \epsilon_x \\ \epsilon_y \\ \epsilon_z \\ \gamma_{yz} \\ \gamma_{zx} \\ \gamma_{xy} \end{Bmatrix} = \begin{bmatrix} \frac{1}{E_1} & \frac{-\nu_{21}}{E_2} & \frac{-\nu_{31}}{E_3} & 0 & 0 & \frac{\nu_{61}}{G_{12}} \\ \frac{-\nu_{12}}{E_1} & \frac{1}{E_2} & \frac{-\nu_{32}}{E_3} & 0 & 0 & \frac{\nu_{62}}{G_{12}} \\ \frac{-\nu_{13}}{E_1} & \frac{-\nu_{23}}{E_2} & \frac{1}{E_3} & 0 & 0 & \frac{\nu_{63}}{G_{12}} \\ 0 & 0 & 0 & \frac{1}{G_{23}} & \frac{\nu_{54}}{G_{31}} & 0 \\ 0 & 0 & 0 & \frac{\nu_{45}}{G_{23}} & \frac{1}{G_{31}} & 0 \\ \frac{\nu_{16}}{E_1} & \frac{\nu_{26}}{E_2} & \frac{\nu_{36}}{E_3} & 0 & 0 & \frac{1}{G_{12}} \end{bmatrix} \begin{Bmatrix} \sigma_x \\ \sigma_y \\ \sigma_z \\ \tau_{yz} \\ \tau_{zx} \\ \tau_{xy} \end{Bmatrix}$$

$$= \begin{bmatrix} \frac{1}{E_1} & \frac{-\nu_{12}}{E_1} & \frac{-\nu_{13}}{E_1} & 0 & 0 & \frac{\nu_{16}}{E_1} \\ \frac{-\nu_{12}}{E_1} & \frac{1}{E_2} & \frac{-\nu_{23}}{E_2} & 0 & 0 & \frac{\nu_{26}}{E_2} \\ \frac{-\nu_{13}}{E_1} & \frac{-\nu_{23}}{E_2} & \frac{1}{E_3} & 0 & 0 & \frac{\nu_{36}}{E_3} \\ 0 & 0 & 0 & \frac{1}{E_4} & \frac{\nu_{45}}{E_4} & 0 \\ 0 & 0 & 0 & \frac{\nu_{45}}{E_4} & \frac{1}{E_5} & 0 \\ \frac{\nu_{16}}{E_1} & \frac{\nu_{26}}{E_2} & \frac{\nu_{36}}{E_3} & 0 & 0 & \frac{1}{E_6} \end{bmatrix} \begin{Bmatrix} \sigma_x \\ \sigma_y \\ \sigma_z \\ \tau_{yz} \\ \tau_{zx} \\ \tau_{xy} \end{Bmatrix} \tag{4.8}$$

where E_4, E_5, and E_6 are shear moduli (G_{ij}). The Poisson's ratio is given by ν_{ij}, which is defined as the ratio of transverse strain in the j-direction to axial strain in the i-direction

$$E_4 = G_{23} \tag{4.9}$$

$$E_5 = G_{31} \tag{4.10}$$

$$E_6 = G_{12} \tag{4.11}$$

As an example, τ_{xy} denotes shear stress in the $x - y$ plane and the corresponding shear modulus is G_{12} in the $x - y$ plane. Also note that

$$\frac{\nu_{ij}}{E_i} = \frac{\nu_{ji}}{E_j} \tag{4.12}$$

It is important to note that ν_{ij} is not equal to ν_{ji} except for isotropic materials ($E_i = E_j$). Again, the total number of engineering constants is 13.

Orthotropic Symmetry

For orthotropic symmetry, there is a further reduction in the number of material constants. The material is assumed to be symmetric with respect to all three orthogonal planes. Now we need nine constants to describe the material. The stiffness matrix

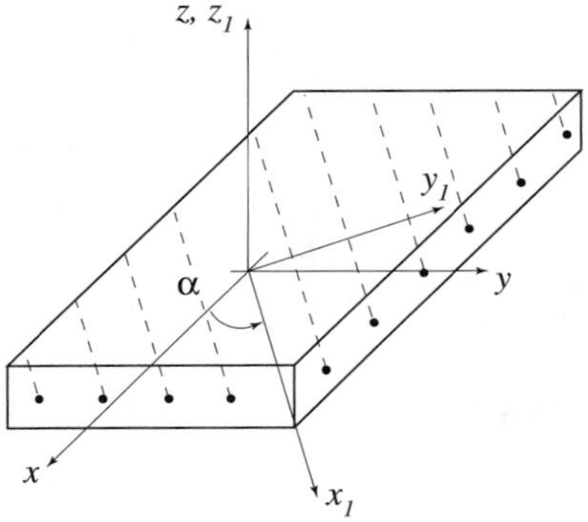

Figure 4.1. Orthotropic material with principal axes different from reference axes.

is given by

$$\begin{Bmatrix} \sigma_x \\ \sigma_y \\ \sigma_z \\ \tau_{yz} \\ \tau_{zx} \\ \tau_{xy} \end{Bmatrix} = \begin{bmatrix} Q_{11} & Q_{12} & Q_{13} & 0 & 0 & 0 \\ Q_{12} & Q_{22} & Q_{23} & 0 & 0 & 0 \\ Q_{13} & Q_{23} & Q_{33} & 0 & 0 & 0 \\ 0 & 0 & 0 & Q_{44} & 0 & 0 \\ 0 & 0 & 0 & 0 & Q_{55} & 0 \\ 0 & 0 & 0 & 0 & 0 & Q_{66} \end{bmatrix} \begin{Bmatrix} \epsilon_x \\ \epsilon_y \\ \epsilon_z \\ \gamma_{yz} \\ \gamma_{zx} \\ \gamma_{xy} \end{Bmatrix} \tag{4.13}$$

and the compliance matrix is

$$\begin{Bmatrix} \epsilon_x \\ \epsilon_y \\ \epsilon_z \\ \gamma_{yz} \\ \gamma_{zx} \\ \gamma_{xy} \end{Bmatrix} = \begin{bmatrix} s_{11} & s_{12} & s_{13} & 0 & 0 & 0 \\ s_{12} & s_{22} & s_{23} & 0 & 0 & 0 \\ s_{13} & s_{23} & s_{33} & 0 & 0 & 0 \\ 0 & 0 & 0 & s_{44} & 0 & 0 \\ 0 & 0 & 0 & 0 & s_{55} & 0 \\ 0 & 0 & 0 & 0 & 0 & s_{66} \end{bmatrix} \begin{Bmatrix} \sigma_x \\ \sigma_y \\ \sigma_z \\ \tau_{yz} \\ \tau_{zx} \\ \tau_{xy} \end{Bmatrix}$$

$$= \begin{bmatrix} \frac{1}{E_1} & \frac{-\nu_{12}}{E_1} & \frac{-\nu_{13}}{E_1} & 0 & 0 & 0 \\ \frac{-\nu_{12}}{E_1} & \frac{1}{E_2} & \frac{-\nu_{23}}{E_2} & 0 & 0 & 0 \\ \frac{-\nu_{13}}{E_1} & \frac{-\nu_{23}}{E_2} & \frac{1}{E_3} & 0 & 0 & 0 \\ 0 & 0 & 0 & \frac{1}{G_{23}} & 0 & 0 \\ 0 & 0 & 0 & 0 & \frac{1}{G_{31}} & 0 \\ 0 & 0 & 0 & 0 & 0 & \frac{1}{G_{12}} \end{bmatrix} \begin{Bmatrix} \sigma_x \\ \sigma_y \\ \sigma_z \\ \tau_{yz} \\ \tau_{zx} \\ \tau_{xy} \end{Bmatrix} \tag{4.14}$$

The nine independent constants are E_1, E_2, E_3, G_{21}, G_{23}, G_{31}, ν_{12}, ν_{23}, and ν_{31}. Note that these equations are only valid if the reference axes coincide with the principal axes of the material. Often, as in the case of composite materials, the principal axes do not coincide with the reference axes. In such a case, a coordinate transformation must be applied to the stress-strain relations. Figure 4.1 shows a composite lamina with fibers oriented at an angle α to the reference axes. As a result, the x- and y-principal axes are oriented at an angle α to the reference axes, and the z-axis is coincident in both coordinate systems. The stresses in the x_1, y_1, z_1 coordinate system (principal axes) are related to the stresses in the x, y, z coordinate system (reference axes) by the following transformation

$$\begin{Bmatrix} \sigma_x \\ \sigma_y \\ \sigma_z \\ \tau_{yz} \\ \tau_{zx} \\ \tau_{xy} \end{Bmatrix}_{(x,y,z)} = T \begin{Bmatrix} \sigma_x \\ \sigma_y \\ \sigma_z \\ \tau_{yz} \\ \tau_{zx} \\ \tau_{xy} \end{Bmatrix}_{(x_1,y_1,z_1)} \tag{4.15}$$

where the subscripts (x, y, z) and (x_1, y_1, z_1) refer to the coordinate systems, and $\boldsymbol{T}$ is the transformation matrix given by

$$\boldsymbol{T} = \begin{bmatrix} \cos^2\alpha & \sin^2\alpha & 0 & 0 & 0 & -2\sin\alpha\cos\alpha \\ \sin^2\alpha & \cos^2\alpha & 0 & 0 & 0 & 2\sin\alpha\cos\alpha \\ 0 & 0 & 1 & 0 & 0 & 0 \\ 0 & 0 & 0 & \cos\alpha & \sin\alpha & 0 \\ 0 & 0 & 0 & -\sin\alpha & \cos\alpha & 0 \\ \sin\alpha\cos\alpha & -\sin\alpha\cos\alpha & 0 & 0 & 0 & \cos^2\alpha - \sin^2\alpha \end{bmatrix} \tag{4.16}$$

Similarly, the transformation between the strains in the two coordinate systems can be derived as

$$\begin{Bmatrix} \epsilon_x \\ \epsilon_y \\ \epsilon_z \\ \gamma_{yz} \\ \gamma_{zx} \\ \gamma_{xy} \end{Bmatrix}_{(x_1,y_1,z_1)} = \boldsymbol{T}^T \begin{Bmatrix} \epsilon_x \\ \epsilon_y \\ \epsilon_z \\ \gamma_{yz} \\ \gamma_{zx} \\ \gamma_{xy} \end{Bmatrix}_{(x,y,z)} \tag{4.17}$$

From Eqs. 4.1, 4.15, and 4.17, it can be seen that

$$\begin{Bmatrix} \sigma_x \\ \sigma_y \\ \sigma_z \\ \tau_{yz} \\ \tau_{zx} \\ \tau_{xy} \end{Bmatrix}_{(x,y,z)} = \boldsymbol{TQT}^T \begin{Bmatrix} \epsilon_x \\ \epsilon_y \\ \epsilon_z \\ \gamma_{yz} \\ \gamma_{zx} \\ \gamma_{xy} \end{Bmatrix}_{(x,y,z)} = \bar{\boldsymbol{Q}} \begin{Bmatrix} \epsilon_x \\ \epsilon_y \\ \epsilon_z \\ \gamma_{yz} \\ \gamma_{zx} \\ \gamma_{xy} \end{Bmatrix}_{(x,y,z)} \tag{4.18}$$

where $\boldsymbol{Q}$ is the stiffness matrix of the lamina along its principal axes and $\bar{\boldsymbol{Q}}$ is the stiffness matrix of the lamina along the reference axes. Because the lamina is orthotropic along its principal axes, we have

$$\boldsymbol{Q} = \begin{bmatrix} Q_{11} & Q_{12} & Q_{13} & 0 & 0 & 0 \\ Q_{12} & Q_{22} & Q_{23} & 0 & 0 & 0 \\ Q_{13} & Q_{23} & Q_{33} & 0 & 0 & 0 \\ 0 & 0 & 0 & Q_{44} & 0 & 0 \\ 0 & 0 & 0 & 0 & Q_{55} & 0 \\ 0 & 0 & 0 & 0 & 0 & Q_{66} \end{bmatrix} \tag{4.19}$$

and from the transformation in Eq. 4.18, we have

$$\bar{\boldsymbol{Q}} = \boldsymbol{T}\boldsymbol{Q}\boldsymbol{T}^T = \begin{bmatrix} \bar{Q}_{11} & \bar{Q}_{12} & \bar{Q}_{13} & 0 & 0 & \bar{Q}_{16} \\ \bar{Q}_{12} & \bar{Q}_{22} & \bar{Q}_{23} & 0 & 0 & \bar{Q}_{26} \\ \bar{Q}_{13} & \bar{Q}_{23} & \bar{Q}_{33} & 0 & 0 & \bar{Q}_{36} \\ 0 & 0 & 0 & \bar{Q}_{44} & \bar{Q}_{45} & 0 \\ 0 & 0 & 0 & \bar{Q}_{45} & \bar{Q}_{55} & 0 \\ \bar{Q}_{16} & \bar{Q}_{26} & \bar{Q}_{36} & 0 & 0 & \bar{Q}_{66} \end{bmatrix} \tag{4.20}$$

where

$$\begin{aligned} \bar{Q}_{11} = {} & Q_{11}\cos^4\alpha - 4Q_{16}\cos^3\alpha\sin\alpha + 2(Q_{12} + 2Q_{66})\cos^2\alpha\sin^2\alpha \\ & - 4Q_{26}\cos\alpha\sin^3\alpha + Q_{22}\sin^4\alpha \end{aligned} \tag{4.21}$$

$$\begin{aligned} \bar{Q}_{12} = {} & Q_{12}\cos^4\alpha + 2(Q_{16} - Q_{26})\cos^3\alpha\sin\alpha \\ & + (Q_{11} + Q_{22} - 4Q_{66})\cos^2\alpha\sin^2\alpha + 2(Q_{26} - Q_{16})\cos\alpha\sin^3\alpha \\ & + Q_{12}\sin^4\alpha \end{aligned} \tag{4.22}$$

$$\bar{Q}_{13} = Q_{13}\cos^2\alpha - 2Q_{36}\cos\alpha\sin\alpha + Q_{23}\sin^2\alpha \tag{4.23}$$

$$\begin{aligned} \bar{Q}_{16} = {} & Q_{16}\cos^4\alpha + (Q_{11} - Q_{12} - 2Q_{66})\cos^3\alpha\sin\alpha \\ & + 3(Q_{26} - Q_{16})\cos^2\alpha\sin^2\alpha \\ & + (2Q_{66} + Q_{12} - Q_{22})\cos\alpha\sin^3\alpha - Q_{26}\sin^4\alpha \end{aligned} \tag{4.24}$$

$$\begin{aligned} \bar{Q}_{22} = {} & Q_{22}\cos^4\alpha + 4Q_{26}\cos^3\alpha\sin\alpha + 2(Q_{12} + 2Q_{66})\cos^2\alpha\sin^2\alpha \\ & + 4Q_{16}\cos\alpha\sin^3\alpha + Q_{11}\sin^4\alpha \end{aligned} \tag{4.25}$$

$$\bar{Q}_{23} = Q_{23}\cos^2\alpha + 2Q_{36}\cos\alpha\sin\alpha + Q_{13}\sin^2\alpha \tag{4.26}$$

$$\begin{aligned} \bar{Q}_{26} = {} & Q_{26}\cos^4\alpha + (Q_{12} - Q_{22} + 2Q_{66})\cos^3\alpha\sin\alpha \\ & + 3(Q_{16} - Q_{26})\cos^2\alpha\sin^2\alpha + (Q_{11} - Q_{12} - 2Q_{66})\cos\alpha\sin^3\alpha \\ & - Q_{16}\sin^4\alpha \end{aligned} \tag{4.27}$$

$$\bar{Q}_{33} = Q_{33} \tag{4.28}$$

$$\bar{Q}_{36} = (Q_{13} - Q_{23})\cos\alpha\sin\alpha + Q_{36}(\cos^2\alpha - \sin^2\alpha) \tag{4.29}$$

$$\bar{Q}_{44} = Q_{44}\cos^2\alpha + 2Q_{45}\cos\alpha\sin\alpha + Q_{55}\sin^2\alpha \tag{4.30}$$

$$\bar{Q}_{45} = Q_{45}(\cos^2\alpha - \sin^2\alpha) + (Q_{55} - Q_{44})\cos\alpha\sin\alpha \tag{4.31}$$

$$\bar{Q}_{55} = Q_{55}\cos^2\alpha + Q_{44}\sin^2\alpha - 2Q_{45}\cos\alpha\sin\alpha \tag{4.32}$$

$$\begin{aligned} \bar{Q}_{66} = {} & 2(Q_{16} - Q_{26})\cos^3\alpha\sin\alpha \\ & + (Q_{11} + Q_{22} - 2Q_{12} - 2Q_{66})\cos^2\alpha\sin^2\alpha \\ & + 2(Q_{26} - Q_{16})\cos\alpha\sin^3\alpha + Q_{66}(\cos^4\alpha + \sin^4\alpha) \end{aligned} \tag{4.33}$$

Transversely Isotropic Symmetry

For a material that is transversely isotropic (assume isotropic in the $y - z$ plane), there is an even further reduction in the number of material constants. Because the $y - z$ plane is isotropic, the index 2 equals the index 3, and index 5 equals index 6.

$$\begin{Bmatrix} \sigma_x \\ \sigma_y \\ \sigma_z \\ \tau_{yz} \\ \tau_{zx} \\ \tau_{xy} \end{Bmatrix} = \begin{bmatrix} Q_{11} & Q_{12} & Q_{12} & 0 & 0 & 0 \\ Q_{12} & Q_{22} & Q_{23} & 0 & 0 & 0 \\ Q_{12} & Q_{23} & Q_{22} & 0 & 0 & 0 \\ 0 & 0 & 0 & \frac{Q_{22}-Q_{23}}{2} & 0 & 0 \\ 0 & 0 & 0 & 0 & Q_{66} & 0 \\ 0 & 0 & 0 & 0 & 0 & Q_{66} \end{bmatrix} \begin{Bmatrix} \epsilon_x \\ \epsilon_y \\ \epsilon_z \\ \gamma_{yz} \\ \gamma_{zx} \\ \gamma_{xy} \end{Bmatrix} \tag{4.34}$$

Now we require only five independent material constants to describe the material. The strain is given by

$$\begin{Bmatrix} \epsilon_x \\ \epsilon_y \\ \epsilon_z \\ \gamma_{yz} \\ \gamma_{zx} \\ \gamma_{xy} \end{Bmatrix} = \begin{bmatrix} \frac{1}{E_1} & \frac{-\nu_{12}}{E_1} & \frac{-\nu_{12}}{E_1} & 0 & 0 & 0 \\ \frac{-\nu_{12}}{E_1} & \frac{1}{E_2} & \frac{-\nu_{23}}{E_2} & 0 & 0 & 0 \\ \frac{-\nu_{12}}{E_1} & \frac{-\nu_{23}}{E_2} & \frac{1}{E_2} & 0 & 0 & 0 \\ 0 & 0 & 0 & \frac{2(1+\nu_{23})}{E_2} & 0 & 0 \\ 0 & 0 & 0 & 0 & \frac{1}{G_{13}} & 0 \\ 0 & 0 & 0 & 0 & 0 & \frac{1}{G_{13}} \end{bmatrix} \begin{Bmatrix} \sigma_x \\ \sigma_y \\ \sigma_z \\ \tau_{yz} \\ \tau_{zx} \\ \tau_{xy} \end{Bmatrix} \tag{4.35}$$

Note that $E_3 = E_2$, $G_{13} = G_{12}$, and $\nu_{12} = \nu_{31}$.

Isotropic Symmetry

For a fully isotropic material, only two independent material constants are required to describe the material. The stress-strain relations for an isotropic material are given by

$$\begin{Bmatrix} \sigma_x \\ \sigma_y \\ \sigma_z \\ \tau_{yz} \\ \tau_{zx} \\ \tau_{xy} \end{Bmatrix} = \begin{bmatrix} Q_{11} & Q_{12} & Q_{12} & 0 & 0 & 0 \\ Q_{12} & Q_{11} & Q_{12} & 0 & 0 & 0 \\ Q_{12} & Q_{12} & Q_{11} & 0 & 0 & 0 \\ 0 & 0 & 0 & \frac{Q_{11}-Q_{12}}{2} & 0 & 0 \\ 0 & 0 & 0 & 0 & \frac{Q_{11}-Q_{12}}{2} & 0 \\ 0 & 0 & 0 & 0 & 0 & \frac{Q_{11}-Q_{12}}{2} \end{bmatrix} \begin{Bmatrix} \epsilon_x \\ \epsilon_y \\ \epsilon_z \\ \gamma_{yz} \\ \gamma_{zx} \\ \gamma_{xy} \end{Bmatrix} \tag{4.36}$$

Note that

$$Q_{11} = Q_{22} = Q_{33}$$

$$Q_{44} = Q_{55} = Q_{66} = \frac{Q_{11} - Q_{22}}{2}$$

$$Q_{12} = Q_{23} = Q_{31}$$

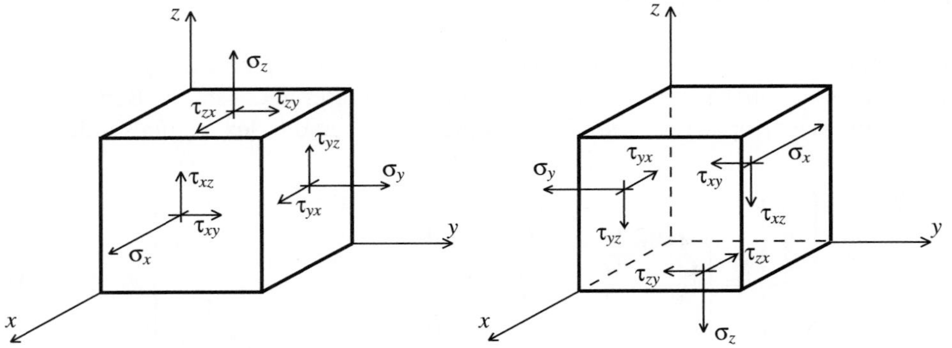

Figure 4.2. Positive stresses acting on a cubic element.

The compliance matrix for an isotropic material is

$$\begin{Bmatrix} \epsilon_x \\ \epsilon_y \\ \epsilon_z \\ \gamma_{yz} \\ \gamma_{zx} \\ \gamma_{xy} \end{Bmatrix} = \begin{bmatrix} s_{11} & s_{12} & s_{12} & 0 & 0 & 0 \\ s_{12} & s_{11} & s_{12} & 0 & 0 & 0 \\ s_{12} & s_{12} & s_{11} & 0 & 0 & 0 \\ 0 & 0 & 0 & 2(s_{11}-s_{12}) & 0 & 0 \\ 0 & 0 & 0 & 0 & 2(s_{11}-s_{12}) & 0 \\ 0 & 0 & 0 & 0 & 0 & 2(s_{11}-s_{12}) \end{bmatrix} \begin{Bmatrix} \sigma_x \\ \sigma_y \\ \sigma_z \\ \tau_{yz} \\ \tau_{zx} \\ \tau_{xy} \end{Bmatrix} \tag{4.37}$$

Rewriting in terms of elastic modulus E and Poisson's ratio ν, we have

$$\begin{Bmatrix} \epsilon_x \\ \epsilon_y \\ \epsilon_z \\ \gamma_{yz} \\ \gamma_{zx} \\ \gamma_{xy} \end{Bmatrix} = \begin{bmatrix} \frac{1}{E} & \frac{-\nu}{E} & \frac{-\nu}{E} & 0 & 0 & 0 \\ \frac{-\nu}{E} & \frac{1}{E} & \frac{-\nu}{E} & 0 & 0 & 0 \\ \frac{-\nu}{E} & \frac{-\nu}{E} & \frac{1}{E} & 0 & 0 & 0 \\ 0 & 0 & 0 & \frac{2(1+\nu)}{E} & 0 & 0 \\ 0 & 0 & 0 & 0 & \frac{2(1+\nu)}{E} & 0 \\ 0 & 0 & 0 & 0 & 0 & \frac{2(1+\nu)}{E} \end{bmatrix} \begin{Bmatrix} \sigma_x \\ \sigma_y \\ \sigma_z \\ \tau_{yz} \\ \tau_{zx} \\ \tau_{xy} \end{Bmatrix} \tag{4.38}$$

4.2 Basic Definitions: Stress, Strains, and Displacements

Most metallic structures are isotropic and linearly elastic below the plastic limit. Although the elastic properties at a unit cell level are anisotropic (different in different directions), their extremely large number (many millions) and random orientation make the material behavior isotropic and homogeneous at the macro-level. External forces can be categorized into two types: surface forces such as aerodynamic pressure, and body or volumetric forces such as inertial forces or magnetic forces. As a result of external forces, the structure experiences two types of internal stresses: normal stresses σ_x, σ_y, and σ_z and shear stresses τ_{xy}, τ_{yz}, and τ_{zx}. The various stresses acting in their positive directions are shown in Fig. 4.2 for a cubic element. For example, σ_y is the direct stress on the plane normal to the y-axis and is assumed positive for tensile stress. Conversely, τ_{xy} is the shear stress along the y-axis on the plane normal to the x-axis. Note that on a plane normal to a negative axis, the sign convention of shear stress changes by 180°. It is well established that $\tau_{xy} = \tau_{yx}$, $\tau_{yz} = \tau_{zy}$, and $\tau_{zx} = \tau_{xz}$. Therefore, six stress components (three normal and three shear stresses) are required to define stress at a point. Let us assume that u, v, and w are the elastic displacements of a point along x, y, and z-axes, respectively (right-hand axes system). This results in six strain components: three direct strains ϵ_x, ϵ_y, and ϵ_z, and three shear strains γ_{xy}, γ_{yz}, and γ_{zx}. With the assumption of small strain, these components are defined (in the nontensor form) as

$$\epsilon_x = \frac{\partial u}{\partial x} \tag{4.39}$$

$$\epsilon_y = \frac{\partial v}{\partial y} \tag{4.40}$$

$$\epsilon_z = \frac{\partial w}{\partial z} \tag{4.41}$$

and

$$\gamma_{xy} = \frac{\partial u}{\partial y} + \frac{\partial v}{\partial x} \tag{4.42}$$

$$\gamma_{yz} = \frac{\partial v}{\partial z} + \frac{\partial w}{\partial y} \tag{4.43}$$

$$\gamma_{zx} = \frac{\partial u}{\partial z} + \frac{\partial w}{\partial x} \tag{4.44}$$

The sign convention for strains is identical to that for stresses. Again, it is well established that $\gamma_{xy} = \gamma_{yx}$, $\gamma_{yz} = \gamma_{zy}$, and $\gamma_{zx} = \gamma_{xz}$. For an isotropic material, stress-strain relations are expressed using Hooke's law

$$\epsilon_x = \frac{1}{E}[\sigma_x - \nu(\sigma_y + \sigma_z)] \tag{4.45}$$

$$\epsilon_y = \frac{1}{E}[\sigma_y - \nu(\sigma_x + \sigma_z)] \tag{4.46}$$

$$\epsilon_z = \frac{1}{E}[\sigma_z - \nu(\sigma_x + \sigma_y)] \tag{4.47}$$

$$\gamma_{xy} = \frac{1}{G}\tau_{xy} \tag{4.48}$$

$$\gamma_{yz} = \frac{1}{G}\tau_{yz} \tag{4.49}$$

$$\gamma_{zx} = \frac{1}{G}\tau_{zx} \tag{4.50}$$

where E is the Young's modulus of the material (N/m^2 or lb/in^2) and ν is the Poisson's ratio. The shear modulus of elasticity G for an isotropic material can be defined in terms of the other two material constants, E and ν, as follows

$$G = \frac{E}{2(1+\nu)} \qquad (\text{N/m}^2 \text{ or lb/in}^2) \tag{4.51}$$

Table 4.1. *Notations for stress and strain components*

	Normal			Shear		
Stress in x, y, z coordinates						
Engineering	σ_x	σ_y	σ_z	τ_{yz}	τ_{zx}	τ_{xy}
Tensorial	σ_{xx}	σ_{yy}	σ_{zz}	σ_{yz}	σ_{zx}	σ_{xy}
Contracted	σ_x	σ_y	σ_z	σ_q	σ_r	σ_s
Stress in 1, 2, 3 coordinates						
Engineering	σ_1	σ_2	σ_3	τ_{23}	τ_{31}	τ_{12}
Tensorial	σ_{11}	σ_{22}	σ_{33}	σ_{23}	σ_{31}	σ_{12}
Contracted	σ_1	σ_2	σ_3	σ_4	σ_5	σ_6
Strain in x, y, z coordinates						
Engineering	ϵ_x	ϵ_y	ϵ_z	γ_{yz}	γ_{zx}	γ_{xy}
Tensorial	ϵ_{xx}	ϵ_{yy}	ϵ_{zz}	ϵ_{yz}	ϵ_{zx}	ϵ_{xy}
Contracted	ϵ_x	ϵ_y	ϵ_z	ϵ_q	ϵ_r	ϵ_s
Strain in 1, 2, 3 coordinates						
Engineering	ϵ_1	ϵ_2	ϵ_3	γ_{23}	γ_{31}	γ_{12}
Tensorial	ϵ_{11}	ϵ_{22}	ϵ_{33}	ϵ_{23}	ϵ_{31}	ϵ_{12}
Contracted	ϵ_1	ϵ_2	ϵ_3	ϵ_4	ϵ_5	ϵ_6

Often, the stress and strain components are written as

$$\begin{Bmatrix} \sigma_x \\ \sigma_y \\ \sigma_z \\ \tau_{yz} \\ \tau_{zx} \\ \tau_{xy} \end{Bmatrix} = \begin{Bmatrix} \sigma_1 \\ \sigma_2 \\ \sigma_3 \\ \sigma_4 \\ \sigma_5 \\ \sigma_6 \end{Bmatrix} \qquad \begin{Bmatrix} \epsilon_x \\ \epsilon_y \\ \epsilon_z \\ \gamma_{yz} \\ \gamma_{zx} \\ \gamma_{xy} \end{Bmatrix} = \begin{Bmatrix} \epsilon_1 \\ \epsilon_2 \\ \epsilon_3 \\ \epsilon_4 \\ \epsilon_5 \\ \epsilon_6 \end{Bmatrix} \tag{4.52}$$

The stress and strain relations in this book are expressed in what is known as the engineering notation. Sometimes in the literature, one comes across tensorial and contracted notations. These are shown in Table 4.1.

The engineering shear strains are twice the tensorial shear strains.

$$\gamma_{yz} = 2\epsilon_{yz} \tag{4.53}$$

$$\gamma_{zx} = 2\epsilon_{zx} \tag{4.54}$$

$$\gamma_{xy} = 2\epsilon_{xy} \tag{4.55}$$

The contracted strains are equal to the engineering strains. The equations for equilibrium of forces acting on a cubic element are obtained as

$$\frac{\partial \sigma_x}{\partial x} + \frac{\partial \tau_{xy}}{\partial y} + \frac{\partial \tau_{zx}}{\partial z} + f_x = 0 \tag{4.56}$$

$$\frac{\partial \tau_{xy}}{\partial x} + \frac{\partial \sigma_y}{\partial y} + \frac{\partial \tau_{yz}}{\partial z} + f_y = 0 \tag{4.57}$$

$$\frac{\partial \tau_{zx}}{\partial x} + \frac{\partial \tau_{yz}}{\partial y} + \frac{\partial \sigma_z}{\partial z} + f_z = 0 \tag{4.58}$$

where f_x, f_y, and f_z are body forces per unit volume, respectively, in x, y, and z directions.

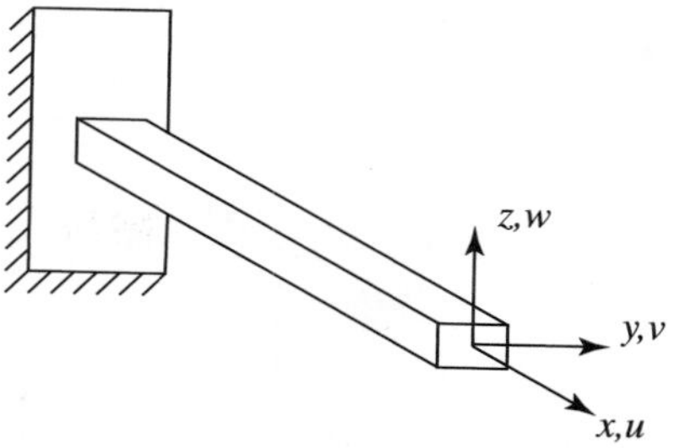

Figure 4.3. Beam with coordinate system and positive deflections.

4.2.1 Beams

A slender beam is a one-dimensional structure with cross-sectional dimensions much smaller than its length. A schematic diagram of a beam with positive deflections u, v, w along the coordinate axes x, y, z is shown in Fig. 4.3. The x-axis is aligned with the beam axis (longitudinal direction), the y-axis is along the width of the beam (lateral direction), and the z-axis is aligned along the thickness direction (transverse direction). Typically, for a structure of length l_b, thickness t_b, and width b_b, to be treated as a beam, $l_b/t_b > 10$ and $l_b/b_b > 10$. For such a beam, the Euler-Bernoulli approximation can be used to develop an engineering theory for beam bending. The neutral axis of the beam is defined as a line passing through the beam cross section that does not undergo any change in length after the beam has undergone a pure bending deformation. The theory assumes that a plane section normal to the neutral axis remains plane and normal to the neutral axis after going through the bending deformation. This means that the transverse shear deformation is negligible as compared to the bending deformation. Because its effect on beam bending is negligible, shear actions are uncoupled from bending. As a result, the effect of shear deformation on the bending response is neglected. For a small deflection approximation, the rotation of the differential element is negligible as compared to vertical deflection. The vertical deflection w due to external transverse load f_z is a function of the axial coordinate x only

$$u(x, y) = u_o(x) - z\frac{dw_o}{dx} \tag{4.59}$$

$$w(x, y) = w_o(x) \tag{4.60}$$

where $u_o(x)$ and $w_o(x)$ are longitudinal and vertical displacements at the neutral axis ($z = 0$) and dw/dx represents rotation of the cross section about the y-axis. For a vertical force distribution, the displacement $v(x)$ in the lateral direction is identically zero. The strain components become

$$\epsilon_y = \epsilon_z = \gamma_{xy} = \gamma_{yz} = \gamma_{zx} = 0 \tag{4.61}$$

$$\epsilon_x(x, z) = \epsilon_x^o(x) - z\frac{d^2w}{dx^2} \tag{4.62}$$

where $\epsilon_x^o = du_o/dx$ is the axial strain at the neutral axis and d^2w/dx^2 is the bending curvature. The sign convention for shear forces and moments acting on a beam element is shown in Fig. 4.4 [1, 2]. A positive bending moment M is defined as one that causes compression on the top fiber of the beam. A positive shear force V results in a clockwise moment acting on the differential element. The bending moment at any cross section is the product of bending or flexural stiffness EI_y and the

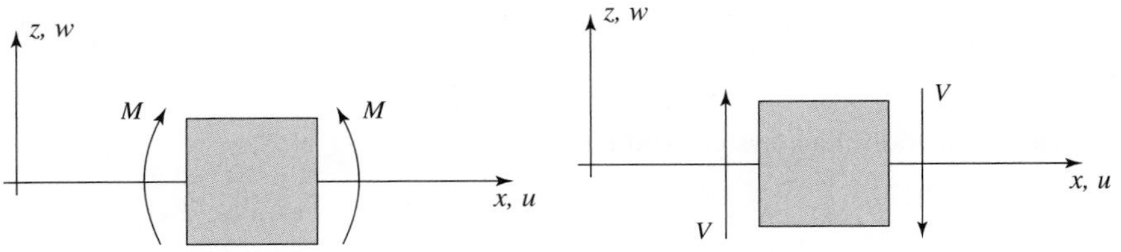

(a) Positive bending moment (b) Positive shear

Figure 4.4. Sign convention for a beam element in the $x - z$ plane.

bending curvature

$$M(x) = EI_y \frac{d^2w}{dx^2} \tag{4.63}$$

where I_y is the area product of inertia about the neutral axis. The flexural stiffness EI_y has the units of N.m^2 or lb.in^2. For a beam with a solid rectangular cross section of thickness t_b and width b_b, the area product of inertia I_y (m^4 or in^4) becomes

$$I_y = \frac{b_b t_b^3}{12} \tag{4.64}$$

The stress components are

$$\sigma_y = \sigma_z = \tau_{xy} = \tau_{yz} = \tau_{zx} = 0 \tag{4.65}$$

$$\sigma_x(x, z) = \sigma_x^o(x) - z\frac{M(x)}{I_y} \tag{4.66}$$

Thus, the axial strain becomes

$$\epsilon_x(x, z) = \epsilon_x^o(x) - z\frac{M(x)}{EI_y} \tag{4.67}$$

The governing equation for a beam undergoing bending, exposed to an external distributed transverse force $f_z(x, t)$ (N/m or lb/in) over its length, and with an axial force $T(x)$ (N or lb) is

$$\frac{\partial^2}{\partial x^2}\left(EI_y \frac{\partial^2 w}{\partial x^2}\right) - \frac{\partial}{\partial x}\left(T\frac{\partial w}{\partial x}\right) + m\frac{\partial^2 w}{\partial t^2} = f_z(x, t) \tag{4.68}$$

where $m(x)$ is the mass per unit length (kg/m or lb.s^2/in^2). This is a partial differential equation with second-order derivatives in time t and fourth-order derivatives in the spatial coordinate x. As a result, two intial conditions and four boundary conditions (two on each end) are required to solve it. For the initial conditions, the displacement $w(x, 0)$ and the velocity $\dot{w}(x, 0)$ must be prescribed. Note that for a beam initially at rest, both $w(x, 0)$ and $\dot{w}(x, 0)$ are set as zero. The boundary conditions involve both kinematic (geometric) and kinetic (force) boundary condtions. Typical boundary conditions are

Clamped Condition

$$u_o(0,t) = 0$$
$$w(0,t) = 0 \rightarrow \text{displacement} = 0$$
$$\frac{\partial w}{\partial x}(0,t) = 0 \rightarrow \text{slope} = 0$$

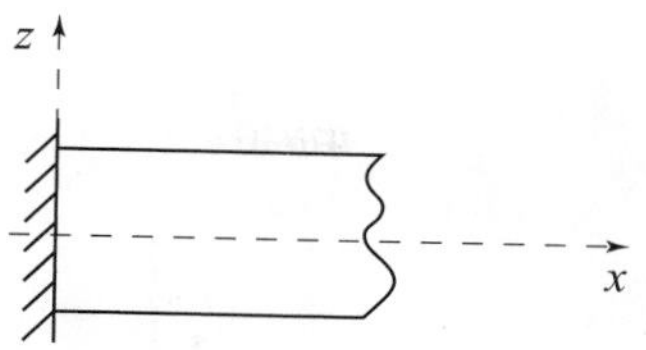

Simply Supported (Hinged or Pinned) Condition

$$u_o(0,t) = 0$$
$$w(0,t) = 0 \rightarrow \text{displacement} = 0$$
$$EI_y \frac{\partial^2 w}{\partial x^2}(0,t) = 0 \rightarrow \text{moment} = 0$$

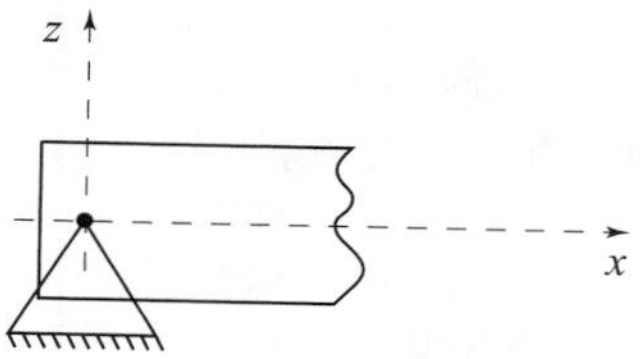

Free Condition

$$EA\frac{\partial u}{\partial x}(0,t) = 0 \rightarrow \text{axial force} = 0$$
$$EI_y \frac{\partial^2 w}{\partial x^2}(0,t) = 0 \rightarrow \text{moment} = 0$$
$$\frac{\partial}{\partial x}\left(EI_y \frac{\partial^2 w}{\partial x^2}\right)(0,t) = 0 \rightarrow \text{shear force} = 0$$

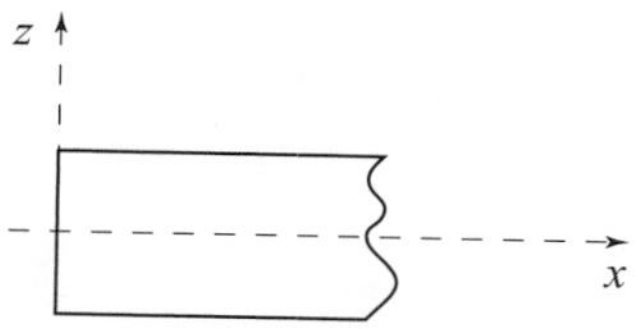

Vertical Spring-Supported Condition (Left End)

$$EA\frac{\partial u}{\partial x}(0,t) = 0 \rightarrow \text{axial force} = 0$$
$$EI_y \frac{\partial^2 w}{\partial x^2}(0,t) = 0 \rightarrow \text{moment} = 0$$
$$\frac{\partial}{\partial x}\left(EI_y \frac{\partial^2 w}{\partial x^2}\right)(0,t) = -kw \rightarrow \text{shear force} = -kw$$

where k is the spring stiffness, (N/m)

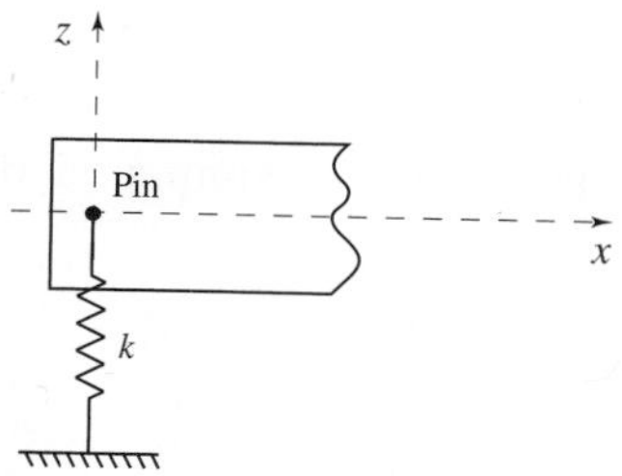

Bending Spring-Supported Condition

$$EA\frac{\partial u}{\partial x}(0,t) = 0 \rightarrow \text{axial force} = 0$$
$$EI_y \frac{\partial^2 w(0,t)}{\partial x^2} = k_\theta \frac{\partial w(0,t)}{\partial x} \rightarrow \text{moment} = k_\theta \frac{\partial w(0,t)}{\partial x}$$
$$\frac{\partial}{\partial x}\left(EI_y \frac{\partial^2 w(0,t)}{\partial x^2}\right) = 0 \rightarrow \text{shear force} = 0$$

where k_θ is the bending spring stiffness, (Nm/rad)

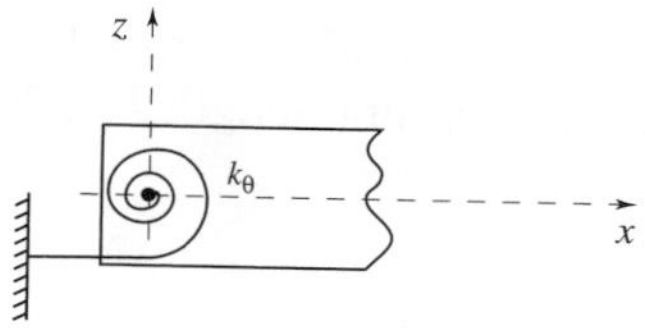

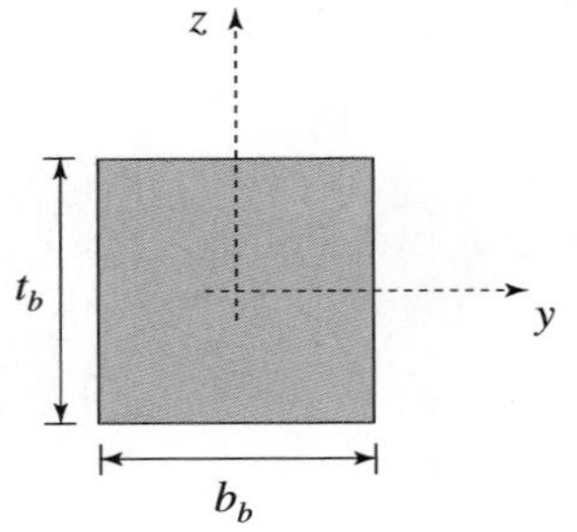

Figure 4.5. Cross section of a uniform rectangular isotropic beam.

The boundary conditions on the other end are defined according to the sign conventions shown in Fig. 4.4. In the derivation of the governing equation for beam bending, the transverse shear strain γ_{zx} is neglected, whereas the shear force V_z (as a result of τ_{zx}) is retained in the equilibrium equation. Let us examine this contradictory assumption. Assume a tip load P acting on a cantilevered beam of length l_b, width b_b, and thickness t_b. This will result in the maximum bending stress at the root surface of the beam

$$\begin{aligned}\sigma_{x_{\max}} &= \frac{M}{I_y}\frac{t_b}{2} = \frac{Pl_b}{I_y}\frac{t_b}{2} \\ &= \frac{6Pl_b}{b_b t_b^2}\end{aligned} \tag{4.69}$$

The distribution of shear stress τ_{zx} is normally assumed parabolic across the beam thickness, and the maximum value occurs at the neutral axis

$$\tau_{zx_{\max}} = \frac{3}{2}\frac{P_z}{b_b t_b} \tag{4.70}$$

The ratio of maximum bending stress to maximum shear stress is

$$\frac{\sigma_{x_{\max}}}{\tau_{zx_{\max}}} = \frac{4l_b}{t_b} \tag{4.71}$$

Because $l_b \gg t_b$, the shear stresses in a slender beam are much smaller than the bending stresses; therefore, we are quite justified in neglecting them.

4.2.2 Transverse Deflection of Uniform Isotropic Beams

Consider a beam having a uniform rectangular cross section with thickness t_b, width b_b, and length L_b (Fig. 4.5). The flexural stiffness is given by

$$EI_b = EI_y = E_b \frac{b_b t_b^3}{12} \tag{4.72}$$

where E_b is the Young's modulus.

(a) Cantilevered Beam: Tip Load

$$w_{tip} = \frac{PL_b^3}{3EI_b}$$

$$M_{root} = PL_b$$

$$w(x) = \frac{PL_b^3}{6EI_b}\left[3\left(\frac{x}{L_b}\right)^2 - \left(\frac{x}{L_b}\right)^3\right]$$

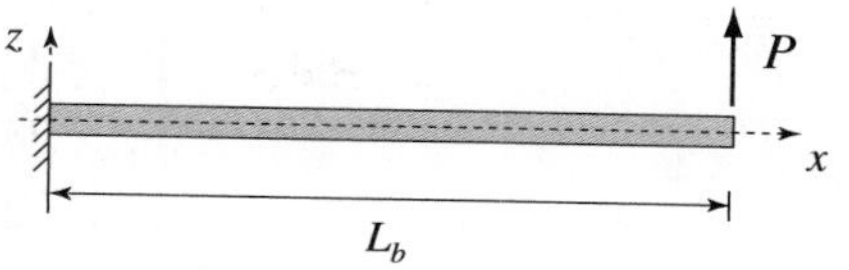

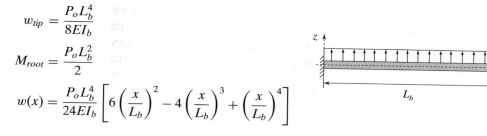

(b) Cantilevered Beam: Uniform Load

$$w_{tip} = \frac{P_o L_b^4}{8EI_b}$$

$$M_{root} = \frac{P_o L_b^2}{2}$$

$$w(x) = \frac{P_o L_b^4}{24EI_b}\left[6\left(\frac{x}{L_b}\right)^2 - 4\left(\frac{x}{L_b}\right)^3 + \left(\frac{x}{L_b}\right)^4\right]$$

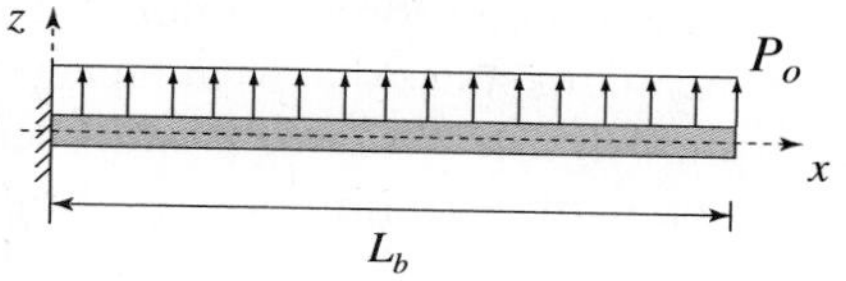

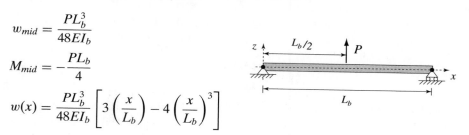

(c) Hinged or Simply Supported: Mid-Point Load

$$w_{mid} = \frac{PL_b^3}{48EI_b}$$

$$M_{mid} = -\frac{PL_b}{4}$$

$$w(x) = \frac{PL_b^3}{48EI_b}\left[3\left(\frac{x}{L_b}\right) - 4\left(\frac{x}{L_b}\right)^3\right]$$

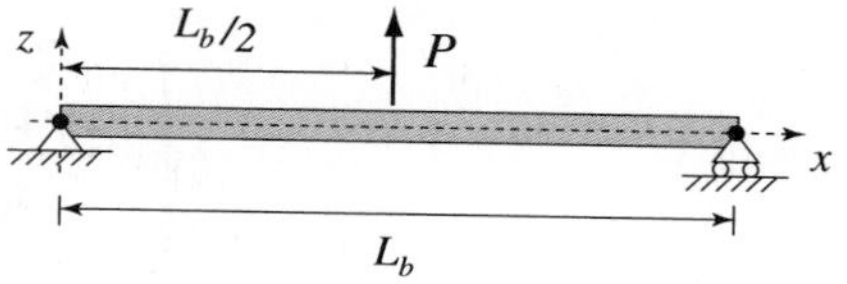

(d) Hinged or Simply Supported: Uniform Load

$$w_{mid} = \frac{5}{384}\frac{P_o L_b^4}{EI_b}$$

$$M_{mid} = -\frac{P_o L_b^2}{8}$$

$$w(x) = \frac{P_o L_b^4}{24EI_b}\left[\left(\frac{x}{L_b}\right) - 2\left(\frac{x}{L_b}\right)^3 + \left(\frac{x}{L_b}\right)^4\right]$$

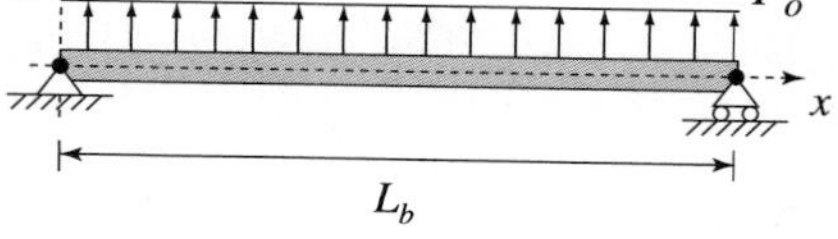

(e) Clamped Both Ends: Mid-point Load

$$w_{mid} = \frac{PL_b^3}{192EI_b}$$

$$M_{mid} = \frac{PL_b}{8}$$

$$w(x) = \frac{PL_b^3}{48EI_b}\left[3\left(\frac{x}{L_b}\right)^2 - 4\left(\frac{x}{L_b}\right)^3\right]$$

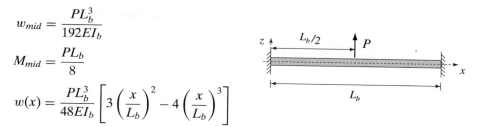

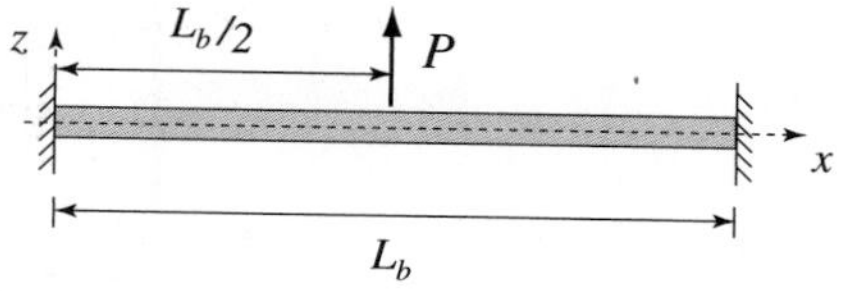

(f) Clamped Both Ends: Uniform Load

$$w_{mid} = \frac{1}{384}\frac{P_o L_b^4}{EI_b}$$

$$M_{mid} = \frac{P_o L_b^2}{12}$$

$$w(x) = \frac{P_o L_b^4}{24EI_b}\left[\left(\frac{x}{L_b}\right)^2 - \left(\frac{x}{L_b}\right)\right]$$

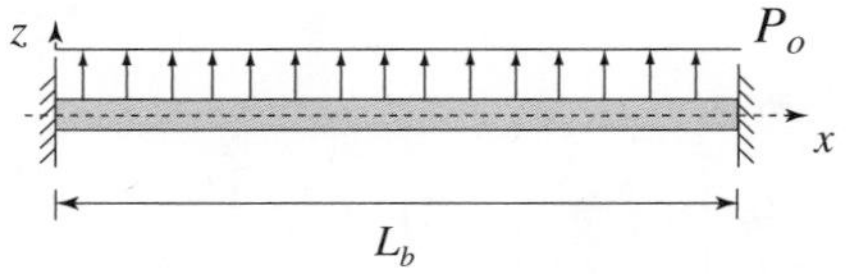

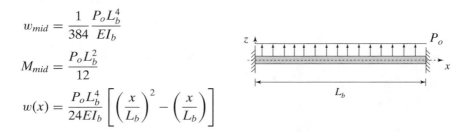

4.3 Simple Blocked-Force Beam Model (Pin Force Model)

The blocked-force method is a simple and physically intuitive approach to estimate beam response due to induced-strain actuation. It is a global and highly approximate model. The actuator is idealized as a line force and, as such, does not include any spanwise variation of stress or strain at the actuator location.

4.3.1 Single Actuator Characteristics

Consider a piezoelectric sheet element of length l_c, width b_c, and thickness t_c attached to an isotropic beam, as shown in Fig. 4.6. If an electric voltage V is applied across an isolated piezo-sheet element along the 3 direction (the direction of polarization), the maximum actuator strain, or free strain, in direction 1 will be

$$\epsilon_{\max} = d_{31}\frac{V}{t_c} = \Lambda \tag{4.73}$$

The piezoelectric sheet actuator axes 1, 2, and 3 are aligned with beam axes x, y, and z, respectively. For convenience, a positive voltage V is assumed to cause a positive strain (extension) along x-axis (direction 1) inducing positive strain in the beam. This may not be strictly true for piezoelectric sheet actuators, but this assumption has no effect on the mathematical formulation of the problem.

The maximum force, or blocked force (zero strain condition), in direction 1 is:

$$F_{max} = d_{31}E_c b_c V = F_{bl} \tag{4.74}$$

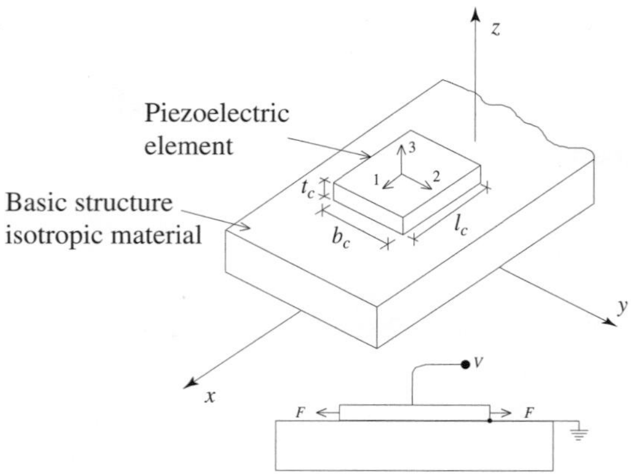

Figure 4.6. Surface-bonded piezo-sheet actuator on a beam.

Figure 4.7. Loadline for a piezoactuator at different excitation voltages.

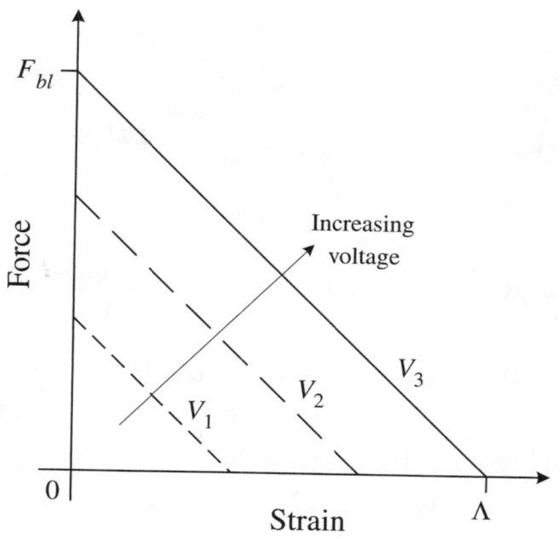

where E_c is the Young's modulus of the piezo (short-circuit condition) and d_{31} is the piezo constant. This relation can be rewritten as

$$F_{bl} = E_c b_c t_c \Lambda = E_c A_c \Lambda \tag{4.75}$$

where the extensional stiffness of the actuator is $E_c A_c$, and A_c is the cross-sectional area of the piezo-sheet. When the piezo is attached to the beam structure, an applied voltage V results in an axial surface force F in the beam. The reactive force in the piezo element will be $-F$. Assuming a sign convention in which tensile stresses and displacements are considered positive, the elastic strain in the piezo, ϵ_c, can be derived from the piezoelectric constitutive relations as

$$\begin{aligned} \epsilon_c &= \frac{\Delta l_c}{l_c} = d_{31}\frac{V}{t_c} - \frac{F}{b_c t_c E_c} \\ &= \Lambda\left(1 - \frac{F}{F_{bl}}\right) \end{aligned} \tag{4.76}$$

The free strain Λ of a piezo-sheet can be measured by attaching a strain gauge on the surface of a free piezo-sheet. The piezo-sheet is then bonded to the surface of a beam (on both the top and bottom surfaces) and the average strain of the beam is measured for an applied voltage. Knowing the properties of the beam, the axial force can be calculated. The actuation force in the beam can be calculated as

$$F = b_b t_b E_b \epsilon_c = E_b A_b \epsilon_c \tag{4.77}$$

where E_b is the Young's modulus of the material of the beam, $E_b A_b$ is extensional stiffness of the beam only, and A_b is the cross-sectional area of the beam. Rewriting the above equation

$$F = F_{bl}\left(1 - \frac{\epsilon}{\Lambda}\right) \tag{4.78}$$

This equation is referred to as the actuator loadline and is plotted in Fig. 4.7.

A few examples illustrate the application of this model.

4.3.2 Dual Actuators: Symmetric Actuation

Consider two identical piezo actuators mounted one on either surface of a beam, as shown in Fig. 4.8. The same voltage applied to the top and bottom actuators will

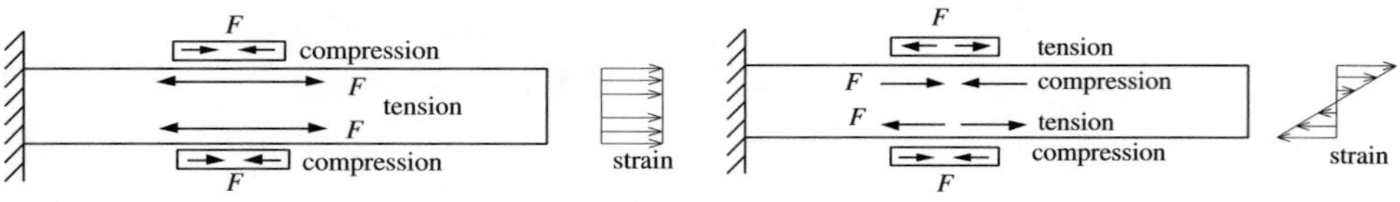

(a) Pure extension (b) Pure bending

Figure 4.8. Beam with two identical piezo actuators.

result in pure extension of the beam (Fig. 4.8(a)), whereas an equal and opposite voltage to the top and bottom actuators will result in pure bending of the beam (Fig. 4.8(b)). Equilibrating forces are produced by the actuators such that the net axial force at any spanwise cross section is zero.

I. Pure-Extension Case

To achieve pure extension in the beam, the same voltage is applied to the top and bottom actuators. Let us imagine that the piezos produce an extensional force F on either surface, resulting in an axial deflection of Δl_b in the beam. The piezo actuators, conversely, will experience equilibrating compressive forces. The axial deflection in the beam is given by

$$\begin{aligned}\Delta l_b &= \frac{2F}{A_b E_b} l_c \\ &= \frac{2F}{E_b b_b t_b} l_c\end{aligned} \tag{4.79}$$

The change in length of each piezo actuator can be found by (from Eq. 4.76)

$$\Delta l_c = \left(\Lambda - \frac{F}{E_c b_c t_c}\right) l_c = \left(d_{31}\frac{V}{t_c} - \frac{F}{E_c b_c t_c}\right) l_c \tag{4.80}$$

For displacement compatibility between the beam and the bonded actuator, $\Delta l_b = \Delta l_c$. Therefore

$$\frac{2F}{E_b b_b t_b} l_c = \left(d_{31}\frac{V}{t_c} - \frac{F}{E_c b_c t_c}\right) l_c \tag{4.81}$$

This results in

$$F = \frac{d_{31}\frac{V}{t_c}}{\frac{2}{E_b b_b t_b} + \frac{1}{E_c b_c t_c}} \tag{4.82}$$

Defining

$EA_c = 2E_c b_c t_c \rightarrow$ extensional stiffness of both actuators, in N (or lb)
$EA_b = E_b b_b t_b \rightarrow$ extensional stiffness of the beam, in N (or lb)

leads to

$$\begin{aligned}F &= \frac{d_{31}V}{2t_c}\frac{EA_b EA_c}{EA_b + EA_c} = \frac{\Lambda}{2}\frac{EA_b EA_c}{EA_b + EA_c} \\ &= F_{bl}\frac{EA_b}{EA_b + EA_c}\end{aligned} \tag{4.83}$$

Table 4.2. *Actuation force and strain limits*

EA_b/EA_c	0.100	1.0	5.000	10.000
F/F_{bl}	0.091	0.5	0.830	0.910
ϵ/Λ	0.910	0.5	0.167	0.091

Total actuation force in the beam is $2F$, generated equally by each actuator. For the case of pure extension, the strain distribution across the beam thickness is uniform (see Fig. 4.8(a)). From Eq. 4.83

$$\epsilon_b = \frac{2F}{EA_b} = \Lambda \frac{EA_c}{EA_b + EA_c} \tag{4.84}$$

Note that this is the same value of strain that would be obtained by considering the blocked force F_{bl} of the piezo actuators acting on the series combination of the actuator and beam stiffnesses, as discussed in Chapter 2. Let us consider two extreme possibilities, as follows

(a) If piezo stiffness $EA_c \gg EA_b$

$$\begin{aligned} F &\approx F_{bl}\frac{EA_b}{EA_c} \approx 0 \\ \epsilon_b &\approx \Lambda \end{aligned} \tag{4.85}$$

The actuation force on the beam approaches zero, whereas the actuation strain approaches the free strain.

(b) If piezo stiffness $EA_c \ll EA_b$

$$\begin{aligned} F &\approx F_{bl} \\ \epsilon_b &\approx \frac{EA_c}{EA_b}\Lambda \approx 0 \end{aligned} \tag{4.86}$$

The actuation strain approaches zero, whereas the actuation force approaches the blocked force.

Table 4.2 illustrates the variation of actuation force and actuation strain with stiffness ratio. It is quite clear that as the beam stiffness becomes more than ten times the actuator stiffness, the actuation strain becomes less than 10% of the free strain of the actuator. Conversely, if the actuator stiffness is more than ten times the beam stiffness, the actuation force in the beam is less than 10% of the blocked force. As discussed in Chapter 2, the maximum work done by the actuators is achieved when the structural impedance of the actuators is equal to the structural impedance of the beam.

II. Pure Bending Case

For a pure bending case, an equal but opposite voltage is applied to the top and bottom actuators. This will result in a pure bending condition with strain varying linearly across the thickness of the beam, as shown in Fig. 4.8(b). The induced bending moment M (positive M causes compression on the top surface of the beam) is caused by the equal but opposite actuation forces exerted by the actuators (positive actuation force F causes tension in the actuator). It is assumed that there is no variation of bending stress along the length of the actuator (i.e., induced moment

M is constant along the length of actuator). To achieve a positive bending moment, a negative field is applied to the top piezo, resulting in a negative actuation strain (and a positive field is applied to the bottom piezo, resulting in a positive actuation strain). Strain on the top surface of the beam where the actuator is attached is given by

$$\epsilon_b^s = -\frac{M}{I_b}\frac{t_b}{2}\frac{1}{E_b} = -\frac{F}{E_bI_b}\frac{t_b^2}{2} \tag{4.87}$$

Because this strain acts over the entire length of the piezo actuator, it results in a net decrease in length on the top surface of the beam, given by

$$\Delta l_b = -\frac{F}{E_bI_b}\frac{t_b^2}{2}l_c \tag{4.88}$$

Because the piezo actuator on the top of the beam has a negative actuation strain, the change in length of the piezo actuator is

$$\Delta l_c = \left(-\Lambda + \frac{F}{E_cA_c}\right)l_c = -\left(d_{31}\frac{V}{t_c} - \frac{F}{E_cb_ct_c}\right)l_c \tag{4.89}$$

For displacement compatibility, $\Delta l_b = \Delta l_c$.

$$\frac{F}{E_bI_b}\frac{t_b^2}{2}l_c = \left(d_{31}\frac{V}{t_c} - \frac{F}{E_cb_ct_c}\right)l_c \tag{4.90}$$

This results in a net actuator force of

$$F = \left(\frac{d_{31}\frac{V}{t_c}}{\frac{t_b^2/2}{E_bI_b} + \frac{1}{E_cb_ct_c}}\right) \tag{4.91}$$

Defining

$$EI_b = E_bI_b = EA_b\frac{t_b^2}{12} \rightarrow \text{bending stiffness of the beam, in (N.m}^2\text{ or lb.in}^2\text{)} \tag{4.92}$$

and

$$EI_c = 2(b_ct_c)\left(\frac{t_b}{2}\right)^2 E_c = EA_c\left(\frac{t_b}{2}\right)^2 \rightarrow \text{bending stiffness of the two actuators} \tag{4.93}$$

The actuation force can be calculated as

$$\begin{aligned} F &= \frac{d_{31}\frac{V}{t_c}}{\frac{t_b^2}{2}\left(\frac{1}{EI_b} + \frac{1}{EI_c}\right)} = (2\frac{d_{31}V}{t_ct_b^2})\left(\frac{EI_bEI_c}{EI_b + EI_c}\right) \\ &= F_{bl}\frac{EI_b}{EI_b + EI_c} \\ &= F_{bl}\frac{EA_b}{EA_b + 3EA_c} \end{aligned} \tag{4.94}$$

and the actuation moment can be calculated as

$$\begin{aligned} M = Ft_b = F_{bl}t_b\left(\frac{EI_b}{EI_b + EI_c}\right) \\ = M_{bl}\left(\frac{EI_b}{EI_b + EI_c}\right) \\ = M_{bl}\left(\frac{EA_b}{EA_b + 3EA_c}\right) \end{aligned} \tag{4.95}$$

where M_{bl} is the blocked moment and is equal to $F_{bl}t_b$. Note that the moment of inertia of the actuators about their own mid-plane is neglected because the piezo-sheets are assumed to be thin. For this pure bending actuation, the beam axial strain varies linearly across the beam thickness.

$$\epsilon_b = -\frac{M}{EI_b}z = -\frac{M_{bl}}{EI_b + EI_c}z \tag{4.96}$$

From Eq. 4.87, the beam top surface strain is

$$\begin{aligned} \epsilon_b^s = -\frac{M_{bl}}{EI_b + EI_c}\frac{t_b}{2} \\ = -\Lambda\frac{EI_c}{EI_b + EI_c} \\ = -\Lambda\frac{3EA_c}{EA_b + 3EA_c} \end{aligned} \tag{4.97}$$

The beam bottom surface strain is

$$\epsilon_b^{-s} = \Lambda\frac{EI_c}{EI_b + EI_c} \tag{4.98}$$

Let us consider two extreme cases, as follows:

(a) If $EI_c \gg EI_b$

$$\begin{aligned} M \approx M_{bl}\left(\frac{EI_b}{EI_c}\right) \approx 0 \\ \epsilon_b^s \approx -\Lambda \text{ (top surface)} \\ \epsilon_b^{-s} \approx \Lambda \text{ (bottom surface)} \end{aligned} \tag{4.99}$$

The actuation moment becomes zero even though the actuation surface strain equals free strain.

(b) if $EI_c \ll EI_b$

$$\begin{aligned} M \approx M_{bl} \\ \epsilon_b^s \approx 0 \end{aligned} \tag{4.100}$$

The actuation surface strain becomes zero because the actuation bending moment equals the blocked moment.

Table 4.3 shows the variation of actuation moment and surface strain (top surface) with bending stiffness ratio. As the beam stiffness increases, the actuation strain decreases. Conversely, if the beam stiffness becomes less than the actuator stiffness, the actuation moment decreases. To find the deflection, consider the cantilevered beam with dual piezo actuators shown in Fig. 4.9(a). Strain relationships on

Table 4.3. *Actuation bending and strain capability*

EI_b/EI_c	0.100	1.0	5.000	10.000
M/M_{bl}	0.091	0.5	0.830	0.910
$-\epsilon_b^s/\Lambda$	0.910	0.5	0.167	0.091

the top surface of the beam are given by

$$\Delta l_c = -\frac{M(t_b/2)}{EI_b} l_c = -\frac{Ft_b}{EI_b}\frac{t_b}{2} l_c = -\frac{6F}{EA_b} l_c \tag{4.101}$$

The bending deflection of the beam can be calculated from the bending moment, which is assumed constant within the length of the beam covered by the piezo. Bending moment $M(x) = Ft_b$.

$$\frac{\partial^2 w}{\partial x^2} = \frac{M}{EI_b} \tag{4.102}$$

Integrating and applying boundary conditions

$$\frac{\partial w}{\partial x} = \frac{M}{EI_b} x + c_1 \tag{4.103}$$

$$\text{At } x = 0, \ \frac{\partial w}{\partial x} = 0, \ \rightarrow c_1 = 0 \tag{4.104}$$

$$w = \frac{M}{EI_b}\frac{x^2}{2} + c_2 \tag{4.105}$$

$$\text{At } x = 0, \ w = 0, \ c_2 = 0 \tag{4.106}$$

$$\begin{aligned} w &= \frac{M}{EI_b}\frac{x^2}{2} = \frac{M_{bl}}{EI_b + EI_c}\frac{x^2}{2} \\ &= \left(\frac{F_{bl}t_b}{EI_b + EI_c}\right)\frac{x^2}{2} \end{aligned} \tag{4.107}$$

Beam bending curvature is nonzero where the piezo actuator is attached to its surface and is assumed uniform along the piezo length. For a cantilevered beam with a piezo actuator attached at the root, beam slope varies linearly along the length of the piezo

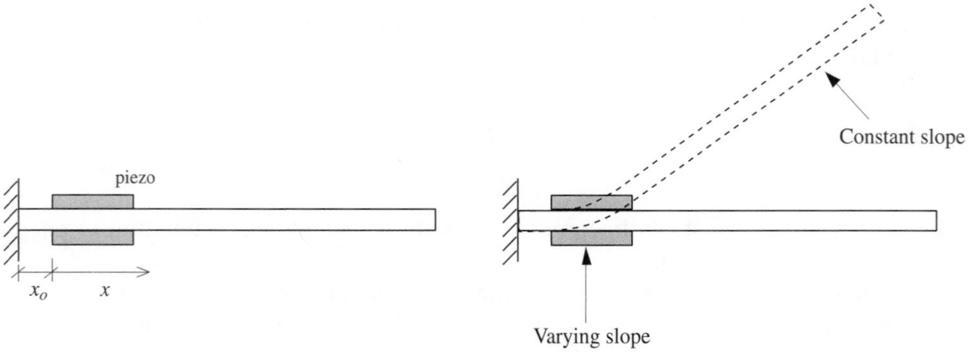

(a) Beam configuration (b) Deflected shape

Figure 4.9. Cantilevered beam in bending with two piezo actuators.

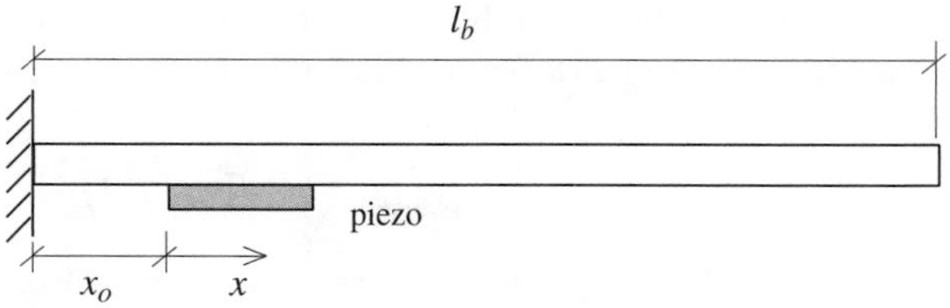

Figure 4.10. Single surface-mounted piezo actuator.

and then remains constant after the edge of the piezo. The deflected shape of the beam is shown in Fig. 4.9(b).

The beam slopes are given by

$$
\begin{aligned}
x < 0 \quad & \frac{\partial w}{\partial x} = 0 \\
0 < x < l_c \quad & \frac{\partial w}{\partial x} = \frac{M_{bl}}{EI_b + EI_c} x \quad \text{within piezo actuators} \\
x > l_c \quad & \frac{\partial w}{\partial x} = \frac{M_{bl}}{EI_b + EI_c} l_c \quad \text{outside piezo actuators}
\end{aligned}
\tag{4.108}
$$

and the beam deflection is given by

$$
\begin{aligned}
x < 0 \quad & w = 0 \\
0 < x < l_c \quad & w = \frac{M_{bl}}{EI_b + EI_c} \frac{x^2}{2} \quad \text{within piezo actuators} \\
x > l_c \quad & w = \frac{M_{bl}}{EI_b + EI_c} l_c \left(x - \frac{l_c}{2} \right) \quad \text{outside piezo actuators}
\end{aligned}
\tag{4.109}
$$

The axial stress in the beam is given by

$$
\begin{aligned}
x < 0 \quad & \sigma_b = 0 \\
0 < x < l_c \quad & \sigma_b = -\frac{M_{bl} E_b}{EI_b + EI_c} z \\
x > l_c \quad & \sigma_b = 0
\end{aligned}
\tag{4.110}
$$

and the strain on the top surface of the beam, ϵ_b^s, is given as

$$
\begin{aligned}
x < 0 \quad & \epsilon_b^s = 0 \\
0 < x < l_c \quad & \epsilon_b^s = -\frac{M_{bl}}{EI_b + EI_c} \frac{t_b}{2} \\
x > l_c \quad & \epsilon_b^s = 0
\end{aligned}
\tag{4.111}
$$

4.3.3 Single Actuator: Asymmetric Actuation

Consider a single piezo actuator surface-mounted on the bottom of a cantilevered beam, as shown in Fig. 4.10. In this case, an electric voltage applied to the piezo actuator will induce both bending and extension of the beam. A positive voltage will induce extension as well as positive bending of the beam.

Following the formulation procedure adopted for dual actuators, the induced actuation bending-extension relations are derived as follows

$$\sigma_b - \sigma_b^o = -\frac{Mz}{I_b} \tag{4.112}$$

where σ_b^o is the axial stress at the neutral axis. Because the thickness of the piezo actuator is small compared to the thickness of the beam, it can be assumed that the neutral axis is at the mid-plane of the beam. The top-surface strain ϵ_b^s and neutral axis strain ϵ_b^o are related to the bending moment by

$$\begin{aligned} \epsilon_b^s - \epsilon_b^o &= -\frac{Mt_b}{2I_b}\frac{1}{E_b} \\ \text{where } M &= F\frac{t_b}{2} \end{aligned} \tag{4.113}$$

The bottom surface strain, ϵ_b^{-s}, is

$$\epsilon_b^{-s} - \epsilon_b^o = \frac{Mt_b}{2I_b}\frac{1}{E_b} \tag{4.114}$$

The mid-plane strain is given by

$$\epsilon_b^o = \frac{F}{b_b t_b E_b} = \frac{F}{E_b A_b} \tag{4.115}$$

Note that in this configuration:

$$\begin{aligned} \text{extensional stiffness of the beam } EA_b &= E_b b_b t_b \\ \text{extensional stiffness of the actuator } EA_{c1} &= E_c b_c t_c \\ \text{bending stiffness of the beam } EI_b &= E_b b_b \frac{t_b^3}{12} \\ &= EA_b \frac{t_b^2}{12} \\ \text{bending stiffness of the actuator } EI_{c1} &= E_c b_c \frac{t_c^3}{12} + E_c b_c t_c \left(\frac{t_c}{2} + \frac{t_b}{2}\right)^2 \\ &\approx EA_{c1}\frac{t_b^2}{4} \\ &\text{(for a comparatively thin piezo)} \end{aligned}$$

On the bottom surface (at the piezo location)

$$\epsilon_b^{-s} = \frac{F(\frac{t_b}{2})^2}{b_b\frac{t_b^3}{12}}\frac{1}{E_b} + \frac{F}{EA_b} = \frac{4F}{EA_b} \tag{4.116}$$

$$\Delta l_b^{-s} = \frac{4F}{EA_b} l_c \tag{4.117}$$

$$\Delta l_c = \left(\Lambda - \frac{F}{EA_{c1}}\right) l_c \tag{4.118}$$

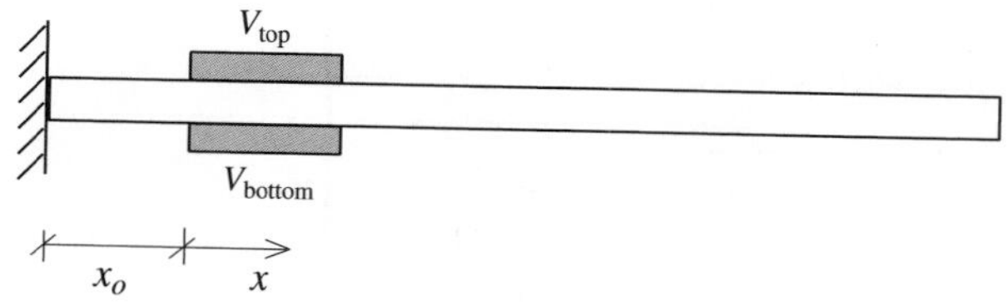

Figure 4.11. Dual surface-mounted actuators with unequal voltage.

Displacement compatibility yields $\Delta l_b^{-s} = \Delta l_c$

$$\frac{4F}{EA_b} = \frac{d_{31}V}{t_c} - \frac{F}{EA_{c1}} \tag{4.119}$$

$$F = \Lambda \frac{EA_b EA_{c1}}{4EA_{c1} + EA_b} \tag{4.120}$$

$$= F_{bl} \frac{EA_b}{4EA_{c1} + EA_b} \tag{4.121}$$

$$= F_{bl} \frac{3EI_b}{4EI_{c1} + 3EI_b} \tag{4.122}$$

where the blocked force $F_{bl} = E_c A_c \Lambda = EA_{c1}\Lambda$. This leads to

$$M = M_{bl} \frac{3EI_b}{3EI_b + 4EI_c} \tag{4.123}$$

where

$$M_{bl} = F_{bl} \frac{t_b}{2} \tag{4.124}$$

4.3.4 Unequal Electric Voltage ($V_{top} \neq V_{bottom}$)

Consider a dual-actuator beam with unequal voltage applied to top and bottom identical actuators, as shown in Fig. 4.11. We resolve this problem into two parts: equal voltages to both the piezos, causing a pure extension, and equal but opposite voltages to top and bottom piezos, causing a pure bending (Fig. 4.12). Then we use superposition to obtain the combined solution due to bending and extension strains. The resolved voltages, shown in Fig. 4.12, can be found by

$$\begin{aligned} V_1 - V_2 &= V_{top} \\ V_1 + V_2 &= V_{bottom} \end{aligned} \tag{4.125}$$

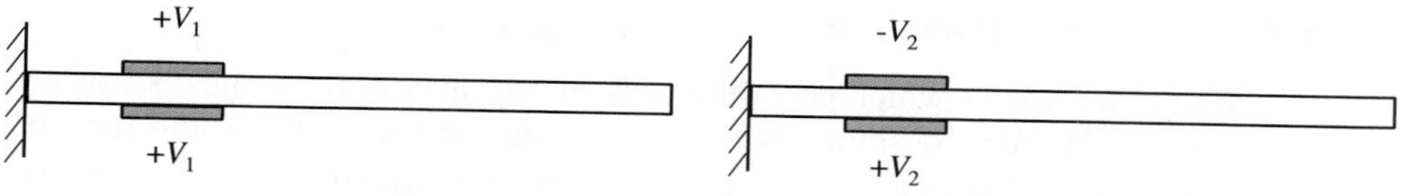

(a) Pure extension (b) Pure bending

Figure 4.12. Resolving piezo actuation into pure extension and pure bending, for unequal actuation voltages.

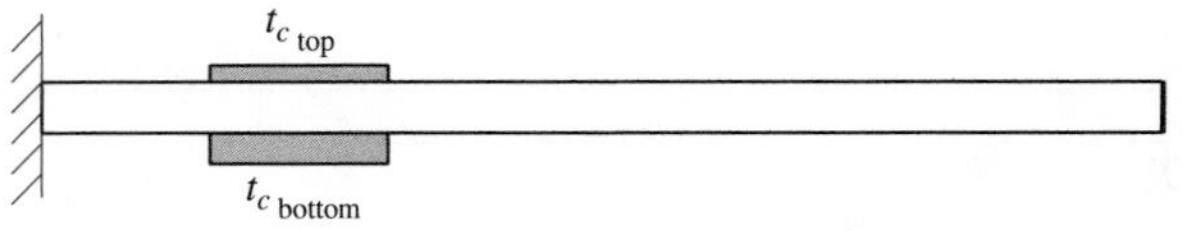

Figure 4.13. Dissimilar thickness of top and bottom piezos.

From these equations

$$V_1 = \frac{V_{\text{bottom}} + V_{\text{top}}}{2}$$
$$V_2 = \frac{V_{\text{bottom}} - V_{\text{top}}}{2} \tag{4.126}$$

Actuation force (extensional) due to V_1 is (from Eq. 4.83)

$$F^e = F_{bl_1} \frac{EA_b}{EA_b + EA_c} \tag{4.127}$$

where

$$F_{bl_1} = EA_c \frac{\Lambda_1}{2} = \frac{d_{31} V_1}{2t_c} EA_c \tag{4.128}$$

Similarly, surface bending actuation force and actuation moment due to V_2 is (from Eqs. 4.94 and 4.95)

$$F^b = F_{bl_2} \frac{EI_b}{EI_b + EI_c} = F_{bl_2} \frac{EA_b}{EA_b + 3EA_c}$$
$$M = M_{bl_2} \frac{EI_b}{EI_b + EI_c}$$
$$= \frac{2d_{31} V_2}{t_b t_c} \frac{EI_b EI_c}{EI_b + EI_c} \tag{4.129}$$

where

$$F_{bl_2} = \frac{d_{31} V_2}{2t_c} EA_c \tag{4.130}$$

The total force on the top surface, F_{top}, is

$$F_{\text{top}} = F^e - F^b \tag{4.131}$$

and the total force on the bottom surface, F_{bottom}, is

$$F_{\text{bottom}} = F^e + F^b \tag{4.132}$$

4.3.5 Dissimilar Actuators: Piezo Thickness ($t_{c_{\text{top}}} \neq t_{c_{\text{bottom}}}$)

This represents a case in which the thickness of top and bottom piezos are not identical (Fig. 4.13). For the same voltage, the actuation force due to the top and bottom piezos will be dissimilar. Proceeding in a manner similar to that for the case of actuation with unequal voltages, the actuation force in the piezos can be resolved into two parts: a force F^b causing pure bending and a force F^e causing pure extension (Fig. 4.14).

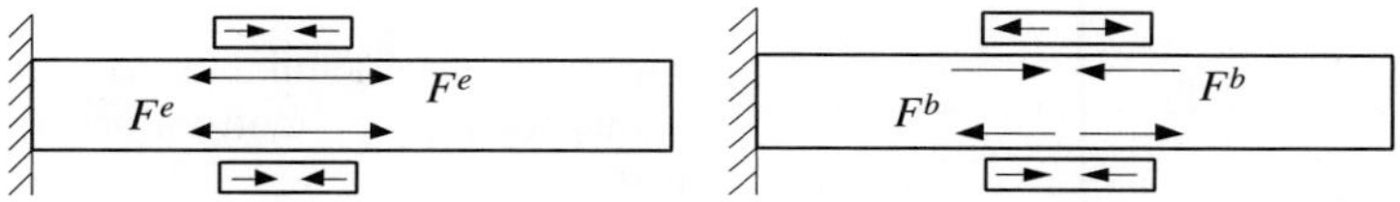

(a) Pure extension (b) Pure bending

Figure 4.14. Resolving piezo actuation into pure extension and pure bending, unequal thickness piezos.

Let us say F_{top} and F_{bottom} represent the actuation forces due to the top and bottom piezos, respectively. Then

$$\begin{aligned} F^e - F^b &= F_{\text{top}} \\ F^e + F^b &= F_{\text{bottom}} \end{aligned} \tag{4.133}$$

This results in

$$\begin{aligned} F^e &= (F_{\text{bottom}} + F_{\text{top}})/2 \\ F^b &= (F_{\text{bottom}} - F_{\text{top}})/2 \end{aligned} \tag{4.134}$$

Because the thicknesses of the two piezos are different, free strains for both piezos will be different for the same applied voltage.

$$\begin{aligned} \text{free strain in the top piezo} \quad \Lambda_{\text{top}} &= d_{31}\frac{V}{t_{c_{\text{top}}}} \\ \text{free strain in the bottom piezo} \quad \Lambda_{\text{bottom}} &= d_{31}\frac{V}{t_{c_{\text{bottom}}}} \end{aligned} \tag{4.135}$$

Displacements at the top surface

$$\begin{aligned} \Delta l_{c_{\text{top}}} &= \left(\Lambda_{\text{top}} - \frac{F_{\text{top}}}{EA_{c_{\text{top}}}}\right) l_c \\ \Delta l_{b_{\text{top}}} &= \left(\frac{2F^e}{EA_b} - \frac{6F^b}{EA_b}\right) l_c \end{aligned} \tag{4.136}$$

Comparing $\Delta l_{c_{\text{top}}} = \Delta l_{b_{\text{top}}}$ and substituting for F^e and F^b from Eq. 4.134 gives

$$F_{\text{top}}\left(\frac{4}{EA_b} + \frac{1}{EA_{c_{\text{top}}}}\right) + F_{\text{bottom}}\left(-\frac{2}{EA_b}\right) = \Lambda_{\text{top}} \tag{4.137}$$

Similarly for the bottom surface

$$F_{\text{top}}\left(-\frac{2}{EA_b}\right) + F_{\text{bottom}}\left(\frac{4}{EA_b} + \frac{1}{EA_{c_{\text{bottom}}}}\right) = \Lambda_{\text{bottom}} \tag{4.138}$$

where

$$\begin{aligned} EA_{c_{\text{top}}} &= b_c t_{c_{\text{top}}} E_c \\ EA_{c_{\text{bottom}}} &= b_c t_{c_{\text{bottom}}} E_c \end{aligned} \tag{4.139}$$

Rewriting these equations

$$\begin{bmatrix} \alpha_1 & \alpha_2 \\ \alpha_2 & \alpha_3 \end{bmatrix} \begin{Bmatrix} F_{\text{top}} \\ F_{\text{bottom}} \end{Bmatrix} = \begin{Bmatrix} \Lambda_{\text{top}} \\ \Lambda_{\text{bottom}} \end{Bmatrix} \tag{4.140}$$

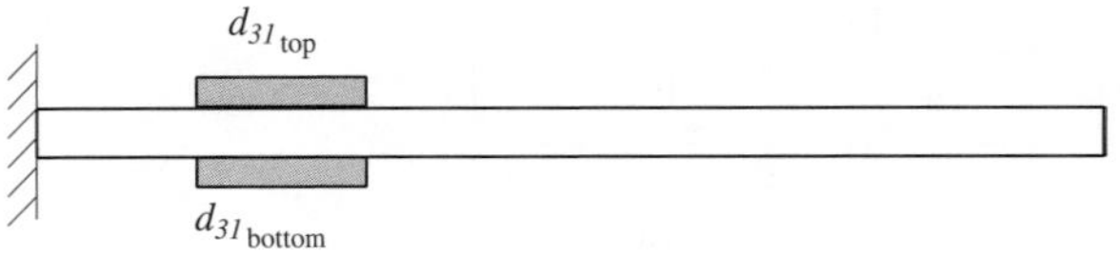

Figure 4.15. Dissimilar piezo constants for top and bottom actuators.

where

$$\alpha_1 = \frac{1}{EA_{c_{\text{top}}}} + \frac{4}{EA_b}$$
$$\alpha_2 = -\frac{2}{EA_b} \tag{4.141}$$
$$\alpha_3 = \frac{1}{EA_{c_{\text{bottom}}}} + \frac{4}{EA_b}$$

Solving the previous equation (Eq. 4.140) gives

$$\begin{Bmatrix} F_{\text{top}} \\ F_{\text{bottom}} \end{Bmatrix} = \frac{1}{\alpha_1\alpha_3 - \alpha_2^2} \begin{bmatrix} \alpha_3 & -\alpha_2 \\ -\alpha_2 & \alpha_1 \end{bmatrix} \begin{Bmatrix} \Lambda_{\text{top}} \\ \Lambda_{\text{bottom}} \end{Bmatrix} \tag{4.142}$$

The final expressions for the forces generated by the piezo actuators are

$$F_{\text{top}} = \frac{1}{\alpha_1\alpha_3 - \alpha_2^2}(\alpha_3\Lambda_{\text{top}} - \alpha_2\Lambda_{\text{bottom}})$$
$$F_{\text{bottom}} = \frac{1}{\alpha_1\alpha_3 - \alpha_2^2}(-\alpha_2\Lambda_{\text{top}} + \alpha_1\Lambda_{\text{bottom}}) \tag{4.143}$$

The actuation force is given by

$$\begin{aligned} F^e &= \frac{F_{\text{top}} + F_{\text{bottom}}}{2} \\ &= \frac{1}{2(\alpha_1\alpha_3 - \alpha_2^2)}\left[\Lambda_{\text{top}}(\alpha_3 - \alpha_2) + \Lambda_{\text{bottom}}(\alpha_1 - \alpha_2)\right] \end{aligned} \tag{4.144}$$

The actuation moment is

$$\begin{aligned} M = F^b t_b &= \frac{F_{\text{bottom}} - F_{\text{top}}}{2} t_b \\ &= \frac{t_b}{2(\alpha_1\alpha_3 - \alpha_2^2)}\left[-\Lambda_{\text{top}}(\alpha_2 + \alpha_3) + \Lambda_{\text{bottom}}(\alpha_1 + \alpha_2)\right] \end{aligned} \tag{4.145}$$

4.3.6 Dissimilar Actuators: Piezo Constants ($d_{31_{\text{top}}} \neq d_{31_{\text{bottom}}}$)

This represents a case in which the top and bottom piezos are not identical in terms of the piezoelectric constant, d_{31}; therefore, their induced strains are different (Fig. 4.15).

Free strain for the top and bottom piezos is given by

$$\Lambda_{\text{top}} = d_{31_{\text{top}}}\frac{V}{t_c}$$
$$\Lambda_{\text{bottom}} = d_{31_{\text{bottom}}}\frac{V}{t_c} \tag{4.146}$$

This case is similar to the case in which the piezo-sheets are of different thicknesses. Using displacement compatibility conditions, actuation forces for the top and bottom piezos can be derived in terms of the free strains. In this case, because the actuator stiffnesses are equal

$$\alpha_1 = \alpha_3 = \frac{4}{EA_b} + \frac{1}{EA_{c1}} \tag{4.147}$$

$$\alpha_2 = -\frac{2}{EA_b} \tag{4.148}$$

The final equations are

$$\begin{bmatrix} \alpha_1 & \alpha_2 \\ \alpha_2 & \alpha_1 \end{bmatrix} \begin{Bmatrix} F_{\text{top}} \\ F_{\text{bottom}} \end{Bmatrix} = \begin{Bmatrix} \Lambda_{\text{top}} \\ \Lambda_{\text{bottom}} \end{Bmatrix} \tag{4.149}$$

where $EA_b = E_b t_b b_b$ and $EA_{c1} = E_c b_c t_c$. This gives

$$F_{\text{top}} = \frac{1}{\alpha_1^2 - \alpha_2^2}\left(\alpha_1 \Lambda_{\text{top}} - \alpha_2 \Lambda_{\text{bottom}}\right) \tag{4.150}$$

$$F_{\text{bottom}} = \frac{1}{\alpha_1^2 - \alpha_2^2}\left(-\alpha_2 \Lambda_{\text{top}} + \alpha_1 \Lambda_{\text{bottom}}\right) \tag{4.151}$$

The actuation force is

$$\begin{aligned} F^e &= \frac{F_{\text{top}} + F_{\text{bottom}}}{2} \\ &= \frac{1}{2(\alpha_1 + \alpha_2)}(\Lambda_{\text{top}} + \Lambda_{\text{bottom}}) \end{aligned} \tag{4.152}$$

and the actuation moment is

$$\begin{aligned} M = F^b t_b &= \frac{t_b(F_{\text{bottom}} - F_{\text{top}})}{2} \\ &= \frac{t_b}{2(\alpha_1 - \alpha_2)}(\Lambda_{\text{bottom}} - \Lambda_{\text{top}}) \end{aligned} \tag{4.153}$$

4.3.7 Worked Example

Two piezo-sheet actuators (PZT-5H & PZT-5A) (length $l_c = 50.8$ mm (2 in), width $b_c = 25.4$ mm (1 in), thickness $t_c = 0.32$ mm (0.0125 in)) are surface-bonded at the top and bottom of a thin aluminum cantilevered beam (length $l_b = 609.6$ mm (24 in), width $b_b = 50.8$ mm (2 in), thickness $t_b = 0.8$ mm (1/32 in)). The configuration is shown in Fig. 4.16 ($x_o = 2$ in).

Material data are as follows:

$$E_c(\text{ PZT-5A and PZT-5H }) = E_b = 72.4 \text{ GPa } (10.5 \times 10^6 \text{ lb/in}^2)$$

$$d_{31}(\text{ PZT-5A }) = -171 \times 10^{-12} \text{ m/V}$$

$$d_{31}(\text{ PZT-5H }) = -274 \times 10^{-12} \text{ m/V}$$

Using the blocked force method,

(a) Show free strain variation in microstrain with voltage for each piezo.
(b) Show variation of piezo strain with axial force F for each piezo.

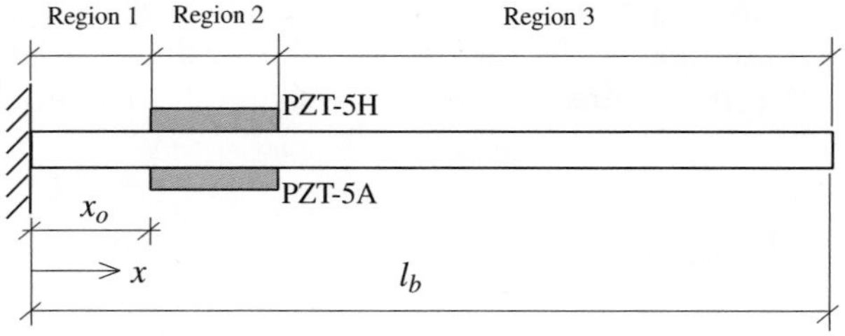

Figure 4.16. Beam with surface bonded piezosheets, split into three regions.

(c) Derive a general bending-extension relation with same field on opposite piezo actuators.
(d) Calculate actuation surface force F and bending moment M for a field of 150 volts to both top and bottom piezos.
(e) Show spanwise distribution of bending slope for this excitation.
(f) Show beam bending displacement distribution for this excitation.
(g) If PZT-5H and PZT-5A elements are replaced with PVDF elements of same size, calculate new surface actuation strain and actuation bending moment (for PVDF $d_{31} = -20 \times 10^{-12}$ m/V and $E_c = 0.2 \times 10^{10}$ N/m^2).

Solutions

(a) The free-strain variation is given by

$$\Lambda_1 = d_{31}\mathbb{E} = d_{31}\frac{V}{t_c}$$

For PZT-5H

$$\Lambda_1 = -274 \times 10^{-12}\frac{V}{0.3175 \times 10^{-3}}$$
$$= -0.863\ V\ \mu\epsilon$$

For PZT-5A

$$\Lambda_2 = -171 \times 10^{-12}\frac{V}{0.3175 \times 10^{-3}}$$
$$= -0.538\ V\ \mu\epsilon$$

(b) The actuator constitutive relation in one-dimension is

$$\epsilon_1 = d_{31}\mathbb{E} + s_{11}^{\mathbb{E}}\sigma_1$$
$$= \Lambda_1 + \frac{F_1}{E_c A_c}$$

The actuator extensional stiffness is given by

$$E_c A_c = E_c t_c b_c = 72.4 \times 10^9 \times 0.3175 \times 10^{-3} \times 1 \times 0.0254 = 0.584 \times 10^6 \text{N}$$

Assuming an applied voltage of 100 volts, for PZT-5H

$$\epsilon_1 = -86.3 + \frac{F_1}{0.584}\ \mu\epsilon$$

$$F_1 = 0.584\epsilon_1 + 50.4 \text{ N}$$

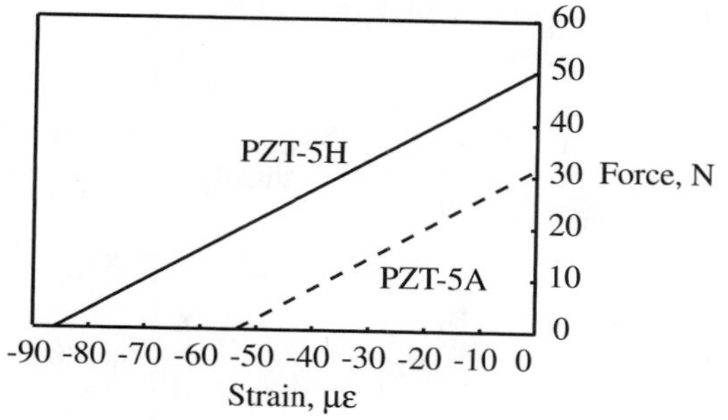

Figure 4.17. Variation of actuator strain with force, PZT-5H and PZT-5A.

The blocked force is given by

$$\epsilon_1 = 0 \rightarrow F_1 = F_{bl1} = 50.4\text{N}$$

and the free strain is

$$\Lambda_1 = -86.3 \ \mu\epsilon$$

For PZT-5A

$$\epsilon_2 = -53.8 + \frac{F_2}{0.584} \ \mu\epsilon$$

$$F_2 = 0.584\epsilon_2 + 31.42 \text{ N}$$

The blocked force is given by

$$\epsilon_2 = 0 \rightarrow F_2 = F_{bl2} = 31.42\text{N}$$

and the free strain is

$$\Lambda_2 = -53.8 \ \mu\epsilon$$

The force-displacement characteristics are shown in Fig. 4.17.

(c) The derivation of beam and actuator strains for the case of a beam with two piezo actuators with different values of d_{31} is discussed in Section 4.3.6.
The actuation force is given by

$$F^e = \frac{1}{2(\alpha_1 + \alpha_2)} \left(\Lambda_{\text{top}} + \Lambda_{\text{bottom}}\right)$$

The actuation moment is

$$M = F^b t_b = \frac{F_{\text{bottom}} - F_{\text{top}}}{2} t_b$$

$$= \frac{t_b}{2(\alpha_1 - \alpha_2)} \left(\Lambda_{\text{bottom}} - \Lambda_{\text{top}}\right)$$

where

$$\alpha_1 = \frac{1}{EA_{c1}} + \frac{4}{EA_b}$$

$$\alpha_2 = -\frac{2}{EA_b}$$

The beam extensional stiffness is given by

$$EA_b = E_b t_b b_b = 72.4 \times 10^9 \times 0.79375 \times 10^{-3} \times 2 \times 0.0254 = 2.92 \times 10^6\text{N}$$

(d) For a voltage of $V = 150$ V

$$\Lambda_{top} = -0.863 \times 150 = -129.45\ \mu\epsilon$$

$$\Lambda_{bottom} = -0.538 \times 150 = -80.7\ \mu\epsilon$$

$$\alpha_1 = \frac{4}{2.92 \times 10^6} + \frac{1}{0.584 \times 10^6} = 3.083 \times 10^{-6}\ 1/\text{N}$$

$$\alpha_2 = -\frac{2}{2.92 \times 10^6} = -0.685 \times 10^{-6}\ 1/\text{N}$$

This gives

$$F^e = -\frac{80.7 + 129.45}{2(3.083 - 0.685)} = -43.82\ \text{N}$$

$$M = 0.79375 \times 10^{-3}\frac{-80.7 + 129.45}{2(3.083 + 0.685)} = 5.13 \times 10^{-3}\ \text{N-m}$$

(e) Assume that $x_1 = 0$ corresponds to the clamped end, where piezo starts

$$\frac{\partial^2 w}{\partial x_1^2} = \frac{M}{EI_b}$$

Integrating

$$\frac{\partial w}{\partial x_1} = \frac{Mx_1}{EI_b} + C$$

$$\text{If } x_1 = 0, \quad \frac{\partial w}{\partial x} = 0 \rightarrow C = 0$$

Integrating again

$$w = \frac{Mx_1^2}{2EI_b} + D$$

$$\text{If } x_1 = 0, \quad w = 0 \rightarrow D = 0$$

Note $x = x_1 + x_0$

The bending stiffness is given by

$$EI_b = E_b b_b \frac{t_b^3}{12} = 72.4 \times 10^9 \times 50.8 \times 10^{-3} \times \frac{(0.79375 \times 10^{-3})^3}{12} = 0.1553\ \text{Nm}^2$$

In Region 1 ($0 < x < x_o$), there is no actuation force or moment.

$$\frac{\partial w}{\partial x} = 0 \quad w = 0$$

In Region 2 ($x_o < x < x_o + l_c$)

$$\frac{\partial w}{\partial x} = \frac{M(x - x_o)}{EI_b} = 0.0335(x - x_o)\ \text{rad}$$

where x and x_o are measured in meters.

$$w = \frac{M(x - x_o)^2}{2EI_b} = 0.0167(x - x_o)^2\ \text{m}$$

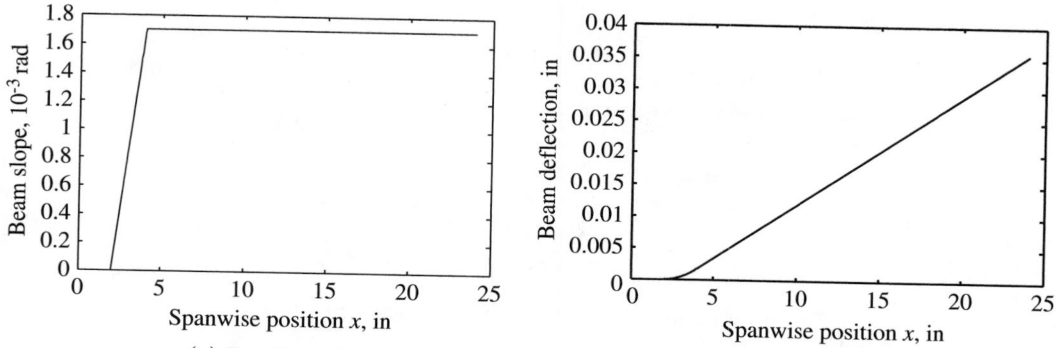

(a) Bending slope (b) Bending displacement

Figure 4.18. Beam deflection with actuators having dissimilar piezoconstants.

In Region 3 ($x_o + l_c < x < l_b$)

$$\frac{\partial w}{\partial x} = \frac{Ml_c}{EI_b} = 1.7 \times 10^{-3} \text{ rad}$$

$$w = \frac{Ml_c^2}{2EI_b} + \frac{Ml_c}{EI_b}(x - l_c - x_o) = 4.31 \times 10^{-5} + 1.7 \times 10^{-3}(x - 0.1016) \text{ m}$$

The tip slope is 1.7×10^{-3} rad and the tip displacement is 0.9067 mm. The slope and bending displacement are plotted in Fig. 4.18

(f) For PVDF, $d_{31} = -20 \times 10^{-12}$ m/V and $E_c = 2$ GPa.
The free-strain variation is given by

$$\begin{aligned}\Lambda &= -20 \times 10^{-12} \frac{V}{0.3175 \times 10^{-3}} \\ &= -0.063\ V\ \mu\epsilon\end{aligned}$$

At 150 V, $\Lambda = -9.5\ \mu\epsilon$.
Because the actuators are now identical, only pure extension will be induced in the beam. The beam-surface strain is given by:

$$\begin{aligned}\epsilon_b &= \frac{\Lambda EA_c}{EA_b + EA_c} \\ &= \Lambda \frac{1}{1 + \frac{E_b b_b t_b}{2E_c b_c t_c}} = \frac{-9.5}{1 + \frac{72.4 \times 0.8}{2 \times 0.32}} \\ &= -0.104\ \mu\epsilon\end{aligned}$$

4.4 Uniform-Strain Model

The simple blocked-force model assumes a perfect transfer of strain between the piezo actuator and the surface of the structure to which it is bonded. In practice, however, this is an idealization because the bond layer between the piezo actuator and the structure has a finite stiffness. Some of the strain generated by the piezo is dissipated in the deformation of the bond layer itself. This phenomenon is also known as "shear lag." An idealized uniform-strain-beam model is used to evaluate the effectiveness of the bond layer in transferring the strain induced by the piezo

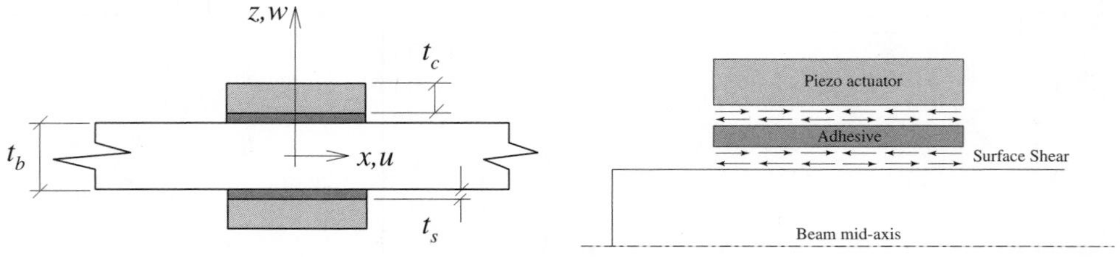

(a) Definition of thickness and coordinate system

(b) Transmission of shear across adhesive layer

Figure 4.19. Beam with symmetric surface-mounted actuators.

actuators to the surface of the beam. The bond layer is assumed to undergo pure shear deformation.

4.4.1 Dual Actuators: Symmetric Actuation

Two identical piezoelectric sheet actuators are bonded to an isotropic beam, one to the top surface and the other to the bottom surface. Between the actuator and the beam-surface, there is a finite-thickness elastic bond layer. Each actuator is assumed to induce a uniform axial strain across its own thickness. Due to bending actuation, there will be a linear distribution of axial strain in the host structure. Conversely, for pure extensional actuation, there will be uniform axial strain in the host structure.

The actuator is constrained by the adhesive, so a shear stress is produced in the adhesive layer. For this analysis, the normal stress in the bond layer is neglected and the beam is subjected to a purely surface shear. The objective is to predict induced strain and induced force due to piezo actuation, including the effects of losses in the bond layer. A schematic of the coordinate system used in this analysis and the shear stress transmitted by the bond layer is shown in Fig. 4.19.

I. Pure Bending Case

Let us first consider pure bending of the beam. This is accomplished by applying an equal but opposite field to the top and bottom actuators. The stress varies linearly across the beam thickness. At the mid-point (neutral axis) of the beam, the axial stress is zero. It is assumed that the axial stress of the actuator σ_c does not vary across its thickness. This also implies a uniform strain across the thickness of the actuator. The forces and moments acting on a differential element of the beam, actuators, and bond layer are shown in Fig. 4.20.

Let t_s, t_c, and t_b denote the adhesive thickness, actuator thickness, and beam thickness, respectively. The strain-displacement relation for the actuator is given by

$$\epsilon_c = \frac{\partial u_c}{\partial x} \tag{4.154}$$

where u_c is the axial deflection of the actuator. On the top surface of the beam

$$\epsilon_b^s = \frac{\partial u_b^s}{\partial x} \tag{4.155}$$

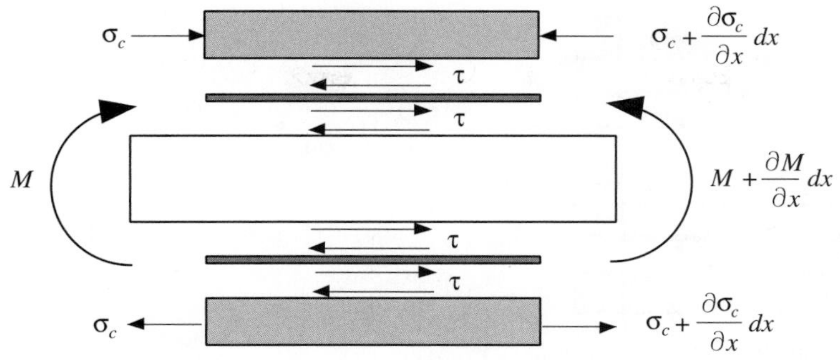

Figure 4.20. Differential element for bending case.

where u_b^s is the axial deflection of the top surface of the beam. Similarly, on the bottom surface of the beam

$$\epsilon_b^{-s} = \frac{\partial u_b^{-s}}{\partial x} \tag{4.156}$$

where u_b^{-s} is the axial deflection of the bottom surface of the beam. These strain-displacement equations relate the normal strain in the x-direction to the u-deflections of the actuator and beam, respectively. Shear strain in the adhesive layer on the bottom side is found by subtracting the deformation of the beam-surface from the deformation of the actuator and dividing by the bond thickness

$$\gamma_{zx} = \frac{1}{t_s}[u_c - u_b^{-s}] = \gamma_s \tag{4.157}$$

Equilibrium of forces on the bottom actuator leads to

$$\left(\sigma_c + \frac{\partial \sigma_c}{\partial x} dx\right) b_c t_c - \sigma_c b_c t_c - \tau b_c \, dx = 0 \tag{4.158}$$

where b_c is the actuator width and τ is the shear stress. Simplification of this expression results in

$$\frac{\partial \sigma_c}{\partial x} - \frac{\tau}{t_c} = 0 \tag{4.159}$$

The same equation is also valid for the top actuator. Equilibrium of bending moments can be written as

$$M + \frac{\partial M}{\partial x} dx - M + \tau t_b b_c \, dx = 0 \tag{4.160}$$

This results in

$$\frac{\partial M}{\partial x} + \tau t_b b_c = 0 \tag{4.161}$$

The stress distribution in the beam can be expressed as

$$\sigma_b(z) = -\frac{M}{I_b} z \tag{4.162}$$

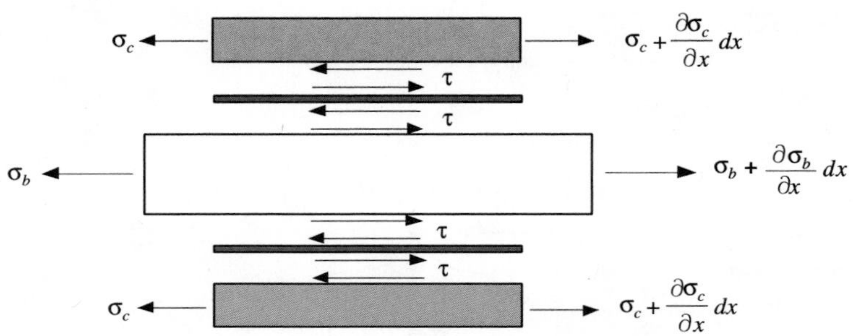

Figure 4.21. Differential element for extensional case.

where I_b is the area product of inertia of the beam cross section about its mid-axis. The stress on the top surface of the beam, σ_b^s, is given by

$$\sigma_b^s = -\frac{M(t_b/2)}{I_b} \tag{4.163}$$

$$= -\frac{M(t_b/2)}{b_b t_b^3/12} = -\frac{6M}{b_b t_b^2} \tag{4.164}$$

where b_b is the beam width. This means that the moment is

$$M = -\frac{b_b t_b^2}{6}\sigma_b^s \tag{4.165}$$

Substituting the relation for M into Eq. 4.161

$$\frac{\partial \sigma_b^s}{\partial x} - \frac{6b_c}{b_b t_b}\tau = 0 \tag{4.166}$$

This is a force equilibrium relation on the top surface of the beam for the pure bending case. Similarly, the stress at the bottom surface of the beam, σ_b^{-s}, can be expressed in terms of the bending moment. This leads to the force equilibrium relation on the bottom surface of the beam for the pure bending case

$$\frac{\partial \sigma_b^{-s}}{\partial x} + \frac{6b_c}{b_b t_b}\tau = 0 \tag{4.167}$$

II. Pure Extension Case

Next, let us consider pure extension of the beam (Fig. 4.21). This is obtained by applying the same voltage to the top and bottom actuators. For this case, σ_b is uniform across the beam thickness. The force equilibrium relation on the top surface of the beam for pure extension becomes

$$\frac{\partial \sigma_b^s}{\partial x} + \frac{2b_c}{b_b t_b}\tau = 0 \tag{4.168}$$

For the bottom surface of the beam, because $\sigma_b^s = \sigma_b^{-s}$

$$\frac{\partial \sigma_b^{-s}}{\partial x} + \frac{2b_c}{b_b t_b}\tau = 0 \tag{4.169}$$

Note that the difference between Eqs. 4.167 and 4.169 is in a term associated with shear force. For the extension case, the factor is 2, and for the pure bending case, the factor is 6. Changing this factor to the variable α gives us the general relation

$$\frac{\partial \sigma_b^{-s}}{\partial x} + \alpha \frac{\tau b_c}{t_b b_b} = 0 \tag{4.170}$$

where α is 2 for the pure extension case and α is 6 for the pure bending case. The force equilibrium in the actuator yields the same equation as Eq. 4.159. Stresses in the actuator, beam, and the bond layer are given by

$$\sigma_c = E_c(\epsilon_c - \Lambda) \tag{4.171}$$

$$\sigma_b^{-s} = E_b \epsilon_b^{-s} \tag{4.172}$$

$$\tau = G_s \gamma_s \tag{4.173}$$

where Λ is the induced strain, or free strain, in the actuator. All together, we have eight equations to define the complete system with eight unknowns. The unknowns are

(a) three stresses: $\sigma_c,\ \sigma_b^{-s},\ \tau$
(b) three strains: $\epsilon_c,\ \epsilon_b^{-s},\ \gamma_s$
(c) two displacements: $u_c,\ u_b^{-s}$

Now, we must combine these equations into something more manageable. Let us consider the case in which the actuator width is the same as the beam width, $b_c = b_b$. The two displacement and three strain-displacement equations can be combined into a compatibility equation. Differentiating Eq. 4.157 with respect to x

$$\frac{\partial \gamma_s}{\partial x} = \frac{1}{t_s}\left[\frac{\partial u_c}{\partial x} - \frac{\partial u_b^{-s}}{\partial x}\right] = \frac{1}{t_s}\left(\epsilon_c - \epsilon_b^{-s}\right) \tag{4.174}$$

Now, we want to combine this compatibility equation (Eq. 4.174) with the equilibrium equations (Eqs. 4.159–4.161). Differentiating Eq. 4.159 with respect to x once gives

$$\frac{\partial^2 \sigma_c}{\partial x^2} - \frac{1}{t_c}\frac{\partial \tau}{\partial x} = 0 \tag{4.175}$$

Similarly, differentiating Eq. 4.170

$$\frac{\partial^2 \sigma_b^{-s}}{\partial x^2} + \frac{\alpha}{t_b}\frac{\partial \tau}{\partial x} = 0 \tag{4.176}$$

Replacing τ in Eq. 4.175 by $G_s \gamma_s$

$$\frac{\partial^2 \sigma_c}{\partial x^2} - \frac{G_s}{t_c}\frac{\partial \gamma_s}{\partial x} = 0 \tag{4.177}$$

Combining this with Eq. 4.174

$$\frac{\partial^2 \sigma_c}{\partial x^2} - \frac{G_s}{t_c t_s}(\epsilon_c - \epsilon_b^{-s}) = 0 \tag{4.178}$$

We can replace σ_c by ϵ_c using Eq. 4.171. We also assume that the induced strain does not vary with x. Simplifying, we obtain

$$\frac{\partial^2 \epsilon_c}{\partial x^2} - \frac{G_s}{E_c t_s t_c}(\epsilon_c - \epsilon_b^{-s}) = 0 \tag{4.179}$$

Carrying out similar steps for the beam equation (Eq. 4.176) yields

$$\frac{\partial^2 \epsilon_b^{-s}}{\partial x^2} + \frac{\alpha G_s}{t_s t_b E_b}(\epsilon_c - \epsilon_b^{-s}) = 0 \tag{4.180}$$

This gives us two governing equations in ϵ_c and ϵ_b^{-s}. These equations can be converted into higher order, uncoupled equations and then solved. It is also possible to solve these equations in an alternate way, without increasing the order of the equations. Subtracting Eq. 4.180 from Eq. 4.179

$$\frac{\partial^2}{\partial x^2}(\epsilon_c - \epsilon_b^{-s}) - \frac{G_s}{t_s E_c}\left(\frac{1}{t_c} + \frac{E_c \alpha}{E_b t_b}\right)(\epsilon_c - \epsilon_b^{-s}) = 0 \tag{4.181}$$

Substituting $\zeta = (\epsilon_c - \epsilon_b^{-s})$ makes this a second-order equation in ζ. Introducing the following nondimensional quantities

$$\bar{x} = \frac{x}{l_c/2} \quad \text{(note that } \bar{x} = 0 \text{ denotes the actuator midpoint)}$$

$$\bar{t}_s = \frac{t_s}{l_c/2}$$

$$\theta_b = \frac{t_b}{t_c}$$

$$\theta_s = \frac{t_s}{t_c}$$

$$\bar{G} = \frac{G_s}{E_c}$$

$$\gamma_b = \frac{E_b}{E_c}$$

Substituting these into Eq. 4.181

$$\frac{\partial^2 \zeta}{\partial \bar{x}^2} - \Gamma^2 \zeta = 0 \tag{4.182}$$

where the shear-lag parameter, Γ, can be defined as

$$\begin{aligned}
\Gamma^2 &= \frac{G_s}{t_s E_c}\left(\frac{1}{t_c} + \frac{E_c \alpha}{E_b t_b}\right)\frac{l_c^2}{4} \\
&= \frac{\bar{G}}{t_s^2}\left(\frac{t_s}{t_c} + \frac{E_c \alpha t_s}{E_b t_b}\right)\frac{l_c^2}{4} \\
&= \frac{\bar{G}}{t_s^2}\left(\theta_s + \frac{\alpha \theta_s}{\gamma_b \theta_b}\right)\frac{l_c^2}{4}
\end{aligned}$$

This term has all of the characteristics of the beam and actuator and represents the shear-lag effects. It becomes larger for higher modulus of the bond layer or for lower bond thickness. In the limiting case, a bond layer of infinite stiffness or zero thickness results in a complete transfer of strain from the actuator to the beam without any losses. Such a case is called a "perfect-bond condition" ($\Gamma \to \infty$).

The solution for Eq. 4.182 can be written as

$$\zeta = A\cosh(\Gamma\bar{x}) + B\sinh(\Gamma\bar{x}) \tag{4.183}$$

Now we have the difference of normal strains, but we want each of these individually. Combining Eqs. 4.179 and 4.183

$$\frac{\partial^2 \epsilon_c}{\partial \bar{x}^2} = \frac{G_s(l_c^2/4)}{t_s t_c E_c}(A\cosh(\Gamma\bar{x}) + B\sinh(\Gamma\bar{x}))$$

leads to

$$\frac{\partial^2 \epsilon_c}{\partial \bar{x}^2} = \frac{\bar{G}}{\bar{t}_s^2}\theta_s(A\cosh(\Gamma\bar{x}) + B\sinh(\Gamma\bar{x})) \tag{4.184}$$

This can be solved as

$$\epsilon_c = C + D\bar{x} + \frac{\bar{G}}{\bar{t}_s^2}\frac{\theta_s}{\Gamma^2}(A\cosh(\Gamma\bar{x}) + B\sinh(\Gamma\bar{x})) \tag{4.185}$$

Combining Eqs. 4.183 and 4.185, we obtain

$$\epsilon_b^{-s} = C + D\bar{x} + A\left[\frac{\bar{G}\theta_s}{\bar{t}_s^2\Gamma^2} - 1\right]\cosh(\Gamma\bar{x}) + B\left[\frac{\bar{G}\theta_s}{\bar{t}_s^2\Gamma^2} - 1\right]\sinh(\Gamma\bar{x}) \tag{4.186}$$

We have four constants that must be determined by the boundary conditions at the edges of the actuator

$$\bar{x} = \pm 1 \rightarrow \epsilon_c = \Lambda \ (\sigma = 0 \text{ no-stress condition})$$
$$\bar{x} = \pm 1 \rightarrow \epsilon_b^{-s} = 0 \ (\text{if no mechanical load})$$

These conditions are used to determine the four constants, as follows

$$C + D + \frac{\bar{G}}{\bar{t}_s^2}\frac{\theta_s}{\Gamma^2}(A\cosh\Gamma + B\sinh\Gamma) = \Lambda$$
$$C - D + \frac{\bar{G}}{\bar{t}_s^2}\frac{\theta_s}{\Gamma^2}(A\cosh\Gamma - B\sinh\Gamma) = \Lambda$$
$$C + D + A\left(\frac{\bar{G}}{\bar{t}_s^2}\frac{\theta_s}{\Gamma^2} - 1\right)\cosh\Gamma + B\left(\frac{\bar{G}}{\bar{t}_s^2}\frac{\theta_s}{\Gamma^2} - 1\right)\sinh\Gamma = 0$$
$$C - D + A\left(\frac{\bar{G}}{\bar{t}_s^2}\frac{\theta_s}{\Gamma^2} - 1\right)\cosh\Gamma - B\left(\frac{\bar{G}}{\bar{t}_s^2}\frac{\theta_s}{\Gamma^2} - 1\right)\sinh\Gamma = 0$$

Solving these equations

$$B = D = 0$$
$$A = \Lambda/\cosh\Gamma$$
$$C = \Lambda\left(1 - \frac{\bar{G}}{\bar{t}_s^2}\frac{\theta_s}{\Gamma^2}\right)$$

Normal strains in the beam and actuator are given as

$$\begin{aligned}\frac{\epsilon_b^{-s}}{\Lambda} &= \frac{\alpha}{\alpha+\Psi} - \frac{\alpha}{(\alpha+\Psi)\cosh\Gamma}\cosh(\Gamma\bar{x}) \\ &= \frac{\alpha}{\alpha+\Psi}\left[1 - \frac{\cosh(\Gamma\bar{x})}{\cosh\Gamma}\right]\end{aligned} \tag{4.187}$$

$$\begin{aligned}\frac{\epsilon_c}{\Lambda} &= -\frac{\Psi}{\alpha+\Psi} + \frac{\Psi}{(\alpha+\Psi)\cosh\Gamma}\cosh(\Gamma\bar{x}) + 1 \\ &= \frac{\alpha}{\alpha+\Psi}\left[1 + \frac{\Psi}{\alpha}\frac{\cosh(\Gamma\bar{x})}{\cosh\Gamma}\right]\end{aligned} \tag{4.188}$$

where

$$\begin{aligned}\Psi &= \text{extensional stiffness ratio} = \frac{E_b t_b}{E_c t_c} \\ &= \frac{\text{extensional stiffness of the beam}}{\text{extensional stiffness of one piezo}}\end{aligned}$$

and $\alpha = 2$ for pure extension and $\alpha = 6$ for pure bending. Shear stress is obtained from Eq. 4.174

$$\gamma_s = \frac{1}{t_s}\int(\epsilon_c - \epsilon_b^{-s})dx \tag{4.189}$$

Using the relations for ϵ_c and ϵ_b^{-s}

$$\frac{\gamma_s}{\Lambda} = \frac{1}{\bar{t}_s\cosh\Gamma}\int_o^{\bar{x}}\cosh(\Gamma\bar{x})\,d\bar{x} = \frac{\sinh(\Gamma\bar{x})}{\Gamma\bar{t}_s\cosh\Gamma} + C_1 \tag{4.190}$$

The constant C_1 is evaluated using the condition

$$\text{at } \bar{x} = 0,\ \gamma_s = 0$$

This gives $C_1 = 0$. Hence

$$\frac{\tau}{\Lambda} = G_s\frac{\gamma_s}{\Lambda} = \frac{G_s\sinh(\Gamma\bar{x})}{\Gamma\bar{t}_s\cosh\Gamma} \tag{4.191}$$

$$\frac{\tau}{E_b} = \frac{\bar{G}\Lambda\sinh(\Gamma\bar{x})}{\gamma_b\Gamma\bar{t}_s\cosh\Gamma} \tag{4.192}$$

As Γ increases, the shear stress becomes more localized at the ends of the piezoelectric sheet.

Finite-Thickness Bond ($\Gamma < 30$)

Pure Extension: ($\alpha = 2$)

$$\epsilon_o = \epsilon_b^{-s}$$

$$\frac{\epsilon_b^{-s}}{\Lambda} = \frac{\partial u_b^{-s}}{\partial x}\frac{1}{\Lambda}$$

$$= \frac{\alpha}{\alpha+\Psi}\left[1 - \frac{\cosh(\Gamma\bar{x})}{\cosh\Gamma}\right]$$

$$\frac{\partial u_b^{-s}}{\partial \bar{x}} = \frac{\alpha\Lambda}{\alpha+\Psi}\frac{l_c}{2}\left[1 - \frac{\cosh(\Gamma\bar{x})}{\cosh\Gamma}\right]$$

$$u_b^{-s} = \frac{\alpha\Lambda}{\alpha+\Psi}\frac{l_c}{2}\left[\bar{x} - \frac{\sinh(\Gamma\bar{x})}{\Gamma\cosh\Gamma}\right] + C_2$$

The constant C_2 is evaluated using the boundary condition

$$\text{at } \bar{x} = -1, \; u = 0$$

$$u_b^{-s}(\bar{x}) = \frac{\alpha\Lambda}{\alpha+\Psi}\left(\frac{l_c}{2}\right)\left[\bar{x} - \frac{\sinh(\Gamma\bar{x})}{\Gamma\cosh\Gamma}\right] + \frac{\alpha\Lambda}{\alpha+\Psi}\left(\frac{l_c}{2}\right)\left[1 - \frac{\tanh\Gamma}{\Gamma}\right]$$

$$= \frac{\alpha\Lambda}{(\alpha+\Psi)}\left[x - \frac{l_c\sinh(\Gamma 2x/l_c)}{2\Gamma\cosh\Gamma}\right] + \frac{\alpha\Lambda}{\alpha+\Psi}\left(\frac{l_c}{2}\right)\left[1 - \frac{\tanh\Gamma}{\Gamma}\right]$$

Pure Bending: ($\alpha = 6$)

$$w''(x) = \frac{\partial^2 w}{\partial x^2} = \frac{2}{t_b}\epsilon_b^{-s}$$

$$= \frac{2\Lambda}{t_b}\frac{\alpha}{\alpha+\Psi}\left(1 - \frac{\cosh(\Gamma\bar{x})}{\cosh\Gamma}\right)$$

$$\frac{\partial w}{\partial \bar{x}} = \frac{2\Lambda}{t_b}\frac{\alpha}{\alpha+\Psi}\left(\frac{l_c}{2}\right)^2\left(\bar{x} - \frac{\sinh(\Gamma\bar{x})}{\Gamma\cosh\Gamma}\right) + C_3$$

The constant C_3 is evaluated using the boundary condition

$$\text{at } \bar{x} = -1, \; \frac{\partial w}{\partial \bar{x}} = 0$$

$$C_3 = \frac{2\Lambda}{t_b}\frac{\alpha}{\alpha+\Psi}\left(\frac{l_c^2}{4}\right)\left(1 - \frac{\tanh\Gamma}{\Gamma}\right)$$

This results in

$$\frac{\partial w}{\partial \bar{x}} = \frac{2\Lambda}{t_b}\frac{\alpha}{\alpha+\Psi}\left(\frac{l_c^2}{4}\right)\left(\bar{x} - \frac{\sinh(\Gamma\bar{x})}{\Gamma\cosh\Gamma}\right) + \frac{2\Lambda}{t_b}\frac{\alpha}{\alpha+\Psi}\left(\frac{l_c^2}{4}\right)\left(1 - \frac{\tanh\Gamma}{\Gamma}\right)$$

$$= \frac{2\Lambda}{t_b}\frac{\alpha}{\alpha+\Psi}\frac{l_c^2}{4}\left[\bar{x} - \frac{\sinh(\Gamma\bar{x})}{\Gamma\cosh\Gamma} + 1 - \frac{\tanh\Gamma}{\Gamma}\right]$$

Table 4.4. *Comparison of strain transfer*

$\frac{\text{Host Structure Stiffness}}{\text{Actuator Stiffness}}$, Ψ	0.1	0.5	1	5	10
Bending, $\frac{\alpha}{\alpha+\Psi}$	0.98	0.92	0.86	0.55	0.38
Extension, $\frac{\alpha}{\alpha+\Psi}$	0.95	0.80	0.67	0.29	0.17

$$\frac{\partial w}{\partial x}(x) = \frac{2\Lambda\alpha}{t_b(\alpha+\Psi)}\left[x - \frac{l_c \sinh(2\Gamma x/l_c)}{2\Gamma\cosh\Gamma} + \frac{l_c}{2} - \frac{l_c}{2}\frac{\tanh\Gamma}{\Gamma}\right]$$

$$= \frac{\Lambda l_c}{t_b}\frac{\alpha}{\alpha+\Psi}\left[\frac{x}{l_c/2} - \frac{\sinh(2\Gamma x/l_c)}{2\Gamma\cosh\Gamma} + 1 - \frac{\tanh\Gamma}{\Gamma}\right]$$

Integrating the slope, we obtain

$$w = \frac{2\Lambda}{t_b}\frac{\alpha}{\alpha+\Psi}\frac{l_c^2}{4}\left(\frac{\bar{x}^2}{2} - \frac{\cosh(\Gamma\bar{x})}{\Gamma^2\cosh\Gamma}\right) + \frac{2\Lambda}{t_b}\frac{\alpha}{\alpha+\Psi}\frac{l_c^2}{4}\left(1 - \frac{\tanh\Gamma}{\Gamma}\right)\bar{x} + C_4$$

The constant C_4 is evaluated using the boundary condition

$$\text{at } \bar{x} = -1, \;\; w = 0$$

This gives

$$C_4 = \frac{2\Lambda}{t_b}\frac{\alpha}{\alpha+\Psi}\frac{l_c^2}{4}\left(\frac{1}{2} + \frac{1}{\Gamma^2} - \frac{\tanh\Gamma}{\Gamma}\right)$$

$$w(x) = \frac{2\Lambda}{t_b}\frac{\alpha}{\alpha+\Psi}\left[\frac{x^2}{2} - \frac{l_c^2/4\cosh(2\Gamma x/l_c)}{\Gamma^2\cosh\Gamma} + \frac{xl_c}{2} - \frac{\tanh\Gamma x l_c}{2\Gamma} + \frac{l_c^2}{8} + \frac{l_c^2}{4\Gamma^2} - \frac{l_c^2\tanh\Gamma}{4\Gamma}\right]$$

Very Thin Bond ($\Gamma > 30$)

This represents a perfectly bonded condition. From Eqs. 4.187 and 4.188

$$\frac{\epsilon_b^{-s}}{\Lambda} = \frac{\epsilon_c}{\Lambda} = \frac{\alpha}{\alpha+\Psi}$$

This means that the induced strain on the surface of a host structure is equal to the actuator strain and it is proportional to the product of the actuation strain, Λ, (which can be generated by the actuation material) and the reciprocal of one plus the stiffness ratio (structural stiffness/actuator stiffness). The second term is a result of the impedance matching (Table 4.4).

The shear-lag parameter must be kept large for efficient transfer of actuation strain to the host structure. As the stiffness of the actuator increases, the strain transfer becomes more effective. However, an extremely large stiffness of the bond layer can cause fracture failure at the edges. Figure 4.22 shows the actuator- and beam strain variation along the actuator length, in the case of pure extension ($\alpha = 2$), for a stiffness ratio of $\Psi = 10$. Three cases of the shear-lag parameter Γ are considered,

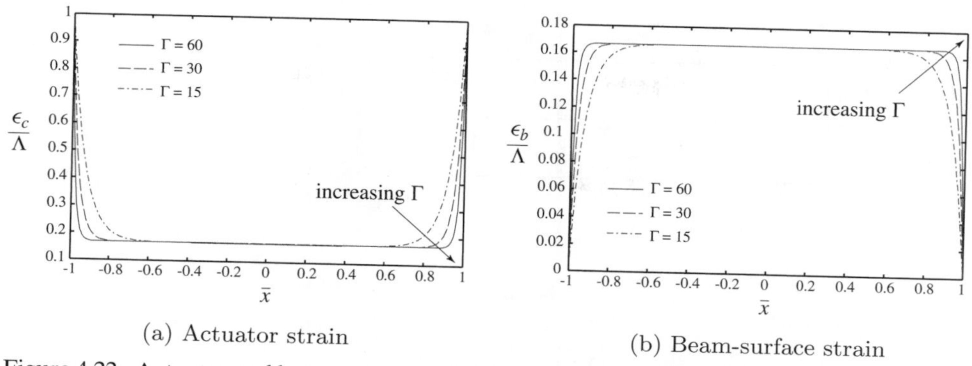

Figure 4.22. Actuator and beam strains on the top surface of the beam, for the pure extension condition ($\alpha = 2$).

and it can be seen that for higher values of Γ, there is a lower loss of strain in the bond layer. As $\Gamma > 30$, the dependence of strain on the value of Γ is less pronounced.

Note that if the actuator width is different from the beam width ($b_c \neq b_b$), the only change in these expressions will be in the definition of Γ^2. In this case

$$\Gamma^2 = \frac{G_s}{t_s E_c}\left(\frac{1}{t_c} + \alpha \frac{E_c}{E_b}\frac{1}{t_b}\frac{b_c}{b_b}\right)\frac{l_c^2}{4} \tag{4.193}$$

However, the governing equation (Eq. 4.182) remains unchanged. The strain distribution results in Eqs. 4.187 and 4.188 are also unchanged.

4.4.2 Single Actuator: Asymmetric Actuation

As shown in Figs. 4.23 and 4.24, a piezo-sheet induced-strain actuator is bonded to the surface of a beam with a finite thickness bond. The governing equations are developed through force and moment equilibrium of the elemental section dx shown in Fig. 4.25. The actuator is assumed to exhibit an axial strain that varies only along its major axis. The neutral axis is assumed to lie at the mid-plane of the beam because the thickness of the actuator is assumed to be small compared to the thickness of the beam. The adhesive layer is assumed to transfer loads only through shear. The strain distribution is assumed to be uniform across the actuator thickness and linear across the beam thickness. In addition, the effect of an actuator of width b_c less than the beam width b_b is considered.

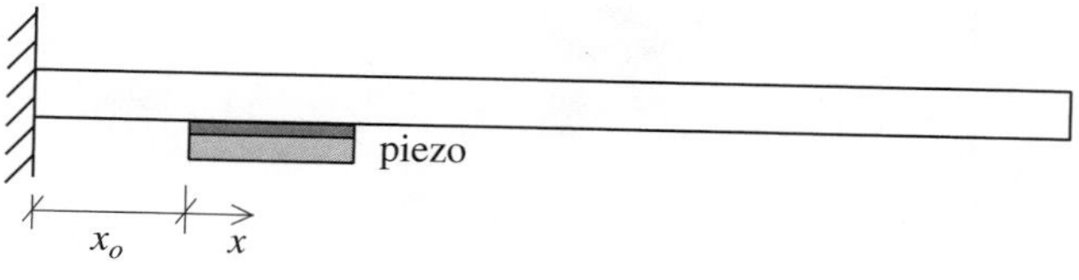

Figure 4.23. Single actuator bonded on bottom surface of beam.

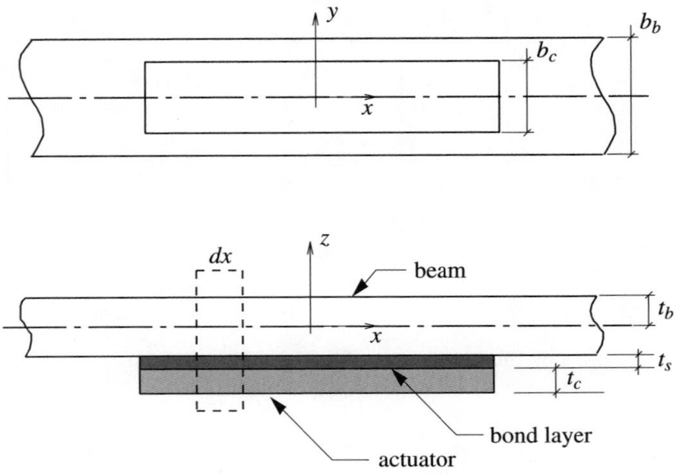

Figure 4.24. Details of the beam geometry.

With the stated assumptions, the equilibrium equations for the differential element of a straight rectangular isotropic beam can be derived as follows:

Equilibrium of piezo element forces

$$\begin{aligned}\sigma_c t_c b_c - t_c b_c \left(\sigma_c + \frac{\partial \sigma_c}{\partial x} dx\right) + \tau b_c dx = 0 \\ \frac{\partial \sigma_c}{\partial x} - \frac{\tau}{t_c} = 0\end{aligned} \tag{4.194}$$

Equilibrium of beam element forces

$$\begin{aligned}\sigma_b^o t_b b_b - t_b b_b \left(\sigma_b^o + \frac{\partial \sigma_b^o}{\partial x} dx\right) - \tau b_c dx = 0 \\ \frac{\partial \sigma_b^o}{\partial x} + \frac{b_c}{b_b t_b} \tau = 0\end{aligned} \tag{4.195}$$

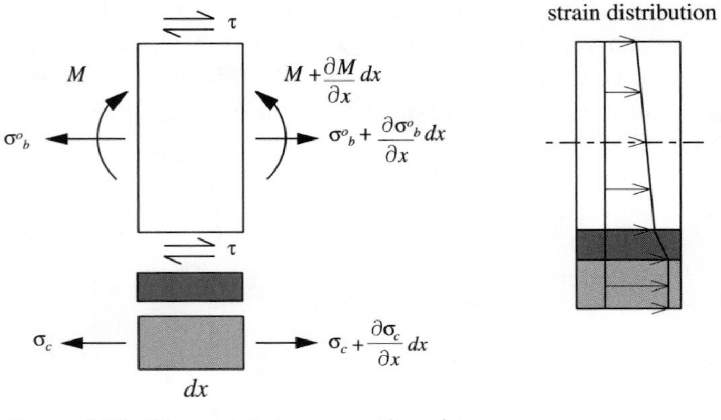

Figure 4.25. Elemental stresses and strains.

where σ_b^o is the axial stress at the mid-axis of the beam ($z = 0$). Equilibrium of beam moments

$$-M + M + \frac{\partial M}{\partial x}dx + \tau b_c \frac{t_b}{2} dx = 0$$

$$\frac{\partial M}{\partial x} + \frac{\tau}{2} b_c t_b = 0$$

$$\sigma_b^{-s} - \sigma_b^o = \frac{M}{I}\frac{t_b}{2}, \quad I = \frac{b_b t_b^3}{12} \tag{4.196}$$

$$M = (\sigma_b^{-s} - \sigma_b^o)\frac{b_b t_b^2}{6}$$

$$\frac{3b_c}{b_b t_b}\tau + \frac{\partial \sigma_b^{-s}}{\partial x} - \frac{\partial \sigma_b^o}{\partial x} = 0$$

where σ_b^{-s} is the axial stress at the bottom surface of the beam. For a one-dimensional system, the strain-displacement relations reduce to

$$\epsilon_c = \frac{\partial u_c}{\partial x} \tag{4.197}$$

$$\epsilon_b^{-s} = \frac{\partial u_b^{-s}}{\partial x} \tag{4.198}$$

$$\epsilon_b^o = \frac{\partial u_b^o}{\partial x} \tag{4.199}$$

$$\gamma_s = \frac{u_c - u_b^{-s}}{t_s} \tag{4.200}$$

where u_c is the uniform axial displacement of the actuator and u_b^{-s} is the axial displacement at the bottom surface of the beam. The other stress-strain relations are

$$\sigma_b^{-s} = E_b \epsilon_b^{-s} \tag{4.201}$$

$$\sigma_b^o = E_b \epsilon_b^o \tag{4.202}$$

$$\tau = G_s \gamma_s \tag{4.203}$$

Substituting Eqs. 4.197 through 4.203 into the equilibrium equations and differentiating with respect to x produces three governing differential equations.

From the actuator force equilibrium

$$\frac{\partial^2 \epsilon_c}{\partial x^2} - \frac{G_s}{E_c t_c t_s}(\epsilon_c - \epsilon_b^{-s}) = 0 \tag{4.204}$$

From the equilibrium of moments

$$\frac{\partial^2 \epsilon_b^{-s}}{\partial x^2} - \frac{\partial^2 \epsilon_b^o}{\partial x^2} + \frac{3b_c G_s}{E_b t_b b_b t_s}(\epsilon_c - \epsilon_b^{-s}) = 0 \tag{4.205}$$

From the equilibrium of axial forces

$$\frac{\partial^2 \epsilon_b^o}{\partial x^2} + \frac{b_c G_s}{E_b t_b b_b t_s}(\epsilon_c - \epsilon_b^{-s}) = 0 \tag{4.206}$$

Combining Eqs. 4.204, 4.205, and 4.206, making a substitution of variables, and non-dimensionalizing with respect to the actuator length reduces the system of equations to a single linear second-order differential equation. From Eqs. 4.205 and 4.206

$$\frac{\partial^2 \epsilon_b^{-s}}{\partial x^2} + \left[\frac{4b_c G_s}{(E_b t_b b_b) t_s}\right](\epsilon_c - \epsilon_b^{-s}) = 0 \tag{4.207}$$

From Eqs. 4.204 and 4.207

$$\frac{\partial^2}{\partial x^2}(\epsilon_c - \epsilon_b^{-s}) - (\epsilon_c - \epsilon_b^{-s})\left[\frac{G_s}{E_c t_c t_s} + \frac{4b_c G_s}{E_b t_b b_b t_s}\right] = 0 \tag{4.208}$$

Assume

$$\zeta = \epsilon_c - \epsilon_b^{-s} \tag{4.209}$$

$$\bar{x} = \frac{x}{l_c/2} \quad (\bar{x} = 0 \text{ indicates the actuator midpoint}) \tag{4.210}$$

$$\alpha = 4 \tag{4.211}$$

This results in

$$\frac{\partial^2 \zeta}{\partial \bar{x}^2} - \Gamma^2 \zeta = 0 \tag{4.212}$$

where

$$\Gamma^2 = \left(\frac{l_c}{2}\right)^2 \frac{b_c G_s}{E_b t_b b_b t_s}\left[4 + \frac{E_b t_b b_b}{E_c t_c b_c}\right] = \frac{l_c^2}{4}\frac{b_c G_s}{EA_b t_s}\left(\alpha + \frac{E_b A_b}{E_c A_c}\right) \tag{4.213}$$

The general solution to this equation is

$$\zeta(\bar{x}) = A\cosh(\Gamma\bar{x}) + B\sinh(\Gamma\bar{x}) \tag{4.214}$$

From actuator equilibrium (Eq. 4.204)

$$\frac{\partial^2 \epsilon_c}{\partial \bar{x}^2} - \frac{G_s}{E_c t_c t_s}\left(\frac{l_c}{2}\right)^2 \zeta = 0 \tag{4.215}$$

$$\frac{\partial^2 \epsilon_c}{\partial \bar{x}^2} - \frac{\psi_s}{\bar{t}_s^2}(A\cosh\Gamma\bar{x} + B\sinh\Gamma\bar{x}) = 0 \tag{4.216}$$

where

$$\psi_s = \frac{G_s b_c t_s}{E_c b_c t_c} = \bar{G}\theta_s \tag{4.217}$$

$$\psi_b = \frac{E_b b_b t_b}{E_c b_c t_c} = \frac{E_b A_b}{E_c A_c} \tag{4.218}$$

From Eq. 4.207

$$\frac{\partial^2 \epsilon_b^{-s}}{\partial \bar{x}^2} = -\frac{\alpha\psi_s}{\psi_b \bar{t}_s^2}(A\cosh(\Gamma\bar{x}) + B\sinh(\Gamma\bar{x})) \tag{4.219}$$

and from axial force equilibrium, Eq. 4.206

$$\frac{\partial^2 \epsilon_b^{o}}{\partial \bar{x}^2} = -\frac{\psi_s}{\psi_b \bar{t}_s^2}(A\cosh(\Gamma\bar{x}) + B\sinh(\Gamma\bar{x})) \tag{4.220}$$

This results in

$$\epsilon_c = C_1 + D_1\bar{x} + \frac{\psi_s}{\bar{t}_s^2\Gamma^2}(A\cosh(\Gamma\bar{x}) + B\sinh(\Gamma\bar{x})) \tag{4.221}$$

$$\epsilon_b^{-s} = C_2 + D_2\bar{x} - \frac{\alpha\psi_s}{\psi_b\bar{t}_s^2\Gamma^2}(A\cosh(\Gamma\bar{x}) + B\sinh(\Gamma\bar{x})) \tag{4.222}$$

$$\epsilon_b^o = C_3 + D_3\bar{x} - \frac{\psi_s}{\psi_b}\frac{1}{\bar{t}_s^2\Gamma^2}(A\cosh(\Gamma\bar{x}) + B\sinh(\Gamma\bar{x})) \tag{4.223}$$

$$\zeta = A\cosh(\Gamma\bar{x}) + B\sinh(\Gamma\bar{x}) \tag{4.224}$$

$$= (C_1 - C_2) + (D_1 - D_2)\bar{x} + \frac{\psi_s}{\bar{t}_s^2\Gamma^2}(1 + \frac{\alpha}{\psi_b})(A\cosh(\Gamma\bar{x}) + B\sinh(\Gamma\bar{x})) \tag{4.225}$$

Comparing this equation with Eq. 4.214, the following relations are obvious

$$C_1 = C_2 \tag{4.226}$$

$$\frac{\psi_s}{\bar{t}_s^2\Gamma^2}\left(1 + \frac{\alpha}{\psi_b}\right) = 1 \tag{4.227}$$

$$D_1 = D_2 \tag{4.228}$$

The boundary conditions are

$$\epsilon_c(\bar{x} = \pm 1) = \Lambda \tag{4.229}$$

$$\epsilon_b^{-s}(\bar{x} = \pm 1) = 0 \tag{4.230}$$

$$\epsilon_b^o(\bar{x} = \pm 1) = 0 \tag{4.231}$$

From these conditions, the unknown constants can be found

$$B = 0 \tag{4.232}$$

$$D_1 = 0 \tag{4.233}$$

$$D_2 = 0 \tag{4.234}$$

$$D_3 = 0 \tag{4.235}$$

$$C_1 = C_2 = \frac{\alpha\psi_s}{\psi_b\bar{t}_s^2\Gamma^2}A\cosh\Gamma \tag{4.236}$$

$$A\left[\frac{\alpha\psi_s}{\psi_b\bar{t}_s^2\Gamma^2} + \frac{\psi_s}{\bar{t}_s^2\Gamma^2}\right]\cosh\Gamma = \Lambda \tag{4.237}$$

$$A = \frac{\Lambda}{\cosh\Gamma} \tag{4.238}$$

$$C_1 = \frac{\alpha\psi_s}{\psi_b\bar{t}_s^2\Gamma^2}\Lambda = \frac{\alpha}{\psi_b}\frac{1}{1 + \frac{\alpha}{\psi_b}}\Lambda = \frac{\alpha}{\alpha + \psi_b}\Lambda$$
$$= C_2 \tag{4.239}$$

$$C_3 = \frac{\psi_s}{\psi_b}\frac{1}{\bar{t}_s^2\Gamma^2}\Lambda = \frac{\Lambda}{\psi_b + \alpha} \tag{4.240}$$

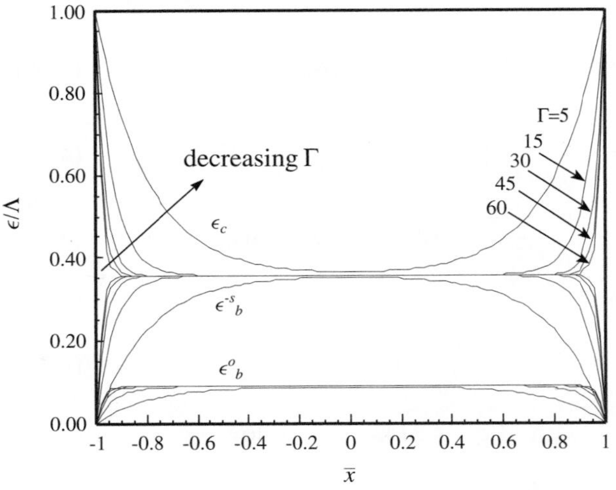

Figure 4.26. Actuator and beam strain distributions.

Substituting these gives the final solutions for the actuator and beam strains.

$$\frac{\epsilon_c(\bar{x})}{\Lambda} = \frac{\alpha}{(\psi_b + \alpha)} + \frac{\psi_b}{(\psi_b + \alpha)\cosh\Gamma}\cosh(\Gamma\bar{x}) \tag{4.241}$$

$$\frac{\epsilon_b^{-s}(\bar{x})}{\Lambda} = \frac{\alpha}{(\psi_b + \alpha)} - \frac{\alpha}{(\psi_b + \alpha)\cosh\Gamma}\cosh(\Gamma\bar{x}) \tag{4.242}$$

$$\frac{\epsilon_b^{o}(\bar{x})}{\Lambda} = \frac{1}{(\psi_b + \alpha)} - \frac{1}{(\psi_b + \alpha)\cosh\Gamma}\cosh(\Gamma\bar{x}) \tag{4.243}$$

Again, note that the value of Γ is directly proportional to shear modulus G_s and is inversely proportional to bond thickness t_s. Figure 4.26 shows the strain distribution in the actuator and in the beam. For large values of Γ, the strain is constant over the span of the actuator and reduces to a perfectly bonded condition, where

$$\frac{\epsilon_c(\bar{x})}{\Lambda} = \frac{\epsilon_b^{-s}(\bar{x})}{\Lambda} = \frac{\alpha}{(\psi_b + \alpha)} \tag{4.244}$$

$$\frac{\epsilon_b^{o}(\bar{x})}{\Lambda} = \frac{1}{(\psi_b + \alpha)} \tag{4.245}$$

Due to a dramatic change of the beam strain near the actuator ends, integration of the strain equations to obtain system deflections and rotations leads to an increasing discrepancy between perfect- and finite-bond conditions. The perfect-bond system response predictions exceed those of the finite-bond equations.

The adhesive shear stress is found as follows

$$\tau = G_s\gamma_s = \frac{G_s}{t_s}\int(\epsilon_c - \epsilon_b^s)dx \tag{4.246}$$

Using Eqs. 4.241 and 4.242

$$\begin{aligned}\tau &= \frac{G_s}{t_s}\int\frac{\cosh(\Gamma\bar{x})}{\cosh\Gamma}d\bar{x}\frac{l_c}{2} \\ &= \frac{G_s}{t_s}\frac{l_c/2}{\cosh\Gamma}\left[\frac{\sinh(\Gamma\bar{x})}{\Gamma}\right]\end{aligned} \tag{4.247}$$

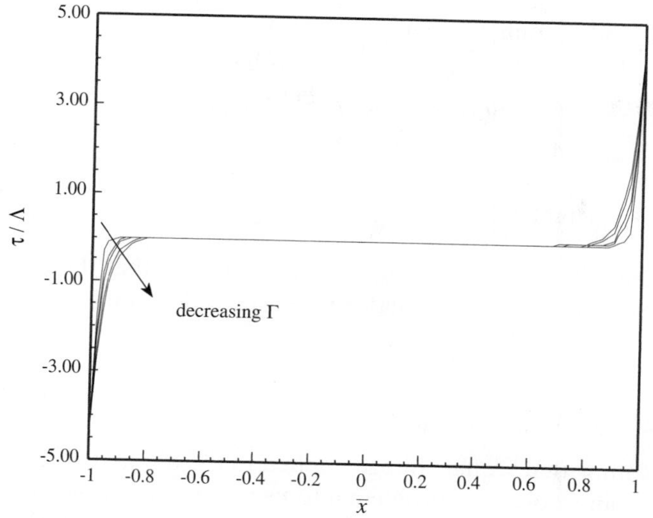

Figure 4.27. Adhesive shear–stress distribution.

As shown in Fig. 4.27, the shear-stress distribution has characteristics similar to the strain distribution near the ends of the actuator. As with the system strains, the rate of load transfer may significantly impact static and dynamic results.

The normalized bending curvature is obtained by using the assumed strain distribution through the beam

$$\epsilon_b(z) = \epsilon_b^o - z\frac{\partial^2 w}{\partial x^2} \tag{4.248}$$

$$\epsilon_b^s = \epsilon_b^o - \frac{t_b}{2}\frac{\partial^2 w}{\partial x^2} \tag{4.249}$$

$$\epsilon_b^{-s} = \epsilon_b^o + \frac{t_b}{2}\frac{\partial^2 w}{\partial x^2} \tag{4.250}$$

$$\frac{\partial^2 w}{\partial x^2} = \kappa = -\frac{2}{t_b}(\epsilon_b^s - \epsilon_b^o) = -\frac{t_b}{2}(\epsilon_b^o - \epsilon_b^{-s}) \tag{4.251}$$

$$= \frac{2}{t_b}\Lambda\left(\frac{\alpha - 1}{\psi_b + \alpha}\right)\left(1 - \frac{\cosh(\Gamma\bar{x})}{\cosh\Gamma}\right) \tag{4.252}$$

$$\frac{\kappa t_b}{2\Lambda}(\bar{x}) = \frac{\alpha - 1}{\psi_b + \alpha}\left(1 - \frac{\cosh(\Gamma\bar{x})}{\cosh\Gamma}\right) \tag{4.253}$$

The bending slope is obtained by integrating the curvature

$$\begin{aligned}\frac{\partial w}{\partial x}(\bar{x}) &= \int_{-1}^{\bar{x}} \frac{\Lambda l_c}{t_b}\frac{\alpha - 1}{\psi_b + \alpha}\left(1 - \frac{\cosh(\Gamma\bar{x})}{\cosh\Gamma}\right) d\bar{x} \\ &= \frac{\Lambda l_c}{t_b}\frac{\alpha - 1}{\psi_b + \alpha}\left[\bar{x} - \frac{\sinh(\Gamma\bar{x})}{\Gamma\cosh\Gamma} + 1 - \frac{\sinh\Gamma}{\Gamma\cosh\Gamma}\right]\end{aligned} \tag{4.254}$$

$$\frac{\partial w}{\partial x}(x) = \frac{\Lambda l_c}{t_b}\frac{\alpha - 1}{\psi_b + \alpha}\left[\frac{x}{l_c/2} - \frac{\sinh(\Gamma 2x/l_c)}{\Gamma \cosh\Gamma} + 1 - \frac{\sinh\Gamma}{\Gamma\cosh\Gamma}\right]$$

$$w = 2\frac{\Lambda}{l_c}\frac{\alpha - 1}{\psi_b + \alpha}\left[\frac{x^2}{2} - \frac{l_c^2}{4}\frac{\cosh(2\Gamma x/l_c)}{\Gamma^2\cosh\Gamma} + \frac{xl_c}{2} - \frac{xl_c}{2}\frac{\tanh\Gamma}{\Gamma} + \frac{l_c^2}{8} + \frac{l_c^2}{4\Gamma^2} - \frac{l_c^2}{4\Gamma}\tanh\Gamma\right] \tag{4.255}$$

Assuming that ψ_b is large, examining the previous equation indicates that the theoretical bending slope achieved with a single actuator ($\alpha = 4$) is half of that for the dual actuator ($\alpha = 6$).

4.4.3 Unequal Electric Voltage ($V_{\text{top}} \neq V_{\text{bottom}}$)

Consider a dual-actuator beam with unequal voltage applied to the top and bottom actuators (see Fig. 4.11). We can resolve this problem into two parts: pure extension and pure bending problems; then use superposition to obtain the combined solution (see Fig. 4.12). The procedure followed is similar to that described in Section 4.3.4, with the only difference being the modeling of the bond layer. The voltages resulting in pure bending and pure extension are (Eq. 4.126)

$$V_1 = \frac{V_{\text{top}} + V_{\text{bottom}}}{2} \tag{4.256}$$

$$V_2 = \frac{V_{\text{bottom}} - V_{\text{top}}}{2} \tag{4.257}$$

For pure extension: $\alpha = \alpha_1 = 2$. The free strain $\Lambda \to \Lambda_1$, corresponding to V_1 and $\Gamma = \Gamma_1$. Therefore, Eqs. 4.241 and 4.242 become

$$\frac{\epsilon_{c1}}{\Lambda_1} = \frac{\alpha_1}{\alpha_1 + \psi_b} + \frac{\psi_b}{\alpha_1 + \psi_b}\frac{\cosh(\Gamma_1 \bar{x})}{\cosh\Gamma_1} \tag{4.258}$$

$$\frac{\epsilon_{b1}^{-s}}{\Lambda_1} = \frac{\alpha_1}{\alpha_1 + \psi_b} - \frac{\alpha_1}{\alpha_1 + \psi_b}\frac{\cosh(\Gamma_1 \bar{x})}{\cosh\Gamma_1} \tag{4.259}$$

For pure bending: $\alpha = \alpha_2 = 6$. The free strain $\Lambda \to \Lambda_2$, corresponding to V_2 and $\Gamma = \Gamma_2$

$$\frac{\epsilon_{c2}}{\Lambda_2} = \frac{\alpha_2}{\alpha_2 + \psi_b} + \frac{\psi_b}{\alpha_2 + \psi_b}\frac{\cosh(\Gamma_2 \bar{x})}{\cosh\Gamma_2} \tag{4.260}$$

$$\frac{\epsilon_{b2}^{-s}}{\Lambda_2} = \frac{\alpha_2}{\alpha_2 + \psi_b} - \frac{\alpha_2}{\alpha_2 + \psi_b}\frac{\cosh(\Gamma_2 \bar{x})}{\cosh\Gamma_2} \tag{4.261}$$

where

$$\Gamma^2 = \frac{l_c^2}{4}\frac{b_c G_s}{E_b A_b t_s}(\alpha + \psi_b) \tag{4.262}$$

The actuator strain is

$$\epsilon_{c_{\text{bottom}}} = \epsilon_{c1} + \epsilon_{c2} \tag{4.263}$$

The beam strain on the bottom surface is

$$\epsilon_b^{-s} = \epsilon_{b1}^{-s} + \epsilon_{b2}^{-s} \tag{4.264}$$

4.4.4 Dissimilar Actuators: Piezo Thickness ($t_{c_{top}} \neq t_{c_{bottom}}$)

This represents a case in which the thicknesses of the top and bottom piezos are not identical (see Fig. 4.13). For the same voltage, the actuation force due to top and bottom piezos will be dissimilar. The actuation force can be resolved into two parts: a force causing pure bending, F^b, and a force causing pure extension, F^e (see Fig. 4.14). The approach followed is similar to that described in Section 4.3.5. Let F_{top} and F_{bottom}, respectively, represent actuation forces due to top and bottom piezos. Then

$$F^e + F^b = F_{bottom} \tag{4.265}$$

$$F^e - F^b = F_{top} \tag{4.266}$$

4.4.5 Dissimilar Actuators: Piezo Constants ($d_{31_{top}} \neq d_{31_{bottom}}$)

This represents a case in which the top and bottom piezos are not identical in terms of induced strain (see Fig. 4.15). The free strain for the top and bottom piezos is given by

$$\Lambda_{top} = d_{31_{top}} \frac{V}{t_c} \tag{4.267}$$

$$\Lambda_{bottom} = d_{31_{bottom}} \frac{V}{t_c} \tag{4.268}$$

By superposing pure bending and pure extension relations, actuation forces for the top and bottom piezos can be derived in terms of free strains (see Fig. 4.14). In this case, the strains on the top and bottom surfaces of the actuator and beam, respectively, are given by

$$\epsilon_{c_{top}} = -\epsilon_c^b + \epsilon_c^e$$

$$\epsilon_b^s = -\epsilon_b^b + \epsilon_b^e$$

$$\epsilon_{c_{bottom}} = \epsilon_c^b + \epsilon_c^e$$

$$\epsilon_b^{-s} = \epsilon_b^b + \epsilon_b^e$$

where the superscript b refers to the quantity resulting from pure bending and the superscript e refers to the quantity resulting from pure extension. Similarly, the free strains on the top and bottom of the beam can also be separated into a component causing pure bending and a component causing pure extension.

$$\Lambda^e = \frac{\Lambda_{top} + \Lambda_{bottom}}{2}$$

$$\Lambda^b = \frac{-\Lambda_{top} + \Lambda_{bottom}}{2}$$

For a perfect-bond condition, the actuator and beam strains can be written as (Eqs. 4.187, 4.188, 4.191)

$$\frac{\epsilon_b^{-s}}{\Lambda} = \frac{\alpha}{\alpha + \Psi} \left(1 - \frac{\cosh(\Gamma \bar{x})}{\cosh \Gamma} \right)$$

$$\frac{\epsilon_c}{\Lambda} = \frac{\alpha}{\alpha + \Psi} \left(1 + \frac{\Psi}{\alpha} \frac{\cosh(\Gamma \bar{x})}{\cosh \Gamma} \right)$$

where

$$\alpha = 2 \quad \text{for pure extension}$$

$$\alpha = 6 \quad \text{for pure bending}$$

$$\Psi = \frac{E_b b_b t_b}{E_c b_c t_c}$$

$$(\Gamma^e)^2 = \frac{G_s}{E_c t_s t_c}\left(1 + \frac{2}{\Psi}\right)\frac{l_c^2}{4} \quad \text{pure extension}$$

$$(\Gamma^b)^2 = \frac{G_s}{E_c t_s t_c}\left(1 + \frac{6}{\Psi}\right)\frac{l_c^2}{4} \quad \text{pure bending}$$

Splitting these equations into pure extension and pure bending, the actuator strains are given by

$$\frac{\epsilon_c^e}{\Lambda^e} = \frac{2}{2+\Psi}\left(1 + \frac{\Psi}{2}\frac{\cosh(\Gamma^e\bar{x})}{\cosh \Gamma^e}\right) \tag{4.269}$$

$$\frac{\epsilon_c^b}{\Lambda^b} = \frac{6}{6+\Psi}\left(1 + \frac{\Psi}{6}\frac{\cosh(\Gamma^b\bar{x})}{\cosh \Gamma^b}\right) \tag{4.270}$$

The beam strains are given by

$$\frac{\epsilon_b^e}{\Lambda^e} = \frac{2}{2+\Psi}\left(1 - \frac{\cosh(\Gamma^e\bar{x})}{\cosh \Gamma^e}\right) \tag{4.271}$$

$$\frac{\epsilon_b^b}{\Lambda^b} = \frac{6}{6+\Psi}\left(1 - \frac{\cosh(\Gamma^b\bar{x})}{\cosh \Gamma^b}\right) \tag{4.272}$$

$$\frac{\tau^e}{\Lambda^e} = \frac{G_s \sinh(\Gamma^e\bar{x})}{\Gamma^e \bar{t}_s \cosh \Gamma^e} \tag{4.273}$$

$$\frac{\tau^b}{\Lambda^b} = \frac{G_s \sinh(\Gamma^b\bar{x})}{\Gamma^b \bar{t}_s \cosh \Gamma^b} \tag{4.274}$$

From these equations, the strains on the top and bottom of the beam, strains in the top and bottom actuators, and shear stress in the top and bottom bond layers can be calculated.

4.4.6 Worked Example

Two piezo-sheet actuators (PZT-5H and PZT-5A) (length $l_c = 50.8$ mm (2 in), width $b_c = 25.4$ mm (1 in), thickness $t_c = 0.3175$ mm (0.0125 in)) are surface-bonded at the top and bottom of a thin aluminum cantilevered beam (length $l_b = 609.6$ mm (24 in), width $b_b = 50.8$ mm (2 in), thickness $t_b = 0.79375$ mm (1/32 in)). The thickness of the bond layer t_s is 0.127 mm (0.005 in) and is assumed uniform. The configuration is shown in Fig. 4.28 ($x_o = 2$ in). Material data are as follows

$$E_c(\text{ PZT-5A and PZT-5H }) = E_b = 72.4 \text{ GPa } (10.5 \times 10^6 \text{ lb/in}^2)$$

$$d_{31}(\text{ PZT-5A }) = -171 \times 10^{-12} \text{ m/V}$$

$$d_{31}(\text{ PZT-5H }) = -274 \times 10^{-12} \text{ m/V}$$

$$\text{Bond-shear modulus } G_s = 965 \times 10^6 \text{ N/m}^2$$

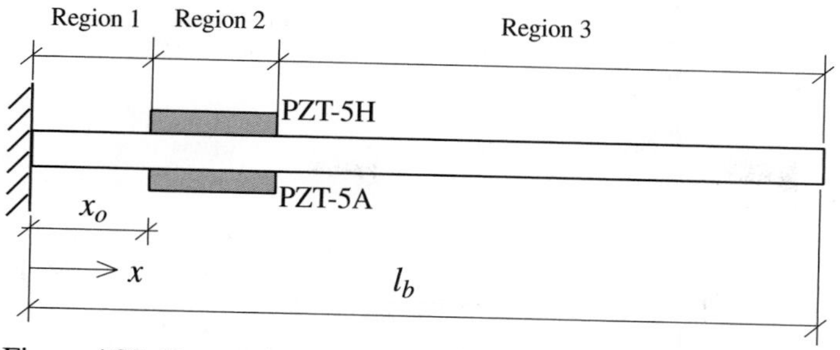

Figure 4.28. Beam with surface-bonded piezosheets, split into three regions.

(a) Using the uniform strain theory, derive general bending as well as extension relations with same field on opposite piezos.
(b) Plot the spanwise variation of beam surface strain and actuator strain for a field of 150 volts to both top and bottom piezos.
(c) Plot the variation of bond shearing force along the piezo span for this field.
(d) Calculate surface actuation force F in lb and bending moment M in in-lb for this excitation for two cases: with bond layer and with perfect bond.
(e) Plot the spanwise distribution of bending slope for this excitation.
(f) Plot the beam bending displacement distribution for this excitation.
(g) If PZT elements are replaced with PVDF elements of same size, calculate the new extensional actuation force and actuation bending moment for 150 volts excitation (for PVDF $d_{31} = -20 \times 10^{-12}$ m/V and $E_c = 0.2 \times 10^{10}$ N/m^2).
(h) Compare the calculated bending slope and displacement distributions with the results from the blocked-force method (see Worked Example in Section 6.5.2).

Solutions

(a),(b),(c) The derivation of the relations for actuator and beam strains is described in Section 4.4.5. The strains are obtained by superposing the bending and extensional strains as given by Eqs. 4.269–4.272. The shear stress is given by Eq. 4.273.
The differential shear force on the top and bottom of the beam is given by

$$dF = \tau b_c dx$$

This gives a shearing force per unit length. The results are shown in Figs. 4.29 and 4.30.

(d) To obtain the total actuation force (force acting on the beam), the shear force is integrated over the length of the actuator, in the region $0 < x < l_c/2$. Integration over the entire actuator length will result in a shear force of zero because the force is an internal force on the structure. The actuation force on the top and bottom can be split into a force producing

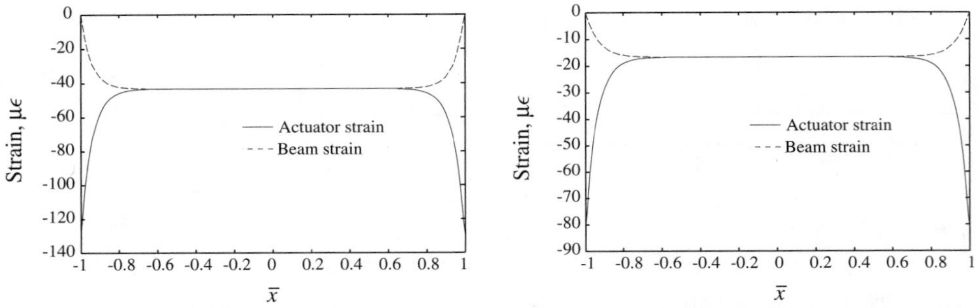

(a) Top of the beam (PZT-5H) (b) Bottom of the beam (PZT-5A)

Figure 4.29. Actuator and beam strains on the top and bottom of the beam.

pure extension, F^e, and a force producing pure bending, F^b.

$$F^e = \int_0^{l_c/2} \tau^e b_c dx = \int_0^1 \tau^e b_c d\bar{x} \frac{l_c}{2}$$

$$= \frac{b_c \Lambda^e G_s l_c}{2\Gamma^e \bar{t}_s \cosh \Gamma^e} \int_0^1 \sinh(\Gamma^e \bar{x})\, d\bar{x}$$

$$= \frac{b_c \Lambda^e G_s l_c}{2\Gamma^{e^2} \bar{t}_s \cosh \Gamma^e} [\cosh \Gamma^e - 1]$$

$$F^b = \frac{b_c \Lambda^b G_s l_c}{2\Gamma^{b^2} \bar{t}_s \cosh \Gamma^b} \left[\cosh \Gamma^b - 1\right]$$

From these equations

$$F^e = -43.8395 \text{ N} \quad F^b = 6.4573 \text{ N}$$

$$F_{\text{top}} = -50.2967 \text{ N} \quad F_{\text{bot}} = -37.3822 \text{ N}$$

$$M = F^b t_b = 6.4573 \times 0.79375 \times 10^{-3} = 5.1255 \times 10^{-3} \text{ Nm}$$

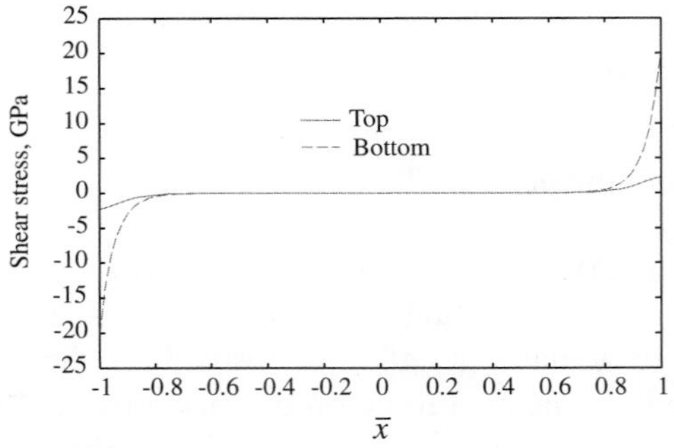

Figure 4.30. Shear stress along the top and bottom of the beam.

For a perfect bond, $\Gamma^e \to \infty$ and $\Gamma^b \to \infty$. The actuator and beam strains are given by

$$\epsilon^e_{c_{\text{perfect}}} = \epsilon^e_{b_{\text{perfect}}} = \frac{2}{2+\Psi}\Lambda^e = -30.03\mu\epsilon$$

$$\epsilon^b_{c_{\text{perfect}}} = \epsilon^b_{b_{\text{perfect}}} = \frac{6}{6+\Psi}\Lambda^b = 13.27\mu\epsilon$$

The actuation force and bending moment for a perfect bond are given by

$$\begin{aligned} F^e_{\text{perfect}} &= \frac{1}{2}\epsilon^e_{b_{\text{perfect}}} E_b b_b t_b \\ &= -30.03 \times 10^{-6} \times 72.4 \times 10^9 \times 0.0508 \times 0.79375 \times 10^{-3}/2 \\ &= -43.8340 \text{ N} \end{aligned}$$

$$\begin{aligned} M_{\text{perfect}} &= -\frac{EI_b}{t_b/2}\epsilon^b_{b_{\text{perfect}}} = -E_b b_b t_b^2 \frac{1}{6+\Psi}\Lambda^b \\ &= -72.4 \times 10^9 \times 0.0508 \times (0.79375 \times 10^{-3})^2 \times \frac{1}{6+\Psi} \times 24.33 \times 10^{-6} \\ &= 5.125 \times 10^{-3} \text{ Nm} \end{aligned}$$

(e),(f) Region 1: Slope and displacement are zero.
Region 2: The curvature at any point along the actuator span is given by

$$\begin{aligned} \frac{\partial^2 w}{\partial x^2}(\bar{x}) &= \frac{\epsilon^b_b}{t_b/2} \\ &= \frac{12\Lambda^b}{t_b(6+\Psi)}\left(1 - \frac{\cosh(\Gamma^b \bar{x})}{\cosh \Gamma^b}\right) \end{aligned}$$

The slope is obtained by integrating the curvature

$$\begin{aligned} \frac{\partial w}{\partial x}(\bar{x}) &= \int_{-1}^{\bar{x}} \frac{12\Lambda^b}{t_b(6+\Psi)}\left(1 - \frac{\cosh(\Gamma^b \bar{x})}{\cosh \Gamma^b}\right) d\bar{x}\, \frac{l_c}{2} \\ &= \frac{12\Lambda^b}{t_b(6+\Psi)}\frac{l_c}{2}\left[\bar{x} - \frac{\sinh(\Gamma^b \bar{x})}{\Gamma^b \cosh \Gamma^b}\right]_{-1}^{\bar{x}} \\ &= \frac{12\Lambda^b}{t_b(6+\Psi)}\frac{l_c}{2}\left[\bar{x} - \frac{\sinh(\Gamma^b \bar{x})}{\Gamma^b \cosh \Gamma^b} + 1 - \frac{\tanh \Gamma^b}{\Gamma^b}\right] \end{aligned}$$

and the displacement is

$$\begin{aligned} w(\bar{x}) &= \int_{-1}^{\bar{x}} \frac{12\Lambda^b}{t_b(6+\Psi)}\frac{l_c}{2}\left[\bar{x} - \frac{\sinh(\Gamma^b \bar{x})}{\Gamma^b \cosh \Gamma^b} + 1 - \frac{\tanh \Gamma^b}{\Gamma^b}\right] d\bar{x}\, \frac{l_c}{2} \\ &= \frac{12\Lambda^b}{t_b(6+\Psi)}\left(\frac{l_c}{2}\right)^2\left[\frac{\bar{x}^2}{2} - \frac{\cosh(\Gamma^b \bar{x})}{\Gamma^{b2} \cosh \Gamma^b} + \bar{x} - \frac{\tanh \Gamma^b}{\Gamma^b}\bar{x}\right]_{-1}^{\bar{x}} \\ &= \frac{12\Lambda^b}{t_b(6+\Psi)}\left(\frac{l_c}{2}\right)^2\left[\frac{\bar{x}^2}{2} - \frac{\cosh(\Gamma^b \bar{x})}{\Gamma^{b2} \cosh \Gamma^b} + \bar{x} - \frac{\tanh \Gamma^b}{\Gamma^b}\bar{x} + \frac{1}{\Gamma^{b2}} + \frac{1}{2} - \frac{\tanh \Gamma^b}{\Gamma^b}\right] \end{aligned}$$

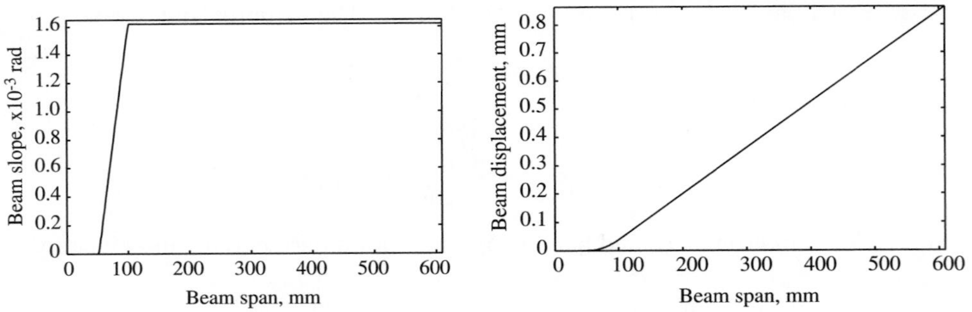

(a) Bending slope (b) Bending displacement

Figure 4.31. Bending slope and bending displacement of the beam.

For a perfect bond, the slope and displacement are given by

$$\frac{\partial w_{\text{perfect}}}{\partial x}(\bar{x}) = \frac{6\Lambda^b l_c}{t_b(6+\Psi)}(\bar{x}+1)$$

$$w_{\text{perfect}}(\bar{x}) = \frac{3\Lambda^b l_c^2}{t_b(6+\Psi)}(\bar{x}^2/2+\bar{x}+1/2)$$

Region 3: The slope remains constant and equal to the value at the end of Region 2.

$$\frac{\partial w}{\partial x} = 1.6203 \times 10^{-3}\text{ rad}$$

The bending displacement increases linearly.

$$w = w\,|_{\bar{x}=1} + (x - l_c)\left.\frac{\partial w}{\partial x}\right|_{\bar{x}=1}$$

The tip displacement $w_{\text{tip}} = 0.8636$ mm. The bending slope and displacement are shown in Fig. 4.31. For a perfect bond, the bending slope is

$$\frac{\partial w_{\text{perfect}}}{\partial x} = \frac{12\Lambda^b l_c}{t_b(6+\Psi)} = 1.6987 \times 10^{-3}\text{ rad}$$

and the tip displacement is

$$w_{\text{perfect}_{\text{tip}}} = \frac{6\Lambda^b l_c^2}{t_b(6+\Psi)} + (l_b - l_c)\frac{12\Lambda^b l_c}{t_b(6+\Psi)}$$
$$= 0.9493\text{ mm}$$

(g) If the actuators are replaced with PVDF, the configuration is symmetric $\rightarrow M = 0$. The actuation force is $F^e = -0.150734$ N. The beam and actuator strains are shown in Fig. 4.32.

(h) The comparison of blocked-force and uniform-strain results is shown in Table 4.5. The uniform-strain theory predicts lower deflections than the blocked-force method.

Table 4.5. *Comparison of blocked-force and uniform-strain theory*

	Blocked-force	Uniform-strain	% Deviation	Uniform-strain – Perfect Bond	% Deviation
Tip slope ($\times 10^{-3}$ rad)	1.7000	1.6203	4.69	1.6987	0.08
Tip displacement (mm)	0.9068	0.8636	4.76	0.9493	4.70

4.5 Euler-Bernoulli Beam Model

The Euler-Bernoulli model is a consistent strain model and generally gives more accurate results for slender beams than the uniform-strain model, especially for thin bond layers. This model considers the beam, adhesive, and actuator as a continuous structure and follows the Euler-Bernoulli assumptions for beam bending. This implies that a plane section normal to the beam axis in the undeformed state remains plane and normal to the beam axis after bending. The effects of transverse shears on bending deformation are neglected. There is a linear distribution of strain in the cross section for both the actuator and the host structure. There is no variation of transverse displacement (w) across the thickness. Using this approach, the deformation of a beam structure is derived with single and dual actuators in the same configurations as in the previous two models. Note that although the previous two models used simplifying assumptions for the strain distribution in the actuator, essentially treating it as a force generator applied to the structure, the Euler-Bernoulli model considers the actuators as an integral part of the structure. The sign convention is defined such that a positive axial force corresponds to tension in the beam, and a positive moment and a positive shear force are as indicated in Fig. 4.4.

4.5.1 Dual Actuators: Symmetric Actuation

Consider two identical piezo-sheet actuators, surface-bonded on either surface of an isotropic beam. Figure 4.33 shows a differential element of the beam and the beam coordinates.

The axial displacement and strain are defined as

$$u(x, z) = u_o(x) - z\frac{\partial w(x)}{\partial x} \tag{4.275}$$

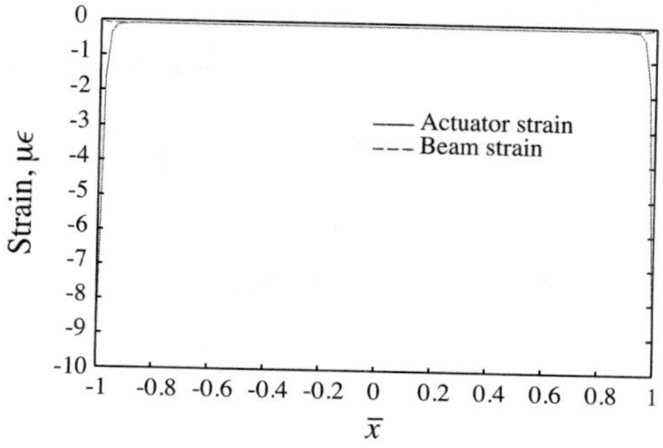

Figure 4.32. Actuator and beam strains on the top of the beam, PVDF actuator.

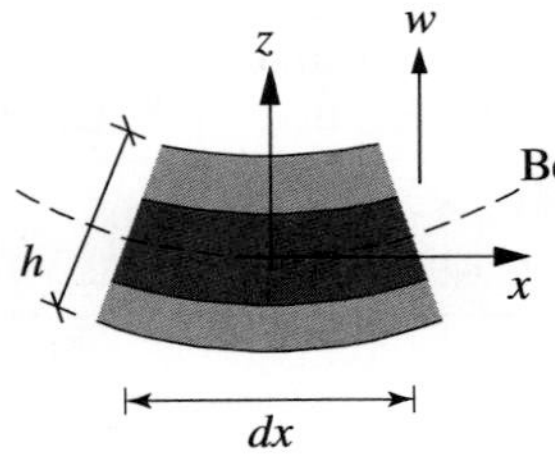

Figure 4.33. Differential beam element and coordinates.

where $u_o(x)$ is the axial displacement at the neutral axis. Thus, the axial strain varies linearly through the thickness according to

$$\begin{aligned}\epsilon(x,z) &= \epsilon_o(x) - z\frac{\partial^2 w(x)}{\partial x^2} \\ &= \epsilon_o(x) - z\kappa\end{aligned} \tag{4.276}$$

where κ is the bending curvature. From the piezoelectric constitutive relations, the stress in the active layer of the beam (i.e., the piezo actuator sheets) is given by

$$\sigma(x,z) = E(x,z)\left[\epsilon(x,z) - \Lambda(x,z)\right] \tag{4.277}$$

For simplicity, this can be written as

$$\sigma(z) = E(z)\left[\epsilon(z) - \Lambda(z)\right] \tag{4.278}$$

The normal (axial) stress for the complete beam in the x-direction is given by Eq. 4.278. Because we are considering the piezo to be an integral part of the beam, the modulus E varies in the z-direction. Note that this equation can also be applied to the passive layers of the beam, by setting $\Lambda(z) = 0$. For a beam of total thickness h (including the actuators), and width $b(z)$, the net axial force is given by

$$\begin{aligned}F &= \int_{-h/2}^{h/2} b(z)\sigma(z)\,dz \\ &= \int_{-h/2}^{h/2} b(z)E(z)[\epsilon_o - z\kappa - \Lambda(z)]\,dz \\ &= \int_{-h/2}^{h/2} \epsilon_o b(z)E(z)\,dz - \kappa\int_{-h/2}^{h/2} b(z)E(z)z\,dz - \int_{-h/2}^{h/2} b(z)E(z)\Lambda(z)\,dz\end{aligned} \tag{4.279}$$

Rewriting

$$F = \epsilon_o EA_{\text{tot}} + \kappa ES_{\text{tot}} - F_\Lambda \qquad \text{(N)} \tag{4.280}$$

$$EA_{\text{tot}} = \int_{-h/2}^{h/2} b(z)E(z)\,dz \qquad \text{(N)} \tag{4.281}$$

$$ES_{\text{tot}} = -\int_{-h/2}^{h/2} b(z)E(z)z\,dz \qquad \text{(N.m)} \tag{4.282}$$

$$F_\Lambda = \int_{-h/2}^{h/2} b(z)E(z)\Lambda(z)\,dz \qquad \text{(N)} \tag{4.283}$$

where F_Λ is the axial force due to induced strain. EA_{tot} is the resultant extensional stiffness and ES_{tot} is an equivalent coupling stiffness. The total moment in the beam is given by

$$\begin{aligned} M &= -\int_{-h/2}^{h/2} zb(z)\sigma(z)\,dz \\ &= -\int_{-h/2}^{h/2} zb(z)E(z)[\epsilon_o - z\kappa - \Lambda(z)]\,dz \\ &= -\int_{-h/2}^{h/2} \epsilon_o zb(z)E(z)\,dz + \kappa\int_{-h/2}^{h/2} b(z)E(z)z^2\,dz + \int_{-h/2}^{h/2} zb(z)E(z)\Lambda(z)\,dz \end{aligned} \tag{4.284}$$

Rewriting

$$M = \epsilon_o ES_{\text{tot}} + \kappa EI_{\text{tot}} - M_\Lambda \qquad \text{(N.m)} \tag{4.285}$$

where

$$EI_{\text{tot}} = \int_{-h/2}^{h/2} b(z)E(z)z^2\,dz \qquad (\text{N.m}^2) \tag{4.286}$$

$$M_\Lambda = -\int_{-h/2}^{h/2} b(z)E(z)\Lambda(z)z\,dz \qquad \text{(N.m)} \tag{4.287}$$

EI_{tot} is the resulting bending stiffness and M_Λ is the bending moment due to induced strain. If the placement of actuators is symmetric, the coupling term ES_{tot} will be zero. If an actuator is attached only on one side, this term will be nonzero, resulting in an extension-bending coupling. Only actuator layers will contribute to the F_Λ and M_Λ terms. The contributions of the passive layers will be zero.

$$F + F_\Lambda = EA_{\text{tot}}\epsilon_o + ES_{\text{tot}}w'' \tag{4.288}$$

Similarly, for the bending moment

$$M + M_\Lambda = ES_{\text{tot}}\epsilon_o + EI_{\text{tot}}w'' \tag{4.289}$$

where the curvature κ is defined as

$$\kappa = w'' = \frac{\partial^2 w}{\partial x^2} \tag{4.290}$$

Combining these into a matrix equation

$$\begin{Bmatrix} F + F_\Lambda \\ M + M_\Lambda \end{Bmatrix} = \begin{bmatrix} EA_{\text{tot}} & ES_{\text{tot}} \\ ES_{\text{tot}} & EI_{\text{tot}} \end{bmatrix} \begin{Bmatrix} \epsilon_o \\ w'' \end{Bmatrix} \tag{4.291}$$

If there is no mechanical load on the structure, $F = 0$ and $M = 0$

$$\begin{Bmatrix} F_\Lambda \\ M_\Lambda \end{Bmatrix} = \begin{bmatrix} EA_{\text{tot}} & ES_{\text{tot}} \\ ES_{\text{tot}} & EI_{\text{tot}} \end{bmatrix} \begin{Bmatrix} \epsilon_o \\ w'' \end{Bmatrix} \tag{4.292}$$

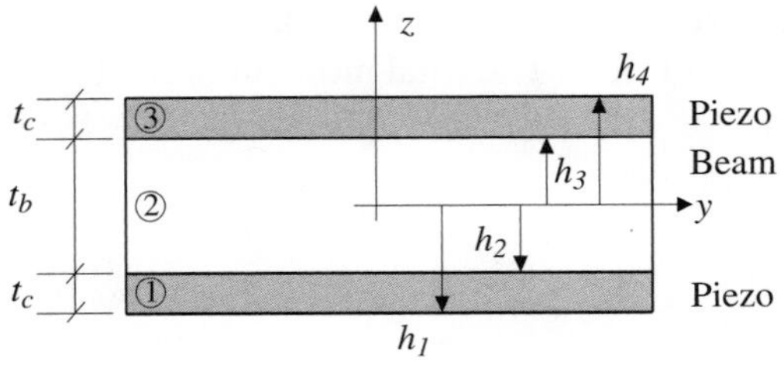

Figure 4.34. Isotropic beam substructure with symmetric surface-bonded actuators.

Let us assume that the beam consists of N layers. Any of these layers can represent an active layer (actuator) or a passive layer (structure or adhesive). Then

$$EA_{\text{tot}} = \sum_{k=1}^{N} b_k E_k (h_{k+1} - h_k) \tag{4.293}$$

$$ES_{\text{tot}} = -\frac{1}{2}\sum_{k=1}^{N} b_k E_k (h_{k+1}^2 - h_k^2) \tag{4.294}$$

$$EI_{\text{tot}} = \frac{1}{3}\sum_{k=1}^{N} b_k E_k (h_{k+1}^3 - h_k^3) \tag{4.295}$$

$$F_\Lambda = \sum_{k=1}^{N} \Lambda_k b_k E_k (h_{k+1} - h_k) \tag{4.296}$$

$$M_\Lambda = -\frac{1}{2}\sum_{k=1}^{N} b_k E_k \Lambda_k (h_{k+1}^2 - h_k^2) \tag{4.297}$$

where h_k is the vertical position of the interface between two different layers. With this approach, it is very easy to incorporate the effects of different widths or thicknesses of each layer, as well as any differences in modulus or free strain of the layers. The properties for each layer can be substituted in these equations without having to re-derive the relations for each configuration.

Consider an isotropic beam with two identical surface-bonded actuators, as in Fig. 4.34. For a symmetric layup, $ES_{\text{tot}} = 0$. The vertical positions of individual layers can be represented as

$$h_1 = -\left(\frac{t_b}{2} + t_c\right) \tag{4.298}$$

$$h_2 = -\frac{t_b}{2} \tag{4.299}$$

$$h_3 = \frac{t_b}{2} \tag{4.300}$$

$$h_4 = \frac{t_b}{2} + t_c \tag{4.301}$$

Figure 4.35. Beam with symmetric surface-bonded actuators in pure extension.

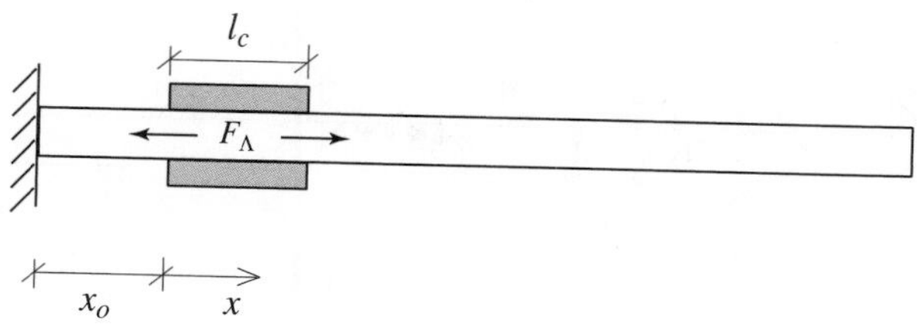

I. Pure Extension Case

The same voltage applied to the top and bottom piezo-sheets causes pure extension, as shown in Fig. 4.35. F_Λ is a uniform extension force induced in the beam due to the piezos, acting on the region $0 \leq x \leq l_c$. From Eqs. 4.293–4.297

$$F_\Lambda = \sum_{k=1}^{3} \Lambda_k b_k E_k (h_{k+1} - h_k) \tag{4.302}$$

$$= E_c b_c \Lambda \left[h_2 - h_1 + h_4 - h_3 \right] \tag{4.303}$$

$$= E_c b_c \Lambda \left[-\frac{t_b}{2} + \left(\frac{t_b}{2} + t_c \right) + \frac{t_b}{2} + t_c - \frac{t_b}{2} \right] \tag{4.304}$$

$$= 2 E_c b_c t_c \Lambda \tag{4.305}$$

$$= EA_c \Lambda \tag{4.306}$$

where EA_c is the extensional stiffness of the two actuators.

$$EA_{\text{tot}} = \sum_{k=1}^{3} E_k b_k (h_{k+1} - h_k) = 2 E_c t_c b_c + E_b b_b t_b \tag{4.307}$$

$$= EA_c + EA_b \tag{4.308}$$

$$M_\Lambda = -\frac{1}{2} \sum_{k=1}^{3} E_k b_k \Lambda_k (h_{k+1}^2 - h_k^2) \tag{4.309}$$

$$= 0 \tag{4.310}$$

The axial strain of the beam is

$$\epsilon_o = \frac{F_\Lambda}{EA_{\text{tot}}} = \frac{2 b_c E_c t_c \Lambda}{EA_b + EA_c} \tag{4.311}$$

$$= \frac{EA_c}{EA_b + EA_c} \Lambda \tag{4.312}$$

The axial displacement can be determined from the strain as

$$\epsilon_o = \frac{\partial u}{\partial x} = \frac{EA_c}{EA_b + EA_c} \Lambda \tag{4.313}$$

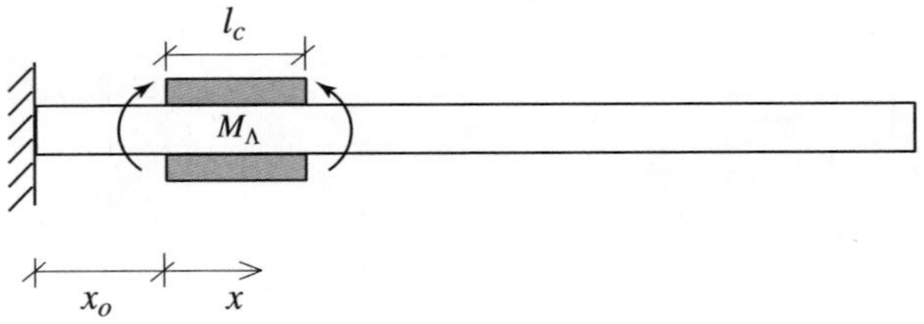

Figure 4.36. Beam with symmetric surface-bonded actuators in pure bending.

from which

$$x < 0 \qquad u = 0$$

$$0 \le x \le l_c \qquad u = \frac{EA_c}{EA_c + EA_b}\Lambda x$$

$$l_c < x \qquad u = \frac{EA_c}{EA_c + EA_b}\Lambda l_c$$

II. Pure Bending Case

For a positive voltage applied to the bottom piezo and a negative voltage applied to the top piezo, a bending deformation results as shown in Fig. 4.36.

$$\Lambda_1 = \Lambda, \quad \Lambda_2 = 0, \quad \Lambda_3 = -\Lambda$$

The induced bending moment M_Λ is uniform over the region where the piezos are attached ($0 \le x \le l_c$).

$$M_\Lambda = -\frac{1}{2}\sum_{k=1}^{3} E_k b_k \Lambda_k (h_{k+1}^2 - h_k^2) \tag{4.314}$$

$$= E_c b_c \Lambda \left[\left(\frac{t_b}{2} + t_c\right)^2 - \frac{t_b^2}{4}\right] = E_c b_c \Lambda t_c (t_b + t_c) \tag{4.315}$$

$$EI_{\text{tot}} = \frac{1}{3}\sum_{k=1}^{3} E_k b_k (h_{k+1}^3 - h_k^3) \tag{4.316}$$

$$= E_c t_c b_c \left(\frac{2}{3}t_c^2 + t_b t_c + \frac{t_b^2}{2}\right) + \frac{E_b b_b t_b^3}{12} \tag{4.317}$$

$$= EI_b + EI_c \tag{4.318}$$

where EI_b is bending stiffness of the beam alone and EI_c is the bending stiffness of the two actuators, which can be written as

$$EI_b = \frac{E_b b_b t_b^3}{12} \tag{4.319}$$

$$EI_c = \frac{2E_c b_c t_c^3}{12} + 2E_c b_c t_c \left(\frac{t_c}{2} + \frac{t_b}{2}\right)^2 \tag{4.320}$$

Note that the first term of this equation is the flexural stiffness of the actuators about their own mid-axis, and the second term is the flexural stiffness of the

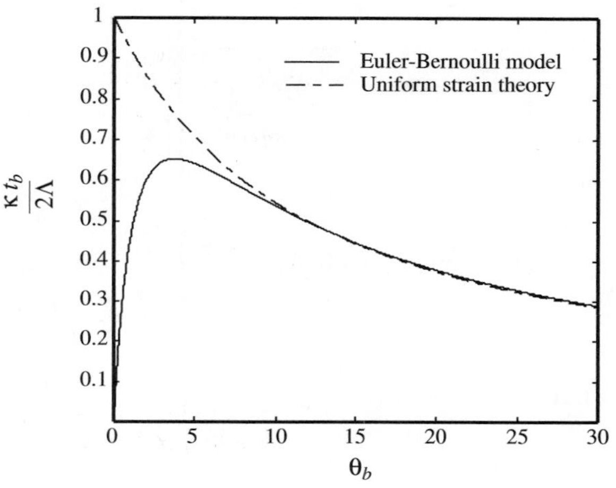

Figure 4.37. Variation of normalized curvature with thickness ratio for the perfect-bond condition.

actuators about the beam mid-axis. In this case, there is no induced axial deformation. Therefore

$$F_\Lambda = 0 \quad \text{and} \quad \epsilon_o = 0$$

The bending deformation can be calculated by

$$M_\Lambda = EI_{\text{tot}} w'' \tag{4.321}$$

The axial strain distribution is

$$\begin{aligned}\epsilon(z) &= -z\frac{M_\Lambda}{EI_{\text{tot}}} \\ &= -\frac{6(1+\frac{1}{\theta_b})\frac{2}{t_b}\Lambda}{(\Psi+6)+\frac{12}{\theta_b}+\frac{8}{\theta_b^2}} z\end{aligned}$$

where

$$\begin{aligned}\Psi &= \text{extensional stiffness ratio} = \frac{E_b t_b b_b}{E_c t_c b_c} \\ &= \frac{E_b A_b}{E_c A_c}\end{aligned}$$

and

$$\theta_b = \frac{\text{beam thickness}}{\text{actuator thickness}} = \frac{t_b}{t_c}$$

The thickness ratio, θ_b, determines whether the strain variation across the piezo element affects the analysis. Figure 4.37 shows the variation of the normalized curvature with the thickness ratio. For thin beams, the uniform-strain model overpredicts strain (curvature). For beams with large thickness ratio ($\theta_b > 8$), induced deformations are identical using both models, away from the edges of the actuator. For induced bending, the Euler-Bernoulli and detailed finite element models predict identical curvatures. The bending slope and deflection can be derived from the

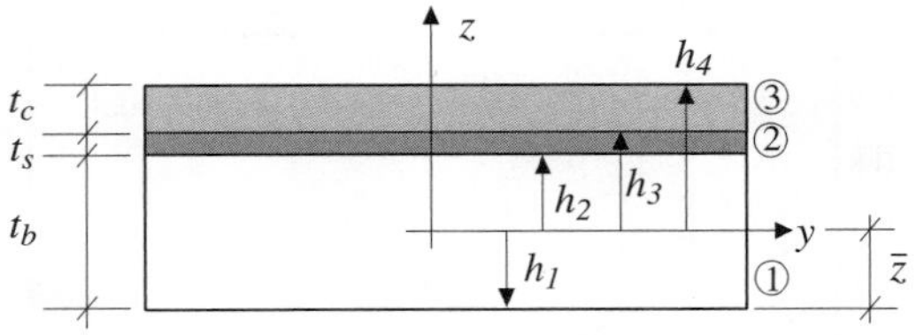

Figure 4.38. Single surface-bonded actuator with bond layer.

induced moment, which gives the beam curvature as

$$\frac{\partial^2 w}{\partial x^2} = \frac{M_\Lambda}{EI_{\text{tot}}} \tag{4.322}$$

where

$$M_\Lambda = E_c b_c t_c (t_b + t_c)\Lambda \tag{4.323}$$

$$EI_{\text{tot}} = EI_b + EI_c \tag{4.324}$$

The bending slope is

$$x < 0 \quad \frac{\partial w}{\partial x} = 0$$

$$0 \leq x \leq l_c \quad \frac{\partial w}{\partial x} = \frac{M_\Lambda}{EI_{\text{tot}}} x$$

$$l_c < x \quad \frac{\partial w}{\partial x} = \frac{M_\Lambda}{EI_{\text{tot}}} l_c$$

The bending deflection is

$$x < 0 \quad w = 0$$

$$0 \leq x \leq l_c \quad w = \frac{M_\Lambda}{EI_{\text{tot}}} \frac{x^2}{2}$$

$$l_c < x \quad w = \frac{M_\Lambda}{EI_{\text{tot}}} l_c (x - l_c/2)$$

4.5.2 Single Actuator: Asymmetric Actuation

In this case, a single piezo actuator is surface-mounted on a cantilevered beam. The cross section of the beam with the positions of individual layers is shown in Fig. 4.38. The effect of the bond layer is also included in this example as another beam layer. An electric field will induce both bending and extension of the beam. Force and moment equilibrium obtained by integration over the cross section provides the governing equations (Eq. 4.291)

$$\begin{Bmatrix} F + F_\Lambda \\ M + M_\Lambda \end{Bmatrix} = \begin{bmatrix} EA_{\text{tot}} & ES_{\text{tot}} \\ ES_{\text{tot}} & EI_{\text{tot}} \end{bmatrix} \begin{Bmatrix} \epsilon_o \\ w'' \end{Bmatrix} \tag{4.325}$$

In the absence of external loads, the deformations of the beam are given by (Eq. 4.292)

$$\begin{Bmatrix} F_\Lambda \\ M_\Lambda \end{Bmatrix} = \begin{bmatrix} EA_{\text{tot}} & ES_{\text{tot}} \\ ES_{\text{tot}} & EI_{\text{tot}} \end{bmatrix} \begin{Bmatrix} \epsilon_o \\ w'' \end{Bmatrix} \tag{4.326}$$

There are two approaches to solve this problem. The first approach involves determination of the neutral axis ($\bar{z}$), to which the locations of the beam and actuators are referenced. In the second approach, the mid-axis of the beam is used as the reference, and coupling terms appear.

Approach I

In the previous examples, to simplify the analysis, the position of the neutral axis was assumed to be at the mid-plane of the beam. However, in the Euler-Bernoulli method, the position of the neutral axis can easily be included in the analysis even if it is not at the mid-plane of the beam. At the neutral axis, $\epsilon_o = 0$ for a pure bending condition. The position of the neutral axis, $\bar{z}$, can be found by

$$\begin{aligned}\bar{z} &= \frac{\int_z E(z)b(z)z\,dz}{\int_z E(z)b(z)dz} \\ &= \frac{E_b b_b t_b(\frac{t_b}{2}) + E_s b_c t_s(t_b + \frac{t_s}{2}) + E_c b_c t_c(t_b + t_s + \frac{t_c}{2})}{E_b b_b t_b + E_s b_c t_s + E_c b_c t_c}\end{aligned} \tag{4.327}$$

where the reference vertical position is taken as the bottom surface of the beam. The cross-sectional properties with respect to the neutral axis are

$$\begin{aligned}EA_{\text{tot}} &= E_b b_b t_b + E_s b_c t_s + E_c b_c t_c \\ ES_{\text{tot}} &= E_b b_b t_b\left(\frac{t_b}{2} - \bar{z}\right) + E_s b_s t_s\left(\frac{t_s}{2} + t_b - \bar{z}\right) + E_c b_c t_c\left(\frac{t_c}{2} + t_b + t_s - \bar{z}\right) \\ &= 0 \\ EI_{\text{tot}} &= \frac{1}{12}E_b b_b t_b^3 + E_b b_b t_b\left(\frac{t_b}{2} - \bar{z}\right)^2 + \frac{1}{12}E_s b_s t_s^3 + E_s b_s t_s\left(\frac{t_s}{2} + t_b - \bar{z}\right)^2 \\ &\quad + \frac{1}{12}E_c b_c t_c^3 + E_c b_c t_c\left(\frac{t_c}{2} + t_s + t_b - \bar{z}\right)^2\end{aligned} \tag{4.328}$$

The induced forces and moments are

$$F_\Lambda = E_c b_c t_c \Lambda \tag{4.329}$$

$$M_\Lambda = -E_c b_c t_c\left(\frac{t_c}{2} + t_b + t_s - \bar{z}\right)\Lambda \tag{4.330}$$

With respect to the neutral axis, if no external forces or moments are present, Eq. 4.292 simplifies to the uncoupled system

$$\begin{bmatrix} EA_{\text{tot}} & 0 \\ 0 & EI_{\text{tot}} \end{bmatrix}\begin{Bmatrix}\epsilon_o \\ \kappa\end{Bmatrix} = \begin{Bmatrix}F_\Lambda \\ M_\Lambda\end{Bmatrix} \tag{4.331}$$

The advantage in writing the equations with respect to the neutral axis is the elimination of coupling between extension and bending.

Approach II

The same solution can be obtained by referring the sectional properties EA_{tot}, ES_{tot}, and EI_{tot} to any vertical location on the beam cross section. In such a case, the coupling term, $ES_{\text{tot}} \neq 0$, and the coupled set of equations (Eq. 4.292) must be solved. It is worth mentioning here that the latter method is usually much simpler in terms of alegbraic manipulatons, despite the presence of the coupling term.

The stiffness terms are given by

$$\begin{aligned}
EA_{\text{tot}} &= E_b b_b t_b + E_s b_c t_s + E_c b_c t_c \\
ES_{\text{tot}} &= E_s b_s t_s \left(\frac{t_s}{2} + \frac{t_b}{2}\right) + E_c b_c t_c \left(\frac{t_c}{2} + t_s + \frac{t_b}{2}\right) \\
EI_{\text{tot}} &= \frac{1}{12} E_b b_b t_b^3 + \frac{1}{12} E_s b_s t_s^3 + E_s b_s t_s \left(\frac{t_s}{2} + \frac{t_b}{2}\right)^2 \\
&\quad + \frac{1}{12} E_c b_c t_c^3 + E_c b_c t_c \left(\frac{t_c}{2} + t_s + \frac{t_b}{2}\right)^2
\end{aligned} \tag{4.332}$$

The forcings are given by

$$\begin{aligned}
F_\Lambda &= E_c b_c t_c \Lambda \\
M_\Lambda &= -E_c b_c t_c \left(\frac{t_c}{2} + \frac{t_b}{2} + t_s\right) \Lambda
\end{aligned} \tag{4.333}$$

The coupled system is

$$\begin{bmatrix} EA_{\text{tot}} & ES_{\text{tot}} \\ ES_{\text{tot}} & EI_{\text{tot}} \end{bmatrix} \begin{Bmatrix} \epsilon_o \\ \kappa \end{Bmatrix} = \begin{Bmatrix} F_\Lambda \\ M_\Lambda \end{Bmatrix} \tag{4.334}$$

4.5.3 Unequal Electric Voltage ($V_{\text{top}} \neq V_{\text{bottom}}$)

Consider a beam with two identical actuators and with an unequal voltage applied to the top and bottom actuators (see Fig. 4.11). We can resolve this problem into two parts: pure extension and pure bending problems, and then use superposition to obtain the composite solution (see Fig. 4.12). This gives

$$\begin{aligned}
V_1 - V_2 &= V_{\text{top}} \\
V_1 + V_2 &= V_{\text{bottom}} \\
V_1 &= \frac{V_{\text{top}} + V_{\text{bottom}}}{2} \text{ (extension)} \\
V_2 &= \frac{V_{\text{bottom}} - V_{\text{top}}}{2} \text{ (bending)}
\end{aligned} \tag{4.335}$$

V_1 produces no bending moment, and it causes an axial induced force given by

$$\begin{aligned}
F_\Lambda &= 2E_c b_c t_c \Lambda_1 = 2E_c A_c d_{31} \frac{V_1}{t_c} \\
&= EA_{\text{tot}} \epsilon_o
\end{aligned} \tag{4.336}$$

V_2 produces no axial force, and it causes an induced bending moment given by

$$\begin{aligned}
M_\Lambda &= E_c b_c t_c (t_c + 2t_s + t_b) \Lambda_2 = E_c A_c (t_c + 2t_s + t_b) d_{31} \frac{V_2}{t_c} \\
&= EI_{\text{tot}} \kappa
\end{aligned} \tag{4.337}$$

The solution is

$$\epsilon_o = \frac{F_\Lambda}{EA_{\text{tot}}} \quad (4.338)$$

$$\kappa = \frac{M_\Lambda}{EI_{\text{tot}}} \quad (4.339)$$

4.5.4 Dissimilar Actuators: Piezo Thickness ($t_{c_{\text{top}}} \neq t_{c_{\text{bottom}}}$)

This represents a case in which the thickness of the top and bottom piezos are not identical (see Fig. 4.13). For the same voltage, the actuation force due to the top and bottom piezos will be dissimilar.

Using the mid-axis as the reference axis, the bending-extension relations are coupled and can be written as

$$\begin{Bmatrix} F_\Lambda \\ M_\Lambda \end{Bmatrix} = \begin{bmatrix} EA_{\text{tot}} & ES_{\text{tot}} \\ ES_{\text{tot}} & EI_{\text{tot}} \end{bmatrix} \begin{Bmatrix} \epsilon_o \\ \kappa \end{Bmatrix} \quad (4.340)$$

where

$$EA_{\text{tot}} = E_c A_{c_{top}} + E_c A_{c_{bottom}} + E_b A_b = EA_{c_{top}} + EA_{c_{bottom}} + EA_b \quad (4.341)$$

$$ES_{\text{tot}} = \frac{1}{2}\left[EA_{c_{bottom}}(t_{c_{bottom}} + t_b) - EA_{c_{top}}(t_{c_{top}} + t_b)\right] \quad (4.342)$$

$$\begin{aligned} EI_{\text{tot}} &= \frac{1}{3}EA_{c_{bottom}}\left[\frac{3}{4}t_b^2 + \frac{3}{2}t_b t_{c_{bottom}} + t_{c_{bottom}}^2\right] + \frac{1}{12}EA_b t_b^2 \\ &\quad + \frac{1}{3}EA_{c_{top}}\left[\frac{3}{4}t_b^2 + \frac{3}{2}t_b t_{c_{top}} + t_{c_{top}}^2\right] \end{aligned} \quad (4.343)$$

$$F_\Lambda = d_{31}E_c b_c(V_{top} + V_{bottom}) = 2d_{31}E_c b_c V \quad (4.344)$$

$$\begin{aligned} M_\Lambda &= \frac{1}{2}E_c b_c d_{31}\left[V_{bottom}(t_{c_{bottom}} + t_b) - V_{top}(t_{c_{top}} + t_b)\right] \\ &= \frac{1}{2}E_c b_c d_{31} V\left[t_{c_{bottom}} - t_{c_{top}}\right] \end{aligned} \quad (4.345)$$

4.5.5 Dissimilar Actuators: Piezo Constants ($d_{31_{\text{top}}} \neq d_{31_{\text{bottom}}}$)

This represents a case in which top and bottom piezos are not identical in terms of induced strain (see Fig. 4.15). Free strain for top and bottom piezos

$$\Lambda_1 = d_{31_{\text{top}}}\frac{V}{t_c} \quad (4.346)$$

$$\Lambda_2 = d_{31_{\text{bottom}}}\frac{V}{t_c} \quad (4.347)$$

Using displacement compatibility conditions, the actuation forces for the top and bottom piezos can be derived in terms of their free strains, and the solution is

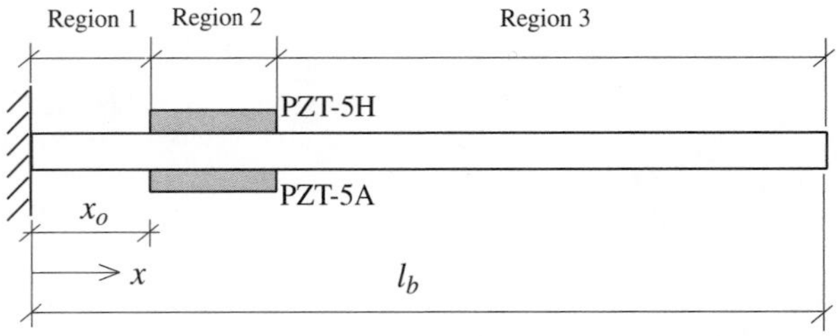

Figure 4.39. Beam with surface-bonded piezo sheets, split into three regions.

similar to Eqs. 4.345 and 4.346.

$$F_\Lambda = E_c b_c t_c(\Lambda_1 + \Lambda_2) \tag{4.348}$$

$$M_\Lambda = \frac{1}{2} E_c b_c t_c(t_c + 2t_s + t_b)(\Lambda_2 - \Lambda_1) \tag{4.349}$$

4.5.6 Worked Example

Two piezo-sheet actuators (PZT-5H and PZT-5A) (length $l_c = 50.8$ mm (2 in), width $b_c = 25.4$ mm (1 in), thickness $t_c = 0.3175$ mm (0.0125 in)) are surface-bonded at the top and bottom of a thin aluminum cantilevered beam (length $l_b = 609.6$ mm (24 in), width $b_b = 50.8$ mm (2 in), and thickness $t_b = 0.79375$ mm ($\frac{1}{32}$ in)). The thickness of the bond layer t_s is 0.127 mm (0.005 in) and is assumed uniform. The configuration is shown in Fig. 4.39 ($x_o = 2''$). Material data are as follows

$$E_c(\text{ PZT-5A and PZT-5H }) = E_b = 72.4 \text{ GPa } (10.5 \times 10^6 \text{ lb/in}^2)$$

$$d_{31}(\text{ PZT-5A }) = -171 \times 10^{-12} \text{ m/V}$$

$$d_{31}(\text{ PZT-5H }) = -274 \times 10^{-12} \text{ m/V}$$

$$\text{Bond shear modulus } G_s = 965 \times 10^6 \text{ N/m}^2$$

(a) Using the Euler-Bernoulli theory, derive general bending as well as extension relations with the same field on opposite piezo actuators.
(b) Plot spanwise variation of beam-surface strain and actuator strain for a field of 150 volts to both top and bottom piezos.
(c) Calculate actuation surface force F in lb and bending moment M in in-lb for this excitation for two cases: with bond layer and with perfect bond.
(d) Show spanwise distribution of bending slope for this excitation.
(e) Show beam bending displacement distribution for this excitation.
(f) If PZT-5H and PZT-5A elements are replaced with PVDF elements of the same size, calculate the new surface actuation strain and actuation bending moment for a field of 150 volts to both top and bottom piezos (for PVDF $d_{31} = -20 \times 10^{-12}$ m/V and $E_c = 0.2 \times 10^{10}$ N/m^2).
(g) Compare the calculated bending slope and displacement distributions with the results from the blocked-force method and the uniform-strain method (see Worked Example in Section 8.9.3).

Solution

(a),(b),(c) The problem can be split into a summation of pure bending and pure extension. The stiffness parameters EA_{tot}, ES_{tot}, and EI_{tot} depend only on the geometry and modulus of the elements of the beam. Because the structure is symmetric about the mid-plane of the beam, $ES_{\text{tot}} = 0$. The problem, therefore, reduces to one of uncoupled bending and extension.

$$EA_{\text{tot}}\epsilon_o = F_\Lambda$$

$$EI_{\text{tot}}\kappa = M_\Lambda$$

where ϵ_o and κ are the mid-plane strain and curvature of the beam, respectively. The total extensional stiffness is given by

$$\begin{aligned}
EA_{\text{tot}} &= \sum_{k=1}^{5} E_k b_k (h_{k+1} - h_k) \\
&= 2E_c b_c t_c + 2E_s b_c t_s + E_b b_b t_b \\
&= 1.1677 \times 10^6 + 16.187 \times 10^3 + 2.919 \times 10^6 \\
&= 4.103 \times 10^6 \text{ N}
\end{aligned}$$

Assume that $E_s = 2(1 + 0.3)G_s$. The total moment of inertia is given by (neglecting the moment of inertia of the acuators and bond layers about their own mid-plane)

$$\begin{aligned}
EI_{\text{tot}} &= \frac{1}{3}\sum_{k=1}^{5} E_k b_k (h_{k+1}^3 - h_k^3) \\
&= 2E_c b_c t_c \left(\frac{t_c}{2} + t_s + \frac{t_b}{2}\right)^2 + \frac{2E_c b_c t_c^3}{12} \\
&\quad + 2E_s b_c t_s \left(\frac{t_s}{2} + \frac{t_b}{2}\right)^2 + \frac{2E_s b_c t_s^3}{12} + E_b b_b \frac{t_b^3}{12} \\
&\approx 0.5441 + 0.1533 \\
&= 0.6974 \text{ Nm}^2
\end{aligned}$$

The induced strains on the top and bottom can be split into strains causing pure extension and pure bending.

$$\Lambda^e = \frac{\Lambda^{\text{top}} + \Lambda^{\text{bot}}}{2}$$

$$\Lambda^b = \frac{-\Lambda^{\text{top}} + \Lambda^{\text{bot}}}{2}$$

Only Λ^e contributes to F_Λ and only Λ^b contributes to M_Λ. As a result, the actuation force is given by (for a voltage of 150 V)

$$\begin{aligned}
F_\Lambda &= \sum_{k=1}^{5} \Lambda_k E_k b_k (h_{k+1} - h_k) \\
&= 2\Lambda^e E_c b_c t_c = -122.75 \text{ N}
\end{aligned}$$

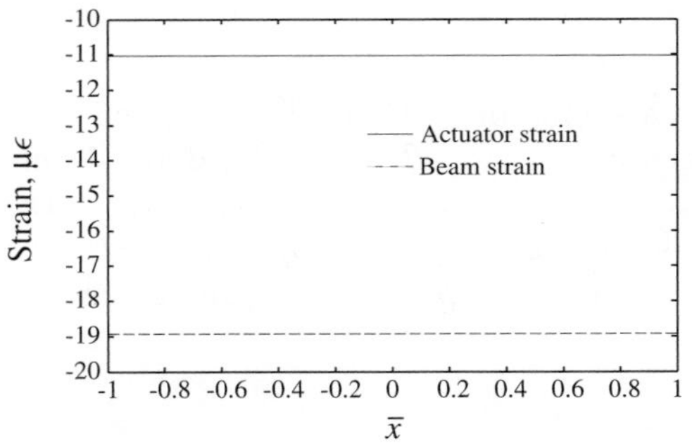

Figure 4.40. Actuator and beam strains on the top of the beam.

Similarly, the actuation moment is given by

$$\begin{aligned} M_\Lambda &= -\frac{1}{2}\sum_{k=1}^{5} \Lambda_k E_k b_k (h_{k+1}^2 - h_k^2) \\ &= \frac{1}{2}\Lambda^b E_c b_c \left[\left(\frac{t_b}{2} + t_s + t_c\right)^2 - \left(\frac{t_b}{2} + t_s\right)^2 \right] 2 \\ &= \Lambda^b E_c b_c t_c \left(t_c + 2t_s + t_b\right) \\ &= 0.0194 \text{ Nm} \end{aligned}$$

The mid-plane strain and bending curvature are

$$\epsilon_o = F_\Lambda / EA_{\text{tot}} = -29.91 \mu\epsilon$$

$$\kappa = M_\Lambda / EI_{\text{tot}} = 0.027835 \text{ 1/m}$$

The strain on the top surface can be obtained by

$$\epsilon^{\text{top}} = \epsilon^b + \epsilon_o = \frac{t_b}{2}\kappa + \epsilon_o$$

This strain is plotted in Fig. 4.40 for both the actuator and the beam. There is no variation in strain along the actuator length. The perfect bond can be modeled by assuming the bond thickness $t_s = 0$ or by assuming a very high modulus of the bond material (e.g., equal to the actuator modulus). However, the expressions for F_Λ and M_Λ contain only the bond thickness, t_s, and are independent of the bond shear modulus. The Euler-Bernoulli method considers only the geometrical effect of the presence of the bond layer and not the loss due to the finite stiffness bond. As a result, the induced force and moment for a perfect bond are the same as the values calculated in part (c). If the bond thickness is assumed to be zero, the induced moment is lower due to the decreased moment arm.

$$F_{\Lambda\text{perfect}} = 2\Lambda^e E_c b_c t_c = -122.75 \text{ N}$$

$$\begin{aligned} M_{\Lambda\text{perfect}} &= \Lambda^b E_c b_c t_c \left(t_c + t_b\right) \\ &= 0.0158 \text{ Nm} \end{aligned}$$

(d),(e) The bending slope and displacement can be obtained by integrating the curvature. The constants of integration disappear because of the cantilevered boundary conditions.

Table 4.6. *Comparision of blocked-force, uniform-strain, and Euler-Bernoulli theory*

	Blocked-force	Uniform-strain		Euler-Bernoulli	
	(Baseline)		% Deviation		% Deviation
Tip slope ($\times 10^{-3}$ rad)	1.7000	1.6203	4.69	1.4060	17.3
Tip displacement (mm)	0.9068	0.8636	4.76	0.7493	17.4

Region 1: Slope and displacement are zero.
Region 2:

$$\frac{\partial w}{\partial x} = \kappa(x - x_o)$$

$$w = \frac{\kappa(x - x_o)^2}{2}$$

Region 3: The slope remains constant and equal to the value at the end of Region 2.

$$\frac{\partial w}{\partial x} = 1.406 \times 10^{-3} \text{ rad}$$

The bending displacement increases linearly.

$$w = w\,|_{\bar{x}=1} + (x - X_o - l_c)\left.\frac{\partial w}{\partial x}\right|_{\bar{x}=1}$$

The tip displacement $w_{\text{tip}} = 0.7493$ mm. The bending slope and displacement are shown in Fig. 4.41.

(f) If the actuators are replaced with PVDF, the configuration is symmetric $\rightarrow M_\Lambda = 0$. The actuation force is $F_\Lambda = -0.3048N$.

(g) The comparison of blocked-force and uniform-strain results is shown in Table 4.6. The uniform-strain theory predicts lower deflections than the blocked-force method.

4.5.7 Bimorph Actuators

A bimorph actuator consists of two identical piezoceramic sheets bonded together. When a voltage is applied, the bimorph actuator undergoes pure bending, resulting in a tip displacement that can be used in a variety of applications. The piezoceramic

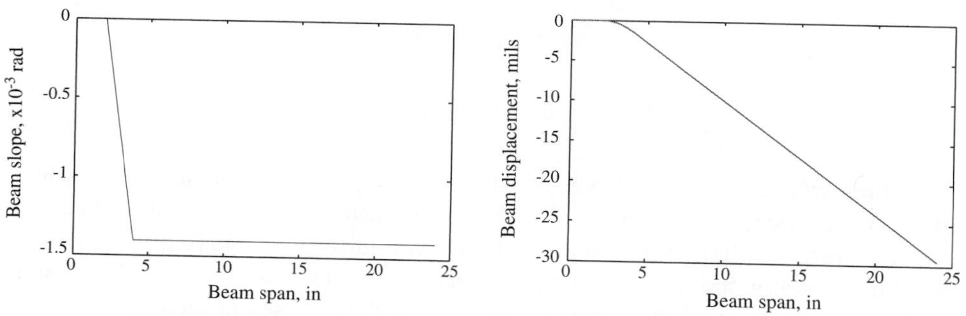

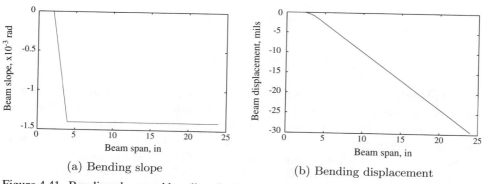

(a) Bending slope (b) Bending displacement

Figure 4.41. Bending slope and bending displacement of the beam.

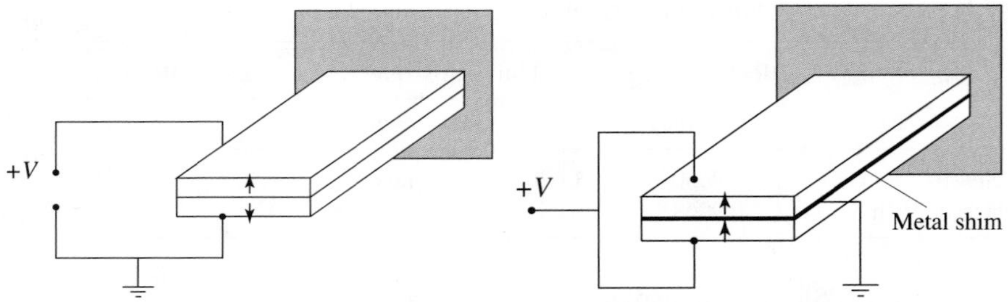

(a) Series configuration (b) Parallel configuration

Figure 4.42. Piezoceramic bimorph actuators. The arrow denotes direction of polarization.

sheets can be bonded in either a series or a parallel configuration. In the series configuration, the piezoceramic sheets are bonded with their polarized axes pointing in opposite directions (Fig. 4.42(a)). If a voltage of $+V$ volts is applied to the top electrode and the bottom electrode is grounded, a pure bending deformation will take place. In this way, the voltage across each piezoceramic sheet is equal to $V/2$ volts. Using the Euler-Bernoulli formulation

$$EI_{\text{tot}} w'' = M_\Lambda$$

$$EI_{\text{tot}} = \frac{1}{3}\sum_{k=1}^{2} b_k E_k (h_{k+1}^3 - h_k^3)$$

$$= \frac{2}{3} E_c b_c t_c^3$$

$$M_\Lambda = -\frac{1}{2}\sum_{k=1}^{2} b_k E_k \Lambda_k (h_{k+1}^2 - h_k^2)$$

As a result of the field, the top piezo-sheet will experience a free strain of $-\Lambda$ and the bottom piezo-sheet will experience a free strain of $+\Lambda$. The induced moment and tip displacement are given by

$$M_\Lambda = E_c b_c t_c^2 \Lambda$$

$$w_{tip} = \frac{M_\Lambda}{EI_{\text{tot}}}\left(\frac{l_c^2}{2}\right)$$

$$= \frac{3}{4}\frac{\Lambda}{t_c} l_c^2$$

$$= \frac{3}{8}\left(\frac{l_c}{t_c}\right)^2 d_{31} V$$

A bimorph actuator can also be constructed in a parallel configuration, in which the piezoceramic sheets are bonded with their polarized axes pointing in the same direction (Fig. 4.42(b)). In this configuration, a common electrode (e.g., a thin metal sheet) must be bonded between the piezoceramic sheets. The common electrode is connected to the ground of the power supply, and the exposed faces of the piezoceramic sheets are connected to $+V$ volts. In this way, one piezo-sheet experiences a positive electric field and the other experiences a negative electric field, resulting in

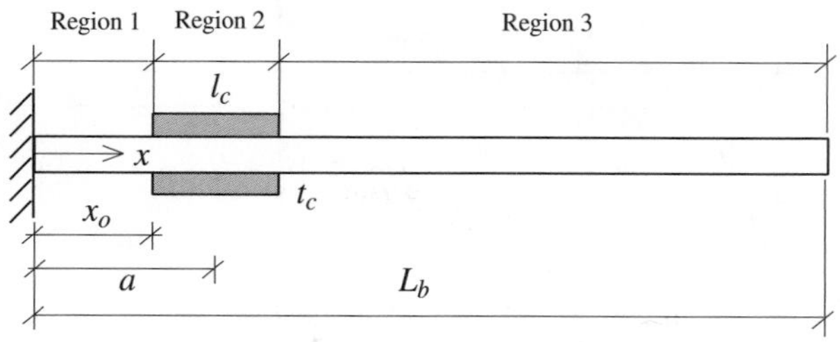

Figure 4.43. Pure bending of cantilevered beam with two piezoceramic actuators, Euler-Bernoulli model.

a pure-bending deformation. Note that the series and parallel configurations refer to the electrical connection of the piezoceramic sheets.

4.5.8 Induced Beam Response Using Euler-Bernoulli Modeling

The induced responses of several beam configurations in pure bending, derived using the Euler-Bernoulli model, are shown here. It is assumed that the length of the piezoceramic-sheet actuator is small compared to the length of the beam. Also, $EI_{\text{tot}} = EI_b = EI$, which is uniform along the length of the beam. Let the mid-point of the piezoceramic sheet be located at a coordinate $a = x_o + l_c/2$.

I. Cantilevered beam (Fig. 4.43)

$$0 \leq x \leq x_o \qquad \frac{\partial w_1}{\partial x} = 0$$

$$w_1 = 0$$

$$x_o \leq x \leq x_o + l_c \qquad \frac{\partial w_2}{\partial x} = \frac{M_\Lambda}{2EI}(2x + l_c - 2a)$$

$$w_2 = \frac{M_\Lambda}{8EI}(2x + l_c - 2a)^2$$

$$x_o + l_c/2 \leq x \leq L_b \qquad \frac{\partial w_3}{\partial x} = \frac{M_\Lambda}{EI} l_c$$

$$w_3 = \frac{M_\Lambda}{EI}(x - a)\, l_c$$

II. Simply-supported beam (Fig. 4.44)

$$0 \leq x \leq x_o \qquad \frac{\partial w_1}{\partial x} = \frac{M_\Lambda}{EI}\frac{l_c}{L_b}(a - L_b)$$

$$w_1 = \frac{M_\Lambda}{EI}\frac{l_c}{L_b}(a - L_b)\, x$$

$$x_o \leq x \leq x_o + l_c \qquad \frac{\partial w_2}{\partial x} = \frac{M_\Lambda}{EI} x + \frac{M_\Lambda}{2EI}\frac{[2a(l_c - L_b) - l_c l_b]}{L_b}$$

$$w_2 = \frac{M_\Lambda}{2EI} x^2 + \frac{M_\Lambda}{2EI}\frac{[2a(l_c - L_b) - l_c l_b]}{L_b} x + \frac{M_\Lambda}{8EI}(2a - l_c)^2$$

$$x_o + l_c/2 \leq x \leq L_b \qquad \frac{\partial w_3}{\partial x} = \frac{M_\Lambda}{EI}\frac{a l_c}{L_b}$$

$$w_3 = -\frac{M_\Lambda}{EI}\left(1 - \frac{x}{L_b}\right) l_c a$$

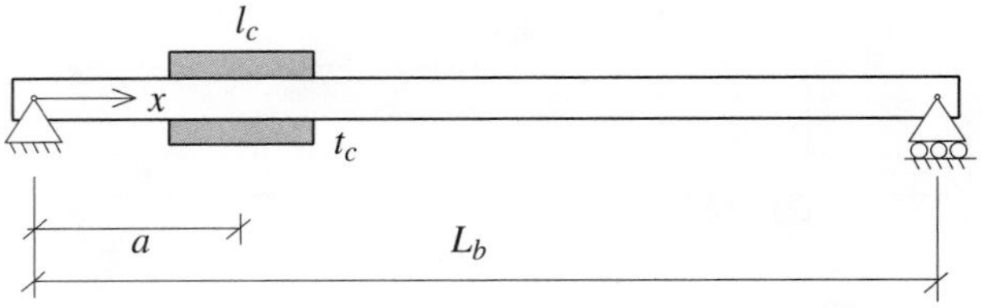

Figure 4.44. Pure bending of simply-supported beam with two piezoceramic actuators, Euler-Bernoulli model.

III. Clamped-clamped beam (Fig. 4.45)

$$0 \le x \le x_o \qquad \frac{\partial w_1}{\partial x} = \frac{M_\Lambda}{EI}\left[\frac{2l_c(3a-2L_b)x}{L_b^2} - \frac{3l_c(2a-L_b)x^2}{L_b^3}\right]$$

$$w_1 = \frac{M_\Lambda}{EI}\left[\frac{l_c(3a-2L_b)x^2}{L_b^2} - \frac{l_c(2a-L_b)x^3}{L_b^3}\right]$$

$$x_o \le x \le x_o + l_c \qquad \frac{\partial w_2}{\partial x} = \frac{M_\Lambda}{EI}\left[\frac{6al_c - L_b(4l_c - L_b)}{L_b^2}x - \frac{3l_c(2a-L_b)}{L_b^3}x^2 - \frac{(2a-l_c)}{2}\right]$$

$$w_2 = \frac{M_\Lambda}{EI}\left[\frac{6al_c - L_b(4l_c - L_b)}{2L_b^2}x^2 - \frac{l_c(2a-L_b)}{L_b^3}x^3 - \frac{(2a-l_c)}{2}x + \frac{(2a-l_c)^2}{8}\right]$$

$$x_o + l_c/2 \le x \le L_b \qquad \frac{\partial w_3}{\partial x} = \frac{M_\Lambda}{EI}\left[\frac{2l_c(3a-2L_b)x}{L_b^2} - \frac{3l_c(2a-L_b)x^2}{L_b^3} + l_c\right]$$

$$w_3 = \frac{M_\Lambda}{EI}\left[\frac{l_c(3a-2L_b)x^2}{L_b^2} - \frac{l_c(2a-L_b)x^3}{L_b^3} + l_c x - al_c\right]$$

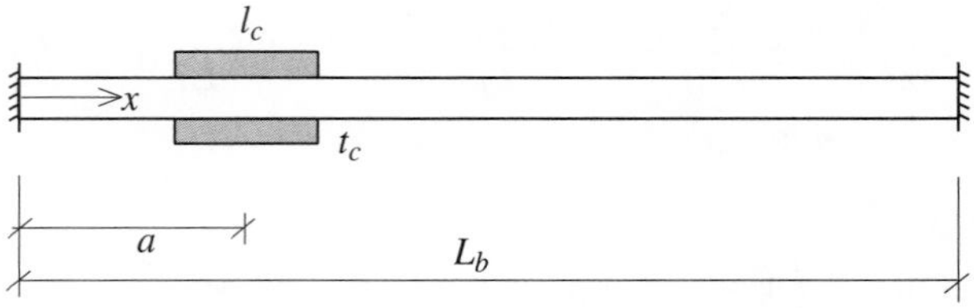

Figure 4.45. Pure bending of clamped-clamped beam with two piezoceramic actuators, Euler-Bernoulli model.

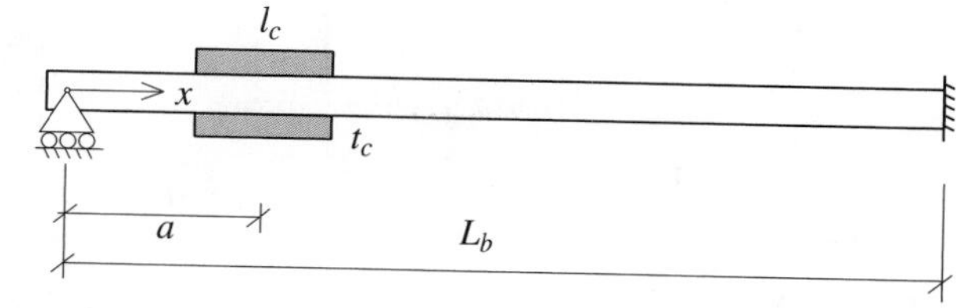

Figure 4.46. Pure bending of simply-supported–clamped beam with two piezoceramic actuators, Euler-Bernoulli model.

IV. Simply-supported–clamped beam (Fig. 4.46)

$$0 \le x \le x_o \qquad \frac{\partial w_1}{\partial x} = \frac{M_\Lambda}{EI}\left[-\frac{3al_cx^2}{2L_b^3} + \frac{l_c(3a-2L_b)}{2L_b}\right]$$

$$w_1 = \frac{M_\Lambda}{EI}\left[-\frac{al_cx^3}{2L_b^3} + \frac{l_c(3a-2L_b)x}{2L_b}\right]$$

$$x_o \le x \le x_o + l_c \qquad \frac{\partial w_2}{\partial x} = \frac{M_\Lambda}{EI}\left[-\frac{3al_cx^2}{2L_b^3} + x + \frac{a(3l_c-2L_b)-l_cL_b}{2L_b}\right]$$

$$w_2 = \frac{M_\Lambda}{EI}\left[-\frac{al_cx^3}{2L_b^3} + \frac{x^2}{2} + \frac{[a(3l_c-2L_b)-l_cL_b]x}{2L_b} + \frac{(2a-l_c)^2}{8}\right]$$

$$x_o + l_c/2 \le x \le L_b \qquad \frac{\partial w_3}{\partial x} = \frac{M_\Lambda}{EI}\left[-\frac{3al_cx^2}{2L_b^3} + \frac{3al_c}{2L_b}\right]$$

$$w_3 = \frac{M_\Lambda}{EI}\left[\frac{al_cx^3}{2L_b^3} + \frac{3al_cx}{2L_b} + al_c\right]$$

4.5.9 Embedded Actuators

The Euler-Bernoulli formulation can be easily used to model the behavior of complex structures, such as those involving embedded actuators. Assume two identical actuators are embedded in an isotropic beam, at an equal distance from the mid-plane, resulting in a symmetric actuation. A very thin bond layer exists between the actuators and the beam such that a perfect-bond assumption may be valid. An equal voltage applied to both actuators results in pure extension, whereas an equal but opposite voltage applied to opposite actuators causes pure bending of the beam. A schematic of two piezo-sheet elements embedded in a beam is shown in Fig. 4.47.

The vertical locations of the layers in the beam are shown in Fig. 4.48, and the distances are defined as follows

$$h_1 = -\frac{t_b}{2}$$
$$h_2 = -(d + t_c)$$
$$h_3 = -d$$
$$h_4 = d$$
$$h_5 = d + t_c$$
$$h_6 = \frac{t_b}{2}$$

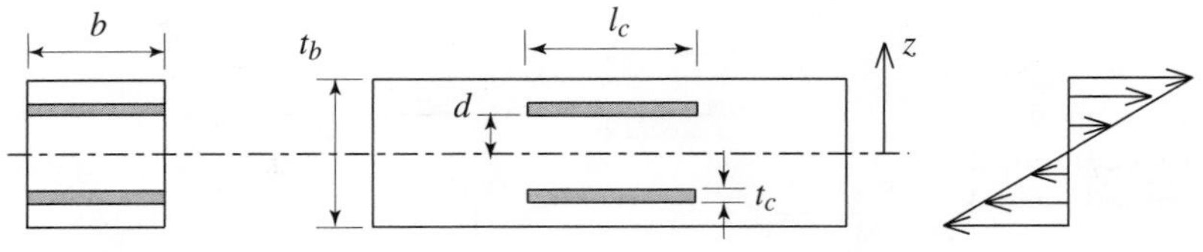

Figure 4.47. Beam substructure with symmetric bending actuators.

Stiffnesses are defined as

$$
\begin{aligned}
EA_{\text{tot}} &= \sum_{k=1}^{5} E_k b_k (h_{k+1} - h_k) \\
&= E_b b[-(d + t_c) - (-t_b/2)] \\
&\quad + E_c b[-d + (d + t_c)] \\
&\quad + E_b b[d - (-d)] \\
&\quad + E_c b[d + t_c - d] \\
&\quad + E_b b[t_b/2 - (d + t_c)] \\
&= E_b b(t_b - 2t_c) + E_c b(2t_c) \\
&= EA_b + EA_c
\end{aligned}
$$

$$
ES_{\text{tot}} = 0 \quad \text{(for a symmetric configuration)}
$$

$$
\begin{aligned}
EI_{\text{tot}} &= \frac{1}{3} \sum_{k=1}^{5} E_k b_k \left(h_{k+1}^3 - h_k^3\right) \\
&= E_b b \frac{t_b^3}{12} + \frac{2}{3}(E_c - E_b) b t_c \left(t_c^2 + 3d^2 + 3dt_c\right)
\end{aligned}
$$

Actuation force and moment can be expressed as

$$
F_\Lambda = \sum_{k=1}^{5} E_k b_k \Lambda_k (h_{k+1} - h_k)
$$

$$
M_\Lambda = -\frac{1}{2} \sum_{k=1}^{5} E_k b_k \Lambda_k \left(h_{k+1}^2 - h_k^2\right)
$$

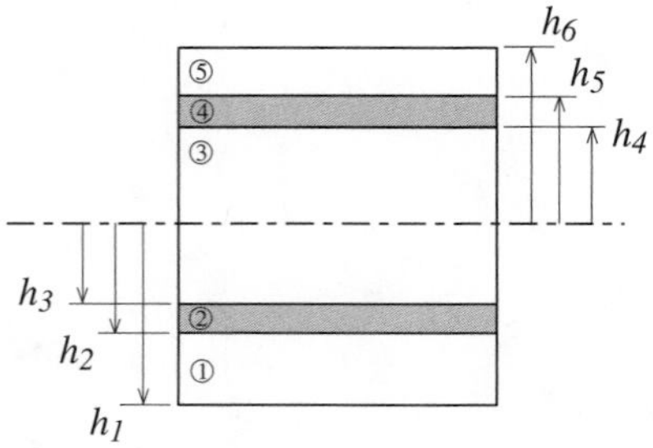

Figure 4.48. Beam cross section with dimensions.

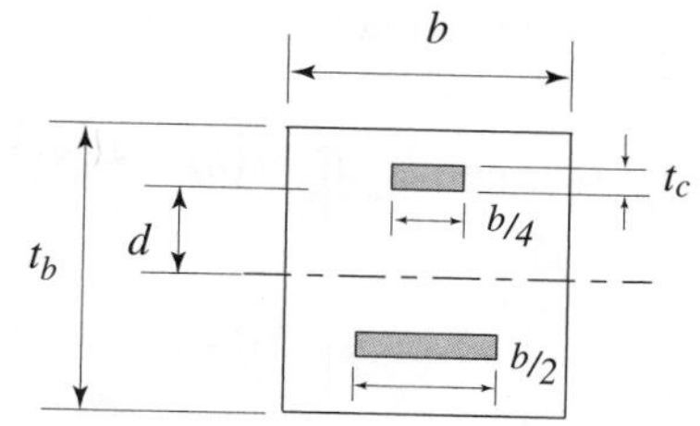

Figure 4.49. Beam with dissimilar embedded piezo-sheets.

For pure extension

$$\Lambda_2 = \Lambda_4 = \Lambda$$
$$F_\Lambda = 2E_c b t_c \Lambda = EA_c \Lambda$$
$$M_\Lambda = 0$$

For pure bending

$$\Lambda_2 = -\Lambda_4 = \Lambda$$
$$F_\Lambda = 0$$
$$M_\Lambda = E_c b \Lambda t_c (2d + t_c)$$

Combining bending-extension relations into the governing matrix equation

$$\begin{Bmatrix} F + F_\Lambda \\ M + M_\Lambda \end{Bmatrix} = \begin{bmatrix} EA_{\text{tot}} & ES_{\text{tot}} \\ ES_{\text{tot}} & EI_{\text{tot}} \end{bmatrix} \begin{Bmatrix} \epsilon_o \\ \frac{\partial^2 w}{\partial x^2} \end{Bmatrix}$$

4.5.10 Worked Example

Using the Euler-Bernoulli theory, derive extension-bending relations for a beam of modulus E_b, with two dissimilar piezo-sheets embedded at a distance d from the mid-plane. The width of the beam is b, and the widths of top and bottom piezos are $b/4$ and $b/2$, respectively. The piezos are of the same thickness and modulus E_c. A cross section of the beam is shown in Fig. 4.49. The same electric field is applied to both piezos.

Solution

The beam can be divided into five layers. The vertical position of each layer relative to the beam mid-plane is given by

$$h_1 = -h_6 = -\frac{t_b}{2}$$
$$h_2 = -h_5 = -(d + t_c)$$
$$h_3 = -h_4 = -d$$

The stiffness coefficients are found from

$$
\begin{aligned}
EA_{\text{tot}} &= \sum_{k=1}^{5} E_k b_k (h_{k+1} - h_k) \\
&= E_b b \left[(h_2 - h_1) + \frac{1}{2}(h_3 - h_2) + (h_4 - h_3) + \frac{3}{4}(h_5 - h_4) + (h_6 - h_5) \right] \\
&\quad + E_c b \left[\frac{1}{2}(h_3 - h_2) + \frac{1}{4}(h_5 - h_4) \right] \\
&= E_b b \left(t_b - \frac{3}{4} t_c \right) + \frac{3}{4} E_c b t_c
\end{aligned}
$$

$$
\begin{aligned}
ES_{\text{tot}} &= -\frac{1}{2} \sum_{k=1}^{5} E_k b_k (h_{k+1}^2 - h_k^2) \\
&= -\frac{1}{2} E_b b \left[(h_2^2 - h_1^2) + \frac{1}{2}(h_3^2 - h_2^2) + (h_4^2 - h_3^2) + \frac{3}{4}(h_5^2 - h_4^2) + (h_6^2 - h_5^2) \right] \\
&\quad - \frac{1}{2} E_c b \left[\frac{1}{2}(h_3^2 - h_2^2) + \frac{1}{4}(h_5^2 - h_4^2) \right] \\
&= - \left[\frac{b}{8}(E_b - E_c) t_c (t_c + 2d) \right]
\end{aligned}
$$

$$
\begin{aligned}
EI_{\text{tot}} &= \frac{1}{3} \sum_{k=1}^{5} E_k b_k (h_{k+1}^3 - h_k^3) \\
&= \frac{1}{3} E_b b \left[(h_2^3 - h_1^3) + \frac{1}{2}(h_3^3 - h_2^3) + (h_4^3 - h_3^3) + \frac{3}{4}(h_5^3 - h_4^3) + (h_6^3 - h_5^3) \right] \\
&\quad + \frac{1}{3} E_c b \left[\frac{1}{2}(h_3^3 - h_2^3) + \frac{1}{4}(h_5^3 - h_4^3) \right] \\
&= \frac{E_b b t_b^3}{12} + \frac{b}{4}(E_c - E_b) t_c \left(t_c^2 + 3d^2 + 3dt_c \right)
\end{aligned}
$$

The induced force and moment are given by

$$
\begin{aligned}
F_\Lambda &= \Lambda_c E_c b \left[\frac{1}{2}(h_3 - h_2) + \frac{1}{4}(h_5 - h_4) \right] \\
&= \frac{3}{4} \Lambda_c E_c b t_c \\
M_\Lambda &= \frac{1}{2} \Lambda_c E_c b \left[\frac{1}{2}(h_3^2 - h_2^2) + \frac{1}{4}(h_5^2 - h_4^2) \right] \\
&= \frac{1}{8} \Lambda_c E_c b t_c (t_c + 2d)
\end{aligned}
$$

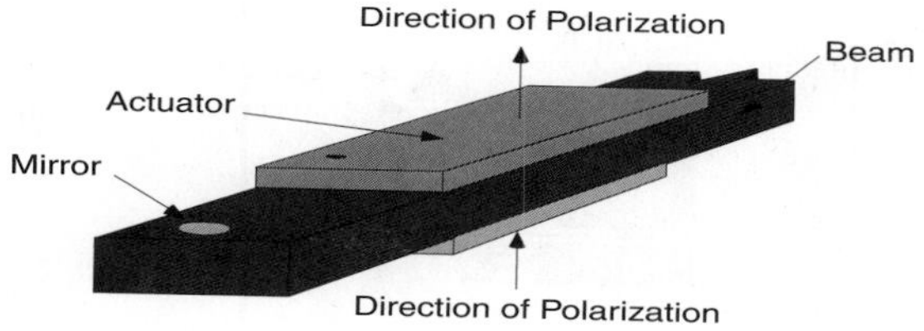

Figure 4.50. Beam setup for tip slope measurement.

The strain and curvature can be found by substituting these stiffnesses and induced forces in the governing equation of the beam.

$$\begin{Bmatrix} F_\Lambda \\ M_\Lambda \end{Bmatrix} = \begin{bmatrix} EA_{\text{tot}} & ES_{\text{tot}} \\ ES_{\text{tot}} & EI_{\text{tot}} \end{bmatrix} \begin{Bmatrix} \epsilon_o \\ \frac{\partial^2 w}{\partial x^2} \end{Bmatrix}$$

4.6 Testing of a Beam with Surface-Mounted Piezoactuators

The static tip bending slope of a beam with surface-mounted piezoactuators can be measured in the laboratory and correlated with the slope predicted by the blocked-force, uniform-strain, and Euler-Bernoulli theories. The free strain of the piezo-actuator can also be directly measured to improve the theoretical correlation.

4.6.1 Actuator Configuration

A sample PZT-5H piezoceramic sheet actuator of thickness 0.01 in (0.254 mm) is connected to a DC power supply. A quarter-bridge strain gage is bonded to the surface of the actuator. The strain gauge is connected to a signal-conditioning unit, from which the voltage is measured by a multimeter. The strain (microstrain) is given by

$$\epsilon = -\frac{4V_R}{\mathbf{GF}(1+2V_R)} \tag{4.350}$$

where **GF** is the gauge factor of the strain gauge (e.g., 2.109) and

$$V_R = \frac{V_{out}}{G.V_{ex}} \tag{4.351}$$

where V_{out} is the output voltage from the signal conditioner, G is the gain of the signal conditioner, and V_{ex} is the bridge-excitation voltage.

4.6.2 Beam Configuration and Wiring of Piezo

To measure the bending slope at the tip of a beam, two piezos are bonded to the beam as shown in Fig. 4.50 (note the direction of polarity). The beam is clamped at one end (not shown) and a mirror is placed at the tip of the beam. The mirror is used in conjunction with a laser to measure the tip slope of the beam on activation of the piezos. Figure 4.51 shows the wiring configuration of the piezos for bending (note that the drawings are not to scale). The black dots on the piezos indicate the positive electrode.

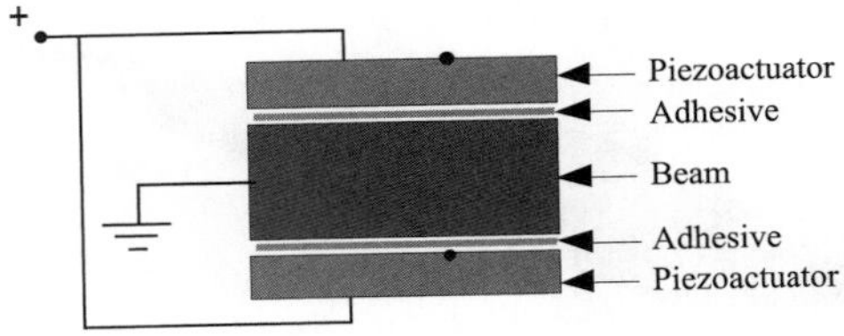

Figure 4.51. Wiring configuration for bending.

4.6.3 Procedure

For piezoactuator free strain:

1. The piezoactuator is cycled by varying the voltage between the maximum and zero until the residual strain stabilizes. The strain gauge is then reset. Care must be taken to cycle the actuator at the same polarity (i.e., zero voltage must not be crossed while cycling).
2. Voltage is applied in increments of 10 V to the maximum allowable voltage to prevent depoling (typically provided by the manufacturer; say, 130 V for PZT-5H). This is performed as follows: Desired voltage is set in the DC power supply, actuator is switched on, output strain is noted, actuator is switched off.
3. Note that at each voltage, the actuator should not be kept on for more than a couple of seconds to minimize error due to drift.
4. After obtaining all of the points for one voltage polarity, the polarity is reversed by switching the connector of the actuator. These steps are repeated for the opposite polarity. Enough datapoints must be measured to obtain a meaningful average.

For a beam with piezoactuators:

1. The beam is tested in bending for voltages of −120 to 120 V in increments of 10 V and the bending slope is measured using the laser-optic system (outlined in Section 4.6.4).
2. Initial position of laser dot is marked.
3. Desired voltages (0–120 V) are applied.
4. Transients are allowed to stop.
5. New position of laser dot is marked.
6. Steps are repeated for all voltages.

4.6.4 Measurement of Tip Slope

The measurement of the tip slope is achieved using a small mirror placed at the tip of the beam so that a laser beam can be reflected across the room. The setup is shown in Fig. 4.52.

The angle between the deflected and undeflected laser light (labeled 2θ in Fig. 4.52) is twice the deflection angle of the beam. This angle is calculated based on the small angle approximation of the tangent function. The distance from the deflected light spot to the undeflected light spot, divided by the distance from the beam to the wall, is the tangent of twice the angle θ. Different angles are calculated by measuring the distance between the deflected and undeflected light for different voltages varying from 0 V to 120 V.

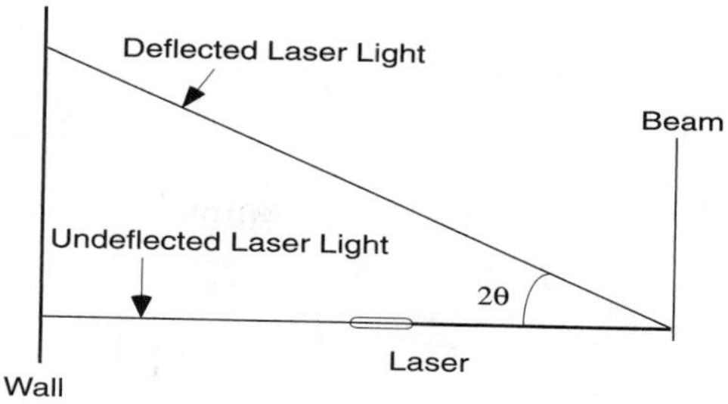

Figure 4.52. Room setup for tip-slope measurement.

4.6.5 Data Processing

1. Distance between deflected and undeflected marks is measured.
2. Beam bending slope is determined.
3. Beam bending slopes (analytical versus experimental) are compared.
4. Typical parameters for the setup are Bond layer: $E_s = 2.509$ GPa, $G_s = 0.965$ GPa, $t_c = 0.01015$ in, $t_b = 0.030$ in, $l_c = 2$ in, $b_c = 1$ in. The remaining parameters can be measured from the experimental setup.

4.7 Extension-Bending-Torsion Beam Model

As an example of an application of the techniques described thus far, this section presents a force-equilibrium formulation of an isotropic rectangular beam with an induced-strain actuator bonded to one surface. Figure 4.53 shows an induced-strain actuator mounted to the surface of a beam with a finite-thickness adhesive layer. In this case, because the actuator axis is offset from the beam axis by an angle β, it induces extension, bending, and twisting of the beam. The uniform-strain theory is used to analyze this structure.

The mechanism that induces torsion in the system is a two-dimensional strain state; however, global beam torsion is adequately represented by a one-dimensional model. Therefore, in keeping the derivation one-dimensional for mathematical simplicity, certain assumptions must be made such that the total state of strain may be sufficiently represented by the state of strain in one axis. Assume that:

(a) The beam may only extend, bend, and twist.
(b) The I_{zz} bending inertia is much greater than I_{yy}.
(c) Chordwise extensional stiffness, EA_y, is much greater than the longitudinal extensional stiffness, EA_x.

Chordwise deflections, v_b, may be neglected. This assumption has the effect of aligning the principal strain axes with the beam axes and setting the transverse principal

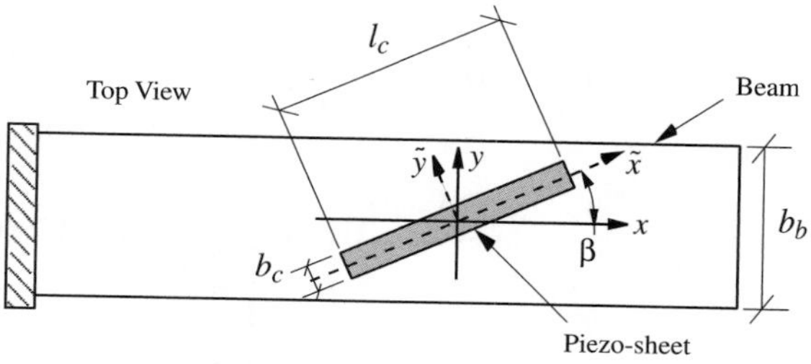

Figure 4.53. Actuator axis offset from beam axis.

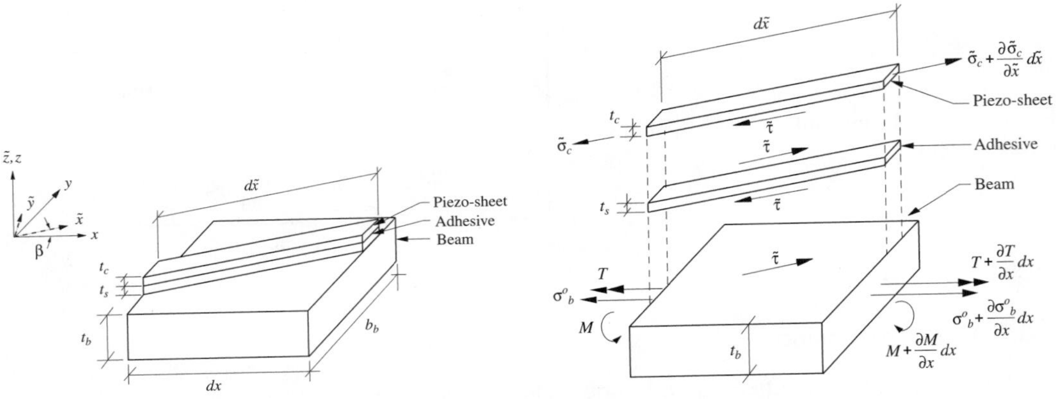

(**a**) Dimensions and co-ordinate system (**b**) Force components

Figure 4.54. Differential element of the beam with piezoactuator and bond layer.

strain identically zero.

$$\boldsymbol{\epsilon_b} = \begin{Bmatrix} \epsilon_b \\ 0 \\ 0 \end{Bmatrix} \tag{4.352}$$

The actuator is assumed to have a high aspect ratio, thereby behaving as a line element and only inducing strain in its longitudinal direction. Similar to the previously presented uniform-strain model, the strain distribution is assumed to be uniform through the piezo thickness and linear through the beam thickness. The adhesive is considered a linear shear layer, which only transfers load in the piezo longitudinal axis direction. A differential element of the beam and actuator section, and the individual components of forces acting on it, are shown in Fig. 4.54. Variables in the coordinate system skewed at an angle β are represented by a tilde.

With the stated assumptions and the geometric relationships, the force and moment equilibrium equations of the differential element are obtained as

$$\frac{\partial \widetilde{\sigma}_c}{\partial \widetilde{x}} - \frac{1}{t_c}\widetilde{\tau} = 0 \qquad \text{(force equilibrium in the actuator in the } \widetilde{x} \text{ direction)} \tag{4.353}$$

$$\frac{\partial \sigma_b^s}{\partial x} + \frac{4b_c}{t_b b_b}\widetilde{\tau} = 0 \qquad \text{(moment equilibrium in the beam about the } y\text{-axis)} \tag{4.354}$$

$$\frac{\partial \sigma_b^o}{\partial x} + \frac{b_c}{t_b b_b}\widetilde{\tau} = 0 \qquad \text{(force equilibrium in the beam in the } x \text{ direction)} \tag{4.355}$$

$$\frac{\partial T}{\partial x} - \frac{b_c t_b \tan(\beta)}{2}\widetilde{\tau} = 0 \qquad \text{(moment equilibrium in the beam about the } x\text{-axis)} \tag{4.356}$$

The strain-displacement relations of the system are

$$\widetilde{\epsilon}_c = \frac{\partial \widetilde{u}_c}{\partial \widetilde{x}} \tag{4.357}$$

$$\epsilon_b^s = \frac{\partial u_b^s}{\partial x} \tag{4.358}$$

$$\epsilon_b^o = \frac{\partial u_b^o}{\partial x} \tag{4.359}$$

$$\widetilde{\gamma}_s = \frac{1}{t_s}(\widetilde{u}_c - \widetilde{u}_b^s) \tag{4.360}$$

Substituting the mechanical strain in the piezo-sheet with $\epsilon_m = (\epsilon_c - \Lambda)$, the stress-strain relations are

$$\widetilde{\sigma}_c = E_c(\widetilde{\epsilon}_c - \Lambda) \tag{4.361}$$

$$\widetilde{\sigma}_b^s = E_b \widetilde{\epsilon}_b^s \tag{4.362}$$

$$\widetilde{\sigma}_b^o = E_b \widetilde{\epsilon}_b^o \tag{4.363}$$

$$\widetilde{\tau} = G_s \widetilde{\gamma}_s \tag{4.364}$$

Given the following transformation relations

$$x = \widetilde{x}\cos(\beta) - \widetilde{y}\sin(\beta) \qquad u = \widetilde{u}\cos(\beta) - \widetilde{v}\sin(\beta) \tag{4.365}$$

$$y = \widetilde{x}\sin(\beta) + \widetilde{y}\cos(\beta) \qquad v = \widetilde{u}\sin(\beta) + \widetilde{v}\cos(\beta) \tag{4.366}$$

$$\widetilde{x} = x\cos(\beta) + y\sin(\beta) \qquad \widetilde{u} = u\cos(\beta) + v\sin(\beta) \tag{4.367}$$

$$\widetilde{y} = -x\sin(\beta) + y\cos(\beta) \qquad \widetilde{v} = -u\sin(\beta) + v\cos(\beta) \tag{4.368}$$

Substituting the strain-displacement and the stress-strain relations into the equilibrium equations, differentiating relative to $\widetilde{x}$, and expanding the resulting equations using the Chain Rule, the governing differential equations relative to the actuator axes are

$$\frac{\partial^2 \widetilde{\epsilon}_c}{\partial \widetilde{x}^2} - \frac{\psi_s}{t_s^2}(\widetilde{\epsilon}_c - \epsilon_b^s \cos^2(\beta)) = 0 \tag{4.369}$$

$$\frac{\partial^2 \epsilon_b^s}{\partial \widetilde{x}^2} + \frac{\alpha \psi_s}{\psi_b}\frac{\cos(\beta)}{t_s^2}(\widetilde{\epsilon}_c - \epsilon_b^s \cos^2(\beta)) = 0 \tag{4.370}$$

$$\frac{\partial^2 \epsilon_b^o}{\partial \widetilde{x}^2} + \frac{\psi_s}{\psi_b}\frac{\cos(\beta)}{t_s^2}(\widetilde{\epsilon}_c - \epsilon_b^s \cos^2(\beta)) = 0 \tag{4.371}$$

$$\frac{\partial^2 T}{\partial \widetilde{x}^2} - \frac{t_b b_c}{2}\frac{t_s G_s \sin(\beta)}{t_s^2}(\widetilde{\epsilon}_c - \epsilon_b^s \cos^2(\beta)) = 0 \tag{4.372}$$

As a result of the stated assumptions, the variations of strains in the y-axis may be written in terms of their variations in the $\widetilde{x}$-axis as

$$\frac{\partial}{\partial y} = 0 \tag{4.373}$$

$$\frac{\partial}{\partial \widetilde{y}} = \frac{\partial}{\partial \widetilde{x}}\tan(\beta) \tag{4.374}$$

Note that the beam extensional strain ϵ_b^o and torque T are uncoupled from the actuator and beam surface strains ($\widetilde{\epsilon}_c$ and ϵ_b^s) and may be found after solving the coupled equations (Eqs. 4.369 and 4.370). Another important observation is that for the given formulation and assumptions, the torsion distribution, actuator extension, beam bending, and beam extension behaviors are all independent of the system torsional rigidity. Multiplying Eq. 4.370 by $\cos^2(\beta)$, subtracting from Eq. 4.369, and making the following substitutions produces a single differential equation for the two coupled equations.

$$\Gamma^2 = \frac{\psi_s}{\bar{t}_s^2}\left(\frac{\psi_b + \alpha\cos^3(\beta)}{\psi_b}\right); \quad \psi_b = \frac{E_b t_b b_b}{E_c t_c b_c}; \quad \psi_s = \frac{G_s t_s b_s}{E_c t_c b_c} \text{ where } b_s = b_c$$
$$\alpha = 4; \quad \bar{t}_s = \frac{t_s}{l_c/2}; \quad \bar{t}_b = \frac{t_b}{l_c/2}; \quad \widetilde{\bar{x}} = \frac{\widetilde{x}}{l_c/2}; \quad \bar{x} = \frac{x}{l_c\cos(\beta)/2} \tag{4.375}$$

$$\zeta = \widetilde{\epsilon}_c - \epsilon_b^s\cos^2(\beta) \tag{4.376}$$

$$\frac{\partial^2\zeta}{\partial\widetilde{\bar{x}}^2} - \Gamma^2\zeta = 0 \tag{4.377}$$

The solution to this equation is given as

$$\zeta = A\cosh(\Gamma\widetilde{\bar{x}}) + B\sinh(\Gamma\widetilde{\bar{x}}) \tag{4.378}$$

Substituting Eq. 4.378 into Eqs. 4.369, 4.371, and 4.372, the actuator strain, beam extensional strain, and beam torque are obtained through direct integration. The beam surface strain is then found by Eq. 4.376.

$$\widetilde{\epsilon}_c(\widetilde{\bar{x}}) = C_1 + D_1\widetilde{\bar{x}} + \frac{\psi_s}{\bar{t}_s^2\Gamma^2}(A\cosh(\Gamma\widetilde{\bar{x}}) + B\sinh(\Gamma\widetilde{\bar{x}})) \tag{4.379}$$

$$\epsilon_b^s(\widetilde{\bar{x}})\cos^2(\beta) = C_1 + D_1\widetilde{\bar{x}} - \left(1 - \frac{\psi_s}{\bar{t}_s^2\Gamma^2}\right)(A\cosh(\Gamma\widetilde{\bar{x}}) + B\sinh(\Gamma\widetilde{\bar{x}})) \tag{4.380}$$

$$\epsilon_b^o(\widetilde{\bar{x}}) = C_2 + D_2\widetilde{\bar{x}} - \frac{\psi_s\cos(\beta)}{\psi_b\bar{t}_s^2\Gamma^2}(A\cosh(\Gamma\widetilde{\bar{x}}) + B\sinh(\Gamma\widetilde{\bar{x}})) \tag{4.381}$$

$$T(\widetilde{\bar{x}}) = C_3 + D_3\widetilde{\bar{x}} + \frac{(G_s b_c t_s)t_b\sin(\beta)}{2\bar{t}_s^2\Gamma^2}(A\cosh(\Gamma\widetilde{\bar{x}}) + B\sinh(\Gamma\widetilde{\bar{x}})) \tag{4.382}$$

Assuming the following stress-free boundary conditions

$$\widetilde{\epsilon}_c(\pm1) = \Lambda \tag{4.383}$$

$$\epsilon_b^o(\pm1) = 0 \tag{4.384}$$

$$\epsilon_b^s(\pm1) = 0 \tag{4.385}$$

$$T(\pm1) = 0 \tag{4.386}$$

and utilizing the geometric relationship $\bar{x} = \tilde{\bar{x}}$, the final solutions are derived as

$$\frac{\tilde{\epsilon}_c(\tilde{\bar{x}})}{\Lambda} = \frac{\alpha\cos^3(\beta)}{\psi_b + \alpha\cos^3(\beta)} + \frac{\psi_b}{\psi_b + \alpha\cos^3(\beta)}\frac{\cosh(\Gamma\tilde{\bar{x}})}{\cosh(\Gamma)} \tag{4.387}$$

$$\frac{\epsilon_b^s(\bar{x})}{\Lambda} = \frac{\alpha\cos(\beta)}{\psi_b + \alpha\cos^3(\beta)}\left(1 - \frac{\cosh(\Gamma\bar{x})}{\cosh(\Gamma)}\right) \tag{4.388}$$

$$\frac{\epsilon_b^o(\bar{x})}{\Lambda} = \frac{\cos(\beta)}{\psi_b + \alpha\cos^3(\beta)}\left(1 - \frac{\cosh(\Gamma\bar{x})}{\cosh(\Gamma)}\right) \tag{4.389}$$

$$\frac{2T(\bar{x})}{\Lambda(E_b b_b t_b)t_b} = \frac{\sin(\beta)}{\psi_b + \alpha\cos^3(\beta)}\left(1 - \frac{\cosh(\Gamma\bar{x})}{\cosh(\Gamma)}\right) \tag{4.390}$$

The normalized bending curvature is obtained as

$$\frac{\kappa t_b}{2\Lambda}(\bar{x}) = \frac{\cos(\beta)(\alpha - 1)}{\psi_b + \alpha\cos^3(\beta)}\left(1 - \frac{\cosh(\Gamma\bar{x})}{\cosh(\Gamma)}\right) \tag{4.391}$$

Integration with respect to x provides the bending slope for a cantilevered beam, $\frac{\partial w}{\partial x} = 0$ at $\bar{x} = -1$, as

$$\frac{\bar{t}_b}{2\Lambda}\frac{\partial w}{\partial x}(\bar{x}) = \frac{\cos^2(\beta)(\alpha - 1)}{\psi_b + \alpha\cos^3(\beta)}\left(\frac{\sin(\Gamma\bar{x}) + \sinh(\Gamma)}{\Gamma\cosh(\Gamma)} - (\bar{x} + 1)\right) \tag{4.392}$$

The twist rate for a rectangular isotropic beam is given by the expression

$$\frac{\partial\phi}{\partial x} = \frac{3T}{G_b b_b t_b^3} \tag{4.393}$$

Integration with respect to x provides the twist angle for a cantilevered beam, $\phi = 0$ at $\bar{x} = -1$, as

$$\frac{\phi}{\Lambda}(\bar{x}) = \frac{3}{4}\frac{l_c}{t_b}\frac{E_b}{G_b}\frac{\cos(\beta)\sin(\beta)}{\psi_b + \alpha\cos^3(\beta)}\left(\frac{\sinh(\Gamma\bar{x}) + \sinh(\Gamma)}{\Gamma\cosh(\Gamma)} - (\bar{x} - 1)\right) \tag{4.394}$$

Examining the solutions for the case in which the actuator is aligned with the beam, $\beta = 0$, the previously derived solutions for this configuration are exactly obtained. For the condition in which the bond layer is infinitely thin, the beam surface strain and actuator strains reduce to

$$\frac{\tilde{\epsilon}_c}{\Lambda} = \frac{\alpha\cos^3(\beta)}{\psi_b + \alpha\cos^3(\beta)} \tag{4.395}$$

$$\frac{\epsilon_b^s}{\Lambda} = \frac{\alpha\cos(\beta)}{\psi_b + \alpha\cos^3(\beta)} \tag{4.396}$$

$$\frac{\epsilon_b^o}{\Lambda} = \frac{\cos(\beta)}{\psi_b + \alpha\cos^3(\beta)} \tag{4.397}$$

The relation between the actuator and beam-surface strains is derived from Eqs. 4.395 and 4.396 as

$$\tilde{\epsilon}_c = \epsilon_b^s\cos^2(\beta) \tag{4.398}$$

The relationship expressed in this equation is exactly the one-dimensional approximation of the compatibility condition at the actuator-beam interface subject to a two-dimensional strain tensor rotation. Figure 4.55 is a plot of the strains versus β

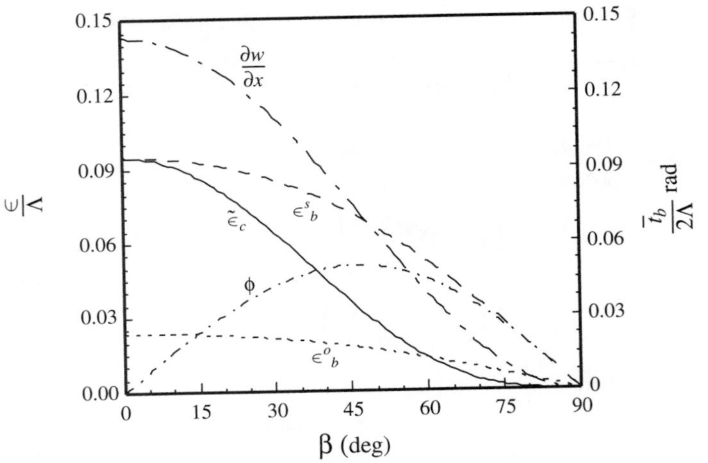

Figure 4.55. Analytical strains and deflections (ψ_b=38).

at $\bar{x} = 0$ and rotations at $\bar{x} = 1$ for a rectangular aluminum beam with one 7.5 mil G-1195 piezoceramic actuator perfectly bonded to one surface [3].

At $\beta = 90°$, the strains all approach zero values due to the one-dimensional assumptions. If two-dimensional strains were considered, the actuator strain would approach the limit compatible with the stiffness in the transverse direction. The longitudinal beam strains, however, would still approach zero. Near $\beta = 45°$, the analysis predicts maximum twist for a fixed actuator length. The limitations of the one-dimensional assumption must be kept in mind when applying this kind of analysis. For more refined modeling, the effects of transverse actuation must be included. Comparison of predicted results with test data (Fig. 4.56) showed that the models were satisfactory in predicting trends for bending slope and twist with different orientation angles. The predicted bending slope deviated significantly from measured values for orientation angles $\beta > 45°$, more so for piezoceramics with moderate aspect ratios. The experimental specimen in this case was a 0.794 mm (1/32 in) thick aluminum beam with three 50 mm × 6.35 mm × 0.19 mm (2 in × 1/4 in × 0.0075 in) piezoceramic sheet actuators bonded along the 406 mm long (16 in) beam in 101 mm (4 in) intervals.

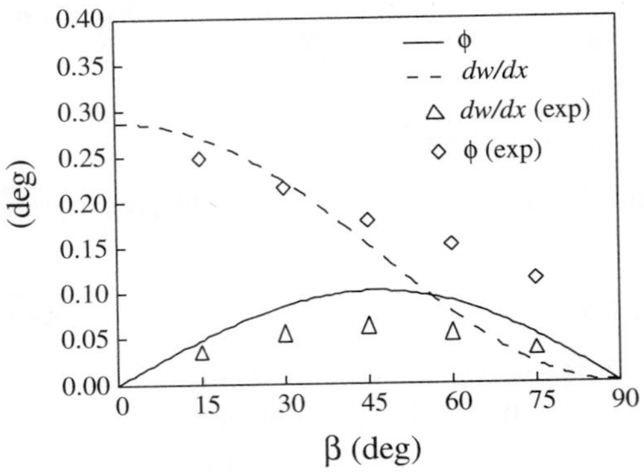

Figure 4.56. Effects of actuator orientation angle β on beam bending slope and twist angle ($\psi_b = 38$).

4.8 Beam Equilibrium Equations

Using the Euler-Bernoulli approach, it is possible to derive the equilibrium equations of a beam. The forces and moments in the beam are given by

$$F + F_\Lambda = EA_{\text{tot}}\epsilon_o + ES_{\text{tot}}\frac{\partial^2 w}{\partial x^2} \tag{4.399}$$

$$M + M_\Lambda = ES_{\text{tot}}\epsilon_o + EI_{\text{tot}}\frac{\partial^2 w}{\partial x^2} \tag{4.400}$$

The equilibrium equations for a one-dimensional structure are

$$\frac{\partial \sigma_x}{\partial x} + \frac{\partial \tau_{zx}}{\partial z} + f_x = 0 \tag{4.401}$$

$$\frac{\partial \tau_{zx}}{\partial x} + \frac{\partial \sigma_z}{\partial z} + f_z = 0 \tag{4.402}$$

where f_x and f_z are body forces in the x and z directions, respectively. Integrating these equations over the beam thickness results in

$$\frac{\partial F}{\partial x} = -p_x \tag{4.403}$$

$$\frac{\partial V}{\partial x} = p_z \tag{4.404}$$

where F and V are the axial force and shear force, respectively, and p_x and p_z are loadings per unit length along the x and z directions. Multiplying the first equilibrium equation by $(-z)$ and integrating over the beam thickness results in

$$\frac{\partial M}{\partial x} - V = 0 \tag{4.405}$$

Substituting in the previous equation

$$\frac{\partial^2 M}{\partial x^2} = p_z \tag{4.406}$$

Therefore, the beam equilibrium equations in x and z directions become

$$u_o \text{ equation: } \frac{\partial}{\partial x}\left[EA_{\text{tot}}\frac{\partial u_o}{\partial x}\right] + \frac{\partial}{\partial x}\left[ES_{\text{tot}}\frac{\partial^2 w}{\partial x^2}\right] - \frac{\partial F_\Lambda}{\partial x} = -p_x \tag{4.407}$$

$$w \text{ equation: } \frac{\partial^2}{\partial x^2}\left[ES_{\text{tot}}\frac{\partial u_o}{\partial x}\right] + \frac{\partial^2}{\partial x^2}\left[EI_{\text{tot}}\frac{\partial^2 w}{\partial x^2}\right] - \frac{\partial^2 M_\Lambda}{\partial x^2} = p_z \tag{4.408}$$

For an isotropic beam without actuators, these equations reduce to

$$\frac{\partial}{\partial x}\left[EA_{\text{tot}}\frac{\partial u_o}{\partial x}\right] = -p_x \tag{4.409}$$

$$\frac{\partial^2}{\partial x^2}\left[EI_{\text{tot}}\frac{\partial^2 w}{\partial x^2}\right] = p_z \tag{4.410}$$

4.9 Energy Principles and Approximate Solutions

The previous sections discussed the modeling of the beam structure with active elements by making several assumptions about the strain distribution in the structure as well as the relative contribution of the various elements to the overall

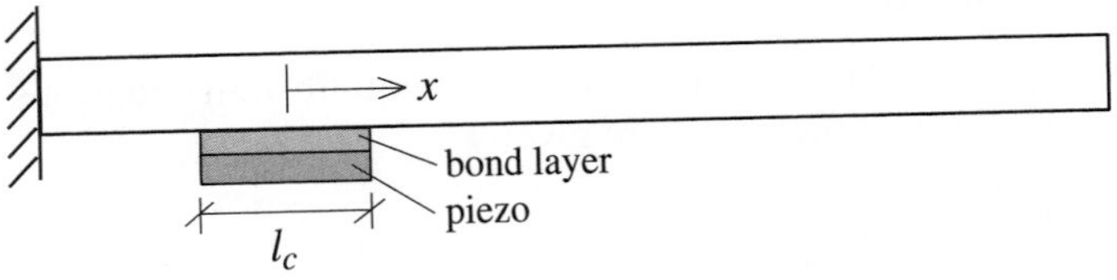

Figure 4.57. Beam with single piezoactuator, energy formulation of uniform-strain model.

deformation of the structure. The governing equations were derived based on a force-balance approach. In some cases (e.g., in the uniform-strain model), this approach can become cumbersome and derivation of the complete set of governing equations can be a tedious process. An alternate approach is to use an energy formulation to derive the governing equations and boundary conditions of the structure. In this approach, the relative energies stored in various elements of the structure can be compared and the assumptions made regarding the importance of each element can be assessed. Because the method does not require keeping track of each force acting in the structure, the derivation is often much simpler than the force-balance method.

In general, one can obtain the exact solution for only a selected few cases, such as a uniform beam under steady loading. This is because an exact solution must satisfy simultaneously all equilibrium equations, boundary conditions, and compatibility relations. Expressing equilibrium equations in terms of displacements inherently satisfies compatibility constraints; however, it is not possible to find a response solution for a generic beam, which satisfies equilibrium equation and all boundary conditions. Therefore, an approximate solution becomes necessary. Approximate methods, such as the Rayleigh-Ritz method and the Galerkin method, are often used in conjunction with energy-based formulations to obtain the solution of complex problems.

The following sections describe the derivation of the uniform-strain model and the Euler-Bernoulli model equations using an energy approach, as well as obtaining the approximate solution of the problem using the Galerkin and the Rayleigh-Ritz methods.

4.9.1 Energy Formulation: Uniform-Strain Model

Using the same basic assumptions made in the force equilibrium formulation, the Principle of Virtual Work readily provides the governing equations and boundary conditions of the system. This can be easily adapted to dynamic equations of motion. However, only the static formulation is presented in this section. It is assumed that the only allowable modes of deformation are actuator extension, adhesive shear, and beam bending and extension. Let us consider a beam with a single actuator bonded to the bottom surface, as shown in Fig. 4.57. The beam extension, bending, and adhesive strain energy relations may be directly written as follows

$$\text{beam extension: } V_b^o = \frac{1}{2}\int_{-\frac{l_c}{2}}^{\frac{l_c}{2}} EA_b \left(\frac{\partial u_b^o}{\partial x}\right)^2 dx \tag{4.411}$$

$$\text{beam bending: } V_b^\kappa = \frac{1}{2}\int_{-\frac{l_c}{2}}^{\frac{l_c}{2}} EI_b \left(\frac{\partial^2 w}{\partial x^2}\right)^2 dx \tag{4.412}$$

$$\text{adhesive shear: } V_s = \frac{1}{2}\int_{-\frac{l_c}{2}}^{\frac{l_c}{2}} G_s A_s (\gamma_s)^2\, dx \tag{4.413}$$

The actuator strain, ϵ_c, consists of the induced-strain term (Λ) and mechanical strain. The strain energy per unit volume of the actuator, ΔV_c, is given by

$$\begin{aligned}\Delta V_c &= \int_o^{\epsilon_m} \sigma d\epsilon_m = \frac{E_c}{2}\epsilon_m^2 \\ &= \frac{E_c}{2}(\epsilon_c - \Lambda)^2\end{aligned} \tag{4.414}$$

Integrating and substituting the strain-displacement relation gives the total strain energy in the actuator

$$\text{actuator extension: } V_c = \frac{1}{2}\int_{-\frac{l_c}{2}}^{\frac{l_c}{2}} E_c A_c \left(\frac{\partial u_c}{\partial x} - \Lambda\right)^2 dx \tag{4.415}$$

Substituting these relations and assuming a solid rectangular beam cross section, the beam bending and adhesive-shear strain energies can be expressed in terms of beam and actuator displacements.

$$I_b = \frac{1}{12}A_b t_b^2 \tag{4.416}$$

$$\sigma_b^{-s} - \sigma_b^o = \frac{M}{I_b}\frac{t_b}{2} \tag{4.417}$$

$$\frac{\partial^2 w}{\partial x^2} = \frac{M}{EI_b} = \frac{2}{t_b}(\epsilon_b^{-s} - \epsilon_b^o) \tag{4.418}$$

$$\gamma_s = \frac{(u_c - u_b^{-s})}{t_s} \tag{4.419}$$

resulting in

$$V_b^\kappa = \frac{1}{2}\int_{-\frac{l_c}{2}}^{\frac{l_c}{2}} \frac{E_b A_b}{3}(\epsilon_b^{-s} - \epsilon_b^o)^2 dx \tag{4.420}$$

$$V_s = \frac{1}{2}\int_{-\frac{l_c}{2}}^{\frac{l_c}{2}} \frac{G_s A_s}{t_s^2}(u_c - u_b^{-s})^2 dx \tag{4.421}$$

The Principle of Virtual Work for static behavior is mathematically stated as

$$\delta W_{\text{ext}} = \delta V \tag{4.422}$$

where δV is the variation in total strain energy and δW_{ext} is the virtual work done by external forces. The total strain energy of the system is given by

$$V = V_b^o + V_b^\kappa + V_s + V_c$$

In the absence of external forces, the virtual work done is zero.

$$\delta W_{\text{ext}} = 0$$

Therefore, summing the first variations of the strain energies and setting to zero

$$\int_{-\frac{l_c}{2}}^{\frac{l_c}{2}} E_cA_c\left(\frac{\partial u_c}{\partial x}-\Lambda\right)\delta\frac{\partial u_c}{\partial x}dx+\int_{-\frac{l_c}{2}}^{\frac{l_c}{2}} E_bA_b\left(\frac{\partial u_b^o}{\partial x}\right)\delta\frac{\partial u_b^o}{\partial x}dx$$

$$+\int_{-\frac{l_c}{2}}^{\frac{l_c}{2}} \frac{E_bA_b}{3}\left(\frac{\partial u_b^{-s}}{\partial x}-\frac{\partial u_b^o}{\partial x}\right)\left(\delta\frac{\partial u_b^{-s}}{\partial x}-\delta\frac{\partial u_b^o}{\partial x}\right)dx$$

$$+\int_{-\frac{l_c}{2}}^{\frac{l_c}{2}} \frac{G_sA_s}{t_s^2}(u_c-u_b^{-s})(\delta u_c-\delta u_b^{-s})dx=0$$

Integrating by parts until no derivatives of variations remain within the integral leads to

$$\left[E_cA_c\left(\frac{\partial u_c}{\partial x}-\Lambda\right)\delta u_c+E_bA_b\left(\frac{\partial u_b^o}{\partial x}\right)\delta u_b^o+\frac{E_bA_b}{3}\left(\frac{\partial u_b^{-s}}{\partial x}-\frac{\partial u_b^o}{\partial x}\right)(\delta u_b^{-s}-\delta u_b^o)\right]_{-\frac{l_c}{2}}^{\frac{l_c}{2}}$$

$$-\int_{-\frac{l_c}{2}}^{\frac{l_c}{2}} E_cA_c\left(\frac{\partial^2 u_c}{\partial x^2}\right)\delta u_c dx-\int_{-\frac{l_c}{2}}^{\frac{l_c}{2}} E_bA_b\left(\frac{\partial^2 u_b^o}{\partial x^2}\right)\delta u_b^o dx$$

$$-\int_{-\frac{l_c}{2}}^{\frac{l_c}{2}} \frac{E_bA_b}{3}\left(\frac{\partial^2 u_b^{-s}}{\partial x^2}-\frac{\partial^2 u_b^o}{\partial x^2}\right)(\delta u_b^{-s}-\delta u_b^o)dx$$

$$+\int_{-\frac{l_c}{2}}^{\frac{l_c}{2}} \frac{G_sA_s}{t_s^2}(u_c-u_b^{-s})(\delta u_c-\delta u_b^{-s})dx=0 \tag{4.423}$$

The governing equations of motion and boundary conditions are obtained by grouping the coefficients of the variations and setting them separately equal to zero.

$$\delta u_c\colon\ \frac{\partial^2 u_c}{\partial x^2}-\frac{G_sA_s}{E_cA_ct_s^2}(u_c-u_b^{-s})=0 \tag{4.424}$$

$$\delta u_b^{-s}\colon\ \frac{\partial^2 u_b^{-s}}{\partial x^2}-\frac{\partial^2 u_b^o}{\partial x^2}+\frac{3G_sA_s}{E_bA_bt_s^2}(u_c-u_b^{-s})=0 \tag{4.425}$$

$$\delta u_b^o\colon\ \frac{\partial^2 u_b^o}{\partial x^2}=\frac{1}{4}\frac{\partial^2 u_b^{-s}}{\partial x^2} \tag{4.426}$$

The boundary conditions are

$$\text{At } x=\pm\frac{l_c}{2},\quad \delta u_c=0 \text{ or } \frac{\partial u_c}{\partial x}=\Lambda \tag{4.427}$$

$$\text{At } x=\pm\frac{l_c}{2},\quad \delta u_b^{-s}=0 \text{ or } \frac{\partial u_b^{-s}}{\partial x}=\frac{\partial u_b^o}{\partial x} \tag{4.428}$$

$$\text{At } x=\pm\frac{l_c}{2},\quad \delta u_b^o=0 \text{ or } \frac{\partial u_b^o}{\partial x}=\frac{1}{4}\frac{\partial u_b^{-s}}{\partial x} \tag{4.429}$$

Based on physical constraints, the only way for Eqs. 4.428 and 4.429 to be satisfied simultaneously is for both strains to be zero at the ends of the actuator. Therefore

$$\text{At } x = \pm\frac{l_c}{2}, \quad \frac{\partial u_b^{-s}}{\partial x} = 0$$
$$\frac{\partial u_b^o}{\partial x} = 0 \qquad (4.430)$$
$$\frac{\partial u_c}{\partial x} = \Lambda$$

The governing equations and boundary conditions are identical to those developed previously with force equilibrium.

4.9.2 Energy Formulation: Euler-Bernoulli Model

The strain energy of a structure is given by

$$V = \frac{1}{2}\int_{\text{volume}} \sigma\epsilon_m dV \qquad (4.431)$$

where the subscript m indicates net mechanical strain and the integration is carried out over the volume of the structure. In other words, only the strain caused by the stress σ contributes to the strain energy of the structure. Let us consider a beam with bonded piezoactuators. As shown previously, the net mechanical strain of the piezoactuator is given by

$$\epsilon_m(x, z) = \epsilon(x, z) - \Lambda(x, z) \qquad (4.432)$$

where ϵ is the total strain of the piezoactuator, x is the coordinate along the beam axis (horizontal direction), and z is the coordinate perpendicular to the beam axis (vertical direction). Note that this equation is valid for any location in the beam, by setting $\Lambda(x, z) = 0$ over the passive volume of the beam. Assuming uniform properties along the width of the beam (in the y direction) and considering an element of the beam of length dx in the x direction, the strain energy of the beam with piezoactuators becomes

$$V = \frac{1}{2}\int_{\text{volume}} \sigma(x, z)\epsilon_m(x, z)dV$$
$$= \frac{1}{2}\int_{\text{volume}} E(z)\left[\epsilon(z) - \Lambda(z)\right]^2 dAdx \qquad (4.433)$$

where E is the local Young's modulus. For a beam deforming as per the Euler-Bernoulli assumption, the total strain at any location at a distance z from the neutral axis of the beam is given by (Eq. 4.276)

$$\epsilon(z) = \epsilon_o - z\kappa \qquad (4.434)$$

where ϵ_o is the axial strain at the neutral axis of the beam and κ is the bending curvature of the beam. Note that ϵ_o and κ are functions of the x-coordinate only and are constant in the beam element under consideration. Substituting for the beam strain (Eq. 4.434) and the actuator strain (Eq. 4.432) in the expression for strain energy (Eq. 4.433)

$$
\begin{aligned}
V &= \frac{1}{2}\int_{\text{volume}} E(z)\,(\epsilon_o - z\kappa - \Lambda)^2\, dAdx \\
&= \frac{1}{2}\int_{\text{volume}} E(z)\left(\epsilon_o^2 + z^2\kappa^2 + \Lambda^2 - 2\epsilon_o z\kappa - 2\epsilon_o\Lambda + 2z\kappa\Lambda\right) dAdx \\
&= \frac{1}{2}\left[EA_{\text{tot}}\epsilon_o^2 + EI_{\text{tot}}\kappa^2 + EA_{\text{tot}}\Lambda^2 + 2\epsilon_o\kappa ES_{\text{tot}} - 2F_\Lambda\epsilon_o - 2M_\Lambda\kappa\right] dx \\
&= \frac{1}{2}\int_{\text{length}} \{\epsilon_o \quad \kappa\}\begin{bmatrix} EA_{\text{tot}} & ES_{\text{tot}} \\ ES_{\text{tot}} & EI_{\text{tot}} \end{bmatrix}\begin{Bmatrix}\epsilon_o \\ \kappa\end{Bmatrix} dx - \int_{\text{length}} \{F_\Lambda \quad M_\Lambda\}\begin{Bmatrix}\epsilon_o \\ \kappa\end{Bmatrix} dx
\end{aligned}
\tag{4.435}
$$

where

$$
\begin{aligned}
EA_{\text{tot}} &= \int_{\text{area}} E(z)dA \\
ES_{\text{tot}} &= -\int_{\text{area}} E(z)zdA \\
EI_{\text{tot}} &= \int_{\text{area}} E(z)z^2dA \\
F_\Lambda &= \int_{\text{area}} E(z)\Lambda dA \\
M_\Lambda &= -\int_{\text{area}} E(z)\Lambda zdA
\end{aligned}
\tag{4.436}
$$

Taking the variation of the strain energy leads to

$$
\delta V = [EA_{\text{tot}}\epsilon_o\delta\epsilon_o + EI_{\text{tot}}\kappa\delta\kappa + ES_{\text{tot}}\kappa\delta\epsilon_o + ES_{\text{tot}}\epsilon_o\delta\kappa - F_\Lambda\delta\epsilon_o - M_\Lambda\Lambda\delta\kappa]\,dx
\tag{4.437}
$$

Because the free strain Λ is a constant, the term Λ^2 does not contribute to the variation in strain energy. Assuming that an external force F and an external moment M are applied to the element, the virtual work done is given by

$$
\delta W_{\text{ext}} = \int_{\text{length}} F\delta\epsilon_o dx + \int_{\text{length}} M\delta\kappa dx
\tag{4.438}
$$

The Principle of Virtual Work states that

$$
\delta V = \delta W_{\text{ext}}
\tag{4.439}
$$

Substituting for δV and δW_{ext} from Eqs. 4.437 and 4.438, and equating the coefficients of $\delta\epsilon_o$ and $\delta\kappa$, leads to two simultaneous equations

$$
EA_{\text{tot}}\epsilon_o + ES_{\text{tot}}\kappa - F_\Lambda = F
\tag{4.440}
$$

$$
ES_{\text{tot}}\epsilon_o + EI_{\text{tot}}\kappa - M_\Lambda = M
\tag{4.441}
$$

Rewriting these equations in matrix form, the Euler-Bernoulli governing equations for the beam with induced-strain actuation are

$$
\begin{bmatrix} EA_{\text{tot}} & ES_{\text{tot}} \\ ES_{\text{tot}} & EI_{\text{tot}} \end{bmatrix}\begin{Bmatrix}\epsilon_o \\ \kappa\end{Bmatrix} = \begin{Bmatrix} F + F_\Lambda \\ M + M_\Lambda \end{Bmatrix}
\tag{4.442}
$$

These equations can be used to estimate the deformations of a beam by an approximate method. Note that because this is a static case, the governing equations are obtained from static force equilibrium, as described in Section 4.5. For a continuous system, the Galerkin method is a widely used method for obtaining the approximate solution.

4.9.3 Galerkin Method

Two of the popular methods used to estimate the approximate solution are the Galerkin and the Rayleigh-Ritz methods. For the Galerkin solution, the response is assumed to be a summation of functions such that each function must separately satisfy all boundary conditions: geometric plus forced-boundary conditions. Expressing strains and curvatures in terms of basic displacements

$$\begin{Bmatrix} \epsilon_o \\ \kappa \end{Bmatrix} = \begin{bmatrix} \frac{\partial}{\partial x} & 0 \\ 0 & \frac{\partial^2}{\partial x^2} \end{bmatrix} \begin{Bmatrix} u_o \\ w \end{Bmatrix} = \mathbb{D}\boldsymbol{U} \tag{4.443}$$

where the operator $\mathbb{D}$ is of order 2×2. Assuming the displacement distribution in terms of a series of functions such as

$$u_o(x) = \sum_{i=1}^{M} \phi_{u_i}(x) q_i$$

$$w(x) = \sum_{j=1}^{N} \phi_{w_j}(x) q_{j+M} \tag{4.444}$$

where ϕ_{u_i} and ϕ_{w_j} are known functions that satisfy all boundary conditions and q_i are unknown coefficients, or generalized coordinates. This means that

$$\boldsymbol{U} = \begin{Bmatrix} u_o \\ w \end{Bmatrix} = \begin{bmatrix} \phi_{u_1} & \phi_{u_2} \cdots & \phi_{uM} & 0 & 0 & 0 \\ 0 & 0 & 0 & \phi_{w_1} & \phi_{w_2} \cdots & \phi_{wN} \end{bmatrix} \begin{Bmatrix} q_1 \\ q_2 \\ \vdots \\ q_{M+N} \end{Bmatrix} = \boldsymbol{\Phi}(x)\boldsymbol{q} \tag{4.445}$$

The size of the matrix $\boldsymbol{\phi}$ is $2 \times (M+N)$ and the size of the vector $\boldsymbol{q}$ is $(M+N) \times 1$. The beam equations are

$$\begin{bmatrix} \frac{\partial}{\partial x}\left(EA_{\text{tot}} \frac{\partial}{\partial x}\right) & \frac{\partial}{\partial x}\left(ES_{\text{tot}} \frac{\partial^2}{\partial x^2}\right) \\ \frac{\partial^2}{\partial x^2}\left(ES_{\text{tot}} \frac{\partial}{\partial x}\right) & \frac{\partial^2}{\partial x^2}\left(EI_{\text{tot}} \frac{\partial^2}{\partial x^2}\right) \end{bmatrix} \begin{Bmatrix} u_o \\ w \end{Bmatrix} - \begin{Bmatrix} \frac{\partial F_\Delta}{\partial x} \\ \frac{\partial^2 M_\Delta}{\partial x^2} \end{Bmatrix} = \begin{Bmatrix} -p_x \\ p_z \end{Bmatrix} \tag{4.446}$$

Substituting the assumed response results in an error function

$$\boldsymbol{\varepsilon}(x) = \begin{bmatrix} \frac{\partial}{\partial x}\left(EA_{\text{tot}} \frac{\partial}{\partial x}\right) & \frac{\partial}{\partial x}\left(ES_{\text{tot}} \frac{\partial^2}{\partial x^2}\right) \\ \frac{\partial^2}{\partial x^2}\left(ES_{\text{tot}} \frac{\partial}{\partial x}\right) & \frac{\partial^2}{\partial x^2}\left(EI_{\text{tot}} \frac{\partial^2}{\partial x^2}\right) \end{bmatrix} \boldsymbol{\Phi q} - \begin{Bmatrix} \frac{\partial F_\Delta}{\partial x} \\ \frac{\partial^2 M_\Delta}{\partial x^2} \end{Bmatrix} - \begin{Bmatrix} -p_x \\ p_z \end{Bmatrix} \tag{4.447}$$

If the assumed solution is an exact solution, this error function will be identically zero. Through the Galerkin method, the error is minimized by orthogonalizing it relative to each assumed function over the complete solution domain. Note that this approach is a specific instance of a more generalized formulation known as the "weighted-residual approach." In the Galerkin method, the weight that is multiplied with the residual $\boldsymbol{\varepsilon}(x)$ is the assumed function itself, and the error is minimized over the complete region (in this case, the length of the beam L_b).

$$\int_0^{L_b} \boldsymbol{\phi}_i^T\ \boldsymbol{\varepsilon}(x)dx = 0 \quad \text{for } i = 1, 2 \ldots M + N \tag{4.448}$$

where the vector $\boldsymbol{\phi}_i$ is the ith mode in the assumed response, corresponding to the ith column of the matrix $\boldsymbol{\phi}$. This results in an $(M + N)$ set of equations that can be concisely put into a matrix form

$$\boldsymbol{K}\boldsymbol{q} = \boldsymbol{Q}_\Lambda \tag{4.449}$$

where the elements of the effective, or generalized, stiffness matrix are

$$K_{ij} = \int_0^{L_b} \boldsymbol{\phi}_i^T \begin{bmatrix} \frac{\partial}{\partial x}\left(EA_{\text{tot}}\frac{\partial}{\partial x}\right) & \frac{\partial}{\partial x}\left(ES_{\text{tot}}\frac{\partial^2}{\partial x^2}\right) \\ \frac{\partial^2}{\partial x^2}\left(ES_{\text{tot}}\frac{\partial}{\partial x}\right) & \frac{\partial^2}{\partial x^2}\left(EI_{\text{tot}}\frac{\partial^2}{\partial x^2}\right) \end{bmatrix} \boldsymbol{\phi}_j\, dx \tag{4.450}$$

and the elements of the generalized forcing vector are

$$Q_{\Lambda_i} = \int_0^{L_b} \boldsymbol{\phi}_i^T \begin{Bmatrix} \frac{\partial F_\Lambda}{\partial x} \\ \frac{\partial^2 M_\Lambda}{\partial x^2} \end{Bmatrix} dx + \int_0^{L_b} \boldsymbol{\phi}_i^T \begin{Bmatrix} -p_x \\ p_z \end{Bmatrix} dx \tag{4.451}$$

The vector $\boldsymbol{\phi}_i$ is the ith mode (column) in the assumed response and is of size 2×1. Generally, it is extremely difficult to choose a $\boldsymbol{\phi}_i(x)$ that satisfies all boundary conditions for all but the simplest problems. Note that the generalized stiffness matrix $\boldsymbol{K}$ has the dimensions of force per unit length (N/m) and the generalized forcing vector $\boldsymbol{Q}_\Lambda$ has the dimensions of force (N).

Consider the case of a symmetric configuration ($ES_{\text{tot}} = 0$) with pure induced extension. This again results in M sets of equations

$$\boldsymbol{K}\boldsymbol{q} = \boldsymbol{Q}_\Lambda \tag{4.452}$$

where the elements of these matrices can be defined as

$$K_{ij} = \int_0^{L_b} \phi_{u_i} \frac{\partial}{\partial x}\left(EA_{\text{tot}}\frac{\partial \phi_{u_j}}{\partial x}\right) dx \tag{4.453}$$

$$Q_{\Lambda_i} = \int_0^{L_b} \phi_{u_i} \frac{\partial F_\Lambda}{\partial x}\, dx \tag{4.454}$$

Note that on integrating by parts and applying the boundary conditions, the generalized force can also be written as

$$Q_{\Lambda_i} = -\int_0^{L_b} F_\Lambda \frac{\partial \phi_{u_i}}{\partial x}\, dx \tag{4.455}$$

This is because $F_\Lambda = 0$ at $x = 0$ and L_b.

Next, consider the case of a symmetric configuration ($ES_{\text{tot}} = 0$) with pure induced bending. This again results in N sets of equations

$$\boldsymbol{K}\boldsymbol{q} = \boldsymbol{Q}_\Lambda \tag{4.456}$$

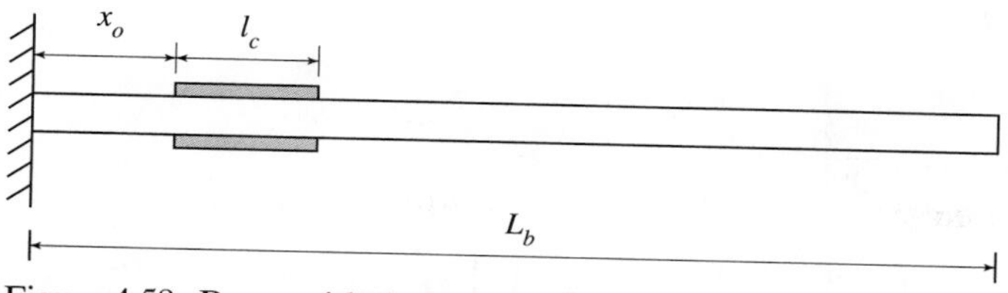

Figure 4.58. Beam with dual piezoactuators.

where the elements of these matrices can be defined as

$$K_{ij} = \int_0^{L_b} \phi_{w_i} \frac{\partial^2}{\partial x^2}\left(EI_{\text{tot}} \frac{\partial^2 \phi_{w_j}}{\partial x^2}\right) dx \tag{4.457}$$

$$Q_{\Lambda_i} = \int_0^{L_b} \phi_{w_i} \frac{\partial^2 M_\Lambda}{\partial x^2} dx \tag{4.458}$$

In this case, on integrating by parts and applying the boundary conditions, the generalized force can also be written as

$$Q_{\Lambda_i} = \int_0^{L_b} M_\Lambda \frac{\partial^2 \phi_{w_i}}{\partial x^2} dx \tag{4.459}$$

4.9.4 Worked Example

Using the Galerkin method, determine the steady-state axial response of a uniform cantilevered beam of length L_b with two identical piezos with the same electric field (Fig. 4.58). Assume a one-term solution as

$$u(x) = \left[\frac{x}{L_b} - \frac{1}{2}\left(\frac{x}{L_b}\right)^2\right] q_1$$

Solution

The assumed shape function is

$$\phi_1 = \frac{x}{L_b} - \frac{1}{2}\left(\frac{x}{L_b}\right)^2$$

At $x = 0$, the boundary condition $\phi_1(0) = 0$ is satisfied.

At $x = L_b$, the boundary condition $\frac{\partial \phi_1(L_b)}{\partial x} = 0$ is satisfied.

Note that for $x_o < x < x_o + l_c$, $EA_{\text{tot}} = EA_b + EA_c$. The governing equation becomes

$$K_{11} q_1 = Q_1$$

where the stiffness is given by

$$K_{11} = \int_0^{L_b} \phi_1 EA_{\text{tot}} \frac{\partial^2 \phi_1}{\partial x^2} \, dx$$

$$= \int_0^{L_b} EA_{\text{tot}} \left(-\frac{1}{L_b^2}\right) \left[\frac{x}{L_b} - \frac{1}{2}\left(\frac{x}{L_b}\right)^2\right] dx$$

$$= -\frac{EA_b}{3L_b} - \frac{EA_c}{L_b^3}\left[\frac{(x_o + l_c)^2 - x_o^2}{2} - \frac{1}{6}\frac{(x_o + l_c)^3 - x_o^3}{L_b}\right]$$

and the forcing is

$$Q_1 = \int_0^{L_b} \phi_{u_1} \frac{\partial F_\Lambda}{\partial x} dx$$

$$= -\int_{x_o}^{x_o + l_c} F_\Lambda \frac{\partial \phi_u}{\partial x} dx$$

$$= -\int_{x_o}^{x_o + l_c} F_\Lambda \left(\frac{1}{L_b} - \frac{x}{L_b^2}\right) dx$$

$$= -\frac{F_\Lambda l_c}{L_b}\left[1 - \frac{l_c + 2x_o}{L_b}\right]$$

The generalized coordinate is obtained from

$$q_1 = \frac{Q_1}{K_{11}}$$

4.9.5 Worked Example

Using the Galerkin method, determine the steady-state bending response of a uniform cantilevered beam of length L_b with two identical piezos with opposite electric fields (see Fig. 4.58). Assume a solution as

$$w(x) = \left[6\left(\frac{x}{L_b}\right)^2 - 4\left(\frac{x}{L_b}\right)^3 + \left(\frac{x}{L_b}\right)^4\right] q_1$$

Note that for $x_o < x < x_o + l_c$, $EI_{\text{tot}} = EI_b + EI_c$; otherwise, $EI_{\text{tot}} = EI_b$.

Solution

The assumed shape function is

$$\phi_1 = 6\left(\frac{x}{L_b}\right)^2 - 4\left(\frac{x}{L_b}\right)^3 + \left(\frac{x}{L_b}\right)^4$$

which satisfies all the boundary conditions

$$\begin{aligned}
\phi_1(0) &= 0 \\
\frac{\partial \phi_1(0)}{\partial x} &= 0 \\
EI_b \frac{\partial^2 \phi_1(L_b)}{\partial x^2} &= 0 \\
EI_b \frac{\partial^3 \phi_1(L_b)}{\partial x^3} &= 0
\end{aligned} \tag{4.460}$$

This assumed shape function results in

$$K_{11} q_1 = Q_1$$

where

$$\begin{aligned}
K_{11} &= \int_0^{L_b} \phi_{w_1} EI_{\text{tot}} \frac{\partial^2 \phi_{w_1}}{\partial x^2} dx \\
&= \frac{24}{L_b^4} \int_0^{L_b} EI \left(6\frac{x^2}{L_b^2} - 4\frac{x^3}{L_b^3} + \frac{x^4}{L_b^4} \right) dx \\
&= \frac{144 EI_{\text{tot}}}{5L_b^3} + \frac{24 EI_c}{L_b^4} \left\{ \frac{3}{L_b^2} \left[(x_o + l_c)^3 - x_o^3 \right] \right. \\
&\quad \left. - \frac{1}{L_b^3} \left[(x_o + l_c)^4 - x_o^4 \right] - \frac{1}{5L_b^4} \left[(x_o + l_c)^5 - x_o^5 \right] \right\}
\end{aligned} \tag{4.461}$$

and

$$\begin{aligned}
Q_1 &= \int_0^{L_b} M_\Lambda \frac{\partial^2 \phi_w}{\partial x^2} dx \\
&= \int_{x_o}^{x_o + l_c} M_\Lambda \left(\frac{12}{L_b^2} - \frac{24x}{L_b^3} + \frac{12x^2}{L_b^4} \right) dx \\
&= M_\Lambda \left\{ \frac{12}{L_b^2} l_c - \frac{12}{L_b^3} \left[(x_o + l_c)^2 - x_o^2 \right] + \frac{4}{L_b^4} \left[(x_o + l_c)^3 - x_o^3 \right] \right\}
\end{aligned} \tag{4.462}$$

from which the generalized coordinate is found as

$$q_1 = \frac{Q_1}{K_{11}} \tag{4.463}$$

4.9.6 Rayleigh-Ritz Method

In the Rayleigh-Ritz method, an assumed solution is directly substituted in the expressions for the energies of the structure. The governing equation of the structure is then obtained using Lagrange's equations. For the Rayleigh-Ritz solution, the response is assumed to be a summation of functions such that each function needs to satisfy only geometric boundary conditions. The Rayleigh-Ritz method is more convenient compared to the Galerkin method because the assumed functions do not need to satisfy forced boundary conditions that are often too involved. Therefore, a larger number of simpler functions is available, from which the approximate solution can be chosen.

$$u(x) = \sum_{i=1}^{M} \phi_{u_i}(x) q_i \tag{4.464}$$

$$w(x) = \sum_{j=1}^{N} \phi_{w_j}(x) q_{j+M} \tag{4.465}$$

$$\boldsymbol{U} = \begin{Bmatrix} u \\ w \end{Bmatrix} = \begin{bmatrix} \phi_{u_1} & \phi_{u_2} \dots & \phi_{u_M} & 0 & 0 & 0 \\ 0 & 0 & 0 & \phi_{w_1} & \phi_{w_2} \dots & \phi_{w_N} \end{bmatrix} \begin{Bmatrix} q_1 \\ q_2 \\ \vdots \\ q_{M+N} \end{Bmatrix} = \boldsymbol{\phi}(x)\boldsymbol{q} \tag{4.466}$$

Similar to the Galerkin method

$$\begin{Bmatrix} \epsilon_o \\ \kappa \end{Bmatrix} = \mathbb{D}\boldsymbol{\phi}\boldsymbol{q} \tag{4.467}$$

For the Rayleigh-Ritz solution, Lagrange's equations are normally used.

$$\frac{\partial}{\partial t}\left(\frac{\partial T}{\partial \dot{q}_i}\right) - \frac{\partial T}{\partial q_i} + \frac{\partial V}{\partial q_i} = Q_i \qquad \text{for } q_i = q_1,\ q_2 \dots q_{M+N} \tag{4.468}$$

where V is the strain energy, T is the kinetic energy, and Q is the generalized force. For a static problem, the kinetic energy T is zero. Substituting for the total strain energy for a beam (Eq. 4.496), this becomes

$$\begin{aligned} V_{\text{total}} &= \frac{1}{2}\int_0^{L_b} \{\mathbb{D}\boldsymbol{\phi}\boldsymbol{q}\}^T \begin{bmatrix} EA_{\text{tot}} & ES_{\text{tot}} \\ ES_{\text{tot}} & EI_{\text{tot}} \end{bmatrix} \{\mathbb{D}\boldsymbol{\phi}\boldsymbol{q}\}\, dx \\ &\quad - \int_0^{L_b} \{F_\Lambda^T\ M_\Lambda^T\}\{\mathbb{D}\boldsymbol{\phi}\boldsymbol{q}\}\, dx \\ &= \frac{1}{2}\boldsymbol{q}^T\boldsymbol{K}\boldsymbol{q} - \boldsymbol{Q}_\Lambda\boldsymbol{q} \end{aligned} \tag{4.469}$$

The size of the generalized stiffness matrix K is $(M+N)\times(M+N)$ and the size of the generalized force vector Q_Λ is $(M+N)\times 1$. The generalized stiffness matrix is defined as

$$K_{ij} = \int_0^{L_b} \{\mathbb{D}\boldsymbol{\phi}_i\}^T \begin{bmatrix} EA_{\text{tot}} & ES_{\text{tot}} \\ ES_{\text{tot}} & EI_{\text{tot}} \end{bmatrix} \{\mathbb{D}\boldsymbol{\phi}_j\}\, dx \tag{4.470}$$

and the generalized force is

$$Q_{\Lambda_i} = \int_0^{L_b} \{F_\Lambda^T M_\Lambda^T\}\{\mathbb{D}\boldsymbol{\phi}_i\}\, dx \tag{4.471}$$

The vector $\boldsymbol{\phi}_i$ is the ith mode in the assumed response and is of size 2×1. The generalized stiffness matrix $\boldsymbol{K}$ has the dimensions of force per unit length (N/m) and the generalized forcing vector $\boldsymbol{Q}_\Lambda$ has the dimensions of force (N).

In the case of a dynamic problem, the kinetic energy T is also included in the total energy of the beam. The kinetic energy of the beam is given by

$$\begin{aligned} T &= \frac{1}{2}\iiint_{\text{volume}} \rho_s\left(\dot{u}^2+\dot{w}^2\right)dxdydz \\ &= \frac{1}{2}\int_0^{L_b}\int_{t_b} \rho_s b\left(\dot{u}^2+\dot{w}^2\right)dxdz \end{aligned} \tag{4.472}$$

where ρ_s is the density of the material of the beam, b is the width of the beam, and t_b is the thickness of the beam. The velocity components at a point (x, z) are

$$\dot{u}(x,z,t) = \dot{u}_o(x,t) - z\frac{\partial}{\partial t}\left(\frac{\partial w}{\partial x}(x,t)\right) \tag{4.473}$$

$$\dot{w}(x,z,t) = \dot{w}(x,t) = \dot{w}(x,t) \tag{4.474}$$

where $\dot{u}_o$ and $\dot{w}$ are the velocity components at the mid-plane. The kinetic energy becomes

$$T = \frac{1}{2}\int_0^{L_b}\left[m_b\left(\dot{u}_o{}^2+\dot{w}^2\right) - 2S_b\left(\dot{u}_o\frac{\partial \dot{w}}{\partial x}\right) + I_b\left(\frac{\partial \dot{w}}{\partial x}\right)^2\right]dx \tag{4.475}$$

This can be rewritten as

$$\begin{aligned} T &= \frac{1}{2}\int_0^{L_b}\left\{\dot{u}_o \quad \dot{w} \quad \tfrac{\partial \dot{w}}{\partial x}\right\}\begin{bmatrix} m_b & 0 & -S_b \\ 0 & m_b & 0 \\ -S_b & 0 & I_b \end{bmatrix}\begin{Bmatrix} \dot{u}_o \\ \dot{w} \\ \frac{\partial^2 w}{\partial t \partial x} \end{Bmatrix}dx \\ &= \frac{1}{2}\int_0^{L_b}\left\{\dot{u}_o \quad \dot{w} \quad \tfrac{\partial^2 w}{\partial t\partial x}\right\}\boldsymbol{m}_s\begin{Bmatrix} \dot{u}_o \\ \dot{w} \\ \frac{\partial^2 w}{\partial t \partial x} \end{Bmatrix}dx \end{aligned} \tag{4.476}$$

where m_b is the mass per unit length (kg/m); S_b is the first mass moment of inertia, per unit length, about the mid-plane (kg); and I_b is the second mass moment of inertia, per unit length, about the mid-plane (kg-m).

$$m_b = \int_{t_b} \rho_s b dz \qquad \text{(kg/m)} \tag{4.477}$$

$$S_b = \int_{t_b} \rho_s b z dz \qquad \text{(kg)} \tag{4.478}$$

$$I_b = \int_{t_b} \rho_s b z^2 dz \qquad \text{(kg-m)} \tag{4.479}$$

For a Rayleigh-Ritz solution

$$\begin{Bmatrix} \dot{u}_o \\ \dot{w} \\ \frac{\partial^2 w}{\partial t\partial x} \end{Bmatrix} = \begin{bmatrix} 1 & 0 \\ 0 & 1 \\ 0 & \frac{\partial}{\partial x} \end{bmatrix}\begin{Bmatrix} \dot{u}_o \\ \dot{w} \end{Bmatrix} = \mathbb{D}_1\begin{Bmatrix} \dot{u}_o \\ \dot{w} \end{Bmatrix} \tag{4.480}$$

Substituting the assumed shape functions for the displacements, the expression for kinetic energy becomes

$$\begin{aligned} T &= \frac{1}{2}\int_0^{L_b} \{\mathbb{D}_1\boldsymbol{\phi}\dot{\boldsymbol{q}}\}^T \boldsymbol{m}_s\{\mathbb{D}_1\boldsymbol{\phi}\dot{\boldsymbol{q}}\}dx \\ &= \frac{1}{2}\dot{\boldsymbol{q}}^T\boldsymbol{M}\dot{\boldsymbol{q}} \end{aligned} \tag{4.481}$$

where $\boldsymbol{M}$ is a generalized mass matrix of size $(M+N)\times(M+N)$ defined as

$$\begin{aligned} M_{ij} &= \int_0^{L_b} \{\mathbb{D}_1\boldsymbol{\phi}_i\}^T \boldsymbol{m}_s\{\mathbb{D}_1\boldsymbol{\phi}_j\}dx \\ &= \int_0^{L_b} \{\mathbb{D}_1\boldsymbol{\phi}_i\}^T \begin{bmatrix} m_b & 0 & -S_b \\ 0 & m_b & 0 \\ -S_b & 0 & I_b \end{bmatrix} \{\mathbb{D}_1\boldsymbol{\phi}_j\}dx \end{aligned}$$

Using Lagrange's equations

$$\boldsymbol{M}\ddot{\boldsymbol{q}} + \boldsymbol{K}\boldsymbol{q} = \boldsymbol{Q}_\Lambda$$

where $\boldsymbol{M}$ and $\boldsymbol{K}$ are of size $(M+N)\times(M+N)$, $\boldsymbol{q}$ is of size $(M+N)\times 1$, and $\boldsymbol{Q}_\Lambda$ is of size $(M+N)\times 1$. The static deflections of the beam are found from

$$\boldsymbol{q} = \boldsymbol{K}^{-1}\boldsymbol{Q}_\Lambda \tag{4.482}$$

$$\boldsymbol{U} = \begin{Bmatrix} u \\ w \end{Bmatrix} = \boldsymbol{\phi}\boldsymbol{q} \tag{4.483}$$

Note that ϕ_{u_i} and ϕ_{w_j} must satisfy at least the geometric boundary conditions.

For an uncoupled beam, where the extension-bending coupling matrix $ES_{\text{tot}} = 0$, the governing equations can be reduced to two sets of uncoupled equations.

$$EA_{\text{tot}}\epsilon_o = F_\Lambda \tag{4.484}$$

$$EI_{\text{tot}}\kappa = M_\Lambda \tag{4.485}$$

Also, the strain energy can be divided into two parts:

extensional strain energy

$$V_{\text{ext}} = \frac{1}{2}\int_0^{L_b} \epsilon_o^T EA_{\text{tot}}\epsilon_o dx \tag{4.486}$$

bending strain energy

$$V_{\text{bend}} = \frac{1}{2}\int_0^{L_b} \kappa^T EI_{\text{tot}}\kappa\, dx \tag{4.487}$$

In the case of a static problem, assuming a symmetric configuration ($ES_{\text{tot}} = 0$) with pure induced extension results in a system of M equations.

$$\boldsymbol{K}\boldsymbol{q} = \boldsymbol{Q}_\Lambda \tag{4.488}$$

where the elements of these matrices can be defined as

$$K_{ij} = \int_0^{L_b} \frac{\partial\phi_{u_i}}{\partial x} EA_{\text{tot}} \frac{\partial\phi_{u_j}}{\partial x}\, dx \tag{4.489}$$

$$Q_{\Lambda_i} = \int_0^{L_b} \phi_{u_i}\frac{\partial F_\Lambda}{\partial x}\, dx \tag{4.490}$$

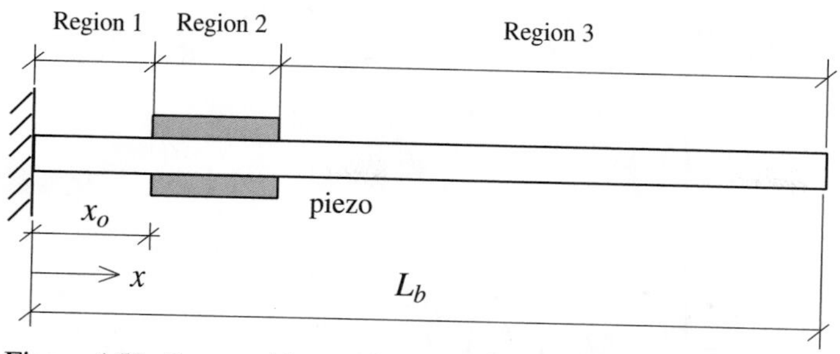

Figure 4.59. Beam with surface-bonded piezoactuators.

Next, consider a symmetric configuration with pure induced bending. The resulting system of N equations can be written as

$$\boldsymbol{Kq} = \boldsymbol{Q}_{\Lambda} \tag{4.491}$$

where the elements of these matrices can be defined as

$$K_{ij} = \int_0^{L_b} \frac{\partial^2 \phi_{w_i}}{\partial x^2} EI_{\text{tot}} \frac{\partial \phi_{w_j}}{\partial x}\, dx \tag{4.492}$$

$$Q_{\Lambda_i} = \int_0^{L_b} F_{\Lambda} \frac{\partial^2 \phi_{w_i}}{\partial x^2}\, dx \tag{4.493}$$

4.9.7 Worked Example

Let us use approximate energy methods to calculate the static response of a cantilevered uniform beam of length L_b, width b, and thickness t_b. (Fig. 4.59). The beam is in a pure bending configuration: identical piezo-sheet actuators of length l_c, thickness t_c, and width b, with opposite polarity.

Solution

Assume the response as a summation of functions such that each function separately satisfies at least the geometric boundary conditions.

$$w(x) = \sum_{i=1}^{N} \phi_{w_i}(x) q_i \tag{4.494}$$

Let us assume a one-term solution for the displacement

$$w(x) = \left(\frac{x}{L_b}\right)^2 q_1 = \phi_1 q_1$$

at $x = 0$,

$$w = 0, \quad \phi_i(0) = 0$$

$$\frac{\partial w}{\partial x} = 0, \quad \frac{\partial \phi_i}{\partial x}(0) = 0$$

This function satisfies the geometric boundary conditions given previously. The bending strain energy is

$$V_{\text{bend}} = \frac{1}{2}\int_0^{L_b} \frac{\partial^2 w}{\partial x^2} EI_{\text{tot}} \frac{\partial^2 w}{\partial x^2} dx - \int_0^{L_b} \frac{\partial^2 w}{\partial x^2} M_{\Lambda} dx$$

Substituting in Lagrange's equation

$$K_{11}q_1 = Q_1$$

the stiffness and forcing function are given by

$$\begin{aligned}
K_{11} &= \int_0^{L_b} EI_{\text{tot}} \left(\frac{\partial \phi_1^2}{\partial x^2}\right)^2 dx \\
&= \int_0^{L_b} EI_{\text{tot}} \left(\frac{2}{L_b^2}\right)^2 dx \\
&= \frac{4}{L_b^4}(EI_b l_b + EI_c l_c) \\
Q_1 &= \int_0^{L_b} M_\Lambda \left(\frac{\partial^2 \phi_1}{\partial x^2}\right) dx \\
&= \int_{x_o}^{x_o+l_c} M_\Lambda \frac{2}{L_b^2} dx \\
&= M_\Lambda \frac{2l_c}{L_b^2}
\end{aligned}$$

where

$$\begin{aligned}
EI_b &= E_b b \frac{t_b^3}{12} \\
EI_c &= E_c b \frac{t_c^3}{6} + E_c b \frac{t_c}{2} (t_c + t_b)^2
\end{aligned}$$

This results in

$$\begin{aligned}
q_1 &= \frac{Q_1}{K_{11}} \\
&= \frac{M_\Lambda l_c L_b^2}{2(EI_b L_b + EI_c l_c)}
\end{aligned}$$

and

$$w(x) = \phi_1 q_1 = \frac{M_\Lambda l_c x^2}{2(EI_b L_b + EI_c l_c)}$$

4.9.8 Worked Example

Using the Rayleigh-Ritz method, determine the steady-state tip response of a beam of length L_b with two identical piezos bonded to the top and bottom surfaces. A sinusoidal field of the same magnitude but opposite polarity is applied to the piezo-sheets (Fig. 4.60).

Assume that

$$w(x,t) = \left(\frac{x}{L_b}\right)^2 q_1$$

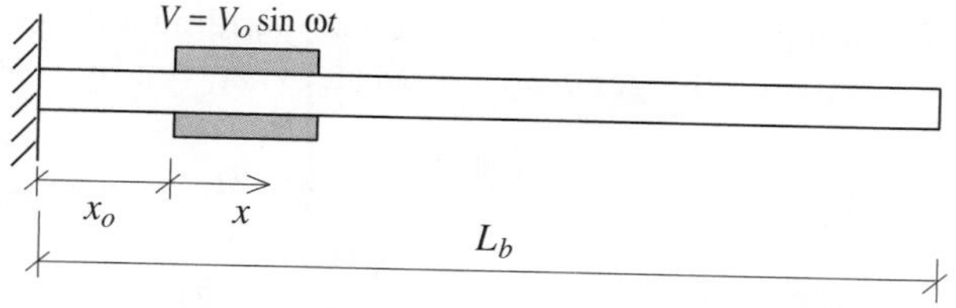

Figure 4.60. Beam with surface-bonded piezoactuators, sinusoidal excitation.

Solution

The actuators are geometrically identical, so $ES_{\text{tot}} = 0$. Because the assumed shape function contains only the w displacement, terms related to u displacements can be ignored. Therefore, from the assumed shape function

$$\phi_1 = \left(\frac{x}{L_b}\right)^2$$

$$\mathbb{D}\phi_1 = \frac{2}{L_b^2}$$

$$\mathbb{D}_1\phi_1 = \begin{Bmatrix} 0 \\ \phi \\ \frac{2x}{L_b^2} \end{Bmatrix}$$

The governing equation becomes

$$M_{11}\ddot{q} + K_{11}q = Q_1$$

where the generalized mass is given by

$$M_{11} = \int_0^{L_b} \begin{Bmatrix} \phi_1 & \frac{\partial \phi_1}{\partial x} \end{Bmatrix} \begin{bmatrix} m & 0 \\ 0 & I \end{bmatrix} \begin{Bmatrix} \phi_1 \\ \frac{\partial \phi_1}{\partial x} \end{Bmatrix} dx$$

Assuming that m_b and m_c are the mass per unit length of the beam and piezo-sheet, respectively, and I_b and I_c are the second mass moments for the beam and actuator, respectively

$$M_{11} = \int_0^{L_b} \left[m\,\phi_1^2 + I\left(\frac{\partial \phi_1}{\partial x}\right)^2 \right] dx$$

$$= \int_0^{L_b} \left(\frac{m_b x^4}{L_b^4} + \frac{4 I_b x^2}{L_b^4} \right) dx + \int_{x_o}^{x_o + l_c} \left(\frac{m_c x^4}{L_b^4} + \frac{4 I_c x^2}{L_b^4} \right) dx$$

$$= \frac{m_b L_b}{5} + \frac{4 I_b}{3 L_b} + \frac{m_c}{5 L_b^4}\left[(x_o + l_c)^5 - x_o^5\right] + \frac{4 I_c}{3 L_b^4}\left[(x_o + l_c)^3 - x_o^3\right]$$

Assuming that EI_b and EI_c are the bending stiffness of the beam and piezo-sheet, respectively, the generalized stiffness is given by

$$
\begin{aligned}
K_{11} &= \int_0^{L_b} \mathbb{D}\phi_1 \, EI_{\text{tot}} \, \mathbb{D}\phi_1 \, dx \\
&= \int_0^{L_b} EI_{\text{tot}} \left(\frac{\partial \phi_1^2}{\partial x^2} \right)^2 dx \\
&= \int_0^{L_b} \frac{4}{L_b^4} EI_b dx + \int_{x_o}^{x_o+l_c} \frac{4}{L_b^4} EI_c dx \\
&= EI_b \frac{4}{L_b^3} + EI_c \frac{4}{L_b^4} l_c
\end{aligned}
$$

The induced moment is

$$
\begin{aligned}
M_\Lambda &= -\frac{1}{2} \sum_{i=1}^{3} E_k b \Lambda_k (h_{k+1}^2 - h_k^2) \\
&= -E_c b \Lambda t_c (t_c + t_b) \\
&= -\frac{E_c b (t_c + t_b)}{t_c} d_{31} V_o \sin \omega t
\end{aligned}
$$

From which the generalized force is derived as

$$
\begin{aligned}
Q_1 &= \int_0^{L_b} M_\Lambda \left(\frac{\partial^2 \phi_1}{\partial x^2} \right) dx \\
&= \int_{x_o}^{x_o+l_c} M_\Lambda \frac{2}{L_b^2} dx \\
&= M_\Lambda \frac{2 l_c}{L_b^2} \\
&= -2 E_c b (t_c + t_b) d_{31} V_o \sin \omega t \frac{l_c}{L_b^2 t_c} \\
&= \bar{Q}_1 \sin \omega t
\end{aligned}
$$

Subsituting $q = q_o \sin \omega t$ in the governing equation

$$
q_o = \frac{\bar{Q}_1}{K_{11} - M_{11} \omega^2}
$$

Therefore, the steady-state tip displacement is given by

$$
w_{\text{tip}}(x, t) = \frac{\bar{Q}_1 \sin \omega t}{K_{11} - M_{11} \omega^2}
$$

4.9.9 Energy Formulation: Dynamic Beam Governing Equation Derived from Hamilton's Principle

The governing equation for the dynamic behavior of a beam can also be derived using Hamilton's Principle. This is a virtual energy principle based on variational calculus. For a dynamic system, it can be written as

$$\delta \int_{t_1}^{t_2} (T - V)dt + \int_{t_1}^{t_2} \delta W_{\text{ext}}\, dt = 0 \tag{4.495}$$

where T is the kinetic energy, V is the strain energy (potential energy), and δW_{ext} is the virtual work done by external forces. These forces include both conservative and nonconservative forces. Taking the variation of the total strain energy

$$\delta V_e = \int_0^{L_b} \{\epsilon_o \quad \kappa\} \begin{bmatrix} EA_{\text{tot}} & ES_{\text{tot}} \\ ES_{\text{tot}} & EI_{\text{tot}} \end{bmatrix} \begin{Bmatrix} \delta\epsilon_o \\ \delta\kappa \end{Bmatrix} dl - \int_0^{L_b} \{F_\Lambda^T \quad M_\Lambda^T\} \begin{Bmatrix} \delta\epsilon_o \\ \delta\kappa \end{Bmatrix} dl \tag{4.496}$$

The kinetic energy is given by

$$\begin{aligned} \delta T &= \iiint_{\text{volume}} \rho_s \left(\dot{u}\delta\dot{u} + \dot{w}\delta\dot{w}\right) dxdydz \\ &= \iiint_{\text{volume}} \rho_s \left[\left(\dot{u}_o - z\frac{\partial^2 w}{\partial t\partial x}\right)\left(\delta\dot{u}_o - z\frac{\partial^2 w}{\partial t\partial x}\right) + \dot{w}\delta\dot{w}\right] dx\, dy\, dz \\ &= \int_0^{L_b} \left\{\dot{u}_o \quad \dot{w} \quad \frac{\partial^2 w}{\partial t\partial x}\right\} \begin{bmatrix} m_b & 0 & -S_b \\ 0 & m_b & 0 \\ -S_b & 0 & I_b \end{bmatrix} \begin{Bmatrix} \delta\dot{u}_o \\ \delta\dot{w} \\ \delta\left(\frac{\partial^2 w}{\partial t\partial x}\right) \end{Bmatrix} dx \end{aligned} \tag{4.497}$$

where ρ_s is the mass density of the material of the beam. The inertia terms are defined as

$$m_b = \int_{t_b} \rho_s b dz \tag{4.498}$$

$$S_b = \int_{t_b} \rho_s b z dz \tag{4.499}$$

$$I_b = \int_{t_b} \rho_s b z^2 dz \tag{4.500}$$

where m_b is the mass per unit length (kg/m); S_b is the first mass moment of inertia, per unit length, about the mid-plane (kg); and I_b is the second mass moment of inertia, per unit length, about the mid-plane (kg-m). If the beam is exposed to a transverse external force $f_z(x, t)$, the virtual work done becomes

$$\delta W_{\text{ext}} = \int_0^{L_b} f_z \delta w dx \tag{4.501}$$

Applying Hamilton's Principle and using integration by parts, we obtain all boundary conditions plus governing equations. Note that the virtual terms are all reduced to δu_o and δw. This results in

$$\delta u_o [\ldots] + \delta w [\ldots] = 0 \tag{4.502}$$

Terms associated with δu_o and δw, respectively, are identically zero.

δu_o:

$$\frac{dF}{dx} - m_b\ddot{u}_o + S_b\frac{d\ddot{w}}{dx} = 0 \tag{4.503}$$

δw:

$$\frac{d^2M}{dx^2} + m_b\ddot{w} + \frac{d}{dx}\left(S_b\ddot{u}_o\right) - \frac{d}{dx}\left(I_b\frac{d\ddot{w}}{dx}\right) = 0 \tag{4.504}$$

Substituting terms for F and M as

$$F = EA_{\text{tot}}\frac{du_o}{dx} + ES_{\text{tot}}\frac{d^2w}{dx^2} - F_\Lambda \tag{4.505}$$

$$M = ES_{\text{tot}}\frac{du_o}{dx} + EI_{\text{tot}}\frac{d^2w}{dx^2} - M_\Lambda \tag{4.506}$$

where F_Λ and M_Λ are the induced force and induced moment, respectively. The governing equations become

δu_o:

$$\frac{d}{dx}\left[EA_{\text{tot}}\frac{du_o}{dx} + ES_{\text{tot}}\frac{d^2w}{dx^2}\right] - m_b\ddot{u}_o + S_b\frac{d\ddot{w}}{dx} = \frac{dF_\Lambda}{dx} \tag{4.507}$$

δw:

$$\frac{d^2}{dx^2}\left[ES_{\text{tot}}\frac{du_o}{dx} + EI_{\text{tot}}\frac{d^2w}{dx^2}\right] + m_b\ddot{w} + \frac{d}{dx}(S_b\ddot{u}_o) - \frac{d}{dx}\left(I_b\frac{d\ddot{w}}{dx}\right) = \frac{d^2M_\Lambda}{dx^2} \tag{4.508}$$

These are coupled partial differential equations with partial derivatives in spatial and temporal coordinates. They require four initial conditions (two for each variable) and six boundary conditions (three at each end). The initial conditions are

$$u_o(x,0) \text{ and } \dot{u}_o(x,0) \quad \rightarrow \quad \text{prescribed}$$

$$w(x,0) \text{ and } \dot{w}(x,0) \quad \rightarrow \quad \text{prescribed}$$

For a system starting from rest, these four initial values are set to zero.

The boundary conditions at each end can be one of the following:

(a) Clamped Condition

$$u_o = 0$$

$$w = 0$$

$$\frac{dw}{dx} = 0$$

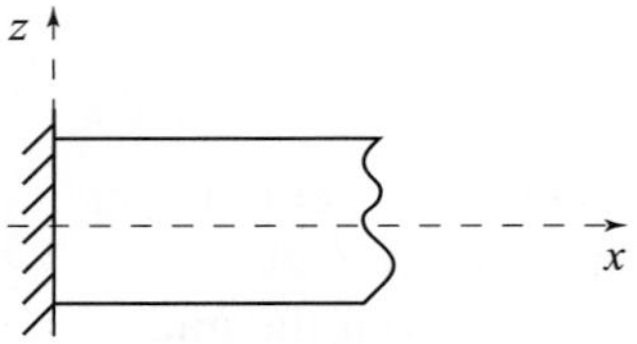

(b) Free Condition

$$\text{axial force } F = EA_{\text{tot}}\frac{\partial u_o}{\partial x} = 0$$

$$\text{bending moment } M = EI_{\text{tot}}\frac{\partial^2 w}{\partial x^2} = 0$$

$$\text{shear force } \frac{dM}{dx} = \frac{\partial}{\partial x}\left(EI_{\text{tot}}\frac{\partial^2 w}{\partial x^2}\right) = 0$$

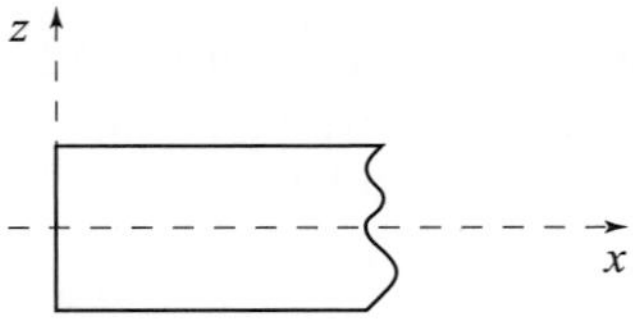

(c) Simply-Supported (Hinged or Pinned) Condition

$$u_o = 0$$

$$w = 0$$

$$M = 0$$

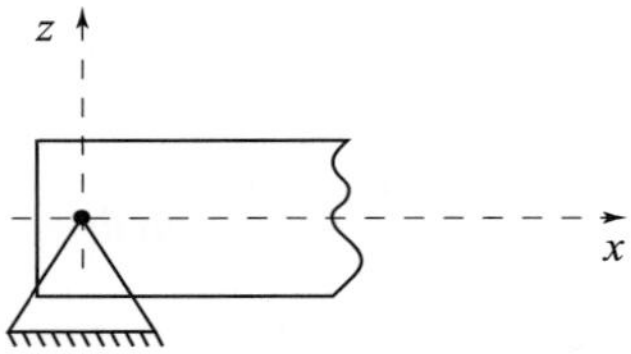

(d) Roller-Supported Condition

$$F = 0$$
$$w = 0$$
$$M = 0$$

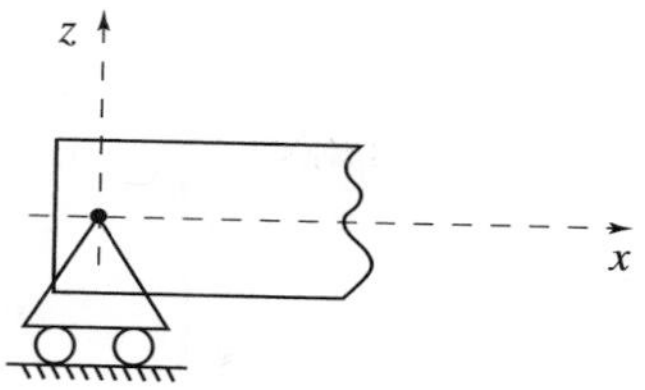

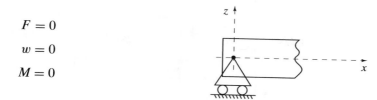

(e) Vertical Spring-Supported Condition (Left End)

$$u_o = 0$$
$$M = 0$$
$$\frac{dM}{dx} = \frac{\partial}{\partial x}\left(EI_{\text{tot}}\frac{\partial^2 w}{\partial x^2}\right) = -kw$$

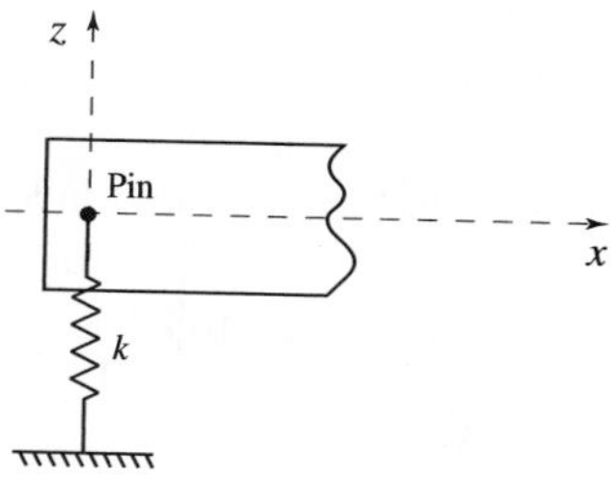

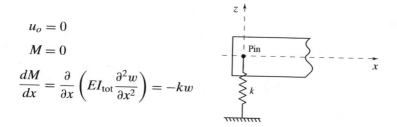

(f) Bending Spring-Supported Condition (Left End)

$$u_o = 0$$
$$w = 0$$
$$M = EI_{\text{tot}}\frac{\partial^2 w}{\partial x^2} = k_\theta \frac{dw}{dx}$$

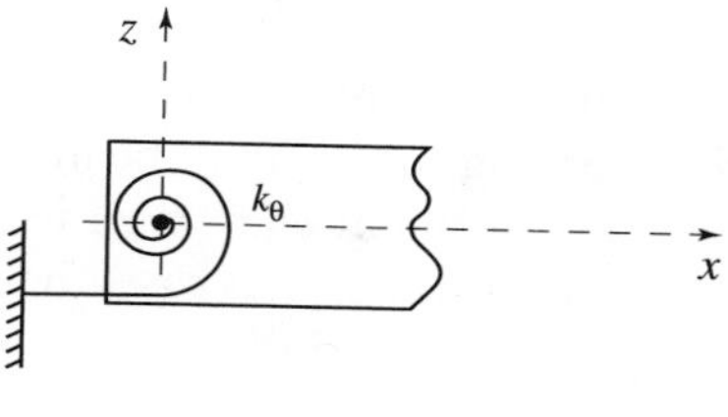

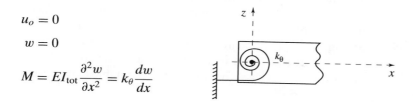

(g) Axial Spring-Supported Condition (Left End)

$$F = k_u u$$
$$M = 0$$
$$\frac{\partial M}{\partial x} = 0$$

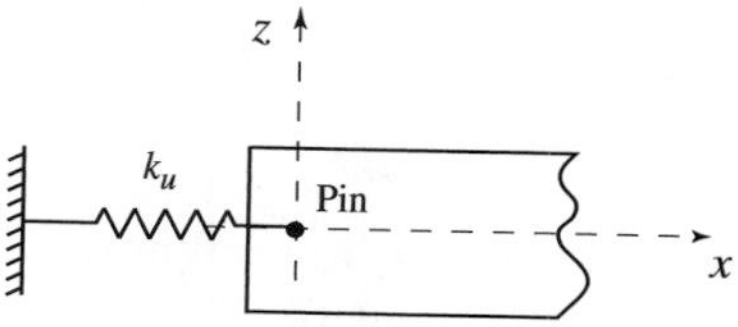

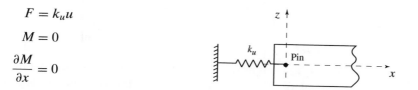

4.10 Finite Element Analysis with Induced-Strain Actuation

This section describes the modeling of structures with induced-strain actuation using the finite element method (FEM). This method is a powerful and convenient technique for modeling the static and dynamic response of a structure. Although the present discussion is focused on a simple beam model, it contains all of the important ingredients that can be easily expanded to more complex beam models.

Consider a beam with two surface-bonded piezo sheet actuators. For the present analysis, we only consider pure bending actuation, but extensional deformation can also be easily incorporated in the derivation. The beam is divided into a finite number of elements connected to one an other by nodes, as shown in Fig. 4.61(a). The properties of the beam are assumed constant in each element. Because the structure and elements are one-dimensional, each node requires two variables to

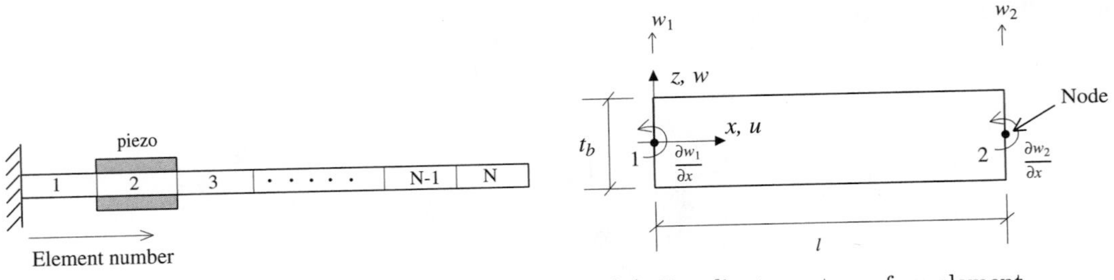

(a) Beam divided into elements (b) Coordinate system of an element

Figure 4.61. Finite element modeling of a beam.

describe its position in the x-z plane. These variables are the displacements in the z direction, represented by w, and the rotation about the y-axis, represented by $\partial w/\partial x$. For small deformations, the axial displacement u due to bending can be neglected, resulting in two degrees of freedom for each node. Equations can be derived for each element that are subsequently assembled to form a mathematical model of the entire beam structure.

4.10.1 Behavior of a Single Element

Consider a beam element of length l, with nodes labeled 1 and 2, as in Fig. 4.61(b). Within this element, the general form of the transverse deflection $w(x)$ must be chosen so that the basic physics of the problem can be adequately represented. Because the governing equation for beam bending contains fourth order derivatives, the transverse displacement $w(x)$ must be at least cubic to ensure that none of the terms identically vanishes. In this case, the minimum acceptable expression for $w(x)$ is

$$w(x) = a_1 + a_2x + a_3x^2 + a_4x^3 = \{1 \quad x \quad x^2 \quad x^3\} \begin{Bmatrix} a_1 \\ a_2 \\ a_3 \\ a_4 \end{Bmatrix} \tag{4.509}$$

where the a_i $(i = 1, 2, 3, 4)$ are unknown coefficients. In the finite element method, this assumed displacement field must be expressed in terms of nodal degrees of freedom; that is, we wish to convert Eq. 4.509 to the following form

$$w(x) = \boldsymbol{H}\boldsymbol{q}_e = \left[H_1(x) \quad H_2(x) \quad H_3(x) \quad H_4(x)\right] \begin{Bmatrix} w_1 \\ \frac{\partial w_1}{\partial x} \\ w_2 \\ \frac{\partial w_2}{\partial x} \end{Bmatrix} \tag{4.510}$$

where the vector $\boldsymbol{q}_e$, of size 4×1 represents the nodal degrees of freedom of the element; w_i and $\partial w_i/\partial x$ represent, respectively, the transverse displacement and rotation at node i; and the functions $H_i(x)$ $(i = 1, 2, 3, 4)$ are the interpolation functions (or shape functions). The required form of the shape functions $H_i(x)$ can be determined by first expressing the coefficients a_i in terms of w_1, $\partial w_1/\partial x$, w_2, and

$\partial w_2/\partial x$, so that Eq. 4.509 satisfies the boundary conditions of the element.

$$\text{at } x = 0: \quad w = w_1 = a_1$$
$$\frac{\partial w}{\partial x} = \frac{\partial w_1}{\partial x} = a_2$$
$$\text{at } x = l: \quad w = w_2 = a_1 + a_2 l + a_3 l^2 + a_4 l^3$$
$$\frac{\partial w}{\partial x} = \frac{\partial w_2}{\partial x} = a_2 + 2a_3 l + 3a_4 l^2$$

Combining the boundary conditions with Eqs. 4.509 and 4.510

$$\boldsymbol{q}_e = \begin{Bmatrix} w_1 \\ \frac{\partial w_1}{\partial x} \\ w_2 \\ \frac{\partial w_2}{\partial x} \end{Bmatrix} = \begin{bmatrix} 1 & 0 & 0 & 0 \\ 0 & 1 & 0 & 0 \\ 1 & l & l^2 & l^3 \\ 0 & 1 & 2l & 3l^2 \end{bmatrix} \begin{Bmatrix} a_1 \\ a_2 \\ a_3 \\ a_4 \end{Bmatrix} \tag{4.511}$$

From this equation

$$\begin{Bmatrix} a_1 \\ a_2 \\ a_3 \\ a_4 \end{Bmatrix} = \begin{bmatrix} 1 & 0 & 0 & 0 \\ 0 & 1 & 0 & 0 \\ -3/l^2 & -2/l & 3/l^2 & -1/l \\ 2/l^3 & 1/l^2 & -2/l^3 & 1/l^2 \end{bmatrix} \begin{Bmatrix} w_1 \\ \frac{\partial w_1}{\partial x} \\ w_2 \\ \frac{\partial w_2}{\partial x} \end{Bmatrix} \tag{4.512}$$

Substituing this in Eq. 4.509 and rewriting leads to

$$\begin{aligned} w(x) &= \boldsymbol{H}\boldsymbol{q}_e \\ &= \{H_1 \quad H_2 \quad H_3 \quad H_4\}\, \boldsymbol{q}_e \end{aligned} \tag{4.513}$$

where the elements of the vector $\boldsymbol{H}$ are given by

$$H_1 = 1 - 3\frac{x^2}{l^2} + 2\frac{x^3}{l^3} \tag{4.514}$$

$$H_2 = x - 2\frac{x^2}{l} + \frac{x^3}{l^2} \tag{4.515}$$

$$H_3 = 3\frac{x^2}{l^2} - 2\frac{x^3}{l^3} \tag{4.516}$$

$$H_4 = -\frac{x^2}{l} + \frac{x^3}{l^2} \tag{4.517}$$

These H_i are called shape functions, and they determine the deflected shape of the element. Neglecting rotational inertia of the element and axial deformation, the kinetic energy of the element, having a mass per unit length m_b, can be written as

$$T_e = \frac{1}{2}\int_0^l m\,(\dot{w})^2\,dx \tag{4.518}$$

Substituting for w from Eq. 4.513, and observing that $\boldsymbol{q}_e$ is independent of x and $\boldsymbol{H}$ is independent of time, the equation for kinetic energy of the element can be simplified to

$$T_e = \frac{1}{2}\dot{\boldsymbol{q}}_e{}^T \int_0^l m\, \boldsymbol{H}^T \boldsymbol{H} dx\, \dot{\boldsymbol{q}}_e = \frac{1}{2}\dot{\boldsymbol{q}}_e{}^T \boldsymbol{M}_e\, \dot{\boldsymbol{q}}_e \tag{4.519}$$

An equivalent elemental mass matrix, $\boldsymbol{M}_e$, can be defined as

$$\boldsymbol{M}_e = \int_0^l m_b\, \boldsymbol{H}^T \boldsymbol{H} dx \tag{4.520}$$

The size of the elemental mass matrix is 4×4 because each element has four degrees of freedom. Similarly, the strain energy of the element can be written as

$$V_e = \frac{1}{2}\int_0^l EI_b \left(\frac{\partial^2 w}{\partial x^2}\right)^2 dx = \frac{1}{2}\boldsymbol{q}_e^T \int_0^l EI_b \frac{\partial^2 \boldsymbol{H}^T}{\partial x^2}\frac{\partial^2 \boldsymbol{H}}{\partial x^2} dx\, \boldsymbol{q}_e = \frac{1}{2}\boldsymbol{q}_e^T \boldsymbol{K}_e\, \boldsymbol{q}_e \tag{4.521}$$

An equivalent elemental stiffness matrix, $\boldsymbol{K}_e$ (of size 4×4), can be defined as

$$\boldsymbol{K}_e = \int_0^l EI_b \frac{\partial^2 \boldsymbol{H}^T}{\partial x^2}\frac{\partial^2 \boldsymbol{H}}{\partial x^2} dx \tag{4.522}$$

The external forces acting on the element, represented by the vector $\boldsymbol{Q}_e$, can be calculated from the expression for virtual work done, $\delta \boldsymbol{W}_{\text{ext}_e}$

$$\delta \boldsymbol{W}_{\text{ext}_e} = \int_0^l (F\delta w + M\frac{\partial \delta w}{\partial x})\, dx = \int_0^l (F\boldsymbol{H} + M\frac{\partial \boldsymbol{H}}{\partial x})\, \delta \boldsymbol{q}_e\, dx = \boldsymbol{Q}_e \delta \boldsymbol{q}_e \tag{4.523}$$

where F is the external force distribution (force per unit length) and M is the external moment distribution (moment per unit length) acting on the element. The vector $\boldsymbol{Q}_e$ gives the force and moment acting on the nodes of the element. From this equation

$$\boldsymbol{Q}_e = \int_0^l (F\boldsymbol{H} + M\frac{\partial \boldsymbol{H}}{\partial x})\, dx \tag{4.524}$$

For a beam element of length l, having a uniform cross section along its length and shape functions given by Eqs. 4.514–4.517, the elemental mass and stiffness matrices can be derived as

$$\boldsymbol{M}_e = \frac{m_b l}{420}\begin{bmatrix} 156 & 22l & 54 & -13l \\ 22l & 4l^2 & 13l & -3l^2 \\ 54 & 13l & 156 & -22l \\ -13l & -3l^2 & -22l & 4l^2 \end{bmatrix} \tag{4.525}$$

$$\boldsymbol{K}_e = \frac{EI_b}{l^3}\begin{bmatrix} 12 & 6l & -12 & 6l \\ 6l & 4l^2 & -6l & 2l^2 \\ -12 & -6l & 12 & -6l \\ 6l & 2l^2 & -6l & 4l^2 \end{bmatrix} \tag{4.526}$$

We now have expressions for the equivalent mass, stiffness, and forcing matrices for each element of the structure. The next step involves assembling the elements to form a global representation of the structure.

4.10.2 Assembly of Global Mass and Stiffness Matrices

The elemental mass and stiffness matrices, $\boldsymbol{M}_e$ and $\boldsymbol{K}_e$, and the forcing vector $\boldsymbol{Q}_e$, are assembled together to form a global mass matrix $\boldsymbol{M}_g$, global stiffness matrix $\boldsymbol{K}_g$, and global forcing vector $\boldsymbol{Q}_g$. The global quantities define the behavior of the entire structure. The assembly process is carried out by considering the total energy of the structure. For a structure divided into N elements, the total kinetic energy T_g and total strain energy V_g are given by

$$T_g = \sum_{i=1}^{N} T_{e_i} \tag{4.527}$$

$$V_g = \sum_{i=1}^{N} V_{e_i} \tag{4.528}$$

where T_{e_i} and V_{e_i} are the kinetic and potential energies of the "i"th element, respectively. By making use of the connectivity of elements in the structure, the total kinetic energy of the structure can be expressed as

$$T_g = \frac{1}{2}\dot{\boldsymbol{q}}_g{}^T M_g \dot{\boldsymbol{q}}_g \tag{4.529}$$

where $\boldsymbol{q}_g$ is the vector of all of the degrees of freedom of the structure. Similarly, the total strain energy and total virtual work are given by

$$V_g = \frac{1}{2}\dot{\boldsymbol{q}}_g{}^T K_g \boldsymbol{q}_g \tag{4.530}$$

$$\delta \boldsymbol{W}_{\text{ext}_g} = \boldsymbol{Q}_g \delta \boldsymbol{q}_g \, dx \tag{4.531}$$

The assembly procedure is carried out based on displacement and force compatibility between two elements at their common node. For example, based on Fig. 4.61(a), for elements 1 and 2

$$w_{2\,\text{element 1}} = w_{1\,\text{element 2}}$$

and

$$\left.\frac{\partial w_2}{\partial x}\right|_{\text{element 1}} = \left.\frac{\partial w_1}{\partial x}\right|_{\text{element 2}}$$

Similar relations can be written for force and moment compatibility, and the entire system can be expressed in terms of global degrees of freedom, $\boldsymbol{q}_g$. Consequently, the entries in the mass, stiffness, and forcing matrices corresponding to the common node are summed up for each element sharing that node. This process is shown schematically in Fig. 4.62, where the shaded areas represent a summation of entries.

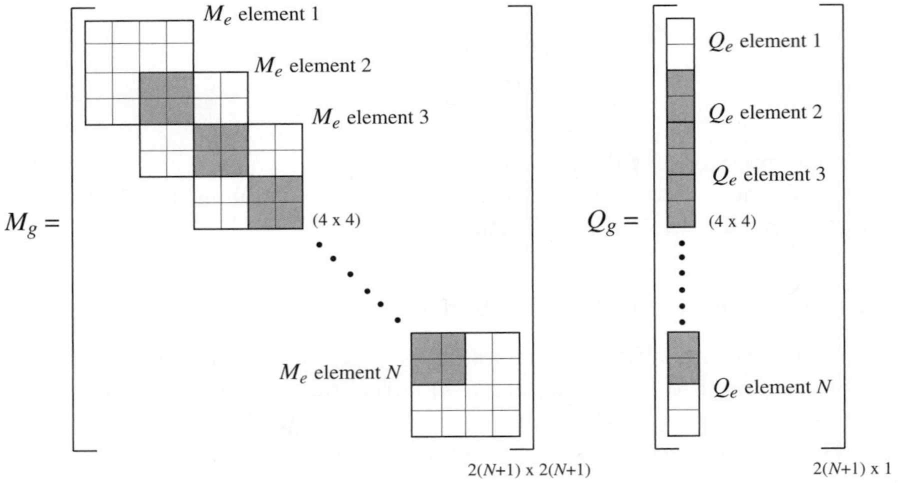

Figure 4.62. Assembly of global matrices.

For a structure with N beam elements, the size of the global mass and stiffness matrices is $2(N+1) \times 2(N+1)$, corresponding to two degrees of freedom per node. The forcing vector is of size $2(N+1) \times 1$.

From Lagrange's equations

$$\frac{d}{dt}\left(\frac{\partial T_g}{\partial \dot{\boldsymbol{q}}_g}\right) + \frac{\partial V_g}{\partial \boldsymbol{q}_g} = \boldsymbol{Q}_g \tag{4.532}$$

Substituting Eqs. 4.529, 4.530, and 4.531 in Lagrange's equation (Eq. 4.532) results in the governing equation for the entire structure

$$\boldsymbol{M}_g\, \ddot{\boldsymbol{q}}_g + \boldsymbol{K}_g\, \boldsymbol{q}_g = \boldsymbol{Q}_g \tag{4.533}$$

This equation can be used to calculate the static and dynamic response of the beam element to an applied external loading. For a static problem, $\ddot{\boldsymbol{q}}_g = 0$, and the mass matrix $\boldsymbol{M}_g$ can be ignored. Note that the process of discretizing the continuous structure with infinite degrees of freedom into a finite number of elements with a finite number of degrees of freedom is equivalent to imposing artificial constraints on the structure. This has the effect of making the mathematical model of the structure somewhat "stiffer" than the real structure, consequently yielding higher natural frequencies and lower deflections than the exact solution. As the number of elements is increased, the constraints on the system decrease, and the FEM solution begins to converge to the exact solution.

4.10.3 Beam Bending with Induced-Strain Actuation

We now examine the effect of induced-strain actuation on the elemental mass and stiffness matrices. Consider a beam bending element as before, of length l, with two piezo-sheet actuators bonded on each surface, as shown in Fig. 4.63. The piezo actuators have a thickness t_c and a mass per unit length m_c. The thickness and mass per unit length of the beam are t_b and m_b, respectively. We consider the case in which the piezo actuators and the beam have the same width, b. The same shape functions as in Eq. 4.513 are used to define the deformation of the element.

Figure 4.63. Beam element with induced-strain actuators.

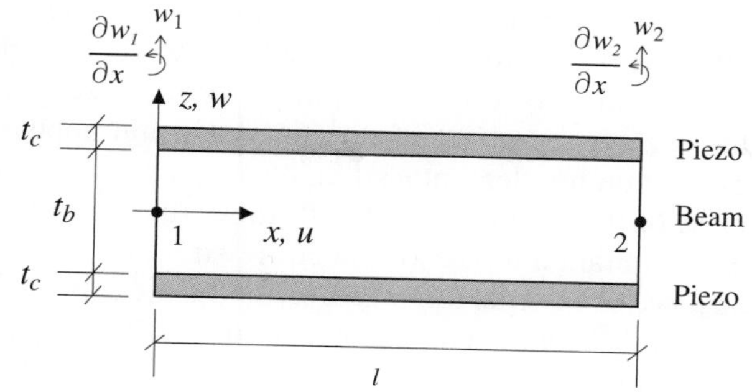

From Eq. 4.519, the kinetic energy of the beam element with piezo actuators is given by

$$\begin{aligned} T &= \frac{1}{2}\dot{\boldsymbol{q}}_e{}^T \int_0^l (m_c + m_b)\boldsymbol{H}^T \boldsymbol{H} dx\, \dot{\boldsymbol{q}}_e \\ &= \frac{1}{2}\dot{\boldsymbol{q}}_e{}^T \boldsymbol{M}_{\text{act}}\, \dot{\boldsymbol{q}}_e \end{aligned} \tag{4.534}$$

where the mass matrix of the actuated element, $\boldsymbol{M}_{\text{act}}$, is defined as

$$\boldsymbol{M}_{\text{act}} = \int_0^l (m_c + m_b)\boldsymbol{H}^T \boldsymbol{H} dx \tag{4.535}$$

The strain energy of the element can be written as

$$V_e = \frac{1}{2}\int_0^l \int_A E(z)\epsilon(z)^2 dA dx \tag{4.536}$$

where A is the cross-sectional area of the element. Only the mechanical strain on the piezo actuators contributes to the potential energy of the element. Therefore, this can be simplified to

$$\begin{aligned} V_e &= \frac{1}{2}\int_0^l 2b\left[\int_0^{t_b/2} E_b\left(z\frac{\partial^2 w}{\partial x^2}\right)^2 dz + \int_{t_b/2}^{t_b/2+t_c} E_c\left(-z\frac{\partial^2 w}{\partial x^2} - \Lambda\right)^2 dz\right] dx \\ &= \frac{1}{2}\boldsymbol{q}_e^T \int_0^l (EI_b + EI_c)\frac{\partial^2 \boldsymbol{H}^T}{\partial x^2}\frac{\partial^2 \boldsymbol{H}}{\partial x^2} dx\, \boldsymbol{q}_e \\ &\quad + E_c\Lambda^2 b l t_c + \int_0^l b E_c \Lambda t_c (t_c + t_b)\frac{\partial^2 \boldsymbol{H}}{\partial x^2}\, dx\, \boldsymbol{q}_e \\ &= \frac{1}{2}\boldsymbol{q}_e^T \boldsymbol{K}_{\text{act}_e} \boldsymbol{q}_e + E_c\Lambda^2 b l t_c - Q_\Lambda \boldsymbol{q}_e \end{aligned} \tag{4.537}$$

where the equivalent stiffness matrix $\boldsymbol{K}_{\text{act}_e}$ is

$$\boldsymbol{K}_{\text{act}_e} = \int_0^l (EI_b + EI_c)\frac{\partial^2 \boldsymbol{H}^T}{\partial x^2}\frac{\partial^2 \boldsymbol{H}}{\partial x^2} dx \tag{4.538}$$

and

$$\begin{aligned} Q_\Lambda &= \int_0^l M_\Lambda \frac{\partial^2 \boldsymbol{H}}{\partial x^2}\, dx \\ M_\Lambda &= E_c t_c b \Lambda (t_b + t_c) \end{aligned} \tag{4.539}$$

Note that because there are no external forces and moments, the virtual work $\delta W_{\text{ext}_e} = 0$, and from the Principle of Virtual Work, we obtain $\delta V_e = 0$. Because the variation of the term containing Λ^2 is zero and the elemental forcing vector is zero, the effect of the induced strain appears as an additional term in the expression for elemental strain energy. The elemental matrices for the stiffness and mass of the beam element with the actuators can be assembled into global matrices to obtain a model of the entire structure with induced-strain actuation. In the absence of external forcing, substituting the expressions for kinetic energy and strain energy into Lagrange's equation (Eq. 5.277) results in the governing equation for the structure

$$\boldsymbol{M}_{\text{act}_g}\, \ddot{\boldsymbol{q}}_g + \boldsymbol{K}_{\text{act}_g}\, \boldsymbol{q}_g = \boldsymbol{Q}_{\text{act}_g} \tag{4.540}$$

where the global forcing vector is obtained by an assembly of vectors for each element given by

$$\boldsymbol{Q}_{\text{act}_e} = -E_c \Lambda t_c \left(t_c + t_b\right) \left[\int_0^l b \frac{\partial^2 \boldsymbol{H}}{\partial x^2}\, dx \right]^T \tag{4.541}$$

The elemental forcing vector is obtained from the elemental strain energy due to the induced-strain term. Therefore, the induced strain effectively appears as a forcing on the system. Each row of the vector $\boldsymbol{Q}_{\text{act}_e}$ represents a forcing corresponding to the particular degree of freedom of each node. Rows 1 and 3 represent forces in the z direction, or shear forces acting on the degrees of freedom w_1 and w_2, whereas rows 2 and 4 represent moments acting on the degrees of freedom $\partial w_1/\partial x$ and $\partial w_2/\partial x$. The response of this structure to induced-strain actuation can be calculated by solving Eq. 4.540. If external forces and moments are present, they will add to the elemental forcing vector and are assembled accordingly into the global forcing vector.

4.10.4 Worked Example

Consider the beam shown in Fig. 4.64, with piezo-sheet actuators, tapered along their width, bonded to both surfaces of the beam. The piezoactuators are identical and are actuated by equal voltages of opposite polarity. Treating the beam as a single element, use the FEM to calculate the mass and stiffness matrices of the beam as well as the actuation force vector. The width of the piezo-sheet is given by

$$b_c = b\left(1 - \frac{x}{l}\right)$$

Solution

The mass per unit length of the actuator, m_c, and of the beam, m_b, are given by

$$m_c(x) = 2\rho_c b t_c \left[1 - \frac{x}{l}\right]$$

$$m_b(x) = \rho_b b t_b$$

where ρ_c and ρ_b are the densities of the actuator and beam material, respectively. Similarly, assuming that $t_c \ll t_b$ and therefore neglecting the moment of inertia of

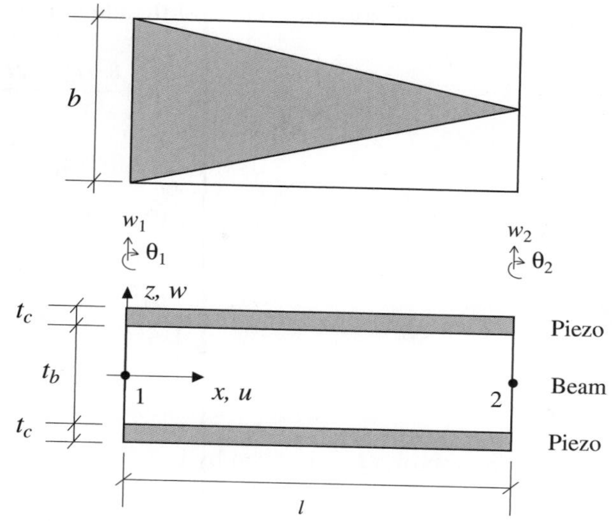

Figure 4.64. Beam with linearly tapered piezoactuators.

the actuators about their own mid-plane, the flexural stiffnesses of the actuator and beam are given by

$$EI_c(x) = 2E_c b t_c \left(\frac{t_c^2}{4} + \frac{t_b^2}{4} + \frac{t_c t_b}{2}\right)\left(1 - \frac{x}{l}\right) = EI_{c_{\text{root}}}\left(1 - \frac{x}{l}\right)$$

$$EI_b(x) = E_b \frac{t_b^3}{12} b = EI_{b_{\text{root}}}$$

where $EI_{c_{\text{root}}}$ and $EI_{b_{\text{root}}}$ are the flexural stiffnesses of the actuator and beam, respectively at the root (left-hand boundary) of the element. From Eq. 4.535, the mass matrix of the element is

$$\boldsymbol{M}_{\text{act}} = \int_0^l (m_c + m_b)\boldsymbol{H}^T \boldsymbol{H} dx$$

$$= \int_0^l b\left[2\rho_c t_c\left(1 - \frac{x}{l}\right) + \rho_b t_b\right] \begin{Bmatrix} H_1 \\ H_2 \\ H_3 \\ H_4 \end{Bmatrix} \{H_1 \quad H_2 \quad H_3 \quad H_4\}\, dx$$

$$= \frac{2\rho_c b t_c}{420}\begin{bmatrix} 120l & 15l^2 & 27l & -7l^2 \\ 15l^2 & 5/2l^3 & 6l^2 & -3/2l^3 \\ 27l & 6l^2 & 36l & -7l^2 \\ -7l^2 & -3/2l^3 & -7l^2 & 3/2l^3 \end{bmatrix}$$

$$+ \frac{\rho_b b t_b}{420}\begin{bmatrix} 156l & 22l^2 & 54l & -13l^2 \\ 22l^2 & 4l^3 & 13l^2 & -3l^3 \\ 54l & 13l^2 & 156l & -22l^2 \\ -13l^2 & -3l^3 & -22l^2 & 4l^3 \end{bmatrix}$$

From Eq. 4.538, the stiffness matrix of the element is

$$\boldsymbol{K}_{\text{act}} = \int_0^l (EI_b + EI_c)\frac{\partial^2 \boldsymbol{H}^T}{\partial x^2}\frac{\partial^2 \boldsymbol{H}}{\partial x^2}dx$$

$$= \frac{EI_{b_{\text{root}}}}{l^3}\begin{bmatrix} 12 & 6l & -12 & 6l \\ 6l & 4l^2 & -6l & 2l^2 \\ -12 & -6l & 12 & -6l \\ 6l & 2l^2 & -6l & 4l^2 \end{bmatrix}$$

$$+ \frac{EI_{c_{\text{root}}}}{l^3}\begin{bmatrix} 6 & 4l & -6 & 2l \\ 4l & 3l^2 & -4l & l^2 \\ -6 & -4l & 6 & -2l \\ 2l & l^2 & -2l & l^2 \end{bmatrix}$$

The forcing vector, $\boldsymbol{Q}_{\text{act}}$, is given by Eq. 4.541

$$\boldsymbol{Q}_{\text{act}} = -E_c\Lambda t_c\,(t_c + t_b)\left[\int_0^l b\frac{\partial^2 \boldsymbol{H}}{\partial x^2}\,dx\right]^T$$

$$= -E_c\Lambda b t_c\,(t_c + t_b)\left[\int_0^l \left(1 - \frac{x}{l}\right)\frac{\partial^2 \boldsymbol{H}}{\partial x^2}\,dx\right]^T$$

$$= -E_c\Lambda b t_c\,(t_c + t_b)\begin{Bmatrix} -1/l \\ -1 \\ 1/l \\ 0 \end{Bmatrix}$$

Here we see an interesting result: The forcing vector has both shear force and bending moment terms. For an actuator of uniform width, only a combination of bending moments or axial forces can be induced. However, by appropriately shaping the actuator, we see that it is possible to also induce shear forces. This has important consequences in applications in which specific types of forcing are required due to boundary conditions or control requirements. Shaping of the active material is also important for sensing applications; for example, when specific modal information is required.

4.11 First-Order Shear Deformation Theory (FSDT) for Beams with Induced-Strain Actuation

Refinements in beam modeling are realized by considering additional terms in the expressions for beam deformation. In general, the deformations can be expanded in a Taylor series with respect to the thickness coordinate z

$$u(x, y, z) = u(x, y, 0) + z\frac{\partial u(x, y, 0)}{\partial z} + \frac{z^2}{2!}\frac{\partial^2 u(x, y, 0)}{\partial z^2} + \cdots \tag{4.542}$$

$$v(x, y, z) = v(x, y, 0) + z\frac{\partial v(x, y, 0)}{\partial z} + \frac{z^2}{2!}\frac{\partial^2 v(x, y, 0)}{\partial z^2} + \cdots \tag{4.543}$$

$$w(x, y, z) = w(x, y, 0) \tag{4.544}$$

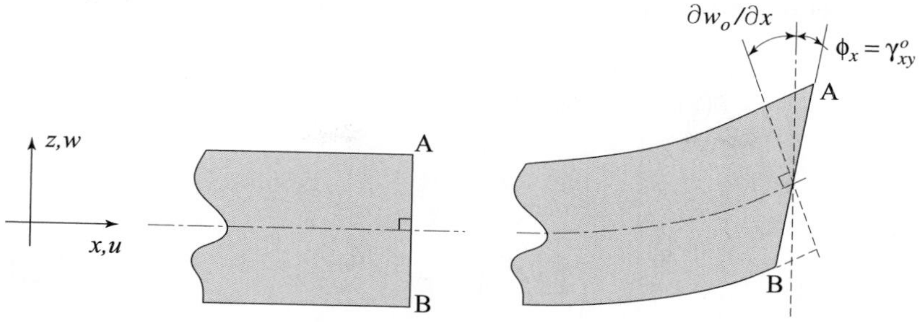

Figure 4.65. Inclusion of uniform transverse shear in beam deformation.

The FSDT for beams (also called the Timoshenko beam theory) retains the first two terms on the right-hand side of Eqs. 4.542 and 4.543. In this theory, the Euler-Bernoulli hypothesis is relaxed and the plane section normal to the neutral axis before deformation does not remain normal to the neutral axis after deformation. This means that the effect of transverse shear is included in the analysis. It is assumed that the transverse shear is uniform across the thickness of the beam. This theory assumes that the primary influence of transverse shear is to introduce an additional rotation of the cross section. The resultant deformation of the cross section can be thought of as a summation of pure Euler-Bernoulli bending and pure transverse shear.

4.11.1 Formulation of the FSDT for a Beam

For a beam under transverse loading, the axial and bending displacements are expressed as

$$u(x, z) = u_o(x) + z\phi_x(x) \tag{4.545}$$

$$w(x, z) = w_o(x) \tag{4.546}$$

Again, u_o and w_o are displacements at the mid-plane ($z = 0$) and ϕ_x is the rotation of the transverse normal about the y-axis (Fig. 4.65). In the case of the Euler-Bernoulli theory

$$\phi_x = -\frac{\partial w}{\partial x} \tag{4.547}$$

For the Timoshenko theory, ϕ_x is an independent variable. Thus, we require three variables (u_o, w, ϕ_x) to determine the strain at any point, where

$$\phi_x = \frac{\partial u}{\partial z} \tag{4.548}$$

$$\begin{Bmatrix} \epsilon_x \\ \gamma_{zx} \end{Bmatrix} = \begin{Bmatrix} \frac{\partial u_o}{\partial x} \\ \frac{\partial w}{\partial x} + \phi_x \end{Bmatrix} - z \begin{Bmatrix} -\frac{\partial \phi_x}{\partial x} \\ 0 \end{Bmatrix} \tag{4.549}$$

The stress in the active layer is

$$\sigma(z) = E(z)(\epsilon(z) - \Lambda(z)) \tag{4.550}$$

E varies in the z-direction, Λ is zero for passive layers. The axial force becomes

$$F = \int_{-h/2}^{h/2} b(z)\sigma(z)dz \tag{4.551}$$

$$= \int_{-h/2}^{h/2} b(z)E(z)\left[\epsilon_x^o - z\left(-\frac{\partial \phi_x}{\partial x}\right) - \Lambda(z)\right] dz \tag{4.552}$$

$$= EA_{\text{tot}}\epsilon_x^o + ES_{\text{tot}}\left(-\frac{\partial \phi_x}{\partial x}\right) - F_\Lambda \tag{4.553}$$

where F_Λ is the axial force due to the induced strain. Similarly, the resultant moment is

$$M = -\int_{-h/2}^{h/2} b(z)\sigma(z)zdz \tag{4.554}$$

$$= -\int_{-h/2}^{h/2} b(z)E(z)z\left[\epsilon_x^o - z\left(-\frac{\partial \phi_x}{\partial x}\right) - \Lambda(z)\right] dz \tag{4.555}$$

$$= ES_{\text{tot}}\epsilon_x^o + EI_{\text{tot}}\left(-\frac{\partial \phi_x}{\partial x}\right) - M_\Lambda \tag{4.556}$$

The resultant transverse shear force is given by

$$V = -\int_{-h/2}^{h/2} b(z)\tau_{zx}(z)dz \tag{4.557}$$

$$= -\int_{-h/2}^{h/2} b(z)G(z)\left(\gamma_{zx} - \Lambda_{zx}\right) dz \tag{4.558}$$

$$= -\int_{-h/2}^{h/2} b(z)G(z)\left[\left(\frac{\partial w}{\partial x} + \phi_x\right) - \Lambda_{zx}\right] dz \tag{4.559}$$

$$= -GA_{\text{tot}}\left(\frac{\partial w}{\partial x} + \phi_x\right) - V_\Lambda \tag{4.560}$$

Assuming that the beam consists of N total layers

$$EA_{\text{tot}} = \sum_{k=1}^{N} b_k E_k (h_{k+1} - h_k) \text{ (N)} \tag{4.561}$$

$$ES_{\text{tot}} = -\frac{1}{2}\sum_{k=1}^{N} b_k E_k (h_{k+1}^2 - h_k^2) \text{ (Nm)} \tag{4.562}$$

$$EI_{\text{tot}} = \frac{1}{3}\sum_{k=1}^{N} b_k E_k (h_{k+1}^3 - h_k^3) \text{ (Nm}^2\text{)} \tag{4.563}$$

$$GA_{\text{tot}} = \sum_{k=1}^{N} b_k G_k (h_{k+1} - h_k) \text{ (N)} \tag{4.564}$$

The forces and moments are given by

$$F_\Lambda = \sum_{k=1}^{N} \Lambda_k b_k E_k (h_{k+1} - h_k) \text{ (N)} \tag{4.565}$$

$$M_\Lambda = -\frac{1}{2}\sum_{k=1}^{N} \Lambda_k b_k E_k (h_{k+1}^2 - h_k^2) \text{ (Nm)} \tag{4.566}$$

$$V_\Lambda = -\sum_{k=1}^{N} \Lambda_{zx} b_k E_k (h_{k+1} - h_k) \text{ (N)} \tag{4.567}$$

Combining these equations

$$\begin{Bmatrix} F + F_\Lambda \\ M + M_\Lambda \\ V + V_\Lambda \end{Bmatrix} = \begin{bmatrix} EA_{\text{tot}} & ES_{\text{tot}} & 0 \\ ES_{\text{tot}} & EI_{\text{tot}} & 0 \\ 0 & 0 & GA_{\text{tot}} \end{bmatrix} \begin{Bmatrix} \epsilon_x^o \\ -\frac{\partial \phi_x}{\partial x} \\ \frac{\partial w}{\partial x} + \phi_x \end{Bmatrix} \tag{4.568}$$

The third equation can be rewritten as

$$\frac{\partial w}{\partial x} + \phi_x = \frac{1}{GA_{\text{tot}}}(V + V_\Lambda) \tag{4.569}$$

$$\frac{\partial \phi_x}{\partial x} = -\frac{\partial^2 w}{\partial x^2} + \frac{\partial}{\partial x}\left[\frac{1}{GA_{\text{tot}}}(V + V_\Lambda)\right] \tag{4.570}$$

The force-displacement relations can be written as

$$\begin{Bmatrix} F + F_\Lambda \\ M + M_\Lambda \end{Bmatrix} = \begin{bmatrix} EA_{\text{tot}} & ES_{\text{tot}} \\ ES_{\text{tot}} & EI_{\text{tot}} \end{bmatrix} \begin{Bmatrix} \epsilon_x^o \\ \frac{\partial^2 w}{\partial x^2} \end{Bmatrix} - \begin{Bmatrix} ES_{\text{tot}} \frac{\partial}{\partial x}\left[\frac{1}{GA_{\text{tot}}}(V + V_\Lambda)\right] \\ EI_{\text{tot}} \frac{\partial}{\partial x}\left[\frac{1}{GA_{\text{tot}}}(V + V_\Lambda)\right] \end{Bmatrix} \tag{4.571}$$

4.11.2 Shear Correction Factor

In the FSDT, transverse shear strains are assumed to be constant through the laminate thickness. As a consequence, a nonzero shear stress appears on the top and bottom surfaces and on the sides of the beam, violating the requirement of zero traction forces on a free surface. In addition, it is well established that for a homogeneous beam under transverse loading, the transverse shear stress varies parabolically through the beam thickness, with the maximum shear stress occurring at the beam neutral axis. For a laminated beam, the distribution of transverse shear stress across the thickness can be more complex. A shear correction factor is often used to make up for the discrepancies in the FSDT formulation.

The corrected transverse shear stresses can be written as [4]

$$V_y = -K \int_{t_b} \tau_{zx}\, b\, dz \tag{4.572}$$

where K is the shear correction factor. The value of K is found by equating the strain energy computed using FSDT to the exact strain energy of the beam. The value of

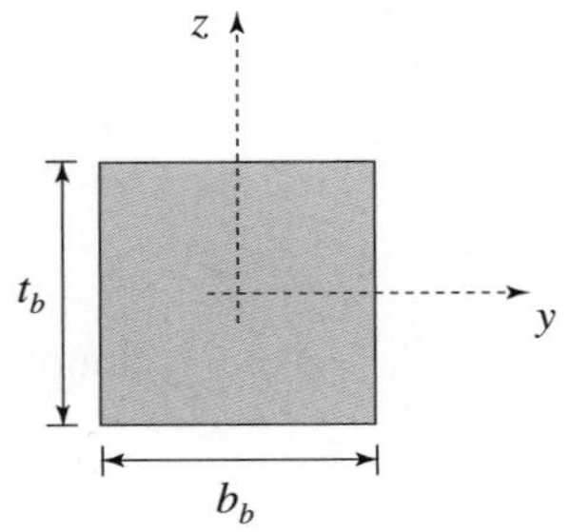

Figure 4.66. Cross section of a uniform rectangular isotropic beam.

K depends on the geometry of the beam cross section and material properties of the laminate.

4.11.3 Transverse Deflection of Uniform Isotropic Beams Including Shear Correction

Consider a beam having a uniform rectangular cross section with thickness t_b, width b_b, and length L_b (Fig. 4.66). The flexural stiffness is given by

$$EI_b = EI_y = E_b \frac{b_b t_b^3}{12} \tag{4.573}$$

where E_b is the Young's modulus. The shear stiffness is given by

$$GA_b = G_b b_b t_b \tag{4.574}$$

where G_b is the shear modulus of the beam. For an isotropic material

$$G_b = \frac{E_b}{2(1+\nu)} \tag{4.575}$$

where ν is the Poisson's ratio.

(a) Cantilevered Beam: Tip Load

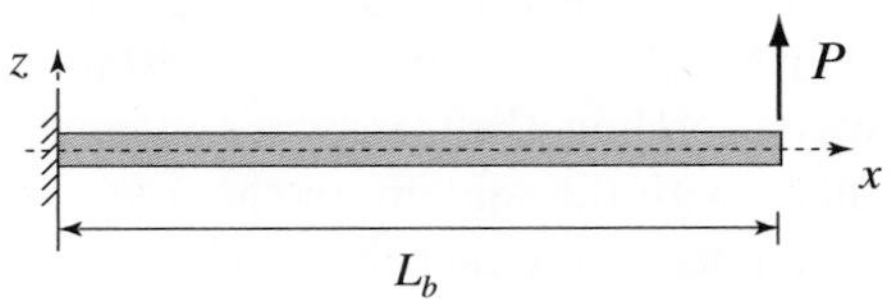

$$w_{tip} = \frac{PL_b^3}{3EI_b} + \frac{PL_b}{KGA_b}$$

$$w(x) = \frac{PL_b^3}{6EI_b}\left[3\left(\frac{x}{L_b}\right)^2 - \left(\frac{x}{L_b}\right)^3\right] + \frac{PL_b}{KGA_b}\left(\frac{x}{L_b}\right)$$

(b) Cantilevered Beam: Uniform Load

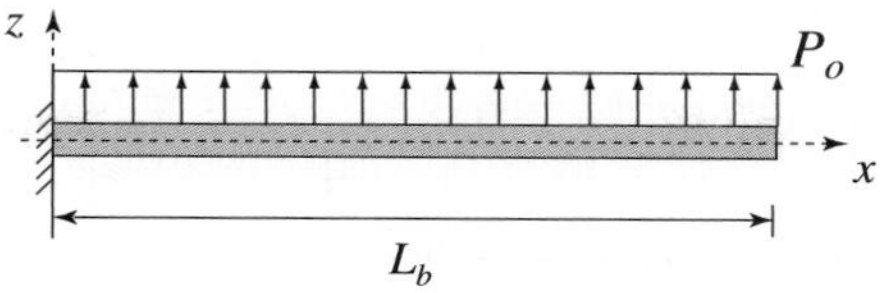

$$w_{tip} = \frac{P_o L_b^4}{8EI_b} + \frac{P_o L_b^2}{2GA_b}$$

$$w(x) = \frac{P_o L_b^4}{24EI_b}\left[6\left(\frac{x}{L_b}\right)^2 - 4\left(\frac{x}{L_b}\right)^3 + \left(\frac{x}{L_b}\right)^4\right] + \frac{P_o L_b^2}{2GA_b}\left[2\left(\frac{x}{L_b}\right) - \left(\frac{x}{L_b}\right)^2\right]$$

(c) Hinged or Simply Supported: Mid-Point Load

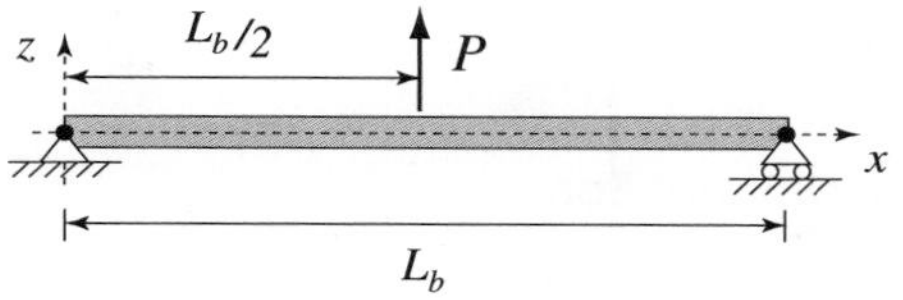

$$w_{mid} = \frac{PL_b^3}{48EI_b} + \frac{PL_b}{4KGA_b}$$

$$w(x) = \frac{PL_b^3}{48EI_b}\left[3\left(\frac{x}{L_b}\right) - 4\left(\frac{x}{L_b}\right)^3\right] + \frac{PL_b}{2KGA_b}\left(\frac{x}{L_b}\right)$$

(d) Hinged or Simply Supported: Uniform Load

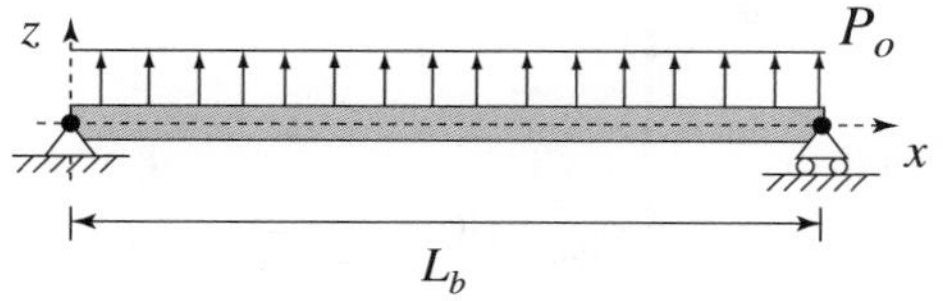

$$w_{mid} = \frac{5}{384}\frac{P_o L_b^4}{EI_b} + \frac{P_o L_b^2}{8GA_b}$$

$$w(x) = \frac{P_o L_b^4}{24EI_b}\left[\left(\frac{x}{L_b}\right) - 2\left(\frac{x}{L_b}\right)^3 + \left(\frac{x}{L_b}\right)^4\right]$$

$$+ \frac{P_o L_b^2}{2GA_b}\left[\left(\frac{x}{L_b}\right) - \left(\frac{x}{L_b}\right)^2\right]$$

(e) Clamped Both Ends: Mid-Point Load

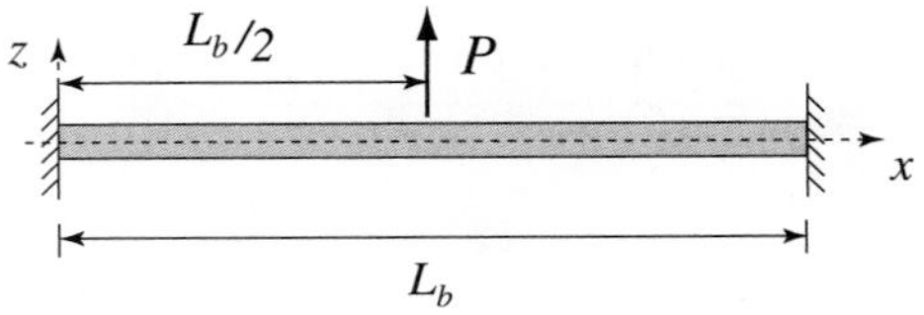

$$w_{mid} = \frac{PL_b^3}{192EI_b} + \frac{PL_b}{4KGA_b}$$

$$w(x) = \frac{PL_b^3}{48EI_b}\left[3\left(\frac{x}{L_b}\right)^2 - 4\left(\frac{x}{L_b}\right)^3\right] + \frac{PL_b}{2KGA_b}\left(\frac{x}{L_b}\right)$$

(f) Clamped Both Ends: Uniform Load

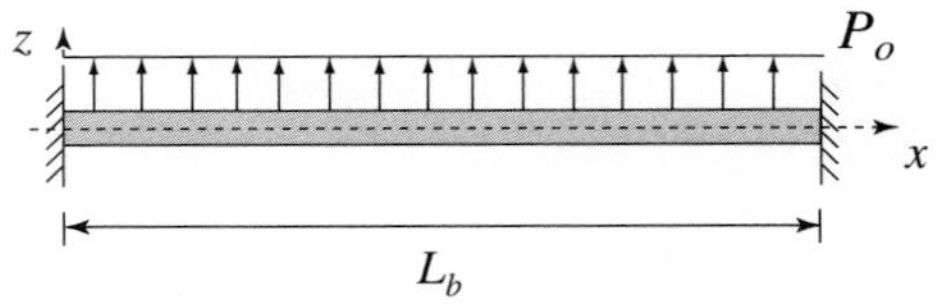

$$w_{mid} = \frac{1}{384}\frac{P_oL_b^4}{EI_b} + \frac{P_oL_b^2}{8GA_b}$$

$$w(x) = \frac{P_oL_b^4}{24EI_b}\left[\left(\frac{x}{L_b}\right)^2 - \left(\frac{x}{L_b}\right)\right] + \frac{P_oL_b^2}{2GA_b}\left[\left(\frac{x}{L_b}\right) - \left(\frac{x}{L_b}\right)^2\right]$$

4.11.4 Induced Beam Response Using Timoshenko Shear Model

The induced responses of several beam configurations in pure bending, derived using the Timoshenko shear model, are shown here. It is assumed that the length of the piezoceramic sheet actuator is small compared to the length of the beam. Also, $EI_{\text{tot}} = EI_b = EI$, which is uniform along the length of the beam. Let the mid-point of the piezoceramic sheet be located at a coordinate $a = x_o + l_c/2$.

I. Cantilevered beam

$$0 \le x \le x_o \qquad w_1 = 0$$

$$\phi_1 = 0$$

$$x_o \le x \le x_o + l_c \qquad w_2 = \frac{M_\Lambda}{8EI}(2x + l_c - 2a)^2 - \frac{M_\Lambda}{KGA}$$

$$\phi_2 = -\frac{M_\Lambda}{2EI}(2x + l_c - 2a)$$

$$x_o + l_c/2 \le x \le L_b \qquad w_3 = \frac{M_\Lambda}{EI}(x - a)\, l_c$$

$$\phi_3 = \frac{M_\Lambda}{EI} l_c$$

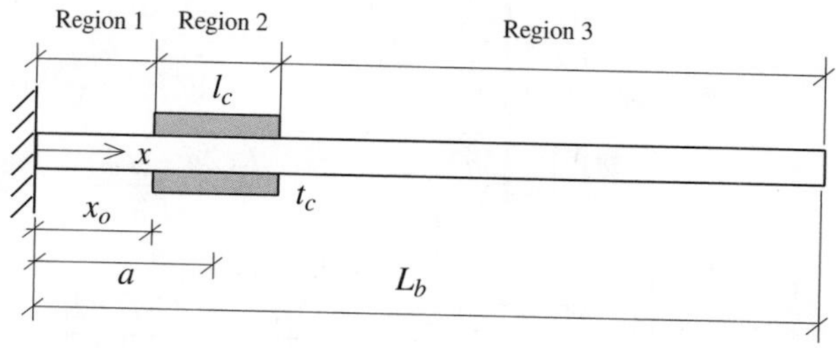

II. Simply-supported beam

$$0 \le x \le x_o \qquad w_1 = \frac{M_\Lambda}{EI} l_c x \left(\frac{a}{L_b} - 1\right)$$

$$\phi_1 = -\frac{M_\Lambda}{EI} l_c \left(\frac{a}{L_b} - 1\right)$$

$$x_o \le x \le x_o + l_c \qquad w_2 = \frac{M_\Lambda}{EI}\left[\frac{1}{2}x^2 + xa\left(\frac{l_c}{L_b} - 1\right) - \frac{1}{2}xl_c + \frac{1}{8}(2a - l_c)^2\right] - \frac{M_\Lambda}{KGA}$$

$$\phi_2 = -\frac{M_\Lambda}{EI}\left[x + a\left(\frac{l_c}{L_b} - 1\right) - \frac{l_c}{2}\right]$$

$$x_o + l_c/2 \le x \le L_b \qquad w_3 = -\frac{M_\Lambda}{EI} a l_c \left(1 - \frac{x}{L_b}\right)$$

$$\phi_3 = \frac{M_\Lambda}{EI} a \frac{l_c}{L_b}$$

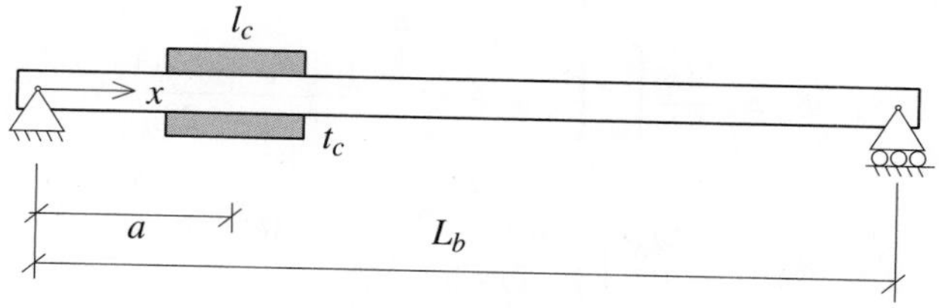

III. Clamped-clamped beam

$$0 \le x \le x_o \qquad w_1 = \frac{M_\Lambda}{EI}\left[\frac{l_c(3a - 2L_b)x^2}{L_b^2} - \frac{l_c(2a - L_b)x^3}{L_b^3}\right]$$

$$\phi_1 = -\frac{M_\Lambda}{EI}\left[\frac{2l_c(3a - 2L_b)x}{L_b^2} - \frac{3l_c(2a - L_b)x^2}{L_b^3}\right]$$

$$x_o \le x \le x_o + l_c \qquad w_2 = \frac{M_\Lambda}{EI}\left[\frac{6al_c - L_b(4l_c - L_b)}{2L_b^2}x^2 - \frac{l_c(2a - L_b)}{L_b^3}x^3 - \frac{(2a - l_c)}{2}x + \frac{(2a - l_c)^2}{8}\right]$$

$$\phi_2 = -\frac{M_\Lambda}{EI}\left[\frac{6al_c - L_b(4l_c - L_b)}{L_b^2}x - \frac{3l_c(2a - L_b)}{L_b^3}x^2 - \frac{(2a - l_c)}{2}\right]$$

$$x_o + l_c/2 \le x \le L_b \qquad w_3 = \frac{M_\Lambda}{EI}\left[\frac{l_c(3a - 2L_b)x^2}{L_b^2} - \frac{l_c(2a - L_b)x^3}{L_b^3} + l_c x - al_c\right]$$

$$\phi_3 = -\frac{M_\Lambda}{EI}\left[\frac{2l_c(3a - 2L_b)x}{L_b^2} - \frac{3l_c(2a - L_b)x^2}{L_b^3} + l_c\right]$$

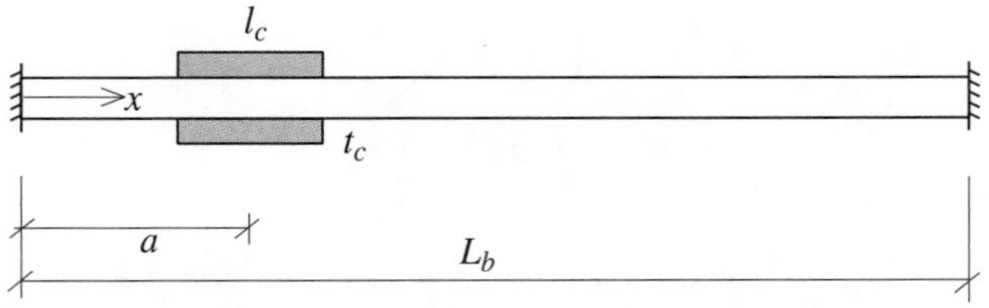

IV. Simply-supported-clamped beam

$$0 \le x \le x_o \qquad w_1 = \frac{M_\Lambda}{EI}\left[-\frac{al_c x^3}{2L_b^3} + \frac{l_c(3a - 2L_b)x}{2L_b}\right]$$

$$\phi_1 = -\frac{M_\Lambda}{EI}\left[-\frac{3al_c x^2}{2L_b^3} + \frac{l_c(3a - 2L_b)}{2L_b}\right]$$

$$x_o \le x \le x_o + l_c \qquad w_2 = \frac{M_\Lambda}{EI}\left[-\frac{al_c x^3}{2L_b^3} + \frac{x^2}{2} + \frac{[a(3l_c - 2L_b) - l_c L_b]x}{2L_b} + \frac{(2a - l_c)^2}{8}\right]$$

$$\phi_2 = -\frac{M_\Lambda}{EI}\left[-\frac{3al_c x^2}{2L_b^3} + x + \frac{a(3l_c - 2L_b) - l_c L_b}{2L_b}\right]$$

$$x_o + l_c/2 \le x \le L_b \quad w_3 = \frac{M_\Lambda}{EI}\left[\frac{al_c x^3}{2L_b^3} + \frac{3al_c x}{2L_b} + al_c\right]$$

$$\phi_3 = -\frac{M_\Lambda}{EI}\left[\frac{3al_c x^2}{2L_b^3} + \frac{3al_c}{2L_b}\right]$$

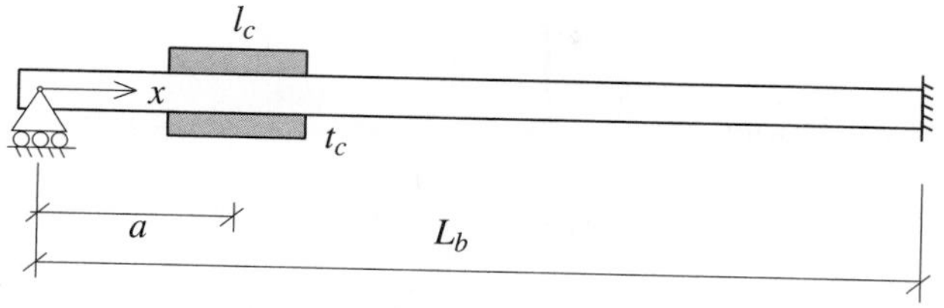

4.11.5 Energy Formulation: FSDT

The governing equations and boundary conditions for a beam modeled using FSDT can also be derived from an energy formulation. Hamilton's principle (see Section 4.9.9) is written as

$$\delta\int_{t_1}^{t_2}(T - V)dt + \int_{t_1}^{t_2}\delta W_{\text{ext}}\,dt = 0 \tag{4.576}$$

where the strain energy is given by

$$\delta V = \iiint_{\text{volume}}(\sigma_x\delta\epsilon_x + \tau_{zx}\delta\gamma_{zx})\,dx\,dy\,dz \tag{4.577}$$

Note that the strain energy due to transverse shear has been included in the total strain energy expression. Substituting for normal and transverse strains, the total strain energy becomes

$$\begin{aligned}\delta V &= \iiint_{\text{volume}}\left[\sigma_x\left(\frac{\partial(\delta u_o)}{\partial x} + z\frac{\partial(\delta\phi_x)}{\partial x}\right) + \tau_{zx}\left(\frac{\partial(\delta w)}{\partial x} + \delta\phi_x\right)\right]dx\,dy\,dz \\ &= \int_0^{L_b}\left[F\frac{\partial(\delta u_o)}{\partial x} - M\frac{\partial(\delta\phi_x)}{\partial x} - V\left(\frac{\partial(\delta w)}{\partial x} + \delta\phi_x\right)\right]dx\end{aligned} \tag{4.578}$$

where

$$F = \int_{area}\sigma_x dydz = \int_{t_b} b\sigma_x dz \tag{4.579}$$

$$M = -\int_{area} z\sigma_x dydz = -\int_{t_b} zb\sigma_x dz \tag{4.580}$$

$$V = -\int_{area}\tau_{zx}dydz = -\int_{t_b} b\tau_{zx}dz \tag{4.581}$$

The kinetic energy is given by

$$\begin{aligned}\delta T &= \iiint_{\text{volume}}\rho_s\left[(\dot{u}_o + z\dot{\phi}_x)(\delta\dot{u}_o + z\delta\dot{\phi}_x) + \dot{w}\delta\dot{w}\right]dx\,dy\,dz \\ &= \int_0^{L_b}\left[m_b(\dot{u}_o\delta\dot{u}_o + \dot{w}\delta\dot{w}) + S_b(\dot{u}_o\delta\dot{\phi}_x + \dot{\phi}_x\delta\dot{u}_o) + I_b\dot{\phi}_x\delta\dot{\phi}_x\right]dx\end{aligned} \tag{4.582}$$

where

$$m_b = \int_{t_b} \rho_s b dz \tag{4.583}$$

$$S_b = \int_{t_b} \rho_s b z dz \tag{4.584}$$

$$I_b = \int_{t_b} \rho_s b z^2 dz \tag{4.585}$$

and the virtual work done is given by

$$\delta W_{\text{ext}} = \int_0^{L_b} f_z \delta w dx \tag{4.586}$$

Substituting in Hamilton's equation and separating out terms related to δu_o, δw, and $\delta\phi_x$ yields the governing equation and boundary conditions.

$$\delta u_o: \quad \frac{\partial F}{\partial x} - m_b\ddot{u}_o - S_b\ddot{\phi}_x = 0 \tag{4.587}$$

$$\delta w: \quad -\frac{\partial V}{\partial x} - m_b\ddot{w} = f_z \tag{4.588}$$

$$\delta\phi_x: \quad -\frac{\partial M}{\partial x} + V - S_b\ddot{u}_o - I_b\ddot{\phi}_x = 0 \tag{4.589}$$

Substituting for the forces and moments

$$F = EA_{\text{tot}}\frac{\partial u_o}{\partial x} - ES_{\text{tot}}\frac{\partial \phi_x}{\partial x} - F_\Lambda \tag{4.590}$$

$$M = ES_{\text{tot}}\frac{\partial u_o}{\partial x} - EI_{\text{tot}}\frac{\partial \phi_x}{\partial x} - M_\Lambda \tag{4.591}$$

$$V = -GA_{\text{tot}}\left(\frac{\partial w}{\partial x} + \phi_x\right) - V_\Lambda \tag{4.592}$$

where

$$F_\Lambda = \int_{\text{area}} E(z)\Lambda dA \qquad \text{(N)} \tag{4.593}$$

$$M_\Lambda = -\int_{\text{area}} E(z)\Lambda z dA \qquad \text{(Nm)} \tag{4.594}$$

$$V_\Lambda = -\int_{\text{area}} E(z)\Lambda_{zx} dA \qquad \text{(N)} \tag{4.595}$$

results in

$$\delta u_o: \quad \frac{\partial}{\partial x}\left[EA_{\text{tot}}\frac{\partial u_o}{\partial x} - ES_{\text{tot}}\frac{\partial \phi_x}{\partial x}\right] - m_b\ddot{u}_o - S_b\ddot{\phi}_x = \frac{\partial F_\Lambda}{\partial x} \tag{4.596}$$

$$\delta w: \quad \frac{\partial}{\partial x}\left[GA_{\text{tot}}\left(\frac{\partial w}{\partial x} + \phi_x\right)\right] - m_b\ddot{w} = -\frac{\partial V_\Lambda}{\partial x} + f_z \tag{4.597}$$

$$\delta\phi_x: \quad \frac{\partial}{\partial x}\left[ES_{\text{tot}}\frac{\partial u_o}{\partial x} - EI_{\text{tot}}\frac{\partial \phi_x}{\partial x}\right] + GA_{\text{tot}}\left[\frac{\partial w}{\partial x} + \phi_x\right] + S_b\ddot{u}_o + I_b\ddot{\phi}_x = 0 \tag{4.598}$$

Typical boundary conditions including shear effects are as follows:

(a) Clamped Condition

$$u_o(0,t) = 0$$
$$w(0,t) = 0 \rightarrow \text{displacement } = 0$$
$$\phi_x(0,t) = 0$$

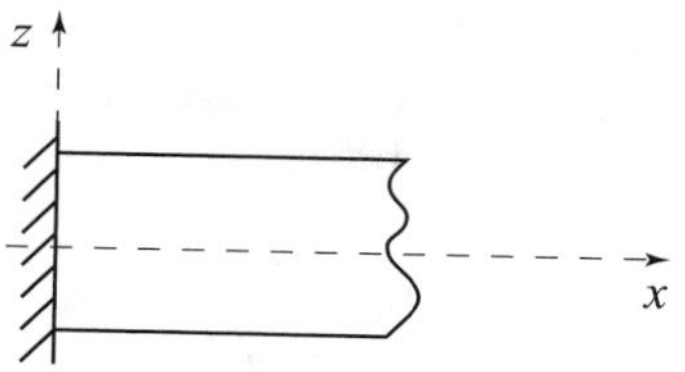

(b) Simply-Supported (Hinged or Pinned) Condition

$$u_o(0,t) = 0$$
$$w(0,t) = 0 \rightarrow \text{displacement } = 0$$
$$M_y(0,t) = 0 \rightarrow \text{moment } = 0$$

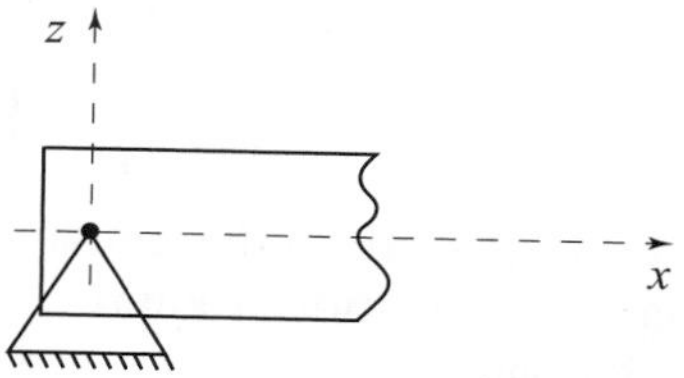

(c) Roller Condition

$$w(0,t) = 0 \rightarrow \text{displacement } = 0$$
$$M_y(0,t) = 0 \rightarrow \text{moment } = 0$$
$$F_x(0,t) = 0 \rightarrow \text{axial force } = 0$$

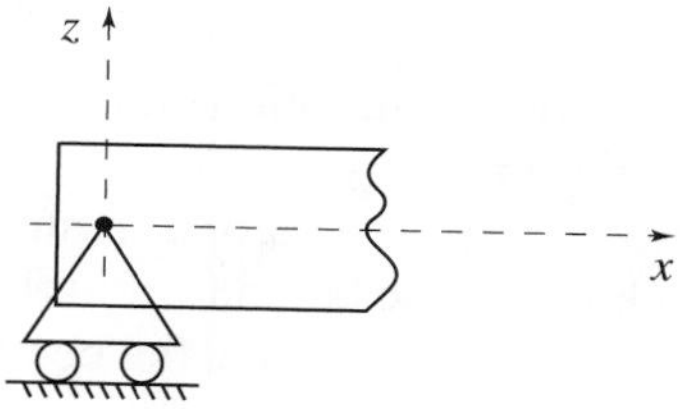

(d) Free Condition

$$F_x(0,t) = 0 \rightarrow \text{axial force } = 0$$
$$M_y(0,t) = 0 \rightarrow \text{moment } = 0$$
$$V_z(0,t) = 0 \rightarrow \text{shear force } = 0$$

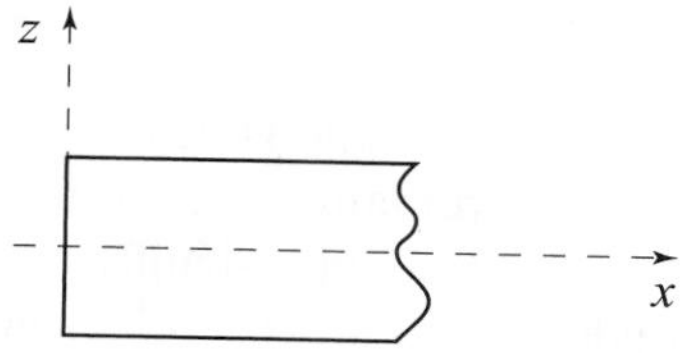

4.12 Layer-Wise Theories

Exact calculation of the force-deflection behavior of a beam requires modeling of the three-dimensional stress-strain behavior. By making certain assumptions regarding the kinematics of deformation, or the transverse stress state, it is possible to simplify this to a one-dimensional problem. Theories based on such simplifications are called equivalent single layer (ESL) theories. The Euler-Bernoulli beam theory and FSDT formulations are two commonly used examples of ESL theories.

ESL theories generally provide fairly accurate predictions of global behavior, especially for thin laminates. However, for an accurate calculation of local stresses at the level of individual laminae, and for thick laminated beams, more refined theories are necessary. The next level of detail is provided by layer-wise theories that model the full three-dimensional behavior at the level of each ply.

In the ESL theories, transverse strains are necessarily assumed to be a continuous function of the beam thickness coordinate. To see how this causes a local variation of transverse force equilibrium, consider two adjacent lamina, labeled k and $k+1$. The assumption of continuous transverse strains implies that

$$\begin{Bmatrix} \gamma_{yz} \\ \gamma_{zx} \\ \epsilon_{zz} \end{Bmatrix}_k = \begin{Bmatrix} \gamma_{yz} \\ \gamma_{zx} \\ \epsilon_{zz} \end{Bmatrix}_{k+1} \tag{4.599}$$

at the interface between the two lamina. In general, the stiffnesses of adjacent lamina are different ($\bar{Q}_k \neq \bar{Q}_{k+1}$), yielding

$$\begin{Bmatrix} \tau_{yz} \\ \tau_{zx} \\ \sigma_{zz} \end{Bmatrix}_k \neq \begin{Bmatrix} \tau_{yz} \\ \tau_{zx} \\ \sigma_{zz} \end{Bmatrix}_{k+1} \tag{4.600}$$

These stresses are called "interlaminar stresses." However, the full three-dimensional elasticity equations require the equilibrium of transverse forces at the interface of the lamina

$$\begin{Bmatrix} \tau_{yz} \\ \tau_{zx} \\ \sigma_{zz} \end{Bmatrix}_k = \begin{Bmatrix} \tau_{yz} \\ \tau_{zx} \\ \sigma_{zz} \end{Bmatrix}_{k+1} \tag{4.601}$$

which contradicts the assumption inherent in ESL theories. To incorporate the continuity of transverse stresses at the interface of the lamina, as given by the previous equation, layer-wise theories assume that the displacements are continuous through the thickness of the laminate, but the transverse strains can be discontinuous at laminar interfaces. For example, the beam displacements can be piece-wise linear in the z direction and for the kth ply are given by

$$u^{(k)}(x, z) = u_o(x) + z\phi_x^{(k)}(x) \tag{4.602}$$

$$w^{(k)}(x, z) = w_o(x) \tag{4.603}$$

where $\phi_x^{(k)}$ represents the rotations of the cross section of the kth layer. Note that for the case of a single-layer laminate, this layer-wise formulation reduces to FSDT. The layer-wise theory results in a significant increase in the number of degrees of freedom of the model. From these displacement relations, the forces and moments in the beam can be found by integrating through the beam thickness, as in the case of the Euler-Bernoulli beam theory and FSDT.

4.13 Review of Beam Modeling

Table 4.7 lists different smart beam models. Crawley and de Luis [5] formulated the uniform-strain model for a beam with surface-bonded piezoceramic sheet actuators (patched and aligned with the beam axis). The model calculated flexural response, including shear-lag effects of the adhesive layer between the piezoceramic actuator and the beam. It was shown that the strain transfer from the piezoceramic actuator to the substructure takes place over a small zone near both edges of the actuator, and there is maximum shear stress in this region. As the adhesive layer becomes thinner and/or stiffer (shear modulus), it approaches a perfect-bond condition (shear concentrated at the two edges of the actuator). The dynamic model was experimentally

Table 4.7. *Comparison of smart beam models*

Modeling Type	Actuators	Piezoelectric Coupling	Beam Type	Validation	Reference
Blocked-Force	Surface and embedded	Uncoupled	Isotropic		
Euler-Bernoulli	Surface and embedded	Uncoupled	Isotropic	Cantilevered aluminum	Crawley and DeLuis [5]
	Straight patches				Park, Walz, and Chopra [3]
	Skewed patches				Park, Walz, and Chopra [3]
					Park and Chopra [7]
Uniform-Strain	Surface and embedded	Uncoupled	Isotropic	Cantilevered aluminum	Crawley and DeLuis [5]
	Straight patches				Park, Walz, and Chopra [3]
	Skewed patches				Park, Walz, and Chopra [3]
					Park and Chopra [7]
Timoshenko (FSDT)	Surface and embedded	Uncoupled	Isotropic		Shen [8]
Vlasov with Chordwise Bending and Shear	Surface-bonded	Uncoupled	Isotropic and composite	Cantilevered composite coupled	Chandra and Chopra [9]
	Straight patches				Bernhard and Chopra [10]
Euler-Bernoulli Coupled	Surface	Coupled	Isotropic	Cantilevered aluminum	Hagood et al. [11]
Layer-Wise Shear Deformation Theory	Surface	Coupled	Isotropic & composite		Robbins and Reddy [12]
					Saravanos et al. [13]

verified for the first two bending modes of a cantilevered aluminum beam. They also presented a uniform-strain model for an isotropic beam with embedded actuators and satisfactorily validated the dynamic response at resonance for aluminum, glass-epoxy, and graphite-epoxy beams. Crawley and Anderson [6] formulated the Euler-Bernoulli model for a beam with surface-bonded or embedded induced-strain actuators (symmetric actuation) and compared it with the uniform-strain model, a finite element model, and experiment. The uniform-strain model was generally found satisfactory except for low beam to actuatorthickness ratios (<4). The Euler-Bernoulli model was quite satisfactory to predict bending and extensional response, even for low-thickness ratios. There is no doubt that for the thickness ratio (beam thickness/actuator thickness) of 1.0 or less (as in the case of bimorphs), a refined model including three-dimensional effects may be needed. Furthermore, the linear model (using linear piezoelectric characteristics) is accurate only for small strains. To predict reliable flexural results with high field conditions, one must include nonlinear field-strain relations.

Im and Atluri [14] developed a nonlinear analysis of a piezoactuated beam with a finite-thickness bond layer, including the effects of transverse shear and axial forces in addition to the bending moment on the beam. Again, it was shown that the maximum shear stress occurs near the two ends of the piezoelectric element and is also a function of externally applied axial and shear forces. Hagood et al. [11] formulated a completely coupled piezoelectric-mechanical model for a beam with surface-bonded actuators. Predicted dynamics were found to be in good agreement with experimental data obtained with a cantilevered aluminum beam. Benjeddow et al. [15] developed a beam finite element model for extension and shear piezoelectric–actuation mechanisms. This is especially suitable for sandwiched beams. The model used the Euler-Bernoulli theory for the surface layers and Timoshenko beam theory for the core. It was shown that the predicted induced deformation was lower with the shear-actuated beam theory.

Park et al. [3] developed coupled bending and extension as well as coupled bending, torsion, and extension analyses for an isotropic beam with isolated surface-bonded actuators. A finite-thickness adhesive layer between the actuator and the beam was included. The convergence point of the Euler-Bernoulli and uniform-strain predictions was shown to be a function of the beam-to-actuator width ratio in addition to the thickness ratio. Satisfactory validation of predicted bending slope with measured values was carried out for several different aluminum beams. Also, Park et al. [7] developed coupled extension, bending, and torsion analysis for an isotropic beam with surface-bonded actuators at an arbitrary orientation β with respect to the beam axis. Piezoceramic actuators were represented as line actuators. Systematic experimental tests with cantilevered aluminum beams were carried out for induced bending and twist at different orientation angles to check the accuracy and limitation of models. It was concluded that the inclusion of the effects of transverse actuation may be necessary to refine the analysis.

Jung et al. [16, 17] made an assessment of the state-of-the-art in modeling thin- and thick-walled composite beams with a view to emphasize the special characteristics of composite materials. The review encompasses modeling nonclassical effects such as out-of-plane warping, warping restraints, and transverse shear. Composite-beam models ranged from simple analytical models to detailed finite element models and some were validated using limited test data from simple tailored specimens [18, 19, 20, 21, 22]. The anisotropic nature of composite materials makes the structural properties direction-dependent. Using special ply layups, structural couplings such as bending-torsion and extension-torsion can be introduced. These couplings can be exploited with induced-strain actuation to actively control aerodynamic shape, in helicopter blades or airplane wings. In modeling a composite beam with induced-strain actuation as a one-dimensional structure, it is important to encompass all of the important effects due to bending and shear deflections, the twist of reference axis of the beam, and the warping deformations of the cross section. Normally, the warping deformations are much smaller than the flexural deformations. This helps to simplify the complexity of inherently three-dimensional problems into two parts: a two-dimensional local deformation field of the cross section that is used to calculate the section properties, and a one-dimensional global deformation field to predict the response of the beam. The first level of idealization of the global deformation includes the Euler-Bernoulli model for bending and the St. Venant model for torsion. In the next level, torsion-related warping, transverse shear strain, and cross-section deformation (in-plane warping) effects are included. For composite thin-walled beams, it is possible to model the shell wall either as a membrane or as a thick laminate, including the effect of transverse shear as well as bending distribution. Chandra et al. [9] developed a formulation for coupled composite thin-walled open- and closed-section beams with distributed induced-strain actuation (surface-mounted or embedded) and then validated the analysis with experimental data. Beam modeling was based on the Vlasov theory in which two-dimensional stress and strain distributions associated with any local plate (shell) element of the beam are reduced to one-dimensional generalized forces and moments. Effects of transverse shear and warping restraints were included. Comparison with experimental data from bending-twist and extension-twist coupled graphite-epoxy–composite solid beams with surface-mounted piezoceramic actuators showed that the inclusion of chordwise (lateral) bending is essential to accurately predict a beam's coupled response. Also, Kaiser [23] carried out a similar type of study with thin-walled, open- and closed-section, coupled composite beams with piezoelectric actuation. Cesnik and Shin [24] developed a refined multicell composite-beam analysis for an active twist rotor with embedded active fiber composite (AFC) actuators. The approach is based on a two-step asymptotic solution: a linear two-dimensional cross-sectional analysis and a global nonlinear one-dimensional analysis. Subsequently, the analysis was successfully validated with test data for different blade configurations and load conditions [25]. Ghiringhelli et al. [26] developed a refined finite-element analysis for anisotropic beams with embedded piezoelectric actuators and successfully compared their results with three-dimensional results. Bernhard et al. [10] developed a Vlasov-type beam analysis for a tailored composite coupled beam with induced-strain actuation. It consisted of a number of spanwise segments with reversed bending-twist couplings for each successive segment. Each segment acts like a bimorph, and the polarity of successive surface-bonded piezoceramics is reversed. Because of the alternating excitation, the beam deflects into a sinusoidal bending wave, whereas the induced twist is additive spanwise. Predictions were validated satisfactorily with test data for several different beam configurations. For accurate predictions, it became necessary to include nonlinear measured characteristics of piezoceramics and modeling of chordwise bending. It is now well established that the effects of transverse shear can be important at both the local and global levels for the response of composite beams because of the low values of shear modulus compared with the direct modulus (G/E ratio).

The effects of transverse shear can be modeled using the Timoshenko beam theory, also called FSDT [27, 8], which assumes a constant transverse shear strain across the cross section. To capture the nonlinear distribution of transverse shear strain across the cross section, higher-order shear deformation theories (HSDTs) are used. These theories, however, are unable to capture accurately a dramatic change of properties at a local ply level. A further refinement to HSDT is the layer-wise shear deformation theory (LWSDT) [28] that models shear distribution separately for each layer. Robbins and Reddy [12] carried out static and dynamic analysis of piezoelectrically actuated beams using LWSDT. Saravanos and Heyliger [13] developed a coupled layer-wise analysis of composite beams with embedded piezoelectric actuators and sensors. It was shown that consistent and more detailed stress distributions, especially near the end of the actuator, are obtained with layer-wise theory. For prediction of higher modes of vibration and/or thicker composite structures, it may be more appropriate to use layer-wise theory.

It is clear from testing of simple isotropic beams with surface-attached piezoelectric elements that the local strain distribution (at or near the actuator) is two-dimensional [7] and, therefore, beam modeling with induced-strain actuation should reflect such a distribution. Simple beam theories often give erroneous results for beams with high actuator-to-beam thickness ratios (as is the case with piezo bimorphs). Detailed three-dimensional models (e.g., FEM models) should be used to establish the strain-actuation mechanism. Most beam theories have either neglected the shearing effect of the bond layer (by assuming perfect-bond condition) or have incorporated a highly approximate shear model (e.g., uniform shear stress within bond thickness); however, test results [29] showed that the bond thickness has a dominant effect on the induced-strain transfer from the actuator to the beam. If the bond layer is important, it may be necessary to include its shearing effect using a HSDT such as LWSDT, which can also help to establish the limits of simple beam models (e.g., uniform-strain and Euler-Bernoulli models). There have been only limited studies on the validation of predictions for composite coupled beams with surface-attached or embedded piezoceramics. These could be expanded to cover more beam configurations and tailored composite couplings for static and dynamic loads. Such studies can be important for shape control of aerospace systems. Most predictions have incorporated linear piezoelectric characteristics that are strictly true only for low electric field conditions. To cover moderate to high electric fields, it is important to include the nonlinear characteristics of piezoelectrics. It will be equally important to examine systematically the effect of piezoelectric-mechanical couplings on actuation strain for a range of isotropic and laminated beams.

This chapter examines several structural models that predict the behavior of different configurations of beams with induced-strain actuators. The existing bending models were expanded to include independent variations in actuator and beam widths. The single-actuator uniform-strain model-governing equations were also formulated using the Principle of Virtual Work as an alternative method that is easily adapted to dynamic applications. A one-dimensional treatment of a strain-actuated beam in coupled extension, bending, and torsion was examined and validated experimentally. The model was found inadequate to predict the structural behavior of the system within acceptable limits. However, because the torsion trend is predicted, analytical accuracy may be improved by integrating a local two-dimensional model of the actuation mechanism with a global one-dimensional system model.

PROBLEMS

1. Two piezo-elements (PZT-5H) (length $l_c = 2$ in, width $b_c = 1$ in, thickness $t_c = 0.0125$ in) of piezoelectric constant $d_{31} = -274 \times 10^{-12}$ m/V are surface-bonded at the top and bottom surfaces of a thin aluminum cantilevered beam (length $L = 24$ in, width $b_b = 2$ in, and thickness $t_b = 0.035$ in). The piezo-elements are bonded 4 in from the root of the beam. During the test, it was discovered that the material constant for the top and bottom piezos were different. Assume the same material modulus for aluminum and piezoceramics as

$$E_b = 10.5 \times 10^6 \text{ lb/in}^2$$
$$E_{ctop} = 9 \times 10^6 \text{ lb/in}^2$$
$$E_{cbottom} = 7 \times 10^6 \text{ lb/in}^2$$

   (a) Show free strain variation in micro-strain with voltage for each piezo.
   (b) Plot the variation of piezo strain with axial blocked force F for each piezo.
   (c) Using block force theory, derive a general bending-extension relation with same field on opposite piezo-elements.
   (d) Calculate actuation surface force F in lb and bending moment M in in-lb for a field of 150 volts to both top and bottom piezos.
   (e) Show spanwise distribution of bending slope for this excitation.
   (f) Show beam bending displacement distribution for this excitation.
   (g) If PZT-5H elements are replaced with PVDF elements of same size, calculate new surface actuation strain and actuation bending moment (for PVDF $d_{31} = -20 \times 10^{-12}$ m/V and $E_c = 0.2 \times 10^{10}$ N/m^2).

2. Two piezo-elements (PZT-5H) (length $l_c = 2$ in, width $b_c = 1$ in), respectively, of thickness $t_c = 0.025$ in and 0.0125 in are surface-bonded at the top and bottom of a thin aluminum cantilevered beam (length $L = 24$ in, width $b_b = 2$ in, thickness $t_b = 0.035$ in). The piezo-elements are bonded 4 in from the root of the beam. Manufacturer-supplied material constants are as follows:

$$d_{31} = -274 \times 10^{-12} \text{ m/V}$$
$$E_c = E_b = 10.5 \times 10^6 \text{ lb/in}^2$$

   (a) Show free-strain variation in micro-strain with voltage for each piezo.
   (b) Plot the variation of piezo strain with axial blocked-force F for each piezo.
   (c) Using blocked-force theory, derive a general bending-extension relation with same field on opposite piezo-elements.
   (d) Calculate actuation surface force F in lb and bending moment M in in-lb for a field of 150 volts to both top and bottom piezos.
   (e) Show spanwise distribution of bending slope for this excitation.
   (f) Show beam bending displacement distribution for this excitation.
   (g) If PZT-5H elements are replaced with PVDF elements of same size, calculate new surface actuation strain and actuation-bending moment (for PVDF $d_{31} = -20 \times 10^{-12}$ m/V and $E_c = 0.2 \times 10^{10}$ N/m^2).

3. Two dissimilar piezoceramic elements (PZT-5H) (length $l_c = 2$ in, thickness $t_c = 0.012$ in), respectively, of width b_c of 1 in and 1/2 in are surface-bonded on the bottom and top of a thin aluminum cantilevered beam (length $L_b = 24$ in,

width $b_b = 1$ in, and thickness $t_b = 0.035$ in). The piezo-elements are bonded 4 in from the root of the beam. Manufacturer-supplied material constants are as follows:

$$d_{31} = -274 \times 10^{-12}\ \text{m/V}$$
$$E_c = E_b = 10.5 \times 10^6\ \text{lb/in}^2$$

(a) Show free-strain variation in micro-strain with voltage for each piezo.
(b) Plot the variation of piezo strain with axial blocked force F for each piezo for an excitation of 150 volts.
(c) Using the blocked-force method, derive a general bending-extension relation with same field on opposite piezo-elements.
(d) Calculate actuation surface force F in lb and bending moment M in in-lb for an excitation of 150 volts to both top and bottom piezos.
(e) Calculate actuation surface force F in lb and bending moment M in in-lb for a field of -150 volts and $+150$ volts, respectively, to top and bottom piezos.
(f) Plot the spanwise distribution of bending slope for the excitation of -150 volts and $+150$ volts, respectively, to top and bottom piezos.
(g) Plot the beam bending displacement for this excitation.
(h) If PZT-5H elements are replaced with PVDF elements of same size, calculate new surface actuation strain and actuation bending moment (for PVDF $d_{31} = -20 \times 10^{-12}$ m/V and $E_c = 0.2 \times 10^{10}$ N/m^2).

4. Two piezo-elements (PZT-5H) (length $l_c = 2$ in, thickness $t_c = 0.012$ in), respectively, of width $b_c = 1$ in and 0.75 in are surface-bonded on the top and bottom of a thin aluminum cantilevered beam (length $L = 24$ in, width $b_b = 2$ in, and thickness $t_b = 0.035$ in) and bond-layer thickness of 0.005 in each on both sides. Manufacturer-supplied material constants are as follows:

$$d_{31} = -274 \times 10^{-12}\ \text{m/V}$$
$$E_c = E_b = 10.5 \times 10^6\ \text{lb/in}^2$$
bond shear modulus $G_s = 965 \times 10^6$ N/m^2

(a) Using uniform-strain theory, derive general bending as well as extension relations with same field on opposite piezo-elements for this dual piezo actuation.
(b) Plot spanwise variation of beam surface strain and actuator strain for a field of 150 volts to both top and bottom piezos.
(c) Show variation of bond shearing force along piezo (top) span for this field.
(d) Calculate actuation surface force F in lb and bending moment M in in-lb for this excitation for two cases: with bond layer and with perfect bond.
(e) If the piezo-elements are bonded 4 in from the root of the beam, show spanwise distribution of bending slope for this excitation.
(f) Show beam bending displacement distribution for this excitation.
(g) If PZT elements are replaced with PVDF elements of same size, calculate new surface actuation strain and actuation-bending moment for 150 volts excitation (for PVDF $d_{31} = -20 \times 10^{-12}$ m/V and $E_c = 0.2 \times 10^{10}$ N/m^2).

5. Two piezo-elements (PZT-5H) (length $l_c = 2$ in, width $b_c = 1$ in, thickness $t_c = 0.0125$ in) of piezoelectric constant $d_{31} = -274 \times 10^{-12}$ are surface-bonded at the top and bottom surfaces of a thin aluminum cantilevered beam (length $L = 24$ in,

width $b_b = 2$ in, thickness $t_b = 0.035$ in). The piezo-elements are bonded 4 in from the root of the beam, and the bondlayer thickness was measured as 0.005 in. During the test, it was discovered that the material constant for the top and bottom piezos were different. The material constants are given as

$E_b = 10.5 \times 10^6$ lb/in^2
$E_{cTOP} = 9 \times 10^6$ lb/in^2
$E_{cBOTTOM} = 7 \times 10^6$ lb/in^2
Bond shear modulus $G_s = 965 \times 10^6$ N/m^2

(a) Using uniform strain theory, derive a general bending-extension relation with same field on opposite piezo-elements.
(b) Calculate actuation surface force F in lb and bending moment M in in-lb for a field of 150 volts to both top and bottom piezos.
(c) Show spanwise distribution of bending slope for this excitation.
(d) Show beam bending displacement distribution for this excitation.
(e) If PZT-5H elements are replaced with PVDF elements of same size, calculate new surface actuation strain and actuation bending moment (For PVDF $d_{31} = -20 \times 10^{-12}$ m/V and $E_c = 0.2 \times 10^{10}$ N/m^2).

6. Two dissimilar piezoceramic elements (PZT-5H) (length $l_c = 2$ in, width $b_c = 1$ in), respectively, of thickness t_c of 0.018 in and 0.012 in are surface-bonded on the top and bottom of a thin aluminum cantilevered beam (length $L_b = 24$ in, width $b_b = 1$ in, thickness $t_b = 0.035$ in). The piezo-elements are bonded 4 in from the root of the beam, and the bond thickness was measured as 0.005 in. Manufacturer-supplied material constants are as follows:

$d_{31} = -274 \times 10^{-12}$ m/V,
$E_c = E_b = 10.5 \times 10^6$ lb/in^2
Bond shear modulus $G_s = 965 \times 10^6$ N/m^2

(a) Using uniform-strain theory, derive a general bending-extension relation with the same field on opposite piezo-elements.
(b) Plot spanwise variation of beam surface strain and actuator strain for a field of 150 volts to both top and bottom piezos.
(c) Calculate actuation surface force F in lb and bending moment M in in-lb for a field of 150 volts to both top and bottom piezos for two cases: with perfect bond and with the effects of the bond layer.
(d) Plot the variation of the bond shearing force along piezo (top) span for this field.
(e) Plot the spanwise distribution of bending slope for this excitation.
(f) Plot the beam bending displacement distribution for this excitation.
(g) If PZT-5H elements are replaced with PVDF elements of same size, calculate new surface actuation strain and actuation bending moment (for PVDF $d_{31} = -20 \times 10^{-12}$ m/V and $E_c = 0.2 \times 10^{10}$ N/m^2).

7. Two piezo-elements (PZT-5H and PZT-5A) (length $l_c = 2$ in, width $b_c = 1$ in, thickness $t_c = 0.012$ in), respectively, of piezoelectric constant d_{31} of -274×10^{-12} and -171×10^{-12} m/volt are surface-bonded at the top and bottom of a thin aluminum cantilevered beam (length $L_b = 24$ in, width $b_b = 2$ in, thickness $t_b = 0.035$ in). The thickness of bond layer t_s is 0.005 in and is assumed uniform. Other manufacturer-supplied material constants are as follows

$E_c = E_b = 10.5 \times 10^6$ lb/in^2, bond shear modulus $G_s = 965 \times 10^6$ N/m^2

(a) Using uniform-strain theory, derive general bending as well as extension relations with same field on opposite piezo-elements for this dual piezo actuation.
(b) Plot spanwise variation of beam surface strain and actuator strain for a field of 150 volts to both top and bottom piezos.
(c) Show the variation of bond shearing force for the top piezo.
(d) Calculate actuation surface force F in lb and bending moment M in in-lb for this excitation for two cases: with bond layer and with perfect bond.
(e) Show spanwise distribution of bending slope for this excitation.
(f) Show beam bending displacement distribution for this excitation.
(g) If PZT-5H and PZT-5A elements are replaced with PVDF elements of same size, calculate new surface actuation strain and actuation bending moment for a field of 150 volts to both top and bottom piezos (for PVDF $d_{31} = -20 \times 10^{-12}$ m/V and $E_c = 0.2 \times 10^{10}$ N/m^2).

8. Two dissimilar piezo-elements (PZT-5H) of width $b_c = 1$ in and thickness $t_c = 0.012$ in, respectively, of length $l_c = 2$ in and 1 in are surface-bonded at top and bottom of a thin aluminum cantilevered beam (length $L_b = 24$ in, width $b_b = 2$ in, and thickness $t_b = 0.035$ in). The thickness of bond layer t_s is 0.005 in and is assumed uniform. Other manufacturer-supplied material constants are as follows

$d_{31} = -274 \times 10^{-12}$ m/V
$E_c = E_b = 10.5 \times 10^6$ lb/in^2
bond shear modulus $G_s = 965 \times 10^6$ N/m^2

(a) Using uniform strain theory, derive general bending as well as extension relations with same field on opposite piezo-elements for this dual piezo actuation.
(b) Plot spanwise variation of beam surface strain and actuator strain for a field of 150 volts to both top and bottom piezos.
(c) Show the variation of bond shearing force for the top piezo.
(d) Calculate actuation surface force F in lb and bending moment M in in-lb for this excitation for two cases: with bond layer and perfect bond.
(e) Show spanwise distribution of bending slope for this excitation.
(f) Show beam bending displacement distribution for this excitation.
(g) If PZT-5H elements are replaced with PVDF elements of same size, calculate new surface actuation strain and actuation bending moment for a field of 150 volts to both top and bottom piezos (For PVDF $d_{31} = -20 \times 10^{-12}$ m/V and $E_c = 0.2 \times 10^{10}$ N/m^2).

9. Two dissimilar piezoceramic elements (PZT-5H) (length $l_c = 2$ in, width $b_c = 1$ in), respectively, of thickness t_c of 0.018 in and 0.012 in are surface-bonded on the top and bottom of a thin aluminum cantilevered beam (length $L_b = 24$ in, width $b_b = 2$ in, and thickness $t_b = 0.035$ in). The piezo-elements are bonded 4 in from the root of the beam. Manufacturer-supplied material constants are as follows

$d_{31} = -274 \times 10^{-12}$ m/V
$E_c = E_b = 10.5 \times 10^6$ lb/in^2

(a) Using the Euler-Bernoulli theory, derive general bending as well as extension relations with same field on opposite piezo-elements for this dual piezo actuation with b_b different from b_c.

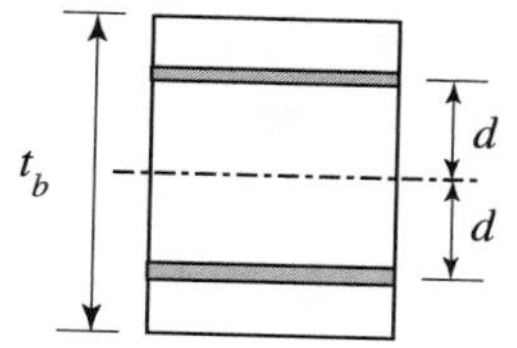

Figure 4.67. Cross section of beam with embedded piezos.

(b) Plot spanwise variation of beam surface strain and actuator strain for a field of 150 volts to both top and bottom piezos.
(c) Calculate actuation surface force F in lb and bending moment M in in-lb for this excitation.
(d) Show spanwise distribution of bending slope for this excitation.
(e) Show beam bending displacement distribution for this excitation.
(f) If PZT-5H and PZT-5A elements are replaced with PVDF elements of same size, calculate new surface actuation strain and actuation bending moment for a field of 150 volts to both top and bottom piezos (for PVDF $d_{31} = -20 \times 10^{-12}$ m/V and $E_c = 0.2 \times 10^{10}$ N/m^2).

10. Two piezo-elements (PZT-5H) (length $l_c = 2$ in, thickness $t_c = 0.012$ in), respectively, of width $b_c = 1$ in and 0.75 in are surface-bonded on the top and bottom of a thin aluminum cantilevered beam (length $L = 24$ in, width $b_b = 2$ in, and thickness $t_b = 0.035$ in). Manufacturer-supplied material constants are as follows

$$d_{31} = -274 \times 10^{-12} \text{ m/V}$$
$$E_c = E_b = 10.5 \times 10^6 \text{ lb/in}^2$$

(a) Using the Euler-Bernoulli theory, derive general bending as well as extension relations with same field on opposite piezo-elements for this dual piezo actuation.
(b) Plot spanwise variation of beam surface strain and actuator strain for a field of 150 volts to both top and bottom piezos.
(c) Calculate actuation surface force F in lb and bending moment M in in-lb for this excitation.
(d) If the piezo-elements are bonded 4 in from the root of the beam, show spanwise distribution of bending slope for this excitation.
(e) Show beam bending displacement distribution for this excitation.
(f) If PZT elements are replaced with PVDF elements of same size, calculate new surface actuation strain and actuation bending moment for 150 volts excitation (for PVDF $d_{31} = -20 \times 10^{-12}$ m/V and $E_c = 0.2 \times 10^{10}$ N/m^2).

11. Using Euler-Bernoulli assumption, derive extension-bending relations for a beam with two dissimilar piezos embedded at a distance d from the mid-axis (Fig. 4.67). The thicknesses of the top and bottom piezos are, respectively, t_{c1} and t_{c2} and they are of the same modulus E_c. The same field is applied to both piezos.
12. Use the Euler-Bernoulli model to calculate the bending displacement at the tip of the beam (Fig. 4.68) for two identical piezos inducing a pure bending actuation ($+V$ field to top and $-V$ to bottom piezo).
13. Two dissimilar piezo-elements (PZT-5H) of width $b_c = 1$ in and thickness $t_c =$ 0.012 in, respectively, of length $l_c = 2$ in and 1 in are surface-bonded at the top and bottom of a thin aluminum cantilevered beam (length $L_b = 24$ in, width

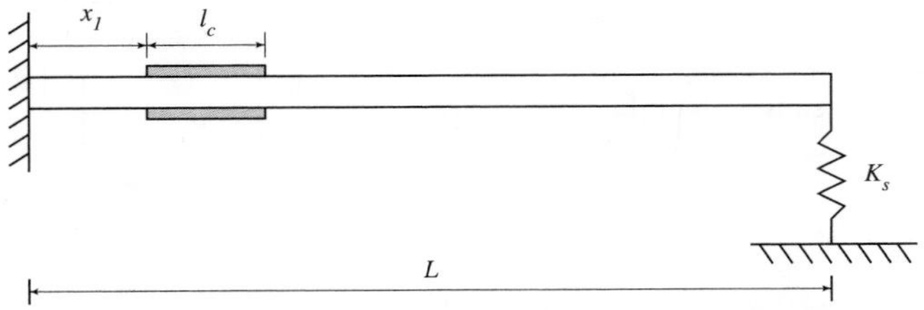

Figure 4.68. Cantilevered beam with surface-mounted piezoactuators and linear spring at the tip.

$b_b = 2$ in, and thickness $t_b = 0.035$ in). Other manufacturer-supplied material constants are as follows

$$d_{31} = -274 \times 10^{-12}\,\text{m/V}$$
$$E_c = E_b = 10.5 \times 10^6\,\text{lb/in}^2$$

(a) Using the Euler-Bernoulli theory, derive general bending as well as extension relations with same field on opposite piezo-elements for this dual piezo actuation.
(b) Plot spanwise variation of beam-surface strain and actuator strain for a field of 150 volts to both top and bottom piezos.
(c) Calculate actuation surface force F in lb and bending moment M in in-lb for this excitation.
(d) Show spanwise distribution of bending slope for this excitation.
(e) Show beam bending displacement distribution for this excitation.
(f) If PZT-5H elements are replaced with PVDF elements of same size, calculate new surface actuation strain and actuation bending moment for a field of 150 volts to both top and bottom piezos (For PVDF $d_{31} = -20 \times 10^{-12}$ m/V and $E_c = 0.2 \times 10^{10}$ N/m^2).

14. Using the Euler-Bernoulli assumption, derive extension-bending relations for a beam with two dissimilar piezos embedded at a distance d from mid-axis (Fig. 4.69). The widths of top and bottom piezos are, respectively, $b/4$ and $b/2$ and they are of the same thickness, length, and modulus E_c. The same field is applied to both piezos.
15. Using the Rayleigh-Ritz method, determine the steady-state tip response of a beam of length L_b with sinusoidal field $V = V_o \sin \omega t$ with two identical piezos but with opposite fields ($+V$ for bottom and $-V$ for top) (Fig. 4.70). Assume a deflection of the form

$$w(x,t) = \left(\frac{x}{L_b}\right)^2 q_1$$

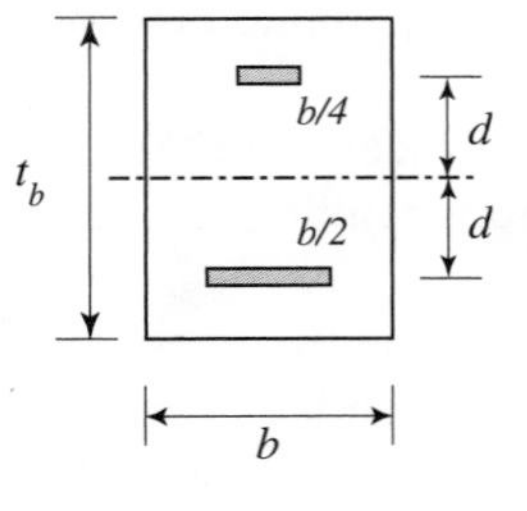

Figure 4.69. Cross section of beam with embedded piezos of different widths.

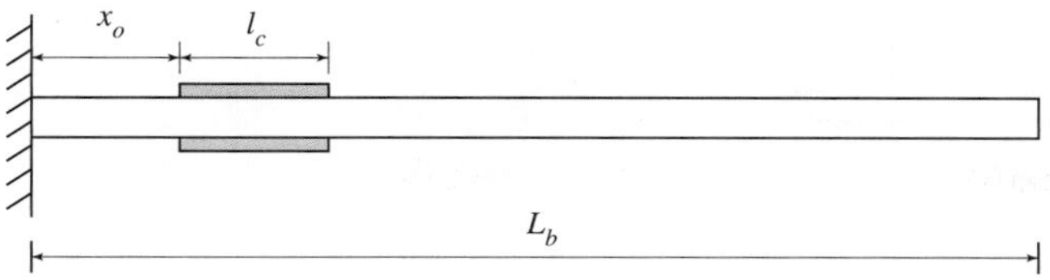

Figure 4.70. Cantilevered beam with surface-mounted piezoactuators.

16. Using the Euler-Bernoulli assumption, derive extension-bending relations for a beam with two dissimilar piezos embedded at a distance d from the mid-axis (similar to Fig. 4.67). The lengths of the top and bottom piezos are, respectively, l_{c1} and l_{c2}, and they are of the same modulus E_c and the same thickness t_c. The same field is applied to both piezos.
17. Two identical piezoceramic elements (PZT-5H) (length $l_c = 2$ in, width $b_c = 1$ in, and thickness $t_c = 0.010$ in) are surface-bonded on the bottom and top of a thin aluminum cantilevered beam as shown in Fig. 4.71 (length $L_b = 24$ in, width $b_b = 1$ in, thickness $t_b = 0.035$ in). Manufacturer-supplied material constants are as follows:

$$d_{31} = -274 \times 10^{-12}\,\text{m/V}$$
$$d_{33} = 593 \times 10^{-12}\,\text{m/V}$$
$$E_c = E_b = 60.6\,\text{GPa (short-circuit)}$$
$$k_{31}^2 = 0.55$$
$$e_{33}^{\sigma} = 30.1 \times 10^{-9}\,\text{F/m}$$
$$k_e = 0.25\,\text{MN/m}$$

   (a) Show free-strain variation in micro-strain with voltage for each piezo.
   (b) Plot the variation of piezo strain with axial blocked force F for each piezo for an excitation of 150 volts.
   (c) For an equal voltage applied to each piezo as shown in Fig. 4.72, what will be the response of the beam? (The dots in the figure represent the poling direction of the piezos.)
   (d) Using the blocked-force method, derive an expression for the tip displacement of the beam.
   (e) Calculate actuation surface force F and tip displacement for an excitation of 150 volts to both top and bottom piezos.
   (f) If PZT-5H elements are replaced with PVDF elements of the same size, calculate new surface actuation strain and tip displacement (for PVDF $d_{31} = -20 \times 10^{-12}$ m/V and $E_c = 0.2 \times 10^{10}$ N/m^2).
18. Two identical piezoceramic elements (PZT-5H) (length $l_c = 2$ in, width $b_c = 1$ in, and thickness $t_c = 0.010$ in) are surface-bonded on the bottom and top of a thin

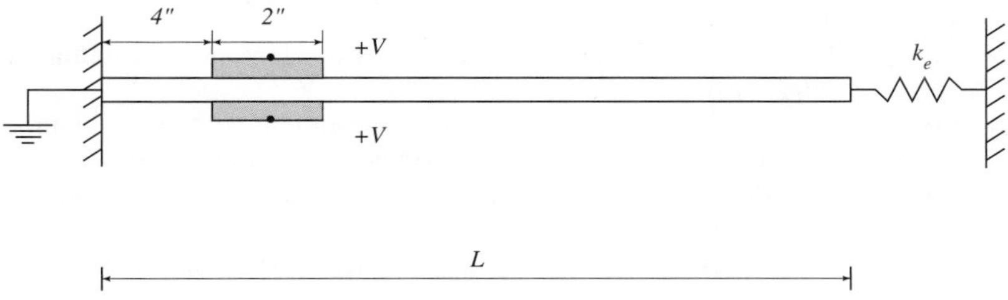

Figure 4.71. Cantilevered beam with surface-mounted piezoactuators and tip spring.

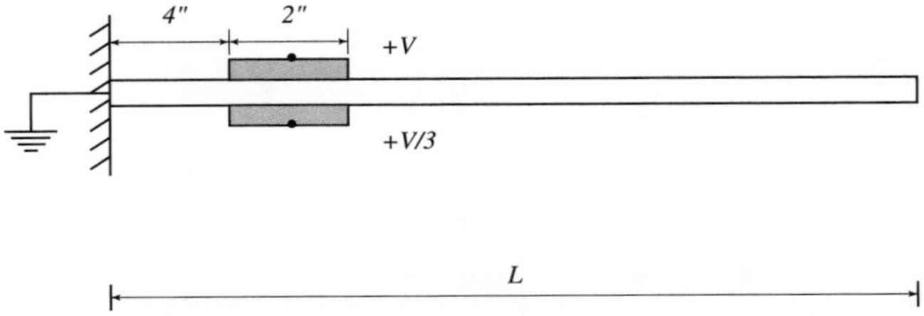

Figure 4.72. Cantilevered beam with surface-mounted piezoactuators.

aluminum cantilevered beam as shown in Fig. 4.72 (length $L_b = 24$ in, width $b_b = 1$ in, thickness $t_b = 0.035$ in). Manufacturer-supplied material constants are as follows:

$$d_{31} = -274 \times 10^{-12}\,\text{m/V}$$
$$d_{33} = 593 \times 10^{-12}\,\text{m/V}$$
$$E_c = E_b = 60.6\,\text{GPa (short-circuit)}$$
$$k_{31}^2 = 0.55$$
$$e_{33}^{\sigma} = 30.1 \times 10^{-9}\,\text{F/m}$$
$$k_e = 0.25\,\text{MN/m}$$

(a) Show free-strain variation in micro-strain with voltage for each piezo.
(b) Plot the variation of piezo strain with axial blocked-force F for each piezo for an excitation of 150 volts.
(c) For $V = 150$ volts applied to the piezos as shown in Fig. 4.72, what will be the response of the beam? (The dots in the figure represent the poling direction of the piezos.)
(d) Using the blocked-force method, derive an expression for the tip displacement of the beam.
(e) Calculate actuation surface force F and tip displacement for this excitation.
(f) If PZT-5H elements are replaced with PVDF elements of the same size, calculate the new surface actuation strain and tip displacement (for PVDF $d_{31} = -20 \times 10^{-12}$ m/V and $E_c = 0.2 \times 10^{10}$ N/m^2).

BIBLIOGRAPHY

[1] R. L. Bisplinghoff, J. W. Mar, and T. H. H. Pian. *Statics of Deformable Solids*. Dover Publications, Inc., New York, 1990.
[2] D. J. Peery. *Aircraft Structures*. McGraw-Hill, 1950.
[3] C. Park, C. Walz, and I. Chopra. Bending torsion models of beams with induced strain actuation. *Smart Materials & Structures*, 5(1):98–113, February 1996.
[4] J. N. Reddy. *Mechanics of Laminated Composite Plates and Shells: Theory and Analysis*. CRC Press, Boca Raton, FL, second edition, 2004.
[5] E. Crawley and J. de Luis. Use of piezoceramic actuators as elements of intelligent structures. *AIAA Journal*, 25(10):1373–1385, October 1987.
[6] E. Crawley and E. Anderson. Detailed models of piezoceramic actuation of beams. *Journal of Intelligent Material Systems and Structures*, 1(1):4–25, January 1990.
[7] C. Park and I. Chopra. Modeling piezoceramic actuation of beams in torsion. *AIAA Journal*, 34(12):2582–2589, December 1996.
[8] M. H. Shen. A new modeling technique for piezoelectric actuated beams. *Computers and Structures*, 57(3):361–366, 1995.

[9] R. Chandra and I. Chopra. Structural modeling of composite beams with induced-strain actuation. *AIAA Journal*, 31(9):1692–1701, September 1993.

[10] A. P. F. Bernhard and I. Chopra. Analysis of bending-torsion coupled actuator for a smart rotor with active blade tips. *Smart Materials and Structures*, 10(1):35–52, February 2001.

[11] N. W. Hagood, W. H. Chung, and A. V. Flotow. Modeling of piezoelectric actuator dynamics for active structural control. *Journal of Intelligent Material Systems and Structures*, 1(3):327–354, 1990.

[12] D. H. Robbins and J. N. Reddy. Analysis of piezoelectrically actuated beams using a layer-wise displacement theory. *Computers and Structures*, 41(2):265–279, 1991.

[13] D. A. Saravanos and P. R. Heyliger. Coupled layerwise analysis of composite beams with embedded piezoelectric sensors and actuators. *Journal of Intelligent Material Systems and Structures*, 6(3):350–363, 1995.

[14] S. Im and S. N. Atluri. Effects of piezo actuator on a finitely deformed beam subjected to general loading. *AIAA Journal*, 27(12):1801–1807, December 1989.

[15] A. Benjeddou, M. A. Trindade, and R. Ohayon. A unified beam finite element model for extension and shear piezoelectric actuation mechanisms. *Journal of Intelligent Material Systems and Structures*, 8(12):1012–1025, December 1997.

[16] S. N. Jung, V. T. Nagaraj, and I. Chopra. Refined structural model for thin- and thick-walled composite rotor blades. *AIAA Journal*, 40(1):105–116, January 2002.

[17] S. N. Jung, V. T. Nagaraj, and I. Chopra. Assessment of composite rotor blade modeling techniques. *Journal of the American Helicopter Society*, 44(3):188–205, July 1999.

[18] O. A. Bauchau and C. H. Hong. Nonlinear composite beam theory. *Journal of Applied Mechanics*, 55(110):156–163, March 1988.

[19] C. E. S. Cesnik and D. S. Hodges. Vabs: A new concept for composite rotor blade cross-sectional modeling. *Journal of the American Helicopter Society*, 42(1):27–38, January 1997.

[20] E. C. Smith and I. Chopra. Formulation and evaluation of an analytical model for composite box-beams. *Journal of the American Helicopter Society*, 36(3):23–35, July 1991.

[21] A. D. Stemple and S. W. Lee. Finite element model for composite beams of arbitrary cross-sectional warping. *AIAA Journal*, 26(12):1512–1520, December 1998.

[22] L. Librescu, L. Meirovitch, and O. Song. Integrated structural tailoring and adaptive materials control for advanced aircraft wings. *Journal of Aircraft*, 33(1):203–213, January-February 1996.

[23] C. Kaiser. Piezothermoelastic behavior of thin-walled composite beams with elastic couplings. Paper # AIAA-2001-1549, *Proceedings of the 42nd AIAA, ASME, ASCE, AHS, and ASC Structures, Structural Dynamics and Materials Conference*, Seattle, WA, April 2001.

[24] C. E. S. Cesnik and S. J. Shin. On the modeling of integrally actuated helicopter blades. *International Journal of Solids and Structures*, 38(10):1765–1789, 2001.

[25] S. J. Shin and C. E. S. Cesnik. Integral twist actuation of helicopter rotor blades for vibration reduction. *AMSL # 01-07, Aero & Astro*, MIT, August 2001.

[26] G. L. Ghiringhelli, P. Masarati, and P. Mantegazza. Characterization of anisotropic non-homogeneous beam sections with embedded piezoelectric materials. *Journal of Intelligent Material Systems and Structures*, 8(10):842–858, October 1997.

[27] S. Raja, K. Rohwer, and M. Rose. Piezothermoelastic modeling and active vibration control of laminated composite beams. *Journal of Intelligent Material Systems and Structures*, 10(11):890–899, November 1999.

[28] J. N. Reddy. A generalization of two-dimensional theories of laminated composite plates. *Communications in Applied Numerical Methods*, 3(3):173–180, 1987.

[29] P. C. Chen and I. Chopra. Hover testing of smart rotor with induced-strain actuation of blade twist. *AIAA Journal*, 35(1):6–16, January 1997.

5 Plate Modeling with Induced-Strain Actuation

The previous chapter discussed the modeling of beam-like structures with induced-strain actuation. Many practical structures can be simplified and analyzed as beams, but such an assumption is not accurate in a large number of other structures, such as fuselage panels in aircraft, low aspect-ratio wings, and large control surfaces. It is possible to treat such structures as plates and perform a simple two-dimensional analysis to estimate their behavior. Some of the theories discussed in the previous chapter can be extended to two-dimensional plate-like structures. This chapter describes the modeling of isotropic and composite plate structures with induced-strain actuation. It will combine both the actuators and substrate into one integrated structure to model its behavior. The discussion focuses on induced-strain actuation by means of piezoceramic sheets, but the general techniques may be equally applicable to other forms of induced-strain actuation.

Plate analysis, including induced-strain actuation, is based on the classical laminated plate theory (CLPT), sometimes referred to as classical laminated theory (CLT). It is an equivalent single layer (ESL) plate theory in which the effects of transverse shear strains are neglected. It is valid for thin plates that have thicknesses of one to two orders of magnitude smaller than their planar dimensions (length and width). In the CLPT formulation, a plane-stress state assumption is used.

5.1 Classical Laminated Plate Theory (CLPT) Formulation without Actuation

A composite laminate consists of a number of laminae or plies, each with different elastic properties. A fiber-reinforced lamina is the fundamental building block of the laminate. The sequence of various orientations of composite laminae is termed the "stacking sequence." A lamina is very strong along the fiber direction and weak in the transverse direction. The stacking sequence and lamina properties help to tailor the stiffness, strength, and coupling between bending, torsion, and extension of the laminate. A macro-mechanical behavior of a lamina is assumed to formulate a linear-elastic analysis. The stress-strain relations for an orthotropic lamina in a plane stress condition are

$$\begin{Bmatrix} \epsilon_1 \\ \epsilon_2 \\ \gamma_{12} \end{Bmatrix} = \begin{bmatrix} S_{11} & S_{12} & 0 \\ S_{12} & S_{22} & 0 \\ 0 & 0 & S_{66} \end{bmatrix} \begin{Bmatrix} \sigma_1 \\ \sigma_2 \\ \tau_{12} \end{Bmatrix} \tag{5.1}$$

Figure 5.1. An orthotropic lamina.

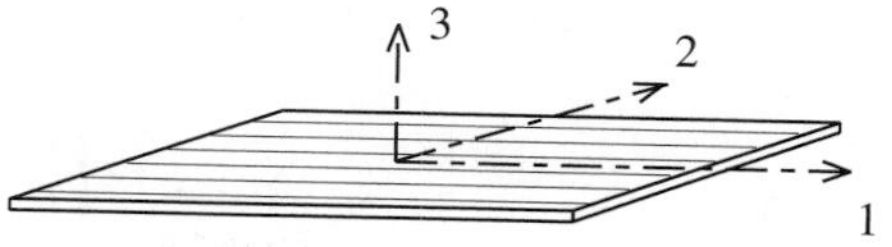

where ϵ_1 and ϵ_2 are normal strains and γ_{12} is the shear strain. Directions 1 and 2 are referred to as principal directions for an orthotropic material. For example, in a composite ply, the fibers are all aligned along Direction 1 (Fig. 5.1). The coefficients of the compliance matrix are defined as

$$S_{11} = \frac{1}{E_1} \tag{5.2}$$

$$S_{12} = -\frac{\nu_{12}}{E_1} = -\frac{\nu_{21}}{E_2} \tag{5.3}$$

$$S_{22} = \frac{1}{E_2} \tag{5.4}$$

$$S_{66} = \frac{1}{G_{12}} \tag{5.5}$$

where E_1 is the longitudinal Young's modulus and E_2 is the transverse Young's modulus. Because the fibers are typically aligned parallel to the 1 axis, E_1 is expected to be much larger than E_2. Typical values of material properties for some commonly available carbon composites (IM7/8552, AS4/3501-6) and fiberglass (E-glass/epoxy, S-glass/epoxy) are shown in Table 5.1.

The units of modulus are N/m^2 or lb/in^2. Sometimes, the moduli are defined in GPa, where G stands for giga (10^9) and Pa (Pascal) means N/m^2. ν_{12} is the longitudinal Poisson's ratio, which is defined as the ratio of the induced strain in the transverse direction due to an imposed longitudinal strain. ν_{21} is the transverse Poisson's ratio, which is defined as the ratio of the induced strain in the longitudinal direction due to an imposed transverse strain. The Poisson's ratio ν_{12} is much larger than ν_{21}. They are related to each other by the following relation

$$\frac{\nu_{12}}{\nu_{21}} = \frac{E_1}{E_2} \tag{5.6}$$

Inverting Eq. 5.1 leads to

$$\begin{Bmatrix} \sigma_1 \\ \sigma_2 \\ \tau_{12} \end{Bmatrix} = \begin{bmatrix} Q_{11} & Q_{12} & 0 \\ Q_{12} & Q_{22} & 0 \\ 0 & 0 & Q_{66} \end{bmatrix} \begin{Bmatrix} \epsilon_1 \\ \epsilon_2 \\ \gamma_{12} \end{Bmatrix} = \boldsymbol{Q} \begin{Bmatrix} \epsilon_1 \\ \epsilon_2 \\ \gamma_{12} \end{Bmatrix} \tag{5.7}$$

Table 5.1. *Material properties of typical composite laminae*

Property	IM7/ 8552	AS4/ 3501–6	E-glass/ epoxy	S-glass/ epoxy	Kevlar 149/ epoxy
Tensile modulus, 0°, E_1 (GPa)	164.00	142.00	39.00	43.00	87.00
Tensile modulus, 90°, E_2 (GPa)	12.00	10.30	8.60	8.90	5.50
Shear modulus, G_{12} (GPa)	11.10	7.20	3.80	4.50	2.20
Poisson's ratio, ν_{12}	0.31	0.27	0.28	0.27	0.34
Specific gravity	1.57	1.58	2.10	2.00	1.38

The coefficients of the reduced stiffness matrix $\boldsymbol{Q}$ are defined as

$$Q_{11} = \frac{S_{22}}{S_{11}S_{22} - S_{12}^2} = \frac{E_1}{1 - \nu_{12}\nu_{21}} \tag{5.8}$$

$$Q_{12} = \frac{S_{12}}{S_{11}S_{22} - S_{12}^2} = \frac{\nu_{12}E_2}{1 - \nu_{12}\nu_{21}} = \frac{\nu_{21}E_1}{1 - \nu_{12}\nu_{21}} \tag{5.9}$$

$$Q_{22} = \frac{S_{11}}{S_{11}S_{22} - S_{12}^2} = \frac{E_2}{1 - \nu_{12}\nu_{21}} \tag{5.10}$$

$$Q_{66} = \frac{1}{S_{66}} = G_{12} \tag{5.11}$$

Rewriting Eq. 5.7,

$$\begin{Bmatrix} \sigma_1 \\ \sigma_2 \\ \tau_{12} \end{Bmatrix} = \begin{bmatrix} \frac{E_1}{(1-\nu_{12}\nu_{21})} & \frac{\nu_{21}E_1}{(1-\nu_{12}\nu_{21})} & 0 \\ \frac{\nu_{21}E_1}{(1-\nu_{12}\nu_{21})} & \frac{E_2}{(1-\nu_{12}\nu_{21})} & 0 \\ 0 & 0 & G_{12} \end{bmatrix} \begin{Bmatrix} \epsilon_1 \\ \epsilon_2 \\ \gamma_{12} \end{Bmatrix} \tag{5.12}$$

Four independent material constants are required to define an orthotropic lamina: E_1, E_2, G_{12}, and ν_{12} or ν_{21}. The units of Q_{ij} are N/m^2 or lb/in^2, whereas the units of S_{ij} are m^2/N or in^2/lb. For isotropic materials, $\nu_{12} = \nu_{21} = \nu$, $E_1 = E_2 = E$ and $G = \frac{E}{2(1+\nu)}$

$$\begin{Bmatrix} \epsilon_1 \\ \epsilon_2 \\ \gamma_{12} \end{Bmatrix} = \frac{1}{E} \begin{bmatrix} 1 & -\nu & 0 \\ -\nu & 1 & 0 \\ 0 & 0 & 2(1+\nu) \end{bmatrix} \begin{Bmatrix} \sigma_1 \\ \sigma_2 \\ \tau_{12} \end{Bmatrix} \tag{5.13}$$

from which,

$$\begin{Bmatrix} \sigma_1 \\ \sigma_2 \\ \tau_{12} \end{Bmatrix} = \frac{E}{1 - \nu^2} \begin{bmatrix} 1 & \nu & 0 \\ \nu & 1 & 0 \\ 0 & 0 & \frac{1-\nu}{2} \end{bmatrix} \begin{Bmatrix} \epsilon_1 \\ \epsilon_2 \\ \gamma_{12} \end{Bmatrix} \tag{5.14}$$

An isotropic material requires two independent material constants (E and ν) to define its behavior.

5.1.1 Stress-Strain Relations for a Lamina at an Arbitrary Orientation

For a lamina with fibers at an arbitrary orientation (Fig. 5.2), the strains can be transformed into the reference coordinate system, as follows

$$\begin{Bmatrix} \epsilon_x \\ \epsilon_y \\ \frac{\gamma_{xy}}{2} \end{Bmatrix} = \begin{bmatrix} \cos^2\alpha & \sin^2\alpha & -2\sin\alpha\cos\alpha \\ \sin^2\alpha & \cos^2\alpha & 2\sin\alpha\cos\alpha \\ \sin\alpha\cos\alpha & -\sin\alpha\cos\alpha & \cos^2\alpha - \sin^2\alpha \end{bmatrix} \begin{Bmatrix} \epsilon_1 \\ \epsilon_2 \\ \frac{\gamma_{12}}{2} \end{Bmatrix} \tag{5.15}$$

where α is the angle of the fibers from the x-axis (+ve in the counterclockwise direction).

The stresses transformed into the reference axes become

$$\begin{Bmatrix} \sigma_x \\ \sigma_y \\ \tau_{xy} \end{Bmatrix} = \begin{bmatrix} \cos^2\alpha & \sin^2\alpha & -2\sin\alpha\cos\alpha \\ \sin^2\alpha & \cos^2\alpha & 2\sin\alpha\cos\alpha \\ \sin\alpha\cos\alpha & -\sin\alpha\cos\alpha & \cos^2\alpha - \sin^2\alpha \end{bmatrix} \begin{Bmatrix} \sigma_1 \\ \sigma_2 \\ \tau_{12} \end{Bmatrix} \tag{5.16}$$

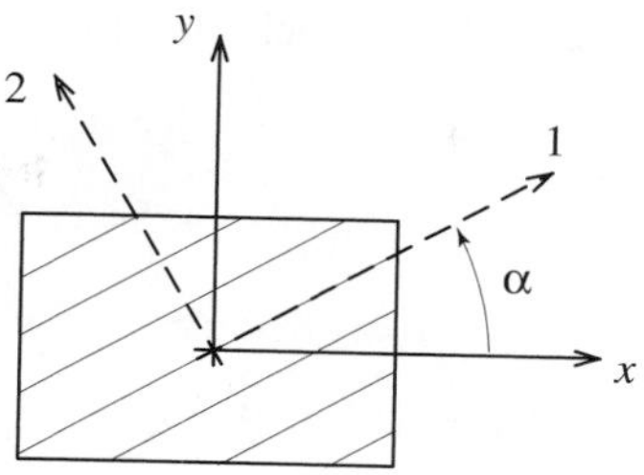

Figure 5.2. Lamina at an arbitrary orientation.

The stress-strain relations become

$$\boldsymbol{\sigma} = \begin{Bmatrix} \sigma_x \\ \sigma_y \\ \tau_{xy} \end{Bmatrix} = \begin{bmatrix} \bar{Q}_{11} & \bar{Q}_{12} & \bar{Q}_{16} \\ \bar{Q}_{12} & \bar{Q}_{22} & \bar{Q}_{26} \\ \bar{Q}_{16} & \bar{Q}_{26} & \bar{Q}_{66} \end{bmatrix} \begin{Bmatrix} \epsilon_x \\ \epsilon_y \\ \gamma_{xy} \end{Bmatrix} = \bar{\boldsymbol{Q}}\boldsymbol{\epsilon} \tag{5.17}$$

where the coefficients of the reduced-stiffness matrix $\bar{\boldsymbol{Q}}$ are defined as

$$\bar{Q}_{11} = Q_{11}\cos^4\alpha + Q_{22}\sin^4\alpha + 2(Q_{12} + 2Q_{66})\sin^2\alpha\cos^2\alpha \tag{5.18}$$

$$\bar{Q}_{22} = Q_{11}\sin^4\alpha + Q_{22}\cos^4\alpha + 2(Q_{12} + 2Q_{66})\sin^2\alpha\cos^2\alpha \tag{5.19}$$

$$\bar{Q}_{12} = (Q_{11} + Q_{22} - 4Q_{66})\sin^2\alpha\cos^2\alpha + Q_{12}(\sin^4\alpha + \cos^4\alpha) \tag{5.20}$$

$$\bar{Q}_{66} = (Q_{11} + Q_{22} - 2Q_{12} - 2Q_{66})\sin^2\alpha\cos^2\alpha + Q_{66}(\sin^4\alpha + \cos^4\alpha) \tag{5.21}$$

$$\bar{Q}_{16} = (Q_{11} - Q_{12} - 2Q_{66})\sin\alpha\cos^3\alpha - (Q_{22} - Q_{12} - 2Q_{66})\sin^3\alpha\cos\alpha \tag{5.22}$$

$$\bar{Q}_{26} = (Q_{11} - Q_{12} - 2Q_{66})\sin^3\alpha\cos\alpha - (Q_{22} - Q_{12} - 2Q_{66})\sin\alpha\cos^3\alpha \tag{5.23}$$

There are still only four independent material constants needed to define the characteristics of a generally orthotropic lamina. Note that for two plies with orientation angles $+\alpha$ and $-\alpha$, the elements of the stiffness matrix are related, as follows

$$\bar{Q}_{11_{+\alpha}} = \bar{Q}_{11_{-\alpha}} \tag{5.24}$$

$$\bar{Q}_{22_{+\alpha}} = \bar{Q}_{22_{-\alpha}} \tag{5.25}$$

$$\bar{Q}_{12_{+\alpha}} = \bar{Q}_{12_{-\alpha}} \tag{5.26}$$

$$\bar{Q}_{66_{+\alpha}} = \bar{Q}_{66_{-\alpha}} \tag{5.27}$$

$$\bar{Q}_{16_{+\alpha}} = -\bar{Q}_{16_{-\alpha}} \tag{5.28}$$

$$\bar{Q}_{26_{+\alpha}} = -\bar{Q}_{26_{-\alpha}} \tag{5.29}$$

Similar relations also hold good for the compliances. In an alternate format, the strains can be expressed in terms of stresses

$$\boldsymbol{\epsilon} = \begin{Bmatrix} \epsilon_x \\ \epsilon_y \\ \gamma_{xy} \end{Bmatrix} = \begin{bmatrix} \bar{S}_{11} & \bar{S}_{12} & \bar{S}_{16} \\ \bar{S}_{12} & \bar{S}_{22} & \bar{S}_{26} \\ \bar{S}_{16} & \bar{S}_{26} & \bar{S}_{66} \end{bmatrix} \begin{Bmatrix} \sigma_x \\ \sigma_y \\ \tau_{xy} \end{Bmatrix} = \bar{\boldsymbol{S}}\boldsymbol{\sigma} \tag{5.30}$$

where the coefficients of the reduced compliance matrix $\bar{\boldsymbol{S}}$ are defined as

$$\bar{S}_{11} = S_{11}\cos^4\alpha + S_{22}\sin^4\alpha + (2S_{12} + S_{66})\sin^2\alpha\cos^2\alpha \tag{5.31}$$

$$\bar{S}_{22} = S_{11}\sin^4\alpha + S_{22}\cos^4\alpha + (2S_{12} + S_{66})\sin^2\alpha\cos^2\alpha \tag{5.32}$$

$$\bar{S}_{12} = (S_{11} + S_{22} - S_{66})\sin^2\alpha\cos^2\alpha + S_{12}(\sin^4\alpha + \cos^4\alpha) \tag{5.33}$$

$$\bar{S}_{66} = 2(2S_{11} + 2S_{22} - 4S_{12} - S_{66})\sin^2\alpha\cos^2\alpha + S_{66}(\sin^4\alpha + \cos^4\alpha) \tag{5.34}$$

$$\bar{S}_{16} = (2S_{11} - 2S_{12} - S_{66})\sin\alpha\cos^3\alpha - (2S_{22} - 2S_{12} - S_{66})\sin^3\alpha\cos\alpha \tag{5.35}$$

$$\bar{S}_{26} = (2S_{11} - 2S_{12} - S_{66})\sin^3\alpha\cos\alpha - (2S_{22} - 2S_{12} - S_{66})\sin\alpha\cos^3\alpha \tag{5.36}$$

5.1.2 Macromechanical Behavior of a Laminate

A laminate consists of two or more laminae bonded together to form an integral structural plate. The stress-strain relations in principal material coordinates for a lamina are

$$\begin{Bmatrix} \sigma_1 \\ \sigma_2 \\ \tau_{12} \end{Bmatrix} = \begin{bmatrix} Q_{11} & Q_{12} & 0 \\ Q_{12} & Q_{22} & 0 \\ 0 & 0 & Q_{66} \end{bmatrix} \begin{Bmatrix} \epsilon_1 \\ \epsilon_2 \\ \gamma_{12} \end{Bmatrix} \tag{5.37}$$

In a different coordinate system, oriented at an angle α to the principal axes, the stress-strain relations become

$$\boldsymbol{\sigma} = \begin{Bmatrix} \sigma_x \\ \sigma_y \\ \tau_{xy} \end{Bmatrix} = \begin{bmatrix} \bar{Q}_{11} & \bar{Q}_{12} & \bar{Q}_{16} \\ \bar{Q}_{12} & \bar{Q}_{22} & \bar{Q}_{26} \\ \bar{Q}_{16} & \bar{Q}_{26} & \bar{Q}_{66} \end{bmatrix} \begin{Bmatrix} \epsilon_x \\ \epsilon_y \\ \gamma_{xy} \end{Bmatrix} = \bar{\boldsymbol{Q}}\boldsymbol{\epsilon} \tag{5.38}$$

For a multilayered laminate, the stress-strain relations are

$$\boldsymbol{\sigma}_k = \bar{\boldsymbol{Q}}_k\boldsymbol{\epsilon}_k \tag{5.39}$$

where the subscript k refers to the 'k'th lamina. The laminate is assumed thin, consisting of uniform layers perfectly bonded together, and undergoing small displacements. When the laminate is extended and bent, a material plane that is initially normal to the mid-surface of the laminate is assumed to remain plane and normal to the mid-surface. This is similar to the assumption made in the bending of Euler-Bernoulli beams and is equivalent to ignoring shear strains in planes perpendicular to the middle surface. This assumption for plates is called the Kirchhoff-Love hypothesis. With this assumption, the strain distribution consists of a linear combination of in-plane extensional strain (constant through thickness) and a bending strain (linearly varying through thickness). The transverse normal effects are neglected.

The variables u, v, and w are laminate displacements in the x, y, and z directions, given by the following (Fig. 5.3)

$$u(x, y, z) = u_o(x, y) - z\frac{\partial w_o}{\partial x}(x, y) \tag{5.40}$$

$$v(x, y, z) = v_o(x, y) - z\frac{\partial w_o}{\partial y}(x, y) \tag{5.41}$$

$$w(x, y, z) = w_o(x, y) \tag{5.42}$$

where u_o, v_o, and w_o are the displacements at the mid-plane or neutral plane ($z = 0$).

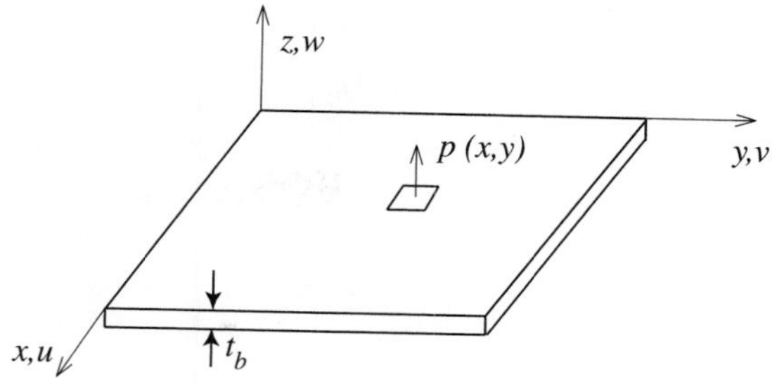

Figure 5.3. Displacements of a thin laminated plate.

By virtue of the Kirchhoff-Love hypothesis

$$\epsilon_z = \gamma_{xz} = \gamma_{yz} = 0 \tag{5.43}$$

$$\epsilon_x = \frac{\partial u}{\partial x} \tag{5.44}$$

$$\epsilon_y = \frac{\partial v}{\partial y} \tag{5.45}$$

$$\gamma_{xy} = \frac{\partial u}{\partial y} + \frac{\partial v}{\partial x} \tag{5.46}$$

Substituting for the displacements (Eqs. 5.40–5.42) in these strain relations, we obtain

$$\epsilon_x = \frac{\partial u_o}{\partial x} - z\frac{\partial^2 w_o}{\partial x^2} \tag{5.47}$$

$$\epsilon_y = \frac{\partial v_o}{\partial y} - z\frac{\partial^2 w_o}{\partial y^2} \tag{5.48}$$

$$\gamma_{xy} = \frac{\partial u_o}{\partial y} + \frac{\partial v_o}{\partial x} - 2z\frac{\partial^2 w_o}{\partial x \partial y} \tag{5.49}$$

These lead to

$$\begin{Bmatrix} \epsilon_x \\ \epsilon_y \\ \gamma_{xy} \end{Bmatrix} = \begin{Bmatrix} \epsilon_x^o \\ \epsilon_y^o \\ \gamma_{xy}^o \end{Bmatrix} - z \begin{Bmatrix} \kappa_x \\ \kappa_y \\ \kappa_{xy} \end{Bmatrix} = \boldsymbol{\epsilon^o} - z\boldsymbol{\kappa} \tag{5.50}$$

Mid-plane strains are given by

$$\boldsymbol{\epsilon^o} = \begin{Bmatrix} \epsilon_x^o \\ \epsilon_y^o \\ \gamma_{xy}^o \end{Bmatrix} = \begin{Bmatrix} \frac{\partial u_o}{\partial x} \\ \frac{\partial v_o}{\partial y} \\ \frac{\partial u_o}{\partial y} + \frac{\partial v_o}{\partial x} \end{Bmatrix} \tag{5.51}$$

and the middle-surface curvatures are

$$\boldsymbol{\kappa} = \begin{Bmatrix} \kappa_x \\ \kappa_y \\ \kappa_{xy} \end{Bmatrix} = \begin{Bmatrix} \frac{\partial^2 w_o}{\partial x^2} \\ \frac{\partial^2 w_o}{\partial y^2} \\ 2\frac{\partial^2 w_o}{\partial x \partial y} \end{Bmatrix} \tag{5.52}$$

If the mid-plane displacements (u_o, v_o, and w_o) are known, the strains at any point (x, y, and z) can be determined. The strains vary linearly through the laminate thickness. Note that not only the transverse strains (ϵ_z, γ_{xz}, γ_{yz}) are zero but also that

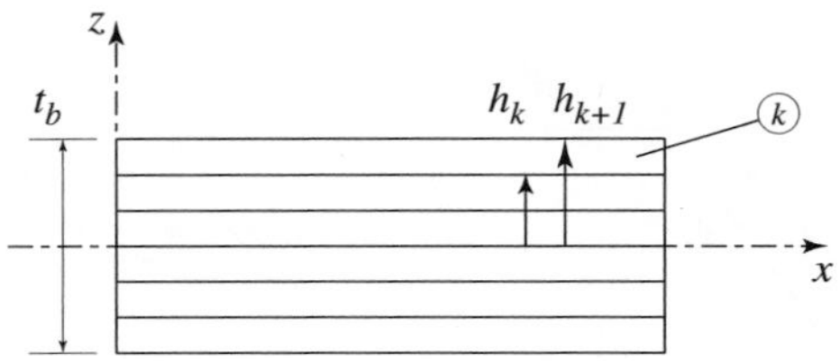

Figure 5.4. Laminate stackup sequence.

the transverse stresses (σ_z, τ_{xz}, τ_{yz}) are either zero or not included in the formulation. For example, the shear stresses τ_{xz} and τ_{yz} are zero, and the transverse normal stress σ_z is not zero identically (Poisson's effect) but does not appear in the virtual strain energy formulation. Thus, this formulation represents a condition of plane stress as well as of plane strain and appears appropriate for very thin laminates.

5.1.3 Resultant Laminate Forces and Moments

Resultant forces and moments on a laminate are obtained by integrating the stresses in each lamina across the laminate thickness, t_b (Fig. 5.4). Note that in plate analysis, the forces and moments are defined with respect to a unit cross-sectional width. Normally, extensional forces and stresses are assumed positive and moments that cause compression on the top fiber are assumed positive.

$$F_x = \int_{t_b} \sigma_x dz \tag{5.53}$$

$$M_x = -\int_{t_b} \sigma_x z\, dz \tag{5.54}$$

This leads to

$$\begin{Bmatrix} F_x \\ F_y \\ F_{xy} \end{Bmatrix} = \int_{t_b} \begin{Bmatrix} \sigma_x \\ \sigma_y \\ \tau_{xy} \end{Bmatrix} dz = \sum_{k=1}^{N} \int_{h_k}^{h_{k+1}} \begin{Bmatrix} \sigma_x \\ \sigma_y \\ \tau_{xy} \end{Bmatrix} dz \tag{5.55}$$

and

$$\begin{Bmatrix} M_x \\ M_y \\ M_{xy} \end{Bmatrix} = -\int_{t_b} \begin{Bmatrix} \sigma_x \\ \sigma_y \\ \tau_{xy} \end{Bmatrix} z\, dz = -\sum_{k=1}^{N} \int_{h_k}^{h_{k+1}} \begin{Bmatrix} \sigma_x \\ \sigma_y \\ \tau_{xy} \end{Bmatrix} z\, dz \tag{5.56}$$

where h_k is the vertical position of the 'k'th layer. Combining these equations gives

$$\begin{Bmatrix} F_x \\ F_y \\ F_{xy} \end{Bmatrix} = \sum_{k=1}^{N} \begin{bmatrix} \bar{Q}_{11} & \bar{Q}_{12} & \bar{Q}_{16} \\ \bar{Q}_{12} & \bar{Q}_{22} & \bar{Q}_{26} \\ \bar{Q}_{16} & \bar{Q}_{26} & \bar{Q}_{66} \end{bmatrix}_k \left(\int_{h_k}^{h_{k+1}} \begin{Bmatrix} \epsilon_x^o \\ \epsilon_y^o \\ \gamma_{xy}^o \end{Bmatrix} dz - \int_{h_k}^{h_{k+1}} \begin{Bmatrix} \kappa_x \\ \kappa_y \\ \kappa_{xy} \end{Bmatrix} z\, dz \right)$$

and

$$\begin{Bmatrix} M_x \\ M_y \\ M_{xy} \end{Bmatrix} = -\sum_{k=1}^{N} \begin{bmatrix} \bar{Q}_{11} & \bar{Q}_{12} & \bar{Q}_{16} \\ \bar{Q}_{12} & \bar{Q}_{22} & \bar{Q}_{26} \\ \bar{Q}_{16} & \bar{Q}_{26} & \bar{Q}_{66} \end{bmatrix}_k \left(\int_{h_k}^{h_{k+1}} \begin{Bmatrix} \epsilon_x^o \\ \epsilon_y^o \\ \gamma_{xy}^o \end{Bmatrix} z\, dz - \int_{h_k}^{h_{k+1}} \begin{Bmatrix} \kappa_x \\ \kappa_y \\ \kappa_{xy} \end{Bmatrix} z^2\, dz \right)$$

(5.57)

These can be rewritten as

$$\boldsymbol{F} = \begin{bmatrix} A_{11} & A_{12} & A_{16} \\ A_{12} & A_{22} & A_{26} \\ A_{16} & A_{26} & A_{66} \end{bmatrix} \begin{Bmatrix} \epsilon_x^o \\ \epsilon_y^o \\ \gamma_{xy}^o \end{Bmatrix} + \begin{bmatrix} B_{11} & B_{12} & B_{16} \\ B_{12} & B_{22} & B_{26} \\ B_{16} & B_{26} & B_{66} \end{bmatrix} \begin{Bmatrix} \kappa_x \\ \kappa_y \\ \kappa_{xy} \end{Bmatrix} \text{ (N/m)}$$

and

$$\boldsymbol{M} = \begin{bmatrix} B_{11} & B_{12} & B_{16} \\ B_{12} & B_{22} & B_{26} \\ B_{16} & B_{26} & B_{66} \end{bmatrix} \begin{Bmatrix} \epsilon_x^o \\ \epsilon_y^o \\ \gamma_{xy}^o \end{Bmatrix} + \begin{bmatrix} D_{11} & D_{12} & D_{16} \\ D_{12} & D_{22} & D_{26} \\ D_{16} & D_{26} & D_{66} \end{bmatrix} \begin{Bmatrix} \kappa_x \\ \kappa_y \\ \kappa_{xy} \end{Bmatrix} \text{ (Nm/m)} \quad (5.58)$$

The coefficients are defined as

$$A_{ij} = \sum_{k=1}^{N} (\bar{Q}_{ij})_k (h_{k+1} - h_k) \rightarrow \text{extensional stiffness (N/m)} \quad (5.59)$$

$$B_{ij} = -\frac{1}{2} \sum_{k=1}^{N} (\bar{Q}_{ij})_k (h_{k+1}^2 - h_k^2) \rightarrow \text{coupling stiffness (N)} \quad (5.60)$$

$$D_{ij} = \frac{1}{3} \sum_{k=1}^{N} (\bar{Q}_{ij})_k (h_{k+1}^3 - h_k^3) \rightarrow \text{bending stiffness (Nm)} \quad (5.61)$$

Note that the strain components ϵ_x^o, ϵ_y^o, and γ_{xy}^o are dimensionless, whereas the units of curvatures κ_x, κ_y, and κ_{xy} are 1/m. The $\boldsymbol{B}$ matrix implies coupling between bending and extension. If a laminate that has a non-zero $\boldsymbol{B}$ is subjected to an extensional stress, it will result in not only extensional deformation but also twisting and bending of the laminate.

Putting together the extension and bending equations

$$\begin{Bmatrix} F_x \\ F_y \\ F_{xy} \\ M_x \\ M_y \\ M_{xy} \end{Bmatrix} = \begin{bmatrix} \begin{bmatrix} A_{11} & A_{12} & A_{16} \\ A_{12} & A_{22} & A_{26} \\ A_{16} & A_{26} & A_{66} \end{bmatrix} & \begin{bmatrix} B_{11} & B_{12} & B_{16} \\ B_{12} & B_{22} & B_{26} \\ B_{16} & B_{26} & B_{66} \end{bmatrix} \\ \begin{bmatrix} B_{11} & B_{12} & B_{16} \\ B_{12} & B_{22} & B_{26} \\ B_{16} & B_{26} & B_{66} \end{bmatrix} & \begin{bmatrix} D_{11} & D_{12} & D_{16} \\ D_{12} & D_{22} & D_{26} \\ D_{16} & D_{26} & D_{66} \end{bmatrix} \end{bmatrix} \begin{Bmatrix} \epsilon_x^o \\ \epsilon_y^o \\ \gamma_{xy}^o \\ \kappa_x \\ \kappa_y \\ \kappa_{xy} \end{Bmatrix} \quad (5.62)$$

Eq. 5.62 can be rewritten in a simpler notation as

$$\begin{Bmatrix} \boldsymbol{F} \\ \boldsymbol{M} \end{Bmatrix} = \begin{bmatrix} \boldsymbol{A} & \boldsymbol{B} \\ \boldsymbol{B} & \boldsymbol{D} \end{bmatrix} \begin{Bmatrix} \boldsymbol{\epsilon^o} \\ \boldsymbol{\kappa} \end{Bmatrix} \quad (5.63)$$

Note that

$$\begin{Bmatrix} \epsilon_x^o \\ \epsilon_y^o \\ \gamma_{xy}^o \\ \kappa_x \\ \kappa_y \\ \kappa_{xy} \end{Bmatrix} = \begin{Bmatrix} \frac{\partial u_o}{\partial x} \\ \frac{\partial v_o}{\partial y} \\ \frac{\partial u_o}{\partial y} + \frac{\partial v_o}{\partial x} \\ \frac{\partial^2 w_o}{\partial x^2} \\ \frac{\partial^2 w_o}{\partial y^2} \\ 2\frac{\partial^2 w_o}{\partial x \partial y} \end{Bmatrix} \quad (5.64)$$

where u_o, v_o, and w_o are the mid-plane (neutral plane) displacements. For uncoupled configurations

$$\boldsymbol{B} = 0 \tag{5.65}$$

which results in

$$\boldsymbol{A\epsilon_o} = \boldsymbol{F} \tag{5.66}$$

and

$$\boldsymbol{D}\boldsymbol{\kappa} = \boldsymbol{M} \tag{5.67}$$

A laminate consists of a number of laminae (plies) laid at arbitrary orientations. Depending on the layup, coupled and uncoupled configurations are generated. In a balanced laminate, for every ply in the $+\alpha^\circ$ direction, there is an identical corresponding ply in the $-\alpha^\circ$ direction. An example is $[45^\circ/-45^\circ/-30^\circ/30^\circ/15^\circ/-15^\circ]$. In a cross-ply laminate, 0° and 90° plies are oriented along x and y directions. Note that there is no distinction between $+0^\circ$ and -0° or between the $+90^\circ$ and -90° plies. In an angle-ply laminate, the plies are oriented at non-zero angles.

The sign conventions used for the forces and moments are shown in Fig. 5.5. The plate equilibrium equations can be obtained from the basic elemental equilibrium equations. For a cubic element, the force equilibrium equations are obtained as

$$\frac{\partial \sigma_x}{\partial x} + \frac{\partial \tau_{xy}}{\partial y} + \frac{\partial \tau_{zx}}{\partial z} + f_x = 0 \tag{5.68}$$

$$\frac{\partial \tau_{xy}}{\partial x} + \frac{\partial \sigma_y}{\partial y} + \frac{\partial \tau_{yz}}{\partial z} + f_y = 0 \tag{5.69}$$

$$\frac{\partial \tau_{zx}}{\partial x} + \frac{\partial \tau_{yz}}{\partial y} + \frac{\partial \sigma_z}{\partial z} + f_z = 0 \tag{5.70}$$

where f_x, f_y, and f_z are the body forces per unit volume, respectively, in the x, y, and z directions. These equations are valid for each element of the plate. The following eight stress resultants can be defined.

In-plane forces per unit length (N/m)

$$F_x = \int_t \sigma_x \, dz \tag{5.71}$$

$$F_y = \int_t \sigma_y \, dz \tag{5.72}$$

$$F_{xy} = \int_t \tau_{xy} \, dz \tag{5.73}$$

Bending moments per unit length (Nm/m)

$$M_x = -\int_t \sigma_x z \, dz \tag{5.74}$$

$$M_y = -\int_t \sigma_y z \, dz \tag{5.75}$$

$$M_{xy} = -\int_t \tau_{xy} z \, dz \tag{5.76}$$

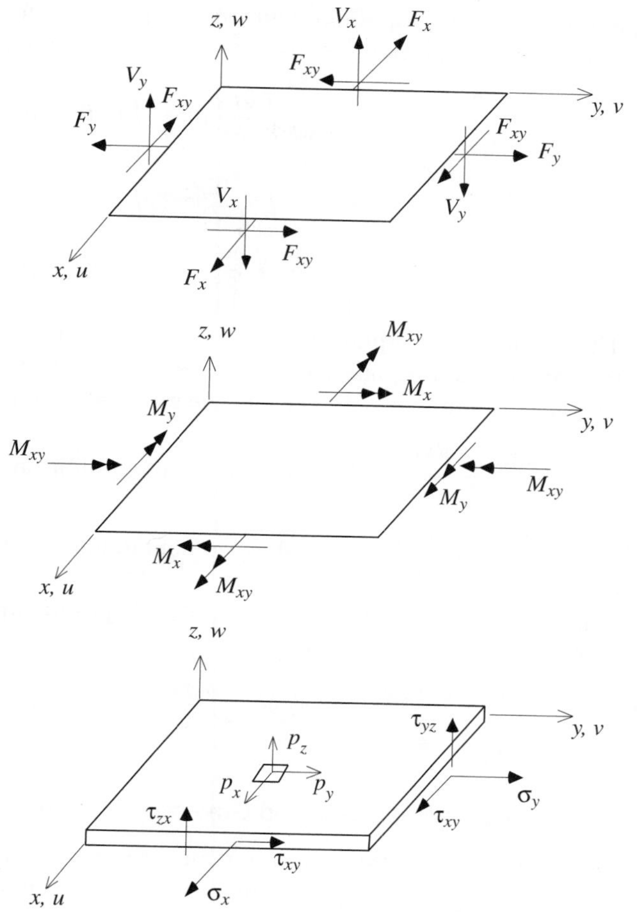

Figure 5.5. Sign convention for forces, stresses, and moments.

Transverse shear forces per unit length (N/m)

$$V_x = -\int_t \tau_{zx}\, dz \tag{5.77}$$

$$V_y = -\int_t \tau_{yz}\, dz \tag{5.78}$$

V_x and V_y are transverse shear forces (per unit length). Positive V_x is defined as pointing in the negative z direction on the $y - z$ plane, and positive V_y is defined as pointing in the negative z direction on the $x - z$ plane. Assuming that the plate is in the $x - y$ plane and integrating the previous equations over the thickness of the plate (integral with respect to the z coordinate) gives the following plate equilibrium equations

$$\frac{\partial F_x}{\partial x} + \frac{\partial F_{xy}}{\partial y} = -p_x \quad \text{(force equilibrium in the } x \text{ direction)} \tag{5.79}$$

$$\frac{\partial F_{xy}}{\partial x} + \frac{\partial F_y}{\partial y} = -p_y \quad \text{(force equilibrium in the } y \text{ direction)} \tag{5.80}$$

$$\frac{\partial V_x}{\partial x} + \frac{\partial V_y}{\partial y} = p_z \quad \text{(force equilibrium in the } z \text{ direction)} \tag{5.81}$$

where F_x, F_y, and F_{xy} are in-plane forces (per unit length). p_x, p_y, and p_z are surface loads (per unit area), respectively, in x, y, and z directions, given by

$$\int_t \frac{\partial \tau_{zx}}{\partial z} dz + \int_t f_x dz = p_x \tag{5.82}$$

$$\int_t \frac{\partial \tau_{yz}}{\partial z} dz + \int_t f_y dz = p_y \tag{5.83}$$

$$\int_t \frac{\partial \sigma_z}{\partial z} dz + \int_t f_z dz = p_z \tag{5.84}$$

The equations for equilibrium of moments can be obtained by multiplying the first two equilibrium equations (Eqs. 5.79 and 5.80) by $-z$ and integrating over the thickness of the plate. As a result, the moment-equilibrium equations are

$$\frac{\partial M_x}{\partial x} + \frac{\partial M_{xy}}{\partial y} - V_x = 0 \quad \text{(moment equilibrium about the } y \text{ axis)} \tag{5.85}$$

$$\frac{\partial M_{xy}}{\partial x} + \frac{\partial M_y}{\partial y} - V_y = 0 \quad \text{(moment equilibrium about the } x \text{ axis)} \tag{5.86}$$

where M_x, M_y, and M_{xy} are moments (per unit length). Combining Eqs. 5.81, 5.85, and 5.86 results in

$$\frac{\partial^2 M_x}{\partial x^2} + \frac{\partial^2 M_y}{\partial y^2} + 2\frac{\partial^2 M_{xy}}{\partial x \partial y} = p_z \tag{5.87}$$

5.1.4 Displacements-Based Governing Equations

Substituting the forces and moments relations from Eq. 5.63 into the equilibrium equations for forces (Eqs. 5.79 and 5.80) and moments (Eq. 5.87), and defining strains in terms of displacements (Eq. 5.50) results in the displacements-based governing equations, given by:

u-equation

$$\begin{aligned}
&A_{11}\frac{\partial^2 u_o}{\partial x^2} + A_{12}\frac{\partial^2 v_o}{\partial x \partial y} + A_{16}\left(2\frac{\partial^2 u_o}{\partial x \partial y} + \frac{\partial^2 v_o}{\partial x^2}\right) + A_{26}\frac{\partial^2 v_o}{\partial y^2} + A_{66}\left(\frac{\partial^2 u_o}{\partial y^2} + \frac{\partial^2 v_o}{\partial x \partial y}\right) \\
&\quad + B_{11}\frac{\partial^3 w_o}{\partial x^3} + B_{12}\frac{\partial^3 w_o}{\partial x \partial y^2} + 3B_{16}\frac{\partial^3 w_o}{\partial x^2 \partial y} + B_{26}\frac{\partial^3 w_o}{\partial y^3} + 2B_{66}\frac{\partial^3 w_o}{\partial x \partial y^2} \\
&\quad = \frac{\partial F_x}{\partial x} + \frac{\partial F_{xy}}{\partial y} - p_x
\end{aligned} \tag{5.88}$$

v-equation

$$\begin{aligned}
&A_{22}\frac{\partial^2 v_o}{\partial y^2} + A_{12}\frac{\partial^2 u_o}{\partial x \partial y} + A_{16}\frac{\partial^2 u_o}{\partial x^2} + A_{26}\left(\frac{\partial^2 u_o}{\partial y^2} + 2\frac{\partial^2 v_o}{\partial x \partial y}\right) + A_{66}\left(\frac{\partial^2 u_o}{\partial x \partial y} + \frac{\partial^2 v_o}{\partial x^2}\right) \\
&\quad + B_{12}\frac{\partial^3 w_o}{\partial x^2 \partial y} + B_{22}\frac{\partial^3 w_o}{\partial y^3} + B_{16}\frac{\partial^3 w_o}{\partial x^3} + 3B_{26}\frac{\partial^3 w_o}{\partial x \partial y^2} + 2B_{66}\frac{\partial^3 w_o}{\partial x^2 \partial y} \\
&\quad = \frac{\partial F_y}{\partial y} + \frac{\partial F_{xy}}{\partial x} - p_y
\end{aligned} \tag{5.89}$$

w-equation

$$\begin{aligned}
&B_{11}\frac{\partial^3 u_o}{\partial x^3} + B_{12}\left(\frac{\partial^3 v_o}{\partial x^2 \partial y} + \frac{\partial^3 u_o}{\partial x \partial y^2}\right) + B_{16}\left(3\frac{\partial^3 u_o}{\partial x^2 \partial y} + \frac{\partial^3 v_o}{\partial x^3}\right) + B_{22}\frac{\partial^3 v_o}{\partial y^3} \\
&+ B_{26}\left(\frac{\partial^3 u_o}{\partial y^3} + 3\frac{\partial^3 v_o}{\partial x \partial y^2}\right) + 2B_{66}\left(\frac{\partial^3 u_o}{\partial x \partial y^2} + \frac{\partial^3 v_o}{\partial x^2 \partial y}\right) \\
&+ D_{11}\frac{\partial^4 w_o}{\partial x^4} + 2D_{12}\frac{\partial^4 w_o}{\partial x^2 \partial y^2} + 4D_{16}\frac{\partial^4 w_o}{\partial x^3 \partial y} + D_{22}\frac{\partial^4 w_o}{\partial y^4} \\
&+ 4D_{26}\frac{\partial^4 w_o}{\partial x \partial y^3} + 4D_{66}\frac{\partial^4 w_o}{\partial x^2 \partial y^2} \\
&= \frac{\partial^2 M_x}{\partial x^2} + \frac{\partial^2 M_y}{\partial y^2} + \frac{\partial^2 M_{xy}}{\partial x \partial y} - p_z
\end{aligned} \tag{5.90}$$

Let us consider an isotropic plate of thickness h with a material of modulus E and Poisson's ration ν. The elements of the stiffness matrices are

$$A_{16} = A_{26} = 0$$

$$A_{11} = A_{22} = \frac{Eh}{1-\nu^2} = A$$

$$A_{12} = \nu A$$

$$A_{66} = \frac{1-\nu}{2}A$$

$$B_{ij} = 0$$

$$D_{11} = D_{22} = \frac{Eh^3}{12(1-\nu^2)} = D$$

$$D_{12} = \nu D$$

$$D_{16} = D_{26} = 0$$

$$D_{66} = \frac{1-\nu}{2}D$$

The forces are given by

$$F_x = A\left(\frac{\partial u_o}{\partial x} + \nu\frac{\partial v_o}{\partial y}\right)$$

$$F_y = A\left(\frac{\partial v_o}{\partial y} + \nu\frac{\partial u_o}{\partial x}\right)$$

$$F_{xy} = \frac{1-\nu}{2}A\left(\frac{\partial u_o}{\partial y} + \frac{\partial v_o}{\partial x}\right)$$

$$V_x = D\left(\frac{\partial^3 w_o}{\partial x^3} + \nu\frac{\partial^3 w_o}{\partial x \partial y^2}\right) + D(1-\nu)\frac{\partial^3 w_o}{\partial x \partial y^2}$$

$$V_y = D\left(\frac{\partial^3 w_o}{\partial y^3} + \nu\frac{\partial^3 w_o}{\partial x^2 \partial y}\right) + D(1-\nu)\frac{\partial^3 w_o}{\partial x^2 \partial y}$$

The moments are given by

$$M_x = D\left(\frac{\partial^2 w}{\partial x^2} + \nu\frac{\partial^2 w}{\partial y^2}\right)$$

$$M_y = D\left(\frac{\partial^2 w}{\partial y^2} + \nu\frac{\partial^2 w}{\partial x^2}\right)$$

$$M_{xy} = D(1-\nu)\frac{\partial^2 w}{\partial x \partial y}$$

Substituting into Eq. 5.87 gives

$$D\left(\frac{\partial^4 w}{\partial x^4} + 2\frac{\partial^4 w}{\partial x^2 \partial y^2} + \frac{\partial^4 w}{\partial y^4}\right) = p_z \qquad (5.91)$$

This is the basic equilibrium equation of plate bending for an isotropic uniform plate with small deflection.

5.1.5 Boundary Conditions

The complete plate problem requires four boundary conditions at each edge: two related to in-plane forces and displacements, and two concerned with out-of-plane displacement and forces. Boundary conditions are broadly classified into geometric (kinematic) and forced boundary conditions. Following is a list of some possible boundary conditions; for example, for a rectangular plate at an edge $x = 0$.

1. Cantilevered (Built-in) Edge: all geometric boundary conditions

$$u_o = 0$$

$$v_o = 0$$

$$w = 0$$

$$\frac{\partial w}{\partial x} = 0$$

2. Free Edge: All forced boundary conditions

$$F_x = 0$$

$$F_{xy} = 0$$

$$M_x = 0$$

$$V_x + \frac{\partial M_{xy}}{\partial y} = 0$$

3. Simply-Supported Edge:
   Case I: In-Plane Motion Freely Permitted. For in-plane response, both are force boundary conditions, and for out-of-plane response, one is force and the second

is a geometric boundary condition

$$F_x = 0$$
$$F_{xy} = 0$$
$$w = 0$$
$$M_x = 0$$

Case II: In-Plane Completely Restrained. Geometric boundary conditions for in-plane displacements

$$u_o = 0$$
$$v_o = 0$$
$$w = 0$$
$$M_x = 0$$

Case III: In-Plane Spring Supported. Linear springs of stiffness k_u and k_v are used to restrain u and v displacements. This results in two in-plane forced-boundary conditions

$$w = 0$$
$$F_x = k_u u$$
$$F_{xy} = k_v v$$
$$M_x = 0$$

This clearly shows that one can have a combination of in-plane and out-of-plane boundary conditions.

4. Bending Spring for Out-of-Plane Displacement and In-Plane Completely Restrained: Bending spring stiffness is k_θ

$$u_o = 0$$
$$v_o = 0$$
$$w = 0$$
$$M_x = k_\theta \frac{\partial w}{\partial x}$$

5. Vertical Spring for Out-of-Plane Displacement and In-Plane Completely Restrained: Linear vertical spring stiffness is k_w. At the root of the plate

$$u_o = 0$$
$$v_o = 0$$
$$M_x = 0$$
$$V_x + \frac{\partial M_{xy}}{\partial y} = -k_w w$$

At the other end of the plate, a similar condition will exist with a difference in sign

$$u_o = 0$$
$$v_o = 0$$
$$M_x = 0$$
$$V_x + \frac{\partial M_{xy}}{\partial y} = k_w w$$

5.2 Plate Theory with Induced-Strain Actuation

Induced-strain actuation can be used to control the extension, bending, and twisting of a plate. With tailored anisotropic plates, control of specific static deformations can be augmented using piezo actuators. Plates with distributed induced-strain actuators have a variety of applications – for example, to control the pointing of precision instruments in space, to control structure-borne noise, and to change aerodynamic shape for vibration reduction, flutter suppression, and gust alleviation.

To develop a consistent plate model with induced strain actuation, the following assumptions are made:

1. Actuators and substrates are integrated as plies of a laminated plate.
2. The displacements in both the actuators and the substrate are defined completely in terms of the deformation of the plate's reference surface.
3. Assumption of thin CLPT is adopted (Kirchhoff-Love hypothesis).

For systems actuated in pure extension, the strains are assumed to be constant across the thickness of the actuators and the plate. For systems actuated in pure bending, strains are assumed to vary linearly through the entire thickness of the structure. The strain in the system therefore depends on the mid-plane strain $\boldsymbol{\epsilon}^o$ and the curvature $\boldsymbol{\kappa}$, as given by

$$\boldsymbol{\epsilon} = \boldsymbol{\epsilon}^o - z\boldsymbol{\kappa} \tag{5.92}$$

The constitutive relation for any ply of a laminated plate with induced-strain actuation is

$$\boldsymbol{\sigma} = \bar{\boldsymbol{Q}}(\boldsymbol{\epsilon} - \boldsymbol{\Lambda}) = \bar{\boldsymbol{Q}}\boldsymbol{\epsilon} - \bar{\boldsymbol{Q}}\boldsymbol{\Lambda} \tag{5.93}$$

where the matrix $\bar{\boldsymbol{Q}}$ is the transformed reduced stiffness matrix of a single ply and the actuation strain vector is

$$\boldsymbol{\Lambda} = \begin{Bmatrix} \Lambda_x \\ \Lambda_y \\ \Lambda_{xy} \end{Bmatrix} \tag{5.94}$$

These are free induced strains consisting of two direct and one shear strain. Integrating through the thickness t of the plate, the forces and moments per unit length of the plate, and the stiffness coefficients can be derived in a manner similar to the passive case. Mechanical forces are

$$F_x = \int_t \sigma_x \, dz, \quad F_y = \int_t \sigma_y \, dz, \quad F_{xy} = \int_t \sigma_{xy} \, dz \tag{5.95}$$

Mechanical moments are

$$M_x = -\int_t \sigma_x z\, dz, \quad M_y = -\int_t \sigma_y z\, dz, \quad M_{xy} = -\int_t \sigma_{xy} z\, dz \tag{5.96}$$

The force vector, $\boldsymbol{F}$, and the moment vector, $\boldsymbol{M}$, can be derived as

$$\begin{aligned} \boldsymbol{F} &= \int_t \bar{\boldsymbol{Q}}(\boldsymbol{\epsilon} - \boldsymbol{\Lambda})\, dz = \int_t \bar{\boldsymbol{Q}}\boldsymbol{\epsilon}^o\, dz - \int_t \bar{\boldsymbol{Q}}\boldsymbol{\kappa}\, z dz - \int_t \bar{\boldsymbol{Q}}\boldsymbol{\Lambda}\, dz \\ &= \boldsymbol{A}\boldsymbol{\epsilon}^o + \boldsymbol{B}\boldsymbol{\kappa} - \boldsymbol{F}_\Lambda \text{ (N/m)} \end{aligned} \tag{5.97}$$

and

$$\begin{aligned} \boldsymbol{M} &= -\int_t \bar{\boldsymbol{Q}}(\boldsymbol{\epsilon} - \boldsymbol{\Lambda})\, z\, dz = \int_t \bar{\boldsymbol{Q}}\boldsymbol{\epsilon}^o\, z dz + \int_t \bar{\boldsymbol{Q}}\boldsymbol{\kappa}\, z^2 dz + \int_t \bar{\boldsymbol{Q}}\boldsymbol{\Lambda}\, z dz \\ &= \boldsymbol{B}\boldsymbol{\epsilon}^o + \boldsymbol{D}\boldsymbol{\kappa} - \boldsymbol{M}_\Lambda \text{ (Nm/m)} \end{aligned} \tag{5.98}$$

From these equations, the stiffness matrices, and the induced force and moment vectors can be derived.

Extensional stiffness:

$$\boldsymbol{A} = \int_t \bar{\boldsymbol{Q}}\, dz \rightarrow A_{ij} = \sum_{k=1}^{N} (\bar{Q}_{ij})_k (h_{k+1} - h_k) \text{ (N/m)} \tag{5.99}$$

Coupling stiffness:

$$\boldsymbol{B} = -\int_t \bar{\boldsymbol{Q}} z dz \rightarrow B_{ij} = -\frac{1}{2}\sum_{k=1}^{N} (\bar{Q}_{ij})_k (h_{k+1}^2 - h_k^2) \text{ (N)} \tag{5.100}$$

Bending stiffness:

$$\boldsymbol{D} = \int_t \bar{\boldsymbol{Q}} z^2 dz \rightarrow D_{ij} = \frac{1}{3}\sum_{k=1}^{N} (\bar{Q}_{ij})_k (h_{k+1}^3 - h_k^3) \text{ (Nm)} \tag{5.101}$$

Induced force vector:

$$\begin{aligned} \boldsymbol{F}_\Lambda &= \int_t \bar{\boldsymbol{Q}}\boldsymbol{\Lambda}\, dz \\ &= \sum_{k=1}^{N} \bar{\boldsymbol{Q}}_k \boldsymbol{\Lambda}_k (h_{k+1} - h_k) \text{ (N/m)} \end{aligned} \tag{5.102}$$

Induced moment vector:

$$\begin{aligned} \boldsymbol{M}_\Lambda &= -\int_t \bar{\boldsymbol{Q}}\boldsymbol{\Lambda} z\, dz \\ &= -\frac{1}{2}\sum_{k=1}^{N} \bar{\boldsymbol{Q}}_k \boldsymbol{\Lambda}_k (h_{k+1}^2 - h_k^2) \text{ (Nm/m)} \end{aligned} \tag{5.103}$$

Both the actuator and the substrate plies contribute to the stiffness in the force and moment equations; however, only the active plies contribute to the forcing functions $\boldsymbol{F}_\Lambda$ and $\boldsymbol{M}_\Lambda$. The total governing equation is

$$\begin{Bmatrix} \boldsymbol{F} \\ \boldsymbol{M} \end{Bmatrix} = \begin{bmatrix} \boldsymbol{A} & \boldsymbol{B} \\ \boldsymbol{B} & \boldsymbol{D} \end{bmatrix} \begin{Bmatrix} \boldsymbol{\epsilon}^o \\ \boldsymbol{\kappa} \end{Bmatrix} - \begin{Bmatrix} \boldsymbol{F}_\Lambda \\ \boldsymbol{M}_\Lambda \end{Bmatrix} \tag{5.104}$$

Expanding the entire set of equations

$$\begin{Bmatrix} F_x \\ F_y \\ F_{xy} \\ M_x \\ M_y \\ M_{xy} \end{Bmatrix} = \begin{bmatrix} \begin{bmatrix} A_{11} & A_{12} & A_{16} \\ A_{12} & A_{22} & A_{26} \\ A_{16} & A_{26} & A_{66} \end{bmatrix} & \begin{bmatrix} B_{11} & B_{12} & B_{16} \\ B_{12} & B_{22} & B_{26} \\ B_{16} & B_{26} & B_{66} \end{bmatrix} \\ \begin{bmatrix} B_{11} & B_{12} & B_{16} \\ B_{12} & B_{22} & B_{26} \\ B_{16} & B_{26} & B_{66} \end{bmatrix} & \begin{bmatrix} D_{11} & D_{12} & D_{16} \\ D_{12} & D_{22} & D_{26} \\ D_{16} & D_{26} & D_{66} \end{bmatrix} \end{bmatrix} \begin{Bmatrix} \epsilon_x^o \\ \epsilon_y^o \\ \gamma_{xy}^o \\ \kappa_x \\ \kappa_y \\ \kappa_{xy} \end{Bmatrix} - \begin{Bmatrix} F_{x_\Lambda} \\ F_{y_\Lambda} \\ F_{xy_\Lambda} \\ M_{x_\Lambda} \\ M_{y_\Lambda} \\ M_{xy_\Lambda} \end{Bmatrix} \tag{5.105}$$

With no external mechanical forces, these equations reduce to

$$\begin{Bmatrix} F_{x_\Lambda} \\ F_{y_\Lambda} \\ F_{xy_\Lambda} \\ M_{x_\Lambda} \\ M_{y_\Lambda} \\ M_{xy_\Lambda} \end{Bmatrix} = \begin{bmatrix} \begin{bmatrix} A_{11} & A_{12} & A_{16} \\ A_{12} & A_{22} & A_{26} \\ A_{16} & A_{26} & A_{66} \end{bmatrix} & \begin{bmatrix} B_{11} & B_{12} & B_{16} \\ B_{12} & B_{22} & B_{26} \\ B_{16} & B_{26} & B_{66} \end{bmatrix} \\ \begin{bmatrix} B_{11} & B_{12} & B_{16} \\ B_{12} & B_{22} & B_{26} \\ B_{16} & B_{26} & B_{66} \end{bmatrix} & \begin{bmatrix} D_{11} & D_{12} & D_{16} \\ D_{12} & D_{22} & D_{26} \\ D_{16} & D_{26} & D_{66} \end{bmatrix} \end{bmatrix} \begin{Bmatrix} \epsilon_x^o \\ \epsilon_y^o \\ \gamma_{xy}^o \\ \kappa_x \\ \kappa_y \\ \kappa_{xy} \end{Bmatrix} \tag{5.106}$$

The vector on the left-hand side represents generalized induced forces and the vector on the right-hand side represents generalized strains. Inverting these equations results in the vector of strains and curvatures in terms of the induced forces and moments. A_{ij} are the in-plane extensional stiffness terms that relate the in-plane induced forces F_{x_Λ}, F_{y_Λ}, and F_{xy_Λ} to the in-plane strains ϵ_x^o, ϵ_y^o, and γ_{xy}^o, and D_{ij} are the bending stiffness terms that relate the induced moments M_{x_Λ}, M_{y_Λ}, and M_{xy_Λ} to the curvatures κ_x, κ_y, and κ_{xy}. Examining these matrices, different types of couplings can be identified, as follows:

(a) **Extension-shear couplings** due to A_{16} and A_{26}: In-plane induced forces F_{x_Λ}, F_{y_Λ} cause shear deformation γ_{xy}^o. Normally, the induced shear force F_{xy_Λ} is zero; however, if it exists, then extensional strains ϵ_x^o and ϵ_y^o are produced.
(b) **Bending-torsion couplings** due to D_{16} and D_{26}: Induced moments M_{x_Λ}, M_{y_Λ} cause twisting (κ_{xy}) of the laminate. Normally, induced twisting M_{xy_Λ} is zero. However, if M_{xy_Λ} exists, these couplings would result in curvatures κ_x and κ_y.
(c) **Extension-torsion couplings** due to B_{16} and B_{26}: Induced forces F_{x_Λ}, F_{y_Λ} cause twisting (κ_{xy}) of the laminate and induced moments M_{x_Λ}, M_{y_Λ} result in shear strain γ_{xy}^o. They are also called bending-shear couplings.
(d) **Extension-bending couplings** due to B_{11} and B_{12}: Induced forces F_{x_Λ} and F_{y_Λ} cause out-of-plane deformation (bending curvatures κ_x and κ_y) and induced moments M_{x_Λ} and M_{y_Λ} cause in-plane deformations in the $x-y$ plane. This is also known as in-plane–out-of-plane coupling.
(e) **Extension-extension couplings** due to A_{12}: The induced force F_{x_Λ} causes deformation in the y direction and induced force F_{y_Λ} causes deformation in the x direction.
(f) **Bending-bending couplings** due to D_{12}: The induced bending moment M_{x_Λ} causes bending deformation (curvature) in the y direction (in plane $y-z$) κ_y and the induced bending moment M_{y_Λ} causes curvature κ_x.

Similarly, the strains can be written in terms of the forces and moments as

$$\begin{Bmatrix} \epsilon_x^o \\ \epsilon_y^o \\ \gamma_{xy}^o \\ \kappa_x \\ \kappa_y \\ \kappa_{xy} \end{Bmatrix} = \begin{bmatrix} \begin{bmatrix} \alpha_{11} & \alpha_{12} & \alpha_{16} \\ \alpha_{12} & \alpha_{22} & \alpha_{26} \\ \alpha_{16} & \alpha_{26} & \alpha_{66} \end{bmatrix} & \begin{bmatrix} \beta_{11} & \beta_{12} & \beta_{16} \\ \beta_{12} & \beta_{22} & \beta_{26} \\ \beta_{16} & \beta_{26} & \beta_{66} \end{bmatrix} \\ \begin{bmatrix} \beta_{11} & \beta_{12} & \beta_{16} \\ \beta_{12} & \beta_{22} & \beta_{26} \\ \beta_{16} & \beta_{26} & \beta_{66} \end{bmatrix} & \begin{bmatrix} \delta_{11} & \delta_{12} & \delta_{16} \\ \delta_{12} & \delta_{22} & \delta_{26} \\ \delta_{16} & \delta_{26} & \delta_{66} \end{bmatrix} \end{bmatrix} \begin{Bmatrix} F_{x_\Lambda} \\ F_{y_\Lambda} \\ F_{xy_\Lambda} \\ M_{x_\Lambda} \\ M_{y_\Lambda} \\ M_{xy_\Lambda} \end{Bmatrix} \tag{5.107}$$

where

$$\begin{bmatrix} \begin{bmatrix} \alpha_{11} & \alpha_{12} & \alpha_{16} \\ \alpha_{12} & \alpha_{22} & \alpha_{26} \\ \alpha_{16} & \alpha_{26} & \alpha_{66} \end{bmatrix} & \begin{bmatrix} \beta_{11} & \beta_{12} & \beta_{16} \\ \beta_{12} & \beta_{22} & \beta_{26} \\ \beta_{16} & \beta_{26} & \beta_{66} \end{bmatrix} \\ \begin{bmatrix} \beta_{11} & \beta_{12} & \beta_{16} \\ \beta_{12} & \beta_{22} & \beta_{26} \\ \beta_{16} & \beta_{26} & \beta_{66} \end{bmatrix} & \begin{bmatrix} \delta_{11} & \delta_{12} & \delta_{16} \\ \delta_{12} & \delta_{22} & \delta_{26} \\ \delta_{16} & \delta_{26} & \delta_{66} \end{bmatrix} \end{bmatrix}$$

$$= \begin{bmatrix} \begin{bmatrix} A_{11} & A_{12} & A_{16} \\ A_{12} & A_{22} & A_{26} \\ A_{16} & A_{26} & A_{66} \end{bmatrix} & \begin{bmatrix} B_{11} & B_{12} & B_{16} \\ B_{12} & B_{22} & B_{26} \\ B_{16} & B_{26} & B_{66} \end{bmatrix} \\ \begin{bmatrix} B_{11} & B_{12} & B_{16} \\ B_{12} & B_{22} & B_{26} \\ B_{16} & B_{26} & B_{66} \end{bmatrix} & \begin{bmatrix} D_{11} & D_{12} & D_{16} \\ D_{12} & D_{22} & D_{26} \\ D_{16} & D_{26} & D_{66} \end{bmatrix} \end{bmatrix}^{-1} \tag{5.108}$$

For uncoupled configurations with no external mechanical forces

$$\boldsymbol{B} = 0$$

$$\boldsymbol{F} = 0$$

$$\boldsymbol{M} = 0$$

resulting in

$$\boldsymbol{A\epsilon^o} = \boldsymbol{F}_\Lambda \quad \text{and} \quad \boldsymbol{D\kappa} = \boldsymbol{M}_\Lambda \tag{5.109}$$

5.2.1 Isotropic Plate: Symmetric Actuation (Extension)

Consider an isotropic plate with identical piezo-sheet actuators bonded to the top and bottom surfaces (Fig. 5.6). The width of the plate is b, the thickness of the plate is t_b, and the thickness of each piezo sheet is t_c. The same voltage applied to both of the actuators causes a pure extension of the plate.

The assumptions are as follows:

1. Free-free isotropic plate.
2. No externally applied loads.
3. Piezo sheet is isotropic in the 1–2 (x, y) plane.

The plate consists of three plies: two are active plies and one is a passive ply. Note that the piezo actuator induces isotropic in-plane strains and cannot induce a

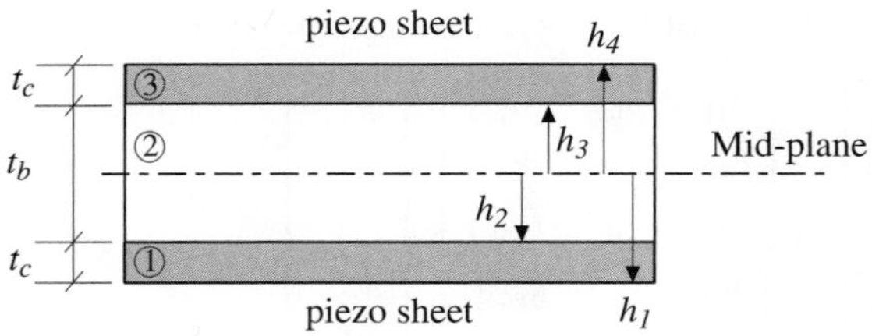

Figure 5.6. Isotropic plate with surface-bonded piezo sheets.

shear strain. Therefore

$$\mathbf{\Lambda}_1 = \begin{Bmatrix} \Lambda \\ \Lambda \\ 0 \end{Bmatrix} \quad \mathbf{\Lambda}_2 = \begin{Bmatrix} 0 \\ 0 \\ 0 \end{Bmatrix} \quad \mathbf{\Lambda}_3 = \begin{Bmatrix} \Lambda \\ \Lambda \\ 0 \end{Bmatrix} \tag{5.110}$$

Assuming that the Poisson's ratios of actuator and substrate are identical, the stresses in each ply are given by (Eq. 5.14)

$$\begin{Bmatrix} \sigma_x \\ \sigma_y \\ \tau_{xy} \end{Bmatrix} = \frac{E}{1-\nu^2} \begin{bmatrix} 1 & \nu & 0 \\ \nu & 1 & 0 \\ 0 & 0 & \frac{1-\nu}{2} \end{bmatrix} \begin{Bmatrix} \epsilon_x^o \\ \epsilon_y^o \\ \gamma_{xy}^o \end{Bmatrix} \tag{5.111}$$

where $E = E_b$ for the plate and $E = E_c$ for the actuator. For an isotropic plate

$$\boldsymbol{B} = 0 \tag{5.112}$$

Therefore, there is no coupling between bending and extension. The resultant induced-strain equation becomes

$$\boldsymbol{A}\boldsymbol{\epsilon}^{\boldsymbol{o}} = \boldsymbol{F}_\Lambda \tag{5.113}$$

The coefficients of matrix $\boldsymbol{A}$ are defined as

$$\begin{aligned} A_{11} &= \sum_{k=1}^{3} (Q_{11})_k (h_{k+1} - h_k) \\ &= \frac{E_c}{1-\nu^2}(-t_b/2 + t_b/2 + t_c) + \frac{E_b}{1-\nu^2}(t_b/2 + t_b/2) + \frac{E_c}{1-\nu^2}(t_b/2 + t_c - t_b/2) \\ &= \frac{E_b t_b}{1-\nu^2} + \frac{2E_c t_c}{1-\nu^2} \end{aligned} \tag{5.114}$$

Similarly, for the other coefficients

$$A_{22} = A_{11} \tag{5.115}$$

$$A_{12} = A_{21} = \frac{\nu}{1-\nu^2}(E_b t_b + 2E_c t_c) \tag{5.116}$$

$$A_{33} = \frac{1-\nu}{2(1-\nu^2)}(E_b t_b + 2E_c t_c) \tag{5.117}$$

The induced force is expressed as

$$\begin{aligned} \boldsymbol{F}_\Lambda &= \sum_{k=1}^{3} \boldsymbol{Q}_k \mathbf{\Lambda}_k (h_{k+1} - h_k) \\ &= \frac{2E_c t_c}{1-\nu^2} \begin{bmatrix} 1 & \nu & 0 \\ \nu & 1 & 0 \\ 0 & 0 & \frac{1-\nu}{2} \end{bmatrix} \begin{Bmatrix} \Lambda \\ \Lambda \\ 0 \end{Bmatrix} = \frac{2E_c t_c \Lambda}{1-\nu} \begin{Bmatrix} 1 \\ 1 \\ 0 \end{Bmatrix} \end{aligned} \tag{5.118}$$

Note that there is no shear actuation and no coupling between extension and shear ($A_{16} = A_{26} = 0$). The shear terms in the equation therefore can be ignored.

$$\gamma^o_{xy} = 0 \tag{5.119}$$

Because the configuration is symmetric, there is no bending.

$$\boldsymbol{M}_\Lambda = 0 \qquad \kappa = 0 \tag{5.120}$$

The extensional strains are calculated by Eq. 5.109. Substituting these terms

$$\begin{bmatrix} A_{11} & A_{12} \\ A_{12} & A_{22} \end{bmatrix} \begin{Bmatrix} \epsilon^o_x \\ \epsilon^o_y \end{Bmatrix} = \frac{2E_c t_c \Lambda}{1-\nu} \begin{Bmatrix} 1 \\ 1 \end{Bmatrix}$$

This reduces to

$$\frac{(E_b t_b + 2E_c t_c)}{1-\nu^2} \begin{bmatrix} 1 & \nu \\ \nu & 1 \end{bmatrix} \begin{Bmatrix} \epsilon^o_x \\ \epsilon^o_y \end{Bmatrix} = \frac{2E_c t_c \Lambda}{1-\nu} \begin{Bmatrix} 1 \\ 1 \end{Bmatrix}$$

from which the extensional strains are

$$\begin{aligned} \begin{Bmatrix} \epsilon^o_x \\ \epsilon^o_y \end{Bmatrix} &= \frac{2E_c t_c \Lambda}{1-\nu} \frac{1-\nu^2}{E_b t_b + 2E_c t_c} \begin{bmatrix} 1 & \nu \\ \nu & 1 \end{bmatrix}^{-1} \begin{Bmatrix} 1 \\ 1 \end{Bmatrix} \\ &= \frac{2E_c t_c \Lambda}{E_b t_b + 2E_c t_c} \begin{Bmatrix} 1 \\ 1 \end{Bmatrix} \end{aligned} \tag{5.121}$$

Defining

$$\alpha_e = 2, \quad \psi = \frac{E_b t_b}{E_c t_c}$$

we obtain

$$\boldsymbol{\epsilon^o} = \begin{Bmatrix} \epsilon^o_x \\ \epsilon^o_y \end{Bmatrix} = \frac{\alpha_e \Lambda}{\alpha_e + \psi} \begin{Bmatrix} 1 \\ 1 \end{Bmatrix} \tag{5.122}$$

As the stiffness of the plate decreases compared to that of the actuator (ψ decreases), the strain transfer from actuator to plate increases.

5.2.2 Isotropic Plate: Antisymmetric Actuation (Bending)

The same assumptions are made as in the previous case (see Section 5.2.1). To induce pure bending, the top and bottom piezos are actuated by equal voltages of opposite polarity. Assume that a positive voltage is applied to the bottom piezo, causing an extensional strain on the bottom surface, and that a negative voltage is applied to the top piezo, causing a compressive strain on the top surface. Because there is no induced shear strain and there is no structural coupling, the shear terms can be ignored completely.

$$\Lambda_{xy} = 0 \quad \text{and} \quad D_{16} = D_{26} = 0 \rightarrow k_{xy} = 0$$

Therefore

$$\boldsymbol{F}_\Lambda = 0 \rightarrow \boldsymbol{\epsilon^o} = 0 \tag{5.123}$$

Actuation strains are

$$\boldsymbol{\Lambda}_1 = \begin{Bmatrix} \Lambda \\ \Lambda \\ 0 \end{Bmatrix} \quad \boldsymbol{\Lambda}_2 = \begin{Bmatrix} 0 \\ 0 \\ 0 \end{Bmatrix} \quad \boldsymbol{\Lambda}_3 = \begin{Bmatrix} -\Lambda \\ -\Lambda \\ 0 \end{Bmatrix} \tag{5.124}$$

In the symmetric case, $\boldsymbol{B} = 0$. The bending equation becomes

$$\boldsymbol{D}\boldsymbol{\kappa} = \boldsymbol{M}_\Lambda \tag{5.125}$$

From Eq. 5.14, the reduced-stiffness matrix is

$$\boldsymbol{Q} = \frac{E}{1-\nu^2}\begin{bmatrix} 1 & \nu \\ \nu & 1 \end{bmatrix} \tag{5.126}$$

where $E = E_b$ for the plate and $E = E_c$ for the actuator plies. Coefficients of the matrix $\boldsymbol{D}$ are

$$\begin{aligned} D_{11} &= \frac{1}{3}\sum_{k=1}^{3}(Q_{11})_k(h_{k+1}^3 - h_k^3) \\ &= \frac{1}{3}\frac{E_c}{(1-\nu^2)}\left[(-t_b/2)^3 - (-t_b/2 - t_c)^3\right] \\ &\quad + \frac{1}{3}\frac{E_b}{(1-\nu^2)}\left[(t_b/2)^3 - (-t_b/2)^3\right] \\ &\quad + \frac{1}{3}\frac{E_c}{(1-\nu^2)}\left[(t_b/2 + t_c)^3 - (t_b/2)^3\right] \\ &= \frac{E_b t_b^3}{12(1-\nu^2)} + \frac{2E_c t_c}{3(1-\nu^2)}\left(\frac{3}{4}t_b^2 + \frac{3}{2}t_b t_c + t_c^2\right) \end{aligned} \tag{5.127}$$

$$\begin{aligned} D_{12} &= \frac{\nu E_b t_b^3}{12(1-\nu^2)} + \frac{2\nu E_c t_c}{3(1-\nu^2)}\left(\frac{3}{4}t_b^2 + \frac{3}{2}t_b t_c + t_c^2\right) \\ &= \nu D_{11} \end{aligned} \tag{5.128}$$

$$D_{22} = D_{11} \tag{5.129}$$

From these equations, $\boldsymbol{D}$ can be written as

$$\boldsymbol{D} = \left[\frac{E_b t_b^3}{12(1-\nu^2)} + \frac{2E_c t_c}{3(1-\nu^2)}\left(\frac{3}{4}t_b^2 + \frac{3}{2}t_b t_c + t_c^2\right)\right]\begin{bmatrix} 1 & \nu \\ \nu & 1 \end{bmatrix} \tag{5.130}$$

The induced moment becomes

$$\begin{aligned} \boldsymbol{M}_\Lambda &= -\frac{1}{2}\sum_{k=1}^{3}\boldsymbol{Q}_k\boldsymbol{\Lambda}_k(h_{k+1}^2 - h_k^2) \\ &= \frac{1}{2}\frac{E_c}{1-\nu^2}\begin{bmatrix} 1 & \nu \\ \nu & 1 \end{bmatrix}\begin{Bmatrix} \Lambda \\ \Lambda \end{Bmatrix} 2\left[(t_b/2 + t_c)^2 - (t_b/2)^2\right] \\ &= \frac{E_c}{1-\nu^2}\begin{bmatrix} 1 & \nu \\ \nu & 1 \end{bmatrix}\begin{Bmatrix} \Lambda \\ \Lambda \end{Bmatrix} t_c(t_b + t_c) \\ &= \frac{E_c t_c(t_b + t_c)\Lambda}{1-\nu}\begin{Bmatrix} 1 \\ 1 \end{Bmatrix} \end{aligned} \tag{5.131}$$

The induced bending equation becomes

$$\left[\frac{E_b t_b^3}{12(1-\nu^2)} + \frac{2E_c t_c}{3(1-\nu^2)}\left(\frac{3}{4}t_b^2 + \frac{3}{2}t_b t_c + t_c^2\right)\right]\begin{bmatrix}1 & \nu\\ \nu & 1\end{bmatrix}\begin{Bmatrix}\kappa_x\\ \kappa_y\end{Bmatrix} = \frac{E_c t_c (t_b + t_c)\Lambda}{1-\nu}\begin{Bmatrix}1\\1\end{Bmatrix} \tag{5.132}$$

from which

$$\begin{Bmatrix}\kappa_x\\ \kappa_y\end{Bmatrix} = \frac{1}{\frac{E_b t_b^3}{12(1-\nu^2)} + \frac{2E_c t_c}{3(1-\nu^2)}\left(\frac{3}{4}t_b^2 + \frac{3}{2}t_b t_c + t_c^2\right)}\frac{E_c t_c}{1-\nu^2}(t_b + t_c)\Lambda\begin{Bmatrix}1\\1\end{Bmatrix} \tag{5.133}$$

defining

$$\theta_b = \frac{t_b}{t_c} \tag{5.134}$$

$$\psi = \frac{E_b t_b}{E_c t_c} \tag{5.135}$$

$$\alpha_b = 6 \tag{5.136}$$

Dividing both top and bottom parts by $\frac{E_c t_c t_b^2}{12(1-\nu^2)}$ gives

$$\begin{aligned}\begin{Bmatrix}\kappa_x\\ \kappa_y\end{Bmatrix} &= \frac{\frac{12}{t_b}(1 + t_c/t_b)\Lambda}{\frac{E_b t_b}{E_c t_c} + 8(\frac{3}{4} + \frac{3}{2}\frac{t_c}{t_b} + \frac{t_c^2}{t_b^2})}\begin{Bmatrix}1\\1\end{Bmatrix}\\ &= \frac{\frac{2\alpha_b}{t_b}(1 + \frac{1}{\theta_b})\Lambda}{\psi + \alpha_b(\frac{4}{3}\frac{1}{\theta_b^2} + \frac{2}{\theta_b} + 1)}\begin{Bmatrix}1\\1\end{Bmatrix}\end{aligned} \tag{5.137}$$

Once again, the bending strain transfer from actuators to plate increases as the plate stiffness decreases with respect to actuator stiffness.

5.2.3 Worked Example

(a) Using laminated plate theory, derive extension-bending equations for a rectangular isotropic plate with a piezo-sheet actuator bonded only on its bottom surface (Fig. 5.7).

(b) Calculate curvature and extension strain at the mid-point of this free-free aluminum plate of size 0.3048 m (12 in) × 0.3048 m (12 in) × 0.79 mm (1/32 in) with a piezo sheet (PZT-5H) of thickness 0.32 mm (0.0125 in) for a voltage of 150 V.

Manufacturer-supplied material constants are as follows

$$E_c = E_b = 72.4 \text{ GPa } (10.5 \times 10^6 \text{ lb/in}^2)$$

$$d_{31} = -274 \times 10^{-12} \text{ m/V}$$

$$\nu_b = \nu_c = 0.3$$

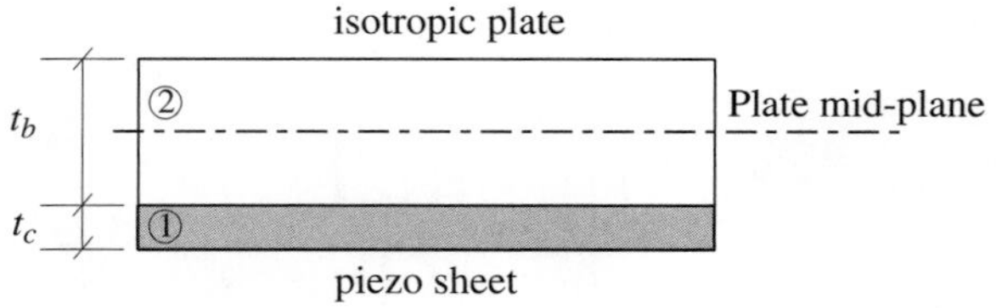

Figure 5.7. Rectangular isotropic plate with one piezo sheet.

Solution

Because the plate is isotropic and there is no induced shear strain, terms related to γ_{xy}^o and κ_{xy} can be ignored.

$\gamma_{xy}^o = 0$ and $\kappa_{xy} = 0$

(a) The governing equation is

$$\begin{bmatrix} \boldsymbol{A} & \boldsymbol{B} \\ \boldsymbol{B} & \boldsymbol{D} \end{bmatrix} \begin{Bmatrix} \boldsymbol{\epsilon^o} \\ \boldsymbol{\kappa} \end{Bmatrix} = \begin{Bmatrix} \boldsymbol{F_\Lambda} \\ \boldsymbol{M_\Lambda} \end{Bmatrix}$$

The stiffnesses are given by

$$\boldsymbol{A} = \sum_{k=1}^{2} \boldsymbol{Q}_k (h_{k+1} - h_k)$$

$$= \frac{1}{1-\nu^2} \begin{bmatrix} 1 & \nu \\ \nu & 1 \end{bmatrix} [E_c\left((-t_b/2) - (-t_b/2 - t_c)\right) + E_b\left(t_b/2 - (-t_b/2)\right)]$$

$$= \frac{E_c t_c + E_b t_b}{1-\nu^2} \begin{bmatrix} 1 & \nu \\ \nu & 1 \end{bmatrix}$$

$$\boldsymbol{B} = -\frac{1}{2} \sum_{k=1}^{2} \boldsymbol{Q}_k (h_{k+1}^2 - h_k^2)$$

$$= -\frac{1}{2(1-\nu^2)} \begin{bmatrix} 1 & \nu \\ \nu & 1 \end{bmatrix} [E_c\left((-t_b/2)^2 - (-t_b/2 - t_c)^2\right)$$

$$+ \; E_b\left((t_b/2)^2 - (-t_b/2)^2\right)]$$

$$= \frac{E_c t_c (t_c + t_b)}{2(1-\nu^2)} \begin{bmatrix} 1 & \nu \\ \nu & 1 \end{bmatrix}$$

$$\boldsymbol{D} = \frac{1}{3} \sum_{k=1}^{2} \boldsymbol{Q}_k (h_{k+1}^3 - h_k^3)$$

$$= \frac{1}{3(1-\nu^2)} \begin{bmatrix} 1 & \nu \\ \nu & 1 \end{bmatrix} [E_b\left((t_b/2)^3 - (-t_b/2)^3\right)$$

$$+ \, E_c\left((-t_b/2)^3 - (-t_b/2 - t_c)^3\right)]$$

$$= \left[\frac{E_b t_b^3}{12(1-\nu^2)} + \frac{E_c t_c}{3(1-\nu^2)} \left(\frac{3}{4} t_b^2 + \frac{3}{2} t_b t_c + t_c^2\right)\right] \begin{bmatrix} 1 & \nu \\ \nu & 1 \end{bmatrix}$$

and the forcing terms are

$$\begin{aligned}
\boldsymbol{F}_\Lambda &= \sum_{k=1}^{2} \boldsymbol{Q}_k \boldsymbol{\Lambda}_k (h_{k+1} - h_k) \\
&= \frac{E_c}{1-\nu^2} \begin{bmatrix} 1 & \nu \\ \nu & 1 \end{bmatrix} \begin{Bmatrix} \Lambda \\ \Lambda \end{Bmatrix} ((-t_b/2) - (-t_b/2 - t_c)) \\
&= \frac{E_c t_c \Lambda}{1-\nu} \begin{Bmatrix} 1 \\ 1 \end{Bmatrix}
\end{aligned}$$

$$\begin{aligned}
\boldsymbol{M}_\Lambda &= -\frac{1}{2} \sum_{k=1}^{2} \boldsymbol{Q}_k \boldsymbol{\Lambda}_k \left(h_{k+1}^2 - h_k^2\right) \\
&= -\frac{E_c}{2(1-\nu^2)} \begin{bmatrix} 1 & \nu \\ \nu & 1 \end{bmatrix} \begin{Bmatrix} \Lambda \\ \Lambda \end{Bmatrix} ((-t_b/2)^2 - (-t_b/2 - t_c)^2) \\
&= \frac{E_c t_c (t_b + t_c) \Lambda}{2(1-\nu)} \begin{Bmatrix} 1 \\ 1 \end{Bmatrix}
\end{aligned}$$

(b) Substituting the given values for the material data, the stiffnesses can be calculated to be

$$\boldsymbol{A} = \begin{bmatrix} 88.4115 & 26.5235 \\ 26.5235 & 88.4115 \end{bmatrix} \times 10^6 \text{ N/m} \quad \boldsymbol{B} = \begin{bmatrix} 0.0140 & 0.0042 \\ 0.0042 & 0.0140 \end{bmatrix} \times 10^6 \text{ N}$$

$$\boldsymbol{D} = \begin{bmatrix} 11.3262 & 3.3979 \\ 3.3979 & 11.3262 \end{bmatrix} \text{ Nm}$$

The induced force and moments are

$$\boldsymbol{F}_\Lambda = \begin{Bmatrix} -4250.9 \\ -4250.9 \end{Bmatrix} \text{ N/m} \quad \boldsymbol{M}_\Lambda = \begin{Bmatrix} -2.3619 \\ -2.3619 \end{Bmatrix} \text{ Nm/m}$$

The strains and curvature of the plate are found from

$$\begin{aligned}
\begin{Bmatrix} \epsilon_x^o \\ \epsilon_y^o \\ \kappa_x \\ \kappa_y \end{Bmatrix} &= \begin{bmatrix} \boldsymbol{A} & \boldsymbol{B} \\ \boldsymbol{B} & \boldsymbol{D} \end{bmatrix}^{-1} \begin{Bmatrix} \boldsymbol{F}_\Lambda \\ \boldsymbol{M}_\Lambda \end{Bmatrix} \\
&= \begin{Bmatrix} -14.34 \times 10^{-6} \\ -14.34 \times 10^{-6} \\ -0.143 \text{ 1/m} \\ -0.143 \text{ 1/m} \end{Bmatrix}
\end{aligned}$$

5.2.4 Single-Layer Specially Orthotropic Plate (Extension)

Consider a free-free plate that consists of a single specially orthotropic ply (with fibers aligned parallel to the x direction). In a specially orthotropic plate, the principal axes of the lamina are aligned with the coordinate axes. Therefore

$$Q_{16} = Q_{26} = 0 \tag{5.138}$$

As a result, the plate does not have any coupling between extension and shear. Two identical piezo actuators are bonded to the top and bottom surfaces, similar to the configuration shown in Fig. 5.6, with the exception that in the present case, the plate is specially orthotropic. Because the structure is symmetric, $\boldsymbol{B} = 0$, and there is no extension-bending coupling. The same assumptions as in Section 5.2.1 are valid. For pure extension, the actuation strains can be expressed as

$$\boldsymbol{\Lambda}_1 = \begin{Bmatrix} \Lambda \\ \Lambda \\ 0 \end{Bmatrix} \quad \boldsymbol{\Lambda}_2 = \begin{Bmatrix} 0 \\ 0 \\ 0 \end{Bmatrix} \quad \boldsymbol{\Lambda}_3 = \begin{Bmatrix} \Lambda \\ \Lambda \\ 0 \end{Bmatrix} \tag{5.139}$$

Note that there is no actuation of the shear components and, because the plate does not have any extension-shear coupling, the shear terms are equal to zero and can be ignored. The stiffness matrix of the plate is given by (Eq. 5.12)

$$Q_{11} = \frac{E_1}{1 - \nu_{12}\nu_{21}} \tag{5.140}$$

$$Q_{22} = \frac{E_2}{1 - \nu_{12}\nu_{22}} \tag{5.141}$$

$$Q_{12} = \frac{\nu_{21}E_1}{1 - \nu_{12}\nu_{21}} = \frac{\nu_{12}E_2}{1 - \nu_{12}\nu_{21}} \tag{5.142}$$

In the absence of external mechanical loading, the induced-strain equation becomes (Eq. 5.104)

$$\begin{bmatrix} A_{11} & A_{12} \\ A_{12} & A_{22} \end{bmatrix} \begin{Bmatrix} \epsilon_x^o \\ \epsilon_y^o \end{Bmatrix} = \begin{Bmatrix} F_{x_\Lambda} \\ F_{y_\Lambda} \end{Bmatrix} \tag{5.143}$$

The coefficients of matrix $\boldsymbol{A}$ are defined as

$$A_{ij} = \sum_{k=1}^{3} (Q_{ij})_k (h_{k+1} - h_k) \tag{5.144}$$

$$A_{11} = \frac{E_1 t_b}{1 - \nu_{12}\nu_{21}} + \frac{2E_c t_c}{1 - \nu^2} \tag{5.145}$$

$$A_{22} = \frac{E_2 t_b}{1 - \nu_{12}\nu_{21}} + \frac{2E_c t_c}{1 - \nu^2} \tag{5.146}$$

$$A_{12} = A_{21} = \frac{\nu_{21}E_1 t_b}{1 - \nu_{12}\nu_{21}} + \frac{\nu}{1 - \nu^2} 2E_c t_c \tag{5.147}$$

The actuation forces are defined as

$$\boldsymbol{F}_\Lambda = \sum_{k=1}^{3} \boldsymbol{Q}_k \boldsymbol{\Lambda}_k (h_{k+1} - h_k) = \frac{2E_c t_c}{1 - \nu^2} \begin{bmatrix} 1 & \nu \\ \nu & 1 \end{bmatrix} \begin{Bmatrix} \Lambda \\ \Lambda \end{Bmatrix} = \frac{2E_c t_c \Lambda}{1 - \nu} \begin{Bmatrix} 1 \\ 1 \end{Bmatrix} \tag{5.148}$$

Substituting in Eq. 5.143 and solving for the strains gives

$$\begin{bmatrix} \frac{E_1 t_b}{1-\nu_{12}\nu_{21}} + \frac{2E_c t_c}{1-\nu^2} & \frac{\nu_{21}E_1 t_b}{1-\nu_{12}\nu_{21}} + \frac{2E_c t_c \nu}{1-\nu^2} \\ \frac{\nu_{21}E_1 t_b}{1-\nu_{12}\nu_{21}} + \frac{2E_c t_c \nu}{1-\nu^2} & \frac{E_2 t_b}{1-\nu_{12}\nu_{21}} + \frac{2E_c t_c}{1-\nu^2} \end{bmatrix} \begin{Bmatrix} \epsilon_x^o \\ \epsilon_y^o \end{Bmatrix} = \frac{2E_c t_c \Lambda}{1 - \nu} \begin{Bmatrix} 1 \\ 1 \end{Bmatrix} \tag{5.149}$$

To simplify this expression, it is assumed that the Poisson's ratios for the actuator (ν) and for the plate (ν_{21}) are equal.

$$\left(\frac{E_1 t_b}{1-\nu_{12}\nu_{21}} + \frac{2E_c t_c}{1-\nu^2}\right)\begin{bmatrix} 1 & \nu \\ \nu & \frac{\frac{E_2 t_b}{1-\nu_{12}\nu_{21}} + \frac{2E_c t_c}{1-\nu^2}}{\frac{E_1 t_b}{1-\nu_{12}\nu_{21}} + \frac{2E_c t_c}{1-\nu^2}} \end{bmatrix}\begin{Bmatrix} \epsilon_x^o \\ \epsilon_y^o \end{Bmatrix} = \frac{2E_c t_c}{1-\nu}\begin{Bmatrix} 1 \\ 1 \end{Bmatrix}\Lambda \tag{5.150}$$

$$\begin{aligned} \begin{Bmatrix} \epsilon_x^o \\ \epsilon_y^o \end{Bmatrix} &= \frac{\frac{2E_c t_c \Lambda}{1-\nu}}{\frac{E_1 t_b}{1-\nu_{12}\nu_{21}} + \frac{2E_c t_c}{1-\nu^2}}\begin{bmatrix} 1 & \nu \\ \nu & \alpha \end{bmatrix}^{-1}\begin{Bmatrix} 1 \\ 1 \end{Bmatrix} \\ &= \frac{\frac{2E_c t_c \Lambda}{1-\nu}}{\frac{E_1 t_b}{1-\nu_{12}\nu_{21}} + \frac{2E_c t_c}{1-\nu^2}}\,\frac{1}{\alpha-\nu^2}\begin{Bmatrix} \alpha-\nu \\ 1-\nu \end{Bmatrix} \\ &= \frac{2E_c t_c \Lambda}{(1-\nu)\left[\frac{E_2 t_b}{1-\nu_{12}\nu_{21}} + \frac{2E_c t_c}{1-\nu^2} - \nu^2\left(\frac{E_1 t_b}{1-\nu_{12}\nu_{21}} + \frac{2E_c t_c}{1-\nu^2}\right)\right]}\begin{Bmatrix} \alpha-\nu \\ 1-\nu \end{Bmatrix} \end{aligned} \tag{5.151}$$

where

$$\alpha = \frac{\frac{E_2 t_b}{1-\nu_{12}\nu_{21}} + \frac{2E_c t_c}{1-\nu^2}}{\frac{E_1 t_b}{1-\nu_{12}\nu_{21}} + \frac{2E_c t_c}{1-\nu^2}} \tag{5.152}$$

5.2.5 Single-Layer Specially Orthotropic Plate (Bending)

Applying a positive voltage to the bottom piezo and a negative voltage to the top piezo results in bending actuation.

$$\mathbf{\Lambda}_1 = \begin{Bmatrix} \Lambda \\ \Lambda \\ 0 \end{Bmatrix} \quad \mathbf{\Lambda}_2 = \begin{Bmatrix} 0 \\ 0 \\ 0 \end{Bmatrix} \quad \mathbf{\Lambda}_3 = \begin{Bmatrix} -\Lambda \\ -\Lambda \\ 0 \end{Bmatrix} \tag{5.153}$$

The actuation equation is

$$\begin{bmatrix} D_{11} & D_{12} \\ D_{12} & D_{22} \end{bmatrix}\begin{Bmatrix} \kappa_x \\ \kappa_y \end{Bmatrix} = \begin{Bmatrix} M_{x_\Lambda} \\ M_{y_\Lambda} \end{Bmatrix} \tag{5.154}$$

where the stiffness coefficients are defined as

$$D_{ij} = \frac{1}{3}\sum_{k=1}^{3}(Q_{ij})_k(h_{k+1}^3 - h_k^3)$$

$$D_{11} = \frac{E_1 t_b^3}{12(1-\nu_{12}\nu_{21})} + \frac{2E_c t_c}{3(1-\nu^2)}\left(\frac{3}{4}t_b^2 + \frac{3}{2}t_b t_c + t_c^2\right) \tag{5.155}$$

$$D_{12} = \frac{\nu_{21} E_1 t_b^3}{12(1-\nu_{12}\nu_{21})} + \frac{2\nu E_c t_c}{3(1-\nu^2)}\left(\frac{3}{4}t_b^2 + \frac{3}{2}t_b t_c + t_c^2\right) \tag{5.156}$$

$$D_{22} = \frac{E_2 t_b^3}{12(1-\nu_{12}\nu_{21})} + \frac{2E_c t_c}{3(1-\nu^2)}\left(\frac{3}{4}t_b^2 + \frac{3}{2}t_b t_c + t_c^2\right) \tag{5.157}$$

The induced moments are given by

$$\begin{Bmatrix} M_{x_\Lambda} \\ M_{y_\Lambda} \end{Bmatrix} = -\frac{1}{2}\sum_{k=1}^{3} \boldsymbol{Q}_k \boldsymbol{\Lambda}_k (h_{k+1}^2 - h_k^2) = \frac{E_c t_c (t_b + t_c)}{(1-\nu)} \Lambda \begin{Bmatrix} 1 \\ 1 \end{Bmatrix} \tag{5.158}$$

$$\begin{Bmatrix} \kappa_x \\ \kappa_y \end{Bmatrix} = \begin{bmatrix} D_{11} & D_{12} \\ D_{12} & D_{22} \end{bmatrix}^{-1} \begin{Bmatrix} M_{x_\Lambda} \\ M_{y_\Lambda} \end{Bmatrix} \tag{5.159}$$

5.2.6 Single-Layer Generally Orthotropic Plate (Extension)

Consider a generally orthotropic plate with two surface-bonded piezo actuators. The geometry of this configuration is similar to that shown in Fig. 5.6, with the exception that in this case, the plate is generally orthotropic. The induced strain vectors are

$$\boldsymbol{\Lambda}_1 = \begin{Bmatrix} \Lambda \\ \Lambda \\ 0 \end{Bmatrix} \quad \boldsymbol{\Lambda}_2 = \begin{Bmatrix} 0 \\ 0 \\ 0 \end{Bmatrix} \quad \boldsymbol{\Lambda}_3 = \begin{Bmatrix} \Lambda \\ \Lambda \\ 0 \end{Bmatrix} \tag{5.160}$$

The stiffness matrix of the plate is given by

$$\bar{\boldsymbol{Q}} = \begin{Bmatrix} \bar{Q}_{11} & \bar{Q}_{12} & \bar{Q}_{16} \\ \bar{Q}_{12} & \bar{Q}_{22} & \bar{Q}_{26} \\ \bar{Q}_{16} & \bar{Q}_{26} & \bar{Q}_{66} \end{Bmatrix} \tag{5.161}$$

The major difference in this case compared to a specially orthotropic plate is the structure of the stiffness matrix $\bar{\boldsymbol{Q}}$. Coupling terms exist in the stiffness matrix because the principal axes of the lamina are not aligned with the coordinate axes. Due to the non-zero extension-shear coupling terms $\bar{Q}_{16}$ and $\bar{Q}_{26}$, there exists a coupling between extension and shear as well as between bending and twist. As a result, even though shear is absent in the induced-strain field, the structure will exhibit shear and twist displacement.

To cause extensional actuation, the same potential is applied to the top and the bottom piezo sheets. Because there is no coupling between bending and extension, the induced-strain actuations are

$$\begin{bmatrix} A_{11} & A_{12} & A_{16} \\ A_{12} & A_{22} & A_{26} \\ A_{16} & A_{26} & A_{66} \end{bmatrix} \begin{Bmatrix} \epsilon_x^o \\ \epsilon_y^o \\ \gamma_{xy}^o \end{Bmatrix} = \begin{Bmatrix} F_{x_\Lambda} \\ F_{y_\Lambda} \\ F_{xy_\Lambda} \end{Bmatrix} \tag{5.162}$$

From Eq. 5.118

$$\begin{Bmatrix} F_{x_\Lambda} \\ F_{y_\Lambda} \\ F_{xy_\Lambda} \end{Bmatrix} = \frac{2E_c t_c \Lambda}{1-\nu} \begin{Bmatrix} 1 \\ 1 \\ 0 \end{Bmatrix} \tag{5.163}$$

also

$$\mathbf{A} = \begin{bmatrix} \bar{Q}_{11} t_b + \frac{2E_c t_c}{1-\nu^2} & \bar{Q}_{12} t_b + \frac{2\nu E_c t_c}{1-\nu^2} & \bar{Q}_{16} t_b \\ \bar{Q}_{12} t_b + \frac{2\nu E_c t_c}{1-\nu^2} & \bar{Q}_{22} t_b + \frac{2E_c t_c}{1-\nu^2} & \bar{Q}_{26} t_b \\ \bar{Q}_{16} t_b & \bar{Q}_{26} t_b & \bar{Q}_{66} t_b + \frac{E_c t_c}{2(1+\nu)} \end{bmatrix} \tag{5.164}$$

The strains in the structure are

$$\begin{Bmatrix} \epsilon_x^o \\ \epsilon_y^o \\ \gamma_{xy}^o \end{Bmatrix} = \frac{2E_c t_c \Lambda}{1-v} \mathbf{A}^{-1} \begin{Bmatrix} 1 \\ 1 \\ 0 \end{Bmatrix} \tag{5.165}$$

From this, it can be seen that the structure exhibits an extension-shear coupling.

5.2.7 Single-Layer Generally Orthotropic Plate (Bending)

Opposite voltage is applied to the top and bottom piezo sheets to create a bending deformation. Again, in this case, there is no coupling between bending and extension. The geometry of the plate and actuators is similar to Fig. 5.6.

$$\mathbf{\Lambda}_1 = \begin{Bmatrix} \Lambda \\ \Lambda \\ 0 \end{Bmatrix} \quad \mathbf{\Lambda}_2 = \begin{Bmatrix} 0 \\ 0 \\ 0 \end{Bmatrix} \quad \mathbf{\Lambda}_3 = \begin{Bmatrix} -\Lambda \\ -\Lambda \\ 0 \end{Bmatrix} \tag{5.166}$$

Induced bending curvatures are found from

$$\begin{bmatrix} D_{11} & D_{12} & D_{16} \\ D_{12} & D_{22} & D_{26} \\ D_{16} & D_{26} & D_{66} \end{bmatrix} \begin{Bmatrix} \kappa_x \\ \kappa_y \\ \kappa_{xy} \end{Bmatrix} = \begin{Bmatrix} M_{x_\Lambda} \\ M_{y_\Lambda} \\ M_{xy_\Lambda} \end{Bmatrix} \tag{5.167}$$

The bending stiffness coefficients are defined as

$$D_{ij} = \frac{1}{3} \sum (\bar{Q}_{ij})_k (h_{k+1}^3 - h_k^3)$$

$$D_{11} = \frac{1}{12}\bar{Q}_{11} t_b^3 + \frac{2E_c t_c}{3(1-\nu^2)} \left(\frac{3}{4} t_b^2 + \frac{3}{2} t_b t_c + t_c^2 \right) \tag{5.168}$$

$$D_{12} = \frac{1}{12}\bar{Q}_{12} t_b^3 + \frac{\nu 2E_c t_c}{3(1-\nu^2)} \left(\frac{3}{4} t_b^2 + \frac{3}{2} t_b t_c + t_c^2 \right) \tag{5.169}$$

$$D_{16} = \frac{1}{12}\bar{Q}_{16} t_b^3 \tag{5.170}$$

$$D_{22} = \frac{1}{12}\bar{Q}_{22} t_b^3 + \frac{2E_c t_c}{3(1-\nu^2)} \left(\frac{3}{4} t_b^2 + \frac{3}{2} t_b t_c + t_c^2 \right) \tag{5.171}$$

$$D_{26} = \frac{1}{12}\bar{Q}_{26} t_b^3 \tag{5.172}$$

$$D_{66} = \frac{1}{12}\bar{Q}_{66} t_b^3 + \frac{E_c t_c}{3(1+\nu)} \left(\frac{3}{4} t_b^2 + \frac{3}{2} t_b t_c + t_c^2 \right) \tag{5.173}$$

where the $\bar{Q}_{ij}$ terms are the stiffness terms of the plate. The induced moments are (Eq. 5.131)

$$\begin{Bmatrix} M_{x_\Lambda} \\ M_{y_\Lambda} \\ M_{xy_\Lambda} \end{Bmatrix} = \frac{E_c t_c (t_b + t_c)\Lambda}{(1-\nu)} \begin{Bmatrix} 1 \\ 1 \\ 0 \end{Bmatrix} \tag{5.174}$$

This results in the bending curvatures,

$$\begin{Bmatrix} \kappa_x \\ \kappa_y \\ \kappa_{xy} \end{Bmatrix} = \frac{E_c t_c (t_b + t_c)\Lambda}{(1-\nu)} \begin{bmatrix} D_{11} & D_{12} & D_{16} \\ D_{12} & D_{22} & D_{26} \\ D_{16} & D_{26} & D_{66} \end{bmatrix}^{-1} \begin{Bmatrix} 1 \\ 1 \\ 0 \end{Bmatrix} \tag{5.175}$$

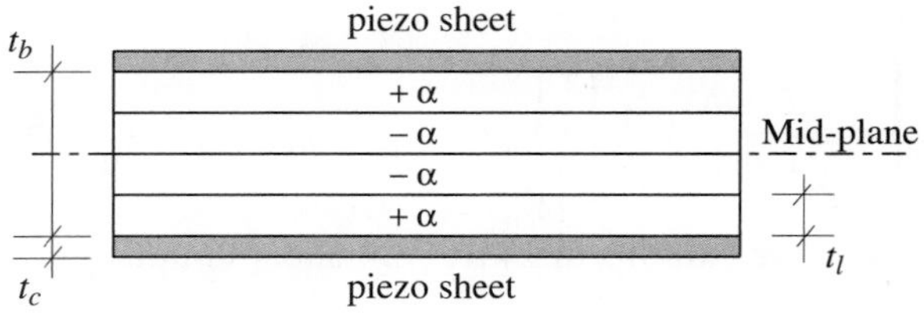

Figure 5.8. Ply layups in a symmetric laminate plate.

Note that a pure bending actuation in this configuration causes a twisting of the plate. The structure exhibits a bending-torsion coupling.

5.2.8 Multilayered Symmetric Laminate Plate

In a symmetric laminate, the ply angles are symmetric with respect to the mid-plane. For example, a five-layered laminate with a layup of $[-\alpha_1^\circ/+\alpha_2^\circ/-\alpha_3^\circ/+\alpha_2^\circ/-\alpha_1^\circ]$ constitutes a symmetric laminate. A three-layered laminate with a layup of $[0^\circ/90^\circ/0^\circ]$ is called a regular symmetric cross-ply laminate. A schematic diagram of the cross section of a four-ply symmetric laminate plate (layup $[+\alpha^\circ/-\alpha^\circ/-\alpha^\circ/+\alpha^\circ]$) with piezo sheets bonded to both surfaces is shown in Fig. 5.8. Such a laminate can also be represented as $[+\alpha^\circ/-\alpha^\circ]_s$. The thickness of each lamina is t_l and the total thickness of the plate is t_b.

The stiffness matrices for a laminate with N plies are defined as

$$A_{ij} = \sum_{k=1}^{N} (\bar{Q}_{ij})_k (h_{k+1} - h_k) \tag{5.176}$$

$$B_{ij} = -\frac{1}{2} \sum_{k=1}^{N} (\bar{Q}_{ij})_k \left(h_{k+1}^2 - h_k^2\right) \tag{5.177}$$

$$D_{ij} = \frac{1}{3} \sum_{k=1}^{N} (\bar{Q}_{ij})_k \left(h_{k+1}^3 - h_k^3\right) \tag{5.178}$$

If the layup and properties of the laminate are symmetric about the middle surface, then $\boldsymbol{B} = 0$. Therefore, there will be no coupling between bending and extension. For purely extensional actuation

$$\begin{bmatrix} A_{11} & A_{12} & A_{16} \\ A_{12} & A_{22} & A_{26} \\ A_{16} & A_{26} & A_{66} \end{bmatrix} \begin{Bmatrix} \epsilon_x^o \\ \epsilon_y^o \\ \gamma_{xy}^o \end{Bmatrix} = \begin{Bmatrix} F_{x_\Lambda} \\ F_{y_\Lambda} \\ F_{xy_\Lambda} \end{Bmatrix} \tag{5.179}$$

where

$$\begin{Bmatrix} F_{x_\Lambda} \\ F_{y_\Lambda} \\ F_{xy_\Lambda} \end{Bmatrix} = \frac{2E_c t_c \Lambda}{1-\nu} \begin{Bmatrix} 1 \\ 1 \\ 0 \end{Bmatrix} \tag{5.180}$$

For purely bending actuation with a voltage $-V$ to the top and $+V$ to the bottom piezo sheet:

$$\begin{bmatrix} D_{11} & D_{12} & D_{16} \\ D_{12} & D_{22} & D_{26} \\ D_{16} & D_{26} & D_{66} \end{bmatrix} \begin{Bmatrix} \kappa_x \\ \kappa_y \\ \kappa_{xy} \end{Bmatrix} = \begin{Bmatrix} M_{x_\Lambda} \\ M_{y_\Lambda} \\ M_{xy_\Lambda} \end{Bmatrix} \tag{5.181}$$

where

$$\begin{Bmatrix} M_{x_\Lambda} \\ M_{y_\Lambda} \\ M_{xy_\Lambda} \end{Bmatrix} = \frac{E_c t_c (t_b + t_c)\Lambda}{1 - \nu} \begin{Bmatrix} 1 \\ 1 \\ 0 \end{Bmatrix} \tag{5.182}$$

Note that there is no piezo-induced twisting of the laminate due to its symmetric nature.

Case I: Symmetric Laminates with Multiple Isotropic Layers

For an isotropic ply,

$$(\bar{Q}_{11})_k = (\bar{Q}_{22})_k = \frac{E_k}{1 - \upsilon_k^2} \tag{5.183}$$

$$(\bar{Q}_{16})_k = (\bar{Q}_{26})_k = 0 \tag{5.184}$$

$$(\bar{Q}_{12})_k = \frac{\upsilon_k E_k}{1 - \upsilon_k^2} \tag{5.185}$$

$$(\bar{Q}_{66})_k = \frac{E_k}{2(1 + \upsilon_k)} \tag{5.186}$$

$$A_{16} = A_{26} = 0 \tag{5.187}$$

$$A_{11} = A_{22} \tag{5.188}$$

$$D_{16} = D_{26} = 0 \tag{5.189}$$

$$D_{11} = D_{22} \tag{5.190}$$

This results in

$$\epsilon_x^o = \epsilon_y^o \tag{5.191}$$

$$\gamma_{xy}^o = 0 \tag{5.192}$$

$$\kappa_x = \kappa_y \tag{5.193}$$

$$\kappa_{xy} = 0 \tag{5.194}$$

Case II: Symmetric Laminates with Multiple Specially Orthotropic Layers

For a specially orthotropic ply,

$$(\bar{Q}_{11})_k = \left(\frac{E_1}{1 - \upsilon_{12}\upsilon_{21}}\right)_k \tag{5.195}$$

$$(\bar{Q}_{12})_k = \left(\frac{\upsilon_{21} E_1}{1 - \upsilon_{12}\upsilon_{21}}\right)_k \tag{5.196}$$

$$(\bar{Q}_{66})_k = (G_{12})_k \tag{5.197}$$

$$(\bar{Q}_{22})_k = \left(\frac{E_2}{1 - \upsilon_{12}\upsilon_{21}}\right)_k \tag{5.198}$$

$$(\bar{Q}_{16})_k = (\bar{Q}_{26})_k = 0 \tag{5.199}$$

This will result in

$$A_{16} = A_{26} = D_{16} = D_{26} = 0 \tag{5.200}$$

and

$$\gamma_{xy}^o = 0 \qquad \kappa_{xy} = 0 \tag{5.201}$$

The behavior of this laminate is very similar to a single-layer specially orthotropic lamina. For purely extensional actuation

$$\begin{bmatrix} A_{11} & A_{12} \\ A_{12} & A_{22} \end{bmatrix} \begin{Bmatrix} \epsilon_x^o \\ \epsilon_y^o \end{Bmatrix} = \begin{Bmatrix} F_{x_\Lambda} \\ F_{y_\Lambda} \end{Bmatrix} \tag{5.202}$$

where

$$\begin{Bmatrix} F_{x_\Lambda} \\ F_{y_\Lambda} \end{Bmatrix} = \frac{2E_c t_c \Lambda}{1-\nu} \begin{Bmatrix} 1 \\ 1 \end{Bmatrix} \tag{5.203}$$

For purely bending actuation

$$\begin{bmatrix} D_{11} & D_{12} \\ D_{12} & D_{22} \end{bmatrix} \begin{Bmatrix} \kappa_x \\ \kappa_y \end{Bmatrix} = \begin{Bmatrix} M_{x_\Lambda} \\ M_{y_\Lambda} \end{Bmatrix} \tag{5.204}$$

where

$$\begin{Bmatrix} M_{x_\Lambda} \\ M_{y_\Lambda} \end{Bmatrix} = \frac{E_c t_c (t_b + t_c)\Lambda}{1-\nu} \begin{Bmatrix} 1 \\ 1 \end{Bmatrix} \tag{5.205}$$

Case III: Symmetric Laminates with Multiple Generally Orthotropic Layers

The behavior of this laminate is very similar to a single-layer generally orthotropic lamina. However, due to symmetry about the mid-plane, $\boldsymbol{B} = 0$. Consequently, there is no coupling between bending and extension. The terms A_{16}, A_{26}, D_{16}, and D_{26} are non-zero for this case. Thus, there is a coupling between normal forces and shearing strain and between twisting moment and bending curvature.

For purely extensional actuation

$$\begin{bmatrix} A_{11} & A_{12} & A_{16} \\ A_{12} & A_{22} & A_{26} \\ A_{16} & A_{26} & A_{66} \end{bmatrix} \begin{Bmatrix} \epsilon_x^o \\ \epsilon_y^o \\ \gamma_{xy}^o \end{Bmatrix} = \frac{2E_c t_c \Lambda}{1-\nu} \begin{Bmatrix} 1 \\ 1 \\ 0 \end{Bmatrix} \tag{5.206}$$

A_{16} and A_{26} are called extension-shear couplings. A normal induced stress results in the shear strain γ_{xy}^o.

For purely bending actuation

$$\begin{bmatrix} D_{11} & D_{12} & D_{16} \\ D_{12} & D_{22} & D_{26} \\ D_{16} & D_{26} & D_{66} \end{bmatrix} \begin{Bmatrix} \kappa_x \\ \kappa_y \\ \kappa_{xy} \end{Bmatrix} = \frac{E_c t_c (t_b + t_c)\Lambda}{1-\nu} \begin{Bmatrix} 1 \\ 1 \\ 0 \end{Bmatrix} \tag{5.207}$$

D_{16} and D_{26} are called bending-twist couplings. An induced moment M_x or M_y causes twisting of the plate.

Case IV: Symmetric Laminates with Multiple Antisymmetric Layers

For every ply with the orientation $+\alpha^\circ$, there is a corresponding ply with the orientation $-\alpha^\circ$ about the same distance from the mid-plane of the plate. The behavior of

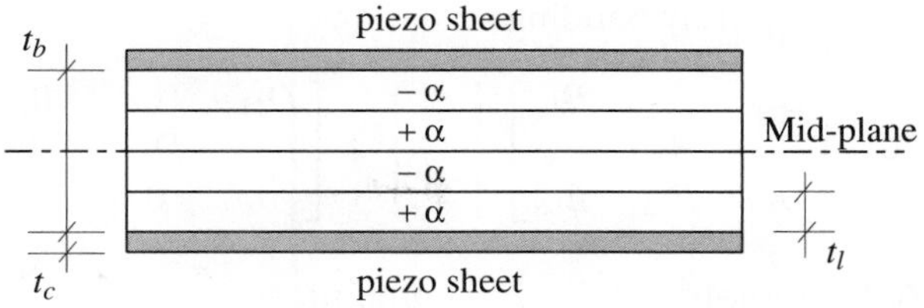

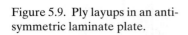

Figure 5.9. Ply layups in an antisymmetric laminate plate.

this laminate is similar to that of a symmetric laminate with generally orthotropic layers. This is also called a balanced laminate (e.g., $[+\alpha^\circ/-\alpha^\circ/-\alpha^\circ/+\alpha^\circ/+\alpha^\circ/-\alpha^\circ/-\alpha^\circ/+\alpha^\circ]$). For a many-layered angle-ply laminate, the values of A_{16}, A_{26}, D_{16}, and D_{26} become quite small as compared to other terms, and the laminate behaves more like a laminate with specially orthotropic layers.

5.2.9 Multilayered Antisymmetric Laminate Plate

In an antisymmetric laminate, the ply angles are antisymmetric with respect to the mid-plane. For example, a four-layered laminate with a layup of $[+\alpha^\circ/-\alpha^\circ/+\alpha^\circ/-\alpha^\circ]$ constitutes an antisymmetric laminate. The behavior of such laminates can be quite different from that of symmetric laminates. An antisymmetric laminate should have an even number of plies. A schematic diagram of the cross section of an antisymmetric laminate is shown in Fig. 5.9.

For plies with angle $+\alpha^\circ$ and $-\alpha^\circ$

$$(\bar{Q}_{11})_{+\alpha^\circ} = (\bar{Q}_{11})_{-\alpha^\circ} \tag{5.208}$$

$$(\bar{Q}_{22})_{+\alpha^\circ} = (\bar{Q}_{22})_{-\alpha^\circ} \tag{5.209}$$

$$(\bar{Q}_{12})_{+\alpha^\circ} = (\bar{Q}_{12})_{-\alpha^\circ} \tag{5.210}$$

$$(\bar{Q}_{66})_{+\alpha^\circ} = (\bar{Q}_{66})_{-\alpha^\circ} \tag{5.211}$$

$$(\bar{Q}_{16})_{+\alpha^\circ} = -(\bar{Q}_{16})_{-\alpha^\circ} \tag{5.212}$$

$$(\bar{Q}_{26})_{+\alpha^\circ} = -(\bar{Q}_{26})_{-\alpha^\circ} \tag{5.213}$$

Therefore, for an antisymmetric laminate with n plies

$$A_{16} = \sum_{k=1}^{n} (\bar{Q}_{16})_k (h_{k+1} - h_k) = 0 \tag{5.214}$$

and

$$D_{16} = \frac{1}{3} \sum_{k=1}^{n} (\bar{Q}_{16})_k \left(h_{k+1}^3 - h_k^3\right) = 0 \tag{5.215}$$

Similarly, $A_{26} = 0$ and $D_{26} = 0$. Thus, for an antisymmetric laminate, the extensional-shear couplings and bending-twist couplings are zero. Force and moment relations are given herein. For purely extensional actuation

$$\begin{bmatrix} A_{11} & A_{12} & 0 \\ A_{12} & A_{22} & 0 \\ 0 & 0 & A_{66} \end{bmatrix} \begin{Bmatrix} \epsilon_x^o \\ \epsilon_y^o \\ \gamma_{xy}^o \end{Bmatrix} + \begin{bmatrix} B_{11} & B_{12} & B_{16} \\ B_{12} & B_{22} & B_{26} \\ B_{16} & B_{26} & B_{66} \end{bmatrix} \begin{Bmatrix} \kappa_x \\ \kappa_y \\ \kappa_{xy} \end{Bmatrix} = \begin{Bmatrix} F_{x_\Lambda} \\ F_{y_\Lambda} \\ F_{xy_\Lambda} \end{Bmatrix} \tag{5.216}$$

For purely bending actuation

$$\begin{bmatrix} B_{11} & B_{12} & B_{16} \\ B_{12} & B_{22} & B_{26} \\ B_{16} & B_{26} & B_{66} \end{bmatrix} \begin{Bmatrix} \epsilon_x^o \\ \epsilon_y^o \\ \gamma_{xy}^o \end{Bmatrix} + \begin{bmatrix} D_{11} & D_{12} & 0 \\ D_{12} & D_{22} & 0 \\ 0 & 0 & D_{66} \end{bmatrix} \begin{Bmatrix} \kappa_x \\ \kappa_y \\ \kappa_{xy} \end{Bmatrix} = \begin{Bmatrix} M_{x_\Lambda} \\ M_{y_\Lambda} \\ M_{xy_\Lambda} \end{Bmatrix} \tag{5.217}$$

Antisymmetric Angle-Ply Laminates

An example of a two-layered antisymmetric laminate is $[+\alpha/-\alpha]$ in which both plies are of equal thickness. For a general layup, if a lamina of $+\alpha^\circ$ orientation is placed at a certain vertical distance on one side of the mid-plane, then an equal-thickness lamina of $-\alpha^\circ$ orientation is placed on the other side at the same vertical distance.

$$B_{ij} = -\frac{1}{2}\sum_{k=1}^{n}(\bar{Q}_{ij})_k\left(h_{k+1}^2 - h_k^2\right)$$

$$B_{11} = -\frac{1}{2}(\bar{Q}_{11})_{+\alpha}t^2 + \frac{1}{2}(\bar{Q}_{11})_{-\alpha}t^2 = 0 \tag{5.218}$$

$$B_{12} = 0 \tag{5.219}$$

$$B_{22} = 0 \tag{5.220}$$

$$B_{66} = 0 \tag{5.221}$$

$$B_{16} = -\frac{1}{2}(\bar{Q}_{16})_{+\alpha}t^2 + \frac{1}{2}(\bar{Q}_{16})_{-\alpha}t^2 = -(\bar{Q}_{16})_{+\alpha}t^2 \tag{5.222}$$

$$B_{26} = -\frac{1}{2}(\bar{Q}_{26})_{+\alpha}t^2 + \frac{1}{2}(\bar{Q}_{26})_{-\alpha}t^2 = -(\bar{Q}_{26})_{+\alpha}t^2 \tag{5.223}$$

where t is the thickness of one ply.

For symmetric induced actuation, the force and moment relations are given herein.

For purely extensional actuation

$$\begin{bmatrix} A_{11} & A_{12} & 0 \\ A_{12} & A_{22} & 0 \\ 0 & 0 & A_{66} \end{bmatrix} \begin{Bmatrix} \epsilon_x^o \\ \epsilon_y^o \\ \gamma_{xy}^o \end{Bmatrix} + \begin{bmatrix} 0 & 0 & B_{16} \\ 0 & 0 & B_{26} \\ B_{16} & B_{26} & 0 \end{bmatrix} \begin{Bmatrix} \kappa_x \\ \kappa_y \\ \kappa_{xy} \end{Bmatrix} = \begin{Bmatrix} F_{x_\Lambda} \\ F_{y_\Lambda} \\ F_{xy_\Lambda} \end{Bmatrix} \tag{5.224}$$

For purely bending actuation

$$\begin{bmatrix} 0 & 0 & B_{16} \\ 0 & 0 & B_{26} \\ B_{16} & B_{26} & 0 \end{bmatrix} \begin{Bmatrix} \epsilon_x^o \\ \epsilon_y^o \\ \gamma_{xy}^o \end{Bmatrix} + \begin{bmatrix} D_{11} & D_{12} & 0 \\ D_{12} & D_{22} & 0 \\ 0 & 0 & D_{66} \end{bmatrix} \begin{Bmatrix} \kappa_x \\ \kappa_y \\ \kappa_{xy} \end{Bmatrix} = \begin{Bmatrix} M_{x_\Lambda} \\ M_{y_\Lambda} \\ M_{xy_\Lambda} \end{Bmatrix} \tag{5.225}$$

From these equations, it can be seen that a normal induced strain (F_{x_Λ} or F_{y_Λ}) causes a twisting of the plate, κ_{xy}, due to the extension-twist couplings B_{16} and B_{26}. In addition, an induced bending moment (M_{x_Λ} or M_{y_Λ}) will also cause a shear deformation γ_{xy}^o. Therefore, these couplings are also called bending-shear couplings.

Antisymmetric Cross-Ply Laminates

These laminates consist of an even number of alternating 0° and 90° plies. For example, a two-ply antisymmetric cross-ply laminate would have the layup [0°/90°],

Figure 5.10. Two-layered cross-ply laminate.

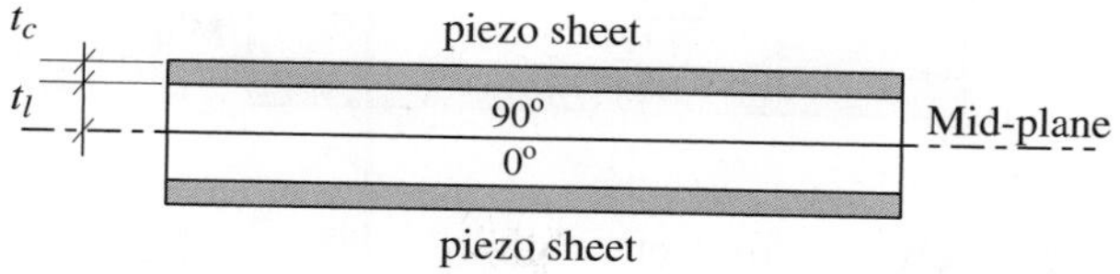

and for a six-ply laminate, the layup would be [0°/90°/90°/0°/0°/90°]. For the 0° ply

$$\bar{\boldsymbol{Q}}_{0^\circ} = \boldsymbol{Q}_{0^\circ} = \begin{bmatrix} Q_{11} & Q_{12} & 0 \\ Q_{12} & Q_{22} & 0 \\ 0 & 0 & Q_{66} \end{bmatrix} \tag{5.226}$$

and for the 90° ply

$$(\bar{Q}_{22})_{90^\circ} = (\bar{Q}_{11})_{0^\circ} \tag{5.227}$$

$$(\bar{Q}_{11})_{90^\circ} = (\bar{Q}_{22})_{0^\circ} \tag{5.228}$$

$$(\bar{Q}_{12})_{90^\circ} = (\bar{Q}_{12})_{0^\circ} \tag{5.229}$$

$$(\bar{Q}_{66})_{90^\circ} = (\bar{Q}_{66})_{0^\circ} \tag{5.230}$$

Therefore

$$\bar{\boldsymbol{Q}}_{90^\circ} = \begin{bmatrix} Q_{22}' & Q_{12} & 0 \\ Q_{12} & Q_{11} & 0 \\ 0 & 0 & Q_{66} \end{bmatrix} \tag{5.231}$$

Let us consider a two-layer cross-ply laminate with a [0°/90°] layup (Fig. 5.10). In this case, the extension and bending stiffness matrices are

$$\mathbf{A} = \begin{bmatrix} A_{11} & A_{12} & 0 \\ A_{12} & A_{22} & 0 \\ 0 & 0 & A_{66} \end{bmatrix} \tag{5.232}$$

and

$$\boldsymbol{D} = \begin{bmatrix} D_{11} & D_{12} & 0 \\ D_{12} & D_{22} & 0 \\ 0 & 0 & D_{66} \end{bmatrix} \tag{5.233}$$

The coupling matrix coefficients are defined as

$$B_{11} = -\frac{1}{2}t_l^2((\bar{Q}_{11})_{90^\circ} - (\bar{Q}_{11})_{0^\circ}) \tag{5.234}$$

$$B_{22} = -\frac{1}{2}t_l^2((\bar{Q}_{22})_{90^\circ} - (\bar{Q}_{22})_{0^\circ}) = -B_{11} \tag{5.235}$$

$$B_{66} = \frac{1}{2}t_l^2((\bar{Q}_{66})_{0^\circ} - (\bar{Q}_{66})_{90^\circ}) = 0 \tag{5.236}$$

$$B_{16} = B_{26} = 0 \tag{5.237}$$

For purely extensional actuation

$$\begin{bmatrix} A_{11} & A_{12} & 0 \\ A_{12} & A_{22} & 0 \\ 0 & 0 & A_{66} \end{bmatrix} \begin{Bmatrix} \epsilon_x^o \\ \epsilon_y^o \\ \gamma_{xy}^o \end{Bmatrix} + \begin{bmatrix} B_{11} & 0 & 0 \\ 0 & -B_{11} & 0 \\ 0 & 0 & 0 \end{bmatrix} \begin{Bmatrix} \kappa_x \\ \kappa_y \\ \kappa_{xy} \end{Bmatrix} = \begin{Bmatrix} F_{x_\Lambda} \\ F_{y_\Lambda} \\ F_{xy_\Lambda} \end{Bmatrix} \tag{5.238}$$

Figure 5.11. A two-ply laminate.

For purely bending actuation

$$\begin{bmatrix} B_{11} & 0 & 0 \\ 0 & -B_{11} & 0 \\ 0 & 0 & 0 \end{bmatrix} \begin{Bmatrix} \epsilon_x^o \\ \epsilon_y^o \\ \gamma_{xy}^o \end{Bmatrix} + \begin{bmatrix} D_{11} & D_{12} & 0 \\ D_{12} & D_{22} & 0 \\ 0 & 0 & D_{66} \end{bmatrix} \begin{Bmatrix} \kappa_x \\ \kappa_y \\ \kappa_{xy} \end{Bmatrix} = \begin{Bmatrix} M_{x_\Lambda} \\ M_{y_\Lambda} \\ M_{xy_\Lambda} \end{Bmatrix} \tag{5.239}$$

From these equations, it can be seen that B_{11} is the in-plane–out-of-plane coupling. Induced forces F_{x_Λ} and F_{y_Λ} cause the bending curvatures κ_x and κ_y, whereas the induced moments M_{x_Λ} and M_{y_Λ} cause the in-plane extensional strains ϵ_x^o and ϵ_y^o.

5.2.10 Summary of Couplings in Plate Stiffness Matrices

To summarize all of the possibilities of coupling in different layups, let us consider a simple two-ply laminate, as shown in Fig. 5.11. The plate stiffness matrices are given by

$$\begin{aligned} \boldsymbol{A} &= \sum_{(k=1)}^{2} \bar{\boldsymbol{Q}}_k (h_{k+1} - h_k) \\ &= \bar{\boldsymbol{Q}}_1 (0 - (-t_l)) + \bar{\boldsymbol{Q}}_2 ((t_l) - 0) \\ &= t_l (\bar{\boldsymbol{Q}}_1 + \bar{\boldsymbol{Q}}_2) \end{aligned} \tag{5.240}$$

where $\bar{\boldsymbol{Q}}_1$ is the reduced-stiffness matrix for ply 1, $\bar{\boldsymbol{Q}}_2$ is the reduced stiffness matrix for ply 2, and t_l is the thickness of each ply

$$\begin{aligned} \boldsymbol{B} &= -\frac{1}{2} \sum_{(k=1)}^{2} \bar{\boldsymbol{Q}}_k (h_{k+1}^2 - h_k^2) \\ &= -\frac{1}{2} [\bar{\boldsymbol{Q}}_1 (0 - (-t_l)^2) + \bar{\boldsymbol{Q}}_2 ((t_l)^2 - 0)] \\ &= -\frac{1}{2} t_l^2 (-\bar{\boldsymbol{Q}}_1 + \bar{\boldsymbol{Q}}_2) \end{aligned} \tag{5.241}$$

$$\begin{aligned} \boldsymbol{D} &= \frac{1}{3} \sum_{(k=1)}^{2} \bar{\boldsymbol{Q}}_k (h_{k+1}^3 - h_k^3) \\ &= \frac{1}{3} [\bar{\boldsymbol{Q}}_1 (0 - (-t_l)^3) + \bar{\boldsymbol{Q}}_2 ((t_l)^3 - 0)] \\ &= \frac{1}{3} t_l^3 (\bar{\boldsymbol{Q}}_1 + \bar{\boldsymbol{Q}}_2) \end{aligned} \tag{5.242}$$

The effect of individual ply stiffness matrices on the plate stiffness matrices can easily be seen from these equations. The results are shown in Table 5.2. The coupling terms are

- $B_{11}, B_{12} \rightarrow$ extension-bending
- $B_{16}, B_{26} \rightarrow$ extension-torsion
- $D_{16}, D_{26} \rightarrow$ bending-torsion

5.2.11 Worked Example

(a) Using laminated plate theory, derive extension-bending equations for a rectangular two-layered angle-ply laminate with a piezo bonded to only its top surface for two cases: symmetric and antisymmetric layups, with $\alpha = 45°$ (Fig. 5.12).

(b) Calculate curvature and extension strain at a mid-point of this free-free laminate of size 0.3048 m (12 in) x 0.3048 m (12 in) x 0.79 mm (1/32 in) with a piezo-sheet (PZT-5H) of thickness 0.3175 mm (0.0125 in) for a voltage of 150 V, for the two cases of symmetric and antisymmetric layups.

Manufacturer-supplied material constants are as follows

$$d_{31} = -274 \times 10^{-12} \text{ m/V}$$

$$E_c = 72.4 \text{ GPa} \ \ (10.5 \times 10^6 \text{ lb/in}^2)$$

$$E_1 = 137.9 \text{ GPa} \ \ (20 \times 10^6 \text{ lb/in}^2)$$

$$E_2 = 14.5 \text{ GPa} \ \ (2.1 \times 10^6 \text{ lb/in}^2)$$

$$G_{12} = 5.86 \text{ GPa} \ (0.85 \times 10^6 \text{ lb/in}^2)$$

$$\nu_c = 0.3$$

$$\nu_{12} = 0.2$$

Solution

(a) With no mechanical force acting on the structure, the governing equation is

$$\begin{bmatrix} \boldsymbol{A} & \boldsymbol{B} \\ \boldsymbol{B} & \boldsymbol{D} \end{bmatrix} \begin{Bmatrix} \boldsymbol{\epsilon}^o \\ \boldsymbol{\kappa} \end{Bmatrix} = \begin{Bmatrix} \boldsymbol{F}_\Lambda \\ \boldsymbol{M}_\Lambda \end{Bmatrix}$$

The stiffness matrices are obtained as follows

$$\begin{aligned} \boldsymbol{A} &= \sum_{k=1}^{3} \boldsymbol{Q}_k (h_{k+1} - h_k) \\ &= \boldsymbol{Q}_c((t_l + t_c) - t_l) + \boldsymbol{Q}_1(0 - (-t_l)) + \boldsymbol{Q}_2(t_l - 0) \\ &= \boldsymbol{Q}_c t_c + t_l(\boldsymbol{Q}_1 + \boldsymbol{Q}_2) \\ &= \boldsymbol{Q}_c t_c + \frac{t_b}{2}(\boldsymbol{Q}_1 + \boldsymbol{Q}_2) \end{aligned}$$

Table 5.2. *Summary of laminated plate stiffness matrices*

Ply	Layup	$\bar{\boldsymbol{Q}}_1$ (N/m^2)	$\bar{\boldsymbol{Q}}_2$ (N/m^2)	$\boldsymbol{A}$ (N/m)	$\boldsymbol{B}$ (N)	$\boldsymbol{D}$ (Nm)
Isotropic	–	$\begin{bmatrix} \bar{Q}_{11} & \bar{Q}_{12} & 0 \\ \bar{Q}_{12} & \bar{Q}_{11} & 0 \\ 0 & 0 & \bar{Q}_{66} \end{bmatrix}$	$\begin{bmatrix} \bar{Q}_{11} & \bar{Q}_{12} & 0 \\ \bar{Q}_{12} & \bar{Q}_{11} & 0 \\ 0 & 0 & \bar{Q}_{66} \end{bmatrix}$	$\begin{bmatrix} A_{11} & A_{12} & 0 \\ A_{12} & A_{11} & 0 \\ 0 & 0 & A_{66} \end{bmatrix}$	0	$\begin{bmatrix} D_{11} & D_{12} & 0 \\ D_{12} & D_{11} & 0 \\ 0 & 0 & D_{66} \end{bmatrix}$
Specially orthotropic	Symmetric	$\begin{bmatrix} \bar{Q}_{11} & \bar{Q}_{12} & 0 \\ \bar{Q}_{12} & \bar{Q}_{22} & 0 \\ 0 & 0 & \bar{Q}_{66} \end{bmatrix}$	$\begin{bmatrix} \bar{Q}_{11} & \bar{Q}_{12} & 0 \\ \bar{Q}_{12} & \bar{Q}_{22} & 0 \\ 0 & 0 & \bar{Q}_{66} \end{bmatrix}$	$\begin{bmatrix} A_{11} & A_{12} & 0 \\ A_{12} & A_{22} & 0 \\ 0 & 0 & A_{66} \end{bmatrix}$	0	$\begin{bmatrix} D_{11} & D_{12} & 0 \\ D_{12} & D_{22} & 0 \\ 0 & 0 & D_{66} \end{bmatrix}$
	Cross-ply	$\begin{bmatrix} \bar{Q}_{11} & \bar{Q}_{12} & 0 \\ \bar{Q}_{12} & \bar{Q}_{22} & 0 \\ 0 & 0 & \bar{Q}_{66} \end{bmatrix}$	$\begin{bmatrix} \bar{Q}_{22} & \bar{Q}_{12} & 0 \\ \bar{Q}_{12} & \bar{Q}_{11} & 0 \\ 0 & 0 & \bar{Q}_{66} \end{bmatrix}$	$\begin{bmatrix} A_{11} & A_{12} & 0 \\ A_{12} & A_{11} & 0 \\ 0 & 0 & A_{66} \end{bmatrix}$	$\begin{bmatrix} B_{11} & 0 & 0 \\ 0 & -B_{11} & 0 \\ 0 & 0 & 0 \end{bmatrix}$	$\begin{bmatrix} D_{11} & D_{12} & 0 \\ D_{12} & D_{11} & 0 \\ 0 & 0 & D_{66} \end{bmatrix}$
Generally orthotropic	Symmetric	$\begin{bmatrix} \bar{Q}_{11} & \bar{Q}_{12} & \bar{Q}_{16} \\ \bar{Q}_{12} & \bar{Q}_{22} & \bar{Q}_{26} \\ \bar{Q}_{16} & \bar{Q}_{26} & \bar{Q}_{66} \end{bmatrix}$	$\begin{bmatrix} \bar{Q}_{11} & \bar{Q}_{12} & \bar{Q}_{16} \\ \bar{Q}_{12} & \bar{Q}_{22} & \bar{Q}_{26} \\ \bar{Q}_{16} & \bar{Q}_{26} & \bar{Q}_{66} \end{bmatrix}$	$\begin{bmatrix} A_{11} & A_{12} & A_{16} \\ A_{12} & A_{22} & A_{26} \\ A_{16} & A_{26} & A_{66} \end{bmatrix}$	0	$\begin{bmatrix} D_{11} & D_{12} & D_{16} \\ D_{12} & D_{22} & D_{26} \\ D_{16} & D_{26} & D_{66} \end{bmatrix}$
	Antisymmetric	$\begin{bmatrix} \bar{Q}_{11} & \bar{Q}_{12} & \bar{Q}_{16} \\ \bar{Q}_{12} & \bar{Q}_{22} & \bar{Q}_{26} \\ \bar{Q}_{16} & \bar{Q}_{26} & \bar{Q}_{66} \end{bmatrix}$	$\begin{bmatrix} \bar{Q}_{11} & \bar{Q}_{12} & -\bar{Q}_{16} \\ \bar{Q}_{12} & \bar{Q}_{22} & -\bar{Q}_{26} \\ -\bar{Q}_{16} & -\bar{Q}_{26} & \bar{Q}_{66} \end{bmatrix}$	$\begin{bmatrix} A_{11} & A_{12} & 0 \\ A_{12} & A_{22} & 0 \\ 0 & 0 & A_{66} \end{bmatrix}$	$\begin{bmatrix} 0 & 0 & B_{16} \\ 0 & 0 & B_{26} \\ B_{16} & B_{26} & 0 \end{bmatrix}$	$\begin{bmatrix} D_{11} & D_{12} & 0 \\ D_{12} & D_{22} & 0 \\ 0 & 0 & D_{66} \end{bmatrix}$

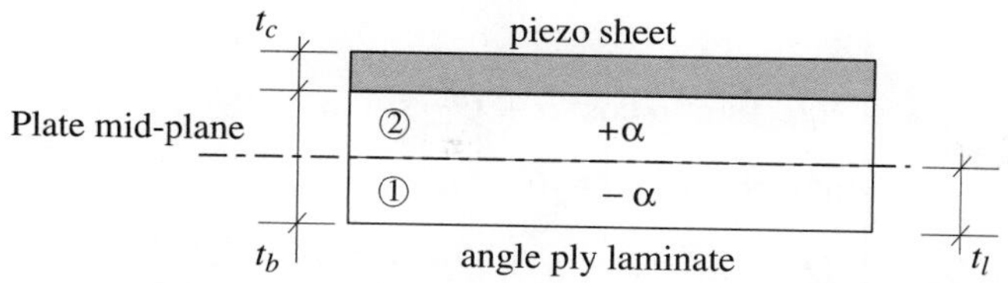

Figure 5.12. Two-layered angle-ply laminate with one piezo sheet.

$$
\begin{aligned}
\boldsymbol{B} &= -\frac{1}{2}\sum_{k=1}^{3} \boldsymbol{Q}_k \left(h_{k+1}^2 - h_k^2\right) \\
&= -\frac{1}{2}\left[\boldsymbol{Q}_c\left((t_l + t_c)^2 - t_l^2\right) + \boldsymbol{Q}_2\left(t_l^2 - 0\right) + \boldsymbol{Q}_1(0 - (-t_l)^2)\right] \\
&= -\frac{1}{2}\boldsymbol{Q}_c t_c (t_c + 2t_l) + \frac{1}{2} t_l^2 (-\boldsymbol{Q}_1 + \boldsymbol{Q}_2) \\
&= -\frac{t_b^2}{8}(-\boldsymbol{Q}_1 + \boldsymbol{Q}_2) + \frac{\boldsymbol{Q}_c t_c}{2}(t_c + t_b)
\end{aligned}
$$

$$
\begin{aligned}
\boldsymbol{D} &= \frac{1}{3}\sum_{k=1}^{3} \boldsymbol{Q}_k (h_{k+1}^3 - h_k^3) \\
&= \frac{1}{3}\left[\boldsymbol{Q}_c\left((t_l + t_c)^3 - t_l^3\right) + \bar{\boldsymbol{Q}}_2\left(t_l^3 - 0\right) + \bar{\boldsymbol{Q}}_1(0 - (-t_l)^3)\right] \\
&= \frac{1}{3}\left[\boldsymbol{Q}_c t_c \left(t_c^2 + 3t_c t_l + 3t_l^2\right) + t_l^3(\bar{\boldsymbol{Q}}_1 + \bar{\boldsymbol{Q}}_2)\right] \\
&= \frac{\boldsymbol{Q}_c t_c}{3}\left(t_c^2 + \frac{3}{2}t_c t_b + \frac{3}{4}t_b^2\right) + \frac{1}{24}t_b^3(\boldsymbol{Q}_1 + \boldsymbol{Q}_2)
\end{aligned}
$$

and the forcing vectors are

$$
\begin{aligned}
\boldsymbol{F}_\Lambda &= \sum_{k=1}^{3} \boldsymbol{Q}_k \boldsymbol{\Lambda}_k (h_{k+1} - h_k) \\
&= \boldsymbol{Q}_c \boldsymbol{\Lambda} t_c \\
&= \frac{E_c t_c \Lambda}{1 - \nu}\begin{Bmatrix} 1 \\ 1 \\ 0 \end{Bmatrix}
\end{aligned}
$$

$$
\begin{aligned}
\boldsymbol{M}_\Lambda &= -\frac{1}{2}\sum_{k=1}^{3} \boldsymbol{Q}_k \boldsymbol{\Lambda}_k \left(h_{k+1}^2 - h_k^2\right) \\
&= -\frac{1}{2}\boldsymbol{Q}_c \boldsymbol{\Lambda} t_c (t_c + 2t_l) \\
&= -\frac{E_c t_c (t_c + t_b)\Lambda}{2(1 - \nu)}\begin{Bmatrix} 1 \\ 1 \\ 0 \end{Bmatrix}
\end{aligned}
$$

(b) For the angle plies

$$\bar{Q}_{11} = Q_{11}\cos^4\alpha + 2(Q_{12} + 2Q_{66})\sin^2\alpha\cos^2\alpha + Q_{22}\sin^4\alpha$$

$$\bar{Q}_{12} = (Q_{11} + Q_{22} - 4Q_{66})\sin^2\alpha\cos^2\alpha + Q_{12}(\sin^4\alpha + \cos^4\alpha)$$

$$\bar{Q}_{22} = Q_{11}\sin^4\alpha + 2(Q_{12} + 2Q_{66})\sin^2\alpha\cos^2\alpha + Q_{22}\cos^4\alpha$$

$$\bar{Q}_{16} = (Q_{11} - Q_{12} - 2Q_{66})\sin\alpha\cos^3\alpha + (Q_{12} - Q_{22} + 2Q_{66})\sin^3\alpha\cos\alpha$$

$$\bar{Q}_{26} = (Q_{11} - Q_{12} - 2Q_{66})\sin^3\alpha\cos\alpha + (Q_{12} - Q_{22} + 2Q_{66})\sin\alpha\cos^3\alpha$$

$$\bar{Q}_{66} = (Q_{11} + Q_{22} - 2Q_{12} - 2Q_{66})\sin^2\alpha\cos^2\alpha + Q_{66}(\sin^4\alpha + \cos^4\alpha)$$

where the stiffness matrix $\boldsymbol{Q}$ with reference to the principal axes is given by (Eq. 5.12)

$$\boldsymbol{Q} = \begin{bmatrix} E_1/(1-\nu_{12}\nu_{21}) & \nu_{21}E_1/(1-\nu_{12}\nu_{21}) & 0 \\ \nu_{21}E_1/(1-\nu_{12}\nu_{21}) & E_2/(1-\nu_{12}\nu_{21}) & 0 \\ 0 & 0 & G_{12} \end{bmatrix}$$

Substituting the given values leads to

$$\boldsymbol{Q} = \begin{bmatrix} 138.4824 & 2.9122 & 0 \\ 2.9122 & 14.5612 & 0 \\ 0 & 0 & 5.8600 \end{bmatrix} \text{GPa}$$

For the lamina with a ply angle $+\alpha = 45^\circ$

$$\bar{\boldsymbol{Q}}_{+\alpha} = \begin{bmatrix} 45.5770 & 33.8570 & 30.9803 \\ 33.8570 & 45.5770 & 30.9803 \\ 30.9803 & 30.9803 & 36.8048 \end{bmatrix} \text{GPa}$$

and for the lamina with a ply angle $-\alpha = -45^\circ$

$$\bar{\boldsymbol{Q}}_{-\alpha} = \begin{bmatrix} 45.5770 & 33.8570 & -30.9803 \\ 33.8570 & 45.5770 & -30.9803 \\ -30.9803 & -30.9803 & 36.8048 \end{bmatrix} \text{GPa}$$

The piezo sheet is isotropic and its stiffness matrix is given by

$$\bar{\boldsymbol{Q}} = \boldsymbol{Q}_c = \frac{E_c}{1-\nu^2}\begin{bmatrix} 1 & \nu & 0 \\ \nu & 1 & 0 \\ 0 & 0 & \frac{1-\nu}{2} \end{bmatrix} = \begin{bmatrix} 79.5604 & 23.8681 & 0 \\ 23.8681 & 79.5604 & 0 \\ 0 & 0 & 27.8462 \end{bmatrix} \text{GPa}$$

Symmetric Layup

For a symmetric layup, $\bar{\boldsymbol{Q}}_1 = \bar{\boldsymbol{Q}}_2$. Substituting this in the previous relations, the stiffness matrices become

$$\boldsymbol{A} = \boldsymbol{Q}_c t_c + \bar{\boldsymbol{Q}}_{+\alpha} t_b$$

$$\boldsymbol{B} = -\frac{1}{2}\boldsymbol{Q}_c t_c (t_c + t_b)$$

$$\boldsymbol{D} = \frac{1}{3}\boldsymbol{Q}_c t_c \left(t_c^2 + \frac{3}{2}t_c t_b + \frac{3}{4}t_b^2\right) + \frac{1}{12}\bar{\boldsymbol{Q}}_{+\alpha} t_b^3$$

The force and moment vectors are the same for the case of symmetric and antisymmetric laminates. Substituting the given values, we obtain

$$\boldsymbol{A} = \begin{bmatrix} 61.4372 & 34.4521 & 24.5906 \\ 34.4521 & 61.4372 & 24.5906 \\ 24.5906 & 24.5906 & 38.0550 \end{bmatrix} \times 10^6 \text{ N/m}$$

$$\boldsymbol{B} = -\begin{bmatrix} 14.0353 & 4.2106 & 0 \\ 4.2106 & 14.0353 & 0 \\ 0 & 0 & 4.9124 \end{bmatrix} \times 10^3 \text{ N}$$

$$\boldsymbol{D} = \begin{bmatrix} 9.9100 & 3.8141 & 1.2911 \\ 3.8141 & 9.9100 & 1.2911 \\ 1.2911 & 1.2911 & 4.3375 \end{bmatrix} \text{ Nm}$$

The induced force and moments are

$$\boldsymbol{F}_\Lambda = \begin{Bmatrix} -4250.9 \\ -4250.9 \\ 0 \end{Bmatrix} \text{ N/m} \quad \boldsymbol{M}_\Lambda = \begin{Bmatrix} 2.3619 \\ 2.3619 \\ 0 \end{Bmatrix} \text{ Nm/m}$$

The strains and curvature of the plate are found from

$$\begin{Bmatrix} \epsilon_x^o \\ \epsilon_y^o \\ \gamma_{xy}^o \\ \kappa_x \\ \kappa_y \\ \kappa_{xy} \end{Bmatrix} = \begin{bmatrix} \boldsymbol{A} & \boldsymbol{B} \\ \boldsymbol{B} & \boldsymbol{D} \end{bmatrix}^{-1} \begin{Bmatrix} \boldsymbol{F}_\Lambda \\ \boldsymbol{M}_\Lambda \end{Bmatrix} = \begin{Bmatrix} -18.55 \times 10^{-6} \\ -18.55 \times 10^{-6} \\ 14.16 \times 10^{-6} \\ 0.1546 \text{ 1/m} \\ 0.1546 \text{ 1/m} \\ -0.076 \text{ 1/m} \end{Bmatrix}$$

Antisymmetric Layup

The relations between the stiffness matrices of antisymmetric angle-ply laminae are given by Eqs. 5.208–5.213.

The coefficients of the stiffness matrix are given by

$$\boldsymbol{A} = \boldsymbol{Q}_c t_c + \frac{t_b}{2}(\bar{\boldsymbol{Q}}_{+\alpha} + \bar{\boldsymbol{Q}}_{-\alpha})$$

$$\boldsymbol{B} = -\left[\frac{\boldsymbol{Q}_c t_c}{2}(t_c + t_b) + \frac{t_b^2}{8}(\bar{\boldsymbol{Q}}_{+\alpha} - \bar{\boldsymbol{Q}}_{-\alpha})\right]$$

$$\boldsymbol{D} = \frac{1}{3}\boldsymbol{Q}_c t_c(t_c^2 + \frac{3}{2}t_c t_b + \frac{3}{4}t_b^2) + \frac{t_b^3}{24}(\bar{\boldsymbol{Q}}_{+\alpha} + \bar{\boldsymbol{Q}}_{-\alpha})$$

Assuming that the bottom ply has an angle $\alpha = -45^\circ$ and substituting the given values, we obtain

$$\boldsymbol{A} = \begin{bmatrix} 61.4372 & 34.4521 & 0 \\ 34.4521 & 61.4372 & 0 \\ 0 & 0 & 38.0550 \end{bmatrix} \times 10^6 \text{ N/m}$$

$$\boldsymbol{B} = -\begin{bmatrix} 14.0353 & 4.2106 & 4.8797 \\ 4.2106 & 14.0353 & 4.8797 \\ 4.8797 & 4.8797 & 4.9124 \end{bmatrix} \times 10^3 \text{ N}$$

$$\boldsymbol{D} = \begin{bmatrix} 9.9100 & 3.8141 & 0 \\ 3.8141 & 9.9100 & 0 \\ 0 & 0 & 4.3375 \end{bmatrix} \text{ Nm}$$

The force and moment vectors are the same as in the case of the symmetric laminate. Solving the governing equation gives the strains and curvature as

$$\begin{Bmatrix} \epsilon_x^o \\ \epsilon_y^o \\ \gamma_{xy}^o \\ \kappa_x \\ \kappa_y \\ \kappa_{xy} \end{Bmatrix} = \begin{bmatrix} \boldsymbol{A} & \boldsymbol{B} \\ \boldsymbol{B} & \boldsymbol{D} \end{bmatrix}^{-1} \begin{Bmatrix} \boldsymbol{F}_\Lambda \\ \boldsymbol{M}_\Lambda \end{Bmatrix} = \begin{Bmatrix} -8.19 \times 10^{-6} \\ -8.19 \times 10^{-6} \\ 51.09 \times 10^{-6} \\ 0.1794 \text{ 1/m} \\ 0.1794 \text{ 1/m} \\ 0.0394 \text{ 1/m} \end{Bmatrix}$$

5.3 Classical Laminated Plate Theory (CLPT) Equations in Terms of Displacements

For a thin laminated plate undergoing small displacements, the plate deflections at any station are expressed in terms of mid-plane displacements as

$$u(x, y, z) = u_o(x, y) - z\frac{\partial w_o}{\partial x}(x, y) \tag{5.243}$$

$$v(x, y, z) = v_o(x, y) - z\frac{\partial w_o}{\partial y}(x, y) \tag{5.244}$$

$$w(x, y, z) = w_o(x, y) \tag{5.245}$$

The equilibrium equations are expressed as

$$\begin{Bmatrix} F_x \\ F_y \\ F_{xy} \\ M_x \\ M_y \\ M_{xy} \end{Bmatrix} = \begin{bmatrix} A_{11} & A_{12} & A_{16} & B_{11} & B_{12} & B_{16} \\ A_{12} & A_{22} & A_{26} & B_{12} & B_{22} & B_{26} \\ A_{16} & A_{26} & A_{66} & B_{16} & B_{26} & B_{66} \\ B_{11} & B_{12} & B_{16} & D_{11} & D_{12} & D_{16} \\ B_{12} & B_{22} & B_{26} & D_{12} & D_{22} & D_{26} \\ B_{16} & B_{26} & B_{66} & D_{16} & D_{26} & D_{66} \end{bmatrix} \begin{Bmatrix} \epsilon_x^o \\ \epsilon_y^o \\ \gamma_{xy}^o \\ \kappa_x \\ \kappa_y \\ \kappa_{xy} \end{Bmatrix} - \begin{Bmatrix} F_{x\Lambda} \\ F_{y\Lambda} \\ F_{xy\Lambda} \\ M_{x\Lambda} \\ M_{y\Lambda} \\ M_{xy\Lambda} \end{Bmatrix} \tag{5.246}$$

The vector on the left-hand side represents mechanical forces and moments, and the vector on the right-hand side represents generalized induced forces and moments. For a thin plate, the strain energy is

$$V_{\text{total}} = \frac{1}{2}\iiint_{volume} \left(\sigma_x\epsilon_x + \sigma_y\epsilon_y + \tau_{xy}\gamma_{xy}\right)\, dx\, dy\, dz \tag{5.247}$$

Including induced-strain actuation, the expression for energy becomes

$$V_{\text{total}} = \iint_{area} \left(\frac{1}{2}\{\boldsymbol{\epsilon}^{oT}\ \boldsymbol{\kappa}^T\} \begin{bmatrix} \boldsymbol{A} & \boldsymbol{B} \\ \boldsymbol{B} & \boldsymbol{D} \end{bmatrix} \begin{Bmatrix} \boldsymbol{\epsilon}^o \\ \boldsymbol{\kappa} \end{Bmatrix} - \{\boldsymbol{F}_\Lambda^T\ \boldsymbol{M}_\Lambda^T\} \begin{Bmatrix} \boldsymbol{\epsilon}^o \\ \boldsymbol{\kappa} \end{Bmatrix} \right) dx\, dy \tag{5.248}$$

where the integration is carried out over the surface of the plate. The force equilibrium equations in the x and y directions are

$$\frac{\partial F_x}{\partial x} + \frac{\partial F_{xy}}{\partial y} = -p_x \tag{5.249}$$

$$\frac{\partial F_{xy}}{\partial x} + \frac{\partial F_y}{\partial y} = -p_y \tag{5.250}$$

and the moment equilibrium equation is

$$\frac{\partial^2 M_x}{\partial x^2} + \frac{\partial^2 M_y}{\partial y^2} + 2\frac{\partial^2 M_{xy}}{\partial x \partial y} = p_z \tag{5.251}$$

where p_x, p_y, and p_z are surface forces per unit area in the x, y, and z direction. The governing equations in terms of displacements for induced actuation only are expressed as

u-equation:

$$\begin{aligned}
&A_{11}\frac{\partial^2 u_o}{\partial x^2} + A_{12}\frac{\partial^2 v_o}{\partial x \partial y} + A_{16}\left(2\frac{\partial^2 u_o}{\partial x \partial y} + \frac{\partial^2 v_o}{\partial x^2}\right) + A_{26}\frac{\partial^2 v_o}{\partial y^2} + A_{66}\left(\frac{\partial^2 u_o}{\partial y^2} + \frac{\partial^2 v_o}{\partial x \partial y}\right) \\
&\quad + B_{11}\frac{\partial^3 w_o}{\partial x^3} + B_{12}\frac{\partial^3 w_o}{\partial x \partial y^2} + 3B_{16}\frac{\partial^3 w_o}{\partial x^2 \partial y} + B_{26}\frac{\partial^3 w_o}{\partial y^3} + 2B_{66}\frac{\partial^3 w_o}{\partial x \partial y^2} \\
&\quad = \frac{\partial\left(F_x + F_{x_\Lambda}\right)}{\partial x} + \frac{\partial\left(F_{xy} + F_{xy_\Lambda}\right)}{\partial y} - p_x
\end{aligned} \tag{5.252}$$

v-equation:

$$\begin{aligned}
&A_{22}\frac{\partial^2 v_o}{\partial y^2} + A_{12}\frac{\partial^2 u_o}{\partial x \partial y} + A_{16}\frac{\partial^2 u_o}{\partial x^2} + A_{26}\left(\frac{\partial^2 u_o}{\partial y^2} + 2\frac{\partial^2 v_o}{\partial x \partial y}\right) + A_{66}\left(\frac{\partial^2 u_o}{\partial x \partial y} + \frac{\partial^2 v_o}{\partial x^2}\right) \\
&\quad + B_{12}\frac{\partial^3 w_o}{\partial x^2 \partial y} + B_{22}\frac{\partial^3 w_o}{\partial y^3} + B_{16}\frac{\partial^3 w_o}{\partial x^3} + 3B_{26}\frac{\partial^3 w_o}{\partial x \partial y^2} + 2B_{66}\frac{\partial^3 w_o}{\partial x^2 \partial y} \\
&\quad = \frac{\partial\left(F_y + F_{y_\Lambda}\right)}{\partial y} + \frac{\partial\left(F_{xy} + F_{xy_\Lambda}\right)}{\partial x} - p_y
\end{aligned} \tag{5.253}$$

w-equation:

$$\begin{aligned}
&B_{11}\frac{\partial^3 u_o}{\partial x^3} + B_{12}\left(\frac{\partial^3 v_o}{\partial x^2 \partial y} + \frac{\partial^3 u_o}{\partial x \partial y^2}\right) + B_{16}\left(3\frac{\partial^3 u_o}{\partial x^2 \partial y} + \frac{\partial^3 v_o}{\partial x^3}\right) + B_{22}\frac{\partial^3 v_o}{\partial y^3} \\
&\quad + B_{26}\left(\frac{\partial^3 u_o}{\partial y^3} + 3\frac{\partial^3 v_o}{\partial x \partial y^2}\right) + 2B_{66}\left(\frac{\partial^3 u_o}{\partial x \partial y^2} + \frac{\partial^3 v_o}{\partial x^2 \partial y}\right) \\
&\quad + D_{11}\frac{\partial^4 w_o}{\partial x^4} + 2D_{12}\frac{\partial^4 w_o}{\partial x^2 \partial y^2} + 4D_{16}\frac{\partial^4 w_o}{\partial x^3 \partial y} + D_{22}\frac{\partial^4 w_o}{\partial y^4} \\
&\quad + 4D_{26}\frac{\partial^4 w_o}{\partial x \partial y^3} + 4D_{66}\frac{\partial^4 w_o}{\partial x^2 \partial y^2} \\
&\quad = \frac{\partial^2\left(M_x + M_{x_\Lambda}\right)}{\partial x^2} + \frac{\partial^2\left(M_y + M_{y_\Lambda}\right)}{\partial y^2} + \frac{\partial^2\left(M_{xy} + M_{xy_\Lambda}\right)}{\partial x \partial y} - p_z
\end{aligned} \tag{5.254}$$

These three governing equations can be concisely put into operator form

$$\begin{bmatrix} \mathbb{D}_{u_1} & \mathbb{D}_{v_1} & \mathbb{D}_{w_1} \\ \mathbb{D}_{u_2} & \mathbb{D}_{v_2} & \mathbb{D}_{w_2} \\ \mathbb{D}_{u_3} & \mathbb{D}_{v_3} & \mathbb{D}_{w_3} \end{bmatrix} \begin{Bmatrix} u \\ v \\ w \end{Bmatrix} = \begin{Bmatrix} \frac{\partial F_{x_\Lambda}}{\partial x} + \frac{\partial F_{xy_\Lambda}}{\partial y} \\ \frac{\partial F_{xy_\Lambda}}{\partial x} + \frac{\partial F_{y_\Lambda}}{\partial y} \\ \frac{\partial^2 M_{x_\Lambda}}{\partial x^2} + 2\frac{\partial^2 M_{xy_\Lambda}}{\partial x \partial y} + \frac{\partial^2 M_{y_\Lambda}}{\partial y^2} \end{Bmatrix} \tag{5.255}$$

where the operators are given by

$$\mathbb{D}_{u_1} = A_{11}\frac{\partial^2}{\partial x^2} + 2A_{16}\frac{\partial^2}{\partial x \partial y} + A_{66}\frac{\partial^2}{\partial y^2}$$

$$\mathbb{D}_{v_1} = A_{12}\frac{\partial^2}{\partial x \partial y} + A_{16}\frac{\partial^2}{\partial x^2} + A_{26}\frac{\partial^2}{\partial y^2} + A_{66}\frac{\partial^2}{\partial x \partial y}$$

$$\mathbb{D}_{w_1} = B_{11}\frac{\partial^3}{\partial x^3} + B_{12}\frac{\partial^3}{\partial x \partial y^2} + 3B_{16}\frac{\partial^3}{\partial x^2 \partial y} + B_{26}\frac{\partial^3}{\partial y^3} + 2B_{66}\frac{\partial^3}{\partial x \partial y^2}$$

$$\mathbb{D}_{u_2} = A_{16}\frac{\partial^2}{\partial x^2} + (A_{12} + A_{66})\frac{\partial^2}{\partial x \partial y} + A_{26}\frac{\partial^2}{\partial y^2}$$

$$\mathbb{D}_{v_2} = 2A_{26}\frac{\partial^2}{\partial x \partial y} + (A_{22} + A_{66})\frac{\partial^2}{\partial y^2}$$

$$\mathbb{D}_{w_2} = B_{16}\frac{\partial^3}{\partial x^3} + (2B_{66} + B_{12})\frac{\partial^3}{\partial x^2 \partial y} + B_{22}\frac{\partial^3}{\partial y^3} + 3B_{26}\frac{\partial^3}{\partial x \partial y^2}$$

$$\mathbb{D}_{u_3} = B_{11}\frac{\partial^3}{\partial x^3} + B_{12}\frac{\partial^3}{\partial x \partial y^2} + 3B_{16}\frac{\partial^3}{\partial x^2 \partial y} + 2B_{66}\frac{\partial^3}{\partial x \partial y^2} + B_{26}\frac{\partial^3}{\partial y^3}$$

$$\mathbb{D}_{v_3} = B_{16}\frac{\partial^3}{\partial x^3} + B_{12}\frac{\partial^3}{\partial x^2 \partial y} + 2B_{66}\frac{\partial^3}{\partial x^2 \partial y} + 3B_{26}\frac{\partial^3}{\partial x \partial y^2} + B_{22}\frac{\partial^3}{\partial y^3}$$

$$\mathbb{D}_{w_3} = D_{11}\frac{\partial^4}{\partial x^4} + (2D_{12} + 4D_{66})\frac{\partial^4}{\partial x^2 \partial y^2} + 4D_{16}\frac{\partial^4}{\partial x^3 \partial y} + 4D_{26}\frac{\partial^4}{\partial x \partial y^3} + D_{22}\frac{\partial^4}{\partial y^4}$$

5.4 Approximate Solutions Using Energy Principles

Laminated plate equations with induced actuation are

$$\begin{bmatrix} \boldsymbol{A} & \boldsymbol{B} \\ \boldsymbol{B} & \boldsymbol{D} \end{bmatrix} \begin{Bmatrix} \epsilon^o \\ \kappa \end{Bmatrix} - \begin{Bmatrix} \boldsymbol{F}_\Lambda \\ \boldsymbol{M}_\Lambda \end{Bmatrix} = \begin{Bmatrix} \boldsymbol{F} \\ \boldsymbol{M} \end{Bmatrix} \tag{5.256}$$

One can obtain an exact solution for these equations only for a few selected cases, such as a uniform laminate with free boundary conditions. Note that in the previous analyses, we did not constrain the plate at its edges.

An exact solution must satisfy all equilibrium equations, boundary conditions, and compatibility relations simultaneously. Expressing equilibrium equations in terms of displacements inherently satisfies compatibility constraints. Again, it is not possible to find a response solution that satisfies the equilibrium equation and all boundary conditions. Therefore, for a generic plate problem, one is forced to estimate an approximate solution. The approximate solution is normally calculated either using energy principles or by using a weighted-residual approach.

The virtual strain energy δV of a deformed body is given by

$$\delta V = \iiint_{volume} \sigma \delta \epsilon \; dx \; dy \; dz \tag{5.257}$$

For a thin plate, the strain energy reduces to

$$V_{\text{total}} = \frac{1}{2}\iiint_{volume} \left(\sigma_x \epsilon_x + \sigma_y \epsilon_y + \tau_{xy}\gamma_{xy}\right) \; dx \; dy \; dz \tag{5.258}$$

With induced-strain actuation, this becomes

$$V_{\text{total}} = \iint_{area} \left(\frac{1}{2} \{\epsilon^{oT} \; \kappa^T\} \begin{bmatrix} \boldsymbol{A} & \boldsymbol{B} \\ \boldsymbol{B} & \boldsymbol{D} \end{bmatrix} \begin{Bmatrix} \epsilon^o \\ \kappa \end{Bmatrix} - \{\boldsymbol{F}_\Lambda^T \; \boldsymbol{M}_\Lambda^T\} \begin{Bmatrix} \epsilon^o \\ \kappa \end{Bmatrix} \right) dx\, dy \tag{5.259}$$

where the integral is evaluated over the area of the plate. For a continuous system, two of the popular approximate methods are the Rayleigh-Ritz method and the Galerkin method.

5.4.1 Galerkin Method

In the Galerkin method, the form of the assumed solution must be chosen in such a way that all boundary conditions (both geometric and forced) are identically satisfied regardless of the values of the undetermined coefficients. This requirement is extremely difficult to satisfy in all but the simplest problems.

The assumed displacement distributions are typically expressed in a series consisting of chosen basis functions with undetermined coefficients

$$u(x, y) = \sum_{i=1}^{M} \phi_{u_i}(x, y) q_i \tag{5.260}$$

$$v(x, y) = \sum_{j=1}^{N} \phi_{v_j}(x, y) q_{j+M} \tag{5.261}$$

$$w(x, y) = \sum_{k=1}^{P} \phi_{w_k}(x, y) q_{k+M+N} \tag{5.262}$$

Each one of these functions, $\phi_{u_i}, \phi_{v_j}, \phi_{w_k}$, must separately satisfy all of the boundary conditions (geometric and forced). In these equations, the q_n $(n = 1, 2, \ldots, M + N + P)$ are undetermined coefficients (with dimensions of length), and $\phi_{u_i}, \phi_{v_j}, \phi_{w_k}$ (dimensionless) are shape functions, respectively, representing longitudinal in-plane, transverse in-plane, and transverse out-of-plane displacement shapes. Eqs. 5.260–5.262 can be written in matrix form as

$$\boldsymbol{U} = \begin{bmatrix} \phi_{u_1} & \phi_{u_2} \ldots & \phi_{uM} & 0 & 0 \ldots & 0 & 0 & 0 \ldots & 0 \\ 0 & 0 \ldots & 0 & \phi_{v_1} & \phi_{v_2} \ldots & \phi_{vN} & 0 & 0 \ldots & 0 \\ 0 & 0 \ldots & 0 & 0 & 0 \ldots & 0 & \phi_{w_1} & \phi_{w_2} \ldots & \phi_{wP} \end{bmatrix} \begin{Bmatrix} q_1 \\ q_2 \\ \vdots \\ q_{M+N+P} \end{Bmatrix}$$

$$= \boldsymbol{\Phi}(x, y) \boldsymbol{q} \tag{5.263}$$

The size of the matrix $\boldsymbol{\Phi}$ is $3 \times (M + N + P)$ and the size of the matrix $\boldsymbol{q}$ is $(M + N + P) \times 1$.

Substituting these assumed-displacement functions into the plate-governing equations expressed in terms of displacements (Eq. 5.255), with no external forces $(F = 0, M = 0)$ results in an error function

$$\boldsymbol{\varepsilon}(x, y) = \begin{bmatrix} \mathbb{D}_{u_1} & \mathbb{D}_{v_1} & \mathbb{D}_{w_1} \\ \mathbb{D}_{u_2} & \mathbb{D}_{v_2} & \mathbb{D}_{w_2} \\ \mathbb{D}_{u_3} & \mathbb{D}_{v_3} & \mathbb{D}_{w_3} \end{bmatrix} \{\boldsymbol{\Phi q}\} - \begin{Bmatrix} \dfrac{\partial F_{x\Lambda}}{\partial x} + \dfrac{\partial F_{xy\Lambda}}{\partial y} \\ \dfrac{\partial F_{xy\Lambda}}{\partial x} + \dfrac{\partial F_{y\Lambda}}{\partial y} \\ \dfrac{\partial^2 M_{x\Lambda}}{\partial x^2} + 2\dfrac{\partial^2 M_{xy\Lambda}}{\partial x \partial y} + \dfrac{\partial^2 M_{y\Lambda}}{\partial y^2} \end{Bmatrix} \tag{5.264}$$

If the assumed solution had been an exact solution, this error function would have been identically zero. Through the Galerkin method, this error is minimized by orthogonalizing it with respect to each assumed function over the entire solution domain.

$$\iint_{area} \boldsymbol{\phi}_j^T \boldsymbol{\varepsilon}(x, y) dx\, dy = 0 \quad \text{for } j = 1, 2 \ldots (M + N + P) \tag{5.265}$$

where the vector $\boldsymbol{\phi}_j$ (of size 3×1) is the jth column in the matrix $\boldsymbol{\Phi}$ and corresponds to the jth mode in the assumed response. The weighted residual minimization approach results in an $(M + N + P)$ set of equations, which can be concisely put into a matrix form

$$\boldsymbol{Kq} = \boldsymbol{Q}_\Lambda \tag{5.266}$$

or

$$K_{ij} q_j = Q_{\Lambda_i} \tag{5.267}$$

where the generalized stiffness matrix is defined as

$$K_{ij} = \iint_{area} \boldsymbol{\phi}_i^T \begin{bmatrix} \mathbb{D}_{u_1} & \mathbb{D}_{v_1} & \mathbb{D}_{w_1} \\ \mathbb{D}_{u_2} & \mathbb{D}_{v_2} & \mathbb{D}_{w_2} \\ \mathbb{D}_{u_3} & \mathbb{D}_{v_3} & \mathbb{D}_{w_3} \end{bmatrix} \boldsymbol{\phi}_j \, dx\, dy \tag{5.268}$$

and the generalized forcing vector is defined as

$$Q_{\Lambda_i} = \iint_{area} \boldsymbol{\phi}_i^T \begin{Bmatrix} \dfrac{\partial F_{x\Lambda}}{\partial x} + \dfrac{\partial F_{xy\Lambda}}{\partial y} \\ \dfrac{\partial F_{xy\Lambda}}{\partial x} + \dfrac{\partial F_{y\Lambda}}{\partial y} \\ \dfrac{\partial^2 M_{x\Lambda}}{\partial x^2} + 2\dfrac{\partial^2 M_{xy\Lambda}}{\partial x \partial y} + \dfrac{\partial^2 M_{y\Lambda}}{\partial y^2} \end{Bmatrix} dx\, dy \tag{5.269}$$

Note that the generalized stiffness matrix $\boldsymbol{K}$ has the dimensions of force per unit length (N/m) and the generalized forcing vector $\boldsymbol{Q}_\Lambda$ has the dimensions of force (N). Normally, the solution monotonically approaches the exact solution as the number of terms in the approximate series is increased. The Galerkin solution understimates the response compared to the exact solution, which means that it overestimates the stiffness. In general, it is extremely difficult to choose $\boldsymbol{\Phi}(x, y)$ that satisfies all boundary conditions.

5.4.2 Rayleigh-Ritz Method

In the Rayleigh-Ritz method, an acceptable trial solution form is much easier to derive because it must satisfy only the geometric boundary conditions. This is because the Rayleigh-Ritz method utilizes energy expressions that incorporate the force boundary conditions as part of the statement of the problem; for example, Lagrange's equations are:

$$\frac{d}{dt}\left(\frac{\partial T}{\partial \dot{q}_i}\right) - \frac{\partial T}{\partial q_i} + \frac{\partial V}{\partial q_i} = Q_i \tag{5.270}$$

where $q_i = q_1,\ q_2\ \ldots\ q_{M+N+P}$

In this equation, V is the strain energy, T is the kinetic energy, and Q_i is the generalized force associated with the undetermined coefficients q_i. For a static problem, the kinetic energy is zero. Assuming a solution of the form

$$u(x, y) = \sum_{i=1}^{M} \phi_{u_i}(x, y) q_i \tag{5.271}$$

$$v(x, y) = \sum_{j=1}^{N} \phi_{v_j}(x, y) q_{j+M} \tag{5.272}$$

$$w(x, y) = \sum_{k=1}^{P} \phi_{w_k}(x, y) q_{k+M+N} \tag{5.273}$$

Expressing strains and curvatures in terms of basic displacements

$$\begin{Bmatrix} \boldsymbol{\epsilon}^o \\ \boldsymbol{\kappa} \end{Bmatrix} = \begin{bmatrix} \frac{\partial}{\partial x} & 0 & 0 \\ 0 & \frac{\partial}{\partial y} & 0 \\ \frac{\partial}{\partial y} & \frac{\partial}{\partial x} & 0 \\ 0 & 0 & \frac{\partial^2}{\partial x^2} \\ 0 & 0 & \frac{\partial^2}{\partial y^2} \\ 0 & 0 & 2\frac{\partial^2}{\partial x \partial y} \end{bmatrix} \begin{Bmatrix} u \\ v \\ w \end{Bmatrix} = \mathbb{D}\boldsymbol{U} \tag{5.274}$$

where the operator $\mathbb{D}$ is of order 6×3. Therefore

$$\begin{Bmatrix} \boldsymbol{\epsilon}^o \\ \boldsymbol{\kappa} \end{Bmatrix} = \mathbb{D}\boldsymbol{U} \tag{5.275}$$

The mid-plane strains and curvatures can be expressed as

$$\begin{Bmatrix} \boldsymbol{\epsilon}^o \\ \boldsymbol{\kappa} \end{Bmatrix} = \mathbb{D}\boldsymbol{\phi}\boldsymbol{q} \tag{5.276}$$

For the static case, Lagrange's equations reduce to

$$\frac{\partial V}{\partial q_i} = Q_i \tag{5.277}$$

The generalized force, Q_i, is found from the virtual work done by external forces

$$\delta W_{\text{out}i} = \iint_{area} \{\boldsymbol{F}^T \; \boldsymbol{M}^T\}\{\mathbb{D}\boldsymbol{\phi}_i\}\delta q_i \, dA = Q_i \delta q_i \tag{5.278}$$

Because the external forces are zero, $\delta W_{\text{out}i} = 0$ and $Q_i = 0$. The total strain energy for a thin laminated plate is (Eq. 5.259)

$$\begin{aligned} V_{\text{total}} &= \frac{1}{2} \iint_{area} \{\mathbb{D}\boldsymbol{\phi q}\}^T \begin{bmatrix} \boldsymbol{A} & \boldsymbol{B} \\ \boldsymbol{B} & \boldsymbol{D} \end{bmatrix} \{\mathbb{D}\boldsymbol{\phi q}\} dA - \iint_{area} \{\boldsymbol{F}_\Lambda^T \; \boldsymbol{M}_\Lambda^T\}\{\mathbb{D}\boldsymbol{\phi q}\} dA \\ &= \frac{1}{2} \sum_{i=1}^{(M+N+P)} \sum_{j=1}^{(M+N+P)} K_{ij} q_i q_j - \sum_{i=1}^{(M+N+P)} Q_{\Lambda_i} q_i \\ &= \frac{1}{2} \boldsymbol{q}^T \boldsymbol{K q} - \boldsymbol{q}^T \boldsymbol{Q}_\Lambda \end{aligned} \tag{5.279}$$

Substituting this equation in Eq. 5.277 leads to

$$\boldsymbol{Kq} - \boldsymbol{Q}_\Lambda = 0 \quad \rightarrow \quad [K_{ij}]\{q_j\} = \{Q_{\Lambda_i}\} \tag{5.280}$$

where $\boldsymbol{K}$ is the generalized stiffness matrix, defined as

$$K_{ij} = \iint_{area} \{\mathbb{D}\boldsymbol{\phi}_i\}^T \begin{bmatrix} \boldsymbol{A} & \boldsymbol{B} \\ \boldsymbol{B} & \boldsymbol{D} \end{bmatrix} \{\mathbb{D}\boldsymbol{\phi}_j\}\, dx\, dy \tag{5.281}$$

and the generalized force is

$$Q_{\Lambda i} = \iint_{area} \{\boldsymbol{F}_\Lambda^T\, \boldsymbol{M}_\Lambda^T\} \{\mathbb{D}\boldsymbol{\phi}_i\}\, dx\, dy \tag{5.282}$$

The vector $\boldsymbol{\phi}_i$ corresponds to the ith mode in the assumed response and is of size 3×1. Using Lagrange's equations

$$\boldsymbol{Kq} = \boldsymbol{Q}_\Lambda \tag{5.283}$$

$\boldsymbol{K}$ is of size $(M+N+P) \times (M+N+P)$, $\boldsymbol{Q}_\Lambda$ is of size $(M+N+P) \times 1$, and $\boldsymbol{q}$ is of size $(M+N+P) \times 1$. The generalized stiffness matrix $\boldsymbol{K}$ has the dimensions of force per unit length (N/m) and the generalized forcing vector $\boldsymbol{Q}_\Lambda$ has the dimensions of force (N).

$$\boldsymbol{q} = \boldsymbol{K}^{-1}\boldsymbol{Q}_\Lambda \tag{5.284}$$

$$\boldsymbol{U} = \begin{Bmatrix} u \\ v \\ w \end{Bmatrix} = \boldsymbol{\phi q} \tag{5.285}$$

Note that ϕ_{u_i}, ϕ_{v_j}, and ϕ_{w_k} need to satisfy only the geometric boundary conditions. Again, like the Galerkin solution, the Rayleigh-Ritz solution underestimates the response and the solution approaches the exact solution as the number of terms in the approximate series is increased. Note that if the assumed response satisfies all of the boundary conditions, then the Rayleigh-Ritz and Galerkin methods result in identical solutions.

5.4.3 Symmetric Laminated Plate Response

Consider a symmetric laminated plate, where the coupling matrix $\boldsymbol{B} = 0$. The governing equations reduce into two sets of uncoupled equations

$$\boldsymbol{A\epsilon^o} = \boldsymbol{F}_\Lambda \tag{5.286}$$

$$\boldsymbol{D\kappa} = \boldsymbol{M}_\Lambda \tag{5.287}$$

The strain energy is divided into two parts: extensional and bending.

Case I: Pure Extension

Extensional strain energy is given by

$$\begin{aligned} V_{\text{ext}} &= \frac{1}{2}\iint_{area} \boldsymbol{\epsilon}^{oT}\boldsymbol{A\epsilon}^o dx\, dy - \iint_{area} \boldsymbol{F}_\Lambda^T \boldsymbol{\epsilon}^o dx\, dy \\ &= \frac{1}{2}\iint_{area} \{\mathbb{D}_{\text{ext}}\boldsymbol{\phi}_{\text{ext}}\boldsymbol{q}_{\text{ext}}\}^T \boldsymbol{A}\{\mathbb{D}_{\text{ext}}\boldsymbol{\phi}_{\text{ext}}\boldsymbol{q}_{\text{ext}}\} dx\, dy \\ &\quad - \iint_{area} \boldsymbol{F}_\Lambda^T\{\mathbb{D}_{\text{ext}}\boldsymbol{\phi}_{\text{ext}}\boldsymbol{q}_{\text{ext}}\} dx\, dy \end{aligned} \tag{5.288}$$

where, neglecting terms related to bending

$$\mathbb{D}_{\text{ext}} = \begin{bmatrix} \frac{\partial}{\partial x} & 0 \\ 0 & \frac{\partial}{\partial y} \\ \frac{\partial}{\partial y} & \frac{\partial}{\partial x} \end{bmatrix} \tag{5.289}$$

$$\boldsymbol{\phi}_{\text{ext}} = \begin{bmatrix} \phi_{u_1} & \phi_{u_2} \dots & \phi_{u_M} & 0 & 0 \dots & 0 \\ 0 & 0 \dots & 0 & \phi_{v_1} & \phi_{v_2} \dots & \phi_{v_N} \end{bmatrix} \tag{5.290}$$

$$\boldsymbol{q}_{\text{ext}} = \begin{Bmatrix} q_1 \\ q_2 \\ \vdots \\ q_{M+N} \end{Bmatrix} \tag{5.291}$$

Eq. 5.288 can be rewritten as

$$\begin{aligned} V_{\text{ext}} &= \frac{1}{2} \iint_{area} \boldsymbol{q}_{\text{ext}}{}^T [\mathbb{D}_{\text{ext}} \boldsymbol{\phi}_{\text{ext}}]^T \boldsymbol{A} [\mathbb{D}_{\text{ext}} \boldsymbol{\phi}_{\text{ext}}] \boldsymbol{q}_{\text{ext}} dx\, dy \\ &\quad - \iint_{area} \boldsymbol{F}_\Lambda^T [\mathbb{D}_{\text{ext}} \boldsymbol{\phi}_{\text{ext}}]\, \boldsymbol{q}_{\text{ext}} dx\, dy \\ &= \frac{1}{2} \sum_{i=1}^{(M+N)} \sum_{j=1}^{(M+N)} K_{ij\,\text{ext}} q_i q_j - \sum_{i=1}^{(M+N)} Q_{\Lambda_i} q_i \end{aligned} \tag{5.292}$$

where $K_{ij\,\text{ext}}$ are elements of a generalized stiffness matrix of size $(M + N) \times (M + N)$, given by

$$K_{ij\,\text{ext}} = \iint_{area} \{\mathbb{D}_{\text{ext}} \boldsymbol{\phi}_{i\text{ext}}\}^T \boldsymbol{A} \left\{\mathbb{D}_{\text{ext}} \boldsymbol{\phi}_{j\,\text{ext}}\right\} dx\, dy \tag{5.293}$$

and the elements of the generalized forcing vector of size $(M + N) \times 1$ are

$$Q_{\Lambda_i} = \iint_{area} \boldsymbol{F}_\Lambda^T \{\mathbb{D}_{\text{ext}} \boldsymbol{\phi}_{i_{\text{ext}}}\} dx\, dy \tag{5.294}$$

This results in the matrix equation

$$\boldsymbol{K q} = \boldsymbol{Q}_\Lambda \tag{5.295}$$

Case II: Pure Bending

Proceeding in a similar manner, the bending strain energy is given by

$$\begin{aligned} V_{\text{bend}} &= \frac{1}{2} \iint_{area} \boldsymbol{\kappa}^T \boldsymbol{D} \boldsymbol{\kappa} dx\, dy - \iint_{area} \boldsymbol{M}_\Lambda^T \boldsymbol{\kappa} dx\, dy \\ &= \frac{1}{2} \iint_{area} \{\mathbb{D}_{\text{bend}} \boldsymbol{\phi}_{\text{bend}} \boldsymbol{q}_{\text{bend}}\}^T \boldsymbol{D} \{\mathbb{D}_{\text{bend}} \boldsymbol{\phi}_{\text{bend}} \boldsymbol{q}_{\text{bend}}\} dx\, dy \\ &\quad - \iint_{area} \boldsymbol{M}_\Lambda^T \{\mathbb{D}_{\text{bend}} \boldsymbol{\phi}_{\text{bend}} \boldsymbol{q}_{\text{bend}}\} dx\, dy \end{aligned} \tag{5.296}$$

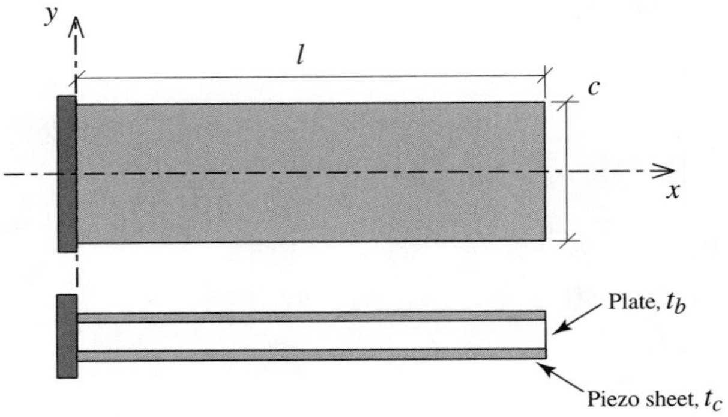

Figure 5.13. Cantilevered plate with piezo sheet actuators.

where, similar to Eqs. 5.274–5.275

$$\mathbb{D}_{\text{bend}} = \begin{Bmatrix} \frac{\partial^2}{\partial x^2} \\ \frac{\partial^2}{\partial y^2} \\ 2\frac{\partial^2}{\partial x \partial y} \end{Bmatrix} \tag{5.297}$$

$$\boldsymbol{\phi}_{\text{bend}} = \{\phi_{w_1} \quad \phi_{w_2} \ldots \quad \phi_{w_P}\} \tag{5.298}$$

$$\boldsymbol{q}_{\text{bend}} = \begin{Bmatrix} q_1 \\ q_2 \\ \vdots \\ q_P \end{Bmatrix} \tag{5.299}$$

Eq. 5.296 can be rewritten as

$$\begin{aligned} V_{\text{bend}} &= \frac{1}{2}\iint_{area} \boldsymbol{q}_{\text{bend}}{}^T \left\{\mathbb{D}_{\text{bend}}\boldsymbol{\phi}_{\text{bend}}\right\}^T \boldsymbol{D} \left\{\mathbb{D}_{\text{bend}}\boldsymbol{\phi}_{\text{bend}}\right\} \boldsymbol{q}_{\text{bend}} dx\, dy \\ &- \iint_{area} \boldsymbol{M}_\Lambda^T \{\mathbb{D}_{\text{bend}}\boldsymbol{\phi}_{\text{bend}}\}\boldsymbol{q}_{\text{bend}} dx\, dy \\ &= \frac{1}{2}\sum_{i=1}^{P}\sum_{j=1}^{P} K_{ij\,\text{bend}} q_i q_j - \sum_{i=1}^{P} Q_{\Lambda_i} q_i \end{aligned} \tag{5.300}$$

where $K_{ij\,\text{bend}}$ are elements of a generalized stiffness matrix of size $P \times P$, given by

$$K_{ij\,\text{bend}} = \iint_{area} \left\{\mathbb{D}_{\text{bend}}\boldsymbol{\phi}_{i_{\text{bend}}}\right\}^T \boldsymbol{D} \left\{\mathbb{D}_{\text{bend}}\boldsymbol{\phi}_{j_{\text{bend}}}\right\} dx\, dy \tag{5.301}$$

and the elements of the generalized forcing vector of size $P \times 1$ are

$$Q_{\Lambda_i} = \iint_{area} \boldsymbol{M}_\Lambda^T \{\mathbb{D}_{\text{bend}}\boldsymbol{\phi}_{i_{\text{bend}}}\} dx\, dy \tag{5.302}$$

5.4.4 Laminated Plate with Induced-Strain Actuation

Let us consider a uniform cantilevered plate of length l and width c, with piezo sheets bonded to both surfaces (Fig. 5.13). The response of the structure is assumed to be a summation of functions such that each function separately satisfies at least geometric boundary conditions.

Pure Extension

For extensional actuation, the same voltage is applied to the top and bottom piezos, which are identical. In this case

$$\boldsymbol{F}_{\Lambda} = \begin{Bmatrix} F_{x_\Lambda} \\ F_{y_\Lambda} \\ 0 \end{Bmatrix} \tag{5.303}$$

$$\boldsymbol{M}_{\Lambda} = \begin{Bmatrix} 0 \\ 0 \\ 0 \end{Bmatrix} \tag{5.304}$$

To simplify the example, let us assume that the stresses and strains in the y direction can be ignored. Then

$$\boldsymbol{F}_{\Lambda} = \begin{Bmatrix} F_{x_\Lambda} \\ 0 \\ 0 \end{Bmatrix} \tag{5.305}$$

$$\boldsymbol{M}_{\Lambda} = \begin{Bmatrix} 0 \\ 0 \\ 0 \end{Bmatrix} \tag{5.306}$$

Let us consider a one-term solution using the Rayleigh-Ritz method, given by the shape functions

$$\begin{aligned} u &= \sum_{i=1}^{N} \phi_{u_i} q_i = \frac{x}{l} q_1 = \phi_{u_1} q_1 \\ v &= 0 \end{aligned} \tag{5.307}$$

At $x = 0$, $u = 0$ and the assumed shape function satisfies this geometric boundary condition. Substituting in Eq. 5.280 and ignoring the terms corresponding to the y direction deformation gives

$$K_{11} q_1 = Q_{\Lambda_1} \tag{5.308}$$

where

$$\begin{aligned} K_{11} &= \int_0^l \int_{-c/2}^{c/2} \left\{ \frac{\partial \phi_{u_1}}{\partial x} \quad 0 \quad \frac{\partial \phi_{u_1}}{\partial y} \right\} \boldsymbol{A} \begin{Bmatrix} \frac{\partial \phi_{u_1}}{\partial x} \\ 0 \\ \frac{\partial \phi_{u_1}}{\partial y} \end{Bmatrix} dx\, dy \\ &= \int_0^l \int_{-c/2}^{c/2} \frac{\partial \phi_{u_1}}{\partial x} A_{11} \frac{\partial \phi_{u_1}}{\partial x} dx\, dy \\ &= \int_0^l \int_{-c/2}^{c/2} \frac{A_{11}}{l^2} dx\, dy \\ &= \frac{A_{11}}{l} c \end{aligned} \tag{5.309}$$

Assuming the piezos are attached to the entire surface of the plate

$$
\begin{aligned}
Q_{\Lambda_1} &= \int_0^l \int_{-c/2}^{c/2} \{F_{x_\Lambda} \quad F_{y_\Lambda} \quad 0\} \begin{Bmatrix} \frac{\partial \phi_{u_1}}{\partial x} \\ 0 \\ \frac{\partial \phi_{u_1}}{\partial y} \end{Bmatrix} dx\, dy \\
&= \int_0^l \int_{-c/2}^{c/2} F_{x_\Lambda} \frac{\partial \phi_{u_1}}{\partial x} dx\, dy \\
&= \int_0^l \int_{-c/2}^{c/2} \frac{F_{x_\Lambda}}{l} dx\, dy \\
&= F_{x_\Lambda} c
\end{aligned} \tag{5.310}
$$

Substituting these expressions for generalized stiffness and generalized force in Eq. 5.308 gives

$$\frac{A_{11}}{l} c q_1 = F_{x_\Lambda} c \tag{5.311}$$

$$q_1 = \frac{F_{x_\Lambda}}{A_{11}} l \tag{5.312}$$

The displacement distribution is

$$u(x, y) = \frac{F_{x_\Lambda}}{A_{11}} x \tag{5.313}$$

where

$$F_{x_\Lambda} = \frac{2 E_c t_c \Lambda}{1 - \nu} \tag{5.314}$$

In a similar way, one can obtain one-term transverse in-plane displacement by assuming

$$v(x, y) = \frac{y}{c} q_1 \tag{5.315}$$

Let us consider a one-term solution using the Galerkin method. Assume a one-term solution as

$$
\begin{aligned}
u_o(x, y) &= \left[\frac{x}{l} - \frac{1}{2} \left(\frac{x}{l} \right)^2 \right] q_1 \\
&= \phi_{u_1}(x) q_1
\end{aligned} \tag{5.316}
$$

$$v_o(x, y) = 0$$

This satisfies both geometric and forced boundary conditions

$$u_o(0, y) = 0$$

$$F_x(x = l) = \frac{Et}{(1 - \nu^2)} \frac{\partial u_o}{\partial x} = 0$$

The governing equation is

$$A_{11} \frac{\partial^2 u_o}{\partial x^2} = \frac{\partial F_{x_\Lambda}}{\partial x} \tag{5.317}$$

Substituting the assumed response

$$\varepsilon(x) = A_{11}\frac{\partial^2 \phi_{u_1}}{\partial x^2} q_1 - \frac{\partial F_{x_\Lambda}}{\partial x} \tag{5.318}$$

Minimizing the error results in

$$\int_0^l \int_{-c/2}^{c/2} \phi_{u_1}\left[A_{11}\frac{\partial^2 u_o}{\partial x^2} - \frac{\partial F_{x_\Lambda}}{\partial x}\right] dx\, dy = 0 \tag{5.319}$$

This gives

$$q_1 = \frac{3}{2}\frac{lF_{x_\Lambda}}{A_{11}} \tag{5.320}$$

where

$$\begin{aligned} F_{x_\Lambda} &= \frac{2E_c t_c \Lambda}{1-\nu} \\ A_{11} &= \frac{E_b t_b + 2E_c t_c}{1-\nu^2} \end{aligned} \tag{5.321}$$

and the displacement is

$$u_o(x, y) = \frac{3}{2}\frac{lF_{x_\Lambda}}{A_{11}}\left[\frac{x}{l} - \frac{1}{2}\left(\frac{x}{l}\right)^2\right] \tag{5.322}$$

Pure Bending

For bending actuation of the cantilevered plate, an opposite voltage is applied to the top and bottom piezo sheets, which are identical in magnitude. Assume that a positive voltage is applied to the bottom piezo and a negative voltage to the top piezo. To keep the example simple, let us also assume that there is no bending in the y direction. Then, the forcing vectors are given by

$$\boldsymbol{F}_\Lambda = \begin{Bmatrix} 0 \\ 0 \\ 0 \end{Bmatrix} \tag{5.323}$$

$$\boldsymbol{M}_\Lambda = \begin{Bmatrix} M_{x_\Lambda} \\ 0 \\ 0 \end{Bmatrix} \tag{5.324}$$

Rayleigh-Ritz Solution

Assume that the w displacement is given by a one-term solution of the form

$$w(x, y) = \frac{x^2}{l^2} q_1 = \phi_{w_1} q_1 \tag{5.325}$$

At $x = 0$, $\phi_{w1} = 0$ and $d\phi_{w1}/dx = 0$ satisfies the geometric boundary condition. Substituting the assumed deflection in Eq. 5.280 gives

$$K_{11} q_1 = Q_{\Lambda_1} \tag{5.326}$$

where the generalized stiffness is

$$\begin{aligned}
K_{11} &= \int_0^l \int_{-c/2}^{c/2} \left\{ \frac{\partial^2 \phi_{w_1}}{\partial x^2} \quad \frac{\partial^2 \phi_{w_1}}{\partial y^2} \quad 2\frac{\partial^2 \phi_{w_1}}{\partial x \partial y} \right\} \boldsymbol{D} \begin{Bmatrix} \frac{\partial^2 \phi_{w_1}}{\partial x^2} \\ \frac{\partial^2 \phi_{w_1}}{\partial y^2} \\ 2\frac{\partial^2 \phi_{w_1}}{\partial x \partial y} \end{Bmatrix} dx\, dy \\
&= \int_0^l \int_{-c/2}^{c/2} \frac{\partial^2 \phi_{w_1}}{\partial x^2} D_{11} \frac{\partial^2 \phi_{w_1}}{\partial x^2} dx\, dy \\
&= \int_0^l \int_{-c/2}^{c/2} \frac{4D_{11}}{l^4} dx\, dy \\
&= \frac{4cD_{11}}{l^3}
\end{aligned} \tag{5.327}$$

The equation for the generalized force becomes

$$\begin{aligned}
Q_{\Lambda_1} &= \int_0^l \int_{-c/2}^{c/2} \{ M_{x_\Lambda} \quad M_{y_\Lambda} \quad 0 \} \begin{Bmatrix} \frac{\partial^2 \phi_{w_1}}{\partial x^2} \\ \frac{\partial^2 \phi_{w_1}}{\partial y^2} \\ 2\frac{\partial^2 \phi_{w_1}}{\partial x \partial y} \end{Bmatrix} dx\, dy \\
&= \int_0^l \int_{-c/2}^{c/2} M_{x_\Lambda} \frac{\partial^2 \phi_{w_1}}{\partial x^2} dx\, dy \\
&= \int_0^l \int_{-c/2}^{c/2} M_{x_\Lambda} \frac{2}{l^2} dx\, dy \\
&= \frac{2M_{x_\Lambda} c}{l}
\end{aligned} \tag{5.328}$$

Substituting these expressions in Eq. 5.326 gives

$$\begin{aligned}
q_1 &= \frac{M_{x_\Lambda} l^2}{2D_{11}} \\
w(x, y) &= \frac{M_{x_\Lambda} x^2}{2D_{11}}
\end{aligned} \tag{5.329}$$

In a similar way, one can calculate the transverse bending displacement using one-term approximation as

$$w(x, y) = \frac{y^2}{c^2} q_1 \tag{5.330}$$

Galerkin Solution

Let us consider a one-term bending solution using the Galerkin method. Neglecting the variation along the y-axis, the one-term response is assumed as

$$\begin{aligned}
w(x, y) &= \left[6\left(\frac{x}{l}\right)^2 - 4\left(\frac{x}{l}\right)^3 + \left(\frac{x}{l}\right)^4 \right] q_1 \\
&= \phi_{w_1}(x) q_1
\end{aligned} \tag{5.331}$$

This satisfies both the geometric and forced boundary conditions.

$$\begin{aligned} w(0, y) &= 0 \\ \left.\frac{\partial w}{\partial x}\right|_{x=0} &= 0 \\ \left.\frac{\partial^2 w}{\partial x^2}\right|_{x=0} &= 0 \\ \left.\frac{\partial^3 w}{\partial x^3}\right|_{x=0} &= 0 \end{aligned} \tag{5.332}$$

The governing equation is

$$D_{11}\frac{\partial^4 w}{\partial x^4} = \frac{\partial^2 M_{x_\Lambda}}{\partial x^2} \tag{5.333}$$

Substituting the assumed response gives an error function

$$\varepsilon(x) = D_{11}\frac{\partial^4 \phi_{w_1}}{\partial x^4}q_1 - \frac{\partial^2 M_{x_\Lambda}}{\partial x^2} \tag{5.334}$$

Minimizing the error results in

$$\int_0^l \int_{-c/2}^{c/2} \phi_{w_1}\left[D_{11}\frac{\partial^4 \phi_{w_1}}{\partial x^4}q_1 - \frac{\partial^2 M_{x_\Lambda}}{\partial x^2}\right] dx\, dy = 0 \tag{5.335}$$

which yields

$$q_1 = \frac{5}{36}\frac{M_{x_\Lambda} l^2}{D_{11}} \tag{5.336}$$

and the displacement is given by

$$w(x, y) = \frac{5}{36}\frac{M_{x_\Lambda} l^2}{D_{11}}\left[6\left(\frac{x}{l}\right)^2 - 4\left(\frac{x}{l}\right)^3 + \left(\frac{x}{l}\right)^4\right] \tag{5.337}$$

5.4.5 Laminated Plate with Antisymmetric Layup: Extension-Torsion Coupling

Let us now examine the induced response of a plate with an antisymmetric layup of generally orthotropic plies. Because an antisymmetric plate exhibits coupling between extension and torsion, we consider the effect of piezo-induced pure extension. To achieve this, identical piezo actuators bonded on the top and bottom of the plate are actuated with the same voltage.

In the absence of external loads, the governing equations for the plate are

$$\boldsymbol{F}_\Lambda = \begin{bmatrix} A_{11} & A_{12} & 0 \\ A_{12} & A_{22} & 0 \\ 0 & 0 & A_{66} \end{bmatrix}\boldsymbol{\epsilon}^o + \begin{bmatrix} 0 & 0 & B_{16} \\ 0 & 0 & B_{26} \\ B_{16} & B_{26} & 0 \end{bmatrix}\boldsymbol{\kappa} \tag{5.338}$$

$$\boldsymbol{M}_\Lambda = \begin{bmatrix} 0 & 0 & B_{16} \\ 0 & 0 & B_{26} \\ B_{16} & B_{26} & 0 \end{bmatrix}\boldsymbol{\epsilon}^o + \begin{bmatrix} D_{11} & D_{12} & 0 \\ D_{12} & D_{22} & 0 \\ 0 & 0 & D_{66} \end{bmatrix}\boldsymbol{\kappa} \tag{5.339}$$

For purely extensional actuation, the induced-force and moment vectors for this configuration are

$$\boldsymbol{F}_\Lambda = \begin{Bmatrix} F_{x_\Lambda} \\ F_{y_\Lambda} \\ 0 \end{Bmatrix} \tag{5.340}$$

$$\boldsymbol{M}_\Lambda = \begin{Bmatrix} 0 \\ 0 \\ 0 \end{Bmatrix} \tag{5.341}$$

To focus on the effect of extension-torsion coupling, the problem is simplified by assuming that the piezo actuator only imparts force in the x direction and by ignoring any stress or deformation in the y direction. Under these assumptions, the induced-force and moment vectors for this configuration are

$$\boldsymbol{F}_\Lambda = \begin{Bmatrix} F_{x_\Lambda} \\ 0 \\ 0 \end{Bmatrix} \tag{5.342}$$

$$\boldsymbol{M}_\Lambda = \begin{Bmatrix} 0 \\ 0 \\ 0 \end{Bmatrix} \tag{5.343}$$

Ignoring the terms corresponding to ϵ_y^o, ϵ_{xy}^o, κ_x, and κ_y, the governing equation, differential operator matrix, and deformation vector simplify to

$$\begin{bmatrix} A_{11} & B_{16} \\ B_{16} & D_{66} \end{bmatrix} \begin{Bmatrix} \epsilon_x^o \\ \kappa_{xy} \end{Bmatrix} = \begin{Bmatrix} F_{x_\Lambda} \\ 0 \end{Bmatrix} \tag{5.344}$$

and

$$\begin{Bmatrix} \epsilon_x^o \\ \kappa_{xy} \end{Bmatrix} = \mathbb{D}\boldsymbol{U} = \begin{bmatrix} \frac{\partial}{\partial x} & 0 \\ 0 & 2\frac{\partial^2}{\partial x \partial y} \end{bmatrix} \begin{Bmatrix} u \\ w \end{Bmatrix} \tag{5.345}$$

The following one-term solutions for u and w identically satisfy the cantilevered boundary conditions at the root of the plate

$$u(x, y) = \frac{x}{l} q_1 = \phi_1 q_1 \tag{5.346}$$

$$w(x, y) = \frac{x^2}{l^2} \frac{y}{c} q_2 = \phi_2 q_2 \tag{5.347}$$

At $x = 0$, the chosen shape functions ensure that the cantilevered boundary conditions are satisfied, regardless of the values of the coefficients q_1 and q_2

$$u = 0 \qquad \phi_1 = 0 \tag{5.348}$$

$$w = 0 \qquad \phi_2 = 0 \tag{5.349}$$

$$\frac{\partial w}{\partial x} = 0 \qquad \frac{\partial \phi_2}{\partial x} = 0 \tag{5.350}$$

From the assumed shape functions, we can write

$$\boldsymbol{\phi} = \begin{bmatrix} \phi_1 & 0 \\ 0 & \phi_2 \end{bmatrix} \tag{5.351}$$

The generalized degrees of freedom q_1 and q_2 are calculated from (Eq. 5.280)

$$\begin{bmatrix} K_{11} & K_{12} \\ K_{12} & K_{22} \end{bmatrix} \begin{Bmatrix} q_1 \\ q_2 \end{Bmatrix} = \begin{Bmatrix} Q_{\Lambda_1} \\ Q_{\Lambda_2} \end{Bmatrix} \tag{5.352}$$

where

$$\begin{aligned} K_{11} &= \int_0^l \int_{-c/2}^{c/2} \{\mathbb{D}\boldsymbol{\phi}_1\}^T \begin{bmatrix} A_{11} & B_{16} \\ B_{16} & D_{66} \end{bmatrix} \{\mathbb{D}\boldsymbol{\phi}_1\}\, dx\, dy \\ &= \int_0^l \int_{-c/2}^{c/2} \{ \tfrac{\partial \phi_1}{\partial x} \quad 0\} \begin{bmatrix} A_{11} & B_{16} \\ B_{16} & D_{66} \end{bmatrix} \begin{Bmatrix} \frac{\partial \phi_1}{\partial x} \\ 0 \end{Bmatrix} dx\, dy \\ &= \int_0^l \int_{-c/2}^{c/2} \frac{\partial \phi_1}{\partial x} A_{11} \frac{\partial \phi_1}{\partial x} dx\, dy \\ &= \int_0^l \int_{-c/2}^{c/2} \frac{1}{l} A_{11} \frac{1}{l} dx\, dy \\ &= \frac{A_{11} c}{l} \end{aligned} \tag{5.353}$$

$$\begin{aligned} K_{12} &= \int_0^l \int_{-c/2}^{c/2} \{\mathbb{D}\boldsymbol{\phi}_1\}^T \begin{bmatrix} A_{11} & B_{16} \\ B_{16} & D_{66} \end{bmatrix} \{\mathbb{D}\boldsymbol{\phi}_2\}\, dx\, dy \\ &= \int_0^l \int_{-c/2}^{c/2} \{ \tfrac{\partial \phi_1}{\partial x} \quad 0\} \begin{bmatrix} A_{11} & B_{16} \\ B_{16} & D_{66} \end{bmatrix} \begin{Bmatrix} 0 \\ 2\frac{\partial^2 \phi_2}{\partial x \partial y} \end{Bmatrix} dx\, dy \\ &= \int_0^l \int_{-c/2}^{c/2} \frac{\partial \phi_1}{\partial x} B_{16}\, 2 \frac{\partial^2 \phi_2}{\partial x \partial y} dx\, dy \\ &= \int_0^l \int_{-c/2}^{c/2} \frac{1}{l} B_{16} \frac{4x}{l^2 c} dx\, dy \\ &= \frac{2B_{16}}{l} \end{aligned} \tag{5.354}$$

$$\begin{aligned} K_{22} &= \int_0^l \int_{-c/2}^{c/2} \{\mathbb{D}\boldsymbol{\phi}_2\}^T \begin{bmatrix} A_{11} & B_{16} \\ B_{16} & D_{66} \end{bmatrix} \{\mathbb{D}\boldsymbol{\phi}_2\}\, dx\, dy \\ &= \int_0^l \int_{-c/2}^{c/2} \left\{ 0 \quad 2\frac{\partial^2 \phi_2}{\partial x \partial y} \right\} \begin{bmatrix} A_{11} & B_{16} \\ B_{16} & D_{66} \end{bmatrix} \begin{Bmatrix} 0 \\ 2\frac{\partial^2 \phi_2}{\partial x \partial y} \end{Bmatrix} dx\, dy \\ &= \int_0^l \int_{-c/2}^{c/2} 2\frac{\partial^2 \phi_2}{\partial x \partial y} D_{66}\, 2 \frac{\partial^2 \phi_2}{\partial x \partial y} dx\, dy \\ &= \int_0^l \int_{-c/2}^{c/2} \frac{4x}{l^2 c} D_{66} \frac{4x}{l^2 c} dx\, dy \\ &= \frac{16 D_{66}}{3lc} \end{aligned} \tag{5.355}$$

Assuming that piezos are attached to the entire surface of the plate, the generalized forces can be found from

$$\begin{aligned} Q_{\Lambda_1} &= \int_0^l \int_{-c/2}^{c/2} \{F_{x_\Lambda} \quad 0\} \{\mathbb{D}\boldsymbol{\phi}_1\}\, dx\, dy \\ &= \int_0^l \int_{-c/2}^{c/2} \{F_{x_\Lambda} \quad 0\} \begin{Bmatrix} \frac{\partial \phi_1}{\partial x} \\ 0 \end{Bmatrix} dx\, dy \\ &= \int_0^l \int_{-c/2}^{c/2} F_{x_\Lambda} \frac{1}{l} dx\, dy \\ &= F_{x_\Lambda} c \end{aligned} \tag{5.356}$$

$$\begin{aligned} Q_{\Lambda_2} &= \int_0^l \int_{-c/2}^{c/2} \{F_{x_\Lambda} \quad 0\} \{\mathbb{D}\boldsymbol{\phi}_2\} dx\, dy \\ &= \int_0^l \int_{-c/2}^{c/2} \{F_{x_\Lambda} \quad 0\} \begin{Bmatrix} 0 \\ 2\frac{\partial^2 \phi_2}{\partial x \partial y} \end{Bmatrix} dx\, dy \\ &= 0 \end{aligned} \tag{5.357}$$

Using these generalized forces, the displacement can be found from

$$\begin{bmatrix} A_{11}c/l & 2B_{16}/l \\ 2B_{16}/l & 16D_{66}/3lc \end{bmatrix} \begin{Bmatrix} q_1 \\ q_2 \end{Bmatrix} = \begin{Bmatrix} F_{x_\Lambda} c \\ 0 \end{Bmatrix} \tag{5.358}$$

Rewriting

$$\begin{Bmatrix} q_1 \\ q_2 \end{Bmatrix} = \frac{1}{\frac{16A_{11}D_{66}}{3l^2} - \frac{4B_{16}^2}{l^2}} \begin{bmatrix} \frac{16D_{66}}{3lc} & \frac{-2B_{16}}{l} \\ \frac{-2B_{16}}{l} & A_{11}\frac{c}{l} \end{bmatrix} \begin{Bmatrix} F_{x_\Lambda} c \\ 0 \end{Bmatrix} \tag{5.359}$$

Eq. 5.359 yields the generalized degrees of freedom as

$$q_1 = \frac{4F_{x_\Lambda} D_{66} l}{4A_{11}D_{66} - 3B_{16}^2} \tag{5.360}$$

$$q_2 = -\frac{3}{2} \cdot \frac{F_{x_\Lambda} B_{16} cl}{4A_{11}D_{66} - 3B_{16}^2} \tag{5.361}$$

This gives the extensional strain in the x direction and the twist rate as

$$\epsilon_x^o = \frac{1}{l} q_1 = \frac{4F_{x_\Lambda} D_{66}}{4A_{11}D_{66} - 3B_{16}^2} \tag{5.362}$$

$$\kappa_{xy} = \frac{4x}{l^2 c} q_2 = -\frac{6F_{x_\Lambda} B_{16} x}{l(4A_{11}D_{66} - 3B_{16}^2)} \tag{5.363}$$

The tip twist of the plate is obtained by setting $x = l$ in the previous equation, yielding

$$\kappa_{xy}^{\text{tip}} = -\frac{6F_{x_\Lambda} B_{16}}{(4A_{11}D_{66} - 3B_{16}^2)} \tag{5.364}$$

5.4.6 Laminated Plate with Symmetric Layup: Bending-Torsion Coupling

Consider a cantilevered plate with a symmetric layup of generally orthotropic plies. Because a plate with this layup exhibits coupling between bending and torsion, it is important to investigate the effect of a piezo-induced pure bending actuation. This is achieved by the application of opposite voltages to identical piezo actuators bonded on the top and bottom surfaces of the plate.

In the absence of external loads, the governing equations for the plate are

$$\boldsymbol{F}_{\Lambda} = \begin{bmatrix} A_{11} & A_{12} & A_{16} \\ A_{12} & A_{22} & A_{26} \\ A_{16} & A_{26} & A_{66} \end{bmatrix} \boldsymbol{\epsilon}^{o} \tag{5.365}$$

$$\boldsymbol{M}_{\Lambda} = \begin{bmatrix} D_{11} & D_{12} & D_{16} \\ D_{12} & D_{22} & D_{26} \\ D_{16} & D_{26} & D_{66} \end{bmatrix} \boldsymbol{\kappa} \tag{5.366}$$

For purely bending actuation, the induced-force and moment vectors for this configuration are

$$\boldsymbol{F}_{\Lambda} = \begin{Bmatrix} 0 \\ 0 \\ 0 \end{Bmatrix} \tag{5.367}$$

$$\boldsymbol{M}_{\Lambda} = \begin{Bmatrix} M_{x_{\Lambda}} \\ M_{y_{\Lambda}} \\ 0 \end{Bmatrix} \tag{5.368}$$

To focus on the effect of bending-torsion coupling, the problem is simplified by assuming that the piezo actuators only impart a moment along the x direction and by ignoring any stress or deformation in the y direction. Because the force and moment equations are uncoupled and the force vector is zero, we can ignore the force-equilibrium equation. Under these assumptions, the induced-moment vector for this configuration is

$$\boldsymbol{M}_{\Lambda} = \begin{Bmatrix} M_{x_{\Lambda}} \\ 0 \\ 0 \end{Bmatrix} \tag{5.369}$$

Ignoring the terms corresponding to $M_{y_{\Lambda}}$ and κ_y, the governing equation, differential operator matrix, and deformation vector simplify to

$$\begin{bmatrix} D_{11} & D_{16} \\ D_{16} & D_{66} \end{bmatrix} \begin{Bmatrix} \kappa_x \\ \kappa_{xy} \end{Bmatrix} = \begin{Bmatrix} M_{x_{\Lambda}} \\ 0 \end{Bmatrix} \tag{5.370}$$

and

$$\begin{Bmatrix} \kappa_x \\ \kappa_{xy} \end{Bmatrix} = \mathbb{D}\boldsymbol{U} = \begin{Bmatrix} \frac{\partial^2}{\partial x^2} \\ 2\frac{\partial^2}{\partial x \partial y} \end{Bmatrix} w \tag{5.371}$$

where w is the out-of-plane deflection. Because the torsional response is important, the out-of-plane deflections along both the x direction and the y direction must be considered. The cantilevered boundary conditions are identically satisfied by the following two-term expansion

$$w(x, y) = \frac{x^2}{l^2} q_1 + \frac{x^2 y}{l^2 c} q_2 = \phi_1 q_1 + \phi_2 q_2 \tag{5.372}$$

Note that w exhibits a linear variation with respect to the y direction in order to accommodate the anticipated twisting. At $x = 0$, the chosen shape functions ensure that the cantilevered boundary conditions are satisfied, regardless of the values of the coefficients q_1 and q_2.

$$w = 0 \qquad \phi_1 = 0 \qquad \phi_2 = 0 \tag{5.373}$$

$$\frac{\partial w}{\partial x} = 0 \qquad \frac{\partial \phi_1}{\partial x} = 0 \qquad \frac{\partial \phi_2}{\partial x} = 0 \tag{5.374}$$

From these shape functions, the deformation vector is given by

$$\begin{Bmatrix} \kappa_x \\ \kappa_{xy} \end{Bmatrix} = \mathbb{D}\boldsymbol{U} = \begin{Bmatrix} \frac{\partial^2}{\partial x^2} \\ 2\frac{\partial^2}{\partial x \partial y} \end{Bmatrix} \{\phi_1 \quad \phi_2\} \begin{Bmatrix} q_1 \\ q_2 \end{Bmatrix} \tag{5.375}$$

The Lagrange's equations reduce to the following form

$$\begin{bmatrix} K_{11} & K_{12} \\ K_{12} & K_{22} \end{bmatrix} \begin{Bmatrix} q_1 \\ q_2 \end{Bmatrix} = \begin{Bmatrix} Q_{\Lambda_1} \\ Q_{\Lambda_2} \end{Bmatrix} \tag{5.376}$$

where

$$\begin{aligned} K_{11} &= \int_0^l \int_{-c/2}^{c/2} \{\mathbb{D}\phi_1\}^T \begin{bmatrix} D_{11} & D_{16} \\ D_{16} & D_{66} \end{bmatrix} \{\mathbb{D}\phi_1\}\, dxdy \\ &= \int_0^l \int_{-c/2}^{c/2} \left\{ \frac{\partial^2 \phi_1}{\partial x^2} \quad 2\frac{\partial^2 \phi_1}{\partial x \partial y} \right\} \begin{bmatrix} D_{11} & D_{16} \\ D_{16} & D_{66} \end{bmatrix} \begin{Bmatrix} \frac{\partial^2 \phi_1}{\partial x^2} \\ 2\frac{\partial^2 \phi_1}{\partial x \partial y} \end{Bmatrix} dx\, dy \\ &= \int_0^l \int_{-c/2}^{c/2} \left\{ \frac{2}{l^2} \quad 0 \right\} \begin{bmatrix} D_{11} & D_{16} \\ D_{16} & D_{66} \end{bmatrix} \begin{Bmatrix} \frac{2}{l^2} \\ 0 \end{Bmatrix} dx\, dy \\ &= \int_0^l \int_{-c/2}^{c/2} \frac{4}{l^4} D_{11} dx\, dy \\ &= \frac{4c}{l^3} D_{11} \end{aligned} \tag{5.377}$$

$$\begin{aligned} K_{12} &= \int_0^l \int_{-c/2}^{c/2} \{\mathbb{D}\phi_2\}^T \begin{bmatrix} D_{11} & D_{16} \\ D_{16} & D_{66} \end{bmatrix} \{\mathbb{D}\phi_1\}\, dxdy \\ &= \int_0^l \int_{-c/2}^{c/2} \left\{ \frac{\partial^2 \phi_2}{\partial x^2} \quad 2\frac{\partial^2 \phi_2}{\partial x \partial y} \right\} \begin{bmatrix} D_{11} & D_{16} \\ D_{16} & D_{66} \end{bmatrix} \begin{Bmatrix} \frac{\partial^2 \phi_1}{\partial x^2} \\ 2\frac{\partial^2 \phi_1}{\partial x \partial y} \end{Bmatrix} dx\, dy \\ &= \int_0^l \int_{-c/2}^{c/2} \left\{ \frac{2y}{l^2 c} \quad \frac{4x}{l^2 c} \right\} \begin{bmatrix} D_{11} & D_{16} \\ D_{16} & D_{66} \end{bmatrix} \begin{Bmatrix} \frac{2}{l^2} \\ 0 \end{Bmatrix} dx\, dy \\ &= \int_0^l \int_{-c/2}^{c/2} \left[\frac{4y}{l^4 c} D_{11} + \frac{8x}{l^4 c} D_{16} \right] dx\, dy \\ &= \frac{4}{l^2} D_{16} \end{aligned} \tag{5.378}$$

$$
\begin{aligned}
K_{22} &= \int_0^l \int_{-c/2}^{c/2} \{\mathbb{D}\phi_2\}^T \begin{bmatrix} D_{11} & D_{16} \\ D_{16} & D_{66} \end{bmatrix} \{\mathbb{D}\phi_2\}\, dxdy \\
&= \int_0^l \int_{-c/2}^{c/2} \left\{ \frac{\partial^2\phi_2}{\partial x^2} \quad 2\frac{\partial^2\phi_2}{\partial x\partial y} \right\} \begin{bmatrix} D_{11} & D_{16} \\ D_{16} & D_{66} \end{bmatrix} \left\{ \begin{matrix} \frac{\partial^2\phi_2}{\partial x^2} \\ 2\frac{\partial^2\phi_2}{\partial x\partial y} \end{matrix} \right\} dx\, dy \\
&= \int_0^l \int_{-c/2}^{c/2} \left\{ \frac{2y}{l^2c} \quad \frac{4x}{l^2c} \right\} \begin{bmatrix} D_{11} & D_{16} \\ D_{16} & D_{66} \end{bmatrix} \left\{ \begin{matrix} \frac{2y}{l^2c} \\ \frac{4x}{l^2c} \end{matrix} \right\} dx\, dy \\
&= \int_0^l \int_{-c/2}^{c/2} \left[\frac{c}{3l^4} D_{11} + \frac{16x^2}{l^4c} D_{66} \right] dx\, dy \\
&= \frac{c}{3l^3} D_{11} + \frac{16}{3lc} D_{66}
\end{aligned}
\tag{5.379}
$$

$$
\begin{aligned}
Q_{\Lambda_1} &= \int_0^l \int_{-c/2}^{c/2} \{M_{x_\Lambda} \quad 0\} \{\mathbb{D}\phi_1\}\, dx\, dy \\
&= \int_0^l \int_{-c/2}^{c/2} \{M_{x_\Lambda} \quad 0\} \left\{ \begin{matrix} \frac{\partial^2\phi_1}{\partial x^2} \\ 2\frac{\partial^2\phi_1}{\partial x\partial y} \end{matrix} \right\} dx\, dy \\
&= \int_0^l \int_{-c/2}^{c/2} \{M_{x_\Lambda} \quad 0\} \left\{ \begin{matrix} \frac{2}{l^2} \\ 0 \end{matrix} \right\} dx\, dy \\
&= \int_0^l \int_{-c/2}^{c/2} \frac{2}{l^2} M_{x_\Lambda}\, dx\, dy \\
&= 2M_{x_\Lambda} \frac{c}{l}
\end{aligned}
\tag{5.380}
$$

$$
\begin{aligned}
Q_{\Lambda_2} &= \int_0^l \int_{-c/2}^{c/2} \{M_{x_\Lambda} \quad 0\} \{\mathbb{D}\phi_2\}\, dx\, dy \\
&= \int_0^l \int_{-c/2}^{c/2} \{M_{x_\Lambda} \quad 0\} \left\{ \begin{matrix} \frac{\partial^2\phi_2}{\partial x^2} \\ 2\frac{\partial^2\phi_2}{\partial x\partial y} \end{matrix} \right\} dx\, dy \\
&= \int_0^l \int_{-c/2}^{c/2} \{M_{x_\Lambda} \quad 0\} \left\{ \begin{matrix} \frac{2y}{l^2c} \\ \frac{4x}{l^2c} \end{matrix} \right\} dx\, dy \\
&= \int_0^l \int_{-c/2}^{c/2} M_{x_\Lambda} \frac{2y}{l^2c}\, dx\, dy \\
&= 0
\end{aligned}
\tag{5.381}
$$

Substituting in Eq. 5.376 gives the solution for the generalized degrees of freedom q_1 and q_2

$$
\begin{bmatrix} K_{11} & K_{12} \\ K_{12} & K_{22} \end{bmatrix} \left\{ \begin{matrix} q_1 \\ q_2 \end{matrix} \right\} = \left\{ \begin{matrix} Q_{\Lambda_1} \\ Q_{\Lambda_2} \end{matrix} \right\}
\tag{5.382}
$$

This reduces to

$$\begin{Bmatrix} q_1 \\ q_2 \end{Bmatrix} = \frac{1}{K_{11}K_{22} - K_{12}^2} \begin{bmatrix} K_{22} & -K_{12} \\ -K_{12} & K_{11} \end{bmatrix} \begin{Bmatrix} Q_{\Lambda_1} \\ Q_{\Lambda_2} \end{Bmatrix} \tag{5.383}$$

Solving this equation yields

$$q_1 = \frac{K_{22}Q_{\Lambda_1}}{K_{11}K_{22} - K_{12}^2} \tag{5.384}$$

$$= \frac{M_{x_\Lambda}\left(l^2c^2D_{11} + 16l^4D_{66}\right)}{2c^2D_{11}^2 + 32l^2D_{11}D_{66} - 24l^2D_{16}^2} \tag{5.385}$$

$$q_2 = -\frac{K_{12}Q_{\Lambda_1}}{K_{11}K_{22} - K_{12}^2} \tag{5.386}$$

$$= \frac{-6l^3cD_{16}M_{x_\Lambda}}{c^2D_{11}^2 + 16l^2D_{11}D_{66} - 12l^2D_{16}^2} \tag{5.387}$$

The bending slope and twist of the plate are

$$\kappa_x = \frac{2q_1}{l^2} + \frac{2y}{l^2c}q_2 \tag{5.388}$$

$$\kappa_{xy} = \frac{4x}{l^2c}q_2 \tag{5.389}$$

From these equations, the bending slope at the tip of the plate is obtained by setting $y = 0$ (mid-chord) and $x = l$.

$$\begin{aligned} \kappa_x^{\text{tip}} &= \left.\frac{\partial^2 w}{\partial x^2}\right|_{y=0} = \frac{2}{l^2}q_1 \\ &= \frac{M_{x_\Lambda}\left(D_{11} + 16(l/c)^2D_{66}\right)}{D_{11}^2 + 16(l/c)^2D_{11}D_{66} - 12(l/c)^2D_{16}^2} \end{aligned} \tag{5.390}$$

Similarly, the twist at the tip of the plate is obtained by setting $x = l$.

$$\begin{aligned} \kappa_{xy}^{\text{tip}} &= 2\left.\frac{\partial^2 w}{\partial x \partial y}\right|_{x=l} = \frac{4}{lc}q_2 \\ &= \frac{-24(l/c)^2D_{16}M_{x_\Lambda}}{D_{11}^2 + 16(l/c)^2D_{11}D_{66} - 12(l/c)^2D_{16}^2} \end{aligned} \tag{5.391}$$

5.4.7 Worked Example

Using laminated plate theory, derive the Rayleigh-Ritz solution for a rectangular cantilevered two-layered cross-ply laminate (Fig. 5.14) with nonidentical piezo sheets (different piezo coefficient d_{31} but same thickness) bonded on either surface for half of the plate length. A PZT-5H sheet is bonded to the top surface and a PZT-5A sheet is bonded to the bottom surface. An equal voltage is applied on each piezo sheet. Although the piezos are stretched across the complete width, neglect the

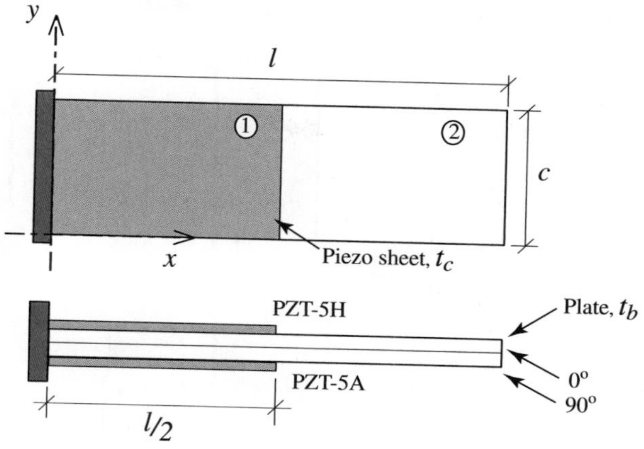

Figure 5.14. Rectangular cross-ply laminate with dissimilar piezo sheets.

influence of lateral strain. Assume the deflection as

$$u(x, y) = \frac{x}{l} q_1$$

$$w(x, y) = \frac{x^2}{l^2} q_2 + \frac{x^2}{l^2} \cdot \frac{y - c/2}{c/2} q_3$$

Note that with an assumed solution of this form, it is possible to represent bending and twisting independently through the coefficients q_2 and q_3, respectively.

Solution

For a cross-ply laminate, from Table 5.2

$$\mathbf{A} = \begin{bmatrix} A_{11} & A_{12} & 0 \\ A_{12} & A_{22} & 0 \\ 0 & 0 & A_{66} \end{bmatrix} \quad \mathbf{B} = \begin{bmatrix} B_{11} & 0 & 0 \\ 0 & -B_{11} & 0 \\ 0 & 0 & 0 \end{bmatrix} \quad \mathbf{D} = \begin{bmatrix} D_{11} & D_{12} & 0 \\ D_{12} & D_{22} & 0 \\ 0 & 0 & D_{66} \end{bmatrix}$$

Because the piezo sheets are isotropic and symmetric with respect to the beam mid-plane, the stiffness matrices in regions 1 and 2 will have the same form as shown here. The assumed deflection and geometric boundary conditions are

$$u(x, y) = \frac{x}{l} q_1, \rightarrow \phi_1 = \frac{x}{l}$$

$$w(x, y) = \frac{x^2}{l^2} q_2 + \frac{x^2}{l^2} \frac{(y - c/2)}{c/2} q_3,$$

$$\rightarrow \phi_2 = \frac{x^2}{l^2}, \quad \phi_3 = \frac{x^2}{l^2} \frac{(y - c/2)}{c/2}$$

At $x = 0$, $w = 0$, $w' = 0$, $u = 0$; the boundary conditions are satisfied. The reduced equation is given by

$$\begin{bmatrix} K_{11} & K_{12} & K_{13} \\ K_{12} & K_{22} & K_{23} \\ K_{13} & K_{23} & K_{33} \end{bmatrix} \begin{Bmatrix} q_1 \\ q_2 \\ q_3 \end{Bmatrix} = \begin{Bmatrix} Q_{\Lambda_1} \\ Q_{\Lambda_2} \\ Q_{\Lambda_3} \end{Bmatrix}$$

The stiffness matrix and forcing vector are derived as follows

$$\mathbb{D} = \begin{bmatrix} \frac{\partial}{\partial x} & 0 & 0 \\ 0 & \frac{\partial}{\partial y} & 0 \\ \frac{\partial}{\partial y} & \frac{\partial}{\partial x} & 0 \\ 0 & 0 & \frac{\partial^2}{\partial x^2} \\ 0 & 0 & \frac{\partial^2}{\partial y^2} \\ 0 & 0 & 2\frac{\partial^2}{\partial x \partial y} \end{bmatrix}, \quad \boldsymbol{\phi}_1 = \begin{Bmatrix} \frac{x}{l} \\ 0 \\ 0 \end{Bmatrix}, \quad \boldsymbol{\phi}_2 = \begin{Bmatrix} 0 \\ 0 \\ \frac{x^2}{l^2}\frac{y}{c} \end{Bmatrix}, \quad \boldsymbol{\phi}_3 = \begin{Bmatrix} 0 \\ 0 \\ \frac{x^2}{l^2}\frac{(y-c/2)}{c/2} \end{Bmatrix}$$

which gives

$$\mathbb{D}\boldsymbol{\phi}_1 = \begin{Bmatrix} \frac{1}{l} \\ 0 \\ 0 \\ 0 \\ 0 \\ 0 \end{Bmatrix}, \quad \mathbb{D}\boldsymbol{\phi}_2 = \begin{Bmatrix} 0 \\ 0 \\ 0 \\ \frac{2}{l^2} \\ 0 \\ 0 \end{Bmatrix}, \quad \mathbb{D}\boldsymbol{\phi}_3 = \begin{Bmatrix} 0 \\ 0 \\ 0 \\ \frac{2}{l^2}\frac{(y-c/2)}{c/2} \\ 0 \\ \frac{8x}{l^2 c} \end{Bmatrix}$$

The terms of the stiffness matrix are

$$\begin{aligned} K_{ij} = & \int_0^{l/2}\int_{-c/2}^{c/2} \{\mathbb{D}\boldsymbol{\phi}_i\}^T \begin{bmatrix} \boldsymbol{A} & \boldsymbol{B} \\ \boldsymbol{B} & \boldsymbol{D} \end{bmatrix}^{(1)} \{\mathbb{D}\boldsymbol{\phi}_j\}\, dx\, dy \\ & + \int_{l/2}^{l}\int_{-c/2}^{c/2} \{\mathbb{D}\boldsymbol{\phi}_i\}^T \begin{bmatrix} \boldsymbol{A} & \boldsymbol{B} \\ \boldsymbol{B} & \boldsymbol{D} \end{bmatrix}^{(2)} \{\mathbb{D}\boldsymbol{\phi}_j\}\, dx\, dy \end{aligned}$$

where the superscripts (1) and (2) refer to the portion of the plate with and without the piezo-sheets, respectively. This gives

$$\begin{aligned} K_{11} = & \int_0^{l/2}\int_{-c/2}^{c/2} \{\mathbb{D}\boldsymbol{\phi}_1\}^T \begin{bmatrix} \boldsymbol{A} & \boldsymbol{B} \\ \boldsymbol{B} & \boldsymbol{D} \end{bmatrix}^{(1)} \{\mathbb{D}\boldsymbol{\phi}_1\}\, dx\, dy \\ & + \int_{l/2}^{l}\int_{-c/2}^{c/2} \{\mathbb{D}\boldsymbol{\phi}_1\}^T \begin{bmatrix} \boldsymbol{A} & \boldsymbol{B} \\ \boldsymbol{B} & \boldsymbol{D} \end{bmatrix}^{(2)} \{\mathbb{D}\boldsymbol{\phi}_1\}\, dx\, dy \\ = & \int_0^{l/2}\int_{-c/2}^{c/2} \frac{1}{l^2} A_{11}^{(1)}\, dx\, dy + \int_{l/2}^{l}\int_{-c/2}^{c/2} \frac{1}{l^2} A_{11}^{(2)}\, dx\, dy \\ = & \frac{c}{2l}\left(A_{11}^{(1)} + A_{11}^{(2)}\right) \end{aligned}$$

$$\begin{aligned} K_{12} = & \int_0^{l/2}\int_{-c/2}^{c/2} \{\mathbb{D}\boldsymbol{\phi}_1\}^T \begin{bmatrix} \boldsymbol{A} & \boldsymbol{B} \\ \boldsymbol{B} & \boldsymbol{D} \end{bmatrix}^{(1)} \{\mathbb{D}\boldsymbol{\phi}_2\}\, dx\, dy \\ & + \int_{l/2}^{l}\int_{-c/2}^{c/2} \{\mathbb{D}\boldsymbol{\phi}_1\}^T \begin{bmatrix} \boldsymbol{A} & \boldsymbol{B} \\ \boldsymbol{B} & \boldsymbol{D} \end{bmatrix}^{(2)} \{\mathbb{D}\boldsymbol{\phi}_2\}\, dx\, dy \\ = & \int_0^{l/2}\int_{-c/2}^{c/2} \frac{2B_{11}^{(1)}}{l^3}\, dx\, dy + \int_{l/2}^{l}\int_{-c/2}^{c/2} \frac{2B_{11}^{(2)}}{l^3}\, dx\, dy \\ = & \frac{c}{l^2}\left(B_{11}^{(1)} + B_{11}^{(2)}\right) \end{aligned}$$

$$K_{22} = \int_0^{l/2} \int_{-c/2}^{c/2} \{\mathbb{D}\boldsymbol{\phi}_2\}^T \begin{bmatrix} \boldsymbol{A} & \boldsymbol{B} \\ \boldsymbol{B} & \boldsymbol{D} \end{bmatrix}^{(1)} \{\mathbb{D}\boldsymbol{\phi}_2\}\, dx\, dy$$

$$+ \int_{l/2}^{l} \int_{-c/2}^{c/2} \{\mathbb{D}\boldsymbol{\phi}_2\}^T \begin{bmatrix} \boldsymbol{A} & \boldsymbol{B} \\ \boldsymbol{B} & \boldsymbol{D} \end{bmatrix}^{(2)} \{\mathbb{D}\boldsymbol{\phi}_2\}\, dx\, dy$$

$$= \int_0^{l/2} \int_{-c/2}^{c/2} \frac{4}{l^4} D_{11}^{(1)}\, dx\, dy + \int_{l/2}^{l} \int_{-c/2}^{c/2} \frac{4}{l^4} D_{11}^{(2)}\, dx\, dy$$

$$= \frac{2c}{l^3} \left(D_{11}^{(1)} + D_{11}^{(2)} \right)$$

$$K_{13} = \int_0^{l/2} \int_{-c/2}^{c/2} \{\mathbb{D}\boldsymbol{\phi}_1\}^T \begin{bmatrix} \boldsymbol{A} & \boldsymbol{B} \\ \boldsymbol{B} & \boldsymbol{D} \end{bmatrix}^{(1)} \{\mathbb{D}\boldsymbol{\phi}_3\}\, dx\, dy$$

$$+ \int_{l/2}^{l} \int_{-c/2}^{c/2} \{\mathbb{D}\boldsymbol{\phi}_1\}^T \begin{bmatrix} \boldsymbol{A} & \boldsymbol{B} \\ \boldsymbol{B} & \boldsymbol{D} \end{bmatrix}^{(2)} \{\mathbb{D}\boldsymbol{\phi}_3\}\, dx\, dy$$

$$= \int_0^{l/2} \int_{-c/2}^{c/2} \frac{B_{11}^{(1)}}{l} \frac{2}{l^2} \frac{(y - c/2)}{c/2}\, dx\, dy$$

$$+ \int_{l/2}^{l} \int_{-c/2}^{c/2} \frac{B_{11}^{(1)}}{l} \frac{2}{l^2} \frac{(y - c/2)}{c/2}\, dx\, dy$$

$$= -\frac{c}{l^2} \left(B_{11}^{(1)} + B_{11}^{(2)} \right)$$

$$K_{23} = \int_0^{l/2} \int_{-c/2}^{c/2} \{\mathbb{D}\boldsymbol{\phi}_2\}^T \begin{bmatrix} \boldsymbol{A} & \boldsymbol{B} \\ \boldsymbol{B} & \boldsymbol{D} \end{bmatrix}^{(1)} \{\mathbb{D}\boldsymbol{\phi}_3\}\, dx\, dy$$

$$+ \int_{l/2}^{l} \int_{-c/2}^{c/2} \{\mathbb{D}\boldsymbol{\phi}_2\}^T \begin{bmatrix} \boldsymbol{A} & \boldsymbol{B} \\ \boldsymbol{B} & \boldsymbol{D} \end{bmatrix}^{(2)} \{\mathbb{D}\boldsymbol{\phi}_3\}\, dx\, dy$$

$$= \int_0^{l/2} \int_{-c/2}^{c/2} \left(\frac{8D_{11}^{(1)} y}{l^4 c} - \frac{4D_{11}^{(1)}}{l^4} \right) dx\, dy$$

$$+ \int_{l/2}^{l} \int_{-c/2}^{c/2} \left(\frac{8D_{11}^{(1)} y}{l^4 c} - \frac{4D_{11}^{(1)}}{l^4} \right) dx\, dy$$

$$= -\frac{2c}{l^3} \left(D_{11}^{(1)} + D_{11}^{(2)} \right)$$

$$K_{33} = \int_0^{l/2} \int_{-c/2}^{c/2} \{\mathbb{D}\boldsymbol{\phi}_3\}^T \begin{bmatrix} \boldsymbol{A} & \boldsymbol{B} \\ \boldsymbol{B} & \boldsymbol{D} \end{bmatrix}^{(1)} \{\mathbb{D}\boldsymbol{\phi}_3\}\, dx\, dy$$

$$+ \int_{l/2}^{l} \int_{-c/2}^{c/2} \{\mathbb{D}\boldsymbol{\phi}_3\}^T \begin{bmatrix} \boldsymbol{A} & \boldsymbol{B} \\ \boldsymbol{B} & \boldsymbol{D} \end{bmatrix}^{(2)} \{\mathbb{D}\boldsymbol{\phi}_3\}\, dx\, dy$$

$$= \int_0^{l/2} \int_{-c/2}^{c/2} \left[\left(\frac{2}{l^2} \frac{(y - c/2)}{c/2} \right)^2 D_{11}^{(1)} + \left(\frac{8x}{l^2 c} \right)^2 \right] dx\, dy$$

$$+ \int_{l/2}^{l} \int_{-c/2}^{c/2} \int_0^{l/2} \int_{-c/2}^{c/2} \left[\left(\frac{2}{l^2} \frac{(y - c/2)}{c/2} \right)^2 D_{11}^{(1)} + \left(\frac{8x}{l^2 c} \right)^2 \right] dx\, dy$$

$$= \frac{8c}{3l^3} \left(D_{11}^{(1)} + D_{11}^{(2)} \right) + \frac{8}{3cl} \left(D_{66}^{(1)} + 7D_{66}^{(2)} \right)$$

We now need to derive the following constants: $A_{11}^{(1)}$, $A_{11}^{(2)}$, $B_{11}^{(1)}$, $B_{11}^{(2)}$, $D_{11}^{(1)}$, $D_{11}^{(2)}$, $D_{66}^{(1)}$, and $D_{66}^{(2)}$

$$
\begin{aligned}
A_{11}^{(1)} &= \sum_{i=1}^{4}(Q_{11})_k(h_{k+1}-h_k)\\
&= (Q_{11})_c\left[-t_b/2-(t_b/2-t_c)\right]+(Q_{11})_{90^\circ}\left[0-(-t_b/2)\right]\\
&\quad +(Q_{11})_{0^\circ}\left[t_b/2-0\right]+(Q_{11})_c\left[t_b/2+t_c-t_b/2\right]\\
&= 2(Q_{11})_c t_c+\frac{t_b}{2}\left((Q_{11})_{0^\circ}+(Q_{11})_{90^\circ}\right)\\
&= 2(Q_{11})_c t_c+\frac{t_b}{2}\left((Q_{11})_{0^\circ}+(Q_{22})_{0^\circ}\right)
\end{aligned}
$$

$$
\begin{aligned}
A_{11}^{(2)} &= \sum_{i=1}^{2}(Q_{11})_k(h_{k+1}-h_k)\\
&= \frac{t_b}{2}\left((Q_{11})_{0^\circ}+(Q_{22})_{0^\circ}\right)
\end{aligned}
$$

$$
\begin{aligned}
B_{11}^{(1)} &= -\frac{1}{2}\sum_{i=1}^{4}(Q_{11})_k\left(h_{k+1}^2-h_k^2\right)\\
&= -\left\{\frac{1}{2}(Q_{11})_c\left[(-t_b/2)^2-(-t_b/2-t_c)^2\right]\right.\\
&\quad +\frac{1}{2}(Q_{11})_{90^\circ}\left[0-(-t_b/2)^2\right]+\frac{1}{2}(Q_{11})_{0^\circ}\left[(t_b/2)^2-0\right]\\
&\quad \left.+\frac{1}{2}(Q_{11})_c\left[(t_b/2+t_c)^2-(t_b/2)^2\right]\right\}\\
&= -\frac{t_b^2}{8}\left((Q_{11})_{0^\circ}-(Q_{11})_{90^\circ}\right)\\
&= -\frac{t_b^2}{8}\left((Q_{11})_{0^\circ}-(Q_{22})_{0^\circ}\right)
\end{aligned}
$$

Because the piezo sheets do not contribute to B_{11}

$$
B_{11}^{(2)} = -\frac{t_b^2}{8}\left((Q_{11})_{0^\circ}-(Q_{22})_{0^\circ}\right)
$$

$$
\begin{aligned}
D_{11}^{(1)} &= \frac{1}{3}\sum_{i=1}^{4}(Q_{11})_k\left(h_{k+1}^3-h_k^3\right)\\
&= \frac{1}{3}(Q_{11})_c\left[(-t_b/2)^3-(-t_b/2-t_c)^3\right]\\
&\quad +\frac{1}{3}(Q_{11})_{90^\circ}\left[0-(-t_b/2)^3\right]+\frac{1}{3}(Q_{11})_{0^\circ}\left[(t_b/2)^3-0\right]\\
&\quad +\frac{1}{3}(Q_{11})_c\left[(t_b/2+t_c)^3-(t_b/2)^3\right]
\end{aligned}
$$

$$= \frac{2}{3}(Q_{11})_c t_c \left(\frac{3}{4}t_b^2 + \frac{3}{2}t_b t_c + t_c^2\right) + \frac{t_b^3}{24}\left((Q_{11})_{0^\circ} + (Q_{11})_{90^\circ}\right)$$

$$= \frac{2}{3}(Q_{11})_c t_c \left(\frac{3}{4}t_b^2 + \frac{3}{2}t_b t_c + t_c^2\right) + \frac{t_b^3}{24}\left((Q_{11})_{0^\circ} + (Q_{22})_{0^\circ}\right)$$

$$D_{11}^{(2)} = \frac{t_b^3}{24}\left((Q_{11})_{0^\circ} + (Q_{22})_{0^\circ}\right)$$

Similarly

$$D_{66}^{(1)} = \frac{2}{3}(Q_{66})_c t_c \left(\frac{3}{4}t_b^2 + \frac{3}{2}t_b t_c + t_c^2\right) + \frac{t_b^3}{12}(Q_{66})_{0^\circ}$$

$$D_{66}^{(2)} = \frac{t_b^3}{12}(Q_{66})_{0^\circ}$$

The generalized forces are given by

$$Q_{\Lambda_1} = \int_0^{l/2}\int_{-c/2}^{c/2} \{\boldsymbol{F}_\Lambda^T\ \boldsymbol{M}_\Lambda^T\}\{\mathbb{D}\ \phi_1\}dx\,dy$$

$$= \int_0^{l/2}\int_{-c/2}^{c/2} F_{x_\Lambda}\frac{1}{l}\,dx\,dy$$

$$= F_{x_\Lambda}\frac{c}{2}$$

$$Q_{\Lambda_2} = \int_0^{l/2}\int_{-c/2}^{c/2} \{\boldsymbol{F}_\Lambda^T\ \boldsymbol{M}_\Lambda^T\}\{\mathbb{D}\ \phi_2\}dx\,dy$$

$$= \int_0^{l/2}\int_{-c/2}^{c/2} M_{x_\Lambda}\frac{2}{l^2}\,dx\,dy$$

$$= M_{x_\Lambda}\frac{c}{l}$$

$$Q_{\Lambda_3} = \int_0^{l/2}\int_{-c/2}^{c/2} \{\boldsymbol{F}_\Lambda^T\ \boldsymbol{M}_\Lambda^T\}\{\mathbb{D}\ \phi_3\}dx\,dy$$

$$= \int_0^{l/2}\int_{-c/2}^{c/2} M_{x_\Lambda}\frac{2}{l^2}\frac{(y - c/2)}{c/2}\,dx\,dy$$

$$= -M_{x_\Lambda}\frac{c}{l}$$

where F_{x_Λ} and M_{x_Λ} are found by

$$\boldsymbol{F}_\Lambda = \begin{Bmatrix} F_{x_\Lambda} \\ F_{y_\Lambda} \\ 0 \end{Bmatrix} = \sum_{i=1}^{4} \boldsymbol{Q}_k \boldsymbol{\Lambda}_k (h_{k+1} - h_k)$$

$$= t_c \boldsymbol{Q}_c (\boldsymbol{\Lambda}_h + \boldsymbol{\Lambda}_a)$$

$$= \frac{E_c t_c (\Lambda_h + \Lambda_a)}{1 - \nu}\begin{Bmatrix} 1 \\ 0 \\ 0 \end{Bmatrix}$$

$$F_{x_\Lambda} = \frac{E_c t_c (\Lambda_h + \Lambda_a)}{1 - \nu}$$

where $\mathbf{\Lambda}_h$ and $\mathbf{\Lambda}_a$ are the free strains of PZT-5H and PZT-5A, respectively. Note that the induced strain in the y direction has been ignored. Similarly, for the induced moment

$$\boldsymbol{M}_\Lambda = \begin{Bmatrix} M_{x_\Lambda} \\ M_{y_\Lambda} \\ 0 \end{Bmatrix} = -\frac{1}{2}\sum_{i=1}^{4} \boldsymbol{Q}_k \mathbf{\Lambda}_k (h_{k+1}^2 - h_k^2)$$

$$= -\frac{1}{2} t_c(t_c + t_b)\boldsymbol{Q}_c(\mathbf{\Lambda}_h - \mathbf{\Lambda}_a)$$

$$= -\frac{1}{2}\frac{E_c t_c(t_c + t_b)(\Lambda_h - \Lambda_a)}{1 - \nu}\begin{Bmatrix} 1 \\ 0 \\ 0 \end{Bmatrix}$$

$$M_{x_\Lambda} = -\frac{1}{2}\frac{E_c t_c(t_c + t_b)(\Lambda_h - \Lambda_a)}{1 - \nu}$$

The generalized forces become

$$Q_{\Lambda_1} = \frac{E_c t_c c}{2(1 - \nu)}(\Lambda_h + \Lambda_a)$$

$$Q_{\Lambda_2} = -\frac{E_c t_c c(t_c + t_b)}{2l(1 - \nu)}(\Lambda_h - \Lambda_a)$$

$$Q_{\Lambda_3} = \frac{E_c t_c c(t_c + t_b)}{2l(1 - \nu)}(\Lambda_h - \Lambda_a)$$

Substituting the expressions for the stiffness matrix and generalized forcing derived previously the governing equation, q_1 and q_2 can be found. This yields the solution for deformations in the x and z directions.

5.4.8 Worked Example

(a) Using laminated plate theory, derive the Rayleigh-Ritz solution for a rectangular cantilevered two-layered antisymmetric laminate with identical piezo sheets bonded on either surface for half of the plate length. An equal voltage is applied on each piezo sheet. Although the piezos are stretched across the complete width, neglect the influence of lateral strain. Assume the deflection as

$$u(x, y) = \frac{x}{l} q_1$$

$$w(x, y) = \frac{x^2}{l^2}\frac{y}{c} q_2 + \frac{x^2}{l^2} \cdot \frac{y - c/2}{c/2} q_3$$

(b) Calculate the tip twist for this two-layered antisymmetric laminated plate (Fig. 5.15) with a ply layup [+30°/−30°]. The size of the plate is 0.3048 m (12 in) × 0.1524 m (6 in) × 0.79375 mm (1/32 in) with piezos of thickness $t_c = 0.3175$ mm (0.0125 in). The voltage applied to the piezos is 100 volts.

Figure 5.15. Rectangular antisymmetric laminate with two identical piezo sheets.

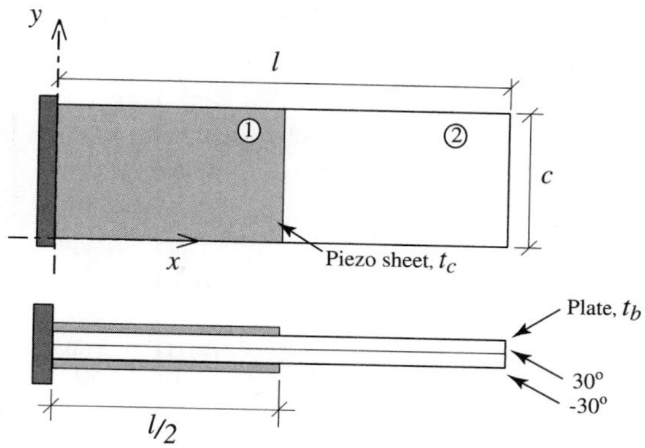

Manufacturer-supplied material constants are as follows

$$\begin{aligned}
d_{31}\text{ (PZT-5H)} &= -274 \times 10^{-12}\text{ m/V}\\
E_c &= 72.4\text{ GPa } (10.5 \times 10^{6}\text{lb/in}^2)\\
E_1 &= 137.9\text{ GPa } (20 \times 10^{6}\text{ lb/in}^2)\\
E_2 &= 14.5\text{ GPa } (2.1 \times 10^{6}\text{ lb/in}^2)\\
G_{12} &= 5.86\text{ GPa } (0.85 \times 10^{6}\text{ lb/in}^2)\\
\nu_c &= 0.3\\
\nu_{12} &= 0.2
\end{aligned}$$

Solution

(a) The plate stiffness matrices for an antisymmetric laminate are given by (see Table 5.2)

$$\boldsymbol{A} = \begin{bmatrix} A_{11} & A_{12} & 0 \\ A_{12} & A_{22} & 0 \\ 0 & 0 & A_{66} \end{bmatrix} \quad \boldsymbol{B} = \begin{bmatrix} 0 & 0 & B_{16} \\ 0 & 0 & B_{26} \\ B_{16} & B_{26} & 0 \end{bmatrix} \quad \boldsymbol{D} = \begin{bmatrix} D_{11} & D_{12} & 0 \\ D_{12} & D_{22} & 0 \\ 0 & 0 & D_{66} \end{bmatrix}$$

We proceed in the same manner as described in Example 5.4.7. The derivation is similar except for the terms involving the $\boldsymbol{B}$ matrix. Therefore, we can write

$$u(x, y) = \frac{x}{l}q_1, \rightarrow \phi_1 = \frac{x}{l}$$

$$w(x, y) = \frac{x^2}{l^2}q_2 + \frac{x^2}{l^2}\frac{(y - c/2)}{c/2}q_3,$$

$$\rightarrow \phi_2 = \frac{x^2}{l^2}, \quad \phi_3 = \frac{x^2}{l^2}\frac{(y - c/2)}{c/2}$$

At $x = 0,\ w = 0,\ w' = 0$; the boundary conditions are satisfied. The governing equation is given by

$$\begin{bmatrix} K_{11} & K_{12} & K_{13} \\ K_{12} & K_{22} & K_{23} \\ K_{13} & K_{23} & K_{33} \end{bmatrix} \begin{Bmatrix} q_1 \\ q_2 \\ q_3 \end{Bmatrix} = \begin{Bmatrix} Q_{\Lambda_1} \\ Q_{\Lambda_2} \\ Q_{\Lambda_3} \end{Bmatrix}$$

where the elements of the stiffness matrix are given by

$$K_{11} = \frac{c}{2l}\left(A_{11}^{(1)} + A_{11}^{(2)}\right)$$
$$K_{22} = \frac{2c}{l^3}\left(D_{11}^{(1)} + D_{11}^{(2)}\right)$$
$$K_{33} = \frac{8c}{3l^3}\left(D_{11}^{(1)} + D_{11}^{(2)}\right) + \frac{8}{3cl}\left(D_{66}^{(1)} + 7D_{66}^{(2)}\right)$$
$$K_{23} = -\frac{2c}{l^3}\left(D_{11}^{(1)} + D_{11}^{(2)}\right) = -K_{22}$$

Because the structure of the $\boldsymbol{B}$ matrix is different from the previous example, the value of K_{12} and K_{13} are given by

$$\begin{aligned} K_{12} &= \int_0^{l/2}\int_{-c/2}^{c/2} \{\mathbb{D}\boldsymbol{\phi}_1\}^T \begin{bmatrix} \boldsymbol{A} & \boldsymbol{B} \\ \boldsymbol{B} & \boldsymbol{D} \end{bmatrix}^{(1)} \{\mathbb{D}\boldsymbol{\phi}_2\}\, dx\, dy \\ &\quad + \int_{l/2}^{l}\int_{-c/2}^{c/2} \{\mathbb{D}\boldsymbol{\phi}_1\}^T \begin{bmatrix} \boldsymbol{A} & \boldsymbol{B} \\ \boldsymbol{B} & \boldsymbol{D} \end{bmatrix}^{(2)} \{\mathbb{D}\boldsymbol{\phi}_2\}\, dx\, dy \\ &= 0 \end{aligned}$$

and

$$\begin{aligned} K_{13} &= \int_0^{l/2}\int_{-c/2}^{c/2} \{\mathbb{D}\boldsymbol{\phi}_1\}^T \begin{bmatrix} \boldsymbol{A} & \boldsymbol{B} \\ \boldsymbol{B} & \boldsymbol{D} \end{bmatrix}^{(1)} \{\mathbb{D}\boldsymbol{\phi}_3\}\, dx\, dy \\ &\quad + \int_{l/2}^{l}\int_{-c/2}^{c/2} \{\mathbb{D}\boldsymbol{\phi}_1\}^T \begin{bmatrix} \boldsymbol{A} & \boldsymbol{B} \\ \boldsymbol{B} & \boldsymbol{D} \end{bmatrix}^{(2)} \{\mathbb{D}\boldsymbol{\phi}_3\}\, dx\, dy \\ &= \int_0^{l/2}\int_{-c/2}^{c/2} \frac{B_{16}^{(1)}}{l}\frac{8x}{l^2 c} dx\, dy + \int_{l/2}^{l}\int_{-c/2}^{c/2} \frac{B_{16}^{(2)}}{l}\frac{8x}{l^2 c} dx\, dy \\ &= \frac{B_{16}^{(1)}}{l} + \frac{3B_{16}^{(2)}}{l} \end{aligned}$$

where

$$\begin{aligned} A_{11}^{(1)} &= 2(Q_{11})_c t_c + \frac{t_b}{2}\left((\bar{Q}_{11})_{+\alpha} + (\bar{Q}_{11})_{-\alpha}\right) \\ &= 2(Q_{11})_c t_c + t_b(\bar{Q}_{11})_{+\alpha} \\ A_{11}^{(2)} &= t_b(\bar{Q}_{11})_{+\alpha} \end{aligned}$$

and

$$D_{11}^{(1)} = \frac{2}{3}(Q_{11})_c t_c \left(\frac{3}{4}t_b^2 + \frac{3}{2}t_b t_c + t_c^2\right) + \frac{t_b^3}{24}\left((\bar{Q}_{11})_{+\alpha} + (\bar{Q}_{11})_{-\alpha}\right)$$

$$= \frac{2}{3}(Q_{11})_c t_c \left(\frac{3}{4}t_b^2 + \frac{3}{2}t_b t_c + t_c^2\right) + \frac{t_b^3}{12}(\bar{Q}_{11})_{+\alpha}$$

$$D_{11}^{(2)} = \frac{t_b^3}{12}(\bar{Q}_{11})_{+\alpha}$$

Similarly

$$D_{66}^{(1)} = \frac{2}{3}(Q_{66})_c t_c \left(\frac{3}{4}t_b^2 + \frac{3}{2}t_b t_c + t_c^2\right) + \frac{t_b^3}{12}(\bar{Q}_{66})_{+\alpha}$$

$$D_{66}^{(2)} = \frac{t_b^3}{12}(\bar{Q}_{66})_{+\alpha}$$

The term B_{16} is found from

$$B_{16}^{(1)} = -\frac{1}{2}\sum_{i=1}^{4}(\bar{Q}_{11})_k(h_{k+1}^2 - h_k^2)$$

$$= -\frac{1}{2}\left[(\bar{Q}_{16})_{+\alpha}\left(\frac{t_b}{2}\right)^2 - (\bar{Q}_{16})_{-\alpha}\left(\frac{t_b}{2}\right)^2\right]$$

$$= -\frac{t_b^2}{4}(\bar{Q}_{16})_{+\alpha}$$

$$B_{16}^{(2)} = B_{16}^{(1)} = -\frac{t_b^2}{4}(\bar{Q}_{16})_{+\alpha}$$

Substituting these relations, the elements of the stiffness matrix are given by

$$K_{11} = \frac{c}{l}\left[(Q_{11})_c t_c + (\bar{Q}_{11})_{+\alpha} t_b\right]$$

$$K_{12} = 0$$

$$K_{13} = -\frac{\bar{Q}_{16})_{+\alpha} t_b^2}{l}$$

$$K_{22} = \frac{2c}{l^3}\left[\frac{2}{3}(Q_{11})_c t_c \left(3/4t_b^2 + 3/2t_b t_c + t_c^2\right) + \frac{t_b^3}{6}(\bar{Q}_{11})_{+\alpha}\right]$$

$$K_{23} = -K_{22}$$

$$K_{33} = \frac{8c}{3l^3}\left[\frac{2}{3}(Q_{11})_c t_c \left(3/4t_b^2 + 3/2t_b t_c + t_c^2\right) + \frac{t_b^3}{6}(\bar{Q}_{11})_{+\alpha}\right]$$

$$= \quad +\frac{8}{3lc}\left[\frac{2}{3}(Q_{66})_c t_c \left(3/4t_b^2 + 3/2t_b t_c + t_c^2\right) + \frac{2t_b^3}{3}(\bar{Q}_{66})_{+\alpha}\right]$$

The generalized forces are given by

$$Q_{\Lambda_1} = \frac{E_c t_c c \Lambda}{1-\nu}$$
$$Q_{\Lambda_2} = 0$$
$$Q_{\Lambda_3} = 0$$

Note that the induced moment is zero. The tip twist is given by

$$\begin{aligned}\theta_{\text{tip}} &= \int_{x=0}^{l} \frac{\partial^2 w}{\partial x \partial y} dx \\ &= \left.\frac{\partial w}{\partial y}\right|_{x=l} \\ &= \frac{2q_3}{c}\end{aligned}$$

(b) Substituting the given material properties, for the lamina with a ply angle $\alpha = 30°$

$$\bar{\boldsymbol{Q}}_{+\alpha} = \begin{bmatrix} 84.2935 & 26.1208 & 40.2292 \\ 26.1208 & 22.3329 & 13.4302 \\ 40.2292 & 13.4302 & 29.0686 \end{bmatrix} \text{ GPa}$$

and for the lamina with a ply angle $\alpha = -30°$

$$\bar{\boldsymbol{Q}}_{-\alpha} = \begin{bmatrix} 84.2935 & 26.1208 & -40.2292 \\ 26.1208 & 22.3329 & -13.4302 \\ -40.2292 & -13.4302 & 29.0686 \end{bmatrix} \text{ GPa}$$

The stiffness matrix of the piezo sheet is

$$\bar{\boldsymbol{Q}} = \boldsymbol{Q}_c = \frac{E_c}{1-\nu^2}\begin{bmatrix} 1 & \nu & 0 \\ \nu & 1 & 0 \\ 0 & 0 & \frac{1-\nu}{2} \end{bmatrix} = \begin{bmatrix} 79.5604 & 23.8681 & 0 \\ 23.8681 & 79.5604 & 0 \\ 0 & 0 & 27.8462 \end{bmatrix} \text{ GPa}$$

The generalized stiffness matrix is

$$\boldsymbol{K} = \begin{bmatrix} 44.95 \times 10^6 & 0 & -83.16 \times 10^3 \\ 0 & 232.55 & -232.55 \\ -83.16 \times 10^3 & -232.55 & 744.13 \end{bmatrix} \text{ N/m}$$

and the generalized forcing is

$$\boldsymbol{Q}_\Lambda = \begin{Bmatrix} -431.878 \\ 0 \\ 0 \end{Bmatrix} \text{ N}$$

Solving the governing equation with these values of generalized stiffness and forcing yields

$$\begin{Bmatrix} q_1 \\ q_2 \\ q_3 \end{Bmatrix} = \begin{bmatrix} K_{11} & K_{12} & K_{13} \\ K_{12} & K_{22} & K_{23} \\ K_{13} & K_{23} & K_{33} \end{bmatrix}^{-1} \begin{Bmatrix} Q_{\Lambda_1} \\ Q_{\Lambda_2} \\ Q_{\Lambda_3} \end{Bmatrix} = \begin{Bmatrix} -13.74 \times 10^{-6} \\ -0.0022 \\ -0.0022 \end{Bmatrix} \text{ m}$$

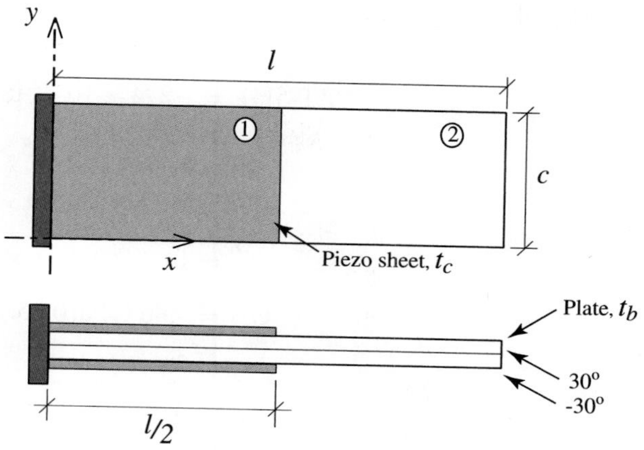

Figure 5.16. Rectangular antisymmetric laminate with two identical piezo sheets.

The tip twist θ_{tip} is found from

$$\theta_{tip} = \frac{2q_3}{c} = -0.0293 \text{ rad} = -1.6797^\circ$$

Note that the out-of-plane displacement at the mid-point of the free end of the plate ($x = l, y = 0$) is

$$w(l, 0) = q_2 - q_3 = 0$$

which indicates that the plate is undergoing twisting without any bending.

5.4.9 Worked Example

(a) Using laminated plate theory, derive the Rayleigh-Ritz solution for a rectangular cantilevered two-layered antisymmetric laminate with identical piezo sheets bonded on either surface for half of the plate length. An equal voltage is applied on each piezo sheet. Although the piezos are stretched across the complete width, neglect the influence of lateral strain. Assume the deflection as

$$u(x, y) = \frac{x}{l} q_1$$

$$w(x, y) = \frac{x^2}{l^2} \frac{y}{c} q_2$$

(b) Calculate the tip twist for this two-layered antisymmetric laminated plate (Fig. 5.16) with a ply layup [+30°/−30°]. The size of the plate is 0.3048 m (12 in) × 0.1524 m (6 in) × 0.79375 mm (1/32 in) with piezos of thickness $t_c = 0.3175$ mm (0.0125 in). The voltage applied to the piezos is 100 volts.

Manufacturer-supplied material constants are as follows

$$
\begin{aligned}
d_{31}\ (\text{PZT-5H}) &= -274 \times 10^{-12}\ \text{m/V} \\
E_c &= 72.4\ \text{GPa}(10.5 \times 10^6 \text{lb/in}^2) \\
E_1 &= 137.9\ \text{GPa}(20 \times 10^6\ \text{lb/in}^2) \\
E_2 &= 14.5\ \text{GPa}(2.1 \times 10^6\ \text{lb/in}^2) \\
G_{12} &= 5.86\ \text{GPa}(0.85 \times 10^6\ \text{lb/in}^2) \\
\nu_c &= 0.3 \\
\nu_{12} &= 0.2
\end{aligned}
$$

Solution

(a) The plate stiffness matrices for an antisymmetric laminate are given by (see Table 5.2)

$$
\boldsymbol{A} = \begin{bmatrix} A_{11} & A_{12} & 0 \\ A_{12} & A_{22} & 0 \\ 0 & 0 & A_{66} \end{bmatrix} \quad \boldsymbol{B} = \begin{bmatrix} 0 & 0 & B_{16} \\ 0 & 0 & B_{26} \\ B_{16} & B_{26} & 0 \end{bmatrix} \quad \boldsymbol{D} = \begin{bmatrix} D_{11} & D_{12} & 0 \\ D_{12} & D_{22} & 0 \\ 0 & 0 & D_{66} \end{bmatrix}
$$

We proceed in the same manner as described in Example 5.4.7. The derivation is similar except for the terms involving the $\boldsymbol{B}$ matrix. Therefore, we can write

$$
u(x,y) = \frac{x}{l} q_1, \ \rightarrow \ \phi_1 = \frac{x}{l}
$$

$$
w(x,y) = \frac{x^2}{l^2}\frac{y}{c} q_2, \ \rightarrow \ \phi_2 = \frac{x^2}{l^2}\frac{y}{c}
$$

At $x = 0,\ w = 0,\ w' = 0$; the boundary conditions are satisfied. The governing equation is given by

$$
\begin{bmatrix} K_{11} & K_{12} \\ K_{12} & K_{22} \end{bmatrix} \begin{Bmatrix} q_1 \\ q_2 \end{Bmatrix} = \begin{Bmatrix} Q_{\Lambda_1} \\ Q_{\Lambda_2} \end{Bmatrix}
$$

where the elements of the stiffness matrix are given by

$$
K_{11} = \frac{c}{2l}\left(A_{11}^{(1)} + A_{11}^{(2)}\right)
$$

$$
K_{22} = \frac{c}{6l^3}\left(D_{11}^{(1)} + D_{11}^{(2)}\right) + \frac{2}{3cl}\left(D_{66}^{(1)} + 7D_{66}^{(2)}\right)
$$

Because the structure of the $\boldsymbol{B}$ matrix is different from the previous example, the value of K_{12} is given by

$$\begin{aligned}
K_{12} &= \int_0^{l/2}\int_{-c/2}^{c/2} \{\mathbb{D}\boldsymbol{\phi}_1\}^T \begin{bmatrix} \boldsymbol{A} & \boldsymbol{B} \\ \boldsymbol{B} & \boldsymbol{D} \end{bmatrix}^{(1)} \{\mathbb{D}\boldsymbol{\phi}_2\}\, dx\, dy \\
&\quad + \int_{l/2}^{l}\int_{-c/2}^{c/2} \{\mathbb{D}\boldsymbol{\phi}_1\}^T \begin{bmatrix} \boldsymbol{A} & \boldsymbol{B} \\ \boldsymbol{B} & \boldsymbol{D} \end{bmatrix}^{(2)} \{\mathbb{D}\boldsymbol{\phi}_2\}\, dx\, dy \\
&= \int_0^{l/2}\int_{-c/2}^{c/2} \left(\frac{B_{16}^{(1)}}{l}\frac{4x}{l^2 c}\right) dx\, dy + \int_{l/2}^{l}\int_{-c/2}^{c/2} \left(\frac{B_{16}^{(2)}}{l}\frac{4x}{l^2 c}\right) dx\, dy \\
&= \frac{1}{2l}\left(B_{16}^{(1)} + 3B_{16}^{(2)}\right)
\end{aligned}$$

where

$$\begin{aligned}
A_{11}^{(1)} &= 2(Q_{11})_c t_c + \frac{t_b}{2}\left((\bar{Q}_{11})_{+\alpha} + (\bar{Q}_{11})_{-\alpha}\right) \\
&= 2(Q_{11})_c t_c + t_b(\bar{Q}_{11})_{+\alpha} \\
A_{11}^{(2)} &= t_b(\bar{Q}_{11})_{+\alpha}
\end{aligned}$$

and

$$\begin{aligned}
D_{11}^{(1)} &= \frac{2}{3}(Q_{11})_c t_c\left(\frac{3}{4}t_b^2 + \frac{3}{2}t_b t_c + t_c^2\right) + \frac{t_b^3}{24}\left((\bar{Q}_{11})_{+\alpha} + (\bar{Q}_{11})_{-\alpha}\right) \\
&= \frac{2}{3}(Q_{11})_c t_c\left(\frac{3}{4}t_b^2 + \frac{3}{2}t_b t_c + t_c^2\right) + \frac{t_b^3}{12}(\bar{Q}_{11})_{+\alpha} \\
D_{11}^{(2)} &= \frac{t_b^3}{12}(\bar{Q}_{11})_{+\alpha}
\end{aligned}$$

Similarly

$$\begin{aligned}
D_{66}^{(1)} &= \frac{2}{3}(Q_{66})_c t_c\left(\frac{3}{4}t_b^2 + \frac{3}{2}t_b t_c + t_c^2\right) + \frac{t_b^3}{12}(\bar{Q}_{66})_{+\alpha} \\
D_{66}^{(2)} &= \frac{t_b^3}{12}(\bar{Q}_{66})_{+\alpha}
\end{aligned}$$

The term B_{16} is found from

$$\begin{aligned}
B_{16}^{(1)} &= -\frac{1}{2}\sum_{i=1}^{4}(\bar{Q}_{11})_k(h_{k+1}^2 - h_k^2) \\
&= -\frac{1}{2}\left[(\bar{Q}_{16})_{+\alpha}\left(\frac{t_b}{2}\right)^2 - (\bar{Q}_{16})_{-\alpha}\left(\frac{t_b}{2}\right)^2\right] \\
&= -\frac{t_b^2}{4}(\bar{Q}_{16})_{+\alpha} \\
B_{16}^{(2)} &= B_{16}^{(1)} = -\frac{t_b^2}{4}(\bar{Q}_{16})_{+\alpha}
\end{aligned}$$

Substituting these relations, the elements of the stiffness matrix are given by

$$K_{11} = \frac{c}{l}\left((Q_{11})_c t_c + (\bar{Q}_{11})_{+\alpha} t_b\right)$$

$$K_{12} = -\frac{t_b^2}{2l}(\bar{Q}_{16})_{+\alpha}$$

$$K_{22} = \frac{c}{9l^3}\left[(Q_{11})_c t_c\left(3/4t_b^2 + 3/2t_b t_c + t_c^2\right) + \frac{t_b^3}{4}(\bar{Q}_{11})_{+\alpha}\right]$$

$$+ \frac{4}{9cl}\left[(Q_{66})_c t_c\left(3/4t_b^2 + 3/2t_b t_c + t_c^2\right) + t_b^3(\bar{Q}_{66})_{+\alpha}\right]$$

The generalized forces are given by

$$Q_{\Lambda_1} = \frac{E_c t_c c\Lambda}{1-\nu}$$

$$Q_{\Lambda_2} = 0$$

Note that the induced moment is zero. The tip twist is given by

$$\theta_{\text{tip}} = \int_{x=0}^{l} \frac{\partial^2 w}{\partial x \partial y} dx = \left.\frac{\partial w}{\partial y}\right|_{x=l} = \frac{q_2}{c}$$

(b) Substituting the given material properties, for the lamina with a ply angle $\alpha = 30^\circ$

$$\bar{\boldsymbol{Q}}_{+\alpha} = \begin{bmatrix} 84.2935 & 26.1208 & 40.2292 \\ 26.1208 & 22.3329 & 13.4302 \\ 40.2292 & 13.4302 & 29.0686 \end{bmatrix} \text{GPa}$$

and for the lamina with a ply angle $\alpha = -30^\circ$

$$\bar{\boldsymbol{Q}}_{-\alpha} = \begin{bmatrix} 84.2935 & 26.1208 & -40.2292 \\ 26.1208 & 22.3329 & -13.4302 \\ -40.2292 & -13.4302 & 29.0686 \end{bmatrix} \text{GPa}$$

The stiffness matrix of the piezo sheet is

$$\bar{\boldsymbol{Q}} = \boldsymbol{Q}_c = \frac{E_c}{1-\nu^2}\begin{bmatrix} 1 & \nu & 0 \\ \nu & 1 & 0 \\ 0 & 0 & \frac{1-\nu}{2} \end{bmatrix} = \begin{bmatrix} 79.5604 & 23.8681 & 0 \\ 23.8681 & 79.5604 & 0 \\ 0 & 0 & 27.8462 \end{bmatrix} \text{GPa}$$

The generalized stiffness matrix is

$$\boldsymbol{K} = \begin{bmatrix} 44.95 \times 10^6 & -41.58 \times 10^3 \\ -41.58 \times 10^3 & 238.93 \end{bmatrix} \text{N/m}$$

and the generalized forcing is

$$\boldsymbol{Q}_\Lambda = \begin{Bmatrix} -431.89 \\ 0 \end{Bmatrix} \text{N}$$

Solving the governing equation with these values of generalized stiffness and forcing yields

$$\begin{Bmatrix} q_1 \\ q_2 \end{Bmatrix} = \begin{bmatrix} K_{11} & K_{12} \\ K_{12} & K_{22} \end{bmatrix}^{-1} \begin{Bmatrix} Q_{\Lambda_1} \\ Q_{\Lambda_2} \end{Bmatrix} = \begin{Bmatrix} -11.45 \times 10^{-6} \\ -0.002 \end{Bmatrix} \quad \text{m}$$

The tip twist θ_{tip} is found from

$$\theta_{tip} = \frac{q_2}{c} = -13.1 \times 10^{-3} \text{ rad} = -0.7493^\circ$$

5.5 Coupling Efficiency

A coupled structure is usually designed to convert one type of motion into another. For example, an extension-torsion coupled beam can be used to convert a linear displacement (e.g., one induced by piezoceramic actuators), into a torsional displacement. A coupling efficiency can be introduced to evaluate the effectiveness of the structure in transforming one type of displacement into another. This also can serve as a performance metric to optimize the design of the structure. The coupling efficiency can be defined in two ways:

1. Displacement-based: A coupling efficiency η_d, based on displacement, can be defined as the ratio of the output displacement or curvature to the force or moment input to the structure. Note that this is, strictly speaking, not an "efficiency" because it is dimensional.
2. Energy-based: A nondimensional coupling efficiency η_e can be defined as the ratio of the strain energy associated with the output displacement to the total strain energy in the structure. In other words, it is a measure of the effectiveness of the transfer of energy between the two types of deformation modes.

Using the Rayleigh-Ritz method in conjunction with the simplest possible assumed modes that capture the structural deformation (see Sections 5.4.5 and 5.4.6), it is possible to obtain a first-order estimate of the coupling efficiency. Note that this estimate depends on the assumed modes. However, because an increasing number of assumed modes will only improve the predictions of structural deformation, the first-order estimate of coupling efficiency serves as a lower bound and can be treated as a worst-case condition.

5.5.1 Extension-Torsion Coupling Efficiency

Consider a plate with extension-torsion coupling, as described in Section 5.4.5. Due to the extension-torsion coupling, a purely extensional induced force results in a torsion of the plate. In this case, the displacement-based coupling efficiency can be defined as the ratio of the tip twist of the plate to the induced force.

$$\eta_d^{ET} = \frac{\kappa_{xy}^{\text{tip}}}{F_{x_\Lambda}} \qquad \text{(rad/N or 1/N)} \tag{5.392}$$

From Eq. 5.364

$$\eta_d^{ET} = -\frac{6B_{16}}{(4A_{11}D_{66} - 3B_{16}^2)} \tag{5.393}$$

Note that this is a function of the coupling stiffness B_{16} as well as the direct extensional and bending stiffnesses, A_{11} and D_{66}, respectively.

Using the second approach, from Section 5.4.5, the torsional strain energy of an extension-torsion coupled plate can be derived as

$$\begin{aligned} V_{\text{torsion}} &= \frac{1}{2}\int_0^l \int_{-c/2}^{c/2} D_{66}\,\kappa_{xy}^2\,dx\,dy \\ &= \frac{1}{2}D_{66}\int_0^l \int_{-c/2}^{c/2} \frac{16x^2}{l^4c^2}\,q_2^2 dx\,dy \\ &= \frac{8}{3}\frac{D_{66}q_2^2}{lc} \end{aligned} \tag{5.394}$$

The total strain energy in the plate is given by

$$\begin{aligned} V_{\text{total}} &= \frac{1}{2}\int_0^l \int_{-c/2}^{c/2} \{\epsilon_x^o\ \kappa_{xy}\} \begin{bmatrix} A_{11} & B_{16} \\ B_{16} & D_{66} \end{bmatrix} \begin{Bmatrix} \epsilon_x^o \\ \kappa_{xy} \end{Bmatrix} dx\,dy \\ &= \frac{1}{2}\int_0^l \int_{-c/2}^{c/2} \left[A_{11}\epsilon_x^{o2} + \kappa_{xy}^2 D_{66} + 2\kappa_{xy}B_{16}\epsilon_x^o\right] dx\,dy \\ &= \frac{1}{2}\int_0^l \int_{-c/2}^{c/2} \left[A_{11}\frac{q_1^2}{l^2} + D_{66}\frac{16x^2}{l^4c^2}q_2^2 + B_{16}\frac{8x}{l^3c}q_1q_2\right] dx\,dy \\ &= \frac{1}{2}A_{11}.\frac{c}{l}q_1^2 + \frac{1}{2}\cdot\frac{16}{3}\cdot\frac{D_{66}}{lc}q_2^2 + \frac{2}{l}B_{16}\,q_1\,q_2 \\ &= V_{\text{extension}} + V_{\text{torsion}} + V_{\text{coupling}} \end{aligned} \tag{5.395}$$

The energy-based extension-torsion coupling efficiency is

$$\begin{aligned} \eta_e^{ET} &= \frac{V_{\text{torsion}}}{V_{\text{total}}} \\ &= \frac{16D_{66}}{3A_{11}c^2\,(q_1/q_2)^2 + 16D_{66} + 12cB_{16}\,(q_1/q_2)} \end{aligned} \tag{5.396}$$

Note that

$$\frac{q_1}{q_2} = -\frac{8}{3}\frac{D_{66}}{B_{16}c} \tag{5.397}$$

Substituting in Eq. 5.396 and simplifying results in

$$\eta_e^{ET} = \frac{3B_{16}^2}{4A_{11}D_{66} - 3B_{16}^2} \tag{5.398}$$

The extension-torsion coupling efficiency can also be defined in terms of an important parameter ψ_{ET}

$$\psi_{ET} = \frac{B_{16}}{\sqrt{A_{11}D_{66}}} \tag{5.399}$$

It can be seen that this extension-torsion coupling parameter is a ratio of the extension-torsion coupling stiffness to the product of the extensional stiffness and

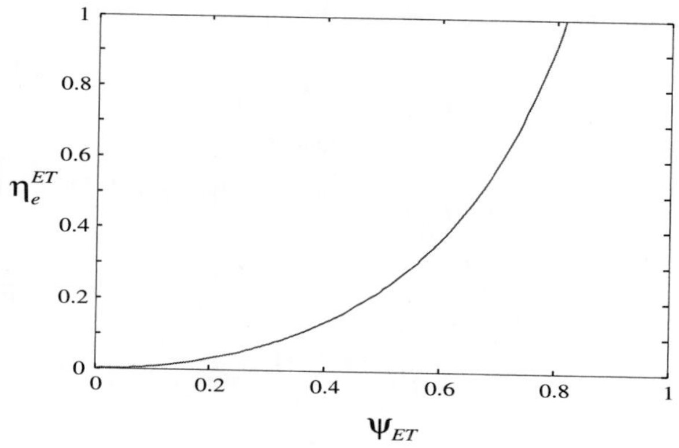

Figure 5.17. Variation of energy-based extension-torsion coupling efficiency η_e^{ET} with coupling parameter ψ_{ET}.

torsional stiffness. Substituting in Eq. 5.398

$$\eta_e^{ET} = \frac{3\psi_{ET}^2}{4 - 3\psi_{ET}^2} \tag{5.400}$$

Note that physically, the efficiency η_e^{ET} cannot be greater than unity, which occurs at $\psi_{ET} = \sqrt{2/3}$. The efficiency monotonically increases with increasing ψ_{ET} in this range and is plotted in Fig. 5.17. For most physical structures, the axial stiffness (A_{11}) is much larger than the other two stiffnesses. As a result, ψ_{ET} is small, which yields a low extension-torsion coupling efficiency.

5.5.2 Bending-Torsion Coupling Efficiency

Consider the bending-torsion coupled plate described in Section 5.4.6. Due to the bending-torsion coupling, a pure induced bending moment results in a twisting of the plate. The bending-torsion coupling efficiency based on displacement can be defined as the ratio of the tip twist of the plate to the induced bending moment on the plate.

$$\eta_d^{BT} = \frac{\kappa_{xy}^{\text{tip}}}{M_{x_\Lambda}} \qquad \text{(rad/(Nm) or 1/(Nm))} \tag{5.401}$$

From Eq. 5.391

$$\eta_d^{BT} = \frac{-24(l/c)^2 D_{16}}{D_{11}^2 + 16(l/c)^2 D_{11} D_{66} - 12(l/c)^2 D_{16}^2} \tag{5.402}$$

Note that this is a function of the coupling stiffness D_{16} as well as the direct extensional and bending stiffnesses, D_{11} and D_{66}, respectively.

Using the second approach, from Section 5.4.6, the torsional strain energy in a bending-torsion coupled plate can be derived as

$$\begin{aligned} V_{\text{torsion}} &= \frac{1}{2}\int_0^l \int_{-c/2}^{c/2} D_{66}\,\kappa_{xy}^2\,dx\,dy \\ &= \frac{1}{2} D_{66} \int_0^l \int_{-c/2}^{c/2} \frac{16x^2}{l^4 c^2}\, q_2^2 dx\,dy \\ &= \frac{8}{3}\frac{D_{66}}{lc} q_2^2 \end{aligned} \tag{5.403}$$

The total strain energy of the plate is

$$
\begin{aligned}
V_{\text{total}} &= \frac{1}{2}\int_0^l \int_{-c/2}^{c/2} \{\kappa_x \quad \kappa_{xy}\} \begin{bmatrix} D_{11} & D_{16} \\ D_{16} & D_{66} \end{bmatrix} \begin{Bmatrix} \kappa_x \\ \kappa_{xy} \end{Bmatrix} dx\,dy \\
&= \frac{1}{2}\int_0^l \int_{-c/2}^{c/2} \left[D_{11}\kappa_x^2 + 2D_{16}\kappa_x\kappa_{xy} + D_{66}\kappa_{xy}^2 \right] dx\,dy \\
&= \frac{1}{2}\int_0^l \int_{-c/2}^{c/2} \left[\frac{4D_{11}}{l^4}q_1^2 + \frac{16D_{16}}{l^4 c}q_1 q_2 x + \frac{8D_{11}}{l^4 c}q_1 q_2 y + \frac{16D_{66}}{l^4 c^2}q_2^2 x^2 \right. \\
&\quad \left. + \frac{4D_{11}}{l^4 c^2}q_2^2 y_2 + \frac{16D_{16}}{l^4 c^2}q_2^2 xy \right] dx\,dy \\
&= 2D_{11}\frac{c}{l^3}\left(q_1^2 + \frac{q_2^2}{12}\right) + \frac{4}{l^2}D_{16}q_1 q_2 + \frac{8}{3lc}D_{66}q_2^2 \\
&= V_{\text{bending}} + V_{\text{coupling}} + V_{\text{torsion}}
\end{aligned}
\tag{5.404}
$$

The energy-based coupling efficiency is

$$
\begin{aligned}
\eta_e^{BT} &= \frac{V_{\text{torsion}}}{V_{\text{total}}} \\
&= \frac{16D_{66}l^2}{D_{11}c^2\left(12(q_1/q_2)^2 + 1\right) + 24D_{16}(q_1/q_2)lc + 16l^2 D_{66}}
\end{aligned}
\tag{5.405}
$$

Note that

$$
\frac{q_1}{q_2} = \frac{-\left(c^2 D_{11} + 16l^2 D_{66}\right)}{12lcD_{16}} \tag{5.406}
$$

Substituting in Eq. 5.405 and simplifying leads to

$$
\eta_e^{BT} = \frac{192l^4 D_{66}D_{16}^2}{c^4 D_{11}^3 + 256l^4 D_{11}D_{66}^2 + 32l^2c^2 D_{11}^2 D_{66} - 12l^2c^2 D_{11}D_{16}^2 - 192l^4 D_{16}^2 D_{66}} \tag{5.407}
$$

The bending-torsion coupling efficiency can be defined in terms of the following parameters

$$
\psi_{BT} = \frac{D_{16}}{\sqrt{D_{11}D_{66}}} \quad \text{(bending-torsion coupling parameter)} \tag{5.408}
$$

$$
K_{BT} = \frac{D_{11}}{D_{66}} \quad \text{(ratio of bending stiffness to torsional stiffness)} \tag{5.409}
$$

$$
A_p = \frac{l}{c} \quad \text{(aspect ratio of the plate)} \tag{5.410}
$$

Substituting in the previous equation

$$
\eta_e^{BT} = \frac{192A_p^4\psi_{BT}^2}{K_{BT}^2 + 256A_p^4 + 32A_p^2 K_{BT} - 12A_p^2\psi_{BT}^2 K_{BT} - 192A_p^4\psi_{BT}^2} \tag{5.411}
$$

5.5.3 Comparison of Extension-Torsion and Bending-Torsion Coupling

The efficiencies of extension-torsion and bending-torsion couplings can be compared by examining the two-ply laminates shown in Fig. 5.18. Note that by changing

Figure 5.18. Extension-torsion and bending-torsion coupled laminates.

the layup as in Figs. 5.18(a) and 5.18(b), only the coupling is affected, leaving other properties of the laminate unchanged. The coupling properties are calculated over the entire range of ply angles $0° < \alpha° < 90°$. The effect of the number of plies in the laminate can be investigated for a symmetric laminate by considering a layup $[(+\alpha°)_n/(+\alpha°)_n]$ and for an antisymmetric laminate by considering a layup $[(+\alpha°)_n/(-\alpha°)_n]$. For example, for a laminate with four plies, $n = 2$. In the symmetric case, the lay-up will be $[+\alpha°/+\alpha°/+\alpha°/+\alpha°]$ and in the antisymmetric case, the layup will be $[+\alpha°/+\alpha°/-\alpha°/-\alpha°]$. Note that the following discussion refers to laminates of this configuration only, and other symmetric and antisymmetric configurations may yield different results.

Figure 5.19 shows the ratio of bending stiffness to torsional stiffness (D_{11}/D_{66}) for the two couplings, as a function of ply angle. Note that this ratio (called K_{ET}) is the same for both couplings. In addition, this ratio does not depend on the aspect ratio of the laminate, A_p, or on the number of plies.

The coupling parameter ψ_{ET} for extension-torsion coupling is plotted in Fig. 5.20 as a function of the ply angle. Note that this parameter also does not depend either on the aspect ratio or the number of plies of the laminate. The sign of the coupling parameter only affects the direction of the induced twist and does not have any major significance. The maximum value of ψ_{ET} is approximately -0.7 at a ply angle of approximately 30°.

The displacement-based extension-torsion coupling efficiency, as a function of the ply angle, is shown in Fig. 5.21. This parameter does not depend on the aspect ratio of the laminate but rather on the number of plies. This plot will serve as a useful tool to design a laminate with appropriate coupling behavior. As expected, with an increasing number of plies, the laminate becomes stiffer and the induced tip twist decreases.

Figure 5.22 shows the energy-based extension-torsion coupling efficiency as a function of the ply angle. This parameter is independent of the number of plies and the plate aspect ratio. The maximum efficiency is approximately 59% at a ply angle of 30°.

The variation of coupling parameter for bending-torsion coupling, ψ_{BT}, with ply angle, is shown in Fig. 5.23. The qualitative behavior is the same as in the case of

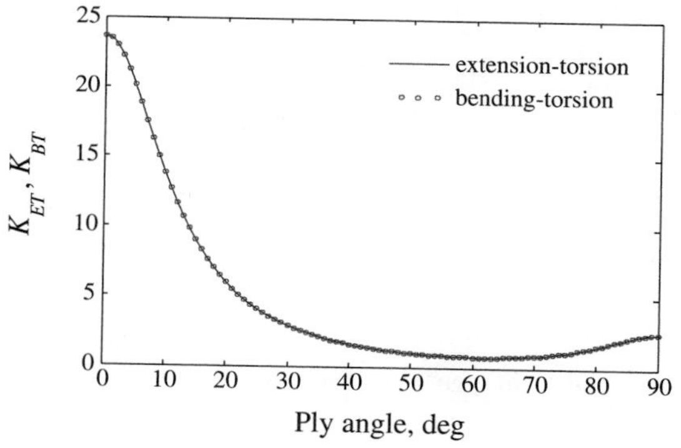

Figure 5.19. Ratio of bending stiffness to torsional stiffness, $A_p = 1$, for two plies.

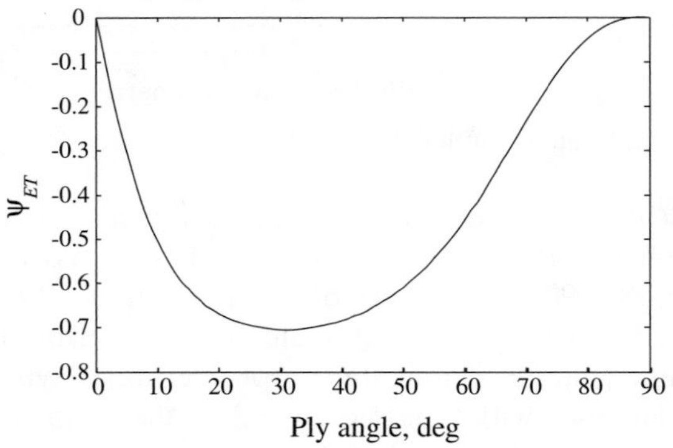

Figure 5.20. Extension-torsion coupling parameter ψ_{ET} for an antisymmetric laminate.

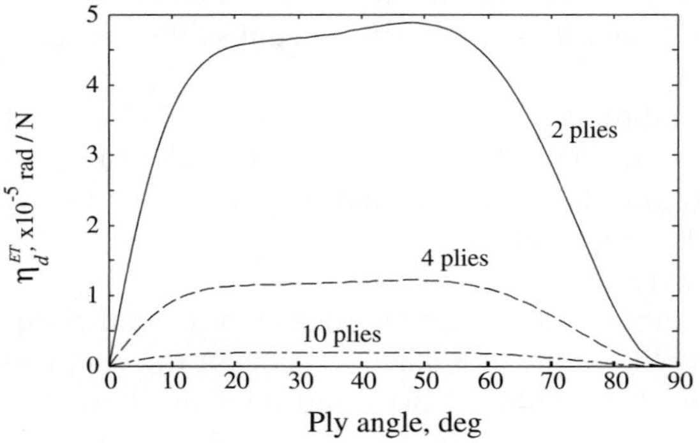

Figure 5.21. Displacement-based extension-torsion coupling efficiency η_d^{ET}, as a function of number of plies.

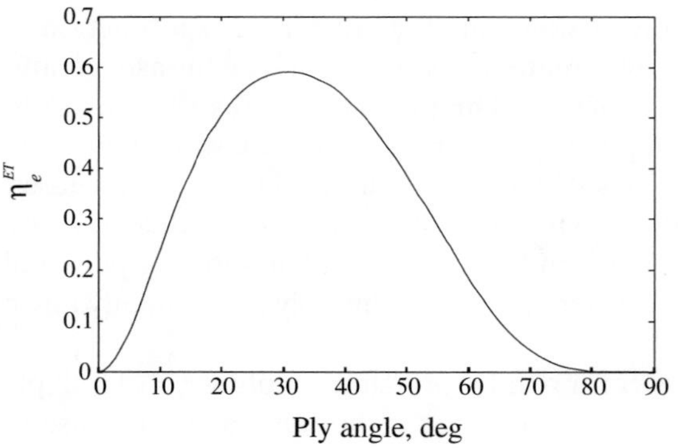

Figure 5.22. Energy-based extension-torsion coupling efficiency η_e^{ET} as a function of ply angle.

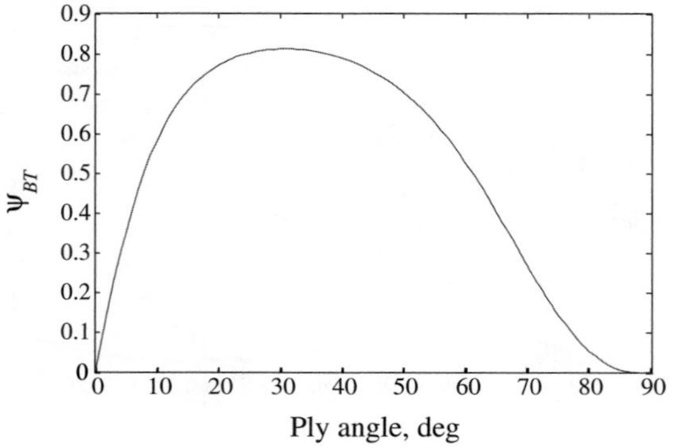

Figure 5.23. Bending-torsion coupling parameter ψ_{BT}, as a function of ply angle.

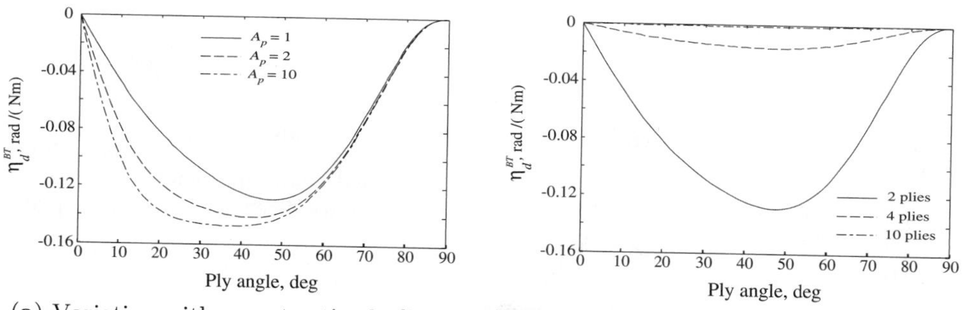

(a) Variation with aspect ratio, 2 plies (b) Variation with number of plies, $A_p = 1$

Figure 5.24. Displacement-based bending-torsion efficiency η_d^{BT}.

extension-torsion coupling, with a maximum efficiency of approximately 81% at a ply angle of 30°. The bending-torsion coupling efficiency based on displacement is shown in Fig. 5.24. It shows that this varies with both the aspect ratio of the plate and the number of plies.

The energy-based bending-torsion coupling efficiency is shown in Fig. 5.25, as a function of the plate aspect ratio. It can be seen that the maximum efficiency increases with plates of increasing aspect ratio, and the optimum ply angle decreases. To compare the relative efficiency of extension-torsion and bending-torsion coupling, we compare the energy-based efficiencies for a plate with $A_p = 1$ and having two plies (Fig. 5.26). It shows that the bending-torsion coupling is more efficient in terms of energy transfer. In addition, the optimum ply angle for bending-torsion coupling is larger than in the case of extension-torsion coupling.

5.6 Classical Laminated Plate Theory (CLPT) with Induced-Strain Actuation for a Dynamic Case

For a thin laminated plate undergoing small displacement motion, the velocity components at a station (x, y, z) are

$$\dot{u}(x, y, z, t) = \dot{u_o}(x, y, t) - z\frac{\partial \dot{w_o}}{\partial x}(x, y, t) \tag{5.412}$$

$$\dot{v}(x, y, z, t) = \dot{v_o}(x, y, t) - z\frac{\partial \dot{w_o}}{\partial y}(x, y, t) \tag{5.413}$$

$$\dot{w}(x, y, z, t) = \dot{w_o}(x, y, t) \tag{5.414}$$

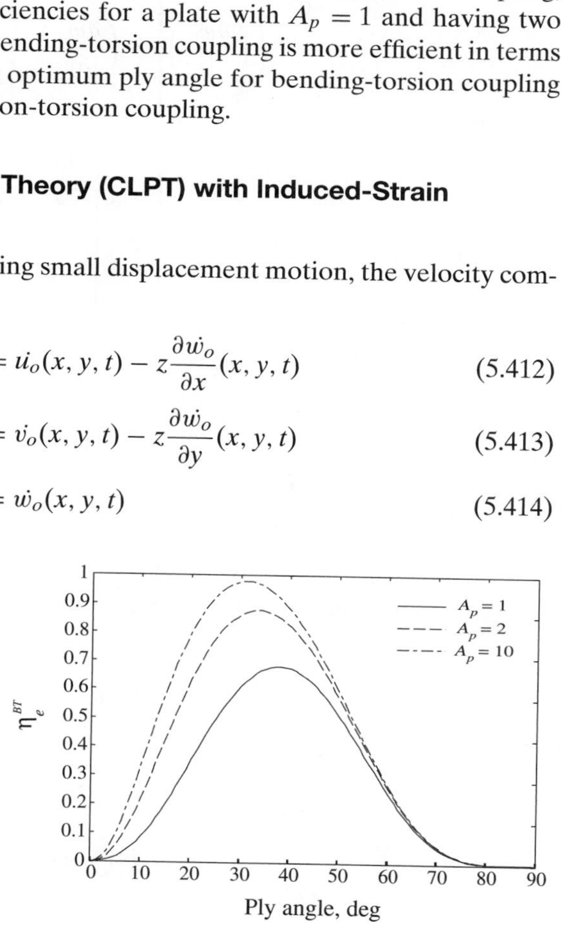

Figure 5.25. Energy-based bending-torsion efficiency η_e^{BT}, as a function of aspect ratio.

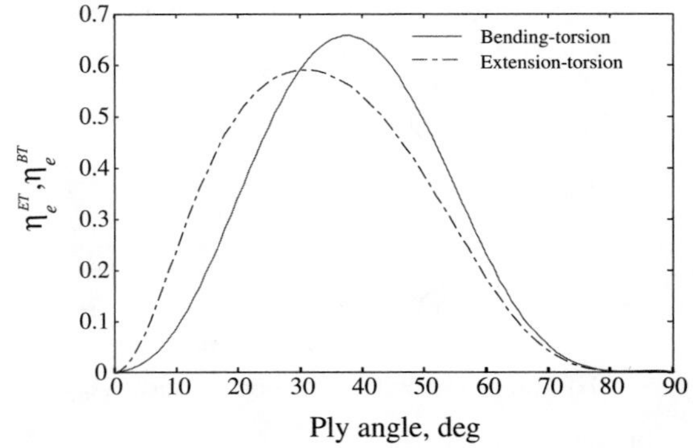

Figure 5.26. Comparison of energy-based extension-torsion and bending-torsion efficiencies, $A_p = 1$, two plies.

where $\dot{u}_o$, $\dot{v}_o$, and $\dot{w}_o$ are the velocity components at the mid-plane. The kinetic energy, T, of an elemental volume dV of the plate is given by

$$T = \frac{1}{2}\iiint_{\text{volume}} \rho_s \left(\dot{u}^2 + \dot{v}^2 + \dot{w}^2\right) dV \tag{5.415}$$

where ρ_s is the mass density. Substituting Eqs. 5.412–5.414

$$\begin{aligned} T = \frac{1}{2}\iint_{\text{area}} \Bigg[& m\left(\dot{u_o}^2 + \dot{v_o}^2 + \dot{w_o}^2\right) - 2S_I\left(\dot{u}_o\frac{\partial \dot{w}_o}{\partial x} + \dot{v}_o\frac{\partial \dot{w}_o}{\partial y}\right) \\ & + I\left(\left(\frac{\partial \dot{w}_o}{\partial x}\right)^2 + \left(\frac{\partial \dot{w}_o}{\partial y}\right)^2\right)\Bigg] dx\,dy \end{aligned} \tag{5.416}$$

This can be rewritten as

$$\begin{aligned} T &= \frac{1}{2}\iint_{\text{area}} \begin{bmatrix} \dot{u}_o & \dot{v}_o & \dot{w}_o & \frac{\partial \dot{w}_o}{\partial x} & \frac{\partial \dot{w}_o}{\partial y} \end{bmatrix} \begin{bmatrix} m & 0 & 0 & -S_I & 0 \\ 0 & m & 0 & 0 & -S_I \\ 0 & 0 & m & 0 & 0 \\ -S_I & 0 & 0 & I & 0 \\ 0 & -S_I & 0 & 0 & I \end{bmatrix} \begin{Bmatrix} \dot{u}_o \\ \dot{v}_o \\ \dot{w}_o \\ \frac{\partial \dot{w}_o}{\partial x} \\ \frac{\partial \dot{w}_o}{\partial y} \end{Bmatrix} dx\,dy \\ &= \frac{1}{2}\iint_{\text{area}} \begin{bmatrix} \dot{u}_o & \dot{v}_o & \dot{w}_o & \frac{\partial \dot{w}_o}{\partial x} & \frac{\partial \dot{w}_o}{\partial y} \end{bmatrix} \boldsymbol{m}_I \begin{Bmatrix} \dot{u}_o \\ \dot{v}_o \\ \dot{w}_o \\ \frac{\partial \dot{w}_o}{\partial x} \\ \frac{\partial \dot{w}_o}{\partial y} \end{Bmatrix} dx\,dy \end{aligned} \tag{5.417}$$

where m is the mass per unit area, S_I is the first mass moment of inertia per unit area about the mid-plane, and I is the second mass moment of inertia per unit area about the mid-plane

$$m = \int_t \rho_s dz \qquad (\text{kg/m}^2) \tag{5.418}$$

$$S_I = \int_t \rho_s z dz \qquad (\text{kg/m}) \tag{5.419}$$

$$I = \int_t \rho_s z^2 dz \qquad (\text{kg}) \tag{5.420}$$

I is also referred to as the rotary inertia term and is often neglected, especially for lower vibration modes. The plate equations are

u-equation:

$$\frac{\partial N_x}{\partial x} + \frac{\partial N_{xy}}{\partial y} = m\frac{\partial^2 u_o}{\partial t^2} - S_I\frac{\partial^2}{\partial t^2}\left(\frac{\partial w_o}{\partial x}\right) \tag{5.421}$$

v-equation:

$$\frac{\partial N_{xy}}{\partial x} + \frac{\partial N_y}{\partial y} = m\frac{\partial^2 v_o}{\partial t^2} - S_I\frac{\partial^2}{\partial t^2}\left(\frac{\partial w_o}{\partial y}\right) \tag{5.422}$$

w-equation:

$$\begin{aligned}\frac{\partial^2 M_x}{\partial x^2} + 2\frac{\partial^2 M_{xy}}{\partial x \partial y} + \frac{\partial^2 M_y}{\partial y^2} + q = m\frac{\partial^2 w_o}{\partial t^2} &+ S_I\frac{\partial^2}{\partial t^2}\left(\frac{\partial u_o}{\partial x} + \frac{\partial v_o}{\partial y}\right)\\ &- I\frac{\partial^2}{\partial t^2}\left(\frac{\partial^2 w_o}{\partial x^2} + \frac{\partial^2 w_o}{\partial y^2}\right)\end{aligned} \tag{5.423}$$

The equation of motion for a laminated plate with induced-strain actuation can be derived as (including inertial forces)

u-equation:

$$\begin{aligned}&A_{11}\frac{\partial^2 u_o}{\partial x^2} + A_{12}\frac{\partial^2 v_o}{\partial x \partial y} + A_{16}\left(2\frac{\partial^2 u_o}{\partial x \partial y} + \frac{\partial^2 v_o}{\partial x^2}\right) + A_{26}\frac{\partial^2 v_o}{\partial y^2} + A_{66}\left(\frac{\partial^2 u_o}{\partial y^2} + \frac{\partial^2 v_o}{\partial x \partial y}\right)\\ &\quad + B_{11}\frac{\partial^3 w_o}{\partial x^3} + B_{12}\frac{\partial^3 w_o}{\partial x \partial y^2} + 3B_{16}\frac{\partial^3 w_o}{\partial x^2 \partial y} + B_{26}\frac{\partial^3 w_o}{\partial y^3} + 2B_{66}\frac{\partial^3 w_o}{\partial x \partial y^2}\\ &\quad = \frac{\partial\left(F_x + F_{x_\Lambda}\right)}{\partial x} + \frac{\partial\left(F_{xy} + F_{xy_\Lambda}\right)}{\partial y} - p_x + m\frac{\partial^2 u_o}{\partial t^2} - S\frac{\partial^2}{\partial t^2}\left(\frac{\partial w_o}{\partial x}\right)\end{aligned} \tag{5.424}$$

v-equation:

$$\begin{aligned}&A_{22}\frac{\partial^2 v_o}{\partial y^2} + A_{12}\frac{\partial^2 u_o}{\partial x \partial y} + A_{16}\frac{\partial^2 u_o}{\partial x^2} + A_{26}\left(\frac{\partial^2 u_o}{\partial y^2} + 2\frac{\partial^2 v_o}{\partial x \partial y}\right) + A_{66}\left(\frac{\partial^2 u_o}{\partial x \partial y} + \frac{\partial^2 v_o}{\partial x^2}\right)\\ &\quad + B_{12}\frac{\partial^3 w_o}{\partial x^2 \partial y} + B_{22}\frac{\partial^3 w_o}{\partial y^3} + B_{16}\frac{\partial^3 w_o}{\partial x^3} + 3B_{26}\frac{\partial^3 w_o}{\partial x \partial y^2} + 2B_{66}\frac{\partial^3 w_o}{\partial x^2 \partial y}\\ &\quad = \frac{\partial\left(F_y + F_{y_\Lambda}\right)}{\partial y} + \frac{\partial\left(F_{xy} + F_{xy_\Lambda}\right)}{\partial x} - p_y + m\frac{\partial^2 v_o}{\partial t^2} - S\frac{\partial^2}{\partial t^2}\left(\frac{\partial w_o}{\partial y}\right)\end{aligned} \tag{5.425}$$

w-equation:

$$
\begin{aligned}
& B_{11}\frac{\partial^3 u_o}{\partial x^3} + B_{12}\left(\frac{\partial^3 v_o}{\partial x^2 \partial y} + \frac{\partial^3 u_o}{\partial x \partial y^2}\right) + B_{16}\left(3\frac{\partial^3 u_o}{\partial x^2 \partial y} + \frac{\partial^3 v_o}{\partial x^3}\right) + B_{22}\frac{\partial^3 v_o}{\partial y^3} \\
& + B_{26}\left(\frac{\partial^3 u_o}{\partial y^3} + 3\frac{\partial^3 v_o}{\partial x \partial y^2}\right) + 2B_{66}\left(\frac{\partial^3 u_o}{\partial x \partial y^2} + \frac{\partial^3 v_o}{\partial x^2 \partial y}\right) \\
& + D_{11}\frac{\partial^4 w_o}{\partial x^4} + 2D_{12}\frac{\partial^4 w_o}{\partial x^2 \partial y^2} + 4D_{16}\frac{\partial^4 w_o}{\partial x^3 \partial y} + D_{22}\frac{\partial^4 w_o}{\partial y^4} \\
& + 4D_{26}\frac{\partial^4 w_o}{\partial x \partial y^3} + 4D_{66}\frac{\partial^4 w_o}{\partial x^2 \partial y^2} \\
& = \frac{\partial^2 (M_x + M_{x_\Lambda})}{\partial x^2} + \frac{\partial^2 (M_y + M_{y_\Lambda})}{\partial y^2} + \frac{\partial^2 (M_{xy} + M_{xy_\Lambda})}{\partial x \partial y} - p_z \\
& + m\frac{\partial^2 w_o}{\partial t^2} + S\frac{\partial^2}{\partial t^2}\left(\frac{\partial u_o}{\partial x} + \frac{\partial v_o}{\partial y}\right) - I\frac{\partial^2}{\partial t^2}\left(\frac{\partial^2 w_o}{\partial x^2} + \frac{\partial^2 w_o}{\partial y^2}\right)
\end{aligned}
\tag{5.426}
$$

For a Rayleigh-Ritz solution

$$
\begin{Bmatrix} \dot{u}_o \\ \dot{v}_o \\ \dot{w}_o \\ \frac{\partial \dot{w}_o}{\partial x} \\ \frac{\partial \dot{w}_o}{\partial y} \end{Bmatrix} = \begin{bmatrix} 1 & 0 & 0 \\ 0 & 1 & 0 \\ 0 & 0 & 1 \\ 0 & 0 & \frac{\partial}{\partial x} \\ 0 & 0 & \frac{\partial}{\partial y} \end{bmatrix} \begin{Bmatrix} \dot{u}_o \\ \dot{v}_o \\ \dot{w}_o \end{Bmatrix} = \mathbb{D}_1 \begin{Bmatrix} \dot{u}_o \\ \dot{v}_o \\ \dot{w}_o \end{Bmatrix} \tag{5.427}
$$

The approximate solution can be expressed in terms of assumed shape functions as

$$
\begin{aligned}
\begin{Bmatrix} \dot{u}_o \\ \dot{v}_o \\ \dot{w}_o \end{Bmatrix} &= \begin{bmatrix} \phi_{u_1} & \phi_{u_2} \ldots & \phi_{u_M} & 0 & 0 \ldots & 0 & 0 & 0 \ldots & 0 \\ 0 & 0 \ldots & 0 & \phi_{v_1} & \phi_{v_2} \ldots & \phi_{v_N} & 0 & 0 \ldots & 0 \\ 0 & 0 \ldots & 0 & 0 & 0 \ldots & 0 & \phi_{w_1} & \phi_{w_2} \ldots & \phi_{w_P} \end{bmatrix} \begin{Bmatrix} \dot{q}_1 \\ \dot{q}_2 \\ \vdots \\ \dot{q}_{M+N+P} \end{Bmatrix} \\
&= \boldsymbol{\phi}(x, y)\dot{\boldsymbol{q}}
\end{aligned}
\tag{5.428}
$$

The expression for kinetic energy becomes

$$
\begin{aligned}
T &= \frac{1}{2}\iint_{\text{area}} \{\mathbb{D}_1 \boldsymbol{\phi}\dot{\boldsymbol{q}}\}^T \boldsymbol{m}_I \{\mathbb{D}_1 \boldsymbol{\phi}\dot{\boldsymbol{q}}\} dxdy \\
&= \frac{1}{2}\dot{\boldsymbol{q}}^T \boldsymbol{M}\dot{\boldsymbol{q}}
\end{aligned}
\tag{5.429}
$$

where $\boldsymbol{M}$ is a generalized mass matrix of size $(M + N + P) \times (M + N + P)$ defined as

$$
M_{ij} = \iint_{\text{area}} \{\mathbb{D}_1 \boldsymbol{\phi}_i\}^T \boldsymbol{m}_I \{\mathbb{D}_1 \boldsymbol{\phi}_j\} dxdy \tag{5.430}
$$

Applying Lagrange's equations results in

$$
\boldsymbol{M}\ddot{\boldsymbol{q}} + \boldsymbol{K}\boldsymbol{q} = \boldsymbol{Q}_\Lambda(t) \tag{5.431}
$$

where the generalized stiffness coefficients are defined as

$$K_{ij} = \iint_{\text{area}} \{\mathbb{D}\boldsymbol{\phi}_i\}^T \begin{bmatrix} \boldsymbol{A} & \boldsymbol{B} \\ \boldsymbol{B} & \boldsymbol{D} \end{bmatrix} \{\mathbb{D}\boldsymbol{\phi}_j\}\, dx\, dy \tag{5.432}$$

and the generalized forces are defined as

$$Q_{\Lambda_i} = \iint_{\text{area}} \{\boldsymbol{F}_\Lambda^T\ \boldsymbol{M}_\Lambda^T\}\{\mathbb{D}\ \phi_i\}\, dx\, dy \tag{5.433}$$

These are now time-varying equations and can be solved for transient and steady-state response.

5.7 Refined Plate Theories

The classical laminated plate theory (CLPT) is based on the Kirchhoff-Love hypothesis, which assumes that straight material lines, initially normal to mid-plane, remain straight, unstretched, and normal to the mid-plane following deformation. This assumption implies that the transverse shear effects are negligible and that the transverse shear stiffness of the plate is infinite. As a result, there is a zero transverse shear strain across the complete thickness. Such a simple theory appears satisfactory for thin laminates (length/thickness >30) and low transverse shear compliance (large shear stiffness). Due to their characteristically low transverse shear stiffness, composite laminates often exhibit more transverse shear effects than homogeneous isotropic plates. As a result, the thinness ratio (thickness/span) becomes even more stringent for composite laminates for accurate prediction of plate response using CLPT theory.

In the CLPT formulation, the in-plane displacements are caused by rotation of the mid-plane normal. However, with the presence of transverse shear strains, the in-plane displacements will be modified due to rotations of cross-sectional planes relative to mid-planes. For refined plate models, different levels of shear modeling are introduced by additional terms in the expression for axial deformation. These correspond to higher-order terms of the Taylor series expansion, which can be written as

$$u = u_o + z\frac{\partial u_o}{\partial z} + \frac{z^2}{2!}\frac{\partial^2 u_o}{\partial z^2} + \frac{z^3}{3!}\frac{\partial^3 u_o}{\partial z^3} + \dots \tag{5.434}$$

$$v = v_o + z\frac{\partial v_o}{\partial z} + \frac{z^2}{2!}\frac{\partial^2 v_o}{\partial z^2} + \frac{z^3}{3!}\frac{\partial^3 v_o}{\partial z^3} + \dots \tag{5.435}$$

$$w = w_o \tag{5.436}$$

where u_o, v_o, and w_o are mid-plane displacements. Neglecting terms of order higher than 3, this generic displacement distribution is usually written as

$$u(x, y, z) = u_o(x, y) + z\phi_x(x, y) + z^2\zeta_x(x, y) + z^3\psi_x(x, y) \tag{5.437}$$

$$v(x, y, z) = v_o(x, y) + z\phi_y(x, y) + z^2\zeta_y(x, y) + z^3\psi_y(x, y) \tag{5.438}$$

$$w(x, y, z) = w_o(x, y) \tag{5.439}$$

where $\phi_x = \partial u_o/\partial z$ and $\phi_y = \partial v_o/\partial z$ represent rotations of the cross section, and ζ_x, ζ_y, ψ_x, and ψ_y correspond to higher-order derivatives in the Taylor series expansion cross-sectional deformation. For the CLPT, ζ_x, ζ_y, ψ_x, and ψ_y are assumed to be zero and ϕ_x, and ϕ_y are gradients of out-of-plane displacements ($\phi_x = -\partial w_o/\partial x$ and $\phi_y = -\partial w_o/\partial y$).

First-order shear deformation plate theory (FSDT) is based on the Reissner-Mindlin plate model and is similar to Timoshenko beam theory. It assumes that straight material lines, initially normal to mid-plane, remain straight and unstretched after deformation but not necessarily normal to the mid-plane. The rotations of these lines are represented by the terms ϕ_x and ϕ_y. For FSDT, ζ_x, ζ_y, ψ_x, and ψ_y are assumed to be zero and the rotations ϕ_x and ϕ_y are assumed constant through the thickness (independent of w_o).

$$u(x, y, z) = u_o(x, y) + z\phi_x(x, y) \tag{5.440}$$

$$v(x, y, z) = v_o(x, y) + z\phi_y(x, y) \tag{5.441}$$

$$w(x, y, z) = w_o(x, y) \tag{5.442}$$

Consequently, the transverse shear strains are assumed uniform through the thickness of the plate, whereas the in-plane displacements vary linearly through the thickness. The FSDT fails to account for changes in shear strains due to the variation of material properties of each layer. To define the local state of displacement, we require five variables that include u_o, v_o, w_o, ϕ_x, and ϕ_y. This theory estimates lower flexural stiffness than that predicted by the CLPT theory. Another anomaly with this theory is that there is non-zero shear strain at the top and bottom free surfaces that violates the physical boundary condition. Normally, a shear correction factor is applied to compensate for non-zero shear strain at free lateral surfaces. However, it is difficult to determine the shear correction factor because it is dependent on lamination and geometric parameters, loading, and boundary conditions. The FSDT theory relaxes somewhat the thinness requirement of the laminate and normally makes more accurate prediction of deformations and curvatures than CLPT theory. However, the FSDT-based finite element models can exhibit spurious shear stiffness (locking) for very thin laminates. Neither CLPT nor Reissner-Mindlin–based include zig-zag form of inplane displacement along the thickness, and neither do they satisfy interlaminar equilibria for the transverse shear. These may become important local effects in multilayered composite plates. Conversely, layerwise displacement theories capture these effects but increase the degrees of freedom.

A higher-order shear deformable theory (HSDT) developed by Reddy [1] models a general distribution of transverse shear strain through the laminate thickness. For HSDT, ζ_x, ζ_y, ψ_x, and ψ_y are assumed non-zero. This represents a cubic variation of in-plane displacements (u, v) through the thickness, resulting in a quadratic variation of shear strain. The form of the displacement distribution is [1]

$$u(x, y, z) = u_o(x, y) + z\phi_x(x, y) + z^3\left(-\frac{4}{3h^2}\right)\left(\phi_x + \frac{\partial w_o}{\partial x}\right) \tag{5.443}$$

$$v(x, y, z) = v_o(x, y) + z\phi_y(x, y) + z^3\left(-\frac{4}{3h^2}\right)\left(\phi_y + \frac{\partial w_o}{\partial y}\right) \tag{5.444}$$

$$w(x, y, z) = w_o(x, y) \tag{5.445}$$

This distribution satisfies the traction-free boundary condition on the top and bottom surfaces but lacks accurate representation of layerwise variation of shear strain due to different material properties of laminae. In general, it is expected that the HSDT should give better prediction of flexural stiffness than FSDT, but this is not assured for all plate configurations [2]. To model the variations of material stiffness from layer to layer, it appears appropriate to use layerwise shear–deformable theory (LWSDT), attributed to Reddy [3] as well as Sun and Whitney [4]. For this theory, the laminate is divided into a number of sublayers that are perfectly bonded. In each layer, the in-plane displacement is to be assumed piece-wise linear along the z direction. There is a significant increase in the degrees of freedom of the model. The plate displacements are given by

$$u^{(k)}(x, y, z) = u_o(x, y) + z\phi_x^{(k)}(x, y) \tag{5.446}$$

$$v^{(k)}(x, y, z) = v_o(x, y) + z\phi_y^{(k)}(x, y) \tag{5.447}$$

$$w^{(k)}(x, y, z) = w_o(x, y) \tag{5.448}$$

where $\phi_x^{(k)}$ and $\phi_x^{(k)}$ represent rotations of the cross section of the kth layer. For the case of a single-layer laminate, LWSDT reduces to FSDT. Even though the shear strain is assumed uniform in each layer, there is a variation from layer to layer. Between different layers, the displacement components are assumed to be continuous, whereas the transverse derivatives of the displacements can be discontinuous. This helps to provide the continuity of transverse stresses at interfaces separating dissimilar materials, as well as a kinematically correct representation of cross-sectional warping, especially associated with the deformation of thick laminates. It does not require any shear correction factor. For an assumption of N layers, it requires $2N + 3$ variables to define the local state of displacement distribution. To cover the detailed three-dimensional behavior of thick laminates, the layerwise theory is further refined to include layerwise expansion for transverse displacement in addition to in-plane displacements, resulting in more dependent variables. This full layerwise theory would provide both discrete-layer transverse shear effects and discrete-layer transverse-normal effect.

5.8 Classical Laminated Plate Theory (CLPT) for Moderately Large Deflections

Most of the CLPT analyses assume small deflections. In this section, the laminated plate is assumed to undergo moderately large deflections. The following terms (i.e., displacement gradients) are of the order of ϵ

$$\frac{\partial u}{\partial x}, \frac{\partial u}{\partial y}, \frac{\partial v}{\partial x}, \frac{\partial v}{\partial y}, \frac{\partial w}{\partial z} \equiv O(\epsilon) \tag{5.449}$$

This means that the rotation angles of transverse normals (i.e., $\partial w/\partial x$ and $\partial w/\partial y$) are moderate (e.g., less than 10–15 deg). As a result, the following terms are of the order of ϵ^2 and should be included in the analysis

$$\left[\frac{\partial w}{\partial x}\right]^2, \quad \left[\frac{\partial w}{\partial y}\right]^2, \quad \frac{\partial w}{\partial x}\frac{\partial w}{\partial y} \tag{5.450}$$

The displacements u, v, and w are defined in terms of the mid-plane displacements ($z = 0$), u_o, v_o, and w_o in the x, y, and z directions, respectively

$$u(x, y, z) = u_o(x, y) + z\phi_x(x, y) \tag{5.451}$$

$$v(x, y, z) = v_o(x, y) + z\phi_y(x, y) \tag{5.452}$$

$$w(x, y, z) = w_o(x, y) \tag{5.453}$$

For the CLPT framework

$$\phi_x(x, y) = -\frac{\partial w_o}{\partial x}(x, y) \tag{5.454}$$

$$\phi_y(x, y) = -\frac{\partial w_o}{\partial y}(x, y) \tag{5.455}$$

The strain-displacement relations for moderate rotations become

$$\epsilon_x = \frac{\partial u}{\partial x} + \frac{1}{2}\left[\frac{\partial w}{\partial x}\right]^2 = \frac{\partial u_o}{\partial x} - z\frac{\partial^2 w_o}{\partial x^2} + \frac{1}{2}\left[\frac{\partial w_o}{\partial x}\right]^2 \tag{5.456}$$

$$\epsilon_y = \frac{\partial v}{\partial y} + \frac{1}{2}\left[\frac{\partial w}{\partial y}\right]^2 = \frac{\partial v_o}{\partial y} - z\frac{\partial^2 w_o}{\partial y^2} + \frac{1}{2}\left[\frac{\partial w_o}{\partial y}\right]^2 \tag{5.457}$$

$$\epsilon_z = \frac{\partial w}{\partial z} = 0 \tag{5.458}$$

$$\gamma_{xy} = \frac{\partial u}{\partial y} + \frac{\partial v}{\partial x} + \frac{\partial w}{\partial x}\frac{\partial w}{\partial y} = \frac{\partial u_o}{\partial y} + \frac{\partial v_o}{\partial x} - 2z\frac{\partial^2 w_o}{\partial x \partial y} + \frac{\partial w_o}{\partial x}\frac{\partial w_o}{\partial y} \tag{5.459}$$

$$\gamma_{yz} = \frac{\partial v}{\partial z} + \frac{\partial w}{\partial y} = 0 \tag{5.460}$$

$$\gamma_{zx} = \frac{\partial u}{\partial z} + \frac{\partial w}{\partial x} = 0 \tag{5.461}$$

These are called the von Kármán nonlinear strains. Once again, the transverse strains ($\epsilon_z, \gamma_{xz}, \gamma_{yz}$) as well as transverse shear stresses (τ_{xz} and τ_{yz}) are identically zero. This leads to

$$\begin{Bmatrix} \epsilon_x \\ \epsilon_y \\ \gamma_{xy} \end{Bmatrix} = \begin{Bmatrix} \epsilon_x^o + \frac{1}{2}(\frac{\partial w_o}{\partial x})^2 \\ \epsilon_y^o + \frac{1}{2}(\frac{\partial w_o}{\partial y})^2 \\ \gamma_{xy}^o + \frac{\partial w_o}{\partial x}\frac{\partial w_o}{\partial y} \end{Bmatrix} - z \begin{Bmatrix} \kappa_x \\ \kappa_y \\ \kappa_{xy} \end{Bmatrix} = \boldsymbol{\epsilon^o} - z\boldsymbol{\kappa} \tag{5.462}$$

Mid-plane (or membrane) strains are given by

$$\boldsymbol{\epsilon^o} = \begin{Bmatrix} \epsilon_x^o \\ \epsilon_y^o \\ \gamma_{xy}^o \end{Bmatrix} = \begin{Bmatrix} \frac{\partial u_o}{\partial x} + \frac{1}{2}(\frac{\partial w_o}{\partial x})^2 \\ \frac{\partial v_o}{\partial y} + \frac{1}{2}(\frac{\partial w_o}{\partial y})^2 \\ \frac{\partial u_o}{\partial y} + \frac{\partial v_o}{\partial x} + \frac{\partial w_o}{\partial x}\frac{\partial w_o}{\partial y} \end{Bmatrix} \tag{5.463}$$

and the middle-surface curvatures (or bending strains) are

$$\boldsymbol{\kappa} = \begin{Bmatrix} \kappa_x \\ \kappa_y \\ \kappa_{xy} \end{Bmatrix} = \begin{Bmatrix} \frac{\partial^2 w_o}{\partial x^2} \\ \frac{\partial^2 w_o}{\partial y^2} \\ 2\frac{\partial^2 w_o}{\partial x \partial y} \end{Bmatrix} \tag{5.464}$$

Again, if mid-plane displacements (u_o, v_o, w_o) are known, the strains at any point (x, y, z) can be determined. Strain components vary linearly through the laminate thickness, and they are independent of material variations through the laminate thickness. The constitutive relation for any ply of a laminated plate with induced-strain actuation is

$$\sigma = \bar{\boldsymbol{Q}}(\epsilon - \boldsymbol{\Lambda}) = \bar{\boldsymbol{Q}}\epsilon - \boldsymbol{Q}_{\Lambda} \tag{5.465}$$

where the matrix $\bar{\boldsymbol{Q}}$ is the transformed reduced stiffness matrix of a single ply and the actuation strain vector is

$$\boldsymbol{\Lambda} = \begin{Bmatrix} \Lambda_x \\ \Lambda_y \\ \Lambda_{xy} \end{Bmatrix} \tag{5.466}$$

$\boldsymbol{Q}_{\Lambda}$ represents an equivalent stress due to actuation for a single ply and represents the forcing on the structure.

Integrating through the thickness t of the plate, the forces and moments per unit length of the plate, and the stiffness coefficients can be derived in a manner similar to the previous case of small deflection. The force vector, $\boldsymbol{F}$, and the moment vector, $\boldsymbol{M}$, can be derived as

$$\begin{aligned} \boldsymbol{F} &= \int_t \bar{\boldsymbol{Q}}(\epsilon - \boldsymbol{\Lambda})\, dz = \int_t \bar{\boldsymbol{Q}}\epsilon^o\, dz - \int_t \bar{\boldsymbol{Q}}\kappa\, z dz - \int_t \bar{\boldsymbol{Q}}\boldsymbol{\Lambda}\, dz \\ &= \boldsymbol{A}\epsilon^o + \boldsymbol{B}\kappa - \boldsymbol{F}_{\Lambda} \text{ (N/m)} \end{aligned} \tag{5.467}$$

and

$$\begin{aligned} \boldsymbol{M} &= -\int_t \bar{\boldsymbol{Q}}(\epsilon - \boldsymbol{\Lambda})\, z\, dz = -\int_t \bar{\boldsymbol{Q}}\epsilon^o\, z dz + \int_t \bar{\boldsymbol{Q}}\kappa\, z^2 dz + \int_t \bar{\boldsymbol{Q}}\boldsymbol{\Lambda}\, z dz \\ &= \boldsymbol{B}\epsilon^o + \boldsymbol{D}\kappa - \boldsymbol{M}_{\Lambda} \text{ (Nm/m)} \end{aligned} \tag{5.468}$$

From these equations, the stiffness matrices and the induced-force and moment vectors can be derived. Extensional stiffness is

$$\boldsymbol{A} = \int_t \bar{\boldsymbol{Q}}\, dz \;\rightarrow\; A_{ij} = \sum_{k=1}^{N} (\bar{Q}_{ij})_k (h_{k+1} - h_k) \text{ (N/m)} \tag{5.469}$$

Coupling stiffness

$$\boldsymbol{B} = -\int_t \bar{\boldsymbol{Q}} z dz \;\rightarrow\; B_{ij} = -\frac{1}{2}\sum_{k=1}^{N} (\bar{Q}_{ij})_k (h_{k+1}^2 - h_k^2) \text{ (N)} \tag{5.470}$$

Bending stiffness

$$\boldsymbol{D} = \int_t \bar{\boldsymbol{Q}} z^2 dz \;\rightarrow\; D_{ij} = \frac{1}{3}\sum_{k=1}^{N} (\bar{Q}_{ij})_k (h_{k+1}^3 - h_k^3) \text{ (Nm)} \tag{5.471}$$

Induced-force vector

$$\begin{aligned} \boldsymbol{F}_{\Lambda} &= \int_t \bar{\boldsymbol{Q}}\boldsymbol{\Lambda}\, dz \\ &= \sum_{k=1}^{N} \bar{\boldsymbol{Q}}_k \boldsymbol{\Lambda}_k (h_{k+1} - h_k) \text{ (N/m)} \end{aligned} \tag{5.472}$$

Induced-moment vector

$$\boldsymbol{M}_\Lambda = -\int_t \bar{\boldsymbol{Q}} \boldsymbol{\Lambda} z \, dz = -\frac{1}{2} \sum_{k=1}^{N} \bar{\boldsymbol{Q}}_k \boldsymbol{\Lambda}_k (h_{k+1}^2 - h_k^2) \quad \text{(Nm/m)} \tag{5.473}$$

Definition of these stiffness terms and induced forces and moments are identical to those defined previously for small-deflection theory. Again, the total governing equations become

$$\begin{Bmatrix} \boldsymbol{F} \\ \boldsymbol{M} \end{Bmatrix} = \begin{bmatrix} \boldsymbol{A} & \boldsymbol{B} \\ \boldsymbol{B} & \boldsymbol{D} \end{bmatrix} \begin{Bmatrix} \boldsymbol{\epsilon^o} \\ \boldsymbol{\kappa} \end{Bmatrix} - \begin{Bmatrix} \boldsymbol{F}_\Lambda \\ \boldsymbol{M}_\Lambda \end{Bmatrix} \tag{5.474}$$

Expanding the entire set of equations

$$\begin{Bmatrix} F_x \\ F_y \\ F_{xy} \\ M_x \\ M_y \\ M_{xy} \end{Bmatrix} = \begin{bmatrix} \begin{bmatrix} A_{11} & A_{12} & A_{16} \\ A_{12} & A_{22} & A_{26} \\ A_{16} & A_{26} & A_{66} \end{bmatrix} & \begin{bmatrix} B_{11} & B_{12} & B_{16} \\ B_{12} & B_{22} & B_{26} \\ B_{16} & B_{26} & B_{66} \end{bmatrix} \\ \begin{bmatrix} B_{11} & B_{12} & B_{16} \\ B_{12} & B_{22} & B_{26} \\ B_{16} & B_{26} & B_{66} \end{bmatrix} & \begin{bmatrix} D_{11} & D_{12} & D_{16} \\ D_{12} & D_{22} & D_{26} \\ D_{16} & D_{26} & D_{66} \end{bmatrix} \end{bmatrix} \begin{Bmatrix} \epsilon_x^o \\ \epsilon_y^o \\ \gamma_{xy}^o \\ \kappa_x \\ \kappa_y \\ \kappa_{xy} \end{Bmatrix} - \begin{Bmatrix} F_{x_\Lambda} \\ F_{y_\Lambda} \\ F_{xy_\Lambda} \\ M_{x_\Lambda} \\ M_{y_\Lambda} \\ M_{xy_\Lambda} \end{Bmatrix} \tag{5.475}$$

The only change from the previous set for small-deflection theory is in the in-plane strains and curvatures vector

$$\begin{Bmatrix} \epsilon_x^o \\ \epsilon_y^o \\ \gamma_{xy}^o \\ \kappa_x \\ \kappa_y \\ \kappa_{xy} \end{Bmatrix} = \begin{Bmatrix} \frac{\partial u_o}{\partial x} + \frac{1}{2}(\frac{\partial w_o}{\partial x})^2 \\ \frac{\partial v_o}{\partial y} + \frac{1}{2}(\frac{\partial w_o}{\partial y})^2 \\ \frac{\partial u_o}{\partial y} + \frac{\partial v_o}{\partial x} + \frac{\partial w_o}{\partial x}\frac{\partial w_o}{\partial y} \\ \frac{\partial^2 w_o}{\partial x^2} \\ \frac{\partial^2 w_o}{\partial y^2} \\ 2\frac{\partial^2 w_o}{\partial x \partial y} \end{Bmatrix} \tag{5.476}$$

With no external mechanical forces, these equations reduce to

$$\begin{Bmatrix} F_{x_\Lambda} \\ F_{y_\Lambda} \\ F_{xy_\Lambda} \\ M_{x_\Lambda} \\ M_{y_\Lambda} \\ M_{xy_\Lambda} \end{Bmatrix} = \begin{bmatrix} \begin{bmatrix} A_{11} & A_{12} & A_{16} \\ A_{12} & A_{22} & A_{26} \\ A_{16} & A_{26} & A_{66} \end{bmatrix} & \begin{bmatrix} B_{11} & B_{12} & B_{16} \\ B_{12} & B_{22} & B_{26} \\ B_{16} & B_{26} & B_{66} \end{bmatrix} \\ \begin{bmatrix} B_{11} & B_{12} & B_{16} \\ B_{12} & B_{22} & B_{26} \\ B_{16} & B_{26} & B_{66} \end{bmatrix} & \begin{bmatrix} D_{11} & D_{12} & D_{16} \\ D_{12} & D_{22} & D_{26} \\ D_{16} & D_{26} & D_{66} \end{bmatrix} \end{bmatrix} \begin{Bmatrix} \frac{\partial u_o}{\partial x} + \frac{1}{2}(\frac{\partial w_o}{\partial x})^2 \\ \frac{\partial v_o}{\partial y} + \frac{1}{2}(\frac{\partial w_o}{\partial y})^2 \\ \frac{\partial u_o}{\partial y} + \frac{\partial v_o}{\partial x} + \frac{\partial w_o}{\partial x}\frac{\partial w_o}{\partial y} \\ \frac{\partial^2 w_o}{\partial x^2} \\ \frac{\partial^2 w_o}{\partial y^2} \\ 2\frac{\partial^2 w_o}{\partial x \partial y} \end{Bmatrix} \tag{5.477}$$

This is called "von Kármán plate analysis." This set of matrices is similar to the previous one for the small-deflection theory. Again, if the displacements at the mid-plane (u_o, v_o, w_o) are known, the strains and curvatures at any point (x, y, z) can be calculated. In general, the strains are a nonlinear function of x and y. These governing equations can be expressed in terms of displacements as given here.

u_o-equation:

$$\begin{aligned}
&A_{11}\left(\frac{\partial^2 u_o}{\partial x^2}+\frac{\partial w_o}{\partial x}\frac{\partial^2 w_o}{\partial x^2}\right)+A_{12}\left(\frac{\partial^2 v_o}{\partial x\partial y}+\frac{\partial w_o}{\partial y}\frac{\partial^2 w_o}{\partial x\partial y}\right)\\
&+A_{16}\left(\frac{\partial^2 u_o}{\partial x\partial y}+\frac{\partial^2 v_o}{\partial x^2}+\frac{\partial^2 w_o}{\partial x^2}\frac{\partial w_o}{\partial y}+\frac{\partial w_o}{\partial x}\frac{\partial^2 w_o}{\partial x\partial y}\right)\\
&+B_{11}\frac{\partial^3 w_o}{\partial x^3}+B_{12}\frac{\partial^3 w_o}{\partial x\partial y^2}+2B_{16}\frac{\partial^3 w_o}{\partial x^2\partial y}\\
&+A_{16}\left(\frac{\partial^2 u_o}{\partial x\partial y}+\frac{\partial w_o}{\partial x}\frac{\partial^2 w_o}{\partial x\partial y}\right)+A_{26}\left(\frac{\partial^2 v_o}{\partial y^2}+\frac{\partial w_o}{\partial y}\frac{\partial^2 w_o}{\partial y^2}\right)\\
&+A_{66}\left(\frac{\partial^2 u_o}{\partial y^2}+\frac{\partial^2 v_o}{\partial x\partial y}+\frac{\partial^2 w_o}{\partial x\partial y}\frac{\partial w_o}{\partial y}+\frac{\partial w_o}{\partial x}\frac{\partial^2 w_o}{\partial y^2}\right)\\
&+B_{16}\frac{\partial^3 w_o}{\partial x^2\partial y}+B_{26}\frac{\partial^3 w_o}{\partial y^3}+2B_{66}\frac{\partial^3 w_o}{\partial x\partial y^2}-\left(\frac{\partial F_x}{\partial x}+\frac{\partial F_{xy}}{\partial y}\right)-\left(\frac{\partial F_{x_\Lambda}}{\partial x}+\frac{\partial F_{xy_\Lambda}}{\partial y}\right)\\
&=m\frac{\partial^2 u_o}{\partial t^2}-S_I\frac{\partial^3 w_o}{\partial x\partial t^2}
\end{aligned} \tag{5.478}$$

v_o-equation:

$$\begin{aligned}
&A_{16}\left(\frac{\partial^2 u_o}{\partial x^2}+\frac{\partial w_o}{\partial x}\frac{\partial^2 w_o}{\partial x^2}\right)+A_{26}\left(\frac{\partial^2 v_o}{\partial x\partial y}+\frac{\partial w_o}{\partial y}\frac{\partial^2 w_o}{\partial x\partial y}\right)\\
&+A_{66}\left(\frac{\partial^2 u_o}{\partial x\partial y}+\frac{\partial^2 v_o}{\partial x^2}+\frac{\partial^2 w_o}{\partial x^2}\frac{\partial w_o}{\partial y}+\frac{\partial w_o}{\partial x}\frac{\partial^2 w_o}{\partial x\partial y}\right)\\
&+B_{16}\frac{\partial^3 w_o}{\partial x^3}+B_{26}\frac{\partial^3 w_o}{\partial x\partial y^2}+2B_{66}\frac{\partial^3 w_o}{\partial x^2\partial y}\\
&+A_{12}\left(\frac{\partial^2 u_o}{\partial x\partial y}+\frac{\partial w_o}{\partial x}\frac{\partial^2 w_o}{\partial x\partial y}\right)+A_{22}\left(\frac{\partial^2 v_o}{\partial y^2}+\frac{\partial w_o}{\partial y}\frac{\partial^2 w_o}{\partial y^2}\right)\\
&+A_{26}\left(\frac{\partial^2 u_o}{\partial y^2}+\frac{\partial^2 v_o}{\partial x\partial y}+\frac{\partial^2 w_o}{\partial x\partial y}\frac{\partial w_o}{\partial y}+\frac{\partial w_o}{\partial x}\frac{\partial^2 w_o}{\partial y^2}\right)\\
&+B_{12}\frac{\partial^3 w_o}{\partial x^2\partial y}+B_{22}\frac{\partial^3 w_o}{\partial y^3}+2B_{26}\frac{\partial^3 w_o}{\partial x\partial y^2}-\left(\frac{\partial F_{xy}}{\partial x}+\frac{\partial F_y}{\partial y}\right)-\left(\frac{\partial F_{xy_\Lambda}}{\partial x}+\frac{\partial F_{y_\Lambda}}{\partial y}\right)\\
&=m\frac{\partial^2 v_o}{\partial t^2}-S_I\frac{\partial^3 w_o}{\partial y\partial t^2}
\end{aligned} \tag{5.479}$$

w_o-equation:

$$
\begin{aligned}
&B_{11}\left(\frac{\partial^3 u_o}{\partial x^3}+\frac{\partial^2 w_o}{\partial x^2}\frac{\partial^2 w_o}{\partial x^2}+\frac{\partial w_o}{\partial x}\frac{\partial^3 w_o}{\partial x^3}\right)\\
&\quad+B_{12}\left(\frac{\partial^3 v_o}{\partial x^2\partial y}+\frac{\partial^2 w_o}{\partial x\partial y}\frac{\partial^2 w_o}{\partial x\partial y}+\frac{\partial w_o}{\partial y}\frac{\partial^3 w_o}{\partial x^2\partial y}\right)\\
&\quad+B_{16}\left(\frac{\partial^3 u_o}{\partial x^2\partial y}+\frac{\partial^3 v_o}{\partial x^3}+\frac{\partial^3 w_o}{\partial x^3}\frac{\partial w_o}{\partial y}+2\frac{\partial^2 w_o}{\partial x^2}\frac{\partial^2 w_o}{\partial x\partial y}+\frac{\partial w_o}{\partial x}\frac{\partial^3 w_o}{\partial x^2\partial y}\right)\\
&\quad+D_{11}\frac{\partial^4 w_o}{\partial x^4}+D_{12}\frac{\partial^4 w_o}{\partial x^2\partial y^2}+2D_{16}\frac{\partial^4 w_o}{\partial x^3\partial y}\\
&\quad+2B_{16}\left(\frac{\partial^3 u_o}{\partial x^2\partial y}+\frac{\partial^2 w_o}{\partial x^2}\frac{\partial^2 w_o}{\partial x\partial y}+\frac{\partial w_o}{\partial x}\frac{\partial^3 w_o}{\partial x^2\partial y}\right)\\
&\quad+2B_{26}\left(\frac{\partial^3 v_o}{\partial x\partial y^2}+\frac{\partial^2 w_o}{\partial x\partial y}\frac{\partial^2 w_o}{\partial y^2}+\frac{\partial w_o}{\partial y}\frac{\partial^3 w_o}{\partial x\partial y^2}\right)\\
&\quad+2B_{66}\left(\frac{\partial^3 u_o}{\partial x\partial y^2}+\frac{\partial^3 v_o}{\partial x^2\partial y}+\frac{\partial^3 w_o}{\partial x^2\partial y}\frac{\partial w_o}{\partial y}+\frac{\partial^2 w_o}{\partial x\partial y}\frac{\partial^2 w_o}{\partial x\partial y}+\frac{\partial^2 w_o}{\partial x^2}\frac{\partial^2 w_o}{\partial y^2}+\frac{\partial w_o}{\partial x}\frac{\partial^3 w_o}{\partial x\partial y^2}\right)\\
&\quad+2D_{16}\frac{\partial^4 w_o}{\partial x^3\partial y}+2D_{26}\frac{\partial^4 w_o}{\partial x\partial y^3}+4D_{66}\frac{\partial^4 w_o}{\partial x^2\partial y^2}\\
&\quad+B_{12}\left(\frac{\partial^3 u_o}{\partial x\partial y^2}+\frac{\partial^2 w_o}{\partial x\partial y}\frac{\partial^2 w_o}{\partial x\partial y}+\frac{\partial w_o}{\partial x}\frac{\partial^3 w_o}{\partial x\partial y^2}\right)\\
&\quad+B_{22}\left(\frac{\partial^3 v_o}{\partial y^3}+\frac{\partial^2 w_o}{\partial y^2}\frac{\partial^2 w_o}{\partial y^2}+\frac{\partial w_o}{\partial y}\frac{\partial^3 w_o}{\partial y^3}\right)\\
&\quad+B_{26}\left(\frac{\partial^3 u_o}{\partial y^3}+\frac{\partial^3 v_o}{\partial x\partial y^2}+\frac{\partial^3 w_o}{\partial x\partial y^2}\frac{\partial w_o}{\partial y}+2\frac{\partial^2 w_o}{\partial x\partial y}\frac{\partial^2 w_o}{\partial y^2}+\frac{\partial w_o}{\partial x}\frac{\partial^3 w_o}{\partial y^3}\right)\\
&\quad+D_{12}\frac{\partial^4 w_o}{\partial x^2\partial y^2}+D_{22}\frac{\partial^4 w_o}{\partial y^4}+2D_{26}\frac{\partial^4 w_o}{\partial x\partial y^3}+q\\
&\quad-\left(\frac{\partial^2 M_x}{\partial x^2}+2\frac{\partial^2 M_{xy}}{\partial x\partial y}+\frac{\partial^2 M_y}{\partial y^2}\right)-\left(\frac{\partial^2 M_{x_\Lambda}}{\partial x^2}+2\frac{\partial^2 M_{xy_\Lambda}}{\partial x\partial y}+\frac{\partial^2 M_{y_\Lambda}}{\partial y^2}\right)\\
&\quad=m\frac{\partial^2 w_o}{\partial t^2}+S_I\frac{\partial^2}{\partial t^2}\left(\frac{\partial u_o}{\partial x}+\frac{\partial v_o}{\partial y}\right)-I\frac{\partial^2}{\partial t^2}\left(\frac{\partial^2 w_o}{\partial x^2}+\frac{\partial^2 w_o}{\partial y^2}\right)
\end{aligned}
\tag{5.480}
$$

5.9 First-Order Shear Deformation Plate Theory (FSDT) with Induced-Strain Actuation

For the FSDT theory, the Kirchhoff-Love hypothesis is relaxed; transverse planes normal to the mid-plane in the undeformed condition do not remain normal to the mid-plane after deformation. This necessitates the inclusion of transverse shear strains in the analysis. However, the assumption of zero transverse-normal strain (ϵ_z) is retained.

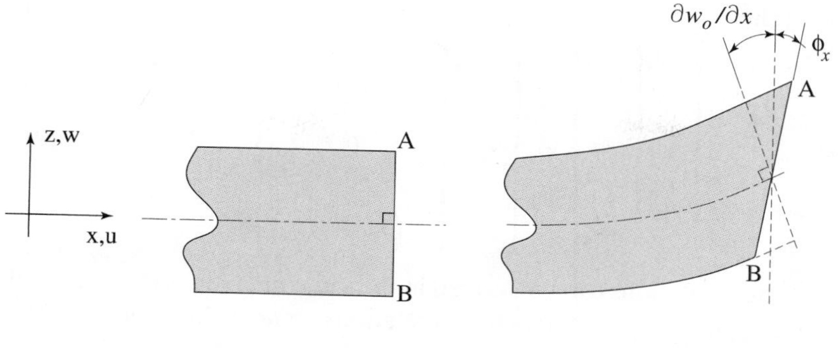

Figure 5.27. Inclusion of first-order shear terms in plate deformation.

The plate displacements are expressed as

$$u(x, y, z) = u_o(x, y) + z\phi_x(x, y) \tag{5.481}$$

$$v(x, y, z) = v_o(x, y) + z\phi_y(x, y) \tag{5.482}$$

$$w(x, y, z) = w_o(x, y) \tag{5.483}$$

Again, u_o, v_o, and w_o are displacements at the neutral plane ($z = 0$) and ϕ_x and ϕ_y are the rotations of the transverse normal plane about the y-axis and x-axis, respectively (Fig. 5.27). Note that the terms ϕ_x and ϕ_y include the effect of both pure bending (as per the Kirchhoff-Love hypothesis) and pure shear. The rotations of the transverse normal plane are given by

$$\phi_x = \frac{\partial u}{\partial z} \tag{5.484}$$

$$\phi_y = \frac{\partial v}{\partial z} \tag{5.485}$$

Now we require five variables (u_o, v_o, w_o, ϕ_x, ϕ_y) to determine the strain at any point. Assuming small displacements, and ignoring the transverse normal strain, the normal and shear strains can be obtained from Eqs. 5.481–5.483 as

$$\epsilon_x = \frac{\partial u}{\partial x} = \epsilon_x^o + z\frac{\partial \phi_x}{\partial x} \tag{5.486}$$

$$\epsilon_y = \frac{\partial v}{\partial y} = \epsilon_y^o + z\frac{\partial \phi_y}{\partial y} \tag{5.487}$$

$$\gamma_{yz} = \frac{\partial w}{\partial y} + \frac{\partial v}{\partial z} = \frac{\partial w_o}{\partial y} + \phi_y \tag{5.488}$$

$$\gamma_{zx} = \frac{\partial w}{\partial x} + \frac{\partial u}{\partial z} = \frac{\partial w_o}{\partial x} + \phi_x \tag{5.489}$$

$$\gamma_{xy} = \frac{\partial u}{\partial y} + \frac{\partial v}{\partial x} = \gamma_{xy}^o + z\left(\frac{\partial \phi_x}{\partial y} + \frac{\partial \phi_y}{\partial x}\right) \tag{5.490}$$

This can be rewritten as

$$\begin{Bmatrix} \epsilon_x \\ \epsilon_y \\ \gamma_{yz} \\ \gamma_{zx} \\ \gamma_{xy} \end{Bmatrix} = \begin{Bmatrix} \epsilon_x^o \\ \epsilon_y^o \\ \frac{\partial w_o}{\partial y} + \phi_y \\ \frac{\partial w_o}{\partial x} + \phi_x \\ \gamma_{xy}^o \end{Bmatrix} - z \begin{Bmatrix} -\frac{\partial \phi_x}{\partial x} \\ -\frac{\partial \phi_y}{\partial y} \\ 0 \\ 0 \\ -\left(\frac{\partial \phi_x}{\partial y} + \frac{\partial \phi_y}{\partial x}\right) \end{Bmatrix} \tag{5.491}$$

The force-displacement relations can be derived in the same manner as in the case of the Kirchhoff-Love hypothesis. The stress-strain relations for an orthotropic material are

$$\boldsymbol{\sigma} = \begin{Bmatrix} \sigma_x \\ \sigma_y \\ \tau_{yz} \\ \tau_{zx} \\ \tau_{xy} \end{Bmatrix} = \begin{bmatrix} \bar{Q}_{11} & \bar{Q}_{12} & 0 & 0 & \bar{Q}_{16} \\ \bar{Q}_{12} & \bar{Q}_{22} & 0 & 0 & \bar{Q}_{26} \\ 0 & 0 & \bar{Q}_{44} & \bar{Q}_{45} & 0 \\ 0 & 0 & \bar{Q}_{45} & \bar{Q}_{55} & 0 \\ \bar{Q}_{16} & \bar{Q}_{26} & 0 & 0 & \bar{Q}_{66} \end{bmatrix} \begin{Bmatrix} \epsilon_x \\ \epsilon_y \\ \gamma_{yz} \\ \gamma_{zx} \\ \gamma_{xy} \end{Bmatrix} = \bar{\boldsymbol{Q}}\boldsymbol{\epsilon} \tag{5.492}$$

For unidirectional plates

$$Q_{44} = G_{23} \tag{5.493}$$

$$Q_{55} = G_{31} \tag{5.494}$$

where G_{23} and G_{31} are the shear moduli in the y-z and z-x planes, respectively. For a lamina at an arbitrary orientation α, the transformed relations become

$$\bar{Q}_{44} = Q_{44}\cos^2\alpha + Q_{55}\sin^2\alpha \tag{5.495}$$

$$\bar{Q}_{45} = (Q_{55} - Q_{44})\sin\alpha\cos\alpha \tag{5.496}$$

$$\bar{Q}_{55} = Q_{44}\sin^2\alpha + Q_{55}\cos^2\alpha \tag{5.497}$$

Again, including the effect of induced-strain actuation, the stress in each lamina can be written as

$$\boldsymbol{\sigma} = \bar{\boldsymbol{Q}}(\boldsymbol{\epsilon} - \boldsymbol{\Lambda}) \tag{5.498}$$

where the actuation strain is

$$\boldsymbol{\Lambda} = \begin{Bmatrix} \Lambda_x \\ \Lambda_y \\ \Lambda_{yz} \\ \Lambda_{zx} \\ \Lambda_{xy} \end{Bmatrix} \tag{5.499}$$

The resultant in-plane forces are

$$\boldsymbol{F} = \begin{Bmatrix} F_x \\ F_y \\ F_{xy} \end{Bmatrix} = \int_t \begin{Bmatrix} \sigma_x \\ \sigma_y \\ \tau_{xy} \end{Bmatrix} dz = \sum_{k=1}^{N} \int_{h_k}^{h_{k+1}} \begin{Bmatrix} \sigma_x \\ \sigma_y \\ \tau_{xy} \end{Bmatrix}_k dz \tag{5.500}$$

The resultant moments are

$$\boldsymbol{M} = \begin{Bmatrix} M_x \\ M_y \\ M_{xy} \end{Bmatrix} = -\int_t \begin{Bmatrix} \sigma_x \\ \sigma_y \\ \tau_{xy} \end{Bmatrix} zdz = -\sum_{k=1}^{N} \int_{h_k}^{h_{k+1}} \begin{Bmatrix} \sigma_x \\ \sigma_y \\ \tau_{xy} \end{Bmatrix}_k zdz \tag{5.501}$$

The resultant transverse shear stresses are

$$\begin{aligned}
\begin{Bmatrix} V_y \\ V_x \end{Bmatrix} &= -\int_t \begin{Bmatrix} \tau_{yz} \\ \tau_{zx} \end{Bmatrix} dz = -\sum_{k=1}^{N} \int_{h_k}^{h_{k+1}} \begin{Bmatrix} \tau_{yz} \\ \tau_{zx} \end{Bmatrix}_k dz \\
&= -\sum_{k=1}^{N} \int_{h_k}^{h_{k+1}} \begin{bmatrix} \bar{Q}_{44} & \bar{Q}_{45} \\ \bar{Q}_{45} & \bar{Q}_{55} \end{bmatrix}_k \begin{Bmatrix} \gamma_{yz} \\ \gamma_{zx} \end{Bmatrix}_k dz \\
&= -\begin{bmatrix} A_{44} & A_{45} \\ A_{45} & A_{55} \end{bmatrix} \begin{Bmatrix} \gamma_{yz} \\ \gamma_{zx} \end{Bmatrix} \\
&= -\begin{bmatrix} A_{44} & A_{45} \\ A_{45} & A_{55} \end{bmatrix} \begin{Bmatrix} \frac{\partial w_o}{\partial y} + \phi_y \\ \frac{\partial w_o}{\partial x} + \phi_x \end{Bmatrix}
\end{aligned} \tag{5.502}$$

This results in the governing equation for the plate with induced-strain actuation, under no external loads

$$\boldsymbol{F}_\Lambda = \begin{Bmatrix} F_{x\Lambda} \\ F_{y\Lambda} \\ F_{xy\Lambda} \end{Bmatrix} = \begin{bmatrix} A_{11} & A_{12} & A_{16} \\ A_{12} & A_{22} & A_{26} \\ A_{16} & A_{26} & A_{66} \end{bmatrix} \begin{Bmatrix} \epsilon_x^o \\ \epsilon_y^o \\ \gamma_{xy}^o \end{Bmatrix} + \begin{bmatrix} B_{11} & B_{12} & B_{16} \\ B_{12} & B_{22} & B_{26} \\ B_{16} & B_{26} & B_{66} \end{bmatrix} \begin{Bmatrix} -\frac{\partial \phi_x}{\partial x} \\ -\frac{\partial \phi_y}{\partial y} \\ -\left(\frac{\partial \phi_x}{\partial y} + \frac{\partial \phi_y}{\partial x}\right) \end{Bmatrix} \tag{5.503}$$

$$\boldsymbol{M}_\Lambda = \begin{Bmatrix} M_{x\Lambda} \\ M_{y\Lambda} \\ M_{xy\Lambda} \end{Bmatrix} = \begin{bmatrix} B_{11} & B_{12} & B_{16} \\ B_{12} & B_{22} & B_{26} \\ B_{16} & B_{26} & B_{66} \end{bmatrix} \begin{Bmatrix} \epsilon_x^o \\ \epsilon_y^o \\ \gamma_{xy}^o \end{Bmatrix} + \begin{bmatrix} D_{11} & D_{12} & D_{16} \\ D_{12} & D_{22} & D_{26} \\ D_{16} & D_{26} & D_{66} \end{bmatrix} \begin{Bmatrix} -\frac{\partial \phi_x}{\partial x} \\ -\frac{\partial \phi_y}{\partial y} \\ -\left(\frac{\partial \phi_x}{\partial y} + \frac{\partial \phi_y}{\partial x}\right) \end{Bmatrix} \tag{5.504}$$

The equation for transverse shears is

$$\boldsymbol{V}_\Lambda = \begin{Bmatrix} V_{y\Lambda} \\ V_{x\Lambda} \end{Bmatrix} = -\begin{bmatrix} A_{44} & A_{45} \\ A_{45} & A_{55} \end{bmatrix} \begin{Bmatrix} \frac{\partial w_o}{\partial y} + \phi_y \\ \frac{\partial w_o}{\partial x} + \phi_x \end{Bmatrix} \tag{5.505}$$

where the stiffness matrices are defined in the same manner as in the case of the CLPT formulation

$$A_{ij} = \sum_{k=1}^{N} (\bar{Q}_{ij})_k (h_{k+1} - h_k) \text{ (N/m)} \tag{5.506}$$

$$B_{ij} = -\frac{1}{2} \sum_{k=1}^{N} (\bar{Q}_{ij})_k (h_{k+1}^2 - h_k^2) \text{ (N)} \tag{5.507}$$

$$D_{ij} = \frac{1}{3} \sum_{k=1}^{N} (\bar{Q}_{ij})_k (h_{k+1}^3 - h_k^3) \text{ (Nm)} \tag{5.508}$$

The induced forces and moments are

$$\boldsymbol{F}_\Lambda = \sum_{k=1}^{N} \begin{bmatrix} \bar{Q}_{11} & \bar{Q}_{12} & \bar{Q}_{16} \\ \bar{Q}_{12} & \bar{Q}_{22} & \bar{Q}_{26} \\ \bar{Q}_{16} & \bar{Q}_{26} & \bar{Q}_{66} \end{bmatrix}_k \begin{Bmatrix} \Lambda_x \\ \Lambda_y \\ \Lambda_{xy} \end{Bmatrix}_k (h_{k+1} - h_k) \text{ (N/m)} \tag{5.509}$$

$$\boldsymbol{M}_\Lambda = -\frac{1}{2} \sum_{k=1}^{N} \begin{bmatrix} \bar{Q}_{11} & \bar{Q}_{12} & \bar{Q}_{16} \\ \bar{Q}_{12} & \bar{Q}_{22} & \bar{Q}_{26} \\ \bar{Q}_{16} & \bar{Q}_{26} & \bar{Q}_{66} \end{bmatrix}_k \begin{Bmatrix} \Lambda_x \\ \Lambda_y \\ \Lambda_{xy} \end{Bmatrix}_k (h_{k+1}^2 - h_k^2) \text{ (Nm/m)} \tag{5.510}$$

$$\boldsymbol{V}_\Lambda = -\sum_{k=1}^{N} \begin{bmatrix} \bar{Q}_{44} & \bar{Q}_{45} \\ \bar{Q}_{45} & \bar{Q}_{55} \end{bmatrix}_k \begin{Bmatrix} \Lambda_{yz} \\ \Lambda_{zx} \end{Bmatrix}_k (h_{k+1} - h_k) \text{ (N/m)} \tag{5.511}$$

5.10 Shear Correction Factors

First-order shear deformation theory assumes constant transverse shears through the laminate thickness. It is well established, from elementary beam theory for homogeneous sections, that the transverse shear stress varies parabolically through the thickness. Also, the transverse stress on the top and bottom free surfaces must be zero. For a uniform isotropic plate, shear stress varies quadratically across the thickness

$$\tau_{zx} = -\frac{3}{2}\frac{V_x}{t_b}\left[1 - \left(\frac{z}{t_b/2}\right)^2\right] \tag{5.512}$$

$$\gamma_{zx} = \frac{\tau_{zx}}{G} \quad \text{(quadratic in } z\text{)} \tag{5.513}$$

The strain $\gamma_{zx} = \phi_x + \partial w/\partial x$ is assumed constant in the z direction, which results in a slight inconsistency. This discrepancy is corrected by applying a correction factor K in computing transverse shear resultants

$$\begin{Bmatrix} V_y \\ V_x \end{Bmatrix} = -K \int_t \begin{Bmatrix} \tau_{yz} \\ \tau_{zx} \end{Bmatrix} dz = -K \sum_{k=1}^{N} \begin{bmatrix} \bar{Q}_{44} & \bar{Q}_{45} \\ \bar{Q}_{45} & \bar{Q}_{55} \end{bmatrix}_k \begin{Bmatrix} \Lambda_{yz} \\ \Lambda_{zx} \end{Bmatrix}_k (h_{k+1} - h_k) \tag{5.514}$$

The correction factor is computed in such a way that the strain energy due to assumed transverse shear stress equals that of true transverse shear stress

$$\text{Work done } W = \int_t \tau_{zx}\gamma_{zx} dz = -V_x \gamma_{zx_A} \tag{5.515}$$

$$\begin{aligned} \gamma_{zx_A} &= -\frac{\int \tau_{zx}\gamma_{zx} dz}{V_x} \\ &= -\frac{1}{V_x G}\int_{-t_b/2}^{t_b/2} \tau_{zx}^2 dx \\ &= -\frac{9}{4}\frac{V_x}{G t_b^3}\int_{-t_b/2}^{t_b/2}\left[1 - 4\frac{z^2}{t_b^2}\right] dz \\ &= -\frac{6}{5}\frac{V_x}{G t_b} \end{aligned} \tag{5.516}$$

$$
\begin{aligned}
V_x &= -\frac{5}{6} G t_b \gamma_{zx_A} \\
\rightarrow K &= \frac{5}{6}
\end{aligned}
\tag{5.517}
$$

The equilibrium equations with shear are

$$
\delta u_o: \quad \frac{\partial F_x}{\partial x} + \frac{\partial F_{xy}}{\partial y} = 0 \tag{5.518}
$$

$$
\delta v_o: \quad \frac{\partial F_{xy}}{\partial x} + \frac{\partial F_y}{\partial y} = 0 \tag{5.519}
$$

$$
\delta w_o: \quad \frac{\partial V_x}{\partial x} + \frac{\partial V_y}{\partial y} = 0 \tag{5.520}
$$

$$
\delta \phi_x: \quad \frac{\partial M_x}{\partial x} + \frac{\partial M_{xy}}{\partial y} + V_x = 0 \tag{5.521}
$$

$$
\delta \phi_y: \quad \frac{\partial M_{xy}}{\partial x} + \frac{\partial M_y}{\partial y} + V_y = 0 \tag{5.522}
$$

u_o-equation:

$$
\begin{aligned}
& A_{11} \frac{\partial^2 u_o}{\partial x^2} + A_{12} \frac{\partial^2 v_o}{\partial x \partial y} + A_{16} \left(\frac{\partial^2 u_o}{\partial x \partial y} + \frac{\partial^2 v_o}{\partial x^2} \right) \\
& \quad - B_{11} \frac{\partial^2 \phi_x}{\partial x^2} - B_{12} \frac{\partial^2 \phi_y}{\partial x \partial y} - B_{16} \left(\frac{\partial^2 \phi_x}{\partial x \partial y} + \frac{\partial^2 \phi_y}{\partial x^2} \right) \\
& \quad + A_{16} \frac{\partial^2 u_o}{\partial x \partial y} + A_{26} \frac{\partial^2 v_o}{\partial y^2} + A_{66} \left(\frac{\partial^2 u_o}{\partial y^2} + \frac{\partial^2 v_o}{\partial x \partial y} \right) \\
& \quad - B_{16} \frac{\partial^2 \phi_x}{\partial x \partial y} - B_{26} \frac{\partial^2 \phi_y}{\partial y^2} - B_{66} \left(\frac{\partial^2 \phi_x}{\partial y^2} + \frac{\partial^2 \phi_y}{\partial x \partial y} \right) \\
& = \frac{\partial F_{x_\Lambda}}{\partial x} + \frac{\partial F_{xy_\Lambda}}{\partial y}
\end{aligned}
\tag{5.523}
$$

v_o-equation:

$$
\begin{aligned}
& A_{16} \frac{\partial^2 u_o}{\partial x^2} + A_{26} \frac{\partial^2 v_o}{\partial x \partial y} + A_{66} \left(\frac{\partial^2 u_o}{\partial x \partial y} + \frac{\partial^2 v_o}{\partial y^2} \right) \\
& \quad - B_{16} \frac{\partial^2 \phi_x}{\partial x^2} - B_{26} \frac{\partial^2 \phi_y}{\partial x \partial y} - B_{66} \left(\frac{\partial^2 \phi_x}{\partial x \partial y} + \frac{\partial^2 \phi_y}{\partial x^2} \right) \\
& \quad + A_{12} \frac{\partial^2 u_o}{\partial x \partial y} + A_{22} \frac{\partial^2 v_o}{\partial y^2} + A_{26} \left(\frac{\partial^2 u_o}{\partial y^2} + \frac{\partial^2 v_o}{\partial x \partial y} \right) \\
& \quad - B_{12} \frac{\partial^2 \phi_x}{\partial x \partial y} - B_{22} \frac{\partial^2 \phi_y}{\partial y^2} - B_{26} \left(\frac{\partial^2 \phi_x}{\partial y^2} + \frac{\partial^2 \phi_y}{\partial x \partial y} \right) \\
& = \frac{\partial F_{xy_\Lambda}}{\partial x} + \frac{\partial F_{y_\Lambda}}{\partial y}
\end{aligned}
\tag{5.524}
$$

w_o-equation:

$$
\begin{aligned}
&KA_{55}\left(\frac{\partial^2 w_o}{\partial x^2}+\frac{\partial \phi_x}{\partial x}\right)+KA_{45}\left(\frac{\partial^2 w_o}{\partial x \partial y}+\frac{\partial \phi_y}{\partial x}\right)\\
&KA_{45}\left(\frac{\partial^2 w_o}{\partial x \partial y}+\frac{\partial \phi_x}{\partial y}\right)+KA_{44}\left(\frac{\partial^2 w_o}{\partial y^2}+\frac{\partial \phi_y}{\partial y}\right)\\
&\quad=-\frac{\partial V_{x_\Lambda}}{\partial x}-\frac{\partial V_{y_\Lambda}}{\partial y}
\end{aligned}
\tag{5.525}
$$

ϕ_x-equation:

$$
\begin{aligned}
&B_{11}\frac{\partial^2 u_o}{\partial x^2}+B_{12}\frac{\partial^2 v_o}{\partial x \partial y}+B_{16}\left(\frac{\partial^2 u_o}{\partial x \partial y}+\frac{\partial^2 v_o}{\partial x^2}\right)\\
&\quad-D_{11}\frac{\partial^2 \phi_x}{\partial x^2}-D_{12}\frac{\partial^2 \phi_y}{\partial x \partial y}-D_{16}\left(\frac{\partial^2 \phi_x}{\partial x \partial y}+\frac{\partial^2 \phi_y}{\partial x^2}\right)\\
&\quad+B_{16}\frac{\partial^2 u_o}{\partial x \partial y}+B_{26}\frac{\partial^2 v_o}{\partial y^2}+B_{66}\left(\frac{\partial^2 u_o}{\partial y^2}+\frac{\partial^2 v_o}{\partial x \partial y}\right)\\
&\quad-D_{16}\frac{\partial^2 \phi_x}{\partial x \partial y}-D_{26}\frac{\partial^2 \phi_y}{\partial y^2}-D_{66}\left(\frac{\partial^2 \phi_x}{\partial y^2}+\frac{\partial^2 \phi_y}{\partial x \partial y}\right)\\
&\quad+KA_{55}\left(\frac{\partial w_o}{\partial x}+\phi_x\right)+KA_{45}\left(\frac{\partial w_o}{\partial y}+\phi_y\right)\\
&=\frac{\partial M_{x_\Lambda}}{\partial x}+\frac{\partial M_{xy_\Lambda}}{\partial y}+V_{x_\Lambda}
\end{aligned}
\tag{5.526}
$$

ϕ_y-equation:

$$
\begin{aligned}
&B_{16}\frac{\partial^2 u_o}{\partial x^2}+B_{26}\frac{\partial^2 v_o}{\partial x \partial y}+B_{66}\left(\frac{\partial^2 u_o}{\partial x \partial y}+\frac{\partial^2 v_o}{\partial x^2}\right)\\
&\quad-D_{16}\frac{\partial^2 \phi_x}{\partial x^2}-D_{26}\frac{\partial^2 \phi_y}{\partial x \partial y}-D_{66}\left(\frac{\partial^2 \phi_x}{\partial x \partial y}+\frac{\partial^2 \phi_y}{\partial x^2}\right)\\
&\quad+B_{12}\frac{\partial^2 u_o}{\partial x \partial y}+B_{22}\frac{\partial^2 v_o}{\partial y^2}+B_{26}\left(\frac{\partial^2 u_o}{\partial y^2}+\frac{\partial^2 v_o}{\partial x \partial y}\right)\\
&\quad-D_{12}\frac{\partial^2 \phi_x}{\partial x \partial y}-D_{22}\frac{\partial^2 \phi_y}{\partial y^2}-D_{26}\left(\frac{\partial^2 \phi_x}{\partial y^2}+\frac{\partial^2 \phi_y}{\partial x \partial y}\right)\\
&\quad+KA_{45}\left(\frac{\partial w_o}{\partial x}+\phi_x\right)+KA_{44}\left(\frac{\partial w_o}{\partial y}+\phi_y\right)\\
&=\frac{\partial M_{xy_\Lambda}}{\partial x}+\frac{\partial M_{y_\Lambda}}{\partial y}+V_{y_\Lambda}
\end{aligned}
\tag{5.527}
$$

Typical boundary conditions including shear are as follows:

Clamped Condition

$$u_o = 0$$
$$v_o = 0$$
$$w_o = 0$$
$$\phi_x = 0$$

Simply-Supported (Hinged or Pinned) Condition

$$u_o = 0$$
$$\frac{\partial v_o}{\partial x} = 0$$
$$w_o = 0$$
$$M_x = 0$$

Roller Condition

$$\frac{\partial v_o}{\partial x} = 0$$
$$w_o = 0$$
$$F_x = 0$$
$$M_x = 0$$

Free Condition

$$N_x = 0$$
$$N_{xy} = 0$$
$$M_x = 0$$
$$V_x = 0$$

5.11 Effect of Laminate Kinematic Assumptions on Global Response

Robbins and Chopra [5] evaluated the importance of accurately accounting for transverse shear strain, transverse normal strain, and discrete layer kinematics on the computed global response of plates actuated by symmetric pairs of surface-mounted piezoceramic sheets acting together to induce global in-plane extension, global in-plane contraction, or global bending in the plate. This study is carried out on a square aluminum plate with a symmetric pair of square piezoceramic actuators that are bonded to the top and bottom surface of the plate (Fig. 5.28). The scope of the present study is restricted to the linear quasi-static global response (in-plane and transverse displacement of the mid-surface) of homogeneous actuated plates two-dimensional computational domain $(0<x<L, 0<y<L)$

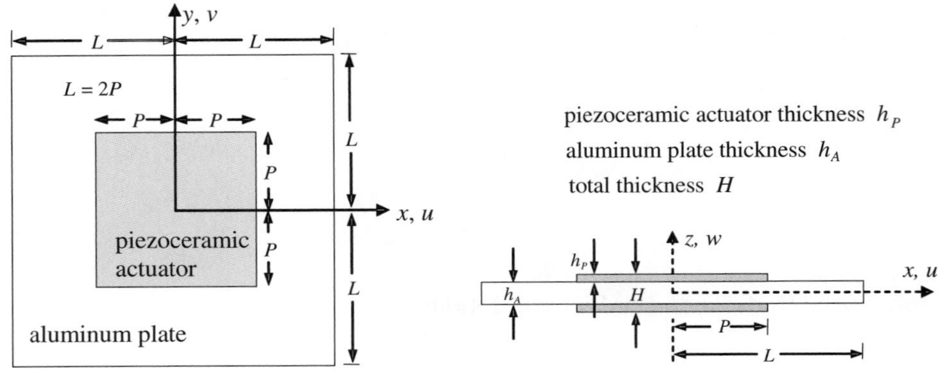

(a) View of specimen in $x - y$ plane (b) View of specimen in $x - z$ plane

Figure 5.28. Two-dimensional computational domain, $0 < x < L, 0 < y < L$. (a) Specimen geometry in $x - y$ plane (fixed at $L/P = 2$), (b) specimen geometry in $x - z$ plane. Relative thickness of material layers is fixed at $h_P/h_A = 0.25$; that is, the thickness of each actuator h_P is 25% of the thickness of the aluminum plate h_A.

covering a wide range of span-to-thickness ratios. Note that the electromechanical coupling is simplified to the form of actuation strain fields that are imposed on the piezoceramic materials.

The global response of the plate is simulated for the cases of induced bending actuation and induced extension or contraction actuation using a series of finite element models that represent a wide range of assumed kinematic complexity. All of the finite element models used in the study are created using a hierarchical, displacement-based, two-dimensional finite element model that is developed specifically for composite laminates. It permits the assumed kinematics of the entire model (or any given element) to be easily changed. The hierarchical model includes the first-order shear deformation model (FSD), a higher-order cubic equivalent single-layer model (ESL3), Type-I layerwise models (LW1), and Type-II layerwise models (LW2) as special cases. Each of the first three models (FSD, ESL3, and LW1) uses a reduced constitutive matrix that is based on the assumption of zero transverse normal stress; however, the models differ significantly in their assumed distribution of transverse shear strain. The FSD and ESL3 models assume transverse shear strain distributions that are C^1 continuous functions (differentiable function whose first derivative is also continuous) of the thickness coordinate (constant and quadratic, respectively), whereas the LW1 model includes discrete-layer transverse shear effects via in-plane displacement components that are C^0 continuous (function is continuous but need not be differentiable) with respect to the thickness coordinate. The LW2 layerwise model utilizes a full three-dimensional constitutive matrix and includes both discrete-layer transverse shear effects and discrete-layer transverse-normal effects by expanding all three displacement components as C^0 continuous functions of the thickness coordinate.

A two-dimensional, hierarchical, displacement-based, variable-kinematic finite element is developed by expressing the total-displacement field as the sum of a low-order primary-displacement field and a higher-order secondary-displacement field. The primary-displacement field is present in all variable-kinematic elements at all times. The individual terms of the secondary-displacement field then serve as relative displacements that can be added to the element's primary field to provide higher-order kinematics as needed. The total-displacement field is expressed as

$$u(x, y, z) = u^{FSD}(x, y, z) + u^{LW}(x, y, z) \tag{5.528}$$

$$v(x, y, z) = v^{FSD}(x, y, z) + v^{LW}(x, y, z) \tag{5.529}$$

$$w(x, y, z) = w^{FSD}(x, y, z) + w^{LW}(x, y, z) \tag{5.530}$$

where u, v, and w are the total-displacement components in the x, y, and z directions, respectively. In this case, the primary-displacement field is provided by u^{FSD}, v^{FSD}, and w^{FSD}, which represent the assumed displacement field for the FSD and is expressed as

$$u^{FSD}(x, y, z) = u_o(x, y) + z\phi_x(x, y) \tag{5.531}$$

$$v^{FSD}(x, y, z) = v_o(x, y) + z\phi_y(x, y) \tag{5.532}$$

$$w^{FSD}(x, y, z) = w_o(x, y) \tag{5.533}$$

where $u_o(x, y)$, $v_o(x, y)$, and $w_o(x, y)$ represent the displacement of points on the plate's mid-surface. The terms $\phi_x(x, y)$ and $\phi_y(x, y)$ represent the rotation of the inextensible transverse-normal fiber in the $x - z$ and $y - z$ planes, respectively. The FSD displacement field includes a rudimentary transverse shear strain that is constant through the thickness of the laminate. Because the FSD displacement field does not explicitly include transverse normal strain, it is intended to be used in conjunction with a reduced constitutive matrix that is based on the assumption of zero transverse normal stress.

The secondary-displacement field consists of u^{LW}, v^{LW}, and w^{LW} and represents the assumed displacement field for a full three-dimensional layerwise theory [3, 6], which is characterized by displacement components that are piece-wise continuous (specifically, C^0 continuous) with respect to the thickness coordinate. The layerwise displacement field is included as an optional, incremental enhancement to the primary-displacement field, so that the element may have full or partial three-dimensional modeling capability when needed. The layerwise field can be expressed as

$$u^{LW}(x, y, z) = U_j(x, y)\varphi_j(z) \tag{5.534}$$

$$v^{LW}(x, y, z) = V_j(x, y)\varphi_j(z) \tag{5.535}$$

$$w^{LW}(x, y, z) = W_j(x, y)\varphi_j(z) \tag{5.536}$$

where the repeated subscript j implies summation over $j = 1, 2, \ldots, n$. The functions $\varphi_j(z)(j = 1, 2, \ldots, n)$ are one-dimensional Lagrangian interpolation functions associated with n nodes distributed through the laminate thickness, located at $z_j(j = 1, 2, \ldots, n)$. Thus, the through-the-thickness variation of the displacement components is defined in terms of a one-dimensional finite-element representation with C^0 continuity of the interpolants. The one-dimensional interpolants $U_j(x, y)$, $V_j(x, y)$, and $W_j(x, y)$ represent additions to the displacement components u_1, u_2, and u_3 on the planes defined by $z = z_j(j = 1, 2, \ldots, n)$.

A hierarchy of three distinctly different types of laminate elements can be obtained from the composite-displacement field of Eqs. 5.528–5.530. The first and simplest type of element is the FSD element. This element is formed using Eqs. 5.531–5.533, while ignoring Eqs. 5.534–5.536. The second type of element is the LW1 element. The LW1 element is formed using Eqs. 5.531, 5.532, 5.533, 5.534, and 5.535, while ignoring Eq. 5.536. Thus, the LW1 element includes discrete-layer transverse shear effects but neglects transverse normal effects and, consequently, uses a reduced-stiffness matrix similar to the FSD element. Due to the inclusion of discrete-layer transverse shear effects, the LW1 element is applicable to thick laminates and often yields results comparable to three-dimensional finite elements while using approximately two-thirds the number of degrees of freedom. The third and most complex element is the LW2 element. The LW2 element is formed using both Eqs. 5.531–5.533 and Eqs. 5.534–5.536. Thus, it is a full three-dimensional layerwise element that explicitly accounts for all six strain components and, consequently, uses a full three-dimensional constitutive matrix. The inclusion of the full layerwise field provides the LW2 element with both discrete-layer transverse shear effects and discrete-layer transverse normal effects. In terms of interpolation capability and number of degrees of freedom, the two-dimensional LW2 element is equivalent to an entire stack of conventional three-dimensional finite elements.

Figure 5.28 shows the geometry of a simple test specimen that is used to study the effect of laminate-kinematic assumptions on the predicted global response of plates that contain surface-bonded actuator pairs. The test specimen is composed of a square aluminum plate and a symmetric pair of square surface-bonded piezoceramic actuators. The aluminum material is characterized by Young's modulus $E = 70$ GPa and Poisson's ratio $\nu = 0.3$, whereas the piezoceramic material is characterized by Young's modulus $E = 63$ GPa, Poisson's ratio $\nu = 0.3$, and piezoelectric constants $d_{31} = d_{32} = 374.537$ pC/N. The length of the aluminum plate ($2L$) is chosen to be twice as large as the length of the piezoceramic actuators ($2P$) to ensure that any local effects associated with the actuator edges will dissipate before reaching the boundary of the aluminum plate. The thickness of each piezoceramic actuator (h_P) is chosen to be one-fourth the thickness of the aluminum plate (h_A). The total thickness of the actuated region is then $H = 2h_P + h_A = 1.5h_A$. The adhesive bond layer between the piezoceramic patch and the aluminum plate is assumed to be sufficiently thin to produce negligible shear lag and, hence, will not be included in the early part of the study.

The edges of the aluminum plate are unconstrained, and the plate is loaded by applying prescribed voltages to the two piezoceramic actuators. The lines $x = 0$ and $y = 0$ represent axes of symmetry; therefore, the computational domain is reduced to one quadrant of the actuated plate $(0 < x < L, 0 < y < L)$. The displacement boundary conditions for the symmetry planes are $u(0, y, z) = 0$ and $v(x, 0, z) = 0$, and the remaining two edges at $x = L$ and $y = L$ are traction-free boundaries. The condition $w(0, 0, 0) = 0$ is also enforced to prevent rigid-body translation in the z direction.

Within the context of the present study, the "global response" of the actuated plate is defined as follows. For the case of equal voltages applied to the piezoceramic actuators, the quasi-static global response is considered to be the distribution of in-plane displacement components on the mid-surface of the actuated plate (i.e., $u(x, y, 0)$ and $v(x, y, 0)$). For the case of opposite voltages applied to the piezoceramic actuators, the quasi-static global response is considered to be the distribution of the

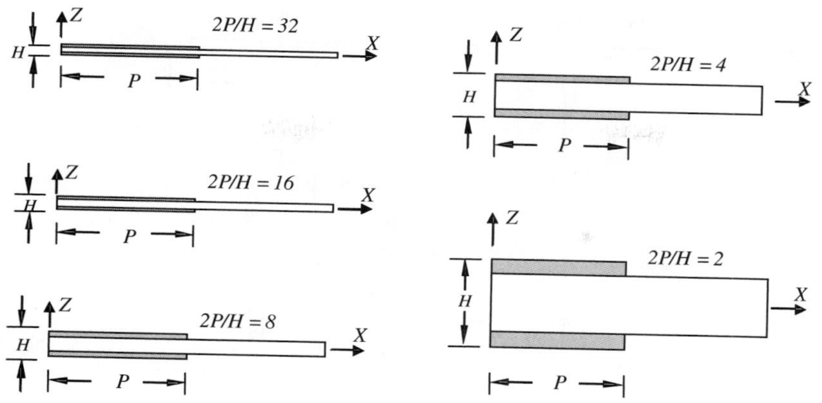

Figure 5.29. View of the $x - z$ plane showing five different levels of span-to-thickness ratio ($2P/H$) used in the actuated-plate problem. In each case, $h_P = (0.25)h_A$.

transverse displacement component on the mid-surface of the actuated plate (i.e., $w(x, y, 0)$).

Figure 5.29 shows five different levels of span-to-thickness ratio ($2P/H$) that are considered in the study: namely, $2P/H = 2$, 4, 8, 16, 32. Note that within the present context, the span-to-thickness ratio ($2P/H$) only describes the geometry of the actuated region of the plate. Thus, the value $2P/H = 2$ does not necessarily imply that the aluminum plate is very thick; it simply implies that the actuator length ($2P$) is only twice as large as the total thickness (H) of the actuated region. In contrast, the value $2P/H = 32$ does indeed represent a relatively thin aluminum plate, regardless of the actual value chosen for the length (P) of the actuators. The span-to-thickness ratio $2P/H$ has a strong influence on the behavior of the actuated plate. The actual load transfer between the piezoceramic actuator and the aluminum plate is known to occur through transverse shear stresses (τ_{xz} and τ_{yz}) and transverse normal stress (σ_z) that are concentrated near the edges of the actuators. These transverse stresses are distributed over a region whose size is approximately two to three times the thickness of the piezoceramic actuator. Consequently, in a specimen with a low value of $2P/H$, these non-zero transverse stresses are present over a higher percentage of the total problem domain than in a specimen with a high value of $2P/H$. Thus, the use of refined models that accurately account for these transverse stresses becomes important as the $2P/H$ ratio decreases.

5.11.1 Effect of Two-Dimensional Mesh Density on the Computed Global Response

Figure 5.30 shows the two-dimensional computational domain as discretized using five different uniform two-dimensional meshes of eight-node, quadratic, quadrilateral elements (2×2 elements, 4×4 elements, 6×6 elements, 12×12 elements, and 24×24 elements). The first part of the study utilizes these five meshes to simply determine the density of the two-dimensional mesh that is required to deliver a well-converged global response with any of the types of laminate models considered in the study. The first ESL model is the first-order shear deformation model (denoted FSD). The second ESL model is a higher-order ESL model similar to that used by Chattopadhyay et al. [7] and Zhou et al. [8]. This ESL model (denoted ESL3) uses

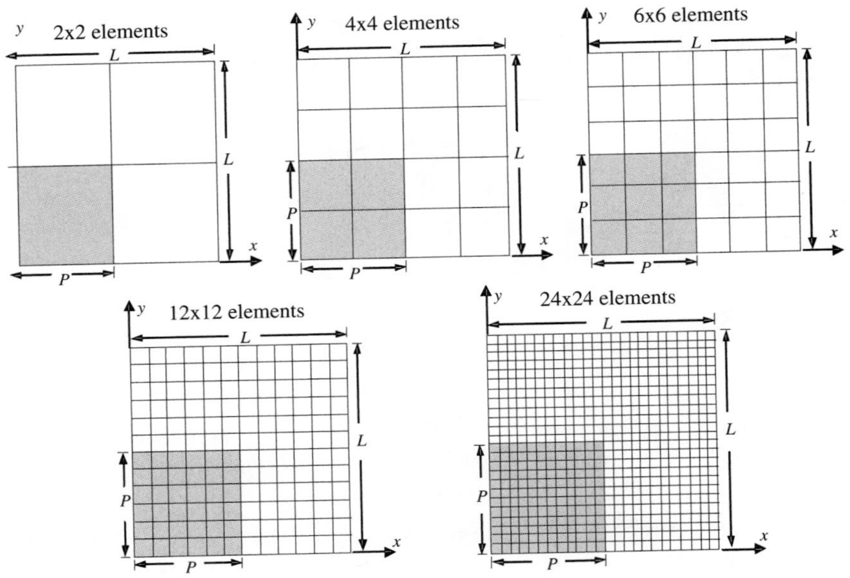

Figure 5.30. Five different levels of uniform two-dimensional mesh density used to assess convergence of the actuated-plate solution. Region of actuator coverage is shown as shaded.

a C^1 continuous cubic expansion of the in-plane displacement components (u and v) through the laminate thickness and uses a reduced constitutive matrix that is based on the assumption of zero transverse normal stress. The last two models are both Type-I layerwise models that enforce the assumption of zero transverse normal stress via a reduced constitutive matrix. The first layerwise model (denoted LW1(1L/2L)) uses one linear layer (1L) for each of the actuators and two linear layers (2L) for the aluminum plate, and the second layerwise model (denoted LW1(1Q/2Q)) uses one quadratic layer (1Q) for each of the actuators and two quadratic layers (2Q) for the aluminum plate.

5.11.2 Pure-Extension Problem (Equal Voltages to Top and Bottom Actuators)

Consider the case in which both actuators are subjected to the same voltage; this causes both actuators to undergo equal extension or contraction, thereby inducing in-plane extension or contraction in the aluminum plate. Specifically, the voltage applied to each piezoceramic actuator is sufficient to provide an electric field strength of 393.7 volts/mm, which in turn is sufficient to induce in-plane normal strains of $\epsilon_{xx} = \epsilon_{yy} = -0.147455 \times 10^{-3}$ in a free actuator. Table 5.3 shows the computed in-plane displacement of the free corner of the aluminum plate (that is, $u(L, L, 0) = v(L, L, 0)$). It shows the results for all four laminate models at all five levels of two-dimensional mesh density; however, only the extreme cases of span-to-thickness ratio are listed ($2P/H = 32$ and 2). The global displacement computed by each of the four laminate models on the 2×2 mesh is noticeably different from the displacement computed by the same laminate model on any of the more refined two-dimensional meshes. However, the global displacements computed by any one laminate model on the 4×4, 6×6, 12×12, and 24×24 two-dimensional meshes show close agreement, thus indicating convergence of the global results. This observation also

Table 5.3. *Effect of two-dimensional mesh density on the normalized in-plane displacement of the free corner, $u(L, L, 0) \times 10^4/P = v(L, L, 0) \times 10^4/P$, for the case of equal voltages applied to both top and bottom actuators*

2P/H	Model Type	Two-Dimensional Mesh	$u(L, L, 0) \times 10^4/P$
32	FSD	2 × 2	−0.19372
32	FSD	4 × 4	−0.20076
32	FSD	6 × 6	−0.20072
32	FSD	12 × 12	−0.20078
32	FSD	24 × 24	−0.20078
32	ESL3	2 × 2	−0.19464
32	ESL3	4 × 4	−0.19896
32	ESL3	6 × 6	−0.19906
32	ESL3	12 × 12	−0.19899
32	ESL3	24 × 24	−0.19898
32	LW1(1L/2L)	2 × 2	−0.19402
32	LW1(1L/2L)	4 × 4	−0.19886
32	LW1(1L/2L)	6 × 6	−0.19915
32	LW1(1L/2L)	12 × 12	−0.19899
32	LW1(1Q/2Q)	2 × 2	−0.19226
32	LW1(1Q/2Q)	4 × 4	−0.19870
32	LW1(1Q/2Q)	6 × 6	−0.19855
32	LW1(1Q/2Q)	12 × 12	−0.19848
2	FSD	2 × 2	−0.19372
2	FSD	4 × 4	−0.20076
2	FSD	6 × 6	−0.20072
2	FSD	12 × 12	−0.20078
2	FSD	24 × 24	−0.20078
2	ESL3	2 × 2	−0.18247
2	ESL3	4 × 4	−0.17011
2	ESL3	6 × 6	−0.17039
2	ESL3	12 × 12	−0.17028
2	ESL3	24 × 24	−0.17024
2	LW1(1L/2L)	2 × 2	−0.18270
2	LW1(1L/2L)	4 × 4	−0.16993
2	LW1(1L/2L)	6 × 6	−0.17029
2	LW1(1L/2L)	12 × 12	−0.17031
2	LW1(1Q/2Q)	2 × 2	−0.15716
2	LW1(1Q/2Q)	4 × 4	−0.16187
2	LW1(1Q/2Q)	6 × 6	−0.16144
2	LW1(1Q/2Q)	12 × 12	−0.16145

Only the two extreme cases of span-to-thickness ratio are shown ($2P/H = 2, 32$). Results are listed for four representative model types [FSD, ESL3, LW1(1L/2L) (one linear layer for each actuator and two linear layers for the aluminum plate), LW1(1Q/2Q)] (one quadratic layer for each actuator and two quadratic layers for the aluminum plate).

applies to the other span-to-thickness ratios not shown in Table 5.3 (e.g., $2P/H = 4$, 8, 16). Based on these results, it is concluded that the 4 × 4 two-dimensional mesh is sufficient for computing the converged global response of all four laminate models. Figure 5.31 shows the distribution of the in-plane displacement components for the thin actuated plate ($2P/H = 32$), where the local coordinate S represents the distance along a diagonal line ($x = y$) that runs from the center of the actuated

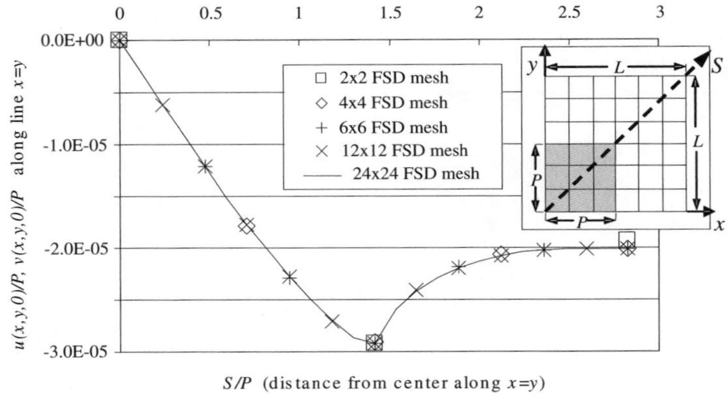

Figure 5.31. Effect of two-dimensional mesh density on the normalized in-plane displacements $u(x, y, 0)/P$ and $v(x, y, 0)/P$ along the diagonal line $x = y$ for the case in which equal contraction is induced in the upper and lower actuators. Results are computed with the FSD model for plate thickness $2P/H = 32$. Corner of actuator is located at $S/P = \sqrt{2}$. Free corner of actuated plate is located at $S/P = 2\sqrt{2}$.

region at $(x, y, z) = (0, 0, 0)$ to the free corner of the aluminum plate at $(x, y, z) = (L, L, 0)$. The results are computed using the FSD model at all five levels of two-dimensional mesh density. The shape of this distribution can be easily understood by considering the in-plane normal strain in the radial direction. Within the actuated region (i.e., $0 < S/P < \sqrt{2}$), the radial normal strain is compressive; thus, the in-plane displacements must have a negative slope. Outside of the actuated region (i.e., $S/P > \sqrt{2}$), the radial normal strain is tensile; thus, the in-plane displacements must have a positive slope. However, the radial normal strain must decrease to zero as the traction-free boundary is approached; thus, the slope of the in-plane displacements must approach zero as S/P approaches $2\sqrt{2}$.

5.11.3 Pure-Bending Problem (Actuators Subjected to Equal but Opposite Voltages)

Consider the case in which both actuators are subjected to opposite voltages of equal magnitude, thus causing the actuator pair to induce pure bending in the aluminum plate. Specifically, the voltage applied to each piezoceramic actuator is sufficient to provide an electric field strength of 393.7 volts/mm, which in turn is sufficient to induce in-plane normal strains of magnitude $\epsilon_{xx} = \epsilon_{yy} = 0.147455 \times 10^{-3}$ in an unconstrained actuator. Table 5.4 shows the computed global response of the actuated plate, as characterized by the transverse displacement of the free corner of the aluminum plate (i.e. $w(L, L, 0)$). The results of the four representative laminate models at all five levels of two-dimensional mesh density are shown; however, only the extreme cases of span-to-thickness ratio are listed ($2P/H = 32$ and 2). The transverse displacement computed by each of the four laminate models on the 2×2 mesh is noticeably different from the displacement computed by the same laminate model on any of the more refined two-dimensional meshes. However, the global displacements computed by any one laminate model on the 4×4, 6×6, 12×12, and 24×24 two-dimensional meshes show close agreement, thus indicating convergence of the computed global results. This observation also applies to the other span-to-thickness ratios (e.g., $2P/H = 4, 8, 16$). Based on these results, it is concluded that the 4×4 two-dimensional mesh is sufficient for computing the converged global response of

Table 5.4. *Effect of two-dimensional mesh density on normalized transverse deflection of the free corner $w(L, L, 0) \times 10^3/P$ during bending actuation*

2P/H	Model Type	Two-Dimensional Mesh	$w(L, L, 0) \times 10^3/P$
32	FSD	2 × 2	7.626
32	FSD	4 × 4	7.595
32	FSD	6 × 6	7.584
32	FSD	12 × 12	7.575
32	FSD	24 × 24	7.573
32	ESL3	2 × 2	7.634
32	ESL3	4 × 4	7.583
32	ESL3	6 × 6	7.559
32	ESL3	12 × 12	7.545
32	ESL3	24 × 24	7.543
32	LW1(1L/2L)	2 × 2	7.632
32	LW1(1L/2L)	4 × 4	7.587
32	LW1(1L/2L)	6 × 6	7.566
32	LW1(1L/2L)	12 × 12	7.550
32	LW1(1Q/2Q)	2 × 2	7.628
32	LW1(1Q/2Q)	4 × 4	7.572
32	LW1(1Q/2Q)	6 × 6	7.546
32	LW1(1Q/2Q)	12 × 12	7.532
2	FSD	2 × 2	0.4663
2	FSD	4 × 4	0.4673
2	FSD	6 × 6	0.4670
2	FSD	12 × 12	0.4667
2	FSD	24 × 24	0.4667
2	ESL3	2 × 2	0.4365
2	ESL3	4 × 4	0.4421
2	ESL3	6 × 6	0.4417
2	ESL3	12 × 12	0.4414
2	ESL3	24 × 24	0.4413
2	LW1(1L/2L)	2 × 2	0.4390
2	LW1(1L/2L)	4 × 4	0.4435
2	LW1(1L/2L)	6 × 6	0.4433
2	LW1(1L/2L)	12 × 12	0.4431
2	LW1(1Q/2Q)	2 × 2	0.4256
2	LW1(1Q/2Q)	4 × 4	0.4282
2	LW1(1Q/2Q)	6 × 6	0.4277
2	LW1(1Q/2Q)	12 × 12	0.4275

Only the two extreme cases of span-to-thickness ratio are shown ($2P/H = 2, 32$). Results are shown for three representative model types [FSD, ESL3, LW1(1L/2L), and LW1(1Q/2Q)].

all four laminate models. Figure 5.32 shows the computed distribution of the transverse deflection $w(x, y, 0)$ for the thin actuated plate ($2P/H = 32$). The deflection computed with the FSD model shows excellent agreement for all five levels of two-dimensional mesh density. Figure 5.33 shows the slope of the plate's mid-surface, dw/dS, computed with the FSD model using all five levels of two-dimensional mesh density for the thin actuated plate ($2P/H = 32$). The slopes dw/dS show excellent agreement for the 4×4, 6×6, 12×12, and 24×24 two-dimensional meshes, and the results computed on the 2×2 mesh are clearly distinguishable from the others.

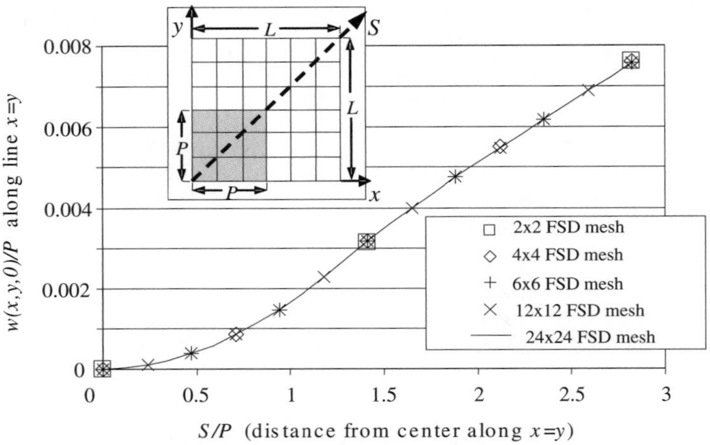

Figure 5.32. Effect of two-dimensional mesh density on the normalized transverse deflection $w(x, y, 0)/P$ along the diagonal line $x = y$ for the case in which opposite voltages are applied to the upper and lower actuators. Results were computed with the FSD model for the thin actuated plate ($2P/H = 32$).

5.12 Effect of Transverse Kinematic Assumptions on Global Response

To determine the effect of various laminate kinematic assumptions on the computed global response of the actuated plate, the problem is solved using the FSD model, the ESL3 model, and 12 different layerwise models. All 14 models utilize the same uniform 6×6 mesh of eight-node, two-dimensional quadratic quadrilateral elements. Note that this two-dimensional mesh is more dense than the 4×4 mesh that was shown to be sufficient in the previous section. However, the increased mesh density is warranted by the need to ensure well-converged results for the wider range of layerwise models that are utilized in this study. For LW1 models, each linear layer admits an independent transverse shear deformation that is constant through the thickness of the layer. For LW2 models, each linear layer admits an independent transverse shear deformation and an independent transverse normal deformation,

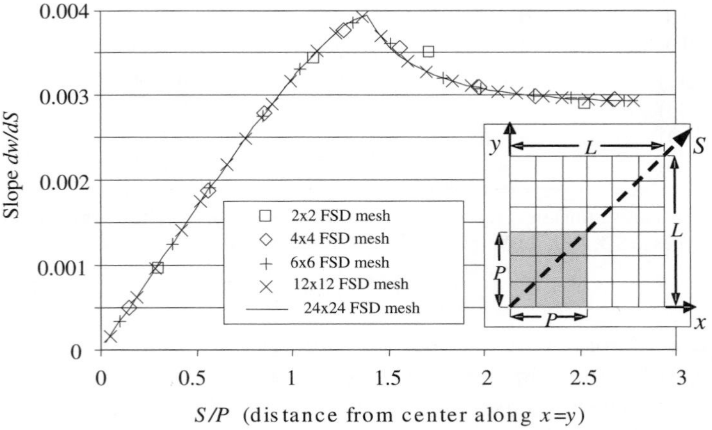

Figure 5.33. Effect of two-dimensional mesh density on the computed slope of the plate's reference surface $dw(x, y, 0)/dS$ along the diagonal line $x = y$ for the case in which opposite voltages are applied to the upper and lower actuators. Results were computed with the FSD model for the thin actuated plate ($2P/H = 32$).

both of which are constant through the thickness of the layer. Using these 14 models (i.e., FSD model, ESL3 model, 6 LW1 models, and 6 LW2 models), the global response of the actuated plate is computed for the cases of induced bending and induced in-plane contraction for five different span-to-thickness ratios ($2P/H = 2, 4, 8, 16, 32$) for a total of 140 different solutions. The computed variables are compared for the series of 14 finite-element solutions that use various levels of refinement in the thickness direction, thus reflecting various levels of laminate-kinematic assumptions.

5.12.1 Case I: Pure-Extension Actuation

Consider the case in which both actuators are subjected to the same voltage, thus causing both actuators to undergo equal extension or contraction, thereby inducing in-plane extension or contraction in the aluminum plate. Specifically, the voltage applied to each piezoceramic actuator is sufficient to provide an electric field strength of 393.7 volts/mm, which in turn is sufficient to induce in-plane normal strains of $\epsilon_{xx} = \epsilon_{yy} = -0.147455 \times 10^{-3}$ in a free actuator. Table 5.5 shows the computed global in-plane response of the actuated plate of the free corner of the aluminum plate, i.e., $u(L, L, 0) = v(L, L, 0)$. It provides a comparison between the FSD model, ESL3 model, and the six LW1 models for all five levels of span-to-thickness ratio and thus shows the effect of including various higher-order representations for the transverse shear deformation while neglecting transverse normal deformation. The FSD model consistently predicts larger global displacements than the ESL3 model and the LW1 models. Thus, the introduction of higher-order transverse shear deformation has the effect of lowering the computed global displacements. To better illustrate this trend, Table 5.5 also shows a computed parameter β that represents the percent decrease in the predicted global response of a higher-order model relative to the FSD model. An examination of the β values in Table 5.5 reveals that the discrepancy between the global responses of the higher-order models and the FSD model is most significant at low values of actuated span-to-thickness ratio $2P/H$. For example, at $2P/H = 2$, the discrepancy between the various higher-order solutions and the FSD solution ranges from 10% to 19%. However, the discrepancy becomes very small as $2P/H$ becomes large; for example, at $2P/H = 32$, the discrepancy only ranges from 0.5% to 1%. This trend is shown graphically in Fig. 5.34, which plots β versus $2P/H$ for some of the higher-order models.

Another general trend that is seen in Table 5.5 and Fig. 5.30 is that for any given level of actuated span-to-thickness ratio $2P/H$, the size of the discrepancy between displacements of the FSD model and a particular higher-order model increases with the level of transverse shear representation that is employed in the higher-order model. For example, increasing the number of numerical layers in the LW1 model causes the predicted global displacements to decrease. Similarly, increasing the polynomial order of the transverse shear deformation in an ESL model causes the predicted global displacements to decrease; for example, the ESL3 model predicts smaller global displacements than the FSD model. In general, an increase in the kinematic order of a laminate model is expected to cause an increase in the overall compliance of the model and, consequently, to result in higher deformations for a given load system. In this sense, these results are somewhat counterintuitive because they clearly show the opposite trend of smaller deformations with increasing kinematic order. To fully explain this counterintuitive behavior, we must consider two separate effects of increasing the kinematics of a model.

Table 5.5. *Normalized in-plane displacement of the free corner, $u(L, L, 0) \times 10^4/P = v(L, L, 0) \times 10^4/P$, caused by applying equal voltages to both upper and lower actuators*

Two-Dimensional Mesh	2P/H	Model Type	Piezo Layers	Aluminum Layers	u(L, L, 0) × 10^4/P and v(L, L, 0) × 10^4/P	β
6 × 6	2	FSD			−0.20072	
6 × 6	2	ESL3			−0.17039	15.1%
6 × 6	2	LW1	1L	1L	−0.17929	10.7%
6 × 6	2	LW1	1L	2L	−0.17029	15.2%
6 × 6	2	LW1	1L	4L	−0.16729	16.7%
6 × 6	2	LW1	2L	1L	−0.17504	12.8%
6 × 6	2	LW1	2L	2L	−0.16645	17.1%
6 × 6	2	LW1	2L	4L	−0.16367	18.5%
6 × 6	4	FSD			−0.20072	
6 × 6	4	ESL3			−0.18595	7.4%
6 × 6	4	LW1	1L	1L	−0.19029	5.2%
6 × 6	4	LW1	1L	2L	−0.18589	7.4%
6 × 6	4	LW1	1L	4L	−0.18439	8.1%
6 × 6	4	LW1	2L	1L	−0.18821	6.2%
6 × 6	4	LW1	2L	2L	−0.18401	8.3%
6 × 6	4	LW1	2L	4L	−0.18263	9.0%
6 × 6	8	FSD			−0.20072	
6 × 6	8	ESL3			−0.19355	3.6%
6 × 6	8	LW1	1L	1L	−0.19558	2.6%
6 × 6	8	LW1	1L	2L	−0.19349	3.6%
6 × 6	8	LW1	1L	4L	−0.19272	4.0%
6 × 6	8	LW1	2L	1L	−0.19458	3.1%
6 × 6	8	LW1	2L	2L	−0.19258	4.1%
6 × 6	8	LW1	2L	4L	−0.19188	4.4%
6 × 6	16	FSD			−0.20072	
6 × 6	16	ESL3			−0.19722	1.7%
6 × 6	16	LW1	1L	1L	−0.19822	1.2%
6 × 6	16	LW1	1L	2L	−0.19725	1.7%
6 × 6	16	LW1	1L	4L	−0.19680	2.0%
6 × 6	16	LW1	2L	1L	−0.19773	1.5%
6 × 6	16	LW1	2L	2L	−0.19683	1.9%
6 × 6	16	LW1	2L	4L	−0.19641	2.1%
6 × 6	32	FSD			−0.20072	
6 × 6	32	ESL3			−0.19906	0.83%
6 × 6	32	LW1	1L	1L	−0.19966	0.53%
6 × 6	32	LW1	1L	2L	−0.19915	0.78%
6 × 6	32	LW1	1L	4L	−0.19888	0.92%
6 × 6	32	LW1	2L	1L	−0.19943	0.64%
6 × 6	32	LW1	2L	2L	−0.19895	0.88%
6 × 6	32	LW1	2L	4L	−0.19870	1.00%

β denotes a higher-order model's percent reduction in $u(L, L, 0)$ compared to the FSD model; i.e., $\beta \equiv 100.[u^{FSD}(L, L, 0) - u^{HOM}(L, L, 0)]/u^{FSD}(L, L, 0)$.

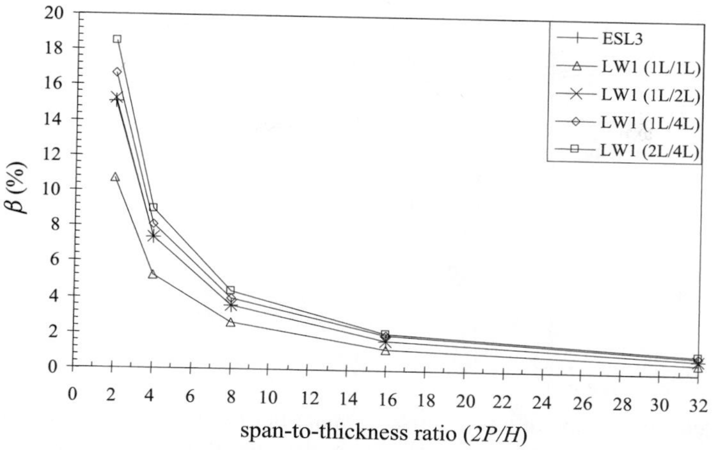

Figure 5.34. Comparison of the computed in-plane displacement of the free corner $u(L, L, 0)$ caused by imposed contraction of both upper and lower actuators. β denotes the higher order model's percent reduction in $u(L, L, 0)$ compared to the FSD model; i.e., $\beta \equiv 100 \cdot [u^{FSD}(L, L, 0) - u^{HOM}(L, L, 0)]/u^{FSD}(L, L, 0)$.

First, let us consider the effect of kinematic order increase on the overall compliance of the model. This is referred to as the global kinematic effect. Unless the laminate is extremely thin, realistic modes of deformation almost always include some level of transverse shear deformation. In this case, the LW1 models and the ESL3 model indeed will be more compliant than the FSD model, and the size of the compliance discrepancy will be directly related to the amount of transverse shear deformation present. Based solely on the existence of the global kinematic effect, one would conclude that the predicted global deformations in the actuated plate should be larger in the LW1 models and the ESL3 model than in the FSD model. However, this is not the case. Despite the fact that the higher-order models are more compliant, their predicted global deformations are lower than the FSD model.

The second effect that must be considered is the effect of kinematic-order increase on the local mechanics of the model in the vicinity of the actuator edges. This is referred to as the local kinematic effect. The actual load transfer between a surface-bonded actuator and the plate substrate occurs through transverse shear stresses (τ_{xz} and τ_{yz}) and transverse normal stress (σ_z) that act across the material interface and tend to be concentrated within a fairly localized region near the edges of the actuators. These transverse stresses cause local transverse shear deformation and local transverse normal deformation, thus allowing a portion of the total actuation energy to be diverted from producing the intended mode of deformation. In other words, an increase in kinematic order allows some of the actuator's energy to produce unwanted or non-useful local deformations. This is the effect that is primarily responsible for the counter intuitive behavior observed in Table 5.5 and Fig. 5.34.

To aid in further discussion of the local kinematic effect, Fig. 5.35 shows the deformed shape of the transverse-normal fiber located at the corner of the actuated region, as predicted by each of the laminate models for the case of equal voltages applied to both actuators (induced global contraction) in a thick plate with a span-to-thickness ratio of $2P/H = 2$. Note that for this particular load case, the in-plane

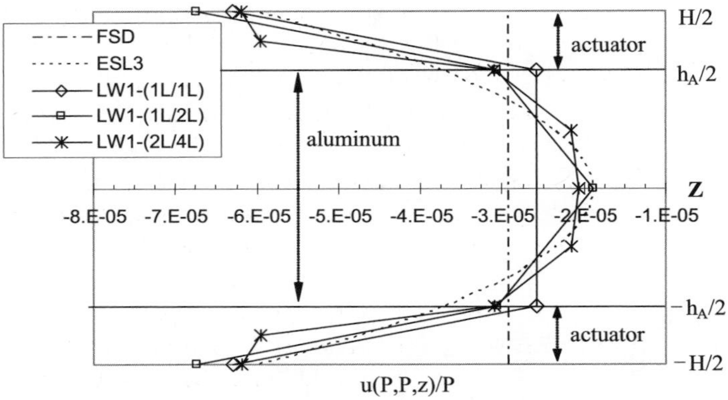

Figure 5.35. Predicted deformed shape of a transverse normal fiber located at the corner of the actuated region for the case of equal voltages applied to both actuators (induced global in-plane contraction). Span-to-thickness ratio is $2P/H = 2$. Results compare the displacement distribution of the FSD model, ESL3 model, and several representative Type-I layer-wise models that include progressively higher-order representation of discrete-layer transverse shear deformation.

displacement distribution must be symmetric about the laminate mid-plane; therefore, the FSD model is unable to make use of its rudimentary transverse shear deformation that is constant through the laminate thickness. As a result, the FSD model does not permit any of the actuation energy to be diverted into local transverse shear deformation. In other words, the FSD model predicts that 100% of the available actuation energy can be directly used to produce the intended effect of in-plane global contraction. Consequently, despite the fact that the FSD model is less compliant than the LW1 layerwise models, the FSD model predicts larger in-plane global deformation for this load case than any of the higher-order models. The ESL3 model uses a C^1-continuous cubic expansion to represent the in-plane displacement components; consequently, the deformed transverse normal fiber is forced to adopt a smooth curved configuration despite the fact that composite laminates do not typically exhibit such smoothness. As seen in Fig. 5.35, the ESL3 model correctly predicts that the transverse shear deformation is highest in the actuator layer and diminishes to zero at the plate's mid-plane. However, the use of a C^1-continuous displacement expansion prevents the ESL3 model from accurately representing the localization of the transverse shear deformation near the actuator/aluminum interface. Consequently, at the actuator/aluminum interface, the ESL3 model underpredicts the transverse shear deformation in the actuator and overpredicts the transverse shear deformation in the aluminum. Among the six LW1 layerwise models, the LW1(1L/1L) model employs the coarsest transverse discretization, using a single linear layer for each of the actuators and a single linear layer for the aluminum plate. The LW1(1L/1L) model is able to exhibit independent transverse shear deformation that is constant within each of the three distinct material layers. For this particular load case, the in-plane displacement must be symmetric about the mid-plane; therefore, the LW1(1L/1L) model does not predict any transverse shear deformation in the aluminum plate. However, both of the piezoceramic actuators are permitted to exhibit gross transverse shear deformation. As shown in Table 5.5, the LW1(1L/1L) model predicts global displacements that are 10.7% smaller than the FSD model for the thick plate ($2P/H = 2$). The next model in the layerwise kinematic hierarchy is the LW1(1L/2L) model, which uses a single linear layer for each of the actuators and two linear layers for the aluminum plate. Compared to the FSD model and the LW1(1L/1L) model, the LW1(1L/2L) model allows the upper and lower halves of the aluminum plate to exhibit independent transverse shear deformation. Thus, the LW1(1L/2L) model permits some of the total actuation energy to be diverted into local transverse shear deformation in both the piezoceramic actuators and the aluminum plate. Consequently, the LW1(1L/2L) model predicts a further reduction in the far-field in-plane displacement: namely, 15.2% less than the FSD model and 5% less than the LW1(1L/1L) model. The final layerwise model is the LW1(2L/4L) model, which uses two linear layers for each of the piezoceramic actuators and four linear layers for the aluminum plate. The LW1(2L/4L) model is better able to represent the locally high transverse shear deformation that is concentrated along the material interface between the aluminum plate and each of the actuators. Thus, the LW1(2L/4L) model is able to divert an even greater portion of the total actuation energy into local transverse shear deformation and, consequently, predicts far-field in-plane displacements that are lower than any of the other models listed in Table 5.5 (18.5% lower than the FSD model).

5.12.2 Case II: Pure-Bending Actuation

Consider the case in which both actuators are subjected to opposite voltages of equal magnitude, thus causing the actuator pair to induce bending in the aluminum plate. Specifically, the voltage applied to each piezoceramic actuator is sufficient to provide an electric field strength of 393.7 volts/mm, which in turn is sufficient to induce free in-plane normal strains of magnitude $\Lambda_x = \Lambda_y = 0.147455 \times 10^{-3}$ in an unconstrained actuator. Table 5.6 shows the computed global-bending response of the actuated plate, as characterized by the transverse displacement of the free corner of the aluminum plate (i.e., $w(L, L, 0)$). It provides a comparison among the FSD model, the ESL3 model, and the six LW1 models for all five levels of span-to-thickness ratio and thus shows the effect of including various higher-order representations for the transverse shear deformation. Again, the FSD model consistently predicts larger global displacements than the ESL3 model and the LW1 models. Thus, the introduction of higher-order transverse shear deformation has the effect of lowering the predicted global-bending response. To better illustrate this trend, Table 5.6 also shows a computed parameter α, which represents the percent decrease in the predicted global response of a higher-order model relative to the FSD model. An examination of the α values reveals that the discrepancy between the global responses of the higher-order models and the FSD model is most significant at low values of actuated span-to-thickness ratio $2P/H$. For example, at $2P/H = 2$, the discrepancy between the various higher-order solutions and the FSD solution ranges from 5% to 8%. However, the discrepancy becomes very small as $2P/H$ becomes larger; for example, at $2P/H = 32$, the discrepancy only ranges from 0.24% to 0.45%. This trend is shown graphically in Fig. 5.36, which plots α versus $2P/H$ for some of the higher-order models. For any given level of actuated span-to-thickness ratio $2P/H$, the size of the discrepancy between displacements of the FSD model and a particular higher-order model increases with the level of transverse shear representation that is employed in the higher-order model. For example, increasing the number of numerical layers in the LW1 model causes the predicted global displacements to decrease. Similarly, increasing the polynomial order of the transverse shear deformation in an

Table 5.6. *Normalized transverse displacement of the free corner $w(L, L, 0) \times 10^3/P$ caused by bending actuation. α denotes the percent reduction in $w(L, L, 0)$ of a higher-order model compared to the FSD model; i.e., $\alpha \equiv 100.[w^{FSD}(L, L, 0) - w^{HOM}(L, L, 0)]/w^{FSD}(L, L, 0)$*

Two-Dimensional Mesh	2P/H	Model Type	Piezo Layers	Aluminum Layers	$w(L, L, 0) \times 10^3/P$	α
6 × 6	2	FSD			0.4670	
6 × 6	2	ESL3			0.4417	5.4%
6 × 6	2	LW1	1L	1L	0.4433	5.1%
6 × 6	2	LW1	1L	2L	0.4433	5.1%
6 × 6	2	LW1	1L	4L	0.4371	6.4%
6 × 6	2	LW1	2L	1L	0.4370	6.4%
6 × 6	2	LW1	2L	2L	0.4370	6.4%
6 × 6	2	LW1	2L	4L	0.4313	7.7%
6 × 6	4	FSD			0.9392	
6 × 6	4	ESL3			0.9108	3.0%
6 × 6	4	LW1	1L	1L	0.9140	2.7%
6 × 6	4	LW1	1L	2L	0.9140	2.7%
6 × 6	4	LW1	1L	4L	0.9062	3.5%
6 × 6	4	LW1	2L	1L	0.9080	3.3%
6 × 6	4	LW1	2L	2L	0.9080	3.3%
6 × 6	4	LW1	2L	4L	0.9007	4.1%
6 × 6	8	FSD			1.8869	
6 × 6	8	ESL3			1.8572	1.6%
6 × 6	8	LW1	1L	1L	1.8612	1.4%
6 × 6	8	LW1	1L	2L	1.8612	1.4%
6 × 6	8	LW1	1L	4L	1.8526	1.8%
6 × 6	8	LW1	2L	1L	1.8556	1.7%
6 × 6	8	LW1	2L	2L	1.8556	1.7%
6 × 6	8	LW1	2L	4L	1.8477	2.1%
6 × 6	16	FSD			3.7847	
6 × 6	16	ESL3			3.7548	0.79%
6 × 6	16	LW1	1L	1L	3.7599	0.65%
6 × 6	16	LW1	1L	2L	3.7599	0.65%
6 × 6	16	LW1	1L	4L	3.7505	0.90%
6 × 6	16	LW1	2L	1L	3.7549	0.79%
6 × 6	16	LW1	2L	2L	3.7549	0.79%
6 × 6	16	LW1	2L	4L	3.7462	1.02%
6 × 6	32	FSD			7.5838	
6 × 6	32	ESL3			7.5594	0.32%
6 × 6	32	LW1	1L	1L	7.5655	0.24%
6 × 6	32	LW1	1L	2L	7.5655	0.24%
6 × 6	32	LW1	1L	4L	7.5542	0.39%
6 × 6	32	LW1	2L	1L	7.5616	0.29%
6 × 6	32	LW1	2L	2L	7.5616	0.29%
6 × 6	32	LW1	2L	4L	7.5508	0.44%

equivalent single-layer model causes the predicted global displacements to decrease; for example, the ESL3 model predicts smaller global displacements than the FSD model.

The inclusion of discrete-layer transverse shear deformation permits some of the total actuation energy to be diverted to the production of local transverse shear

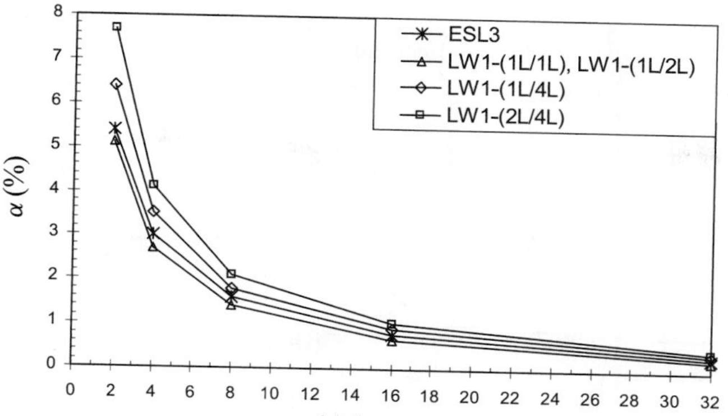

Figure 5.36. Comparison of the predicted transverse displacement of the free corner $w(L, L, 0)$ caused by bending actuation. α denotes the percent reduction in $w(L, L, 0)$ of a higher-order model compared to the FSD model; i.e., $\alpha \equiv 100.[w^{FSD}(L, L, 0) - w^{HOM}(L, L, 0)]/w^{FSD}(L, L, 0)$.

deformation, thus diminishing the amount of actuation energy available for the intended purpose of producing global bending. This local shear deformation can be seen in Figs. 5.37 and 5.38. Figure 5.37 shows the deformed shape of the transverse normal fiber located at the corner of the actuated region for the case of actuation bending in a thick plate with a span-to-thickness ratio of $2P/H = 2$, as predicted by the FSD, ESL3, LW1(1L/2L), and LW1(2L/4L) models. Figure 5.38 shows the distribution of transverse shear strain through the laminate thickness at the two-dimensional reduced Gauss–point located closest to the corner of the actuated region, as predicted by the FSD, ESL3, LW1(1L/2L), and LW1(2L/4L) models. Note that for this particular load case, the transverse shear-strain distribution must be symmetric about the laminate mid-plane. Because the FSD model uses a transverse shear strain that is constant through the entire laminate thickness, it is unable to resolve any of the localized transverse shear that occurs at the actuator/aluminum interface.

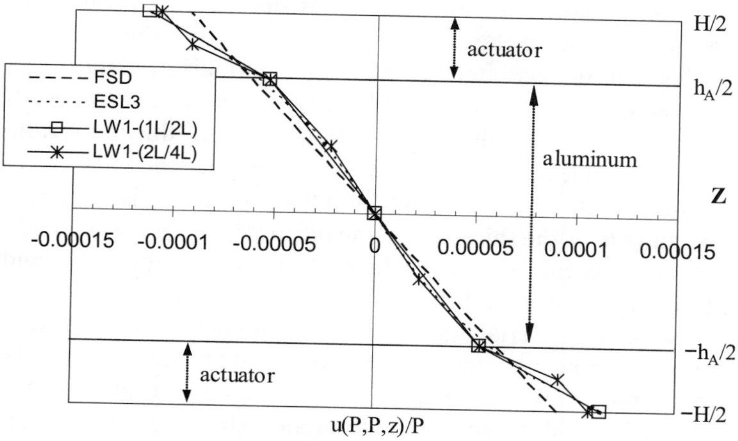

Figure 5.37. Thickness distribution of in-plane displacement at the corner of the actuated region for the case of bending actuation. Span-to-thickness ratio is $2P/H = 2$. Results compare the FSD distribution with two representative Type-I layerwise distributions, which include progressively higher-order representation of discrete-layer transverse shear deformation.

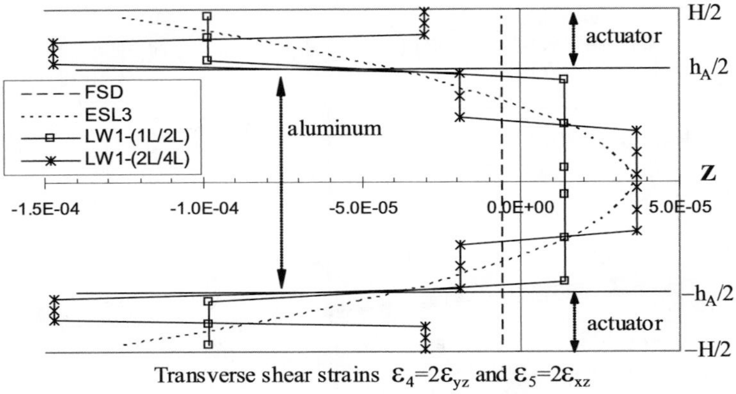

Figure 5.38. Thickness distribution of transverse shear strain at the two-dimensional reduced Gauss point closest to the corner of the actuated region for the case of bending actuation. Span-to-thickness ratio is $2P/H = 2$. Results compare the FSD distribution with two representative Type-I layerwise distributions, which include progressively higher-order representation of discrete-layer transverse shear deformation.

Consequently, the FSD model exhibits very little transverse shear deformation near the actuator edges, which leaves more of the actuation energy to be devoted to producing global bending. In contrast, Fig. 5.38 shows that the ESL3 model and both of the LW1 models are able to represent the localized transverse shear concentration, thus permitting some of the actuation energy to be diverted away from the intended purpose of producing global bending. Consequently, these higher-order models predict smaller global-bending deformation than the FSD model.

5.13 Effect of Finite Thickness Adhesive Bond Layer

In all of the previous solutions for the actuated plate, the adhesive bond layer was neglected; thus, the actuators were assumed to be in a perfect-bond condition. The objective of this section is to check whether the trends observed previously for the perfect-bond case remain valid in the presence of a compliant adhesive layer of finite thickness. To this end, the actuated-plate model is modified by adding an adhesive layer between each of the actuators and the aluminum plate. The resulting actuated plate is simulated at five different levels of span-to-thickness ratio ($2P/H = 2, 4, 8, 16, 32$). At each level, five different adhesive-layer thicknesses are considered: namely, $h_G/h_P = 0.00, 0.02, 0.04, 0.08$, and 0.12, where h_G is the thickness of the adhesive layer and h_P is the thickness of the piezoceramic actuator. Thus, the zero adhesive-layer thickness corresponds to the perfect-bonding condition, and the thickest adhesive layer is chosen to be 12% of the thickness of the actuator. The stiffness of the isotropic adhesive material is assumed to be one-tenth the stiffness of the piezoceramic material. To show the effect of various laminate kinematic assumptions in the presence of a finite-thickness adhesive bond, each configuration of the actuated plate is solved with five different laminate models that represent different portions of the kinematic assumption spectrum. These models include the FSD model, the ESL3 model, a LW1(1L/1L/2L) model, a LW1(2L/1L/4L) model, and a LW2(2L/1L/4L) model. The naming convention for the layerwise models indicates both the type of layerwise model (e.g., LW1 or LW2) and the level of transverse

Table 5.7. *Normalized in-plane displacement of the free corner, $u(L, L, 0) \times 10^4/P = v(L, L, 0) \times 10^4/P$, caused by applying equal voltages to both upper and lower actuators. Results are listed for actuated plates with five different span-to-thickness ratios and five different adhesive-layer thicknesses*

		(Adhesive Layer Thickness)/(Actuator Thickness)				
2P/H	Model Type	0.00	0.02	0.04	0.04	0.12
2	FSD	−0.2007	−0.2006	−0.2005	−0.2003	−0.2002
2	ESL3	−0.1704	−0.1696	−0.1688	−0.1672	−0.1656
2	LW1(1L/1L/2L)	−0.1703	−0.1588	−0.1516	−0.1402	−0.1310
2	LW1(2L/1L/4L)	−0.1637	−0.1541	−0.1475	−0.1369	−0.1281
2	LW2(2L/1L/4L)	−0.1608	−0.1505	−0.1439	−0.1332	−0.1244
4	FSD	−0.2007	−0.2006	−0.2005	−0.2003	−0.2002
4	ESL3	−0.1860	−0.1855	−0.1851	−0.1842	−0.1833
4	LW1(1L/1L/2L)	−0.1859	−0.1802	−0.1766	−0.1709	−0.1660
4	LW1(2L/1L/4L)	−0.1826	−0.1779	−0.1746	−0.1691	−0.1645
4	LW2(2L/1L/4L)	−0.1810	−0.1759	−0.1726	−0.1672	−0.1626
8	FSD	−0.2007	−0.2006	−0.2005	−0.2003	−0.2002
8	ESL3	−0.1936	−0.1933	−0.1930	−0.1925	−0.1920
8	LW1(1L/1L/2L)	−0.1935	−0.1907	−0.1889	−0.1860	−0.1836
8	LW1(2L/1L/4L)	−0.1919	−0.1896	−0.1879	−0.1851	−0.1828
8	LW2(2L/1L/4L)	−0.1909	−0.1882	−0.1867	−0.1840	−0.1817
16	FSD	−0.2007	−0.2006	−0.2005	−0.2003	−0.2002
16	ESL3	−0.1972	−0.1970	−0.1969	−0.1965	−0.1962
16	LW1(1L/1L/2L)	−0.1973	−0.1959	−0.1949	−0.1934	−0.1921
16	LW1(2L/1L/4L)	−0.1964	−0.1953	−0.1944	−0.1929	−0.1917
16	LW2(2L/1L/4L)	−0.1954	−0.1939	−0.1932	−0.1919	−0.1908
32	FSD	−0.2007	−0.2006	−0.2005	−0.2003	−0.2002
32	ESL3	−0.1991	−0.1989	−0.1988	−0.1985	−0.1983
32	LW1(1L/1L/2L)	−0.1991	−0.1985	−0.1979	−0.1970	−0.1963
32	LW1(2L/1L/4L)	−0.1987	−0.1981	−0.1976	−0.1968	−0.1960
32	LW2(2L/1L/4L)	−0.1972	−0.1961	−0.1958	−0.1951	−0.1946

discretization used in each layerwise model; for example, the label (2L/1L/4L) indicates the use of two linear layers (2L) per actuator, one linear layer (1L) per adhesive bond, and four linear layers (4L) for the aluminum plate.

5.13.1 Case I: Pure-Extension Actuation

Table 5.7 and Fig. 5.39 show the computed global response for the case of equal voltages applied to the top and bottom actuators. Specifically, the voltage applied to each piezoceramic actuator is sufficient to provide an electric field strength of 393.7 volts/mm, which in turn is sufficient to induce in-plane normal strains of $\Lambda_x = \Lambda_y = -0.147455 \times 10^{-3}$ in a free actuator. Table 5.7 contains 125 different solutions that represent a combination of model type, adhesive-layer thickness, and

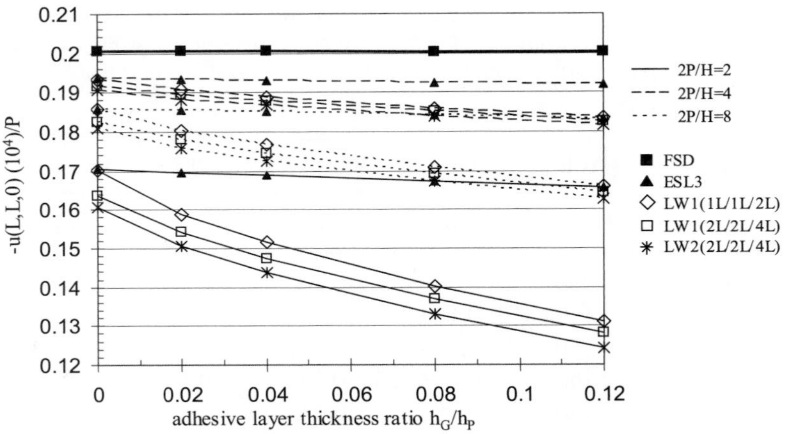

Figure 5.39. Normalized in-plane displacement of the free corner, $u(L, L, 0) \times 10^4/P = v(L, L, 0) \times 10^4/P$, caused by applying equal voltages to both upper and lower actuators. Line type indicates actuated span-to-thickness ratio ($2P/H = 2, 4, 8$), and symbol types indicate different laminate models.

span-to-thickness ratio. Figure 5.39 displays these results graphically for the three thickest plate configurations ($2P/H = 2, 4, 8$) and is included to aid interpretation of the results.

The FSD global response is completely unaffected by the span-to-thickness ratio and is only mildly affected by the thickness of the adhesive layer. For this particular problem, the insensitivity exhibited by the FSD model is caused by the fact that the in-plane displacement components must be symmetric with respect to the laminate mid-plane; therefore, the FSD model is unable to make use of its rudimentary transverse shear deformation. Consequently, the local kinematic effect is completely undetected by the FSD model. The higher-order ESL model (ESL3) predicts smaller global displacements than the FSD model for all levels of span-to-thickness ratio and adhesive thickness ratio. Furthermore, for the perfectly bonded configurations ($h_G/h_P = 0$), the ESL3 model predicts global displacements that are comparable to the low-order layerwise model (LW1(1L/1L/2L)). However, for plate configurations with finite-thickness adhesive layers ($h_G/h_P = 0.02$, 0.04, 0.08, and 0.12), the ESL3 model predicts global displacements that are larger than the three layerwise models. In fact, the ESL3 results are relatively insensitive to changes in the adhesive thickness ratio; for example, even at the lowest span-to-thickness ratio ($2P/H = 2$), the ESL3 model displacements decrease by only 2.8% as the adhesive-thickness ratio increases from $h_G/h_P = 0$ to 0.12.

For plate configurations with finite-thickness adhesive layers ($h_G/h_P = 0.02$, 0.04, 0.08, and 0.12), the layerwise models collectively predict smaller global responses than the FSD model and the ESL3 model. This observation applies to all combinations of span-to-thickness ratio and adhesive-layer thickness. Furthermore, unlike the ESL models (FSD and ESL3), each of the layerwise models predicts a significant decline in the global response as the thickness of the adhesive layer is progressively increased. As seen in Fig. 5.39, this decline is more pronounced for low values of span-to-thickness ratio $2P/H$. For example, at a span-to-thickness ratio of $2P/H = 2$, each of the layerwise models predicts that the global response decreases by approximately 22% as the adhesive-layer-thickness ratio changes from $h_G/h_P = 0.0$ to 0.12. In contrast, at the high span-to-thickness ratio of $2P/H = 32$, each of the layerwise models predicts that the global response decreases by approximately 1% as the adhesive-layer-thickness ratio changes from $h_G/h_P = 0.0$ to 0.12. These observations are entirely consistent with the local kinematic effect, where the presence of the relatively compliant adhesive layer readily permits additional actuation energy to be diverted to the production of local transverse shear strain and local transverse normal strain in the adhesive layer. Also, it is observed that for any particular combination of span-to-thickness ratio and adhesive-layer thickness, the LW2(2L/1L/4L) model predicts a slightly smaller global response than the LW1(2L/1L/4L), which in turn predicts a slightly smaller global response than the LW1(1L/1L/2L) model. In other words, the predicted global response decreases as the kinematic order increases. Again, this observation is entirely consistent with the local kinematic effect. Thus, the trends observed previously for the case of perfect bonding are also exhibited in the presence of a finite-thickness, compliant-adhesive layer. In fact, the presence of a finite-thickness adhesive layer actually causes the trends to be more pronounced.

5.13.2 Case II: Pure-Bending Actuation

Consider the case in which both actuators are subjected to opposite voltages of equal magnitude, thus causing the actuator pair to induce bending in the aluminum plate. Specifically, the voltage applied to each piezoceramic actuator is sufficient to provide an electrical field strength of 393.7 volts/mm, which in turn is sufficient to induce in-plane normal strains of magnitude $\Lambda_x = \Lambda_y = 0.147455 \times 10^{-3}$ in an unconstrained actuator. Table 5.8 shows the computed global response for this case, which contains 125 different solutions for a combinations of model type, adhesive-thickness ratio, and span-to-thickness ratio. Figure 5.40 displays these results graphically for the four thickest plate configurations ($2P/H = 2, 4, 8, 16$) and is included to aid in interpreting the results. Again, for any particular combination of span-to-thickness ratio and finite-adhesive-layer thickness, the layer-wise models collectively predict smaller global responses than the FSD model and the ESL3 model. However, the discrepancy observed between the layer-wise models and the ESL models (FSD and ESL3) is smaller for the present case of induced bending than for the previous case of induced in-plane contraction. This last observed trend is caused by the fact that the global-kinematic effect is significant for the case of induced bending, but the global-kinematic effect opposes and partially cancels the local-kinematic effect. For any particular combination of model type and span-to-thickness ratio, Fig. 5.40 clearly shows that the computed global response decreases as the thickness of the adhesive layer increases. This particular effect is more pronounced in the layerwise models than in the ESL models (FSD and ESL3). Furthermore, this effect is more pronounced in plates with low span-to-thickness ratios than in plates with high span-to-thickness ratios.

5.14 Strain Energy Distribution

Robbins and Chopra [9] examined the distribution of strain energy in the various component materials of actuated plates and investigated the manner in which

Table 5.8. *Normalized transverse displacement of the free corner $w(L, L, 0) \times 10^3/P$ caused by bending actuation. Results are listed for actuated plates with five different levels of total span-to-thickness ratio and five different adhesive-layer-thickness ratios (h_G/h_P = 0.0, 0.2, 0.4, 0.8, 0.12)*

		(Adhesive Layer Thickness)/(Actuator Thickness)				
2P/H	Model Type	0.00	0.02	0.04	0.04	0.12
2	FSD	0.4670	0.4664	0.4653	0.4632	0.4611
2	ESL3	0.4417	0.4317	0.4317	0.4318	0.4319
2	LW1(1L/1L/2L)	0.4433	0.4236	0.4115	0.3920	0.3759
2	LW1(2L/1L/4L)	0.4313	0.4150	0.4039	0.3857	0.3704
2	LW2(2L/1L/4L)	0.4240	0.4053	0.3938	0.3749	0.3590
4	FSD	0.9392	0.9379	0.9359	0.9317	0.9275
4	ESL3	0.9108	0.8948	0.8949	0.8950	0.8951
4	LW1(1L/1L/2L)	0.9140	0.8944	0.8820	0.8619	0.8450
4	LW1(2L/1L/4L)	0.9007	0.8845	0.8731	0.8544	0.8382
4	LW2(2L/1L/4L)	0.8900	0.8713	0.8598	0.8407	0.8242
8	FSD	1.8869	1.8839	1.8799	1.8718	1.8634
8	ESL3	1.8572	1.8293	1.8293	1.8295	1.8296
8	LW1(1L/1L/2L)	1.8612	1.8407	1.8266	1.8033	1.7830
8	LW1(2L/1L/4L)	1.8477	1.8305	1.8174	1.7953	1.7757
8	LW2(2L/1L/4L)	1.8270	1.8065	1.7943	1.7727	1.7534
16	FSD	3.7847	3.7781	3.7703	3.7542	3.7375
16	ESL3	3.7548	3.7031	3.7032	3.7033	3.7035
16	LW1(1L/1L/2L)	3.7599	3.7373	3.7190	3.6874	3.6589
16	LW1(2L/1L/4L)	3.7462	3.7265	3.7097	3.6795	3.6518
16	LW2(2L/1L/4L)	3.6994	3.6725	3.6583	3.6313	3.6055
32	FSD	7.5838	7.5703	7.5548	7.5229	7.4899
32	ESL3	7.5594	7.4606	7.4607	7.4609	7.4611
32	LW1(1L/1L/2L)	7.5655	7.5383	7.5134	7.4658	7.4202
32	LW1(2L/1L/4L)	7.5508	7.5256	7.5018	7.4560	7.4115
32	LW2(2L/1L/4L)	7.4464	7.3967	7.3768	7.3371	7.2974

the strain energy distribution is influenced by the actuated span-to-thickness ratio, the thickness of the adhesive bond layer, and the effect of modeling choices (e.g., kinematic assumptions and mesh density). Again, the focus problem consisted of a square aluminum plate with a single symmetric pair of surface-mounted piezoceramic actuators, which were used to produce in-plane extension or bending in the aluminum plate. The behavior of the actuated plate was examined over a range of plate thicknesses and adhesive–bond layer thicknesses using a series of finite element models that feature different levels of kinematic complexity and different levels of two-dimensional mesh density. The study confirmed the existence and quantified the magnitude of the local-kinematic effect, whereby a portion of the available actuation energy is diverted to the production of localized transverse shear deformation and transverse normal deformation, thus reducing the amount of actuation energy available to produce in-plane deformation in the structural substrate.

The relevant strain energy quantities that are computed from the results of each simulation include the total strain energy U_{total}, the in-plane strain energy U_{2D} (or strain energy associated with the in-plane strain components ϵ_1, ϵ_2, and ϵ_6), the transverse shear strain energy U_{ts} (or strain energy associated with the transverse

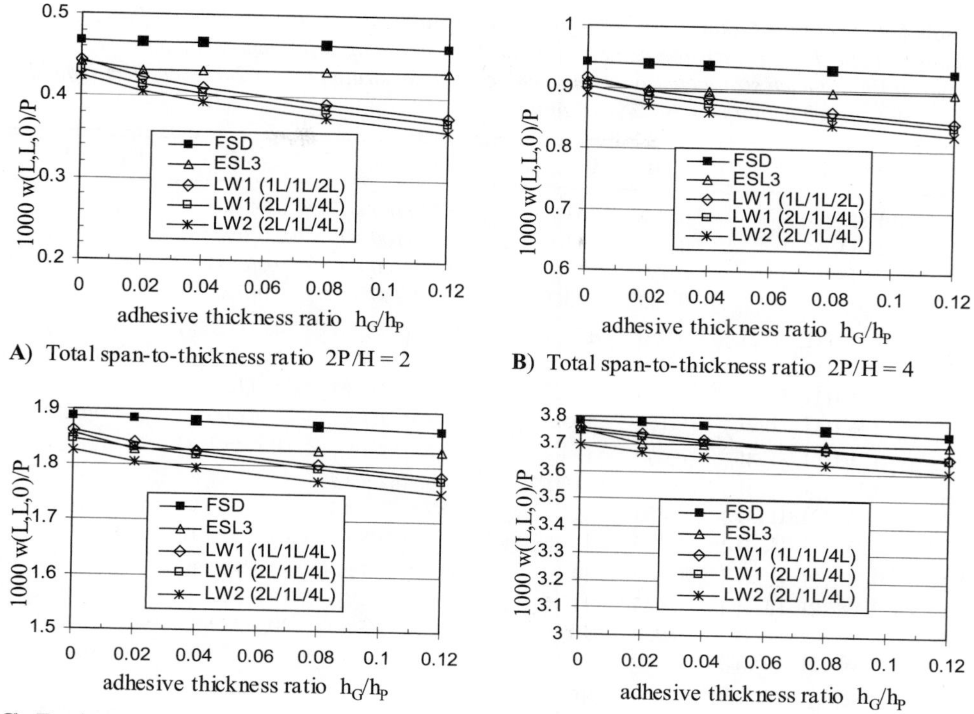

A) Total span-to-thickness ratio 2P/H = 2

B) Total span-to-thickness ratio 2P/H = 4

C) Total span-to-thickness ratio 2P/H = 8

D) Total span-to-thickness ratio 2P/H = 16

Figure 5.40. Normalized transverse displacement of the free corner $w(L, L, 0) \times 10^3/P$ caused by bending actuation. Results are shown for actuated plates with four different levels of total span-to-thickness ratio ($2P/H = 2, 4, 8, 16$) and five different adhesive layer thickness ratios ($h_G/h_P = 0.0, 0.2, 0.4, 0.8, 0.12$).

shear strain components ϵ_4 and ϵ_5), and the transverse normal strain energy U_{tn} (or strain energy associated with the transverse normal strain component ϵ_3). These quantities are defined in Eqs. 5.537–5.540.

$$U_{total} \equiv \int_V \left(\int \sigma_\alpha d\varepsilon_\alpha \right) dv \text{ (implied summation on } \alpha \text{ from 1 to 6)} \tag{5.537}$$

$$U_{2D} \equiv \int_V \left(\int \sigma_1 d\varepsilon_1 \right) dv + \int_V \left(\int \sigma_2 d\varepsilon_2 \right) dv + \int_V \left(\int \sigma_6 d\varepsilon_6 \right) dv \tag{5.538}$$

$$U_{ts} \equiv \int_V \left(\int \sigma_4 d\varepsilon_4 \right) dv + \int_V \left(\int \sigma_5 d\varepsilon_5 \right) dv \tag{5.539}$$

$$U_{tn} \equiv \int_V \left(\int \sigma_3 d\varepsilon_3 \right) dv \tag{5.540}$$

In each case, the computed strain energy values are computed for the entire actuated plate and thus include the contribution from all three component materials. Because the thickness dimension of the plate varies, each of the strain energy quantities is expressed on a per–unit volume basis by dividing by the total volume of the actuated plate. Tables 5.9 and 5.10 list the strain energy values for all configurations

Table 5.9. *Computed total strain energy and strain energy components (in J/m³) for entire actuated plate during extension actuation. Quantities in parentheses denote the percent contribution of each strain energy component to the total strain energy. Imposed electric field strength for each actuator is 393.7 volts/mm*

2P/H	Model	h_A/H	U_{total}/Vol	U_{2D}/Vol	U_{ts}/Vol	U_{tn}/Vol
2	FSD	0.000	163.1	163.1*(100%)*	0.0*(0%)*	0.0*(0%)*
2	FSD	0.025	163.1	163.1*(100%)*	0.0*(0%)*	0.0*(0%)*
2	FSD	0.050	163.2	163.2*(100%)*	0.0*(0%)*	0.0*(0%)*
2	FSD	0.100	163.2	163.2*(100%)*	0.0*(0%)*	0.0*(0%)*
2	LW1(1L/1L/2L)	0.000	132.0	117.3*(88.9%)*	14.6*(11.1%)*	0.0*(0%)*
2	LW1(1L/1L/2L)	0.025	122.9	104.1*(84.7%)*	18.8*(15.3%)*	0.0*(0%)*
2	LW1(1L/1L/2L)	0.050	115.8	93.9*(81.1%)*	21.8*(18.9%)*	0.0*(0%)*
2	LW1(1L/1L/2L)	0.100	105.3	79.4*(75.4%)*	25.9*(24.6%)*	0.0*(0%)*
2	LW1(2L/1L/4L)	0.000	128.8	112.7*(87.5%)*	16.1*(12.5%)*	0.0*(0%)*
2	LW1(2L/1L/4L)	0.025	119.1	98.7*(82.9%)*	20.4*(17.1%)*	0.0*(0%)*
2	LW1(2L/1L/4L)	0.050	112.6	89.5*(79.5%)*	23.1*(20.5%)*	0.0*(0%)*
2	LW1(2L/1L/4L)	0.100	102.8	76.1*(74%)*	26.7*(26%)*	0.0*(0%)*
2	LW2(2L/1L/4L)	0.000	125.7	110.3*(87.7%)*	13.6*(10.8%)*	1.8*(1.5%)*
2	LW2(2L/1L/4L)	0.025	115.5	95.8*(82.9%)*	17.6*(15.3%)*	2.1*(1.8%)*
2	LW2(2L/1L/4L)	0.050	109.0	86.6*(79.5%)*	20.2*(18.5%)*	2.2*(2.0%)*
2	LW2(2L/1L/4L)	0.100	99.1	73.1*(73.8%)*	23.6*(23.9%)*	2.3*(2.3%)*
4	FSD	0.000	163.1	163.1*(100%)*	0.0*(0%)*	0.0*(0%)*
4	FSD	0.025	163.1	163.1*(100%)*	0.0*(0%)*	0.0*(0%)*
4	FSD	0.050	163.2	163.2*(100%)*	0.0*(0%)*	0.0*(0%)*
4	FSD	0.100	163.2	163.2*(100%)*	0.0*(0%)*	0.0*(0%)*
4	LW1(1L/1L/2L)	0.000	147.3	139.7*(94.8%)*	7.7*(5.2%)*	0.0*(0%)*
4	LW1(1L/1L/2L)	0.025	142.8	132.9*(93.1%)*	9.9*(6.9%)*	0.0*(0%)*
4	LW1(1L/1L/2L)	0.050	138.9	127.1*(91.5%)*	11.8*(8.5%)*	0.0*(0%)*
4	LW1(1L/1L/2L)	0.100	133.1	118.5*(89%)*	14.6*(11.0%)*	0.0*(0%)*
4	LW1(2L/1L/4L)	0.000	145.8	137.4*(94.2%)*	8.4*(5.8%)*	0.0*(0%)*
4	LW1(2L/1L/4L)	0.025	140.8	129.9*(92.3%)*	10.8*(7.7%)*	0.0*(0%)*
4	LW1(2L/1L/4L)	0.050	137.2	124.6*(90.8%)*	12.6*(9.2%)*	0.0*(0%)*
4	LW1(2L/1L/4L)	0.100	131.7	116.5*(88.5%)*	15.2*(11.6%)*	0.0*(0%)*
4	LW2(2L/1L/4L)	0.000	144.8	136.8*(94.5%)*	7.5*(5.2%)*	0.49*(0.3%)*
4	LW2(2L/1L/4L)	0.025	139.6	129.2*(92.5%)*	9.9*(7.1%)*	0.56*(0.4%)*
4	LW2(2L/1L/4L)	0.050	135.9	123.7*(91.1%)*	11.5*(8.5%)*	0.62*(0.5%)*
4	LW2(2L/1L/4L)	0.100	130.2	115.5*(88.7%)*	14.0*(10.8%)*	0.72*(0.6%)*
8	FSD	0.000	163.1	163.1*(100%)*	0.0*(0%)*	0.0*(0%)*
8	FSD	0.025	163.1	163.1*(100%)*	0.0*(0%)*	0.0*(0%)*
8	FSD	0.050	163.2	163.2*(100%)*	0.0*(0%)*	0.0*(0%)*
8	FSD	0.100	163.2	163.2*(100%)*	0.0*(0%)*	0.0*(0%)*
8	LW1(1L/1L/2L)	0.000	155.8	152.4*(97.8%)*	3.5*(2.2%)*	0.0*(0%)*
8	LW1(1L/1L/2L)	0.025	153.7	149.1*(97.1%)*	4.5*(2.9%)*	0.0*(0%)*
8	LW1(1L/1L/2L)	0.050	151.6	145.9*(96.3%)*	5.6*(3.7%)*	0.0*(0%)*
8	LW1(1L/1L/2L)	0.100	148.4	141.1*(95.1%)*	7.3*(4.9%)*	0.0*(0%)*
8	LW1(2L/1L/4L)	0.000	155.1	151.3*(97.5%)*	3.8*(2.5%)*	0.0*(0%)*
8	LW1(2L/1L/4L)	0.025	152.6	147.6*(96.7%)*	5.1*(3.3%)*	0.0*(0%)*
8	LW1(2L/1L/4L)	0.050	150.7	144.6*(96.0%)*	6.1*(4.0%)*	0.0*(0%)*
8	LW1(2L/1L/4L)	0.100	147.6	140.0*(94.8%)*	7.6*(5.2%)*	0.0*(0%)*
8	LW2(2L/1L/4L)	0.000	155.2	151.6*(97.7%)*	3.7*(2.4%)*	−0.07*(0.04%)*
8	LW2(2L/1L/4L)	0.025	152.6	147.8*(96.8%)*	4.9*(3.2%)*	−0.05*(0.03%)*
8	LW2(2L/1L/4L)	0.050	150.6	144.7*(96.1%)*	5.9*(3.9%)*	−0.04*(0.02%)*

2P/H	Model	h_A/H	U_{total}/Vol	U_{2D}/Vol	U_{ts}/Vol	U_{tn}/Vol
8	LW2(2L/1L/4L)	0.100	147.5	140.0*(95.0%)*	7.5*(5.1%)*	0.00*(0%)*
16	FSD	0.000	163.1	163.1*(100%)*	0.0*(0%)*	0.0*(0%)*
16	FSD	0.025	163.1	163.1*(100%)*	0.0*(0%)*	0.0*(0%)*
16	FSD	0.050	163.2	163.2*(100%)*	0.0*(0%)*	0.0*(0%)*
16	FSD	0.100	163.2	163.2*(100%)*	0.0*(0%)*	0.0*(0%)*
16	LW1(1L/1L/2L)	0.000	160.4	159.1*(99.2%)*	1.2*(0%)*	0.0*(0%)*
16	LW1(1L/1L/2L)	0.025	159.5	157.8*(99.0%)*	1.6*(1%)*	0.0*(0%)*
16	LW1(1L/1L/2L)	0.050	158.5	156.4*(98.7%)*	2.1*(1%)*	0.0*(0%)*
16	LW1(1L/1L/2L)	0.100	156.8	153.8*(98.1%)*	3.0*(1%)*	0.0*(0%)*
16	LW1(2L/1L/4L)	0.000	160.1	158.7*(99.2%)*	1.4*(0%)*	0.0*(0%)*
16	LW1(2L/1L/4L)	0.025	159.0	157.1*(98.2%)*	1.9*(1%)*	0.0*(0%)*
16	LW1(2L/1L/4L)	0.050	158.0	155.7*(98.5%)*	2.3*(1%)*	0.0*(0%)*
16	LW1(2L/1L/4L)	0.100	156.5	153.3*(98.0%)*	3.1*(1%)*	0.0*(0%)*
16	LW2(2L/1L/4L)	0.000	160.4	159.2*(99.2%)*	1.4*(0.9%)*	−0.13*(0.08%)*
16	LW2(2L/1L/4L)	0.025	159.3	157.5*(98.9%)*	1.9*(1.2%)*	−0.13*(0.08%)*
16	LW2(2L/1L/4L)	0.050	158.3	156.1*(98.6%)*	2.4*(1.5%)*	−0.12*(0.08%)*
16	LW2(2L/1L/4L)	0.100	156.7	153.6*(98.0%)*	3.2*(2.0%)*	−0.11*(0.07%)*

and models. Figures 5.41 and 5.42 show the total strain energy, the in-plane strain energy, and the transverse shear strain energy stored in the actuated plate for the respective load cases of extension actuation and bending actuation. The FSD model results are observed to be completely insensitive to both the actuated span-to-thickness ratio ($2P/H$) and adhesive thickness ratio (h_A/H). Furthermore, for all of the thickness configurations tested, the FSD model predicts higher levels of total strain energy and in-plane strain energy than any of the layerwise models. All of these observed behaviors are caused by the FSD model's use of a transverse shear strain that is required to be constant through the thickness of the laminate, thus preventing the FSD model from detecting any of the localized discrete-layer transverse shear strain that occurs near the actuator edges. As a result, the actuator is subjected to an artificially elevated degree of elastic constraint, which increases the total amount of strain energy produced during actuation. Furthermore, this increased strain energy is manifested predominantly as in-plane strain energy because the FSD model's assumed kinematics do not allow the production of localized discrete-layer transverse shear deformation.

Next, consider the behavior predicted by the layerwise models. Part A of Figs. 5.41 and 5.42 shows that all of the layerwise models predict that the total strain energy (per unit volume) decreases significantly as (1) the actuated span-to-thickness ratio ($2P/H$) decreases, and/or (2) the adhesive thickness ratio (h_A/H) increases. In both load cases, the decrease in total strain energy is due to the decreased level of elastic constraint that is exerted on the piezoceramic patch, thus permitting the piezoceramic patch to deform without accumulating as much stress. Because the piezoceramic patch is surface-mounted, only one of its six surfaces is elastically constrained. As the actuated span-to-thickness ratio ($2P/H$) decreases, this constrained surface accounts for a smaller percentage of the patch's total surface area; thus, its level of constraint is effectively lowered. Furthermore, as the adhesive thickness ratio is increased, the relatively compliant adhesive results in lower elastic constraint forces exerted on the patch; that is, the patch becomes more able to deform the compliant adhesive without deforming the relatively stiff substrate.

Table 5.10. *Computed total strain energy and strain energy components (in J/m³) for entire actuated plate during bending actuation. Quantities in parentheses denote the percent contribution of each strain energy component to the total strain energy. Imposed electric field strength for each actuator is 393.7 volts/mm*

2P/H	Model	h_A/H	U_{total}/Vol	U_{2D}/Vol	U_{ts}/Vol	U_{tn}/Vol
2	FSD	0.000	86.9	86.8(*99.9%*)	0.1(*0.1%*)	0.0(*0%*)
2	FSD	0.025	86.0	85.9(*99.9%*)	0.1(*0.1%*)	0.0(*0%*)
2	FSD	0.050	85.2	85.1(*99.9%*)	0.1(*0.1%*)	0.0(*0%*)
2	FSD	0.100	83.5	83.3(*99.9%*)	0.1(*0.1%*)	0.0(*0%*)
2	LW1(1L/1L/2L)	0.000	77.3	72.5(*93.7%*)	4.9(*6.3%*)	0.0(*0%*)
2	LW1(1L/1L/2L)	0.025	73.3	66.8(*91.2%*)	6.4(*8.8%*)	0.0(*0%*)
2	LW1(1L/1L/2L)	0.050	69.4	61.5(*88.6%*)	7.9(*11.4%*)	0.0(*0%*)
2	LW1(1L/1L/2L)	0.100	63.6	53.9(*84.7%*)	9.7(*15.3%*)	0.0(*0%*)
2	LW1(2L/1L/4L)	0.000	75.8	70.2(*92.6%*)	5.6(*7.4%*)	0.0(*0%*)
2	LW1(2L/1L/4L)	0.025	70.4	62.6(*89.0%*)	7.8(*11.0%*)	0.0(*0%*)
2	LW1(2L/1L/4L)	0.050	66.9	57.9(*86.6%*)	9.0(*13.4%*)	0.0(*0%*)
2	LW1(2L/1L/4L)	0.100	61.6	51.1(*82.8%*)	10.6(*17.2%*)	0.0(*0%*)
2	LW2(2L/1L/4L)	0.000	76.9	71.6(*93.1%*)	5.3(*6.9%*)	0.02(*0.03%*)
2	LW2(2L/1L/4L)	0.025	70.9	63.3(*89.3%*)	7.4(*10.4%*)	0.19(*0.26%*)
2	LW2(2L/1L/4L)	0.050	67.2	58.3(*86.9%*)	8.6(*12.7%*)	0.27(*0.41%*)
2	LW2(2L/1L/4L)	0.100	61.6	51.1(*83.0%*)	10.1(*16.3%*)	0.38(*0.62%*)
4	FSD	0.000	87.1	87.0(*99.9%*)	0.1(*0.1%*)	0.0(*0%*)
4	FSD	0.025	86.2	86.1(*99.9%*)	0.1(*0.1%*)	0.0(*0%*)
4	FSD	0.050	85.3	85.2(*99.9%*)	0.1(*0.1%*)	0.0(*0%*)
4	FSD	0.100	83.6	83.5(*99.9%*)	0.1(*0.1%*)	0.0(*0%*)
4	LW1(1L/1L/2L)	0.000	82.4	80.0(*97.0%*)	2.5(*3.0%*)	0.0(*0%*)
4	LW1(1L/1L/2L)	0.025	80.1	77.0(*96.1%*)	3.1(*3.9%*)	0.0(*0%*)
4	LW1(1L/1L/2L)	0.050	77.6	73.6(*94.9%*)	4.0(*5.1%*)	0.0(*0%*)
4	LW1(1L/1L/2L)	0.100	73.7	68.6(*93.1%*)	5.1(*6.9%*)	0.0(*0%*)
4	LW1(2L/1L/4L)	0.000	81.7	78.9(*96.6%*)	2.8(*3.4%*)	0.0(*0%*)
4	LW1(2L/1L/4L)	0.025	78.6	74.8(*95.1%*)	3.9(*4.9%*)	0.0(*0%*)
4	LW1(2L/1L/4L)	0.050	76.3	71.7(*94.0%*)	4.6(*6.0%*)	0.0(*0%*)
4	LW1(2L/1L/4L)	0.100	72.7	67.1(*92.3%*)	5.6(*7.7%*)	0.0(*0%*)
4	LW2(2L/1L/4L)	0.000	82.9	80.3(*96.8%*)	2.8(*3.3%*)	−0.12(*0.15%*)
4	LW2(2L/1L/4L)	0.025	79.6	75.8(*95.3%*)	3.8(*4.8%*)	−0.07(*0.09%*)
4	LW2(2L/1L/4L)	0.050	77.1	72.6(*94.2%*)	4.5(*5.9%*)	−0.03(*0.04%*)
4	LW2(2L/1L/4L)	0.100	73.3	67.8(*92.5%*)	5.5(*7.5%*)	0.02(*0.03%*)
8	FSD	0.000	87.3	87.2(*95%*)	0.1(*0.1%*)	0.0(*0%*)
8	FSD	0.025	86.4	86.3(*95%*)	0.1(*0.1%*)	0.0(*0%*)
8	FSD	0.050	85.5	85.4(*95%*)	0.1(*0.1%*)	0.0(*0%*)
8	FSD	0.100	83.8	83.7(*95%*)	0.1(*0.1%*)	0.0(*0%*)
8	LW1(1L/1L/2L)	0.000	85.3	84.3(*98.8%*)	1.0(*1.2%*)	0.0(*0%*)
8	LW1(1L/1L/2L)	0.025	83.9	82.6(*98.5%*)	1.3(*1.5%*)	0.0(*0%*)
8	LW1(1L/1L/2L)	0.050	82.1	80.4(*97.9%*)	1.7(*2.1%*)	0.0(*0%*)
8	LW1(1L/1L/2L)	0.100	79.2	76.9(*97.1%*)	2.3(*2.9%*)	0.0(*0%*)
8	LW1(2L/1L/4L)	0.000	85.0	83.9(*98.6%*)	1.2(*1.4%*)	0.0(*0%*)
8	LW1(2L/1L/4L)	0.025	83.1	81.5(*98.0%*)	1.7(*2.0%*)	0.0(*0%*)
8	LW1(2L/1L/4L)	0.050	81.5	79.5(*97.5%*)	2.0(*2.5%*)	0.0(*0%*)
8	LW1(2L/1L/4L)	0.100	78.7	76.1(*96.7%*)	2.6(*3.3%*)	0.0(*0%*)
8	LW2(2L/1L/4L)	0.000	86.2	85.1(*98.7%*)	1.2(*1.4%*)	−0.12(*0.11%*)
8	LW2(2L/1L/4L)	0.025	84.2	82.6(*98.1%*)	1.7(*2.0%*)	−0.11(*0.09%*)

2P/H	Model	h_A/H	U_{total}/Vol	U_{2D}/Vol	U_{ts}/Vol	U_{tn}/Vol
8	LW2(2L/1L/4L)	0.050	82.5	80.5*(97.6%)*	2.1*(2.5%)*	−0.10*(0.09%)*
8	LW2(2L/1L/4L)	0.100	79.6	77.0*(96.8%)*	2.7*(3.3%)*	−0.08*(0.07%)*
16	FSD	0.000	87.5	87.4*(99.9%)*	0.1*(0.1%)*	0.0*(0%)*
16	FSD	0.025	86.6	86.5*(99.9%)*	0.1*(0.1%)*	0.0*(0%)*
16	FSD	0.050	85.7	85.6*(99.9%)*	0.1*(0.1%)*	0.0*(0%)*
16	FSD	0.100	84.0	83.9*(99.9%)*	0.1*(0.1%)*	0.0*(0%)*
16	LW1(1L/1L/2L)	0.000	86.9	86.5*(99.6%)*	0.4*(0.4%)*	0.0*(0%)*
16	LW1(1L/1L/2L)	0.025	85.8	85.3*(99.5%)*	0.4*(0.5%)*	0.0*(0%)*
16	LW1(1L/1L/2L)	0.050	84.5	83.9*(99.3%)*	0.6*(0.7%)*	0.0*(0%)*
16	LW1(1L/1L/2L)	0.100	82.3	81.4*(99.0%)*	0.9*(1.0%)*	0.0*(0%)*
16	LW1(2L/1L/4L)	0.000	86.8	86.3*(99.5%)*	0.4*(0.5%)*	0.0*(0%)*
16	LW1(2L/1L/4L)	0.025	85.5	84.9*(99.3%)*	0.6*(0.7%)*	0.0*(0%)*
16	LW1(2L/1L/4L)	0.050	84.3	83.5*(99.1%)*	0.7*(0.9%)*	0.0*(0%)*
16	LW1(2L/1L/4L)	0.100	82.0	81.1*(98.8%)*	1.0*(1.2%)*	0.0*(0%)*
16	LW2(2L/1L/4L)	0.000	87.9	87.5*(99.6%)*	0.4*(0.5%)*	−0.11*(0.13%)*
16	LW2(2L/1L/4L)	0.025	86.5	86.0*(99.4%)*	0.6*(0.7%)*	−0.11*(0.12%)*
16	LW2(2L/1L/4L)	0.050	85.3	84.7*(99.2%)*	0.8*(0.9%)*	−0.10*(0.12%)*
16	LW2(2L/1L/4L)	0.100	83.0	82.1*(98.9%)*	1.01.2%)	−0.10*(0.12%)*

Part B of Figs. 5.41 and 5.42 shows that the in-plane strain energy (per unit volume) decreases even more significantly than the total strain energy. In practical terms, this is important because the actuator is usually intended to induce in-plane normal deformation in the substrate. The dramatic decrease in the in-plane strain energy can be explained as follows. As the actuated span-to-thickness ratio decreases, or as the adhesive thickness ratio increases, the actuator tends to produce an

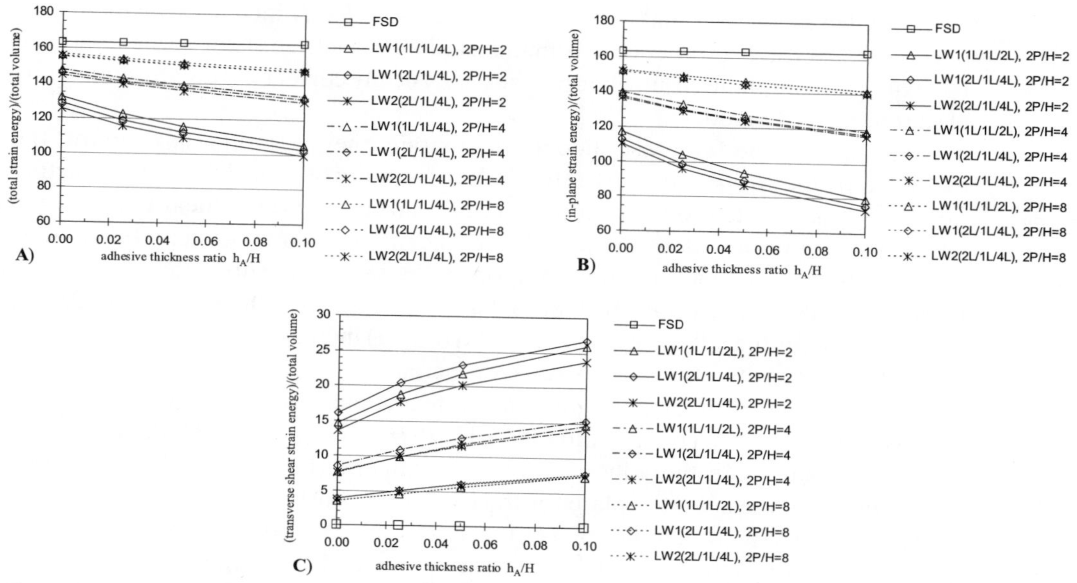

Figure 5.41. Extension actuation results showing the effect of actuated span-to-thickness ratio $2P/H$ and adhesive thickness ratio h_A/H on (A) total strain energy, (B) in-plane strain energy, and (C) transverse–shear strain energy. Energy density is expressed in J/m^3.

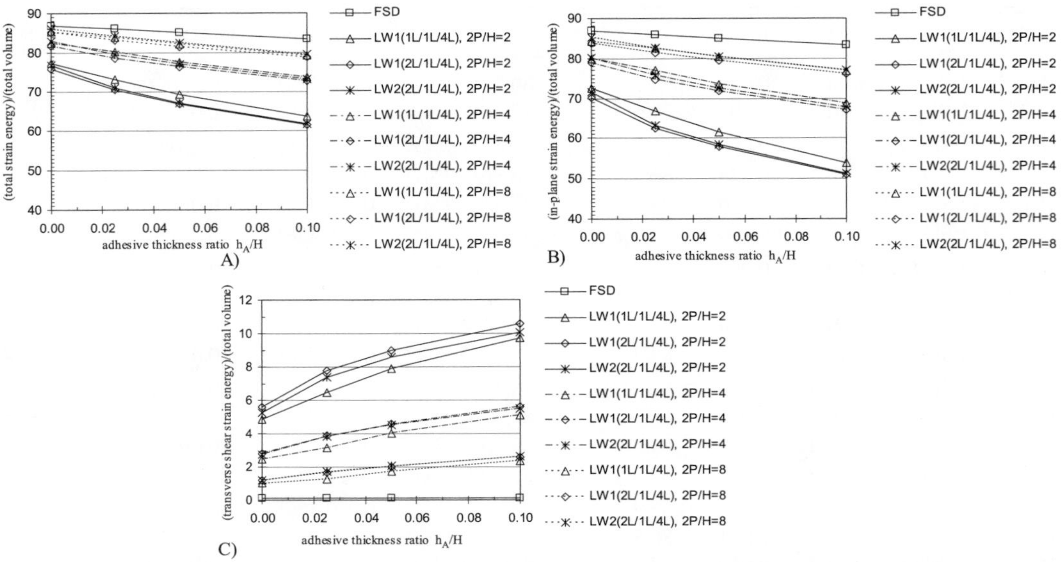

Figure 5.42. Bending actuation results showing the effect of actuated span-to-thickness ratio $2P/H$ and adhesive thickness ratio h_A/H on (A) total strain energy, (B) in-plane strain energy, and (C) transverse–shear strain energy. Energy density is expressed in J/m^3.

increasing amount of localized transverse-shear deformation. This can be seen in Part C of Figs. 5.41 and 5.42, which shows that the transverse shear strain energy increases significantly as the actuated span-to-thickness ratio ($2P/H$) decreases and/or the adhesive thickness ratio increases. In summary, as $2P/H$ decreases or as h_A/H increases, the total strain energy decreases due to a reduction in the degree of elastic constraint exerted on the patch. This reduction in total strain energy is compounded by the fact that as $2P/H$ decreases or as h_A/H increases, an increasing portion of the total strain energy is manifested as transverse shear strain energy at the expense of in-plane strain energy. These physically correct trends are completely absent in the FSD model.

All three of the layerwise models yield strain energy results that confirm the existence of the local kinematic effect, whereby a portion of the available actuation energy is diverted to the production of localized transverse shear deformation and transverse normal deformation, thus reducing the amount of actuation energy available to produce in-plane deformation in the structural substrate. Each of the layerwise models clearly predicts that as the actuated span-to-thickness ratio ($2P/H$) decreases and/or the adhesive thickness ratio (h_A/H) increases, the transverse shear-strain energy density of the adhesive layer increases at the expense of the in-plane strain energy density of the aluminum substrate and piezoceramic actuators. For the thickest actuated plate configuration tested ($2P/H = 2$, $h_A/H = 0.1$), the transverse shear-strain energy in the adhesive layer accounts for 15% and 11% of the total strain energy of the actuated plate for the cases of extension actuation and bending actuation, respectively. Considering that the adhesive layer only accounts for 1% of the volume of the actuated plate, the strain energy density of the adhesive layer is seen to be much higher than that of the other constituent materials. The results also clearly show that the primary destination of energy diverted by the local kinematic effect is local transverse shear deformation as opposed to local transverse normal deformation. This is determined by examining the strain energy results from comparable LW1 and LW2 layerwise models, where it is observed that the transverse shear strain energy predicted by the LW1 and LW2 models shows good agreement and is at least an order of magnitude larger than the transverse normal strain energy predicted by the LW2 model. Even for the thickest actuated plate configuration tested ($2P/H = 2$, $h_A/H = 0.1$), the transverse shear strain energy is approximately 10 and 25 times as large as the transverse normal strain energy for the respective load cases of extension actuation and bending actuation.

For actuated plates with relatively high span-to-thickness ratios, the boundary-layer region (where local transverse shear strains are significant) occupies only a small portion of the total computational domain. In such cases, it is impractical to use a two-dimensional mesh with sufficient refinement to permit a smooth, nonoscillating transverse shear strain distribution in the adhesive bond layer because this requires a minimum of two or three elements across the width of the boundary layer. However, the strain energy contributions from each constituent material were consistently predicted by each of the layerwise models over a wide range of two-dimensional mesh densities and were shown to converge at two-dimensional mesh densities that are far below that required to accurately depict the local transverse shear strain distribution. Even using coarse two-dimensional meshes where the element size is considerably larger than the width of the boundary-layer region, the layerwise models were able to correctly distinguish the magnitude and mode of the dominant strain energy form in each constituent material.

5.15 Review of Plate Modeling

The modeling of laminated composite structural components can be broadly classified into two basic categories according to the kinematics assumed in each case. The first category of models, known as "equivalent single-layer" or ESL models, are identified by the use of a displacement field that exhibits C^1 continuity with respect to the laminate-thickness coordinate. This means that the displacement components and their thickness derivatives are continuous through the entire laminate thickness. This assumption results in a high computation efficiency because of only a few evaluations of functions. Many ESL models of actuated plates are adaptations of the CLPT or CLT, which is based on Kirchhoff-Love hypothesis (transverse normal material fibers remain straight and normal to the curved mid-plane). The CLPT is strictly valid only for very thin laminates.

Table 5.11 lists different smart-plate models that can be found in the literature. Crawley and Lazarus [10] systematically developed the CLPT formulation and a Rayleigh-Ritz analysis for anisotropic plates and validated it with data obtained by testing cantilevered aluminum and composite plates with surface-bonded piezoceramic actuators (fully attached on the top and bottom surfaces). Nonlinear piezo characteristics (d_{31} with field) were measured experimentally and included in the analysis using an iterative approach. Results demonstrated the validity of the analysis for selected plate configurations and showed the potential for shape control with induced-strain actuation. Also, Lee [11, 12] developed a CLPT formulation for composite plates using linear actuation characteristics of piezoelectric laminae. A limited validation study was carried out with test data obtained from a thin composite plate actuated with piezoelectric polymer film (PVDF and PVF2). Wang and Rogers [13] applied CLPT to determine the equivalent force and moment induced by

Table 5.11. *Comparison of smart plate models*

Modeling Type	Actuators	Piezoelectric Coupling	Plate Type	Validation	Reference
CLPT	Surface-bonded full surface	Uncoupled	Composite, nonlinear piezo characteristics	Cantilevered aluminum and composite	Crawley and Lazarus [10]
	Surface-bonded patches		Composite, linear piezo characteristics	Cantilevered composite	Lee [12]
Modified CLPT with transverse shear	Surface and embedded Discrete patches	Uncoupled	Composite, nonlinear piezo characteristics	Cantilevered composite	Hong and Chopra [15]
Reissner-Mindlin FSDT	Surface and embedded Discrete patches	Coupled	Nonlinear Karman analysis, thick isotropic		Carrera [17]
LWSDT	Surface and embedded Piezoply	Coupled	Isotropic and composite		Mitchell and Reddy [18] Robbins and Reddy [19] Chattopadhyay et al. [7] Zhou et al. [8]
Higher-order three-dimensional thick-plate theory	Surface and embedded Piezoply	Coupled	Isotropic, thick		Sun and Whitney [4] Batra and Vidoli [20] Ha et al. [21]

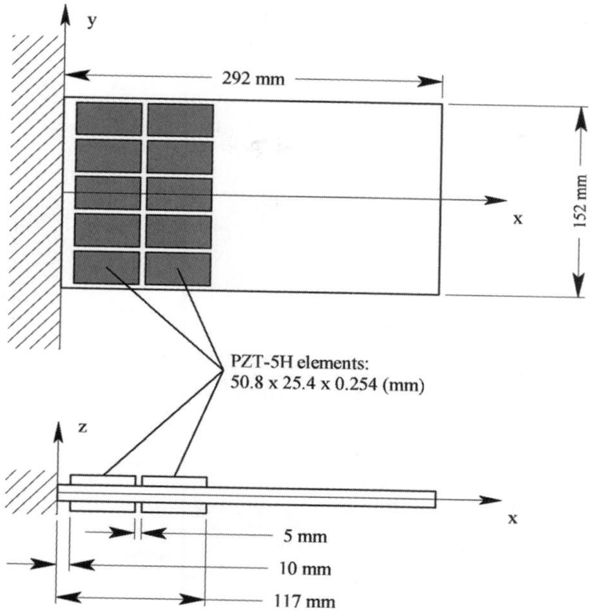

Figure 5.43. Cantilevered plate with surface-mounted piezoceramic sheet actuators.

a finite-length surface-attached piezoelectric actuator to a laminate. They used linear characteristics of piezoelectrics and developed a simplified analysis to calculate bending and extension of the plate. Within the CLPT framework, the piezoelectric sheet is assumed to be an integrated ply of the laminated plate. For thicker plates, the FSDT provides ESL representation to estimate gross macroscopic shear deformation behavior. Lin et al. [14] developed a FSD finite element model of piezoelectrically actuated plates.

Hong et al. [15] developed a consistent finite-element formulation for coupled composite plates including modeling of transverse shear and nonlinear piezoelectric characteristics. The analysis is applicable to a generic anisotropic plate with a number of piezoactuators of arbitrary size, surface-bonded or embedded at arbitrary locations. Composite cantilevered plates with extension-twist and bending-twist couplings with two rows of surface-bonded piezoceramics on both top and bottom surfaces were tested extensively and data were used to validate analysis (Fig. 5.43). Predictions agreed satisfactorily with test data for most configurations, the exception being strongly bending–twist coupled plates, where the predicted spanwise bending was overestimated by up to 20% (Fig. 5.44). The use of an iterative procedure with the incorporation of nonlinear piezoelectric characteristics (as suggested by other researchers) was found to be unnecessary. Heyliger [16] obtained exact solutions for some idealized plate configurations.

Higher-order ESL models with full thermo-electro mechanical coupling were formulated for laminated plates by Chattopadhyay et al. [7] for static analysis and by Zhou et al. [8] for dynamic analysis. In these studies, the in-plane displacement components were assumed to be cubic functions of the thickness coordinate, and the transverse normal effects were neglected through the assumption of zero transverse normal stress ($\sigma_z = 0$). For undamaged, relatively thin, homogeneous plates, the assumption of C^1 thickness continuity for the displacement field is generally considered adequate. However, for composite laminates, where adjacent material

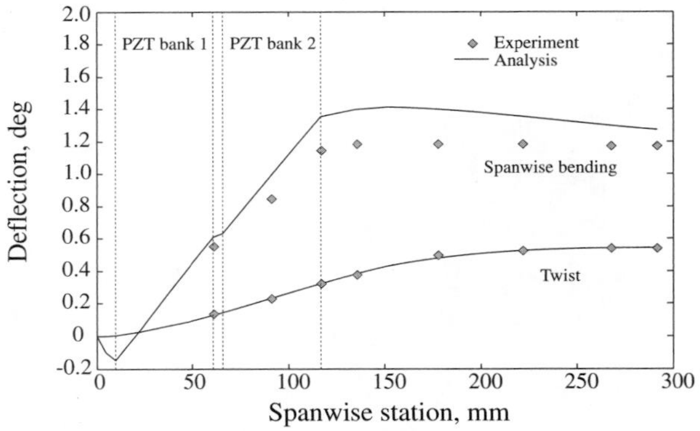

(a) $[+30_2/0]_s$

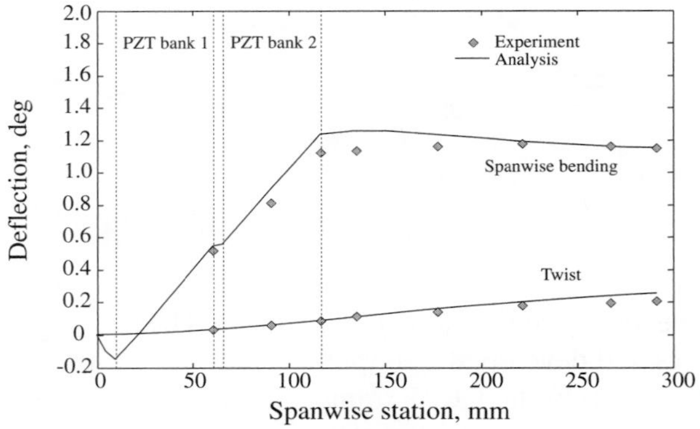

(b) $[0/\pm 45]_s$

Figure 5.44. Spanwise bending and twist distribution at the mid-chord of a composite coupled plate due to piezo-bending excitation at 100 volts.

layers are likely to be quite different in material characteristics, this assumption (of C^1 continuity) is overly restrictive and prevents the transverse normal fibers from exhibiting localized kinking at the material interface. In fact, it will result in a loss of transverse stress equilibrium at layer interfaces. This warping and kinking is most noticeable within very thick laminates or near geometric or material discontinuities (e.g., free edges) and near damaged areas. ESL models are not expected to provide accurate solutions for such cases.

The second category of models, known as "discrete-layer" models (or layer-wise models) are identified by the use of a displacement field that exhibits only C^0 continuity with respect to the laminate thickness coordinate. This means that the displacement components are continuous through the entire laminate thickness, but their thickness derivatives can be discontinuous along the thickness direction (commensurate with the level of transverse discretization employed). In a layer-wise model, the laminate thickness is divided ino a contiguous set of numerical layers. The displacement field is then separately defined within each assumed layer in such a way that the displacement components maintain continuity across interlayer boundaries, whereas their thickness derivatives are not required to be continuous across the interlayer boundaries. It is important that the assumed number of layers is greater than or equal to the number of distinct material laminae. The layer-wise model is capable of representing a variable displacement field across the thickness, appropriately representing the kinking and warping of transverse normal fibers in a multilayer laminate. This modeling capability becomes important in an actuated plate due to the presence of adhesive bond layers. It should be kept in mind that full three-dimensional finite element models can be classified as discrete layer-wise models provided more than one three-dimensional element is used to discretize the laminate thickness dimension. Although layer-wise models are capable of providing accurate solutions to problems that exhibit significant three-dimensional stress and strain fields, they are computationally too expensive for use in most practical problems. Even though the layer-wise models utilize a two-dimensional data structure (similar to two-dimensional ESL models), each node contains a large number of degrees of freedom. Thus, a layer-wise finite-element model produces a global system of equations that is comparable in size to a full three-dimensional finite element model.

References [22]–[23] reviewed several refined plate theories for induced-strain actuation. Bisegna et al. [24, 25] developed a Reissner-Mindlin-type finite-element formulation (i.e., locking-free quadrangular elements) for the analysis of a plate with surface-bonded thin piezoelectric sheet actuators. It was shown to be suitable and effective for some vibration control analyses. Carrera [17] extended the Reissner-Mindlin plate model to the multilayered structures through the inclusion of both the zigzag effect in-plane displacements and the interlamina equilibrium of transverse shear stresses. The theory is normally denoted by an acronym RMZC (Reissner-Mindlin Zigzag Continuity). For the calculation of results, a quadratic distribution of the voltage field along the thickness direction and von Karman–type nonlinear plate analysis were used factoring in the effect of electro-mechanical coupling. It was shown that RMZC effects become important for thick anisotropic plates, especially in the evaluation of transverse shear stresses. Kapuria and Achary [26] developed a coupled zigzag theory for hybrid piezoelectric plates under thermomechanical loads, in which the deflection is approximated as a combination of a global uniform term across the thickness and local piece-wise quadratic variations across sublayers to account for the transverse normal strain. The inplane displacements are approximated as a combination of a third-order global variation across the thickness and a piece-wise linear variation across layers.

Yu and Hodges [27] developed a variational-asymptotic analysis using the Reissner-Mindlin formulation to solve laminated composite plates under mechanical, thermal, and electrical loads. Through this approach, they split the three-dimensional problem into two parts: (1) a nonlinear two-dimensional global plate analysis, and (2) a linear analysis through the thickness to provide two-dimensional generalized constitutive model and recovery relations. Edery-Azulay and Abramovich [28] presented a Reissner-Mindlin theory for plates, developed for laminated plates with continuous piezoelectric layers. The formulation included Levy's solution for plates with two opposite simply-supported edges.

Mitchell and Reddy [18] used LWSDT to model smart-composite laminates with embedded piezoelectric sheets using linear piezoelectric characteristics. Also, this model included the coupling between mechanical deformation and electrostatic charge equations. Robbins and Reddy [29] incorporated a layer-wise composite plate model using an induced-strain approach to approximate the piezoelectric effect. They demonstrated that the resulting layer-wise plate model produced laminate solutions that were equivalent to three-dimensional finite-element solutions, provided that comparable levels of discretization were used. However, the layer-wise model is based on an efficient two-dimensional data structure, which permits the finite-element equations to be computed and assembled more quickly than the three-dimensional model.

Robbins and Reddy [19] formulated a linear global-local analysis based on layer-wise shear-deformation theory to determine local shear fields and global response in surface-mounted piezoelectric-actuated plates. Using a variable-order finite-element discretization, interlaminar stresses in the adhesive layer were determined. It was shown that the highest transverse normal stress occurs at the interface between the bond layer and surface near the free edges that may be the likely source of debonding. Chattopadhyay et al. [8, 7] used LWSDT to calculate static and dynamic response of composite plates with surface-bonded piezoelectric actuators using a completely coupled thermo-piezoelectric mechanical model. Most researchers have neglected these coupling effects. They have shown that to accurately model the behavior of smart composite laminates, it is important to model the transverse shear of each layer using LWSDT and incorporate piezoelectric-mechanical two-way coupling effects. Heyliger et al. [30] and Saravanos et al. [31] developed layer-wise plate models with full electromechanical coupling. Vel and Batra [32] developed a three-dimensional analytical solution using Eshelby-Stroh formalism to calculate the static response of thick multilayered piezoelectric plates. Only linear piezoelectric characteristics are incorporated. Using a three-dimensional mixed-variational principle [33], Batra and Vidoli [20] derived higher-order (kth order) anisotropic homogeneous piezoelectric plate theory. The electric potential, mechanical displacement, and in-plane stresses were expressed as a finite series of order k in the thickness coordinate using Legendre polynomials as the basis functions. The boundary conditions on the top and bottom surfaces were exactly satisfied. Results were obtained for bending of cantilevered thick plates with surface-bonded PZT sheets. It was shown that the seventh-order plate theory captured well the boundary-layer effects near the free and clamped edges. Kulkarni and Bajoria [34] developed a geometrically nonlinear analysis of piezolaminated composite plates and shells using higher-order shear-deformation theory in conjugation with the von Karman hypothesis. The finite-element formulation was based on energy principles and linear piezoelectric characteristics were used. When there is an abrupt change of material properties of the layers and the thickness of the sandwich plate, higher-order shear deformation theory appears more appropriate and shows considerable deviation from first-order shear deformation analysis. Ha et al. [21] used a three-dimensional composite brick element to analyze the static and dynamic response of a laminated plate with distributed piezoceramic actuators. Even though such an analysis can increase the computational involvement enormously, it provides the flexibility to analyze generic plate configurations, including thick plates with surface-bonded or embedded-patched actuators.

Most current plate analyses assume a perfect-bond condition between actuator and bond surface (i.e., neglect shearing effect of adhesive). This assumption is too restrictive and therefore requires a careful assessment, especially for discrete actuators. Furthermore, simple plate theories such as CLPT are routinely used to analyze plate structures. It should be important to examine its limits for different plate configurations and actuation fields with the help of either higher-order shear-deformation theories (e.g., LWSDT) or detailed FEMs (e.g., three-dimensional solid elements). There have been limited studies to validate predictions using experimental test data for coupled composite plates with surface-bonded or embedded piezoelectric elements. These studies should be expanded to cover a range of plate configurations, including strongly coupled bending-torsion coupled plates.

BIBLIOGRAPHY

[1] J. N. Reddy. A simple higher-order theory for laminated composite plates. *Journal of Applied Mechanics, Transactions of ASME*, 51(4):745–752, 1984.

[2] A. Nosier, R. K. Kapania, and J. N. Reddy. Free vibration analysis of laminated plates using a layerwise theory. *AIAA Journal*, 31(12):2335–2346, December 1993.

[3] J. N. Reddy. A generalization of two-dimensional theories of laminated composite plates. *Communications in Applied Numerical Methods*, 3(3):173–180, 1987.

[4] C. T. Sun and J. M. Whitney. Theories for the dynamic response of laminated plates. *AIAA Journal*, 11(2):178–183, February 1973.

[5] D. H. Robbins and I. Chopra. The effect of laminate kinematic assumptions on the global response of actuated plates. *Journal of Intelligent Material Systems and Structures*, 17(4): 273–299, 2006.

[6] J. N. Reddy. On the generalization of displacement-based laminate theories. *Applied Mechanics Reviews*, 42(11):S213–S222, 1989.

[7] A. Chattopadhyay, J. Li, and H. Gu. Coupled thermo-piezoelectric-mechanical model for smart composite laminate. *AIAA Journal*, 37(12):1633–1638, December 1999.

[8] X. Zhou, A. Chattopadhyay, and H. Gu. Dynamic response of smart composites using a coupled thermo-piezoelectric-mechanical model. *AIAA Journal*, 38(10):1939–1948, October 2000.

[9] D. H. Robbins and I. Chopra. Quantifying the local kinematic effect in actuated plates via strain-energy distribution. *Journal of Intelligent Material Systems and Structures*, 18(6):569–589, 2007.

[10] E. F. Crawley and K. B. Lazarus. Induced-strain actuation of isotropic and anisotropic plates. *AIAA Journal*, 29(6):944–951, June 1991.

[11] C. K. Lee. In piezoelectric laminates: Theory and experiments for distributed sensors and actuators, pages 75–168. *Intelligent Structural Systems*, edited by H. S. Tzou and G. L. Anderson, Kluwer Academic Publishers, 1992.

[12] C. K. Lee. Theory of laminated piezoelectric plates for the design of distributed sensors/actuators: Part I: Governing equations and reciprocal relationships. *Journal of the Acoustical Society of America*, 87(3):1144–1158, 1990.

[13] B. T. Wang and C. A. Rogers. Laminate plate theory for spatially distributed induced-strain actuators. *Journal of Composite Materials*, 25(4):433–452, April 1991.

[14] C. C. Lin, C. Y. Hsu, and H. N. Huang. Finite element analysis on deflection control of plates with piezoelectric actuators. *Composite Structures*, 35(4):423–433, 1996.

[15] C. H. Hong and I. Chopra. Modeling and validation of induced-strain actuation of composite coupled plates. *AIAA Journal*, 37(3):372–377, March 1999.

[16] P. Heyliger. Exact solutions for simply-supported laminated piezoelectric plates. *Journal of Applied Mechanics*, 64(2):299–306, 1997.

[17] E. Carrera. An improved Reissner-Mindlin type model for the electromechanical analysis of multi-layered plates including piezo-layers. *Journal of Intelligent Material Systems and Structures*, 8(3):232–248, March 1997.

[18] J. A. Mitchell and J. N. Reddy. A refined hybrid plate theory for composite laminates of piezoelectric laminae. *International Journal of Solids and Structures*, 32(16):2345–2367, August 1995.

[19] D. H. Robbins and J. N. Reddy. An efficient computational model for the stress analysis of smart plate structures. *Smart Materials and Structures*, 5(3):353–360, 1996.

[20] R. C. Batra and S. Vidoli. Higher-order piezoelectric plate theory derived from a three-dimensional variational principle. *AIAA Journal*, 40(1):91–104, January 2002.

[21] S. K. Ha, C. Keilers, and F. K. Chang. Finite element analysis of composite structures containing distributed piezoelectric sensors and actuators. *AIAA Journal*, 30(3):772–780, March 1992.

[22] S. V. Gopinathan, V. V. Varadan, and V. K. Varadan. A review and critique of theories for piezoelectric laminates. *Smart Materials and Structures*, 9(1):24–48, February 2000.

[23] Y. Y. Yu. Some recent advances in linear and nonlinear dynamical modeling of elastic and piezoelectric plates. *Journal of Intelligent Material Systems and Structures*, 6(2): 237–254, March 1995.

[24] P. Bisegna and G. Carusa. Mindlin-type finite elements for piezoelectric sandwich plates. *Journal of Intelligent Material Systems and Structures*, 11(1):14–25, January 2000.

[25] P. Bisegna and F. Maceri. A consistent theory of thin piezoelectric plates. *Journal of Intelligent Material Systems and Structures*, 7(4):372–389, July 1996.

[26] S. Kapuria and G. G. S. Achary. Electromechanically coupled zigzag third-order theory for thermally loaded hybrid piezoelectric plates. *AIAA Journal*, 44(1):160–170, 2006.

[27] W. Yu and D. H. Hodges. A simple thermopiezoelastic model for smart composite plates with accurate stress recovery. *Smart Materials and Structures*, 13(4):926–938, August 2004.

[28] L. Edery-Azulay and H. Abramovich. A reliable plain solution for rectangular plates with piezoceramic patches. *Journal of Intelligent Material Systems and Structures*, 18(5): 419–433, May 2007.

[29] D. H. Robbins and J. N. Reddy. Modelling of thick composites using a layerwise laminate theory. *International Journal for Numerical Methods in Engineering*, 36(4):655–677, 1993.

[30] P. Heyliger, G. Ramirez, and D. A. Saravanos. Coupled discrete-layer finite elements for laminated piezoelectric plates. *Communications in Numerical Methods in Engineering*, 10(12):971–981, 1994.

[31] D. A. Saravanos, P. R. Heyliger, and D. A. Hopkins. Layerwise mechanics and finite element for the dynamic analysis of piezoelectric composite plates. *International Journal of Solids and Structures*, 34(3):359–378, 1997.

[32] S. S. Vel and R. C. Batra. Three-dimensional analytical solution for hybrid multilayered piezoelectric plates. *Journal of Applied Mechanics, Transactions of the ASME*, 67(3): 558–567, September 2000.

[33] J. S. Yang and R. C. Batra. Mixed variational principles in nonlinear piezoelectricity. *International Journal of Nonlinear Mechanics*, 30(5):719–726, 1995.

[34] S. A. Kulkarni and K. M. Bajoria. Large deformation analysis of piezolaminated smart structures using higher-order shear-deformation theory. *Smart Materials and Structures*, 16(5):1506–1516, 2007.

6 Magnetostrictives and Electrostrictives

Magnetostrictives and electrostrictives are active materials that exhibit magneto-mechanical and electromechanical coupling, respectively. These materials undergo a change in dimensions in response to an applied magnetic or electric field. A common property of both materials is that the induced strain depends only on the magnitude of the applied field and is independent of its polarity. In other words, it can be said that the induced strain has a quadratic dependence on the applied field. It is this behavior that differentiates electrostriction from the piezoelectric effect, which is also caused by an electric field. This chapter discusses the basic mechanisms behind magnetostriction and electrostriction, and it describes how these materials are used to construct practical actuators and sensors. The behavior of magnetic shape memory alloys (SMAs) is also described.

6.1 Magnetostriction

A ferromagnetic material placed in a magnetic field generally undergoes a change in shape [1]. The internal structure of a ferromagnetic material consists of randomly oriented magnetic domains. When a magnetic field is applied, the domains rotate to align themselves along the field, causing a change in the material dimensions. This phenomenon is known as "magnetostriction." The effect is small in most materials but is measurable (on the order of microstrain) in ferromagnetic materials. Some materials, such as Terfenol-D, exhibit magnetostrictive strains on the order of 2000 microstrain (2000×10^{-6}). Such materials can be used as both solid-state actuators and magnetic-field sensors. Magnetostrictive materials are available in the form of rods, thin films, and powder. The material is usually supplied by manufacturers ready to assemble into devices, without the need for any processing; however, some manufacturers also provide complete actuator assemblies, including the active material, magnetic field generators, and housing. Note that because the material is brittle, any machining operations such as threading, drilling, soldering, and welding should be avoided. Magnetostrictive materials are now being used for a wide range of applications that include active vibration and noise control systems, machine tools, servo-valves, hybrid motors, sonar devices and tomography, automotive brake systems, micro-positioners, particulate-actuators and sensors, ultrasonic cleaning, machining, welding, micropositioning and sensors [2, 3, 4].

Table 6.1. *Maximum magnetic field induced strain*

Material	Magnetostriction ($\times 10^{-6}$)
Iron	20
Nickel	−40
Cobalt	−60
Alfenol 13	40
NiCo	186
Galfenol	300
$TbFe_2$	1750
Terfenol-D	2000
$SmFe_2$	−1560

James Prescott Joule first discovered the magnetostrictive effect in nickel in 1842. Later, cobalt, iron, and their alloys were shown to have significant magnetostrictive effects similar to those of nickel. The maximum strains were on the order of 50 ppm (parts per million, 0.005%). Table 6.1 shows the magnetostriction of different materials. Note that nickel has a negative magnetostrictive constant, which means that a decrease in dimension occurs in the presence of a magnetic field along that dimension. The early applications of magnetostriction, using nickel and other magnetostrictive materials, date back to the first half of the twentieth century. These applications include telephone recievers, hydrophones, oscillators, torquemeters, and sonar devices. These early applications were limited because of the low saturation strains of the materials (less than 100 ppm). The discovery of Terfenol-D, with its large magnetostriction, expanded the range of applications.

Magnetostrictive nickel-based alloys (magnetostriction $\approxeq$50 ppm) were employed in building transducers for sonar devices applications in World War II. In the early 1960s, there was a discovery of "giant magnetostriction" in the rare earth elements Terbium and Dysprosium. Even though one could obtain large induced strain (1000 microstrain), it could be achieved only at cryogenic temperatures. Because of this temperature requirement, this discovery had limited applications. In the early 1970s, researchers at the Naval Ordnance Laboratories (NOL), later known as the Naval Surface Warfare Center (NSWC), began developing giant magnetostrictive alloys at room temperature with the lanthanide elements. One such alloy was Terfenol-D, developed by Arthur Clark [5] and his coworkers. Terfenol-D is an intermetallic alloy of Terbium, Dysprosium, and Iron, ($Tb_xDy_{1-x}Fe_y$) that is produced as a near-single crystal. The value of x varies from 0.27 to 0.3 and y varies from 1.92 to 2.0. Small changes in x and y can have a major influence on the alloy's magnetic, magnetostrictive, and elastic properties. For example, a small decrease in y below 2.0 reduces brittleness significantly but also decreases the maximum strain capability. Increasing x above 0.27 improves magnetostriction at lower fields and results in more efficient energy transduction. These findings demonstrate that modifications in the stoichiometry of Terfenol-D can have a significant influence on its properties. The material characteristics of Terfenol-D are nonlinear functions of mechanical, magnetic, and thermal operating conditions. For example, the Young's modulus changes with applied stress.

Butler et al. [6, 7] provided an introduction to the magnetostrictive materials and especially to ETREMA's Terfenol-D (*Ter* for Terbium, *Fe* for Ferrous, *NOL* for

Naval Ordnance Laboratory, and *D* for Dysprosium). The maximum strain produced by Terfenol-D (on the order of 2000×10^{-6} or 0.2% in a magnetic field of 10 kA/m or 2kOe) is almost twice the maximum strain produced by piezoceramics. The material coupling factor, k^2 (the ability to convert magnetic energy into mechanical energy) of Terfenol-D, is of the same level ($\approxeq$50%) as that of piezoceramics. However, the magnetic permeability and mechanical stiffness of Terfenol-D are generally low. For example, the Young's modulus of Terfenol-D is about half that of a typical piezoceramic. Terfenol-D is available in a variety of forms that include thin films, powder material, and monolithic rods, which is the most common. In 1978, Clark and coworkers developed a new magnetostrictive material based on amorphous metal, called metglas (metallic glass), in the form of thin ribbons. This material has an extremely high coupling factor ($k^2 = 0.85$), which makes it a prime candidate for sensor applications.

The manufacturing of Terfenol-D is carried out by melting the material and then casting and directionally solidifying it to provide the unidirectional crystalline microstructure needed to produce large strains. Two common manufacturing techniques are the free-standing zone melt (FSZM) and the modified Bridgman (MB) methods. These methods are known as directional-solidification methods and are described in detail along with other methods of production in Refs. [8, 9, 10, 11, and 12]. Today, advanced crystalline magnetostrictive materials are also being manufactured using crystal growth techniques to obtain directional solidificiation along the longitudinal axis of the rod, including precision laminations. The crystal growth process requires a high degree of purity of terbium, dysprosium, and iron. To improve material characteristics, heat treatment and magnetic annealing are used.

The main drawback of Terfenol-D is its low tensile strength and extremely brittle nature. These limitations make it difficult to design actuators with complex shapes, optimized for specific applications. The design space available to magnetostrictive materials has recently been expanded by the development of a new class of magnetostrictive alloys called iron-gallium alloys (or Galfenol) by researchers at the NSWC [13]. These alloys exhibit moderate magnetostriction (350×10^{-6}) under very low magnetic fields ($\approxeq$100 Oe), have very low hysteresis, and demonstrate high tensile strength ($\approxeq$500 MPa) and limited variation in magneto-mechanical properties for temperatures between −20°C and 80°C [14, 15, 16]. In addition, Galfenol is highly ductile, machinable, and weldable. The behavior of different alloy compositions under a variety of operating conditions is currently under investigation.

As a result of the magnetostrictive, or Joule, effect, an application of a magnetic field results in a longitudinal extensional strain accompanied by a transverse compressive strain with a negligible change in net volume. A converse effect also exists, which is a change in the magnetization of the material in response to deformations. This is called the "Villari effect." The Joule effect is used in actuators, whereas the Villari effect is used in sensors. The Joule effect transforms magnetic energy into mechanical energy, whereas the Villari effect transforms mechanical energy into magnetic energy. Using a helical magnetic field around the magnetostrictive material, a twisting action can be produced that is called the "Wiedemann effect." The inverse effect, in which application of torque results in a change of magnetization, is called the "Matteusi effect." Due to the bidirectional exchange of energy, magnetostrictive materials can be used as both actuators and sensors. Above the Curie temperature (specific material characteristic), the materials lose their magnetostrictive property and become paramagnetic.

For an actuator, an electrical coil, usually in the shape of a solenoid, is used to convert electrical energy into magnetic energy, and a Terfenol-D rod is used to convert the magnetic energy into mechanical energy. For a sensor, the strain in the sensing element (Terfenol-D) changes its magnetization as well as the magnetic energy in the solenoid. Thus, the sensor converts mechanical energy into magnetic energy, which can be measured using either a Hall probe or a sensing coil. In this way, magnetostrictive materials can deform due to induced strain in a magnetic field (actuation mode) or change their magnetiziation state when mechanically deformed (sensing mode). Also, magnetostrictive materials change their stiffness under an external magnetic field, often called the "ΔE" effect. For example, the Young's modulus of Terfenol-D becomes higher under a DC magnetic field than when under no magnetic field. The stiffness of magnetostrictive materials also depends on whether these materials are operated under mechanically free conditions (zero external load), mechanically clamped conditions (zero strain), or a combination of the two. The material is typically stiffer when mechanically clamped than when allowed to strain freely. At magnetic saturation, an intrinsic or uncoupled stiffness is achieved.

The design of the magnetic circuit is crucial to obtaining good performance in terms of uniformity of the magnetic field, maximum field intensity, and so on. In addition, the weight of the magnetic field generator, which includes the coils of the electrical conductor and the magnetic flux paths, is often the largest fraction of the total weight of the actuator. A good design of the magnetic field generator can therefore significantly increase the overall power efficiency of the system, in terms of weight as well as volume. For example, to minimize eddy current losses, laminated magnetostrictive cores and slit permanent magnets are used. For most applications, the magnetostrictive material is a monolithic, grain-oriented Terfenol-D ($Tb_{0.3}Dy_{0.7}Fe_{1.92}$) rod, which is manufactured such that a large number of magnetic moments are oriented normal to the longitudinal axis of the rod. A compressive bias stress further improves the alignment of magnetic moments, as well as minimizes the tensile stresses that are applied to the brittle Terfenol-D rod. For a zero-bias magnetic field, the oscillatory response of the rod takes place at twice the excitation frequency (frequency of the magnetic field). To achieve a bidirectional dynamic response of the Terfenol-D rod, a DC magnetic bias is applied by including a permanent magnet in the circuit or by applying a DC current in the magnetic coil. In such a case, the output response occurs at the same frequency as the excitation field.

6.2 Review of Basic Concepts in Magnetism

The phenomenon of magnetism has been well documented and remains one of the cornerstones of modern science. As such, it is not possible to provide a comprehensive background of magnetism without filling several volumes. Magnetic quantities are expressed in several different systems of units and can often be confusing. However, a one-to-one correspondence exists between electrical and magnetic quantities, and the behavior of electrical and magnetic circuits are analogous. Although detailed discussions of electromagnetism can be found in standard textbooks, a brief review of some basic definitions and concepts in magnetism is useful before discussing the

Figure 6.1. Magnetic field induced by a current element.

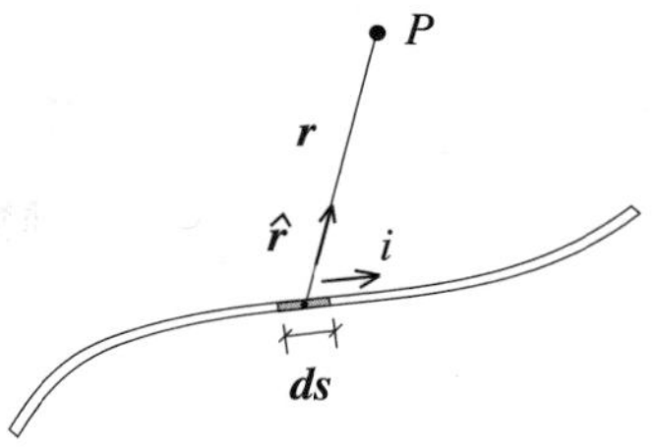

magnetostrictive effect and methods for actuation of magnetostrictive and magnetic SMA elements.

6.2.1 Magnetic Field *B* and the Biot-Savart Law

A basic quantity used in the discussion of magnetism is the magnetic field, $\boldsymbol{B}$. This is a vector quantity and is also referred to as the magnetic induction or the magnetic flux density. In the International System of Units (SI) system, the unit of $\boldsymbol{B}$ is the Tesla (T), or N/(A.m). The field required to exert a force of 1 N on a charge of 1 Coulomb (C) moving at a velocity of 1 m/sec is defined as 1 Tesla. A physical feel of the magnitude of a 1-Tesla field can be obtained by noting that the magnetic field of the Earth near its surface is approximately 0.5×10^{-4} T [17]. Permanent magnets for laboratory use are commonly available with fields of up to 2.5 T.

Fundamentally, magnetic fields are generated as a result of the motion of electrical charges (discovered by Hans Christian Oersted in 1819). Even in the case of permanent magnets, the origin of the magnetic field can be traced to the motion of electrons within the material. In practice, a magnetic field can be produced by a current carrying coil. The magnitude and direction of the magnetic field can be conveniently controlled by the magnitude and direction of the applied current. At a point P, a conductor element of length $\boldsymbol{ds}$ carrying a steady current vector i (Amperes) generates a magnetic field in free space given by the Biot-Savart Law (Fig. 6.1)

$$d\boldsymbol{B} = \frac{\mu_o}{4\pi} \frac{i\, \boldsymbol{ds} \times \hat{\boldsymbol{r}}}{r^2} \tag{6.1}$$

where r is the magnitude of the distance of the point P from the elemental conductor, and $\hat{\boldsymbol{r}}$ is the unit vector pointing from the element to P. Note that the symbols in boldface, such as the magnetic field $d\boldsymbol{B}$, are vector quantities. The Biot-Savart law is a fundamental relation of electromagnetism and can be used to calculate the magnetic field around a current carrying conductor of any given geometry. Numerical methods are often used to obtain the solution for complex geometries, including nonlinear effects [1].

The constant μ_o is called the "permeability of free space" and is given by

$$\mu_o = 4\pi \times 10^{-7} \text{ T.m/A or N/A}^2 \text{ or H/m} \tag{6.2}$$

where symbol H represents Henry, and it is the SI-derived unit.

Note that as a result of the vector cross product, the magnetic field $d\boldsymbol{B}$ lies in a plane perpendicular to the elemental conductor $\boldsymbol{ds}$. Along the elemental conductor, $d\boldsymbol{B}$ becomes zero. The magnetic field is maximum in a plane perpendicular to the

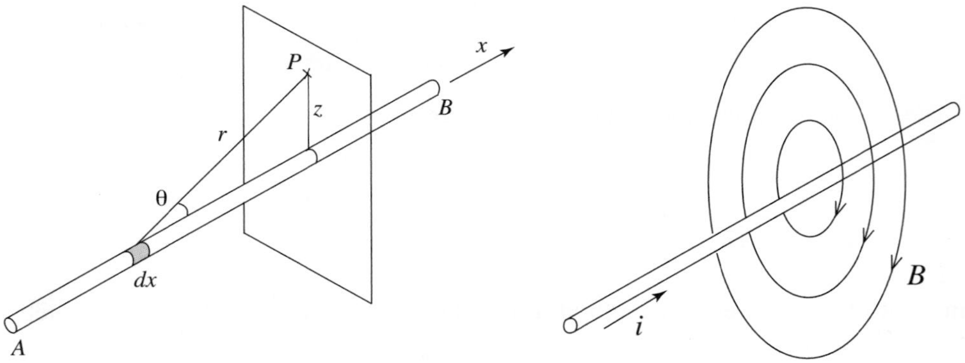

(a) Application of Biot-Savart law (b) Direction of magnetic field

Figure 6.2. Magnetic field due to a straight current carrying conductor.

elemental conductor and passing through the element. Closed-form solutions are possible for a select few simple cases. For more general configurations, a numerical approach is necessary.

6.2.2 Current Carrying Conductors

Let us examine the magnetic field produced by current carrying conductors of three commonly used geometries. This provides useful insight into the design of magnetic field generation circuits, which are extremely important in the construction of actuators and sensors. First, we consider a finite straight conductor; second, a single circular coil; and third, a solenoid. Of these, the solenoid is one of the easiest and most widely used methods to obtain a uniform magnetic field.

Finite Straight Conductor

Consider a straight conductor of finite length lying along the x-axis, carrying a current i (Fig. 6.2(a)). Lines of constant magnetic field are given by concentric circles centered on the conductor, lying in a plane perpendicular to the axis of the conductor, and their direction is determined by the right-hand rule (Fig. 6.2(b)). The magnetic field at the point P can be found by applying the Biot-Savart law to elemental lengths of the current carrying conductor and integrating along the length of the conductor between the ends A and B (Fig. 6.3). The magnetic field due to the elemental length of conductor $\boldsymbol{ds}$ is given by

$$d\boldsymbol{B} = \frac{\mu_o}{4\pi}\frac{i\,\boldsymbol{ds} \times \hat{\boldsymbol{r}}}{r^2} \tag{6.3}$$

From the figure

$$-s = z\cot\theta \tag{6.4}$$

$$ds = \frac{z}{\sin^2\theta}d\theta \tag{6.5}$$

Figure 6.3. Biot-Savart law applied to a finite straight conductor.

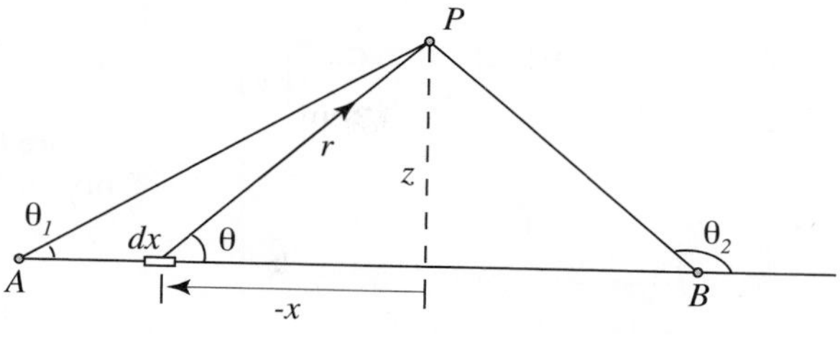

Substituting in Eq. 6.3 and integrating along the length yields the magnetic field at point P

$$B = \frac{\mu_o i}{4\pi z}\int_{\theta_1}^{\theta_2} \sin\theta d\theta$$
$$= \frac{\mu_o i}{4\pi z}(\cos\theta_1 - \cos\theta_2) \qquad (6.6)$$

From this expression, the magnetic field at point P due to an infinitely long current carrying conductor can be found by setting $\theta_1 = 0$ and $\theta_2 = \pi$, yielding

$$B = \frac{\mu_o i}{2\pi z} \qquad (6.7)$$

Circular Coil

Let us consider a single-turn circular coil of radius R, carrying a current i, and lying in the x-z plane (Fig. 6.4). At a point P on the y-axis of the coil, at a distance l from the center, the magnetic field is given by

$$d\mathbf{B} = \frac{\mu_o}{4\pi}\frac{i\,\mathbf{ds} \times \hat{\mathbf{r}}}{r^2} \qquad (6.8)$$

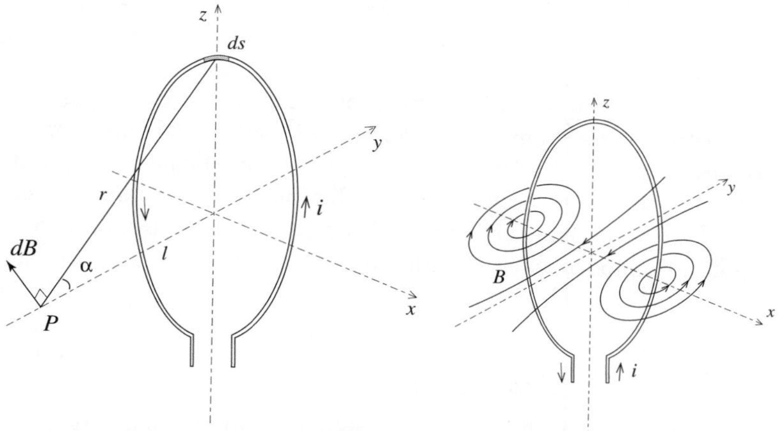

(a) Application of Biot-Savart law (b) Direction of magnetic field

Figure 6.4. Magnetic field due to a current carrying loop.

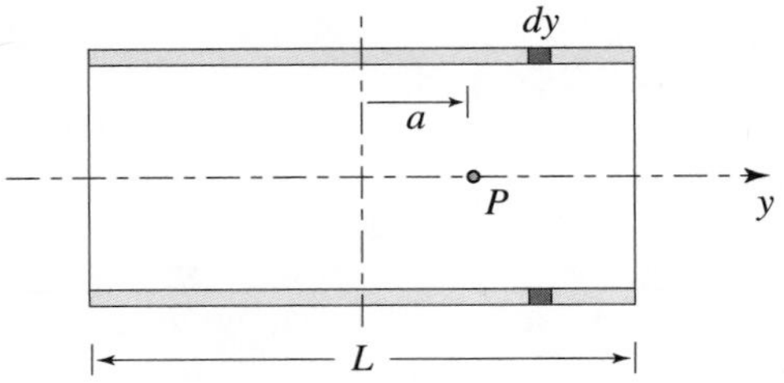

Figure 6.5. Calculation of field in cross section of current carrying solenoid.

The magnetic field along the y-axis becomes

$$B_y = \int_0^{2\pi R} dB \sin\alpha \tag{6.9}$$

$$= \int_0^{2\pi R} \frac{\mu_o}{4\pi} i \frac{ds}{r^2} \sin\alpha \tag{6.10}$$

$$= \frac{\mu_o}{4\pi} i \frac{\sin\alpha}{r^2} \int_0^{2\pi R} ds \tag{6.11}$$

$$= \frac{\mu_o}{2} \frac{iR}{r^2} \sin\alpha \tag{6.12}$$

$$= \frac{\mu_o}{2} \frac{iR^2}{(l^2 + R^2)^{3/2}} \tag{6.13}$$

At the center of the coil, where $\alpha = 90^\circ$ (or $l = 0$), the magnetic field becomes

$$B = \frac{\mu_o}{2} \frac{i}{R} \tag{6.14}$$

Note that the lines of constant magnetic field are almost parallel close to the center of the loop (see Fig. 6.4(b)). Therefore, by stacking together a large number of current carrying loops, a uniform magnetic field can be obtained along their central axis. Such an arrangement of current carrying loops is called a "solenoid."

Solenoid

A solenoid is typically built by winding a large number of helical turns of insulated wire around a straight central axis. Consider a solenoid of length L and diameter D, with N turns of wire and carrying a current i (Fig. 6.5). The magnetic field at a point P, at a distance a from the center of the solenoid, can be found in a similar manner as for the case of a single circular coil. The solenoid can be treated as a summation of individual current carrying loops, each of width dy. The number of turns n_y in an element dy is

$$n_y = \frac{N}{L} dy \tag{6.15}$$

where each turn carries the current i. The magnetic field at point P is given by

$$dB = \frac{\mu_o}{2} \frac{iR^2}{\left[(y-a)^2 + R^2\right]^{3/2}} \frac{N}{L} dy \tag{6.16}$$

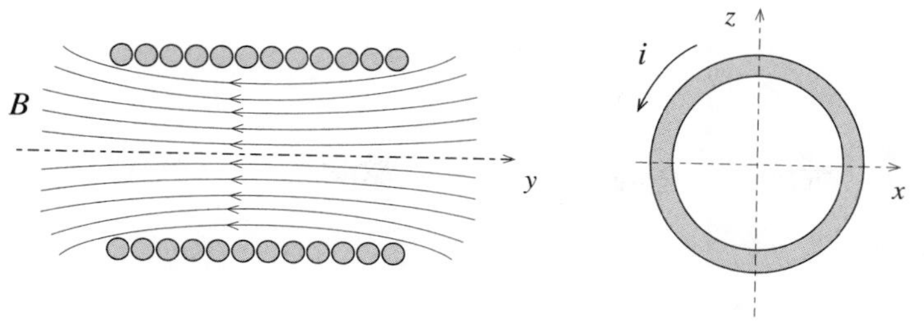

Figure 6.6. Magnetic field due to a current carrying solenoid.

The net effect due to all of the elements is

$$B = \int_{-L/2}^{L/2} \frac{\mu_o}{2} \frac{iR^2}{[(y-a)^2 + R^2]^{3/2}} \frac{N}{L} dy$$
$$= \frac{\mu_o N i}{L} \left[\frac{L+2a}{2\,[D^2 + (L+2a)^2]^{1/2}} + \frac{L-2a}{2\,[D^2 + (L-2a)^2]^{1/2}} \right] \tag{6.17}$$

At the center ($a = 0$) of a long solenoid ($L \gg D$ or $L \to \infty$), this reduces to

$$B = \frac{\mu_o N i}{L} = \mu_o n i \tag{6.18}$$

where n is the number of turns per unit length. Note that the magnetic field is independent of the radius of the solenoid. This relation is true for a long thin solenoid. For a thick solenoid of finite length, the magnetic field distribution is more involved and is additionally a function of the internal and external radii of the solenoid. One approach is to directly solve the Biot-Savart relation with imposed constraints. The other approach is to use the superposition of analytical solutions for current carrying loops; the solution at any point in space is then the vector sum of the contributions from each loop.

As the length of the solenoid increases, the magnetic field near the center of the solenoid becomes more uniform (Fig. 6.6). Usually, it is assumed that the field is uniform inside the solenoid over a large part of its length; however, this assumption breaks down near the ends of the solenoid. An empirical factor, sometimes called the "fringing factor," can be used to quantify the fraction of the solenoid length over which a uniform field exists. As a rule of thumb, a fringing factor of 10% is adequate for most solenoid applications. Physically, this means that the magnetic field within the solenoid can be assumed constant except within a distance of 10% of the total solenoid length from the edges. Sometimes to increase the magnitude of the magnetic field, the coil is wound around a core of high permeability. In such a case, if the coil is wound around a core material of permeability μ_c, the quantity μ_o in Eq. 6.18 is replaced by μ_c. The magnetic field generated by a permanent bar magnet emerges from its north pole (field source) and ends on its south pole (field sink) (Fig. 6.7). This field pattern is similar to that induced by a solenoid (external to the solenoid).

An important property of the solenoid is its inductance L_s. For a solenoid of length L, cross-sectional area A_x, with N turns wound around a core of permeability μ_c, the inductance can be derived as

$$L_s = \mu_c N^2 A_x / L \qquad \text{(H)} \tag{6.19}$$

where the inductance is expressed in terms of Henry (H).

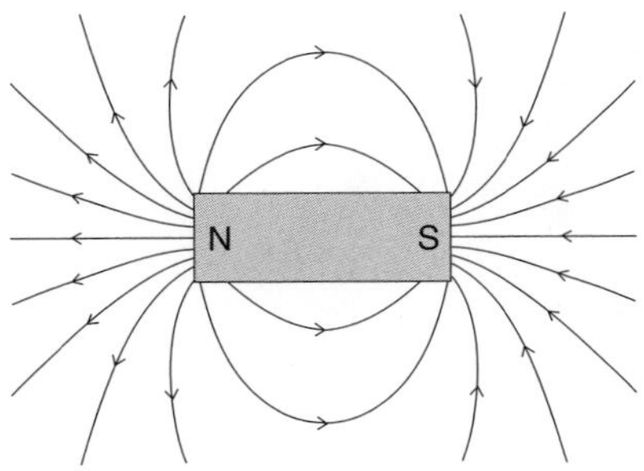

Figure 6.7. Lines of force produced by a bar magnet.

6.2.3 Magnetic Flux Φ and Magnetic Field Intensity *H*

The "magnetic flux," Φ, is defined as

$$\Phi = \int \boldsymbol{B}.d\boldsymbol{A} \tag{6.20}$$

with units T.m^2 (Nm/A). This quantity gives an idea of the total magnetic field in a given region.

From Eq. 6.3, it can be seen that the magnetic field also depends on the permeability of the material surrounding the current carrying conductor. However, it is possible to define a quantity called the "magnetic field intensity," $\boldsymbol{H}$, that is independent of the material surrounding the current carrying coil and depends linearly on the current alone. For this reason, it is often convenient to express the strength of the magnetic field in terms of $\boldsymbol{H}$ rather than $\boldsymbol{B}$.

The magnetic field is related to the magnetic field intensity through the permeability of the material, μ. The magnetic permeability μ has the units of Henry/meter (H/m) or Tesla meter/Ampere (Tm/A). In centimeter-gram-second (CGS) units, it is expressed as Gauss/Oersted (G/Oe). In general, the permeability is a function of field intensity, stress level, temperature, and magnetic history. A highly permeable material is one in which a large magnetic flux is induced.

$$\boldsymbol{B} = \mu \boldsymbol{H} \tag{6.21}$$

Consequently, the units of $\boldsymbol{H}$ are A/m. It can also be defined in the following way: a straight conductor in free space, of infinite length, carrying a current of 1A, generates a tangential magnetic field intensity of $1/2\pi$ A/m at a distance of 1m. Notice that there are several different types of nomenclature for the quantities $\boldsymbol{B}$ and $\boldsymbol{H}$.

In practical applications, magnetic field is generated by means of a specific configuration of coils carrying current. The shape of the coils is dictated by the geometry of the required field and is normally designed to obtain a uniform field over a region of interest. Two common coil configurations are the solenoid and the toroid. The field intensity inside a solenoid having turns/m, carrying a current of i amps is given by (from Eqs. 6.18 and 6.21)

$$H = n\, i \tag{6.22}$$

From this equation, it can be seen that the units of H can also be expressed as A.turns/m. Note that one A.turn/m is equal to $4\pi \times 10^{-3}$ Oe. The field intensity inside a toroid having turns/m, carrying a current i, and with a radius r is

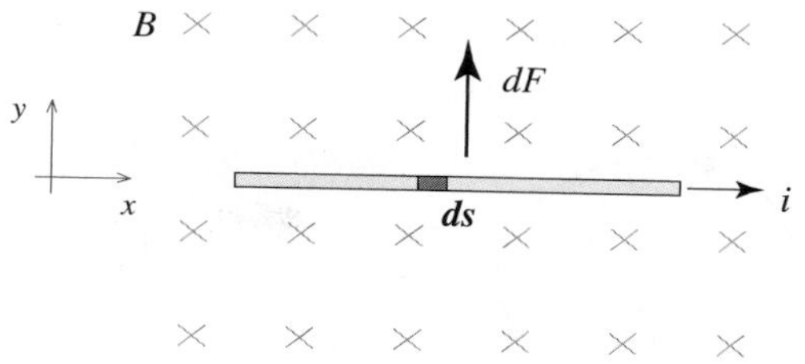

Figure 6.8. Force on a current carrying conductor.

given by

$$H = \frac{n\,i}{2\pi r} \quad \text{A/m} \tag{6.23}$$

6.2.4 Interaction of a Current Carrying Conductor and a Magnetic Field

The effect of the quantity $\boldsymbol{B}$ can be understood from the Lorentz force law. This gives the force $\boldsymbol{F}$ on a charge q, moving at a velocity $\boldsymbol{v}$, in a magnetic field $\boldsymbol{B}$ as

$$\boldsymbol{F} = q\boldsymbol{v} \times \boldsymbol{B} \tag{6.24}$$

Note that the direction of the induced force is perpendicular to both the magnetic field as well as the velocity of the charge. From Eq. 6.24, we can see that a conductor of length $\boldsymbol{ds}$, carrying a current $\boldsymbol{i}$ in a magnetic field $\boldsymbol{B}$ experiences a force $d\boldsymbol{F}$, given by

$$d\boldsymbol{F} = i\,\boldsymbol{ds} \times \boldsymbol{B} \tag{6.25}$$

This effect is shown in Fig. 6.8. A conductor of length $\boldsymbol{ds}$, carrying a current i, is placed in a uniform magnetic field $\boldsymbol{B}$. The conductor is along the x axis, and the magnetic field is directed along the negative z-axis (shown by the "×" marks). The resultant force F on the conductor is along the positive y-axis (into the page).

From Eq. 6.25, a magnetic field of 1 Tesla can be defined as that in which one Coulomb of charge experiences a force of 1 Newton, when it is moving normal to the magnetic field at a velocity of 1 meter per second. Because 1 Ampere is defined as 1 Coulomb/second

$$1T = 1\frac{\text{N}}{\text{C-m/s}} = 1\frac{\text{N}}{\text{A-m}} \tag{6.26}$$

A loop of current in an external magnetic field experiences a net torque but no net force. The torque on a loop enclosing an area $\boldsymbol{A}_x$ and carrying a current i, in a magnetic field $\boldsymbol{B}$, is given by

$$\boldsymbol{\tau} = i\,\boldsymbol{A}_x \times \boldsymbol{B} \tag{6.27}$$

where the direction of the area vector $\boldsymbol{A}_x$ is given by the right-hand rule applied to the current carrying loop. The quantity $i\boldsymbol{A}_x$ is defined as the magnetic moment vector $\mathfrak{M}$ and is in the same direction as the area vector of the loop (Fig. 6.9). From the definition, it can be seen that the units of magnetic moment are A.m^2.

The concept of magnetic moment is useful to calculate the forces acting on magnetic elements, and it is applicable to both current loops and permanent magnets. The magnetic moment $\mathfrak{M}$ of a bar magnet of length l with a flux Φ at its center

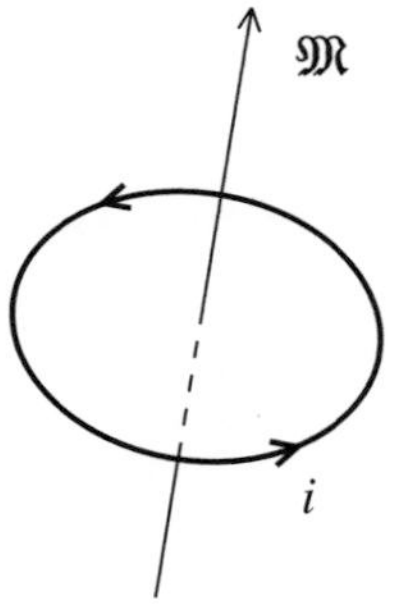

Figure 6.9. Magnetic moment of a current carrying loop.

is given by

$$\mathfrak{M} = \Phi \frac{l}{\mu_o} \tag{6.28}$$

Just as a current carrying conductor produces a magnetic field, a changing magnetic field induces a voltage in a conductor. The magnitude of this voltage is given by Faraday's law. This law states that the voltage induced V_i, in a coil of N turns, is related to the magnetic flux Φ by

$$V_i = -N \frac{d\Phi}{dt} \tag{6.29}$$

The negative sign is due to the law of conservation of energy, which states that the effect of the voltage produced is to oppose the change in magnetic field. This statement is also known as Lenz's law. The generation of a voltage in response to a changing magnetic field is often used to measure the magnetic flux and, consequently, to measure stress or strain using a magnetostrictive material as a sensor.

6.2.5 Magnetization M, Permeability μ, and the $B-H$ Curve

The magnetic field inside a given material is often treated as originating from a collection of small current loops or, equivalently, a collection of magnetic moments. Any material has a large number of randomly oriented magnetic moments on the atomic level. In the absence of an external magnetic field, the random orientation of the magnetic moments in the material leads to a net zero magnetic moment, as shown in Fig. 10(a). When the material is placed in an external magnetic field, the magnetic moments in the material reorient themselves preferentially along the external magnetic field, resulting in a net internal magnetic field (Fig. 10(b)). There is a magnetic phase transition from a disordered paramagnetic state to an ordered ferromagnetic state. The material in this latter state is said to be magnetized. The transition to ferromagnetism is accompanied by a change in shape, referred to as "magnetostriction." A region in which the magnetic moments are oriented in the

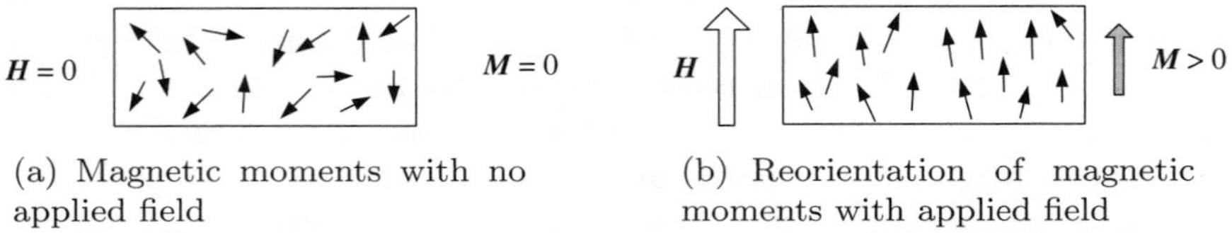

(a) Magnetic moments with no applied field

(b) Reorientation of magnetic moments with applied field

Figure 6.10. Effect of external magnetic field on a ferromagnetic material.

same direction is called a "magnetic domain." The net magnetic field in the material is the vector sum of the externally applied magnetic field and the internal magnetic field. The internal field can increase the net field, as in the case of ferromagnets, or decrease the net field, as in the case of diamagnets. The magnetic state of the material can be quantified in terms of a magnetization vector, $\boldsymbol{M}$. This is defined as the magnetic moment per unit volume of the material. For a volume of material ΔV with a net magnetic moment $\Delta \mathfrak{M}$

$$\boldsymbol{M} = \frac{\Delta \mathfrak{M}}{\Delta V} \tag{6.30}$$

The units of $\boldsymbol{M}$ are A/m. The total magnetic field in a material placed in an external magnetic field $\boldsymbol{B}_o$ is given by

$$\boldsymbol{B} = \boldsymbol{B}_o + \mu_o \boldsymbol{M} \tag{6.31}$$

where $\mu_o \boldsymbol{M}$ is the contribution arising from the orientation of magnetic moments inside the material (internal magnetization). Note that this equation is a vector addition because the direction of the net magnetic moment of the material may not be the same as that of the external magnetic field. In linear isotropic materials, the magnetization $\boldsymbol{M}$ is proportional to the magnetic field intensity $\boldsymbol{H}$ by a factor called the "magnetic susceptibility," χ_m. Substituting for the magnetic field from Eq. 6.21

$$\boldsymbol{B} = \mu_o \boldsymbol{H} + \mu_o \boldsymbol{M} \tag{6.32}$$

$$= \mu_o(1 + \chi_m)\boldsymbol{H} = \mu_o \mu_r \boldsymbol{H} \tag{6.33}$$

$$= \mu \boldsymbol{H} \tag{6.34}$$

where $\mu_r = 1 + \chi_m$ is the relative permeability and μ is the permeability of the material. The relative permeability of air is approximately equal to that of free space, $\mu_r \approxeq 1$. The value of μ describes the behavior of the material in response to an applied magnetic field. Based on the values of μ, the material is classified as diamagnetic ($\mu < \mu_o$), paramagnetic ($\mu > \mu_o$), or ferromagnetic ($\mu \gg \mu_o$ by several orders of magnitude). The magnetic susceptibility χ_m is small and negative for diamagnetic materials, and their magnetic response opposes the applied field. Examples are copper, silver, gold, and berylium. For paramagnetics, the value of χ_m is small and positive. Examples are aluminum, platinum, and manganese. For ferromagnetics, the value of χ_m is large and positive. Examples are iron, cobalt, and nickel. Among ferromagnetics, supermalloy (nickel-iron-molybdenum alloy) has a maximum relative permeability on the order of 10^6. This means that a solenoid wrapped around a supermalloy core will induce a magnetic flux 10^6 times that induced in free space.

When a ferromagnetic material is heated beyond a temperature called the Curie temperature, it undergoes a transition to a paramagnetic state. In addition, in the case of ferromagnetic materials, μ is not a constant but rather depends on the field. However, in the case of diamagnetic and paramagnetic materials, μ is constant over a large range of applied field. Such materials are referred to as linear. In a similar way, the susceptibility of the material χ_m may not be a constant.

A ferromagnetic material contains a large number of magnetic domains that are randomly oriented in an unmagnetized sample. The magnetic domains are easily aligned by external fields, resulting in a net magnetization in the material. The magnetization is partly retained even on the removal of the external field, as internal stresses prevent some of the domains from returning to their original orientation.

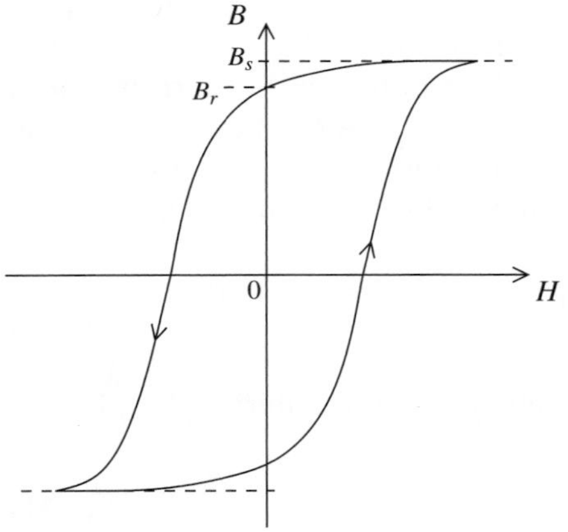

Figure 6.11. Typical magnetic field B versus applied magnetic field H for a ferromagnetic material.

This leads to a magnetic hysteresis in the material. The hysteretic behavior can be shown on a B–H diagram (Fig. 6.11) that describes the behavior of a material when exposed to a cyclically varying magnetic-field intensity. It can be seen that the magnetic field does not go to zero when the applied field intensity is zero. The value of the magnetic field, which persists after the applied field intensity becomes zero, is called remanent field B_r. In addition to the remanent field, it can be seen that the magnetic field saturates (at a value B_s) after a certain value of applied field intensity. At this point, all the magnetic moments in the material are aligned with the applied field intensity, and no further rearrangement is possible. In practice, saturation occurs over the region of the B–H curve, where the magnitude of B begins to "level off" for an increasing field intensity, H. As more flux is forced into the same cross-sectional area of the ferromagnetic material, fewer domains are available within that area to align with the additional field intensity. It is important to note that saturation only occurs in ferromagnetic materials. For pure iron, B_r is about 2 T and B_s is 2.15 T.

Note that at any point, the slope of the B–H curve gives the permeability of the material. Typically, when the variation in H is large and it passes through zero (changes sign), the resulting hysteresis curve is called a major loop.

$$\mu = \frac{\partial B}{\partial H} \tag{6.35}$$

When the variation in H is small and the magnitude of H increases and decreases without changing polarity, the resulting curve is called a "minor loop." The minor loops are completely enclosed within the major loop. It can be seen that the value of μ does not remain constant but instead decreases as the material reaches saturation. Physically, the magnetic permeability can be considered as a measure of the material's acceptance of magnetic flux. For purified iron, the value of μ_r, at a magnetic field of 2 T, is 5000.

An important property of the B–H curve is that the area enclosed by the curve is equal to the work done in one cycle of magnetizing and demagnetizing the material or, equivalently, it is equal to the stored magentic energy per unit volume, V_m. The origin of this hysteretic loss can be attributed to the work done in reorienting the magnetic moments in the material. The work done can be expressed as

$$V_m = \frac{1}{2} \int B dH \tag{6.36}$$

Table 6.2. *Demagnetizing factors for simple geometries, from Ref. [18]*

Geometry	l/d	N_d
Toroid	–	0
Cylinder	∞	0
Cylinder	20	0.006
Cylinder	10	0.017
Cylinder	5	0.040
Cylinder	1	0.27
Sphere	–	0.333

6.2.6 Demagnetization

The magnetic field between two poles can be calculated in terms of the pole strength, the distance between the poles, and the permeability of the material between the poles. The magnetic field is caused only by the presence of the two poles (Fig. 6.12(a)). However, when a ferromagnetic material is introduced between the poles so that it does not completely occupy the volume between them, magnetic poles are induced in the material. As a result, the material produces its own magnetic field, which alters the original magnetic field (Fig. 6.12(b)). This effect is called "demagnetization," because the magnetization induced in the ferromagnetic material tends to decrease the original field inside the material. The demagnetizing field strength, H_d, is given by

$$H_d = -N_d M \tag{6.37}$$

where N_d is a demagnetization factor ranging from zero to one, and M is the magnetization in the material. Substituting this expression in Eq. 6.32, the magnetic field inside the material, B_{int}, is given by

$$B_{int} = \mu_o(M - N_d M) = \mu_o M(1 - N_d) \tag{6.38}$$

The demagnetization factor is determined by the geometry of the magnetic material. Only for an ellipsoidal body, a uniform magnetization causes a uniform demagnetizing field. In this case, an exact expression for N_d can be derived [18]. For other shapes, experimentally determined values of N_d are used. In practice, a general rule of thumb is that the higher the aspect ratio (i.e., ratio of length to diameter, l/d) of the specimen, the lower the demagnetization field tends to be. For very high – aspect ratio specimens, the demagnetization field is often neglected. Table 6.2 shows

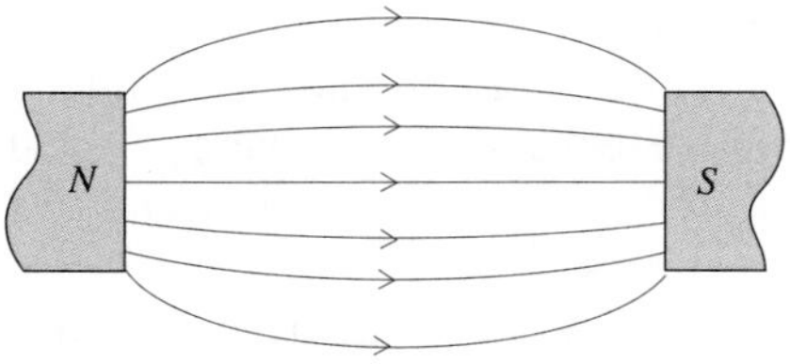

(a) Magnetic field due to two poles

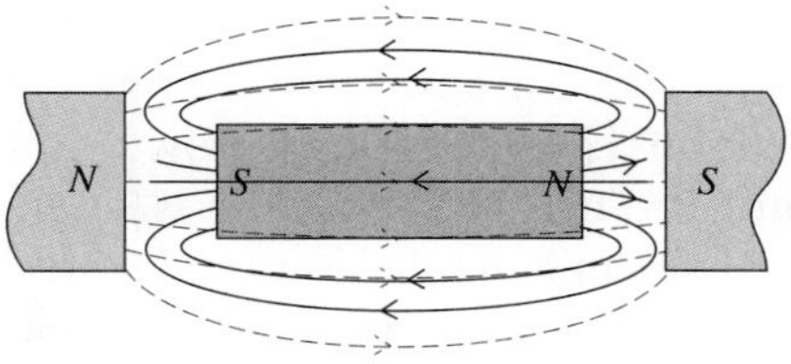

(b) Effect of introducing a ferromagnetic material

Figure 6.12. Demagnetization effect.

the demagnetizing factors for some simple geometries with various aspect ratios. Note that cuboidal specimens can be treated as cylinders of equivalent volume. The goal in magnetostrictive transducer design is to include low-reluctance flux return paths in the magnetic circuit. The magnetic circuit should direct the magnetic flux into the sample with minimal leakage into the surrounding air and with a small demagnetization effect in the core.

6.2.7 Electrical Impedance

The electrical impedance of a long, thin solenoid is approximately represented as (under a harmonic applied voltage)

$$Z(\omega) = \frac{V}{i} = R + j\omega\mathbb{L} \tag{6.39}$$

where V is the voltage across the solenoid, i is the current flowing through it, R is the resistance of the solenoid coil windings, ω is the frequency of the applied voltage, and $\mathbb{L}$ is the inductance of the solenoid. The magnetic field intensity is given by

$$H = \frac{N}{L} i \tag{6.40}$$

where N is the number of turns of the coil and L is the length of the coil. The inductance is given by

$$\mathbb{L} = \mu_o \frac{N^2 A}{L} \tag{6.41}$$

where A is the area enclosed in the coil and μ_o is the magnetic permeability. The total electrical impedance is the sum of the blocked electrical impedance Z_e (mechanically blocked condition) and the motional or mobility impedance Z_m (transduction of mechanical energy to electrical energy). Z_m is calculated from the difference between Z and Z_e.

$$Z = Z_e + Z_m \tag{6.42}$$

6.2.8 Systems of Units

Several different systems of units are used to represent magnetic quantities. The most commonly used systems are the SI system, the MKSA system (meter-kilogram-second-ampere), and the CGSM system (centimeter-gram-second-magnetic). In the SI system, there are two conventions: the Sommerfeld convention and the Kennelly convention [1]. The SI system provides a more intuitive feel of the underlying physical quantities. The CGSM system is often encountered in older literature, and the MKSA system is similar to the SI system but sometimes differs in the definition of magnetization. The remainder of this chapter uses the SI system. The correspondence between important magnetic quantities in the SI system and CGSM system, along with their dimensions in the SI system, is shown in Table 6.3. Note that the field equation changes depending on the system of units.

Table 6.3. *Systems of magnetic units*

Quantity	Symbol	SI	CGS	SI Dimensions	Conversion $\frac{\text{SI value}}{\text{CGS value}}$
Magnetic field, magnetic induction, magnetic flux density	$\boldsymbol{B}$	Tesla (T), Wb/m^2	Gauss (G)	$MA^{-1}T^{-2}$	10^{-4}
Magnetic field intensity, Magnetic field strength	$\boldsymbol{H}$	A/m, A.Turns/m	Oersted (Oe)	AL^{-1}	79.58
Magnetic flux	Φ	Weber (Wb)	Maxwell (Mx), G.cm^2	$ML^2A^{-1}T^{-2}$	10^{-8}
Magnetization	$\boldsymbol{M}$	A/m	emu/cc, G	AL^{-1}	1000
Field equation		$\boldsymbol{B} = \mu_o(\boldsymbol{H} + \boldsymbol{M})$	$\boldsymbol{B} = \boldsymbol{H} + 4\pi\boldsymbol{M}$	–	–
Magnetic moment	$\mathfrak{M}$	A.m^2, Wb.m	emu, erg/G	AL^2	–
Magnetomotive force	*mmf*	A.Turn	Gilbert(Gb)	A	–
Magnetic permeability	μ	H/m	–	$MLA^{-2}T^{-2}$	$4\pi \times 10^{-7}$
Inductance	L	Henry (H)	second^2/centimeter (abhenry)	$ML^2A^{-2}T^{-2}$	–
Reluctance	$\mathbb{R}$	1/H	Gb/Mx	$M^{-1}L^{-2}A^2T^2$	7.96×10^7

The conversion factors among different systems of units are as follows

$$1\ \text{Tesla} = 1\frac{\text{Volt Second}}{\text{Meter}^2}\ \text{or}\ 1\frac{\text{Newton}}{\text{Ampere Meter}} \tag{6.43}$$

$$1\ \text{Gauss} = 1\frac{\text{Maxwell}}{\text{Centimeter}^2} = 10^{-4}\frac{\text{Weber}}{\text{Meter}^2} = 10^{-4}\ \text{T} \tag{6.44}$$

$$1\ \text{Oersted} = \frac{1000}{4\pi}\frac{\text{Ampere}}{\text{Meter}} = 79.58\frac{\text{A}}{\text{m}} \tag{6.45}$$

$$1\ \text{Weber} = 10^8\ \text{Maxwell} \tag{6.46}$$

$$1\frac{\text{emu}}{\text{cm}^3} = 1000\frac{\text{A}}{\text{m}} \tag{6.47}$$

$$1\ \text{Henry} = \frac{\text{Volt Second}}{\text{Ampere}} \tag{6.48}$$

6.2.9 Magnetic Circuits

There is a close analogy between electrical and magnetic phenomena. For example, whereas a current carrying conductor induces a magnetic field around it, a flow of current will be induced in a conductor if it is placed in a time-varying magnetic field. In practical applications, a magnetic field is usually generated by a current carrying conductor of a specific geometry. The magnetic field is directed to and focused on a region of interest by a flux path constructed out of a material with a high magnetic permeability. The combination of the magnetic field producing coil, the flux path, and the region of interest is referred to as a "magnetic circuit." Often, it is desired to design a magnetic circuit to produce a specified magnetic field over a region of interest; for example, on a volume of magnetostrictive material. Conversely, given a magnetic circuit of a known geometry with a known electric current, it may be required to calculate the magnetic field produced. Magnetic circuits can be conveniently analyzed by considering an equivalent electric circuit.

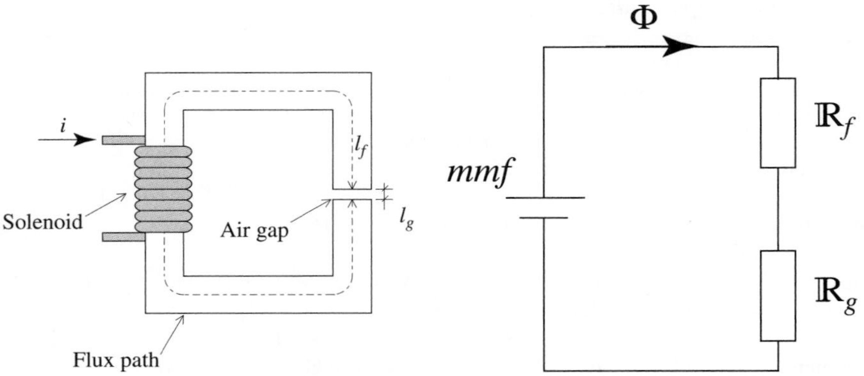

(a) Schematic of magnetic circuit (b) Equivalent circuit

Figure 6.13. Analysis of a magnetic circuit.

Let us consider the simple magnetic circuit shown in Fig. 6.13(a). The circuit consists of a coil of N turns, carrying a current i, and wound over a core, or flux path. The flux path has a length l_f, permeability μ_f, and a constant cross-sectional area A_f (we neglect the effect of the sharp corners). The flux path is broken by a small air gap of height l_g. We also neglect the fringing effect around the edges of the air gap. The goal of this magnetic circuit is to produce a uniform magnetic field across the air gap. Because the magnetic circuit contains interfaces between several materials with different permeabilities, such as the flux path and air gap, it is important to know how the magnetic field and magnetic field strength behave at an interface. These relations can be obtained by using Gauss's law and Ampere's law. It can be shown from Gauss's law that across an interface of two materials with different permeabilities, the component of $\boldsymbol{B}$ normal to the interface is continuous. From Ampere's law, it can be shown that the component of $\boldsymbol{H}$ tangential to the interface is continuous.

Applying Ampere's law along the entire circuit

$$Ni = H_f l_f + H_g l_g \tag{6.49}$$

where the quantities with subscripts 'g' and 'f' refer to the air gap and flux path, respectively. A simple equivalent circuit can be constructed by recognizing the analogy between electric and magnetic quantities. The quantity Ni is called the magnetomotive force (mmf) and is analogous to the voltage (electromotive force) in an electric circuit. Because the normal component of the magnetic field is constant at each interface ($B_f = B_g = B$) and the cross-sectional area of the magnetic circuit is uniform, it follows that the magnetic flux, Φ, is constant at any cross section of the magnetic circuit. Therefore, it can be seen that the magnetic flux is analogous to current in an electric circuit.

$$\Phi_f = \Phi_g = \Phi \tag{6.50}$$

The total mmf can be written as (from Eq. 6.49)

$$\begin{aligned} mmf &= \frac{Bl_f}{\mu_f} + \frac{Bl_g}{\mu_g} \\ &= \frac{\Phi l_f}{\mu_f A_x} + \frac{\Phi l_g}{\mu_g A_x} \\ &= \mathbb{R}_f \Phi + \mathbb{R}_g \Phi \end{aligned} \tag{6.51}$$

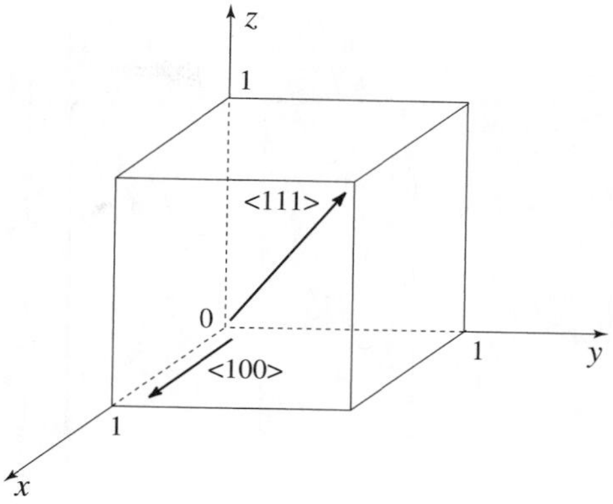

Figure 6.14. Definition of direction vectors in a unit cell.

where the quantity $\mathbb{R}$ is defined as the reluctance. For a volume of material of length L, cross-sectional area A_x, and permeability μ_c, the reluctance is given by

$$\mathbb{R} = \frac{L}{\mu_c A_x} \tag{6.52}$$

Therefore, in a magnetic circuit, the magnetomotive force is given by the product of the magnetic flux and the reluctance

$$mmf = \mathbb{R}\Phi \tag{6.53}$$

By recognizing the similarity between Eq. 6.53 and Ohm's law in electricity, it follows that the reluctance is analogous to electrical resistance. An equivalent circuit can now be constructed as shown in Fig. 6.13(b). Eq. 6.53 is sometimes referred to as Ohm's law for magnetism. For a core constructed out of a typical low-carbon steel, $\mu_f \approxeq 1000\mu_g$, and almost all of the magnetomotive force appears across the air gap. Note that in the previous discussion, it is assumed that the permeability is independent of B. In reality, the discussion is valid at each point on the B-H curve of the material.

6.3 Mechanism of Magnetostriction

Magnetostrictive materials transduce or transform magnetic energy to mechancial energy and vice versa. As a magnetostrictive material is magnetized, it exhibits a change in length. Conversely, if an external force is applied, it produces strain in the magnetostrictive material, which in turn changes the magnetic state of the material. The phenomenon of magnetostriction is closely linked to the presence of magnetic anisotropy and the alignment of magnetic domains in the material.

6.3.1 Definition of Crystal Axes and Magnetic Anisotropy

References to direction vectors with respect to crystal axes are often found in literature discussing the microstructure and properties of materials. Especially in the case of magnetostrictive materials, these direction vectors help in understanding fundamental phenomena. The definition of crystal axes and direction vectors is shown in Fig. 6.14. Consider the edges of a cubic unit cell oriented along the x, y, and z axes. The sides of the cube are of unit length. The direction vectors are assumed to start from the origin and end at a point with coordinates specified by vertices of

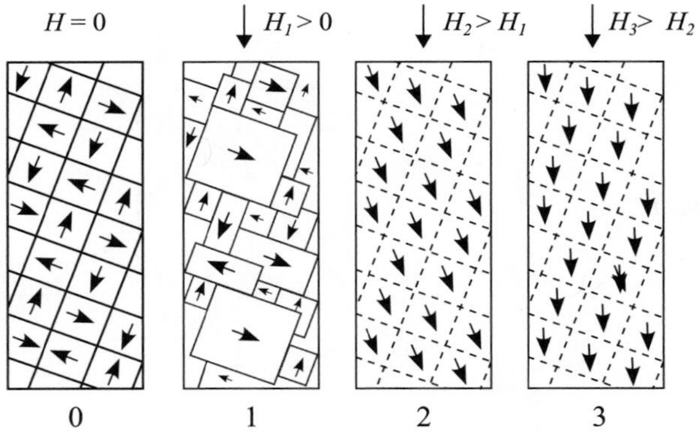

Figure 6.15. Progression of magnetization with applied field.

the cube. They are denoted by enclosing the *xyz* coordinates of the endpoint within square brackets. For example, the direction vector defines the lower side of the cube on the *x*-axis is [100]. Similarly, vectors pointing in the opposite direction can be defined, such as $[\bar{1}00]$, where the $\bar{1}$ refers to the coordinate $x = -1$. The entire set of directions is denoted by a single dimension enclosed in angular brackets, such as $< 100 >$. Similarly, a plane is denoted by the *xyz* coordinates enclosed in round brackets, and the entire set of planes is denoted by the *xyz* coordinates enclosed in curly brackets.

A number of material properties, such as elastic, electric, and magnetic properties, depend on the direction along which they are measured with respect to the orientation of a unit cell. Typically, unit cells are dispersed with random orientations throughout a volume of material, resulting in isotropic macromechanical behavior. In some cases, in which a number of unit cells, or domains, are aligned in a particular direction, the macromechanical behavior can be anisotropic. A concept that is crucial to understanding the phenomenon of magnetostriction is magnetic anisotropy, which is the major cause of the preferential orientation of magnetization along specific directions in a unit cell. Magnetic anisotropy is said to exist when the internal energy of a material depends on the orientation of its spontaneous magnetization with respect to its crystallographic axes [19]. It follows that the overall energy of the material is minimized if the magnetization is oriented along specific directions, which the system naturally prefers. These preferred directions are sometimes referred to as magnetically "easy" directions (or axes), and they depend on the geometry of the unit cell. For tetragonal and hexagonal materials, the easy axis is typically along the *c* axis of the unit cell.

Let us consider Terfenol-D material with a stoichiometry of $Tb_{0.27}Dy_{0.73}Fe_{1.95}$. Normally, it is produced as a monolithic cubic crystal using the FSZM process. It has a positive magnetostriction coefficient and exhibits magnetostrictive anisotropy. Figure 6.15 shows the progression from the demagnetized state to magnetization saturation as an increasing magnetic field is applied in the $[11\bar{2}]$ direction. Stage 0 represents the initial demagnetized state of Terfenol-D. The magnetic domain vectors are randomly oriented and the total magnetization is nearly zero. On the application of a magnetic field H_1, the magnetic domains start to align with the applied field. Stage 1 shows an early alignment with low magnetic field, in which the domains start regrouping (growing and shrinking), whereas the orientation of magnetization within the domains is unchanged. As the applied magnetic field is increased, Stage 2 is reached, where the orientations of magnetization change and

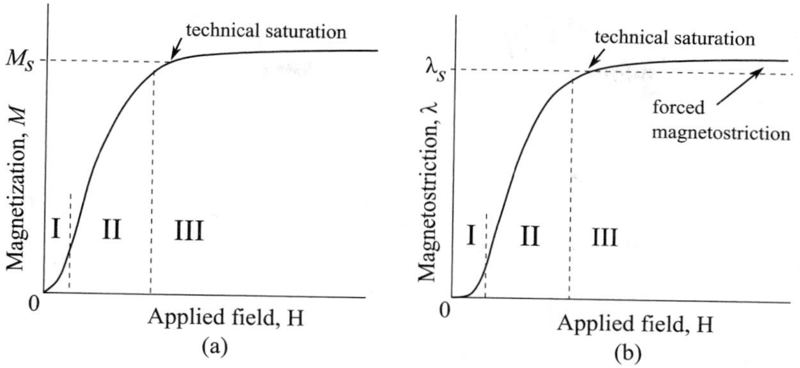

Figure 6.16. Simulated variation with applied field of (a) magnetization and (b) magnetostriction, adapted from Ref. [20].

the majority of domains are aligned along the $[11\bar{1}]$ axis. On the application of a large magnetic field, Stage 3 is reached, in which further reorientation of the magnetization occurs and the majority of domains are aligned along the $[11\bar{2}]$ axis. This stage corresponds to the magnetization saturation of the material. The process is explained further in Fig. 6.16. It can be seen that magnetostriction is a nonlinear process. Terfenol-D undergoes positive strain (extension) along the direction of the applied magnetic field until magnetic saturation is reached. However, in the direction transverse to the applied field, the strain is negative (compressive) with typically one half the magnitude of that in the axial direction (no net change in volume).

6.3.2 Origin of the Magnetostrictive Effect

When the material is above its Curie temperature (i.e., around 380°C for Terfenol-D), it exists in a paramagnetic state and is composed of unordered magnetic moments in random orientations. On cooling below the Curie temperature, the material becomes ferromagnetic, and the magnetic moments become ordered over small volumes. A volume in which all of the magnetic moments are parallel is called a "domain." At this stage, each domain has a spontaneous magnetization due to the ordering of the magnetic moments. However, because the domains are randomly oriented, the net magnetization of the material is zero.

The formation of domains is accompanied by a spontaneous deformation of the crystal lattice in the direction of domain magnetization. This change in dimension leads to an overall change in dimension of the material, which is called the "spontaneous magnetostriction." A schematic of this effect, simplified to one dimension, is shown in Fig. 6.17(a). The unordered material in the paramagnetic state can be represented as spherical volumes. When magnetic domains form in the material, each volume undergoes a strain e along its axes of magnetization. Because the magnetic domains are randomly oriented, the overall material strain, when resolved into components along the reference axes (e.g., the x-axis), is $e/3$. This can be easily explained by the following argument: because the domains are randomly oriented and the material is isotropic, the strain e can occur along each of the three reference axes with equal probability. Therefore, the effective strain along any of the reference axes is $e/3$. The phenomenon of change in dimension due to a transition from a

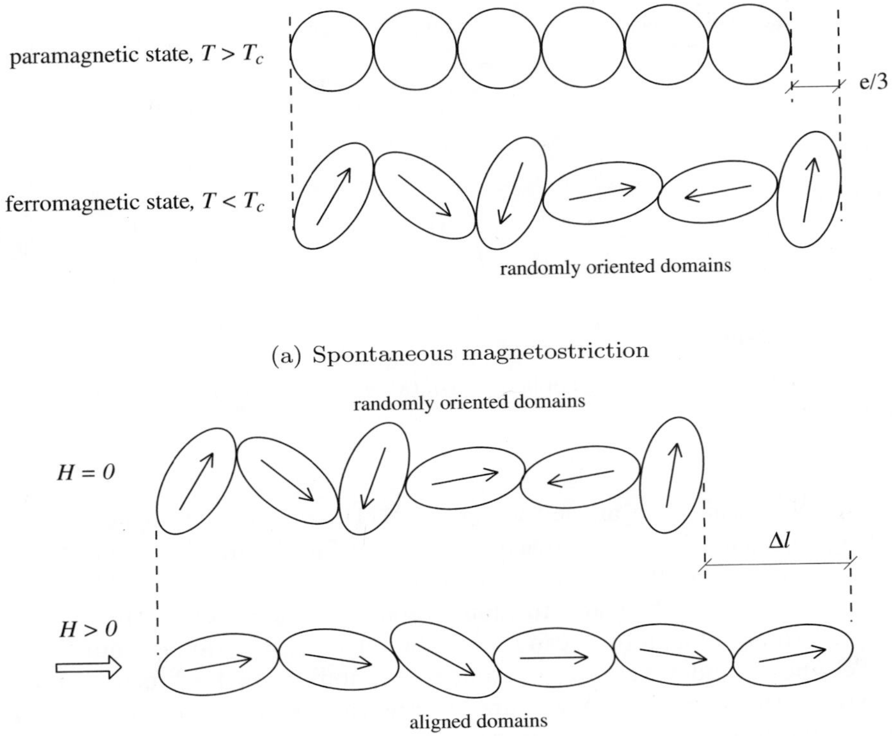

(a) Spontaneous magnetostriction

(b) Field induced magnetostriction

Figure 6.17. Schematic of magnetostrictive effect.

paramagnetic state to a ferromagnetic state is called spontaneous magnetostriction, and the strain associated with it ($e/3$) is a material-dependent constant.

Due to the magnetic anisotropy of the material, when an external magnetic field H is applied, the individual domain magnetization vectors tend to orient as closely as possible along the direction of the applied field. Because all of the domains are then oriented in a specific direction, the material becomes magnetized. In addition, due to the geometry of the domains, the rotation of the domain magnetizations results in an effective change in dimension of the material in addition to the spontaneous magnetostriction. The change in dimension of the material in response to an external magnetic field is called field-induced magnetostriction. Because the spontaneous magnetization is constant, the field-induced magnetization is the most important quantity with respect to typical engineering applications, such as magnetostrictive actuators. A schematic of the field-induced magnetostriction in one dimension is shown in Fig. 6.17(b), where the external field H causes a net change in length ΔL. As the magnetization of the material reaches the saturation magnetization M_s, the material reaches its saturation magnetostriction λ_{max}, which is the maximum achievable magnetostrictive strain. From the figure, it can be seen that $\lambda_{max} = e$. Therefore, the maximum achievable field-induced magnetostriction λ_s is given by

$$\lambda_s = \lambda_{max} - \frac{e}{3} = \frac{2}{3}e \qquad (6.54)$$

The magnetostriction is defined as

$$\lambda = \frac{\Delta L}{L} \tag{6.55}$$

where ΔL is the change in length from the original length L, and it is different from the total strain that includes both elastic and magnetostrictive components. Thus, in ferromagnetic materials, there can be two types of magnetostriction: (1) spontaneous magnetostriction due to alignment of domains on cooling through the Curie temperature, and (2) field-induced magnetostriction arising from the reorientation of magnetic moments due to the applied magnetic field.

Because magnetostriction involves motion on a molecular level, the mechanical response to the applied field is very fast (i.e., bandwidth on the order of kHz). The phenomenon described here is also known as "Joule magnetostriction" and occurs with a net zero change in volume. In reality, this is an approximation but, for all practical purposes, it can be assumed that the volume of the material remains constant and the transverse dimensions change appropriately. Furthermore, note that to obtain extensional strain in the longitudinal direction, magnetic flux lines must be arranged parallel to the longitudinal axis of the magnetostrictive specimen.

In practice, for anisotropic materials, the value of saturation magnetostriction along each crystal axis is different [21]. For example, in cubic materials (e.g., Terfenol-D), there are two independent constants, λ_{100} and λ_{111}, which define the saturation magnetostrictions along the [100] and [111] directions, respectively [22, 23]. The expression for the saturation magnetization in a single domain along any arbitrary angle is then given by

$$\begin{aligned}\lambda_s = &\frac{3}{2}\lambda_{100}\left(\alpha_1^2\beta_1^2 + \alpha_2^2\beta_2^2 + \alpha_3^2\beta_3^2 - \frac{1}{3}\right) \\ &+ 3\lambda_{111}\left(\alpha_1\alpha_2\beta_1\beta_2 + \alpha_2\alpha_3\beta_2\beta_3 + \alpha_3\alpha_1\beta_3\beta_1\right)\end{aligned} \tag{6.56}$$

where α_1, α_2, and α_3 are the direction cosines of the domain magnetization (magnetic moments) with respect to the reference coordinate system (denoted as 123, or xyz). β_1, β_2, and β_3 are the direction cosines of the axes along which the magnetostriction is measured with respect to the reference coordinate system. In a polycrystalline material with randomly oriented domains, the strain is assumed to be evenly distributed in all directions. Therefore, the strain in a particular direction is obtained as an average quantity. In such a case, the saturation magnetostriction measured along the direction of the external field is given by

$$\lambda_s = \frac{2}{5}\lambda_{100} + \frac{3}{5}\lambda_{111} \tag{6.57}$$

Note that in the previous equation, if the material is isotropic ($\lambda_{100} = \lambda_{111}$), the components of saturation magnetostriction along the $< 100 >$ and $< 111 >$ directions would add up to unity.

6.3.3 Effect of Magnetic Field Polarity

Figure 6.18 shows the effect of a change in polarity of the external magnetic field. Because the magnetic field has only the effect of reorienting the domains, it can be seen that the effective change in length is the same irrespective of the polarity of the

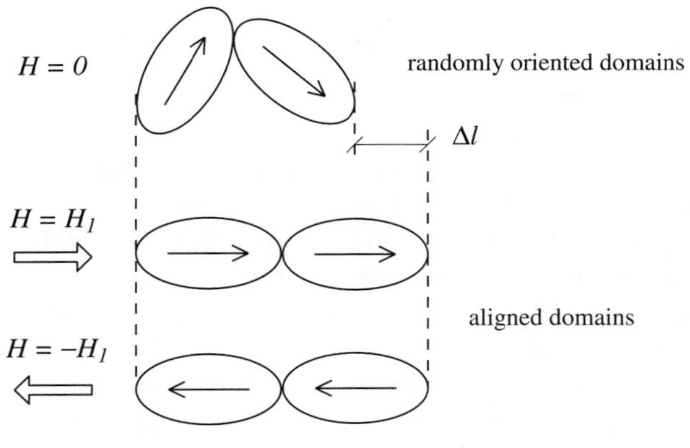

Figure 6.18. Independence of strain on field polarity.

applied field. Such behavior is characteristic of electrostriction as well as magnetostriction and is the main difference from induced strain due to the piezoelectric effect. Consequently, the strain is often treated as having a quadratic dependence on the external magnetic field.

The quadratic dependence of magnetostrictive strain on the applied field can also be shown mathematically [24]. Assuming that a magnetic field is applied at an angle θ to the magnetization vector of a domain, the magnetostriction along the direction of the applied field is given by

$$\lambda = \frac{3}{2}\lambda_s \left(\frac{M}{M_s}\right)^2 \tag{6.58}$$

where M is the component of magnetization along the direction of the applied field. Although this equation assumes that the reorientation of the domain occurs purely by rotation, it captures the correct qualitative trend for all cases.

A schematic of a general quasi-static strain-field curve can be seen in Fig. 6.19(a), and the behavior for Terfenol-D is shown in Fig. 6.19(b). The curve is symmetric for positive and negative magnetic fields, and it saturates at high values of field. The slope of the curve is relatively constant at moderate values of field. An important consequence of the "quadratic" nature of the strain response is that it is not possible to obtain a bipolar output strain with a bipolar input magnetic field. However, a bipolar output strain can be obtained by operating around a bias point, as shown in Fig. 6.19(a). The bias point is chosen to be the mid-point of the linear region of the curve. A steady magnetic field H_b is applied to the material, resulting in a constant bias strain ϵ_b. A bipolar field superimposed on the steady field H_b will result in a bipolar output strain about the constant strain ϵ_b, as shown by the arrow in Fig. 6.19(a). The bias field can be introduced by means of a permanent magnet or by applying a DC current in the magnetic coil. For optimum performance, it may be important to tune the bias field, which can be accomplished by a combination of permanent magnet and DC current. Also, for zero magnetic bias, a sinusoidal current input at a discrete frequency would result in a sinusoidal magnetic field at this frequency, which in turn would result in a magnetostrictive strain at twice the input frequency. However, with a bias magnetic field (large enough to result in a purely unipolar total magnetic field), the output response is at the same frequency as the excitation frequency.

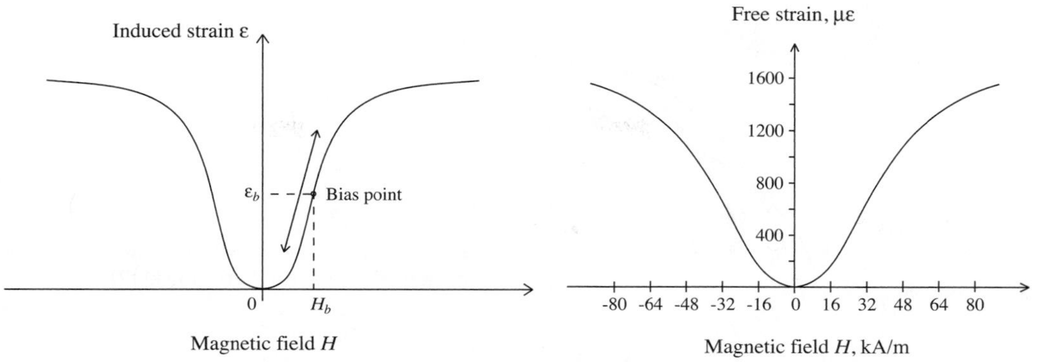

(a) Schematic of strain response (b) Typical values for Terfenol-D

Figure 6.19. Induced strain in response to an applied magnetic field.

6.3.4 Effect of External Stresses

Because the orientation of the domains depends on both the magnetic field and internal stresses, it follows that an externally applied stress also has the effect of reorienting the domains. For example, with the application of a compressive pre-stress, most of the domains are oriented normal to the direction of the stress. This occurs due to the inherent asymmetry of the domains and can be understood by looking at the geometry of the effect, as shown in Fig. 6.20. The material undergoes a decrease in length of Δl_c as a result of the compressive stress σ_1. On the application of an external magnetic field, the domains reorient along the direction of the applied field, and the material elongates by an amount Δl_h. It can be seen that if the material is given an initial compressive pre-stress, the recoverable strain is larger than in the

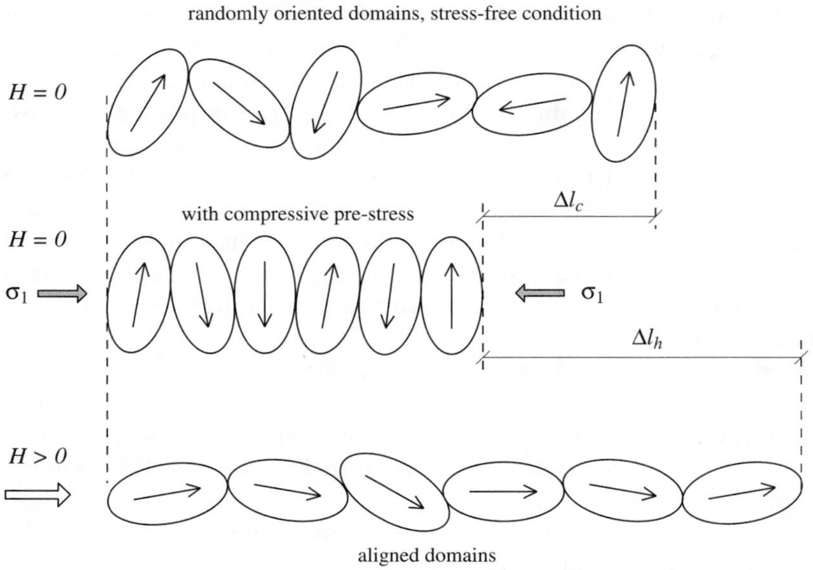

Figure 6.20. Effect of compressive stress on the induced strain.

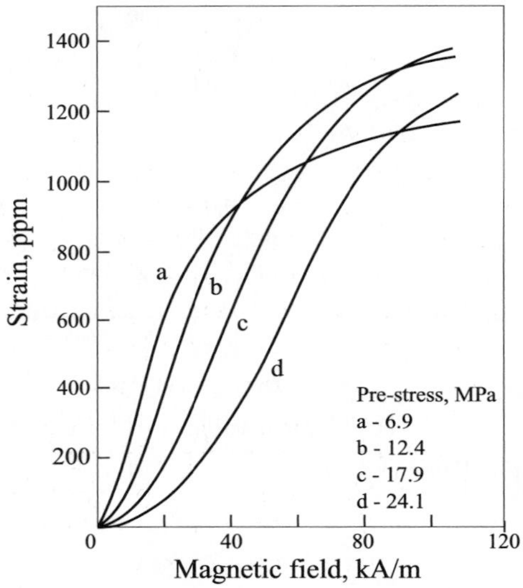

Figure 6.21. Magnetostriction with various pre-stress levels for Terfenol-D [7].

case of zero compressive pre-stress. However, at high values of compressive pre-stress, the material is unable to respond by the same extent to the applied magnetic field, and the induced strain starts to decrease. Therefore, the best performance can be achieved by operating the material at a moderate value of compressive pre-stress. In addition, because magnetostrictive materials (especially Terfenol-D) are brittle in tension (tensile strength ≊28 MPa, compressive strength ≊700 MPa), they are normally placed under a mechanical compressive bias stress to ensure their mechanical integrity during operation.

The angle by which a magnetic domain rotates in response to an applied external magnetic field is a balance among several different energies of the system. In the most simple terms, this can be understood by considering the torque equilibrium on each magnetic domain in the material. The external magnetic field exerts a torque on each domain that tends to orient it along the direction of the field. Internal stresses are created in the material as a result of the distortion in internal structure that accompanies the rotation of the domains. The result of the internal stresses, or elastic forces, is to exert a restoring torque on the domain, tending to oppose any change in its orientation. Therefore, for small values of external field, the final angle of rotation of the domains is a balance between the magnetic forces and the internal elastic forces. The resulting motion of the domain walls is sometimes termed as "reversible" because the change in orientation of the domain magnetizations is small and the internal stresses return the domains to their original orientation on removal of the external field.

Figure 6.21 shows the effect of compressive pre-stress on the induced strain due to applied magnetic field for Terfenol-D. As the pre-stress is increased, a larger applied magnetic field is necessary to reach magnetostriction saturation. It is clear that the pre-stress impacts elastic and magnetic properties as well as coupling between mechanical and magnetic states.

As the magnetic field is increased, the internal forces are overcome and the magnetization vector of the domains switches to another magnetically easy axis that is better oriented with the external field. When the external field is subsequently

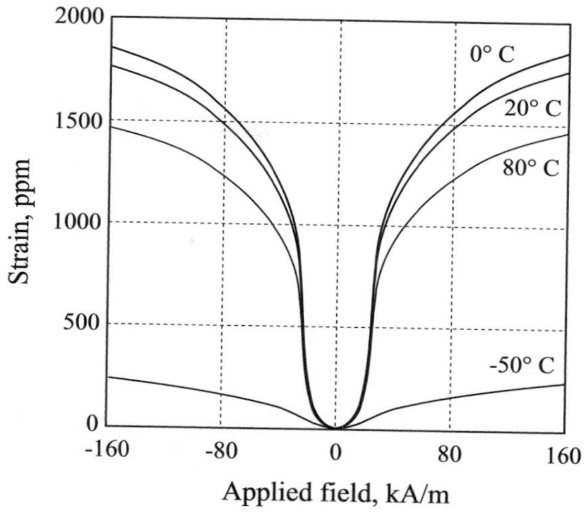

Figure 6.22. Temperature dependence of magnetostriction with applied field for Terfenol-D.

removed, the magnetization vector remains oriented along the new axis, unless a large external field of the opposite polarity is applied. Hence, the resulting motion of the domain walls is termed "irreversible." The reversible and irreversible domain-wall motion is the main cause of the hysteresis observed in the B–H curve of the material. Of the two effects, the irreversible domain wall motion is dominant.

6.3.5 Effect of Temperature

Temperature also has a significant effect on the performance of magnetostrictive materials. Generally, magnetostriction decreases with an increase in temperature and ultimately becomes zero at the Curie temperature. For example, in the case of Terfenol-D, at a compressive pre-stress of 13.3 MPa and a magnetic field of 2000 Oe, the magnitude of magnetostriction changes from 200 $\mu\epsilon$ at $-50°$C to 1740 $\mu\epsilon$ at 0° [25]. The optimal temperature for the operation of Terfenol-D is around room temperature. This behavior is related to the change in the magnetically easy axes from the $< 100 >$ direction at low temperatures to the $< 111 >$ direction at higher temperatures. Because the values of magnetostriction along each direction are different, this translates to a change in the net magnetostriction of the material with temperature. The dependence of magnetostriction on temperature can be captured in constitutive models by incorporating higher-order interaction terms [26].

Figure 6.22 presents the effect of temperature on magnetostriction of Terfenol-D ($Tb_{0.27}Dy_{0.73}Fe_{1.95}$), optimized for use in ambient conditions. The magnetostriction decreases with an increase in temperature (above ambient temperature) and ultimately becomes zero at the Curie temperature. For example, there is a reduction of 20% in saturation strain at 80°C compared to the strain at 0°C. Furthermore, there is a degradation of magnetostriction at negative temperatures (below ambient temperature).

Note that the coefficient of thermal expansion of the material is around 12×10^{-6}/°C, which is small compared to the magnetostriction [7]. The sensitivity of magnetostriction to temperature is an important factor to be considered during the design of actuators. Typically, the magnetic field is generated by a current-carrying coil, which generates Ohmic heating. In addition, eddy current losses in the

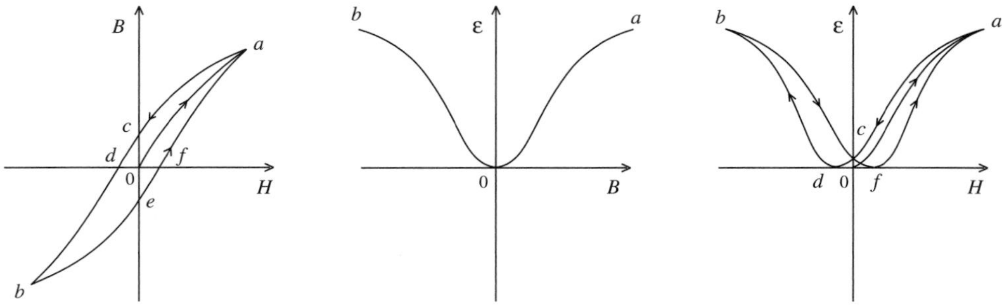

Figure 6.23. Hysteretic behavior for magnetostriction.

flux return path that surrounds the magnetostrictive material also contribute to an increase in temperature. During the design process, care must be taken to minimize the current required and the eddy current losses, as well as to incorporate some means of heat dissipation so that the temperature rise is restricted to acceptable levels.

6.3.6 Strain Hysteresis

The induced-strain curves shown in Fig. 6.19 represent the quasi-static behavior of the material. If the magnetic field is varied sinusoidally about a zero mean value, a hysteretic behavior of the induced strain is observed. However, the hysteresis exhibited by magnetostrictive and electrostrictive materials is much smaller than in the case of piezoceramics and is often ignored. The origin of the hysteresis in case of magnetostriction is largely due to the hysteresis inherent in the *B-H* curve of the material, as shown in Fig. 6.23. However, the strain-*B* curve shows little hysteresis. Again, the strain versus magnetic field shows a highly nonlinear behavior.

The origin 0 represents the original nonmagnetized state of the material. As the magnetic field *H* is increased along curve 0*a*, the magnetic induction *B* also increases until magnetic saturation is reached at point *a*. A further increase in *H* does not increase either the magnetic induction or the strain ϵ. Decreasing the field would result in the curve *acdb*. At zero field, there is a residual strain due to the residual, or remanent magnetic induction (given by *c*). Further decreasing the field will bring the material to a zero strain state and then increase the strain again. Similar behavior can be observed for an increase in field, along the curve *befa*. Note that the strain goes to zero at the points *d* and *f*, where the magnetic induction is zero. The strain is positive and nonzero at the points of remanent magnetic induction, *c* and *e*, even though the field is zero.

The *B-H* curve clearly shows that the magnetic permeability μ is a nonlinear function of the magnetic field *H* and the time history (during the initial cycles). The minimum value of permeability occurs at points *a* and *b* (magnetic saturation), and the maximum values occur at points *d* and *f* (remnant magnetic induction). The strain versus field ($\epsilon - H$) curve, known as a butterfly curve, shows the hysteretic nature of the material. Hysteresis can be visualized as a result of internal friction as the domains attempt to rotate to align with the magnetic field. The stoichiometry of the material is key to changing the aspect ratio of the hysteresis loop. For example, Terfenol-D, with stoichiometry $Tb_{0.27}Dy_{0.73}Fe_{1.97}$, exhibits

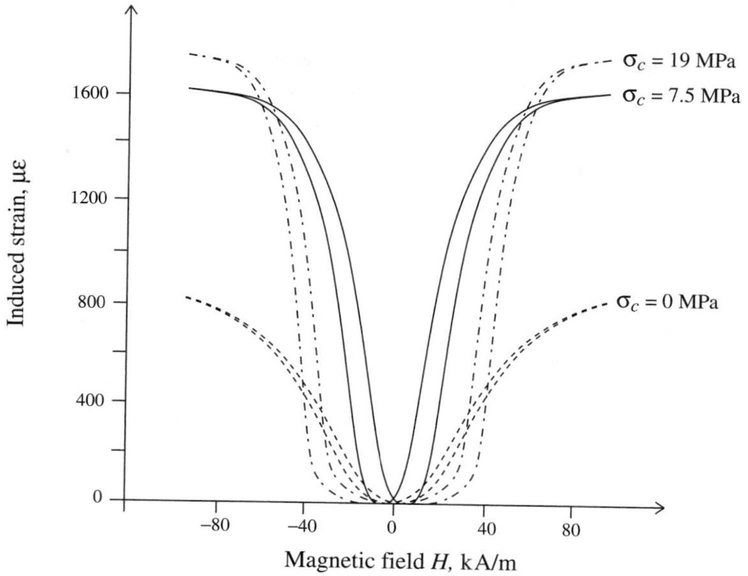

Figure 6.24. Effect of compressive stress.

significantly less hysteresis as well as reduced maximum strain compared to $Tb_{0.3}Dy_{0.7}Fe_{1.98}$. Eddy currents are induced in Terfenol-D due to AC magnetic field, which in turn produce a magnetic flux that resists the applied AC field. Eddy currents are electric currents induced within a conductor by a changing magnetic field. Eddy currents reduce effective permeability and increase power losses due to Ohmic heating. Often, the effects of eddy currents are minimized by laminating the rods.

Figure 6.24 shows a typical set of hysteretic curves for Terfenol-D at different values of compressive pre-stress. As discussed in Section 6.3.4, it can be seen that the maximum induced strain in the case of a compressive pre-stress σ_c of 7.5 MPa is larger than in the case of no pre-stress. Additionally, the effect of an even larger pre-stress, σ_c of 19 MPa, is not as pronounced as in the lower pre-stress curve. There is a marginal increase in maximum induced strain but also lower induced strains over most of the operating region. Terfenol-D is also extremely brittle, and has a very low tensile yield strength compared to its compressive yield strength. A compressive pre-stress will decrease the magnitude of any tensile forces seen by the active material during operation and, therefore, protect the material from failure. It can be concluded that it is beneficial to operate a magnetostrictive element under a compressive pre-stress.

6.4 Constitutive Relations

Reliable constitutive models are important for the design and development of actuators and sensors, to understand scaling effects, and for proper assessment of input power and field requirements. Due to the coupling between physical parameters such as input current, magnetic field, and output displacement, a comprehensive model must account for interactions on several levels. Typically, for a magnetostrictive actuator consisting of a Terfenol-D rod (magnetostrictive core) and a current carrying coil, the development of a model can be divided into the following four steps:

1. Electromagnetic equations are used to find the field generated in the magnetostrictive material as a function of the applied current. The distribution of the field in the material and in the flux return path is a function of the geometry of the actuator, flux return path, and current carrying coil, as well as the magnetic permeabilities of different materials in the circuit.
2. The magnetization in the material is calculated based on the generated field. A number of different models have been proposed, capturing various aspects of the M–H or B–H curve. For example, the Jiles-Atherton model [27, 28, 29] is a well-known model that can capture ferromagnetic hysteresis. This model requires five experimentally determined parameters to define the state of the material. Other more detailed models include the effects of minor loops [30, 31] and time-varying magnetic fields [32, 33].
3. From the magnetization, the magnetostriction in the material is calculated. This can be based on models of varying complexity. The simplest model is the quadratic model discussed in Section 6.3.3 (Eq. 6.58). However, this model does not account for hysteresis in the M–λ curve. More refined models can be obtained by expressing the magnetostriction as a series expansion of even powers of the magnetization [34], by deriving expressions for the magnetostriction based on the energy equation [35, 36], or by examining the rotation of magnetic dipoles on a micromagnetic scale [37, 38].
4. Once the magnetostriction and the magnetization in the material are calculated, the interaction between the material and the external load or external structure can be calculated [39]. In its simplest form, the coupled actuator-structure problem can be treated as an arrangement of springs, each representing the stiffness of a specific part of the system. A more complex representation could be to treat the system as a continuum, set up equations using the force-balance method, and solve the resultant set of partial differential equations (similar to the wave equation).

Some of the models that have been reported in the literature are discussed herein, focusing on the magnetomechanical aspects of the material behavior. Note that a similar approach can be followed in the case of magnetostrictive sensors, the only difference being that the input quantity will be a stress and the output will be a voltage or current. Most of the engineering models are phenomenological, which fit experimental behavior of the bulk material to physically based laws. It is important to consider the coupling between magnetic and electric fields (electromagnetic coupling), interaction between magnetic and elastic state of the material (magnetomechanical coupling), interaction of magnetization and thermal effects (thermomagnetic), and coupling between the thermal and elastic effects (thermoelastic). For some effects, such as magnetomechanical coupling, there is a two-way coupling between the magnetic and elastic states. To model a magnetostricitve transducer, it is esential to model the effects of electrical, magnetic, and elastic components. Thermal effects can also be significant for dynamic cases due to ohmic heating, eddy current losses, and magnetomechanical hysteresis.

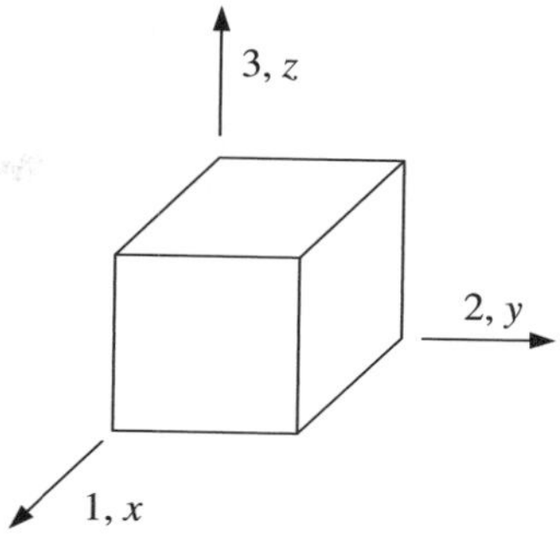

Figure 6.25. Coordinate system for the constitutive relations.

6.4.1 Linear Piezomagnetic Equations

Even though magnetostrictive transduction is intrinsically nonlinear and hysteretic, the quasi-steady linearized piezomagnetic representations provide insight on the performance, especially at low excitation levels. The linear piezomagnetic equations are the simplest representation of the interaction between the magnetic field and the mechanical response of the material. From the qualitative discussion on the mechanism of magnetostriction in Section 6.3, it can be seen that the induced strain is a nonlinear function of the applied magnetic field, which can be approximately expressed as a quadratic function of the field. However, for moderate values of applied field, or for operation about a bias point, the induced strain can be assumed to vary linearly with the field. In this region, linear constitutive relations can be written to model the behavior of the material. These relations are analogous to the piezoelectric constitutive relations and are sometimes known as piezomagnetic equations. However, in this case, they represent a coupling between magnetic and elastic quantities. The linear coupled magnetomechanical constitutive relations for a magnetostrictive material at a constant temperature (coordinate axes shown in Fig. 6.25) are

$$\boldsymbol{\epsilon} = \boldsymbol{s}^H \boldsymbol{\sigma} + \boldsymbol{d}\boldsymbol{H} \tag{6.59}$$

$$\boldsymbol{B} = \boldsymbol{d}^* \boldsymbol{\sigma} + \boldsymbol{\mu}^\sigma \boldsymbol{H} \tag{6.60}$$

where $\boldsymbol{d}$ (m/A) is the matrix of magnetostrictive constants that correspond to the slope of the linear region of the ϵ-H curve, and μ^σ (H/m or Tm/A) is the permeability of the material at constant stress, corresponding to the slope of the B–H curve in the first quadrant. $\boldsymbol{s}^H$ (m^2/N) is the compliance matrix of the material at constant magnetic field. In this case, the elastic modulus can be measured with coil leads in open-circuit condition (zero current) or with a means of maintaining a constant current in the coil. The magnetic field vector $\boldsymbol{H}$ consists of three components (H_1, H_2, and H_3) with units of A/m, or Oersted. The strain of a magnetostrictive element consists of two parts, one due to mechanical stress and the second due to applied magnetic field. The magnetic induction B also consists of two parts, one due to mechanical stress and the second due to applied magnetic field. The strain in the material is given by $\boldsymbol{\epsilon}$ (dimensionless) and the mechanical stress is given by $\boldsymbol{\sigma}$ (N/m^2). The units of $\boldsymbol{d}^*$ is Tm^2/N and this term is equivalent to $\boldsymbol{d}$ (m/A).

Note the similarity between the form of Eqs. 6.59 and 6.60 and the piezoelectric constitutive relations. The magnetostrictive relations can be obtained from the piezoelectric relations by replacing the electric field $\mathbb{E}$ with the magnetic field $\boldsymbol{H}$, the electric permittivity $\boldsymbol{e}^{\sigma}$ with the magnetic permeability $\boldsymbol{\mu}^{\sigma}$, and the electric displacement $\boldsymbol{D}$ with the magnetic induction $\boldsymbol{B}$. These relations are applicable for small changes in stress and applied field. The first-order temperature and frequency effects, hysteresis effects, and higher-order coupling among temperature, stress, and applied field are neglected.

Note that, in general, the strain and magnetic induction can be obtained by differentiating the total energy of the material with respect to various quantities, giving

$$\epsilon = \epsilon_o + \left(\frac{\partial \epsilon}{\partial \sigma}\right)_{H,T} \sigma + \left(\frac{\partial \epsilon}{\partial H}\right)_{\sigma,T} H + \left(\frac{\partial \epsilon}{\partial T}\right)_{\sigma,H} T + \text{higher-order terms} \tag{6.61}$$

$$B = B_o + \left(\frac{\partial B}{\partial \sigma}\right)_{H,T} \sigma + \left(\frac{\partial B}{\partial H}\right)_{\sigma,T} H + \left(\frac{\partial B}{\partial T}\right)_{\sigma,H} T + \text{higher-order terms} \tag{6.62}$$

where the subscripts denote that those quantities are being kept constant and T refers to the temperature. For reversible magnetostriction, it can be shown that [40]

$$d^* = \left(\frac{\partial B}{\partial \sigma}\right)_{H,T} \equiv \left(\frac{\partial \epsilon}{\partial H}\right)_{\sigma,T} = d \tag{6.63}$$

Neglecting the higher-order terms and ignoring the temperature term, the previous equations reduce to the familiar piezomagnetic equations for small variations. The constants in these equations can be experimentally determined and are related to each other [41]. The equations are useful for representing a magnetostrictive material operating in its linear region, such as when it is exposed to a low-level AC field superimposed on a steady bias field. Commercial finite element packages (e.g., ANSYS, PZFLEX, and ATILA) typically use the piezomagnetic equations to solve coupled structural-magnetostrictive problems. To address the nonlinear behavior of magnetostrictives (as seen in Fig. 6.23), the higher-order terms become important. Furthermore, these equations do not model the hysteretic behavior.

For Terfenol-D, it has been theoretically proven that the coefficient matrices in the piezomagnetic equations can be expanded as

$$\begin{Bmatrix} \epsilon_1 \\ \epsilon_2 \\ \epsilon_3 \\ \gamma_{23} \\ \gamma_{31} \\ \gamma_{12} \end{Bmatrix} = \begin{bmatrix} s_{11}^H & s_{12}^H & s_{13}^H & 0 & 0 & 0 \\ s_{12}^H & s_{11}^H & s_{13}^H & 0 & 0 & 0 \\ s_{13}^H & s_{13}^H & s_{33}^H & 0 & 0 & 0 \\ 0 & 0 & 0 & s_{44}^H & 0 & 0 \\ 0 & 0 & 0 & 0 & s_{44}^H & 0 \\ 0 & 0 & 0 & 0 & 0 & s_{66}^H \end{bmatrix} \begin{Bmatrix} \sigma_1 \\ \sigma_2 \\ \sigma_3 \\ \tau_{23} \\ \tau_{31} \\ \tau_{12} \end{Bmatrix} + \begin{bmatrix} 0 & 0 & d_{31} \\ 0 & 0 & d_{31} \\ 0 & 0 & d_{33} \\ 0 & d_{15} & 0 \\ d_{15} & 0 & 0 \\ 0 & 0 & 0 \end{bmatrix} \begin{Bmatrix} H_1 \\ H_2 \\ H_3 \end{Bmatrix} \tag{6.64}$$

$$\begin{Bmatrix} B_1 \\ B_2 \\ B_3 \end{Bmatrix} = \begin{bmatrix} 0 & 0 & 0 & 0 & d_{15}^* & 0 \\ 0 & 0 & 0 & d_{15}^* & 0 & 0 \\ d_{31}^* & d_{31}^* & d_{33}^* & 0 & 0 & 0 \end{bmatrix} \begin{Bmatrix} \sigma_1 \\ \sigma_2 \\ \sigma_3 \\ \tau_{23} \\ \tau_{31} \\ \tau_{12} \end{Bmatrix} + \begin{bmatrix} \mu_{11}^\sigma & 0 & 0 \\ 0 & \mu_{11}^\sigma & 0 \\ 0 & 0 & \mu_{33}^\sigma \end{bmatrix} \begin{Bmatrix} H_1 \\ H_2 \\ H_3 \end{Bmatrix} \tag{6.65}$$

It can be seen that the material is transversely isotropic (isotropic in the 1–2 plane) in terms of both its elastic and magnetic properties. The structure of the matrices is identical to that of a piezoelectric material. In the actuator equation, the coefficients $\boldsymbol{d}$ represent the change in strain per unit change in magnetic field at a constant stress. Alternatively, in the sensor equation, the coefficients d^* represent the change in magnetic induction due to a unit change in stress at a constant magnetic field. Again, it is typically assumed that the coefficients $d = d^*$.

Note that the coefficient matrices $\boldsymbol{s}^H$, $\boldsymbol{d}$, and $\boldsymbol{\mu}^\sigma$ can be dependent on the level of pre-stress, applied bias magnetic field, and driving field amplitude to represent the actual nonlinear material behavior more accurately. This dependence can be characterized experimentally [42, 43, 44] to identify the pre-stress and bias field required for optimum performance of the magnetostrictive actuator. For example, Moffet et al. [42] reported sets of experiments to measure the effect of driving field amplitude (from 8 kA/m to 160 kA/m) and pre-stress (from 7 MPa to 63 MPa), at an optimum bias field, on the properties of Terfenol-D. These constants are also a function of the applied stress and magnetic field; however, a linear assumption is valid in cases of small variations in inputs. The nonlinear behavior may be advantageous in some applications. Pratt et al. [45] exploited the nonlinear transduction of nonbiased Terfenol-D actuators to design an autoparametric vibration absorber. To include nonlinear effects approximately, one can either include higher-order terms in constitutive relations or use a lookup table (from experimentally measured values) for coefficients.

6.4.2 Refined Magnetostrictive Models

Based on the linearized constitutive equations (Eqs. 6.59 and 6.60), several models of magnetostrictive behavior of varying complexity have been proposed. Engdahl and Svensson [46] presented a simple, uncoupled finite difference analysis to predict the steady response of a magnetostrictive rod due to applied sinusoidal magnetic field using linear material characteristics. Kvarnsjo and Engdahl [47] developed a two-dimensional finite difference transient analysis in response to a magnetic field using nonlinear material characteristics. The finite-difference methods are less versatile to deal with structures constituting dissimilar materials such as the case with smart structures. Claeyssen et al. [48] developed a three-dimensional, coupled, linear finite element analysis to establish the effective dynamic coupling constants of a magnetostrictive actuator using an empirical representation of material characteristics.

Sherman and Butler [49] developed a nonlinear constitutive model for Terfenol-D by expanding the stress and magnetic field in terms of strain and applied magnetic field intensity for higher-order longitudinal components. These are described for a constant temperature and no hysteresis.

$$\begin{aligned}\sigma = c_1\epsilon + c_2\epsilon^2 + c_3\epsilon^3 + c_4\epsilon^4 - e_1H - e_2H^2 - e_3H^3 - e_4H^4 - 2c_a\epsilon H \\ - 3c_b\epsilon^2H - 3c_c\epsilon H^2 - 4c_d\epsilon^3H - 6c_e\epsilon^2H^2 - 4c_f\epsilon H^3\end{aligned} \tag{6.66}$$

$$\begin{aligned}B = e_1\epsilon + c_a\epsilon^2 + c_b\epsilon^3 + c_d\epsilon^4 + \mu_1H + \mu_2H^2 + \mu_3H^3 + \mu_4H^4 + 2e_2\epsilon H \\ + 3c_c\epsilon^2H + 3e_3\epsilon H^2 + 4c_e\epsilon^3H + 6c_f\epsilon^2H^2 + 4e_4\epsilon H^3\end{aligned} \tag{6.67}$$

where e_i, c_i, and μ_i are coefficients. For the unbiased condition, stress and strain are even functions of the magnetic field intensity H. Thus, the coefficients c_a, c_b, c_d, c_f, e_1, and e_3 will be identically zero.

Roberts et al. [50] developed nonlinear equations including higher-order interactions of stress, magnetic field, and temperature

$$\epsilon_{kl} = \sigma_{ij} s_{ijkl}^{H,T} + \Delta T \alpha_{kl}^{H} + \frac{1}{2} H_n H_m d_{klmn}^{T} + \frac{1}{2} H_n H_p \sigma_{ij} s_{klijnp}^{T} + \frac{1}{2} H_m H_n \Delta T \alpha_{klmn}$$

$$B_m = H_n \mu_{mn}^{T,\sigma} + \Delta T P_m^{\sigma} + \sigma_{ij} H_n d_{ijnm}^{T} + \Delta T \sigma_{ij} H_n \alpha_{ijnm} + \frac{1}{2} \sigma_{ij} \sigma_{kl} H_n s_{ijklnm}^{T} \tag{6.68}$$

where d_{ijnm} is the pyromagnetic parameter relating temperature and magnetization P_m, s_{ijklnm} is the elastic compliance, and α_{ijnm} is the coefficient of thermal expansion.

Jiles [51] developed a magnetomechanical model including elastic effects. This model provides a representation of the bidirectional coupling between the magnetic and elastic states. The model appears to accurately represent the magnetic hysteresis in the material. Anjanappa et al. [52, 53] presented a simple one-dimensional model to simulate the quasi-static response of a magnetostrictive mini-actuator (they developed) due to the applied magnetic field. Pradhan et al. [54] developed the first-order shear deformation theory (FSDT) to study the vibration control of laminated composite plates with embedded magnetostrictive layers. The effects of material properties and placement of magnetostrictive layers on vibration suppression were examined. It was found that the maximum suppression is obtained when the magnetostrictive layers were relatively thin and placed far away from the neutral axis.

6.4.3 Preisach Model

The Preisach model is normally used to describe a hysteretic process. It empirically fits the input u_p and output f_p of experimental data. The major drawback of this model is its lack of insight into the underlying physical mechanisms. The Preisach model utilizes kernels $\gamma_{\alpha\beta}$ in conjunction with the weighting function $W_p(\alpha_p, \beta_p)$ to fit the experimental data

$$f_p(t) = \iint_{\alpha_p > \beta_p} W_p(\alpha_p, \beta_p) \gamma_{\alpha,\beta} u_p(t) d\alpha_p d\beta_p \tag{6.69}$$

A continuous distributed system can be assembled with shifted values of α_p and β_p as long as α_p is larger than β_p. For example, the input quantity can be magnetization and the output quantity can be strain. For two input variables, u_p and v_p, this can be written as

$$f_p(t) = \iint_{\alpha_p > \beta_p} W_p(\alpha_p, \beta_p) \gamma_{\alpha,\beta} u_p(t) d\alpha_p d\beta_p + \iint_{\alpha_p > \beta_p} V_p(\alpha_p, \beta_p) \gamma_{\alpha,\beta} v_p(t) d\alpha_p d\beta_p \tag{6.70}$$

where V_p is the weighting function for v_p. For a magnetostrictive actuator, u_p and v_p can be magnetization and stress and the output can be strain.

Carman and Mitrovic [55] formulated a coupled, one-dimensional, nonlinear finite element analysis for a magnetostrictive actuator. Interactions among magnetization, stress, and temperature were included in the model, which showed good agreement with test data at high pre-loads. Duenas et al. [26] developed a more comprehensive constitutive model of magnetostrictive material that includes magnetization hysteresis ($M\!-\!H$ loop) and thermal effects. Although the model accounted for the quadratic variation of magnetostrictive strain with applied field, it did not capture saturation effects. One of the main features of this model is that it is developed in terms of magnetic field intensity as the dependent variable, as opposed to magnetic induction as in the piezomagnetic equations. Dapino et al. [34] developed a coupled nonlinear and hysteretic magnetomechanical model for magnetostrictives. The magnetostrictive effect is modeled by taking into account the Jiles-Atherton model of ferromagnetic hysteresis in combination with a quartic magnetostriction law ($\lambda = k_1 M^2 + k_2 M^4$). This model provides a representation of the bi-directional coupling between the magnetic and elastic states. The model appears to accurately represent the magnetic hysteresis in the material. Sablik and Jiles [56, 57] included magnetic hysteresis effects predicted using the Jiles-Atherton model along with a model for magnetostriction.

6.4.4 Energy Methods

The coupled magnetomechanical equations are often derived using energy methods. In its simplest form, the energy of the material per unit volume is written as a sum of the energy due to internal effects, external magnetic field, and elastic deformation. A brief description of this method is useful in understanding the physical basis behind the constitutive equations.

Consider a domain of magnetostrictive material exposed to a magnetic field as well as a stress σ. Let the orientation of the domain magnetization of the material with respect to the reference axes (taken to be the axes of the unit cell) be defined by the direction cosines α_1, α_2, and α_3 and the orientation of the stress with respect to the reference axis be defined by the direction cosines γ_1, γ_2, and γ_3. The total energy E_{tot} of the magnetostrictive material per unit volume can be expressed as [1, 18, 24, 58]

$$E_{tot} = E_o + E_a + E_{me} + E_m + E_e \tag{6.71}$$

where each energy term is explained as follows:

(1) E_o, long-range coupling energy or exchange energy: This accounts for the effect of the interaction between aligned magnetic moments over large distances (compared to the size of the unit cell). This term is constant in a given domain. Therefore, if only one domain is considered, this term is usually neglected. The exchange energy is given by

$$E_o = \frac{1}{2}\alpha_w M_s^2 \tag{6.72}$$

where α_w is the mean long-range coupling factor and M_s is the magnetization of the domain (equal to the saturation magnetization).

(2) E_a, anisotropy energy: This term is related to the dependence of the magnetic energy on the crystal symmetry of the material. Hence, it is also sometimes known as magnetocrystalline anisotropy. Minimization of this energy along specific directions is the reason for the preferential orientation of magnetic moments

in a unit cell. The anisotropy energy for a cubic unit cell (e.g., iron, nickel, or Terfenol-D) is given by

$$E_a = K + K_1\left(\alpha_1^2\alpha_2^2 + \alpha_2^2\alpha_3^2 + \alpha_3^2\alpha_1^2\right) + K_2\left(\alpha_1^2\alpha_2^2\alpha_3^2\right) + \text{ higher-order terms} \tag{6.73}$$

where K, K_1, and K_2 are the material-dependent anisotropy constants. The constant K is rarely used because, typically, the change in energy or derivative of energy is the quantity of most interest. Note that odd powers of the direction cosines do not appear in the equation because a change in sign of the direction cosine does not result in a change in orientation.

(3) E_{me}, magnetoelastic coupling energy: The coupling between the applied magnetic field and the magnetostrictive strain in the material is captured by this term. For a cubic unit cell, this is given by

$$\begin{aligned} E_{me} = &-\frac{3}{2}\lambda_{100}\sigma\left(\alpha_1^2\gamma_1^2 + \alpha_2^2\gamma_2^2 + \alpha_3^2\gamma_3^2 - \frac{1}{3}\right) \\ &- 3\lambda_{111}\sigma\left(\alpha_1\alpha_2\gamma_1\gamma_2 + \alpha_2\alpha_3\gamma_2\gamma_3 + \alpha_1\alpha_3\gamma_1\gamma_3\right) \end{aligned} \tag{6.74}$$

where λ_{100} and λ_{111} are the magnetostrictions in the $< 100 >$ and $< 111 >$ directions, respectively.

(4) E_m, magnetic energy: This is the energy that is required to magnetize the material. Over one cycle, this is also the energy lost due to hysteresis in the B–H curve (the area enclosed in the loop). For a single domain of magnetization $\boldsymbol{M}_s$ in a magnetic field $\boldsymbol{H}$, the energy is given by

$$E_m = -\mu_o \int \boldsymbol{H}.d\boldsymbol{M}_s \tag{6.75}$$

where μ_o is the permeability of the material.

(5) E_e, elastic energy: Also known as the strain energy, this term captures the effect of the deformation caused by a stress field. For a cubic crystal, we have

$$E_e = \frac{1}{2}c_{11}\left(\epsilon_x^2 + \epsilon_y^2 + \epsilon_z^2\right) + \frac{1}{2}c_{44}\left(\gamma_{yz}^2 + \gamma_{zx}^2 + \gamma_{xy}^2\right) + c_{12}\left(\epsilon_y\epsilon_z + \epsilon_z\epsilon_x + \epsilon_x\epsilon_y\right) \tag{6.76}$$

where the ϵ quantities are the strains in the material along the reference axes.

The total energy as described here is typically written in terms of a potential function (e.g., the Gibbs potential) and minimized with respect to a particular quantity, such as the strains in the material. This yields a set of coupled constitutive equations for the material. Note that an equilibrium state is reached between the elastic and magnetoelastic energies, which determines the final strain values of the material.

6.5 Material Properties

The properties of magnetostrictive materials vary widely with their composition, external magnetic field level, bias stress, and temperature. Typical material properties that are variable include Young's modulus, magnetostrictive strain, magnetic permeability, and saturation magnetization. Examples of experimental data showing these variations, as well as simple physical explanations, are discussed in the preceding sections. Of special interest is the variation of Young's modulus and magnetic permeability with magnetic and mechanical boundary conditions. These phenomena can be quantified in terms of the "Delta-E effect" and magnetostrictive coupling factor, and they are discussed herein.

The Young's modulus and permeability are two of the most important properties of the material in terms of actuator and sensor applications. For example, the capability of the material to operate as an actuator can be evaluated in terms of its blocked force F_{bl}. This is the maximum force that the actuator is capable of producing under quasi-static conditions

$$F_{bl} = E_3^B A \epsilon_3^s \tag{6.77}$$

where A is the cross-sectional area of the actuator and ϵ_3^s is the saturation strain, or maximum free strain. Note that this output capability is evaluated in terms of the Young's modulus (E_3^B). Another important parameter is the actuator's free strain, or magnetostrictive strain, which depends on the coefficient d_{33}. The value of d_{33} varies significantly with magnetic field, magnetic bias, stress distribution in the material, and frequency. As an example, for a constant magnetic induction with the elastic modulus $E_3^B = 45$ GPa and saturation strain $\epsilon_3^s = 1600\mu\epsilon$, the blocked force for a Terfenol-D rod of diameter 24.5 mm is 34 kN.

The permeability of Terfenol-D is about 5–10 times that of free space. This is quite small compared to ferromagnetic materials such as low carbon steel and is a key parameter in the design of the magnetic flux path. The saturation magnetization M_s is the magnetization of a single domain and is dependent on the atomic configuration of the material. For Terfenol-D, M_s is approximtely 0.79 mA/m (milliampere/meter). The magnetomechanical coupling k^2 represents the fraction of magnetic energy that can be converted to mechanical energy per cycle. Again, there is a considerable variation of k^2 with operating conditions. The value of k for Terfenol-D is around 0.7, whereas its value is about 0.3 for nickel.

Important magnetic and elastic coefficients for a magnetostrictive material designated as M5, biased at 60 kA/m at a pre-stress of 20 MPa [48], are shown in Table 6.4 and important parameters for a Terfenol-D rod are shown in Table 6.5. Note that as a result of the conservation of total volume during magnetostriction, $d_{31} = -d_{33}/2$. In addition, it can be seen that a range of values is given for the Young's modulus and magnetic permeability of the Terfenol-D rod because they are dependent on other parameters.

6.5.1 Magnetomechanical Coupling

Let us consider a one-dimensional rod with a magnetic field aligned along the longitudinal axis (axis-3). The constitutive equations (Eqs. 6.59 and 6.60) can be written as

$$\epsilon_3 = s_{33}^H \sigma_3 + d_{33} H_3 \tag{6.78}$$

$$B_3 = d_{33}^* \sigma_3 + \mu_{33}^\sigma H_3 \tag{6.79}$$

Recall that the superscript ϵ means constant or zero strain, or clamped boundary conditions, and the superscript σ denotes free boundary conditions (zero stress, or constant stress). The superscript B means zero induction or a short-circuit coil. The superscript H means an open-circuit condition. From the equation for ϵ_3

$$\sigma_3 = \frac{\epsilon_3}{s_{33}^H} - \frac{d_{33}}{s_{33}^H} H_3 \tag{6.80}$$

Table 6.4. *Material properties for M5, 60 kA/m bias, 20 MPa pre-stress*

d_{31}, ($\times 10^{-9}$ m/A)	−5.2
d_{33}, ($\times 10^{-9}$ m/A)	10.4
d_{15}, ($\times 10^{-9}$ m/A)	28.0
s^H_{11}, ($\times 10^{-12}$ m²/N)	27.0
s^H_{33}, ($\times 10^{-12}$ m²/N)	42.0
s^H_{44}, ($\times 10^{-12}$ m²/N)	167.0
s^H_{66}, ($\times 10^{-12}$ m²/N)	63.0
s^H_{12}, ($\times 10^{-12}$ m²/N)	−4.3
s^H_{13}, ($\times 10^{-12}$ m²/N)	−19.0
E_1, (GPa)	37.04
E_3, (GPa)	23.81
ν_{12}	0.1593
ν_{31}	0.4524
μ^σ_{11}	6.9
μ^σ_{33}	4.4
k_{31}	0.43
k_{33}	0.69
k_{15}	0.74

Substituting this in the equation for induction

$$\begin{aligned} B_3 &= d^*_{33}\left(\frac{\epsilon_3}{s^H_{33}} - \frac{d_{33}}{s^H_{33}}H_3\right) + \mu^\sigma_{33}H_3 \\ &= \frac{d_{33}}{s^H_{33}}\epsilon_3 + \left(\mu^\sigma_{33} - \frac{d^2_{33}}{s^H_{33}}\right)H_3 \\ &= \frac{d_{33}}{s^H_{33}}\epsilon_3 + \mu^\epsilon_{33}H_3 \end{aligned} \tag{6.81}$$

Table 6.5. *Nominal properties for a Terfenol-D rod*

Nominal Composition	–	$Tb_{0.3}Dy_{0.7}Fe_{1.92}$
Maximum field induced magnetostriction, $\mu\epsilon$	Λ_s	1740
Young's modulus, constant field, MPa	E^H	35–50
Young's modulus, constant induction, MPa	E^B	40–65
Magnetic permeability, constant stress, Tm/A	μ^σ	$3–10 \times 10^{-6}$
Relative permeability	μ_r	5–10
Saturation magnetization, A/m	M_s	0.79×10^6
Magnetostrictive coefficient, m/A	d	$3–20 \times 10^{-9}$
Magnetomechanical coupling factor	k	0.7–0.75
Density, kg/m³	ρ	9250
Resistivity, Ωm	ϱ	60×10^{-8}
Coefficient of thermal expansion, ppm/°C	α_T	12
Compressive strength, MPa	–	≈700
Tensile strength, MPa	–	≈28
Curie temperature, °C	T_c	380

assuming $d_{33}^* = d_{33}$. It can be seen that two values of magnetic permeability can be defined, one at constant stress and another at constant strain

$$\mu_{33}^{\epsilon} = \mu_{33}^{\sigma} - \frac{d_{33}^2}{s_{33}^H} = \mu_{33}^{\sigma} - d_{33}^2 E_3^H \tag{6.82}$$

where E_3^H is the Young's modulus of the material in the 3 direction at constant magnetic field. It is clear that $\mu_{33}^{\sigma} > \mu_{33}^{\epsilon}$. A magnetomechanical coupling factor, or coupling coeffecient k_{33}, can be defined as

$$k_{33}^2 = \frac{d_{33}^2}{s_{33}^H \mu_{33}^{\sigma}} = \frac{d_{33}^2 E_3^H}{\mu_{33}^{\sigma}} \tag{6.83}$$

This results in

$$\mu_{33}^{\epsilon} = \mu_{33}^{\sigma} \left(1 - k_{33}^2\right) \tag{6.84}$$

Note that the value of k_{33} (or in any other direction) is such that $0 \le k \le 1$, although in reality, no material can have $k = 1$. It is possible to define other values of k to represent other directions, such as k_{11} in the 1-direction. In general, d_{33}, E_3^H, and μ_{33}^{σ} vary with magnetic field strength H; and, hence, k_{33} also varies with H. The magnetomechanical coupling factor k is also referred to as the figure of merit of the actuator because k^2 identifies the fraction of magnetic energy that is converted to mechanical energy and vice versa. Improvements in manufacturing techniques have helped increase the coupling factor k to close to 0.7 for Terfenol-D in the longitudinal direction. However, due to magnetic, mechanical, and thermal losses, this factor is reduced for the complete actuator system. These losses can be minimized by careful design of the magnetic path and by incorporating laminated material in the magnetic flux path.

As in the case of the permeabilities, there are also two values of material Young's modulus; one at constant magnetic field E^H and another at constant magnetic induction E^B. It can be seen that the magnetomechanical coupling relations for a magnetostrictive material are fundamentally similar to the electromechanical coupling relations for a piezoelectric material. Rewriting the constitutive relations to eliminate the magnetic field H

$$H_3 = \frac{B_3}{\mu_{33}^{\sigma}} - \frac{d_{33}}{\mu_{33}^{\sigma}} \sigma_3 \tag{6.85}$$

Substituting in the strain equation

$$\begin{aligned} \epsilon_3 &= s_{33}^H \sigma_3 + d_{33} \left(\frac{B_3}{\mu_{33}^{\sigma}} - \frac{d_{33}}{\mu_{33}^{\sigma}} \sigma_3 \right) \\ &= \sigma_3 \left(s_{33}^H - \frac{d_{33}^2}{\mu_{33}^{\sigma}} \right) + \frac{d_{33}}{\mu_{33}^{\sigma}} B_3 \\ &= \sigma_3 s_{33}^H \left(1 - k_{33}^2\right) + \frac{d_{33}}{\mu_{33}^{\sigma}} B_3 \end{aligned} \tag{6.86}$$

Therefore, the relationship between the Young's modulus at constant field and Young's modulus at constant induction is given by

$$E_3^H = E_3^B \left(1 - k_{33}^2\right) \tag{6.87}$$

This means that $E_3^B > E_3^H$. It is clear that the Young's modulus under constant field condition is reduced due to the magnetomechanical coupling factor. In terms of a transducer, where the magnetic field is created by an electric current, the condition of constant magnetic field is equivalent to a condition of constant electric current passing through the coils of the field generator (e.g., a solenoid). Let us consider a case of zero mechanical stress (unloaded rod)

$$\epsilon_3 = d_{33}H_3 \tag{6.88}$$

$$B_3 = \mu_{33}^{\sigma}H_3 \tag{6.89}$$

It can be seen that the magnetostrictive constant d_{33} is the local slope of the ϵ_3-versus-H_3 curve, and the material permeability μ_{33}^{σ} is the slope of the B-H curve (see Fig. 6.23). The magnetomechanical coupling factor is also related to the energy-conversion efficiency of the material. Specifically, it can be shown that the coupling factor is related to the ratio of the elastic energy to the magnetic energy stored in the material

$$k_{33}^2 = \frac{U_{\text{elastic}}}{U_{\text{magnetic}}} \tag{6.90}$$

In one-dimension

$$U_{\text{magnetic}} = \frac{1}{2}\mu_{33}H_3^2 \tag{6.91}$$

$$U_{\text{elastic}} = \frac{1}{2}E_3\epsilon_3^2 \tag{6.92}$$

The maximum magnetic energy for a given field strength takes place when material permeability μ_{33} is maximum, and it occurs at $\mu_{33} = \mu_{33}^{\sigma}$. The elastic energy for a given strain is maximum when E_3 is maximum, and it happens for a case of constant induction E_3^B. Ignoring eddy current losses and material damping, the ratio of the difference between the maximum and minimum elastic energies to the maximum elastic energy is

$$\frac{\frac{1}{2}E_3^B\epsilon_3^2 - \frac{1}{2}E_3^H\epsilon_3^2}{\frac{1}{2}E_3^B\epsilon_3^2} = k_{33}^2 \tag{6.93}$$

A value of zero for the coupling coefficient corresponds to no transduction and unity corresponds to perfect transduction. A typical value is $k \approx 0.7$ for Terfenol-D, indicating that only 50% of the total stored magnetic energy is converted to mechanical energy. This value is quite comparable to the corresponding value for a piezoceramic material. Materials with a high coupling factor are especially preferred as sensors.

Values of k_{33} that have been determined experimentally in Terfenol-D with a stoichiometry of $Tb_{0.27}Dy_{0.73}Fe_{1.95}$ vary from 0.7 to 0.8 (Table 6.6). For $k = 0.707$

$$k^2 = 0.5 \tag{6.94}$$

$$\mu_{33}^{\sigma} = 2\mu_{33}^{\epsilon} \tag{6.95}$$

$$E_3^B = 2E_3^H \tag{6.96}$$

Table 6.6. *Maximum magnetomechanical coupling along longitudinal axis and magnetostriction along easy axis*

Alloy	k_{33}	λ_{100}, $\times 10^{-6}$
Ni	0.31	−23
$TbFe_2$	0.35	2450
$Tb_{0.5}Dy_{0.5}Fe_2$	0.51	1840
$SmFe_2$	0.35	−2100
$Tb_{0.27}Dy_{0.73}Fe_2$	0.74	1620

In addition, the previous discussion shows that the speed of sound in the material c is dependent on its magnetic boundary conditions

$$c^H = \sqrt{\frac{E_3^H}{\rho}} \tag{6.97}$$

$$c^B = \sqrt{\frac{E_3^B}{\rho}} \tag{6.98}$$

where ρ is the material density. Also note that d_{33} has two definitions

$$d_{33} = \left.\frac{d\epsilon_3}{dH_3}\right|_{\sigma} \text{ and} \tag{6.99}$$

$$d_{33}^* = \left.\frac{dB_3}{d\sigma_3}\right|_{H} \tag{6.100}$$

6.5.2 Worked Example

Consider a Terfenol-D rod of length 2 in (50.8 mm) and diameter 0.25 in (6.35 mm), surrounded by a solenoid of 1000 turns with a current of 2 amperes passing through it. Calculate the flux density in the rod, the change in the rod length, and the inductance of the solenoid with the Terfenol-D rod. Assume that the rod is long enough to neglect the demagnetization effects at the ends. Use the following data

$$d_{33} = 20 \times 10^{-9} \text{ m/A or } 1.6 \times 10^{-6} \text{ Oe}^{-1}$$

$$\mu_{33}^{\sigma} = 11.56 \times 10^{-6} \text{ W/A.m or H/m}$$

$$k_{33} = 0.72$$

Solution

The magnetic field intensity H is calculated as

$$H_3 = ni = \frac{1000 \times 2}{0.0508} = 39.37 \text{ kA/m} = 495 \text{ Oe}$$

For an unloaded rod, $\sigma_3 = 0$. Therefore, the magnetic field induced strain is

$$\epsilon_3 = d_{33}H_3 = 20 \times 10^{-9} \times 39370 = 787\mu\epsilon$$

The change in length of the rod, Δl, is given by

$$\Delta l = 787 \times 10^{-6} \times 50.8 = 0.04 \text{ mm} = 0.00157 \text{ in}$$

The flux density is calculated as

$$B_3 = \mu_{33}^{\sigma} H_3 = 11.56 \times 10^{-6} \times 39370 = 0.455 \text{ T or } 4.55 \text{ kiloGauss}$$

$$\Phi = B_3 \times A = 0.455 \times \pi \times (6.35 \times 10^{-3})^2/4 = 14.41 \times 10^{-6} \text{ W} = 1.441 \text{ Maxwells}$$

The inductance of the solenoid in the stress-free condition, L_f, is given by

$$L_f = \mu_{33}^{\sigma} N^2 A/l = 11.56 \times 10^{-6} \times 1000^2 \times (6.35 \times 10^{-3})^2/4/0.0508 = 7.21 \text{ mH}$$

If the Terfenol-D rod is clamped, the inductance of the solenoid is L_o

$$L_o = L_f \left(1 - k_{33}^2\right) = 3.47 \text{ mH}$$

6.5.3 Delta-E Effect

From the previous discussion, it can be seen that due to the magnetomechanical coupling, the Young's modulus of the magnetostrictive material depends on the magnetic boundary conditions (constant induction or constant field). Based on these conditions, it can change by a significant amount, depending on the value of the magnetomechanical coupling factor. In addition to this variation, the Young's modulus of a magnetostrictive material also depends on the bias stress and the magnetic field. The dependence of the Young's modulus on the applied magnetic field is termed the ΔE effect, and it is a consequence of the geometry of the domains as well as the inherent anisotropy of the unit cells. In simple terms, as more domains get oriented along a particular direction, the modulus of the material in that direction changes.

For a given bias stress, the modulus changes for magnetizations between zero and the saturation magnetization. The ΔE effect is defined as the change in the Young's modulus between the magnetically saturated and unsaturated states, divided by the Young's modulus at the unsaturated state [59]

$$\Delta \text{E} = \frac{E_s - E_o}{E_o} \tag{6.101}$$

where E_s and E_o are the Young's moduli at the saturated and unsaturated states, respectively. In the same way, a ΔE can be defined between two values of magnetic field, H_1 and H_2. In this case, the change in Young's modulus is represented as

$$\Delta \text{E}_{H_2 H_1} = \frac{E_{H_2} - E_{H_1}}{E_{H_1}} \tag{6.102}$$

where E_{H_1} and E_{H_2} are the Young's moduli at magnetic fields H_1 and H_2, respectively. Typically, as the magnetic field increases, the Young's modulus decreases.

Changes in the modulus of elasticity with magnetization were observed in materials such as iron and nickel as early as the beginning of the past century [21]. However, for these materials, the changes in modulus are small (0.4% to 18%). In Terfenol-D, a ΔE effect of up to 161% was measured on the application of a magnetic field of strength 342 kA/m at zero bias stress [59]. At a lower temperature

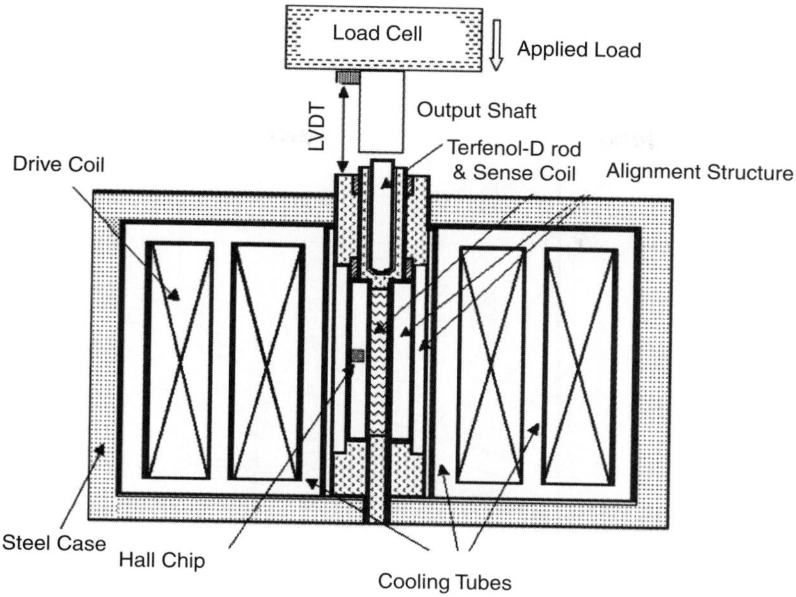

Figure 6.26. Schematic of water-cooled transducer, from Ref. [62].

(−196°C) and a bias stress of 20 MPa, a ΔE effect of 680% has been reported for Terfenol-D [60]. A typical value of ΔE is around 150% for Terfenol-D at room temperature.

The ΔE effect can have a significant effect on the performance of magnetostrictive actuators and sensors. This variation also introduces nonlinearities in the input/output relationship, which are often perceived as undesirable. At the same time, this effect can be used to make novel devices; for example, a resonator in which the natural frequency can be tuned by adjusting a bias magnetic field [61]. In the development of this device, comprehensive quasi-static tests were performed to characterize the Young's modulus and damping ratio of Terfenol-D under controlled thermal, magnetic, and mechanical loading conditions. A ΔE approaching 266% was demonstrated with bias magnetic field levels of up to 61 kA/m. The damping ratio was found to increase, and the Young's modulus was found to decrease, with increasing magnetic field.

The bias stress also has a significant effect on the change in modulus. For example, in a study to characterize the blocked force of a Terfenol-D actuator with varying bias stress and bias fields [62], the Young's modulus was observed to increase monotonically with applied field at a bias stress of 0 MPa. However, at a bias stress of 6.9 MPa, the Young's modulus increased with increasing field but began to level off or even decrease at high field levels (>1000 Oe). As such, the minimum elastic modulus occurs for a combination of high compressive stress and high applied magnetic field. Thus, with an appropriate selection of bias compressive stress, the desired modulus can be obtained with a minimum external magnetic field. Therefore, a comprehensive set of experimental data is essential to characterize the material at all operating conditions. Empirical relations can be extracted from the data and subsequently used in design tools.

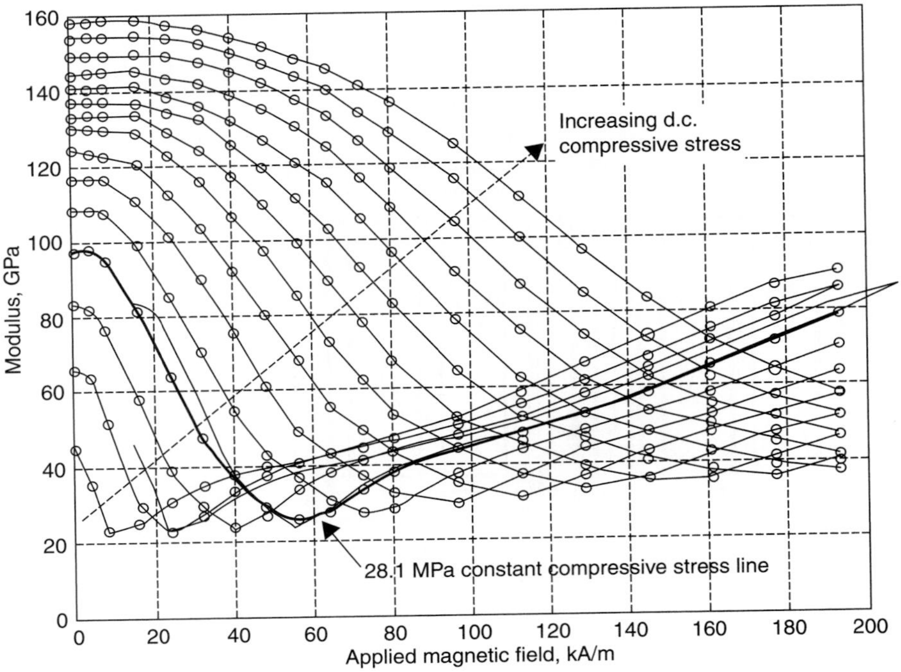

Figure 6.27. Modulus at constant compressive stress values (6.9–103.5 MPa) for constant applied fields (0–193.2 kA/m), from Ref. [61].

To characterize Terfenol-D under controlled quasi-static environments (thermal, mechanical, and magnetic), Kellogg [62] built a special water-cooled apparatus (Fig. 6.26). The material specimen is placed at the center, surrounded by a sense coil for measuring the magnetic induction. A Hall effect sensor is used to measure the applied magnetic field. Water cooling tubes surround the cylindrical solenoid, and loads are transmitted from the specimen to a load cell.

Figure 6.27 presents the modulus of Terfenol-D for a range of compressive stress values (6.9 to 103.5 MPa in steps of 6.9 MPa) and applied fields. The modulus values were extracted from stress-strain curves. Considering a representative 28.1 MPa case (highlighted), as the field is increased from zero to 50 kA/m, the modulus decreases from 72 GPa to 14 GPa; thereafter, the modulus increases with further increase in magnetic field. It becomes 69 GPa at a maximum field of 193.2 kA/m. In this case, the $\Delta E_{H_2H_1}$ is determined as -414% with $H_1 = 0$ kA/m and $H_2 = 50$ kA/m. At high compressive loading with a low or zero magnetic field, the modulus becomes 112 GPa. It is apparent that the largest changes in modulus may be achieved using the appropriate pre-stress at a low magnetic field.

6.5.4 Magnetostrictive Composites

A composite material can be produced by solidifying Terfenol-D powder in a matrix. The finished composite then exhibits magnetostrictive properties. These materials have numerous advantages over pure magnetostrictive materials. They can be cast into complex shapes, and they show a much higher machinability and ductility compared to Terfenol-D. Therefore, they are more mechanically robust under harsh operating conditions. Another major advantage is that there are no eddy current losses because the matrix is electrically nonconducting. In addition, the composite is lighter than the magnetostrictive material itself. The main disadvantage of magnetostrictive composites is their strain capability. The volume fraction of magnetostrictive material in the composite is typically 10%–40%. As a result, the overall strain of the composite is less than that of the magnetostrictive material itself [26, 53].

Wu and Anjanappa [53] and Krishnamurthy et al. [63] developed a simple rule-of-mixture model to calculate the response of magnetostrictive particulate composites. Flatau et al. [64] discussed magnetostrictive particle composites in terms of the underlying physical processes that occur during fabrication, material characterization, design considerations, and structural health sensing.

6.6 Magnetostrictive Actuators

Typically, a magnetostrictive actuator consists of a multilayer solenoid for magnetic field generation, a magnetic flux return path for routing the magnetic flux into the magnetostrictive element, a permanent magnet to provide a DC bias magnetic field, and a mechanical preload mechanism. This design can help to achieve maximum bidirectional transduction of energy between the magnetic and elastic states. For actuation, the direct effect (Joule effect) is used where the externally applied magnetic field induces magnetization in the material causing a measurable change in strain. The inverse effect (Villari effect) is used in sensors where the mechanical energy is transformed into magnetic energy.

For optimum performance in an actuator, both permanent magnets and DC currents in the solenoid are often used. The precompression mechanism helps to expand the range of magnetostriction by increasing the population of magnetic moments normal to the rod axis. Also, the precompression helps to safeguard the brittle magnetostrictive material from tensile stresses.

Applications of magnetostrictors can be broadly classified into the following three categories based on the operating frequency:

1. Low-frequency, high-power sonar applications: The initial impetus for the development of Terfenol-D was for underwater applications (for the U.S. Navy). The goal was to develop a small size sonar system that could radiate high acoustic–power signatures. For example, flextensional transducers were developed in the 1930s based on PZT ceramics. These PZT actuators were natural candidates to be replaced by Terfenol-D actuators to improve their low-frequency vibration characteristics. Technological challenges associated with flextensionals include stress-induced fatigue, high deep-water hydrostatic pressure, and compactness. Sonar designs include flexing of oval-shaped shells, piston-type actuation (Tonpilz-type transducers), and ring-type actuation.
2. Motion generation against external loads: The goal is to generate actuation forces and motion against external loads with compact devices over a range of operating frequencies. The motion can be linear or rotational. Applications include active vibration control, micropositioners, valve controls, and active struts. Stroke is amplified by either using a long active element or incorporating a motion amplification mechanism.

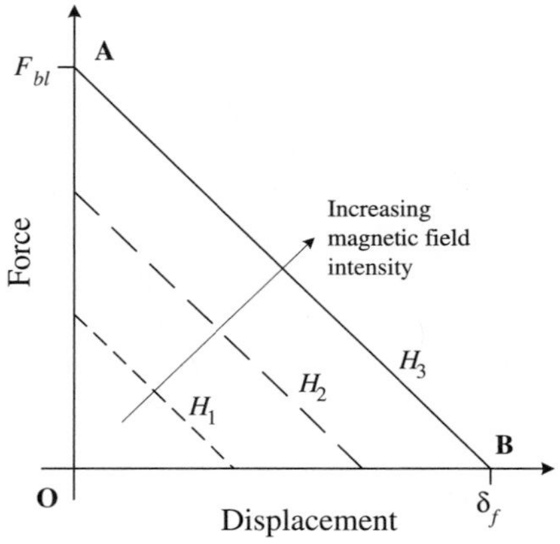

Figure 6.28. Actuator force as a function of displacement.

3. Ultrasonic applications: This involves high-frequency wave generation (above 20 kHz) for a broad range of applications that include industrial machining, welding, cleaning, and medical applications. The power losses due to eddy currents are proportional to the square of operating frequency, and they become a key factor in the design of these devices. There has been a growing interest in the development of ultrasonic motors for commercial and medical applications.

Several issues are crucial to the efficient operation of a magnetostrictive actuator. In addition, the design of an actuator capable of generating oscillatory output displacement is complicated by several factors:

(1) generation of the required oscillatory magnetic field
(2) providing a bias magnetic field and compressive pre-stress
(3) power supply to the device

Most commercially available magnetostrictive actuators contain Terfenol-D as the active material. A few important material properties of Terfenol-D are listed in Table 6.5.

Two of the key characteristics of any actuator are the maximum free stroke (free displacement) δ_f and the blocked force F_{bl}. Both of these parameters are a function of the magnetic field intensity. When expressed as a linear relation, they can be plotted as shown in Fig. 6.28. Such a plot is typically known as the actuator load line. As the output force level increases, the output displacement capability decreases. F_{bl} is the maximum force capability of the actuator and the output displacement at this force is zero. Let us assume that the stiffness of the longitudinal rod is k_T N/m, given by

$$k_T = \frac{E_c^H A_c}{L} \tag{6.103}$$

where E_c^H is the Young's modulus of the actuator at constant induction (constant current), A_c is the cross-sectional area, and L is the total length of the actuators. The axial displacement of the rod is

$$w = \epsilon_{33} L \tag{6.104}$$

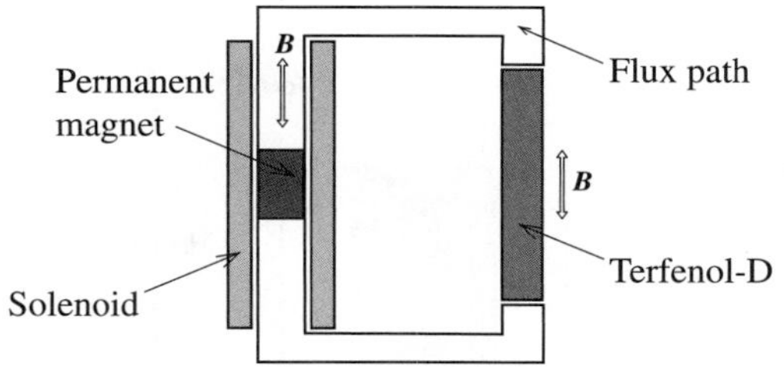

Figure 6.29. Magnetic field generation.

and the actuation force is

$$F = \frac{E_c^H A_c}{L}(\delta_f - \epsilon_{33} L) \tag{6.105}$$

$$\frac{F}{F_{bl}} = \left(1 - \frac{\epsilon_{33} L}{\delta_f}\right) \tag{6.106}$$

To increase the effectiveness of output dynamic strains, mechanical resonance is exploited. Eddy current losses and device-specific magnetomechanical and electrical resonances limit the operating bandwidth of magnetostrictive devices to the low ultrasonic regimes (below 100 kHz).

6.6.1 Generation of the Magnetic Field

The magnetic field required for inducing strain in the active material is usually generated by means of a current carrying coil. This is typically in the shape of a solenoid, resulting in a relatively uniform magnetic field over a long length. A core of highly permeable material is often used to direct and concentrate the magnetic field where desired, forming a flux path. The magnetic field is concentrated in this flux path because of its much larger permeability compared to air. The bias magnetic field can be produced either by a constant DC current in the solenoid or by placing a permanent magnet in the flux path. The latter method is preferable due to its much lower power requirements and decreased heating effects.

A schematic of a Terfenol-D rod placed in a magnetic field generator is shown in Fig. 6.29. In this example, the bias magnetic field is created by a permanent magnet. The weight and volume associated with the coil windings, magnetic core, and other elements of the flux path result in a decrease in the energy density of the actuator. It is difficult to design a compact field generation system because of the problems associated with saturation of the magnetic core and dissipation of the heat produced in the coil windings. Consequently, in applications with stringent constraints on allowable weight penalty or available volume, actuators based on piezoelectric or electrostrictive materials may be preferred despite their lower induced strains.

6.6.2 Construction of a Typical Actuator

A cross section of a typical magnetostrictive actuator is shown in Fig. 6.30. The main components of the device can be seen: active material (Terfenol-D rod), field generation system (solenoid, flux return), an output piece to transmit the induced strain from the active material, and a pre-load spring to exert a compressive pre-load on the active material.

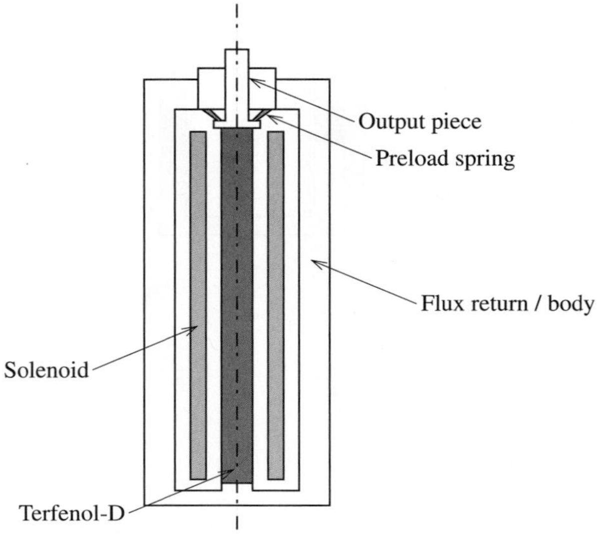

Figure 6.30. Cross section of a typical magnetostrictive actuator.

Such linear actuators with strokes of up to 250 μm are commercially available. They can be assembled directly into a structure with only a source of electrical power needed to complete the installation. Magnetostrictive actuators producing other types of output displacements can be created using active material of different shapes; however, the same basic configuration is applicable in each case.

6.6.3 Measurement of Magnetic Field

At this point, it is worth discussing the measurement of the magnetic field or the magnetic state of the magnetostrictive material. The constitutive relations of the material are given in terms of the magnetic field. Whereas the magnetic flux is constant at any point in a magnetic circuit, the magnetic field depends on the permeability of the material, which is a function of the stress in the material as well as the magnitude of the magnetic induction and the magnetization in the material. Hence, it is often more convenient to measure the magnetic induction and to express the behavior of the material in terms of magnetic induction. We discuss two commonly used measurement techniques in the following subsections.

Hall Effect Sensor

The Hall effect is a widely used method of measuring magnetic induction. A schematic of this effect is shown in Fig. 6.31. The sensor consists of a thin sheet of conducting material (shaded area) placed in a plane perpendicular to the magnetic induction (which is directed along the negative z direction). A constant current i_s is passed across the length of the sheet (along the x-axis). The magnetic induction produces a voltage V_h across the width of the sheet (along the y-axis) as a result of the Hall effect. V_h is a linear function of the magnetic induction, which can be calculated from the measured voltage after calibration. This sensor is introduced to the magnetic circuit and measures the magnetic induction perpendicular to its plane. Hall sensors of very small dimensions (i.e., thickness 0.01 in, sensing area

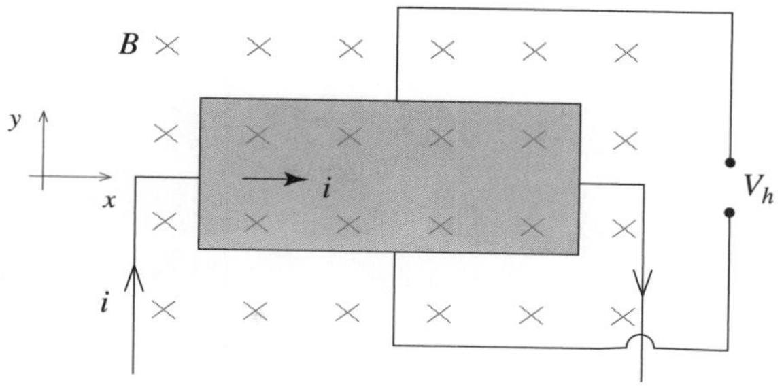

Figure 6.31. Operating principle of a Hall-effect sensor.

0.03 in × 0.06 in) are commercially available, which facilitates their installation in thin air gaps.

Sense Coil

This is based on Faraday's law. The sense coil is a solenoid of known turns, wound with thin wire directly onto the sample. It measures the magnetic induction along its axis. Therefore, it is well suited to measure the magnetic induction produced by a solenoid, and it is often wound around its core (e.g., around a magnetostrictive rod). From Faraday's law (Eq. 6.29), we know that the voltage produced is proportional to the product of the number of turns in the coil and the rate of change of magnetic flux.

6.6.4 DC Bias Field

To operate the actuator in a linear range, a DC bias magnetic field is needed. This would result in output response at the same frequency as the input field frequency. The bias field can be introduced either by means of a permanent magnet or by passing a DC current through the magnetic coil. A simple circuit to achieve this is shown in Fig. 6.32. The blocking capacitor C isolates the AC source from the DC power supply, V_{DC}, because of its infinite impedance at DC. Similarly, the blocking inductance isolates the DC power supply from the AC source because its impedance increases with frequency. In this way, the magnetostrictive actuator can be excited with an alternating current superimposed on a DC current, without complex electronics to protect the AC and DC power supplies. Note that the value of the blocking capacitor must be chosen such that it presents an impedance high enough to block the DC current while being low enough to let the AC current pass

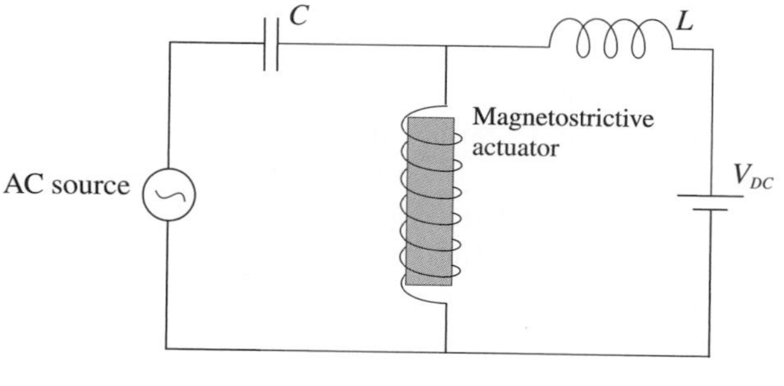

Figure 6.32. Simple circuit to apply a DC bias field to the magnetostrictive element.

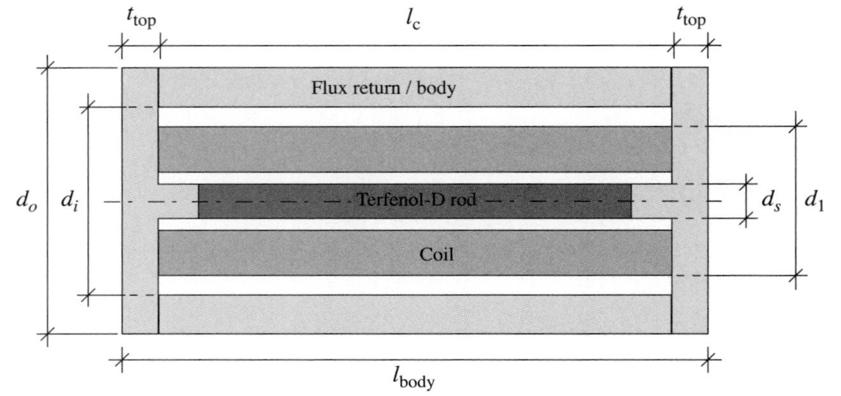

Figure 6.33. Cylindrical magnetic field generator for a magnetostrictive actuator.

through. Similarly, the value of the blocking inductor must be chosen such that it presents an impedance high enough to block the AC current while being low enough to let the DC current pass through.

6.6.5 Design of the Magnetic Field Generator for a Magnetostrictive Actuator

The generation of the magnetic field is an important aspect in the design of a practical magnetostrictive actuator. The magnetic field can be generated either by a combination of permanent magnets and a current carrying coil or entirely by a current carrying coil. For a specific value of magnetic field acting in the active material, it is possible to design several different current carrying coils with varying parameters such as wire thickness and number of turns. Let us examine the design of an optimum magnetic field generator for a typical magnetostrictive actuator as described in Section 6.6.2, where the entire magnetic field is generated by the current in the coil.

A schematic of a typical magnetic field generator is shown in Fig. 6.33. The geometry of the flux return path is cylindrical because the magnetostrictive rod is cylindrical. The coil is wound around the rod and induces a magnetic field along its longitudinal axis. To simplify the analysis, the return spring and output piece have been included as part of the two endcaps. Therefore, an idealized flux return path is considered, consisting of the body and the two endcaps.

The major geometrical parameters of the magnetic field generator are determined by the dimensions of the magnetostrictive rod. The main requirement is to ensure that the magnetic field acting on the bar is as uniform as possible. A general algorithm for design of the field generator is described as follows:

1. Determine the required magnetic field in the magnetostrictive material. For example, if a strain of 1000 $\mu\epsilon$ is required, the magnetic field in the material, H_s, can be obtained from the λ-H curves of the material. The permeability of the material, μ_s, at that operating condition can be obtained from the B-H curves. From this, the magnetic induction in the material can be calculated as $B_s = \mu_s H_s$.
2. Determine the length of the solenoid l_c. The length of the solenoid is based on the length of the magnetostrictive rod, l_s. A fringing factor F of approximately

20% is added to this length to ensure that the field in the material is close to uniform.

$$l_c = l_s + Fl_s = 1.2l_s \tag{6.107}$$

3. Determine the dimensions of the actuator body. The actuator body provides the flux return path. It is important to ensure that the dimensions of the flux path are large enough to ensure that the material is well within its magnetic saturation limits, so that its reluctance is as low as possible. In many cases, the outer diameter of the actuator is dictated by geometrical constraints of the final application. Therefore, we assume that the outer diameter d_o is a fixed parameter with input as a requirement. We also assume that the thickness of the top and bottom endcaps are equal. The unknown quantities are the inner diameter of the body (d_i) and the thickness of the endcaps (t_{top}). Once these two parameters are found, the overall length of the actuator, l_{body}, is given by

$$l_{\text{body}} = l_c + 2t_{\text{top}} \tag{6.108}$$

The magnetic flux at any point in the magnetic circuit is a constant and is equal to the flux in the magnetostrictive rod. For an efficient design, we should ensure that no part of the magnetic circuit is saturated. The dimensions d_i and t_{top} are evaluated with the condition that the flux through critical points on the top and bottom endcaps (Φ_e), as well as through the cylindrical body (Φ_b), are 80% of that required to drive the material to saturation. This can be obtained by solving the equations

$$\Phi_e = B_s A_s = 0.8 B_{\text{sat},1018} \pi d_s t_{\text{top}} \tag{6.109}$$

$$\Phi_b = B_s A_s = 0.8 B_{\text{sat},1018} \frac{\pi(d_o^2 - d_i^2)}{4} \tag{6.110}$$

where d_s and A_s are the diameter and cross-sectional area, respectively, of the magnetostrictive rod. $B_{\text{sat},1018}$ is the saturation–magnetic induction of the material of the body, which we have chosen as 1018 low-carbon steel. Assuming $B_{\text{sat},1018} = 1.5T$ (which is a conservative estimate), we can obtain the values of t_{top} and d_i.

However, the requirement that the actuator body be adequately stiff (to ensure that it does not undergo large elastic deformations) usually dictates much larger values of t_{top} and d_i than the magnetic induction saturation condition. In general, an actuator body designed to meet strength and stiffness specifications is more than adequate to provide an effective flux return path.

4. Estimate the required *mmf*. This estimate is based on Ohm's law for magnetism in which the *mmf* is equal to the flux in the circuit multiplied by the sum of the reluctances in the circuit. This law is only valid for cases in which there is a well-defined magnetic path. Furthermore, the reluctance of a magnetic material changes with the concentration of flux. Although this implies that Ohm's law for magnetism is nonlinear, it will still provide a useful estimate of the *mmf*. In this initial calculation, assume that the *mmf* required is dependent on the reluctances of the air gap and sample only. The effect of the flux return path is neglected for now but will be added in a later step to serve as a check on the initial estimate. This assumption is only valid if the permeability of the flux return is large and its reluctance is small compared to that of the magnetostrictive rod

and the air gap. The *mmf* is given by

$$mmf = H_s l_s \frac{\mathbb{R}_c + \mathbb{R}_s}{\mathbb{R}_s} = N_{tot} i_w \tag{6.111}$$

where $\mathbb{R}_c$ and $\mathbb{R}_s$ are the reluctances of the magnetic circuit (excluding the magnetostrictive rod) and the magnetostrictive rod, respectively. N_{tot} is the total number of turns in the coil and i_w is the current passing through the coil. Typically, $\mathbb{R}_c >> \mathbb{R}_s$ if 1018 steel is used as the body material. However, due to unavoidable air gaps and flux leakage, the reluctance of the magnetic circuit is considerably increased. Because the relative permeability of Galfenol is much higher than Terfenol-D (by approximately two orders of magnitude), the reluctance of a Galfenol rod will be much less than the reluctance of a Terfenol-D rod of the same dimensions. Therefore, in a practical actuator, it is found that for a Galfenol rod, $\mathbb{R}_c \approxeq \mathbb{R}_s$ and for a Terfenol-D rod, $\mathbb{R}_c \ll \mathbb{R}_s$. For design purposes, the following empirical expressions (motivated by experiments on magnetic transducers) are used

$$N_{tot}\, i_w = 1.05\, H_s\, l_s \text{ for Terfenol-D} \tag{6.112}$$

$$N_{tot}\, i_w = 2.0\, H_s\, l_s \text{ for Galfenol} \tag{6.113}$$

5. Calculate the coil geometry. For a chosen wire gauge, the number of turns per layer N_t and the number of layers in the coil N_l can be determined. The product of these two gives the total number of turns, N_{tot}. We assume that the coil is wound up to a diameter d_1. For a wire of diameter d_w

$$N_t = \frac{l_c}{d_w} \tag{6.114}$$

$$N_l = \frac{d_1 - d_s}{2d_w} \tag{6.115}$$

In these equations, any imperfections in winding the wire are neglected. This assumption becomes more accurate as the wire diameter decreases.

6. Determine the electrical properties of the circuit. Now that the geometry of the coil has been determined, the length of the wire in the coil, l_w, the resistance of the coil, and the inductance of the coil can be calculated. The length of the wire is given by

$$l_w = \pi \frac{d_1 + d_s}{2} N_{tot} \tag{6.116}$$

For a wire of cross-sectional area A_w, with a resistivity ϱ_w, the resistance of the coil R_w and the inductance of the coil L_w are

$$R_w = \frac{\varrho_w l_w}{A_w} \tag{6.117}$$

$$L_w = \frac{\mu_s \pi d_s^2 N_{tot}^2}{4 l_c} \tag{6.118}$$

7. Determine the magnitude of the voltage and the power required by the coil. The current flowing in the coil, i_w, is determined by dividing the *mmf* by the assumed number of turns.

$$i_w = \frac{mmf}{N_{tot}} \tag{6.119}$$

From the required current and the calculated coil impedance, the voltage V_w, and the power required P_w, at a given operating frequency ω, can now be determined.

$$V_w = i_w\sqrt{R_w^2 + \omega^2 L_w^2} \tag{6.120}$$

$$P_w = i_w^2\sqrt{R_w^2 + \omega^2 L_w^2} \tag{6.121}$$

Note that the heat produced in the coil, P_d, is due purely to Ohmic heating, and is given by

$$\begin{aligned} P_d = i_w^2 R_w &= \frac{(mmf)^2}{N_{\text{tot}}^2}\frac{\rho_w l_w}{A_w} \\ &= 4\rho_w (mmf)^2 \frac{(d_1 + d_s)}{(d_1 - d_s)l_c} \end{aligned} \tag{6.122}$$

From this equation, it can be seen that the minimum dissipated power is obtained when $d_1 \gg d_s$. Because the maximum value of the coil diameter is constrained by the inner diameter of the actuator body, it follows that for minimum power dissipation, the entire volume inside the actuator body must be used to wind the coil. Therefore, for minimum dissipated power, $d_1 = d_i$. Also note that the dissipated power is independent of the wire diameter. Similarly, by substituting for R_w and L_w, it can be seen that the total power is also independent of the wire diameter. However, the required voltage increases with decreasing wire diameter.

8. Determine the mass of the magnetic field generator. Provided that the density of the material used in the body and in the wire is known, the mass of the magnetic field generator can be calculated. The mass of the body M_b and of the coil M_w are given by

$$M_b = \rho_b\left[\frac{\pi d_o^2}{4}.2.t_{\text{top}} + \frac{\pi(d_o^2 - d_i^2)}{4}l_c\right] \tag{6.123}$$

$$M_w = \rho_w l_w A_w = \frac{\rho_w}{16}\pi^2\left(d_1^2 - d_s^2\right)l_c \tag{6.124}$$

where ρ_c is the density of the material of the body and ρ_w is the density of the material of the wire. The mass of the rods transferring strain from the magnetostrictive rod is neglected compared to the mass of the body. The total mass of the magnetic field generator is $M_{\text{tot}} = M_b + M_w$. Note that the coil mass is also independent of the wire diameter.

9. The coil impedance, required power, and total mass can be calculated for different values of wire gauge. Based on constraints such as maximum driving voltage, the final value of wire gauge can be chosen. Alternately, a winding ratio W_r can be defined as

$$W_r = (d_1 - d_s)/(d_i - d_s) \tag{6.125}$$

This ratio quantifies the fraction of the actuator body that is filled with the coil. For $W_r = 1$, the entire actuator body is filled by the coil windings. These calculations can be performed for different values of $0 \le W_r \le 1$ to choose a coil with a lower mass at the cost of an increase in dissipated power.

6.6.6 Worked Example: Design of a Magnetic Field Generator for a Magnetostrictive Actuator

A numerical example illustrates the design procedure. Consider a Terfenol-D rod of diameter 6.35 mm (0.25 in) and length 50.8 mm (2 in). A strain of 1000 $\mu\epsilon$ is required from the actuator. Let us investigate the possible dimensions of the coil and actuator body to obtain the specified strain. The data required for the calculation are summarized as follows:

(a) General data
- Required strain: 1000 $\mu\epsilon$
- Fringing factor F: 0.2
- Design operating frequency: 500 Hz
- Maximum operating voltage: 100 V

(b) Terfenol-D rod
- Length l_s: 50.8 mm
- Diameter d_s: 6.35 mm

(c) Body
- Material: 1018 steel
- Density: 7850 kg/m^3
- Relative permeability μ_b: 1000
- Saturation magnetic induction $B_{sat,1018}$: 1.5 T
- Outer diameter d_o: 38.1 mm
- Minimum body wall thickness: 4.00 mm
- Minimum endcap thickness: 6.35 mm

(d) Coil wire
- Density: 8906 kg/m^3
- Resistivity: 1.72×10^{-8} ohm-m

Solution

Based on the required strain and the λ-H curves of Terfenol-D, a magnetic field of $H_s = 60$ kA/m is required (magnetic induction $B_s = 0.8$ T). This yields a required *mmf* of

$$mmf = N_{\text{tot}} i_w = 1.05 H_s l_s = 3200.4 \text{ A-turns}$$

For the given rod length and fringing factor, the length of the coil is

$$l_c = (1 + F) l_s = 60.96 \text{ mm}$$

The body dimensions are first calculated on the basis of preventing magnetic saturation in the body. For the given saturation–magnetic induction of steel, $B_{sat,1018} = 1.5$ T, and the magnetic induction in the Terfenol-D rod of $B_s = 0.8$ T, the thickness of the endcaps and the inner diameter of the body are calculated as

$$t_{\text{top}} = \frac{B_s A_s}{0.8 \pi d_s B_{sat,1018}} = 1.06 \text{ mm}$$

$$d_i = \sqrt{d_o^2 - \frac{4 B_s A_s}{0.8 \pi B_{sat,1018}}} = 37.7 \text{ mm}$$

Table 6.7. *Parameters as a function of wire gauge for a coil with $W_r = 1$ (body entirely filled by the coil)*

AWG	14	16	18	20	22	24	26	28	30
Wire diameter, mm	1.629	1.291	1.025	0.812	0.644	0.510	0.405	0.322	0.255
N_t, turns/layer	37.400	47.200	59.400	75.100	94.600	119.500	150.400	189.500	239.500
N_l, layers	7.300	9.200	11.600	14.600	18.400	23.300	29.300	36.900	46.600
l_w, m	15.621	24.855	39.419	62.914	99.914	159.169	252.283	400.450	639.501
R_w, ohms	0.129	0.327	0.823	2.096	5.287	13.417	33.708	84.927	216.588
L_w, mH	0.730	1.850	4.640	11.820	29.820	75.680	190.120	479.020	1221.640
i_w, A	11.730	7.372	4.649	2.913	1.834	1.151	0.726	0.458	0.287
V_w, V	26.905	42.810	67.894	108.359	172.087	274.145	434.519	689.713	1101.443

This gives a body wall thickness of $(d_o - d_i)/2 = 0.18$ mm. It can be seen that both the endcap thickness and the body wall thickness are much smaller than the minimum requirements for the application (most probably based on body stiffness specifications). Therefore, the minimum requirements are chosen for these dimensions.

Let us first design a coil for minimum dissipated power. This requires a winding ratio $W_r = 1$. Assuming a wire diameter, the number of turns in the coil is calculated. From this, the resistance and inductance of the coil are found. The current required is found by dividing the *mmf* by the total number of turns and is used to calculate the voltage required. A table of these parameters can be generated for different values of wire diameter (Table 6.7). Based on the requirement that the voltage should be less than 100 V, we choose an 18-AWG wire for the coil. The power dissipated in the coil and the mass of the coil are calculated as

$$P_d = 4\rho_w (mmf)^2 \frac{(d_1 + d_s)}{(d_1 - d_s) l_c} = 17.78 \text{ W}$$

$$M_w = \frac{\rho_w}{16} \pi^2 \left(d_1^2 - d_s^2\right) l_c = 290 \text{ gm}$$

The mass of the body is

$$M_b = \rho_b \left[\frac{\pi d_o^2}{4} .2.t_{\text{top}} + \frac{\pi (d_o^2 - d_i^2)}{4} l_c \right] = 319 \text{ gm}$$

If a lighter coil is required, the winding ratio can be decreased and the entire calculation can be repeated as described previously. Figure 6.34 shows the variation of the dissipated power and the coil mass as a function of the winding ratio. From these curves, it can be seen that the dissipated power does not change much at high winding ratios compared to that at low winding ratios. However, the coil mass shows an approximately linear dependence on winding ratio. Therefore, by choosing a high winding ratio ($W_r < 1$), we can obtain a significant saving in coil mass at the cost of a small increase in dissipated power. We choose a winding ratio of 80%, which results in a coil mass of 202 gm and a power dissipation of 19.33 W. The decrease in winding ratio results in a different voltage requirement, based on which a 20-AWG wire is chosen for the coil.

The resulting values of actuator dimensions, power dissipated, mass, required current, and required voltage are shown in Table 6.8. The number of layers and number of turns can be rounded off to the next highest integer value. Note that the total mass of the coil and body is 520 gm, whereas the mass of the Terfenol-D rod alone is approximately 15 gm. This gives an idea of the extra mass of the components

Table 6.8. *Parameters of Terfenol-D actuator designed to produce* $1000\mu\epsilon$

Body Dimensions	
d_o, mm	38.1
d_i, mm	30.1
t_{top}, mm	6.35
l_{body}, mm	73.7
Coil geometry	
Wire gauge, AWG	20
Wire diameter, mm	0.81
d_1, mm	25.35
l_c, mm	61.0
N_t, turns	75.1
N_l, turns	11.7
Mass of components	
Coil, gm	319
Body, gm	202
Total, gm	520
Electrical quantities	
Current required, A	3.64
Voltage required, V	86.7
Power dissipated, W	19.33

required to apply the magnetic field to the magnetostrictive element. This extra mass results in a large decrease in the overall energy density of the actuator compared to the energy density of the active material itself.

6.6.7 Power Consumption and Eddy Current Losses

The power consumption of a magnetostrictive actuator can be calculated by knowing the effective impedance of the field generation system. The effective impedance consists of the electrical impedance of the coils as well as a component due to the mechanical impedance of the combination of the actuator and load. The mechanical impedance represents the output work of the actuator, and the electrical impedance

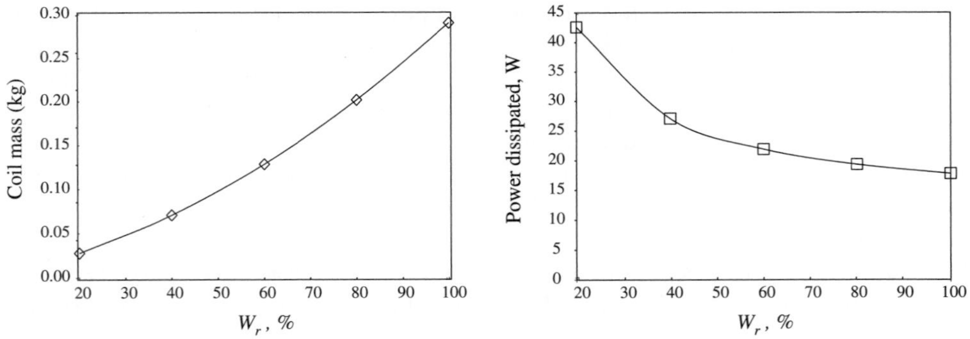

(a) Mass of the coil (b) Power dissipated in the coil

Figure 6.34. Power as a function of wire gauge ($N_{tot} = 50$), for a specified magnetic field.

represents the power required purely to generate the magnetic field. As in the case of piezoelectric actuators, the power supply requirements are determined primarily by the electrical impedance of the actuator.

The electrical impedance of a magnetostrictive actuator appears as a large inductance due to the coils and magnetic core. In addition to the inductance, the coil's windings also have a finite resistance. To maintain the required field in the device, coil currents on the order of a few amperes are not uncommon. Due to the large inductance of the actuator, the power supply must be capable of operating at high voltages to maintain these large currents, especially at high actuation frequencies.

The electrical impedance, Z, of the actuator at an actuation frequency ω can be written as

$$Z = R_l + j\omega L \tag{6.126}$$

where R_l is the resistance of the coil windings and L is the inductance of the coil. The resistance of the coil windings results in Ohmic heating losses in the coil given by

$$P_{\text{heat}} = i_c^2 R_l \tag{6.127}$$

where i_c is the current passing through the coil, given by the sum of the current i_a required to create the actuation magnetic field and i_b, the steady current required to create the bias magnetic field.

$$i_c = i_a + i_b \tag{6.128}$$

Because the heat generated is proportional to the square of the current, the advantage of using a permanent magnet to generate the bias magnetic field is evident. In such a case, $i_b = 0$ and the DC power requirements and heat generation are considerably reduced. In the case of dynamic actuation, the magnitude of the voltage required, V_l, can be written as

$$V_l = i_a\sqrt{R_l^2 + \omega^2 L^2} \tag{6.129}$$

It can be seen that the required voltage increases with the actuation frequency. Careful attention must be given to the design of power supplies for magnetostrictive actuators operating in the high-frequency range ($\approx$1 kHz). In addition to the Ohmic losses in the coil windings, another major source of loss is due to eddy currents. As a result of the alternating magnetic induction, and due to the conductive nature of the magnetic material, eddy current loops are set up in the material. This is schematically shown in Fig. 6.35. The eddy currents result in an energy loss, primarily through Ohmic heating as a result of the material's resistivity. The eddy currents also induce a magnetic induction in a direction opposite to the applied magnetic induction, resulting in a higher required power to achieve the same induced strain. A simple and commonly used remedy is to laminate the magnetic material, separating each laminate by a layer of nonconducting material. This has the effect of breaking up the eddy currents into much smaller loops and greatly attenuates their effect, as shown in Fig. 6.35. The eddy current losses can be accounted for by using a complex permeability in the calculations [65, 7].

Consider a lamina of thickness h, width b, and length l (Fig. 6.36(a)). The lamina is placed in a sinusoidally varying magnetic induction B, aligned perpendicular to

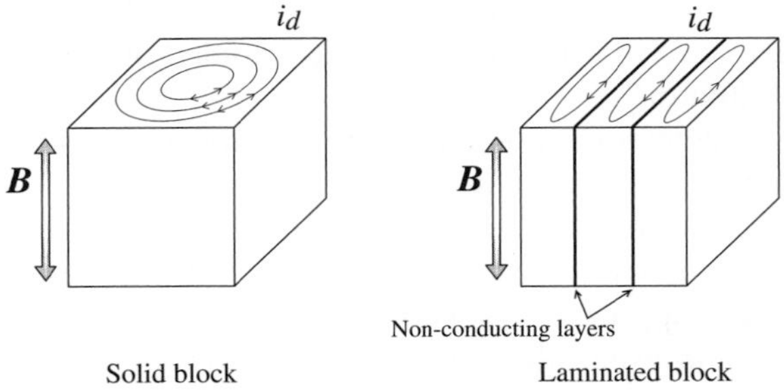

Figure 6.35. Eddy currents set up due to an alternating magnetic field.

the x-y plane. The magnetic induction can be written as

$$B = B_o \sin \omega t \tag{6.130}$$

where B_o is the amplitude and ω is the circular frequency (rad/s). The changing magnetic induction produces a voltage in the laminate that results in a flow of current. Consider an elemental piece of the lamina of thickness h, width dz, and length dx. A current element of thickness dy can be constructed as shown in Fig. 6.36(b), which is part of a current loop extending across the entire cross section of the lamina (Fig. 6.36(a)). The resistance of the element is given by

$$dR = \varrho \frac{l}{A} = 2\varrho \frac{dx}{dydz} \tag{6.131}$$

where ϱ is the resistivity of the laminate material. The voltage produced is

$$dV = -\frac{d\Phi}{dt} = -2\ y\ dx\ B_o \omega \cos \omega t \tag{6.132}$$

Therefore, the power dissipated in the element as a function of time is given by

$$dP(t) = \frac{(dV)^2}{dR} = \frac{4y^2 B_o^2 \omega^2 \cos^2 \omega t}{2\varrho}\ dy\ dx\ dz \tag{6.133}$$

The average power dissipated, dP_d, is given by

$$\begin{aligned} dP_d &= \frac{\omega}{2\pi} \int_0^{2\pi/\omega} dP(t)\ dt \\ &= \frac{B_o^2 \omega^2}{3\varrho} \frac{h^3}{8}\ dz\ dx \end{aligned} \tag{6.134}$$

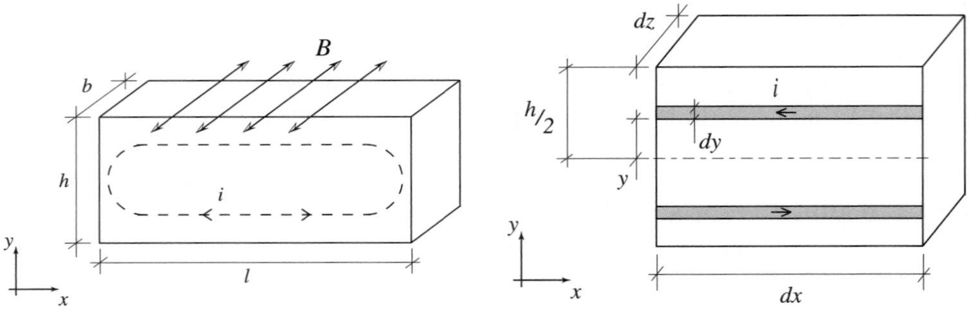

(a) Eddy current in a lamina (b) Element of lamina and current loop

Figure 6.36. Modeling of eddy-current losses.

The power dissipated per unit volume is found by dividing this power by the volume of the element, $h\,dx\,dz$. The resulting expression for eddy current losses per unit volume is

$$P_d = \frac{B_o^2 \omega^2 h^2}{24\varrho} \tag{6.135}$$

It can be seen that the eddy current losses are inversely proportional to the resistivity of the lamina material and directly proportional to the square of the lamina thickness. Therefore, an effective way to minimize eddy current losses is to construct a flux return path by stacking many thin laminae (typical thickness can be on the order of 0.5 mm). Note that this analysis assumes that the eddy currents do not affect the flux density and that the permeability of the lamina material is constant.

6.6.8 Magnetostrictive Particulate Actuators

It may be possible to embed micron size (50-300 μm) magnetostrictive particles in a host structure without any significant effect on the mechanical integrity of the structure. These particles can then be excited using a remote magnetic field. Anjanappa and Wu [53] uniformly distributed Terfenol-D particles in a polymeric host material, which were magnetically oriented in a specific direction, by applying a magnetic field during fabrication. In practice, it is difficult to achieve a perfect orientation of particles; hence, the attainable induced strain is comparatively less. It was shown that the volume fraction, orientation field, modulus of elasticity of the matrix, and pre-stress have an important role in defining the performance of particulate composites.

6.7 Magnetostrictive Sensors

Magnetostrictive sensors take advantage of the coupling between the elastic and magnetic states of a material to measure motion, stress, and magnetic field. From the second constitutive equation (Eq. 6.60), it can be seen that a change in mechanical stress results in a change in the magnetic induction in the material, which can be sensed. This forms the principle behind magnetostrictive sensing.

Calkins et al. [66] and Dapino et al. [4] provided an overview of magnetostrictive-sensor technology. Sensors are classified into three categories: passive, active, and hybrid. Passive sensors are based on the Villari effect and measure changes in the magnetic flux due to an externally imposed stress, by means of a coil surrounding the sensor. Active sensors use an internal excitation of the material (e.g., with a coil) and measure the change in permeability (often with another coil) due to an external forcing. Hybrid or combined sensors rely on the use of a magnetostrictive element to actively excite another material (e.g., an optical fiber) that allows measurement of change in its properties due to external inputs. Many different sensors based on their applications have been investigated and contrasted with conventional sensors in terms of sensitivity and implementation issues.

Flatau et al. [67] developed a high-bandwidth tuned vibration absorber using a Terfenol-D actuator and showed a significant change of modulus from demagnetized state to magnetic saturation. Simple experiments were conducted to demonstrate proof of the concept. Pratt and Flatau [68] developed a self-sensing magnetostrictive actuator and formulated an analysis of the noncontact nature of sensing using magnetostrictives.

Overall, there is no sufficiently detailed database for magnetostrictive sensors for a wide range of test conditions. More in-depth investigations are needed to understand the behavior of magnetostrictive materials under a wide range of controlled operating conditions. For modeling, the least well-defined component is the magnetic state of the magnetostrictive core, which is a function of operating conditions. It is important to develop reliable modeling of magnetization using either micro-magnetic representation of material, the Preisach model, or the ferromagnetic hysteresis model. There is a need to develop a three-dimensional constitutive model of magnetostrictive materials that includes nonlinear thermal effects, magnetization saturation, eddy current losses, pre-stress, hysteretic behaviors, and dynamic effects and then systematically validate it with test data.

6.7.1 Worked Example

Consider the Terfenol-D rod of Example 6.5.2, used as a sensor. Derive an expression for the output voltage and current developed by the sensor in response to a mechanical stress.

Solution

Let us consider the case when this Terfenol-D rod is biased and a sinusoidal force is applied at one end, given by

$$F = F_o \sin \omega t$$

This will generate an output voltage at the open leads of the coil (Faraday's law)

$$V = -N\frac{d\phi}{dt}$$

For an open-circuit coil, $i = 0$; hence, $H = 0$. This results in

$$\epsilon_3 = s_{33}^H \sigma_3$$

$$B_3 = d_{33}\sigma_3$$

where

$$s_{33}^H = \frac{1}{E^H} \quad (E^H \text{ is the open-circuit Young's modulus})$$

$$E^H = 2.85 \times 10^{10} \text{ N/m}^2$$

$$s_{33}^H = 0.377 \times 10^{-10} \text{ m}^2\text{/N}$$

$$\phi = BA_x = d_{33}\sigma_3 A_x$$

The output voltage is

$$\begin{aligned} V &= -Nd_{33}A_x\frac{d\sigma_3}{dt} \\ &= -Nd_{33}\frac{dF}{dt} \\ &= -\omega d_{33}NF_o \cos \omega t \\ &= -2\pi f d_{33}NF_o \cos \omega t \end{aligned}$$

where f is the frequency of the sinusoidal force in Hz. This can be written as

$$V = -\omega d_{33} n l A_x \sigma_o \cos \omega t$$

where l is the length of the rod and n is the number of turns per unit length. The Terfenol-D rod can also be used to sense strain. Because the magnetic field in the rod is zero, the strain in the rod can be calculated using the constant field Young's modulus of the material, $E_3^H = 26.5$ GPa. Therefore, the voltage generated by a sinusoidally varying longitudinal strain $\epsilon_3 = \epsilon_o \sin \omega t$ is

$$V_s = -\omega\frac{d_{33}}{s_{33}^H} NA\epsilon_o \cos \omega t$$

The Terfenol-D rod can be used to measure the current in the coil due to a sinusoidal forcing. Now we need to close the circuit and include a current-sensing resistor in series with the coil. For the short-circuit condition, $B = 0$. This results in

$$H = -\frac{d\sigma}{\mu^\sigma} = ni$$

The current generated by the mechanical stress σ is

$$i = -\frac{d\sigma}{\mu^\sigma n}$$

Typically, a sensing constant g is defined as

$$g = \frac{d}{\mu^\sigma}$$

The current i is then $i = \frac{-g}{n}\sigma$, which is proportional to σ and inversely proportional to n.

6.8 Iron-Gallium Alloys

Iron-gallium (FeGa) alloys, also known as Galfenol, were developed at the NSWC by Clark et al. [13]. These alloys exhibit moderate magnetostriction (350 ppm) under very low magnetic fields (≊100 Oe or 8 kA/m) and have very low hysteresis. They also have a high tensile strength (≊500 MPa) and limited dependence of magnetomechanical properties on temperatures between −20°C and 80°C [15]. In comparison, the tensile strength of Terfenol-D is about 30 MPa and that of PZT-5H is about 75 MPa.

The machinability, ductility, weldability, high Curie temperature (675°C), and low raw-material cost make FeGa an attractive low-cost actuator and a sensor material well suited to applications in harsh environments, including high shock loads, capable of being easily integrated with a structure and functioning as a load-bearing member. Its corrosion resistance, fatigue properties, and the stability of its properties over time are likely to make it a reliable engineering material. Furthermore, FeGa demonstrates potential for microscale actuation and sensing applications. The bias field required is also low ($\approxeq$ 10 times smaller than in the case of Terfenol-D) and may be achieved with a small permanent magnet.

The effect of alloying iron with other third-group elements, aluminum, and beryllium has also been investigated [69]. Whereas both FeGa and FeAl alloys exhibit similar trends up to 25 atomic % of Ga or Al, the magnetostriction ($\lambda 100$) of FeGa is more than twice that of FeAl, occurring at $\approxeq$19 atomic % Ga or Al. Iron-beryllium (FeBe) alloys, investigated up to 11 atomic % Be, show magnetostriction similar to FeGa, but the high toxicity of Be makes FeBe alloys difficult to process. Furthermore, limited studies to date have shown that ternary alloys of Fe and Ga with nickel, molybdenum, tin, aluminum, and cobalt at best do not significantly improve its magnetostrictive properties and have a detrimental effect at some critical compositions [70, 71].

Atulasimha and Flatau [72] reviewed the state-of-the-art in Galfenol alloys. They also provided an overview of the typical experimental behavior of single-crystal Galfenol in actuation and sensing modes [73]. Figure 6.37(a) shows the strain as a function of magnetic field for 24.7% (atomic) Ga content single-crystal Galfenol. It can be seen that the characteristics of Galfenol improve under compressive pre-stress. As expected, the application of compressive stress helps to orient all of the magnetic moments perpendicular to the axis of the Galfenol rod (in the stress direction) in the demagnetized state. The application of magnetic field causes the magnetic moments to reorient parallel to the rod axis. Beyond an optimum compressive stress, a saturation in the maximum magnetostrictive strain takes place. With an increase in the Ga content, there is a decrease in the maximum strain. The effect of compressive stress on magnetic induction is shown in Fig. 6.37(b). At low fields, the slope of the *B–H* curves is small, representing a state of low magnetic permeability. As the slope becomes nearly zero, a saturation in magnetization is reached. The Galfenol is a highly anisotropic material, and it is expected that its magnetoelastic behavior along various crystallographic directions will be different.

For sensing applications, the interaction between the transducer's magnetic circuit and the magnetostrictive element becomes quite important. This is because Galfenol has a sufficiently high magnetic permeability such that its reluctance becomes comparable to that of the magnetic circuit. Hence, only a part of the magnetomotive force (MMF) generated by the coils is dropped across the magnetostrictive element. In sensing applications, a constant drive current, or MMF, is applied to the transducer. On the application of stress, a change in the magnetomechanical state of the Galfenol element occurs, which in turn produces a large change in the magnetic field (H) in the transducer. To maintain a constant magnetic field through the Galfenol element, a feedback loop is introduced to vary the drive current to compensate for variation in the sample reluctance. The field through the sample is monitored by a Hall effect sensor. Figure 6.38 shows the variation of magnetic induction with stress applied in the (100) axis for a range of bias fields. There is more hysteresis in the sensing behavior than in the actuation behavior.

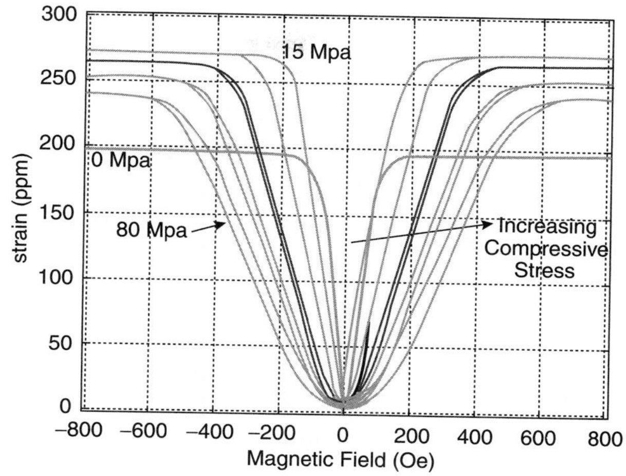

(a) Magnetostriction

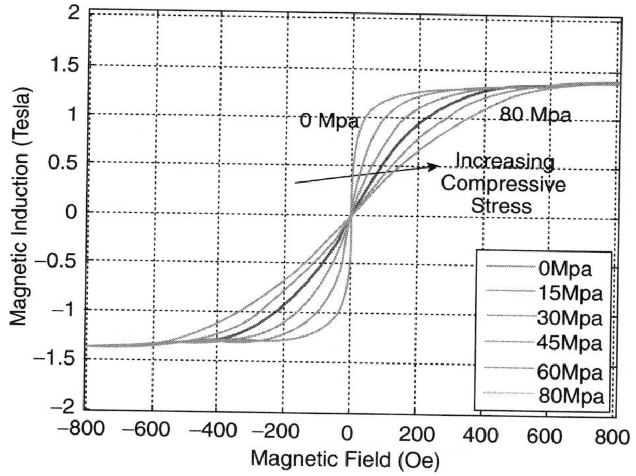

(b) Magnetic induction

Figure 6.37. Effect of compressive pre-stress on experimental behavior of furnace-cooled, 24.7 at. % single-crystal FeGa, from Ref. [73].

6.9 Magnetic Shape Memory Alloys

Magnetic shape-memory alloys (MSMAs), also known as ferromagnetic shape memory alloys (FSMAs), are recently discovered smart materials that display a magnetically driven shape-memory effect (SME). Initial development started as early as in 1996, when Ullakko et al. [74] demonstrated a 0.2% magnetic field induced strain in a sample of single-crystal Ni_2MnGa. Later more than 10% magnetic field induced strain was measured in Ni-Mn-Ga by Sozinov et al. [75]. Several materials such as Ni_2MnGa_2, Co_2MnGa, FePt, CoNi, and FeNiCoTi exhibit this type of behavior, but the largest magnetic field induced strain was achieved in Ni-Mn-Ga alloys [76], which as a result, remain the leading material in this class. Several modes of deformation are possible. Initial studies reported axial strains [74] and, subsequently, macroscopic bending deformations were reported [77]. Several actuator designs based on linear, bending, and torsional deformations have been proposed, and some models of linear actuators are commercially available (e.g., ADAPTAMAT [78]). Much of the following discussion on MSMAs is based on the characteristics of this material.

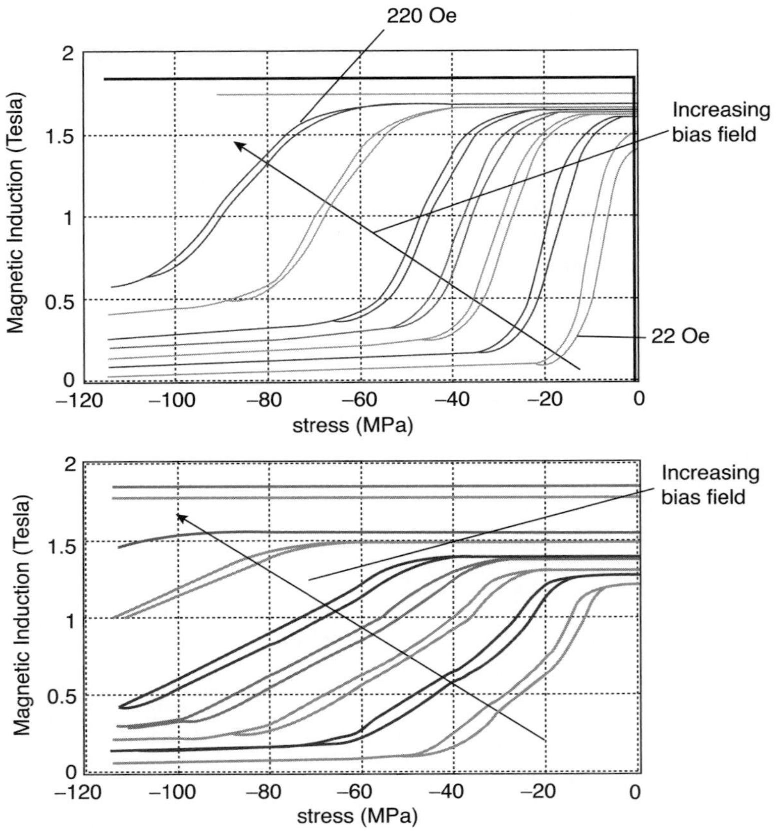

Figure 6.38. Magnetic induction as a function of applied stress and bias field, for (100) oriented 19 at. % Ga, furnace-cooled, single-crystal FeGa, from Ref. [73].

6.9.1 Basic Mechanism

At high temperatures, Ni-Mn-Ga has a cubic lattice structure (in the austenite state). On cooling, the material transforms to martensite. This transition temperature is typically around 20°C–35°C, although the exact transition temperature is highly dependent on the alloy composition. The maximum transition temperature achieved so far is 70°C [79]. In the martensite phase, the material has a tetragonal unit cell, with a long axis (*a*-axis) and a shorter axis (*c*-axis). In addition, the martensitic phase is ferromagnetic and has two twin variants. The magnetic field induced deformation exhibited by these materials is the direct result of the rearrangement of these martensitic twin variants [80].

In a typical ferromagnetic material like Fe, the direction of magnetization can be easily aligned with an external magnetic field. However, the MSMA exhibits a large magnetocrystalline anisotropy (larger than conventional magnetostrictive materials), which means that the axis of magnetization is rigidly fixed in each unit cell. This axis of magnetization is aligned parallel to the *c*-axis, which is the "easy" axis. Consquently, in a magnetic field, the entire unit cell tends to rotate such that its easy axis is aligned with the external field. Because the entire unit cell tends to change its orientation, the process of alignment with an external magnetic field results in transformation of the material from one twin variant to another, accompanied by a

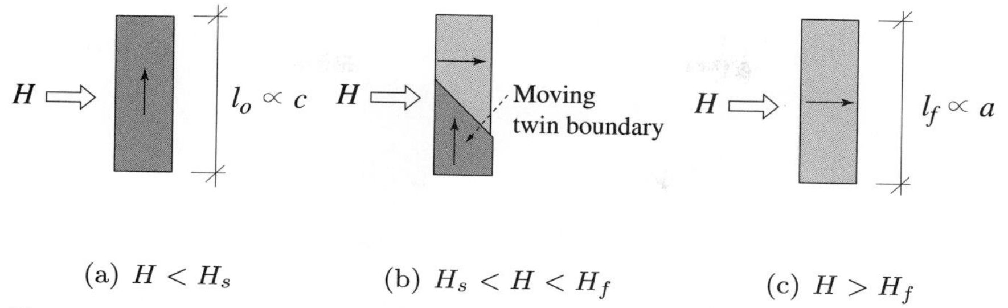

Figure 6.39. MSMA bar in an external magnetic field (zero stress).

change in dimensions. In a similar manner, the orientation of the unit cells can be affected by the application of an external stress.

6.9.2 Effect of an External Magnetic Field

The effect of an external magnetic field on a MSMA sample is shown schematically in Fig. 6.39. A bar of MSMA is placed in a magnetic field H, acting perpendicular to the length of the bar. Initially, the entire bar consists of a single twin variant of MSMA, in which all the c-axes (easy axes) are aligned parallel to the length of the bar (Fig. 6.39(a)). The direction of the easy axis is shown by the vertical arrow on the bar. The initial length of the bar, l_o, is proportional to the length of the c-axis. As the magnetic field is increased above a critical value H_s, the material begins to transform into the twin variant in which the c-axis is aligned parallel to the applied magnetic field. Because the easy axis is aligned with the applied field, this variant is called the "field-preferred" variant. The boundary between the regions in which the two twin variants exist appears as an inclined twin boundary.

As the magnetic field is further increased, a larger fraction of the original twin variant transforms into the field-preferred variant, resulting in a motion of the twin boundary. In physical samples of MSMA, these twin boundaries can be clearly seen with the naked eye, and their motion can be observed by moving a sufficiently strong permanent magnet near the bar.

When the field reaches a value H_f, the entire bar exists in the field-preferred variant, that is, with the c-axis aligned parallel to the applied field (Fig. 6.39(c)). The final length of the bar after the transformation, l_f, is proportional to the length of the a-axis. Because the a-axis is longer than the c-axis, we can see that the total length of the bar has increased by the amount a/c, and the total field induced strain is given by $1 - c/a$. For a typical $c/a = 0.94$ [77], it follows that the maximum achievable magnetic field induced strain is 6%. Note that once the material has transformed entirely into the field-preferred variant, any subsequent change in the magnetic field produces no change in the dimensions of the material. Consequently, if the magnetic field is now set to zero, the deformed shape remains unchanged.

6.9.3 Effect of an External Stress

The twin boundaries can also be moved by the application of a mechanical stress (Fig. 6.40). Consider the bar in Fig. 6.40(a), which is the same as in Fig. 6.39(c), in which the entire sample is in the field-preferred variant. Now let the field be turned to zero, and let a compressive stress be applied along the length of the bar. Because

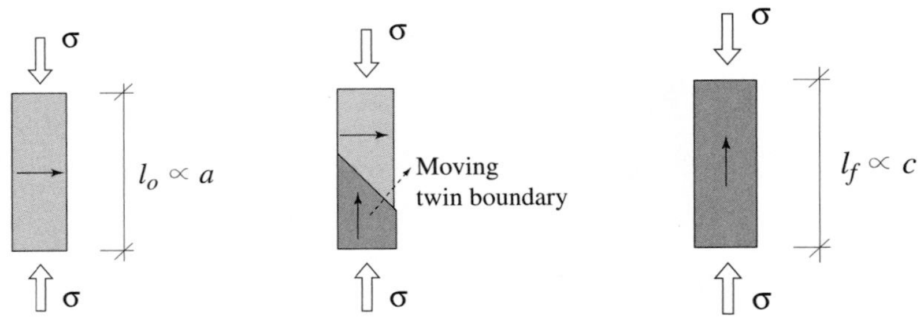

(a) $\sigma < \sigma_s$ (b) $\sigma_s < \sigma < \sigma_f$ (c) $\sigma > \sigma_f$

Figure 6.40. MSMA bar under an external compressive stress (zero magnetic field).

the c-axis is shorter than the a-axis, the compressive stress tends to move the unit cells such that their long a-axes are aligned perpendicular to the applied compressive stress. Therefore, this twin variant is called the "stress-preferred" variant.

At low values of compressive stress, only elastic deformation occurs. As the compressive stress increases beyond a critical stress σ_s, the material begins to transform into the stress-preferred twin variant. The twin boundary begins to move as the stress is increased, until a final critical stress σ_f is reached, at which point the entire material exists in the stress-preferred variant. It can be seen that the length of the bar has now returned to its original dimensions, as in Fig. 6.39(a). The application of compressive stress reverses the effects of the applied magnetic field. Typically, to move the twin boundaries, the minimum field required is around 0.2 T and the minimum stress is about 2–3 MPa.

6.9.4 Behavior under a Combination of Magnetic Field and Compressive Stress

The entire process of applying a magnetic field cycle followed by a compressive stress on a sample of MSMA, as described previously, is shown in Fig. 6.41. The sample is initially fully in the stress-preferred twin variant. As the magnetic field is increased above a critical value H_s, the material starts to deform. The free strain reaches a value of approximately 6% at a magnetic field H_f and then saturates. Decreasing the magnetic field to zero has no subsequent effect on the dimensions of the sample. The sample is returned to its original length by the application of an external compressive stress. It is important to note that both the magnetic field and the compressive stress have critical values that define the beginning and the end of the change in dimensions of the material.

In a similar way, let us examine the response of the MSMA bar to a compressive stress cycle followed by the application of a magnetic field. This behavior is shown in Fig. 6.42. The sample is initially entirely in the field-preferred variant. As the compressive stress is increased, the material deforms elastically. The elastic modulus at this point is called E_f (field-preferred) and is approximately 450 MPa. When the compressive stress reaches a value σ_s, the material begins to transform into the stress-preferred variant. During the transformation, which corresponds to the nearly horizontal portion of the stress-strain curve, the material has a very low stiffness. At the stress value σ_f, the transformation is complete, and the sample exists entirely in the stress-preferred variant. A subsequent increase in compressive stress results

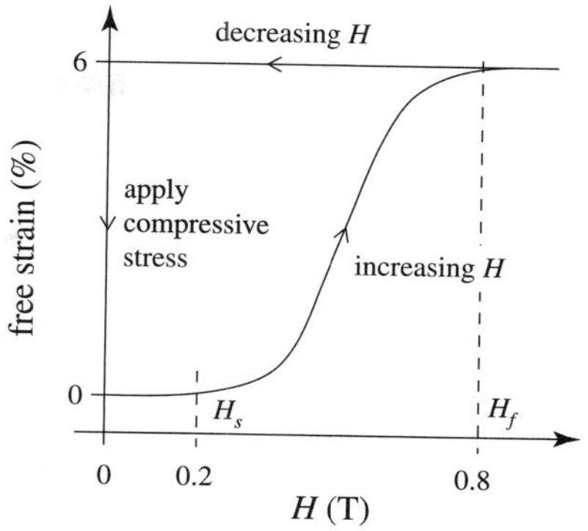

Figure 6.41. Induced strain of an MSMA in response to an applied magnetic field at zero stress, followed by the application of a compressive stress at zero field.

in an elastic deformation of the stress-preferred variant. At this stage, the material has an elastic modulus E_s (stress-preferred) that is approximately 850 MPa. Note that E_s is nearly double E_f. When the compressive stress is removed, the material recovers the elastic deformation of the stress-preferred variant and retains a residual strain ϵ_L. The strain ϵ_L can be found by extrapolating the elastic deformation curve of the stress-preferred variant to the zero stress axis. This strain can be completely recovered by applying a magnetic field $H > H_f$, which converts the material back into the field-preferred variant. Because all deformation occurring at a stress greater than σ_f is elastic in nature, ϵ_L is the maximum recoverable strain of the material and is a constant that depends on the material composition. Based on this discussion, it is obvious that ϵ_L is also the maximum magnetic field induced free strain of the material.

As the unit cells reorient in response to the applied stress, the permeability of the material also changes significantly. This is due to the large magnetic anisotropy in the unit cell itself. The effect of the compressive stress on the induced strain is also similar to that observed in the case of conventional magnetostrictive materials. As the compressive stress increases, the induced strain increases, reaches a maximum, and then decreases. The reason for this behavior is also similar to that in the case of magnetostrictive materials; that is, the compressive stress tends to transform the material into the stress-preferred variant and, therefore, a larger change in length is achievable on transformation to the field-preferred variant. The optimum value

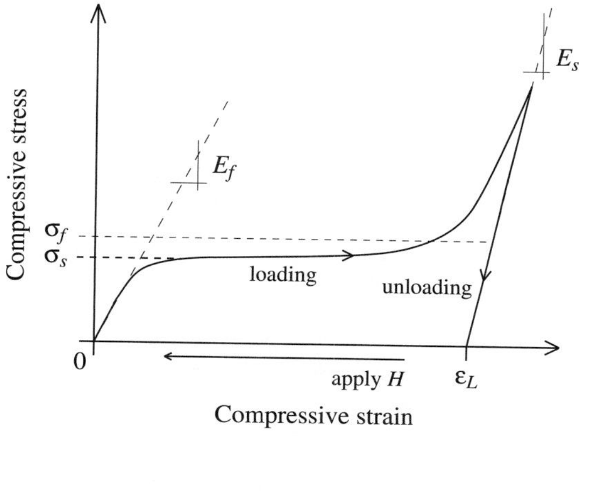

Figure 6.42. Stress-strain behavior of an MSMA at zero magnetic field.

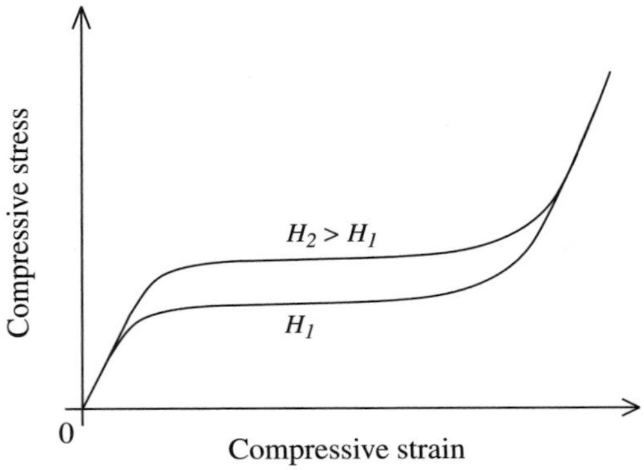

Figure 6.43. Effect of magnetic field on stress-strain behavior of an MSMA.

of compressive stress, in order to obtain the maximum induced strain, has been observed to be around 1–1.5 MPa [78].

Figure 6.43 shows the stress-strain curve of the MSMA at different values of magnetic field. Because the unit cells tend to align along the magnetic field, it is more difficult to reorient them by applying a compressive stress. As a result, the critical stresses σ_s and σ_f are higher in the case of the higher applied field. However, the stiffness of the two twin variants are unaffected by the magnetic field.

The response of the material to a compressive stress cycle at a non-zero magnetic field is shown in Fig. 6.44. The magnetic field has the effect of causing some recovery of the strain on unloading. If the applied magnetic field is larger than H_f, the entire

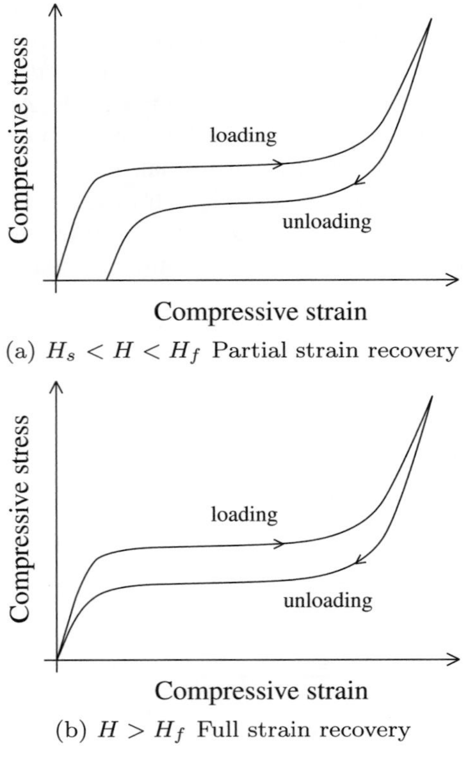

Figure 6.44. Response of MSMA to a compressive stress cycle in a non-zero magnetic field.

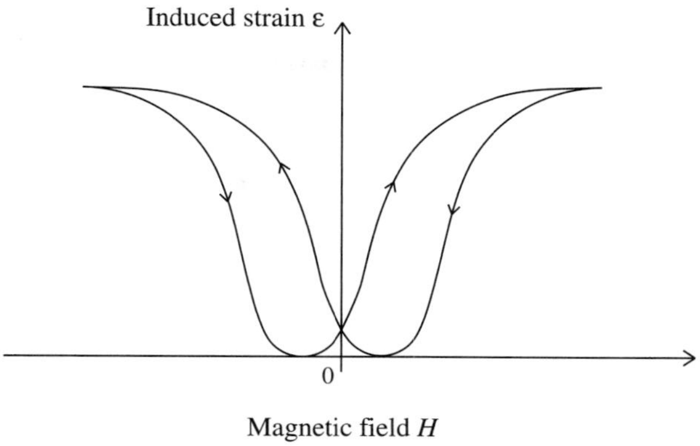

Figure 6.45. Induced-strain response of an MSMA under a sinusoidal magnetic field, at a non-zero compressive stress.

strain can be recovered. If the magnetic field is less than H_f, only partial recovery is possible (i.e., pseudoelastic behavior).

6.9.5 Dynamic Response

Because the transformation between twin variants does not depend on any heat transfer (as in the case of SMAs) and is purely a reorientation of unit cells, it is a fast process. As a result, the material exhibits a bandwidth on the order of kHz. Coupled with the large achievable strains, this high bandwidth makes MSMAs attractive as actuator materials.

To obtain a bidirectional induced strain, it is essential to apply a compressive stress to return the material to its original dimensions when the magnetic field is zero. In practice, this can be realized by actuating the material against a mechanical spring. Figure 6.45 shows a schematic of the strain induced in an MSMA bar by a sinusoidal magnetic field, under a constant compressive stress. Because the process of alignment of the unit cells along the direction of the external magnetic field is similar to the alignment of domains in the case of a magnetostrictive material, it can be seen that the induced strain is independent of the polarity of the applied magnetic field. Therefore, the induced-strain response can be approximated as a quadratic function of the applied magnetic field. Similar to the case of a magnetostrictive element, the MSMA can be actuated in two ways:

1. Constant bias magnetic field superimposed on a bipolar magnetic field: The induced strain is bi-directional but has a non-zero mean value. The output strain can be almost linear and has the same frequency as the actuation.
2. Purely bipolar magnetic field: In this case, the induced strain is highly nonlinear. The strain is bi-directional with a non-zero mean but occurs at a frequency double that of the actuation, due to the quadratic dependence of induced strain on magnetic field.

6.9.6 Comparison with SMAs

Comparing the stress-strain behavior of an MSMA with that of an SMA makes it clear why the material is called a "magnetic" SMA. For reference, the stress-strain curve of a conventional SMA in shape memory mode is shown in Fig. 6.46(a). A detailed discussion of this behavior is in Chapter 3. Also shown in Fig. 6.46(b) is the

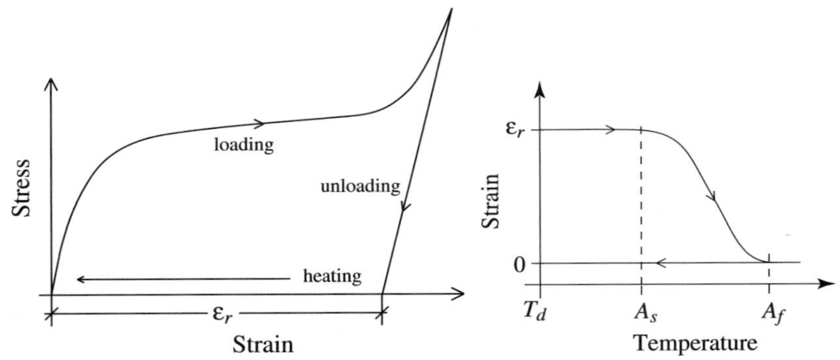

(a) Stress-strain behavior, $T < M_f$ (b) Strain-temperature behavior at zero stress

Figure 6.46. Strain variation of a SMA.

strain-temperature behavior of a pre-strained SMA under no load (free recovery). T_d is the initial temperature, or room temperature.

The stress-strain curve for an MSMA sample undergoing a loading and unloading cycle at zero magnetic field is shown in Fig. 6.42 and is discussed in Section 6.9.3. The residual strain can be completely recovered by the application of a magnetic field. This behavior is identical to that of an SMA in the fully martensite phase, except that the strain recovery occurs on the application of a magnetic field in the case of the MSMA, as opposed to a temperature rise as in the case of the SMA.

The analogy between the effects of temperature and magnetic field becomes more obvious on comparison of the strain-field behavior of an MSMA with the strain-temperature behavior of an SMA, each under no external stress. This is also referred to as the "free recovery behavior." Let us consider an MSMA material that has been imparted a residual strain ϵ_r. As the magnetic field is increased from zero, the material begins to transform to the field-preferred variant, thereby recovering the residual strain and returning to its original dimensions. The transformation begins at a field H_s and is complete at a field H_f (Fig. 6.41). This behavior corresponds to the temperature-induced free recovery of an SMA (Fig. 6.46(b)), with the quantities H_s and H_f corresponding to the austenite start and finish temperatures A_s and A_f, respectively.

The strain behavior of the MSMA, when exposed to a compressive stress cycle in the presence of a non-zero magnetic field (Fig. 6.44), is also similar to the pseudoelastic behavior of an SMA, shown in Fig. 6.47. For this reason, the phenomenon in the case of MSMAs is referred to as "magnetic pseudoelasticity."

Note that there is also a qualitative similarity between the shapes of the stress-strain curves of the MSMA and the SMA, as well as between the shapes of the

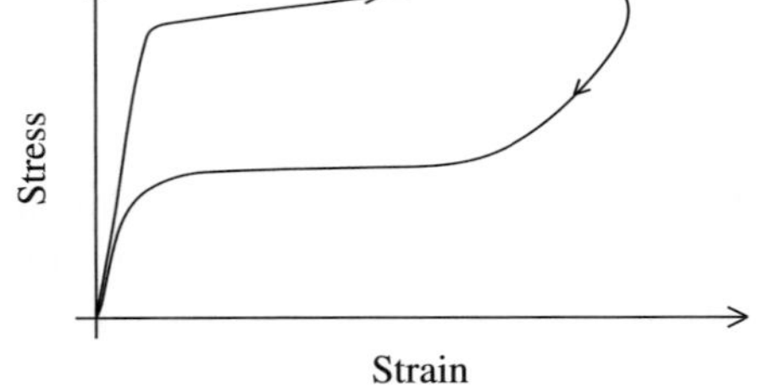

Figure 6.47. Pseudoelastic behavior of an SMA, $T > A_f$.

strain-field curve of the MSMA and the strain-temperature curve of the SMA. By comparing the stress-strain curves, free recovery, and pseudoelasticity, we can conclude that the behavior of SMAs and fully martensitic MSMAs is analogous. There is a one-to-one correspondence between martensite (SMA) and the field-preferred twin variant (MSMA), between austenite (SMA) and the stress-preferred twin variant (MSMA), and between temperature (SMA) and the magnetic field (MSMA).

6.9.7 Experimental Behavior

The procedure for measuring the properties of an MSMA sample is described in this section. Two types of tests are performed on the material, either to measure the stress-strain behavior of the material at a constant magnetic field or to measure the field-strain behavior at a constant stress. The properties measured by this technique can be used in a constitutive model of MSMA behavior, described in Section 6.9.8.

The samples used in the experiments are two single-crystal martensite NiMnGa rods, obtained from Adaptamat (Helsinki, Finland). The specimen dimensions were $2 \times 3 \times 16$ mm. In addition, the magnetic easy axis (c-axis) is oriented perpendicular to the direction of the long axis. Therefore, magnetic strain is induced when a field is applied perpendicular to the long axis of the specimen. The density of the material was measured to be 8.36 gm/cm^3.

The experimental setups for the constant stress and constant magnetic field tests were designed around similar magnetic field generators. The field generators consisted of laminated, transformer-steel core electromagnets capable of producing inductive fields on the order of 1.0 Tesla. The core consisted of two E-shaped halves, joined together by an aluminum frame. Two copper wire coils were wound around the center arm of the E-frames. The ends of the center arms were tapered to concentrate the magnetic flux and the NiMnGa specimen was situated between them.

It is important to point out that it is difficult to experimentally measure the magnetic field H in the MSMA because of the varying permeability of the material. However, it is easy to measure the magnetic induction B, by means of a Hall effect sensor. For this reason, the experimental behavior of the MSMA is often quantified in terms of the applied magnetic induction.

Due to the tapering of the magnetic poles of the E-frame, and by keeping the cross-sectional area of the poles significantly larger than the area of the sample, the measured variation in applied magnetic induction across the face of the poles was less than 2%.

Constant Magnetic Field Testing Apparatus

For the constant magnetic field tests, the NiMnGa specimen was gripped by a stationary and a moveable push rod, holding the specimen parallel to the electromagnet poles. Axial loads were applied to the specimen by an advance screw behind the moveable push rod. A 10-lb load cell, mounted between the moveable push rod and the specimen, was used to acquire force data. The accuracy of the load cell was within 0.0045 N. Actuator deflections and strains were measured by a laser sensor, accurate to within 0.01 mm. Magnetic measurements were taken by a Gauss meter. Power to the coils of the electromagnet was provided by a 20V/10A DC power supply. A photograph of the constant field test rig is shown in Fig. 6.48.

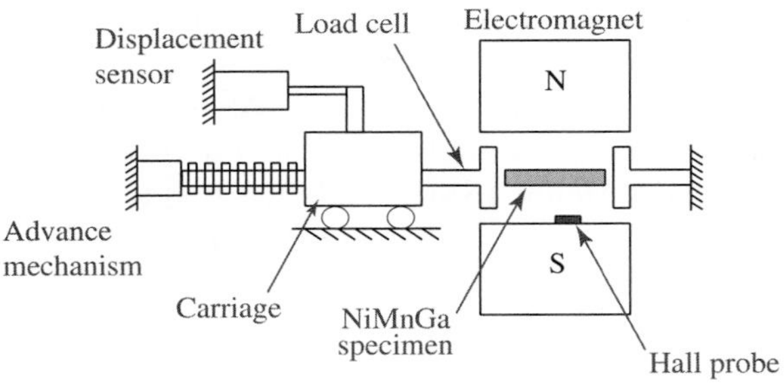

(a) Schematic of apparatus

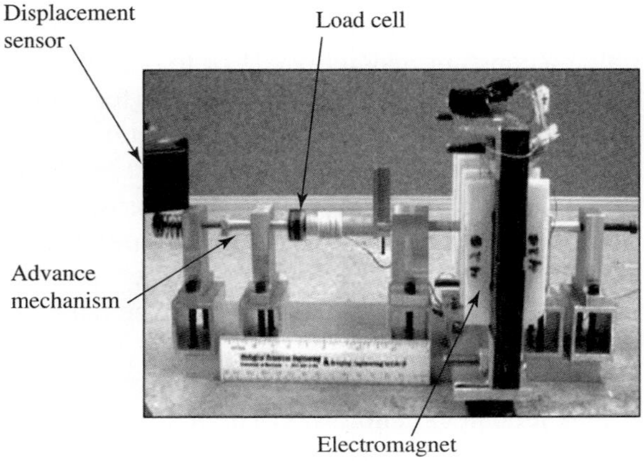

(b) Picture of the test setup

Figure 6.48. Constant magnetic field testing.

Constant Stress Testing Apparatus

For the constant stress tests, a setup similar to the constant field apparatus was developed. The main difference between the two setups is that the specimen is oriented vertically in the constant-stress test setup and horizontally in the constant-field test setup. The NiMnGa specimen was glued into grips between the poles of the electromagnet. The specimen is supported by a stationary, lower rod so that strain is restricted to one direction. In the direction of strain, the specimen acts against a rod attached to a low-friction, linear bearing. Another rod at the other end of the bearing connects the bearing-pushrod combination to a linear potentiometer and weight pan. Strains are measured by the linear potentiometer, accurate to within 0.002 mm, and the level of constant stress is regulated by adding and subtracting weights to and from the weight pan. Applied magnetic field measurements are taken by Hall effect sensors located in the air gap between the pole and NiMnGa bar. The electromagnet in this rig is powered by two 30V/5A power supplies connected in series. A rack of capacitors was connected in parallel with the coils to obtain a desired RC time constant in the electrical circuit. A high RC constant is necessary to have a slow decay in the magnetic field when the power is removed so that the quasi-static behavior of the material may be observed. A photograph of the constant stress test rig is shown in Fig. 6.49.

Experimentally measured magnetic SME and magnetic pseudoelastic behavior is shown in Fig. 6.50.

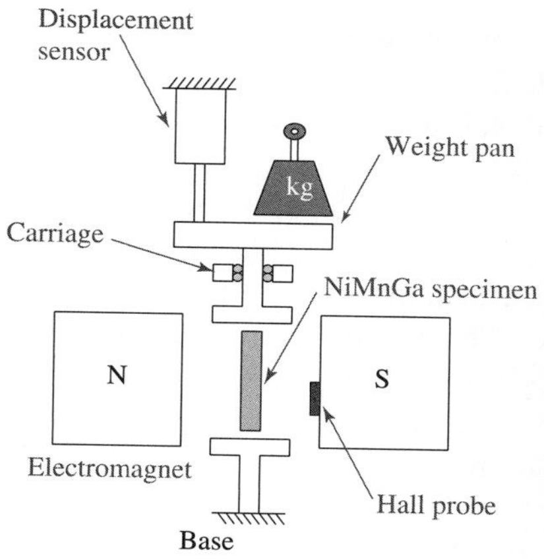

(a) Schematic of apparatus

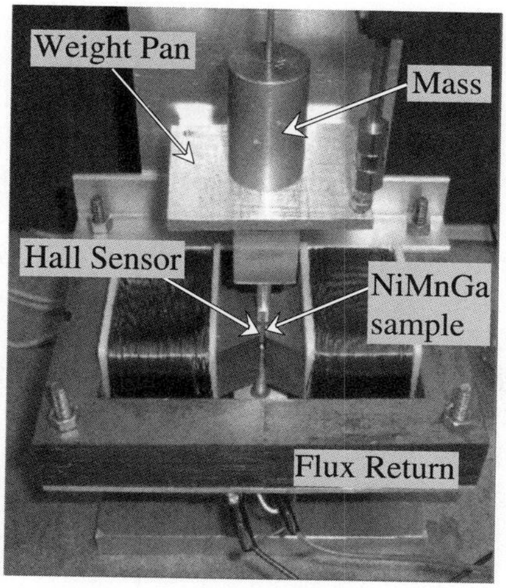

(b) Picture of the test setup

Figure 6.49. Constant stress testing.

6.9.8 MSMA Constitutive Modeling

Several models of varying degrees of complexity have been proposed to predict MSMA behavior. These range from free energy-based models, representing the MSMA as containing two twin variants separated by a mobile twin boundary, to

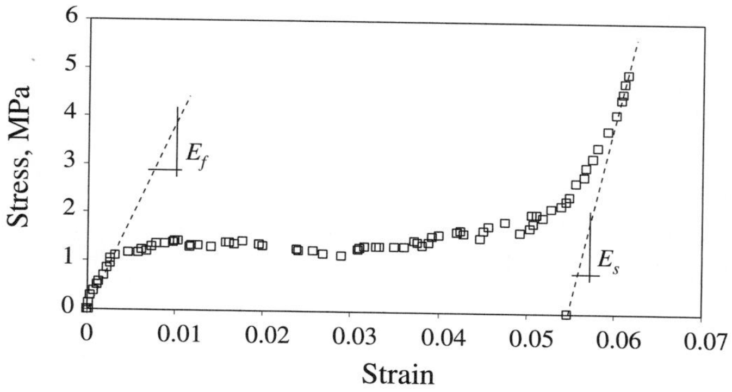

(a) Magnetic shape memory effect at 0 T

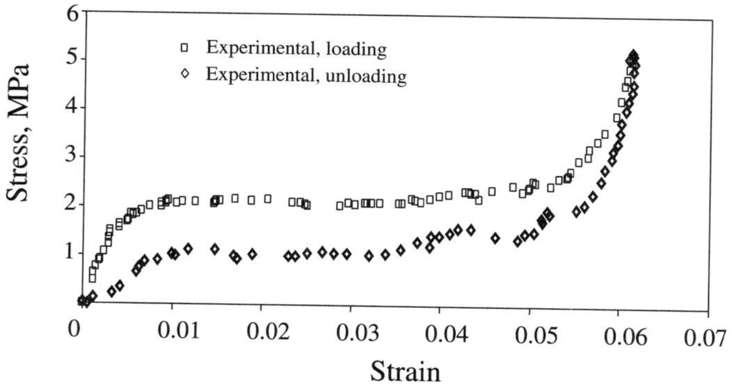

(b) Magnetic pseudoelasticity at 0.6 T

Figure 6.50. Experimental behavior of an MSMA sample.

macroscopic phenomenological models. Free energy-based micromechanics models can predict the local stress and strain states of the material but are difficult to implement from an engineering perspective. In contrast, phenomenological constitutive models are easy to implement and can satisfactorily predict the macroscopic behavior of the material. Here, we outline the development of a quasi-static phenomenological model developed by Couch et al. [81].

Because of the analogous behavior of MSMAs and SMAs, as discussed in Section 6.9.6, it is possible to adapt SMA constitutive models to MSMAs. Note that the assumption here is that the MSMAs operate only at temperatures low enough to ensure that they are in the fully martensite phase. The Tanaka model is adapted to model MSMA behavior by replacing all quantities describing the martensite phase of SMAs with the field-preferred variant of MSMAs and the austenite phase of SMAs with the stress-preferred variant of MSMAs. The martensite volume fraction ξ is replaced by a quantity called the "stress-preferred volume fraction," ξ_σ, that varies from 0 to 1. The constitutive equation can be written as

$$\sigma - \sigma_o = E(\xi_\sigma)(\epsilon - \epsilon_o) + \Omega_s(\xi_\sigma - \xi_{\sigma o}) + \lambda(H - H_o) \tag{6.136}$$

In this equation, Ω_s is a constant related to the transformation from one twin variant to another and λ is a constant related to magnetostriction. Quantities with the subscript o refer to initial conditions. This magnetostriction term refers to the change in dimensions of each unit cell in response to an applied magnetic field. In the case of MSMAs, we assume that the induced strain is predominantly a result of the change in geometry caused by transformation between twin variants and not due to change in the dimensions of each unit cell. Therefore, the term related to λ is comparatively small and can be neglected. This term is similar to the coefficient of thermal expansion of SMAs.

The constant Ω_s can be found by starting from a set of initial conditions and then applying a combination of stresses and magnetic fields on the sample. Consider an MSMA sample in the fully field-preferred variant, under zero compressive stress and zero magnetic field, and with no initial strain. Therefore, the initial conditions are

$$\sigma_o = 0$$

$$\epsilon_o = 0$$

$$\xi_{\sigma o} = 0$$

$$H_o = 0$$

A compressive stress is then applied on the sample to convert it completely into the stress-preferred variant and then removed. The strain in the material is now ϵ_L. The variables at this point are

$$\sigma = 0$$

$$\epsilon = \epsilon_L$$

$$\xi_\sigma = 1$$

$$H = 0$$

Table 6.9. *Parameters used in the constitutive model*

H_s	1.0	kOe
H_f	3.5	kOe
σ_s	0.284	MPa
σ_f	0.902	MPa
C_s	0.452	MPa/kOe
C_f	0.488	MPa/kOe
E_s	820	MPa
E_f	450	MPa
ϵ_L	5.5	%

Substituting these initial and final conditions in the constitutive relation (Eq. 6.136), we obtain

$$\Omega_s = -E(\xi_\sigma)\epsilon_L \tag{6.137}$$

The final form of the constitutive relation becomes

$$\sigma - \sigma_o = E(\xi_\sigma)(\epsilon - \epsilon_o) - E(\xi_\sigma)\epsilon_L(\xi_\sigma - \xi_{\sigma o}) \tag{6.138}$$

where the magnetostrictive term has been neglected. The stress-preferred volume fraction is a function of the applied magnetic field. This function can be defined in different ways depending on the state of the material, similar to the procedure followed in the case of SMA modeling. The model is characterized by nine experimentally determined constants. These are as follows:

(1) Material parameters: Maximum free strain, ϵ_L; stress-preferred variant Young's modulus, E_s; field-preferred variant Young's modulus, E_f
(2) Critical stresses and fields: σ_s, σ_f, H_s, and H_f
(3) Stress-influence coefficients: $C_s = 1/(dH_s/d\sigma)$ and $C_f = 1/(dH_f/d\sigma)$

These parameters are obtained from experimental testing of the MSMA. The material properties are obtained from constant magnetic field stress-strain tests. The remaining constants are determined by varying the magnetic field at a constant stress. Typical values of the constants used in the model are shown in Table 6.9. Correlation of the constitutive model with some experimental data is shown in Fig. 6.51.

6.9.9 Linear Actuator

The behavior of the material in response to a magnetic field and a compressive stress can be used to construct a bi-directional linear actuator. A schematic of such an actuator is shown in Fig. 6.52.

The basic construction is similar to that of a magnetostrictive actuator. A magnetic field is applied to the active material by means of a field generator, consisting of a current carrying solenoid coil and a highly permeable flux path. To reduce the power requirements, a permament magnet (shown in the diagram by the poles N and S) can be incorporated in the flux path to provide a constant-bias magnetic field. The active material deforms a spring that serves to return the material to its original dimensions after the magnetic field goes to zero.

The major difference between the MSMA actuator and a conventional magnetostrictive actuator is the configuration of the magnetic field generator. In the case of a

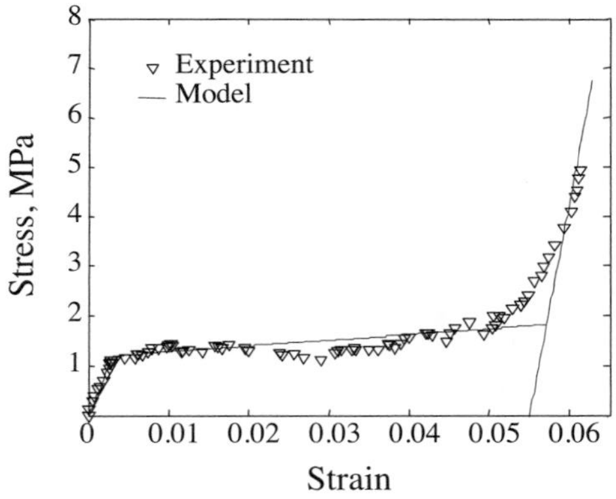

(a) Magnetic shape memory effect at 0 T

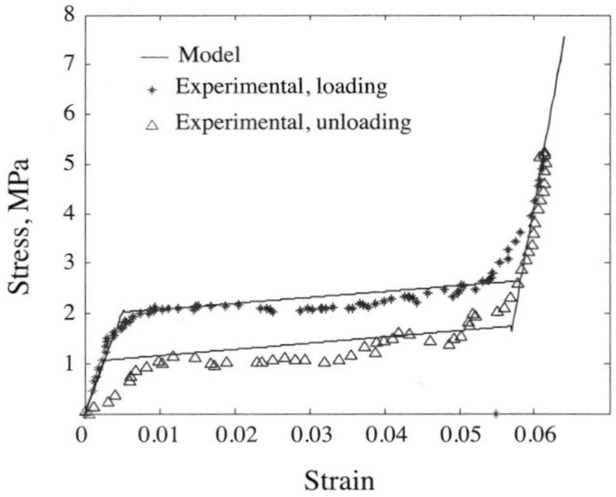

(b) Magnetic pseudoelasticity at 0.6 T

Figure 6.51. Validation of quasi-static constitutive model of an MSMA.

magnetostrictive actuator, the applied field is along the length of the active material and is parallel to the output displacement. However, in the case of the MSMA actuator, the applied field is perpendicular to both the length of the active material and the output displacement. Therefore, the design of the magnetic field generating circuit is different for the two types of active materials.

Although the MSMA has a high bandwidth, the useful bandwidth of the actuator is often limited by the time taken for the spring to return the active material to its original length. In other words, the dynamics of the return spring and the external load can have a significant effect on the output of the actuator. As a consequence, if the actuator is excited by a sinusoidal current, the output displacement will follow a sinusoidal waveform only when the MSMA is expanding (pushing) and not during the return stroke. One solution to this problem might be to have two actuators operating antagonistically against the same load.

6.9.10 Design of the Magnetic Field Generator (E-Frame)

A schematic of an E-frame magnetic field generator is shown in Fig. 6.53(a). This geometry is well suited for MSMA actuators because of the cuboidal shape of the active element. The coil is wound on the middle arms of the E-frame, and the MSMA bar is placed between the faces of the middle arms. The mass of the actuator is evenly

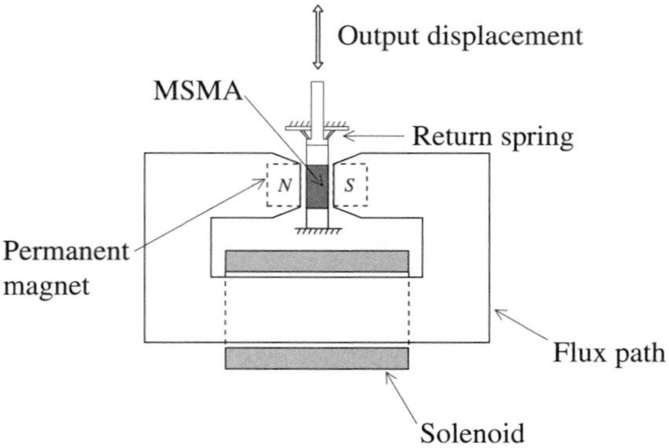

Figure 6.52. MSMA linear actuator.

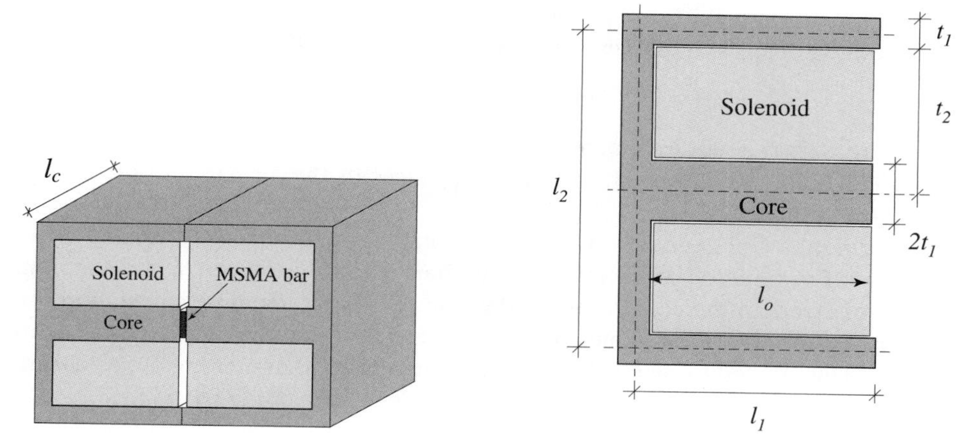

(a) E-frame with MSMA sample (b) Cross-section of one half of the E-frame

Figure 6.53. E-frame magnetic field generator for an MSMA linear actuator.

distributed on both sides of the MSMA bar. Note that the output displacement occurs perpendicular to the plane of the E-frame.

The major geometrical parameters of the E-frame are determined by the dimensions of the MSMA bar. The main requirement is to ensure that the magnetic field acting on the bar is as uniform as possible. A general algorithm for design of the coil is described here. All important dimensions are shown in Fig. 6.53(b).

1. Calculate the cross-sectional area of the flux return path. The parameters t_1 and l_c should match the corresponding dimensions of the active material plus an additional 10–20% to account for fringing in the air gap. This can be represented by a "fringing factor" F that is expressed as a percentage of the length of the active material. Including the fringing factor will ensure that the field across the poles of the field generator is close to uniform. For an MSMA bar of width w_s and length l_s

$$t_1 = w_s(1 + F)/2 \tag{6.139}$$

$$l_c = l_s(1 + F) \tag{6.140}$$

2. Determine the length of the solenoid l_o. The length of the solenoid is based on the dimension l_1, which can be chosen based on other specifications such as overall actuator dimensions.

$$l_o = k(l_1 - t_1/2) \tag{6.141}$$

Assume that k is between 0.6 and 0.9 to account for the gap between the poles or tapering of the central arms. The gap between the faces of the middle arms of the E-frame is fixed by the thickness of the MSMA bar, t_s, with as little air gap as possible.

3. Estimate the required *mmf*. Because the geometry of the flux return path is not known at this stage, in this initial calculation, we assume that the *mmf* required is dependent on the reluctances of the air gap and sample only. The effect of the flux return path is neglected for now but will be added in a later step to serve as a check on our initial estimate. This assumption is only valid if the permeability of

the flux return is large and its reluctance is small compared to that of the sample and the air gap. The *mmf* is given by

$$mmf = B_a A_c(\mathbb{R}_a + \mathbb{R}_s) \tag{6.142}$$

where B_a is the required magnetic induction, A_c is the core cross section (given by $l_c \times 2t_1$), and $\mathbb{R}_a$ and $\mathbb{R}_s$ are the reluctances of the air gap and the MSMA sample, respectively.

4. Assume the coil geometry. Choose the total number of turns, N_{tot}, and a wire gauge. From these parameters, the number of turns per layer N_t and the number of layers in the coil N_l can be determined. The remaining calculations are performed for several values of N_{tot} so that the optimum value can be identified. This procedure is much simpler to implement than finding a closed-form solution for the optimum number of turns. For a wire of diameter d_w

$$N_t = \frac{l_o}{d_w} \tag{6.143}$$

$$N_l = \frac{N_{\text{tot}}}{2N_t} = \frac{t_2 - t_1}{d_w} \tag{6.144}$$

In these equations, any imperfections in winding the wire are neglected. This assumption becomes more accurate as the wire diameter decreases. Also note that N_t and N_l are the number of turns and number of layers in the coil on only half of the E-frame. The other half of the E-frame will have an identical coil, and the total number of turns of both of the coils is N_{tot}.

5. Find the remaining dimensions of the flux return path. To minimize the mass of the flux return, it is assumed that the coil will fill of all the empty space in the E-frame. Using the value of N_l, the remaining dimensions of the E-frame, t_2 and l_2, are calculated

$$t_2 = N_l d_w + t_1 \tag{6.145}$$

$$l_2 = 3t_1 + 2N_l d_w \tag{6.146}$$

6. Determine the electrical properties of the circuit. Now that the geometry of the E-frame has been determined, the total length of the wire in the coil (both halves of the E-frame), l_w, and the resistance of the coil can be calculated. In addition, because the geometry of both the coil and the E-frame are now known, the inductance of the coil can be calculated. The total length of the wire is given by

$$l_w = N_{\text{tot}}(2l_c + 4t_2) \tag{6.147}$$

For a wire of cross-sectional area A_w, with a resistivity ϱ_w, and a flux return path of permeability μ_c, the resistance of the coil R_w and the inductance of the coil L_w are

$$R_w = \frac{N_{\text{tot}}\varrho_w}{A_w}(2l_c + 4t_2) \tag{6.148}$$

$$L_w = \frac{N_{\text{tot}}^2 \mu_c l_c t_1}{l_2 + 4l_1} \tag{6.149}$$

7. Determine the magnitude of the power required by the coil. The current flowing in the coil, i_w, is determined by dividing the *mmf* by the assumed number of

turns

$$i_w = \frac{mmf}{N_{\text{tot}}} \tag{6.150}$$

From the required current and the calculated coil impedance, the power required, P_w, at a given operating frequency can now be determined

$$P_w = i_w^2\sqrt{R_w^2 + L_w^2} \tag{6.151}$$

Note that the heat produced in the coil, P_d, is purely due to Ohmic heating and is given by

$$P_d = i_w^2 R_w \tag{6.152}$$

8. Determine the mass of the magnetic field generator. Provided that the density of the material used in the flux return and in the wire is known, the mass of the magnetic field generator can be calculated. The mass of the flux return M_f and coil M_w are given by

$$M_f = 2\rho_f l_c\left[4t_1\left(l_1 - \frac{t_1}{2}\right) + t_1\left(l_2 + t_1\right)\right] \tag{6.153}$$

$$M_w = \rho_w l_w A_w \tag{6.154}$$

where ρ_f is the density of the material of the flux return and ρ_w is the density of the material of the wire. The total mass of the magnetic field generator is $M_{\text{tot}} = M_f + M_w$.

9. Determine the actual mmf produced. Now that all of the parameters are known, the reluctance of the flux return path can be determined and included in the mmf calculation. If it is small, then our initial assumption is valid and the reluctance of the flux return can be neglected. If the reluctance is large, then it must be included in the calculation of mmf. Comparing the predicted mmf to the required mmf gives an indication of the accuracy of the coil design. Because the number of turns, N_{tot}, is fixed for each design, the current in the coil must be increased to compensate for the increase in mmf. Once the true current in the coil is known, it can be substituted into the power equation and the actual required power of the coil can be determined.
10. The coil impedance, required power, and total mass are calculated for each assumed value of N_{tot}. From these calculations, based on important requirements such as maximum driving frequency (minimum L_w), minimum total power, minimum heating of the coil (minimum R_w), and minimum total mass, the optimum coil geometry can be chosen.

6.9.11 Worked Example: Design of a Magnetic Field Generator (E-Frame)

A numerical example illustrates the design procedure. Consider an MSMA sample of length 17 mm, width 3 mm, and thickness 2 mm. A uniform magnetic induction of 1 Tesla is required over the width of the sample (across the thickness). Let us investigate the possible dimensions of the coil and flux return path. The data required for the calculation are summarized as follows:

(a) General data
- Required magnetic induction B_a: 1 T

- Air gap between sample and poles of the field generator t_g: 1 mm
- Fringing factor F: 0.2
- Design operating frequency: 100 Hz

(b) MSMA sample
- Length l_s: 17 mm
- Width w_s: 3 mm
- Thickness t_s: 2 mm
- Relative permeability μ_s: 1.5

(c) Flux return
- Material: Steel
- Density: 7850 kg/m^3
- Relative permeability μ_c: 1000
- Saturation magnetic induction $B_{sat,c}$: 2 T
- E-frame parameter L_1: 40 mm

(d) Coil wire
- Density: 8906 kg/m^3
- Resistivity: 1.72×10^{-8} ohm-m

Solution

Based on the given length and width of the sample and the assumed fringing factor, the E-frame parameters t_1 and l_c can be calculated (see Fig. 6.53(b)).

$$t_1 = w_s(1 + F)/2 = 18 \text{ mm}$$

$$l_c = (1 + F)l_s = 20.4 \text{ mm}$$

Assuming $k = 0.9$, the parameter l_o is given by

$$l_o = k(l_1 - t_1/2) = 35.2 \text{ mm}$$

The reluctances of the air gap and MSMA sample are

$$\mathbb{R}_a = \frac{t_g}{\mu_o l_s w_s} = 1560342.579 \text{ A-turn/Wb}$$

$$\mathbb{R}_s = \frac{t_s}{\mu_o \mu_s l_s w_s} = 2080456.772 \text{ A-turn/Wb}$$

The *mmf* is given by

$$mmf = 2B_a l_c t_1(\mathbb{R}_a + \mathbb{R}_s) = 267.38 \text{ A-turns}$$

At this point, the number of turns in the coil has to be assumed. The other parameters are then calculated as described in Steps 4–10 in Section 6.9.10. A set of parameters calculated for a coil with 50 turns for different wire gauges is shown in Table 6.10. The number of turns per layer and number of layers are found based on the E-frame parameters calculated previously. For some of the thinner wires, it can be seen that the number of layers is less than 1. This is because the coil is assumed to cover the entire length l_o. In such a case, keeping in mind physically realizable limits, either the number of turns per layer can be reduced so that one layer can be wound, or the dimensions l_1 can be changed.

From this point, the geometric parameters t_2, l_2, and l_w; the electrical parameters R_w and L_w; and the mass properties are calculated. It can be seen that for different

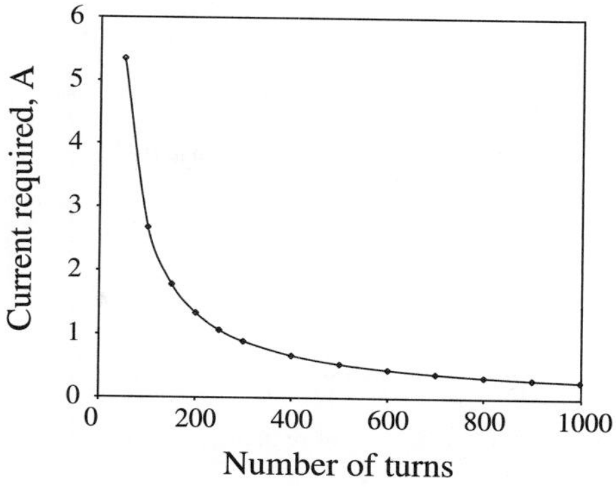

Figure 6.54. Current required to achieve specified magnetic induction in the MSMA sample as a function of number of turns in the coil.

wire diameters, the coil resistance varies much more than the coil inductance. This has a significant effect on the heat generated in the coil, P_d. The total required power P_w does not vary much with the chosen wire diameter. To check the effect of the flux-return core on the magnetic circuit, its reluctance, $\mathbb{R}_{\text{core}}$, can be calculated based on the dimensions of the E-frame. The actual required *mmf*-, mmf_{act}, is calculated by including $\mathbb{R}_{\text{core}}$ in the total reluctance. It can be seen that $\mathbb{R}_{\text{core}}$ is small compared to $\mathbb{R}_a$ and $\mathbb{R}_s$, and its effect on the required *mmf* can be neglected.

Such a table of parameters can be generated for several values of total turns in the coil, and the important quantities can be plotted as a function of wire gauge and number of turns. Based on specific criteria, an acceptable coil geometry can be chosen. Because the current required is constant for an assumed number of turns, a plot of current required as a function of total number of turns is shown in Fig. 6.54. It can be seen that there is a large increase in the required current as the number of turns decreases. Figure 6.55(a) shows the heat generated in a 50-turn coil as a function of the wire gauge. In general, to reduce the amount of heat generated, it is better to use a larger diameter wire (smaller gauge) and more number of turns. However, it can be seen that the difference in heat produced by the 500-turn coil and 1000-turn coil is small. The total power required to achieve the specified magnetic induction of 1 T is shown in Fig. 6.55(b). The total power decreases with increasing number of turns and increasing wire diameter, especially for the larger diameter wires.

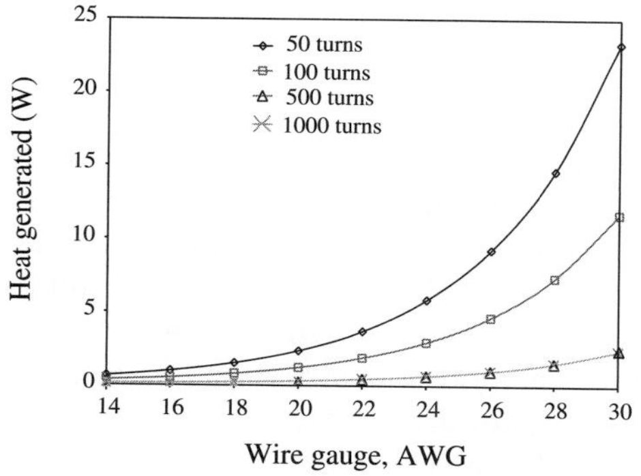

(a) Heat generated in the coil

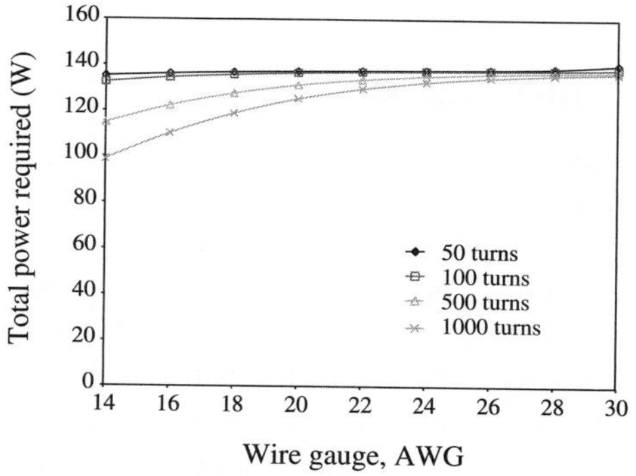

(b) Total power required

Figure 6.55. Power as a function of wire gauge and total number of turns to achieve a specified magnetic field.

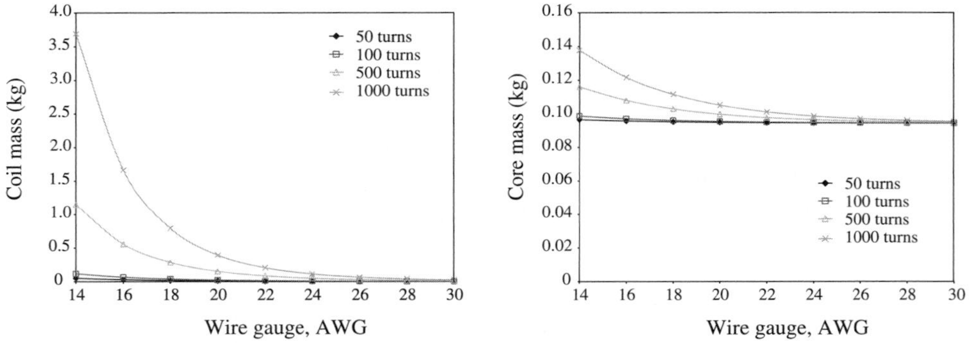

(**a**) Mass of the coil (**b**) Mass of the flux return core

Figure 6.56. Mass of the magnetic field generator as a function of wire gauge and total number of turns to achieve a specified magnetic field.

Figure 6.56 shows the mass of the magnetic field generator as a function of wire gauge. The mass of the flux return core remains relatively constant, especially for a lower number of turns, but the mass of the coil shows a large increase at higher wire diameters and number of turns.

It can be seen that there is a tradeoff between lower power and lower mass. Based on these plots and other operating considerations, such as maximum available power and maximum allowable mass, optimum coil parameters can be chosen. This will determine the geometry of the flux return core. The heat generated in the coil is often a limiting factor because of the poor thermal conductivity of the coil and core materials. Although significant flexibility exists in the design, we choose a 500-turn, 22-AWG coil based on low total mass and low heat generated as driving factors. The resulting values of E-frame dimensions, power, mass, and current are shown in Table 6.11. The number of layers and number of turns can be rounded off to the next highest integer value.

It is interesting to note that while the mass of the active material is approximately 0.9 gm, the total mass of the magnetic field generator is 184.4 gm. This results in a large decrease in the overall actuator power density compared to the capability of the active material and is one of the most important challenges in developing an effective actuator.

6.10 Electrostrictives

Electrostrictive materials undergo deformations under the influence of an electric field. However, the phenomenon of electrostriction is fundamentally different from the converse piezoelectric effect. In electrostrictive materials, unlike piezoelectrics, the unit cells are centrosymmetric, and the change in dimensions is not the result of a modification of the structure of the material but instead is inherent to the material itself. The basic mechanism is a separation of charged ions in the unit cell of the material. The phenomenon exists in all of the materials; however, the magnitude of electrostriction is negligible in most cases. Some materials that have a large polarization, such as as relaxor ferroelectrics, undergo large electrostrictive strains when an electric field is applied (on the order of 0.1% strain). Electrostrictives, like piezoelectrics, belong to a class of ionic crystals called ferroelectrics. They consist of

Table 6.10. *Parameters as a function of wire gauge for a 50-turn coil*

AWG	14	16	18	20	22	24	26	28	30
Wire diameter, mm	1.629	1.291	1.025	0.812	0.644	0.510	0.405	0.322	0.255
N_t, turns/layer	21.600	27.300	34.300	43.400	54.600	69.000	86.800	109.400	138.200
N_l, layers	1.200	0.900	0.700	0.600	0.500	0.400	0.300	0.200	0.200
t_2, mm	3.685	2.985	2.547	2.268	2.095	1.985	1.917	1.874	1.846
l_2, mm	9.170	7.769	6.894	6.336	5.989	5.770	5.633	5.547	5.492
l_w, m	2.777	2.637	2.549	2.494	2.459	2.437	2.423	2.415	2.409
R_w, ohms	0.023	0.035	0.053	0.083	0.130	0.205	0.324	0.512	0.816
L_w, mH	6.819	6.876	6.912	6.935	6.950	6.959	6.965	6.968	6.971
P_{act}, W	135.176	136.199	136.851	137.282	137.577	137.833	138.161	138.789	140.263
P_{heat}, W	0.657	0.993	1.522	2.376	3.721	5.875	9.259	14.645	23.334
Mass of core, kg	0.096	0.096	0.095	0.095	0.095	0.095	0.094	0.094	0.094
Mass of coil, kg	0.052	0.031	0.019	0.011	0.007	0.004	0.003	0.002	0.001
Total mass, kg	0.148	0.126	0.114	0.106	0.102	0.099	0.097	0.096	0.095
$\mathbb{R}_{\text{core}}$, A-turn/Wb	183308	181790	180841	180237	179861	179624	179476	179382	179322
mmf_{act}, A-turns	280.842	280.731	280.661	280.617	280.589	280.572	280.561	280.554	280.550

domains that have a uniform, permanent, reorientable polarization. These domains are randomly oriented, resulting in a net zero–bulk polarization. On the application of an electric field, these domains reorient, resulting in a change in the overall dimensions of the material sample. Electrostriction is a coupled electromechanical effect and induced strain is a quadratic function of the applied field. Under this category of materials, lead magnesium niobate (PMN) ceramics have sufficiently

Table 6.11. *Parameters of an E-frame magnetic field generator designed to produce 1 T*

Flux return core dimensions	
l_1, mm	40.00
l_2, mm	11.29
t_1, mm	1.80
t_2, mm	4.75
l_o, mm	35.20
l_c, mm	20.40
Coil geometry	
Wire gauge, AWG	22.000
Wire diameter, mm	0.644
N_{tot}, turns	500.000
N_t, turns	54.600
N_l, turns	4.600
Mass of components	
Coil, gm	86.7
Core, gm	97.7
Total, gm	184.4
Electrical quantities	
Current required, A	0.535
Power required, W	133.600
Heat generated, W	0.450

Table 6.12. *Characteristics of* $(1-x)$*PMN-*(x)*PT with optimized processing conditions, at 1 KHz, from Ref. [83]*

Ceramic	Density (kg/m^3)	Average Grain Size (μm)	T_c (°C)	Stress-Free Relative Permittivity e_r	tan δ
0.9PMN-0.1PT	7980	2.07	45	10713	0.083
0.8PMN-0.2PT	7940	2.02	100	2883	0.079
0.7PMN-0.3PT	7860	1.72	150	1976	0.045
0.6PMN-0.4PT	7830	1.93	210	1909	0.031
0.5PMN-0.5PT	7780	2.11	260	1375	0.022

large dielectric permittivity that help to generate significant polarization and, hence, large induced strains. These ceramics are often defined as $(1-x)$PMN-(x)PT, where x normally varies from 0.1 to 0.5. Superior characteristics are obtained when PMN is doped with lead titanate (PT) in low ratios, such as 0.9PMN-0.1PT [82].

These materials are often fabricated from calcined powders by a sintering process. Table 6.12 shows typical material characteristics of these ceramics, at room temperature and stress-free conditions, measured using X-ray diffraction and scanning electron microscopy [83]. The relative permittivity undergoes a large decrease from 10713 for 0.9PMN-0.1PT to 2883 for 0.8PMN-0.2PT. In addition, there is a large drop in the loss factor, tan δ. This may be attributed to the transition temperature T_c being closer to room temperature for 0.9PMN-0.1PT.

The variation of strain with electric field is approximately quadratic (independent of polarity of field). At a sufficiently high field, the induced strain becomes saturated, as shown in Fig. 6.57. Unlike piezoelectrics, uncharged electrostrictives are isotropic and are not poled. With an application of field, the materials become instantly polarized and become anisotropic. For example, the transverse material stiffness of PMN-PT decreases by about 20% as the electric field becomes 1300 V/mm. On the removal of field, the materials become depolarized. An electric field produces an extensional strain in the direction of field and contraction in the transverse direction. If the field is reversed, the domains reverse direction, but it again induces an extensional strain in the direction of field (thickness direction). To produce an oscillatory (bidirectional) strain, it becomes necessary to apply a bias DC field. Electrostrictives are primarily used as actuators in a wide range of

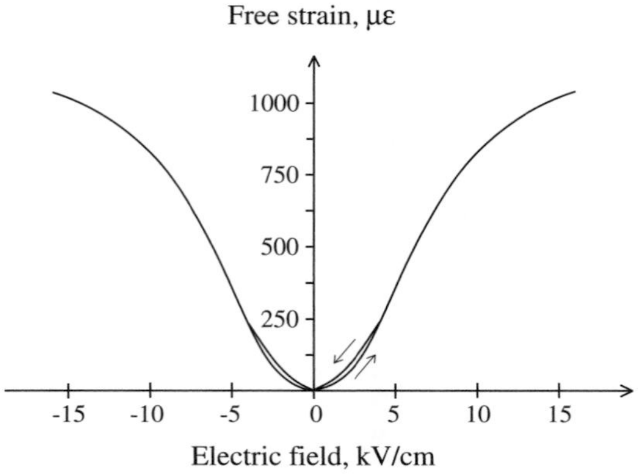

Figure 6.57. Typical induced-strain curve of a PMN-based ceramic.

applications [84, 85]. Because no permanent polarization is needed for electrostrictives, these are not subjected to electric aging. They are characterized by very low hysteresis (less than 1%) but are very sensitive to surrounding temperature [86]. PMN and doped derivatives also have high relative permittivities (20,000–30,000) and high electrostrictive coefficients. In addition, doping has the effect of changing the operating temperature range of the material (where the electromechanical performance is maximum). For example, PMN-15 (0.9PMN-0.1PT), PMN-38 (0.85PMN-0.15PT), and PMN-85 have operating temperature ranges of 0°C–30°C, 10°C–50°C, and 75°C–90°C, respectively. In general, the higher-temperature materials have higher coupling coefficients.

In the absence of an electric field, the material is not polarized. As a result, an application of stress does not change the electric displacement. However, a change in electric permittivity of the material does occur. Hence, electrostrictives are not normally used as sensors. Because these materials are sensitive to temperature (variation within 10°C), most applications of electrostrictives are focused on operations underwater or in vivo, ranging from ultrasonic motors to medical probes. Because of the nonhysteretic nature of this material, it is used in micro-positioners.

A number of differences between piezoelectrics and electrostrictors are noted. Piezoelectric actuators and sensors must be initially polarized, and they also suffer from the problems of depolarization (in the presence of high negative fields) and aging (decrease in polarization with time). Furthermore, they show significant hysteresis leading to large energy losses in dynamic applications. Conversely because electrostrictors do not require polarization, they do not suffer from aging. They also show far less hysteresis (less than 1%), even at high frequencies, which is important for dynamic applications. Electromechanical behavior is linear for piezoelectrics and quadratic for electrostrictives. On application of stress, piezoelectric materials exhibit spontaneous change of polarization, whereas compressive stress on electrostrictive materials only results in a change in strain levels. Because of the absence of remnant polarization, electrostrictives return to the non-field zero-strain state and, hence, are quite suitable for optical positioning. Unlike piezoelectrics, the electromechanical response of electrostrictives does not deteriorate under severe operating conditions. Piezoelectrics are far less sensitive to temperature variations. Dielectric constants of piezoelectrics are smaller than those of electrostrictive materials and, hence, piezoelectric exhibits faster response. These differences result in different areas of application of the two materials.

Like piezoelectric ceramics, electrostrictive ceramics are compact, deliver small but accurate displacements, and are less prone to overheating. They find applications in sonar transducers, precision machine tools, dot-matrix printers, and ultrasonic motors. Above the Curie temperature, these materials undergo a transition from a ferroelectric (polar) state to a paraelectric (nonpolar) state, and the spontaneous polarization vanishes. Electrostrictives without a DC bias field cannot be used as stress sensors. However, with the presence of a DC bias voltage, a change in polarization induced by mechanical stress can be measured. The coupling coefficients (k_{ij}) determine the fraction of stored electrical energy that can be converted into mechanical work. These coefficients provide a direct figure of merit to compare ferroelectric (piezoelectric and electrostrictive) devices. These coefficients also provide a measure of the efficiency of a sensor in terms of the fraction of stored strain energy that can be converted into electrical energy. It has been observed that the coupling coefficients of electrostrictives are lower than those of piezoelectrics [87].

In general, for electrostrictives, the induced strain versus electric field is quadratic for low field, becomes close to linear for moderate field, and saturates at high field values. However, such a saturation is not observed in the strain versus polarization curves. This shows that the saturation nonlinearity is primarily due to electrical phenomena involving the polarization and electric field [82] and is not an electromechanical phenomenon. The characteristics of electrostrictive ceramics are sensitive to operating conditions such as electric field magnitude, excitation frequency, and ambient temperature. Significant heating of electrostrictive ceramics occurs when subjected to a high-frequency, high-magnitude excitation field.

6.10.1 Constitutive Relations

The phenomenology-based macroscopic behavior of an electrostrictive material can be derived using a parametric Gibbs elastic–free-energy function. The material constitutive relations were developed by Devonshire [88] by expressing electrostriction as a quadratic function of dielectric polarization. The direct effect is obtained as

$$Q_{ijkl} = \frac{1}{2}\frac{\partial^2 \epsilon_{ij}}{\partial P_k \partial P_l} \tag{6.155}$$

and the converse effect is

$$Q_{klij} = \frac{1}{2}\frac{\partial^2 \mathbb{E}_k}{\partial \sigma_{ij} \partial P_l} \tag{6.156}$$

where Q_{ijkl} is defined as the electrostriction coefficient. P_k are the components of polarization of the dielectric, ϵ_{ij} are the strain components, σ_{ij} are the stress components, and $\mathbb{E}_k$ is the applied electric field. For engineering applications, it is more convenient to express electrostriction relations in terms of electric field rather than polarization. In terms of the electric field, the direct-effect coefficient can be written as

$$\tilde{m}_{ijkl} = \frac{1}{2}\frac{\partial^2 \epsilon_{ij}}{\partial \mathbb{E}_k \partial \mathbb{E}_l} \tag{6.157}$$

Similarly, the converse effect coefficient becomes

$$\tilde{m}_{klij} = \frac{1}{2}\frac{\partial^2 D_k}{\partial \sigma_{ij} \partial \mathbb{E}_l} = \frac{1}{2}\frac{\partial e_{kl}}{\partial \sigma_{ij}} \tag{6.158}$$

where $\tilde{m}_{ijkl}$ is the electrostriction coefficient, D_k is the electric displacement, and e_{kl} is the dielectric permittivity. Neglecting temperature effects and higher-order terms, the full elastic Gibbs free energy for an electrostrictive material with a crystal center of symmetry is [84]

$$\begin{aligned}\Delta G = & -\frac{1}{2}e_{mn}\mathbb{E}_m\mathbb{E}_n - \frac{1}{4}e_{mnpq}\mathbb{E}_m\mathbb{E}_n\mathbb{E}_p\mathbb{E}_q - \frac{1}{6}e_{mnpqrs}\mathbb{E}_m\mathbb{E}_n\mathbb{E}_p\mathbb{E}_q\mathbb{E}_r\mathbb{E}_s \\ & - \frac{1}{2}s_{ijkl}\sigma_{ij}\sigma_{kl} - \tilde{m}_{mnij}\mathbb{E}_m\mathbb{E}_n\sigma_{ij} - r_{mnijkl}\mathbb{E}_m\mathbb{E}_n\sigma_{ij}\sigma_{kl}\end{aligned} \tag{6.159}$$

From the energy expression, the electrical displacement and strain can be derived as

$$\left(\frac{\partial \Delta G}{\partial \mathbb{E}_m}\right)_\sigma = -D_m \tag{6.160}$$

$$\left(\frac{\partial \Delta G}{\partial \sigma_{ij}}\right)_\mathbb{E} = -\epsilon_{ij} \tag{6.161}$$

This results in

$$D_m = e^\sigma_{mn}\mathbb{E}_n + e^\sigma_{mnpq}\mathbb{E}_n\mathbb{E}_p\mathbb{E}_q + e^\sigma_{mnpqrs}\mathbb{E}_n\mathbb{E}_p\mathbb{E}_q\mathbb{E}_r\mathbb{E}_s + 2\tilde{m}_{mnij}\mathbb{E}_n\sigma_{ij} + 2r_{mnijkl}\mathbb{E}_n\sigma_{ij}\sigma_{kl} \tag{6.162}$$

$$\epsilon_{ij} = s^\mathbb{E}_{ijkl}\sigma_{kl} + \tilde{m}_{ijmn}\mathbb{E}_m\mathbb{E}_n + 2r_{ijmnkl}\mathbb{E}_m\mathbb{E}_n\sigma_{kl} \tag{6.163}$$

where e_{mn}, e_{mnpq}, and e_{mnpqrs} are, respectively, the second-order, fourth-order, and sixth-order dielectric permittivities; r_{mnijkl} is the sixth-order elastostriction tensor; and s_{ijkl} is the compliance tensor. These equations represent the nonlinear electrostriction relations at constant temperature. The higher-order nonlinear terms can take into account saturation effects at high fields. For example, the elastostriction tensor is a correction factor for the compliance of the material under an applied electric field. It can also be treated as a correction to the electrostrictive constant under applied mechanical stress.

Neglecting the elastostriction tensor, as well as higher-order dielectric terms, the constitutive equations at constant temperature are derived from these equations as

$$D_m = e^\sigma_{mn}\mathbb{E}_n + 2\tilde{m}_{mnij}\mathbb{E}_n\sigma_{ij} \tag{6.164}$$

$$\epsilon_{ij} = \tilde{m}_{ijmn}\mathbb{E}_m\mathbb{E}_n + s^\mathbb{E}_{ijkl}\sigma_{kl} \tag{6.165}$$

For simplicity, these can be rewritten in matrix form as

$$\boldsymbol{D} = e^\sigma\mathbb{E} + 2\boldsymbol{m}\boldsymbol{\sigma} \quad \text{(direct effect)} \tag{6.166}$$

$$\boldsymbol{\epsilon} = \boldsymbol{m}^T\mathbb{E} + \boldsymbol{s}^\mathbb{E}\boldsymbol{\sigma} \quad \text{(converse effect)} \tag{6.167}$$

where $\boldsymbol{D}$ is the electric displacement (C/m^2), $\mathbb{E}$ is the electric field (V/m), $\boldsymbol{s}^\mathbb{E}$ is the material compliance at constant electric field (m^2/N), $\boldsymbol{m}$ is the electrostrictive coupling matrix similar to the piezoelectric coefficient $\boldsymbol{d}$ (m/V or C/N), and e^σ is the dielectric permittivity (C/Vm). The direct effect relates the electric displacement to the external stress and is used in sensor applications. The converse effect relates the induced strain to the applied electric field and is used in actuators. Note that although these constitutive relations appear to be linear, the coupling matrix $\boldsymbol{m}$ contains an electric field term. Therefore, these equations can only be considered as linearized about a given operating point. In other words, the electrostrictive constitutive relations can be considered linear for small changes in electric field about a given value of electric field. In addition, note the presence of the factor of 2 in the equation relating charge due to stress and the absence of this factor in the equation for strain due to field. This arises from the fact that the electromechanical coupling energy term is proportional to the product of the mechanical stress and two electric field components.

As a result of the cubic unit cell of PMN, it can be shown that a number of coefficients are identically equal to zero [84]. Consequently, from Eq. 6.165, the electrostriction term can be expanded as

$$\begin{Bmatrix} \epsilon_1 \\ \epsilon_2 \\ \epsilon_3 \\ \gamma_{23} \\ \gamma_{31} \\ \gamma_{12} \end{Bmatrix} = \begin{bmatrix} m_{11} & m_{12} & m_{12} & 0 & 0 & 0 \\ m_{12} & m_{11} & m_{12} & 0 & 0 & 0 \\ m_{12} & m_{12} & m_{11} & 0 & 0 & 0 \\ 0 & 0 & 0 & m_{44} & 0 & 0 \\ 0 & 0 & 0 & 0 & m_{44} & 0 \\ 0 & 0 & 0 & 0 & 0 & m_{44} \end{bmatrix} \begin{Bmatrix} \mathbb{E}_1^2 \\ \mathbb{E}_2^2 \\ \mathbb{E}_3^2 \\ \mathbb{E}_2\mathbb{E}_3 \\ \mathbb{E}_3\mathbb{E}_1 \\ \mathbb{E}_1\mathbb{E}_2 \end{Bmatrix} \tag{6.168}$$

From this expression, the electrostrictive coupling matrix in Eq. 6.166 becomes

$$\boldsymbol{m} = \begin{bmatrix} m_{11}\mathbb{E}_1 & m_{12}\mathbb{E}_1 & m_{12}\mathbb{E}_1 & 0 & \frac{m_{44}}{2}\mathbb{E}_3 & \frac{m_{44}}{2}\mathbb{E}_2 \\ m_{12}\mathbb{E}_2 & m_{11}\mathbb{E}_2 & m_{12}\mathbb{E}_2 & \frac{m_{44}}{2}\mathbb{E}_3 & 0 & \frac{m_{44}}{2}\mathbb{E}_1 \\ m_{12}\mathbb{E}_3 & m_{12}\mathbb{E}_3 & m_{11}\mathbb{E}_3 & \frac{m_{44}}{2}\mathbb{E}_2 & \frac{m_{44}}{2}\mathbb{E}_1 & 0 \end{bmatrix} \tag{6.169}$$

where m_{ij} are the electrostrictive coefficients with units m^2/V^2. Again, note that the quadratic nonlinearity appears in the electrostrictive coupling matrix as an additional electric field term. The electrostrictive sensor equation (direct effect) can be expanded as

$$\begin{Bmatrix} D_1 \\ D_2 \\ D_3 \end{Bmatrix} = \begin{bmatrix} e_{11}^{\sigma} & 0 & 0 \\ 0 & e_{11}^{\sigma} & 0 \\ 0 & 0 & e_{11}^{\sigma} \end{bmatrix} \begin{Bmatrix} \mathbb{E}_1 \\ \mathbb{E}_2 \\ \mathbb{E}_3 \end{Bmatrix}$$

$$+2 \begin{bmatrix} m_{11}\mathbb{E}_1 & m_{12}\mathbb{E}_1 & m_{12}\mathbb{E}_1 & 0 & \frac{m_{44}}{2}\mathbb{E}_3 & \frac{m_{44}}{2}\mathbb{E}_2 \\ m_{12}\mathbb{E}_2 & m_{11}\mathbb{E}_2 & m_{12}\mathbb{E}_2 & \frac{m_{44}}{2}\mathbb{E}_3 & 0 & \frac{m_{44}}{2}\mathbb{E}_1 \\ m_{12}\mathbb{E}_3 & m_{12}\mathbb{E}_3 & m_{11}\mathbb{E}_3 & \frac{m_{44}}{2}\mathbb{E}_2 & \frac{m_{44}}{2}\mathbb{E}_1 & 0 \end{bmatrix} \begin{Bmatrix} \sigma_1 \\ \sigma_2 \\ \sigma_3 \\ \tau_{23} \\ \tau_{31} \\ \tau_{12} \end{Bmatrix} \tag{6.170}$$

The electrostrictive actuator equation (converse effect) can be expanded as

$$\begin{Bmatrix} \epsilon_1 \\ \epsilon_2 \\ \epsilon_3 \\ \gamma_{23} \\ \gamma_{31} \\ \gamma_{12} \end{Bmatrix} = \begin{bmatrix} m_{11}\mathbb{E}_1 & m_{12}\mathbb{E}_2 & m_{12}\mathbb{E}_3 \\ m_{12}\mathbb{E}_1 & m_{11}\mathbb{E}_2 & m_{12}\mathbb{E}_3 \\ m_{12}\mathbb{E}_1 & m_{12}\mathbb{E}_2 & m_{11}\mathbb{E}_3 \\ 0 & \frac{m_{44}}{2}\mathbb{E}_3 & \frac{m_{44}}{2}\mathbb{E}_2 \\ \frac{m_{44}}{2}\mathbb{E}_3 & 0 & \frac{m_{44}}{2}\mathbb{E}_1 \\ \frac{m_{44}}{2}\mathbb{E}_2 & \frac{m_{44}}{2}\mathbb{E}_1 & 0 \end{bmatrix} \begin{Bmatrix} \mathbb{E}_1 \\ \mathbb{E}_2 \\ \mathbb{E}_3 \end{Bmatrix}$$

$$+ \begin{bmatrix} s_{11}^{\mathbb{E}} & s_{12}^{\mathbb{E}} & s_{12}^{\mathbb{E}} & 0 & 0 & 0 \\ s_{12}^{\mathbb{E}} & s_{11}^{\mathbb{E}} & s_{12}^{\mathbb{E}} & 0 & 0 & 0 \\ s_{12}^{\mathbb{E}} & s_{12}^{\mathbb{E}} & s_{11}^{\mathbb{E}} & 0 & 0 & 0 \\ 0 & 0 & 0 & s_{44}^{\mathbb{E}} & 0 & 0 \\ 0 & 0 & 0 & 0 & s_{44}^{\mathbb{E}} & 0 \\ 0 & 0 & 0 & 0 & 0 & s_{44}^{\mathbb{E}} \end{bmatrix} \begin{Bmatrix} \sigma_1 \\ \sigma_2 \\ \sigma_3 \\ \tau_{23} \\ \tau_{31} \\ \tau_{12} \end{Bmatrix} \tag{6.171}$$

Note that the compliance matrix is isotropic. Some electrostrictives such as 0.9PMN-0.1PT exhibit a significant change of the apparent Young's modulus as a function of the electric field. The change can be more than 50% in the direction of the field. However, the variation of modulus in a plane normal to the field is shown to be much smaller (around 6%) [89]. These equations lead to an induced strain that is proportional to the square of the applied field. For example, for a field applied only along the 1-direction, in the absence of external stresses, the strain in the 1-direction is given by

$$\epsilon_1 = m_{11}\mathbb{E}_1^2 \tag{6.172}$$

This representation is useful for low electric fields but does not take into account the saturation of induced strain at high fields. A hyperbolic tangent model [90, 91] can be used to include this effect. In this case, the strain in this equation becomes

$$\epsilon_1 = \frac{1}{k_r^2} m_{11} \tanh^2(k_r \mathbb{E}_1) \tag{6.173}$$

where k_r is a relaxation factor that represents the point of saturation. It is important to note that the coefficient matrices described here are dependent on the applied electric field and mechanical stress, resulting in deviations from the quadratic or hyperbolic–tangent variation assumptions.

A significant amount of research has gone into the modeling and experimental investigation of electrostrictive materials. Hom and Shankar [92, 93] presented a fully coupled, two-dimensional, quasi-static finite element analysis for electroceramics and applied it to electrostrictive stack actuators. This formulation incorporates the effect of body forces of dielectric origin but ignores the body moments of dielectric origin. Fripp and Hagood [94] presented a set of constitutive equations for electrostrictive materials and developed a dynamic analysis for an electromechanical system with distributed electrostrictive couplings. A Rayleigh-Ritz analysis was formulated for a cantilevered beam actuated with surface-bonded electrostrictive wafers and satisfactorily validated for static and dynamic response with experimental test data. Piquet and Forsythe [95] covered nonlinear modeling of PMN materials. Pablo and Petitjean [82] carried out stress-free electric behavior (in transverse direction) of electrostrictive patches experimentally at a macro level for a range of excitation fields, frequencies, and temperatures.

In summary, electrostrictive materials are well suited to precise positioning applications in a laboratory environment due to their high stroke and stiffness. In generic applications, special attention must be paid to the design of an appropriate control system to compensate for the large temperature sensitivity of electrostrictives and their inherent nonlinearity. The main advantage of these materials is their low drift and low hysteresis, resulting in low self-heating during dynamic actuation. The electrostrictive effect is a quadratic effect, so it depends on the square of the voltage. The lack of polarization in electrostrictives means that there is no depoling field. This fundamental difference has also some consequences on the electric behavior of the ceramic. Indeed, this symmetry implies that there is less hysteresis loss than in a piezoelectric.

The following sections discuss the major characteristics of electrostriction, illustrated by experiments performed on a commercially available sheet sample of PMN ceramic. The test specimens used in these experiments were sheet samples of BM600 ceramic (composition 0.9PMN-0.1PT) with a size of 2 in $\times$ 1 in and

Table 6.13. *Small-signal characteristics of BM600 electrostrictive ceramic*

Physical Properties	
Density, g/cc	7.80
Young's modulus, GPa	100
K_{33} (at 25°C, 1 kHz)	22,000
$\tan\delta$ (at 25°C)	0.08
Static d_{33}, pC/N (at 25°C and 4.0 kV/cm)	1800
Dynamic d_{33}, pC/N (at 38°C and 4.5 kV/cm)	650
Dynamic d_{31}, pC/N (at 38°C and 4.0 kV/cm)	290
k_{33} (at 6 kV/cm)	0.55
k_{31} (at 5 kV/cm)	0.25

thickness 0.01 in (Sensortech Technology). Important properties of this material, as given by the manufacturer, are listed in Table 6.13. The experimental setups and test procedures are similar to those used for testing of piezoceramic actuators, as described in Chapter 2.

6.10.2 Behavior under Static Excitation Fields

Under static excitation fields, two important phenomena can be observed. These are the static free strain and the strain drift. The measurement of these two quantities using a BM600 sheet element is described in this section.

Induced Strain under Static Excitation Fields

The strain of the electrostrictive sheet element in response to a static excitation field can be measured using a strain gauge bonded to the surface of the sample. The excitation field is varied from −13.78 kV/cm (−350 V) to +25 kV/cm (+635 V) in uniform increments. The limitations on excitation field were dictated by practical considerations, such as arcing between the high-voltage electrodes. This voltage range is considered sufficient for practical applications. Each value of the excitation field is maintained for a given amount of time, at the end of which the strain reading is recorded. An average of three separate measurements are recorded. As a result, a quasi-static variation of induced strain with excitation field is obtained. Because the hysteresis of the material is small compared to that in the case of piezoelectric materials, a quasi-static hysteresis curve can be measured. During this test, the following important points must be kept in mind:

(1) Despite low inherent hysteresis in the material, the measurements of the induced strain for positive and for negative excitation fields must be taken separately. This is because the remnant strain for positive excitation fields and for negative excitation fields can be different. Either the positive or negative polarity is chosen first, and the material is cycled until the residual strain stabilizes. The measurements are then made by increasing the excitation field to the required value, holding the field constant for a short period of time, recording the induced-strain value, and then increasing the excitation field to the next higher value. This is repeated for increasing values of the excitation field until the maximum field

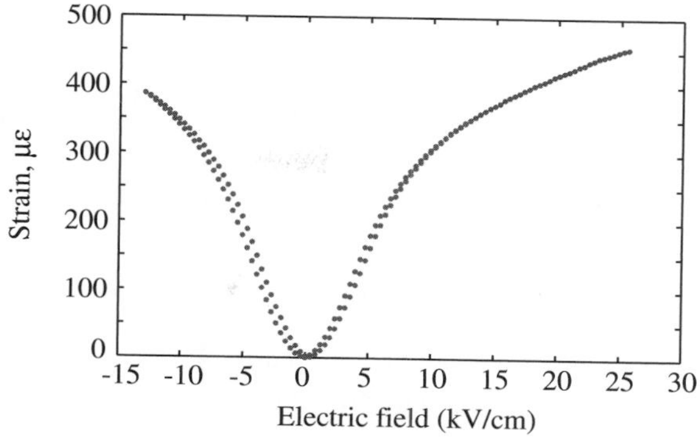

Figure 6.58. Static free strain of PMN sample.

is reached. The process is then repeated while decreasing the excitation field in uniform steps until the applied field is zero. In this manner, all datapoints are recorded for the chosen polarity. The procedure is repeated for the opposite polarity of excitation field, yielding a quasi-static hysteresis curve of induced strain versus excitation field.

(2) When a static field is applied, the strain induced in the material is not constant with time. A slow increase in the induced strain with time is observed. This phenomenon is called "drift." To have consistent results, the measurement is taken after maintaining the excitation field for the same period of time at each measurement point. For example, the excitation field can be switched on, and after the chosen time (e.g., 2 seconds), the induced strain is recorded and the excitation field is switched off.

The quasi-static hysteresis behavior measured as described here is shown in Fig. 6.58. The most important feature that can be observed is that the induced strain does not depend on the polarity of the applied excitation field. This is sometimes referred to as a "quadratic" response because at low values of excitation field, the induced strain is proportional to the square of the applied excitation field. Another important observation is that the amount of hysteresis is small, especially compared to the case of piezoelectric ceramics.

The induced strain of the BM600 electrostrictive sample is compared to that of a piezoelectric ceramic (composition PZT-5H) in Fig. 6.59. Note that the hysteresis in the induced strain is not shown. The most important difference in the behavior is the independence of electrostrictive-induced strain on the polarity of applied excitation

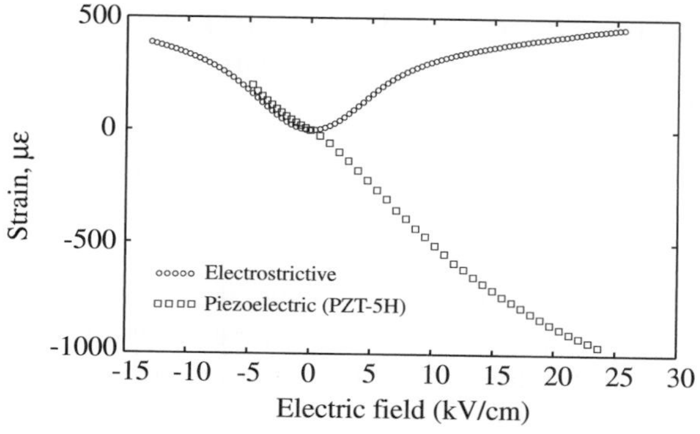

Figure 6.59. Comparison of PZT and PMN static free strain.

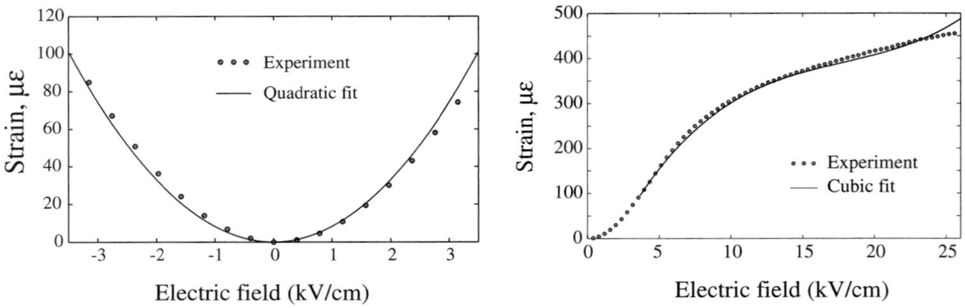

(a) Low excitation field (b) High excitation field

Figure 6.60. Empirical model of PMN static free strain.

field. In addition, the absence of a depoling field in the case of the electrostrictive enables the application of much lower negative fields as compared to the piezoelectric. However, for positive excitation fields, the maximum free strain is higher in the case of the piezoelectric material.

At low excitation fields (magnitude less than 3.5 kV/cm), the induced strain Λ can be represented as a function of the excitation field E by a quadratic relation; for higher fields (up to 27 kV/cm), the induced strain can be represented as a cubic function of the excitation field. Figure 6.60 shows the induced strain of the electrostrictive material at low and high excitation fields as well as the empirical model.

$$\Lambda = \begin{cases} 8.258E^2 & 0 \leq E < 3.5\ \text{kV/cm} \\ 0.076E^3 - 3.903E^2 + 74.5E - 124.06 & 3.5\ \text{kV/cm} \leq E < 27\ \text{kV/cm} \end{cases} \tag{6.174}$$

where Λ is in microstrain. It can be seen that the measured data are closely represented by the quadratic relation for $|E| \leq 3.5$ kV/cm.

Drift in Induced Strain

Under a static excitation field, the strain induced in the electrostrictive does not remain constant but rather exhibits a slow increase with time. This phenomenon is called drift and is similar to the drift phenomenon observed in piezoelectric ceramics. The increase in strain with time can typically be represented by an exponential relationship. The drift in induced strain at four different voltages, along with the empirical fit, is shown in Fig. 6.61. For reference, a voltage of 50 V applied to the sample corresponds to a field of approximately 2 kV/cm. The increase in strain $\Delta\epsilon$ can be expressed as

$$\Delta\epsilon = \epsilon_o \left(1 + \gamma \ln \frac{t}{0.1}\right) \tag{6.175}$$

where ϵ_o is the strain 0.1 second after the excitation field was applied, t is the time elapsed in seconds, and γ is a time constant. In the present case, the value of γ is 1.8%. It is interesting to note that the drift in the case of electrostrictive is similar to that of the piezoelectric.

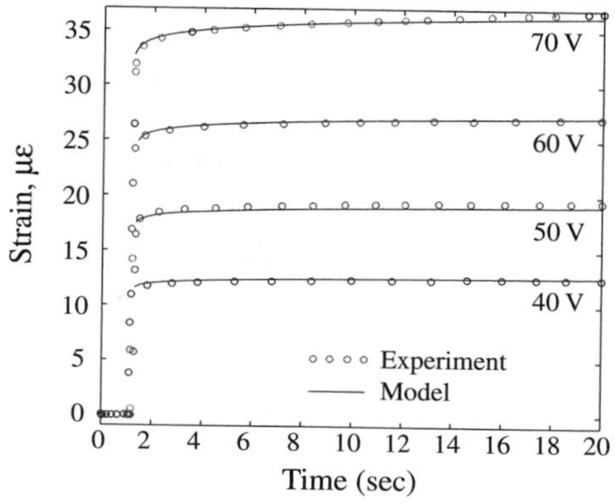

Figure 6.61. Drift of PMN free strain.

6.10.3 Behavior under Dynamic Excitation Fields

In a number of practical applications, the actuators must be operated under a dynamic excitation field, producing a dynamic strain. Under such conditions, the hysteresis of the induced strain and the electrical impedance of the material are important considerations.

Strain Hysteresis

The hysteresis of the induced strain of the material is shown in Fig. 6.62 for two different amplitudes of the excitation field, at a frequency of 1 Hz. The quadratic nature of the induced-strain response is clearly visible. Note that the strain values are generally smaller than in the case of a typical PZT-5H sheet. The hysteresis of the induced strain for three different excitation frequencies at the same peak excitation voltage is shown in Fig. 6.63. The hysteresis does not appear to vary strongly with excitation frequency.

It has been observed that for PMN ceramics at room temperature, the hysteresis is low and the response is approximately linear for most of the applied field [84]. As the temperature is reduced, the hysteresis and the total strain increase. At the same time, the field at which strain saturation occurs decreases with lower temperature. Under a high-cyclic electric field, some heating of the material takes place, which is a function of the excitation frequency. Induced strain is more dependent on the equivalent temperature than excitation frequency. Frequency affects the strain indirectly through an increase of material temperature.

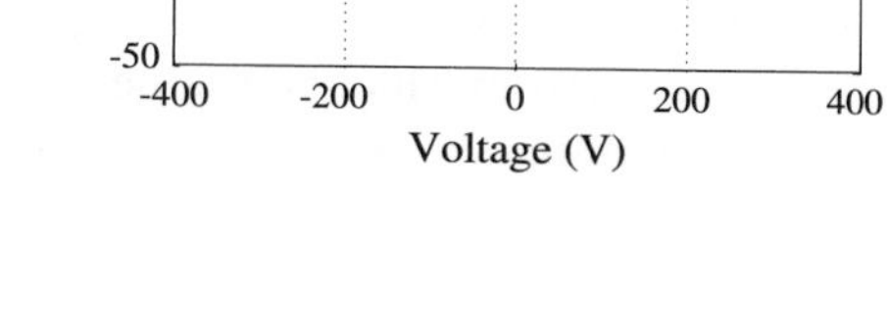

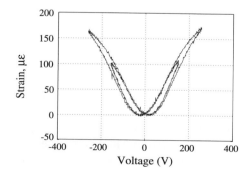

Figure 6.62. PMN strain hysteresis as a function of applied voltage at 1 Hz.

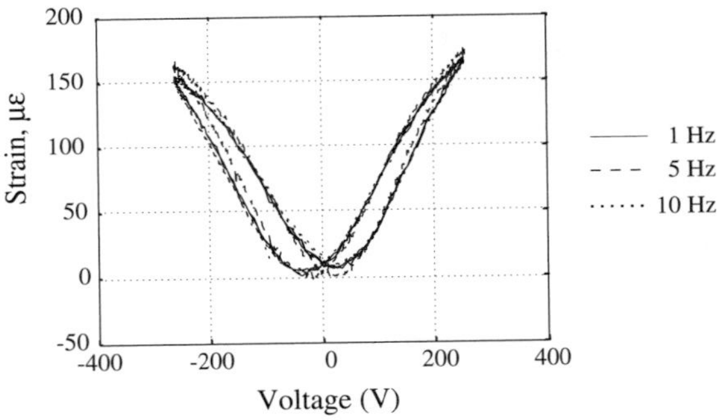

Figure 6.63. PMN strain hysteresis as a function of frequency.

Electrical Impedance

The electrical impedance of the electrostrictive sample is found by exciting it with a swept sine wave. The voltage and current are measured at excitation frequency, from which the electrical impedance can be calculated. As in the case of piezoceramics, the electrical impedance of the electrostrictive material is primarily capacitive in nature. For excitation frequencies much lower than resonance, the impedance can be expressed as an ideal capacitor in series with a resistance. The resistance models the losses in the material, which occur due to the motion of the dipoles in response to the applied electric field. These losses are typically quantified by a constant called the "dissipation factor."

An electrostrictive sheet behaves like a parallel-plate capacitor. For a sheet of thickness t, with electrodes of area A, the capacitance C is given by

$$C = \frac{eA}{t} \tag{6.176}$$

As in the case of a piezoceramic sheet, the electric permittivity e is given by

$$e = K_3 e_o - j \tan\delta \tag{6.177}$$

where e_o is the permittivity of free space or vacuum, K_3 is the relative permittivity of the material (for electrodes in the 1-2 plane), and $\tan\delta$ is the dissipation factor. In practice, the value of $\tan\delta$ is usually much less than unity. Simplifying these equations, the electrical impedance (for harmonic excitation at a circular frequency ω) can be expressed as

$$\begin{aligned} Z &= \frac{1}{j\omega C} \\ &= \frac{1}{j\omega C_o} + \frac{\tan\delta}{\omega C_o} \end{aligned} \tag{6.178}$$

where C_o represents the ideal capacitance of the sample

$$C_o = \frac{K_3 e_o A}{t} \tag{6.179}$$

Note that, in general, as in the case of piezoceramics, the values of K_3 and $\tan\delta$ can depend on the magnitude as well as the frequency of the excitation field. From the electrical impedance measurements, the real and imaginary parts of the impedance can be found, from which the value of K_3 and $\tan\delta$ can be calculated using Eq. 6.178, at each operating condition.

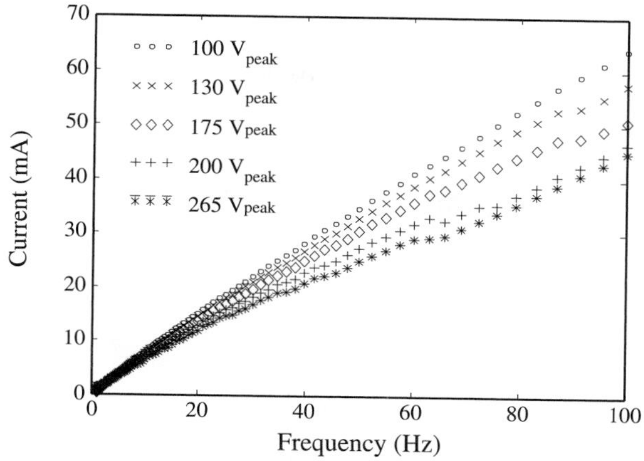

Figure 6.64. Current drawn by PMN sample.

The current drawn by the electrostrictive sample as a function of frequency is shown in Fig. 6.64 for several excitation voltage amplitudes. At lower excitation voltages, the current increases linearly with frequency, which shows that the impedance is dominated by the capacitive term. At higher excitation voltages, the resistive part of the impedance becomes significant. This is shown more clearly in Fig. 6.65, which shows the variation of relative permittivity, and in Fig. 6.66, which shows the variation of dissipation factor. The solid lines in these figures are empirical relations that were fit to these measurements. The relative permittivity is given by

$$K_3 = af^2 + bf + c \tag{6.180}$$

where f is the excitation frequency in Hertz and a, b, and c are functions of the amplitude of the excitation voltage V

$$a = -1.945 \times 10^{-3}V + 0.898 \tag{6.181}$$

$$b = 0.36669V - 170.59 \tag{6.182}$$

$$c = 0.6116V^2 - 340.14V + 56.421 \times 10^3 \tag{6.183}$$

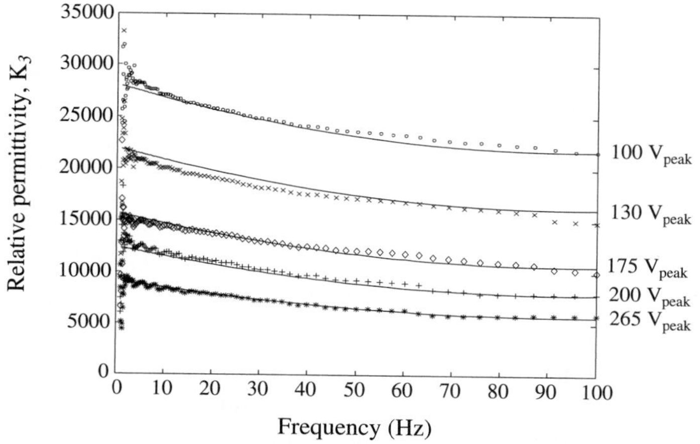

Figure 6.65. Relative permittivity as a function of excitation voltage and frequency.

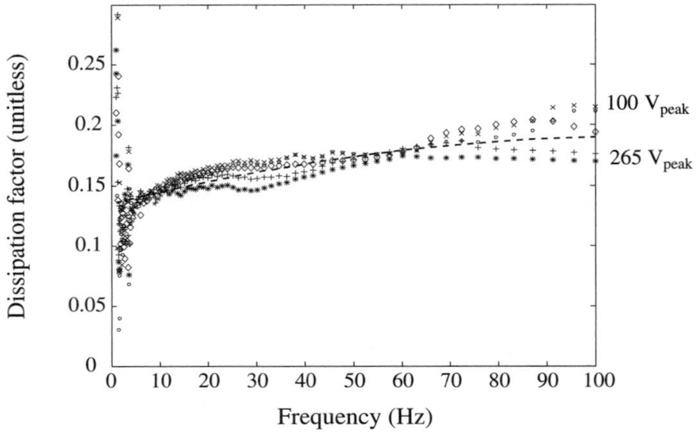

Figure 6.66. Dissipation factor as a function of excitation voltage and frequency.

The dissipation factor is given by a quadratic function of the frequency

$$\tan\delta = -4.5235 \times 10^{-6} f^2 + 0.001 f + 0.13556 \tag{6.184}$$

Using these empirical relations, the electrical impedance and, therefore, the power consumed by the electrostrictive actuators, can be predicted for any excitation voltage and frequency in the range described previously. Two important conclusions can be made from the data in Figs. 6.65 and 6.66. First, the relative permittivity and, therefore, the effective capacitance of the actuator, is much higher for the electrostrictive material than in the case of the piezoceramic. Second, the dissipation factor for the electrostrictive material is less than for the piezoceramic and also is independent of the magnitude of the excitation frequency. These are extremely important considerations when choosing an appropriate actuator material for a given application.

6.10.4 Effect of Temperature

Temperature has a strong effect on the characteristics of electrostrictives. The permittivity of the material is significantly changed with temperature. In addition, relaxor ferroelectrics such as PMN do not have a well-defined Curie temperature. Conversely, there exists a range of temperatures at which the material exists in a mix of both paraelectric and ferroelectric states. Using the modified Curie-Weiss law, one can obtain the tangent permittivity e^*_{11} at any temperature

$$e^*_{11} = \frac{\partial D_1}{\partial \mathbb{E}_1} \tag{6.185}$$

which is given by

$$\frac{1}{e^*_{11}} = \frac{1}{\left(e^*_{11}\right)_M} \exp\left[\frac{(T - T_M)^2}{2\delta^2}\right] \tag{6.186}$$

where $\left(e^*_{11}\right)_M$ is the maximum permittivity, δ is a parameter governing local Curie temperature, and T_M is the temperature at maximum permittivity. For a nonzero bias field, the permittivity may not be maximum at mean Curie temperature. Using a Taylor series, Blackwood and Ealey [84] simplified this relation, valid for small $(T - T_M)/\sqrt{2}\delta$, to

$$\frac{1}{e_{11}^*} = \frac{1}{\left(e_{11}^*\right)_M} + \frac{(T - T_M)^2}{2\delta^2 \left(e_{11}^*\right)_M} \tag{6.187}$$

6.11 Polarization

The net polarization of a ferroelectric material consists of three fundamental mechanisms: the electronic polarization, the ionic polarization, and the dipole-orientation polarization. A dielectric material between electrodes forms a capacitor that can store charge. On application of an electric field, the center of positive charge of the ionic crystal is attracted to the cathode and the center of negative charge to the anode due to electrostatic attraction. This process, called "polarization," induces electric dipoles within the material. The stored electric charge per unit area is called the electric displacement D

$$D = e_o\mathbb{E} + P \tag{6.188}$$

where e_o is the permittivity of free space (F/m) and P is the polarization. This can be rewritten as

$$D = Ke_o\mathbb{E} \tag{6.189}$$

where K is the relative permittivity, which is also referred to as the "dielectric constant." On the application of an electric field, the spontaneous polarization of the dielectric material can be reversed for a ferroelectric material. Note that not all dielectric materials are ferroelectric.

Barium titanate $BaTiO_3$ is a ferroelectric material. At high temperature (above Curie temperature T_c), it exists in a paraelectric phase and there is no spontaneous polarization. Below the transition temperature, it develops spontaneous polarization. Above the transition temperature, the crystal structure is of cubic symmetry; below the transition temperature, the crystal structure becomes tetragonal symmetric (slightly elongated). The spontaneous polarization P_s and spontaneous strain ϵ_s are related as

$$\epsilon_s = QP_s^2 \tag{6.190}$$

where Q is the electrostrictive coefficient. Thus, the spontaneous strain due to spontaneous polarization as a result of an applied electric field decreases with increasing temperature and becomes zero at and beyond the transition temperature.

Spontaneous polarization decreases with temperature and becomes zero at the Curie temperature T_c, whereas electric permittivity e diverges at Curie temperature. The inverse of permittivity appears linear and is referred to as Curie-Weiss law

$$e = \frac{C}{T - T_o} \tag{6.191}$$

where C is the Curie-Weiss constant and T_o is the Curie-Weiss temperature (slightly lower than T_c). For capacitor dielectrics, the maximum dielectric constant occurs around the Curie temperature T_c. For pyroelectric transducers, spontaneous polarization below T_c is important. Thus, the material behaves as an electrostrictive for temperatures above the transition temperature.

6.12 Young's Modulus

Electrostrictive ceramics such as 0.9PMN-0.1PT show a large reduction of the apparent Young's modulus as a function of electric field (more than 50%), in the direction of the applied field. However, the variation of the elastic modulus in the direction perpendicular to the applied electric field can be comparatively small (less than 6%) [89]. It was found that the Young's modulus decreases from 120 GPa to 50 GPa when a 1 MV/m static electric field is applied at room temperature. However, with the application of a uniaxial compressive stress of 30 MPa parallel to the electric field, the Young's modulus increases to 60 GPa.

6.13 Summary and Conclusions

Magnetosrictives and electrostictives are active materials that exhibit quadratic induced strain characteristics with respect to applied magnetic/electric field. This behavior differentiates these materials from piezoelectrics that exhibit predominantly linear induced strain characteristics with field. Typically, for magnetostrictives and electrostrictives, the induced strain with field varies quadratically at low-field values, becomes close to linear for moderate-field values, and saturates at high-field values. Even though magnetostriction was discovered in nickel a long time ago (in 1842 by Joule), the maximum induced-strain levels were quite low for most practical applications. With the discovery of Terfenol-D with its large magnetostriction, in the 1970s, it is now used in a wide range of applications that include machine tools, servo valves, sonar, ultrasonic cleaning, load sensors, and micropositioners. Terfenol-D is an alloy of terbium, dysprosium, and iron ($Tb_xDy_{1-x}\,Fe_y$) and is normally produced as a near-single crystal. The value of x varies from 0.27 to 0.3 and y varies from 1.92 to 2.0. Small changes in x and y (stoichiometry) can have a major influence on the material magnetization, magnetostriction, and elastic characteristics, which are a nonlinear function of magnetic, mechanical (stress), and thermal operating conditions. The converse effect in magnetostrictives was discovered in the early 1900s by Villari and is often referred to as the Villari effect. It is a change in magnetization of the material in response to its deformation. Thus, the Joule effect transforms magnetic energy into mechanical energy and the Villari effect transforms mechanical energy into magnetic energy. Because of the bi-directional exchange of energy, magnetostrictives can be used both as actuators and sensors. Magnetostrictive materials elongate in the direction of an applied magnetic field and contract in the direction normal to the field, such that the net change in volume is nearly invariant. The maximum induced strain in Terfenol-D is on the order of 0.2% (2000×10^{-6}), which is about twice that of piezoelectrics; however, its stiffness is about 40% lower than that of piezoelectrics. There is a significant change in stiffness characteristics of the material with magnetization and bias stress, called the ΔE effect. The major drawbacks of Terfenol-D are brittleness and low tensile strength. Normally, a mechanical compressive pre-stress is used to improve the performance of the transducer.

A key element of a magnetostrictive transducer is the need for a magnetic coil (solenoid) that transforms electric energy into magnetic induction for actuator operation and mechanical energy into the magnetization state for sensor operation. It becomes important to design an appropriate magnetic coil for a specific application to achieve a uniform magnetic field of high intensity. The magnetic field generator consists of electrical conducting coils, magnetic flux paths, and a mechanical preload mechanism; these together result in a significant weight and volume of the complete transducer. To increase the overall efficiency of the magnetostrictive transducer, laminated magnetic cores and permanent magnets are often used. Because of the requirement of magnetic field generation, magnetostrictive transducers are usually heavy and bulky in comparison to piezoelectric and electrostrictive counterparts. For a precise constitutive model of a magnetostrictive transducer, it is important to consider appropriately electro-magnetic coupling (between magnetic and electric fields), magnetomechanical coupling (between magnetic and elastic states), thermomagnetic coupling (between magnetization and thermal states), and thermoelastic coupling (between thermal and elastic states). Accurate comprehensive models of magnetostrictives covering all couplings of different states are not readily available. Most of the engineering models are phenomenological, fitting the experimental data of the bulk material into physics-based laws. A simple and commonly used constitutive model is the linear piezomagnetic model. Despite the actual behavior being intrinsically nonlinear and hysteretic, this quasi-steady linearized macromechanics model is quite insightful and amenable for inclusion in engineering analyses. This model represents interaction between the magnetic field and mechanical stress about a bias point for moderate field amplitudes. A major barrier for use of magnetostrictive actuators in aerospace applications is its bulky magnetic field generator. It is important to creatively develop lightweight, compact magnetic coil and flux paths. Many industrial applications require robust ductile materials with high tensile strength. Also, there is a need for detailed material characteristics for a wide range of static and dynamic operating conditions. Because of low tensile strength (30 MPa) and brittleness, the magnetostrictives are not easily machinable and cannot be fabricated in complex shapes. A new class of iron-gallium alloy called Galfenol, which has high tensile strength (500 MPa), has emerged. This material exhibits moderate magnetostriction (350×10^{-6}) under a very low magnetic field (100 Oe) but is highly ductile, machinable, and weldable. It has very low hysteresis and a high Curie temperature. Again, the low-cost Galfenol can be used both as actuators and sensors. To exploit this material in engineering applications, one requires material characteristics for a wide range of static and dynamic operating conditions as well as simplified macromechanic constitutive models that can easily be included in engineering analyses.

Most of the active materials, such as piezoelectrics, magnetostrictives, and electrostrictives, inherently possess very low maximum induced strain (0.1–0.2%) over a range of frequencies (in kHz). Conversely, SMAs such as Nitinol have a very large induced strain (6–8%) but at an extremely low frequency (less than 1 Hz). Recently, a new class of nickel-manganese-gallium alloy called ferromagnetic SMA has emerged, which displays a large induced strain (up to 6%) at a high frequency (kHz). On the application of magnetic field, the original twin variant transforms into the field-preferred variant, resulting in motion of the twin boundary causing elongation normal to the field. A compressive stress along the length of the specimen can cause the stress-preferred twin variant to reverse the effects of magnetic field. At a critical compressive stress, the saturation of the stress-preferred twin variant is reached and, beyond this stress, it results in elastic strain. If there is initially no magnetic field, the removal of compressive stress results in a residual strain, which is completely recoverable on the application of magnetic field normal to the stress direction. To achieve dynamic (bi-directional) induced strain, an initial compressive stress is needed. The induced-strain response is a quadratic function of the applied magnetic field (similar to magnetostriction). Two major drawbacks of this material are very low stiffness (one to two orders lower than standard SMA) and the requirement of a bulky magnetic coil. For most practical applications, one needs higher stiffness material than the current magnetic SMAs.

Relaxor ferroelectrics undergo large electrostriction strains under the application of electric field. Under this category of materials, lead-magnesium-niobate (PMN) ceramics display a large electrostriction (on the order of 0.1%). To improve their characteristics, PMNs are normally doped with lead titanate (PT) in a small volume fraction. Unlike piezoelectrics, electrostrictives do not require initial polarization and are isotropic under zero electric field. Under the application of electric field, these materials are instantly polarized and become anisotropic. Electrostrictives undergo a quadratic variation in induced strain with field, extension in the direction of field, and contraction in the transverse direction. To produce dynamic (bidirectional strain), it becomes necessary to apply a bias DC field. These materials are stiffer and have lower hysteresis than piezoelectrics, but they are very sensitive to temperature. Electrostrictives are primarily used as actuators, especially in underwater and in vivo applications where the change in temperature is expected to be small. These materials provide small but accurate displacement and they do not age with time. Overall, there is a general lack of detailed static and dynamic material characteristics for a range of operating conditions. It is also important to develop simplified macromechanic–constitutive-material models that can easily be included in engineering analyses.

BIBLIOGRAPHY

[1] D. Jiles. *Introduction to Magnetism and Magnetic Materials, 2nd Edition*. Chapman and Hall, 1998.

[2] F. V. Hunt. *Electroacoustics: The Analysis of Transduction and its Historical Background*. American Institute of Physics for the Acoustical Society of America, 1953.

[3] M. Goodfriend, K. Shoop, and T. Hansen. Applications of magnetostrictive Terfenol-D. *Proceedings of Actuator 94, 4th International Conference on New Actuators*, Bremen, Germany, 1994.

[4] M. J. Dapino, F. T. Calkins, and A. B. Flatau. Magnetostrictive devices. *Wiley Encyclopedia of Electrical and Electronics Engineering*, edited by J. G. Webster. John Wiley and Sons, Inc., 1999.

[5] A. E. Clark. *Magnetostrictive rare earth Fe_2 compounds. Ferromagnetic materials*, edited by E. P. Wohlfarth. North-Holland Pub., 1980.

[6] J. L. Butler, S. C. Butler, and A. E. Clark. Unidirectional magnetostrictive piezoelectric hybrid transducer. *Journal of the Acoustical Society of America*, 88(1):7–11, July 1990.

[7] J. L. Butler. *Application Manual for the Design of ETREMA Terfenol-D Magnetostrictive Transducers*. ETREMA Products, Edge Technologies, 1988.

[8] O. D. McMasters. Method of forming magnetostrictive rods from rare earth-iron alloys. Technical Report, U.S. Patent No. 4,609,402, September 1986.

[9] E. D. Gibson, J. D. Verhoeven, F. A. Schmidt, and O. D. McMasters. Method of forming magnetostrictive rods from rare earth-iron alloys. Technical Report, U.S. Patent No. 4,770,704, September 1988.

[10] J. D. Snodgrass and O. D. McMasters. Optimized Terfenol-D manufacturing processes. Technical Report, ETREMA Products Inc., Preprint, 1997.

[11] J. D. Verhoeven, E. D. Gibson, O. D. McMasters, and H. H. Baker. The growth of single crystal Terfenol-D crystals. *Metallurgical Transactions A*, 18A, 1987.

[12] J. D. Verhoeven, E. D. Gibson, O. D. McMasters, and J. E. Ostenson. Directional solidification and heat treatment of Terfenol-D magnetostrictive materials. *Metallurgical Transactions A*, 21(8):2249–2255, 1990.

[13] A. E. Clark, M. Wun-Fogle, J. B. Restorff, and T. A. Lograsso. Magnetic and magnetostrictive properties of Galfenol alloys under large compressive stresses. *Proceedings of the International Symposium on Smart Materials: Fundamentals and System Applications, Pacific Rim Conference on Advanced Materials and Processing (PRICM-4)*, Honolulu, Hawaii, December 2001.

[14] R. A. Kellogg, A. B. Flatau, A. E. Clark, M. Wun-Fogle, and T. A. Lograsso. Quasi-static transduction characterization of Galfenol. *Journal of Intelligent Material Systems and Structures*, 16(6):471–479, June 2005.

[15] R. A. Kellogg, A. M. Russell, T. A. Lograsso, A. B. Flatau, A. E. Clark, and M. Wun-Fogle. Tensile properties of magnetostrictive Iron-Gallium alloys. *Acta Materialia*, 52 (17):5043–5050, October 2004.

[16] A. E. Clark, J. B. Restorff, M. Wun-Fogle, T. A. Lograsso, and D. L. Schlagel. Magnetostrictive properties of body-centered cubic Fe-Ga and Fe-Ga-Al alloys. *IEEE Transactions on Magnetics*, 36(5):3238–3240, September 2000.

[17] R. A. Serway. *Physics for Scientists and Engineers*. Saunders College Publishing, 1983.

[18] S. Chikazumi. *Physics of Magnetism*. John Wiley and Sons, Inc., New York, 1964.

[19] K. H. J. Buschow and F. R. De Boer. *Physics of magnetism and magnetic materials*. Kluwer Academic Press, New York, 2003.

[20] M. J. Dapino. Nonlinear and hysteretic magnetomechanical model for magnetostrictive transducers. Ph.D. Dissertation, Iowa State University, 1999.

[21] E. W. Lee. Magnetostriction and magnetomechanical effects. *Reports on Progress in Physics*, 18:184–220, 1955.

[22] A. V. Andreev and K. H. J. Buschow. *Handbook of Magnetic materials*, Volume 8. Elsevier Science Publishers, Amsterdam, 1995.

[23] D. Gignoux. *Material Science and Technology*, Edited by R. W. Cahn, P. Haasen and E. J. Kran, volume 3A. VCH Verlag, Weinheim, 1992.

[24] B. D. Cullity. *Introduction to Magnetic Materials*. Addison-Wesley, Reading, MA, 1972.

[25] A. E. Clark. High power rare earth magnetostrictive materials. In *Proceedings of Recent Advances in Adaptive and Sensory Materials and Their Applications*, Technomic Publishing Co., Inc., Lancaster, PA, 1992.

[26] T. A. Duenas, L. Hsu, and G. P. Carman. Magnetostrictive composite material systems analytical/experimental. *Symposium on Advances in Smart Materials–Fundamentals Applications*, Boston, MA, 1996.

[27] D. C. Jiles and D. L. Atherton. Ferromagnetic hysteresis. *IEEE Transactions on Magnetics*, 19(5):2183–2185, 1983.

[28] D. C. Jiles and D. L. Atherton. Theory of ferromagnetic hysteresis. *Journal of Applied Physics*, 55(6):2115–2120, 1984.

[29] D. C. Jiles and D. L. Atherton. Theory of ferromagnetic hysteresis. *Journal of Magnetism and Magnetic Materials*, 61:48–60, 1986.

[30] K. H. Carpenter. A differential equation approach to minor loops in the Jiles-Atherton hysteresis model. *IEEE Transactions on Magnetics*, 27(6):4404–4406, 1991.

[31] D. C. Jiles. A self-consistent generalized model for the calculation of minor loop excursions in the theory of hysteresis. *IEEE Transactions on Magnetics*, 28(5):2602–2604, 1992.

[32] D. C. Jiles. Frequency dependence of hysteresis curves in conducting magnetic materials. *Journal of Applied Physics*, 76(10):5849–5855, 1994.

[33] D. C. Jiles. Modelling the effects of eddy current losses on frequency dependent hysteresis in electrically conducting media. *IEEE Transactions on Magnetics*, 30(6):4326–4328, 1994.

[34] M. J. Dapino, R. Smith, L. E. Faidley, and A. B. Flatau. A coupled structural magnetic strain and stress model for magnetostrictive transducers. *Journal of Intelligent Material Systems and Structures*, 11(2):135–152, February 2000.

[35] F. Delince, A. Genon, J. M. Gillard, H. Hedia, W. Legros, and A. Nicolet. Numerical computation of the magnetostriction coefficient in ferromagnetic materials. *Journal of Applied Physics*, 69(8):5794–5796, 1991.
[36] V. Agayan. Thermodynamic model of ideal magnetostriction. *Physica Scripta*, 54:514–521, 1996.
[37] R. D. James and D. Kinderlehrer. Theory of magnetostriction with applications to $Tb_xDy_{1-x}Fe_2$. *Philosophical Magazine B*, 68(2):237–274, 1993.
[38] A. E. Clark, H. T. Savage, and M. L. Spano. Effect of stress on the magnetostriction and magnetization of single crystal $Tb_{0.27}Dy_{0.73}Fe_2$. *IEEE Transactions on Magnetics*, 20(5):1443–1445, 1984.
[39] F. Claeyssen, N. Lhermet, R. Le. Letty and P. Bouchilloux. Design and construction of a resonant magnetostrictive motor. *IEEE Transactions on Magnetics*, 32(5):4749–4751, 1996.
[40] R. M. Bozorth. *Ferromagnetism*. Van Nostrand, New York, 1951.
[41] H. W. Katz. *Solid-State Magnetic and Dielectric Devices*. John Wiley and Sons, Inc., New York, 1959.
[42] M. Moffet, A. E. Clark, M. Wun-Fogle, J. Linberg, J. Teter, and E. McLaughlin. Characterization of Terfenol-D for magnetostrictive transducers. *Journal of the Acoustical Society of America*, 89(3):1448–1455, 1991.
[43] F. T. Calkins, M. J. Dapino, and A. B. Flatau. Effect of pre-stress on the dynamic performance of a Terfenol-D transducer. *Proceedings of the SPIE Symposium on Smart Structures and Materials*, 3041:293–304, 1997.
[44] R. Greenough, A. Jenner, M. Schulze, and A. Wilkinson. The properties and applications of magnetostrictive rare-earth compounds. *Journal of Magnetism and Magnetic Materials*, 101:75–80, 1991.
[45] J. R. Pratt, S. C. Oueini, and A. H. Nayfeh. Terfenol-D nonlinear vibration absorber. *Journal of Intelligent Material Systems and Structures*, 10(1):29–35, January 1999.
[46] G. Engdahl and L. Svensson. Simulation of the magnetostrictive performance of Terfenol-D in mechanical devices. *Journal of Applied Physics*, 63(8):3924–3926, 1988.
[47] L. Kvarnsjo and G. Engdahl. Nonlinear 2-D transient modeling of Terfenol-D rods. *IEEE Transactions on Magnetics*, 27(6):5349–5351, 1991.
[48] F. Claeyssen, R. Bossut, and D. Boucher. Modeling and characterization of the magnetostrictive coupling. In B. F. Hamonic, O. B. Wilson, and J.-N. Decarpigny, editors, *Proceedings of the International Workshop on Power Transducers for Sonics and Ultrasonics*, Toulon, France, pages 132–151. Springer-Verlag, June 1990.
[49] C. H. Sherman and J. L. Butler. Analysis of harmonic distortion in electroacoustic transducers. *The Journal of the Acoustical Society of America*, 98(3):1596–1611, 1995.
[50] M. M. Roberts, M. Mitrovic, and G. P. Carman. Nonlinear behavior of coupled magnetostrictive material systems analytical/experimental. *Proceedings of the SPIE Smart Structures and Materials Symposium*, 2441:341–354, 1995.
[51] D. C. Jiles. Theory of the magnetomechanical effect. *Journal of Physics D: Applied Physics*, 28:1537–1546, 1995.
[52] M. Anjanappa and J. Bi. A theoretical and experimental study of magnetostrictive mini-actuators. *Smart Materials and Structures*, 3(2):83–91, 1994.
[53] M. Anjanappa and Y. Wu. Magnetostrictive particulate actuators: Configuration, modeling and characterization. *Smart Materials and Structures*, 6(4):393–402, 1997.
[54] S. C. Pradhan, Y. T. Ng, K. Y. Lam, and J. N. Reddy. Control of laminated composite plates using magnetostrictive layers. *Smart Materials and Structures*, 10(4):657–667, 2001.
[55] G. P. Carman and M. Mitrovic. Nonlinear constitutive relations for magnetostrictive materials with applications to 1-D problems. *Journal of Intelligent Material Systems and Structures*, 6(5):673–683, 1995.
[56] M. J. Sablik and D. C. Jiles. A model of hysteresis in magnetostriction. *Journal of Applied Physics*, 64(10):5402–5404, 1988.
[57] M. J. Sablik and D. C. Jiles. Coupled magnetoelastic theory of magnetic and magnetostrictive hysteresis. *IEEE Transactions on Magnetics*, 29(4):2113–2123, 1993.

[58] E. T. Lacheisserie. Magnetoelastic coupling in materials with spherical symmetry. In W. Gorzkowski, M. Gutowski, H. K. Lachowicz, and H. Szymczak, editors, *Proceedings of the Fifth International Conference on Physics of Magnetic Materials*, Madralin, Poland, pages 164–203. World Scientific Publishing Co., Singapore, October 1990.

[59] A. E. Clark and H. T. Savage. Giant magnetically induced changes in the elastic moduli in $Tb_{0.3}Dy_{0.7}Fe_2$. *IEEE Transactions on Sonics and Ultrasonics*, 22(1):50–52, 1975.

[60] A. E. Clark, J. B. Restorff, and M. Wun-Fogle. Magnetoelastic coupling and Delta-E effect in Tb_xDy_{1-x} single crystals. *Journal of Applied Physics*, 73:6150–6152, May 1993.

[61] R. Kellogg and A. B. Flatau. Wide-band tunable mechanical resonator employing the ΔE effect of Terfenol-D. *Journal of Intelligent Material Systems and Structures*, 15(5): 355–368, May 2004.

[62] R. Kellogg and A. B. Flatau. Blocked-force characteristics of Terfenol-D transducers. *Journal of Intelligent Material Systems and Structures*, 15(2):117–128, February 2004.

[63] A. V. Krishnamurty, M. Anjanappa, and Y. Wu. Use of magnetostrictive particle actuators for vibration attenuation of flexible beams. *Journal of Sound and Vibration*, 206(2): 133–149, 1997.

[64] A. B. Flatau, M. J. Dapino, and F. T. Calkins. *Comprehensive Composite Materials Handbook*, edited by A. Kelly and C. Zweben, volume 5, chapter Magnetostrictive Composites, pages 563–574. Elsevier Science, 2000.

[65] R. L. Stoll. *The Analysis of Eddy Currents*. Clarendon Press, Oxford, 1974.

[66] F. T. Calkins, A. B. Flatau, and M. J. Dapino. Overview of magnetostrictive sensor technology. *Paper# AIAA-1999-1551, Proceedings of the 40th AIAA, ASME, ASCE, AHS, and ASC Structures, Structural Dynamics and Materials Conference*, St. Louis, MO, April 1999.

[67] A. B. Flatau, M. J. Dapino, and F. T. Calkins. High bandwidth tunability in a smart vibration absorber. *Journal of Intelligent Material Systems and Structures*, 11(12):923–929, December 2000.

[68] J. Pratt and A. B. Flatau. Development and analysis of a self-sensing magnetostrictive actuator design. *Journal of Intelligent Material Systems and Structures*, 6(5):639–648, 1995.

[69] A. E. Clark, M. Wun-Fogle, J. B. Restorff, T. A. Lograsso, and G. Petculescu. Magnetostriction and elasticity of body-centered cubic $Fe_{100-x}Be_x$ alloys. *Journal of Applied Physics*, 95(11):6942–6944, 2004.

[70] J. B. Restorff, M. Wun-Fogle, A. E. Clark, T. A. Lograsso, A. R. Ross, and D. L. Schlagel. Magnetostriction of ternary Fe-Ga-X alloys (X = Ni,Mo,Sn,Al). *Journal of Applied Physics*, 91(10):8225, 2002.

[71] L. Dai, J. Cullen, M. Wuttig, E. Quandt, and T. Lograsso. Magnetism, elasticity, and magnetostriction of FeCoGa alloys. *Journal of Applied Physics*, 93(10):8267–8269, 2003.

[72] J. Atulasimha and A. B. Flatau. A review of magnetostrictive iron-gallium alloys. *Smart Materials and Structures*, 20(4):043001–15, pp., 2011.

[73] J. Atulasimha and A. B. Flatau. Experimental actuation and sensing behavior of single-crystal iron-gallium alloys. *Journal of Intelligent Material Systems and Structures*, 19(12): 1371–1381, 2008.

[74] K. Ullakko, J. K. Huang, C. Kantner, R. C. O'Handley, and V. V. Kokorin. Large magnetic-field-induced strains in Ni_2MnGa single crystals. *Applied Physics Letters*, 69 (13):1966–1968, September 1996.

[75] A. Sozinov, A. A. Likhachev, N. Lamska, and K. Ullakko. Giant magnetic-field-induced strain in NiMnGa seven-layered martensitic phase. *Applied Physics Letters*, 80(10): 1746–1748, March 2002.

[76] A. A. Likhachev and K. Ullakko. Magnetic-field-controlled twin boundary motion and giant magneto-mechanical effects in Ni-Mn-Ga shape memory alloys. *Physics Letters A*, 275(1–2):142–151, 2000.

[77] S. J. Murray, M. A. Marioni, A. M. Kukla, J. Robinson, R. C. O'Handley, and S. M. Allen. Large field-induced strain in single crystalline Ni-Mn-Ga ferromagnetic shape memory alloy. *Journal of Applied Physics*, 87(9):5774–5776, May 2000.

[78] J. Tellinen, I. Suorsa, A. Jääskeläinen, I. Aaltio, and K. Ullakko. Basic properties of magnetic shape memory actuators. *8th International Conference ACTUATOR 2002*, Bremen, Germany, June 2002.

[79] K. Ullakko, Y. Ezer, A. Sozinov, G. Kimmel, P. Yakovenko, and V. K. Lindroos. Magnetic-field-induced strains in polycrystalline Ni-Mn-Ga at room temperature. *Scripta Materialia*, 44(3):475–480, March 2001.

[80] K. Ullakko, A. Likhachev, O. Heczko, A. Sozinov, T. Jokinen, K. Forsman, and I. Aaltio. Magnetic shape memory (MSM) – A new way to generate motion in electromechanical devices. *ICEM 2000*, pages 1195–1199, August 2000.

[81] R. N. Couch, J. Sirohi, and I. Chopra. Development of a quasi-static model of NiMnGa magnetic shape memory alloy. *Journal of Intelligent Material Systems and Structures*, 18(6):611–622, 2007.

[82] F. Pablo and B. Petitjean. Characterization of 0.9PMN-0.1PT patches of active vibration control of plate host structures. *Journal of Intelligent Material Systems and Structures*, 11(11):857–867, November 2000.

[83] R. Yimnirun, M. Unruan, Y. Laosiritaworn, and S. Ananta. Change of dielectric properties of ceramics in lead magnesium niobate-lead titanate systems with compressive stress. *Journal of Physics D: Applied Physics*, 39(14):3097–3102, 2006.

[84] G. H. Blackwood and M. A. Ealey. Electrostrictive behavior in lead magnesium niobate (PMN) actuators. I. Materials perspective. *Smart Materials and Structures*, 2(2):124–133, 1993.

[85] K. Uchino. Electrostrictive actuators. *Ceramic Bulletin*, 65:647–652, 1986.

[86] D. Damjanovic and R. E. Newnham. Electrostrictive and piezoelectric materials for actuator applications. *Journal of Intelligent Material Systems and Structures*, 3(2):190–208, 1992.

[87] C. L. Hom, S. M. Pilgrim, N. Shankar, K. Bridger, M. Massuda, and S. R. Winzer. Calculation of quasi-static electromechanical coupling coefficients for electrostrictive ceramic materials. *IEEE Transactions on Ultrasonics, Ferroelectrics and Frequency Control*, 41(4):542 -551, July 1994.

[88] A. F. Devonshire. Theory of ferroelectrics. *Philosophical Magazine*, 3:85–130, 1954.

[89] J. Scortesse, J. F. Manceau, F. Bastien, M. Lejeune, S. Kurutcharry, and M. Oudjedi. Apparent Young's modulus in PMN-PT electrostrictive ceramics. *The European Physical Journal of Applied Physics*, 14(3):155–158, 2001.

[90] C. Namboodri. Experimental investigation and modeling of the electrostrictive relaxor ferroelectric lead magnesium niobate-lead titanate. Master's thesis, Department of Mechanical Engineering, Virginia Polytechnic Institute and State University, 1992.

[91] M. Fripp and N. Hagood. Comparison of electrostrictive and piezoceramic actuators for vibration suppression. *Proceedings of the SPIE Smart Structures and Materials Symposium*, 2443:334–348, 1995.

[92] C. L. Hom and N. Shankar. A fully coupled constitutive model for electrostrictive ceramic materials. *Journal of Intelligent Material Systems and Structures*, 5(6):795–801, November 1994.

[93] C. L. Hom and N. Shankar. A constitutive model for relaxor ferroelectrics. *Proceedings of the SPIE Smart Structures and Materials Symposium*, 3667:134–144, 1999.

[94] M. Fripp and N. Hagood. Distributed structural actuation with electrostrictives. *Journal of Sound and Vibration*, 203(1):11–40, 1997.

[95] J. C. Piquette and S. E. Forsythe. A nonlinear material model of lead magnesium niobate (PMN). *Journal of the Acoustical Society of America*, 101(1):289–296, 1997.

7 Electrorheological and Magnetorheological Fluids

The previous chapters discuss the properties and behavior of active materials that existed in a solid state. These materials exhibited changes in properties and physical dimensions when subjected to an electric, magnetic, or thermal field. A special class of fluids exists that change their rheological properties on the application of an electric or a magnetic field. These controllable fluids can generally be grouped under one of two categories: electrorheological (ER) fluids and magnetorheological (MR) fluids. An electric field causes a change in the viscosity of ER fluids, and a magnetic field causes a similar change in MR fluids. The change in viscosity can be used in a variety of applications, such as controllable dampers, clutches, suspension shock absorbers, valves, brakes, prosthetic devices, traversing mechanisms, torque transfer devices, engine mounts, and robotic arms. Other applications such as electropolishing do not rely directly on the change in viscosity but rather on the ability to change properties of the fluid locally.

Most mechanical dampers consist of fixed damping that is designed as a compromise between a range of operating conditions. As a result, these devices do not provide an optimum level of damping for any specific operating environment. Using ER/MR fluid dampers, variable damping levels can be obtained, and the system performance can be optimized over a wide range of operating conditions. In such dampers, the resistance to flow and, consequently, the energy dissipation, can be modulated through the applied electric or magnetic field. The quasi-steady flow characteristics are nonlinear functions of many variables, including the effects of fluid inertia and compressibility for dynamic conditions. For practical considerations, the gap across which an electric or magnetic field is applied needs to be small and uniform. The application of field results primarily in an increase in the static yield stress of the fluid. Typical examples of modern ER fluids are alumino-silicates in silicone oil, silica spheres in mineral oil, and polymer particles in chlorinated hydrocarbon oil. These fluids undergo a change from a viscous fluid to an almost solid gel under the application of an electric field. Recently, interest in MR fluids has grown and activities in ER fluids have waned because of the superior characteristics of MR fluids.

The ER effect was first observed in 1947 by Willis Winslow [1, 2], who discovered that the application of a large electric field across an organic suspension caused the fluid to solidify. Winslow experimented with a variety of solid particulates including starch, stone, lime, gypsum, carbon, and silica, dispersed in various insulating oils such as mineral oil, paraffin, and kerosene, to demonstrate ER effects. Subsequent research led to a patent describing ER fluid couplings [3]. At approximately the same time (1948), the MR effect was observed by Jacob Rabinow [4, 5, 6]. However, the early studies pointed out many shortcomings of these fluids that include their abrasiveness, chemical instability, and rapid deterioration of properties, which prevented their widespread application. Significant advances toward improving the properties of MR fluids took place in the 1980s, and many applications were demonstrated.

Although the ER/MR fluid is an active material in the sense that its properties such as viscosity, elasticity, and plasticity change, within the order of milliseconds, in response to an applied electric or magnetic field, it is not capable of directly generating any actuation force. This is in contrast to active materials such as piezoelectrics, electro-magnetostrictives, and SMAs, which can be used as force generators in actuators. Therefore, devices based on ER/MR fluids are referred to as "semi-active" devices.

A perceived barrier for a successful MR device is the settling of ferrous particles in the carrier fluid. As early as 1949, Rabinow dispelled this myth and demonstrated complete suspension stability for most MR fluid devices (e.g., dampers and rotary brakes) because of their high mixing characteristics. The motion of the piston in an MR damper rapidly moves fluid through orifices resulting in swirl and eddy motions, which in turn vigorously mix the suspension. Except for seismic dampers (which remain quiescent for a long period), suspension stability is not an important consideration. For many devices, it is important to have a low zero-field viscosity and high yield stress under field activation.

Even though one can produce ER/MR fluids in the laboratory, it is necessary to use a reliable manufacturing process to obtain fluids with repeatable characteristics. During the early development of MR fluid devices, a critical issue of "in-use-thickening" was encountered, in which the fluid viscosity increases under the application of high stress and high shear rate over an extended period of time (more than thousands of cycles). The viscosity increase is believed to result from spalling of the surface layer of the iron particles in the MR fluid. As a result of this spalling, the zero-field stress increases significantly, rendering the device (e.g., a damper) unsuitable for semi-active vibration control. There is no doubt that the zero-field characteristics of the fluid are important for the successful functioning of the device. Good MR fluids should not show any measurable in-use thickening until more than 10 million cycles.

Another issue is the deterioration of the fluid, especially silicone oil based fluid, over a period of time. This can happen because of cross-linking if silicone fluid is exposed to high temperatures or to ionizing radiation for an extended period of time. Again, this results in thickening of the fluid (increase in viscosity). The thickening generally depends on the shear rate, temperature, and duration of the applied stress. For practical applications of ER/MR fluids durability and life can be considered overwhelming barriers to commerical success, compared to material characteristics such as yield strength and suspension stability. Furthermore, characteristics such as force-velocity (damping) and force-displacement (stiffness) are highly nonlinear and are functions of a number of variables, including the size of the device.

7.1 Fundamental Composition and Behavior of ER/MR Fluids

ER and MR fluids are similar in terms of their composition and behavior. ER fluids change their properties in response to an electric field, whereas MR fluids respond to a magnetic field. ER and MR fluids are, however, different in terms of their density, yield stress, and other mechanical parameters.

7.1.1 Compostion of ER/MR Fluids

Both ER and MR fluids consist of a colloidal suspension of particles in a carrier fluid. In the case of ER fluids, the particles are micron-sized dielectric particles and could be cornstarch or an alumino-silicate compound. The carrier fluid is electrically nonconducting and could be mineral oil, silicone oil, or paraffin oil. On the application of an electric field, the particles become charged and experience electrostatic forces. ER fluids require a high electric field (in the range of 8 kV/mm). The response time is on the order of 1 ms (bandwidth of less than 1 kHz). The electric field causes the suspended particles to form chains linking the electrodes (in the direction of the applied field) and, as a result, increases the resistance to flow of fluid (i.e., increases the viscosity of the fluid). In the case of MR fluids, the properties of the carrier fluid are similar to those of ER fluids. However, the particles must be a ferromagnetic material. On the application of a magnetic field, the particles attract each other due to magnetic induction. The size of the particles in both cases is on the order of 10 microns. There exists a class of fluids called ferrofluids that are also composed of a suspension of magnetic particles in a carrier fluid. However, in the case of ferrofluids, the particle size is on the order of nanometers. On the application of a magnetic field, ferrofluids experience a net body force but do not exhibit any change in rheological properties. In both ER and MR fluids, surfactants (i.e., compounds that lower surface tension) are used to achieve high particle-volume fractions and, hence, high variations in rheological properties, as well as to minimize sedimentation.

7.1.2 Viscosity

The dynamic viscosity μ of a Newtonian fluid is defined as the ratio between the shear stress τ and the shear strain rate in the fluid, $\dot{\gamma}$. This relationship can be expressed as

$$\begin{aligned} \tau &= \mu \frac{\partial u}{\partial y} \\ &= \mu \dot{\gamma} \end{aligned} \tag{7.1}$$

where u is the velocity of the fluid and y is a spatial coordinate perpendicular to the flow of the fluid. These quantities are seen in the schematic diagram in Fig. 7.1, which shows the velocity profile of a fluid flow past a stationary wall. The viscosity μ is also called the dynamic viscosity and has the units Pa.s. (Pascal-second). In general, the viscosity is defined as the variation of shear stress with shear strain rate, which can be written as

$$\mu = \frac{\partial \tau}{\partial \dot{\gamma}} \tag{7.2}$$

There are two basic methods of measuring viscosity. One is using a Couette cell (Fig. 7.2(a)) and the other is using Poiseulle flow (Fig. 7.2(b)). In the case of the Couette cell, the fluid is sheared between two coaxial cylinders, one rotating and the other stationary, yielding a linear velocity profile. In the case of the Poiseulle flow, the fluid is made to flow through a passage, yielding a velocity profile that

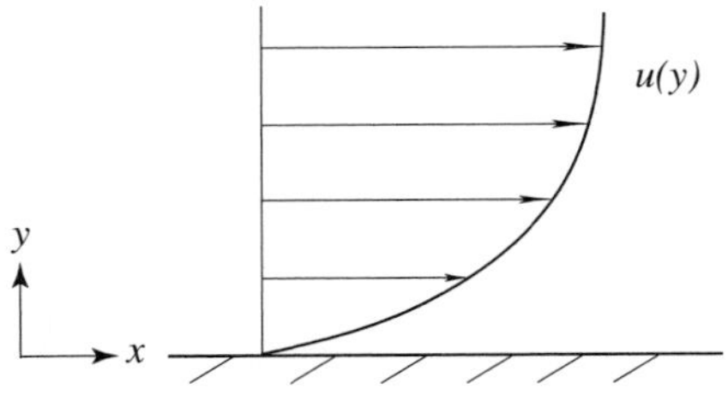

Figure 7.1. Velocity profile of a fluid flowing past a stationary wall.

is parabolic. A third method is to shear the fluid between two parallel plates, one rotating with respect to the other (Fig. 7.2(c)). This technique is similar to the Couette cell; however, the shear strain rate is not a constant across the area of the plates. Several differences exist between these techniques [7]. ER/MR fluids usually exhibit shear thinning, which means that the viscosity decreases with increasing shear rate. When an electric or magnetic field is applied to the fluid, a large increase in viscosity occurs. This change in viscosity is rapid (on the order of milliseconds) and is reversed on removal of the field. As the shear stress is increased under an applied field, the viscosity reverts back to its original zero-field value at a particular value of shear stress. This sudden decrease in viscosity is called "yielding." The yield shear stress and the plastic, or post-yield, viscosity of the fluid is one of its most important characteristics. The plastic viscosity is usually assumed to be constant (i.e., equal to the zero-field viscosity), or a weak function of the applied field. For both ER and MR fluids, the value of plastic viscosity is in the range of 0.2–0.3 Pa.s.

To enhance the ER effect, a small amount of water (about 5% by volume) is often added to the fluid, which helps to bond together the suspended particles. However, the presence of water can limit the temperature range of applications. Conversely, MR fluids are less sensitive to a small addition of water.

7.1.3 Origin of the Change in Viscosity

ER and MR fluids exhibit similar rheological properties [8]. The change in viscosity also occurs in a similar way for the two types of fluids. In the absence of an electric or magnetic field, the particles are randomly distributed throughout the carrier fluid, and they are free to move about (Fig. 7.3(a)). The viscosity of the fluid, in this case, is a function of the viscosity of the carrier fluid and the concentration of dispersed particles. In the case of an ER fluid, when an electric field is applied, the particles

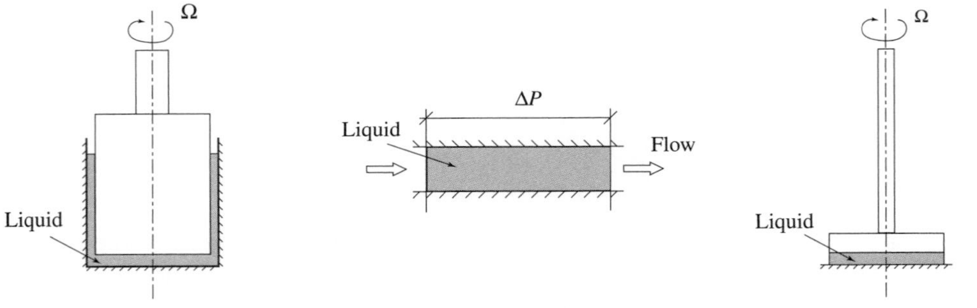

(a) Couette cell (b) Poiseulle flow (c) Parallel plate

Figure 7.2. Viscosity measurement techniques.

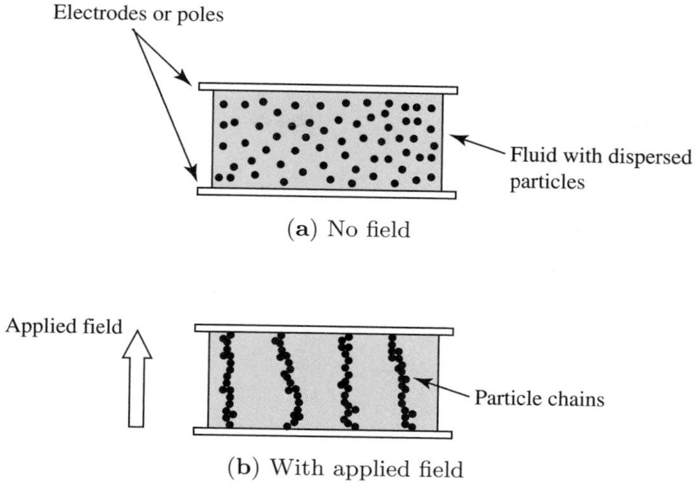

Figure 7.3. Effect of applied field on the dispersed particles.

become polarized and attract each other due to electrostatic forces. As a result, chains of particles form in the fluid between the electrodes, as shown in Fig. 7.3(b). In the absence of a field, the fluid can freely flow across the electrodes in response to an applied pressure gradient or can be sheared by a relative motion of the electrodes. On the application of the field, the fluid flow across the electrodes is impeded by the particle chains. A larger pressure gradient is required to break the chains and maintain the flow of the fluid. As a result, a larger force is required on the electrodes to produce a relative motion between them. The forming and breaking of the chains results in a significant change in the viscosity of the fluid. The yield stress can be defined as the shear stress at which the particle chains begin to break. It should be kept in mind that the chain formations may be influenced by the flow field.

Similarly, in the case of an MR fluid, the application of a magnetic field causes chains of the magnetic particles to form along the applied magnetic field. The particles attract each other by magnetic induction, and the fluid at this point exhibits a much larger viscosity than in the case of zero applied field. A yield stress can be defined, similar to the case of ER fluids, corresponding to the breaking of chain structures in the fluid. An optical photomicrograph of the chain formation in an MR fluid is shown in Fig. 7.4 [9, 10]. This photograph was taken under a microscope, with a magnetic field applied in the plane of the paper, as shown. The dark stripes are chains of magnetic particles aligned along the direction of the applied magnetic field, and the clear region is the carrier fluid.

Figure 7.4. Formation of chains in an MR fluid parallel to the applied magnetic field, adapted from Dimock et al. [9].

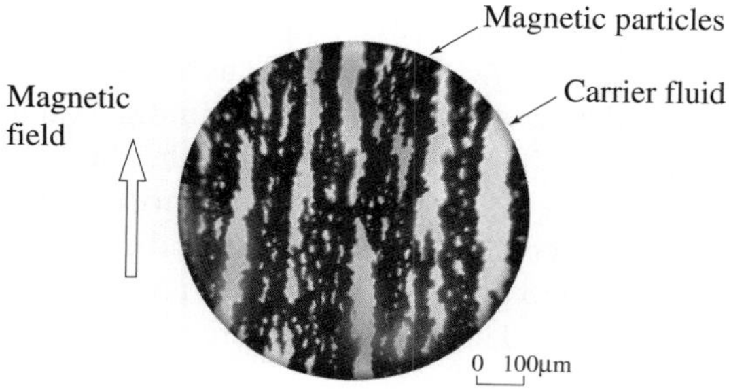

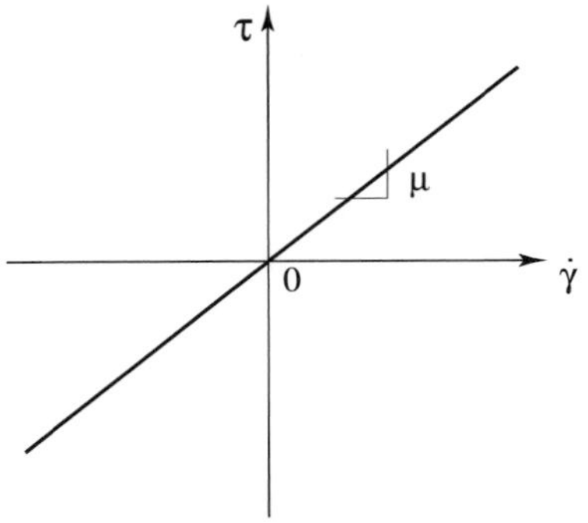

(a) No field

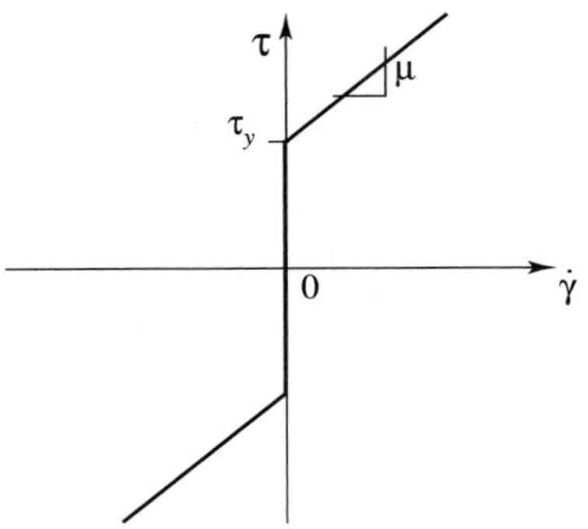

(b) With applied field

Figure 7.5. Simplified yield behavior.

A rudimentary ER fluid can be created by mixing a cup of corn starch with a cup of mineral oil to obtain a uniform suspension and then carefully removing the air bubbles. Similarly, a rudimentary MR fluid can be created by mixing a cup of iron filings with a cup of hydraulic oil. Commercial compositions are similar, with extra chemicals added to generally improve the properties of the fluid; for example, to prevent the particles from agglomerating.

7.1.4 Yield Behavior

As the shear stress in the fluid is increased, the particle chains start deforming. When a certain value of shear stress is reached, the chains begin to break. The value of shear stress at which this occurs is called the yield stress of the fluid. After the yield point, or in the post-yield condition, an equilibrium exists between the breaking and reforming of particle chains. The viscosity of the fluid in this state, also called the plastic viscosity, is much lower than the viscosity of the fluid with unbroken chains.

This behavior is represented simplistically in Fig. 7.5, which shows the variation of shear stress τ in the fluid with shear strain rate $\dot{\gamma}$, at zero applied field and at a constant field. At zero applied field, the fluid behaves like a Newtonian fluid, with a constant viscosity given by the slope of the curve. When a field is applied to the fluid, the behavior becomes discontinuous. Initially, at low values of shear stress, the particle chains are unbroken, and the applied shear stress can be sustained without any flow in the fluid (zero strain rate). In this condition, the fluid essentially behaves as a solid. This phenomenon is used in some devices to operate the fluid as a valve in a flow circuit. The flow of fluid can be stopped by simply applying an external field, without introducing a mechanical valve into the flow.

As the shear stress is increased to the yield stress τ_y, the particle chains start breaking and the fluid yields. Thereafter, a finite fluid flow can be maintained, and the fluid behaves as a Newtonian fluid. The slope of the curve in the post-yield region is equal to the plastic viscosity of the fluid. It is generally assumed that the plastic viscosity is equal to the viscosity at no field. From this simple description of the fluid behavior, it can be seen that the fluid exhibits two distinct values of viscosity, in the pre-yield and post-yield regions. In this case, the yield stress is the static yield stress $\tau_{y,s}$ of the fluid, defined as the stress required to cause fluid flow from a state of zero strain rate.

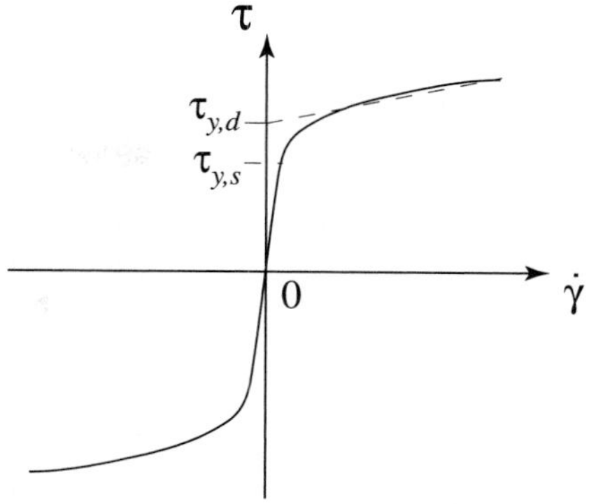

Figure 7.6. Static and dynamic yield stress.

In reality, however, the behavior of the fluid is more complex. The fluid does not become completely solid on the application of a field but rather has a finite viscosity. This corresponds to the fact that the particle chains deform before breaking. After the particle chains are broken, the viscosity decreases and may vary with flow rate. A quasi-steady approximation of this behavior is schematically represented in Fig. 7.6. A dynamic yield stress $\tau_{y,d}$ can be defined as the y-intercept of the straight line fit to the shear strain versus strain rate curve. The dynamic yield stress is approximately twice the static yield stress and can be attributed to friction effects between the dispersed particles.

In the pre-yield region, the fluid can be treated as elastic or viscoelastic, with the dominant deformation mechanism being the stretching of particle chains with occasional breaking. In the post-yield region, an equilibrium is reached between the chains breaking and reforming. In this region, the fluid can be treated as a viscous Newtonian fluid [11, 12, 13]. A description of the yield behavior and the actual structural processes occurring during chain formation and rupture can be found in several references [14, 15, 16, 17] .

As the field is increased, the yield stress also increases, as shown in Fig. 7.7. In the case of ER fluids subjected to an electric field E, the yield stress can be expressed as

$$\tau_y = E^n \tag{7.3}$$

where the exponent n ranges from 1.2 to 2.5 depending on the consistency of the suspension [18]. In general, it can be assumed that the dynamic yield stress of ER

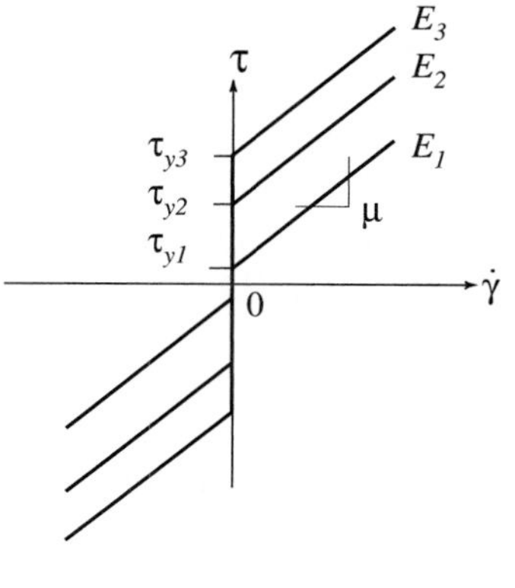

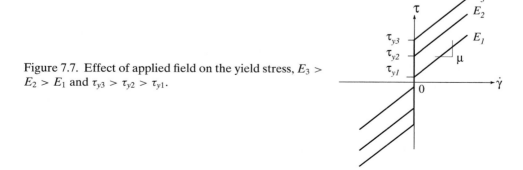

Figure 7.7. Effect of applied field on the yield stress, $E_3 > E_2 > E_1$ and $\tau_{y3} > \tau_{y2} > \tau_{y1}$.

and MR fluids exhibits a quadratic dependence on the applied electric or magnetic field [15, 19].

The maximum yield stress in the case of ER fluids is in the range of 2–5 kPa, whereas it is an order of magnitude higher for MR fluids, around 50–100 kPa. The maximum yield stress also depends on the maximum field that can be applied to the fluid. In the case of ER fluids, the maximum field is limited by dielectric breakdown of the carrier fluid; in the case of MR fluids, the maximum field is limited by saturation of the magnetic field in the dispersed particles.

7.1.5 Temperature Dependence

The dependence of the structural properties on temperature is an important factor to be considered in the design of devices such as automobile dampers. Both the viscosity and the yield stress of ER/MR fluids change with temperature, depending on the composition. Klass and Martinek [19] showed that the viscosity of ER fluids increases with temperature. However, the properties of MR fluids are stable over a wider range than ER fluids [20]. For example, over a temperature range of −25°C to +125°C, the dynamic yield stress of ER fluids decreases by 70% and the plastic viscosity decreases by 95%. This temperature range is considered the allowable operating temperature range for ER fluids. In the case of MR fluids, the yield stress decreases by approximately 10% and the plastic viscosity decreases by 5% over a temperature range of −40°C to +150°C.

7.1.6 Dynamic Behavior and Long-Term Effects

The behavior of the fluid under dynamic conditions (e.g. with a time-dependent shear rate) is important in many applications. Generally speaking, this time dependence could be a harmonic function or it could be close to an impulse, as in the case of dampers subjected to shock loads.

Many investigations have been conducted into the dynamic behavior of controllable fluids. For example, Gamota and Filisko [12] and Ehrgott and Masri [21] studied the dynamic response of an ER fluid subjected to oscillatory shear strains using specific device geometries. Experiments performed up to a frequency of 50 Hz showed that the response consisted of three parts. For small strain rates, where the shear stress is much below the yield stress, the behavior is linear and the material behaves as a viscoelastic – that is, a combination of an elastic solid (spring) and a viscous liquid (dashpot). At high strain rates, where most of the cycle is in the post-yield region, the behavior resembles that of a viscous fluid. Near the yield point, the material is highly nonlinear and the response is a combination of the viscoelastic and plastic response. Over a complete cycle, a hysteresis is exhibited in the plot of shear stress versus shear strain rate. For simple models, this hysteresis can be neglected, but more accurate rheological models include some sort of approximation for this behavior. A schematic of the dynamic shear stress versus shear strain rate curve is shown in Fig. 7.8.

In the case of step excitations, such as the sudden application of an electric or magnetic field, it is generally observed that the response of both ER and MR fluids is in the millisecond range [22, 23, 24]. A study of the time response of ER and MR dampers by Choi et al. [25] suggests that the time delay is smaller for ER fluid devices.

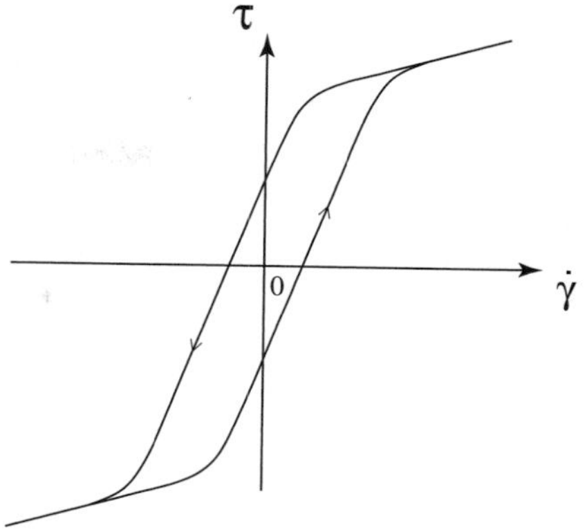

Figure 7.8. Dynamic rheological behavior.

The response of MR fluids subjected to a large number of cycles, such as in dampers, is a subject of great interest. It has been observed that when an MR fluid composition is activated and subjected to a large number of cycles, the zero-field viscosity of the fluid increases. This phenomenon is called "In-Use-Thickening" [26]. The origin of this phenomenon is attributed to the long-term stress exerted on the dispersed particles, which cause nanometer-sized pieces to separate from the micron-sized dispersed particles. The nanometer-sized particles then remain suspended in the carrier fluid, greatly increasing the viscosity of the fluid. Another issue of concern regarding the long-term behavior of ER and MR fluids is the settling of the dispersed particles. When stored for a period of time, which can be as short as a few days, the particles in the fluid tend to settle to the bottom of the container, destroying the properties of the fluid. The fluid then has to be thoroughly mixed before use in order to redistribute the particles evenly throughout the carrier fluid, after which the fluid regains its original properties. This problem is more severe in the case of MR fluids due to their heavier ferrous particles.

7.1.7 Comparison of ER and MR Fluids

ER and MR fluids were discovered around the same time. However, most of the initial research was focused on ER fluids. This is mainly because devices based on ER fluids have a simple geometry and are easy to construct. ER fluids can be easily developed in the laboratory. Recently, more interest has focused on MR fluid based devices. This interest is fueled by commercial applications requiring a more stable fluid with higher yield stress. The yield stress of MR fluids is an order of magnitude higher than ER fluids. MR fluids are also much more tolerant to impurities [27] and can be operated off a low-voltage power supply (≈28 V DC). This low voltage is much safer to work with as compared to the high voltage (≈3 kV) required for ER fluid devices. MR fluids are also stable over a wider temperature range (−40°C to 150°C) than ER fluids (−25°C to 125°C). The dynamic response characteristics are similar for the two types of fluids.

However, the design of MR fluid devices is complicated by the requirement of an efficient magnetic circuit. The entire magnetic circuit, including current carrying coils and flux return path, must be carefully designed and incorporated into the device. The high currents passing through the coil cause heating, which must be dissipated satisfactorily. As the device becomes smaller and gains more complex

Table 7.1. *Comparison of the properties of ER and MR fluids*

	ER Fluids	MR Fluids
Required voltage	2–10 kV	2–25 V
Required current	1–10 mA	1–2 A
Maximum yield stress	2–5 kPa (at 3–5 kV/mm)	50–100 kPa (at 150–250 kA/m)
Maximum field	4 kV/mm	250 kA/m
Volume factor (μ/τ_y^2)	10^{-7} s/Pa to 10^{-8} s/Pa	10^{-10} s/Pa to 10^{-11} s/Pa
Specific gravity	1–2.5	3–4
Temperature range	−25°C to 125°C	−40°C to 150°C
Device and actuation geometry	Simple	Complex

geometry, it becomes easier to create an electric field compared to a magnetic field. MR fluids are also much heavier than ER fluids, as a result of the high density of the ferro-magnetic particles. This is another factor that must be considered in weight-critical applications.

A volume factor (μ/τ_y^2) can be defined for the fluid that is directly proportional to the size of the device. This quantity is three orders of magnitude larger for MR fluids than for ER fluids. A comparison of the properties of ER and MR fluids is summarized in Table 7.1.

7.2 Modeling of ER/MR Fluid Behavior and Device Performance

Several phenomenological models of varying complexity have been proposed by different researchers to predict the performance of ER/MR fluid devices. The modeling approaches fall into one of the two following categories:

1. Apply a specific fluid model to the device geometry in question. The model is chosen based on a qualitative representation of the fluid behavior. The parameters in the model are adjusted so that the predicted performance of the device matches experimental data.
2. The device as a whole is treated as a "black box" and a model is fit to the behavior relating input and output quantities.

The first approach requires the application of specific fluid models. Most of these models are quasi-steady, piecewise continuous approximations to the rheological behavior of the fluid. The performance of many devices such as dampers [16, 28, 29, 30], clutches, brakes, and valves [31, 32, 18, 16, 33, 34] has been evaluated using these models.

Examples of the second approach can be found in studies performed by Stanway et al. [30, 35] in which an ER damper is considered as a Coulomb element in parallel with a viscous damper. Figure 7.9 shows various simplified phenomenological models to represent ER and MR devices. The model parameters such as spring stiffness, dashpot damping, and Coulomb friction are identified from test data usually using a sinusoidal input forcing.

The coeffiecients of these models are typically extracted from experimental data. Ehrgott and Masri [36, 37] expressed the restoring force of an ER device as a function of velocity and displacement using Chebyshev polynomials. Extraction of these coefficients from experimental data can be computationally demanding. In

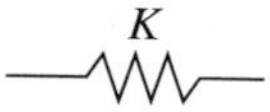

(a) Elastic spring with stiffness coefficient K

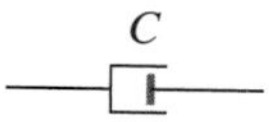

(b) Viscous dashpot with damping coefficient C

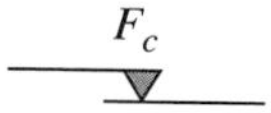

(c) Coulomb damping with friction force F_c

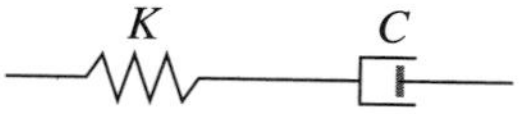

(d) Maxwell model with spring and damper in series arrangement

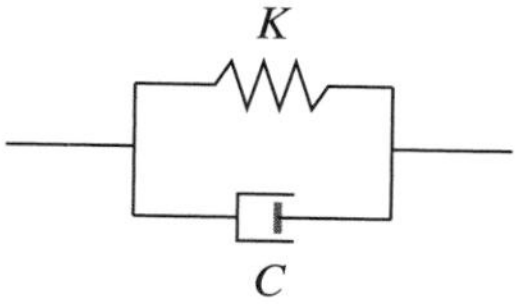

(e) Kelvin-Voigt model with spring and damper in parallel arrangement

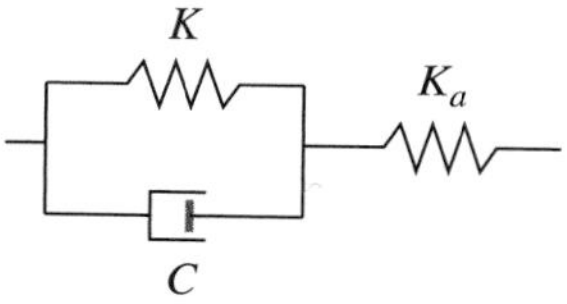

(f) Zener model with three parameter arrangement

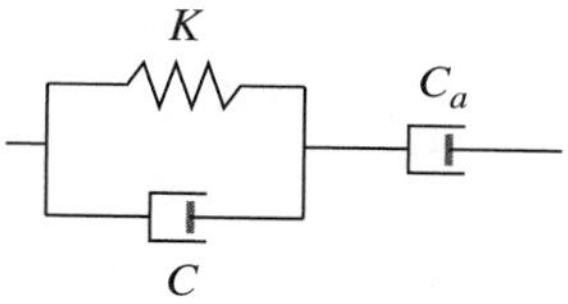

(g) Three parameter model with viscous damping

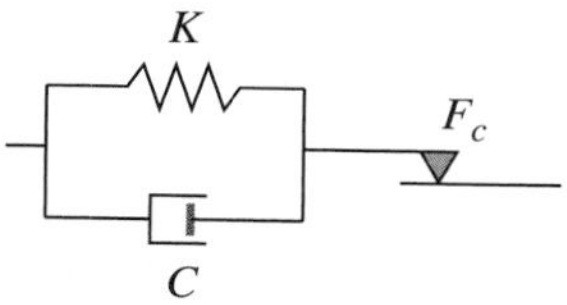

(h) Three parameter model with Coulomb damping

Figure 7.9. Various simplified representations of phenomenological models of ER and MR devices.

a simpler approach, the ER device can be modeled purely as a viscous damper by matching the damping coefficient to experimental data.

There are a number of phenomenological models for ER/MR fluids that are used by researchers [38]. Several quasi-steady phenomenological fluid models are discussed in the following subsections, with emphasis on a few important models.

7.2.1 Equivalent Viscous Damping

This is a basic model intended to capture only the damping of the fluid. The rheological behavior, which is the variation of shear stress with shear strain rate in the fluid, is not captured. The energy dissipated by the device in one cycle is equated to that dissipated by an equivalent viscous damper. The nonlinear variation of shear stress with shear strain rate is effectively linearized. Consequently, the model can be treated as a device performance model, not a model of the fluid behavior itself.

Assuming a harmonic excitation at a frequency ω and amplitude X_o given by

$$x(t) = X_o \cos(\omega t) \tag{7.4}$$

The energy dissipated in one cycle ΔW by a system with damping coefficient c_{eq} can be derived as

$$\Delta W = \int_{\text{cycle}} F dx = \int_0^{2\pi/\omega} F\dot{x}dt \tag{7.5}$$

where the damping force F is given by

$$F = c_{eq}\dot{x} = -c_{eq}X_o\omega\sin(\omega t) \tag{7.6}$$

This leads to

$$\begin{aligned}\Delta W &= \int_0^{2\pi/\omega} c_{eq}\,\dot{x}\,\dot{x}\,dt \\ &= c_{eq}X_o^2\omega^2\int_0^{2\pi/\omega}\sin^2(\omega t)dt \\ &= \pi c_{eq}X_0^2\omega\end{aligned} \tag{7.7}$$

The energy dissipated by damping in the device, ΔW_{DE}, is calculated from experimental data as the area under the force-displacement curve. Equating the two energies yields the equivalent viscous damping coefficient as

$$c_{eq} = \frac{\Delta W_{DE}}{\pi X_0^2\omega} \tag{7.8}$$

It can be seen that the equivalent viscous damping model is useful only for quantifying the damping properties of an ER/MR fluid damper. The model does not capture the rheological behavior of the fluid.

7.2.2 Bingham Plastic Model

The Bingham plastic model is an idealized model of fluid behavior. In this model, when a shear stress is applied, the fluid behaves as a solid until a specific yield stress is reached. At stress levels higher than the yield stress, the fluid behaves like a Newtonian fluid with constant viscosity. Above the yield point, the stress in the fluid can be expressed as

$$\tau = \tau_y\,sgn\,(\dot{\gamma}) + \mu\dot{\gamma} \qquad \tau > \tau_y \tag{7.9}$$

where τ and τ_y are the shear stress and yield shear stress, respectively; $\dot{\gamma}$ is the shear strain rate; and μ is the viscosity of the fluid. A schematic of this behavior is shown in Fig. 7.10.

The behavior of an ER/MR fluid device can be modeled using the Bingham plastic model by representing the force in the device as a summation of a viscous force and a frictional force. This is equivalent to modeling the device as a parallel combination of a dashpot and a Coulomb friction element, as shown in Fig 7.11. The damping force in the device is therefore given by

$$F = F_c\,sgn\,(\dot{x}) + c_o\dot{x} \qquad F > F_c \tag{7.10}$$

where c_o is the viscous damping coefficient and F_c is the Coulomb friction force.

The yield stress, τ_y, and therefore, F_c, is dependent on the applied field. It should be noted that the Bingham plastic model is an idealized model and treats the fluid as a solid before the yield point. Therefore, although high strain rate behavior is

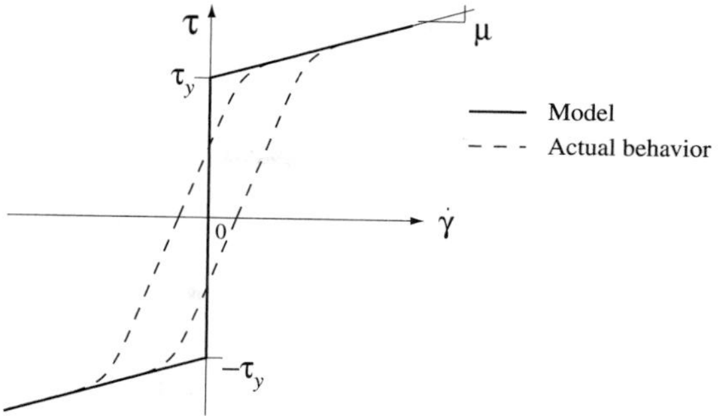

Figure 7.10. Bingham plastic model.

captured with good accuracy, the model is not accurate for low strain rate behavior. The overall damping is captured well despite the fact that the rheological behavior is idealized.

7.2.3 Herschel-Bulkley Model

The Herschel-Bulkley model focuses on capturing the shear thinning and shear thickening effects of the fluid. This is achieved by representing the shear stress as a power law of the the shear strain rate [39, 40]. The shear stress in the fluid is given by

$$\tau = \tau_y \, sgn(\dot{\gamma}) + K\dot{\gamma}^n \qquad \text{for } \tau > \tau_y \tag{7.11}$$

where K is a constant. The exponent n defines the properties of the fluid in the post-yield region (Fig. 7.12) and is called the flow behavior index. In the case in which $n > 1$, the fluid is said to exhibit shear thickening, and when $n < 1$, the fluid exhibits shear thinning. Note that the Herschel-Bulkley model reduces to the Bingham plastic model when $n = 1$. In the post-yield condition, an apparent viscosity can be expressed as (Eq. 7.2)

$$\begin{aligned} \mu_a &= \frac{\partial \tau}{\partial \dot{\gamma}} \\ &= nK\dot{\gamma}^{n-1} \end{aligned} \tag{7.12}$$

7.2.4 Biviscous Model

In the biviscous model, the fluid in the pre-yield region is treated as having a finite viscosity that is much larger than the viscosity in the post-yield region (Fig. 7.13). The fluid has two specific values of viscosity, depending on the strain rate. This behavior

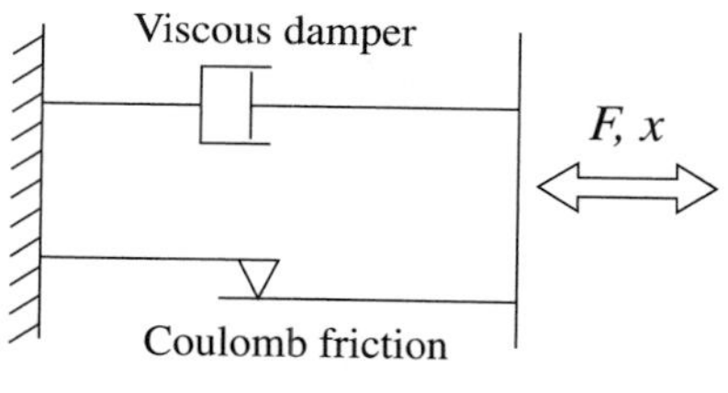

Figure 7.11. Bingham plastic model represented by mechanical elements.

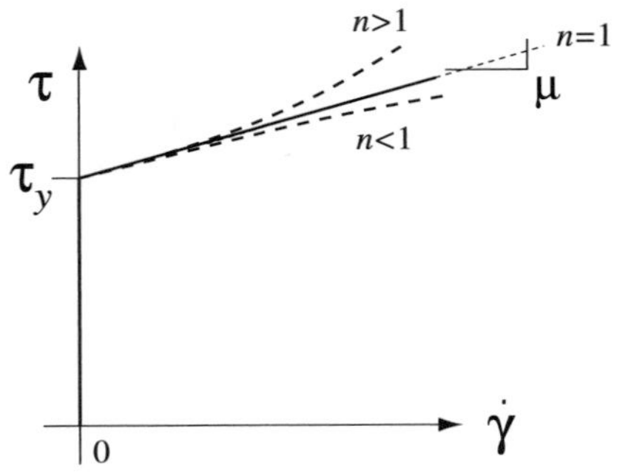

Figure 7.12. Herschel-Bulkley model.

can be represented as

$$\begin{aligned} \tau &= \mu_{pr}\dot{\gamma} & \tau < \tau_y \\ &= \mu_{po}\dot{\gamma} + \tau_y & \tau \geq \tau_y \end{aligned} \tag{7.13}$$

Alternatively, the equations can be expressed in terms of the yield strain rate as follows

$$\begin{aligned} \tau &= \mu_{pr}\dot{\gamma} & \text{for } \dot{\gamma} < \dot{\gamma}_y \\ &= \mu_{po}\dot{\gamma} + \tau_y & \text{for } \dot{\gamma} \geq \dot{\gamma}_y \end{aligned} \tag{7.14}$$

where the yield strain rate is defined as

$$\dot{\gamma}_y = \frac{\tau_y}{\mu_{pr} - \mu_{po}} \tag{7.15}$$

Note that the pre-yield viscosity, μ_{pr}, is much higher than the post-yield viscosity, μ_{po}. This model captures both low strain rate and high strain rate behavior. Damping is also represented well, but the hysteresis in the shear stress versus strain rate behavior is not captured. Note that the biviscous model reduces to the Bingham plastic model if the pre-yield viscosity is set to infinity.

7.2.5 Hysteretic Biviscous

This model extends the biviscous model to capture the dynamic pre-yield hysteresis (Fig. 7.14). Four parameters are required for this model: the pre-yield and post-yield

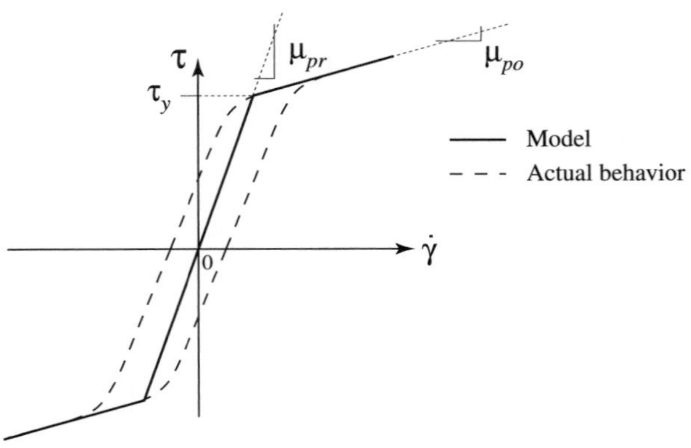

Figure 7.13. Biviscous model.

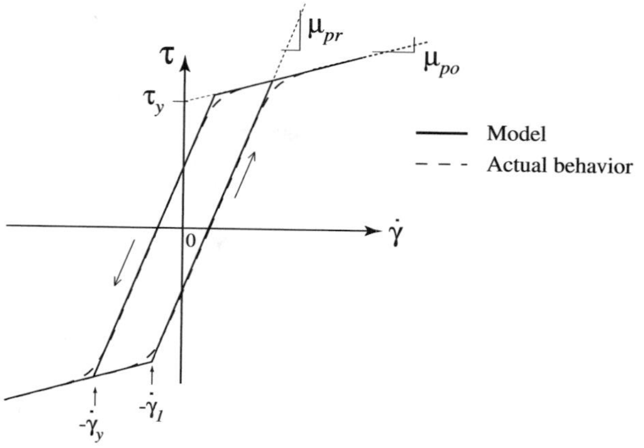

Figure 7.14. Hysteretic biviscous model.

viscosities μ_{pr} and μ_{po}, the yield stress τ_y, and the yield strain rate, $\dot{\gamma}_y$. The shear stress versus strain rate behavior can be written as follows

If $\dot{\gamma} > 0$,

$$\tau(\dot{\gamma}) = \begin{cases} \mu_{po}\dot{\gamma} - \tau_y & \text{for } \dot{\gamma} < -\dot{\gamma}_1 \\ \mu_{pr}(\dot{\gamma} - \dot{\gamma}_o) & \text{for } -\dot{\gamma}_1 \leq \dot{\gamma} < \dot{\gamma}_y \\ \mu_{po}\dot{\gamma} + \tau_y & \text{for } \dot{\gamma} \geq \dot{\gamma}_y \end{cases} \tag{7.16}$$

If $\dot{\gamma} \leq 0$

$$\tau(\dot{\gamma}) = \begin{cases} \mu_{po}\dot{\gamma} + \tau_y & \text{for } \dot{\gamma} > \dot{\gamma}_1 \\ \mu_{pr}(\dot{\gamma} - \dot{\gamma}_o) & \text{for } -\dot{\gamma}_y \leq \dot{\gamma} < \dot{\gamma}_1 \\ \mu_{po}\dot{\gamma} + \tau_y & \text{for } \dot{\gamma} \leq -\dot{\gamma}_y \end{cases} \tag{7.17}$$

7.2.6 Other Models

Many models exist based on the fundamental Bingham plastic model, incorporating additional stiffness and viscous damping elements, along with the basic combination of viscous damper and Coulomb friction elements described previously. Examples of such models are as follows:

1. Extended Bingham model [12]. Viscoelastic element added in series to Bingham plastic model to capture pre-yield behavior.
2. Three-element method [17]. Introduces a nonlinear spring in parallel with the elements of the Bingham plastic model. Static and dynamic coefficients are incorporated in terms of Coulomb friction. A hyperbolic tangent function is incorporated instead of the $sgn(\dot{x})$ to simplify numerical calculations.
3. Bing Max model [28, 41, 42]. This model has a series combination of a spring and a dashpot, in parallel with a Coulomb friction element.
4. Nonlinear viscoelastic-plastic model [43]. The fluid behavior is separated into pre-yield and post-yield regions. In the pre-yield region, the fluid is represented by a three-element model (a damper in series with a parallel combination of a spring and a damper), behaving like a viscoelastic material. In the post-yield region, the fluid is represented by a viscous damping element.

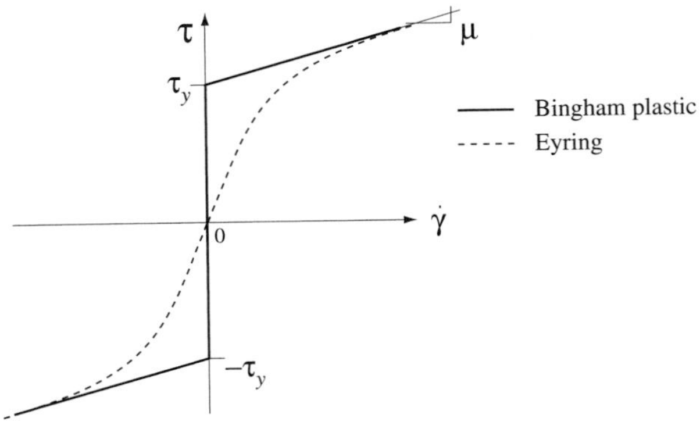

Figure 7.15. Comparison of Eyring and Bingham plastic models.

5. Eyring model [44]. The Eyring model was proposed to address the large change in shear stress occurring in piecewise continuous models such as the Bingham plastic model. For example, at the yield shear stress in the Bingham plastic model, the ER/MR fluid abruptly changes from exhibiting fluid-like behavior to a rigid solid at zero shear rate. In the Eyring model, the ER/MR fluid is no longer treated as piecewise continuous and has a smooth transition from low shear rate to high shear rate behavior. The shear stress is given by

$$\tau = \tau_o \sinh^{-1}(\lambda\dot{\gamma}) \tag{7.18}$$

where τ_o and λ are two constants that determine the rheological behavior of the ER/MR fluid. Figure 7.15 shows a comparison of Eyring and Bingham plastic models.

7.3 ER and MR Fluid Dampers

The most common damping mechanism in modern systems is the fluid-filled viscous damper. Such dampers are widespread in many applications ranging from complex mechanical systems such as automobile and motorcycle suspensions, to aircraft landing gear, to simple systems such as doors and artillery pieces.

A typical viscous damper basically consists of an oil-filled cylinder in which slides a loose-fitting piston. The upper and lower chambers of the cylinder are connected by the annular gap around the loose-fitting piston. Motion of the piston inside the cylinder forces the fluid between the two chambers through the annular gap. The geometry of this flowpath determines its resistance to the flow of fluid, which in turn determines the amount of damping. Hence, for a given geometry, the damping coefficient is a constant.

In many applications, it is desirable to have different damping coefficients depending on the operating condition of the system. For example, in automotive suspensions, low damping is desirable to isolate the passengers from a bumpy road, whereas high damping is required to improve handling of the vehicle. Conventional automotive dampers are designed to provide a compromise between a comfortable ride and good handling. The degree of this trade-off depends on the type of vehicle such as a passenger car or a sports car. The dampers are often designed with a complicated network of passages, springs, bypass channels, and check valves that provide different flow resistances and, therefore, different damping coefficients, depending on the speed of the vehicle [45].

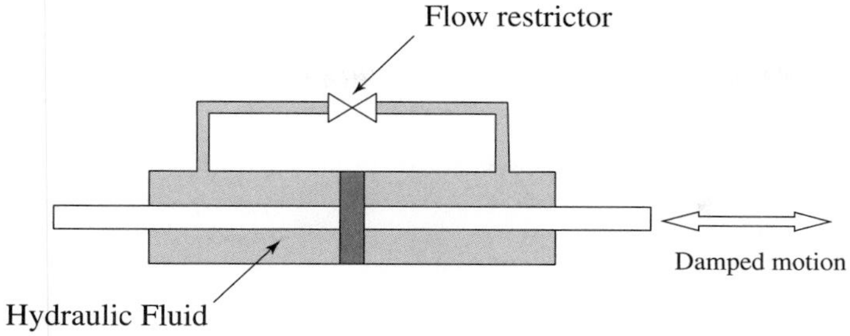

(a) Conventional passive damper

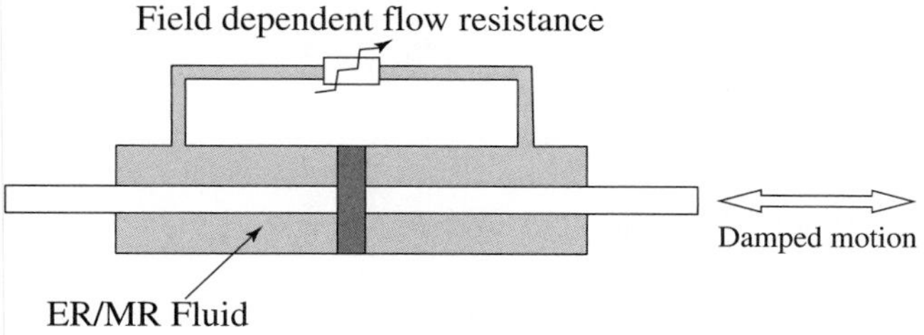

(b) Controllable damper with ER/MR fluid

Figure 7.16. Passive and semi-active damping.

The more expensive shock absorbers provide a larger variation in damping by using more complicated mechanisms. However, even such variable dampers have some disavantages. High-performance adjustable dampers are expensive, mechanically complex, and require time-consuming maintenance. In addition, even the most complicated mechanical dampers provide only a fixed number of damping coefficients that are permanently set by the design.

Dampers utilizing ER/MR fluids overcome these drawbacks. The viscosity of the fluid and, hence, the damping coefficient can be controlled by the application of an electric or magnetic field. In this way, control of the damping is possible over a wide range, with infinite resolution using a device of simple geometry with few moving parts. A schematic of the controllable damping concept is illustrated in Fig. 7.16. The conventional passive damper has a flow restrictor of fixed geometry. As a result, the damping coefficient is a constant. In the ER/MR fluid damper, the flow restriction can be controlled by the applied field. Consequently, the damping coefficient can be varied at any time, even during the application of loads on the damper.

Three types of basic damping mechanisms can be utilized in the construction of ER/MR dampers, as follows:

1. Flow mode (Poiseulle flow) (Fig. 7.17(a)): The ER/MR fluid is made to flow through a passage, across which the field can be applied. In this case, the electrodes are stationary. This is also the configuration typically utilized to construct ER/MR valves.
2. Shear mode (Couette flow) (Fig. 7.17(b)): The ER/MR fluid is enclosed between two electrodes, or magnetic poles. One of the electrodes or poles is kept fixed, whereas the other undergoes displacement and is connected to the system that

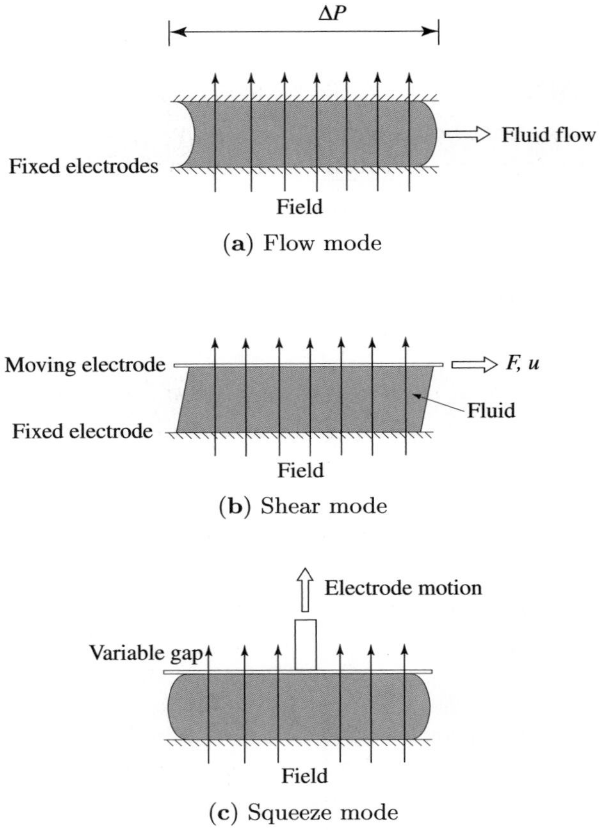

Figure 7.17. Controllable fluid-damper operating modes.

requires damping. Relative displacement between the two electrodes, or magnetic poles, results in shearing of the fluid while maintaining a constant gap between them. The strength of the applied field is not expected to change with motion.

3. Squeeze mode (Fig. 7.17(c)): The ER/MR fluid is enclosed between two electrodes or magnetic poles that undergo relative motion along the direction of the field. The field strength varies with the displacement of the electrodes.

Figure 7.18 shows schematic cross sections of damper configurations operating in the flow mode, shear mode, and mixed mode. Note that the mixed mode is a combination of the flow mode and shear mode. In Fig. 7.18(a), a flow passage is formed by an annular gap in the piston head. An electric or magnetic field is applied across this gap to activate the fluid. The part of the flow passage over which the field is applied is referred to as the "active region." The motion of the piston forces fluid through the flow passage. The inner and outer walls of the flow passage translate with the same velocity and can be considered stationary with respect to the flow of fluid. Therefore, this damper operates in flow mode. In Fig. 7.18(b), the fluid is contained in an annular gap between the stationary outer shell of the damper and the movable inner shaft, across which an electric or magnetic field can be applied. As the motion of the inner shaft results in shearing of the fluid in the active region, this damper operates in shear mode. Mixed-mode operation involves a combination of flow mode as well as shear mode. In Fig. 7.18(c), a flow passage is formed by the

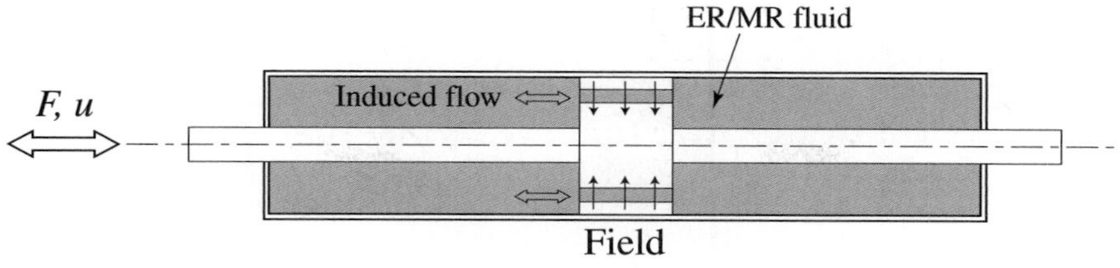

(a) Flow mode

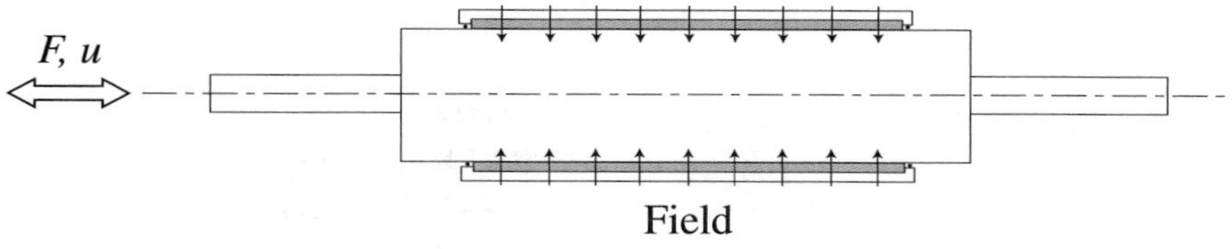

(b) Shear mode

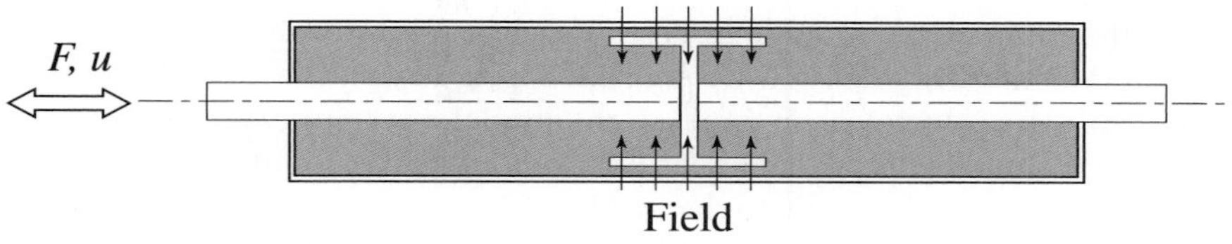

(c) Mixed mode

Figure 7.18. ER/MR fluid damper configurations.

annular gap between the piston and the outer shell of the cylinder. This area also constitutes the active region of the damper. The fluid in the active region is sheared by the motion of the piston. In addition, the motion of the piston forces fluid from one side of the piston to the other through the annular gap. Therefore, a combination of flow mode and shear mode operation results.

From the damper configurations in Fig. 7.18, the simplicity of construction of ER dampers is evident. It is easy to obtain a uniform electric field between two concentric cylindrical surfaces by connecting each surface to one terminal of a high-voltage power supply. However, in the case of MR fluids, a uniform magnetic field must be generated by carefully placing a current carrying coil inside the device such that the field lines are perpendicular to the direction of shear strain in the fluid. One way to achieve this is by using a bypass construction as shown in Fig. 7.19. The current carrying coil is wound on a bobbin core made of highly permeable material. The outer cylinder is also made of the same material and forms a flux return path. The magnetic field is uniformly concentrated along the radial direction in the region at the ends of the bobbin. The fluid flows through this annulus, which is the active region of the device.

In the case of both ER and MR fluids, the height of the gap across which the field is applied is an important parameter. A small gap enables the required actuation field to be achieved with smaller values of applied voltage or current. Also, a smaller gap has a higher flow resistance in the inactive state (zero applied field).

A schematic of a mixed mode ER damper designed and tested by Kamath et al. [16] is shown in Fig. 7.20. This damper was used to develop and validate quasi-steady damper models using idealized Bingham plastic fluid behavior. The fluid

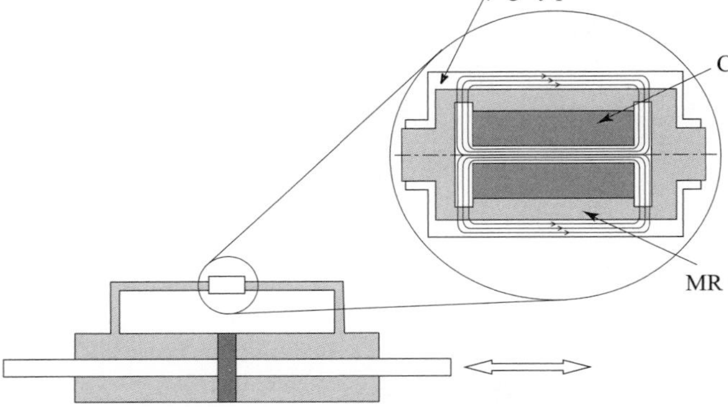

Figure 7.19. MR bypass damper.

used in this damper was VersaFlo ER-100 [46]. The force-velocity curves that were measured for this damper, along with theoretical predictions, are shown in Fig. 7.21. It can be seen that the Bingham plastic analysis captures well the overall trend of the experimental data. However, in general, the model does not capture the exact behavior, which includes several nonlinear phenomena. To account for this, either nonlinear corrections must be made to the idealized Bingham plastic model or other nonlinear fluid models must be used. In addition, it was noted that slight adjustments in the published material data yielded a much-improved correlation. This could be due to aging of the ER fluid or variation in properties between batches of fluid due to the manufacturing processes. Therefore, it is necessary to carefully measure the field-dependent fluid properties before using them in an analysis.

7.4 Modeling of ER/MR Fluid Dampers

The performance of ER/MR fluid dampers can be modeled using the phenomenological approaches described in Section 7.2. The complete equations for three-dimensional states of stress and field are very involved because they contain nonlinear terms and anisotropic properties. For practical purposes, it is instructive to use a simple approach. The application of a fluid model to a device of a specific geometry is described herein. To illustrate the modeling procedure, we examine the

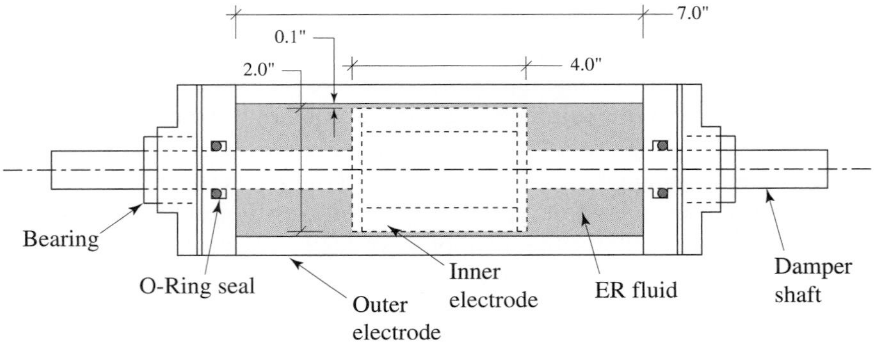

Figure 7.20. Schematic of an mixed mode ER fluid damper, adapted from Kamath et al. [16].

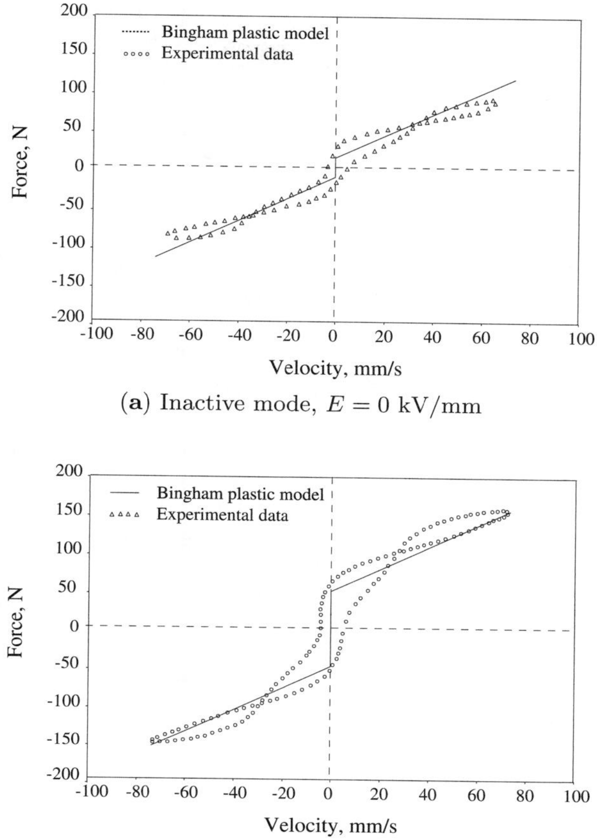

(a) Inactive mode, $E = 0$ kV/mm

(b) Active mode, $E = 3$ kV/mm

Figure 7.21. Comparison of experimental damper data with Bingham plastic model predictions, at electric fields of $E = 0$ kV/mm and $E = 3$ kV/mm, adapted from Kamath et al. [16].

behavior of a damper in the shear mode as well as in the flow mode, using a Bingham plastic fluid model. We first consider the simplest flow geometry, which is a passage of rectangular cross section, and then explore an annular flow passage, which is more suited for many practical engineering applications. The flow passage of rectangular cross section is formed by the gap between two parallel plates that also act as the electrodes (in the case of ER fluids) or magnetic poles (in the case of MR fluids) for the application of a field. An annular flow passage can be formed by the gap between two concentric cylinders that also act as the electrodes or magnetic poles. The behavior of dampers in the squeeze mode is not considered here; discussion of this aspect has been provided by Stanway et al. [47, 48].

7.4.1 Rectangular Flow Passage

Let us consider the behavior of the fluid in a passage of rectangular cross section. The fluid is enclosed between two parallel plates that also form the electrodes or magnetic poles. An electric or magnetic field is applied across the height of the passage d. The length over which the field is applied, or the active length, is L, and the width of the passage is b. A schematic of this flow passage is shown in Fig. 7.22. It can be assumed that a uniform field exists across the height of the passage, over an area $L \times b$. The fluid enclosed in this volume forms a simple active fluid element.

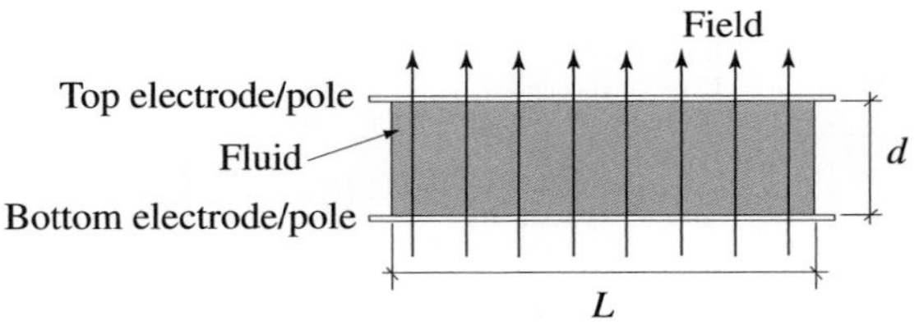

Figure 7.22. ER/MR fluid in rectangular flow passage.

Consider the force equilibrium on a rectangular fluid element of length dx, height dy, and width b as shown in Fig. 7.23. The force equilibrium equation can be written as

$$-m\ddot{x} + P\,dy\,b \;-\; \tau\,dx\,b \;-\; \left(P + \frac{\partial P}{\partial x}\,dx\right)dy\,b \;+\; \left(\tau + \frac{\partial \tau}{\partial y}\,dy\right)\,dx\,b \;=\; 0 \quad (7.19)$$

where P is the fluid pressure, τ is the shear stress, and m is the mass of the fluid element given by

$$m = \rho\,dy\,dx\,b \quad (7.20)$$

where ρ is the mass density of the fluid. Substituting in Eq. 8.171, we obtain

$$-\rho\,\frac{\partial u}{\partial t} - \frac{\partial P}{\partial x} + \frac{\partial \tau}{\partial y} = 0 \quad (7.21)$$

where u is the axial velocity ($\partial x/\partial t$). Assuming a quasi-steady flow

$$\frac{\partial u}{\partial t} = 0 \quad (7.22)$$

The governing equation reduces to

$$\frac{\partial \tau}{\partial y} = \frac{\partial P}{\partial x} \quad (7.23)$$

We examine the behavior of a damper using this active fluid element operating in two modes: shear mode and flow mode.

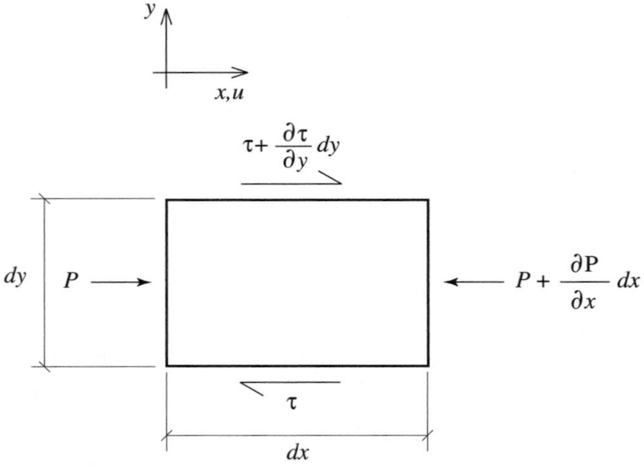

Figure 7.23. Differential fluid element, rectangular flow passage.

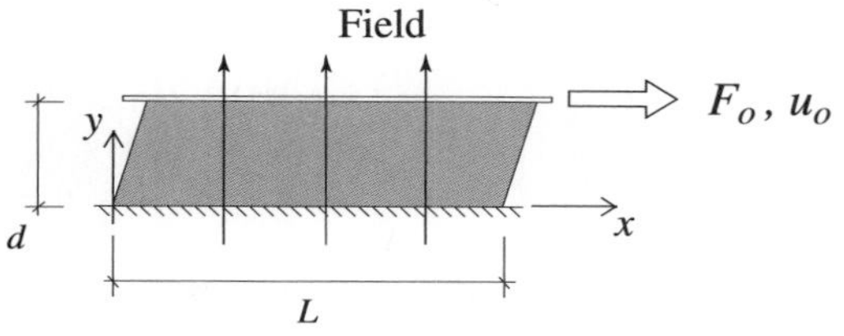

Figure 7.24. Rectangular flow passage: shear mode operation.

Shear Mode

A shear mode damper can be constructed with the rectangular flow geometry shown in Fig. 7.22 by moving the upper plate with respect to the lower one, while maintaining a constant gap d between them. Assume that a force F_o acts on the upper plate, moving it with a constant velocity u_o. A schematic of this configuration is shown in Fig. 7.24. In this case, the pressure gradient is

$$\frac{\partial P}{\partial x} = 0 \tag{7.24}$$

The governing equation reduces to

$$\frac{\partial \tau}{\partial y} = 0 \tag{7.25}$$

(a) Solution under zero applied field

When no field is applied, the fluid behaves like a Newtonian fluid. The shear stress is given by (Eq. 7.1)

$$\tau = \mu \frac{\partial u}{\partial y} \tag{7.26}$$

where μ is the dynamic viscosity of the fluid. Substituting in Eq 7.25, we obtain

$$\mu \frac{\partial^2 u}{\partial y^2} = 0 \tag{7.27}$$

Integrating twice leads to

$$u(y) = A\,y + B \tag{7.28}$$

The constants A and B are determined from the boundary conditions

$$\begin{cases} u(0) = 0 \\ u(d) = u_o \end{cases} \quad \Longrightarrow \quad \begin{cases} B = 0 \\ A = u_o/d \end{cases} \tag{7.29}$$

The velocity profile is given by

$$u(y) = \frac{u_o}{d}\, y \tag{7.30}$$

and the shear stress is

$$\tau(y) = \mu \frac{\partial u}{\partial y} = \mu \frac{u_o}{d} \tag{7.31}$$

The force on the upper plate required to move it with the velocity u_o is given by

$$F_o = \tau(d)\, L\, b = \mu \frac{u_o}{d} L\, b \tag{7.32}$$

This can be equated to the equivalent damping force, yielding an effective damping coefficient (inactive state) c^o_{eq}.

$$F_o = c^o_{eq}\, u_o \quad \Longrightarrow \quad c^o_{eq} = \frac{\mu\, L\, b}{d} = \mu\, \Gamma \tag{7.33}$$

where Γ is a parameter that depends only on the geometry of the flow passage.

(b) Solution under non-zero applied field

When a field is applied across the gap, the fluid is modeled as a Bingham plastic. The shear stress is given by

$$\tau(y) = \tau_y + \mu \frac{\partial u}{\partial y} \tag{7.34}$$

The velocity profile is calculated from the governing Eq. 7.25. Because τ_y is independent of y, and the boundary conditions are the same, the velocity profile is the same as before

$$u(y) = \frac{u_o}{d}\, y \tag{7.35}$$

The shear stress is given by

$$\tau(y) = \tau_y + \mu \frac{u_o}{d} \tag{7.36}$$

and the force in the damper is

$$\begin{aligned} F_o &= \tau(d)\, L\, b \\ &= \left(\tau_y + \mu \frac{u_o}{d}\right) L\, b \\ &= \left(\frac{\tau_y\, d}{\mu\, u_o} + 1\right) \mu \frac{u_o}{d} L\, b \\ &= c^a_{eq}\, u_o \end{aligned} \tag{7.37}$$

where c^a_{eq} is the effective damping coefficient in the active state, defined as

$$c^a_{eq} = \mu \Gamma (1 + \mathsf{Bi}) \tag{7.38}$$

The quantity Bi is called the Bingham number and is a nondimensional quantity relating the yield stress to the viscous stress. Introducing nondimensional quantities in the analysis, such as the Bingham number and other parameters based on the damper geometry, enables the performance of different types and sizes of devices to be compared on the same basis. Note that if the velocity u_o is high, then the Bingham number is small and, consequently, the increase in damping coefficient on activation of the fluid is small. It can be concluded that when an activated fluid is subjected to high velocities, because the Bingham number is small, the fluid tends to behave more like a Newtonian fluid than like a Bingham plastic. Therefore, the displacement

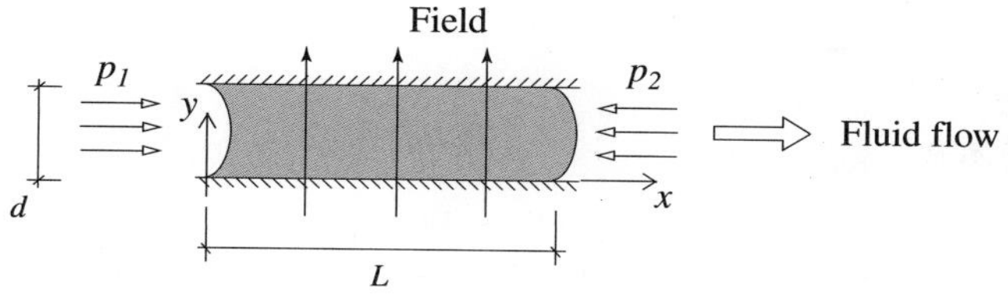

Figure 7.25. Rectangular flow passage: flow mode operation.

amplitude and operating frequency are also important parameters in characterizing the performance of a damper. The expression for the Bingham number is

$$\mathsf{Bi} = \frac{\tau_y}{\mu u_o / d} = \frac{\text{yield stress}}{\text{viscous stress}} \tag{7.39}$$

It can be seen that the Bingham number depends on the yield stress and viscosity of the fluid, as well as on the gap height and the velocity of motion. The smaller the gap, the smaller the Bingham number. Note that for a Newtonian fluid, the Bingham number is zero. The equivalent active damping coefficient, c_{eq}^a, can be written as (from Eq. 7.38)

$$c_{eq}^a = c_{eq}^o(1 + \mathsf{Bi}) \tag{7.40}$$

We see that the damping coefficient in the active state has increased by the amount Bi. Therefore, Bi defines the amount of active damping in the device. To create the largest change in damping on the application of a field, the ratio of active damping coefficient to inactive damping coefficient must be high. Therefore

$$\frac{c_{eq}^a}{c_{eq}^o} \gg 1 \quad \rightarrow \quad \mathsf{Bi} \gg 1 \quad \rightarrow \tau_y \gg \frac{\mu u_o}{d} \tag{7.41}$$

This means that the yield stress must be much higher than the viscous stress. Because u_o is based on the application and d is based on the geometry of the device, the ideal controllable fluid should have a high yield stress τ_y and a low dynamic viscosity μ.

Flow Mode

A flow mode damper can be constructed with the rectangular flow geometry shown in Fig. 7.22 by holding both of the plates fixed and creating a fluid flow between them. A schematic of this configuration is shown in Fig. 7.25. The fluid flow is caused by the difference in pressures p_1 and p_2 at the ends of the flow passage. In this case, the pressure gradient is related to the applied differential pressure ΔP across the active length (assumed constant over the entire active length). Note that $\Delta P = p_1 - p_2$ is the pressure drop across the length of the gap. The pressure gradient is given by

$$\frac{\partial P}{\partial x} = -\frac{\Delta P}{L} = \frac{p_2 - p_1}{L} \tag{7.42}$$

It is assumed that the location under consideration is sufficiently far away from the ends of the flow passage such that the flow profile is fully developed. The governing

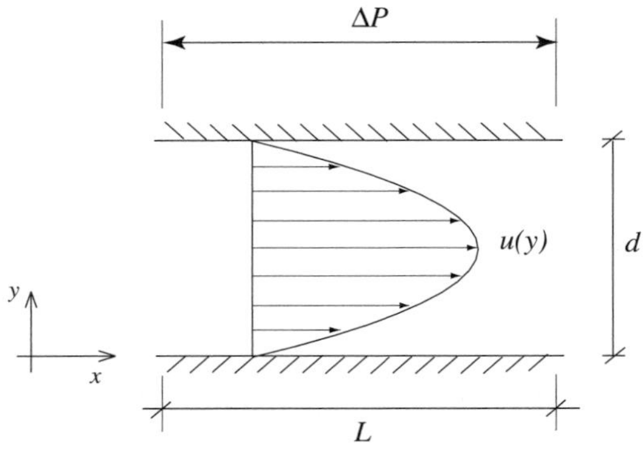

Figure 7.26. Flow profile of the fluid in the flow mode, no field applied (rectangular cross section).

equation becomes

$$\frac{\partial \tau}{\partial y} = \frac{\partial P}{\partial x} = -\frac{\Delta P}{L} \tag{7.43}$$

(a) Solution under zero applied field

In the inactive state, the fluid behavior is Newtonian. The governing equation becomes

$$\mu \frac{\partial^2 u}{\partial y^2} = -\frac{\Delta P}{L} \tag{7.44}$$

Integrating twice yields

$$u(y) = -\frac{\Delta P}{2\mu L} y^2 + Cy + D \tag{7.45}$$

The constants C and D are determined from the boundary conditions

$$\begin{cases} u(0) = 0 \\ u(d) = 0 \end{cases} \implies \begin{cases} D = 0 \\ C = \frac{d}{2}\frac{\Delta P}{\mu L} \end{cases} \tag{7.46}$$

Substituting these constants into Eq. 7.45, the velocity profile of the flow across the gap can be written as

$$\begin{aligned} u(y) &= -\frac{\Delta P}{2\mu L} y^2 + \frac{\Delta P d}{2\mu L} y \\ &= \frac{\Delta P}{2\mu L} y(d - y) \end{aligned} \tag{7.47}$$

It can be seen that the velocity profile is parabolic (shown in Fig. 7.26). By symmetry, it is evident that the velocity is maximum at the center of the gap

$$\begin{aligned} u(d/2) &= u_o \\ &= \frac{\Delta P}{2\mu L}\frac{d}{2}\frac{d}{2} \\ &= \frac{\Delta P d^2}{8\mu L} \end{aligned} \tag{7.48}$$

The velocity profile can also be conveniently expressed in nondimensional form

$$\bar{u}(\bar{y}) = 4\bar{y}(1 - \bar{y}) \tag{7.49}$$

where

$$\bar{y} = \frac{y}{d} \quad \text{and} \quad \bar{u} = \frac{u}{u_o} \tag{7.50}$$

The shear stress in the gap is

$$\begin{aligned} \tau(y) &= \mu \frac{\partial u}{\partial y} = \mu \left(-\frac{\Delta P y}{\mu L} + \frac{\Delta P d}{2\mu L} \right) \\ &= \frac{\Delta P}{L} \left(\frac{d}{2} - y \right) \end{aligned} \tag{7.51}$$

The force required to maintain the flow velocity in the passage, which is basically the damping force in the device, is given by the product of the differential pressure and the cross-sectional area. We can assume that the flow is created by a piston with the same cross section as the flow passage, moving with a constant velocity u_o. The force required to move the piston is F_o. Because the velocity profile across the gap is parabolic, a mean velocity u_m can be defined that is constant across the gap and that yields the same volumetric flow as the parabolic profile. The volumetric flow Q is given by

$$\begin{aligned} Q &= \int_{y=0}^{d} u(y) b dy = b \int_0^d \frac{\Delta P}{2\mu L} y(d - y) dy \\ &= \frac{\Delta P b}{2\mu L} \left[\frac{dy^2}{2} - \frac{y^3}{3} \right]_0^d \\ &= \frac{\Delta P b d^3}{12\mu L} \end{aligned} \tag{7.52}$$

The volumetric flow can also be expressed in terms of the mean velocity, u_m, as follows

$$Q = u_m b d \tag{7.53}$$

From Eqs. 7.52 and 7.53, the mean velocity is

$$u_m = \frac{\Delta P d^2}{12\mu L} \tag{7.54}$$

The damping coefficient of the fluid element can be found from the force and velocity of the piston. The differential pressure is related to the force on the piston by

$$\Delta P = \frac{F}{bd} \tag{7.55}$$

which yields

$$F = \frac{12\mu b L}{d} u_m \tag{7.56}$$

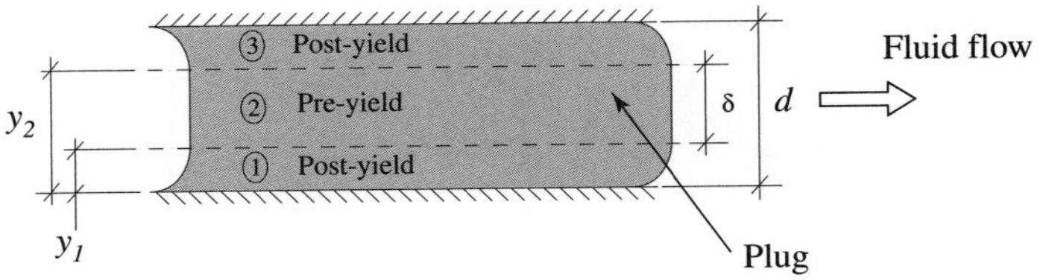

Figure 7.27. Flow profile of the fluid in the flow mode, under an external field (rectangular cross section), Bingham plastic model.

The damping coefficient of the fluid element under zero applied field, c^o_{eq}, can be found from the previous equation as

$$\begin{aligned} c^o_{eq} &= \frac{F}{u_m} \\ &= \frac{12\mu Lb}{d} \end{aligned} \tag{7.57}$$

It is seen that the damping coefficient depends on the geometry of the damper and the viscosity of the fluid.

(b) Solution under non-zero applied field

When a field is applied, the velocity profile of the fluid changes depending on the local shear stress. The flow velocity profile in the Newtonian case is parabolic and the shear stress at the middle of the gap is zero. Therefore, around this region, the fluid is in the pre-yield condition. Near the walls of the passage, the shear stresses may be higher than the yield stress, resulting in post-yield fluid behavior. Treating the fluid as a Bingham plastic, it can be seen that in the pre-yield region, the fluid behaves like a solid and, therefore, has a constant translational velocity around the center of the gap. Near the walls, the fluid behavior is Newtonian, with a parabolic velocity profile. The resulting flow profile across the height of the gap can be considered as a solid plug around the center of the gap, being carried along in a Newtonian fluid. This flow profile is depicted in Fig. 7.27. The flow is divided into three regions: regions 1 and 3 are the post-yield regions and region 2 is the pre-yield region. The thickness of the plug in the center of the gap is δ.

To find the flow profile in the gap and the effective damping coefficient, each of the three regions is treated separately. Substituting the expressions for shear stress in each region, we see that the governing equation for all three regions reduces to Eq. (7.44)

$$\mu \frac{\partial^2 u}{\partial y^2} = -\frac{\Delta P}{L} \tag{7.58}$$

and the location of each region is

$$y_1 = \frac{d-\delta}{2} \tag{7.59}$$

$$y_2 = \frac{d+\delta}{2} \tag{7.60}$$

Region 1

Integrating the previous governing equation twice leads to

$$u_1(y) = -\frac{\Delta P}{2\mu L}y^2 + C_1 y + C_2 \tag{7.61}$$

The boundary conditions in this case are

$$u_1(0) = 0 \tag{7.62}$$

$$\left.\frac{\partial u_1}{\partial y}\right|_{y=y_1} = 0 \tag{7.63}$$

Whereas the first boundary condition is a result of the no-slip condition at the wall, the second boundary condition occurs because there can be no discontinuity in the flow profile. Substituting and solving yields the constants

$$C_2 = 0 \tag{7.64}$$

$$-\frac{\Delta P}{\mu L}y_1 + C_1 = 0 \implies C_1 = \frac{\Delta P y_1}{\mu L} \tag{7.65}$$

Therefore, the velocity profile in region 1 is given by

$$\begin{aligned} u_1(y) &= -\frac{\Delta P}{2\mu L}y^2 + \frac{\Delta P y_1}{\mu L}y \\ &= \frac{\Delta P}{2\mu L}\, y\,(2y_1 - y) \\ &= \frac{\Delta P}{2\mu L}\, y\,(d - \delta - y) \end{aligned} \tag{7.66}$$

Region 3

Integrating the governing equation twice leads to

$$u_3(y) = -\frac{\Delta P}{2\mu L}y^2 + C_3 y + C_4 \tag{7.67}$$

The boundary conditions in this case are

$$u_3(d) = 0 \tag{7.68}$$

$$\left.\frac{\partial u_3}{\partial y}\right|_{y_2} = 0 \tag{7.69}$$

These boundary conditions are similar to that of the previous case. Substituting and solving yields the constants

$$-\frac{\Delta P}{\mu L}y_2 + C_3 = 0 \implies C_3 = \frac{\Delta P y_2}{\mu L} \tag{7.70}$$

$$-\frac{\Delta P}{2\mu L}d^2 + \frac{\Delta P}{\mu L}d\, y_2 + C_4 = 0 \implies C_4 = \frac{\Delta P}{2\mu L}\, d\,(d - 2y_2) \tag{7.71}$$

Therefore, the velocity profile in region 3 is given by

$$\begin{aligned} u_3(y) &= -\frac{\Delta P}{2\mu L}y^2 + \frac{\Delta P}{\mu L}yy_2 + \frac{\Delta P}{2\mu L}\left(d^2 - 2d\,y_2\right) \\ &= \frac{\Delta P}{2\mu L}\left[(d^2 - y^2) - 2y_2(d - y)\right] \\ &= \frac{\Delta P}{2\mu L}(d - y)(y - \delta) \end{aligned} \tag{7.72}$$

Note that this result can also be obtained from the symmetry of the flow

$$u_3(y) = u_1(d - y) \tag{7.73}$$

Applying this relation to Eq. 7.66 results in Eq. 7.72.

Region 2

The velocity is constant in region 2, given by the velocity at the locations y_1 and y_2. Let us call the velocity of the fluid in region 2 the plug velocity, u_p. Then we can write

$$\begin{aligned} u_1(y_1) &= u_p \\ u_3(y_2) &= u_p \end{aligned} \tag{7.74}$$

Substituting in Eq. 7.66, we obtain

$$\begin{aligned} u_p = u_1(y_1) &= \frac{\Delta P}{2\mu L}y_1^2 \\ &= \frac{\Delta P(d - \delta)^2}{8\mu L} \end{aligned} \tag{7.75}$$

As a check

$$\begin{aligned} u_3(y_2) &= \frac{\Delta P}{2\mu L}(d - y_2)(y_2 - \delta) \\ &= \frac{\Delta P}{8\mu L}(d - \delta)^2 \\ &= u_p \end{aligned} \tag{7.76}$$

Note that the solution of the governing flow equation (Eq. 7.43) in all three regions involves a total of five constants: C_1, C_2, C_3, C_4, and δ. The boundary conditions in regions 1 and 3 (Eqs. 7.62, 7.63, 7.68, and 7.69) provide four equations. The condition of equal flow velocities at the locations y_1 and y_2 (Eq. 7.74) does not provide any additional information because y_1 and y_2 are fixed by the assumption that the flow profile is symmetric about the center of the flow passage. Therefore, an additional condition is required to find the thickness of the plug, δ. This can be found by solving for the shear stress at the boundary of region 2. The governing equation (Eq. 7.43) in region 2 is written as

$$\frac{\partial \tau_2}{\partial y} = -\frac{\Delta P}{L} \tag{7.77}$$

Integrating this equation yields

$$\tau_2(y) = -\frac{\Delta P}{L} y + C_5 \tag{7.78}$$

The constants δ and C_5 can be found from the following boundary conditions

$$\tau_2(y_1) = \tau_y \tag{7.79}$$

$$\tau_2(y_2) = -\tau_y \tag{7.80}$$

Substitution in Eq. 7.78 results in an expression for C_5

$$C_5 = \frac{\Delta P}{2L}(y_1 + y_2) = \frac{\Delta P}{L} d \tag{7.81}$$

Therefore, the shear stress in region 2 is given by

$$\begin{aligned} \tau_2(y) &= -\frac{\Delta P}{L} y + \frac{\Delta P}{2L} d \\ &= \frac{\Delta P}{2L}(d - 2y) \end{aligned} \tag{7.82}$$

The plug thickness can be found by substituting the constant C_5 in the first boundary condition (Eq. 7.79)

$$-\frac{\Delta P}{L} y_1 + \frac{\Delta P}{2L} d = \tau_y \implies \delta = \tau_y \frac{2L}{\Delta P} \tag{7.83}$$

It is convenient to nondimensionalize the plug thickness by the height of the gap

$$\bar{\delta} = \frac{\delta}{d} = \frac{\tau_y 2L}{\Delta P d} \tag{7.84}$$

The value of $\bar{\delta}$ defines the state of flow through the gap.

(1) $\bar{\delta} = 0$: The flow is purely Newtonian.
(2) $\bar{\delta} = 1$: The gap is completely blocked and there is no flow of fluid. Given a specific fluid, the differential pressure below which the flow passage remains blocked can be derived as

$$\Delta P \leq \frac{2\tau_y L}{d} \tag{7.85}$$

Alternatively, to sustain a specified pressure differential without allowing any flow, a fluid can be chosen with a yield stress such that

$$\tau_y \geq \frac{\Delta P d}{2L} \tag{7.86}$$

To calculate the effective damping coefficient of the activated fluid element, it is necessary to find a mean flow velocity, u_m, by finding the total volumetric flow Q through the passage

$$\begin{aligned} Q &= \int_{y=0}^{d} u(y)\, b\, dy \\ &= 2\, Q_1 + Q_2 \end{aligned} \tag{7.87}$$

where Q_1 and Q_2 is the volumetric flow through region 1 and region 2, respectively, given by (from Eqs. 7.66 and 7.75)

$$Q_1 = b\int_0^{y_1} \frac{\Delta P}{2\mu L}\left(2yy_1 - y^2\right) dy = \frac{\Delta P b}{24\mu L}(d-\delta)^3 \tag{7.88}$$

$$Q_2 = b\int_{y_1}^{y_2} u_p\, dy = \frac{\Delta P b}{8\mu L}(d-\delta)^2\delta \tag{7.89}$$

Note that $Q_3 = Q_1$. The total volumetric flow is given by

$$\begin{aligned} Q &= u_m b\, d \\ &= \frac{\Delta P b}{12\mu L}(d-\delta)^3 + \frac{\Delta P b}{8\mu L}(d-\delta)^2\delta \\ &= \frac{\Delta P b}{12\mu L}(d-\delta)^2\left(d + \frac{\delta}{2}\right) \\ &= \frac{\Delta P b\, d^3}{12\mu L}(1-\bar{\delta})^2\left(1 + \frac{\bar{\delta}}{2}\right) \end{aligned} \tag{7.90}$$

From this equation, the mean velocity can be extracted as

$$u_m = \frac{\Delta P d^2}{12\mu L}(1-\bar{\delta})^2\left(1 + \frac{\bar{\delta}}{2}\right) \tag{7.91}$$

The damping coefficient c_{eq}^a in the active state is given by

$$c_{eq}^a = \frac{F_a}{u_m} \tag{7.92}$$

where F_a is the force required to move the piston when the fluid is activated, given by

$$F_a = \Delta P\, bd \tag{7.93}$$

From these equations, the active damping coefficient is

$$c_{eq}^a = \frac{12\mu L b d}{d^2(1-\bar{\delta})^2(1+\bar{\delta}/2)} = \frac{c_{eq}^o}{(1-\bar{\delta})^2(1+\bar{\delta}/2)} \tag{7.94}$$

The ratio of the damping coefficient in the active state to the damping coefficient in the inactive state, as a function of different plug thicknesses, is shown in Fig. 7.28. It can be seen that this ratio increases steeply as the plug thickness increases. For a plug thickness of around 0.6, the damping coefficient increases by an order of magnitude from the inactive to the active state.

The ratio of the damping coefficients in the active and inactive states can also be expressed in terms of the Bingham number. The Bingham number is defined as

$$\begin{aligned} \text{Bi} &= \frac{\tau_y d}{\mu u_m} \\ &= \frac{\tau_y d}{\mu}\frac{12\mu L}{\Delta P d^2(1-\bar{\delta})^2(1+\bar{\delta}/2)} \end{aligned} \tag{7.95}$$

From the definition of plug thickness (Eq. 7.84)

$$\text{Bi} = \frac{6\bar{\delta}}{(1-\bar{\delta})^2(1+\bar{\delta}/2)} \tag{7.96}$$

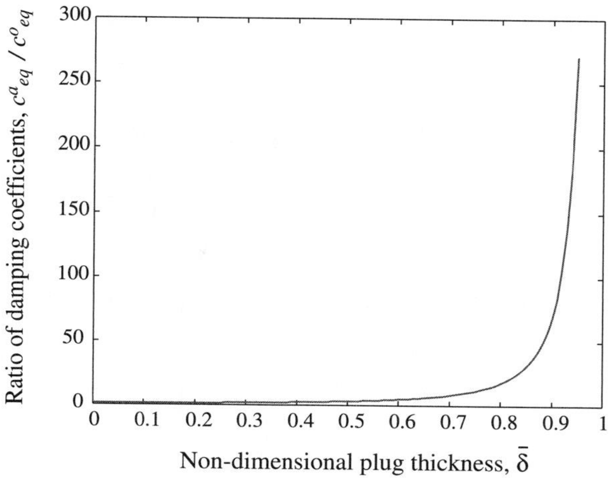

Figure 7.28. Variation of the ratio of damping coefficients with plug thickness.

which yields the ratio of damping coefficients as

$$\frac{c^a_{eq}}{c^o_{eq}} = \frac{\mathrm{Bi}}{6\bar{\delta}} \tag{7.97}$$

It is interesting to note that using the Bingham plastic model, it is possible to obtain a value of $\bar{\delta} = 1$, meaning fully blocked flow. This would yield a damping coefficient of infinity, which is not realistic. Using the biviscous fluid model would alleviate this problem because of the finite pre-yield viscosity.

Piston Area and Flow Passage Area

Often, for flow mode dampers, the cross-sectional area of the piston head (A_p) may not be the same as the cross-sectional area of the flow passage (A_d). An example of such a case is a bypass damper (shown in Fig. 7.29). In this case, the damping coefficient calculated from the force and velocity in the flow passage is different from the damping coefficient with respect to the force and velocity of the piston. The volume of fluid displaced by the piston head is given by

$$Q_p = A_p u_p \tag{7.98}$$

where u_p is the velocity of the piston head. The effective damping coefficient of the bypass damper, c_{eq}, is defined with respect to the piston velocity and the force on the piston, F_p

$$F_p = c_{eq} u_p \tag{7.99}$$

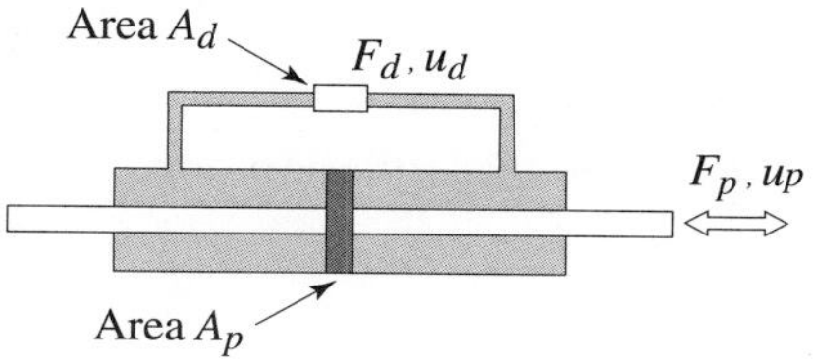

Figure 7.29. Equivalent damping coefficient of a bypass damper.

From conservation of mass of the fluid, we obtain

$$A_p u_p = A_d u_d \tag{7.100}$$

where u_d is the mean flow velocity in the flow passage. By equality of pressures

$$\frac{F_p}{A_p} = \frac{F_d}{A_d} \tag{7.101}$$

where F_d is the force that would be exerted on a piston having the same area as the flow passage. The damping coefficient with respect to the flow passage, $c_{eq,d}$, is defined as

$$c_{eq,d} = \frac{F_d}{u_d} \tag{7.102}$$

From these equations, we see that

$$\begin{aligned} c_{eq} &= \frac{A_p^2}{A_d^2}\frac{F_d}{u_d} \\ &= \frac{A_p^2}{A_d^2}c_{eq,d} \end{aligned} \tag{7.103}$$

7.4.2 Worked Example: Herschel-Bulkley Fluid Model

Derive the inactive and active damping coefficients for a flow mode damper with a rectangular flow passage, using the Herschel-Bulkley fluid model

$$\tau = \tau_y \, sgn(\dot{\gamma}) + K\dot{\gamma}^n \qquad \text{for } \tau > \tau_y \tag{7.104}$$

Solution

Proceeding as in Section 7.4.1, fluid flowing through the rectangular passage can be divided into three regions. The governing equation for all three regions is given by Eq. (7.44)

$$\frac{\partial \tau}{\partial y} = -\frac{\Delta P}{L} \tag{7.105}$$

On the application of a non-zero field, the shear stress in the fluid is expressed by the Herschel-Bulkley model as

$$\tau = \tau_y + K\left(\frac{\partial u}{\partial y}\right)^n \tag{7.106}$$

Substituting in the governing equation leads to

$$\frac{\partial}{\partial y}\left[K\left(\frac{\partial u}{\partial y}\right)^n\right] = -\frac{\Delta P}{L} \tag{7.107}$$

Integrating this equation once

$$\frac{\partial u}{\partial y} = \left[-\frac{\Delta P}{KL}y + C_1\right]^{\frac{1}{n}} \tag{7.108}$$

Integrating again, we obtain

$$u(y) = -\frac{nKL}{(1+n)\Delta P}\left[-\frac{\Delta P}{KL}y + C_1\right]^{\frac{n+1}{n}} + C_2 \tag{7.109}$$

The flow can be divided into three distinct regions, as shown in Fig. 7.27.

Region 1

In this region (post-yield), the shear stress is given by

$$\tau_1 = \tau_y + K\left(\frac{\partial u_1}{\partial y}\right)^n \tag{7.110}$$

and the boundary conditions are

$$\begin{aligned} u_1(0) &= 0 \\ \left.\frac{\partial u_1}{\partial y}\right|_{y=y_1} &= 0 \end{aligned} \tag{7.111}$$

This yields the constants C_1 and C_2 as

$$\begin{aligned} C_1 &= \frac{\Delta P}{KL}y_1 \\ C_2 &= \frac{n}{n+1}\left(\frac{\Delta P}{KL}\right)^{\frac{1}{n}} y_1^{\frac{n+1}{n}} \end{aligned} \tag{7.112}$$

Therefore, the velocity distribution in region 1 becomes

$$u_1(y) = -\frac{n}{n+1}\left(\frac{\Delta P}{KL}\right)^{\frac{1}{n}}\left[(y_1 - y)^{\frac{n+1}{n}} - y_1^{\frac{n+1}{n}}\right] \tag{7.113}$$

Note that if we substitute $n = 1$ in the previous equation, we obtain the same expression for flow velocity as in the case of the Bingham plastic analysis (Eq. 7.66).

Region 3

This is also a post-yield region. The velocity profile can be obtained by symmetry, using the relation

$$u_3(y) = u_1(d - y) \tag{7.114}$$

which results in

$$u_3(y) = -\frac{n}{n+1}\left(\frac{\Delta P}{KL}\right)^{\frac{1}{n}}\left[(y - y_2)^{\frac{n+1}{n}} - (d - y_2)^{\frac{n+1}{n}}\right] \tag{7.115}$$

Region 2

In this region, the fluid is in a pre-yield condition. The fluid has a uniform velocity given by

$$\begin{aligned} u_2(y) &= u_1(y_1) = u_3(y_2) \\ &= \frac{n}{n+1}\left(\frac{\Delta P}{KL}\right)^{\frac{1}{n}}\left(\frac{d-\delta}{2}\right)^{\frac{n+1}{n}} \end{aligned} \tag{7.116}$$

The shear stress profile can be obtained from the following relation

$$\frac{\partial \tau_2}{\partial y} = -\frac{\Delta P}{L} \tag{7.117}$$

Integrating, we obtain

$$\tau_2(y) = -\frac{\Delta P}{L} y + C_3 \tag{7.118}$$

Note that this is the same expression as that obtained in the case of the Bingham plastic model (Eq. 7.78). The boundary conditions in this case are also the same and are given by

$$\begin{aligned} \tau_2(y_1) &= \tau_y \\ \tau_2(y_2) &= -\tau_y \end{aligned} \tag{7.119}$$

Substituting these relations, we obtain

$$\begin{aligned} C_3 &= \frac{\Delta P d}{2L} \\ \delta &= \tau_y \frac{2L}{\Delta P} \end{aligned} \tag{7.120}$$

which are the same relations as in the case of the Bingham plastic model. The shear stress profiles in the three regions are

$$\begin{aligned} \tau_1(y) &= \tau_y + \frac{\Delta P}{L}(y_1 - y) \\ \tau_2(y) &= \frac{\Delta P}{2L}(d - 2y) \\ \tau_3(y) &= -\tau_y - \frac{\Delta P}{L}(y - y_2) \end{aligned} \tag{7.121}$$

The total volume flux can be determined by the summation of volume fluxes from the three regions

$$\begin{aligned} Q &= Q_1 + Q_2 + Q_3 = 2Q_1 + Q_2 \\ &= b\left[2\int_0^{y_1} u_1 dy + \int_{y_1}^{y_2} u_2 dy\right] \\ &= 2b\int_0^{y_1}\left(-\frac{n}{n+1}\right)\left(\frac{\Delta P}{KL}\right)^{\frac{1}{n}}\left[(y_1 - y)^{\frac{n+1}{n}} - y_1^{\frac{n+1}{n}}\right] dy \\ &\quad + b\int_{y_1}^{y_2}\left(\frac{n}{n+1}\right)\left(\frac{\Delta P}{KL}\right)^{\frac{1}{n}}\left(\frac{d-\delta}{2}\right)^{\frac{n+1}{n}} dy \\ &= bn\left(\frac{\Delta P}{KL}\right)^{\frac{1}{n}}\left(\frac{d-\delta}{2}\right)^{\frac{n+1}{n}} \frac{n(d+\delta)+d}{(2n+1)(n+1)} \end{aligned} \tag{7.122}$$

From which the mean flow velocity is derived as

$$u_m = \frac{n}{d}\left(\frac{\Delta P}{KL}\right)^{\frac{1}{n}}\left(\frac{d-\delta}{2}\right)^{\frac{n+1}{n}} \frac{n(d+\delta)+d}{(2n+1)(n+1)} \tag{7.123}$$

Nondimensionalizing the plug thickness with respect to the height of the flow passage

$$u_m = n\left(\frac{\Delta P}{KL}\right)^{\frac{1}{n}}\left(\frac{1-\bar{\delta}}{2}\right)^{\frac{n+1}{n}}\frac{n(1+\bar{\delta})+1}{(2n+1)(n+1)}d^{\frac{n+1}{n}} \tag{7.124}$$

The pressure differential ΔP can be obtained in terms of u_m as

$$\Delta P = \frac{KLd^n u_m^n}{n^n}\left(\frac{2}{d-\delta}\right)^{n+1}\left[\frac{(2n+1)(n+1)}{n(d+\delta)+d}\right]^n \tag{7.125}$$

From this equation, the active damping coefficient is found to be

$$\begin{aligned} c_{eq}^a &= \frac{F}{u_m} = \frac{\Delta Pbd}{u_m} \\ &= u_m^{n-1}\frac{KLbd^{n+1}}{n^n}\left(\frac{2}{d-\delta}\right)^{n+1}\left[\frac{(2n+1)(n+1)}{n(d+\delta)+d}\right]^n \\ &= u_m^{n-1}\frac{KLb}{n^n d^n}\left(\frac{2}{1-\bar{\delta}}\right)^{n+1}\left[\frac{(2n+1)(n+1)}{n(1+\bar{\delta})+1}\right]^n \end{aligned} \tag{7.126}$$

Note that the damping coefficient in this case depends on the mean flow velocity u_m. As a result, the active and inactive damping coefficients can only be compared at constant u_m, which translates to the condition of equal flow rate in both the active and inactive cases. The inactive damping coefficient c_{eq}^o can be easily determined by setting the plug thickness to zero ($\bar{\delta} = 0$) in the previous equation. This leads to

$$c_{eq}^o = u_m^{n-1}\frac{KLb}{n^n d^n}2^{n+1}(2n+1)^n \tag{7.127}$$

The ratio of active and inactive damping coefficients is

$$\frac{c_{eq}^a}{c_{eq}^o} = \frac{1}{(1-\bar{\delta})^{n+1}}\left(\frac{1}{1+\frac{n\bar{\delta}}{n+1}}\right)^n \tag{7.128}$$

7.4.3 Worked Example: Bingham Biplastic Fluid Model

The Bingham plastic model assumes constant post-yield viscosity for all shear strain rates. However, in reality, the post-yield viscosity of ER/MR fluids can vary with shear strain rate, exhibiting shear thinning or shear thickening. The Bingham biplastic model has been proposed to capture this behavior [9]. In this model, as shown in Fig. 7.30, the post-yield behavior is approximated by two regions of different viscosity, one for low shear strain rate and the other for high shear strain rate. The value of the shear strain rate at which the viscosity changes is independent of field strength. Below a shear rate of $\dot{\gamma}_t$, the viscosity is μ_o; above this shear rate, the viscosity is μ_1. For shear thinning, $\mu_1 < \mu_o$, whereas for shear thickening, $\mu_1 > \mu_o$.

$$\begin{aligned} \tau &= \tau_y\, sgn(\dot{\gamma}) + \mu_o\dot{\gamma} \qquad \text{for } 0 < |\dot{\gamma}| < \dot{\gamma}_t \\ &= \left[\tau_y + (\mu_o - \mu_1)\dot{\gamma}_t\right] sgn(\dot{\gamma}) + \mu_1\dot{\gamma} \qquad \text{for } |\dot{\gamma}| > \dot{\gamma}_t \end{aligned} \tag{7.129}$$

Using this model, derive an expression for the active damping coefficient, for a flow mode damper with a rectangular flow passage.

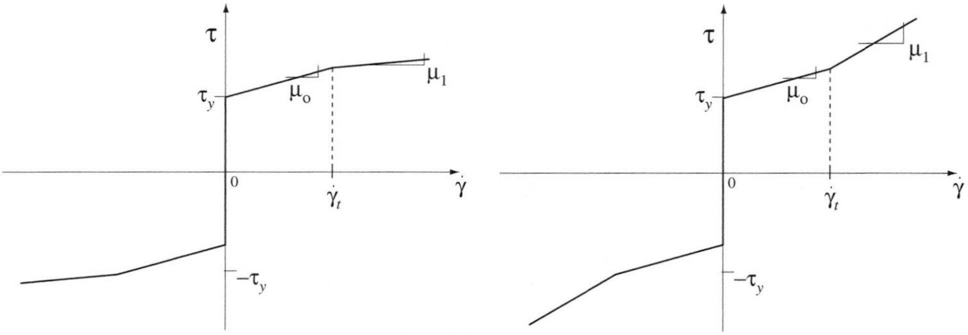

(**a**) Shear thinning (**b**) Shear thickening

Figure 7.30. Shear thinning and shear thickening represented by the Bingham biplastic model.

Solution

Consider the quasi-static flow through a rectangular flow passage of width b, height d, and length L. When the fluid is activated, a plug forms in the center of the flow passage similar to the case of the Bingham-plastic model. However, in the Bingham biplastic model, two distinct viscosities exist in the post-yield region. Therefore, the flow profile must be divided into five regions, compared to three regions in the case of the Bingham plastic model. These five regions are shown in Fig 7.31. Regions 1, 2, 4, and 5 represent the post-yield condition and region 3 represents the pre-yield central plug. The y-axis is assumed to originate from the mid-axis of the gap. Because the flow profile is symmetric, only half of the gap can be considered for analysis.

As shown previously for a rectangular flow passage, the force equilibrium equation is reduced to

$$\frac{\partial \tau}{\partial y} = \frac{\partial P}{\partial x} = -\frac{\Delta P}{L} \tag{7.130}$$

where ΔP is the pressure drop along the length of the passage. Integrating the previous equation leads to

$$\tau = -\frac{\Delta P}{L} y + C_1 \tag{7.131}$$

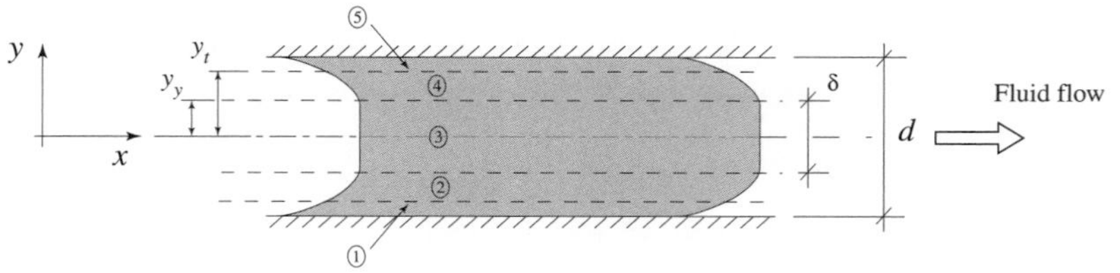

Figure 7.31. Flow profile of the fluid in the flow mode, under an external field (rectangular cross section), Bingham biplastic model.

The boundary conditions are

$$\begin{aligned} &\text{At } y = 0, \quad \tau = 0 \quad \Longrightarrow \quad C_1 = 0 \\ &\text{At } y = y_y, \quad \tau = \tau_y \quad \Longrightarrow \quad y_y = \frac{\delta}{2} = -\frac{\tau_y L}{\Delta P} \end{aligned} \tag{7.132}$$

where δ is the pre-yield plug thickness. In addition, at $y = y_t$, $\tau = \tau_y + \mu_o \dot{\gamma}$. This results in

$$y_t = -\frac{\mu_o \dot{\gamma}_t L}{\Delta P} - \frac{\tau_y L}{\Delta P} \tag{7.133}$$

It is clear that y_t does not depend on the second viscosity μ_1.

Region 5

$u_5(y)$ is the flow velocity in region 5. The shear stress is given by

$$\begin{aligned} \tau(y) &= -\left[\tau_y + (\mu_o - \mu_1)\dot{\gamma}_t\right] + \mu_1 \frac{\partial u_5}{\partial y} \\ &= -\frac{\Delta P}{L} \end{aligned} \tag{7.134}$$

From which the strain rate is

$$\frac{\partial u_5}{\partial y} = \frac{1}{\mu_1}\left[\tau_y + (\mu_o - \mu_1)\dot{\gamma}_t\right] - \frac{1}{\mu_1}\frac{\Delta P}{L} y \tag{7.135}$$

Integrating this expression yields

$$u_5(y) = \frac{1}{\mu_1}\left[\tau_y + (\mu_o - \mu_1)\dot{\gamma}_t\right] y + \frac{1}{\mu_1}\frac{(-\Delta P)}{L}\frac{y^2}{2} + C_1 \tag{7.136}$$

The constant C_1 can be found from the no-slip boundary condition on the upper wall of the flow passage

$$u_5(d/2) = 0 \tag{7.137}$$

Applying this boundary condition and substituting for the constant C_1 yields

$$u_5(y) = \frac{y - d/2}{\mu_1}\left[\tau_y + (\mu_o - \mu_1)\dot{\gamma}_t + (y + d/2)\frac{(-\Delta P)}{2L}\right] \tag{7.138}$$

Region 4

The shear stress in region 4 is given by

$$\begin{aligned} \tau(y) &= -\tau_y + \mu_o \frac{\partial u_y}{\partial y} \\ &= -\frac{\Delta P}{L} y \end{aligned} \tag{7.139}$$

From which the velocity distribution in region 4, $u_4(y)$, can be found as

$$u_4(y) = \frac{\tau_y}{\mu_o} y - \frac{\Delta P}{\mu_o L}\frac{y^2}{2} + C_2 \tag{7.140}$$

Flow continuity between regions 4 and 5 requires that

$$u_4(y_t) = u_5(y_t) \tag{7.141}$$

Applying this flow continuity condition at y_t results in

$$\begin{aligned} u_4(y) = &\frac{y - y_t}{\mu_o}\left[\tau_y + \frac{(-\Delta P)}{2L}(y + y_t)\right] \\ &+ \frac{y_t - d/2}{\mu_1}\left[\tau_y + (\mu_o - \mu_1)\dot{\gamma}_t + (y + d/2)\frac{(-\Delta P)}{2L}\right] \end{aligned} \tag{7.142}$$

Substituting for y_t from Eq. 7.133

$$\begin{aligned} u_4(y) = &\frac{1}{\mu_o}\left[\tau_y + \frac{(-\Delta P)}{2L}y\right]y \\ &+ \frac{1}{2\mu_1 L}\left\{\frac{\Delta P}{4}d^2 - dL\left[\tau_y + (\mu_o - \mu_1)\dot{\gamma}_t\right]\right. \\ &\left. + \frac{L^2}{\Delta P}\left(\tau_y + \dot{\gamma}_t\mu_o\right)^2\left(1 - \frac{\mu_1}{\mu_o}\right)\right\} \end{aligned} \tag{7.143}$$

Plug Region 3

The flow velocity $u_3(y)$ is uniform in this region and is equal to the pre-yield plug velocity u_p. From flow continuity at the interface of regions 3 and 4

$$u_3(y) = u_4(y_y) \tag{7.144}$$

where

$$y_y = -\frac{\delta}{2} = -\frac{\tau_y L}{\Delta P} \tag{7.145}$$

Substituting in Eq. 7.143 results in

$$\begin{aligned} u_3(y) = &\frac{\tau_y^2 L}{2\mu_o \Delta P} + \frac{1}{2\mu_1 L}\left\{\frac{\Delta P}{4}d^2 - dL\left[\tau_y + (\mu_o - \mu_1)\dot{\gamma}_t\right]\right. \\ &\left. + \frac{L^2}{\Delta P}\left(\tau_y + \dot{\gamma}_t\mu_o\right)^2\left(1 - \frac{\mu_1}{\mu_o}\right)\right\} \end{aligned} \tag{7.146}$$

Let us define the low shear strain rate plug thickness as $\delta_t = 2y_t$. Introducing the nondimensional plug thicknesses $\bar{\delta} = \delta/d$ and $\bar{\delta}_t = \delta_t/d$ and a nondimensional height $\bar{y} = y/d$, we obtain

$$\begin{aligned} u_1(y) &= \frac{\Delta P d^2}{8L}\left[\frac{(1 - \bar{\delta}_t)^2}{\mu_1} + \frac{(\bar{\delta}_t - \bar{\delta}_y)(2 - \bar{\delta}_t - \bar{\delta}_y)}{\mu_o}\right] \\ u_2(y) &= \frac{\Delta P d^2}{8L}\left[\frac{(1 - \bar{\delta}_t)^2}{\mu_1} - \frac{4\bar{y}(\bar{y} - \bar{\delta}_y) + 2(\bar{\delta}_t - \bar{\delta}_y) + \bar{\delta}_t^2}{\mu_o}\right] \\ u_3(y) &= \frac{\Delta P d^2}{8L}\left[\frac{(1 - 2\bar{y})(2\bar{y} + 1 - 2\bar{\delta}_t)}{\mu_1} + \frac{2(1 - 2\bar{y})(\bar{\delta}_t - \bar{\delta}_y)}{\mu_o}\right] \end{aligned} \tag{7.147}$$

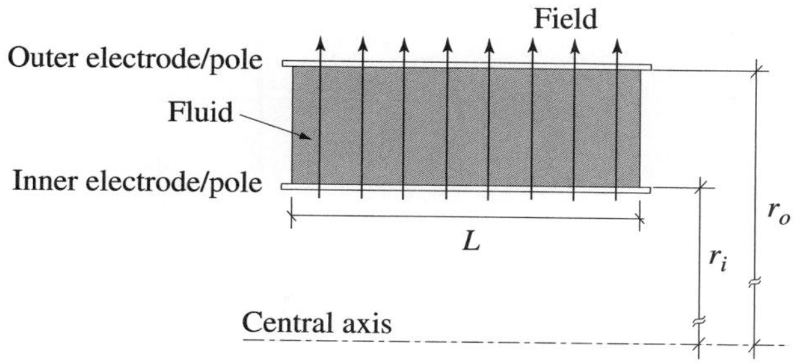

Figure 7.32. ER/MR fluid in annular flow passage.

The total volumetric flow rate is given by

$$
\begin{aligned}
Q &= Q_4 + 2Q_5 + 2Q_6 \\
&= 2b\int_0^{y_y} u_4(y)\,dy + 2b\int_{y_y}^{y_t} u_5(y)\,dy + 2b\int_{y_t}^{d} u_6(y)\,dy \\
&= \frac{\Delta P b d^3}{12\mu_o L}\left[(1-\bar{\delta}_y)^2\left(1+\frac{\bar{\delta}_y}{2}\right) - (1-\bar{\delta}_t)^2\left(1+\frac{\bar{\delta}_t}{2}\right)\left(1-\frac{\mu_o}{\mu_1}\right)\right]
\end{aligned}
\tag{7.148}
$$

Assuming the fluid is being forced through the flow passage by a piston of area A_p at a mean velocity u_m, the flow rate is given by $Q = A_p u_m$. The force on the piston is $F_p = \Delta P A_p$. Therefore, the active damping coefficient c_{eq}^a is given by

$$c_{eq}^a = \frac{12\mu_o L A_p^2}{bd^3}\,\frac{1}{(1-\bar{\delta}_y)^2(1+\bar{\delta}_y/2) - (1-\bar{\delta}_t)^2(1+\bar{\delta}_t/2)(1-\mu_o/\mu_1)} \tag{7.149}$$

Expressing this equation in terms of the inactive Newtonian damping coefficient c_{eq}^o (Eq. 7.57) yields the increase in damping due to the application of the field

$$\frac{c_{eq}^a}{c_{eq}^o} = \frac{1}{(1-\bar{\delta}_y)^2(1+\bar{\delta}_y/2) - (1-\bar{\delta}_t)^2(1+\bar{\delta}_t/2)(1-\mu_o/\mu_1)} \tag{7.150}$$

7.4.4 Annular Flow Passage

Let us consider a damper with an annular active region. The annulus is formed by the gap between two concentric cylinders that form two electrodes, or magnetic poles. The annular gap is filled with the controllable fluid, and the applied field acts over an axial length L. The radius of the inner cylinder is r_i and the radius of the outer cylinder is r_o. Typically, the width of the annular gap, $d = r_o - r_i$, is small compared to the radius of the inner cylinder. In such a case, the electric or magnetic field can be assumed to be uniform across the gap, which considerably simplifies the analysis. A diagram of this configuration is shown in Fig. 7.32.

The governing equation for the fluid in the annulus can be derived by considering force equilibrium on an annular fluid element, as shown in Fig. 7.33. The mass of the fluid element dm is given by

$$dm = 2\pi r\,dr\,dx\,\rho \tag{7.151}$$

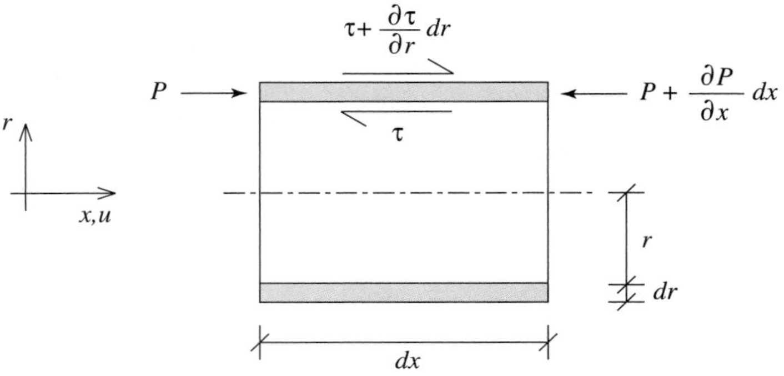

Figure 7.33. Differential fluid element, annular flow passage.

The force balance on the element can be written as

$$- dm\frac{\partial u}{\partial t} + 2\pi r\, dr\, P + \left(\tau + \frac{\partial \tau}{\partial r}dr\right) 2\pi(r + dr)\, dx$$

$$- \tau 2\pi r\, dx - \left(P + \frac{\partial P}{\partial x}dx\right) 2\pi r\, dr = 0 \quad (7.152)$$

Substituting for the elemental mass from Eq. 7.151, the governing equation can be derived as

$$-\rho\frac{\partial u}{\partial t} + \frac{\tau}{r} + \frac{\partial \tau}{\partial r} - \frac{\partial P}{\partial x} = 0 \qquad (7.153)$$

Assuming steady incompressible flow, $\partial u/\partial t = 0$. The governing equation reduces to

$$\frac{\partial \tau}{\partial r} + \frac{\tau}{r} = \frac{\partial P}{\partial x} \qquad (7.154)$$

The pressure gradient $\partial P/\partial x$ is defined in the same way as for the parallel plate case, by Eq. 7.42. It is important to note that the electric field across the annular gap is not uniform because of the curvature of the surfaces. However, if the gap is assumed small compared to the radius of curvature, it can be assumed that the electric field is uniform across the gap. We can now examine the shear mode and the flow mode cases separately.

Shear Mode

A shear mode damper can be constructed using the annular geometry by holding the outer cylinder fixed and moving the inner cylinder in the axial direction. Let us assume that a force F_o is acting on the inner cylinder, which is moving with a velocity u_o (Fig. 7.34). A one-dimensional axisymmetric model is sufficient for this analysis. Again, the pressure gradient is zero and the governing equation becomes

$$\frac{\tau}{r} + \frac{\partial \tau}{\partial r} = 0 \qquad (7.155)$$

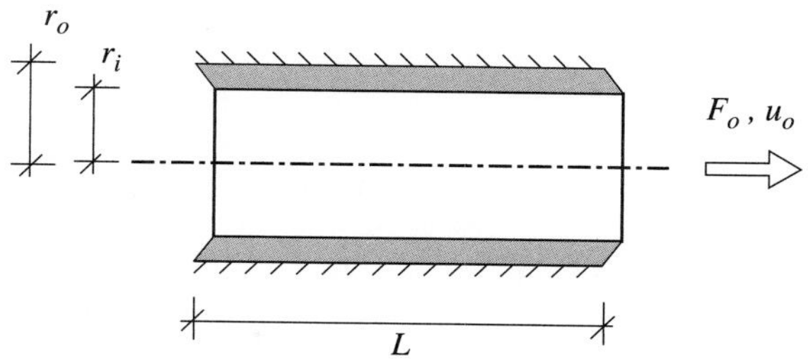

Figure 7.34. Damper with annular gap operating in shear mode.

(a) Solution under zero applied field

When no field is applied, the fluid behaves like a Newtonian fluid. The shear stress is then given by Eq. 7.1, with the radial variable r replacing the Cartesian variable y.

$$\tau(r) = \mu \frac{\partial u}{\partial r} \tag{7.156}$$

Substituting this into the governing equation (Eq. 7.154), with the pressure gradient for shear mode, gives

$$\mu \frac{\partial^2 u}{\partial r^2} + \frac{\mu}{r} \frac{\partial u}{\partial r} = 0 \tag{7.157}$$

This can be rewritten as

$$\frac{\partial}{\partial r}\left(r \frac{\partial u}{\partial r}\right) = 0 \tag{7.158}$$

Integrating this equation twice with respect to r yields

$$u(r) = A \ln r + B \tag{7.159}$$

The constants A and B are determined from the boundary conditions

$$\begin{aligned} u(r_i) = u_o &\implies A = \frac{u_o}{\ln(r_i/r_o)} \\ u(r_o) = 0 &\implies B = -\frac{u_o \ln r_o}{\ln(r_i/r_o)} \end{aligned} \tag{7.160}$$

Therefore, the velocity profile across the gap is given by

$$u(r) = \frac{u_o}{\ln(r_i/r_o)} \ln(r/r_o) \tag{7.161}$$

and the shear stress in the fluid is

$$\tau(r) = -\frac{\mu u_o}{r \ln(r_o/r_i)} \tag{7.162}$$

Note that the shear strain rate, $\partial u/\partial r$, is negative in the annular gap because the inner cylinder is moved while the outer cylinder remains at rest. As a result, the shear stress has a negative sign, which can be ignored when calculating the damping force. Following the same procedure as in Section 7.4.1, the damping coefficient is found by the ratio of the force and velocity. The force is given by (dropping the negative sign)

$$\begin{aligned} F_o &= 2\pi r_i \, L \, \tau(r_i) \\ &= \frac{2\pi L \mu u_o}{\ln(r_o/r_i)} \end{aligned} \tag{7.163}$$

The damping force is

$$F_o = c_{eq}^o \, u_o \tag{7.164}$$

where c_{eq}^o is the damping coefficient in the inactive state. Equating these two expressions for F_o, the damping coefficient is obtained as

$$c_{eq}^o = \frac{2\pi L \mu}{\ln(r_o/r_i)} = \Gamma \mu \tag{7.165}$$

where, similar to the case of the rectangular flow duct, Γ is a parameter that depends only on the geometry of the device

$$\Gamma = \frac{2\pi L}{\ln(r_o/r_i)} \tag{7.166}$$

(b) Solution under non-zero applied field

Once the activation field is applied, the fluid is modeled as a Bingham plastic. In the pre-yield region, the fluid is modeled as a Newtonian fluid as described previously. In the post-yield region, the shear stress is given by (Eq. 7.9)

$$\begin{aligned} \tau &= \tau_y \, sgn(\dot{\gamma}) + \mu \dot{\gamma} \\ &= -\tau_y + \mu \frac{\partial u}{\partial r} \end{aligned} \tag{7.167}$$

Note that a negative sign appears because the shear stress in the annular gap is negative. Substituting this in the governing equation (Eq. 7.154)

$$\mu \frac{\partial^2 u}{\partial r^2} + \frac{\mu}{r} \frac{\partial u}{\partial r} = \frac{\tau_y}{r} \tag{7.168}$$

which can be rewritten as

$$\frac{\partial}{\partial r} \left(r \frac{\partial u}{\partial r} \right) = \frac{\tau_y}{\mu} \tag{7.169}$$

Integrating this twice leads to

$$u(r) = \frac{\tau_y}{\mu} r + C \ln r + D \tag{7.170}$$

Applying the same boundary conditions as before (Eq. 7.160), the constants are obtained as

$$\begin{aligned} C &= -\frac{1}{\ln(r_o/r_i)} \left[u_o + \frac{\tau_y}{\mu} (r_o - r_i) \right] \\ D &= \frac{u_o}{\ln(r_o/r_i)} + \frac{\tau_y}{\mu} \left[-r_o + \frac{r_o - r_i}{\ln(r_o/r_i)} \right] \end{aligned} \tag{7.171}$$

The fluid velocity across the gap is

$$u(r) = -\frac{\tau_y}{\mu}(r_o - r) + \frac{u_o + \tau_y/\mu(r_o - r_i)}{\ln(r_o/r_i)} \ln(r_o/r) \tag{7.172}$$

and the shear stress is given by

$$\begin{aligned} \tau(r) &= -\frac{\mu}{r\ln(r_o/r_i)}\left[u_o + \frac{\tau_y}{\mu}(r_o - r_i)\right] \\ &= -\frac{\mu u_o}{r\ln(r_o/r_i)}\left[1 + \frac{\tau_y d}{\mu u_o}\right] \\ &= -\frac{\mu u_o}{r\ln(r_o/r_i)}(1 + \mathsf{Bi}) \end{aligned} \tag{7.173}$$

where the Bingham number, Bi, is defined as before (Eq. 7.39). The force on the inner cylinder is (ignoring the negative sign)

$$\begin{aligned} F_o &= 2\pi r_i L\tau(r_i) \\ &= \mu\,(1 + \mathsf{Bi})\,\Gamma u_o \end{aligned} \tag{7.174}$$

from which the equivalent damping coefficient in the active state is found to be

$$\begin{aligned} c^a_{eq} &= \mu\,(1 + \mathsf{Bi})\,\Gamma \\ &= c^o_{eq}\,(1 + \mathsf{Bi}) \end{aligned} \tag{7.175}$$

Flow Mode

In the flow mode, a pressure differential forces fluid through the gap between the two cylinders. The configuration is similar to that of a parallel plate (see Section 7.4.1), except that in this case the electrodes (or magnetic poles) are cylindrical.

The governing equation is given by Eq. 7.154, with the pressure gradient being the same as in the case of the rectangular flow passage (Eq. 7.42)

$$\frac{\partial \tau}{\partial r} + \frac{\tau}{r} = -\frac{\Delta P}{L} \tag{7.176}$$

When no field is applied, the fluid behavior is Newtonian. Substituting for the shear stress (Eq. 7.156) yields the governing equation for fluid flow through the annulus

$$\mu\frac{\partial^2 u}{\partial r^2} + \frac{\mu}{r}\frac{\partial u}{\partial r} = -\frac{\Delta P}{L} \tag{7.177}$$

This can be rewritten as

$$\frac{\partial}{\partial r}\left(r\frac{\partial u}{\partial r}\right) = -\frac{\Delta P}{\mu L}r \tag{7.178}$$

Integrating twice yields

$$u(r) = -\frac{\Delta P}{\mu L}\frac{r^2}{4} + C_1 \ln r + C_2 \tag{7.179}$$

The constants are determined from the boundary conditions, which in this case are the no-slip conditions at the wall

$$\begin{aligned} u(r_i) &= 0 \\ u(r_o) &= 0 \end{aligned} \tag{7.180}$$

from which the constants can be derived as

$$C_1 = \frac{\Delta P}{4\mu L} \frac{\left(r_i^2 - r_o^2\right)}{\ln(r_i/r_o)}$$
$$C_2 = \frac{\Delta P}{4\mu L} r_i^2 - \frac{\Delta P}{4\mu L} \frac{\left(r_i^2 - r_o^2\right)}{\ln(r_i/r_o)} \ln r_i \tag{7.181}$$

Substituting the two constants, the velocity profile in the annulus becomes

$$u(r) = \frac{\Delta P}{4\mu L} \left[-r^2 + r_i^2 \frac{\ln(r_o/r)}{\ln(r_o/r_i)} + r_o^2 \frac{\ln(r/r_i)}{\ln(r_o/r_i)} \right] \tag{7.182}$$

This represents a paraboloid enclosed in the annulus. The maximum velocity is no longer in the center of the gap; however, as the gap thickness becomes small in comparison to the inner radius, the flow profile becomes more symmetric with respect to the center of the gap, and the flow profile can be approximated by that of a rectangular cross-section flow passage. To find an equivalent constant flow velocity u_m, the volumetric flow through the annulus, Q, is calculated

$$\begin{aligned} Q &= \int_{r_i}^{r_o} u(r)\, 2\pi\, r\, dr \\ &= 2\pi \frac{\Delta P}{4\mu L} \int_{r_i}^{r_o} \left(r_i^2 r - r^3 + \frac{\left(r_i^2 - r_o^2\right)}{\ln(r_i/r_o)} r \ln(r/r_i) \right) dr \\ &= \frac{\Delta P \pi}{8\mu L} \left[r_o^4 - r_i^4 + \frac{\left(r_i^2 - r_o^2\right)^2}{\ln(r_i/r_o)} \right] \end{aligned} \tag{7.183}$$

This volumetric flow is equated to that resulting from the equivalent constant flow

$$Q = u_m\, \pi\, \left(r_o^2 - r_i^2\right) \tag{7.184}$$

Following the same procedure as in the case of the rectangular flow passage (see Section 7.4.1), the effective damping coefficient, c_{eq}^o can be derived

$$\begin{aligned} u_m &= \frac{\Delta P}{8\mu L} \left[r_o^2 + r_i^2 + \frac{r_i^2 - r_o^2}{\ln(r_o/r_i)} \right] \\ &= \frac{F}{\pi\left(r_o^2 - r_i^2\right) 8\mu L} \left[r_o^2 + r_i^2 + \frac{r_i^2 - r_o^2}{\ln(r_o/r_i)} \right] \\ &= \frac{F}{c_{eq}^o} \end{aligned} \tag{7.185}$$

$$c_{eq}^o = \frac{8\pi\mu L}{\frac{r_o^2 + r_i^2}{r_o^2 - r_i^2} + \frac{1}{\ln(r_o/r_i)}} \tag{7.186}$$

When a field is applied and the fluid is treated as a Bingham plastic, the flow through the annulus can be divided into three regions, similar to the case of the rectangular cross-section flow passage. In this case, an annular plug forms in the flow passage. Due to the geometry of the flow passage, the flow profile is not symmetric across the gap as in the case of the rectangular cross-section flow passage. The procedure for finding the velocity profile, shear stresses, and equivalent damping is the same as that outlined in the case of the rectangular cross-section flow passage (see Section. 7.4.1).

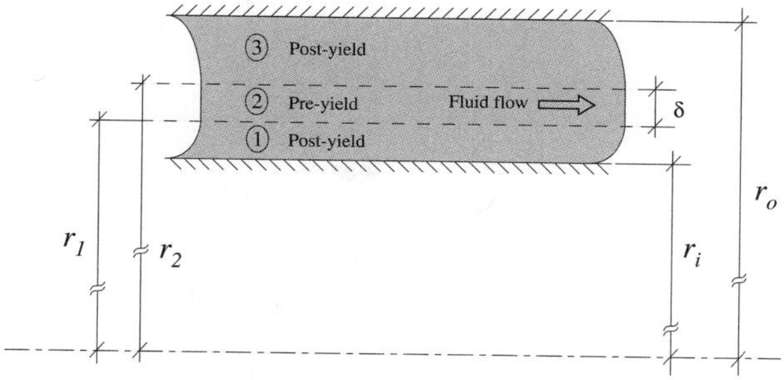

Figure 7.35. Flow profile in the annulus, with an external field applied.

The flow is divided into three regions, as shown in Fig. 7.35. The flow in each region is treated separately, as follows:

Region 1

The governing equation in regions 1 and 3 (post-yield) is given by (Eq. 7.176)

$$\frac{\partial \tau}{\partial r} + \frac{\tau}{r} = -\frac{\Delta P}{L} \tag{7.187}$$

Using the Bingham plastic model, the shear stress in the post-yield region is given by (Eq. 7.9)

$$\tau(r) = \mu \frac{\partial u}{\partial r} + \tau_y \tag{7.188}$$

Substituting this equation in the governing equation (Eq. 7.188), we obtain

$$\mu \frac{\partial^2 u}{\partial r^2} + \frac{\mu}{r}\frac{\partial u}{\partial r} + \frac{\tau_y}{r} = -\frac{\Delta P}{L} \tag{7.189}$$

This can be rewritten as

$$\frac{\partial}{\partial r}\left(r \frac{\partial u}{\partial r}\right) = -\frac{\Delta P}{\mu L} r - \frac{\tau_y}{\mu} \tag{7.190}$$

Integrating the previous equation twice leads to

$$u_1(r) = -\frac{\Delta P}{4\mu L} r^2 - \frac{\tau_y}{\mu} r + C_3 \ln r + C_4 \tag{7.191}$$

The constants are determined from the boundary conditions

$$u(r_i) = 0 \tag{7.192}$$

$$\frac{\partial u}{\partial r}(r_1) = 0 \tag{7.193}$$

Substituting these boundary conditions leads to

$$-\frac{\Delta P}{4\mu L} r_i^2 - \frac{\tau_y}{\mu} r_i + C_3 \ln r_i + C_4 = 0 \tag{7.194}$$

$$-\frac{\Delta P r_1^2}{2L} - \tau_y r_1 + \mu C_3 = 0 \tag{7.195}$$

from which we obtain the constants

$$C_3 = \frac{\Delta P}{2\mu L} r_1^2 + \frac{\tau_y}{\mu} r_1 \tag{7.196}$$

$$C_4 = \frac{\Delta P}{4\mu L} r_i^2 + \frac{\tau_y}{\mu} r_i - \left(\frac{\Delta P r_1^2}{2\mu L} + \frac{\tau_y r_1}{\mu} \right) \ln r_i \tag{7.197}$$

Substituting the constants, the velocity profile in region 1 is given by

$$u_1(r) = \frac{\Delta P}{4\mu L} \left[-r^2 + r_i^2 + 2r_1^2 \ln(r/r_i) \right] + \frac{\tau_y}{\mu} \left[-r + r_i + r_1 \ln(r/r_i) \right] \tag{7.198}$$

Region 3

The governing equation in this region is the same as in the case of region 1. Integrating the governing equation yields an expression for flow velocity

$$u_3(r) = -\frac{\Delta P}{4\mu L} r^2 - \frac{\tau_y}{\mu} r + C_5 \ln r + C_6 \tag{7.199}$$

The constants are determined from the boundary conditions

$$u_3(r_o) = 0 \tag{7.200}$$

$$\left. \frac{\partial u_3}{\partial r} \right|_{r=r_2} = 0 \tag{7.201}$$

Substituting these boundary conditions leads to

$$-\frac{\Delta P}{4\mu L} r_o^2 - \frac{\tau_y}{\mu} r_o + C_5 \ln r_o + C_6 = 0 \tag{7.202}$$

$$-\frac{\Delta P r_2^2}{2L} - \tau_y r_2 + \mu C_5 = 0 \tag{7.203}$$

from which we obtain the constants

$$C_5 = \frac{\Delta P}{2\mu L} r_2^2 + \frac{\tau_y}{\mu} r_2 \tag{7.204}$$

$$C_6 = \frac{\Delta P}{4\mu L} r_o^2 + \frac{\tau_y}{\mu} r_o - \left(\frac{\Delta P r_2^2}{2\mu L} + \frac{\tau_y r_2}{\mu} \right) \ln r_o \tag{7.205}$$

Substituting the constants, the velocity profile in region 3 is given by

$$u_3(r) = \frac{\Delta P}{4\mu L} \left[-r^2 + r_o^2 + 2r_2^2 \ln(r/r_o) \right] + \frac{\tau_y}{\mu} \left[-r + r_o + r_2 \ln(r/r_o) \right] \tag{7.206}$$

Region 2

The fluid in region 2 has a constant flow velocity, u_p. The plug thickness δ can be found from the shear stress conditions at the boundaries of the plug. The governing equation for shear stress in region 2 is (Eq. 7.176)

$$\frac{\partial \tau}{\partial r} + \frac{\tau}{r} = -\frac{\Delta P}{L} \tag{7.207}$$

This can be rewritten as

$$\frac{\partial}{\partial r}(\tau r) = -\frac{\Delta P}{L} r \tag{7.208}$$

Integrating this equation leads to an expression for the shear stresses as a function of radial position

$$\tau(r) = -\frac{\Delta P}{2L} r + \frac{C_o}{r} \tag{7.209}$$

The boundary conditions for the shear stress are given by the yield condition at the edges of the plug

$$\tau(r_1) = \tau_y \tag{7.210}$$

$$\tau(r_2) = -\tau_y \tag{7.211}$$

This leads to

$$\tau(r_1) = -\frac{\Delta P}{2L} r_1 + \frac{C_o}{r_1} = \tau_y \tag{7.212}$$

$$\tau(r_2) = -\frac{\Delta P}{2L} r_2 + \frac{C_o}{r_2} = -\tau_y \tag{7.213}$$

from which the constant C_o and the plug thickness δ can be found

$$C_o = \frac{\Delta P}{2L} r_1 r_2 \tag{7.214}$$

$$\tau_y = \frac{\Delta P}{2L} \delta \tag{7.215}$$

Note that the expression for the plug thickness is the same as in the case of the rectangular cross-section flow passage. However, unlike the case of the rectangular cross-section flow passage, the flow profile in the annular gap is not symmetric. The plug velocity u_p and the location of the plug (r_1 and r_2) must be found by equating the flow velocities at the boundaries of the post-yield and pre-yield regions

$$\begin{aligned} u_p = u_1(r_1) = {} & \frac{\Delta P}{4\mu L}\left[r_i^2 - r_1^2 + 2r_1^2 \ln(r_1/r_i)\right] \\ & + \frac{\tau_y}{\mu}\left[-r_1 + r_i + r_1 \ln(r_1/r_i)\right] \end{aligned} \tag{7.216}$$

$$\begin{aligned} u_p = u_3(r_2) = {} & \frac{\Delta P}{4\mu L}\left[-r_2^2 + r_o^2 + 2r_2^2 \ln(r_2/r_o)\right] \\ & + \frac{\tau_y}{\mu}\left[-r_2 + r_o + r_2 \ln(r_2/r_o)\right] \end{aligned} \tag{7.217}$$

The equivalent damping can be found by calculating the total volumetric flux Q through the annulus and finding an equivalent constant velocity

$$\begin{aligned} Q &= \int_{r=r_i}^{r=r_o} 2\pi\, r\, u(r) dr \\ &= 2\pi \int_{r=r_i}^{r=r_1} u_1(r) r\, dr + 2\pi \int_{r=r_1}^{r=r_2} u_2(r) r\, dr + 2\pi \int_{r=r_2}^{r=r_o} u_3(r) r\, dr \end{aligned} \tag{7.218}$$

It can be seen that the expressions become tedious to manipulate. Because the annular gap d is much smaller than the inner radius r_i, the annular flow passage can

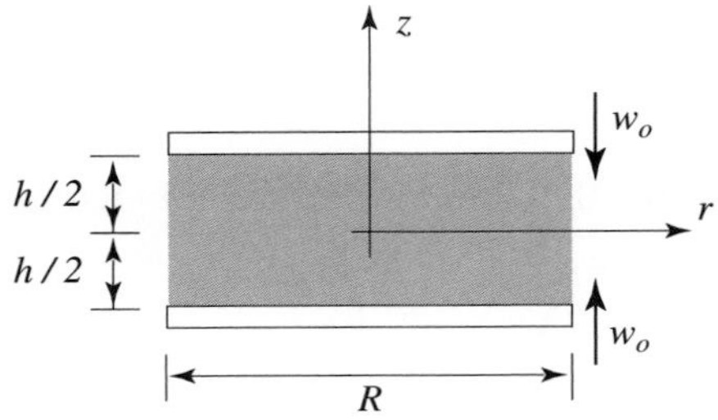

Figure 7.36. Geometry of fluid in squeeze mode.

be approximated as a rectangular flow passage between two parallel conductors. The width b of the equivalent rectangular passage can be defined in terms of the mean circumference of the annulus as

$$b = 2\pi \left(R + \frac{d}{2} \right) \tag{7.219}$$

As a result of this approximation, the flow profile between the conductors becomes symmetric and a simplified analysis, as described in Section 7.4.1, can be applied. Note that this assumption is sufficiently accurate only if $d/r_i \ll 1$. A detailed analysis of the errors introduced by such an approximation is described by Atkin et al. [49] and Yoo and Wereley [50].

7.4.5 Squeeze Mode

The electric or magnetic field is applied across a narrow gap in which the fluid is situated, and the field strength is assumed constant across the gap. In the squeeze mode, the motion of a channel wall is in the normal direction, and the fluid is forced to flow along the channel. The volume of the channel is reduced due to the motion of the channel wall, and the fluid is subjected to shear due to its motion in the radial direction. A discussion of the modeling of the fluid in such a mode is provided by Nilsson et al. [51]. At a high field strength, particle chains are formed between the walls, which try to prevent radial flow. In a pure shear flow mode, there is no net change of flow and the volume is constant. Thus, the stiffness in squeeze mode is expected to be an order of magnitude larger than that in a pure shear mode.

Consider two circular plates of radius R, arranged symmetrically at a distance of $h/2$ with respect to a mid-plane, as shown in Figure 7.36. The plates are moving at a velocity w_o toward each other. The volume of fluid displaced in time Δt is given by

$$\Delta V = 2\pi r^2 w_o \Delta t \tag{7.220}$$

Assuming the fluid is incompressible, the radial fluid velocity can be estimated from the continuity equation

$$\Delta V = 2\pi r h \Delta t u(r) \tag{7.221}$$

where $u(r)$ is the radial velocity of the fluid. From these two equations, we obtain

$$u(r) = \frac{w_o}{h} r \tag{7.222}$$

Applying Bernoulli's equation (energy conservation)

$$p(r) + \frac{1}{2}\rho [u(r)]^2 = p(R) + \frac{1}{2}\rho [u(R)]^2 \tag{7.223}$$

where $p(r)$ is the pressure in the fluid. Assuming a free outlet at the edge of the plates, $p(R) = 0$. This yields

$$p(r) = \frac{1}{2}\rho \frac{w_o^2}{h^2}(R^2 - r^2) \tag{7.224}$$

Neglecting inertial effects, the fluid equilibrium equations become

$$-\frac{\partial p}{\partial r} + \mu \left[\frac{\partial^2 u}{\partial r^2} + \frac{1}{r}\frac{\partial u}{\partial r} + \frac{u}{r^2} + \frac{\partial^2 u}{\partial z^2} \right] = 0 \tag{7.225}$$

$$-\frac{\partial p}{\partial z} + \mu \left[\frac{\partial^2 w}{\partial r^2} + \frac{1}{r}\frac{\partial w}{\partial r} + \frac{w}{r^2} + \frac{\partial^2 w}{\partial z^2} \right] = 0 \tag{7.226}$$

Assuming that the fluid is Newtonian, the continuity equation can be written as

$$\frac{1}{r}\frac{\partial}{\partial r}(ru) + \frac{\partial w}{\partial z} = 0 \tag{7.227}$$

The boundary conditions in this case are

$$\text{At } z = \pm h/2, \quad u = 0, \quad w = \pm w_o \tag{7.228}$$

$$\text{At } r = R, \quad p = 0 \tag{7.229}$$

The solution for velocities and pressures is obtained as

$$u = u_o \frac{4r}{h^2 R}\left(\frac{h^2}{4} - z^2\right) \tag{7.230}$$

$$w = -w_o \left(\frac{3z}{h} - \frac{4z^3}{h^3}\right) \tag{7.231}$$

$$p = \frac{p_o}{R^2}(R^2 - r^2) \tag{7.232}$$

$$= \frac{w_o}{h^2}(R^2 - r^2)\left(\frac{1}{2}\rho w_o + 6\frac{\mu}{h}\right) \tag{7.233}$$

where

$$u_o = w_o \frac{3R}{2h} \tag{7.234}$$

The total force from a single-sided squeeze mode cell is

$$F_z = \frac{\pi R^2 w_o}{h^2}\left(\frac{\rho w_o}{4} + \frac{3\mu}{h}\right) \tag{7.235}$$

The first term dominates when the density ρ and velocity w_o are large and the gap is moderate to large. The force increases rapidly with increasing radius and decreasing gap.

7.5 Summary and Conclusions

ER/MR fluids are a special class of fluids that dramatically change their rheological characteristics on the application of electric/magnetic field, with response times on the order of milliseconds. When there is no field, the suspended particles are randomly distributed in the nonconducting fluid; in the presence of a field, they form chains. The change of rheological property (viscosity) is used in a range of applications such as controllable dampers, shock absorbers, valves, brakes, prosthetic devices, and engine mounts. Even though both of these smart fluids were discovered about the same time (1947–1948), most of the early applications were focused on ER fluids because of their ready availability and ease in implementation. In the 1980s, MR fluids became available commercially (by Lord Corporation); their applications have grown rapidly and surpassed those of ER fluids since then. A characteristic of smart fluids is their yield stress, which is an order of magnitude higher for MR fluids than ER fluids (i.e., 50–100 kPa for MR and 2–5 kPa for ER). The MR devices are operated off a low-voltage power supply (2–25 V and 1–2 A), whereas the ER devices require a high-voltage power supply (2–10 kV and 1–10 mA). Conversely, MR fluids are heavier than ER fluids and also require a complex magnetic field generator consisting of electrical conducting coils and magnetic flux paths.

Two types of models are used to characterize ER/MR dampers: first–principle-based models and phenomenology-based models. The first-category models are based on fundamental fluid mechanics principles. These are complex and less amenable to incorporate in engineering analyses. In the second category of models, one of the three basic flow mechanisms is often used: shear mode (Couette flow), flow mode (Poiseulle flow), or squeeze mode. One of the widely adopted rheological idealizations of fluid is the Bingham-plastic model. Other refined representations include the Herschel-Bulkley model, the biviscous model, the extended Bingham model, and the Bing-Max model. Most of these phenomenological-based engineering models fit experimental data of the bulk fluid into simple representations. One of the major drawbacks of these ER/MR damper models is the widely adopted quasi-steady approximation, in which the dynamic effects are neglected.

It is important to include dynamic effects in these models and examine their performance for a range of operating conditions. For MR dampers, another challenge is to design a compact and lightweight magnetic field generator that includes coil and magnetic flux paths. Magnetic particles are susceptible to sedimentation with time and it is important to optimize the size of suspended particles so that there is no possibility of sedimentation over a long period. For aerospace applications, the challenge is to develop lightweight, compact, highly effective adaptive dampers that can replace expensive, fixed-damping elastomeric dampers.

BIBLIOGRAPHY

[1] W. M. Winslow. Method and means for translating electrical impulses into mechanical force. U.S. Patent 2,417,850, 1947.

[2] W. M. Winslow. Induced fibration of suspensions. *Journal of Applied Physics*, 20:1137–1140, 1949.

[3] W. M. Winslow. Field responsive fluid couplings. U.S. Patent 2,886,151, 1959.

[4] J. Rabinow. The magnetic fluid clutch. *AIEE Transactions*, 67:1308–1315, 1948.

[5] J. Rabinow. Magnetic fluid clutch. *National Bureau of Standards Technical News Bulletin*, 32(4):54–60, 1948.

[6] J. Rabinow. Magnetic fluid torque and force transmitting device. U.S. Patent 2,575,360, 1951.

[7] K. Shimada, H. Nishida, and T. Fujita. Differences in steady charactersitics and response time of ERF on rotational flow between rotating disk and concentric cylinder. *International Journal of Modern Physics B*, 15(6–7):1050–1056, 2001.

[8] K. D. Weiss, J. D. Carlson, and D. A. Nixon. Viscoelastic properties of magneto- and electro-rheological fluids. *Journal of Intelligent Material Systems and Structures*, 5:772–775, 1994.

[9] G. A. Dimock, J.-H. Yoo, and N. M. Wereley. Quasi-steady Bingham biplastic analysis of electrorheological and magnetorheological dampers. *Journal of Intelligent Material Systems and Structures*, 13(9):549–559, 2002.

[10] P. Poddar, J. L. Wilson, H. Srikanth, J.-H. Yoo, N. M. Wereley, S. Kotha, L. Barghouty, and R. Radhakrishnan. Nanocomposite magneto-rheological fluids with uniformly dispersed Fe nanoparticles. *Journal of Nanoscience and Nanotechnology*, 4(1/2):192–196, 2004.

[11] Y. Choi, A. F. Sprecher, and H. Conrad. Vibration characteristics of a composite beam containing an electrorheological fluid. *Journal of Intelligent Material Systems and Structures*, 1:91–104, 1990.

[12] D. R. Gamota and F. E. Filisko. Dynamic mechanical studies of electrorheological materials: Moderate frequencies. *Journal of Rheology*, 35:399–425, 1991.

[13] M. R. Jolly, J. D. Carlson, and B. C. Munoz. A model of the behavior of magnetorheological materials. *Smart Materials and Structures*, 5:607–614, 1996.

[14] H. Block and J. P. Kelly. Electro-rheology. *Journal of Physics D: Applied Physics*, 21: 1661–1667, 1988.

[15] D. W. Felt, M. Hagenbuchle, J. Liu, and J. Richard. Rheology of a magnetorheological fluid. *Journal of Intelligent Material Systems and Structures*, 7(5):589–593, 1996.

[16] G. M. Kamath, M. K. Hurt, and N. M. Wereley. Analysis and testing of Bingham plastic behavior in semi-active electrorheological fluid dampers. *Smart Materials and Structures*, 5:576–590, 1996.

[17] J. A. Powell. Modelling the oscillatory response of an electrorheological fluid. *Smart Materials and Structures*, 3:416–438, 1994.

[18] H. P. Gavin, R. D. Hanson, and F. E. Filisko. Electrorheological dampers, Part I: Analysis and design. *Journal of Applied Mechanics*, 63:669–675, 1996.

[19] D. L. Klass and T. W. Martinek. Electroviscous fluids. I. Rheological properties. *Journal of Applied Physics*, 38(1):67–74, 1967.

[20] K. D. Weiss and T. G. Duclos. Controllable fluids: Temperature dependence of post-yield properties. *International Journal of Modern Physics B*, 8(20&21):3015–3032, 1994.

[21] R. C. Ehrgott and S. F. Masri. Experimental characterisation of an electrorheological material subjected to oscillatory shear strains. *Journal of Vibration and Acoustics*, 116: 53–60, 1994.

[22] W. S. Yen and P. J. Achron. A study of the dynamic behavior of an electrorheological fluid. *Journal of Rheology*, 35:1375–1384, 1991.

[23] H. Lee and S. B. Choi. Dynamic properties of an ER fluid under shear and flow modes. *Material and Design*, 23(1):69–76, 2002.

[24] O. Ashour and C. Rogers. Magnetorheological fluids: Materials, characterization and devices. *Journal of Intelligent Material Systems and Structures*, 7:123–130, 1996.

[25] Y-T. Choi and N. M. Wereley. Comparative analysis of the time response of electrorheological and magnetorheological dampers using nondimensional parameters. *Journal of Intelligent Materials Systems and Structures*, 13(7):443–451, 2002.

[26] J. D. Carlson. What makes a good MR fluid? In *8th International Conference on ER Fluids and MR Fluids Suspensions*, Nice, 9–13 July 2001.

[27] K. D. Weiss, T. G. Duclos, J. D. Carlson, M. J. Chrzan, and A. J. Margida. High-strength magneto- and electro-rheological fluids. *Society of Automotive Engineering Transactions, SAE Paper No. 932451*, pages 425–430, 1993.

[28] N. Makris, S. A. Burton, and D. P. Taylor. Electrorheological damper with annular ducts for seismic protection applications. *Smart Materials and Structures*, 5:551–564, 1996.

[29] B. F. Spencer Jr., S. J. Dyke, M. K. Sain, and J. D. Carlson. Phenomenological model of a magnetorheological damper. *Journal of Engineering Mechanics*, 123:230–238, 1997.

[30] R. Stanway, J. L. Sproston, and N. G. Stevens. Non-linear modeling of an electrorheological vibration damper. *Journal of Electrostatics*, 20:167–184, 1987.

[31] D. A. Brooks. Design and development of flow based electrorheological devices. *International Journal of Modern Physics B*, 6:2705–2730, 1992.

[32] T. G. Duclos. Design of devices using electrorheological fluids. *Society of Automotive Engineering Transactions, Sec. 2, SAE Paper No. 881134*, 97:2532–2536, 1988.

[33] D. J. Peel, W. A. Bullough, and R. Stanway. Dynamic modeling of an ER vibration damper for vehicle suspension applications. *Smart Materials and Structures*, 5(5):591–606, 1996.

[34] R. Stanway, J. L. Sproston, and A. K. El-Wahed. Application of electrorheological fluids in vibration control: A survey. *Smart Materials and Structures*, 5:464–482, 1996.

[35] R. Stanway, J. Sproston, and R. Firoozian. Identification of the damping law of an electro-rheological fluid: A sequential filtering approach. *ASME Journal of Dynamic Systems, Measurement and Control*, 111:91–96, March 1989.

[36] R. C. Ehrgott and S. F. Masri. Modeling the oscillatory dynamic behavior of electrorheological materials. *Smart Materials and Structures*, 1:275–285, 1992.

[37] S. F. Masri, R. Kumar, and R. C. Ehrgott. Modeling and control of an electrorheological device for structural control applications. *Smart Materials and Structures*, 4(1A):A121–A123, 1995.

[38] T. Butz and O. von Stryk. Modelling and simulation of electro- and magnetorheological fluid dampers. *Zeitschrift fur Angewandte Mathematik und Mechanik*, 82(1):3–20, 2002.

[39] D. Y. Lee, Y.-T. Choi, and N. M. Wereley. Performance analysis of ER/MR impact damper systems using Hershcel-Bulkley model. *Journal of Intelligent Material Systems and Structures*, 13:525–531, 2002.

[40] D. Y. Lee and N. M. Wereley. Quasi-steady Herschel-Bulkley analysis of electro- and magneto-rheological flow mode dampers. *Journal of Intelligent Material Systems and Structures*, 10:761–769, 1999.

[41] B. Bird, R. Armstrong, and O. Hassager. *Dynamics of Polymeric Fluids*. John Wiley and Sons, New York, 1987.

[42] S. A. Burton, N. Makris, I. Konstantopoulos, and P. J. Antsaklis. Modeling the response of ER damper: phenomenology and emulation. *Journal of Engineering Mechanics*, 122: 897–906, 1996.

[43] G. M. Kamath and N. M. Wereley. A non-linear viscoelastic-plastic model for electrorheological fluids. *Smart Materials and Structures*, 6:351–359, 1997.

[44] Y.-T. Choi, L. Bitman, and N. M. Wereley. Nondimensional analysis of electrorheological dampers using an Eyring constitutive relationship. *Journal of Intelligent Material Systems and Structures*, 16:383–394, May 2005.

[45] J. C. Dixon. *The Shock Absorber Handbook*. Society of Automotive Engineers, Inc., Warrendale, PA, 1999.

[46] *VersaFlo Fluids Product Information, ER-100 Fluid Form PI01-ER100A*. Lord Corporation, Cary, NC, 1996.

[47] A. K. El Wahed, J. L. Sproston, and R. Stanway. The rheological characteristics of electrorheological fluids in dynamic squeeze. *Journal of Intelligent Materials Systems and Structures*, 13(10):655–660, 2002.

[48] A. K. El Wahed, J. L. Sproston, and R. Stanway. The performance of an electrorheological fluid in dynamic squeeze flow: The influence of solid phase size. *Journal of Colloid and Interface Science*, 211(2):264–280, 1999.

[49] R. J. Atkin, X. Shi, and W. A. Bullough. Solutions of the constitutive equations for the flow of an electrorheological fluid in radial configurations. *Journal of Rheology*, 35(7): 1441–1461, 1991.

[50] J. H. Yoo and N. M. Wereley. Approximating annular duct flow in ER/MR dampers using a rectangular duct. *Proceedings of FEDSM'03, 4th ASME/JSME Joint Fluids Engineering Conference* (#FEDSM03/45034), July 2003.

[51] M. Nilsson and N. G. Ohlson. An electrorheological fluid in squeeze mode. *Journal of Intelligent Material Systems and Structures*, 11(7):545–554, 2000.

8 Applications of Active Materials in Integrated Systems

Applications of smart structures technology to various physical systems are primarily focused on actively controlling vibration, performance, noise, and stability. Applications range from space systems to fixed-wing and rotary-wing aircraft, automotive, civil structures, marine systems, machine tools, and medical devices. Early applications of smart structures technology were focused toward space systems to actively control vibration of large space structures [1] as well as for precision pointing in space (e.g., telescope, and mirrors [2]). The scope and potential of smart structures applications for aeronautical systems have subsequently expanded. Embedded or surface-bonded smart material actuators on an airplane wing or helicopter blade can induce alteration of twist/camber of airfoil (shape change), which in turn can cause variation of lift distribution and may help to control static and dynamic aeroelastic problems. For fixed-wing aircraft, applications cover active control of flutter [3, 4, 5, 6, 7], static divergence [8, 9], panel flutter [10], performance enhancement [11], and interior structure-borne noise [12]. Compared to fixed-wing aircraft, helicopters appear to show the most potential for a major payoff with the application of smart structures technology. Given the broad scope of smart structures applications, developments in the field of rotorcraft are highlighted in a subsequent section. Although most current applications are focused on the minimization of helicopter vibration, there are other potential applications such as interior/exterior noise reduction, aerodynamic performance enhancement that includes stall alleviation, aeromechanical stability augmentation, rotor tracking, handling qualities improvement, rotor head health monitoring, and rotor primary controls implementation (e.g., swashplateless rotors) [13]. For aerospace systems, two types of actuation concepts have been incorporated. One approach uses active materials directly, surface-bonded or embedded, to actively twist or control the camber of primary lifting surfaces. Another approach actively controls auxiliary lifting devices such as leading-/trailing-edge flaps using smart material actuators, which in turn twist the primary lifting surface.

8.1 Summary of Applications

A key element in any smart structures application is its actuation mechanism. There are many important factors that must be taken into account in the selection of the actuation mechanism. These include maximum free strain/displacement (or stroke), maximum blocked force, permissible bandwidth, compactness

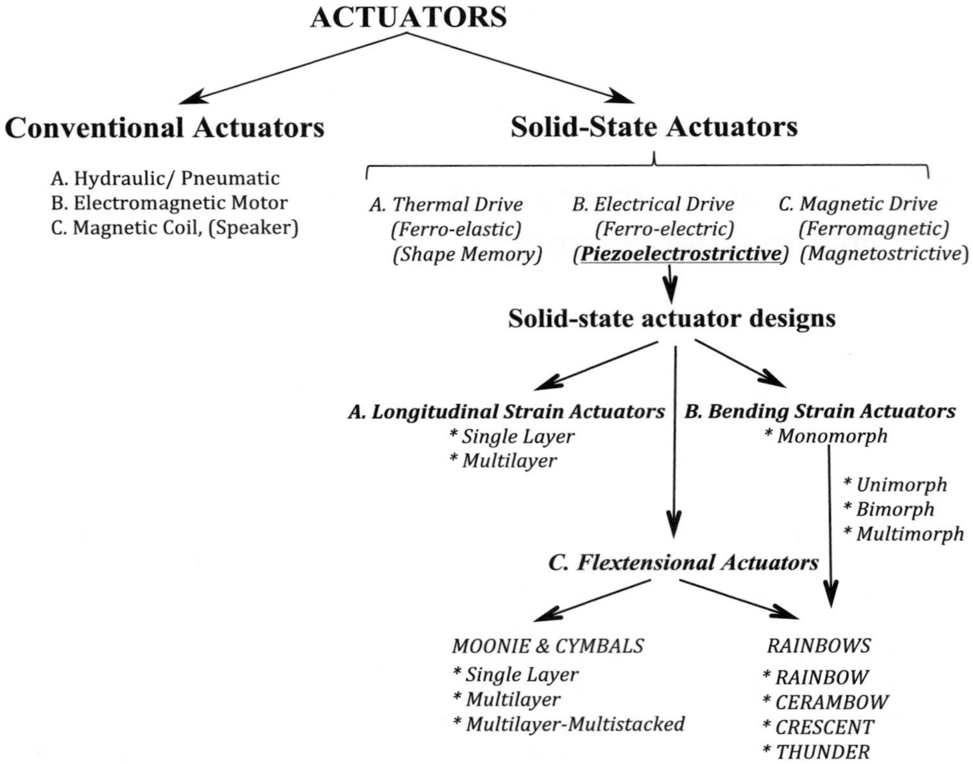

Figure 8.1. Classification of actuators, adapted from Dogan et al. [14].

(integration issue), specific energy requirement (weight issue), maximum field/current requirement (power transfer issue), safety and operational needs (associated electronics), material integrity/longevity (fatigue life over 200 million cycles or performance degradation with time) and operational sustainability (centrifugal/aerodynamic forces), tolerance to environment (temperature/humidity), cost, and technical maturity. Depending on a specific application, any one of these issues can be a dominating factor during the selection of an actuator. Displacement actuators can be generally classified into two categories: conventional displacement actuators and solid-state actuators [14] (Fig. 8.1).

The smart material actuators are referred to as solid-state actuators.The most common actuators are monolithic sheet actuators, bender actuators, and stack actuators. Overall, the specific power density of smart material actuators is higher than that of conventional actuators such as electromagnetic, hydraulic and pneumatic actuators. Other disadvantages of conventional actuators are large space requirements (size), weight, and long response times. Also, conventional displacement actuators are often inadequate for precise positioning. The good points of conventional actuators are their low cost, high force and stroke, easy commercial availability, and proven and well-understood technology. Conventional displacement actuators can be categorized into three types: hydraulic actuators, servo- or stepper-motors, and electrodynamic actuators. Hydraulic actuators operate with oil pressure acting on pistons in cylinders. The principal disadvantages are the large volume and bandwidth requirements. The servo- or stepper-motors convert rotary motion (from an electric motor) into linear displacement through a gearbox or screw mechanism. Mechanical backlash can be a major problem and is sometimes avoided with the use of ball screws. The electrodynamic actuators, such as voice coils and shakers, use magnetic coils and springs. The major issues are efficiency, maximum force output, and bandwidth. Even though conventional displacement actuators can achieve large displacements, they have a lower specific actuation energy, are bulkier, and become less precise as their scale decreases compared to solid-state actuators. Conventional displacement actuators typically feature a number of moving parts that can increase issues related to reliability and maintainability.

The applications of smart structures technology are summarized in the following sections.

8.1.1 Space Systems

Large space structures consist of multi-member, lightweight, flexible trusses that support precision equipment, including telescopes and mirrors. Applications in space systems include active vibration control of large space structures, adaptive geometric control of large truss configurations, precision pointing of telescopes and mirrors, structural integrity monitoring, condition-based maintenance, and active structural-acoustic control to mitigate interior noise.

8.1.2 Fixed-Wing Aircraft and Rotorcraft

A fixed-wing aircraft consists of many flexible structures such as wings, ailerons, flaps, fins, and elevators, which undergo coupled aeroservoelastic deformations. These deformations impact the performance and controllability of the vehicle. Applications of smart structures include active vibration control, gust alleviation, flutter and divergence stability augmentation, increasing panel flutter stability, interior structure-borne noise control, shape control for performance enhancement, and structural integrity monitoring.

For jet engines, smart structures technology (specifically SMA actuators) is used to develop adaptive variable geometry chevrons (engine nozzle surfaces) and inlets to optimize acoustics and performance for multiple flight conditions.

As compared to fixed-wing aircraft, rotorcraft suffer from severe vibratory loads, increased vulnerability to aeromechanical instabilities, excessive noise signature, poor flight stability characteristics, poor aerodynamic performance, and a restricted flight envelope. The primary source of all of these problems is the main rotor, which operates in an unsteady and complex aerodynamic environment. There is a wide range of potential applications of smart structures technology in rotorcraft, which include active vibration control, aeromechanical stability augmentation, handling qualities enhancement, external acoustics suppression, stall alleviation, rotor performance enhancement, in-flight rotor tracking, structure-borne interior-noise control, rotor head and drive train structural health monitoring, and primary rotor control toward development of a swashplateless rotor.

8.1.3 Civil Structures

Civil structures include buildings, bridges, water and gas pipelines, chimneys, and dams. Applications of smart structures technology in civil structures include active vibration and motion suppression, earthquake mitigation, and structural health monitoring. These active material components and systems could be installed in new structures or could also be retrofitted in existing structures. The goal is to increase the overall safety, life-cycle cost, occupant comfort, and life of the structures.

8.1.4 Machine Tools

There have been increasing applications of smart structures technology in machine tools to improve their performance. These applications include the active control of vibratory motion of the cutting tools for precision machining, adaptive tools for high-speed glass cutting, smart papermill graders, intelligent presses for sheet-metal forming, active tension control in wire electro-discharge machining, adaptable high-speed traversing mechanisms, microscopic polishing, and smart compact grippers.

8.1.5 Automotive

Because of low awareness, lack of reliable material database, cost, and reliability concerns, smart structures technology has not widely penetrated the automotive industry at this time. One of the areas in which this technology has started appearing in a few makes of production vehicles is magnetorheological suspensions. Other potential applications include active control of vibration and noise, active suspension and engine mounts, controllable clutching and braking mechanisms, and haptic joystick controllers.

8.1.6 Marine Systems

Applications in marine systems include active control of machinery vibration, structural acoustic control, radiated noise control, shape/flow control to increase maneuverability, biomimetic active hydrofoils, mini-underwater propulsors, and health monitoring/condition-based maintenance. Affordability, design simplicity, and stroke and actuation authority as well as robustness are key factors in these applications.

8.1.7 Medical Systems

There is a wide range of applications of smart structures technology in the medical field. Many applications require soft materials with large strain capability. Precise control, compactness, low weight, and durability are key factors for the application of smart structures technology in medical systems. Applications include compact adaptable dampers in prosthetic devices, artificial muscles, variable-resistance rehabilitation exercise machines and haptic devices, artificial hands and fingers, artificial anal sphincters and urethral valves, robotic eyes with intelligent orbital prostheses, telerobotic surgical systems, robots for rehabilitation therapy, tools for minimally invasive surgery (MIS), novel therapeutic approaches for cancer, swimming micro-robots, recoverable eyeglass frames, active palpation sensors for detection of prostatic cancer and hypertrophy, orthopedic implants, orthodontic treatments, and tissue fixators.

8.1.8 Electronic Equipment

Many electronic equipments are being built using smart structures technology. The key factors are cost, expanded capability, power requirements, complexity, durability, and precision control. Applications include ultrasonic motors, large size LCD televisions, high-capacity CD-ROM devices, active antennas, and precision sensors.

8.1.9 Rail

There have been applications of smart structures technology in rail systems. For example, in high-speed trains, vibration is a major issue that affects ride quality, stability, and maintenance cost of the tracks. Key factors for applications are robustness, durability, and maintenance cost. Applications include active suspensions to control vibration and structure-borne noise, and active buffers.

8.1.10 Robots

Applications of smart structures technology in robotic systems, especially at the small and miniature scales, are growing rapidly. Key factors for applications are stroke and actuation authority, robustness, maintenance cost, power requirements, precise control, and durability. Applications include the robotic gripper, miniature stepping robot, and high-speed robotic manipulator.

8.1.11 Energy Harvesting

Using low-power, efficient micro-electronics, compact energy harvesting systems are being built using smart structures technology. Using the direct piezoelectric effect, energy harvesters are being developed to take advantage of the vibratory motions induced by wind; mechanical systems (machinery); human shoes during walking; and moving platforms such as airplane wings, automobiles, ships, and rails. The key challenge is to develop efficient acquisition and storage of the input low-level energy.

8.2 Solid-State Actuation and Stroke Amplification

For most applications, there is a need for compact, moderate force, moderate bandwidth (less than 100 Hz), and moderate to large displacement actuators. Actuators based on piezoceramic mechanisms show great promise for aerospace applications, primarily due to their high energy density and wide bandwidth. The high energy density allows these actuators to meet the severe volumetric and weight constraints imposed by a large number of applications (e.g. the on-blade actuators in a smart helicopter rotor). The high bandwidth of the actuators is essential for achieving the desired authority in vibration- and noise-control applications. In addition, the low number of moving parts involved in such "solid-state" actuators decreases complexity and operational wear, and increases reliability of the system.

Monolithic PZT (piezoceramic) sheet actuators are available commercially in a variety of sizes and shapes. One of the most common types is in the form of thin rectangular sheets (Fig. 8.3(a)). Let us take a PZT-5H sheet actuator of size 50.8 mm $\times$ 25.4 mm $\times$0.3048 mm; its maximum free displacement is about 0.00685 mm at a permissible voltage of 150 volts (field $\mathbb{E} = 492$ V/mm) and the corresponding blocked force is approximately 70 N. If we increase the thickness of the sheet actuator, this will not affect its maximum free strain; however, the applied voltage must be increased accordingly to achieve the same electric field (V/t_c). The maximum blocked force will be a linear function of thickness. As a result, it is not expected to

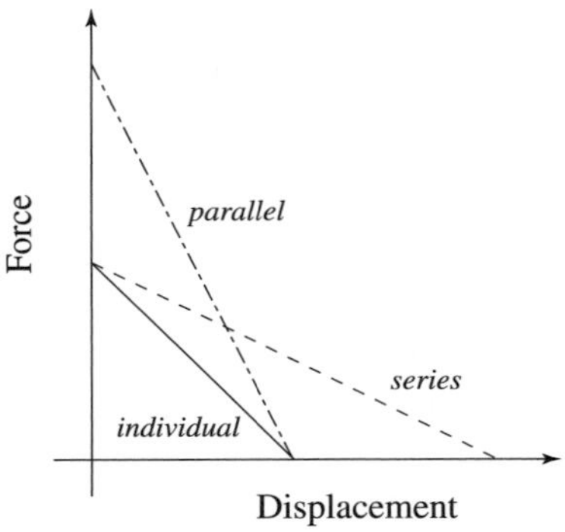

Figure 8.2. Force-deflection characteristic of smart material actuator.

use thick monolithic actuators in actual applications due to the small displacements and extremely high voltage requirements. Even though the maximum displacement is a function of the length of the actuator, there is a limit on the maximum length; a larger length may result in buckling of the actuator. Typically, sheet actuators are either surface-bonded to the structure or embedded as a laminated structure. To overcome the drawback of low displacement of these actuators, many different stroke amplification mechanisms have been investigated. These can be divided into the following two basic categories:

(1) amplification by means of special geometry or arrangement of the active material
(2) amplification by external leverage mechanisms

8.2.1 Amplification by Means of Special Geometry or Arrangement of the Active Material

Individual piezoelectric sheet actuators can be combined in series to obtain higher actuation displacement. The actuation force is, however, unaffected and, also, there is a limit on increasing the length of thin sheet actuators (buckling constraint). The actuation force can be increased by placing actuators in parallel (Fig. 8.2). This, however, does not change the maximum displacement.

Another approach to increase the actuation displacement is by building piezoelectric bimorphs (Fig. 8.3(b)). A bimorph or bending actuator consists of two or more even layers of piezoelectric sheets bonded on either side of a thin metallic shim (main load-carrying member). By applying an opposite potential to the top and bottom sheets, a pure bending actuation is generated. In a cantilevered arrangement, the tip displacement can be used for actuation of a system. With piezobimorphs, one can obtain displacements from 5 to 10 mils and forces up to 0.5 lb. Using more layers can increase the actuation force, but the displacement is reduced.

To increase actuation force, multilayered actuators such as piezostacks can be used (Fig. 8.3(c)). Piezostacks consist of a large number of thin piezoelectric sheets stacked in a series arrangement, separated by electrodes. Piezostacks make use of induced strain in thickness direction (d_{33} actuation). These devices induce small free displacements but much larger actuation force than sheet actuators. Nominal performance of piezostack actuators range from free displacement of 15 to 250 μm, blocked forces of up to 1000 lbs, and frequencies of up to 20 kHz. One can obtain a similar type of actuation with a bulk piezoelectric actuator; however, the electric voltage requirement becomes impractical. Combined with suitable external

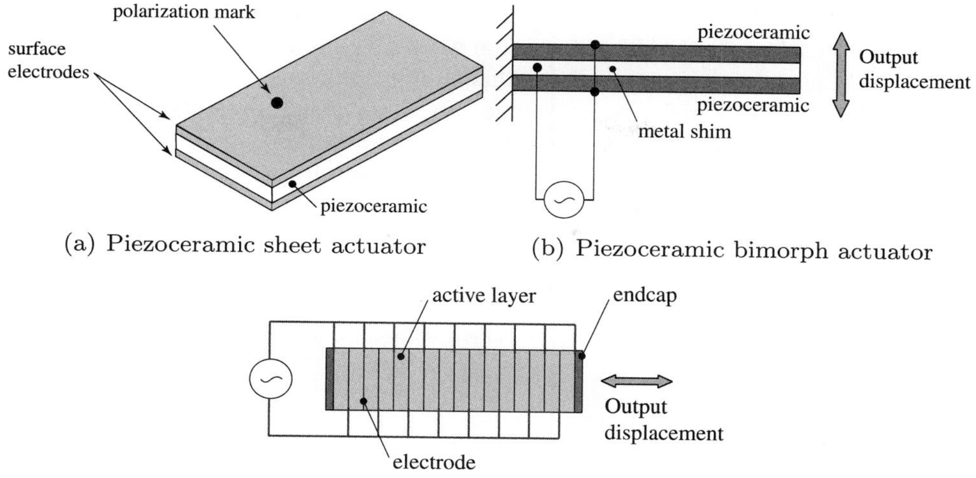

(a) Piezoceramic sheet actuator (b) Piezoceramic bimorph actuator

(c) Piezoceramic stack actuator

Figure 8.3. Piezoceramic actuators.

amplification mechanisms, piezostacks have been used in a wide range of applications. There have been several studies to characterize the electromechanical behavior of piezostacks [15], [16], [17], and [18]. For example, Lee et al. [15] evaluated the characteristics of 11 different stack actuators including maximum free strain, maximum blocked force, operating voltage, and energy density (Table 8.1). These actuators were tested systematically using specially built test apparatus under different field levels, operating frequencies, and pre-loads. Because the commercially available piezostacks are different in shape and in size, a strain-force index consisting of the product of maximum strain and normalized blocked force (blocked force divided by cross-sectional area) can be defined and used to compare different actuators. Graphically, this is equivalent to twice the area enclosed by the force-strain curve of the piezostack. Note that similar performance metrics can be defined for other active material actuators. Another performance metric is the energy density,

Table 8.1. *Maximum operation voltage, free strain, and blocked force for static excitation testing*

Piezostack Part/Material No.	Max. Voltage *V*	Strain *μ-strain*	Blocked Force (BF) *N*	Normalized BF *MPa*	Energy Density *J/kg* [19]
MM 8M (70018)	360	254	570	5.50	0.12
MM 5H (70023-1)	200	449	449	4.35	0.17
MM 4S (70023-2)	360	497	636	6.13	0.26
PI P-804.10	100	1035	5042	38.30	3.41
PI PAH-018.102	1000	1358	6697	50.87	5.85
XI RE0410L	100	468	423	27.04	1.07
XI PZ0410L	100	910	311	18.76	1.45
EDO 100P-1 (98)	800	838	685	10.48	0.74
EDO 100P-1 (69)	800	472	222	3.46	0.14
SU 15C (H5D)	150	940	1184	39.19	3.12
SU 15C (5D)	150	1110	1219	40.34	3.79

which is equal to the maximum work that can be extracted from the actuator divided by the mass of the actuator. The maximum work is equal to half the area under the force-stroke curve of the actuator. Note that although the energy density of the active material itself is quite high, the energy density of the actuator can be considerably lower due to the addition of various mechanisms for amplification and conversion of the stroke of the active material into the desired output.

The thickness of each piezoceramic layer varies from 0.002 in to 0.040 in. The total axial deflection of the piezostack is proportional to the applied field

$$\Delta L = nd_{33}V_{p-p} \tag{8.1}$$

where ΔL is the axial displacement, d_{33} is the piezoelectric coefficient, n is the total number of layers, and V_{p-p} is the peak-to-peak operating voltage. The blocked force F_{bl} is

$$F_{bl} = \Delta L K_p \tag{8.2}$$

where K_p is the stiffness of the piezostack. If $E_{33}^{\mathbb{E}}$ is the Young's modulus of the material in the polarized direction at constant field (short-circuited electrodes), A is the cross-sectional area of the piezostack, and L is its length

$$F_{bl} = nd_{33}V_{p-p}E_{33}^{\mathbb{E}}\frac{A}{L} \tag{8.3}$$

In an actual piezostack, there are losses due to the bond layers, which is usually accounted for in terms of a constant C_n. The actual blocked force is given by

$$F_a = C_n nd_{33}V_{p-p}E_{33}^{\mathbb{E}}\frac{A}{L} \tag{8.4}$$

PZT and PMN stack actuators are typically fabricated by one of two approaches. In the first approach, thin layers of active material are assembled and glued together using an adhesive. The modulus of the adhesive (typically 4–5 GPa) is much lower than the modulus of the active ceramic layer (typically 70–90 GPa). This leads to an effective reduction of stack stiffness. In the second approach, the thin layers of active material and the electrodes are assembled together and fired together (co-fired) in the processing oven in conjunction with a high isostatic pressure (HIP) process. In this process, the electrodes and ceramic material are processed together, the wafer thickness is typically thinner, and the electrodes extend only partially through the ceramic. This process ensures stiff stacks; note that stacks are weak in tension irrespective of the fabrication approach. A polymeric or elastomeric wrapping is normally applied around the stack as a protective layer and stiff, insulated endcaps (metallic or ceramic) are attached to both ends. Three major characteristics of the stack actuators are maximum free stroke (at maximum applied voltage), maximum blocked force, and maximum applied voltage (which depends on the thickness of the individual layers).

In a pre-stressed stack, the stack is enclosed in a casing with a pre-stress mechanism (Fig. 8.4). The casing not only protects the stack against mechanical impact and damage from the environment, it also provides the possibility of applying a pre-stress on the stack to enable it to sustain tensile forces. The mechanical compressive stress generally improves the performance (stiffness and stroke) of piezoceramic actuators and provides bi-directional operation. The goal of the pre-stress mechanism is to obtain a high compressive stress on the stack by incorporating a spring in parallel,

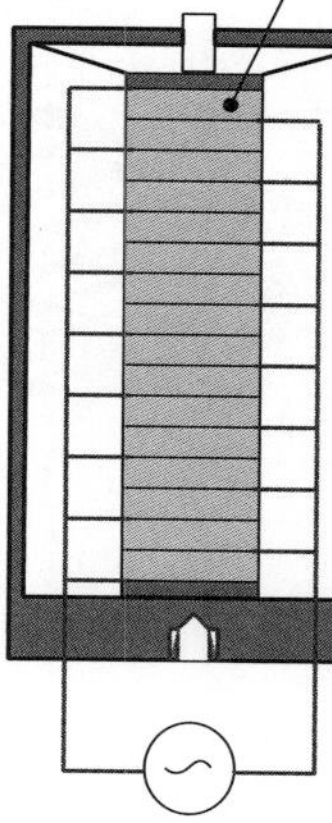

Figure 8.4. Pre-stressed stack actuator.

with a stiffness as low as possible. If the spring stiffness is too high, the stroke of the actuator will be reduced. Typically, the compressive pre-stress is on the order of 10%–20% of the blocked stress. This also results in a limit on the tensile stress that the actuator can sustain. In case it becomes necessary to change the pre-stress (e.g., to accommodate increased tensile stresses in dynamic cases), one can adjust the pre-stress mechanism using a mechanical screw.

Piezoelectric actuators exhibit self-heating due to dielectric dissipation in the material. This typically increases with frequency and amplitude. Because of the low thermal conductivity of PZT and poor heat radiation in the case of enclosed stack actuators, self-heating becomes a serious issue at high-frequency operation. A stack built using thin ceramic layers, densified by high pressure, and sintered at a high temperature results in a low-voltage actuator. Conversely, hard sintered ceramic plates or ceramic layers individually bonded together often result in higher voltage requirements. Because of the higher thickness of the active layers in a high-voltage stack, there is a better insulation stability compared to that of low-voltage stacks. It takes a longer time for the electrochemical degradation of the insulation/ceramic interface across the electrodes.

Several types of piezoelectric actuators with special geometry have been developed to enhance the output stroke. Some of these are discussed in more detail in the following sections. Because of the possible geometric and density variations of the different actuators, a more appropriate parameter is the specific energy density (energy density per unit weight). For magnetostrictive actuators, it is appropriate to also include the weight of the excitation coils for comparison. A large free strain is another preferred characteristic. A large material induced strain reduces the stroke amplification requirement, which in turn improves the overall efficiency of the actuation system. To compare different types of actuators, the maximum strain is referred to as half peak-to-peak (HPP) strain. Depending on the application, the bandwidth (frequency range) of the material is another important index. Because of this consideration, SMAs, despite their high specific energy density, are restricted to static applications only. Ceramics are brittle and suffer from fatigue issues. Materials that are quadratically dependent on the applied field may be difficult to integrate into a linear control system. Because of significant variation in temperature in a specific application, sensitivity to temperature can be a major cause of concern. As a result of

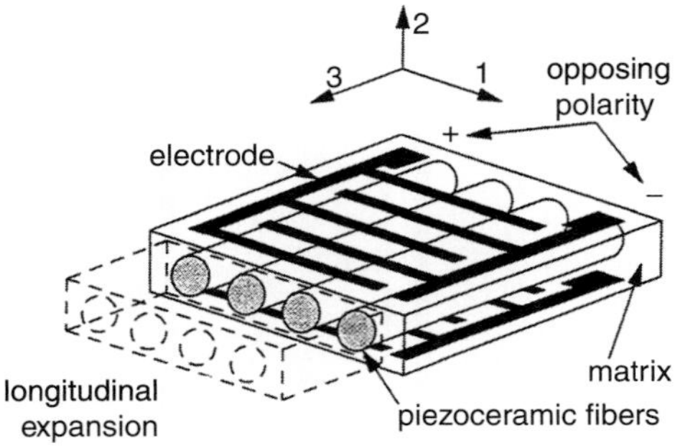

Figure 8.5. Schematic of active fiber composite.

this concern, electrostrictives are restricted to a few specific applications. In practical applications, cost can also be a critical factor. For most applications, it is preferred to use actuators with low-voltage requirements.

Active Fiber Composites (AFC)/Macro Fiber Composites (MFC)

One major development in piezoceramics has been the emergence of active fiber composites (AFCs) and macro fiber composites (MFCs), in which active piezoceramic fibers are embedded in a matrix. The piezo fibers are actuated in the d_{33} mode using interdigitated electrodes. For example, the piezo active fibers were used successfully in the development of an active twist rotor [20, 21, 22, 23].

The AFC material typically consists of 0.25 mm (or 250 μm) diameter continuous PZT-5A fibers aligned in an epoxy matrix to provide in-plane actuation, which fill up to 90% of the width (Fig. 8.5). The fibers are manufactured through an extrusion process using soft PZT-5A powder. The strength and toughness (brittleness) characteristics of the composite are significantly enhanced with the incorporation of a polymer matrix that surrounds the fibers. The fibers embedded in the matrix are sandwiched between two layers of polyimide film that have a conductive interdigitated-electrode pattern printed on the inner surface. The interdigitated pattern creates an effective width for the fibers to achieve d_{33} effect (electric field applied along the direction of fibers; i.e., actuation direction). Because the fibers are of fine scale and embedded in the polymer matrix, they conform to the shape of irregular structures. The combination of interdigitated electrodes and ceramic fibers offers an enhanced toughness (ductility) and damage tolerance. It is also possible to tailor multiple composites that can induce torsional actuation. Applications of AFCs include vibration control, shape control, and structural health monitoring. AFCs were initially developed at the Massachusetts Institute of Technology by Hagood and Bent [24, 25]. Since the initial development, there have been significant advancements in AFCs, which include fiber manufacture, matrix materials, electrode design, manufacturing technique, and modeling. There are other approaches to manufacture PZT fibers, which include sol-gel, extrusion, and viscous suspension spinning process. Wilkie et al. [26] developed a MFC that incorporated fibers of rectangular cross section (smaller than AFC fibers). Again, the uniaxial piezoceramic fibers are embedded in the polymer matrix in conjunction with interdigitated electrodes. This approach was expected to increase the contact area between PZT fibers and the interdigitated electrodes. To lower the cost, the PZT

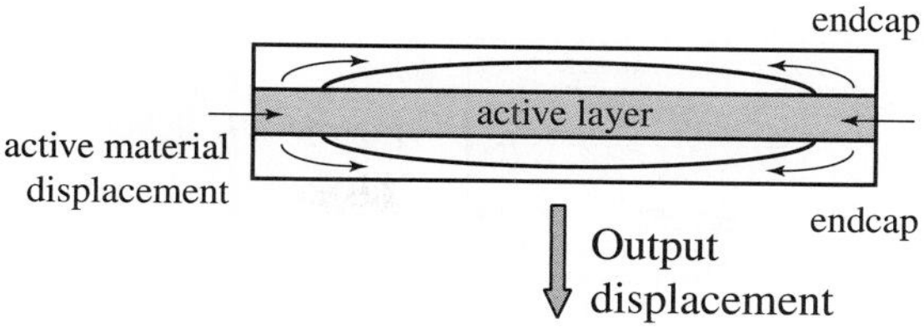

Figure 8.6. Schematic of Moonie actuator.

fibers were built by dicing monolithic PZT wafers. Williams et al. [27, 28] examined the mechanical properties of a MFC using classical laminated plate theory and measured the nonlinear actuation characteristics under various loads. Ruggerio et al. [29] used MFCs as both actuators and sensors to determine the dynamic behavior of an inflatable satellite structure. Park and Kim [30] investigated the introduction of single-crystal piezoelectric fibers instead of polycrystalline-piezoceramic fibers, and they estimated the variation of their mechanical and electromechanical properties.

Bowen et al. [31] manufactured AFCs by viscous plastic processing, which creates a highly viscous material composed of ceramic powder particles (PZT-5A) dispersed in a polymer (polyvinyl butyral) and solvent gel structure. These are mixed under high shear force. Then, green fibers of a diameter of approximately 250 μm are extruded from a die. Subsequently, the fibers are embedded in a lead-rich zirconia sand bed and sintered to 6000°C. The macrostructure and microstructure characterization showed the homogeneous structure of fibers with a control of size, microstructure, and composition.

Typically, the stress-strain characteristics of AFC and MFC are quite nonlinear. The disadvantages of these actuators are high cost, difficulty of processing and handling during fabrication, and high-voltage requirement (e.g., 3 kV peak-to-peak).

Specially-Designed Flextensional Actuators

In the early 1990s, different versions of flextensional transducers emerged. These are Moonie, Cymbal, RAINBOW, and THUNDER actuators. In these actuators, the radial displacement of the active piezoceramic material disk is transformed into axial displacement (normal to disk) by flexing or bending the structure. As a result, the stroke is amplified and the actuators exhibit large displacement.

Moonie Actuator

Newnham et al. [32, 33] devised a compact version of a flextensional actuator called the Moonie actuator. The name "Moonie" comes directly from "moon-shaped" spaces between the metal endcaps and the piezoceramic sheet. The basic composite circular configuration of the piezoceramic-metal caps is shown in Fig. 8.6.

Each metal cap has a varying thickness with a shallow crescent-shaped cavity on the inner surface and is bonded to the active diskmaterial around the circumference. The two metal endcaps serve as a stroke amplifier (flextensional) to transform the lateral motion of the piezoceramic (d_{31} effect) into a large axial displacement normal to the endcaps. Additionally, the "d_{33} effect" is superposed to increase the net axial displacement. The active component can be electroded PZT-5A, PMN-PT, or multilayer piezoceramic disk, and the endcaps can be machined from brass, phosphor

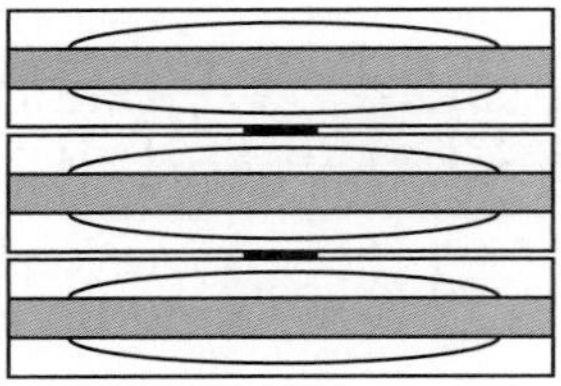

Figure 8.7. Schematic of three Moonie actuators stacked in series.

bronze, or acrylic. Note that the voltage is a function of piezoceramic thickness. Stroke increases exponentially with an increase in cavity diameter, increases linearly with an increase in cavity depth, and is inversely proportional to the endcap thickness. The response time of a Moonie actuator becomes larger with an increase of endcap compliance and cavity diameter [14]. A key element in the design is the bond layer between the endcaps and the ceramic driving disk, which undergoes severe shear stress. The Moonie actuator shows a larger generative force than a bimorph actuator and a higher displacement than a stack actuator. By stacking together Moonie actuators in series, more displacement can be obtained (Fig. 8.7). The ceramic element is kept primarily under compressive stress. Moonie actuators have also been used as hydrophones and transceivers.

Cymbal

An improved version of the Moonie actuator was developed by Dogan et al. [34] as a Cymbal actuator with higher efficiency, more displacement, and larger generative force. With a new design of endcap, the stress concentration at the bond layer was eliminated. The newly shaped endcap looked more like the musical instrument Cymbal and, hence, was named after it. The Cymbal cap is thinner than the Moonie cap and can be easily mass-produced using a punch/die fabrication scheme. Whereas the displacement of the Moonie actuator is produced through a flexural action of the caps, the displacement in the Cymbal is a mix of both flexural and rotational motions (Fig. 8.8). As a result, the output displacement is further amplified. The modulus of elasticity of endcaps is an important design parameter for the Cymbal actuator. The Moonie and the Cymbal appear to show potential for application in the automotive and aviation industries. Also, they can be used as micro-positioners, a role requiring small size and quick response. Other applications include optical scanners and high-density memory storage drives.

Most failures in multilayer Cymbal piezocomposites are caused due to inhomogeneous stresses. PZT actuators are susceptible to fracture failure under tensile stress. Often, PZT actuators are pre-loaded with compressive stresses that need to be below the values that can cause depolarization and microcracking. Ochoa et al. [35]

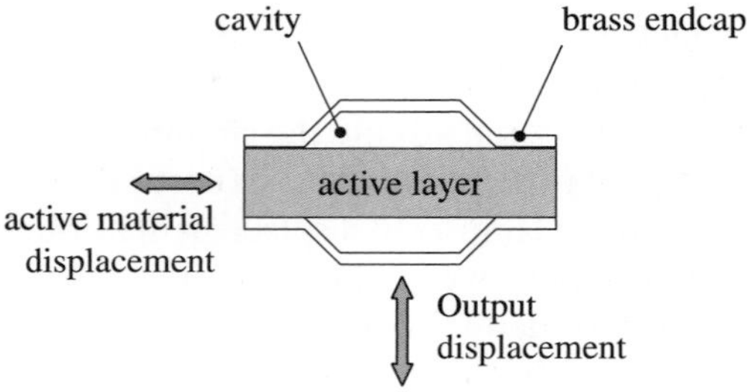

Figure 8.8. Schematic of Cymbal actuator.

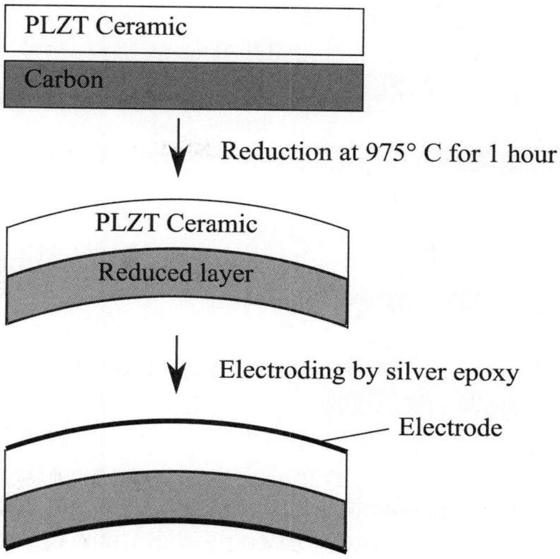

Figure 8.9. Fabrication of the RAINBOW actuator.

examined the depolarization of Cymbal piezocomposites by measuring the electric charge generated during the application of compressive load. The depolarization was found to be more severe in Cymbal actuators than in PZT disks.

RAINBOW (Reduced and Internally Biased Oxide Wafers)

These actuators are constructed by bonding a piezoceramic layer and a chemically reduced layer [36, 37]. These actuators are also categorized as monomorph actuators. Because of the thermal expansion mismatch between the reduced layer and the parent layer, a curvature is formed on cooling, giving the actuator a dome shape with oxide layer in compression. It is a pre-stressed, monolithic, axial-mode bender and, because of its dome or saddle-shaped configuration, it is able to produce more displacement and a moderate blocked force. The RAINBOW actuator is dome-shaped (circular) with the piezoelectric layer on the convex side (Fig. 8.9).

Applying an electric field across the piezoelectric layer results in an increase or decrease in the curvature of the actuator. The RAINBOW disks are typically of 0.5 mm or less thickness and can range in diameter from 1 to 10 cm. Typically, they show an actuation displacement of up to 1 mm, blocked force of up to 500 N, and actuation frequency of less than 10 kHz. By stacking RAINBOWS in a clamshell configuration, it is possible to obtain a larger stroke, which is proportional to the number of actuators. Materials used for the RAINBOW may include PZT, PMN, PLZT (lead lanthanum zirconate titanate), PBZT (lead barium zirconate titanate), and PSZT (lead stannate zirconate titanate). Because a part of the actuator is in compression while the other part is in tension, the RAINBOW has a long-term material integrity problem due to degradation of the interface between the oxide and the reduced layer. Hence, a RAINBOW actuator often shows degraded performance under cyclic loads. Li and Haertling [38] characterized PLZT RAINBOW ceramic actuators in the dome mode for a range of sizes, thickness ratios, and sizes of electroded area. The sensitivity study showed a progressive decrease of performance with increasing frequency below 5 Hz, and then a flat response up to the resonance frequency. Using thin-plate theory in conjunction with nonlinear

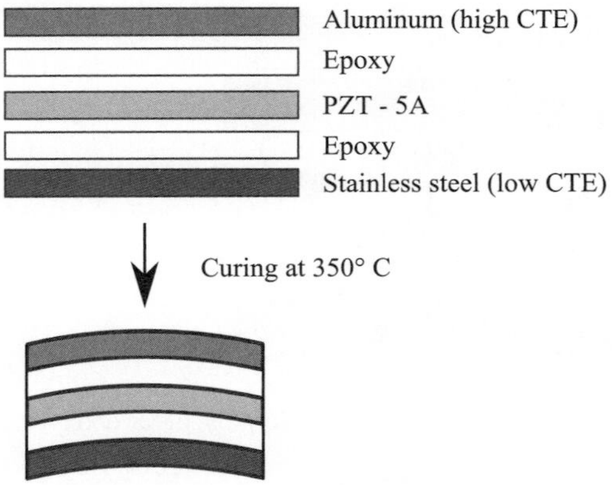

Figure 8.10. Fabrication of the THUNDER actuator.

strain-displacement relations, Hyer and Jilani [39] carried out a modal analysis of RAINBOW actuators to predict quasi-static response with applied field. They identified key geometric parameters such as radius-to-thickness ratio and reduced layer thickness to total thickness ratio, which impact the performance of actuators. These transducers appear to show potential for application in aero-acoustic cancellation, pumps, and switches.

THUNDER (Thin Layer Unimorph Ferroelectric Driver and Sensor)

The THUNDER actuator is a unimorph-type actuator initially developed by NASA (Langley) [40, 41]. It is a curved device composed of three layers: a metallic layer (typically aluminum) on the top, bonded to a pre-stressed piezoceramic layer using high-performance epoxy (LaRC-SI), and a metallic layer (typically steel) on the bottom surface (Fig. 8.10).

Both initial curvature and pre-stressing are introduced during the manufacturing process. Because of the difference in the coefficients of thermal expansion and Young's modulus between the conductor and piezoceramic layers, the composite actuator during the cooling process deforms to a shallow dome shape. Due to pre-stressing, the piezoceramic sheet is in a state of compression, whereas the substrate is in a state of tension, which enhances the induced deflection capability of the actuator. Displacement is achieved via the induced d_{31} contraction effect, which tends to flatten the actuator. When the voltage is released, the actuator tends to return to its natural dome shape. A commercial version of THUNDER consists of a stainless-steel substrate, a piezoceramic layer, and an aluminum top layer. It can be mounted as a cantilevered or simply-supported configuration. An actuator of length 1.0 inch and width of 0.5 inch can generate displacement of 10 mils for a cantilevered configuration and a blocked force of 8 lbs in a simply-supported configuration. As compared with THUNDER actuators, RAINBOW actuators show 10%–25% lower displacement and THUNDERs are relatively more rugged. To increase the blocked force, multiple THUNDER elements can be stacked together in parallel. Both THUNDER and RAINBOW show the largest displacement at the center of the dome. Marouze and Cheng [42] developed a hybrid isolation system using both the passive and active effects of THUNDER actuators. A simple prototype was built using three THUNDER actuators, demonstrating successful active vibration control over a range of frequencies.

Kim et al. [43] fabricated and characterized a THUNDER actuator. Five sets of actuators with different dimensions were built, showing that the large residual stresses within PZT layers developed during the fabrication process result in significant nonlinear electromagnetic coupling. The severity of the residual stresses and ensuing nonlinear response increased with higher substrate/piezoelectric thickness ratio and, to a lesser extent, with decreasing in-plane dimensions.

LIPCA (Lightweight Piezoelectric Composite Actuator)

LIPCA is a variant of the THUNDER actuator, in which some or all parts of the metallic layer are replaced with fiber-reinforced composite layers to reduce weight. In LIPCA, the adhesive layer is not needed [44]. Syaifuddin et al. [45] used LIPCAs to actuate a flapping-wing mechanism, through a four-bar linkage system. It was successfully tested up to a frequency of 9 Hz, where the maximum flapping amplitude was obtained. Lim et al. [46] built a small bird-like flapping wing in which the trailing-edges are actuated by LIPCA. Because of the material nonlinearity of the piezoceramic wafer in the LIPCA, the measured displacements were found to be larger than those predicted based on linear theory, especially for high field (more than 150 V).

The flextensional actuators described herein are all referred to as solid-state actuators. Dogan et al. [14] provided a comparison of their characteristics (except THUNDER), given in Table 8.2. Because of wide variation of geometric and operating conditions, a fair comparison of these actuators is difficult. Choosing similar dimensions, Dogan et al. [14] provided an assessment of these actuators, as shown in Table 8.3.

Let us consider PZT-5A as an active material for all actuators. The larger axial displacement of the RAINBOW than that of the Moonie or the Cymbal may be due to the difference in the R/t (radius/thickness) ratio. If identical dimensions are used, the displacements will be similar [47]. To achieve a positive longitudinal displacement, the applied field will be in the opposite direction to the polarization in the RAINBOWs and THUNDERs but in the same direction as the polarization in the Moonie and the Cymbal designs. The axial displacement is approximately represented as

$$\delta \approx \pm \frac{d_{31}\mathbb{E}l^2}{2t} \tag{8.5}$$

where $\mathbb{E}$ is the applied electrical field (V/m), l is the length or diameter, and t is the thickness for RAINBOW or THUNDER design or the cavity height for the Moonie or Cymbal design. All of these flextensional actuators provide moderate generative force and displacement values, and their actuation capabilities lie between multilayer stacks and bimorph actuators. Aimmanee and Hyer [48] carried out an analysis to predict the residual stresses of rectangular-shaped THUNDER actuators using the Rayleigh-Ritz approach. Because of the large out-of-plane deformations that take place during cooling, geometric nonlinearities are included in the analysis. It was shown that the geometric nonlinearities that are a function of the actuator shape have an important role in the actuation response.

Another actuator similar to the Cymbal uses bimorph-based displacement (Fig. 8.11). This architecture combines both bending and flextensional features to produce output displacement [49].

Table 8.2. *Comparison of solid-state actuator designs, from Ref. [14]*

Features	Multilayer	Bimorph	RAINBOW	Cymbal	Moonie
Dimensions, mm	5 × 5 × 12.7 (L × W × T)	12.7 × 10 × 1 (L × W × T)	$\phi = 12.7$ T = 0.5 mm	$\phi = 12.7$ T = 1.7 mm	$\phi = 12.7$ T = 1.7 mm
Driving voltage, V	100	100	450	100	100
Displacement, μm	10	35	20	40	20
Displacement direction	Positive	Positive	Negative	Positive	Positive
Contact surface, mm^2	25	1	1	3	1
Generative force, N	900	0.5–1	1–3	15–100	3
Position dependence of displacement	No	Maximum at the tip	Maximum at the center	Maximum at the center but more diffuse	Maximum at the center
Stability under loading	Very high	Very low	Low	High	Low
Response time, μs	1–5	100	100	5–50	5–50
Fabrication method co-firing at 1200°C	Type casting and with metal shim	Bonding ceramic element at 950°C	Reducing ceramic with metal endcaps	Bonding ceramic element with metal endcaps	Bonding ceramic element
Cost	High	Medium	Medium	Low	Low

Table 8.3. *Performance of various flextensional composite actuators, from Ref. [14]*

Feature	Moonie	Cymbal	RAINBOW	THUNDER
Dimensions (mm)	ϕ25.4 Disk	ϕ25.4 Disk	ϕ25.4 Disk	25.4 r Square
PZT	PZT-5A	PZT-5A	PZT-5A	PZT-5A
Applied field, kV/mm	Unipolar 1.0	Unipolar 1.0	Bipolar ±0.65	Unipolar 1.0
Thickness of PZT, mm	0.500	0.500	0.380	0.325
Displacement, μm	50	80	88	60

C-Block

This consists of a semicircular piezoelectric bender, poled in the radial direction and activated in the circumferential direction by a voltage applied across the thickness. Basically, it is a multilayered, curved bimorph (Fig. 8.12). The induced piezoelectric strain causes a bending action in each individual C-block, similar to a straight bender (bimorph), flexing the entire architecture. Individual C-blocks can generate more than twice the force of a straight bender with a slight reduction in deflection. Individual C-blocks can be combined in series to increase the total axial deflection without changing the force capability. Also, C-blocks can be stacked in parallel to increase the force output without any loss of stroke [50, 51]. Thus, it is possible to tailor the performance of a solid-state actuator to an application within a constrained volume. Changing the piezoelectric material (piezoelectric constant and stiffness) results in a change in the performance of the actuator. As expected, the change in width of a rectangular straight actuator only changes its force capability linearly, whereas the C-block will increase the output force capability cubically and decrease the deflection quadratically. To validate the force-deflection behavior of C-block actuator arrays, Moskalik and Brei [52] built prototypes using PZT-5H and PVDF, testing each prototype across a range of voltages to obtain the force-displacement behavior. Through a numerical study, authors showed that a tailored C-block actuator produces the largest specific-energy index among other actuators such as bimorphs, RAINBOWs, THUNDERs, Cymbals, Moonies, and leveraged stacks.

8.2.2 Amplification by External Leverage Mechanisms

For many practical applications, it becomes necessary to amplify the small stroke of actuators using external mechanisms. Amplification mechanisms in general may involve many moving parts that contribute to actuation losses and degrade rapidly

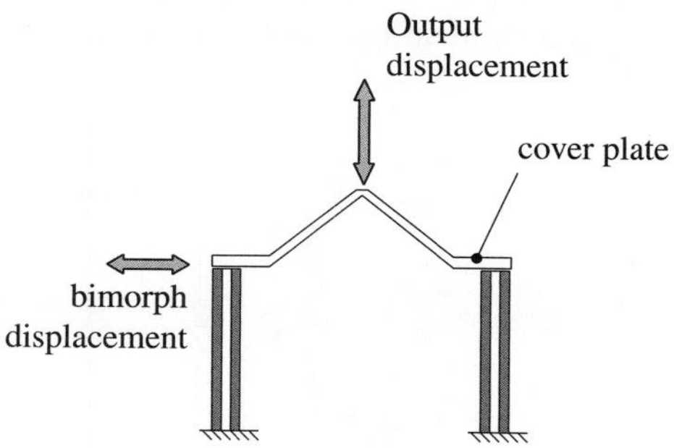

Figure 8.11. Bimorph-based double-amplifier actuator.

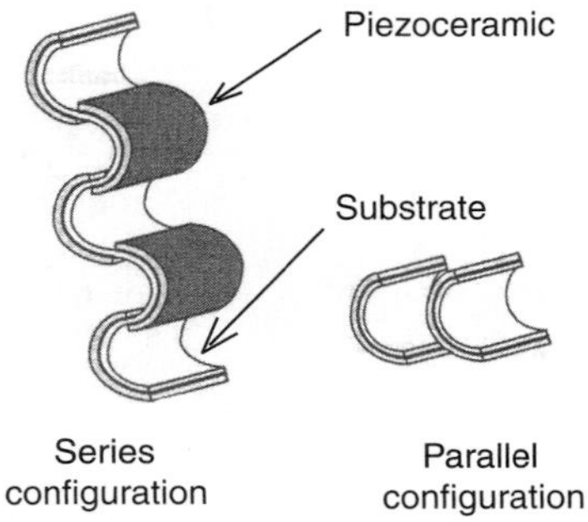

Figure 8.12. C-block actuator.

under high loading. In addition, the maximum practical amplification ratio is on the order of 15–20. At larger amplification ratios, the stiffness of the amplification mechanism itself becomes a serious issue. Studies have shown that mechanical amplification leads to a 60%–80% reduction in the overall energy density of the device, compared to that of the base active material [53]. Several issues important to the design of mechanical stroke amplifiers, such as the positioning of the hinge/fulcrum and the effect of the finite stiffness of the lever arms, is discussed.

Amplification mechanisms can generally be divided into two categories: fluidic and mechanical. Typically, the fluidic approach uses two cylinders of different diameters to give the desired stroke amplification [54, 55, 56], as shown in Fig. 8.13. This approach can provide higher amplifications than general mechanical amplifiers. However, the fundamental limitations – that is, the finite stiffness related to the compressibility of the working fluid and flexibility of the hydraulic chambers [57], as well as frictional losses due to fluid viscosity, are the same.

Several single-stage mechanical amplification devices that include lever-fulcrum mechanisms and triangular frame mechanisms have been built [58, 59, 19]. In comparison with the fluidic system, the mechanical lever-fulcrum stroke amplifier is a simple, lightweight, and compact actuation mechanism. From the stiffness point of view, the triangular frame system is more efficient than the lever-fulcrum system because its structural members experience mostly extensional (axial) loads in contrast to bending loads for lever-fulcrum amplifiers. Significantly increasing the stiffness of the lever-fulcrum assembly will result in a large weight penalty.

Mechanical amplification devices trade force with displacement but have a detrimental effect on power transfer efficiency and energy density [19, 53], especially at high amplification factors. These mechanisms consist of a framework of stiff passive members that are interconnected by hinges to provide a mechanical advantage [15, 59, 19]. To achieve amplification factors higher than about 10, multistage amplification is incorporated, as in the L-L amplification mechanism [60]. Any stroke amplification mechanism can be represented by its linkage equivalence. In

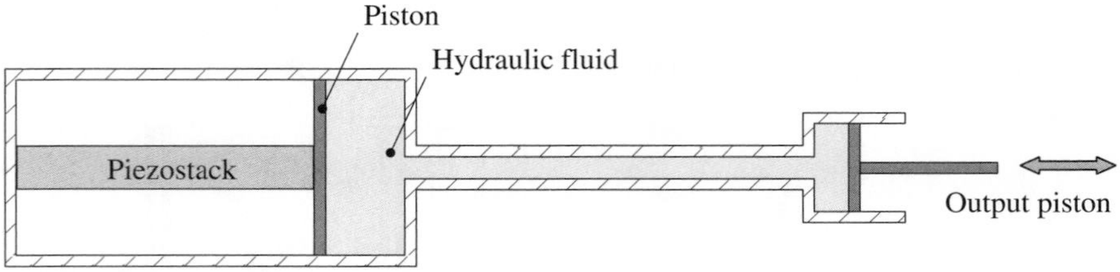

Figure 8.13. Hydraulic amplification system.

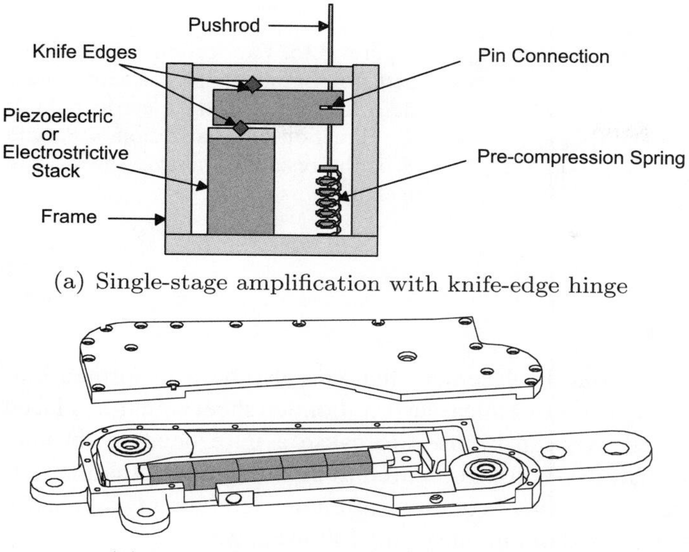

(a) Single-stage amplification with knife-edge hinge

(b) Two-stage amplification with flexures

Figure 8.14. Actuators with mechanical stroke amplification.

this way, it is possible to identify whether or not the stroke amplification mechanism is practically realizable. Using Gruebler's equation, the feasibility can be checked

$$F = 3(n-1) - 2f_1 - f_2 \tag{8.6}$$

where n is the number of links, f_1 is the number of pin joints, and f_2 is the number of roll-slide contact joints. To have a single-degree-of-freedom actuator stroke, the actuator would be equivalent to either a four-bar linkage with

$$F = 3(4-1) - 2 \times 4 - 0 = 1 \tag{8.7}$$

or a six-bar linkage with

$$F = 3(6-1) - 2 \times 7 - 0 = 1 \tag{8.8}$$

Large mechanical amplification using a compact leverage system often leads to substantial losses at hinges and slippage at knife edges. Figure 8.14(a) shows a single-stage mechanical linkage amplification system. Because of mechanical losses and slippage at the knife edges, the measured stroke was far less than predicted.

To overcome losses due to the finite play inherent in pin-jointed amplification mechanisms, flexure hinges [61] or fully compliant mechanisms are used. A double L-arm lever amplification with flexures is shown in Fig. 8.14(b). The mechanical losses in this device were less than in the single-stage system. However, the approach utilizing flexures requires careful design to optimize the mechanism. In addition, the actuation efficiency is reduced due to the strain energy stored in the flexures. Frecker and Canfield [62] formulated a systematic topology-optimization approach to the design of compliant-mechanical amplifiers for piezoceramic stack actuators. In this approach, any direction of force and motion transmission from the active material can be chosen. This methodology shows potential to build devices in which precise motions are important [63, 64].

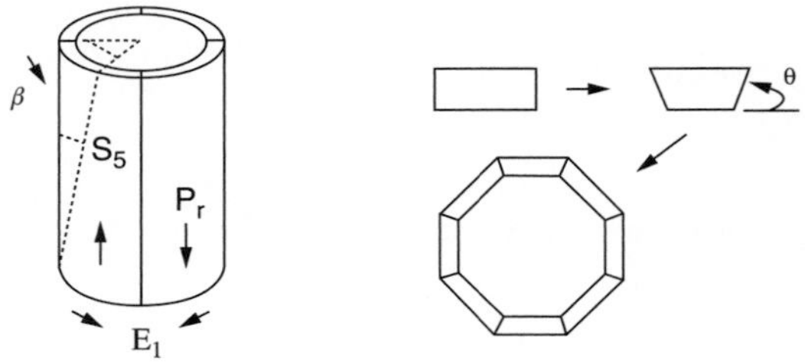

Figure 8.15. Fabrication of torsional actuator based on piezoelectric shear deformation. Driving electric field is E_1, direction of polarization is P_r, and S_5 is the resultant induced shear, from Ref. [67].

8.2.3 Torsional Actuators

Actuators normally provide axial displacement that can also be transformed into bending actuation. For example, two identical surface-bonded sheet actuators placed equidistant from the neutral axis can be used to cause a pure bending through application of opposite actuation strains. Compared to pure bending actuation, it is more challenging to cause pure torsion actuation (often needed in many aerospace applications). It is typically carried out in one of the following ways.

Specially Arranged Conductors

Glazounov et al. [65, 66, 67] developed a torsional tube actuator using piezoelectric d_{15} shear coupling. The tube consists of an even number of piezoelectric cylindrical segments, poled in radial or axial direction, and bonded together in a circumferential direction with sandwiched conductors between segments (at joints) to form a circular tube. On the application of field (normal to the polarized direction), shear strain is created in the circumferential direction, resulting in twisting of the tube. The objective is to take advantage of a comparatively large value of piezoelectric d_{15} shear coupling and induce directly a large torsional displacement and a large blocked torque. Using a tubular structure, the shear strain is converted into angular displacement.

Case I: Polarization in Axial Direction

In this case, the polarization direction (1-axis) is along the length of the tube, and the direction of polarization alternates between adjacent segments. The segments are connected in parallel to provide coherent shear strain γ_{31} due to an applied field $\mathbb{E}_1$, as shown in Fig. 8.15.

Let us consider a tube of internal radius R_{in}, outer radius R_{out}, and length l_c. For a tube of this geometry

$$\text{Shear strain} \quad \gamma_{31} = d_{15}\mathbb{E}_1 \tag{8.9}$$

$$\text{Compliance} \quad s_{55} = \frac{2(1+\nu_{31})}{E_3} = \frac{1}{G_{31}} \tag{8.10}$$

$$\text{Shear modulus} \quad G_{31} = \frac{E_3}{2(1+\nu_{31})} \tag{8.11}$$

$$\text{Angular displacement} \quad = \frac{l_c}{R_{out}} d_{15}\mathbb{E}_1 \tag{8.12}$$

The angular displacement is amplified by the ratio of the length of the tube and its outer radius (i.e., l_c/R_{out}). The blocked torque can be calculated as

$$\text{Shear stress} \quad \tau_{31} = \gamma_{31} G_{31} \tag{8.13}$$

$$\text{Blocked torque} \quad T_{bl} = \pi(R_{out}^2 - R_{in}^2) R_{out} G_{31} d_{15} \mathbb{E}_1 \tag{8.14}$$

The blocked torque is independent of the length of the tube. Thus, it is possible to change the amplification factor l_c/R_{out} by changing the length without any loss of the blocked torque (unlike many other actuators). It has also been pointed out that the performance of the actuator does not degrade under external torque load and that a pre-stress in the radial direction improves the mechanical strength of the actuator. Also, in most PZT actuators, the d_{15} shear coupling coefficient has the largest value. For example, the value of d_{15}/d_{31} for PZT-5H is 2.7. Also, it is a direct transformation of shear strain into angular displacement. Replacing a cylindrical shape with a polygonal shape is expected to have a small influence on its performance. A major drawback with this actuator is initial poling with a large electric field. Typically, the piezoelectric shear coefficient d_{15} is quite nonlinear with respect to applied electrical field [68, 69].

Glazounov et al. [67] used a continuous poling technique to initially pole along the length of a long cylindrical PZT segment. In this technique, the segment is secured by a holder, and a couple of electrodes made out of conducting rubber are applied to the surface of the segment. These electrodes are separated (e.g., by a distance of 1.5 cm), and a high DC field of 20 kV/cm is applied. By slowly moving the rubber electrodes along the length of the segment, the specimen is poled. After poling, the segments are bonded together using a conducting high shear stiffness adhesive (e.g., silver-filled epoxy MB-10HT/S from Master Bond). Thakkar and Ganguli [70] and Centolanza and Smith [71, 72, 73] examined the application of this d_{15}-based torsional actuator in a helicopter rotor system to actuate a trailing-edge flap to actively control vibration.

Case II: Polarization in Radial Direction

The polarization direction (1-axis) for each segment is along the radial direction of the tube, and the direction of polarization alternates between adjacent segments. The segments are connected in parallel to provide coherent shear strain γ_{31} due to an applied field $\mathbb{E}_1$, as shown in Fig. 8.15.

$$\text{Shear strain} \quad \gamma_{31} = d_{15} \mathbb{E}_1 \tag{8.15}$$

$$\text{Compliance} \quad s_{55} = \frac{2(1+\nu_{31})}{E_3} = \frac{1}{G_{31}} \tag{8.16}$$

$$\text{Shear modulus} \quad G_{31} = \frac{E_3}{2(1+\nu_{31})} \tag{8.17}$$

$$\text{Shear stress} \quad \tau_{31} = \gamma_{31} G_{31} \tag{8.18}$$

$$\text{Blocked torque} \quad T_{bl} = 2\pi R_{out}^2 l_c G_{31} d_{15} \mathbb{E}_1 \tag{8.19}$$

The angular displacement is proportional to the shear itself, which is a function of electric field $\mathbb{E}_1$, and is independent of the tube length. The blocked torque is a function of the tube length and tube outer radius.

Case III: Polarization in Radial Direction and Stepper Motor

In this type of stepper motor, the angular motion produced by the tubular torsional actuator at resonance condition is accumulated in one direction using a direct coupling mechanism between the stator and rotor. A clutch drives the motor by locking it. Due to direct coupling, there is no energy loss in the frictional contact. The locking mechanism permits smooth motion in either a continuous or stepwise manner with a precise control over angular positioning. The drawbacks of frictional contacts are eliminated.

Coupling to Structure or External Mechanism

It is possible to convert the linear displacement of the actuators into angular displacement using a simple mechanism. However, mechanical conversion can significantly reduce the effectiveness of the device due to play in the linkages. Therefore, the conversion of the displacement output is often achieved by using structural couplings of the base structure on which the actuators are mounted. Bothwell et al. [74] used extension-torsion coupling of a thin-walled composite tube to convert the linear motion of a magnetostrictive actuator into a torsional displacement to actuate a trailing-edge flap. Bernhard et al. [75] used bending-torsion coupling of a composite beam in conjunction with surface-bonded piezoelectric elements to convert the bending of the beam into a tip twist, which was used to actuate a rotor blade tip. Giurgiutiu and Rogers [17] used the twist-warping concept of thin-wall open-section tube to convert linear motion of PZT stacks into rotary motion. This large-amplitude rotary induced-strain (LARIS) actuator was built using a 28-mm diameter, 1.2-m long open tube in conjunction with a PZT stack actuator, and a maximum twist of 8° was measured in the free condition. The main issue to be considered in the case of structural coupling is the coupling efficiency. This efficiency can be defined as the ratio of the energy output to the total energy input to the structure. Normally, bending-torsion coupling is more efficient than extension-torsion coupling because of the large extensional stiffness of the structure.

Specially Arranged Actuators

By bonding specially cut piezoceramic sheet elements at $\pm\theta$ degree orientation on the top and bottom surfaces of an uncoupled beam, respectively, a pure twist can be caused by in-phase excitation. It is also possible to induce a pure bending of the beam by out-of-phase excitation of the top and bottom banks of the piezo elements (see Section 8.6 for more details).

To obtain maximum twist, θ should be 45° and the piezoceramic elements should be of high aspect ratio (length/width > 4). Chen et al. [76] built a Froude-scaled rotor blade with surface-bonded piezoceramic sheets. To increase actuation authority, two-layered piezoceramic sheets were used.

The bond layer has an important role in induced twist of the beam. The maximum torsional and bending deflection increased by 60% and 90%, respectively, when the bond thickness was reduced from 0.020 in to 0.0025 in. A minimum bond-layer thickness results in the most efficient shear transfer, which in turn results in

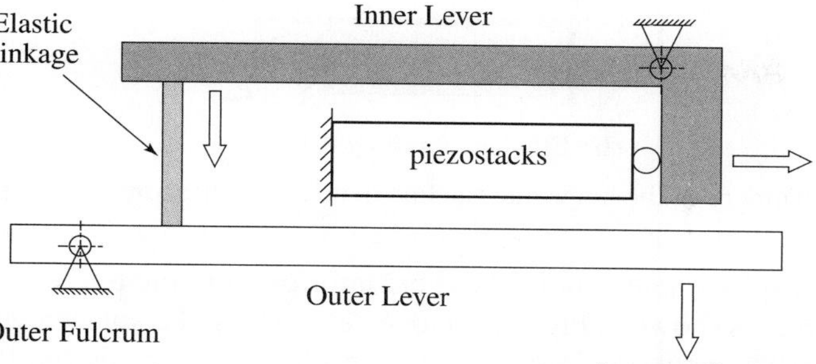

Figure 8.16. Schematic of L-L amplification mechanism.

maximum torsional and bending response. Test results also showed that increasing the actuator spacing reduced the structural stiffness of the beam as well as the nonlinear interference effect. For example, the beam tip twist was reduced by 38% when the spacing between the two piezoelectric elements was reduced from 1.5 in to 0.1 in.

8.3 Double-Lever (L-L) Actuator

To illustrate the challenges involved in high-amplification–ratio lever mechanisms, we explore the design of the L-L actuator, intended as a stroke amplifier for a piezostack driven trailing-edge flap in an active helicopter rotor blade. The L-L amplification mechanism is designed to have a high amplification factor with a low level of energy loss. Figure 8.16 shows a schematic of the L-L amplification mechanism. It is a combination of two lever-fulcrums and an elastic linkage. The stroke of the piezostacks is amplified by the inner lever with a low amplification factor (≤ 6) and then amplified again by another lever-fulcrum (outer lever). The two lever-fulcrums are connected in series using an elastic linkage, which transmits forces axially from one lever to the other. In addition, the flexure of the elastic linkage applies a returning force as well as a pre-load to the piezostacks. In comparison to a rigid axial member with pin joints on either end, the elastic linkage does not suffer from any losses due to play in the pin joints. For this reason, flexural joints such as the elastic linkage are preferred over mechanical joints in mechanisms amplifying the small stroke of active materials.

By connecting the two lever-fulcrum mechanisms in series, it is possible to obtain a high amplification while allowing a moderate actuation loss. The advantages of an L-L actuator are a planar structure, with the potential of further increasing the amplification factor, an embedded spring mechanism for piezostack pre-load, and ease of conducting structural analysis and optimization.

8.3.1 Positioning of the Hinges

The line of action of the force is known to be one of the major issues that affect the output performance. The line of action depends directly on the location of the hinge. Two cases are considered for the design of an L-L amplification mechanism.

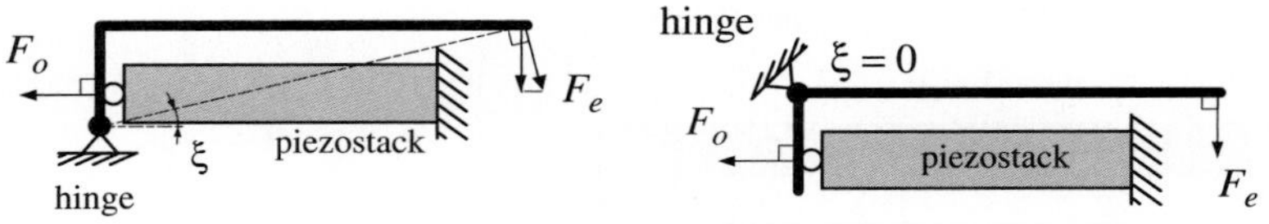

(a) Loss of force (b) Efficient force transfer

Figure 8.17. Effect of the alignment of the hinge and the line of action on the force transfer.

In the first case, a loss of displacement occurs at the actuator output, due to the line of action of the output force, as shown in Fig. 8.17. However, no loss of displacement occurs in the second case. The actuator force output in both of these cases can be expressed as

$$F_e = \frac{F_o}{\text{(Displacement Gain)}} \cdot \cos \xi \tag{8.20}$$

where F_e is the output force, F_o is the piezostack force, and ξ is the angle between the line of action and the actuator output direction. If the line of action is not perpendicular to the lever ($\xi \neq 0$), the actuator output has a loss factor of $(1 - \cos \xi)$. A similar situation happens at the interface between piezostack and lever, depicted in Fig. 8.18, and results in a loss of displacement due to misalignment of the hinge. The effective piezostack displacement for the amplification mechanism can be written as

$$(u_o)_{eff} = (u_o) \cdot \cos \zeta \tag{8.21}$$

where u_o and $(u_o)_{eff}$ are the actual and effective displacement of the piezostack, respectively. For nonzero ζ, the amplification mechanism loses the actuation stroke by a factor of $(1 - \cos \zeta)$. Therefore, the hinge location that satisfies both $\xi = 0$ and $\zeta = 0$ results in the most efficient force and displacement transfer, and this configuration is preferred in the design of an L-L amplification mechanism.

8.3.2 Actuation Efficiency: Stiffness of the Actuator, Support, and Linkages

The finite stiffness of the actuator, support structure, and linkages results in loss of energy transmitted to the output load. This energy loss is due to the strain energy stored in the flexible structure. The effect of the finite stiffness of the active material is discussed in Chapter 2 from the point of view of impedance matching. We revisit

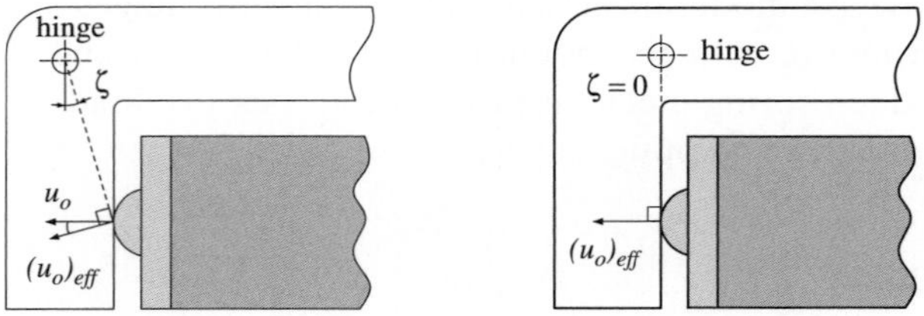

(a) Loss of displacement (b) Efficient displacement transfer

Figure 8.18. Effect of the alignment of the hinge and the piezostack on the displacement transfer.

Figure 8.19. Effect of the finite stiffness of the actuator.

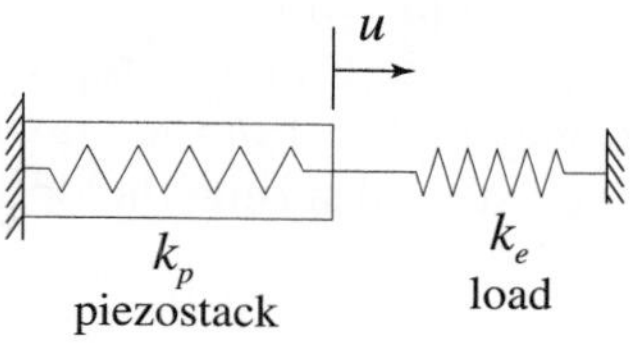

this analysis herein as a starting point for the analysis of the effect of flexible support and flexible linkages.

Effect of Actuator Stiffness

Consider a piezostack actuator of stiffness k_p on a rigid support acting against an external spring load of stiffness k_e (Fig. 8.19). The displacement of the piezostack, u, is given by

$$u = \delta_f - \frac{F}{k_p} \tag{8.22}$$

where δ_f is the free displacement of the piezostack and F is the force generated by the piezostack. Rewriting in terms of the external stiffness

$$\begin{aligned} u &= \delta_f - \frac{k_e u}{k_p} \\ &= \delta_f - ru \end{aligned} \tag{8.23}$$

where the ratio of the external stiffness to the piezostack stiffness is defined as

$$r = \frac{k_e}{k_p} \tag{8.24}$$

Simplifying the previous equation, the displacement can be written as

$$u = \frac{\delta_f}{1+r} \tag{8.25}$$

The energy transferred to the external stiffness is

$$\begin{aligned} U_e &= \frac{1}{2} k_e u^2 \\ &= \frac{1}{2} \frac{r}{(1+r)^2} \delta_f^2 k_p \\ &= \frac{r}{(1+r)^2} U_{max} \end{aligned} \tag{8.26}$$

where U_{max} is a measure of the maximum output energy capability of the piezostack and is equal to the area under its force-stroke curve. The efficiency of the actuator, η, can be defined as

$$\eta = \frac{U_e}{U_{max}} = \frac{r}{(1+r)^2} \tag{8.27}$$

$r = 0$ represents a free condition and the actuator efficiency is zero, whereas r approaching infinity represents a blocked condition and again the actuator efficiency is zero. For the maximum efficiency, and the maximum energy transferred to the external spring, the optimum value of stiffness ratio, r_{opt}, can be found as

$$\frac{\partial U_e}{\partial r} = 0 \quad \rightarrow \quad r_{opt} = 1 \tag{8.28}$$

This yields the condition that $k_e = k_p$ and corresponds to the impedance-matched condition. Note that in this condition, the maximum energy that can be transferred to the external spring and the maximum efficiency are

$$\begin{aligned} U_{e_{max}} &= \frac{U_{max}}{2} \\ \eta_{max} &= \frac{1}{4} \end{aligned} \tag{8.29}$$

We can conclude that the maximum energy that can be transferred to an external load occurs at the impedance-matched condition and is equal to half the area under the force-stroke curve of the actuator. At this point, the efficiency of the actuator is a maximum and is equal to 1/4. Note that this discussion assumes that the external load and the force transmission mechanism are linear. It may be possible to obtain better energy transfer using a nonlinear transmission between the actuator and the external load [77]. In addition, the previous analysis assumes that the support is rigid. We now consider the effect of support flexibility on the efficiency of energy transfer.

Effect of Finite Support Stiffness

In reality, the actuator support also has a finite stiffness. This can be incorporated in the analysis by considering the stiffness of the support to be acting in series with that of the actuator itself. A conceptual diagram of this scenario is shown in Fig. 8.22. Let us consider an elastic support structure with a stiffness k_s. Following the same procedure as before, the deflection u is given by

$$\begin{aligned} u &= \delta_f - \frac{F}{k_p} - \frac{F}{k_s} \\ &= \delta_f - \frac{uk_e}{k_p} - \frac{uk_e}{k_s} \end{aligned} \tag{8.30}$$

Defining the ratio of support stiffness to actuator stiffness as the support stiffness ratio r_s

$$r_s = \frac{k_s}{k_p} \tag{8.31}$$

we obtain

$$u = \frac{\delta_f}{1 + r(1 + 1/r_s)} \tag{8.32}$$

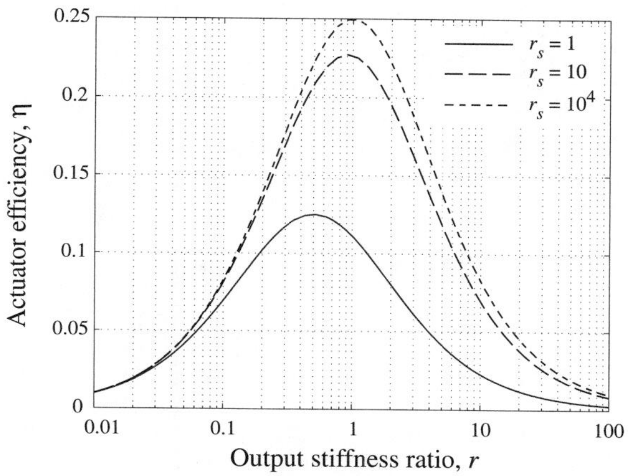

Figure 8.20. Variation of actuator efficiency with output stiffness ratio and support stiffness ratio.

and the output energy is

$$\begin{aligned} U_e &= \frac{1}{2} k_e u^2 \\ &= \frac{r}{[1 + r(1 + 1/r_s)]^2} U_{max} \end{aligned} \tag{8.33}$$

The actuator efficiency is

$$\eta = \frac{r}{[1 + r(1 + 1/r_s)]^2} \tag{8.34}$$

For maximum energy transferred to the output or the maximum actuator efficiency

$$\frac{\partial \eta}{\partial r} = 0 \quad \rightarrow \quad r_{opt} = \frac{1}{1 + 1/r_s} \tag{8.35}$$

This gives the maximum efficiency as

$$\eta_{max} = \frac{1}{4} \frac{r_s}{1 + r_s} \tag{8.36}$$

The variation of actuator efficiency with output stiffness ratio, for several values of the support stiffness ratio, is shown in Fig. 8.20. The case in which $r_s = 10^4$ corresponds to the ideal case with an infinitely rigid support. In the case in which $r_s = 10$, the support is 10 times stiffer than the actuator, and this condition is close to a rigid support. For this case, the maximum energy transfer takes place near $r = 1$. Also, the output deflection is lower with a softer support. This can be seen in Fig. 8.21, which shows the variation of the ratio of output displacement to the maximum actuator displacement (u/δ_f) as a function of output-stiffness ratio and support stiffness ratio. Again, it is seen that the case of $r_s = 10$ can be considered almost rigid.

Note that this expression reduces to the case of the rigid support in the limit $r_s \rightarrow \infty$. The actuation efficiency depends on both output stiffness ratio r and support-stiffness ratio r_s. If $r_s = 10$, it represents a case in which the support stiffness is 10 times the actuator stiffness and it is quite close to the rigid support case. For a flexible-support case (e.g., $r_s = 1$), there is not only a reduction of actuator efficiency but also a reduction of r at which maximum efficiency takes place. For this case, there is a reduction in maximum output energy of 50%. It is clear that now half

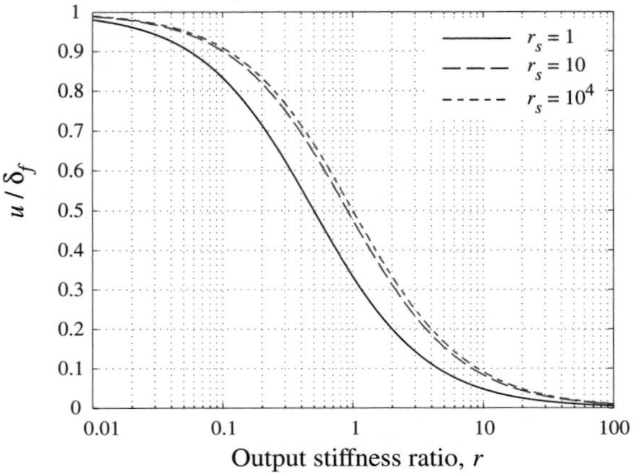

Figure 8.21. Variation of output displacement with output stiffness ratio and support stiffness ratio.

of the available energy is stored in the support system. To improve the actuation efficiency, it is important to increase the stiffness of the supporting structure that, in turn, results in an increase in the weight of the system. Therefore, another important and practical index of efficiency should be to consider the mass of supporting and active structures. Let us define an active material energy-density ratio as

$$\eta_{mass} = \frac{U_e}{U_{max}} \frac{M_{act}}{M_{tot}} \tag{8.37}$$

where M_{act} is the mass of the actuator and M_{tot} is the total mass of the structure including the frame, supports, and active systems. This efficiency helps to evaluate different actuation mechanisms, especially under static conditions, when the weight penalty is an important factor.

Effect of Finite Stiffness of the Linkages

The finite stiffness of the linkages in the amplification mechanism results in a degradation of the output stroke. This loss appears as an increase in strain energy stored in the linkages of the mechanism. Ideally, an inert frame mechanism, such as a lever and fulcrum, can be the most efficient stroke amplifier as long as the stiffness of the frame remains much higher than that of the active material. However, in practice, the amplification factor of these mechanisms is limited to a moderate value (≈20) because the deformation in the linkages increases due to the larger loads resulting from the higher amplification factors. This effect can be easily understood by analyzing the kinematics of a single-lever amplification mechanism, as shown in Fig. 8.23. Consider a piezostack of stiffness k_a acting against a spring of stiffness k_e through an infinitely stiff lever with lengths l_1 and l_2. Let the deflections of the piezostack and

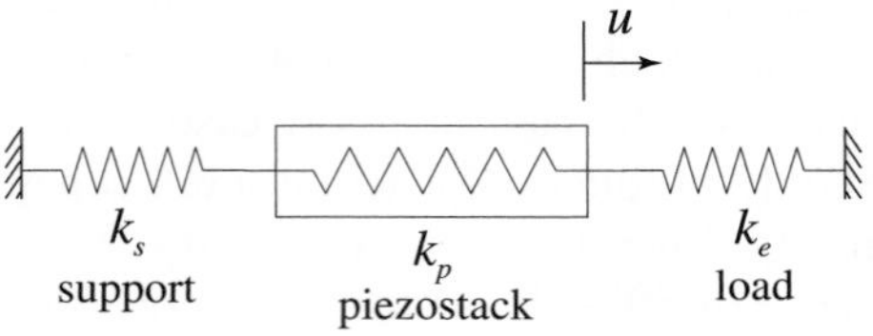

Figure 8.22. Effect of the finite stiffness of the support.

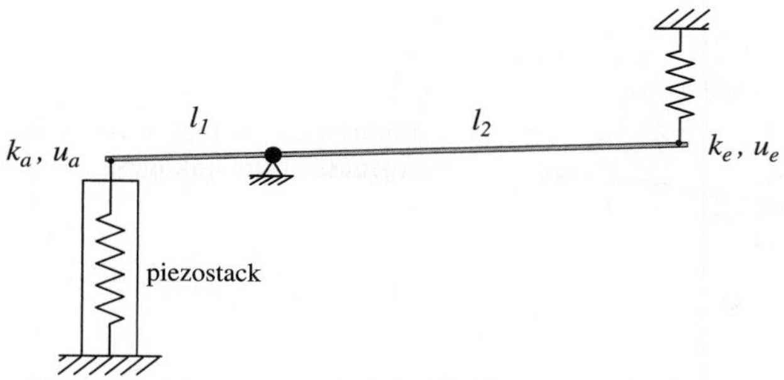

Figure 8.23. Single-lever amplification mechanism.

the spring be u_a and u_e, respectively. The piezostack displacement is given by

$$u_a = u_{free} - \frac{F_a}{k_a} \tag{8.38}$$

where F_a is the force acting on the piezostack and u_{free} is the free displacement of the piezostack. The deflection of the spring is given by

$$\begin{aligned} u_e &= u_a \frac{l_2}{l_1} = \frac{l_2}{l_1}\left[u_{free} - \frac{F_a}{k_a}\right] = \frac{l_2}{l_1}\left[u_{free} - \frac{k_e}{k_a}\frac{l_2}{l_1}u_e\right] \\ &= \frac{G u_{free}}{1 + rG^2} \end{aligned} \tag{8.39}$$

where we define the amplification ratio by G and the stiffness ratio by r, as

$$G = \frac{l_2}{l_1} \tag{8.40}$$

$$r = \frac{k_e}{k_a} \tag{8.41}$$

The energy stored in the spring is

$$\begin{aligned} U_o &= \frac{1}{2} k_e u_e^2 \\ &= \frac{1}{2}\frac{G^2}{(1 + rG^2)^2} r K_a u_{free}^2 \end{aligned} \tag{8.42}$$

Therefore, the actuation energy efficiency becomes

$$\begin{aligned} \eta &= \frac{\frac{1}{2} k_e u_e^2}{\frac{1}{2} k_a u_{free}^2} \\ &= \frac{rG^2}{(1 + rG^2)^2} \end{aligned} \tag{8.43}$$

For maximum energy efficiency

$$\frac{\partial \eta}{\partial r} = 0 \quad \rightarrow \quad r_{opt} = \frac{1}{G^2} \tag{8.44}$$

This gives

$$\eta_{opt} = \frac{1}{4} \tag{8.45}$$

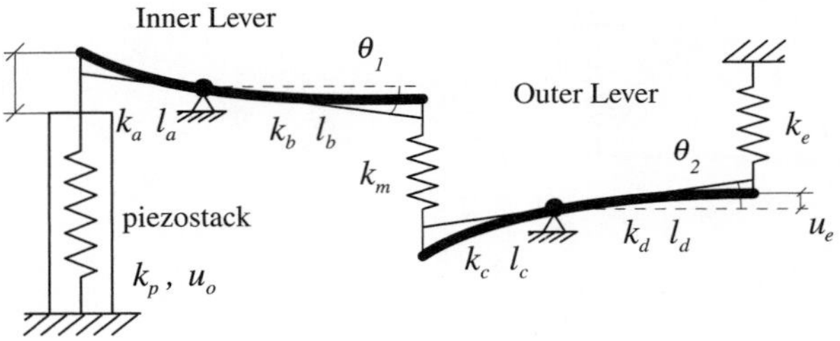

Figure 8.24. Effect of the finite stiffness of the linkages.

The maximum value of energy efficiency can be 1/4. Because $G > 1$, the maximum energy transfer occurs when the output stiffness is lower than the actuator stiffness. The optimal value of actuator stiffness increases rapidly with amplification factor G. To include the effect of mass, the active material energy density ratio is defined as

$$\begin{aligned}\eta_{mass} &= \frac{U_o}{U_{max}} \frac{M_{act}}{M_{tot}} \\ &= \frac{rG^2}{(1+rG^2)^2} \frac{M_{act}}{M_{tot}}\end{aligned} \tag{8.46}$$

We now examine the efficiency of the the L-L actuator. A schematic diagram of the actuator, with the linkages modeled as elastic beams, is shown in Fig. 8.24.

The piezostack is assumed to have free actuation displacement u_o and internal stiffness k_p. The elastic linkage has a stiffness k_m, which is equivalent to $(EA)/L$. The external or load stiffness is assumed as k_e. The bending stiffnesses of the levers are denoted as k_a, k_b, k_c, and k_d. The displacement at the piezostack can then be expressed as

$$\theta_1 \, l_a = u_o - \frac{F_p}{k_p} - \frac{F_p}{k_a} \tag{8.47}$$

$$u_e = \theta_2 \, l_d - \frac{F_e}{k_d} \tag{8.48}$$

$$F_e = k_e \, u_e \tag{8.49}$$

$$\frac{F_m}{k_m} = l_b \, \theta_1 - \frac{F_m}{k_b} - (l_c \, \theta_2 + \frac{F_m}{k_c}) \tag{8.50}$$

where F_p is the force induced by the piezostack and F_e is the force acting on the external stiffness. The displacement at the elastic linkage is

$$\frac{F_m}{k_m} = l_b \, \theta_1 - l_c \, \theta_2 \tag{8.51}$$

where F_m is the force (compressive) on the elastic linkage. By applying the force equilibrium

$$F_p = \frac{l_b}{l_a} F_m = G_1 \, F_m = G_1 \, G_2 \, F_e \tag{8.52}$$

$$F_m = \frac{l_d}{l_c} F_e = G_2 \, F_e \tag{8.53}$$

where G_1, G_2 is the kinematic gain of the inner lever and outer lever, respectively. Rearranging these equations and eliminating $\theta_1, \theta_2, F_p, F_m$, and F_e yields

$$\frac{u_e}{u_o} = \frac{G_1 G_2}{1 + G_e^*} \tag{8.54}$$

where

$$G_e^* = \frac{k_e}{k_d} + G_2^2\left(\frac{k_e}{k_m} + \frac{k_e}{k_b} + \frac{k_e}{k_c}\right) + (G_1 G_2)^2\left(\frac{k_e}{k_p} + \frac{k_e}{k_a}\right) \tag{8.55}$$

High efficiency in the amplification mechanism can be accomplished by minimizing G_e^*. Because k_e is a given parameter, it is required to maximize k_a, k_b, k_c, and k_d. The stiffness of elastic linkage k_m cannot be simply maximized because it should allow small bending displacement. Therefore, it can be seen that there is a practical limit to the amount of amplification possible. Note that the support is assumed to be rigid in the previous analysis. However, the effect of support stiffness can be easily incorporated, as shown in Section 8.3.2.

8.4 Energy Density

One way to compare different smart actuators is using a specific-energy index, defined as

$$\eta_e = \frac{1}{2}\frac{u_{free}F_{bl}}{W} \tag{8.56}$$

where u_{free} is the free displacement, F_{bl} is the blocked force, and W is the weight of the actuator. It can be expressed for a sheet actuator of length l_c, width b_c and thickness t_c, Young's modulus E_c, and weight density ρ_s (lb/in^3 or N/m^3) as

$$\eta_e = \frac{1}{2}\frac{E_c\Lambda^2}{l_c\rho_s} \tag{8.57}$$

where Λ is the free strain. To compare different type of actuators, Λ is the strain amplitude or HPP strain. In this way, linear piezoelectric actuators can be compared with quadratic electrostrictive/magnetostrictive actuators.

The larger this number, the lighter the actuator is. It is clear that a large free strain (i.e., stroke) is a key to increase actuator efficiency. Also, the larger the stroke, the lower will be the requirement for amplification in specific applications. Normally, stroke amplification decreases the overall efficiency of an actuation system. Magnetostrictive materials require a comparatively heavy solenoid coil to actuate the material. Accounting for solenoid weight in the calculation will substantially reduce the effective energy efficiency index. Also, the resultant energy efficiency index of a stack is lower than the value based on individual material sheet properties, primarily because of losses due to bond layers. Additional weight due to endcaps and electrodes further lowers the energy index.

Another way of defining efficiency is by the ratio of the output energy to the maximum strain energy of the actuator. Suppose the actuator deflects an external load of stiffness k_{ext} by a distance u; then, the energy efficiency of the system is given by

$$\eta = \frac{(1/2)k_{ext}u^2}{(1/2)k_{act}u_{free}^2} \tag{8.58}$$

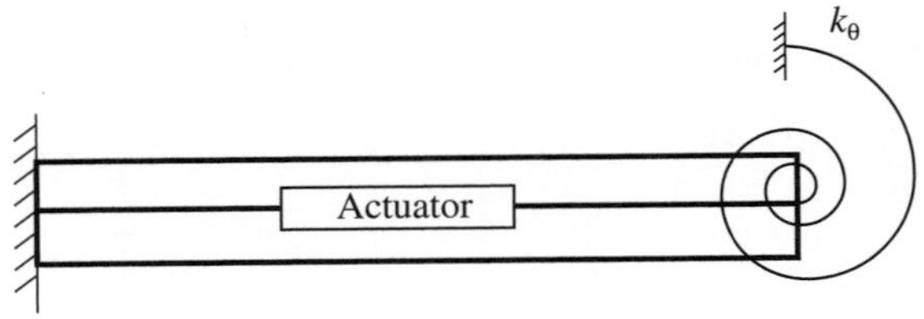

Figure 8.25. Magnetostrictive actuator driving an extension-torsion–coupled composite tube connected to a torsional spring.

where k_{act} and u_{free} are the stiffness and free displacement of the actuator, respectively.

One of the major applications of smart structures technologies is active vibration control of a flexible structure. Specific applications include automobiles (in the chassis from engine and tires), helicopters (in the airframe, rotor-induced), aircraft (in the airframe due to gust and engine-induced), ships (in the cabin, both marine-engine and waves-induced), and machine tools (imbalance of rotating shaft). In an active vibration control system, force inputs from actuators are used to suppress vibration based on real-time measurements from sensors. The controllers provide input signals to actuators to minimize a performance function such as a weighted sum of vibration amplitudes at selected stations. There is no doubt that there are differing requirements from actuators for a specific application to control vibration. Two key characteristics of an actuator are free displacement (stroke) and blocked force. For active cancellation of vibratory forces, the actuator must have the capacity to generate sufficient force and displacement to overcome the vibratory source. In a simplified single-input/single-output case, the free displacement of the actuator must be equal to or larger than the free displacement of the structure (at point of contact; i.e., the source); the blocked force should be more than the internal stiffness times the free displacement. Brennan et al. [78] carried out a set of vibration control experiments using five actuators: three were piezoceramic-based and the other two were magnetostrictive and electrodynamic. The first three actuators were a cylindrical high-force PZT actuator (40 mm diameter, 15 mm height), a high-displacement hydraulic PZT actuator (95 mm diameter, 58 mm height), and a high-displacement PZT RAINBOW actuator (50 mm diameter, 2 mm height). The fourth actuator used Terfenol-D (25 mm diameter, 65 mm height) and the fifth one was an electrodynamic device (66 mm-diameter tweeter, 20 mm height). Using the energy index as a figure of merit, the actuators were ranked as PZT tube, magnetostrictive, RAINBOW, hydraulic PZT, and tweeter.

8.4.1 Worked Example

A magnetostrictive actuator in conjunction with an extension-torsion coupled composite tube is used to actuate a trailing-edge flap, as shown in Fig. 8.25. Assume that the flap acts as a linear torsional spring k_θ (in-lb/rad) and the actuator stiffness is $k_a = E_c A_c / l_c$ (lb/in). Given are magnetostrictive free displacement u_{free} and blocked force F_{bl}. Calculate the actuation energy efficiency η_e. Assume that F is the actuation force, T is the flap torque, u is the axial deflection, and θ is the beam tip twist.

$$\begin{bmatrix} k_{11} & k_{12} \\ k_{12} & k_{22} \end{bmatrix} \begin{Bmatrix} u \\ \theta \end{Bmatrix} = \begin{Bmatrix} F \\ T \end{Bmatrix} \tag{8.59}$$

$$\eta = \frac{(1/2) k_{ext} \theta^2}{1/2 k_{act} u_{free}^2} \tag{8.60}$$

Solution

The output energy is

$$U_o = \frac{1}{2} k_\theta \theta^2$$

The maximum actuation energy is

$$U_{max} = \frac{1}{2} \frac{E_c A_c}{l_c} u_{free}^2$$

From the geometry of the actuator, the force can be written as

$$k_{11} u + k_{12} \theta = F$$

$$u = u_{free} - \frac{F}{k_a}$$

from which

$$F = (u_{free} - u) k_a$$

Substituting in the actuator force equation

$$k_{11} u + k_{12} \theta = (u_{free} - u) k_a$$

$$u = -\frac{k_{12}}{k_{11} + k_a} \theta + \frac{k_a}{k_{11} + k_a} u_{free}$$

Similarly, we can write the actuator torque as

$$k_{12} u + k_{22} \theta = -k_\theta \theta$$

$$u = -\frac{k_{22} + k_\theta}{k_{12}} \theta$$

from which we obtain

$$\theta = \frac{\frac{k_a}{k_{11}+k_a} u_{free}}{\frac{k_{12}}{k_{11}+k_a} - \frac{k_{22}+k_\theta}{k_{12}}}$$

The actuation energy efficiency is

$$\eta = \frac{k_\theta}{k_a} \frac{\left[\frac{k_a}{k_{11}+k_a}\right]^2}{\left[\frac{k_{12}}{k_{11}+k_a} - \frac{k_{22}+k_\theta}{k_{12}}\right]^2}$$

Let us define the following nondimensional quantities

$$\bar{k}_{12} = \frac{k_{12}}{(k_{11} k_{22})^{1/2}}$$

$$\bar{k}_\theta = \frac{k_\theta}{k_{22}}$$

$$\bar{k}_a = \frac{k_a}{k_{11}}$$

Substituting these in the expression for energy efficiency leads to

$$\eta = \frac{\bar{k}_\theta \bar{k}_a \bar{k}_{12}^2}{\left[\bar{k}_{12}^2 - (1+\bar{k}_\theta)(1+\bar{k}_a)\right]^2}$$

Note that the units of k_{11} is lb/in (or N/m), k_{22} is in-lb/rad (or m-N/rad), and k_{12} is lb (or N). The $\bar{k}_{12}$ is a coupling coefficient with its value varying from -1.0 to $+1.0$. The parameter $\bar{k}_\theta$ defines how stiff the external load is in comparison to the direct stiffness of the coupling mechanism.

8.5 Stroke Amplification Using Frequency Rectification: The Piezoelectric Hybrid Hydraulic Actuator

Frequency rectification is a method of increasing the stroke of an active material without the need for a mechanical amplification mechanism. The concept consists of a mechanical system that accepts a low-magnitude, high-frequency, oscillatory displacement from an active material and converts it into a larger magnitude displacement at a lower frequency. The large bandwidth capability of the active material is traded off into an increase in output displacement. Hence, this concept is suitable for materials like piezoceramics, magnetostrictives, and electrostrictives and is not applicable to low-bandwidth materials like SMAs. The mechanical system operates by rectifying the oscillatory input displacement from the active material. In this regard, the system can be thought of as a "mechanical diode." In addition, the mechanical system sums up each cycle of the rectified displacement, resulting in a steadily increasing output displacement. Because the output moves by a small step for each cycle of the input displacement, this concept is also referred to as "stepwise" or "step-and-repeat" actuation. The output can be converted again into an oscillatory displacement, as required by the application, although at a much lower frequency than the input displacement. The rectification and summation can be achieved by the following broad classes of mechanical systems:

1. Friction based – clamp and release: In this type of mechanism, the rectification and summation of the input displacement is achieved by clamping elements on the output of the device. When the clamping elements are actuated, they grip onto a fixed surface by means of frictional force. The clamping elements are actuated in the appropriate phase in relation to the input displacement to yield a rectified and summed output.
2. Friction based – traveling wave: The active material is arranged in such a way that it generates a traveling wave when actuated with the appropriate phasing. The out-of-plane displacement (crests of the traveling wave) result in the active material contacting a fixed surface and creating tangential frictional forces. Due to the friction generated, the active material assembly moves in the opposite direction.
3. One-way mechanical elements: Roller clutches are an example of mechanical elements that allow motion in only one direction. The oscillatory motion of the active material is coupled to the input stage of the roller clutch, and the output is rectified and summed.
4. Hydraulic elements: Check valves in a hydraulic fluid allow flow in only one direction. An oscillatory flow-rate input to the check valves will result in a cumulative unidirectional fluid flow.

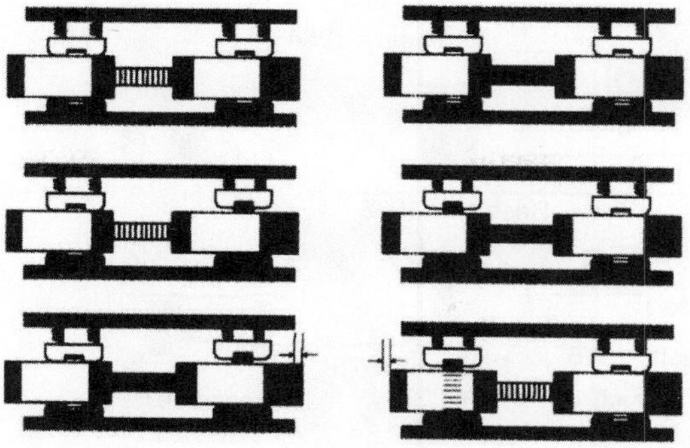
Starting position: both ends clamped
Step 1: release clamp 2
Step 2: Extend the pusher (leader indicates one step)
Step 3: reapply clamp 2
Step 4: release clamp 1
Step 5: Contract the pusher

Figure 8.26. Schematic of the translational operation of an inchworm motor (H3C), from Ref. [81].

Because the first three concepts rely on mechanical contact or friction, they are susceptible to wear. This issue is considerably alleviated in the case of a hydraulic system with check valves, and such devices could demonstrate advantages in terms of reliability and service life. Examples of each of these concepts are given herein, followed by a detailed discussion of a hybrid hydraulic actuator. Descriptions of the fundamental concept of frequency-rectified devices, as well as examples of several designs, can be found in Refs. [79] and [80].

8.5.1 Inchworm Motors

Inchworm motors utilize three or more active elements to achieve frequency rectification. One group provides the motive power and the other group acts as a brake and alternately clamps the ends of the motive piezostack elements to a stationary rail or shaft. This results in a net displacement in only one direction. By actuating the motive stacks at a high frequency, the entire assembly can achieve large linear velocities. The clamping can also be performed by passive mechanical elements. A schematic of the operation of a linear inchworm motor is shown in Fig. 8.26. This concept has been investigated in one form or another since the 1960s. The majority of the early concepts were focused on precision positioning applications. A comprehensive review of the historical development of inchworm-type actuators was published by Galante et al. [81] (Fig. 8.27). They also developed a compact inchworm motor operated by three piezostacks for a shape control application. The device measured 60 × 40 × 20 mm and was capable of a no-load output velocity of 0.6 cm/s and a blocked force in excess of 40 N. A holding force of greater than 200 N was achieved by careful design of the clamping mechanism. This aspect of the design is often the most crucial and the most susceptible to wear because the holding force is based on friction on the clamping elements. To address this issue, Park et al. [82] developed an inchworm motor with clamping ability enhanced by the use of MEMS micro-ridges.

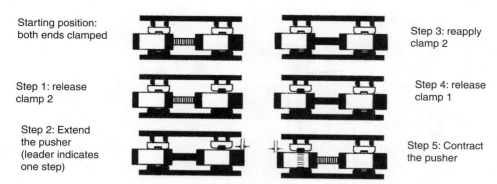

Based on the inchworm concept, rotary output motion can also be achieved. In such rotary motors, the clamping elements engage a cylindrical stator and the motive elements, which are mounted on a rotor, to provide a torque about a central shaft [83]. Another rotary motor concept is based on rectifying the angular displacements created by a set of piezoelectric bimorphs vibrating at resonance using a roller

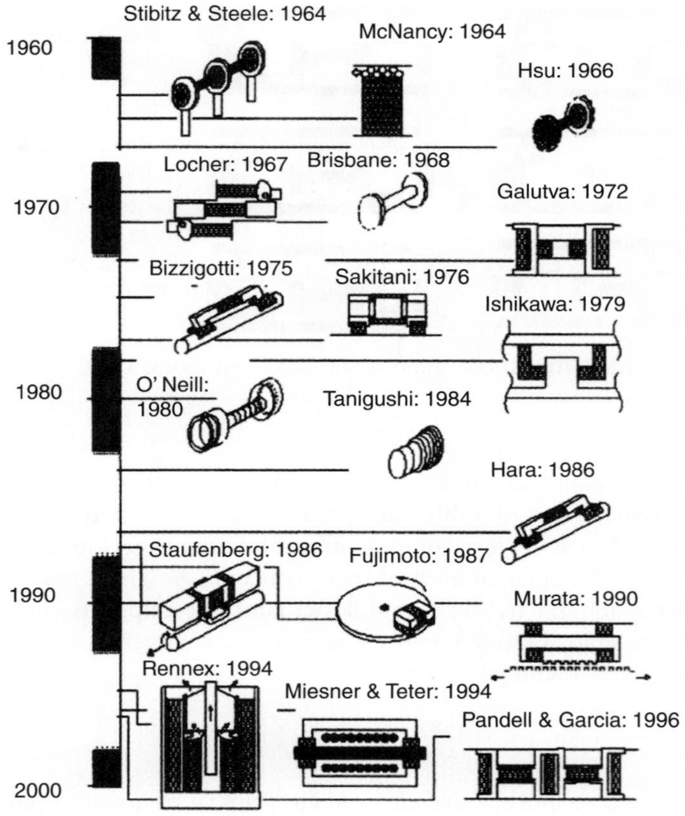

Figure 8.27. Illustrated timeline of inchworm history, from Ref. [81].

clutch [84, 85, 86]. Because the roller clutch is a passive frequency-rectification device, the construction of this device is much simpler than devices requiring active clamping by secondary piezoelectric actuators.

8.5.2 Ultrasonic Piezoelectric Motors

In contrast to inchworm-type motors, which require active clamping elements, ultrasonic piezoelectric motors are based on the passive generation of frictional forces. These motors are based on a traveling wave created in a ring of piezoelectric material. The ring is sandwiched between two fixed stator rings and presses against them. The motion of the traveling wave in the piezoelectric ring creates a frictional force between it and the stator, which results in a relative motion. In an alternate type of design, the longitudinal vibrations of a piezoelectric element are converted to rotary motion by means of specially shaped or angled mechanical links in contact with the output shaft. However, devices based on this design can only operate unidirectionally and were only seen in early research in this area. A detailed review and description of various types of piezoelectric motors can be found in the references [79, 87, 88, 89].

Because the piezoelectric element is normally driven at resonance to obtain maximum displacement, the operation frequency is linked to the physical dimensions of the device. The majority of such motors were developed for applications in small devices (e.g., in focusing drives for camera lenses). Consequently, these motors operate at very high frequencies, typically above 20 kHz. At these ultrasonic frequencies, the motors have the added advantage of being practically noiseless. Several different types of this motor exist: traveling-wave motors, standing-wave ultrasonic motors, multimode ultrasonic motors, and hybrid-transducer ultrasonic motors. All of these concepts are based on the same fundamental principle. Although the majority of applications are based on precision positioning, ultrasonic motors have also been used as actuators to control surfaces in an unmanned aerial vehicle as part of the Smart Wing program [5, 90].

Piezoelectric motors can have a mechanical efficiency as high as 50%, although typical values are between 20% and 30%. The friction-based design and tight manufacturing tolerances required make them suited only for relatively low-power applications, on the order of less than 100 W. Their main advantages compared to conventional DC motors are that they typically operate at a high torque and low rotational speed, they are noiseless, and they can be used in environments where magnetic fields are undesirable.

Frank et al. [86] designed and tested a rotary motor driven by resonant piezoelectric bimorphs. Tip masses were attached to the bimorphs, which were radially arranged around a central hub. A rotary clutch was installed between the central hub and an output shaft. Actuation of the bimorphs at their resonant bending mode resulted in an oscillatory rotation of the central hub. The rotary clutch rectified this oscillatory displacement into a constant velocity output. In this manner, the inertial forces produced by the resonant masses were transmitted to the load on the output shaft. The device was successful and several prototypes were fabricated for a flow control application. A mathematical model was developed that was used for optimization of the design. The final prototype demonstrated a power density of 10.5 W/kg, with a stall torque of 0.048 N.m and a no-load rotational speed of 366 RPM, and the bimorph drive frequency was around 894 Hz. Again, the low output speed without the need for a gearbox is one of the main attractions of this concept over a conventional DC motor.

8.5.3 Hybrid Hydraulic Actuation Concept

A common feature of all of the actuation mechanisms described previously is their reliance on friction to achieve relative motion. As a result of this friction, the surfaces in contact undergo significant wear, which limits the useful lifetime of the actuator. These actuators also require tight tolerances to generate sufficient friction and to achieve efficient frequency rectification. The wear of the surfaces adversely affects these tolerances and, therefore, affects the performance of the actuators. In addition, the blocked force of the actuator depends on achieving maximum friction between the stationary and moving parts. Although the frequency-rectification principle promises to simultaneously enable large output force and stroke, it is desirable to eliminate the reliance on friction. One concept that realizes this is the hybrid hydraulic actuator. A description of the working principle, followed by the construction and performance of a specific device, is given here to illustrate the concept of this actuator.

In a hybrid hydraulic actuator, an active material actuator is excited at a high frequency, displacing a hydraulic fluid in a constrained volume referred to as the pumping chamber, through a displacement rectification device. The rectification device is a set of unidirectional flow valves that allows the fluid to flow in only one direction. These valves are usually passive, but some current research efforts are exploring actively controlled valves. The active material actuator, pumping chamber, and valves form a solid-state hydraulic pump with no moving parts. The pump is coupled to a conventional hydraulic cylinder through a manifold. The fluid pressurized by the pump is utilized to transmit power to the hydraulic cylinder, resulting in a localized, self-contained hydraulic actuation system. The entire system, consisting of active material actuator, pumping mechanism, valves, manifold, and output hydraulic cylinder, is referred to as a hybrid hydraulic actuator. In principle, the pump can be actuated by any active material that has a high-stiffness and high-frequency response. To avoid confusion in the remainder of this discussion, the entire assembly is referred to as the "device" and the active material is referred to as the "actuator."

Several researchers have developed different versions of the hybrid hydraulic actuator. There has been considerable interest in hybrid actuation systems as potential actuators for a variety of aerospace [91, 92, 93] and automotive applications [94]. Several prototype piezohydraulic actuators have been designed and tested, developing an output power in the ranges of tens of watts. A promising application of this technology is in the area of control surface actuation for aerospace vehicles. Conventionally, control surfaces on aircraft are actuated by hydraulic actuators that are supplied with high-pressure fluid from a centralized pump. The weight of the associated hydraulic hoses, fittings, and hydraulic fluid contained in the system can be significant. The conventional actuators can be replaced with multiple localized piezohydraulic actuators. Because several can be located at one control surface and each has its own self-contained hydraulic circuit, the overall system can be more redundant and weight efficient than a conventional centralized hydraulic actuation system.

Konishi et al. [95, 96] developed a piezoelectric hybrid hydraulic actuator using a piezostack of length 55.5 mm and diameter 22 mm. The operating voltage of this piezostack was -100V to $+500$V, and its free strain and blocked force were 60 μm and 10.8 kN, respectively. This piezoelectric pump was excited at 300 Hz and delivered an output power of approximately 34 W. Mathematical models were developed and the possibility of using fluid resonance to increase the output power was investigated [97, 98].

A magnetostrictive water pump, which is conceptually similar to a pump driven by a piezoelectric stack, was developed by Gerver et al. [99]. This pump operates at a relatively low pressure, on the order of 34.5 kPa (5 psi), and makes use of an additional hydraulic stroke amplification scheme to increase the flow rate. A flow rate of 15 ml/sec at an output pressure of 34.5 kPa (5 psi) was reported.

Among early studies, Nasser et al. [100] presented a piezohydraulic actuation system that made use of the compressibility of the working fluid to eliminate accumulators and four-way valves. Use of active bi-directional valves was envisaged to control output actuator motion. One of the main goals of this work was to determine the effects of the friction of the actuator on its performance. The piezostack driving this pump had a free stroke of 100 μm and a capacitance of 40 μF. The system was run at 10 Hz at an input voltage amplitude of 150 V, and it demonstrated an overall amplification factor of 1.42. Because this device made use of commercially available solenoid valves as active check valves, operation of the device was limited to low pumping frequencies. As a result, it functions more as a hydraulic amplification device than a frequency-rectification device.

Mauck and Lynch [91, 101] investigated a system consisting of a pump driven by a high-voltage piezostack of length 10.2 cm and cross-sectional area 1.9 cm × 1.9 cm. Several versions of designs were investigated, with accumulators incorporated into later designs. The final device achieved a blocked force of 61 lbs and an output actuator velocity of 7 cm/sec. The large current requirements and heating of the piezostack limited the pumping frequency of the system to 60 Hz. The overall performance of the system was analytically examined in terms of actuation efficiencies, and the effect of the viscosity of the hydraulic fluid was experimentally determined. A lumped-parameter theoretical model of this system was developed.

Anderson et al. [102] described the development of a compact piezohydraulic actuator for potential application as a control surface actuator on a UAV such as the X-45A (UCAV). A maximum output power of 42 W was measured, with the piezostacks being driven at 750 Hz. It was concluded that the concept was promising, but substantial improvement in performance was necessary before the technology could compete with conventional electromechanical actuators. Cavitation in the hydraulic circuit was determined to be a major factor limiting the maximum pumping frequency.

Sirohi et al. [93] developed a piezohydraulic actuator that operated at a maximum pumping frequency of 600 Hz. The tested prototype pump weighs 300 gm, and the actuator had an output blocked force of 70.6 N (16 lbs), with an unloaded velocity of 140 mm/sec (5.5 in/sec). An improved version of this device was tested with piezoelectric, magnetostrictive, and electrostrictive driving elements, and their relative performance and efficiencies were compared [103]. This hybrid actuator system was extensively tested in a closed system, and a detailed description of the design, development, and performance of this system follows in a subsequent section.

A magnetostrictive hydraulic pump was developed by Bridger et al. [104], with the goal of achieving a power output of 400 W, with a 20.7 MPa (3000 psi) operating pressure and a no-load flow rate of 57.35 cm^3/sec (3.5 in^3/sec). Two designs were investigated, one with a clamped active element pushing against a piston and another with a Tonpilz-type active material resonant driver displacing a piston. Limitations were encountered with the passive check valves, and compliance in the pumping chamber made it difficult to achieve the desired operating pressure of 3000 psi. The Tonpilz design was lighter and was designed to operate at a resonant frequency of 2 kHz; however, the required 3000 psi pressure was achieved only at very high pumping frequencies.

A hybrid hydraulic actuator based on an SMA thin-film bubble was also developed [105] that had the same fundamental operational principle as a piezohydraulic actuator. In this device, the piezostack-diaphragm assembly was replaced by a SMA membrane that displaced the hydraulic fluid in the pumping chamber. The goals of this device were to achieve a power density of 100 W/kg, with an output force of 100 N, stroke of 4 mm, and no-load output velocity of 50 mm/s. The SMA film intrinsically was expected to have an energy density of 40 kW/kg, which is at least an order of magnitude larger than other active materials. The final device comprised 10 SMA membranes working in parallel, each pumping hydraulic fluid at a frequency of 100 Hz. Each membrane was 8 μm thick and 1 cm in diameter. The pump was coupled to a hydraulic cylinder, and at 50 Hz pumping frequency, a velocity of 5.4 mm/s and an output force of 100 N was measured. The SMA membranes were activated by passing short bursts of high current through them, followed by relatively long cooling periods. The pump also incorporated an array of MEMS check valves operating in parallel. Each valve was a single flap or reed type design with dimensions on the order of 600 μm. The valves were designed using a trade study based on FEM analysis, with a minimum flow resistance and a first natural frequency of greater than 20 kHz.

Active valves based on piezoelectric unimorph disks were tested in a piezoelectric pump by Lee et al. [106]. A structural optimization was performed to maximize the volume of displaced fluid and to minimize the weight of the pump housing. Simulations indicated valve operation up to 15 kHz. A maximum power density of 12 W/kg was measured; however, the device was not extensively tested in a closed hydraulic circuit. A piezohydraulic pump utilizing proprietary check valves was tested by Tieck et al. [107]. The power output of the pump was measured as 46 W while operating at a bias pressure of 3.45 MPa (500 psi) and a pumping frequency of 1 kHz, in an open hydraulic circuit.

At this point, it is worth mentioning that significant research has been focused in the biomedical field on piezoelectrically driven micropumps [108, 109]. The main goal of these devices is to move small, precise quantities of fluid from one location to another. Shoji et al. [108] compiled an extensive review of microflow devices. One microflow concept was developed as a hybrid energy-harvesting transducer [110]. Some of these designs also dispense with mechanical valves and achieve flow directionality by means of appropriately designing the geometry of the inlet and outlet ports [111, 112, 113]. Several micropumps utilize piezoelectric unimorphs or bimorphs as their actuators, which are of relatively low stiffness but have a large free displacement. Other micropump concepts include peristaltic pumps [114] and resonantly driven pumps [115]. However, most of these devices operate on the micro-scale and are too small to be considered as actuators that produce significant mechanical work output.

A schematic of a hybrid hydraulic actuator driven by piezoelectric stack actuators [93] is shown in Fig. 8.28. The frequency rectification is achieved by passive mechanical valves that have a high natural frequency. The solenoid valves that are mounted in the manifold operate at a much lower frequency than the active material, and they serve to control the direction of output motion. The accumulator is present to enable easy filling of the hydraulic circuit and, more important, to maintain a positive bias pressure on the hydraulic fluid as well as on the piezostacks. The bias pressure reduces the possibility of cavitation in the hydraulic fluid and decreases the effect of entrained air on the fluid compressibility. In addition, the bias pressure provides a steady compressive pre-load to the piezostacks.

8.5.4 Operating Principles

The hybrid hydraulic actuator operates by displacing a small volume of hydraulic fluid during each stroke of the active material. A schematic diagram of the hydraulic pump driven by a volume of active material is shown in Fig. 8.28(a). The active material is excited by an oscillatory electric or magnetic field, resulting in an oscillatory displacement of the piston. The diaphragm acts as a seal, preventing hydraulic fluid from entering the body of the pump, and also functions as a return spring for the piston. A small volume of hydraulic fluid is displaced by the piston during each stroke of the active material. Whereas the displacement of the active material is

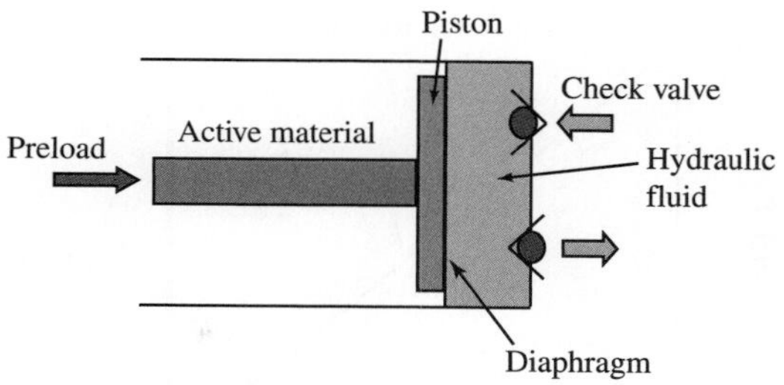

(a) Piezoelectric pump

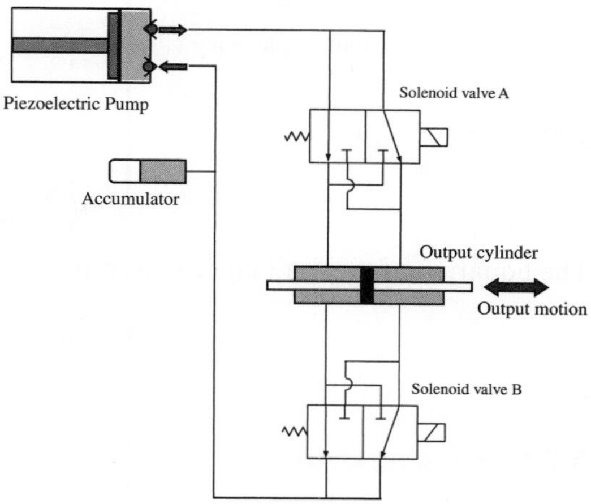

(b) Hybrid actuator

Figure 8.28. Schematic of piezo-hydraulic hybrid actuator.

bi-directional, the flow of the hydraulic fluid is rectified by mechanical check valves. This results in a unidirectional flow of pressurized hydraulic fluid.

The volume of fluid displaced in each cycle depends on a number of factors, such as stiffness of the actuator, compressibility of the fluid, and impedance of the hydraulic circuit. The mechanism of pumping can be better understood by looking closely at the energy transfer between the active material and the hydraulic fluid. We now focus on the operation of a hydraulic pump driven by piezostack actuators. However, the basic operating principle remains the same for any type of active material driving the hydraulic pump.

8.5.5 Active Material Load Line

The force and displacement characteristics of an active material actuator are typically defined by its load line. In general, the relationship between force and displacement may not be linear. However, it is convenient to approximate it by a straight line, especially for a preliminary analysis. A typical load line for a piezostack actuator is shown in Fig. 8.29. The force F_o and displacement δ_o of the actuator are related by

$$\delta_o = \delta_f \left(1 - \frac{F_o}{F_{bl}}\right) \tag{8.61}$$

where F_{bl} is the blocked force and δ_f is the free displacement of the actuator. Note that the slope of the load line is equal to the stiffness of the actuator, k_{act}, and is

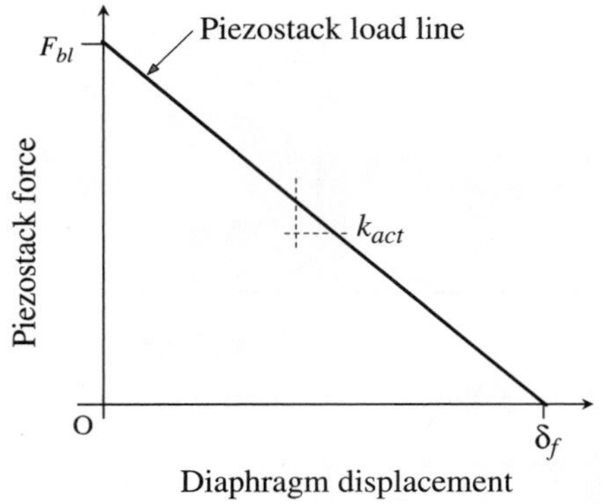

Figure 8.29. Load line for a piezostack actuator.

given by

$$k_{act} = \frac{F_{bl}}{\delta_f} \tag{8.62}$$

The equation of the load line can also be written as

$$\delta_o = \delta_f - \frac{F_o}{k_{act}} \tag{8.63}$$

8.5.6 Pumping Cycle

A simplified diagram of the working cycle of a piezohydraulic hybrid actuator, with an ideal incompressible hydraulic fluid, is shown in Fig. 8.30. This figure plots the force on the piezostack as a function of the piston displacement in the pump. Note that the force on the piezostack is equal to the product of the pressure in the pumping chamber and the area of the piston, less the force required to deflect the diaphragm. In addition, the pressure of the hydraulic fluid is equal to the ratio of the external force on the device and the area of the output cylinder.

It is assumed that the piston is always in contact with the active material, which is an accurate assumption at low frequencies. The actual kinematics of this motion is a function of the design of each particular pump. The part of the curve labeled **OA-AB** corresponds to the compression stroke of the pump. In this part of the

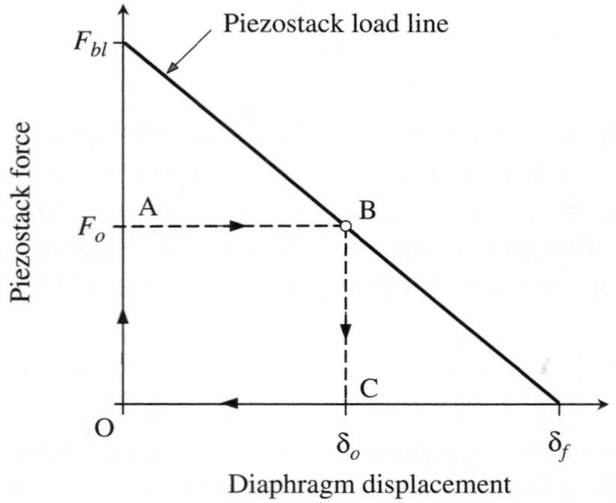

Figure 8.30. Pumping cycle for an ideal incompressible hydraulic fluid.

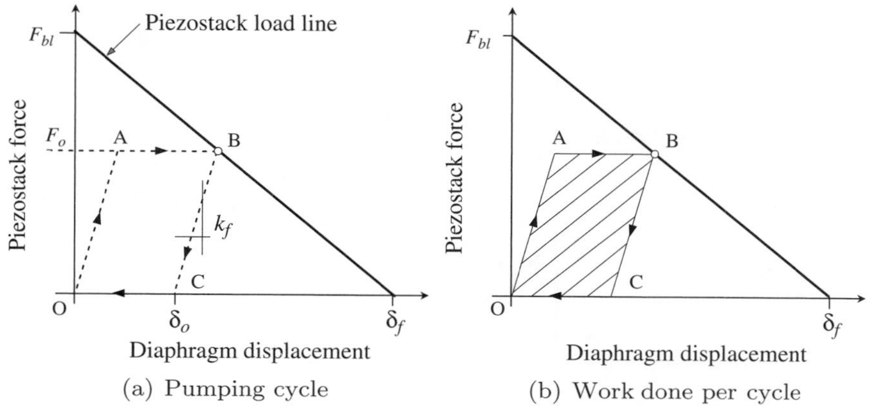

Figure 8.31. Force, displacement, and work done during one pumping cycle, for a hydraulic fluid of finite stiffness.

cycle, the pumping chamber pressure increases instantaneously as the piston starts displacing, to the point **A**. At this point, the force on the piezostack, F_o, corresponds to the external load on the device. The outlet check valve opens, and the piston continues to displace fluid till the point **B**, defined by the load line of the piezostack. Similarly, as the piezostack starts contracting, the force on the piezostack drops instantaneously to zero (point **C**) and then stays constant until the piston returns to its original position at point **O**. The entire loop **OABCO** comprises one pumping cycle. Note that the instantaneous increase and decrease of piezostack force is due to the incompressible nature of the hydraulic fluid. However, in reality, the fluid is compressible, and the stiffness of the fluid contained in the pumping chamber is comparable to the stiffness of the piezostack. Therefore, a more realistic representation of the pumping cycle is as shown in Fig. 8.31(a).

As the piston starts displacing fluid from the point **O**, the piezostack force increases linearly along the line **OA**. The slope of this line is given by the effective stiffness of the fluid in the pumping chamber, k_f. At the point **A**, the outlet check valve opens and the piston moves to the point **B**, which is a point defined by the load line of the piezostack. Useful external work is done in the segment **AB**, during which the output load undergoes some displacement. The segment **BC** is the return stroke of the piezostacks, where the pumping chamber pressure falls back to the original value and the piezostacks return to their initial length. The work done by the piezostack on the external load per cycle is given by the shaded area **OABCO** in Fig. 8.31(b).

The limits of operation of the device are defined by the points **B** and **C** in Fig. 8.32. Two limiting cases of pumping cycles are shown. The pumping cycle corresponding to the blocked condition of the device is given by **OABAO**. The point **B** is the point of intersection of the fluid stiffness line and the piezostack load line. In the blocked condition, the external load on the device is so large that all of the work done by the piezostack goes into compression of the fluid in the pumping chamber, and the net work output of the device is zero. The corresponding force at the output hydraulic cylinder is the blocked force of the hybrid actuator, and $F_o + F_{crack}$ corresponds to the force on the piezostack at this condition. Note that this force is less than the blocked force of the piezostack. The force F_{crack} is defined as the force on the

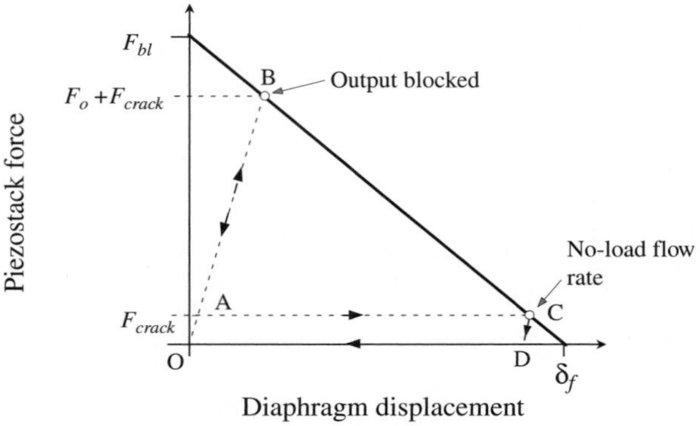

Figure 8.32. Limits of operation of the piezostack driven pump.

piezostack corresponding to the minimum pressure required to open the mechanical check valve, which is usually called the cracking pressure of the valve.

The pumping cycle corresponding to the unloaded condition of the device is given by **OACDO**. At point **C**, because the external load is zero, the only force on the piezostacks is due to the cracking pressure of the check valves. In the unloaded condition, the maximum flow rate, or no-load flow rate, of the device is achieved. The output work of the device is zero because the external load in this condition is zero. Maximum work output of the actuator is achieved at an intermediate point that can be determined from a tradeoff study.

It is worth mentioning here that Fig. 8.31(a) also contains some level of idealization. In reality, the segments **AB** and **OC** will have a non-zero slope, corresponding to the effective stiffness of the working fluid in the hydraulic circuit when the outlet check valve is open. However, because the volume of fluid in the circuit is large compared to the volume of fluid in the pumping chamber, the stiffness of the fluid in the circuit will be much higher than that of the segments **OA** and **BC**. Hence, this slope is approximated to be zero. For the remainder of this discussion, the segments **AB** and **OC** are assumed to be horizontal.

8.5.7 Energy Transfer

The performance of the actuator is highly sensitive to the dimensions of both the pumping chamber and the output hydraulic cylinder. The viscosity of the fluid, the diameter of the tubing, and the modulus of elasticity of the tubing material are also important for the compliance and frictional losses that occur in the device, as well as for its frequency response. At high pumping frequencies, the performance of the device is dominated by the dynamics of the hydraulic circuit. However, for an initial assessment, it is convenient to neglect any frequency dynamics of the system and perform a quasi-static analysis of the energy transferred between the active material and the hydraulic fluid. As a result of this quasi-static pumping assumption, the volumetric flow rate of the pump, Q, at any given pumping frequency is equal to the product of the volumetric displacement of the pump per cycle, Δ_{pump}, and the pumping frequency, f_{pump}

$$Q = \Delta_{pump} f_{pump} \tag{8.64}$$

The large flexibility in trading off force and stroke afforded by the use of hydraulics results in many possible combinations of pumping chamber and output

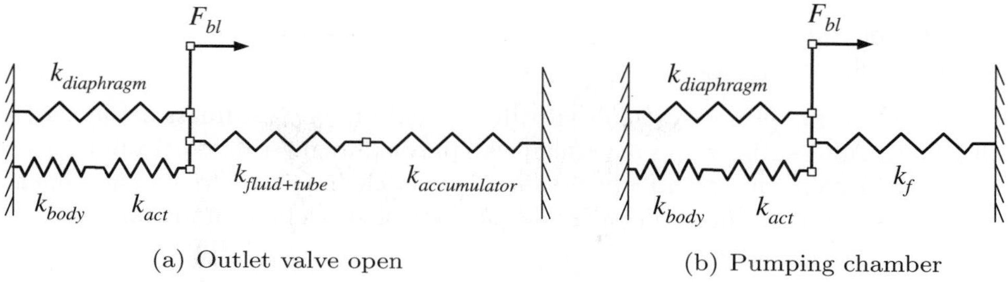

Figure 8.33. Simplified system model.

cylinder dimensions. The final design of the hybrid actuator is dependent on its output requirements. Due to this strong coupling between the pump and output cylinder, all subsequent discussions of the performance of the device are with respect to its force and stroke as opposed to the pressure and flow rate generated by the pump alone.

Because the ultimate goal is to maximize output power density for a device of known external dimensions, the output power is considered to be the primary performance metric. The major variables on which the output power depends are as follows:

1. pumping chamber diameter, d_{cham}
2. output cylinder diameter, d_{out}
3. pumping chamber height, Δ_{gap}
4. piezostack characteristics: blocked force F_{bl} and free displacement δ_f
5. fluid compressibility, β

As a result of the quasi-static pumping assumption (Eq. 8.64), to maximize the output power, it is sufficient to maximize the work done by the device per pumping cycle. This is given by the area **OABCO** in Fig. 8.31(b), which outlines a typical pumping cycle. The work done per cycle is given by

$$\Delta W_{cyc} = F_o \delta_o \tag{8.65}$$

where F_o and δ_o are related by the equation of the piezostack load line (Eq. 8.61). To calculate the area under the curve, it is necessary to derive expressions for the locations of the points **A**, **B**, and **C** in Fig. 8.31(a). This involves calculating the fluid stiffness k_f. Because the present analysis is quasi-static, the mass terms can be ignored and the overall system can be treated as an arrangement of springs, as shown in Fig. 8.33(a), yielding a static solution for deformations of the system. Note that Fig. 8.33(a) is the equivalent system representation when the outlet valve is open, and the output hydraulic cylinder is assumed to be clamped. The effective stiffness obtained from this configuration will give the slope of the segments **OA** and **BC**. The stiffness of the fluid in this case can be thought of as similar to the finite stiffness of the linkages in a mechanical amplification system. The equivalent system when both valves are closed, which leads to the stiffness k_f, is shown in Fig. 8.33(b). This stiffness is only due to the fluid in the pumping chamber.

The stiffness elements that comprise the device with output cylinder clamped and inlet valve closed are as follows:

1. pump body – k_{body}
2. piezostack internal stiffness – k_{act}

3. diaphragm – $k_{diaphragm}$
4. fluid and tubing – $k_{fluid+tube}$
5. accumulator – $k_{accumulator}$

Note that the piezostack internal stiffness and the body stiffness act in series, and the resulting stiffness acts in parallel with the diaphragm stiffness, resulting in an actuation element with an effective stiffness k_p. Each element is treated as a linear mechanical spring element with a force-deflection characteristic of the form

$$\Delta F = k_{spring} \Delta x \tag{8.66}$$

where ΔF is the change in force in the spring, k_{spring} is the spring constant, and Δx is the change in length of the spring. The stiffness of a fluid element can be expressed in terms of its compressibility as [116, 117]

$$\Delta P_e = \beta_{eff} \frac{\Delta \mathbb{V}_e}{\mathbb{V}_e} \tag{8.67}$$

where ΔP_e is the change in pressure in the element. In practice, the tubing exhibits some degree of compliance as well, and this can be accounted for by replacing the fluid bulk modulus with an effective bulk modulus, β_{eff}, that is derived considering the change in volume of the tubing resulting from a pressure rise. $\mathbb{V}_e$ and $\Delta \mathbb{V}_e$ are the initial volume and change in volume of the element, respectively. The pumping chamber and the tubing in the hydraulic circuit can be treated as cylindrical volumes of fluid. Considering a cylindrical volume of fluid with a cross-sectional area A_e, and assuming that the change in volume is caused purely due to a change in length Δx of the cylinder, we have

$$\begin{aligned} \Delta P_e &= \frac{\Delta F}{A_e} = k_{spring} \frac{\Delta x}{A_e} \\ &= \beta_{eff} \frac{\Delta x A_e}{\mathbb{V}_e} \end{aligned} \tag{8.68}$$

The effective stiffness of the fluid volume is

$$k_{spring} = \frac{\beta_{eff} A_e^2}{\mathbb{V}_e} \tag{8.69}$$

The fluid in the pumping chamber can be represented as a cylindrical volume of fluid with a length Δ_{gap} and a cross-sectional area $A_p = \pi d_{cham}^2/4$. From this equation, the stiffness of the fluid in the pumping chamber is given by

$$k_f = \beta_{eff} \frac{A_p}{\Delta_{gap}} \tag{8.70}$$

From Fig. 8.31(a), and Eq. 8.63

$$\begin{aligned} \delta_o &= \delta_f - \frac{F_o}{k_p} - \frac{F_o}{k_f} \\ &= \delta_f - F_o \left[\frac{1}{k_p} + \frac{1}{k_f} \right] \end{aligned} \tag{8.71}$$

Defining an effective stiffness, k_{eff}, the output displacement can be written as

$$\delta_o = \delta_f - \frac{F_o}{k_{eff}} \tag{8.72}$$

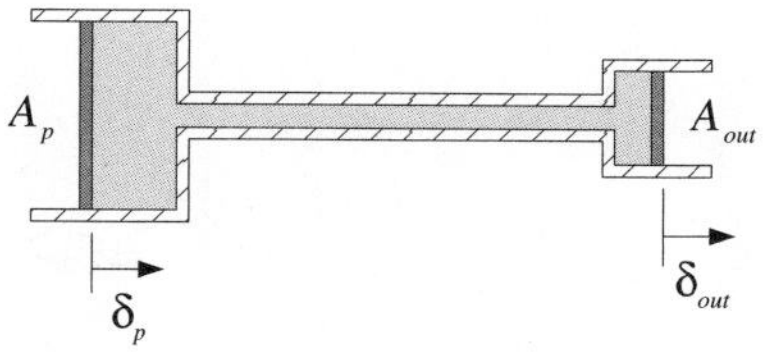

Figure 8.34. Relation between the pumping chamber area and output cylinder area.

Note that k_{eff} is the effective stiffness of the piezostack and the pumping chamber. For a typical set of system parameters, the stiffness of the accumulator is usually small compared to the stiffness of the fluid and the tubing, which is the dominant factor in the design. The diaphragm stiffness is typically designed to be around 10% of the piezostack stiffness. Note that the stiffness of the pump body, k_{body}, should be much larger than the piezostack stiffness to minimize energy loss in the pump body. In practice, a good rule of thumb is to design the stiffness of the body to be at least a factor of 10 higher than the stiffness of the piezostack.

8.5.8 Work Done Per Cycle

An effective load line of the piezostack is used in the analysis to account for the diaphragm stiffness. This is obtained by scaling the original piezostack load line to account for the combined stiffness of the piezostack-diaphragm assembly. From Eqs. 8.72 and 8.65, the work output of the device per cycle is

$$\Delta W_{cyc} = F_o \left[\delta_f - \frac{F_o}{k_{eff}} \right] \tag{8.73}$$

This work can be expressed in terms of the fluid pressure, or piezostack force, and the piston displacement. Figure 8.34 shows a schematic of the fluid column between the pumping chamber and the output cylinder during the output stroke of the pump. The pumping chamber has a cross-sectional area A_p and a piston displacement δ_p, whereas the output cylinder has a cross-sectional area A_{out} and a displacement δ_{out}. The area ratio is given by

$$A_R = \frac{A_{out}}{A_p} \tag{8.74}$$

The work output per cycle can be rewritten in terms of the force (F_{out}) and displacement (δ_{out}) of the output cylinder as

$$\Delta W_{cyc} = \frac{F_{out}}{A_R} \left[\delta_f - \frac{F_{out}}{A_R k_{eff}} \right] \tag{8.75}$$

For a given working fluid and tubing, β_{eff} is fixed, and A_p may be constrained by the pump geometry. In such a case, the only parameter that the designer is free to choose is the pumping chamber height, Δ_{gap}. To increase the power output of the device, it is desirable to maximize the stiffness of the fluid in the pumping chamber. This can be accomplished by either increasing the pumping chamber diameter or decreasing the pumping chamber height. However, if there are no other geometrical restrictions, it is more effective to increase the pumping chamber diameter than to decrease the height because the fluid stiffness depends on the square of the diameter and is inversely proportional to the height. In addition, the pumping-chamber height

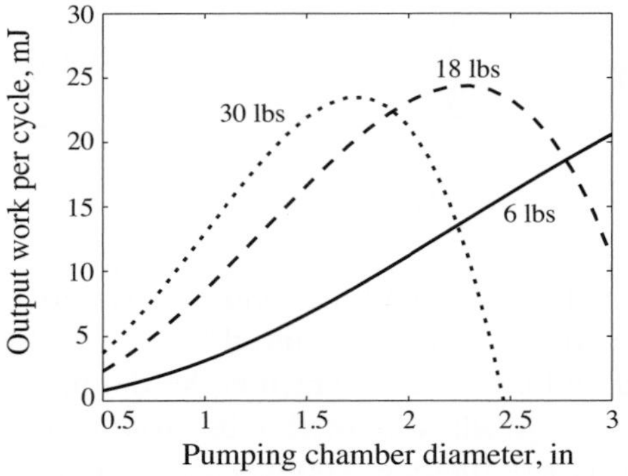

(a) Work output as a function of pumping chamber diameter, $\Delta_{gap} = 0.05''$

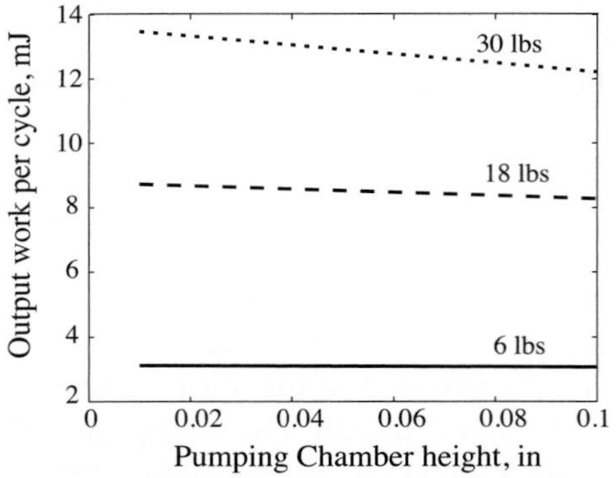

(b) Work output as a function of pumping chamber height, $d_{cham} = 1''$

Figure 8.35. Dependence of work output per cycle on pumping chamber geometry.

cannot be decreased indefinitely because the mechanical check valves require a finite clearance to function properly. Substituting the expression for fluid stiffness (Eq. 8.70) in Eq. 8.75, the expression for output work per cycle is

$$\Delta W_{cyc} = \frac{F_{out}}{A_R}\delta_f - \frac{F_{out}^2}{A_R^2 k_p} - \frac{F_{out}^2 \Delta_{gap}}{A_R^2 \beta_{eff} A_p} \tag{8.76}$$

8.5.9 Maximum Output Work

The dimensions of the output cylinder are typically fixed by the geometric constraints of the intended application of the device. Let us choose a commercially available 0.5 in bore diameter hydraulic cylinder as the output cylinder. This fixes the value of A_{out}, and the only parameters that remain to be fixed are the diameter and height of the pumping chamber. Note here that although maximum work output is the primary goal of the device, the specific application may require a certain output displacement at a certain bandwidth. This requirement directly translates into the flow rate of the piezoelectric pump. Therefore, in addition to sizing the parameters of the device for maximum output work per cycle, the constraint of achieving the required output displacement must also be included in the design process.

Plots of the variation of output work with A_{out} and Δ_{gap} are shown in Fig. 8.35. In Fig. 8.35(a), the work output per cycle is plotted as a function of the pumping chamber diameter for various external loads, with a pumping chamber height of 0.05 in. The maximum work output per cycle is achieved at a certain value of the pumping chamber diameter, and this maximum value decreases with external load. Figure 8.35(b) shows the work output per cycle as a function of pumping chamber height for a pumping chamber diameter of 1 in. The maximum work output increases monotonically (at a slower rate) with decreasing pumping chamber height. This is to be expected because a smaller pumping chamber height increases the stiffness of the fluid and essentially provides a direct energy transfer to the output load. A value of the pumping chamber height between 0.02 in and 0.05 in can be chosen depending on other factors, such as machinability and operating clearances for the mechanical check valves. The conditions for maximum output work per cycle can be obtained by differentiating Eq. 8.76 with respect to the two parameters, A_p and Δ_{gap}

$$\frac{\partial(\Delta W_{cyc})}{\partial A_p} \Rightarrow A_p = F_b \frac{A_{out}}{2F_{out}} - \Delta_{gap} \frac{k_p}{2\beta_{eff}} \tag{8.77}$$

If $\beta_{eff} \to \infty$ in the previous equation, then the value of A_p reduces to an impedance-matched condition; the resulting piezostack force is at the midpoint of the piezostack load line. This is as expected because the maximum work output is known to occur at an impedance-matched condition.

Based on the previous discussion, taking into consideration any geometric constraints on the overall size of the device, an optimum pump geometry can be arrived at for a given external load. Note that the fluid compressibility has a significant effect on the performance of the device. The actual compressibility depends on several factors, such as the system bias pressure, the amount of air entrained, and the flexibility of the tubing, and can be as low as 10% of the reference value [117]. Specifically, entrained air has a strong influence on the effective fluid bulk modulus and care must be taken to ensure that there are no air bubbles in the hydraulic circuit. At the same time, the application of a bias pressure on the hydraulic circuit significantly decreases the effect of entrained air on fluid compressibility and decreases the possibility of cavitation in the hydraulic fluid.

8.5.10 Prototype Actuator

We now describe the construction of a prototype hybrid hydraulic actuator driven by piezoelectric stacks. The device is constructed with the same configuration as shown in the schematic in Fig. 8.28(b). The function of each part of the circuit is as follows:

- **Piezoelectric pump:** This serves as a source of pressurized hydraulic fluid. Electrical energy is converted to mechanical energy by piezoelectric stack actuators. The piezoelectric pump achieves the same function as a conventional hydraulic pump, with a much simpler design and almost no moving parts. The hydraulic fluid transmits mechanical energy from the piezoelectric pump to the output cylinder.
- **Output cylinder:** This is a conventional hydraulic cylinder that can be connected to an external load. Work is done on the load by extracting energy from the pressurized hydraulic fluid.
- **Solenoid valves:** These change the direction of the actuator output by redirecting the hydraulic fluid into the appropriate sides of the output cylinder. At present, these valves are conventional electromagnetic valves; however, they can be miniaturized to yield a decrease in the overall volume of the actuator.
- **Accumulator:** The accumulator consists of a volume of the hydraulic circuit that is separated from a volume of air or gas by a rubber diaphragm. The accumulator provides a convenient means of filling the device with the hydraulic fluid. In addition, a bias pressure can be applied to the hydraulic circuit by filling the accumulator with pressurized gas. The bias pressure serves three important functions. First, it reduces the effect of entrained air on the compressibility of the hydraulic fluid. Second, it reduces the possibility of cavitation occurring in the hydraulic circuit. Third, it ensures that the piezoelectric stacks are always subjected to a compressive stress, thereby maintaining their structural integrity.

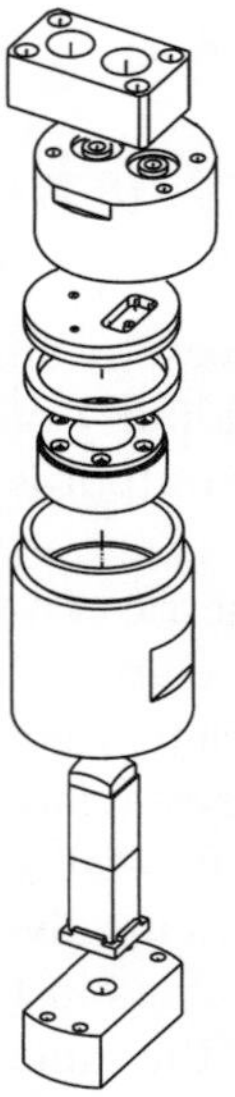

Figure 8.36. Exploded view of piezoelectric pump assembly.

- **Manifold:** The hydraulic circuit that goes between the pump, cylinder, valves, and accumulator can all be housed in a compact manifold. A special manifold can also be constructed to allow unidirectional testing of the device. In addition to housing the hydraulic circuit, the manifold is used to properly mount the device to work against an external load.

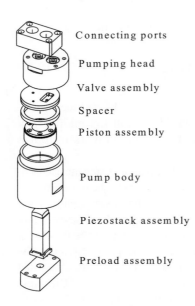

An exploded view of the piezoelectric pump is shown in Fig. 8.36. The main components of the piezoelectric pump are the piezostack assembly, piston assembly, pump body, pumping head, and pre-load assembly. The piezostack assembly consists of two commercially available low-voltage piezostacks (i.e., Model P-804.10 and Physik Instrumente [118]), that are bonded together, end to end. One end of the piezostack assembly is bonded to a pre-load mechanism and the other end is pushed up against a piston-diaphragm assembly. The pre-load assembly consists of a fine thread screw and a locking nut, and it serves to adjust the position of the piezostack assembly relative to the pump body as well as to provide a compressive pre-load to the piezostacks. The piston-diaphragm assembly consists of a steel piston, which has a tight running fit with the bore of the pump body and is clamped to a 0.002 in thick C-1095 spring-steel diaphragm. The diaphragm seals the pump body from the hydraulic fluid in the pumping chamber, and the piston serves to constrain the deflected shape of the diaphragm to remain flat over most of its surface, thereby maximizing the swept volume of the pump per cycle. Whereas one face of the pumping chamber is formed by the movable piston, the other face is formed by the pumping head, which contains the valve assembly. The valve assembly contains two passive check valves that are formed by thin (0.002 in) C-1095 spring-steel reeds sandwiched between two aluminum disks. The natural frequency of the steel reeds in air is designed to be above 3 kHz; therefore, it can be assumed that the reeds do not interfere with the pumping dynamics for pumping frequencies of at least 1.5 kHz. A schematic of the piston-diaphragm assembly and the valve assembly is shown in Fig. 8.37.

The temperature rise in the piezostacks caused by high-frequency operations can be minimized by surrounding the piezostacks with a thermally conductive

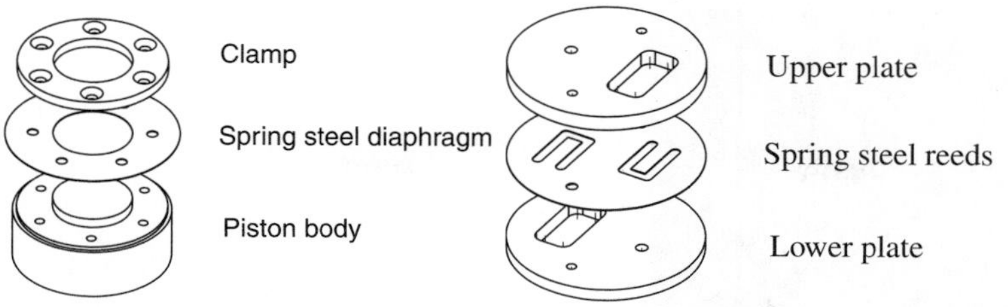

(a) Piston-diaphragm assembly (b) Reed valve assembly

Figure 8.37. Schematic of piston and reed valve assemblies in the piezoelectric pump.

compound that conducts heat away from the piezostacks and into the pump body [93]. To facilitate the heat conduction, the pump body is constructed of aluminum. The body of the pump has an outer diameter of 1.25 in, a length of 4 in, and has a mass of 300 g. The total length of the pump body can be increased as necessary to accommodate longer piezostacks. The pump is coupled to a commercially available hydraulic cylinder, with a bore of 7/16 in and a shaft diameter of 3/16 in, through a custom-built manifold. The manifold also houses the accumulator. A list of all of the important parameters of the current design is in Table 8.4. The piezostack data, shown for each piezostack, are taken from Ref. [15]. The parts that comprise the pump assembly are shown in Fig. 8.38(a), and Fig. 8.38(b) shows a picture of the assembled prototype piezoelectric hybrid hydraulic actuator, incorporating the custom manifold and the output cylinder. It can be seen that the system is completely self-contained and only requires electrical power input to the piezostacks. The entire device can be mounted where required, and power can be harnessed from the output shaft.

Table 8.4. *Prototype device parameters*

Piezostack – Model P-804.10		
Number of Piezostacks	2	
Length	0.3937	in
Width	0.3937	in
Height	0.7087	in
Blocked Force (0–100 V)	1133	lbs
Free Displacement (0-100 V)	$\approxeq$ 0.5	mil
Maximum Voltage	120	V
Minimum Voltage	–24	V
Capacitance	$\approxeq$ 7	μF
Hydraulic Fluid – MIL-H-5606F		
Density	0.859	g/cc
Kinematic Viscosity	15	centistokes
Reference Bulk Modulus β_{ref}	260,000	psi
Pumping Chamber		
Diameter	1	in
Height	0.050	in
Output Actuator – Double Rod		
Bore Diameter	0.4375	in
Shaft Diameter	0.1875	in
Stroke	2	in

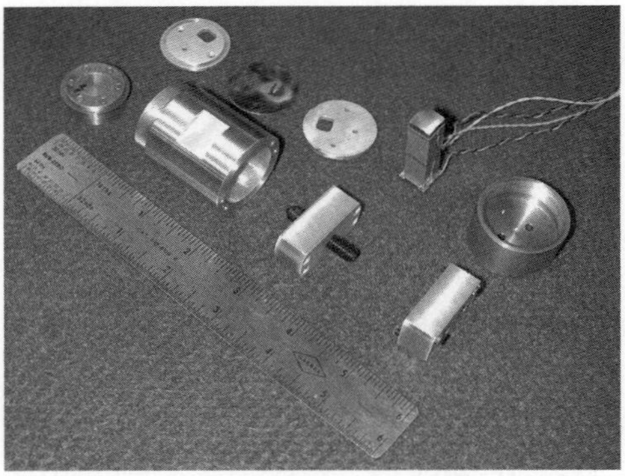

(a) Pump assembly components

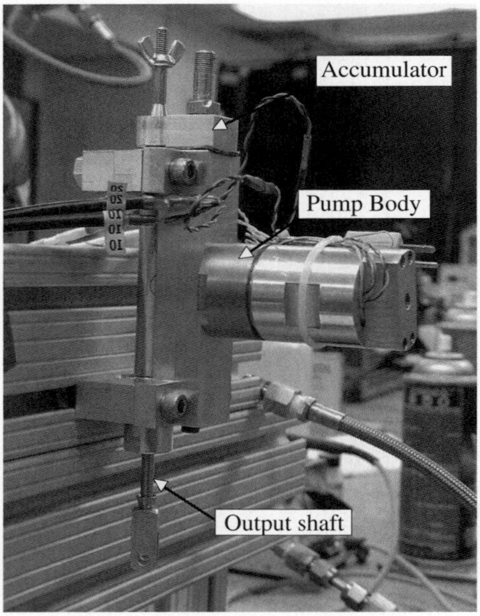

(b) Assembled hybrid actuator

Figure 8.38. Prototype piezoelectric hybrid hydraulic actuator.

8.5.11 Experimental Testing

Experiments are performed on the device to investigate its frequency response, power output, and efficiency. These experiments can provide data to validate analytical models of the device and can also provide insight on the effect of various system parameters on the performance of the device. The main objective is typically to maximize the power density of the device. The experiments involve operation of the device in either the unidirectional or the bi-directional mode.

Unidirectional Testing

To minimize the number of variables in the system and to decrease its overall complexity, it is convenient to perform initial testing of the device in the unidirectional mode. In this mode, the directional-control solenoid valves (valves **A** and **B** in

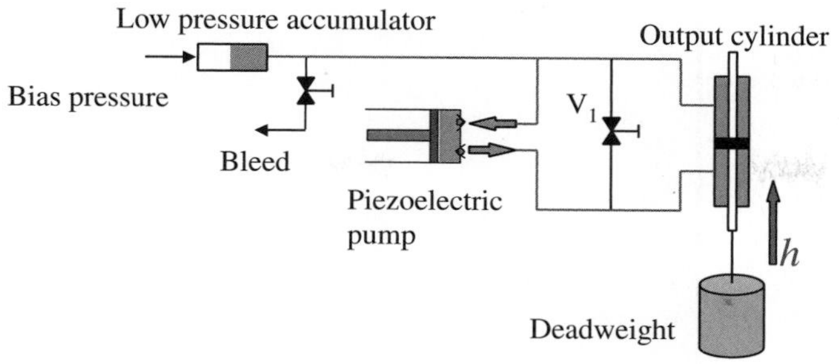

Figure 8.39. Schematic of unidirectional test setup.

Fig. 8.28(b)) are eliminated from the hydraulic circuit. A special manifold is constructed that includes a manually operated return valve. When the return valve is open, both sides of the output cylinder are connected, and the output shaft can be manually moved to any position. By eliminating the solenoid valves from the circuit, the experiments can focus on the pumping dynamics as well as the interaction between the external load and the piezoelectric pump. Different constant external loads are simulated by hanging deadweights from the output of the device. The power output of the device is calculated by measuring the velocity with which the deadweight is lifted.

A schematic of the unidirectional test setup is shown in Fig. 8.39. The entire hydraulic circuit, including the accumulator, is machined into an aluminum manifold. The valve V_1 is the manually operated return valve. The testing procedure is as follows:

(1) The device is first filled with hydraulic fluid. This is accomplished by connecting a vacuum pump and a reservoir of hydraulic fluid, through a two-way adapter, to the valve labeled "Bleed." The air in the device is drawn out using the vacuum pump, and the two-way adapter is then connected to the hydraulic fluid reservoir. Atmospheric pressure forces the hydraulic fluid into the evacuated hydraulic circuit.
(2) The bleed valve is closed and the accumulator is charged with nitrogen to the desired bias pressure.
(3) The output shaft is manually positioned at its lowest point, and the desired deadweight is hung from it.
(4) The return valve V_1 is manually closed and the device is now ready for actuation.
(5) The piezostacks are excited at the desired frequency, and the displacement of the output shaft is measured by a linear potentiometer. The output velocity is obtained from the slope of the output displacement-versus-time curve. The excitation is maintained until the output shaft reaches the end of its travel.
(6) The return valve V_1 is manually opened, and the output shaft is returned to its lowest position, bringing the device to the state at step (3). Steps (3) through (6) are repeated for different values of deadweight and excitation frequency.

From the unidirectional testing, the output velocity of the device can be plotted as a function of pumping frequency as well as external load. Note that the output velocity, multiplied by the cross-sectional area of the output cylinder (which is constant), gives the flow rate of the piezoelectric pump. Therefore, the dependence of flow rate on pumping frequency and external load can be determined in this way.

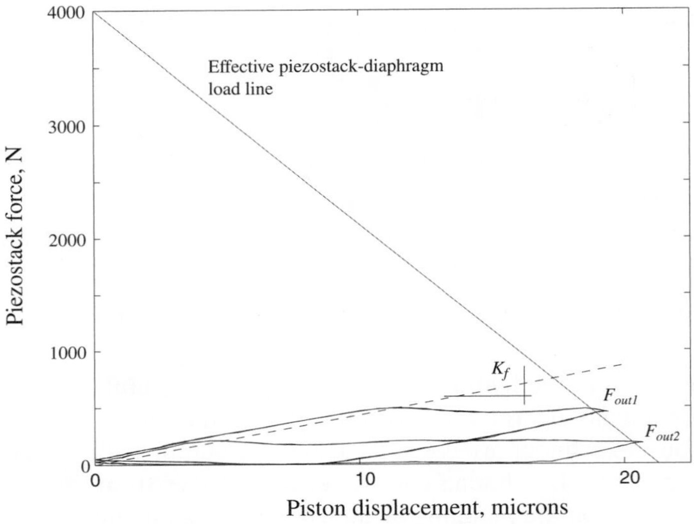

Figure 8.40. Measured piezostack force-displacement curves, 50 Hz pumping frequency.

Unidirectional Performance

Preliminary tests were performed with a special pumping head incorporating a pressure transducer. Figure 8.40 shows the force on the piezostack, obtained by multiplying the pressure measured in the pumping chamber by the piston area, as a function of the piston displacement. Two separate loops are shown, each one for a given value of output load (F_{out1} and F_{out2}), corresponding to different deadweights in Fig. 8.39 ($F_{out2} > F_{out1}$).

Superimposed on the same plot is the effective load line of the piezostack-diaphragm combination. Due to their combined stiffness, they show a lower blocked force and free displacement than the piezostack alone. It can be seen that the force-displacement curves are parallelograms, similar to the theoretical variation described in Fig. 8.31(a). Also note that the extents of the parallelograms are defined by the effective load line. The fluid stiffness, k_f, is also marked in the figure. The importance of the fluid stiffness can be observed from the large area under the effective load line that remains unused by the pumping cycle. Note that these curves were measured at a pumping frequency of 50 Hz. At higher pumping frequencies, the dynamics of the check valves and hydraulic fluid circuit result in shapes that can deviate considerably from parallelograms.

The fluid stiffness can be significantly affected by bias pressure, especially in the presence of entrained air. Figure 8.41 shows the effect of bias pressure on the no-load output velocity of a piezoelectric pump driven by two piezostacks. In this particular case, the pump is coupled to the manifold through two lengths of 6 in long tube. Note that the higher bias pressure results in increased output velocity. In addition, the most important characteristic of this plot is the highly nonlinear variation of the no-load velocity with pumping frequency. This nonlinear variation is primarily caused by the coupled dynamics of the hydraulic circuit and the piezostacks, and it is accentuated by the presence of long tubes from the piezoelectric pump. A systematic experimental study of the effect of the stiffness of the diaphragm, reed valves, and accumulator on the performance of the device was performed by Sirohi et al. [119],

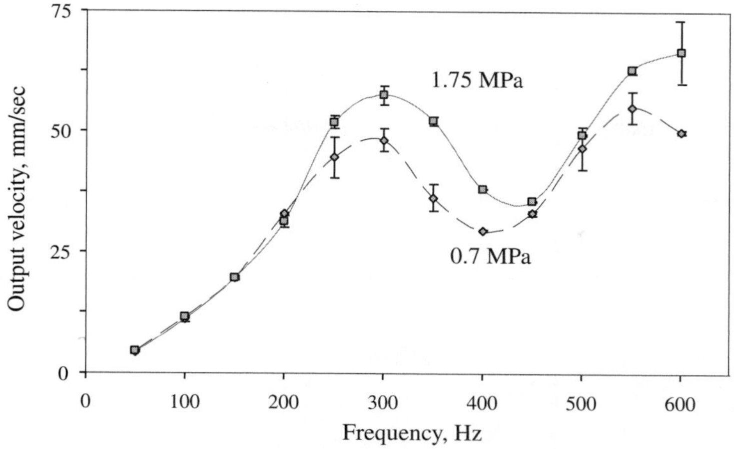

Figure 8.41. No-load output velocity as a function of pumping frequency, for different bias pressures.

where it was found that the accumulator stiffness is the dominant factor affecting the dynamics of the device.

Figure 8.42 shows the no-load output velocity of the unidirectional actuator with a pump driven by three piezostacks. In this case, the pump is directly assembled on the manifold, and a higher resonant frequency results due to the lower mass of the hydraulic fluid compared to the case in Fig. 8.41. The predictions of no-load output velocity obtained from a linear analysis (see Section 8.5.7) are shown as the dotted line. It can be seen that the linear analysis shows satisfactory agreement with measurements up to a pumping frequency of around 300 Hz, above which the quasi-steady pumping assumption breaks down. The spread in the measured velocity increases around the point of resonance of the curve and is especially sensitive to the amount of bias pressure and entrained air in the fluid.

By measuring the output velocity for different values of deadweight (output load), the load line of the device can be obtained at each pumping frequency. An example of the load lines at pumping frequencies of 100 Hz, 200 Hz, and 300 Hz for

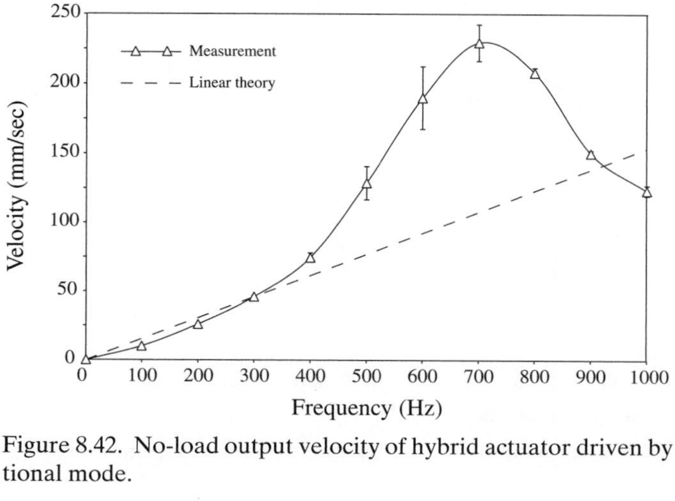

Figure 8.42. No-load output velocity of hybrid actuator driven by three piezostacks, unidirectional mode.

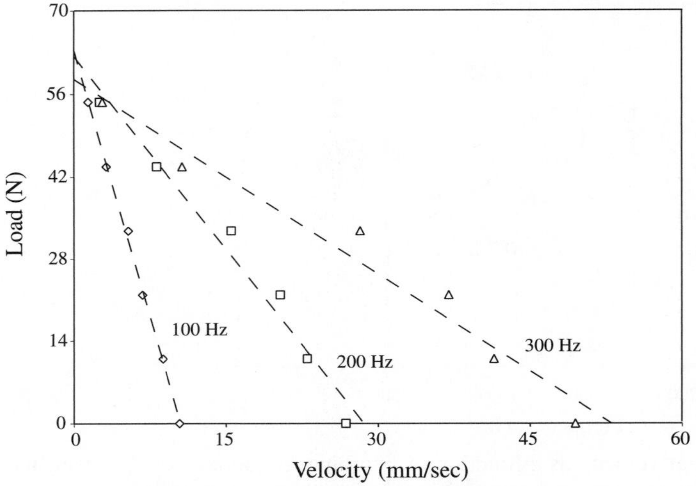

Figure 8.43. Variation of output load and output velocity of hybrid actuator driven by three piezostacks, unidirectional mode.

the device with three piezostacks is shown in Fig. 8.43. The measured datapoints are shown, along with a straight-line fit. As the pumping frequency increases, the spread in the datapoints increases because the output deadweight is subjected to increasing inertial loads at each pumping cycle. It can be seen that the blocked force is relatively unaffected by the pumping frequency. From these measurements, the output power of the device as a function of pumping frequency is plotted in Fig. 8.44.

Because the piezostacks represent a highly capacitive load, the reactive power consumption is very high compared to the active power. As a result, the efficiency of the device with respect to the apparent input electrical power is poor, on the order of 5%. A study of the power output and efficiency of a hybrid actuator driven by piezostacks, a magnetostrictive rod, and electrostrictive stacks was performed by

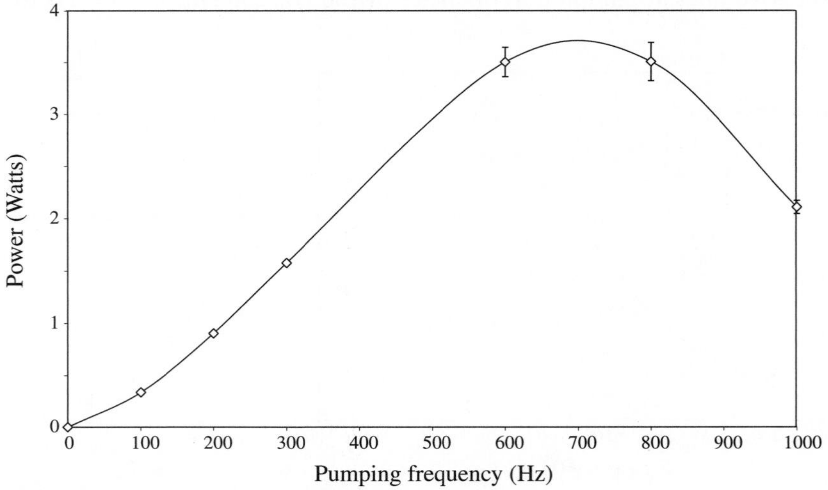

Figure 8.44. Output power of hybrid actuator driven by three piezostacks, unidirectional mode.

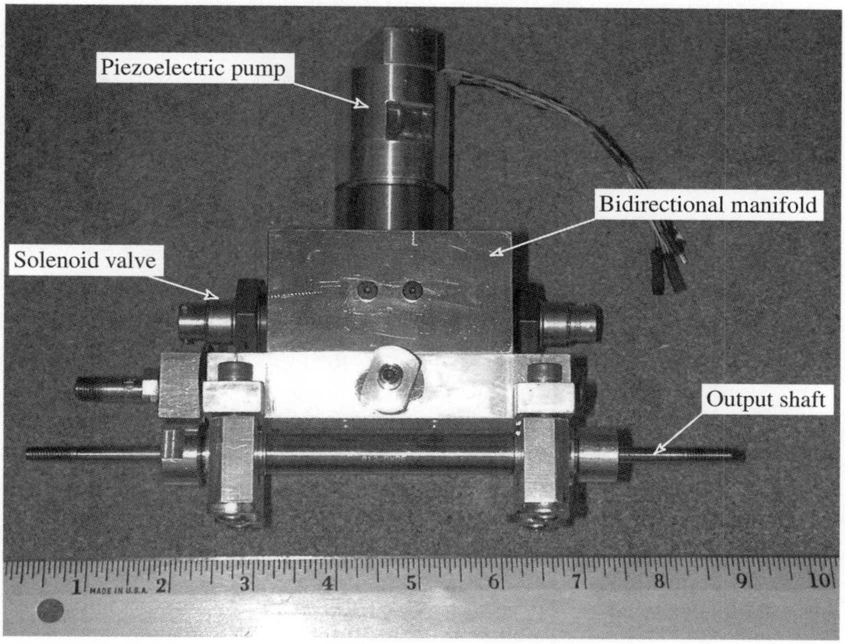

Figure 8.45. Bi-directional piezoelectric hybrid actuator.

John et al. [103]. It was found that the electrostrictive stack resulted in the highest efficiency and output power, primarily due to its much higher free strain at the same stiffness. In general, the free strain and stiffness of the active material driver of the hybrid actuator are the dominant factors in its power output.

Bi-Directional Testing

For a practical application, the hybrid actuator must be capable of bi-directional output. To achieve this, an additional manifold incorporating the solenoid valves (valves **A** and **B** in Fig. 8.28(b)) was assembled between the piezoelectric pump and the output actuator. Figure 8.45 shows the assembled bi-directional hybrid actuator with the piezoelectric pump. Note that the addition of bi-directional capability significantly increases the overall mass and volume of the device. In addition, the increased length of the hydraulic circuit significantly affects its dynamics. The large volume of the manifold is a consequence of the size of currently available solenoid valves. Some research efforts have explored alternate solutions to this issue, such as active valves utilizing MEMS technology, piezoelectric actuation, or ER and MR fluids. Although some of these active valve concepts are aimed at replacing the passive the check valves of the active material driven pump, it may be possible to combine the functions of the check valve and directional-control valve by appropriately adjusting the phasing of valve actuation, resulting in an extremely compact device.

The solenoid valves in the bi-directional device shown in Fig. 8.45 are actuated in an on-off manner by a square wave at different frequencies. Because the flow rate generated by the piezoelectric pump is constant at a given pumping frequency, the product of the stroke and frequency of the output displacement will be a constant. This is shown as a rectangular hyperbola in Fig. 8.46, along with measured values for

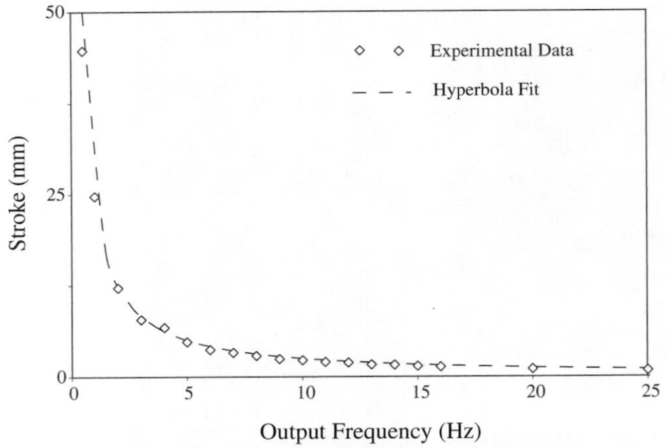

Figure 8.46. Output displacement of a bi-directional device, at a constant pumping frequency.

the bi-directional device driven by three piezostacks. The close agreement indicates that leakage of hydraulic fluid between the high- and low-pressure sides of the cylinder is not a significant issue.

8.5.12 Modeling Approaches

Testing of the prototype actuator revealed a highly nonlinear variation of the flow rate with pumping frequency. The quasi-static pumping assumption described in Section 8.5.7 only approximates the flow rate at low pumping frequencies, less than around 100 Hz. To improve the performance of these devices, accurate modeling of the behavior of the device as a function of pumping frequency is essential. Modeling the behavior of the hybrid hydraulic actuator is complicated by several factors. The system is inherently nonlinear due to the presence of the check valves in the hydraulic circuit, rod seals in the output actuator, and friction in the piston. The hydraulic circuit itself is geometrically complex and contains numerous turns and restrictions in the flow passage. The combined effects of fluid inertia, viscosity, and compressibility become more important at higher pumping frequencies, necessitating the modeling of the hydraulic circuit as a continuous system. A further complexity in the modeling of the hydraulic circuit is the presence of elements of greatly differing stiffnesses close to each other, such as the accumulator and the pumping chamber.

Several researchers have presented models for the performance of the device, including lumped-parameter descriptions of the hydraulic circuit, computational fluid dynamics (CFD)-based calculations of the flow impedance, distributed parameter models of the hydraulic fluid, and solution of the fluid governing differential equations in the time domain.

Tang et al. [56] developed a frequency domain model of a piezoelectrically driven hydraulic amplification device for vibration control. However, this model did not include the effects of viscosity. Simple fluid models have also been developed to predict the performance of fluid micropumps by Ullmann et al. [112] and other researchers.

Konishi et al. [98, 94] developed time-domain solutions to directly solve the coupled governing differential equations of the piezostack and the hydraulic fluid.

This approach used a simplified viscous model that correlated well with their measurements at a single pumping frequency of 300 Hz.

Nasser and Leo [120] investigated piezohydraulic and piezopneumatic pumps and identified upper bounds for their mechanical and electrical efficiencies. It was shown that piezohydraulic pumps are inherently more efficient due to the lower compliance of the working fluid. The pumping cycle was divided into four parts: namely, intake, compression, expansion, and exhaust, which were modeled separately and then assembled together. The steady electrical and mechanical states of the system were evaluated after each part of the pumping cycle to obtain expressions for the energy transfer.

A coupled piezostack-hydraulic circuit lumped-parameter model was developed by Oates et al. [101] incorporating check-valve resistance and fluid compliance but neglecting fluid inertia and check-valve dynamics. The pressure losses in the pumping chamber were calculated using CFD. The system equations were assembled and solved in the state-space form, which enabled simulation of the stepwise output displacement of the device. The model showed good agreement with measurements up to a pumping frequency of 60 Hz.

Cadou et al. [121] developed a quasi-static model including the effects of fluid inertia as well as a refined model for impulsively started flow, which is what happens when the check valves open. The stiffness of the fluid in the pumping chamber and the output hydraulic line were included to obtain a detailed representation of the quasi-static pumping cycle. A static force balance approach was adopted on each component of the hydraulic circuit, the intake and output strokes were modeled separately, and the entire system was solved iteratively. The model was found to agree with experimental data up to a pumping frequency of 150 Hz, and it showed the same qualitative trend at higher frequencies. An important conclusion of the analysis was that the inertial forces dominated the viscous forces at higher pumping frequencies.

Tan et al. [122] developed a model of a piezohydraulic actuator with active valves incorporating both an incompressible viscous fluid representation as well as a compressible fluid representation. The pumping cycle was divided into four parts, each of which was modeled separately and subsequently combined into a complete cycle. Based on the dimensions of their device, laminar fluid flow was assumed for pumping frequencies less than 960 Hz. The fluid energy equations for laminar flow were directly incorporated in the incompressible fluid representation and were used with appropriate assumptions in the compressible fluid representation. Differences between the compressible and incompressible fluid representations became apparent at pumping frequencies above 100 Hz and at high output forces, indicating that compressibility effects were important. The compressible fluid model showed good agreement with measurements up to a pumping frequency of 200 Hz.

John et al. [123] used a commercial CFD software to perform two-dimensional and three-dimensional simulations of the steady flow through the pumping chamber of a hybrid actuator. It was found that three-dimensional effects such as vortex rings in the flow channels are major sources of pressure loss. The effect of pumping chamber height as well as the geometry and location of the output ports was investigated. A quadratic variation of pressure loss with flow rate was derived based on continuity and momentum equations and validated using the computed results. Analytical expressions for the scaling of the pressure loss with pumping frequency were also derived.

Due to the discretization introduced by the lumped-parameter models, they do not represent the system accurately at high pumping frequencies. The dynamics of the system, specifically the first natural frequency of the system, affects the frequency up to which the quasi-steady analysis can give satisfactory predictions. As the pumping frequency approaches the first natural frequency, refined models are required to accurately predict the performance of the device. The effects of fluid inertia, compressibility, and viscosity can be incorporated completely only by representing the fluid as a continuum. Sirohi et al. [124] developed a transmission-line model that solved the equations of motion in the frequency domain. A transfer matrix type approach was adopted, with each fluid line treated as a transmission line and represented by a transfer matrix. Such an approach makes it easy to add additional fluid elements to the system and to change the properties of specific elements. The fluid equations were coupled with the structural elements and the entire system was solved in the frequency domain up to a pumping frequency of 1kHz. The first resonant mode correlated well with measurements for three different geometries of the hydraulic circuit; however, nonlinearities in the system resulted in poor correlation of higher modes.

The implementation of the quasi-static, lumped-parameter/state-space, transmission-line, and CFD approaches are described in the following subsections.

Quasi-Static Analysis

The quasi-static analysis is described in Section 8.5.7 as a means to estimate the work output of the actuator. A linear variation of flow rate with pumping frequency is assumed. This method can be used to obtain an approximate estimate of the blocked force of the device and the no-load cylinder velocity, and it can be used as a preliminary design tool.

The blocked force, or the maximum output force, of the device F_{max} is obtained when $\delta_o = 0$. Applying this condition to Eq. 8.71 leads to

$$F_{max} = \frac{\delta_f}{1/k_p + 1/k_f} \tag{8.78}$$

The no-load output velocity, v_{max}, can be calculated from the volumetric flow rate of the pump, given by the product of pumping frequency and the swept volume per cycle

$$v_{max} = \frac{A_p \delta_f f_{pump}}{A_{out}} \tag{8.79}$$

For the pumping chamber geometry given in Table 8.4, based on a measured piston deflection $\delta_p = 0.0008$ in and a pumping frequency of 50 Hz, the no-load output velocity is 5.3 mm/sec. It should be noted that because the analysis neglects fluid compressibility and flow resistance, this estimate represents an upper bound of the achievable no-load cylinder velocity. The correlation between the quasi-static theory and no-load output velocity is shown in Fig. 8.42, and the correlation between output force-velocity measurements of a different configuration is shown in Fig. 8.47. Note that the first natural frequency for both of these cases was above 600 Hz.

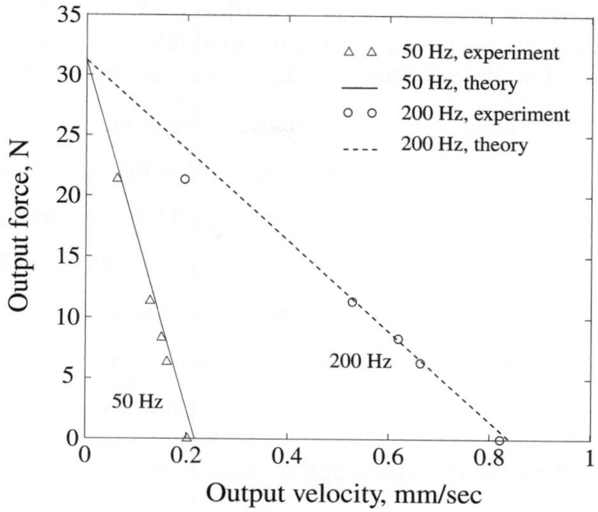

Figure 8.47. Quasi-static theory compared with measured performance of the hybrid actuator at a pumping frequency of 50 Hz and 200 Hz.

Lumped-Parameter State Space Method

This is a simple method capable of quickly predicting the overall actuator performance based on a given set of input parameters. The opening and closing of the check valves can be easily modeled using this method by incorporating a flow resistance based on the pressure gradients across the valve. The overall system is represented by a set of states that are related to each other by equations of continuity and force balance. The components of the actuator as well as the fluid are represented using a lumped-parameter formulation, and the equations are solved using the state-space method.

States of the Device

Consider the schematic diagram of the device as shown in Fig. 8.48. The output hydraulic actuator, or cylinder, is shown connected to an external load that has a mass M_{out} and a spring of stiffness k_{out}. The goal of the model is to predict the output cylinder displacement x_{out} in response to the excitation voltage V_p applied to the piezostacks. However, by setting up the system in state-space form, the time response of any of the other states of the system can be calculated without rewriting the equations of the system. Only one-way operation of the device is modeled, with

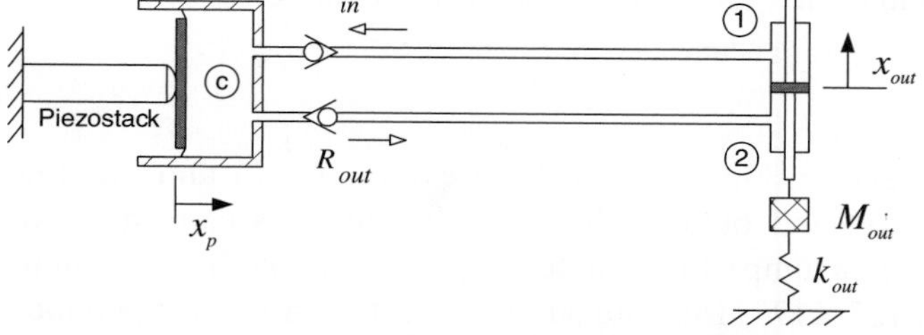

Figure 8.48. Schematic diagram of the piezohydraulic actuator, lumped-parameter model.

fluid flowing out through the outlet check valve R_{out} into the lower end of the output hydraulic actuator, causing the load mass to move upward.

The state variables of the system are (Fig. 8.48)

$$\begin{aligned}
&x_1 \rightarrow x_p \quad \text{displacement of the pump piston} \\
&x_2 \rightarrow x_{out} \quad \text{displacement of the output hydraulic actuator} \\
&P_2 \quad \text{pressure in the upper part of the output hydraulic actuator} \\
&P_1 \quad \text{pressure in the lower part of the output hydraulic actuator} \\
&P_c \quad \text{pressure in the pumping chamber} \\
&x_3 \rightarrow \dot{x}_p \quad \text{velocity of the pump piston} \\
&x_4 \rightarrow \dot{x}_{out} \quad \text{velocity of the output hydraulic actuator}
\end{aligned} \tag{8.80}$$

Setting Up the State-Space Equations

From these definitions, the state-space equations can be set up as follows

$$\dot{x}_1 = x_3 \tag{8.81}$$

$$\dot{x}_2 = x_4 \tag{8.82}$$

By equating forces acting on the pumping piston

$$m_p\ddot{x}_p + b_p\dot{x}_p + k_px_p = c_vV_p - P_ca_p \tag{8.83}$$

where m_p, b_p, and k_p are the effective mass, damping, and stiffness of the piezostack assembly, respectively. From energy considerations, it can be shown that one-third of the mass of the piezostack also needs to be added to the mass of the piston to obtain the effective mass of the combination, m_p. k_p is the effective stiffness of the diaphragm and piezostack, in parallel. a_p is the area of the pumping piston and V_p is the voltage applied to the piezostack. The constant c_v is an effective piezoelectric coefficient expressed in terms of the blocked force of the piezostack and the voltage at which the piezostack is excited to obtain the specified blocked force (V_{max}).

$$c_v = \frac{F_b}{V_{max}} \tag{8.84}$$

Similarly, by equating forces acting on the output hydraulic actuator piston

$$M_{out}\ddot{x}_{out} + B_{out}\dot{x}_{out} + k_{out}x_{out} = (P_2 - P_1)a_{out} - M_{out}g - f_{friction} \tag{8.85}$$

where B_{out} is the damping at the output hydraulic actuator, g is the acceleration due to gravity, and $f_{friction}$ is the frictional force in the output actuator as a result of the tight fit between the piston and the inner bore of the output cylinder.

Fluid Impedance

The next step is to find a relationship between the pressures P_1, P_2 and the flow rate of fluid through the inlet and outlet valves. To do this, it is convenient to approximate the fluid circuit as a lumped-parameter system and derive an equivalent fluid impedance [125, 126, 127, 117, 116]. The similarity between fluid impedance and electrical impedance can also be utilized to solve the hydraulic system as an equivalent electrical network. For example, Bourouina et al. [128] used an electrical equivalent network to model a fluid micropump. Because the analogy extends to mechanical systems, a lumped-parameter mass-spring–damper model can also be used to model such a system. A simple lumped-parameter representation for the impedance of a fluid line is [117]

$$Z_f = \frac{\Delta P}{Q} = R_f + \frac{1}{sC_f} + sL_f \tag{8.86}$$

where R_f, C_f, and L_f are the resistance, capacitance, and inductance of the fluid line, respectively. ΔP is the pressure differential across the fluid line, causing a volumetric flow rate of Q to occur. Note that the pressure differential, ΔP, has the units N/m^2; the volumetric flow rate, Q, has the units m^3/s; and the fluid impedance, Z_f, has the units of $Pa.s/m^3$. Typically, the flow resistance, R_f, is given by an empirical formula. A more refined value of R_f can be calculated using CFD [101, 123]. In the present discussion, however, the flow resistance of the pumping chamber is neglected and only the flow resistance of the check valves and tubing is considered. Assuming laminar flow, a simple expression for flow resistance of a tube of inner diameter D and length L, with a fluid of dynamic viscosity μ, is given by [129, 117, 116]

$$R_f = \frac{128\mu L}{\pi D^4} \tag{8.87}$$

The fluid resistance term represents the viscous losses in the fluid and depends directly on the fluid dynamic viscosity. The fluid capacitance, C_f, is a measure of the compliance of the fluid, or the energy stored in the form of volumetric deformation. This has the units m^3/Pa and, for a tube of length L, cross-sectional area A, filled with a fluid of effective bulk modulus β, is given by [117, 116]

$$C_f = \frac{AL}{\beta} \tag{8.88}$$

The last term in the fluid impedance, Z_f, is the fluid inductance, which is a measure of the inertia, or mass of the fluid. Fluid inductance has the units $Pa.s^2/m^3$, and for a fluid in a tube of length L and cross-sectional area A, it can be expressed as

$$L_f = \frac{\rho L}{A} \tag{8.89}$$

Using the expression for fluid impedance, along with the continuity equation, a relationship between the flow rates and pressures in the pumping circuit can be derived. The continuity equation applied to the pumping chamber leads to

$$a_p \dot{x}_p + C_c \dot{P}_c + Q_{out} - Q_{in} = 0 \tag{8.90}$$

Substituting for the flow rate exiting the pumping chamber, Q_{out}, and the flow rate entering the pumping chamber, Q_{in}, this can be simplified to

$$\dot{P}_c = \frac{P_1}{C_c(R_1 + R_{in})} + \frac{P_2}{C_c(R_2 + R_{out})} + \frac{P_c}{C_c}\left[-\frac{1}{R_1 + R_{in}} - \frac{1}{R_2 + R_{out}}\right] + \frac{a_p}{C_c}x_3 \tag{8.91}$$

Similarly, applying the continuity equation to the upper and lower sides of the output hydraulic actuator

$$\dot{P}_1 = -\frac{P_1}{C_1(R_1 + R_{in})} + \frac{P_c}{C_1(R_1 + R_{in})} + \frac{a_{out}}{C_1}x_4 \tag{8.92}$$

$$\dot{P}_2 = \frac{P_2}{C_2(R_2 + R_{out})} - \frac{P_c}{C_2(R_2 + R_{out})} - \frac{a_{out}}{C_2}x_4 \tag{8.93}$$

where a_{out} is the cross-sectional area of the output hydraulic actuator. C_1 and C_2 are calculated using Eq. 8.88. R_1 and R_2 are the flow resistances of the inlet tubing and outlet tubing, respectively, and R_{in} and R_{out} are the flow resistances of the inlet check valve and outlet check valve, respectively.

Assembly of the State-Space Matrices

Eqs. 8.81–8.93 can be combined into the familiar state-space matrices

$$\begin{aligned}\dot{\tilde{y}} &= \mathbf{A}\tilde{y} + \mathbf{B}\tilde{u} \\ \tilde{y} &= \mathbf{C}\tilde{x} + \mathbf{D}\tilde{u}\end{aligned} \tag{8.94}$$

The terms in these equations are as follows

$$\tilde{y} = \begin{Bmatrix} x_1 \\ x_2 \\ P_2 \\ P_1 \\ P_c \\ x_3 \\ x_4 \end{Bmatrix} \tag{8.95}$$

$$\mathbf{A} = \begin{bmatrix} 0 & 0 & 0 & 0 & 0 & 1 & 0 \\ 0 & 0 & 0 & 0 & 0 & 0 & 1 \\ 0 & 0 & -\frac{1}{C_2(R_2+R_{out})} & 0 & \frac{1}{C_2(R_2+R_{out})} & 0 & -\frac{a_{out}}{C_2} \\ 0 & 0 & 0 & -\frac{1}{C_1(R_1+R_{in})} & \frac{1}{C_1(R_1+R_{in})} & 0 & \frac{a_{out}}{C_1} \\ 0 & 0 & \frac{1}{C_c(R_2+R_{out})} & \frac{1}{C_c(R_1+R_{in})} & -\frac{1}{C_c(R_1+R_{in})} - \frac{1}{C_c(R_2+R_{out})} & \frac{a_p}{C_c} & 0 \\ -\frac{k_p}{m_p} & 0 & 0 & 0 & -\frac{a_p}{m_p} & -\frac{b_p}{m_p} & 0 \\ 0 & -\frac{k_{out}}{M_{out}} & -\frac{a_{out}}{M_{out}} & \frac{a_{out}}{M_{out}} & 0 & 0 & -\frac{B_{out}}{M_{out}} \end{bmatrix} \tag{8.96}$$

$$\mathbf{B} = \begin{bmatrix} 0 & 0 \\ 0 & 0 \\ 0 & 0 \\ 0 & 0 \\ 0 & 0 \\ \frac{c_v}{m_p} & 0 \\ 0 & \frac{-M_{out}g - f_{friction}}{M_{out}} \end{bmatrix} \tag{8.97}$$

$$\tilde{u} = \begin{Bmatrix} V_p \\ 1 \end{Bmatrix} \tag{8.98}$$

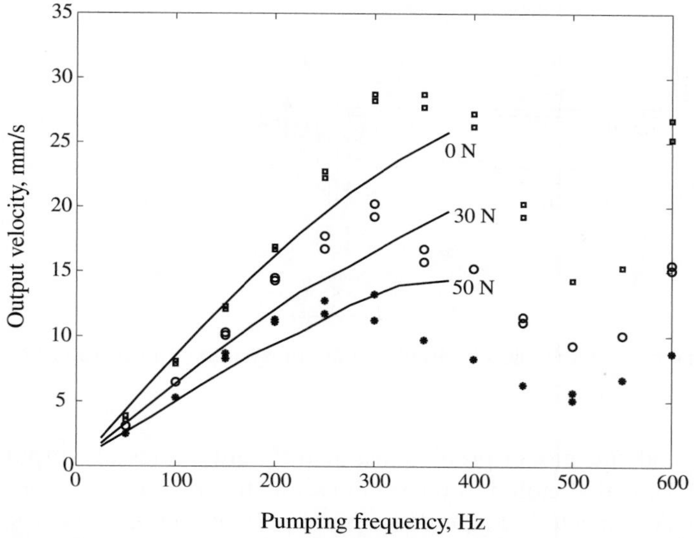

Figure 8.49. Correlation of a lumped-parameter state-space model with experimental data, at different values of output force.

The **C** matrix can be set depending on the states of interest. In the present case, the cylinder displacement is the output, which makes the **C** matrix

$$\mathbf{C} = \begin{bmatrix} 0 & 1 & 0 & 0 & 0 & 0 & 0 \end{bmatrix} \tag{8.99}$$

The **D** matrix is taken as zero

$$\mathbf{D} = \begin{bmatrix} 0 & 0 \end{bmatrix} \tag{8.100}$$

Numerical Solution

For specified initial conditions, these state-space equations are solved over a given time period. The directionality of the check valves and their opening and closing are modeled by means of the values of R_{in} and R_{out}. Based on the pressures P_c, P_1, and P_2, the values of check-valve resistance are assigned. If the valve is open, the check-valve resistance is assigned a specific value (determined through experiment, analysis, or CFD); if the valve is closed, the check-valve resistance is assigned a value of infinity. The changing value of valve resistance means that the **A** matrix is not constant with time. Other entries in the **A** matrix are also dependent on the state vector; for example, the fluid capacitances of the upper and lower parts of the output hydraulic cylinder, C_1 and C_2, depend on the position of the output piston. However, to solve the system, it is assumed invariant for short periods of time. In each of these short periods, the state-space system is solved and the values of the various states are computed. Using the state vector computed in the previous time step as the initial condition, a new **A** matrix is calculated and the system is solved for the next time period. In this manner, the solution is calculated over the entire time period of interest. Once the output displacement is obtained over a certain time period, the slope of the curve determines the output velocity.

Figure 8.49 shows the correlation between the measured output velocities from a certain configuration of the device (i.e., specific values of tubing length and

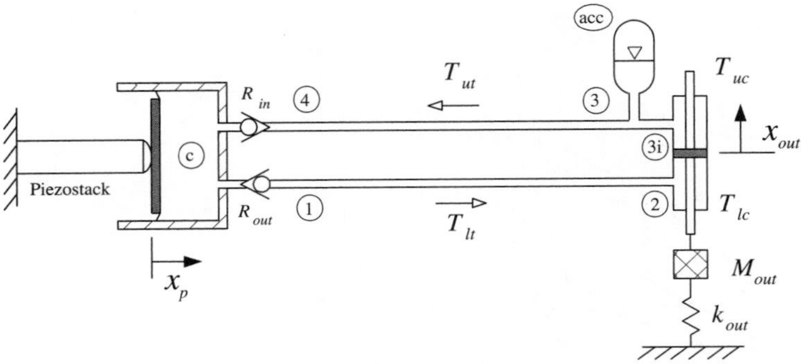

Figure 8.50. Schematic diagram of the piezohydraulic actuator system, transmission-line approach.

check-valve resistance) and the model predictions, at different values of output force. The experimental data are denoted by the symbols, and the model predictions are the continuous lines. As expected, the predictions are poor at high frequencies. However, at frequencies up to approximately 250 Hz, the theory can predict the behavior with sufficient accuracy. It can be concluded that the model is sufficient for the purpose of parameter optimization and to size the device to achieve required performance specifications.

8.5.13 Transmission-Line Approach

To accurately capture the dynamics of the hydraulic circuit, it must be treated as a continuous system. The entire piezohydraulic actuator is divided into sections, as in Fig. 8.50. A mathematical model for each section of the fluid circuit is developed in terms of transfer matrices. These are subsequently combined with models for the valves, pump diaphragm, and active material stack to yield an overall coupled fluid-structural model for the hybrid system that relates the mechanical output (displacement) of the output cylinder to the electrical input to the active material [124]. The advantage of this approach is that any additional elements in the hydraulic circuit can be easily added on without affecting the rest of the formulation. It is assumed that in the frequency range of interest, up to 1kHz, the check valves do not affect the overall response of the device.

Accordingly, the rectification effect of the check valves is neglected in the model, resulting in a sinusoidal displacement of the output shaft in response to a sinusoidal voltage applied to the active material. The analytical model is formulated in the frequency domain, and it is expected that the variation of the output shaft velocity with pumping frequency is the same as the variation of output shaft displacement with frequency without flow rectification.

Fluid Transfer Matrix Model

A fluid in a tube, such as in the tube between points "1" and "2" in Fig. 8.50, can be represented in terms of the pressure (P) and volumetric flow rate (Q) quantities at each end, defined by P_1, Q_1, and P_2, Q_2, respectively. An accurate model of such a fluid transmission line can be developed by treating it as a distributed-parameter system [130, 116].

Starting from the basic fluid equations of continuity, momentum, and energy, the relationship between the pressure and flow rate variables at the ends of the fluid line can be derived in terms of a transfer matrix T_{12} as

$$\begin{Bmatrix} P_2 \\ Q_2 \end{Bmatrix} = \mathbf{T}_{12} \begin{Bmatrix} P_1 \\ Q_1 \end{Bmatrix} \tag{8.101}$$

where

$$\mathbf{T}_{12} = \begin{bmatrix} \cosh\Gamma & -Z_c \sinh\Gamma \\ -\dfrac{1}{Z_c}\sinh\Gamma & \cosh\Gamma \end{bmatrix} \tag{8.102}$$

The behavior of the fluid line is governed by two quantities: the characteristic impedance Z_c and the propagation parameter Γ. For an inviscid fluid, these quantities are given by

$$\Gamma = \bar{D}_c = \frac{1}{\omega_c}\frac{d}{dt} \tag{8.103}$$

$$Z_c = Z_o = \frac{\rho c_o}{\pi r^2} \tag{8.104}$$

where $\omega_c = c_o/l$ is the characteristic frequency of the fluid line and the speed of sound in the fluid, c_o, is given by

$$c_o = \sqrt{\frac{\beta}{\rho}} \tag{8.105}$$

Fluid viscosity can be incorporated in two ways: one could use a linear friction model using a friction factor calculated from the Hagen-Poiseuille flow theory, or a dissipative model derived using the energy equation [131]. This results in different expressions for Γ and Z_c compared to the inviscid case, whereas the basic transfer matrix (Eq. 8.101) between pressure and flow quantities remains the same. The exact solution for liquids with frequency-dependent viscous dissipation yields expressions for Γ and Z_c in terms of a ratio of Bessel functions, B_r, as given here [132, 133, 134, 135]

$$\Gamma = \bar{D}_c \left[\frac{1}{1 - B_r}\right]^{1/2} \tag{8.106}$$

$$Z_c = Z_o \left[\frac{1}{1 - B_r}\right]^{1/2} \tag{8.107}$$

The Bessel function ratio, B_r, can be expressed as a first-order square root approximation [136]

$$B_r = \frac{1}{\sqrt{1 + 2\bar{D}_v}} \tag{8.108}$$

where the operator $\bar{D}_v$ is defined in terms of the viscous frequency, ω_v, by

$$\bar{D}_v = \frac{1}{\omega_v}\frac{d}{dt} \tag{8.109}$$

and

$$\omega_v = \frac{8\nu}{d^2} \tag{8.110}$$

Eqs. 8.101–8.110 constitute a comprehensive model of the fluid lines, incorporating the effects of fluid inertia, compressibility, and viscous dissipation.

Frequency Response of the Device

The frequency response of the device is calculated by assuming a harmonic excitation at a frequency ω, resulting in the following substitution for the operator D

$$D = j\omega \tag{8.111}$$

This substitution greatly simplifies the fluid line equations. The complete system model is now obtained by combining the fluid line equations with the governing equations of the active material stack, output cylinder, and continuity equation for the pumping chamber. Pressure and flow rate continuity relations are applied between different elements of the system.

Force equilibrium on the active material, assuming a piezostack, gives

$$c_v V - P_c a_p = m_p \ddot{x}_p + b_p \dot{x}_p + k_p x_p \tag{8.112}$$

The continuity equation for the pumping chamber can be written as

$$C_c \dot{P}_c = a_p \dot{x}_p + Q_4 - Q_1 \tag{8.113}$$

where the fluid capacitance of the pumping chamber, C_c, is given by [116, 117]

$$C_c = \frac{a_p \Delta_{gap}}{\beta} \tag{8.114}$$

Assuming the output mechanical load to be lumped together with the output piston, force equilibrium on the output piston gives

$$(P_{lp} - P_{up}) a_{out} = m_{out} \ddot{x}_{out} + b_{out} \dot{x}_{out} + k_{out} x_{out} \tag{8.115}$$

At the check valves

$$P_c - P_1 = R_{out} Q_1 \tag{8.116}$$

$$P_4 - P_c = R_{in} Q_4 \tag{8.117}$$

From Eqs. 8.101–8.115

$$\begin{Bmatrix} P_1 \\ Q_1 \end{Bmatrix} = \mathbf{A} \begin{Bmatrix} P_{lp} \\ a_{out} \dot{x}_{out} \end{Bmatrix} \tag{8.118}$$

$$\begin{Bmatrix} P_4 \\ Q_4 \end{Bmatrix} = \mathbf{B} \begin{Bmatrix} P_{up} \\ a_{out} \dot{x}_{out} \end{Bmatrix} \tag{8.119}$$

where the matrices **A** and **B** are given by

$$\mathbf{A} = (\mathbf{T}_{lt}.\mathbf{T}_{lc})^{-1} = \begin{bmatrix} A_{11} & A_{12} \\ A_{21} & A_{22} \end{bmatrix} \tag{8.120}$$

$$\boldsymbol{B} = \mathbf{T}_{ut}.\mathbf{T}_{acc}.\mathbf{T}_{uc} = \begin{bmatrix} B_{11} & B_{12} \\ B_{21} & B_{22} \end{bmatrix} \tag{8.121}$$

The complete set of equations can be simplified and expressed in terms of an equivalent mass-spring–damper system, in which the vector $\{q\}$ contains the system variables and is given by

$$\{q\} = \begin{Bmatrix} x_p \\ x_{out} \\ P_c \\ P_1 \\ P_4 \\ P_{lp} \\ P_{up} \end{Bmatrix} \tag{8.122}$$

The forcing vector, $\{F\}$, contains the electrical input to the piezostack

$$\{F\} = \begin{Bmatrix} c_v V \\ 0 \\ 0 \\ 0 \\ 0 \\ 0 \\ 0 \end{Bmatrix} \tag{8.123}$$

For harmonic forcing at a frequency ω, the solution to this system of equations can be written as

$$\{q\} = \left(-\omega^2 \mathbf{M} + j\omega \mathbf{C} + \mathbf{K}\right)^{-1} \{F\} \tag{8.124}$$

Because the system is assumed to be linear and check-valve dynamics are ignored, constant values of R_{out} and R_{in} are assumed. By setting $R_{in} = \infty$ and $R_{out} = 0$, one valve is permanently closed and the other is permanently open. This represents the dynamics of the real device with check valves during half of the pumping cycle, the other half being symmetric.

Eq. 8.124 is solved to obtain the frequency response of the system variables in the vector $\{q\}$. Because the hyperbolic sines and cosines are calculated exactly, the calculated frequency response accurately represents an infinite number of modes, which is a significant advantage over typical lumped-parameter methods. The response of the system is calculated up to pumping frequencies of 1kHz for correlation with experiments.

Model Predictions

The effect of the length of the fluid line on the frequency response of the output displacement x_{out}, for a fluid of kinematic viscosity $\nu = 2$ cSt, is shown in Fig. 8.51. It can be seen that a large increase in output displacement is obtained at the resonant

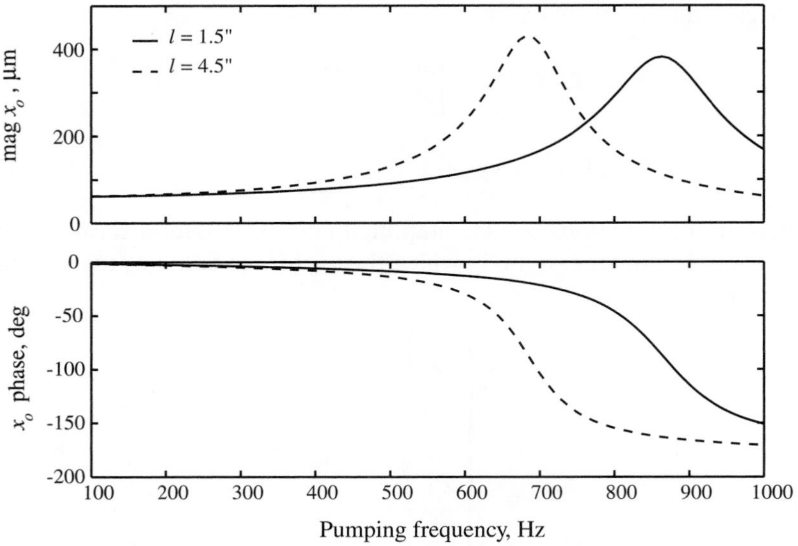

Figure 8.51. Predicted variation of output displacement x_o with tubing length, $\nu = 2$ cSt.

condition and that the resonant frequency is strongly dependent on the length of the fluid line. A correspondingly large increase in the pumping chamber pressure is also observed.

It can be seen that the frequency response is dominated by the dynamics of the hydraulic circuit, and this offers a powerful method of increasing the output power of the device. Although similar concepts have been investigated in the past [96, 137], investigation of any possible applications requires a refined prediction tool. The transmission-line approach may address this issue by allowing a more accurate representation of the device behavior at high frequencies, including the effect of fluid viscosity.

Correlation with Experimental Data

To obtain experimental data to validate the previous analysis, a special pump configuration was developed. The reed valve assembly was replaced by a steel plate with one hole aligned with one of the ports of the pump. This resulted in one check valve being permanently open ($R_{in} = \infty$) and the other one being permanently closed ($R_{out} = 0$), which is the case treated in the analysis.

A schematic of the experimental setup is shown in Fig. 8.52. It can be seen that one of the check valves is permanently closed and the other one is permanently open. The output displacement is measured by a laser vibrometer or a laser displacement sensor. As a result of the elimination of the check valves, a sinusoidal voltage applied to the piezostacks results in a sinusoidal output displacement.

A swept sinusoid, from a frequency of 50 Hz to 1kHz, is input to a power amplifier, which actuates the piezostacks. The actuating waveform is offset by a DC value equal to the amplitude of the sinusoid. This ensures that the piezostacks are only actuated by a positive voltage and minimizes the possibility of piezostack failure due to tensile stresses. At each frequency of actuation, the magnitude and phase of the displacement of the output shaft are measured. Although the actual

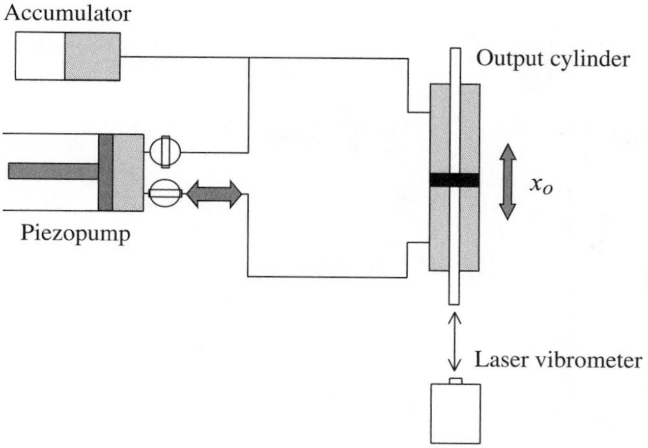

Figure 8.52. Schematic of the experimental setup to validate transmission-line analysis.

device can be actuated with voltages from 0 to 100V, the voltage amplitude for the present testing was conducted at 12.5V and 25V due to amplifier current limitations. Figure 8.53 shows a comparison of the predicted and measured frequency response functions of the output displacement, for 4.5 in long tubes between the pump and the output cylinder. In general, it can be seen that the analysis underpredicts the first natural frequency by 10–15%. Below the resonant peak, the magnitude of the response is accurately predicted. However, at frequencies higher than the resonance, there is a significant underprediction of the response. Regarding the phase of the response, there is an underprediction (10° to 30°) below resonance and a mixed variation above the resonance condition.

To understand the causes of the discrepancy between analysis and experiment, it is useful to look closely at the time-domain signal from the vibrometer, which is directly proportional to the velocity of the output shaft. Based on the assumptions regarding the linear behavior of the system, a purely sinusoidal waveform is expected.

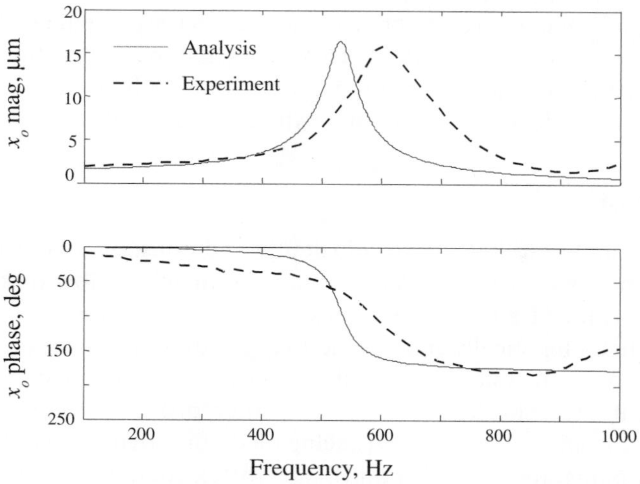

Figure 8.53. Comparison of experimental and analytical frequency response functions, tubing length = 4.5 inches.

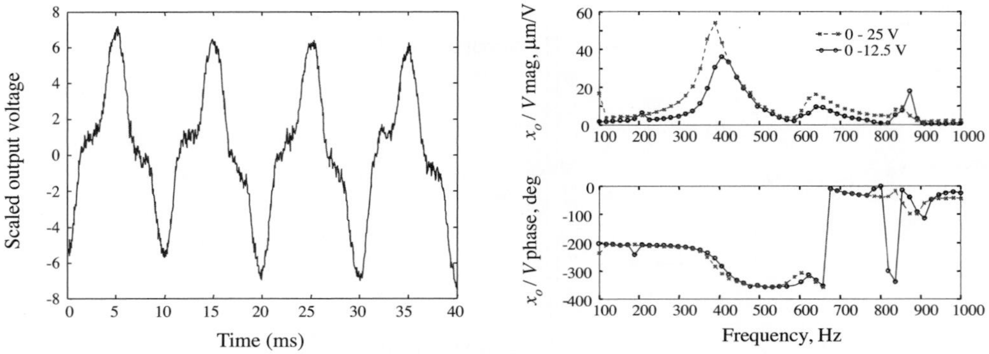

(a) Vibrometer output signal, 100 Hz pumping frequency, 25 V actuation

(b) Effect of actuation voltage on the transfer function

Figure 8.54. Nonlinear behavior of the hybrid hydraulic actuator.

Figure 8.54(a) shows the signal from the vibrometer at a pumping frequency of 100 Hz, at an actuation voltage of amplitude 25 V. Although the driving voltage is sinusoidal, it can be clearly seen that the output waveforms are not purely sinusoidal. The most notable feature is the presence of a discontinuous region around each zero crossing, with a lower slope than the neighboring regions. Because the scaled voltage signal is proportional to velocity, the zero-crossing region corresponds to the time period around which the output shaft achieves its maximum displacement. At this time, the output shaft changes direction and momentarily achieves zero velocity. This discontinuous region can be attributed to static friction in the rod seals around the output shaft, resulting in a stick-slip type of behavior.

Another consequence of the assumption of a linear system is that the transfer function between the output displacement and input voltage should be independent of the magnitude of the input voltage. To verify the accuracy of this assumption, the frequency response of the output displacement was measured as described previously at actuation voltages of amplitude 12.5 V and 25 V. Figure 8.54(b) shows the comparison between the two frequency response functions. It can be seen that at the higher actuation voltage of 25 V, the first resonant peak moves to a lower frequency compared to the case of the 12.5 V actuation voltage. This shows that the assumption of linearity in the actuator system is an approximation and explains, at least in part, the discrepancy between the analytical predictions and experimental results.

8.6 Smart Helicopter Rotor

Helicopters have the unique ability to both hover efficiently and cruise in forward flight. This is achieved by means of a large diameter main rotor with a low disk loading. However, the main rotor of a helicopter is also the source of a variety of problems. Helicopters are characteristically susceptible to high vibratory loads and noise levels. The rotor is the key subsystem, setting limits on vehicle performance, handling qualities, and reliability. The flow field on the rotor disk is extremely complex and may involve transonic flow on the advancing blade tips, dynamic stall, and reversed flow on the retreating side of the disk and blade-vortex interactions. The primary source of helicopter vibration is the main rotor that transmits large vibratory forces and moments to the fuselage. For an N-bladed rotor, the N/rev, $N+1$/rev, and $N-1$/rev vibratory blade loads (in the rotating frame) are transmitted to the fuselage through the hub as dominant N/rev forces and moments.

The high vibration levels limit helicopter performance [138] and reduce the structural life of components [139], leading to increased maintenance and operating costs. In addition, the high vibration and noise levels lead to pilot fatigue and passenger discomfort, and they have been recognized as major barriers to public acceptance of rotorcraft for the short-haul commuter transport role [140]. Even in military applications, high vibration levels are undesirable from the point of view of crew fatigue and errors in target tracking and firing systems. Decreasing noise levels has also become a priority in modern rotorcraft. The civilian rotorcraft sector has seen the introduction of increasingly stringent noise requirements for rotorcraft flying in and around airports and residential areas. Military rotorcraft would like to generate as little noise as possible in order to increase stealth and improve battlefield survivability.

Significant research is directed toward realizing the goal of jet-smooth-and-quiet rotary-wing flight [13] to improve the cost effectiveness and to achieve wider community acceptance of rotorcraft. Extensive research has been focused on active and passive methods of vibration control. Passive methods suffer from several inherent disadvantages such as a large weight penalty and poor off-design performance. Active control strategies show more promise for controlling vibration levels over a wide range of flight conditions. However, a helicopter rotor blade presents an extremely challenging environment for conventional actuators due to the severe volumetric constraints, small allowable weight penalty, large centrifugal forces, and complexity of power/pressure transfer from the fixed-frame to the rotating-frame (electric/pneumatic slip rings).

In active vibration control, the blade pitch is excited at higher harmonics of rotational speed, generating new unsteady forces that cancel the vibratory forces at their source. Blade feathering/twisting on the order of 1 to 2 degrees at higher harmonics is needed to suppress vibration. The higher harmonic control (HHC) system incorporates excitation of the swashplate at N/rev with servo-actuators (typically hydraulic). It has been found to be a viable concept to suppress vibration and may incur a lower weight penalty than a passive system. The power requirements of the servo-actuators can become substantial at extreme flight conditions where vibrations are likely to be highest. Also, the swashplate can be excited only at integer multiples of N/rev. Using individual blade control (IBC), the blades can be excited at any pitch using actuators in the rotating frame. However, with hydraulic actuators in the rotating frame, one faces the complexity of hydraulic slip rings.

Advances in active materials and smart structures technology have introduced the possibility of designing compact, lightweight actuators that can be integrated in the blade structure to deflect a control surface or change the blade geometry. In this manner, the airloads on the rotor blade could be affected in an active control scheme.

Many actuation mechanisms have been proposed, in both model-scale and full-scale versions. Most of these mechanisms are based on piezoceramic actuators, which provide the benefits of high energy density and high bandwidth. Additionally, the coupled electromechanical behavior of piezoceramics enables the use of these materials as sensors as well as actuators. This property creates numerous possibilities; for example, self-sensing actuators for collocated control and high-sensitivity embedded sensors to sense strain in the rotor blade. The rotor with on-blade actuators and

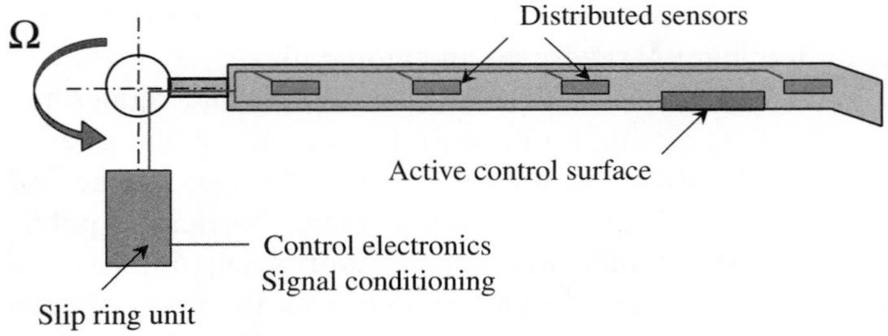

Figure 8.55. Smart rotor concept.

sensors, together with real-time control algorithms, results in a smart rotor system (Fig. 8.55). In contrast, a rotor with only actuators is referred to as an active rotor.

Recent interest has also been focused on the concept of a swashplateless rotor, in which the bulky and mechanically complex swashplate of the main rotor could be replaced by much more compact fly-by-wire type control systems. Primary control of the vehicle could be achieved by means of on-blade actuators that deflect elevon-like control surfaces [141] or create shape changes in the rotor blades. These actuators would operate off electrical power that would be transmitted from the fixed frame to the rotating frame through an electrical slip-ring unit. Because the basic hardware requirements are the same, a smart rotor could be designed to achieve the objectives of both active vibration control and primary flight control. Several model-scale as well as full-scale active rotors actuated by smart materials are described next.

8.6.1 Model-Scale Active Rotors

The first active-twist rotor, using direct twist actuation, was developed by Chen and Chopra [76, 142]. The rotor blade incorporated dual-layer monolithic piezo patch elements embedded at $+45^\circ$ under the upper-surface skin and -45° under the lower-surface skin of the rotor blade (Fig. 8.56). The high aspect ratio (length = 2 in, width = 1/4 in) piezo elements extended from approximately 17.5% to 70% chord, and the ratio of the piezo to fiberglass skin thickness was on the order of 4:1. With both the upper and lower piezo elements excited in phase, a net shear strain is induced in the skin, which in turn causes a pure twisting of the blade. Similarly, it is possible to

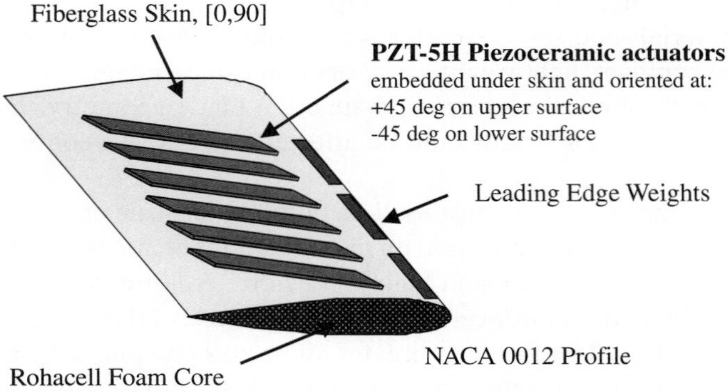

Figure 8.56. Active twist rotor using piezoceramic sheet actuators, from Ref. [13].

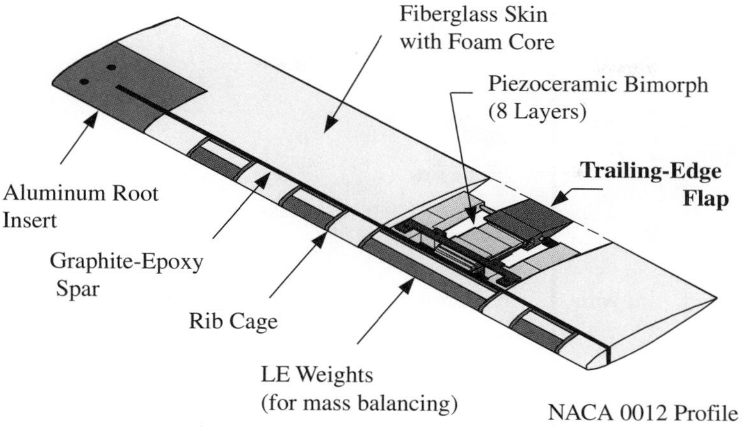

Figure 8.57. Mach-scaled active rotor blade with trailing-edge flap actuated by piezo-bender elements, from Ref. [13].

induce pure bending by an out-of-phase excitation. The total number of actuators per blade ranges from 24 (single layer, 1.5 in apart) to 120 (dual layer, 0.75 in apart). A 1.83 m (6 ft) diameter Froude-scale active-twist rotor was tested on the hover stand and in the wind tunnel. The tip Mach number of the reduced-speed rotor was 0.28. Blade tip twist amplitudes of $\pm 0.25^\circ$ were achieved (excitation field -560 to 1110 V/mm, excitation frequency below the torsional resonance frequency of 4.75/rev). Open-loop wind tunnel tests demonstrated that despite the low blade-tip twist amplitudes, it was possible to appreciably alter the rotor vibratory hub loads.

Another active twist concept involves the integration of active piezo-fiber plies into the composite blade structure. Interdigitated electrodes are deposited on the active plies to utilize the larger d_{33} effect of the piezoelectric material. Active piezoceramic fibers replace the conventional graphite or glass fibers in a resin matrix, creating an active composite ply. When cured in a $+45^\circ/-45^\circ$ orientation on the blade, actuation of these active layers results in a linear twist along the blade section. The piezofiber concept was originally developed by Hagood et al. [24] and has subsequently been commercialized. A two-bladed 1/6th Mach-scale model of the CH-47 rotor with active piezofibers was tested and a tip twist of $\pm 0.4^\circ$ was measured at full rotor speed and 8° collective, with a mass penalty of 16%.

Extensive research has been conducted on Froude- and Mach-scaled rotor models (1/8 scale) with trailing-edge flaps actuated by piezo-bender elements (bimorphs). Koratkar and Chopra [143] tested Froude-scaled and Mach-scaled, four-bladed model rotors on the hover stand followed by testing of a Mach-scaled model in the wind tunnel (Fig. 8.57). In the wind tunnel, the rotor was tested at rotational speeds of up to 1800 RPM and an advance ratio of 0.3 and a collective pitch setting of 6°. A micro-thrust bearing was needed to attach the flap to the blade to reduce the frictional torque under high centrifugal forces. A schematic of an eight-layered piezo-bimorph is shown in Fig. 8.58. These tapered bimorphs were built with a decreasing thickness from root to tip to increase the actuation authority and weight efficiency. A mechanical leverage arrangement between the actuator and flap was incorporated using a rod-cusp arrangement (Fig. 8.59).

The piezo-bender actuators were excited at 90 Vrms with a 3:1 bias – the positive half-cycle of the actuating sinusoidal waveform was amplified by a factor of three,

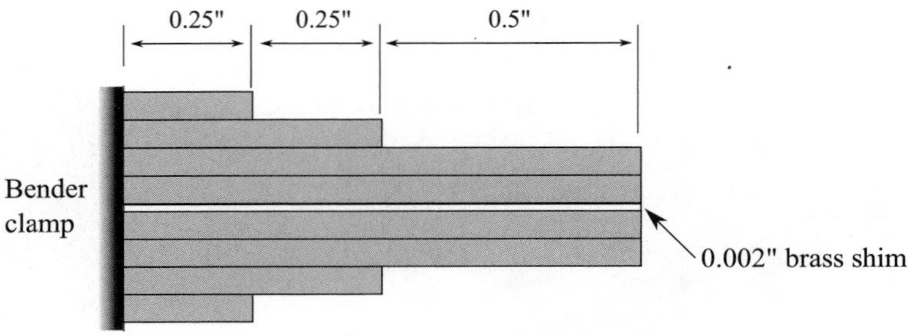

Figure 8.58. Eight-layered, 1 in wide, tapered piezo-bimorph (each PZT layer is 0.0075 in thick).

whereas the negative half-cycle was not amplified (Fig. 8.60). This enabled a higher excitation field to be applied in the direction of poling, without depoling the piezoelectric material), up to frequencies of 5/rev and generated a deflection on the order of $\pm 4^\circ$. Open-loop and closed-loop tests in the wind tunnel demonstrated the control authority of the actuation system and the ability to minimize vibratory hub loads for a range of flight conditions (Fig. 8.61). Fulton and Ormiston [144, 145] successfully tested an improved bimorph flap on a reduced-speed rotor (tip Mach 0.27, diameter 2.23 m). The 12% span, 10% chord flap was centered at 75% radius and was driven by two 38.1 mm wide piezo-bimorphs. At full speed, open-loop flap deflections of $\pm 7.5^\circ$ were achieved at an excitation of ± 610 V/mm. The test program clearly demonstrated the ability of the trailing-edge flap to alter the blade torsion and flap bending loads. However, this concept does not easily lend itself to scaling up to a full-scale rotor because of the large weight penalty.

Bernhard and Chopra [146] developed a novel actuation mechanism consisting of a bending-torsion coupled composite beam with piezoceramic sheet actuators bonded on its surface (Fig. 8.62). The beam is divided into a number of spanwise segments with reversed bending–torsion couplings for each successive segment. Over each beam segment, identical piezoceramic actuators are bonded on the top and bottom surfaces, resulting in equivalent bimorph units. The polarity is reversed for successive piezo elements. This composite beam is located spanwise at the quarter-chord of the blade profile. When the piezoceramic actuators are actuated in a bending configuration, the total bending in the beam cancels out and the total twist adds up

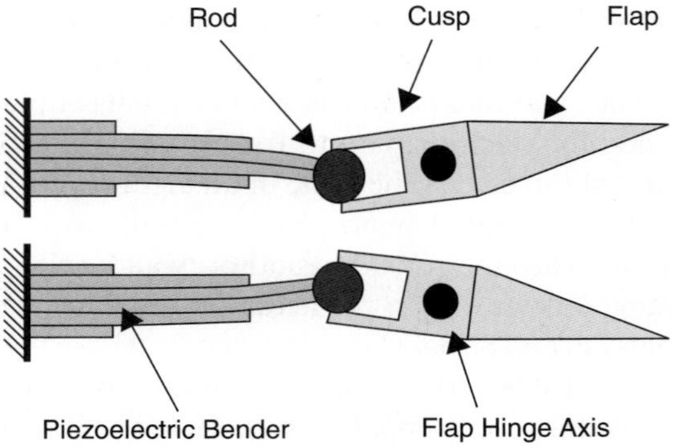

Figure 8.59. Piezo-bimorph flap actuation system, from Ref. [13].

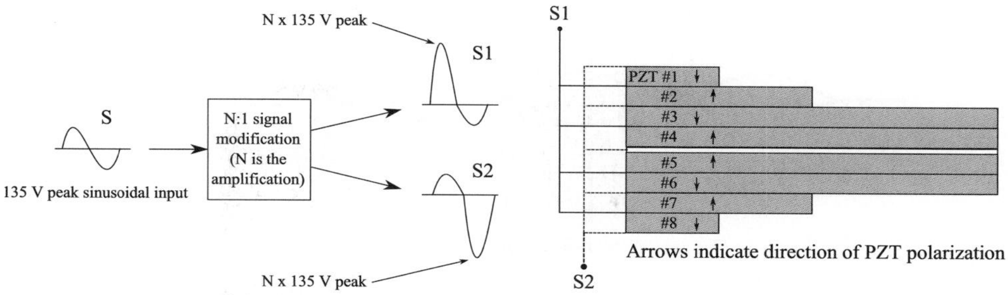

Figure 8.60. AC bias circuit used to power the piezo-bimorphs.

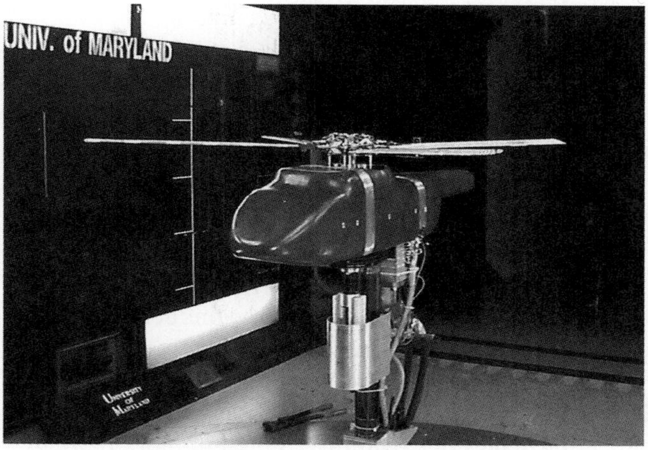

Figure 8.61. Mach-scale rotor model with piezo-bimorph–actuated trailing-edge flaps in the Glenn L. Martin wind tunnel, from Ref. [13].

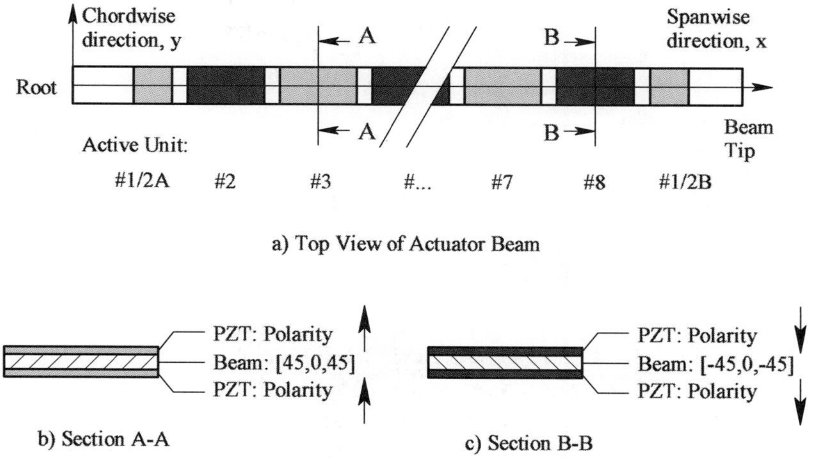

Figure 8.62. Composite bending-torsion coupled beam with piezo actuators, from Ref. [13].

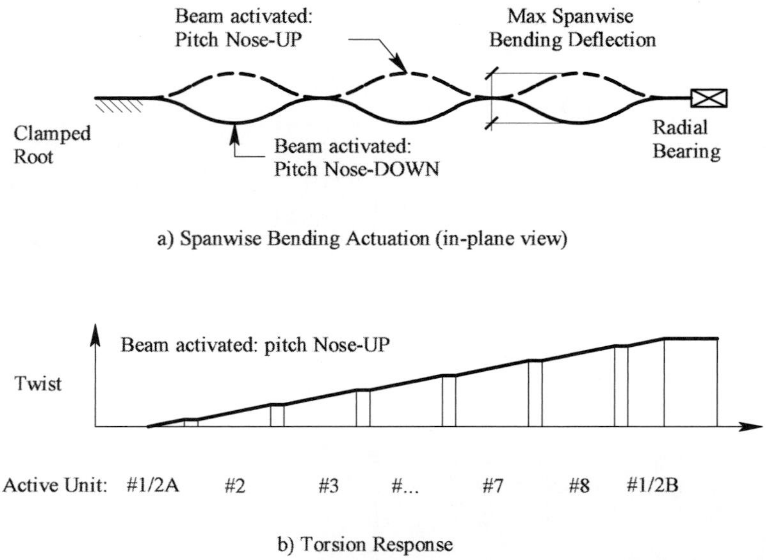

Figure 8.63. Schematic of composite bending-torsion–coupled beam actuator mechanics, from Ref. [13].

(Fig. 8.63). The resulting tip twist was used to actuate trailing-edge flaps as well as an all-moving, 10%-span blade tip (Smart Active Blade Tip, SABT), as shown schematically in Fig. 8.64. Froude- and Mach-scaled tests were conducted on active rotors with this actuation mechanism. In Mach-scaled hover tests, at 2000 RPM, 2° collective, and an actuation voltage of 125 Vrms, the measured tip deflection at the first four rotor harmonics was between ±1.7° and ±2.8°, increasing to ±5.3° at 5/rev due to resonance amplification. The tip activation resulted in more than 50% variation in the steady rotor thrust levels at 8° collective.

Bothwell et al. [74] researched the concept of actuating trailing-edge flaps by means of an extension-torsion coupled composite tube, with an internal piezostack

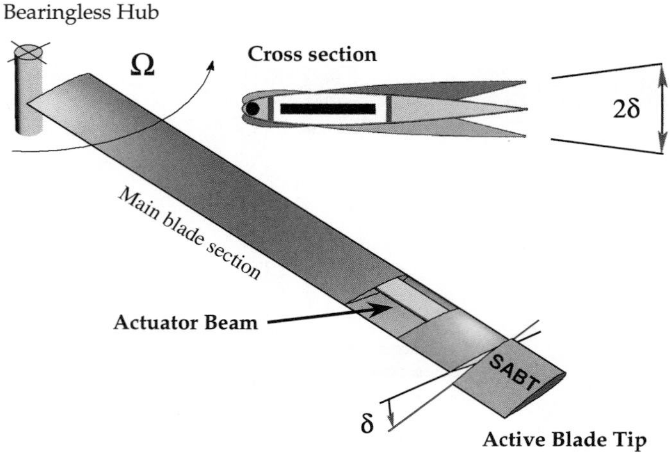

Figure 8.64. Mach-scaled rotor blade with Smart Active Blade Tip (SABT), from Ref. [13].

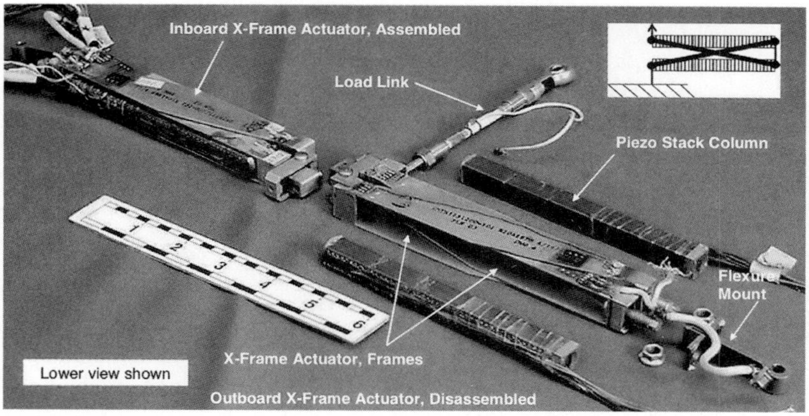

Figure 8.65. X-frame piezostack amplifier, from Ref. [147].

or magnetostrictive actuator. As a result of the composite coupling, the tube extends and twists in response to extension of the actuator. Based on experimental success, it was proposed to stack multiple tubes in series to generate sufficient twist to deflect a trailing-edge flap.

Modifications of the conventional straight bimorph have also been proposed, such as the C-block actuator [52]. The C-block has a greater stroke capability than a conventional straight bender at the cost of reduced force output. A blade section incorporating a 10% chord trailing-edge flap with a 50% pivot overhang for aerodynamic balancing was wind tunnel tested. With a driving voltage of 55% (of the maximum level), flap deflections of $\pm 5^\circ$ to $\pm 9^\circ$ were measured; however, the dynamic pressure was less than 3% of full-scale dynamic pressure.

The X-frame-actuator concept was developed by Prechtl and Hall [19]. The X-frame is a piezo-stack amplification mechanism that uses stroke amplification via shallow angles. The actuator was integrated into a 1/6th Mach-scale Boeing CH-47 (Chinook) blade and tested in hover. The flap is a slotted servo-flap with a 11.5% span, 20% chord, and aerodynamic overhang 27.5% of flap chord. At the operating speed (tip Mach number 0.63) and 8° collective, flap deflections of $\pm 3.9^\circ$ were achieved.

8.6.2 Full-Scale Active Rotors

The baseline theoretical and experimental work validated in the model scale is slowly transitioning into full-scale applications. A full-scale active-flap rotor was developed by McDonnel–Douglas/Boeing for an MD900 Explorer helicopter, which is an eight-seat utility helicopter with a maximum take-off weight of 6250 lbs, with a five-bladed, 34-foot-diameter bearingless rotor. The trailing-edge flap was driven by a bi-directional version of the X-frame actuator. The flap had a span of 3 feet, a chord of 3.5 inches, and was located at 83% radial position. The actuator was scaled up from the model scale to meet full-scale requirements [147]. Two X-frames were coupled together to obtain a positive force during both extension and retraction of the flap actuator push rod. The dual X-frame actuator was capable of a blocked force of 80 lbs and a free displacement of approximately 100 mils. A dual X-frame actuator undergoing benchtop testing is shown in Fig. 8.65. The rotor was successfully tested

Figure 8.66. Full-scale BK117 rotor blade with trailing-edge flaps and piezoactuators, from Ref. [154].

in the 40 ft × 80 ft wind tunnel, and results showed reductions up to 6 dB in blade vortex interaction and in-plane noise, as well as a reduction in vibratory hub loads of up to 80%. Rotor performance was affected by 2/rev flap inputs [148].

Early studies by Eurocopter for a full-scale active trailing-edge flap used a piezostack actuator with a shallow-angle flextensional amplification mechanism [149, 150, 151, 152]. The same actuator was also proposed for a leading-edge droop concept, delay dynamic stall [153]. More recently, Eurocopter investigated and selected two candidate technologies for vibration cancellation and BVI noise suppression by means of IBC using trailing-edge flaps [154]. The first approach used DWARF piezoceramic actuators driving a 15% chord flap integrated in modified BK117/ATR rotor blades (Fig. 8.66). The second approach utilized COCE electromagnetic actuators driving a 25% chord flap in a modified Dauphin blade section.

A bi-directional flap actuator driven by piezostacks, based on lever-arm amplification was developed by Lee and Chopra [15] (Fig. 8.67). This actuator was designed to meet the requirements of a trailing-edge flap on the blade section of a full-scale MD900 Explorer helicopter. The actuator was driven by five piezoceramic stack elements driven at a peak-to-peak voltage of 120 V and achieved a blocked force of approximately 9 lbs with a free stroke of approximately 75 mils. This actuator was

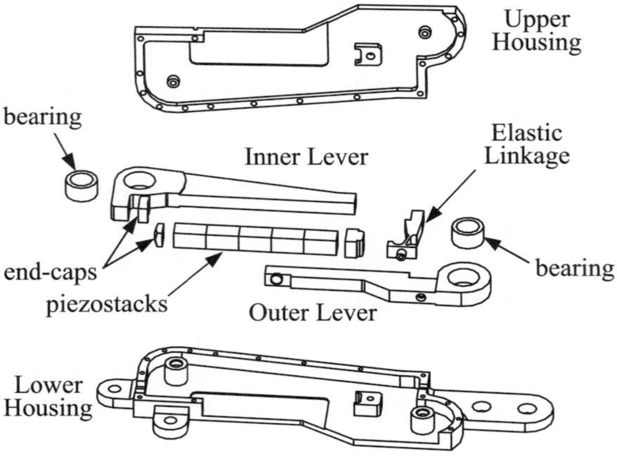

Figure 8.67. L-L amplification actuator.

tested in a vacuum chamber spin test and in a wind tunnel (nonrotating condition), at frequencies of up to 5/rev. Spin testing results showed less than 10% degradation of actuator deflection at 710gs of centrifugal acceleration. In the wind tunnel, peak-to-peak flap deflections of up to 12° were measured at free-stream velocities of 120 ft/sec and 12° collective.

A shear-mode piezoelectric tube actuator was developed to drive a trailing-edge flap [72, 65]. This actuation mechanism utilizes the d_{15} effect of the base piezoelectric material, which is the largest piezoelectric coupling effect. Design studies were conducted for a Boeing MD 900 helicopter with a plain trailing-edge flap. It was estimated that flap deflections of ±2.5° could be achieved at full speed. This was based on a 6% span flap, with a 25% chord, driven by a tube with an outer and inner diameter of 17.8 and 11.4 mm, respectively, and a length of 203 mm (corresponding to 4% of rotor radius). Using a spring to simulate aerodynamics, deflections of ±1.5° were measured in a bench-top test at 75% of the maximum electrical field, on the order of 1kV.

A trailing-edge flap actuator based on MSMA as the active material was designed to provide primary flight control authority on a search and rescue helicopter [155]. Two trailing-edge flaps were used on each blade, and flap deflections on the order of ±5° at hinge moments of approximately 3 lb-ft were required for trim. Two permanent magnets were used in conjunction with two magnetic coils that provided a differential magnetic field of ±100 kA/m. The total actuator weight, including the housing, was 1.9 lbs, of which 0.798 lbs was the active MSMA material. The power requirement was 210 W, at a current of 4 A, which corresponds to approximately 0.2% of the total installed continuous power of the vehicle. This design study clearly demonstrates the feasibility and attractiveness of the swashplateless concept.

8.6.3 Adaptive Controllers for Smart Rotors

A number of control approaches were published in the literature for different types of smart structures and systems. Helicopter rotors have certain unique features in this regard; for example, time-periodic equations of motion and nonlinear aeroelastic response. We focus on adaptive controllers for smart rotors in this section.

It is well established that the vibration and dynamic loads in a helicopter can be alleviated using higher harmonic pitch controls. Most frequently, the helicopter rotor system is expressed in the frequency domain through a transfer function T relating the input control harmonics to the output response harmonics of loads or stresses. This simple linear quasi-static model (Fig. 8.68) is expressed as

$$Z = Z_o + T\theta + \nu \tag{8.125}$$

where the response vector Z consists of sine and cosine components of stresses and the vibration level, in either the rotating frame or the fixed frame. The input control vector θ consists of sine and cosine components of higher-harmonic pitch, in either the rotating frame or the fixed frame. The transfer function matrix T and the uncontrolled response Z_o depend on flight conditions; for example, forward speed, rotor thrust, and rotational speed. The measurement noise ν is expected to be random in nature.

These controllers are broadly classified into two categories: open loop and closed loop. For open loop controllers, there is no direct feedback of response; for the closed

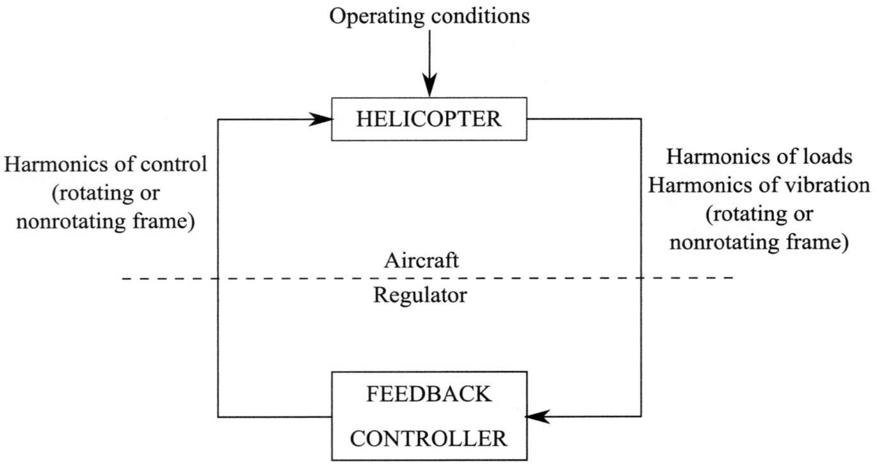

Figure 8.68. Multicyclic control of helicopter vibration.

loop controllers, there is a feedback of measured response. Two types of models are used to represent the control system, as follows:

(1) Local model: The model is linear about the current control value.

$$Z_n = Z_{n-1} + T_{n-1}\left(\theta_n - \theta_{n-1}\right) \tag{8.126}$$

(2) Global model: The model is linear for the composite range of the control.

$$Z_n = Z_o + T\theta_n \tag{8.127}$$

The n characterize the time step, $t_n = n\Delta t$, where Δt is assumed to be large enough so that any transient has settled (e.g., one or two rotor revolutions). The local model is more general and is quite applicable for nonlinear conditions because T is linearized about the current control value and the range of $\Delta\theta_n = \theta_n - \theta_{n-1}$ is assumed to be small. For the local model, the T matrix is identified in each time-cycle (n). For the global model, there are three possibilities: (1) identification of the T matrix only, (2) identification of the uncontrolled response vector Z_o only, and (3) identification of both the T matrix and the vector Z_o. For the open loop controllers, the input controls are based on the uncontrolled response Z_o, whereas for the closed loop controllers, the input controls for any time-cycle n are based on the measured response from the previous cycle Z_{n-1}. The controllers are further subdivided into two categories: (1) off-line identification in which the characteristics of the control system (T and Z_o) are identified initially and are assumed invariant; and (2) on-line identification in which the characteristics of the control system is updated continuously in each time-cycle. The first category of controllers is applicable to the global model and the control gains are fixed. The second category of controllers is applicable to both global and local models, and the control gains vary with time.

The quadratic performance function J is typically expressed as

$$J = Z_n^T W_z Z_n + \theta_n^T W_\theta \theta_n + \Delta\theta_n^T W_{\Delta\theta} \Delta\theta_n \tag{8.128}$$

where W_z, W_θ, and $W_{\Delta\theta}$ are the weighting matrices for the response, pitch controls, and pitch control rates, respectively. These matrices are typically diagonal. For example, the elements of the W_z matrix provide relative weighting to the response (vibration) harmonics, and setting any one of these diagonal elements to zero results in unconstraining the corresponding vibration harmonic. The weighting matrix W_θ constrains the amplitude of input control harmonics, which in turn helps to regulate the stroke of the actuators. The weighting matrix $W_{\Delta\theta}$ constrains the control rate, which in turn helps to reduce control (actuators) excursions. This is important from the actuator hardware point of view and generally stabilizes the feedback system. A large value on the diagonal of $W_{\Delta\theta}$ results in control sluggishness.

Let us consider a simple case

$$W_z = I \qquad \text{(unit matrix)} \tag{8.129}$$

$$W_\theta = 0 \tag{8.130}$$

$$W_{\Delta\theta} = \tau I \tag{8.131}$$

where τ is the time constant of the control lag. If this has a large value, it slows down convergence to the optimal control condition. The optimal control input is obtained from the minimization of the performance function J with respect to each control harmonic

$$\frac{\partial J}{\partial \theta_n} = 0 \tag{8.132}$$

for each component θ_n. This provides

$$\theta_n = \theta_{n-1} + C Z_{n-1} \tag{8.133}$$

where the gain matrix C is given by

$$C = -D T^T W_z \tag{8.134}$$

and

$$D = T^T W_z T + W_{\Delta\theta}^{-1} \tag{8.135}$$

For the global model, the input control solution can take the following form

$$\theta_n = C Z_o + C_{\Delta\theta} \theta_{n-1} \tag{8.136}$$

where

$$C_{\Delta\theta} = D W_{\Delta\theta} \tag{8.137}$$

This form is applicable to the open loop case and the input controls are functions of the uncontrolled response Z_o. For the closed loop case, the optimal controls are obtained using the first form and for the nth time cycle, these are a function of the feedback of response from the previous cycle Z_{n-1}.

Off-Line Identification: The helicopter model (T matrix) and response (Z) are identified off-line using a succession of input and output measurements. This is a key part of the study for both off-line and on-line identification control schemes. For on-line identification, a good initial estimate of model characteristics helps enormously in stability and convergence to the steady-state solutions. The input control vectors used for off-line identification are generally randomly selected, and criteria should be such that the generated output responses are within stress limits. The larger the output response amplitude, the smaller will be the influence of measurement noise.

A set of N measurements is made using a prescribed schedule of independent control inputs (random), and the T matrix is identified using the least-squared-error method

$$T = Z\theta^T(\theta\theta^T)^{-1} \tag{8.138}$$

where $\theta = [\theta_1, \theta_2, \ldots \theta_N]$ is the vector of control inputs. The minimum number of measurements N must be at least equal to the number of control input harmonics; typically, N must be 2–3 times this minimum value.

On-Line Identification: The T matrix and response Z are continuously updated in each cycle of time. This is normally carried out using a Kalman filter estimation. Let us consider that there are j measurement response harmonics and m control harmonics; then, the dimension of transfer function T is $j \times m$. In any particular time cycle n, we take j concurrent response measurements, and we want to identify the T matrix with $j \times m$ elements. It is not possible to identify the elements of the T matrix directly because the number of unknown quantities far exceeds the number of measurements. Through the Kalman filter, the T matrix is divided into j states (number of columns); a prior estimate of each state is made at the time of measurement, and then the estimation is updated using the current measurement. The jth measurement vector can be written as

$$Z_{j_n} = \theta_n^T t_{j_n} + \nu_{j_n} \tag{8.139}$$

where t_j is the jth row of the T matrix and ν_j is the measurement noise, which is typically assumed to be Gaussian with zero mean. The variation of state t_j is assumed to be

$$t_{j_{n+1}} = t_{j_n} + u_{j_n} \tag{8.140}$$

where u_j is the process noise, which is also assumed to be Gaussian with zero mean. The variances of noise levels are defined as

$$E(\nu_n, \nu_i) = r_n \delta_{ni} \tag{8.141}$$

$$E(u_n, u_i) = Q_n \delta_{ni} \tag{8.142}$$

where r and Q are the covariances of measurement noise and process noise, respectively. Using a Kalman filter solution, an estimate of t_j at the nth time cycle is based on current measurements and an estimate of t_j at the $(n-1)$th time cycle. The Kalman filter gives a minimum error variance solution

$$\hat{t}_{j_n} = \hat{t}_{j_{n-1}} + K_n(Z_n - \theta_n^T \hat{t}_{j_{n-1}}) \tag{8.143}$$

where

$$M_n = P_{n-1} + Q_{n-1} \tag{8.144}$$

$$P_n = M_n - M_n\theta_n\theta_n^T M_n/(r_n + \theta_n^T M_n \theta_n) \tag{8.145}$$

$$K_n = P_n\theta_n/r_n \tag{8.146}$$

where M_n is the covariance of the error in the estimate of t_n before the measurement and P_n is the covariance of the error in the estimate t_n after the measurement. The elements of Q represent the variation of the actual t from the estimated one. For changing flight conditions, t is expected to vary rapidly; hence, Q should be large. However, a large value of Q can slow the convergence process. One has to try a few different values of Q to obtain the proper value that makes the results acceptable. The r_n represents the measurement error due to the sensor. Again, a meaningful value must be assigned to r_n. To start the process, a large value on the diagonal matrix of P is assigned initially. For details, see Chopra and McCloud [156] and Johnson [157].

Johnson [157] classified these frequency-domain higher-harmonic controllers into four types. The first is the invariant open-loop controller, in which the model properties (transfer functions) are identified off-line and input controls are based on the uncontrolled response. The second type is the invariant closed-loop controller, in which the model properties are identified off-line and the input controls are based on feedback of the measured output. The third type is the open-loop adaptive controller, in which the control inputs are based on identified uncontrolled vibration rather than on measured outputs. This type of controller can use on-line identification of uncontrolled vibration only or can use on-line identification of both the uncontrolled vibration and the transfer function. The fourth type is the closed-loop adaptive controller, in which the model properties are identified on-line and the controls are based on feedback of measured output. This type of controller can use both local-linear and global-linear models.

8.7 SMA Actuated Tracking Tab for a Helicopter Rotor

An untracked rotor system is a common source of large 1/rev vibrations in helicopters. Small dissimilarities in structural or aerodynamic properties of the blades created during the manufacturing process, or occurring as a result of wear, result in the rotor system going out of track (i.e., the motion of each blade tip lies in a different plane). The masses of the blade are closely matched by adding small masses at specific locations on the blade. However, the aerodynamic properties of the blade can only be modified by small variations in the root pitch of the blade by means of adjustments in the lengths of the pitch links or by special devices known as tracking tabs. These are small aerodynamic surfaces located at the trailing-edge of the blade, at approximately 75% of the blade span. Small differences in aerodynamic loads are created by mechanically bending the tracking tabs to specific angular positions, specified as needed by the blade manufacturer. The conventional procedure for rotor tracking is a ground-based method requiring manual adjustment of tracking tabs and pitch links. The difference in tip-path plane between the rotor blades is measured while the rotor is spinning. The rotor is then stopped and tracking tabs and/or pitch links are manually adjusted. Because a small change in the length of the pitch link can result in a large change in overall aerodynamic loads of the blade, fine adjustment of the tracking is often performed using the tracking tabs alone. This procedure is repeated on a trial-and-error basis until the rotor is tracked to a sufficient accuracy. The conventional rotor tracking procedure is both time consuming and expensive. Large savings in maintenance costs can be achieved by automating the tracking procedure.

A few researchers [158, 159] examined methods to replace the current manual tracking procedure with an on-blade tracking mechanism capable of deflecting a tracking tab in-flight. It is anticipated that an on-blade tracking system would effectively replace manual tracking operations and thus reduce operation errors, helicopter downtime, maintenance time, and associated costs. Actuators based on high energy density smart materials are ideally suited to this application because of severe volumetric constraints and low allowable weight penalties associated with mechanisms mounted on a rotor blade.

Actuators based on SMAs are particularly relevant to this application. The relatively large output force and stroke capability offered by SMAs enables the design of simple actuation mechanisms. This is in contrast to complicated designs requiring gear reduction or motion amplification when employing conventional materials or other active materials, such as piezoelectrics. Low actuation voltages, low costs, and a reduced number of moving parts are additional advantages associated with an SMA actuator. Because tracking operations need not be conducted at high frequencies, the low bandwidth of SMAs is more than adequate for this application.

Tab actuation systems implementing SMAs, which have been built and/or tested in the past, have demonstrated the anticipated advantages of these materials. These actuation systems may be broadly classified under two types, based on their deflection mechanism, as follows:

1. Torsional tubes/rods – Actuators implementing torsional tubes/rods develop rotational strains and moments that are directly transmitted to the tracking tab. A preliminary study exploring the feasibility of on-blade SMA torsional actuators [160] and an experimental bi-directional actuator consisting of two antagonistic SMA torsional tubes [161] have been reported. Due to their large thermal inertia, torsional actuators require external heating and cooling systems for activation. The large volume occupied by these heating and cooling elements may impede the integration of the entire system into the confined space in a rotor blade section.
2. Wire actuators – The extensional strains of SMA wires are translated into a rotational motion of the tracking tab. Tab actuators employing SMA wires have been designed for operation in a hydrofoil [162] and for rotor blade tracking [163, 158]. In contrast to torsion tubes, wires demonstrate a much smaller thermal inertia. This property permits faster thermal actuation of SMA wires. Additionally, internal resistive heating of the wires eliminates the need for bulky external heating mechanisms. Consequently, SMA wire based actuators can be easily integrated into the blade section.

The design, analysis, and testing of a tracking tab actuated by SMA wires [159] is described in the following sections.

8.7.1 Actuator Design Goals

The parameters for designing the tracking tab actuator evolved from the angular deflections and loads that were estimated to be experienced by the tracking tab during operation on the rotor blade. Previous studies [160, 161] report a quantitative estimate of the structural and environmental conditions that the actuator must operate in, sized for a Boeing MD900 helicopter (weight 6,250 lbs). Based on these studies, the goals for the actuator were determined and are summarized in Table 8.5.

As far as possible, the actuator should conform to a weight of less than 1 lb, tab deflection of $\pm 5^\circ$, angular resolution of $\pm 0.1^\circ$, output and braking moments (to overcome hinge moments due to aerodynamic and rotating frame loads) of 4.0 in-lbs, and a duty cycle of 20 cycles/hr. It was planned to test the mechanism integrated

Table 8.5. *Tracking tab actuator goals*

Parameters	Goals
Actuator stroke	$\pm 5^o$
Resolution	$\pm 0.1^o$
Braking moment	4.0 in-lb
Actuator weight	<1 lb
Actuator dimensions	10 in × 8 in × 1 in
Duty cycle	20 cycles per hour
Temperature range	−60 to 160 F

in a 12 in chord NACA 0012 blade section and, therefore, the entire actuator must be capable of meeting geometric design requirements imposed by space limitations of the blade profile. These spatial constraints were established to be a thickness dimension of 1.4 in at the quarter-chord section and 0.8 in at the location of the hinge tube. The system should be capable of withstanding aerodynamic and rotating frame loads expected to be encountered near the 75% radius of the blade. However, for the model tested in the laboratory, the focus was on the behavior of the SMA actuator, and only testing under aerodynamic loads in a wind tunnel was planned. Consequently, the mechanism was not designed to operate under centrifugal loads. Temperature, force (moment), and position (angle) sensors must be located on the tab assembly, providing feedback to a position control mechanism. Additionally, the tab deflection must be sustained under a power-off condition.

8.7.2 Construction and Operating Principle

The basis of the actuation system is the antagonistic operation of two sets of SMA wires. Figure 8.69 schematically illustrates the principle of operation. The upper and lower wires are fixed rigidly at one end and are connected to a rotating hinge tube at the other. They are both given an equal tensile pre-strain and are insulated from one another, both thermally and electrically. The wires are resistively heated and convectively cooled. To deflect the tab upward, a current is passed through the upper set of wires. This results in heating of the upper wires, which then undergo a phase transformation and try to recover their pre-strain. Consequently, the upper wires contract by a certain length while extending the lower set of wires by an equal length. This action results in a rotation of the hinge tube and an upward deflection of the tab. Deflection of the tab in the opposite direction is accomplished by passing a current through the bottom set of wires alone.

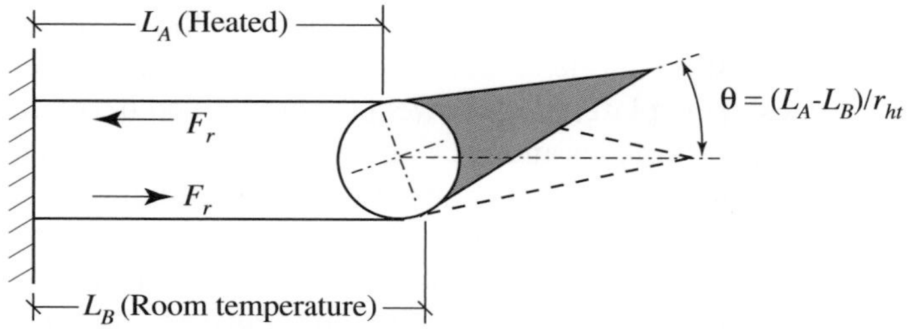

Figure 8.69. Schematic of operation of bi-directional SMA wire actuator.

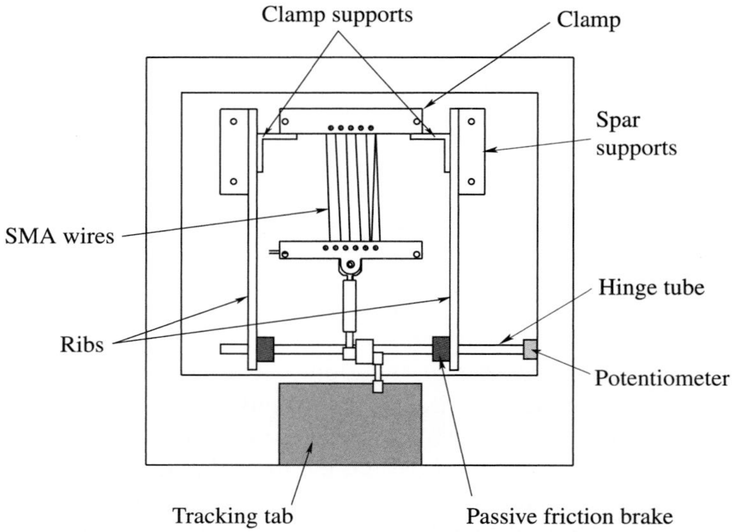

Figure 8.70. Tab actuation system components.

The deflection of the tab is given by

$$\theta = \left(\frac{L_o \epsilon_r}{r_{ht}}\right) \tag{8.147}$$

where L_o is the initial length of the wire segments, also referred to as the length of the wires; ϵ_r is the strain in the wires; and r_{ht} is the radius of the hinge tube.

The primary elements that comprise this actuator are the SMA wires and clamping mechanism, a pre-strain mechanism, a passive friction brake to maintain position, and a position controller. A schematic diagram of the various components of the actuation system is in Fig. 8.70.

SMA Wires and Clamping Mechanism

The SMA wires used were of commercially procured Nitinol (Ni-51%, Ti-49%) material. The diameter, length, and number of wire segments used are discussed in Section 8.7.5 describing the parametric design of the actuator. The clamping mechanism assembly was designed to restrain one end of the SMA wires near the main spar of the blade while allowing the other end to freely translate along the chord of the blade section. The assembly consists two pairs of stainless steel clamps, of which one pair is rigidly bolted to the main spar. The linear motion of the aft clamp is transmitted to the hinge tube, resulting in deflection of the tab. Each clamp pair comprises two 1/8 in thick stainless steel plates, with dowel pins embedded between them. The wire is wound around the dowel pins and back again in a manner such that each wire could be fixed at one end of the clamp. The purpose of this configuration is to effect force multiplication in the system due to the multiple SMA wire segments acting in parallel. The use of one SMA wire wound around the dowel pins results in an equal distribution of the tension in each segment, minimizing transverse loading on the mechanism. Figure 8.71 shows schematic details of this clamping mechanism.

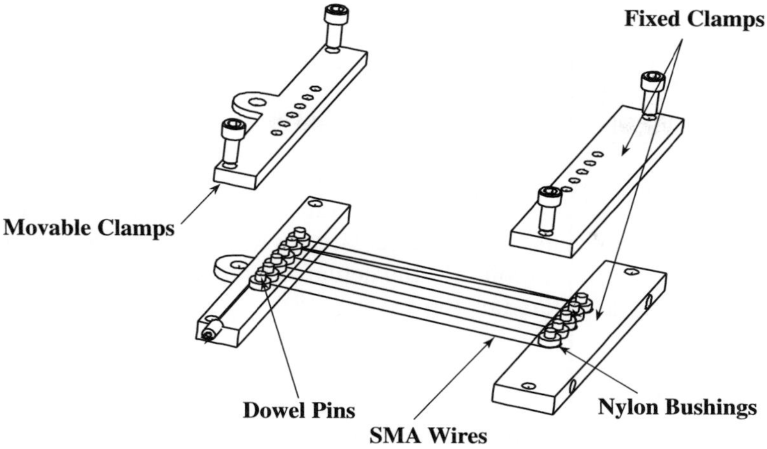

Figure 8.71. Clamping mechanism.

Pre-Strain Mechanism

The clamp output motion is transmitted to the tab through a pair of linkages. These consist of a pair of oppositely threaded rod-ends, connected to a threaded turnbuckle. The rod-ends are attached to the movable clamp at one end and the rotating hinge tube at the other end. These turnbuckles have multiple roles in this design. In addition to transmitting the linear motion of the wires to the tab, they provide a convenient method of pre-straining the wires after assembly by rotating the turnbuckles through a set number of turns.

Locking Mechanism

A locking device is necessary to maintain the tab position without further supply of power once the desired tracking position is acquired. The recovery of a small amount of elastic strain on unloading the SMA wires also necessitates the presence of a locking device to prevent a change in the angle once the heat activation is stopped.

The main specification for the lock is that it must allow for rotation in both directions as well as hold the hinge tube in position without slipping, under external loading. Several active friction brake designs employing piezostacks, electrostrictives, and SMAs were experimented with; eventually, a passive friction brake was selected as the final design. The passive brake consists of a shaft collar, rigidly mounted on the rib and around the shaft. A torque wrench is used to tighten the collar to the required frictional braking moment. For actuation moments exceeding this braking moment, the hinge tube undergoes rotation.

Position Feedback Controller

Closed-loop control was required to demonstrate the capability of the actuator to accurately deflect a tracking tab to the commanded input position. A closed-loop PID controller was implemented using a LabView™ Virtual Instrument, which performed functions of both data acquisition and control. The inputs to the controller

Table 8.6. *PID control gains*

Proportional Gain, K_P	1.280
Derivative Gain, K_D	8.000
Integral Gain, K_I	0.051

were the tab position measured by a rotary potentiometer and the desired tab position, or setpoint. The error signal, V_{err}, was calculated as the difference between the measured tab position and the setpoint. In the actuated state, the output of the controller was a voltage, $u(t)$, as defined by the classical PID control law [164]

$$U_{PID} = K_P V_{err}(t) + K_D \dot{V}_{err}(t) + K_I \int V_{err}(t)dt \qquad (8.148)$$

where K_P, K_I, and K_D are the proportional, integral, and derivative gains, respectively. These gains were determined for the present system by the Ziegler-Nichols method [164] and are tabulated in Table 8.6. A small deadband Δ_{dead} was introduced in the controller, such that

$$u(t) = 0 \quad \text{if} \quad |V_{err}| \leq \Delta_{dead} \qquad (8.149)$$

$$u(t) = U_{PID} \quad \text{if} \quad |V_{err}| > \Delta_{dead} \qquad (8.150)$$

The deadband ensures that the output control voltage is zero when the tab position reaches the desired position within the acceptable error margin. The output voltage serves as an input to a power Metal Oxide Semiconductor Field Effect Transistor (MOSFET) driver that is connected to both sets of SMA wires. The sign of the output voltage determines which set of wires were to be actuated. Note that for the sake of simplicity of the driving electronics, the control voltage $u(t)$ is the gate-source voltage, V_{GS} of the output MOSFETS to which the SMA wires were connected.

8.7.3 Blade Section Assembly

A NACA 0012 blade section of 12 in span and 12 in chord section was fabricated. The actuator was mounted into this blade section. The fabricated blade consists of a foam core, trailing-edge tab, and actuator assembly with spar and ribs to provide structural integrity. Teflon spring bushings are embedded at the 72% chordwise position and provide mounting points for the rotating hinge tube.

The tracking tab is embedded in the planform of the blade section and has a dimension of 4 in span and 3.4 in chord. This is in contrast to existing tracking tab designs where the metal tab is typically 12–18 inches in span and projects out of the blade nominal planform. The motive for selection of the present configuration was primarily ease of fabrication. Figure 8.72 shows some of the important features of the actuator assembly installed in the NACA 0012 blade profile.

8.7.4 Modeling of the Device

A mathematical model of the device is developed implementing the thermo-mechanical response of the SMA wires under applied stress and temperature. From the theoretical model of the device, a parametric design study can be performed to

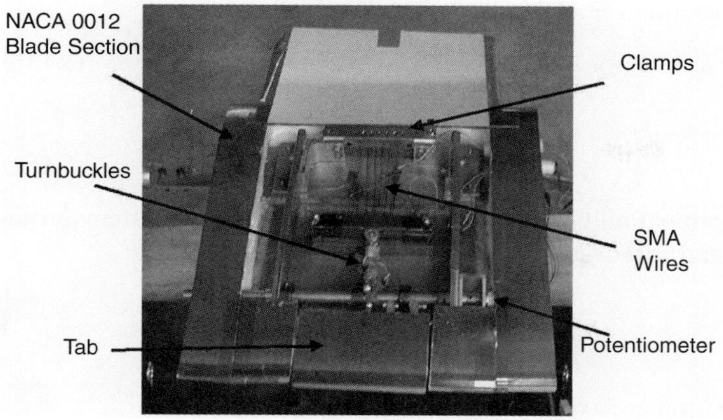

Figure 8.72. Tracking tab system installed in NACA 0012 blade profile.

determine the optimum length of SMA wires and the number of segments required to achieve the force and stroke specifications.

The force acting on the active, or heated, SMA wire as a function of its displacement is shown in Fig. 8.73(a). This diagram depicts the physical constraints imposed on the active wire. During the initial part of activation, marked by the line **OA**, the wire acts only against the friction brake. At the point **A**, the brake starts slipping and force is transmitted to the inactive, or cold wire. The portion of the curve **AB** is the force-displacement characteristic in the martensite region of the inactive SMA wire, which behaves like a nonlinear spring attached to the active wire. The corresponding stress-strain behavior of the wires is shown in Fig. 8.73(b).

For the sake of brevity in this discussion, the upper, heated wires are referred to as wire A and the lower wires, kept at room temperature, are referred to as wire B. For this analysis, the equations of motion are coupled with the transformation kinetic equations based on the Brinson model [165], applied to a one-dimensional SMA wire.

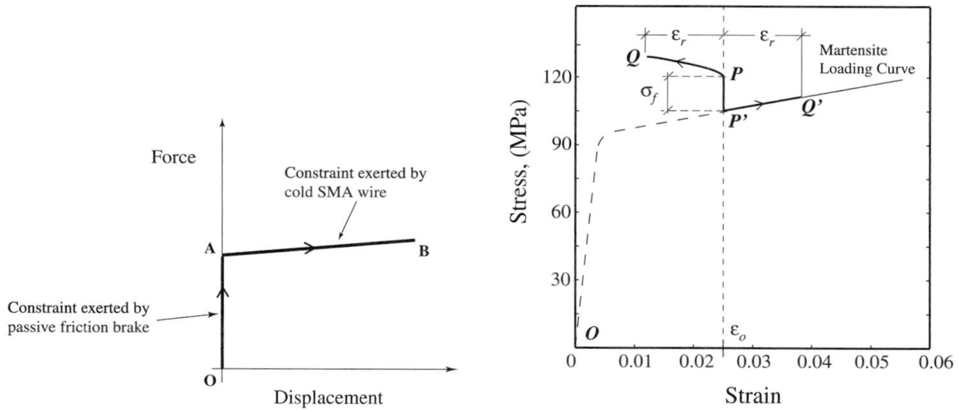

(a) Constraints imposed on the active SMA wire upon heat activation

(b) Stress-strain behavior

Figure 8.73. Modeling of the SMA wire.

The constitutive equation for an SMA wire is given by

$$\sigma - \sigma_o = E(\xi)\epsilon - E(\xi_o)\epsilon_o + \Omega(\xi)\xi_s - \Omega(\xi_o)\xi_{so} \tag{8.151}$$

where, by definition

$$\Omega(\xi) = -\epsilon_L E(\xi) \tag{8.152}$$

The stress- and strain-compatibility conditions define the states of the system during actuation. For wire A and wire B, these are given as

$$\begin{aligned} x^A &= -x^B \\ F^A &= F^B + F_F \end{aligned} \tag{8.153}$$

The basic steps of actuation are explained as follows:

1. Pre-strain Step: The pre-straining method specifies the initial conditions of the SMA wires, prior to heating. This is illustrated on the stress-strain curve in Fig. 8.73(b) as the dotted line (**OP'**). Both wires are pre-strained equally to ensure symmetric operation. The initial conditions are defined as

$$\epsilon_o^A = \epsilon_o^B = \epsilon_o \tag{8.154}$$

$$\sigma_o^A = \sigma_o^B = \sigma(\epsilon_o) \tag{8.155}$$

The following approximation is made in the model to define the initial volume fraction of the material

$$\xi_{SO} = \frac{\epsilon_o}{\epsilon_L} \tag{8.156}$$

2. Heating – Zero Tab Deflection Condition: Wire A is heated ($M \rightarrow A$ transformation) and undergoes constrained recovery until the stress in wire A overcomes the stress σ_f corresponding to the frictional moment (path **P'P** in Fig. 8.73) and is mathematically given by Eq. 8.151. In this state, there is no change in stress or strain of wire B (point **P'**) because the following stress condition holds

$$\sigma^A - \sigma^B < \sigma_f \tag{8.157}$$

3. Heating – Tab Deflection Condition: As the temperature of the wire rises, the transformation to austenite introduces stress in wire A to a level sufficient to overcome the frictional force. The actuating wire A is now able to exert a stress on wire B represented by

$$\sigma^A = \sigma_f + \sigma^B \tag{8.158}$$

The strain developed in each wire is symmetric with respect to the pre-strain ϵ_o

$$\begin{aligned} \epsilon^B &= \epsilon_o^B + \epsilon_r \\ \epsilon^A &= \epsilon_o^A - \epsilon_r \end{aligned} \tag{8.159}$$

The fundamental difference in the state of the two wires is represented in the differing volume fractions of the two wires

$$\begin{aligned} \xi^B &= 1 \\ \xi^A &= \xi(T^A, \sigma^A, \epsilon^A) \rightarrow 0 \end{aligned} \tag{8.160}$$

The transformation kinetic equations as developed by the Brinson model define the martensite fraction for the two wires. Newton-Raphson's iterative technique is applied to solve Eqs. 8.151–8.159 simultaneously with the transformation equations with the objective of determining the thermomechanical parameters for the system.

The moment at the shaft in the counterclockwise direction is

$$\tau = \left(\sigma^A - \sigma^B\right) \frac{\pi d_o^2}{4} N_{wire} r_{ht} \tag{8.161}$$

where N_{wire} is the number of wires or, in this particular configuration, the number of wire segments acting in parallel. The strain recovery condition results in a deflection of the tab given by Eq. 8.147.

8.7.5 Parametric Studies and Actuator Design

Using the model described previously, a parametric study can be conducted to determine the optimum length and diameter of SMA wires and the number of segments, based on the given constraints of the system. The results of this parametric study can be used as a design tool. In addition to the specified force and stroke requirements, several physical constraints exist that must be considered in the design of the actuation system. The ultimate objective is to integrate the actuator assembly into a 12 in chord NACA 0012 blade section and to achieve a deflection of the $\pm 5°$ at a wind speed of 120 ft/sec (Mach number 0.107) at an angle of attack of 15°.

The actuator parameters identified are classified by their influence on either the angular deflection or the actuation moment. The influence of actuator parameters on the output can be seen from Eqs. 8.147 and 8.161. The initial length of wire (L_o), radius of hinge tube (r_{ht}), and maximum recoverable strain (ϵ_r) directly influence the angular deflection (θ). The maximum recoverable strain (ϵ_r) is, in turn, a function of the pre-strain imparted to the wires (ϵ_o). The parameters affecting the actuation moment (τ) are the diameter of the wires (d_o), radius of hinge tube (r_{ht}), and number of wire segments (N_{wire}).

The basis for selection of various design parameters is discussed in the following subsections.

Output Goals

The tab actuator goals are described in Section 8.7.1 and summarized in Table 8.5. The maximum aerodynamic hinge moment acting on the tracking tab can be estimated using a simple quasi-static model [159]. A quasi-steady analysis is considered sufficient for this application because tab deflections occur at a frequency of less than 1 Hz. Because only testing in a wind tunnel is planned, centrifugal loads need not be included in the calculations. From the predicted hinge moments, the actuator stroke/force capability is calculated for the design configuration described in Section 8.7.2. The expression for the total hinge moment H for a simple blade section with a plain flap is given by

$$H = \frac{1}{2}\rho V^2 c_f^2 l_f \left[C_{l_\alpha} \frac{dC_h}{dC_l} \left(\alpha_o + \frac{\Delta\alpha}{\Delta\delta}\delta \right) + \frac{dC_h}{d\delta}\delta \right] \tag{8.162}$$

where c_f is the flap chord, l_f is the flap length, and δ is the flap deflection angle. $\frac{dC_h}{dC_l}$ and $\frac{dC_h}{d\delta}$ are obtained as a function of the ratio of flap chord to total chord, $(\frac{c_f}{c})$ [166]. Note that Eq. 8.162 represents a steady hinge moment necessary to maintain a given tab deflection angle. From this equation, a hinge moment of 0.85 in-lbs is calculated for maximum loading conditions described previously. Because a power-off hold is required at the maximum loading condition, the braking moment τ_f is set equal to the maximum hinge moment. This requirement sets the braking moment to be 0.85 in-lbs.

Material Constraints

The deflection angle is a function of the recoverable strain, which in turn is a function of the pre-strain imparted. The material itself imposes a restriction on the maximum pre-strain that can be applied. For the SMA wire selected, this pre-strain is set at 2.5% and a wire diameter is 15 mil. The wire diameter is selected based on the availability of material data (characterized in-house by Prahlad and Chopra [167]).

Geometric Constraints

The volume inside the blade section places severe constraints on the dimensions of the actuator. The hinge tube is located at the 72% chord location of the 12 in chord NACA 0012 blade section. A spar is located between 10% and 30% chord. A major constraint on the actuator size is imposed along the thickness direction, which ranges from 1.2 in to 0.85 in depending on the chordwise location. As a result, the hinge tube radius (r_{ht}) is limited by the space available at the 72% chord location, which fixes this parameter at 0.35 in The wire length (L_o), although constrained by the available chordwise dimensions, does allow a certain margin of variation between 3.4 and 3.7 in. The number of wire segments (N_{wire}) may vary over a fairly large range (2–20 wires for the present configuration). Consequently, this forms an important control parameter.

8.7.6 Results of Parametric Studies

The effect of varying the key parameters, wire length (L_o), and number of wires (N_{wire}) is discussed herein. The influence of varying these parameters is then quantified in terms of wind speed. Although the braking moment is set by the power-off hold requirements, it is interesting to explore the effect of that parameter as well.

1. Frictional braking moment τ_f: This has a direct impact on the actuator output. It is observed that the required actuation moment increases as the frictional moment to be overcome increases, and the range of available angular deflection decreases. This trend is shown in Figure 8.74, where the effect of increasing external loading moment is plotted.
2. Number of wires, N_{wire}: Figure 8.75 shows the influence of increasing the number of wires on the actuator output characteristics. The other control parameter, wire length (L_o), is kept constant during calculations for this specific case. From Eqs. 8.147 and 8.161, it is evident that an increase in N_{wire} increases the maximum actuation moment, whereas it has no effect on maximum angular deflection. The wind speeds that the actuator could operate at are shown in Fig. 8.75. Note that

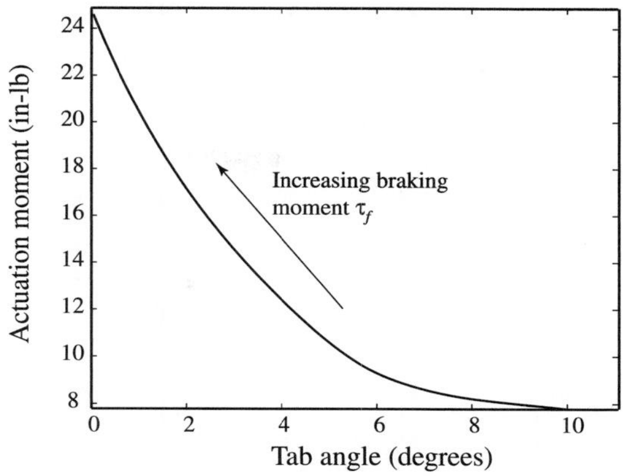

Figure 8.74. Actuator output for increasing brake friction (zero wind speed, number of wires = 12).

the actuator must deflect the tab under both the air loads and the braking moments that are required to overcome the air loads. As a result, the force capability of the actuator must be at least twice that required for overcoming the air loads. The horizontal lines in Fig. 8.75 define the maximum wind speeds for effective operation of the passive brake, including the effect of the required braking moments. From this figure, it can be concluded that tab actuation under higher wind speeds (higher air loads) is possible by increasing the control parameter (N_{wire}).

3. Length of wire, L_o: In Fig. 8.76, the effect of varying the wire length L_o is examined while maintaining the number of wires (N_{wire}) constant. The increase in L_o increases maximum angular deflection, whereas it does not influence the maximum actuation moment. This trend is quantified in terms of actuation wind speeds in which the actuator can operate.

It is worth noting that based on the parametric plots and the fact that there is a limited scope for varying the parameter L_o, the number of wires N_{wire} becomes a key parameter in the design of the actuator. The final set of parameters selected is tabulated in Table 8.7. An increase in the number of wires is possible by simply scaling up in the spanwise direction without increasing the chordwise or thickness dimensions. Therefore, to achieve the force-deflection requirements of a full-scale

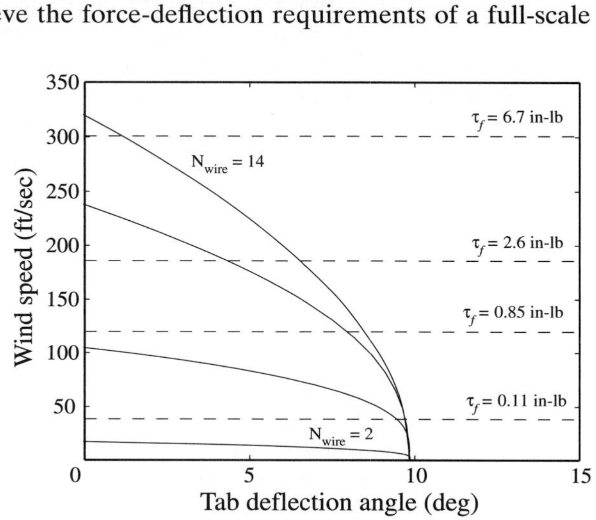

Figure 8.75. Influence of number of wires, N_{wire}, on actuator characteristics (wire length = 3.6 in).

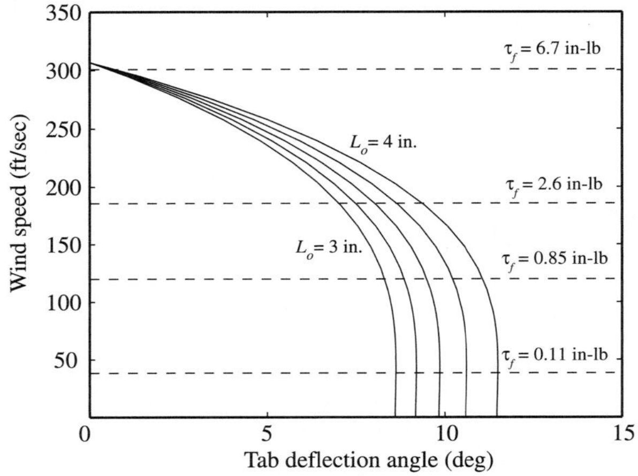

Figure 8.76. Influence of length of the wire, L_o, on actuator characteristics (number of wires = 12).

blade section, the present actuator dimensions need only be increased in the spanwise direction.

8.7.7 Testing and Performance of the System

Open-loop experiments were conducted on the system to validate the analytical model. K-type thermocouples were used to measure the temperature of the two sets of wires. The tab deflection angle was measured using a three-quarter-turn 10kΩ potentiometer, and the force was measured by means of strain gauges bonded to the turnbuckles. In these tests, the wires were heated using a Hewlett Packard 6642-A DC power supply, set at an output current of 3 amps. The heating rate was maintained at approximately 0.5° F/sec. This corresponded to a strain rate of 0.0004 /sec during the heat activation cycle, and the opposing wire was maintained at room temperature. Experimental results were recorded over a period of 600 seconds. The power requirements of the actuator were calculated based on the maximum voltage and current requirements over the complete testing time and were found to vary over a range of 3–4.5 W.

Figure 8.77(a) shows a comparison of predicted wire temperatures with test data for zero braking moment. Figure 8.77(b) shows the predicted output moment as a function of tab deflection for different braking moments, compared to experimentally measured values. To estimate the braking moments, the setting on the friction brake was calibrated with a torque wrench. For each set of tests, the friction imparted by the brake was incremented by adjusting the screw according to the

Table 8.7. *Design parameters for constructing actuator*

Length of SMA wire	L_o	3.6 in
Diameter of wire	d_o	0.015 in
Radius of hinge tube	r_{ht}	0.35 in
Braking moment	τ_f	0.85 in-lb
Range of tab deflection	θ_{max}	$\pm 5^o$
Number of wires	N_{wire}	12
Input power	P	3–4.5 Watts

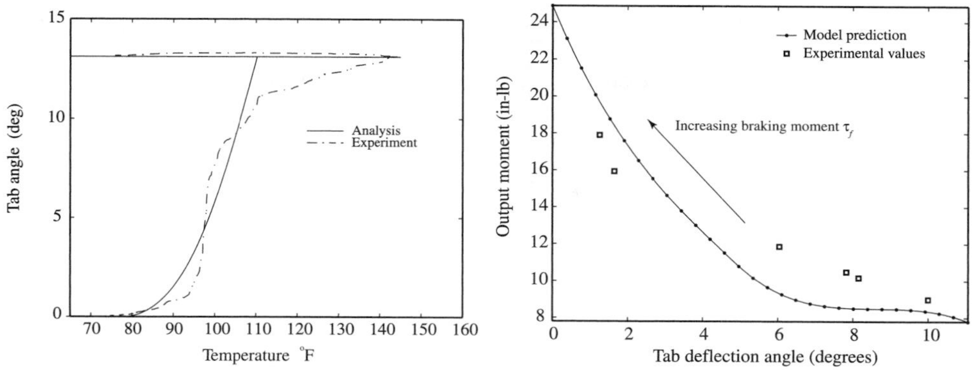

(a) Wire temperature variation (b) Output moment and tab deflection

Figure 8.77. Validation of analytical model.

calibration. The maximum deflections and moments achieved were then compared with the analytical results as shown in Fig. 8.77(b).

It is observed that the model captures the actuator behavior fairly well. A reason for the discrepancies between the model and experimental data could be the assumption that the brake is a quasi-static friction-generating element. In reality, static and dynamic friction are different, and careful testing is required to develop precise models. This might account for the overprediction of strains at higher frictional moments.

Closed-loop tests were performed to assess the capability of the actuator to accurately deflect the tracking tab in response to commanded inputs. The position control system discussed in Section 8.7.2 was implemented to test closed-loop performance both on the bench-top and in an open-jet wind tunnel.

The control system was implemented on a Windows-NT-based Pentium III, 450-MHz computer equipped with a National Instruments, PCI-6031E, 16-bit DAQ card. The controller was programmed using LABVIEW^{TM} 5.1. Five input channels to the DAQ measured temperature and angular rotation of the tab, and the actuating signals to the wires were sent through two output channels. The sampling period of the DAQ system was selected to be 1.5 seconds, which was determined to be adequate for these quasi-steady tests.

The bench-top tests were conducted under zero load and under simulated external loads. For the simulated loading case, a deadweight was mounted at the tip of the tab, perpendicular to the blade chord. The maximum moment imparted was 0.85 in-lbs, corresponding to the maximum loads anticipated in the open-jet wind tunnel (at 120 ft/sec and angles of attack $\alpha = 0^\circ$ and $\alpha = 15^\circ$).

Figure 8.78 plots steady-state error in tab angle for all of the wind tunnel test cases. The horizontal dashed lines indicate the deadband in the controller. Steady-state error was less than 0.05° for all of the tested wind-speed/angle-of-attack cases. Note that although the system is highly nonlinear, the PID controller achieved good results for all tested loading conditions with a constant set of control gains. However, if better tracking performance is desired at specific loading conditions, the control gains may be changed as a function of the operating condition to yield optimum overall performance.

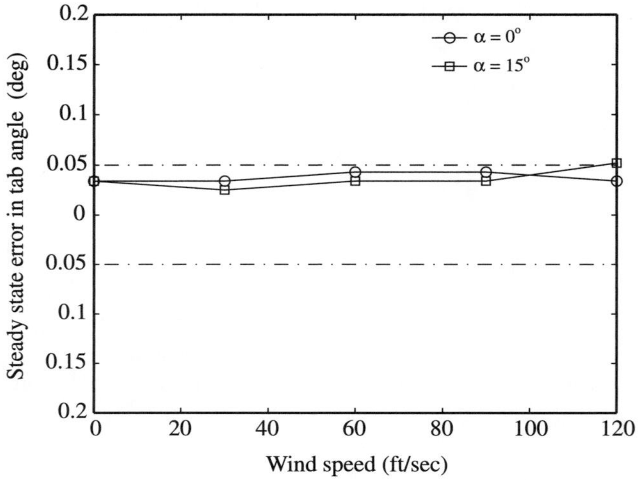

Figure 8.78. Influence of wind speed and angle of attack on actuator steady-state error for tracking input of 5°.

Figure 8.79 plots time histories of the tracking response at a wind speed of 120 ft/sec and an angle of attack of $\alpha = 15°$. The commanded signal θ_{SET} is indicated by the dashed line in these plots, and the tab response θ_{TAB} is the solid line. Similar to this time trace, it was observed that the system exhibited zero overshoot for all cases, regardless of loading. There exists, however, a definite tradeoff in the overall closed-loop response characteristics of the system. This is evident when evaluating the excellent overshoot characteristics in conjunction with large rise and settling times, which are on the order of 200 seconds for almost all wind loading conditions. It is possible to reduce these characteristic times, but this will result in a degradation in output overshoot behavior. The closed-loop system response

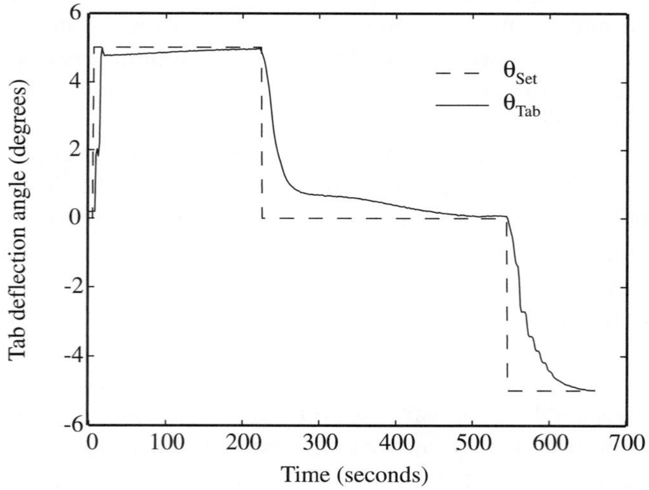

Figure 8.79. Time trace of the tracking response for tab up and down inputs of 5°, at a wind speed of 120 ft/sec and $\alpha = 15°$.

may be controlled by tuning the PID control system gains according to the desired closed-loop characteristics.

8.8 Tuning of Composite Beams

Embedding SMAs into composite structures offers the capability to tune the properties of the structure. This capability has been used in a variety of applications to enhance the functionality of the structure. SMA wires have been used to alter the natural frequencies of composite structures in several studies [168, 169, 170]. Epps and Chandra [169] presented an experimental-theoretical study on the active tuning of graphite-epoxy rectangular-solid-section beams with SMA wires inserted in embedded sleeves and showed a 22% increase in the first natural frequency using one 20 mil diameter wire. The volume fraction of SMA wires in this configuration was 2%. Good correlation between theory and experiment was achieved in this study. Note that the SMA wires were separately clamped, and they act as an elastic foundation for the parent beam. On heating, the spring stiffness of the elastic foundation increases and, as a result, the natural frequency changes. Baz, Imam, and McCoy [171] also conducted a study on the active vibration control of flexible beams. Experiments were conducted on flexible beams with SMA wires mechanically constrained on the exterior of the structure. The recovery force due to mechanically constrained, heated, pre-strained SMA wires was used to demonstrate active vibration control. In such an application, external access to the substructure becomes essential. For many aerospace structures like rotor blades, it may not be possible to use this configuration.

In addition to the possibility of tuning the dynamic properties of the structure, SMA-embedded composites also offer advantages such as structural damping augmentation [172], controlling the buckling in a thin structure [173], structural acoustic transmission control [174], and delay in the fracture of composites due to fatigue and low-velocity ballistic impact [175]. When combined with the advantages of structural tailoring offered by composites, embedded SMAs provide enhanced flexibility in design. In another study, Baz et al. [176] inserted SMA wires into flexible beams with sleeves to control their buckling and vibration behavior. As a typical example of a structure with embedded SMAs, the fabrication, testing, and analysis of a composite beam with embedded SMA wires is described in this section.

8.8.1 Fabrication of Composite Beams with SMA in Embedded Sleeves

Figure 8.80 shows a schematic of the mold and lay-up for fabrication of a composite beam with SMA wires embedded in sleeves. The function of the sleeves is to transmit the recovery force generated by the SMA wires to the ends of the beam. This is equivalent to applying an external axial force, F_r, to the beam (Fig. 8.81), resulting in a change in its bending frequencies.

The sleeves are formed by silica tubes, which can withstand the high curing temperature of the composite material and have a low coefficient of friction. Steel wires are inserted in the sleeves during curing of the beams to maintain their inner diameter. After the material is cured, the steel wires are replaced by pre-strained SMA wires that are clamped to the ends of the beam with appropriate fixtures. The

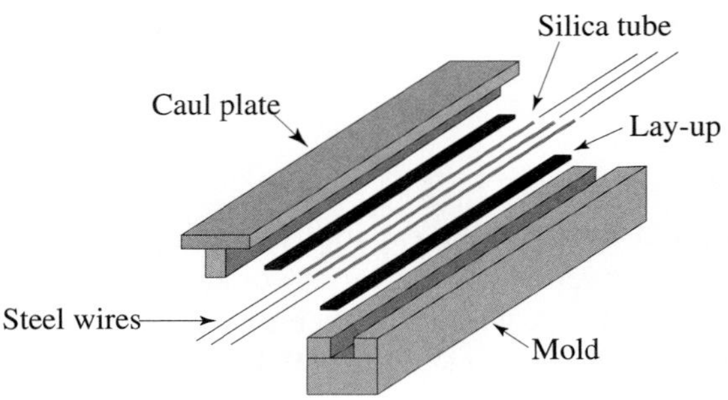

Figure 8.80. Schematic of mold for fabrication of composite beam with embedded sleeves.

resulting graphite-epoxy beam of solid rectangular cross sections with embedded SMA wires is shown in Fig. 8.82.

8.8.2 Dynamic Testing of Composite Beams with SMA Wires

The composite beams were tested for their bending frequencies under clamped-clamped boundary conditions. This setup was enclosed by a plexiglas chamber to minimize external temperature effects on the wire. Figure 8.83 shows the setup for a clamped-clamped beam test. For this test, two piezoceramic elements bonded on opposite faces at the root of the beam were excited with equal but opposite voltage to induce a bending moment. Strain gauges located on the piezoelements and the beam surface were used to measure the structural response. Figure 8.84 shows a cross section of the beam with the piezo-elements bonded to the beam. Natural frequencies of the beam were obtained by exciting the piezo-elements with a swept sine wave and examining the resulting strain on the beam as a function of frequency.

8.8.3 Free Vibration Analysis of Composite Beams with SMA Wires

In the case of a composite beam with SMA wires inserted in it via sleeves, the activated SMA wire behaves like an elastic foundation that is represented by a series of springs. The governing equation of a uniform composite beam on an elastic foundation undergoing transverse bending vibration is

$$m\frac{\partial^2 w}{\partial t^2} + EI\frac{\partial^4 w}{dx^4} + k(x)w = 0 \tag{8.163}$$

where m = mass per unit length of beam
EI = bending stiffness of beam
w = transverse displacement of beam
$k(x)$ = spring constant of activated SMA wire

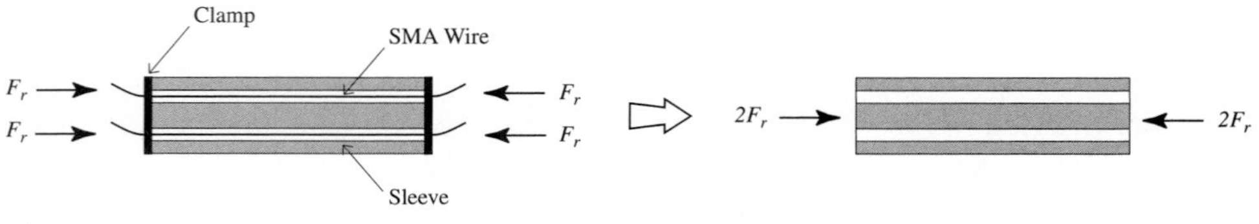

Figure 8.81. SMA recovery force acting as an external axial force.

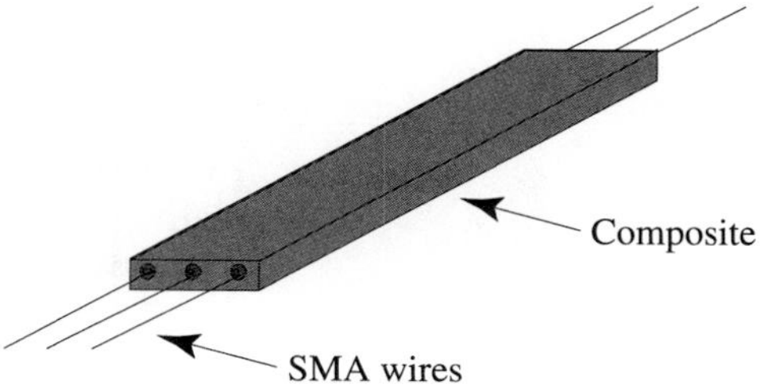

Figure 8.82. Composite beam with embedded SMA wires.

This equation is solved using the Galerkin method, and the transverse displacement w is assumed as

$$w(x,\ t) = \sum_{i=1}^{n} W_i(t)\phi_i(x) \tag{8.164}$$

where ϕ_i are beam functions. For a clamped-clamped beam, the beam functions ϕ_i are given as

$$\phi_i = \sinh\frac{\lambda_i x}{l} - \sin\frac{\lambda_i x}{l} - \alpha_i\left(\cosh\frac{\lambda_i x}{l} - \cos\frac{\lambda_i x}{l}\right) \tag{8.165}$$

where

$$\alpha_i = \frac{\sinh\frac{\lambda_i x}{l} - \sin\frac{\lambda_i x}{l}}{\cosh\frac{\lambda_i x}{1} - \cos\frac{\lambda_i x}{l}} \tag{8.166}$$

and $\lambda_1 = 4.730041$, $\lambda_2 = 7.853205$, $\lambda_3 = 10.995608$, and so on. Applying the Galerkin method, the following differential equation in the time domain is obtained

$$\boldsymbol{M}\ddot{W} + \boldsymbol{K}\boldsymbol{W} = 0 \tag{8.167}$$

where

$$\boldsymbol{W} = \{W_1,\ W_2,\ W_3 \ldots W_n\}^T$$

$$k_{ij} = \frac{EI}{l^4}\lambda_i^4 + \frac{I_{ii}}{l};\quad k_{ij} = 0 \text{ for } i \neq j$$

$$I_{ii} = \int_0^1 k(x)\phi_i^2\, dx$$

$$M_{ii} = m,\ M_{ij} = 0,\ \text{for } i \neq j$$

$$l = \text{ length of beam}$$

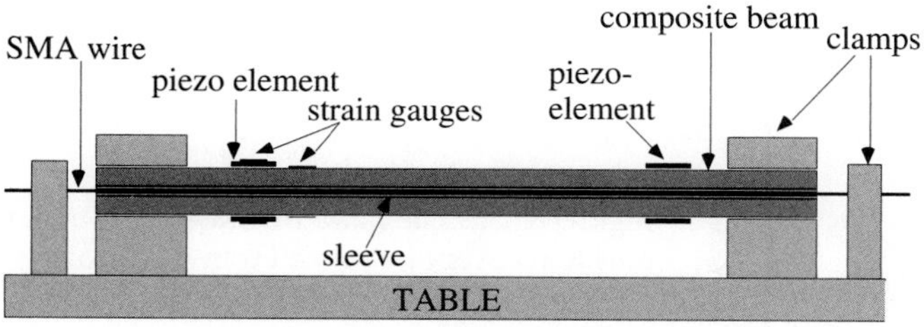

Figure 8.83. Vibration testing of composite beam with embedded SMA wires.

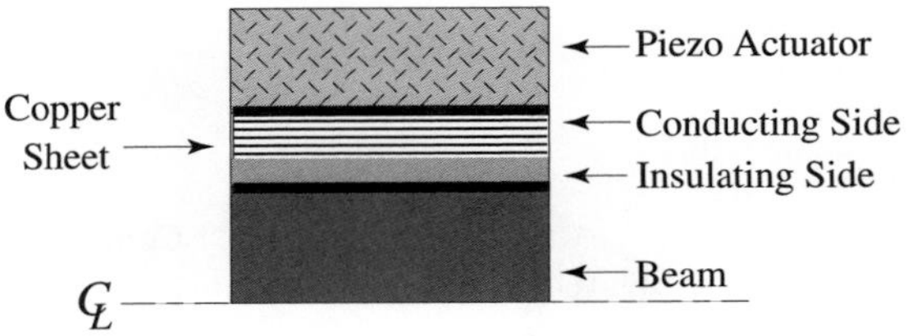

Figure 8.84. Cross section of a composite beam with copper sheet and piezo actuator.

The natural frequencies are obtained as

$$\omega_i = \sqrt{\left[\frac{EI}{m}\left(\frac{\lambda_i}{l}\right)^4 + \frac{I_{ii}}{ml}\right]} \tag{8.168}$$

The spring constant $k(x)$ is derived in the following section.

8.8.4 Calculation of the Spring Coefficient of SMA Wire under Tension

The governing equation of a SMA wire under tension, subjected to transverse-load q (Fig. 8.85), is:

$$Fw,_{xx} + q\delta(x - \xi) = 0 \tag{8.169}$$

where F = recovery force in the wire
q = transverse force
$\delta(x - \xi)$ = Dirac delta function centered at $x = \xi$
Integrating Eq. 8.169 twice with respect to x

$$Fw + qr(x - \xi) + Ax + B = 0 \tag{8.170}$$

where contants A and B are determined by enforcing the boundary conditions and $r(x - \xi)$ is the unit ramp function. The boundary conditions at clamped ends are

$$w(x = 0) = w(x = l) = 0 \tag{8.171}$$

Using the boundary conditions (Eq. 8.171) in Eq. 8.169, the deflection w is

$$w = \frac{q}{F}\left[\frac{x}{l}r(l - \xi) - r(x - \xi)\right] \tag{8.172}$$

The spring constant k per unit length of the beam is defined as

$$k(\xi) = \frac{q}{w(x = \xi)l} \tag{8.173}$$

Using Eq. 8.172, the spring constant becomes

$$k(\xi) = \frac{F}{\xi r(l - \xi)} \tag{8.174}$$

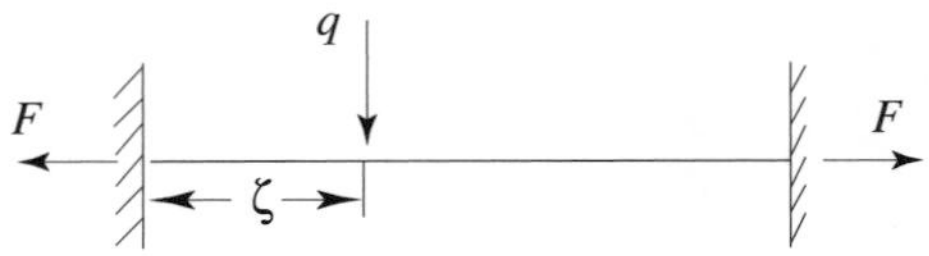

Figure 8.85. Schematic of SMA wire under tensile recovery force with transverse load.

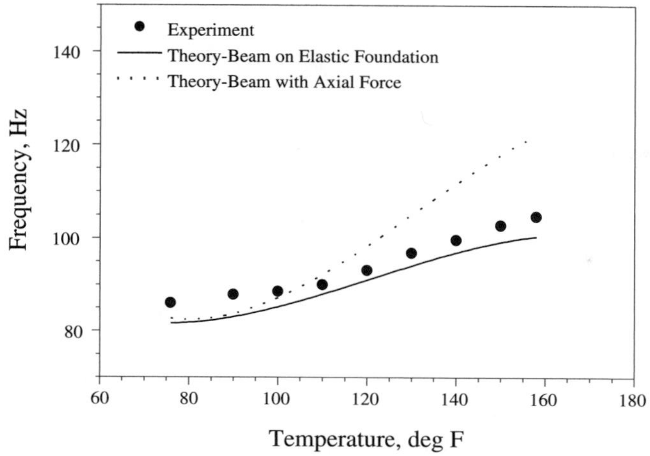

Figure 8.86. Fundamental bending frequency of clamped-clamped graphite-epoxy beam activated with one 20 mil SMA wire, beam length 18 in, width 0.25 in, and thickness 0.068 in.

Zhang et al. [177] built E-glass composite beams with integrated Nitinol wires with a fiber fraction of 10%, and tested the beams to failure in both martensite phase (room temperature) and austenite phase (75°C). The SMA-embedded composite beam showed significant increase in the strain energy absorption prior to failure, which in turn increased the fracture toughness and crashworthiness of the structure. At room temperature, energy absorption was increased by 50% from a baseline non-SMA structure, whereas at elevated temperatures (austenite phase), the increase was as high as 600%. However, the stiffness and failure stress of the composite with reinforced-SMA in the austenite phase were significantly lower than those for the baseline beam as well as those of reinforced-SMA beams at room temperature.

Recently, there have been growing investigations related to hybrid composites embedded with SMA wires. Examples are finite element analysis [178], adhesion characteristics between SMA wires and composite (fiber/matrix interface) [179], and using heavily cold-worked, ultra-thin wires ($A_s > 180°C$) in conjunction with low-temperature heat treatment [180]. Xu et al. [180] used a resin with a curing temperature of 180°, embedded ultra-thin (50μm) NiTi wires in a SMA prepreg sheet and removed the restriction of any special jigs and fixtures.

8.8.5 Correlation with Test Data

The natural frequencies of a composite beam with SMA wires inserted into sleeves embedded in the beam depend on the beam parameters and the SMA characteristics. The beam parameters are length, thickness, width, material, and boundary conditions. The SMA wire parameters are recovery force in each wire (which in turn depends on pre-strain, mechanical properties, and temperature) and number of wires. Figure 8.86 shows the first bending frequency of a graphite-epoxy composite beam activated by one 20 mil diameter SMA wire. The dimensions of this beam are as follows: clamped length = 18.0 in, width = 0.25 in, and thickness = 68.0 mils.

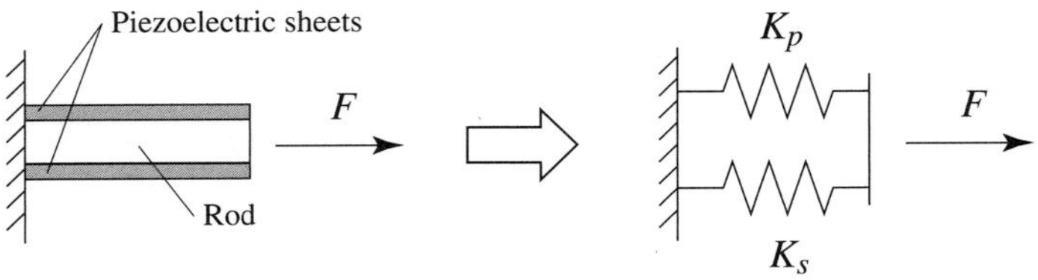

Figure 8.87. Effective stiffness of rod with attached piezoelectric sheets.

The increase in the fundamental frequency due to 100% SMA activation (temperature = 160°F) is 21.8%. The agreement between theory and experiment is within 5%. Note that the experimentally obtained recovery force is used in the prediction of the frequency. It is important to note that the prediction of frequency correlates with the experimental results within acceptable limits only when the beam-on-elastic foundation analysis is used.

8.9 Shunted Piezoelectrics

We have seen that due to its electromechanical coupling, a piezoelectric material behaves as a transducer between electrical energy and mechanical energy. If one form of energy is input to the material, it is partly converted into the other form of energy. The ratio of the energy output to the energy input, or the fraction of input energy that is transduced, is given by the electromechanical coupling factor of the material. The coupling factor is a material constant, and it depends on the permittivity of the piezoelectric (at constant stress), the compliance of the piezoelectric (at constant field), and the piezoelectric coefficient. In some cases, it may be possible to change the effective coupling factor by attaching the piezoelectric material to a structure having an appropriate stiffness. As a consequence of this coupling, the electric properties of a sample of piezoelectric material (primarily the dielectric constant) depend on its mechanical boundary conditions. Conversely, the mechanical properties (stiffness and damping) depend on the electrical boundary conditions. Note that it is not possible to affect the mass of a piezoelectric element by changing either the electrical or mechanical boundary conditions.

In a typical structure incorporating piezoelectric material, the stiffness of the piezoelectric elements acts in parallel with the stiffness of the base structure. Hence, the total stiffness of the structure is given by the sum of the stiffnesses of the piezoelectric element and the base structure. For example, Fig. 8.87 shows a prismatic rod of length L and cross-sectional area A_s made of a material with Young's modulus E_s. Two piezoelectric sheets are bonded on the top and bottom of the rod. The Young's modulus of the piezoelectric material is E_p and the total cross-sectional area of the piezoelectric sheets is A_p. An axial force F is applied to the structure. The effective stiffness K_{eff} is given by the parallel combination of the stiffnesses of the piezoelectric sheets (K_p) and the rod (K_s) as

$$K_{eff} = K_s + K_p = \frac{E_s A_s}{L} + \frac{E_p A_p}{L} \tag{8.175}$$

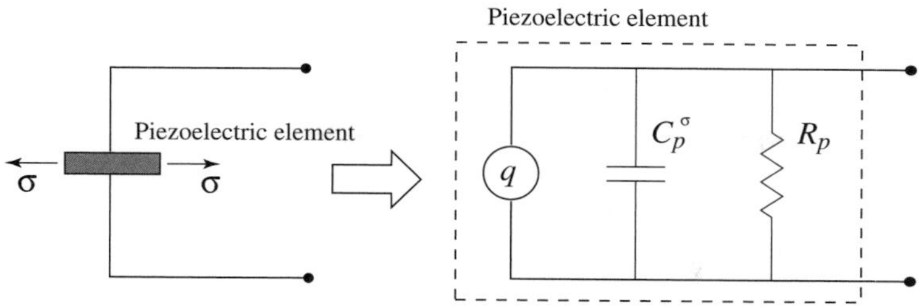

Figure 8.88. Equivalent circuit of a piezoelectric element under a uniaxial stress σ.

By changing the electrical boundary conditions of the piezoelectric sheets, it is possible to change the effective stiffness and damping of the structure. A simple way to change the electrical boundary conditions in a passive manner is to connect an impedance across the electrodes of the piezoelectric element. Because the impedance appears in parallel with the impedance of the piezoelectric element, it is called a "shunt impedance." Hence, this technique is known as "piezoelectric shunting." Numerous researchers have described different methods and applications of passive as well as semi-passive piezoelectric shunting. The technique was first described by Forward [181]. Hagood et al. [182] gave a detailed description of the use of passive electrical shunt networks in conjunction with piezoelectric elements for damping augmentation. They showed that the piezoelectric element with a resistive shunt behaved like a viscoelastic material. They also discussed resonant shunt circuits, compared them to conventional proof mass damper systems, and derived methods to choose the optimum parameters of the circuit. The analysis was validated by experiments on a cantilevered beam with bonded piezoelectric sheets. Several reviews of developments in piezoelectric shunting and its applications have been published. A comprehensive review of shunted piezoelectric materials for vibration damping and control is given by Lesieutre [183], where four basic types of shunt circuits are discussed: inductive, resistive, capacitive, and switched. A switched shunt, in its simplest form, consists of a fast-acting switch that opens or closes to convert the electric boundary conditions of the piezoelectric element from short-circuit to open-circuit. The energy transfer from the piezoelectric can be affected in this manner by actively controlling the switch. This type of shunt circuit is not discussed further in this chapter. Tang et al. [184] discuss semi-active damping techniques using piezoelectric shunt networks. They also describe active-passive techniques in which the piezoelectric element is simultaneously used as an actuator and as a passive damper. Ahmadian et al. [185] describe vibration suppression using actively controlled piezoelectric elements with positive position feedback control techniques, as well as using passive electrical shunts.

8.9.1 Principle of Operation

A simple equivalent circuit of a piezoelectric element in the sensor mode [186] under a uniaxial stress σ is shown in Fig. 8.88. The piezoelectric element can be treated as a charge generator (q) in parallel with a capacitance (C_p^σ) and a leakage resistance (R_p). Because R_p is typically very large, we can ignore it in the present discussion. Let an external impedance Z_{sh} be connected between the electrodes of the piezoelectric element, as shown in Fig. 8.89. It can be seen that the shunt impedance acts in parallel

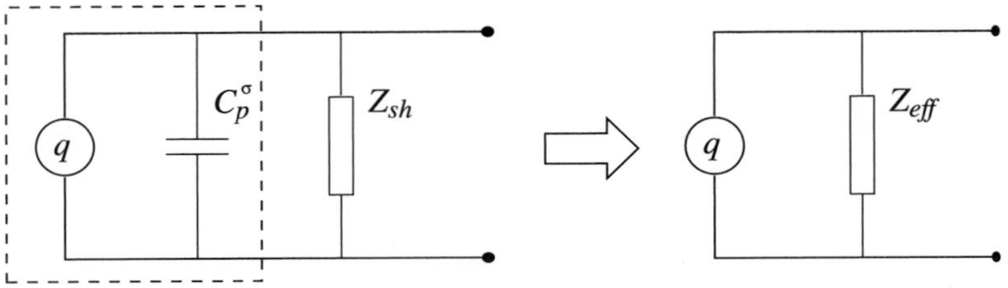

Figure 8.89. Effective impedance of a shunted piezoelectric element.

with the impedance of the piezoelectric element Z_p. The effective impedance Z_{eff} is given by

$$Z_{eff} = \frac{Z_p Z_{sh}}{Z_p + Z_{sh}} \tag{8.176}$$

Defining a nondimensional effective impedance ratio $\bar{Z}_{eff}$ and ignoring R_p (assuming it is infinite), we obtain

$$\begin{aligned} \bar{Z}_{eff} &= \frac{Z_{eff}}{Z_p} \\ &= \frac{j\omega C_p^\sigma Z_{sh}}{1 + j\omega C_p^\sigma Z_{sh}} \end{aligned} \tag{8.177}$$

where a harmonic excitation at a frequency ω rad/s is assumed. Although the general form of the equations can be derived using a Laplace transform, the remainder of this discussion considers the special case of a harmonic excitation.

Because the shunt impedance changes the effective impedance of the piezoelectric element, the electrical boundary conditions are changed. Physically, the charge generated by the piezoelectric flows through the shunt impedance, changing the voltage across the electrodes. Depending on whether the shunt impedance is resistive, capacitive, or inductive, this manifests itself as a change in the stiffness and damping of the piezoelectric element. If the piezoelectric element is attached to a structure, the stiffness and damping characteristics of the structure are affected. For example, if the shunt impedance is resistive, energy dissipation occurs due to Ohmic heating, resulting in an increase in damping of the structure. If the shunt impedance is purely capacitive or inductive, there is no energy loss, and hence no change in damping. The effect in this case can only appear as a change in the effective stiffness of the structure. Therefore, we can conclude that to add damping to the structure, the shunt impedance must have a resistive component.

To further explore the effect of the shunt impedance, let us examine the constitutive relations of the piezoelectric. Consider a piezoelectric sheet element with electrodes parallel to the 1–2 planes and poled along the 3-direction, as shown in Fig. 8.90. The constitutive relations for this element are

$$\begin{Bmatrix} \boldsymbol{\epsilon} \\ \boldsymbol{D} \end{Bmatrix} = \begin{bmatrix} \boldsymbol{s}^{\mathbb{E}} & \boldsymbol{d} \\ \boldsymbol{d}^c & \boldsymbol{e}^\sigma \end{bmatrix} \begin{Bmatrix} \boldsymbol{\sigma} \\ \mathbb{E} \end{Bmatrix} \tag{8.178}$$

These relations can be rewritten in terms of the applied voltage as

$$\begin{Bmatrix} \boldsymbol{\epsilon} \\ \boldsymbol{D} \end{Bmatrix} = \begin{bmatrix} \boldsymbol{s}^{\mathbb{E}} & \boldsymbol{dL}^{-1} \\ \boldsymbol{d}^c & \boldsymbol{e}^\sigma \boldsymbol{L}^{-1} \end{bmatrix} \begin{Bmatrix} \boldsymbol{\sigma} \\ \boldsymbol{V} \end{Bmatrix} \tag{8.179}$$

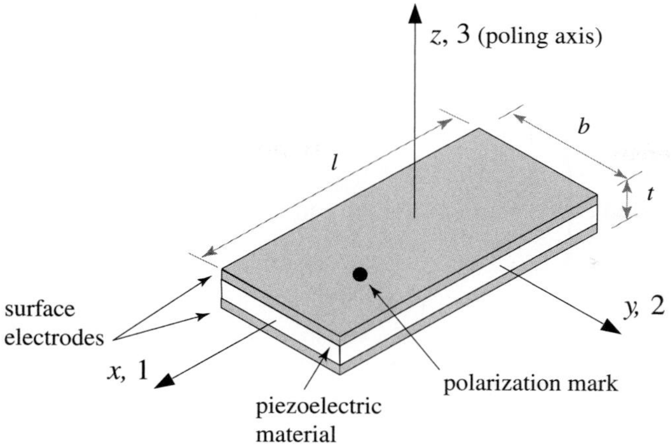

Figure 8.90. Piezoelectric-sheet element.

where the matrix $\boldsymbol{L}$ (size 3×3) is a diagonal matrix of the lengths of the piezoelectric element along the 1-, 2-, and 3-directions.

$$\boldsymbol{L} = \begin{bmatrix} L_1 & 0 & 0 \\ 0 & L_2 & 0 \\ 0 & 0 & L_3 \end{bmatrix} = \begin{bmatrix} l & 0 & 0 \\ 0 & b & 0 \\ 0 & 0 & t \end{bmatrix} \tag{8.180}$$

Recalling that the charge $\boldsymbol{q}$ generated by the piezoelectric element and the current $\boldsymbol{i}$ are given by

$$\boldsymbol{q} = \int_A \boldsymbol{D} d\boldsymbol{A} \tag{8.181}$$

$$\boldsymbol{i} = \dot{\boldsymbol{q}} \tag{8.182}$$

and assuming a harmonic excitation at a frequency ω rad/s, the constitutive relations can be written in terms of the current as

$$\begin{Bmatrix} \boldsymbol{\epsilon} \\ \boldsymbol{i} \end{Bmatrix} = \begin{bmatrix} \boldsymbol{s}^{\mathbb{E}} & \boldsymbol{dL}^{-1} \\ j\omega \boldsymbol{A}\boldsymbol{d}^{c} & j\omega \boldsymbol{A}\boldsymbol{e}^{\sigma}\boldsymbol{L}^{-1} \end{bmatrix} \begin{Bmatrix} \boldsymbol{\sigma} \\ \boldsymbol{V} \end{Bmatrix} \tag{8.183}$$

where the matrix $\boldsymbol{A}$ (size 3×3) is a diagonal matrix of areas perpendicular to the 1-, 2-, and 3-directions, respectively, given by

$$\boldsymbol{A} = \begin{bmatrix} A_1 & 0 & 0 \\ 0 & A_2 & 0 \\ 0 & 0 & A_3 \end{bmatrix} = \begin{bmatrix} bt & 0 & 0 \\ 0 & lt & 0 \\ 0 & 0 & lb \end{bmatrix} \tag{8.184}$$

Note that for the piezoelectric element under consideration, only a voltage V_3 (along the 3-direction) can be applied. Considering a one-dimensional case (stress applied along the 1-direction), the constitutive equation can be written as

$$\begin{Bmatrix} \epsilon_1 \\ i_3 \end{Bmatrix} = \begin{bmatrix} s_{11}^{\mathbb{E}} & d_{31}/t \\ j\omega A_3 d_{31} & j\omega A_3 e_{33}^{\sigma}/t \end{bmatrix} \begin{Bmatrix} \sigma_1 \\ V_3 \end{Bmatrix} \tag{8.185}$$

Recall that the capacitance of the piezoelectric sheet element, C_p^{σ}, is given by

$$C_p^{\sigma} = \frac{e_{33}^{\sigma} A_3}{t} \tag{8.186}$$

where the superscript σ indicates a constant stress condition. Therefore, the constitutive relation can be rewritten as (dropping the directional subscripts for V and A)

$$\begin{Bmatrix} \epsilon_1 \\ i \end{Bmatrix} = \begin{bmatrix} s_{11}^{\mathbb{E}} & d_{31}/t \\ j\omega A d_{31} & 1/Z_p \end{bmatrix} \begin{Bmatrix} \sigma_1 \\ V \end{Bmatrix} \tag{8.187}$$

In the case of the shunted piezoelectric element, the impedance of the piezoelectric is replaced by the effective impedance, yielding

$$\begin{Bmatrix} \epsilon_1 \\ i \end{Bmatrix} = \begin{bmatrix} s_{11}^{\mathbb{E}} & d_{31}/t \\ j\omega A d_{31} & 1/Z_{eff} \end{bmatrix} \begin{Bmatrix} \sigma_1 \\ V \end{Bmatrix} \tag{8.188}$$

Eliminating the voltage V from this equation, we obtain

$$V = Z_{eff}\, i - Z_{eff}\, j\omega A d_{31}\sigma_1 \tag{8.189}$$

and

$$\begin{aligned} \epsilon &= \sigma\left(s_{11}^{\mathbb{E}} - Z_{eff}\frac{j\omega A d_{31}^2}{t}\right) + \frac{Z_{eff}\, d_{31}}{t} i \\ &= s_{11}^{i}\sigma + \frac{Z_{eff}\, d_{31}}{t} i \end{aligned} \tag{8.190}$$

where s_{11}^{i} is the compliance at constant-current, or open-circuit compliance. A physical way to understand this is by recalling that the impedance of a constant-current source is infinite; hence, the constant-current condition corresponds to an open-circuit condition. The open-circuit compliance of the shunted piezoelectric element can be simplified as

$$\begin{aligned} s_{11}^{i} &= s_{11}^{\mathbb{E}} - Z_{eff}\frac{j\omega A d_{31}^2}{t} \\ &= s_{11}^{\mathbb{E}}\left(1 - \frac{Z_{eff}}{Z_p}k_{31}^2\right) \\ &= s_{11}^{E}\left(1 - \frac{k_{31}^2}{1+\alpha}\right) \end{aligned} \tag{8.191}$$

where k_{31}^2 is the electromechanical coupling coefficient (typically ≈ 0.4) and α is the ratio of the impedance of the piezoelectric element to the impedance of the shunt.

$$\alpha = \frac{Z_p}{Z_{sh}} \tag{8.192}$$

From Eq. 8.177

$$\bar{Z}_{eff} = \frac{1}{1+\alpha} \tag{8.193}$$

Note that $\bar{Z}_{eff}$ and α can be complex numbers, depending on the constituents of the shunt impedance. Different authors use either $\bar{Z}_{eff}$ or α to represent the shunt impedance. In the present discussion, we use α because it gives a direct feel of the magnitude of shunt impedance compared to the impedance of the piezoelectric element. From the compliance (Eq. 8.191), the effect of the shunt impedance on the stiffness of the piezoelectric element can be derived as [187]

$$K^* = K^{\mathbb{E}}\left(1 + \frac{k_{31}^2}{1 + \alpha - k_{31}^2}\right) \tag{8.194}$$

where K^* is the effective stiffness of the shunted piezoelectric element and $K^{\mathbb{E}}$ is the short-circuit stiffness of the unshunted piezoelectric element. Some authors represent the effective stiffness in terms of the open-circuit stiffness (K^D) of the piezoelectric element as

$$K^{*D} = K^D\left(1 + \frac{k_{31}^2}{1 + \alpha - k_{31}^2}\right) = \frac{K^{\mathbb{E}}}{1 - k_{31}^2}\left(1 + \frac{k_{31}^2}{1 + \alpha - k_{31}^2}\right) \tag{8.195}$$

In the present discussion, we represent the effective stiffness in terms of the short-circuit stiffness, $K^{\mathbb{E}}$, as in Eq. 8.194. Let us also define an effective stiffness ratio (similar to a mechanical impedance ratio) as

$$\bar{K} = \frac{K^*}{K^{\mathbb{E}}} \tag{8.196}$$

Note that Eq. 8.190 represents the strain in the piezoelectric element in response to an applied stress as well as a current passing through it. Therefore, the effect of the shunt impedance on the compliance appears while the piezoelectric element is being actuated as well as in the passive case ($i = 0$).

8.9.2 Types of Shunt Circuits

The value of the shunt impedance sets upper and lower bounds for the effective stiffness ratio. These limits are between $Z_{sh} = 0$ if the electrodes of the piezoelectric element are short-circuited, and $Z_{sh} = \infty$ if the electrodes are open-circuited; that is, no shunt impedance is connected between them. Note that the present discussion concerns only the stiffness of the piezoelectric element. When analyzing a structure incorporating piezolectric elements, one way to model the effect of the shunt impedance is by appropriately changing the modal stiffness, as described by Hagood et al. [182]. The shunt circuit can be purely capacitive, purely resistive, purely inductive, or a combination of the three types. Based on the type of shunt circuit, the effective stiffness of the piezoelectric element can exhibit different characteristics. Let us first consider a general case and then examine special cases separately.

General Case of Shunt Impedance

Consider a shunt impedance consisting of a resistance R_{sh} in series with a reactance X_{sh}. The shunt impedance is given by

$$Z_{sh} = R_{sh} + jX_{sh} \tag{8.197}$$

Neglecting the resistance R_p and assuming harmonic excitation, the impedance of the piezoelectric element is given by

$$Z_p = \frac{1}{j\omega C_p^{\sigma}} \tag{8.198}$$

The impedance ratio becomes

$$\alpha = \frac{Z_p}{Z_{sh}} = \frac{1}{j\omega C_p^\sigma (R_{sh} + jX_{sh})} = \frac{1}{j\omega R_{sh} C_p^\sigma - \omega X_{sh} C_p^\sigma} \tag{8.199}$$

Substituting this result in Eqs. 8.194 and 8.196, we obtain

$$\begin{aligned} \bar{K} &= \frac{j\omega R_{sh} C_p^\sigma - \omega X_{sh} C_p^\sigma + 1}{\left(1 - k_{31}^2\right)\left(j\omega R_{sh} C_p^\sigma - \omega X_{sh} C_p^\sigma\right) + 1} \\ &= \frac{1}{1 - k_{31}^2} \cdot \frac{j\omega R_{sh} C_p^\epsilon - \omega X_{sh} C_p^\epsilon + 1 - k_{31}^2}{j\omega R_{sh} C_p^\epsilon - \omega X_{sh} C_p^\epsilon + 1} \end{aligned} \tag{8.200}$$

The factor $(1 - k_{31}^2)$ is being retained for ease of conversion between open-circuit and short-circuit stiffness ratios and to maintain consistency with existing literature. In the previous equation, we have made use of the relationship between the capacitance of the piezoelectric element at constant stress C_p^σ and its capacitance at constant strain C_p^ϵ

$$C_p^\epsilon = \left(1 - k_{31}^2\right) C_p^\sigma \tag{8.201}$$

Let us make the following substitutions

$$\rho = \omega R_{sh} C_p^\epsilon \tag{8.202}$$

$$\lambda = \omega X_{sh} C_p^\epsilon \tag{8.203}$$

where ρ is called the nondimensional frequency or nondimensional resistance because the quantity $R_{sh} C_p^\epsilon$ has the dimensions of time. The effective stiffness ratio becomes

$$\begin{aligned} \bar{K} &= \frac{1}{1 - k_{31}^2} \cdot \frac{1 - k_{31}^2 - \lambda + j\rho}{1 - \lambda + j\rho} \\ &= \frac{1}{1 - k_{31}^2} \cdot \left[1 - \frac{(1 - \lambda) k_{31}^2}{(1 - \lambda)^2 + \rho^2}\right]\left[1 + j\frac{k_{31}^2 \rho}{(1 - \lambda)^2 - k_{31}^2 (1 - \lambda) + \rho^2}\right] \\ &= \frac{1}{1 - k_{31}^2} E'(1 + j\eta) \end{aligned} \tag{8.204}$$

It can be seen that the addition of the shunt impedance makes the piezoelectric element behave like a viscoelastic material. The quantity E' is typically known as the storage modulus, and η is called the loss factor.

$$E' = 1 - \frac{(1 - \lambda) k_{31}^2}{(1 - \lambda)^2 + \rho^2} \tag{8.205}$$

$$\eta = \frac{k_{31}^2 \rho}{(1 - \lambda)^2 - k_{31}^2 (1 - \lambda) + \rho^2} \tag{8.206}$$

The condition for maximum loss factor can be found by differentiating Eq. 8.206 with respect to ρ, as follows

$$\frac{\partial \eta}{\partial \rho} = (1 - \lambda)^2 - k_{31}^2 (1 - \lambda) - \rho^2 \tag{8.207}$$

Setting the previous equation to zero yields the value of ρ for maximum loss factor as

$$\rho\Big|_{\eta_{max}} = \sqrt{(1-\lambda)\left(1-\lambda-k_{31}^2\right)} \tag{8.208}$$

which yields the value of maximum loss factor as

$$\eta_{max} = \frac{k_{31}^2}{2\sqrt{(1-\lambda)(1-\lambda-k_{31}^2)}} \tag{8.209}$$

From the generic expressions for storage modulus and loss factor given previously, we can derive the expressions for special cases of shunt impedance.

Resistive Shunt

In the case of a purely resistive shunt, the impedance is given by

$$Z_{sh} = R_{sh} \tag{8.210}$$

Comparing this to the generic expression for shunt impedance (Eq. 8.197)

$$X_{sh} = 0 \tag{8.211}$$

From Eqs. 8.202 and 8.203, we obtain

$$\rho = \omega R_{sh} C_p^{\epsilon} \tag{8.212}$$

$$\lambda = 0 \tag{8.213}$$

Substituting this in Eqs. 8.205 and 8.206, we obtain the storage modulus and loss factor for a purely resistive shunt as

$$E' = 1 - \frac{k_{31}^2}{1+\rho^2} \tag{8.214}$$

$$\eta = \frac{\rho k_{31}^2}{\left(1-k_{31}^2\right)+\rho^2} \tag{8.215}$$

from which the effective stiffness ratio is

$$\bar{K} = \frac{1}{1-k_{31}^2} \cdot \left[1 - \frac{k_{31}^2}{1+\rho^2}\right]\left[1 + j\frac{\rho k_{31}^2}{\left(1-k_{31}^2\right)+\rho^2}\right] \tag{8.216}$$

It can be seen that the resistive shunt effectively adds structural damping to the system through a non-zero loss factor η. In physical terms, the energy dissipated in the resistance due to Ohmic heating appears as a damping in the system. The condition for maximum damping, which corresponds to the maximum achievable value of η, can be calculated using Eqs. 8.208 and 8.209 by setting $\lambda = 0$. Assuming $k_{31}^2 = 0.4$, this yields

$$\rho\Big|_{\eta_{max}} = \sqrt{1-k_{31}^2} = 0.7746$$

$$\eta_{max} = \frac{k_{31}^2}{2\sqrt{1-k_{31}^2}} = 0.2582 \tag{8.217}$$

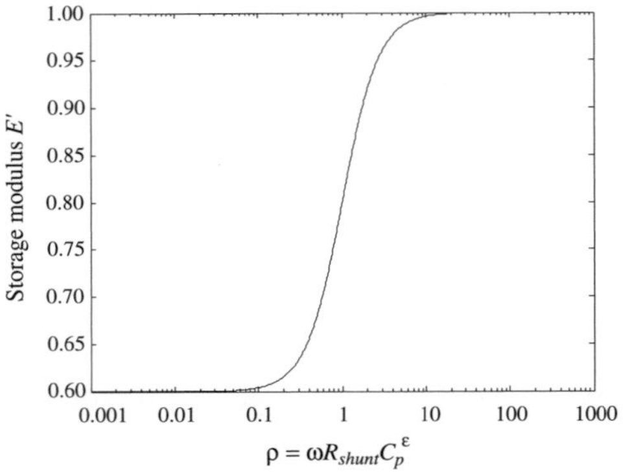

(a) Storage modulus E'

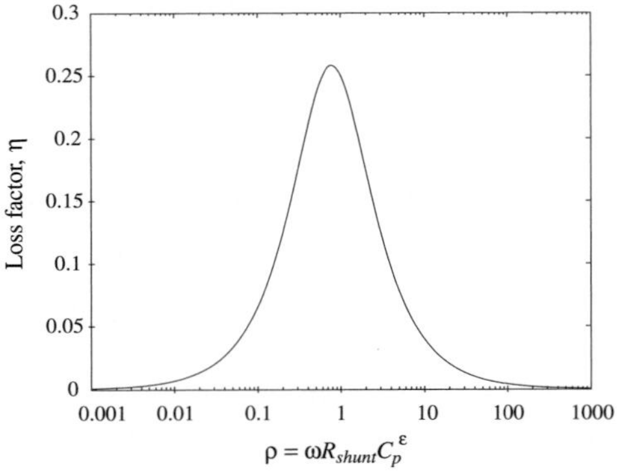

(b) Loss factor η

Figure 8.91. Storage modulus and loss factor of resistively shunted piezoelectric element as a function of dimensionless frequency ($k_{31}^2 = 0.4$).

Figure 8.91 shows the variation of storage modulus and loss factor of a resistively shunted piezoelectric element. The short-circuit condition is realized as $\rho \to 0$ ($R_{sh} = 0$) and yields a storage modulus $E' = 0.6$, which when divided by the factor $(1 - k_{31}^2)$ corresponds to $\bar{K} = 1$. Note that the loss factor increases because the coupling coefficient increases. This is to be expected because a larger coupling coefficient implies that a larger fraction of the input mechanical energy is converted to electrical energy, which in turn can dissipate through the shunt resistance.

To maximize the loss factor, the value of R_{sh} must be chosen based on the operating frequency. In the case of a steady-state forced response, the energy dissipated during one cycle (ΔE_{cyc}) is given by [188]

$$\Delta E_{cyc} = A^2 \eta k^E E' \pi = A^2 k^E \pi \frac{k_{31}^2 \rho}{(1 - k_{31}^2)(1 + \rho^2)} \tag{8.218}$$

where A is the amplitude of motion. Note that when $\rho = 0$ ($R_{sh} = 0$), $\Delta E_{cyc} = 0$.

Capacitive Shunt

Let the shunt circuit consist of a pure capacitance C_{sh} in series with a resistance R_{sh} (RC shunt). In this case, the shunt impedance is given by

$$Z_{sh} = R_{sh} + \frac{1}{j\omega C_{sh}} \tag{8.219}$$

Comparing this to the generic expression for shunt impedance (Eq. 8.197), we get

$$X_{sh} = -\frac{1}{\omega C_{sh}} \tag{8.220}$$

From Eqs. 8.202 and 8.203, we obtain

$$\rho = \omega R_{sh} C_p^{\epsilon} \tag{8.221}$$

$$\lambda = -\frac{C_p^{\epsilon}}{C_{sh}} \tag{8.222}$$

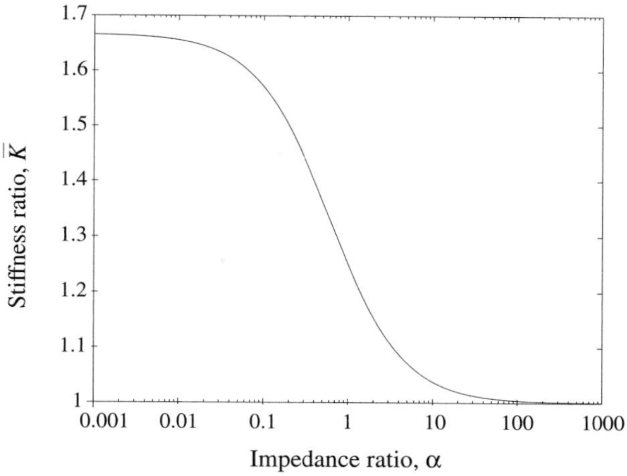

Figure 8.92. Variation of effective stiffness ratio with shunt impedance ratio for a purely capacitive shunt ($k_{31}^2 = 0.4$).

Let us first examine the case of a purely capacitive shunt. Setting $R_{sh} = 0$ and substituting in Eqs. 8.205 and 8.206

$$E' = 1 - \frac{k_{31}^2}{1 + C_p^{\epsilon}/C_{sh}} = 1 + \frac{k_{31}^2}{1 + C_{sh}/C_p^{\sigma} - k_{31}^2} \tag{8.223}$$

$$\eta = 0 \tag{8.224}$$

It can be seen that the purely capacitive shunt results in a change in the effective stiffness of the piezoelectric element. If the piezoelectric element is bonded to a structure, the stiffness of the structure is affected. The loss factor is zero, indicating that there is no damping in the piezoelectric element. This result can be expected because there is no resistance in the circuit and, therefore, no dissipative element in the system.

Davis and Lesieutre [187] derived Eq. 8.194 and described the use of capacitive shunting to change the stiffness of a tunable passive vibration absorber consisting of an inertial mass mounted on a capacitively shunted piezoelectric element. Variation in the shunt impedance results in a change in the stiffness of the piezoelectric element and, therefore, a change in the natural frequency of the absorber. In this way, a small mass of piezoelectric material can be used to absorb vibrations of a larger structure over a range of frequencies. They plotted the effect of capacitive shunt impedance on effective stiffness, as shown in Fig. 8.92. For an assumed value of $k_{31}^2 = 0.4$, it can be seen that the effective stiffness ratio $\bar{K}$ varies between an upper limit of 1.6667 ($1/\sqrt{(1 - k_{31}^2)}$) when the shunt capacitance $C_{sh} \ll C_p^{\sigma}$ (tending toward an open-circuit condition) and a lower limit of 1, when the shunt capacitance $C_{sh} \gg C_p^{\sigma}$, (tending toward a closed-circuit condition). Recall that although it is possible to realize a significant change in the stiffness of the piezoelectric element, the change in effective stiffness of a structure incorporating piezoelectric elements could be considerably lower, depending on the geometry and amount of piezoelectric material.

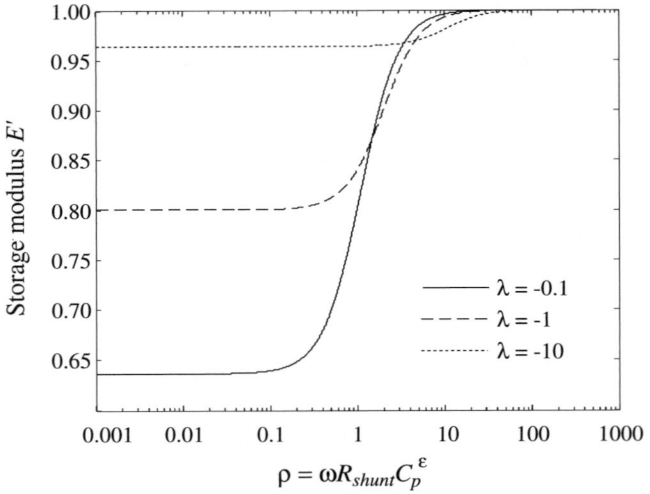

(a) Storage modulus E'

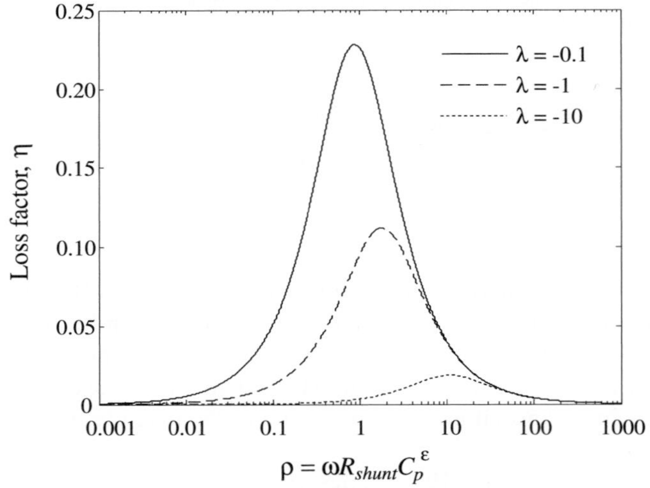

(b) Loss factor η

Figure 8.93. Storage modulus and loss factor of RC shunted piezoelectric element as a function of dimensionless frequency ($k_{31}^2 = 0.4$).

Let us now see what happens when a non-zero resistance is included in the shunt circuit. The effective storage modulus and loss factor become

$$E' = 1 - \frac{\left(1 + C_p^\epsilon / C_{sh}\right) k_{31}^2}{\left(1 + C_p^\epsilon / C_{sh}\right)^2 + \left(\omega R_{sh} C_p^\epsilon\right)^2} \tag{8.225}$$

$$\eta = \frac{k_{31}^2 \omega R_{sh} C_p^\epsilon}{\left(1 + C_p^\epsilon / C_{sh}\right)^2 - k_{31}^2 \left(1 + C_p^\epsilon / C_{sh}\right) + \left(\omega R_{sh} C_p^\epsilon\right)^2} \tag{8.226}$$

The storage modulus and loss factor are plotted as a function of ρ in Fig. 8.93 for different ratios of shunt capacitance to piezoelectric capacitance ($\lambda = C_p^\epsilon / C_{sh}$). It can be seen that the loss factor reaches a maximum at a particular value of ρ, and that the maximum loss factor increases with increasing shunt capacitance for a given piezoelectric element. Also note that the maximum storage modulus is equal to unity (because $\lambda < 1$) and that the change in storage modulus increases with increasing shunt capacitance. The maximum loss factor can be found by substituting for λ in Eqs. 8.229 and 8.209. For example, with a shunt capacitance equal to 10 times the piezoelectric constant-strain capacitance ($\lambda = -0.1$), assuming $k_{31}^2 = 0.4$

$$\begin{aligned} \rho|_{\eta_{max}} &= 0.8775 \\ \eta_{max} &= 0.2279 \end{aligned} \tag{8.227}$$

In this discussion, $\lambda < 0$ because capacitance is a positive number. As a result, because $k_{31}^2 < 1$, a solution for $\rho|_{\eta_{max}}$ always exists for all values of λ. As the value of the shunt capacitance becomes very large, or $\lambda \to 0$, the maximum value of loss factor tends toward a limit of (Eq. 8.209)

$$\eta_{max}(\text{as } \lambda \to 0) = \frac{k_{31}^2}{2\sqrt{1 - k_{31}^2}} \tag{8.228}$$

which is the same as in the case of a purely resistive shunt. This represents the upper limit of loss factor that can be achieved with a positive shunt capacitance. It is interesting to examine what happens if the shunt capacitance is negative. In practice, this can be achieved using a negative impedance converter, which is an active circuit based on an operational amplifier. For a negative shunt capacitance, $\lambda > 0$. For a solution to exist for Eq. 8.208, it can be seen that $1 - k_{31}^2 > \lambda$ or $\lambda > 1$. As λ approaches these limits, η tends to infinity and the value of ρ at which the loss factor is maximum tends to zero. At the same time, the minimum value of the storage modulus tends to zero.

For other values of λ, the variation of loss factor with ρ does not exhibit an extremum because there is no real solution for ρ. However, η goes to infinity at $\rho = \sqrt{-(1-\lambda)(1-\lambda-k_{31}^2)}$. Another way of looking at this is by differentiating the expression for loss factor (Eq. 8.206) with respect to λ to find the optimum value of λ.

$$-\left(k_{31}^2\rho\right)\left[2(1-\lambda)(-1) + k_{31}^2\right] = 0 \quad \to \quad \lambda\Big|_{\eta_{max}} = 1 - \frac{k_{31}^2}{2} \tag{8.229}$$

Note that this optimum value of λ is always positive, which means it can only occur with a negative shunt capacitance. An inductive shunt can also lead to this condition, as discussed later in this chapter. Substituting this value of λ in the expression for loss factor, we obtain

$$\eta_{max} = \frac{k_{31}^2\rho}{\rho^2 - k_{31}^2/4} \tag{8.230}$$

As $\rho \to k_{31}^2/2$, $\eta_{max} \to \infty$. Therefore, using a negative capacitance, it is possible to achieve extremely large values of loss factor. This is shown in Fig. 8.94, which compares the storage modulus and loss factor of a positive shunt capacitance ($\lambda = -0.5$) and a negative shunt capacitance ($\lambda = 0.5$). Figure 8.95 shows the effect of increasing the value of λ as it approaches 0.6, in the case of a negative shunt capacitance.

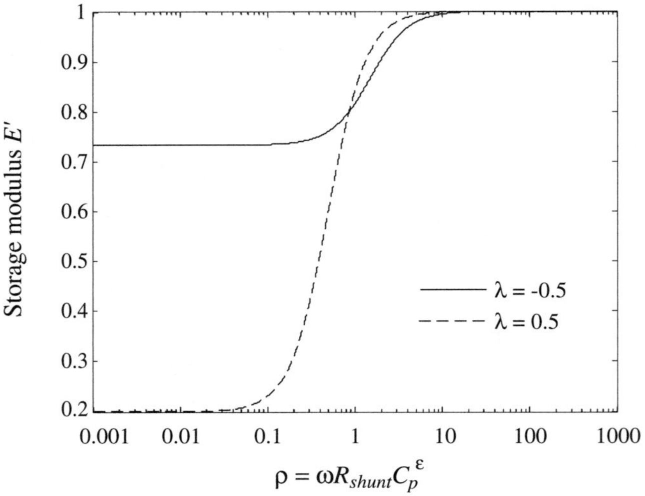

(a) Storage modulus E'

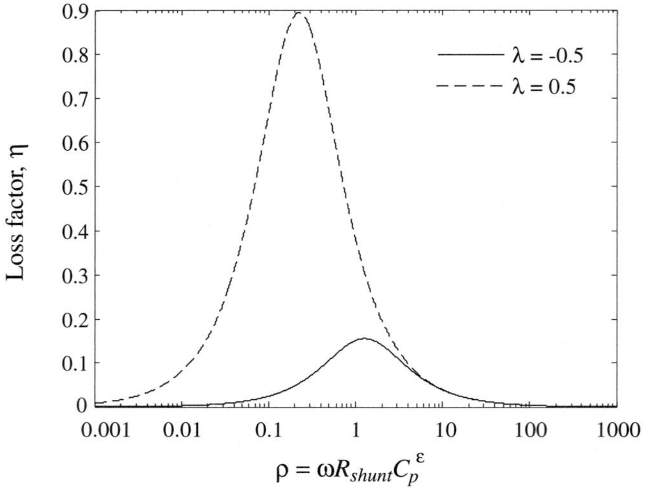

(b) Loss factor η

Figure 8.94. Comparison of storage modulus and loss factor of RC shunted piezoelectric element as a function of dimensionless frequency, for positive and negative shunt capacitance ($k_{31}^2 = 0.4$).

Inductive Shunting

Let the shunt circuit consist of a pure inductance L_{sh} in series with the resistance R_{sh} (RL shunt). In this case, assuming harmonic excitation, the shunt impedance is given by

$$Z_{sh} = R_{sh} + j\omega L_{sh} \tag{8.231}$$

Comparing this to the generic expression for shunt impedance (Eq. 8.197), we obtain

$$X_{sh} = \omega L_{sh} \tag{8.232}$$

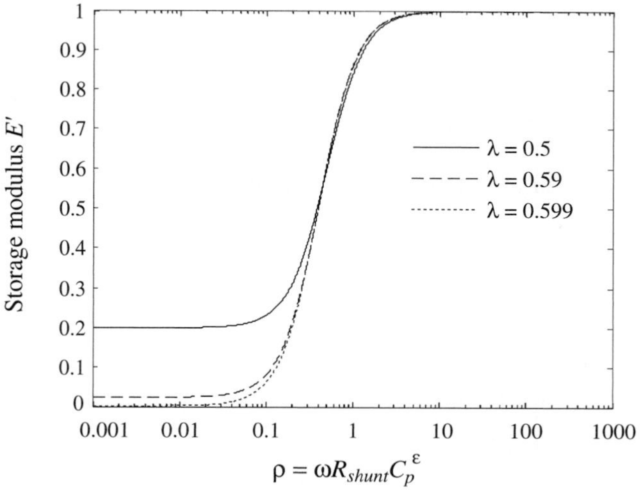

(a) Storage modulus E'

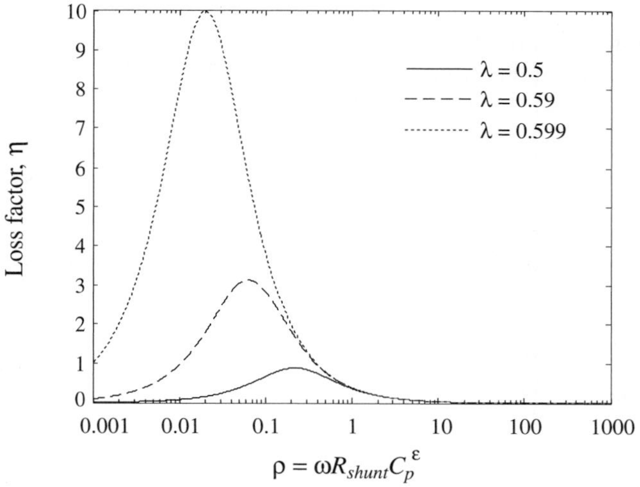

(b) Loss factor η

Figure 8.95. Effect of negative shunt capacitance on storage modulus and loss factor of RC shunted piezoelectric element ($k_{31}^2 = 0.4$).

From Eqs. 8.202 and 8.203, we obtain

$$\rho = \omega R_{sh} C_p^{\epsilon} \tag{8.233}$$

$$\lambda = \omega^2 L_{sh} C_p^{\epsilon} \tag{8.234}$$

It can be seen that $\lambda > 0$, similar to the case of a negative shunt capacitance. A quantity $\bar{L}$ can be defined such that

$$\bar{L} = \frac{L}{R_{sh}^2 C_p^{\epsilon}} \rightarrow \omega^2 L_{sh} C_p^{\epsilon} = \rho^2 \bar{L} \tag{8.235}$$

The physical meaning of the quantity $\bar{L}$ can be understood by considering the definition of quality factor Q of the RLC circuit formed by the shunt impedance and the piezoelectric element.

$$Q = \frac{1}{R_{sh}}\sqrt{\frac{L_{sh}}{C_p^{\epsilon}}} = \sqrt{\bar{L}} \tag{8.236}$$

Recall that the Q factor is directly related to the damping factor ζ of the RLC circuit. Specifically

$$Q = \frac{1}{2\zeta} \tag{8.237}$$

Therefore, it is convenient to use the quantity $\bar{L}$ because it gives an indication of the amount of damping in the circuit. It is also useful to remember one more relation between $\bar{L}$ and ρ

$$\bar{L} = \frac{\omega^2}{\omega_e^2}\frac{1}{\rho^2} \tag{8.238}$$

where ω_e^2 is the resonant frequency of the LC circuit formed by the shunt inductance and the constant strain capacitance of the piezoelectric element. These relations help to obtain a physical understanding of the parameters involved.

$$\omega_e^2 = \frac{1}{L_{sh}C_p^{\epsilon}} \tag{8.239}$$

Substituting for ρ and λ in Eqs. 8.205 and 8.206, the effective storage modulus and loss factor become

$$E' = 1 - \frac{k_{31}^2\left(1-\rho^2\bar{L}\right)}{\left(1-\rho^2\bar{L}\right)^2+\rho^2} \tag{8.240}$$

$$= 1 - \frac{k_{31}^2\left(1-\omega^2 L_{sh}C_p^{\epsilon}\right)}{\left(1-\omega^2 L_{sh}C_p^{\epsilon}\right)^2+\left(\omega R_{sh}C_p^{\epsilon}\right)^2} \tag{8.241}$$

$$\eta = \frac{\rho k_{31}^2}{\left(1-\rho^2\bar{L}\right)^2 - k_{31}^2\left(1-\rho^2\bar{L}\right)+\rho^2} \tag{8.242}$$

$$= \frac{\omega R_{sh}C_p^{\epsilon}k_{31}^2}{\left(1-\omega^2 L_{sh}C_p^{\epsilon}\right)^2 - k_{31}^2\left(1-\omega^2 L_{sh}C_p^{\epsilon}\right)+\left(\omega R_{sh}C_p^{\epsilon}\right)^2} \tag{8.243}$$

The storage modulus and loss factor are plotted in Fig. 8.96 for different values of $\bar{L}$. The case of $\bar{L} = 0$ reduces to a purely resistive shunt, as plotted in Fig. 8.91. Note that the resonant frequency in the case of an inductive shunt with a series resistance is different from the case of a purely inductive shunt due to the presence of damping in the system. Because λ is positive, the condition for maximum loss factor is the same as in the case of a negative capacitance (Eq. 8.229)

$$\lambda\Big|_{\eta_{max}} = 1 - \frac{k_{31}^2}{2} \tag{8.244}$$

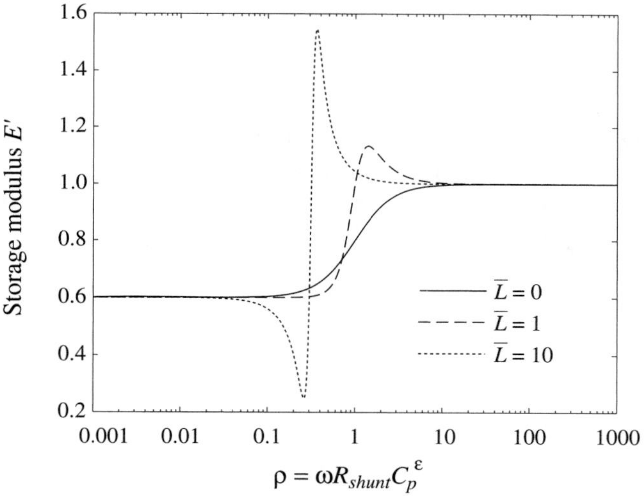

(a) Storage modulus E'

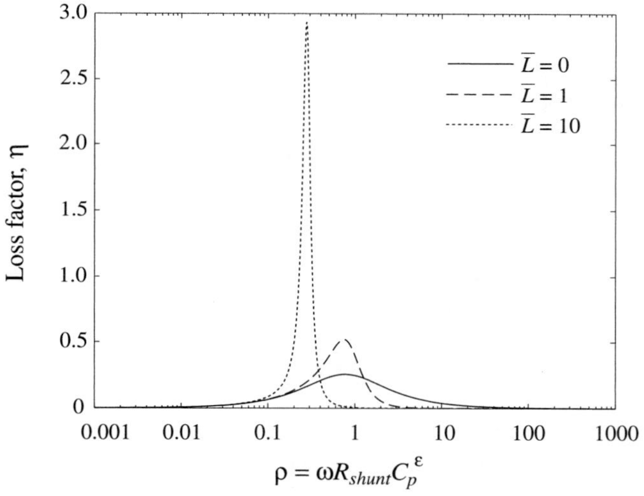

(b) Loss factor η

Figure 8.96. Storage modulus and loss factor of an RL shunted piezoelectric element as a function of dimensionless frequency ($k_{31}^2 = 0.4$), for different values of inductance.

From which the maximum loss factor is obtained as

$$\eta_{max} = \frac{k_{31}^2 \rho}{\rho^2 - k_{31}^2/4} \tag{8.245}$$

In practice, the resonant frequency of the shunt circuit would be tuned to occur at the same frequency as the structural mode that is to be damped. From Fig. 8.96, it can be seen that compared to the case of a purely resistive shunt, the loss factor is higher at low frequencies and is reduced at higher frequencies. The large values loss factor achievable is the primary advantage of using an inductive element in the shunt circuit. In comparison, the maximum loss factor of an RC shunt is the same as that of a purely resistive shunt and is limited to 0.2279. However, a purely capacitive shunt can be used to change the effective stiffness without adding any damping to the structure.

8.9.3 Worked Example

Design a series RC shunt circuit for a piezoelectric sheet to achieve a maximum loss factor of 0.25. Plot the variation of required shunt resistance over a range of operating frequency from 10 Hz to 1 kHz.

Material data are as follows

$$\begin{aligned} k_{31}^2 &= 0.4 \\ \text{length } l_c &= 50.8\,\text{mm (2 in)} \\ \text{width } b_c &= 25.4\,\text{mm (1 in)} \\ \text{thickness } t_c &= 0.3175\,\text{mm (0.0125 in)} \\ \text{Relative permittivity } K_{31}^\sigma &= 3400 \end{aligned}$$

Solution

The capacitance of the piezoelectric at constant stress is given by

$$C_{31}^\sigma = \frac{K_{31}^\sigma \epsilon_o l_c b_c}{t_c} = \frac{3400 \times 8.854 \times 10^{-12} \times 0.0508 \times 0.0254}{0.0003175}$$
$$= 122.34 \text{ nF}$$

The constant strain capacitance can be calculated from this value

$$C_p^\epsilon = \left(1 - k_{31}^2\right) C_{31}^\sigma = 73.4 \text{ nF}$$

The condition for the maximum loss factor is given by Eq. 8.209. Because we are given the desired value of the maximum loss factor, we can rewrite this equation in terms of λ as

$$\lambda^2 + \left(k_{31}^2 - 2\right)\lambda + 1 - k_{31}^2 = \frac{k_{31}^4}{4\eta_{max}^2}$$

Solving this quadratic equation, we obtain $\lambda = 1.6246$ or $\lambda = -0.0246$. Let us choose the negative value because we are using a positive shunt capacitance. The shunt capacitance is given by

$$C_{sh} = -\frac{C_p^\epsilon}{-0.0246} = 2.984\ \mu\text{F}$$

The value of ρ at which the maximum loss factor is achieved is given by Eq. 8.208. Substituting for C_p^ϵ, we can calculate the value of shunt resistance required to obtain the maximum loss factor at each operating frequency in the range of interest. The result is plotted in Fig. 8.97. It can be seen that the required resistance becomes very large at lower frequencies.

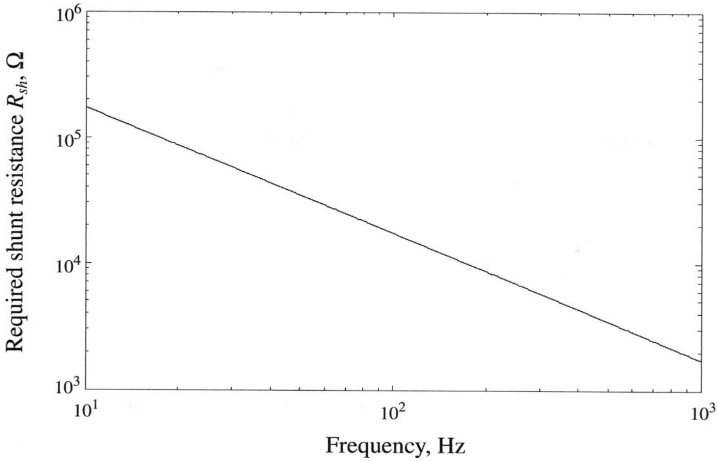

Figure 8.97. Resistance required for maximum loss factor, as a function of operating frequency ($k_{31}^2 = 0.4$, $\eta_{max} = 0.25$).

8.9.4 Worked Example

Design a series RL shunt circuit for a piezoelectric sheet to achieve a maximum loss factor of 0.25. Plot the variation of required shunt resistance and shunt inductance over a range of operating frequency from 10 Hz to 1 kHz.

Material data are as follows

$$k_{31}^2 = 0.4$$
$$\text{length } l_c = 50.8\,\text{mm (2 in)}$$
$$\text{width } b_c = 25.4\,\text{mm(1 in)}$$
$$\text{thickness } t_c = 0.3175\,\text{mm (0.0125 in)}$$
$$\text{Relative permittivity } K_{31}^{\sigma} = 3400$$

Solution

Proceeding similarly to the previous worked example, the constant-strain capacitance of the piezoelectric sheet is calculated as

$$C_p^{\epsilon} = 73.4\text{ nF}$$

The condition for the maximum loss factor is given by Eq. 8.229 as

$$\lambda\Big|_{\eta_{max}} = 1 - \frac{k_{31}^2}{2}$$

From this equation, we obtain the value of λ to achieve the maximum loss factor at each frequency of interest

$$\lambda = 1 - \frac{0.4}{2} = 0.8$$

Using this value of λ, we can calculate the required shunt inductance L_{sh} at each operating frequency. The maximum value of loss factor desired is specified as 0.25. Substituting in the expression for maximum loss factor (Eq. 8.230), we obtain a

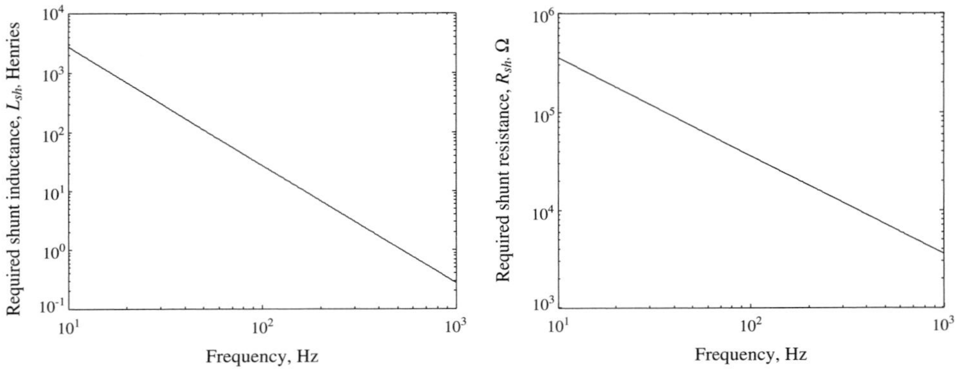

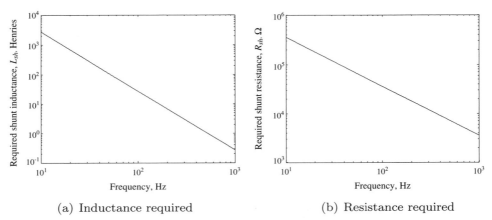

(a) Inductance required (b) Resistance required

Figure 8.98. Inductance and resistance required for maximum loss factor, as a function of operating frequency ($k_{31}^2 = 0.4$, $\eta_{max} = 0.25$).

quadratic equation for ρ

$$\eta_{max}\rho^2 - k_{31}^2\rho - \eta_{max}\frac{k_{31}^2}{4} = 0$$

Solving this equation, we obtain $\rho = 1.6602$ and $\rho = -0.0602$. We ignore the negative root because it is not physical. From the positive root, we can calculate the shunt resistance required

$$R_{sh} = \frac{\rho}{\omega C_{sh}^{\epsilon}}$$

The required shunt inductance and shunt resistance are plotted as a function of operating frequency in Fig. 8.98. It can be seen that the value of inductance required to tune the circuit at low frequencies becomes too large to be realized by a practical physical inductor. In these cases, an active circuit based on an operational amplifier can be used to simulate an inductance of the appropriate value. Such circuits fall under a category of circuits called gyrators and are widely used in active filter design [189, 190, 191]. Note that whereas gyrators have the effect of inverting a physical impedance (converting a capacitance to an effective inductance), negative impedance converters have the effect of creating the negative of a physical impedance (e.g., a negative capacitance). The real advantage in using an RL shunt is that a much higher loss factor can be achieved compared to an RC shunt with a physical capacitance.

8.9.5 Worked Example

Two piezoelectric sheets are bonded to the top and bottom of an aluminum beam, as shown in Fig. 8.99. The piezoelectric sheets are connected in parallel, and the polarity is indicated by the dots. The tip of the beam is subjected to a unit sinusoidal

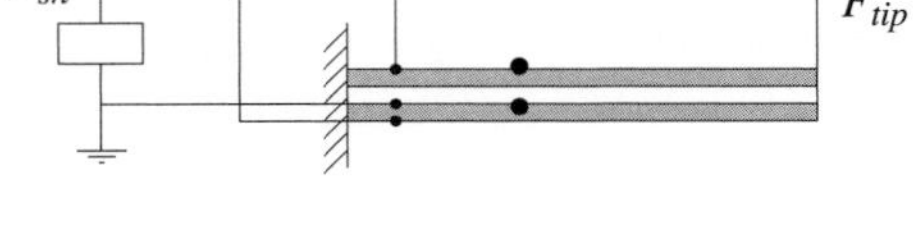

Figure 8.99. Beam with shunted piezoelectric sheets subjected to a tip force.

forcing. Using a finite element approach, calculate the tip deflection of the beam over the range 50 Hz to 1500 Hz (capturing the first two bending modes) for the following cases:

(1) no shunt (short-circuited electrodes)
(2) purely resistive shunt
(3) RL shunt

Material data are as follows

$$k_{31}^2 = 0.4$$
$$\text{piezo length } l_c = 50.8\,\text{mm}\,(2\text{ in})$$
$$\text{piezo width } b_c = 25.4\,\text{mm}\,(1\text{ in})$$
$$\text{piezo thickness } t_c = 0.254\,\text{mm}\,(0.01\text{ in})$$
$$\text{beam thickness } t_b = 0.254\,\text{mm}\,(0.01\text{ in})$$
$$\text{relative permittivity } K_{31}^{\sigma} = 3400$$
$$\text{Young's modulus of aluminum } = 70\text{ GPa}$$
$$\text{Young's modulus of piezoelectric (constant field) } = 70\text{ GPa}$$
$$\text{density of aluminum} = 2700\,\text{kg/m}^3$$
$$\text{density of piezoelectric} = 7600\,\text{kg/m}^3$$

Solution

The finite element formulation is used to develop a model for a beam with bonded piezoelectric elements. The governing equation of the beam is obtained as

$$\boldsymbol{M}_g\, \ddot{\boldsymbol{q}}_g + \boldsymbol{K}_g\, \boldsymbol{q}_g = \boldsymbol{Q}_g \tag{8.246}$$

In the case of the shunted piezoelectric, the effective modulus E_{sh} is given by

$$E_{sh} = \frac{E^E}{1 - k_{31}^2} E'(1 + j\eta)$$

Due to the geometry of the problem, we are only concerned with the modulus in the "1"-direction. The storage modulus and loss factor are calculated based on the shunt impedance from Eqs. 8.205 and 8.206

$$E' = 1 - \frac{(1-\lambda)k_{31}^2}{(1-\lambda)^2 + \rho^2}$$

$$\eta = \frac{k_{31}^2 \rho}{(1-\lambda)^2 - k_{31}^2(1-\lambda) + \rho^2}$$

The effective modulus of the piezoelectric elements is used to calculate the global stiffness matrix K_g in the finite element formulation. Note that the matrix K_g can now be complex as well as frequency-dependent. For the present problem, because there is no applied voltage, the forcing due to induced strain is zero.

(1) No shunt: In this case, the storage modulus is $E' = 1 - k_{31}^2$ and loss factor is $\eta = 0$. The effective stiffness reduces to the short-circuit stiffness of the piezoelectric.

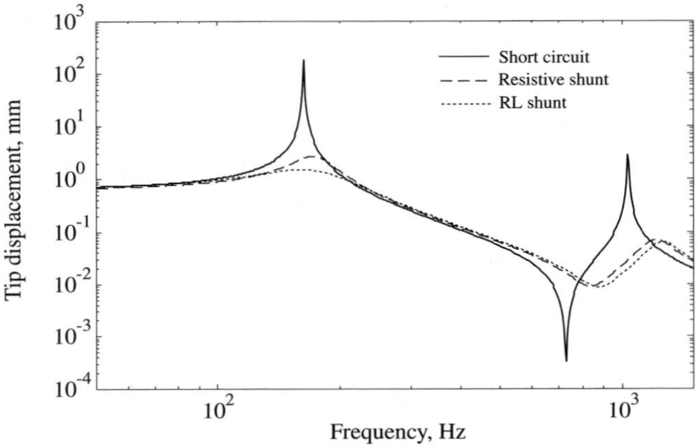

Figure 8.100. Response of beam with shunted piezoelectric sheets subjected to a tip force, for different shunt impedances.

The stiffness matrix of the structure is real and constant. The governing equation is solved and the tip displacement is calculated and plotted in Fig. 8.100. The first two modes occur at 163 Hz and 1034 Hz, respectively.

(2) Let us target the first mode occurring at 163 Hz. The constant strain capacitance of the piezoelectric is

$$C_p^{\epsilon} = 2\left(1 - k_{31}^2\right) \times 3400 \times 8.854 \times 10^{-12} \times l_c \times b_c/t_c = 183.5 \text{ nF}$$

The condition for maximum damping (Eq. 8.217) yields $\rho = 0.7746$ and a loss factor of $\eta = 0.2582$. At the frequency of the first mode, the required shunt resistance is

$$R_{sh} = \frac{0.7746}{2 \times \pi \times 163 \times 183.5 \times 10^{-9}} = 4.1214 \ \text{k}\Omega$$

The effect of this shunt resistance is plotted in Fig. 8.100. It can be seen that the response is damped and the peaks are shifted to the right due to the increased stiffness of the piezoelectric elements.

(3) The condition for the maximum loss factor is given by Eq. 8.229 as

$$\lambda = 1 - \frac{k_{31}^2}{2} = 0.8$$

At the frequency of the first mode, this results in a shunt inductance of

$$L_{sh} = \frac{0.8}{(2 \times \pi \times 163)^2 \times 183.5 \times 10^{-9}} = 4.1562 \ \text{H}$$

Let us assume that this large value of inductance is achieved using an electronic pseudo-inductor. Let us also use the same value of resistance used for the case of the purely resistive shunt. This yields a value of $\rho = 0.7746$ at the frequency of the first mode and a loss factor of $\eta = 0.6197$ (from Eq. 8.230). The results are plotted in Fig. 8.100.

8.10 Energy Harvesting

In recent years, there have been rapid developments in the area of structural health monitoring for large civil structures using wireless sensor networks. A number of sensor nodes are installed over a structure (e.g.) a bridge [192, 193]. Each sensor node collects local information such as vibration amplitude or strain, and either stores this information locally or transmits it to a central base station. In this way, the state or health of the bridge can be monitored. Applications such as these require a source of energy at each node. Typically, this has been accomplished by a battery pack. However, the need for periodic replacement of the batteries, especially at locations that are difficult to access, has spurred the development of methods to locally generate the required power. One of the most popular approaches is to harvest the energy inherent in the ambient vibrations of the structure. The conversion of the mechancial energy into electrical energy can be accomplished by different types of transducers (e.g., electrostatic, electromagnetic, and piezoelectric). The devices based on piezoelectric materials are attractive due to their solid-state nature and the high volumetric density of harvested power [194]. Piezoelectric energy harvesters are finding application in a variety of areas with similar requirements, such as MEMS devices and wearable electronics.

The previous section on shunted piezoelectrics described how the transduction of mechanical energy to electrical energy by piezoelectric materials can be used to enhance the damping of a structure. It follows that this electrical energy, instead of being dissipated, can be accumulated and used to power other devices. This concept forms the basis of energy harvesting using piezoelectric materials. Energy harvested in this manner also increases the effective damping of the system, similar to the effect of shunted piezoelectrics. In the case of energy harvesting, the goal is to accumulate the energy, whereas in the case of shunt damping, the goal is to dissipate as much of the energy as possible.

8.10.1 Vibration-Based Energy Harvesters

Piezoelectric materials have found wide application as low-power generators. In the majority of these applications, the piezoelectric material extracts energy from ambient structural vibrations by operating as a base-excited oscillator. Due to the inherently low energy in structural vibrations, these devices are limited to relatively low-power outputs, in the range of 1–1000 μW. Sodano et al. [195] provided an overview of several studies related to piezoelectric energy harvesting, including devices based on impact, wearable energy harvesting devices based on motion of the human body, and devices designed to power wireless sensors. They also discussed methods to accumulate the harvested energy using rechargeable batteries, capacitors, or flyback converters. Although most of the energy harvesting devices are based on cantilever beams, other geometries such as annular piezoelectric unimorphs/bimorphs have also been explored [196].

Piezoelectric energy harvesters have also been investigated at the MEMS-scale, to power autonomous sensors. duToit et al. [194] described the design of a MEMS-scale piezoelectric energy harvester based on a unimorph piezoelectric cantilever beam with a proof mass. They compared the power density of electrostatic, electromechanical, and piezoelectric vibration-based energy harvesters and concluded that the piezoelectric devices have the highest power density based on volume. In addition to a low number of moving parts, this accounts for the popularity of piezoelectric energy harvesters.

There have also been numerous studies on optimizing the power-conditioning and storage electronics. This forms an important part of the overall device, especially for MEMS-scale systems. Several designs have been proposed. For example, Ottman et al. [197] designed optimal power-conditioning electronics for a vibration energy harvester using a step-down converter. Wickenheiser and Garcia [198] investigated the conditions for maximum power generated by a vibration-based energy harvester connected to four different circuits. These circuits were a simple resistive load, a standard rectifier, and parallel and series-switching circuits. It was concluded that the active switching circuits are advantageous for systems with low electromechanical coupling, and this advantage decreases as the coupling increases. Therefore, passive harvesting circuits may be adequate for systems with high electromechanical coupling.

8.10.2 Wind-Based Energy Harvesters

Structures with piezoelectric elements have also been used to harvest energy from other sources, such as wind. For example, the energy harvester developed by Tan and Panda [199] is based on vibrations excited in a piezoelectric bimorph when exposed to wind. The device developed by Wang and Ko [200] generates on the order of 0.2W in response to flow-induced pressure fluctuations. Robbins et al. [201] investigated the use of flexible, flag-like, piezoelectric sheets to generate power while flapping in an incident wind. The energy that can be harvested using these approaches is comparable to that of a vibration-based device. By exploiting structures with aeroelastic instabilities, it is possible to extract significantly higher amounts of energy from the wind. Bryant and Garcia [202] developed a device to harvest energy from flutter, using a piezoelectric bimorph with a flap at its tip. Linear and nonlinear models were developed to predict the performance of the device. The device generated an output power on the order of 2mW. Sirohi and Mahadik [203] investigated wind energy harvesting using a beam with piezoelectric sheets attached to a tip body with a D-shaped cross section. Wind-induced galloping of the tip body resulted in oscillatory bending of the beam, and the maximum power generated was measured to be on the order of 0.5 mW.

8.10.3 Modeling of Piezoelectric Energy Harvesters

An analytical model incorporating the electromechanical coupling of the piezoelectric material must be derived to predict the behavior of the system. Such a model can be derived by directly coupling the constitutive relations of the piezoelectric and the structure (see Erturk and Inman [204]), representing the system in terms of an equivalent electric circuit (see Elvin et al. [205]), or by using an energy-based variational formulation.

In the energy-based formulation, the basic approach is to formulate a variational indicator incorporating the kinetic energy, potential energy, and nonconservative virtual work on the system. The potential energy and nonconservative virtual work include contributions from both mechanical (strain energy) and electrical (stored charge) terms. There are several ways to represent the potential energy due to electrical and mechanical contributions, depending on the choice of independent variables. Mason [206] lists these different representations of energy in differential form, along with the corresponding independent variables. Two of these representations are convenient for modeling structures with electromechanical coupling. These are the internal energy U and the electric enthalpy H_2, given by

$$U(\boldsymbol{\epsilon}, \boldsymbol{D}) = \frac{1}{2}\int_{V_s} \boldsymbol{\sigma}^T \boldsymbol{\epsilon} dV_s + \frac{1}{2}\int_{V_s} \mathbb{E}^T \boldsymbol{D} dV_s \tag{8.247}$$

where $\boldsymbol{\epsilon}$ is the strain vector, $\boldsymbol{\sigma}$ is the stress vector, $\boldsymbol{D}$ is the electric displacement vector, $\mathbb{E}$ is the electric field vector, and V_s is the volume of the structure. Note that the internal energy must be expressed as a function of independent variables corresponding to displacement and charge, which are, in this case, $\boldsymbol{\epsilon}$ and $\boldsymbol{D}$. The electric enthalpy is given by

$$H_2(\boldsymbol{\epsilon}, \mathbb{E}) = \frac{1}{2}\int_{V_s} \boldsymbol{\sigma}^T \boldsymbol{\epsilon} dV_s - \frac{1}{2}\int_{V_s} \mathbb{E}^T \boldsymbol{D} dV_s \tag{8.248}$$

Note that the electric enthalpy must be expressed as a function of independent variables corresponding to displacement and electric field, which are, in this case, $\boldsymbol{\epsilon}$ and $\mathbb{E}$.

Based on the choice of either the internal energy or the electric enthalpy to represent the potential energy of the structure, it is possible to formulate a variational indicator in two ways [207]. In one approach, the variational indicator ($V.I.$) is written as

$$V.I. = \int_{t_1}^{t_2} [\delta(T - V - W_e) + \sum_i f_i \delta w_i + \sum_j \mathbb{V}_j \, \delta q_j] \, dt \tag{8.249}$$

$$= \int_{t_1}^{t_2} [\delta(T - U) + \sum_i f_i \delta w_i + \sum_j \mathbb{V}_j \, \delta q_j] \, dt \tag{8.250}$$

where T is the kinetic energy of the structure, V is the strain energy, and W_e is the electrical energy. The summations represent the virtual work done by all nonconservative mechanical and electrical elements in the system. In the present case, f_i are the transverse forces applied to the beam, w_i are the transverse displacements, $\mathbb{V}_j$ is the voltage drop across the nonconservative electrical elements (e.g., a load resistance across the electrodes of the piezoelectric sheets), and q_j is the electric charge. Several researchers have adapted this approach to model the electromechanical coupling in structures with piezoelectric material [208, 195, 194, 205].

The other approach makes use of the electric enthalpy and flux linkage to formulate the variational indicator as [209]

$$V.I. = \int_{t_1}^{t_2} [\delta(T - V + W_e^*) + \sum_i f_i \delta w_i + \sum_j i_j \, \delta\lambda_j] \, dt \tag{8.251}$$

$$= \int_{t_1}^{t_2} [\delta(T - H_2) + \sum_i f_i \delta w_i + \sum_j i_j \, \delta\lambda_j] \, dt \tag{8.252}$$

where W_e^* is the electrical co-energy and i_j are the currents flowing through the dissipative electrical elements in the system. The λ_j are the flux linkages, which are

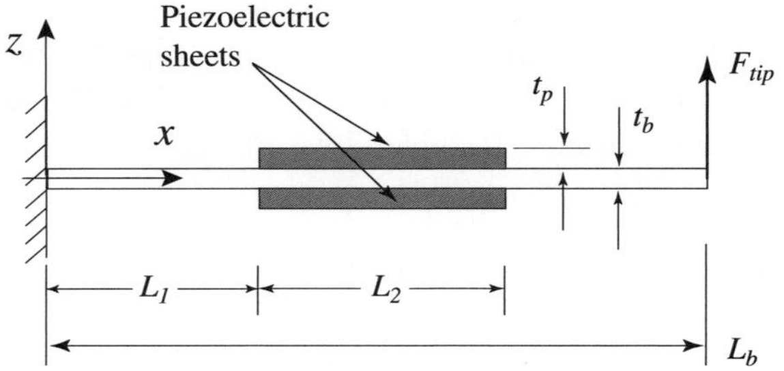

Figure 8.101. Schematic of energy harvester: cantilever beam with piezoelectric sheets.

related to voltages by

$$\mathbb{V} = \dot{\lambda} \tag{8.253}$$

It can be shown that the approaches based on the two variational indicators are equivalent because the internal energy and electric enthalpy are related to each other by a Legendre transformation.

Applying Hamilton's principle, integrating by parts, and setting the variations at t_1 and t_2 equal to zero, we obtain the governing equations of the system. Let us use the formulation based on internal energy (Eq. 8.249) to derive a model of an energy harvester consisting of an aluminum cantilever beam with surface-bonded piezoelectric sheets (Fig. 8.101). An oscillatory force F_{tip} acting at the tip of the beam acts as a source of energy. The piezoelectric sheets are connected in parallel such that the beam is a common electrical ground and the charges induced by the opposite polarity strains on the top and bottom surfaces add up. The electrical energy is dissipated across a load resistance connected between the electrodes of the piezoelectric sheets (Fig. 8.102).

Let the piezoelectric sheets have electrodes parallel to the 1–2 planes and poled along the 3-direction, as shown in Fig. 8.90. The piezoelectric sheets are attached so that their 1-axis is along the length of the beam (x direction) and the 3-axis is along the thickness of the beam (z direction). The constitutive relations for these sheet elements are then given by Eq. 8.178. These relations can be rearranged in terms of the strain and electric displacement as

$$\begin{Bmatrix} \sigma \\ \mathbb{E} \end{Bmatrix} = \begin{bmatrix} \boldsymbol{c}^D & -\boldsymbol{h}^T \\ -\boldsymbol{h} & \boldsymbol{\beta}^{\epsilon} \end{bmatrix} \begin{Bmatrix} \epsilon \\ \boldsymbol{D} \end{Bmatrix} \tag{8.254}$$

In the case of the cantilever beam in the present example, strains along the y direction can be ignored and a one-dimensional representation can be used to model the device. Reducing Eq. 8.254 to one dimension and substituting the relevant

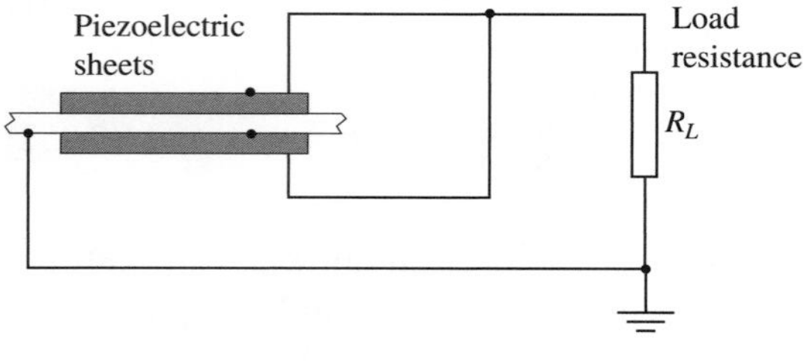

Figure 8.102. Schematic of energy harvester circuit with load resistance.

piezoelectric constants from Eq 8.178 yields

$$\begin{Bmatrix} \sigma_{11} \\ \mathbb{E}_3 \end{Bmatrix} = \begin{bmatrix} Y_{11}^D & -\frac{1}{d_{31}'} \\ -\frac{1}{d_{31}'} & \frac{1}{e_{33}^{\epsilon}} \end{bmatrix} \begin{Bmatrix} \epsilon_{11} \\ D_3 \end{Bmatrix} \tag{8.255}$$

where

$$Y_{11}^D = \frac{Y_{11}^{\mathbb{E}}}{1-k_{31}^2} = \frac{1}{s_{11}^{\mathbb{E}}\left(1-k_{31}^2\right)} \tag{8.256}$$

$$d_{31}' = \frac{d_{31}\left(1-k_{31}^2\right)}{k_{31}^2} \tag{8.257}$$

$$e_{33}^{\epsilon} = e^{\sigma}\left(1-k_{31}^2\right) \tag{8.258}$$

The superscripts D and ϵ refer to quantities measured at constant electric displacement and constant strain, respectively. The quantity Y_{11} is the Young's modulus of the piezoelectric material, and the electromechanical coupling factor of the piezoelectric sheets is defined as

$$k_{31}^2 = \frac{d_{31}^2 Y_{11}^{\mathbb{E}}}{e_{33}^{\sigma}} \tag{8.259}$$

It is convenient to model the coupled behavior of the piezoelectric sheets in this way because it is relatively simple to measure the constants $Y_{11}^{\mathbb{E}}$, d_{31}, and e_{33}^{σ}. Substituting these quantities into Eq. 8.247 yields the internal energy of the device as

$$U = \frac{1}{2}\int_{V_s} \sigma_{11}\epsilon_{11} dV_s + \frac{1}{2}\int_{V_s} D_3\mathbb{E}_3 dV_s \tag{8.260}$$

$$= \frac{1}{2}\int_{V_s} \left(Y_{11}^D\epsilon_{11}^2 + \frac{D_3^2}{e_{33}^{\epsilon}}\right) dV_s - \int_{V_s} \frac{D_3\epsilon_{11}}{d_{31}'} dV_s \tag{8.261}$$

The integration is performed over the volume of the entire structure, taking care to set the appropriate material constants over the piezoelectric elements and the aluminum beam. Applying the Euler-Bernoulli assumption to the aluminum beam results in the longitudinal strain given by

$$\epsilon_{11} = -zw'' \tag{8.262}$$

where w is the transverse deflection of the beam and z is the coordinate along the beam thickness, measured from the neutral axis. The assumed modes method is typically used in the derivation of equations of motion. A superposition of assumed shape functions $\phi(x)$ and generalized displacement coordinates $r(t)$ can be used to represent the transverse deflection as

$$w(x,t) = \sum_{i=1}^{N} \phi(x) r(t) = \boldsymbol{\phi r} \tag{8.263}$$

Typically, the most accurate results are obtained when exact solutions (mode shapes) to the same structure with identical boundary conditions are used as the assumed shape functions. However, simple polynomials that satisfy the same geometric boundary conditions often give satisfactory results. The longitudinal strain is

$$\epsilon_{11} = -z\boldsymbol{\phi}''\boldsymbol{r} \tag{8.264}$$

Similarly, the electric displacement can be represented as a summation of assumed functions $\psi(z)$ and generalized charge coordinates $q(t)$ as

$$D(z,t) = \sum_{j=1}^{M} \psi(z)q(t) = \boldsymbol{\psi}\boldsymbol{q} \tag{8.265}$$

Note that if the electric field across the piezoelectric sheets is assumed to be a constant

$$\psi = \frac{1}{A_p} \tag{8.266}$$

where A_p is the area of the electrodes on the piezoelectric sheets. In this case, the electric displacement is given by

$$D = \frac{q}{A_p} \tag{8.267}$$

where q is the physical charge generated by the piezoelectric sheets. The assumption of constant electric field across the piezoelectric sheets is sufficiently accurate for most practical purposes. Substituting the strain and electric displacement into Eq. 8.260, the internal energy can be written as

$$U = \frac{1}{2}\int_{V_s}\left(\boldsymbol{r}^T\boldsymbol{\phi}''^T Y_{11}^D \boldsymbol{\phi}''\boldsymbol{r}z^2 + \boldsymbol{q}^T\boldsymbol{\psi}^T\frac{1}{e_{33}^{\epsilon}}\boldsymbol{\psi}\boldsymbol{q}\right)dV_s + \int_{V_s}\boldsymbol{r}^T\boldsymbol{\phi}^T d'_{31}z\boldsymbol{\psi}\boldsymbol{q}dV_s \tag{8.268}$$

$$= \frac{1}{2}\boldsymbol{r}^T\boldsymbol{K}\boldsymbol{r} + \frac{1}{2}\boldsymbol{q}^T\boldsymbol{C}'_p\boldsymbol{q} + \boldsymbol{r}^T\boldsymbol{\Theta}\boldsymbol{q} \tag{8.269}$$

where the stiffness matrix $\boldsymbol{K}$ (size $N \times N$) and the coupling matrix $\boldsymbol{\Theta}$ (size $N \times M$) are given by

$$\boldsymbol{K} = \int_{V_s} Y_{11}^D z^2 \boldsymbol{\phi}''^T\boldsymbol{\phi}'' dV_s \tag{8.270}$$

$$\boldsymbol{\Theta} = \int_{V_s}\frac{z\boldsymbol{\phi}''^T}{d'_{31}}\boldsymbol{\psi}dV_s \tag{8.271}$$

For a uniform electric field across the piezoelectric sheets

$$\boldsymbol{C}'_p = \int_{V_p}\boldsymbol{\psi}^T\frac{1}{e_{33}^{\epsilon}}\boldsymbol{\psi}dV_p = \frac{t_p}{A_p e_{33}^{\epsilon}} = \frac{1}{C_p} \tag{8.272}$$

$$\boldsymbol{\Theta} = \int_{V_s}\frac{z\boldsymbol{\phi}''^T}{d'_{31}A_p}dV_s \tag{8.273}$$

The subscripts b and p denote quantities corresponding to the beam and the piezoelectric sheets, respectively. Note that the appropriate Young's modulus must be used when integrating over the volume of the aluminum beam; that is, Y_{11}^D must be substituted by Y_b. Similarly, over the volume of the beam, $d'_{31} = 0$.

The term C_p is recognized as the capacitance of the piezoelectric sheets (at constant strain). The kinetic energy of the structure is given by

$$T = \frac{1}{2}\int_{V_s} \rho \dot{w}^2 dV_s \tag{8.274}$$

$$= \frac{1}{2}\dot{\boldsymbol{r}}^T \boldsymbol{M}\dot{\boldsymbol{r}} \tag{8.275}$$

where the mass matrix is

$$\boldsymbol{M} = \int_{V_s} \rho \boldsymbol{\phi}^T \boldsymbol{\phi} dV_s \tag{8.276}$$

In the case of the energy harvesting device under consideration, the nonconservative mechanical virtual work arises only due to the force F_{tip} acting on the tip of the beam. The nonconservative electrical virtual work is the energy dissipated by the load resistance.

Substituting the internal energy (Eq. 8.268), kinetic energy (Eq. 8.274), and nonconservative virtual works into Eq. 8.249 and setting the variational indicator to zero

$$V.I. = \int_{t_1}^{t_2} [\delta(T - U) + F_{tip}\delta w(L_b) + \mathbb{V}\delta q]\, dt = 0 \tag{8.277}$$

yields the equations of motion of the energy harvesting device as (assuming constant electric field across the piezoelectric sheets)

$$\boldsymbol{M}\ddot{\boldsymbol{r}} + \boldsymbol{K}\boldsymbol{r} + \boldsymbol{\Theta}q = F_{tip}\boldsymbol{\phi}^T(L_b) \tag{8.278}$$

$$\boldsymbol{\Theta}^T\boldsymbol{r} + \frac{1}{C_p}q - \mathbb{V} = 0 \tag{8.279}$$

In the case of a load resistance connected between the electrodes of the piezoelectric sheets, the voltage drop is given by

$$\mathbb{V} = -R_L i = -R_L \dot{q} \tag{8.280}$$

Any mechanical damping in the structure can also be incorporated into the model in terms of a proportional damping matrix [195] $\boldsymbol{C}$ given by

$$\boldsymbol{C} = \alpha\boldsymbol{M} + \beta\boldsymbol{K} \tag{8.281}$$

where the constants α and β are determined from experiments, typically an impulse response or equivalent test with an appropriate electrical boundary condition for the piezoelectric sheets. The modal damping can be written as

$$\zeta_i = \frac{\alpha}{2\omega_i} + \frac{\beta\omega_i}{2} \qquad (i = 1, 2 \ldots N) \tag{8.282}$$

In this equation, ω_k is the natural frequency of the kth mode, ζ_k is the modal damping, and N is the number of modes (equal to the dimension of the mass and stiffness matrices). An additional damping is introduced due to the energy dissipation in the internal resistance of the piezoelectric sheets, R_i. This resistance is a function of the dissipation factor (expressed as $\tan\delta$) of the piezoelectric material. For small

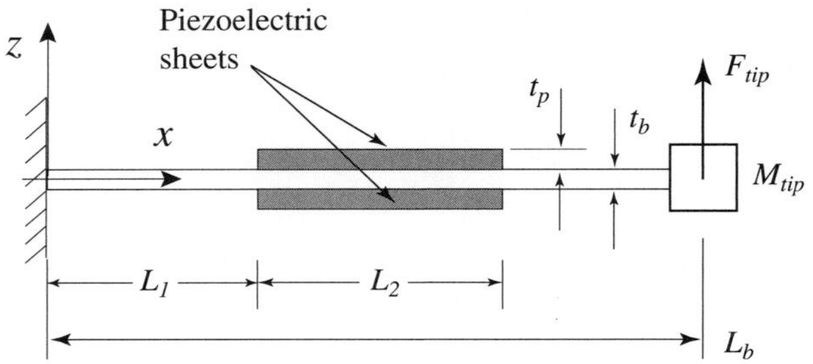

Figure 8.103. Cantilever beam with surface-bonded piezoelectric sheets and tip mass.

values of $\tan\delta$, the internal resistance can be written as [210]

$$R_i = \frac{\tan\delta}{\omega C_p} \tag{8.283}$$

where ω is the frequency of voltage across the electrodes of the piezoelectric sheet. At large values of electric field, the dissipation factor and other piezoelectric constants become nonlinear functions of the electric field.

The equations of motion can be written in the state-space form by defining a state vector containing the generalized displacement (consisting of N assumed modes), generalized velocity, and charge.

$$\boldsymbol{x} = \begin{Bmatrix} \boldsymbol{r} \\ \dot{\boldsymbol{r}} \\ q \end{Bmatrix} \tag{8.284}$$

The equations of motion (Eqs. 8.278 and 8.279) can then be written as

$$\dot{\boldsymbol{x}} = \begin{bmatrix} \boldsymbol{0}_{(N\times N)} & \boldsymbol{I}_{(N\times N)} & \boldsymbol{0}_{(N\times 1)} \\ -\boldsymbol{M}^{-1}\boldsymbol{K} & -\boldsymbol{M}^{-1}\boldsymbol{C} & -\boldsymbol{M}^{-1}\boldsymbol{\Theta} \\ -\frac{1}{R_L}\boldsymbol{\Theta}^T & \boldsymbol{0}_{(1\times N)} & -\frac{1}{R_L C_p} \end{bmatrix} \boldsymbol{x} + \begin{Bmatrix} \boldsymbol{0}_{(N\times 1)} \\ \boldsymbol{M}^{-1}\boldsymbol{\phi}^T(L_b) \\ 0 \end{Bmatrix} F_{tip} \tag{8.285}$$

These equations can be solved using standard time-marching algorithms to find the voltage developed and power dissipated by the load resistance.

8.10.4 Worked Example

Consider a cantilever beam with surface-bonded piezoelectric sheets and a tip mass as shown in Fig. 8.103. The two piezoelectric sheets are connected in parallel with a load resistance, as in Fig. 8.102. Assume a uniform electric field across the piezoelectric sheets and a beam transverse displacement given by

$$w(x,t) = \phi_1 r_1 = \left(\frac{x}{L_b}\right)^3 r_1 \tag{8.286}$$

Derive the equations of motion of the system. For an oscillatory tip force of unit amplitude at a frequency of 100 Hz, plot the voltage generated and power dissipated for a range of load resistances. Neglect structural damping.

The parameters of the piezoelectric sheets and the aluminum beam are listed in Table 8.8.

Table 8.8. *Parameters of example energy harvesting device*

Property	Symbol	Value
Piezoelectric		
Strain coefficient (pC/N)	d_{31}	−320
Young's modulus (GPa)	$Y_{11}^{\mathbb{E}}$	62
Dielectric constant (nF/m)	e_{33}^{σ}	33.65
Density (kg/m^3)	ρ_p	7800
Thickness (mm)	t_p	0.1905
Length (mm)	$L_2 - L_1$	50.8
Width (mm)	b_p	25.4
Offset from beam root (mm)	L_1	6.35
Beam		
Young's modulus (GPa)	Y_b	69
Density (kg/m^3)	ρ_b	2700
Thickness (mm)	t_b	0.79375
Length (mm)	L_b	152.4
Tip mass (kg)	M_{tip}	0.010

Solution

The equations of motion of the system are given by Eq. 8.285. The stiffness and coupling matrices are given by Eqs. 8.270 and 8.271. These are obtained by substituting the assumed deflection and electric displacement. In the present case, these are scalars.

$$K = \int_{V_b} Y_b(\phi_1'')^2 z^2 dV_b + \int_{V_p} Y_{11}^D(\phi_1'')^2 z^2 dV_p \tag{8.287}$$

$$= \frac{Y_b b_b t_b^3}{L_b^3} + \frac{8Y_{11}^D b_p \left(L_2^3 - L_1^3\right)}{L_b^6}\left[\left(\frac{t_b}{2} + t_p\right)^3 - \left(\frac{t_b}{2}\right)^3\right] \tag{8.288}$$

$$\Theta = \int_{V_p} \frac{z\phi_1}{d_{31}' A_p} dV_p \tag{8.289}$$

$$= \frac{3d_{31} Y_{11}^{\mathbb{E}} b_p}{L_b^3 C_p t_p} t_p (t_b + t_p)\left(L_2^2 - L_1^2\right) \tag{8.290}$$

Because of the additional tip mass M_{tip}, we need to derive an expression for the kinetic energy of the structure, from which we can obtain the appropriate mass matrix. The kinetic energy is now given by

$$T = \frac{1}{2}\int_{V_s} \rho \dot{w}^2 dV_s + \frac{1}{2} M_{tip} (\dot{w}(L_b))^2 \tag{8.291}$$

$$= \frac{1}{2}\dot{\boldsymbol{r}}^T \boldsymbol{M}\dot{\boldsymbol{r}} \tag{8.292}$$

from which the mass matrix can be written as

$$\boldsymbol{M} = \int_{V_s} \rho \boldsymbol{\phi}^T \boldsymbol{\phi} dV_s + \boldsymbol{\phi}^T (L_b) M_{tip} \boldsymbol{\phi}(\boldsymbol{L_b}) \tag{8.293}$$

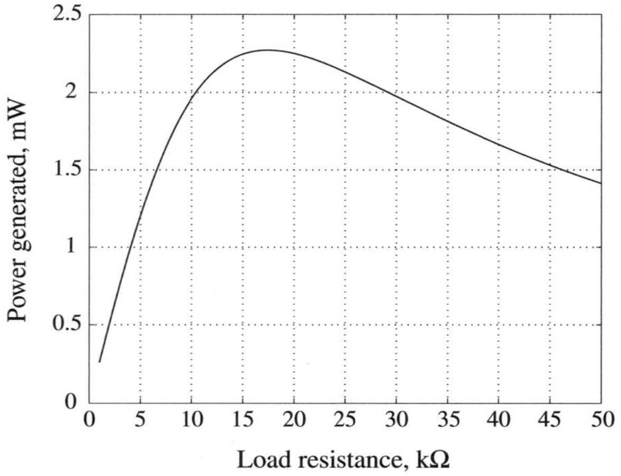

Figure 8.104. Output power as a function of load resistance at a forcing frequency 100 Hz.

Substituting the assumed displacement, we get a scalar equivalent mass of the system

$$M = \int_{V_b} \rho_b \phi_1^2 dV_b + \int_{V_p} \rho_p \phi^2 dV_p + M_{tip}\phi_1^2(L_b) \tag{8.294}$$

$$= \frac{\rho_b b_b t_b L_b}{7} + \frac{2\rho_p b_p t_p \left(L_2^7 - L_1^7\right)}{7L_b^6} + M_{tip} \tag{8.295}$$

The final equations become

$$\begin{Bmatrix} \dot{r}_1 \\ \ddot{r}_1 \\ q \end{Bmatrix} = \begin{bmatrix} 0 & 1 & 0 \\ -K/M & 0 & -\Theta/M \\ -\Theta/R_L & 0 & -1/(R_L C_p) \end{bmatrix} \begin{Bmatrix} r_1 \\ \dot{r}_1 \\ q \end{Bmatrix} + \begin{Bmatrix} 0 \\ 1/M \\ 0 \end{Bmatrix} F_{tip} \tag{8.296}$$

The values of the constants can be obtained by substituting the values given in Table 8.8. The results of solving the system of equations for a forcing of unit magnitude are shown in Figs. 8.104–8.106.

It is seen that the output power reaches a maximum for a specific value of load resistance. This corresponds to the impedance-matched condition, in which the load resistance is equal to the output impedance of the piezoelectric sheets. The voltage generated is seen to increase with increasing load resistance and asymptote to a constant value. Similarly, the tip displacement increases and asymptotes to a constant value. This corresponds to the changing stiffness of the piezoelectric sheets based on the load resistance connected across their electrodes.

8.10.5 Worked Example

Consider a cantilever beam with surface-bonded piezoelectric sheets and a tip force as shown in Fig. 8.101. The two piezoelectric sheets are connected in parallel with a load resistance, as in Fig. 8.102. Assume a uniform electric field across the piezoelectric sheets and derive the equations of motion using the electric enthalpy approach. Neglect structural damping.

Assuming a one-term expression for the displacement, write the equations of the system in state-space form.

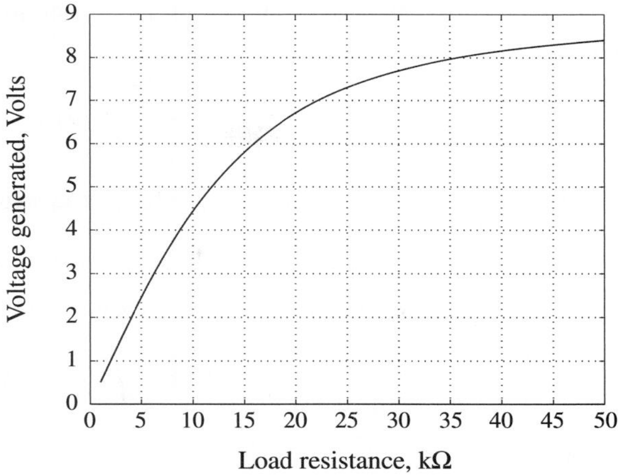

Figure 8.105. Magnitude of voltage generated as a function of load resistance at a forcing frequency 100 Hz.

Solution

The variational indicator in this case is written as in Eq. 8.251. Ignoring strains along the y direction and using a one-dimensional representation to model the device, the piezoelectric constitutive relations can be written as

$$\begin{Bmatrix} D_3 \\ \sigma_{11} \end{Bmatrix} = \begin{bmatrix} e^{\epsilon}_{33} & d_{31} Y^{\mathbb{E}}_{11} \\ -d_{31} Y^{\mathbb{E}}_{11} & Y^{\mathbb{E}}_{11} \end{bmatrix} \begin{Bmatrix} \mathbb{E}_3 \\ \epsilon_{11} \end{Bmatrix} \tag{8.297}$$

Using these relations, the electric enthalpy is written as

$$H_2 = \frac{1}{2}\int_{V_s} \sigma_{11}\epsilon_{11} dV_s - \frac{1}{2}\int_{V_s} D_3 \mathbb{E}_3 dV_s \tag{8.298}$$

$$= \frac{1}{2}\int_{V_s} \left(Y^{\mathbb{E}}_{11}\epsilon^2_{11} - e^{\epsilon}_{33}\mathbb{E}^2_3 - 2d_{31} Y^{\mathbb{E}}_{11}\mathbb{E}_3\epsilon_{11} \right) dV_s \tag{8.299}$$

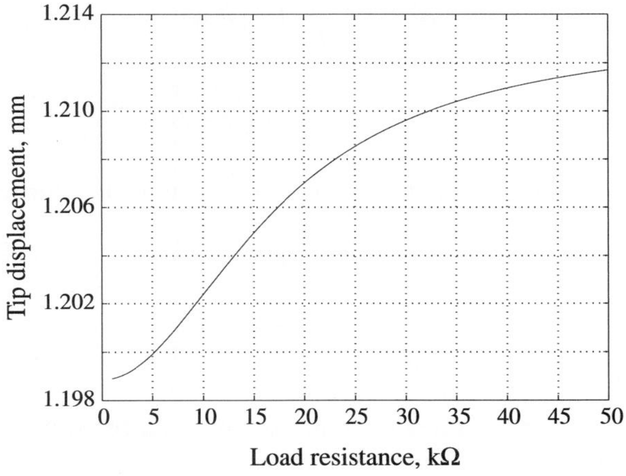

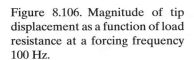

Figure 8.106. Magnitude of tip displacement as a function of load resistance at a forcing frequency 100 Hz.

The transverse displacement is given by a summation of assumed shape functions as

$$w(x,t) = \sum_{i=1}^{N} \phi(x)r(t) = \boldsymbol{\phi r} \tag{8.300}$$

From the Euler-Bernoulli assumption, the strain is

$$\epsilon_{11} = -\boldsymbol{\phi}''\boldsymbol{r}z \tag{8.301}$$

The kinetic energy of the structure is given by

$$T = \frac{1}{2}\int_{V_s} \rho \dot{w}^2 dV_s \tag{8.302}$$

which gives

$$T - H_2 = \frac{1}{2}\int_{V_s} \left(\rho \dot{\boldsymbol{r}}^T \boldsymbol{\phi}^T \boldsymbol{\phi} \dot{\boldsymbol{r}} - Y_{11}^{\mathbb{E}} z^2 \boldsymbol{r}^T \boldsymbol{\phi}''^T \boldsymbol{\phi}'' \boldsymbol{r} + e_{33}^{\epsilon} \mathbb{E}_3^2 - 2d_{31} Y_{11}^{\mathbb{E}} \mathbb{E}_3 z \boldsymbol{\phi}'' \boldsymbol{r}\right) dV_s \tag{8.303}$$

Taking the variation of this expression yields

$$\begin{aligned}\delta(T - H_2) = \int_{V_s} \Big(&\rho \boldsymbol{\delta} \dot{\boldsymbol{r}}^T \boldsymbol{\phi}^T \boldsymbol{\phi} \dot{\boldsymbol{r}} - Y_{11}^{\mathbb{E}} z^2 \boldsymbol{\delta r}^T \boldsymbol{\phi}''^T \boldsymbol{\phi}'' \boldsymbol{r} + e_{33}^{\epsilon} \mathbb{E}_3 \delta \mathbb{E}_3 \\ &- d_{31} Y_{11}^{\mathbb{E}} z \boldsymbol{\phi}'' \boldsymbol{r} \delta \mathbb{E}_3 - d_{31} Y_{11}^{\mathbb{E}} z \mathbb{E}_3 \boldsymbol{\delta r}^T \boldsymbol{\phi}''^T \Big) dV_s\end{aligned} \tag{8.304}$$

The voltage is related to the gradient of the electric field. Assuming a uniform electric field across the piezoelectric sheets

$$V = -\mathbb{E}_3 t_p \tag{8.305}$$

from which we obtain an expression for the variation of electric field in terms of flux linkage as

$$\delta \mathbb{E}_3 = -\frac{\delta \dot{\lambda}}{t_p} \tag{8.306}$$

Substituting this in Eq. 8.304 yields

$$\begin{aligned}\delta(T - H_2) = \frac{1}{2}\int_{V_s} \Bigg(&\rho \boldsymbol{\delta} \dot{\boldsymbol{r}}^T \boldsymbol{\phi}^T \boldsymbol{\phi} \dot{\boldsymbol{r}} - Y_{11}^{\mathbb{E}} z^2 \boldsymbol{\delta r}^T \boldsymbol{\phi}''^T \boldsymbol{\phi}'' \boldsymbol{r} + \frac{e_{33}^{\epsilon}}{t_p^2} \dot{\lambda} \delta \dot{\lambda} \\ &+ \boldsymbol{\phi}'' \boldsymbol{r} \frac{d_{31} Y_{11}^{\mathbb{E}} z}{t_p} \delta \dot{\lambda} + \frac{d_{31} Y_{11}^{\mathbb{E}} z}{t_p} \boldsymbol{\delta r}^T \boldsymbol{\phi}''^T \delta \dot{\lambda} \Bigg) dV_s\end{aligned} \tag{8.307}$$

The mechanical and electric virtual work terms are given by

$$\sum_i f_i \delta w_i = F_{tip} \delta w_{tip} = F_{tip} \boldsymbol{\delta r}^T \boldsymbol{\phi}^T (L_b) \tag{8.308}$$

$$\sum_j i_j \, \delta \lambda_j = -\frac{V}{R_L} \delta \lambda = -\frac{1}{R_L} \dot{\lambda} \delta \lambda \tag{8.309}$$

where the voltage drop across the load resistance has been substituted in terms of the current flowing through it.

These expressions are substituted into the variational indicator (Eq. 8.251), which is then set equal to zero

$$V.I. = \int_{t_1}^{t_2} \left[\delta\dot{\boldsymbol{r}}^T \boldsymbol{M}\dot{\boldsymbol{r}} + C_p \dot{\lambda}\delta\dot{\lambda} + \boldsymbol{\Theta}\boldsymbol{r}\delta\dot{\lambda} - \delta\boldsymbol{r}^T \boldsymbol{K}\boldsymbol{r} + \delta\boldsymbol{r}^T \boldsymbol{\Theta}\dot{\lambda} \right.$$
$$\left. + F_{tip}\delta\boldsymbol{r}^T \boldsymbol{\phi}^T(L_b) - \frac{1}{R_L}\dot{\lambda}\delta\lambda \right] dt \tag{8.310}$$
$$= 0$$

where the mass, stiffness, and coupling matrices are given by

$$\boldsymbol{M} = \int_{V_s} \rho\boldsymbol{\phi}^T\boldsymbol{\phi}dV_s \tag{8.311}$$

$$\boldsymbol{K} = \int_{V_s} Y_{11}^{\mathbb{E}} z^2 \boldsymbol{\phi}''^T \boldsymbol{\phi}'' dV_s \tag{8.312}$$

$$\boldsymbol{\Theta} = \int_{V_s} \frac{d_{31}Y_{11}^{\mathbb{E}} z}{t_p} \boldsymbol{\phi}''^T dV_s \tag{8.313}$$

The capacitance of the piezoelectric sheets at constant strain is

$$C_p = \int_{V_s} \frac{e_{33}^{\epsilon}}{t_p^2} dV_s = \frac{e_{33}^{\epsilon} A_p}{t_p} \tag{8.314}$$

The coupling matrix defined using the electric enthalpy, $\boldsymbol{\Theta}_{H_2}$, is related to the coupling matrix defined using the internal energy approach, $\boldsymbol{\Theta}_U$. Assuming constant properties over the volume of the piezoelectric material

$$\boldsymbol{\Theta}_U = \int_{V_s} \frac{z\boldsymbol{\phi}''^T}{A_p d_{31}'} dV_s = \int_{V_s} \frac{d_{31}Y_{11}^{\mathbb{E}} z}{\left(1 - k_{31}^2\right) e_{33}^{\sigma} A_p} \boldsymbol{\phi}''^T dV_s$$
$$= \frac{d_{31}Y_{11}^{\mathbb{E}} t_p}{\left(1 - k_{31}^2\right) e_{33}^{\sigma} t_p A_p} \int_{V_s} z\boldsymbol{\phi}''^T dV_s \tag{8.315}$$
$$= \boldsymbol{\Theta}_{\boldsymbol{H_2}} \frac{1}{C_p}$$

Integrating Eq. 8.310 by parts, setting the variations at t_1 and t_2 to zero, and collecting coefficients of $\delta\boldsymbol{r}^T$ and $\delta\lambda$ results in the equations of motion

$$\boldsymbol{M}\ddot{\boldsymbol{r}} + \boldsymbol{K}\boldsymbol{r} - \boldsymbol{\Theta}V = F_{tip}\boldsymbol{\phi}^T(L_b) \tag{8.316}$$

$$C_p\dot{V} + \boldsymbol{\Theta}^T\dot{\boldsymbol{r}} + \frac{V}{R_L} = 0 \tag{8.317}$$

These equations can be put in the state-space form

$$\begin{Bmatrix} \dot{\boldsymbol{r}} \\ \ddot{\boldsymbol{r}} \\ \dot{V} \end{Bmatrix} = \begin{bmatrix} \boldsymbol{0}_{(N\times N)} & \boldsymbol{I}_{(N\times N)} & \boldsymbol{0}_{(N\times 1)} \\ -\boldsymbol{M}^{-1}\boldsymbol{K} & \boldsymbol{0}_{(N\times N)} & \boldsymbol{M}^{-1}\boldsymbol{\Theta} \\ \boldsymbol{0}_{(1\times N)} & -\frac{1}{C_p}\boldsymbol{\Theta}^T & -\frac{1}{R_L C_p} \end{bmatrix} \begin{Bmatrix} \boldsymbol{r} \\ \dot{\boldsymbol{r}} \\ V \end{Bmatrix} + \begin{Bmatrix} \boldsymbol{0}_{(N\times 1)} \\ \boldsymbol{M}^{-1}\boldsymbol{\phi}^T(L_b) \\ 0 \end{Bmatrix} F_{tip} \tag{8.318}$$

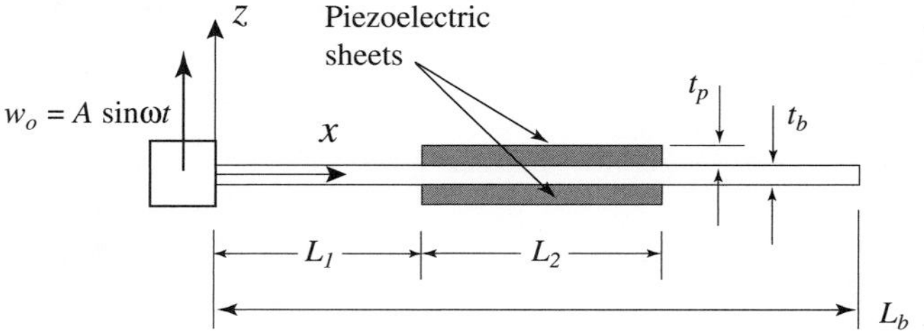

Figure 8.107. Cantilever beam with surface-bonded piezoelectric sheets and base excitation.

Let us assume the displacement to be given by

$$w(x,t) = \phi_1 r_1 = \left(\frac{x}{L_b}\right)^3 r_1 \tag{8.319}$$

The equivalent mass, stiffness, and coupling matrix of the system are

$$M = \int_{V_b} \rho_b \phi_1^2 dV_b + \int_{V_p} \rho_p \phi_1^2 dV_p \tag{8.320}$$

$$= \frac{\rho_b b_b t_b L_b}{7} + \frac{2\rho_p b_p t_p \left(L_2^7 - L_1^7\right)}{7L_b^6} \tag{8.321}$$

$$K = \int_{V_b} Y_b (\phi_1'')^2 z^2 dV_b + \int_{V_p} Y_{11}^{\mathbb{E}} (\phi_1'')^2 z^2 dV_p \tag{8.322}$$

$$= \frac{Y_b b_b t_b^3}{L_b^3} + \frac{8Y_{11}^{\mathbb{E}} b_p \left(L_2^3 - L_1^3\right)}{L_b^6}\left[\left(\frac{t_b}{2} + t_p\right)^3 - \left(\frac{t_b}{2}\right)^3\right] \tag{8.323}$$

$$\Theta = \int_{V_p} \frac{d_{31} Y_{11}^{\mathbb{E}} z \phi_1''}{t_p} dV_p \tag{8.324}$$

$$= 3d_{31} Y_{11}^{\mathbb{E}} b_p (t_b + t_p) \frac{\left(L_2^2 - L_1^2\right)}{L_b^3} \tag{8.325}$$

The final equations in state-space form become

$$\begin{Bmatrix} \dot{r}_1 \\ \ddot{r}_1 \\ V \end{Bmatrix} \begin{bmatrix} 0 & 1 & 0 \\ -K/M & 0 & \Theta/M \\ 0 & -\Theta/C_p & -1/(R_L C_p) \end{bmatrix} \begin{Bmatrix} r_1 \\ \dot{r}_1 \\ V \end{Bmatrix} + \begin{Bmatrix} 0 \\ 1/M \\ 0 \end{Bmatrix} F_{tip} \tag{8.326}$$

8.10.6 Worked Example

Consider a cantilever beam with surface-bonded piezoelectric sheets, as shown in Fig. 8.107, being excited by a harmonic base motion. This is a schematic of a typical vibration-based energy harvester. The two piezoelectric sheets are connected in parallel with a load resistance, as in Fig. 8.102. Derive the equations of motion of the system assuming a uniform electric field across the piezoelectric sheets and neglecting structural damping. Assume the beam transverse displacement to be given by

$$w(x,t) = \phi_1 r_1 = \left(\frac{x}{L_b}\right)^3 r_1 \tag{8.327}$$

Solution

In this case, the mass, stiffness, and coupling terms remain the same as in Eqs. 8.276, 8.270, and 8.271.

$$M = \int_{V_b} \rho_b \phi_1^2 dV_b + \int_{V_p} \rho_p \phi_1^2 dV_p \tag{8.328}$$

$$= \frac{\rho_b b_b t_b L_b}{7} + \frac{2\rho_p b_p t_p \left(L_2^7 - L_1^7\right)}{7L_b^6} \tag{8.329}$$

$$K = \int_{V_p} Y_{11}^D z^2 (\phi_1'')^2 dV_p + \int_{V_b} Y_b z^2 (\phi_1'')^2 dV_b \tag{8.330}$$

$$= \frac{Y_b b_b t_b^3}{L_b^3} + \frac{8Y_{11}^{\mathbb{E}} b_p \left(L_2^3 - L_1^3\right)}{L_b^6}\left[\left(\frac{t_b}{2} + t_p\right)^3 - \left(\frac{t_b}{2}\right)^3\right] \tag{8.331}$$

$$\Theta = \int_{V_p} \frac{z\phi_1}{d_{31}' A_p} dV_p \tag{8.332}$$

$$= \frac{3d_{31} Y_{11}^{\mathbb{E}} b_p}{L_b^3 C_p t_p} t_p (t_b + t_p)\left(L_2^2 - L_1^2\right) \tag{8.333}$$

In this case, the mechanical virtual work is due to the inertial force acting on the structure.

$$\delta W_f = \int_{V_s} f(x,t)\delta w \tag{8.334}$$

$$= -\left(\int_{V_b} \rho_b b_b t_b \delta w dV_b + 2\int_{V_p} \rho_p b_p t_p \delta w dV_p\right)\ddot{w}_o \tag{8.335}$$

$$= \omega^2 A \delta \boldsymbol{r}^T \left(\int_{V_b} \rho_b b_b t_b \boldsymbol{\phi}^T dV_b + 2\int_{V_p} \rho_p b_p t_p \boldsymbol{\phi}^T dV_p\right) \tag{8.336}$$

$$= \delta \boldsymbol{r}^T \boldsymbol{F_a} \tag{8.337}$$

Substituting the assumed displacement gives

$$F_a = \omega^2 A \left[\rho_b b_b t_b \int_0^{L_b} \left(\frac{x}{L_b}\right)^3 dx + 2\rho_p b_p t_p \int_{L_1}^{L_2} \left(\frac{x}{L_b}\right)^3 dx\right] \tag{8.338}$$

$$= \omega^2 A \left[\frac{\rho_b b_b t_b L_b}{4} + \frac{\rho_p b_p t_p}{2} \frac{\left(L_2^4 - L_1^4\right)}{L_b^3}\right] \tag{8.339}$$

The final equations of the system are

$$\begin{Bmatrix} \dot{r}_1 \\ \ddot{r}_1 \\ q \end{Bmatrix} = \begin{bmatrix} 0 & 1 & 0 \\ -K/M & 0 & -\Theta/M \\ -\Theta/R_L & 0 & -1/(R_L C_p) \end{bmatrix} \begin{Bmatrix} r_1 \\ \dot{r}_1 \\ q \end{Bmatrix} + \begin{Bmatrix} 0 \\ 1/M \\ 0 \end{Bmatrix} F_a \tag{8.340}$$

8.11 Constrained Layer Damping

Passive surface treatments are extensively used to increase damping of flexible structures such as plain and sandwich plates, beams, blades, and other dynamic systems.

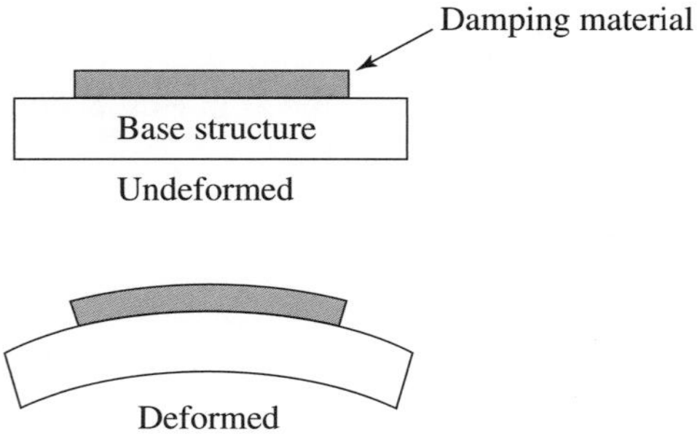

Figure 8.108. Passive surface damping treatment (unconstrained).

These passive treatments can be broadly classified into two categories: extensional (unconstrained) and shear (constrained). The unconstrained layer treatment consists of a simple layer of high-damping viscoelastic material, firmly bonded to the elastic baseline structure, as in Fig. 8.108. As the surface vibrates in bending, the treatment deforms cyclically in compression and tension, and energy is dissipated. Because there is a negligible shear deformation in viscoelastic material, a very low level of energy dissipation takes place. This is not an effective way of damping augmentation. In the constrained layer treatment, a stiff constrained layer is added to the top surface of the highly damped viscoelastic layer that is firmly bonded to the baseline structure at its bottom surface, as in Fig. 8.109. The flexural modulus of constrained layer is of the same order as that of the baseline structure. During bending motion of the base structure, the viscoelastic layer is forced to deform in shear mode. During dynamic motion, the energy is dissipated in viscoelastic material. Increasing the thickness of the damping material layer can increase damping, but it will also increase the weight penalty, which is a critical issue especially for aerospace systems. This approach of damping augmentation in a structure is simple, reliable, and less expensive but often is of limited effectiveness. Early works on constrained layer damping can be attributed to DiTaranto [211] Mead and Markus [212], who worked on sandwich beams with viscoelastic cores for axial and bending vibration of beams. Following these studies, there have been numerous investigations on constrained layer damping for plates and beams [213, 214].

The constrained layer analysis is based on the following assumptions: (1) the constrained layer bends in transverse direction as an integral part of the base layer, (2) the viscoelastic layer undergoes pure shear deformation, and (3) the viscoelastic layer does not undergo a change in thickness during deformation. These assumptions appear quite satisfactory as long as the viscoelastic layer is comparatively thin.

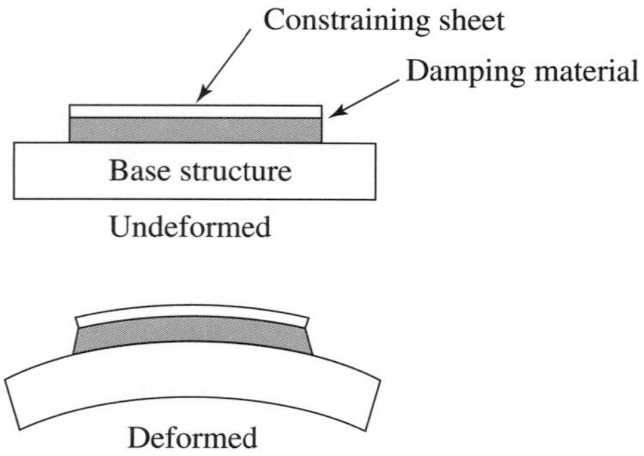

Figure 8.109. Passive surface damping treatment (constrained).

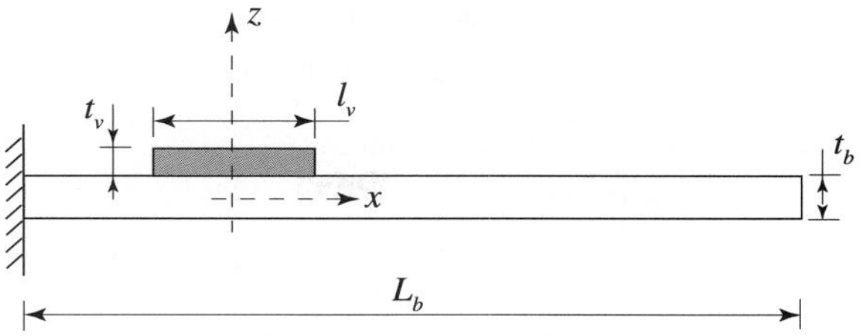

(a) Beam geometry

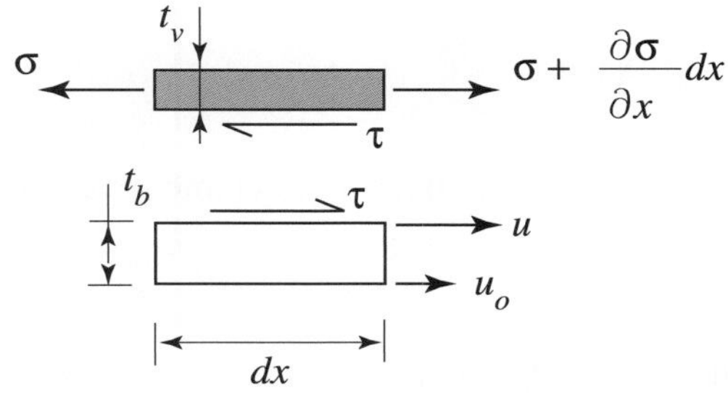

(b) Free body diagram of beam element

Figure 8.110. Beam with unconstrained viscoelastic layer.

Let us consider a cantilevered beam of length L_b, thickness t_b, and Young's modulus E_b with an unconstrained viscoelastic layer of length l_v, thickness t_v, and Young's modulus E_v. Let us draw a free body diagram of the treated element (Fig. 8.110). Equilibrium of forces gives

$$\frac{\partial \sigma}{\partial x} t_v = \tau \tag{8.341}$$

$$\sigma = E_v \frac{\partial u}{\partial x} \tag{8.342}$$

Shear strain in the baseline beam is based on the assumption that displacement varies linearly with thickness as

$$\gamma_b = \frac{u - u_o}{t_b} \tag{8.343}$$

$$\tau = G\left(\frac{u - u_o}{t_b}\right) \tag{8.344}$$

where u_o is the displacement of the beam surface and G is the shear modulus. Substituting, we obtain

$$E_v t_v \frac{\partial^2 u}{\partial x^2} = G\left(\frac{u - u_o}{t_b}\right) \tag{8.345}$$

which gives

$$\frac{E_v t_v t_b}{G} \frac{\partial^2 u}{\partial x^2} - u = -\frac{u_o}{t_b} \tag{8.346}$$

Let us define

$$\left(\frac{E_v t_v t_b}{G}\right)^{\frac{1}{2}} = B \quad \text{and} \quad \epsilon_o = \frac{u_o}{t_b} \tag{8.347}$$

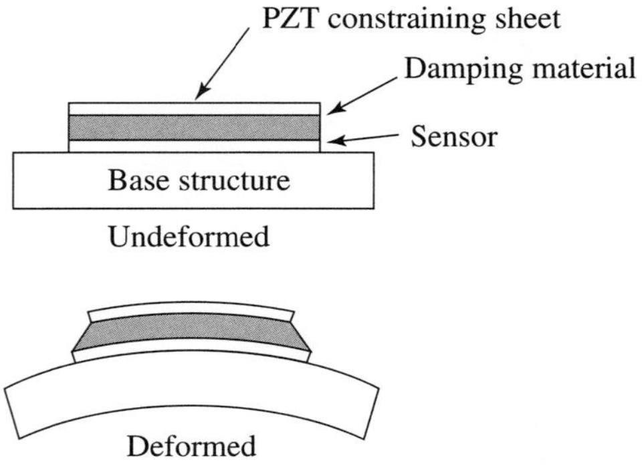

Figure 8.111. Active constrained layer damping treatment.

The governing equation of the element becomes

$$B^2 \frac{\partial^2 u}{\partial x^2} - u = -\epsilon_o \tag{8.348}$$

with the boundary conditions

$$\text{at } x = \pm \frac{l_v}{2} \quad E_v t_v \frac{\partial u}{\partial x} = 0 \quad \text{i.e.,} \quad \frac{\partial u}{\partial x} = 0 \tag{8.349}$$

The solution to the governing equation (Eq. 8.348) is given by

$$u(x) = \epsilon_o \left[x - B \frac{\sinh(x/B)}{\cosh(l_v/2B)} \right] \tag{8.350}$$

where ϵ_o is the strain of the beam surface. The energy dissipated in the damping layer per cycle is

$$\Delta W = \int_{-l_v/2}^{l_v/2} \pi t_b G'' \gamma^2 dx \tag{8.351}$$

from which the loss coefficient is found to be

$$\eta = \frac{\Delta W}{\frac{1}{2}\epsilon_o^2 E_v t_v l_c} \tag{8.352}$$

8.11.1 Active Constrained Layer Damping

In active surface treatments, the constraining surface for the high-damping viscoelastic material layer is an active material sheet such as PZT (Fig. 8.111). This helps to increase the shearing strain in viscoelastic material and thereby augment the damping of the passive layer. Even though this technique is more involved than the passive constrained layer damping approach, it is far more effective to increase specific damping (damping/weight) of the system. This system inherently has both the capabilities of active and passive constrained layer damping. Active capability enhances the damping augmentation, whereas the passive layer introduces robustness and reliability of the system. Early efforts in this area are due to Plump and Hubbard [215], Baz and Ro [216], Van Nostrand and Inman [217], and Shen [218] for active constrained layer damping of beam and plates. Liao and Wang [219] enhanced the active constrained layer damping augmentation using a new configuration with

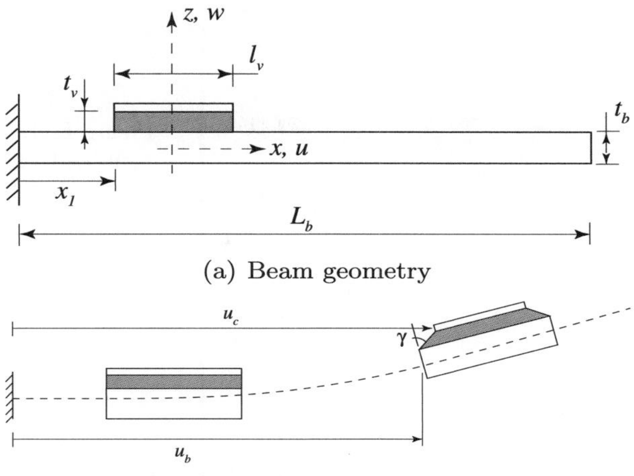

(a) Beam geometry

(b) Displacement of beam element

Figure 8.112. Beam with constrained viscoelastic layer.

edge elements. This helps to increase the active action transmissibility while retaining passive damping capability.

Let us consider a beam of length L_b with a constrained layer treatment of thickness t_v over a length l_v (Fig. 8.112). The kinematic beam relations are

$$\text{beam:} \quad \epsilon_b = \frac{\partial u_b}{\partial x} - z\frac{\partial^2 w}{\partial x^2} \tag{8.353}$$

$$\text{constrained layer:} \quad \epsilon_c = \frac{\partial u_c}{\partial x} - z\frac{\partial^2 w}{\partial x^2} \tag{8.354}$$

The shear strain in the viscoelastic is

$$\gamma = \frac{1}{t_v}\left[d\frac{\partial w}{\partial x} + u_c - u_b\right] \tag{8.355}$$

where u_b and u_c are the axial displacements of the host beam and constraint layer, respectively, at the neutral axis, or mid-axis, and d is the distance between the neutral axis of the beam and the mid-plane of the constraint layer.

$$d = \frac{t_b}{2} + t_v + \frac{t_c}{2} = t_v + \frac{t_c + t_b}{2} \tag{8.356}$$

Assuming that the constraint layer has a negligible effect on the neutral axis, and neglecting the kinetic energy in the axial direction, the strain energy and kinetic energy of the beam are

$$U_b = \frac{1}{2}\int_0^{L_b} E_b I_b \left(\frac{\partial^2 w}{\partial x^2}\right)^2 dx + \frac{1}{2}\int_0^{L_b} E_b A_b \left(\frac{\partial u_b}{\partial x}\right)^2 dx \tag{8.357}$$

$$T_b = \frac{1}{2}\int_0^{L_b} m_b \left(\frac{\partial w}{\partial t}\right)^2 dx \tag{8.358}$$

where m_b is the mass per unit length of the beam, $E_b I_b$ is the flexural stiffness, and $E_b A_b$ is the extensional stiffness. For the viscoelastic layer, the strain energy is assumed to be entirely due to the shear strain. The strain energy and kinetic energy

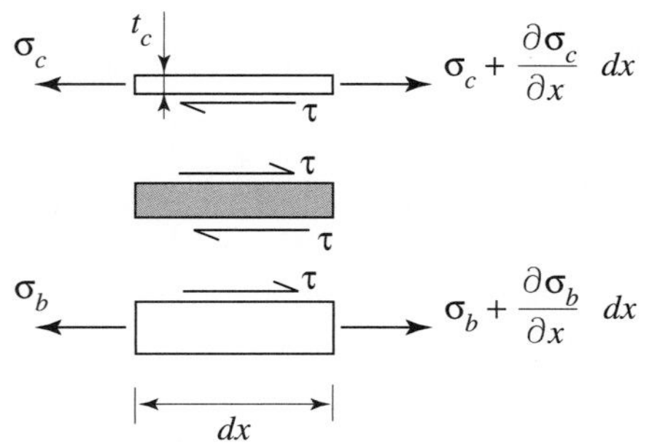

Figure 8.113. Free body diagram of beam element with constrained damping layer.

are given by

$$U_v = \frac{1}{2}\int_{x_1}^{x_1+l_v} GA_v \left[\frac{t_b}{2t_v}\frac{\partial w}{\partial x} + \frac{u_c - u_b}{t_v}\right]^2 dx \tag{8.359}$$

$$T_v = \frac{1}{2}\int_{x_1}^{x_1+l_v} m_v \left(\frac{\partial w}{\partial t}\right)^2 dx \tag{8.360}$$

where GA_v is the shear stiffness, and m_v is the mass per unit length of the viscoelastic material only. The strain energy and kinetic energy of the constraining layer (or active layer) are

$$U_c = \frac{1}{2}\int_{x_1}^{x_1+l_c} E_c I_c \left(\frac{\partial^2 w}{\partial x^2}\right)^2 dx + \frac{1}{2}\int_{x_1}^{x_1+l_c} E_c A_c \left(\frac{\partial u_c}{\partial x}\right)^2 dx \tag{8.361}$$

$$T_c = \frac{1}{2}\int_{x_1}^{x_1+l_c} m_c \left(\frac{\partial w}{\partial t}\right)^2 dx \tag{8.362}$$

where $E_c I_c$ is the flexural stiffness, $E_c A_c$ is the extensional stiffness, and m_c is the mass per unit length. Because the inertial forces in the x direction are neglected, the static equilibrium in the axial direction can be uncoupled. From the free body diagram of the beam element (Fig. 8.113), we obtain the equilibrium equation for the beam as

$$t_b \frac{\partial \sigma_b}{\partial x} + \tau = 0 \tag{8.363}$$

$$\frac{\partial}{\partial x}\left[t_b E_b \frac{\partial u_b}{\partial x}\right] + G\gamma = 0 \tag{8.364}$$

$$E_b t_b \frac{\partial^2 u_b}{\partial x^2} + \frac{G}{t_v}\left[d\frac{\partial w}{\partial x} + u_c - u_b\right] = 0 \tag{8.365}$$

from which we get

$$u_c = u_b - d\frac{\partial w}{\partial x} + \left(\frac{E_b t_b}{G} t_v\right)\frac{\partial^2 u_b}{\partial x^2} \tag{8.366}$$

The equilibrium equation for the constrained layer can be obtained as

$$t_c \frac{\partial \sigma_c}{\partial x} - \tau = 0 \tag{8.367}$$

$$t_c \frac{\partial}{\partial x}\left[E_c \frac{\partial u_c}{\partial x}\right] - \frac{G}{t_v}\left[d\frac{\partial w}{\partial x} + u_c - u_b\right] = 0 \tag{8.368}$$

$$\frac{E_c t_c}{G} t_v \frac{\partial^2 u_c}{\partial x^2} - u_c + u_b - d\frac{\partial w}{\partial x} = 0 \tag{8.369}$$

Define

$$C_b = \frac{E_b t_b t_v}{G} \tag{8.370}$$

$$C_c = \frac{E_c t_c t_v}{G} \tag{8.371}$$

Baz and Ro [220] examined bending vibration control of flat plates using patches of active constrained layer damping. Each patch consists of a viscoelastic damping layer sandwiched between two piezoelectric layers, one used as a sensor and a second one as an actuator. Numerical simulation is carried out using the finite element approach and results are validated experimentally by testing aluminum plate by treating it with two patches (viscoelastic and PVDF piezo films). Active constrained layer treatment was found to be far more effective to attenuate vibration amplitudes than passive treatment: a three-fold increase in damping augmentation.

Huang et al. [221] compared results from three configurations of pure active control by surface-attached piezoceramics, passive constrained layer damping, and active constrained layer damping treatments. In these studies, the total thickness of the damping treatment is restricted. A complex modulus approach is used to model damping to carry out steady-state analysis. It was shown that the active constrained layer damping treatment provides far superior vibration suppression (i.e., damping augmentation) than passive damping treatment, and it even outperforms pure active control for low-gain applications. From open-loop studies, it is possible to determine the optimal size of active constrained layer treatment; the closed-loop studies provide the optimal control gains, thereby ensure robustness of operation [222]. Shields et al. [223] presented a theoretical (FEM) and an experimental demonstration of the effectiveness of active control of sound radiation from a plate into an acoustic cavity using patches of active piezoelectric damping composites. Each patch consists of piezoelectric fibers embedded across the thickness of a viscoelastic matrix. This study demonstrated the effectiveness of active control of sound and low-frequency structural vibration. Chantalakhana and Stanway [224] addressed the suppression of vibrations of a clamped-clamped plate using active constrained layer damping, both numerically and experimentally. An active modal controller was implemented using the reduced-order model (i.e., FEM transformed to state-space format). It was shown that the control of the first two vibration modes (bending and torsion) could be achieved using only a single actuator and a single sensor. Overall, the best configuration is two actuators and two sensors. Some investigators examined active constrainted layer damping augmentation for cylindrical shell structures [225], rings [226], and arc-type shells [227].

8.12 Interior Noise Control

Interior noise control in automotive, fixed-wing aircraft, and rotorcraft systems is an important issue. For example, the contributions to noise spectra in a helicopter are the main rotor, tail rotor, and engine system in the frequency range of 50–500 Hz and gear trains in the main transmission for frequencies above 500 Hz. To understand the phenomenon, one can examine a simple case of transmission of noise into an enclosure with flexible walls. For a flexible structure, interior noise and structural vibration are coupled phenomena. The sound and vibration energy is propagated in the interior cabin through two modes: structure-borne transmission and direct air radiation. Passive techniques are widely used to control interior noise, but these normally result in a significant weight penalty. A possible paradigm for noise control in a three-dimensional enclosure such as helicopter cabin consists of an active control approach in the low-frequency range (below 500 Hz) and a passive or a combined active/passive approach in the high-frequency range (above 500 Hz). Passive approaches include stiffening the structure, isolating the structure, damping augmentation, and soundproofing treatments (insulation and absorption). Passive techniques are less effective in the low-frequency range because of the relatively large wavelength of acoustic signature as compared to the thickness of the treatment. Hence, the active control schemes using either secondary force inputs or external-acoustic sources may be used to cancel noise in an enclosure. Among secondary force inputs, electrodynamic shakers and piezoelectric actuators are adopted. Sampath and Balachandran [228] described an analytical formulation for active control of noise in a three-dimensional enclosure using piezoelectric actuators. They defined three different performance functions to evaluate the effectiveness of the system. For external acoustic sources, speakers can be used to cancel a specific noise source. The nature of acoustics in an enclosure is characterized by a parameter called the "Schroeder frequency," which identifies a transition boundary between a low modal density frequency range and a high modal density frequency range. From an active-control point of view, sound fields below the Schroeder frequency are important. For example, in rotorcraft cabins, the Schroeder frequency is in the range of 80–100 Hz.

BIBLIOGRAPHY

[1] E. Crawley. Intelligent structures for aerospace: A technology overview and assessment. *AIAA Journal*, 32(8):1689–1699, August 1994.

[2] B. K. Wada, J. L. Fanson, and E. F. Crawley. Adaptive structures. *Journal of Intelligent Material Systems and Structures*, 1(2):157–174, April 1990.

[3] C. Y. Lin, E. F. Crawley, and J. Heeg. Open- and closed-loop results of a strain-actuated active aeroelastic wing. *Journal of Aircraft*, 33(5):987–994, September–October 1996.

[4] J. Kudva, K. Appa, C. Martin, P. Jardine, and G. Sendeckji. Design, fabrication and testing of the DARPA/WL "smart wing" wind tunnel model. *Paper # AIAA-1997-1198, Proceedings of the 38th AIAA/ASME/ASCE/AHS/ASC Structures, Structural Dynamics, and Materials Conference*, Kissimmee, FL, April 1997.

[5] J. Kudva, C. A. Martin, L. B. Scherer, A. P. Jardine, A. M. R. McGowan, R. C. Lake, G. P. Sendeckyj, and B. P. Sanders. Overview of the DARPA/AFRL/NASA Smart Wing program. *Proceedings of the SPIE Smart Structures and Materials Symposium*, 3674:230–236, 1999.

[6] J. Becker, H. W. Schroeder, K. W. Dittrich, E. J. Bauer, and H. Zippold. Advanced aircraft structures program: An overview. *Proceedings of the SPIE Smart Structures and Materials Symposium*, 3674:2–21, 1999.

[7] J. K. Durr, U. Herold-Schmidt, and H. W. Zaglauer. On the integration of piezoceramic actuators in composite structures for aerospace applications. *Journal of Intelligent Material Systems and Structures*, 10(11):880–889, November 1999.

[8] S. M. Ehlers and T. A. Weisshaar. Static aeroelastic control of an adaptive lifting surface. *Journal of Aircraft*, 30(4):534–540, July–August 1993.

[9] K. B. Lazarus, E. F. Crawley, and J. D. Bohlmann. Static aeroelastic control using strain actuated adaptive structures. *Journal of Intelligent Material Systems and Structures*, 2(3):386–410, July 1991.

[10] K. D. Frampton, R. L. Clark, and E. H. Dowell. Active control of panel flutter with piezoelectric transducers. *Journal of Aircraft*, 33(4):768–774, July–August 1996.

[11] T. Bein, H. Hanselka, and E. Breitbach. An adaptive spoiler to control the transonic shock. *Smart Materials and Structures*, 9(2):141–148, April 2000.

[12] C. R. Fuller, C. H. Hansen, and S. D. Snyder. Experiments on active control of sound radiation from a panel using a piezoceramic actuator. *Journal of the Acoustical Society of America*, 91(6):3313–3320, 1992.

[13] I. Chopra. Status of application of smart structures technology to rotorcraft systems. *Journal of the American Helicopter Society*, 45(4):228–252, October 2000.

[14] A. Dogan, J. Tressler, and R. E. Newnham. Solid-state ceramic actuator designs. *AIAA Journal*, 39(7):1354–1362, July 2001.

[15] T. Lee and I. Chopra. Design of peizostack-driven trailing-edge flap actuator for helicopter rotors. *Smart Materials and Structures*, 10(1):15–24, February 2001.

[16] V. Giurgiutiu, C. A. Rogers, and Z. Chaudhary. Energy-based comparison of solid-state induced-strain actuators. *Journal of Intelligent Material Systems and Structures*, 7(1):4–14, January 1996.

[17] V. Giurgiutiu and C. A. Rogers. Large amplitude rotary induced-strain (laris). *Journal of Intelligent Material Systems and Structures*, 8(1):41–50, January 1997.

[18] M. Mitrovic, G. P. Carman, and F. K. Straub. Electromechanical characterization of piezoelectric stack actuators. *Proceedings of the SPIE Smart Structures and Materials Symposium*, 3668:586–603, 1999.

[19] E. F. Prechtl and S. R. Hall. Design of a high efficiency, large stroke, electromechanical actuator. *Smart Materials and Structures*, 8(1):13–30, February 1999.

[20] J. P. Rodgers and N. W. Hagood. Preliminary Mach-scale hover testing of an integral twist-actuated rotor blade. *Proceedings of the SPIE Smart Structures and Materials Symposium*, 3329:291–308, 1998.

[21] J. P. Rodgers and N. W. Hagood. Design and manufacture of an integral twist-actuated rotor blade. *Paper # AIAA-1997-1264, Proceedings of the 38th AIAA/ASME/ASCE/AHS/ASC Structures, Structural Dynamics and Materials Conference*, Kissimmee, FL, April 1997.

[22] R. C. Derham and N. W. Hagood. Rotor design using smart materials to actively twist blades. *Proceedings of the 52nd Annual Forum of the American Helicopter Society*, June 1996.

[23] K. W. Wilkie, M. L. Wilbur, P. H. Mirick, C. E. S. Cesnik, and S. J. Shin. Aeroelastic analysis of the NASA/Army/MIT active twist rotor. *55th Annual Forum of the American Helicopter Society*, May 1999.

[24] A. A. Bent and N. W. Hagood. Anisotropic actuation with piezoelectric fiber composites. *Journal of Intelligent Material Systems and Structures*, 6(3):338–349, 1995.

[25] A. A. Bent and N. W. Hagood. Piezoelectric fiber composites with interdigitated electrodes. *Journal of Intelligent Material Systems and Structures*, 8(11):903–919, 1997.

[26] W. K. Wilkie, R. G. Bryant, J. W. High, R. L. Fox, R. F. Hellbaum, A. Jalink, B. D. Little, and P. H. Mirick. Low-cost piezocomposite actuator for structural control application. *Proceedings of the SPIE Smart Structures and Materials Symposium*, 3991:323–334, 2000.

[27] R. B. Williams, B. W. Grimsley, D. J. Inman, and W. K. Wilkie. Manufacturing and mechanics-based characterization of macro fiber composite actuators. *Proceedings,*

2002 ASME International Adaptive Structures and Materials Systems Symposium, pages 17–22, New Orleans, LA, 2002.

[28] R. Williams, D. Inman, and W. Wilkie. Nonlinear actuation properties of macro fiber composite actuators. *Proceedings, ASME International Mechanical Engineering Congress*, pages 15–21. Washington, DC, 2003.

[29] E. Ruggiero, G. H. Park, D. Inman, and J. Wright. Multi-input, multi-output modal testing techniques for a Gossamer structure. *Proceedings, ASME International Adaptive Structures Symposium*, pages 17–22. New Orleans, LA, 2002.

[30] J. S. Park and J. H. Kim. Analytical development of single crystal macro fiber composite actuators for active twist rotor blades. *Smart Materials and Structures*, 14(4):745–753, 2005.

[31] C. R. Bowen, R. Stevens, L. J. Nelson, A. C. Dent, G. Dolman, B. Su, T. W. Button, M. G. Cain, and M. Stewart. Manufacture and characterization of high activity piezoelectric fibres. *Smart Materials and Structures*, 15(2):295–301, 2006.

[32] R. E. Newnham, Q. C. Xu, and S. Yoshikawa. Transformed stress direction acoustic transducer. U.S. Patent No. 4,999,819, 1991.

[33] K. Onitsuka, A. Dogan, J. F. Tressler, Q. Xu, S. Yoshikawa, and R. E. Newnham. Metal-ceramic composite transducer, the Moonie. *Journal of Intelligent Material Systems and Structures*, 6(4):447–455, July 1995.

[34] A. Dogan, K. Uchino, and R. E. Newnham. Composite piezoelectric transducer with truncated conical endcaps – Cymbal. *IEEE Transactions on Ultrasonics, Ferroelectrics and Frequency Control*, 44(3):597–605, 1997.

[35] P. Ochoa, J. de Frutos, and J. F. Fernandez. Electromechanical characterization of Cymbal piezocomposites. *Smart Materials and Structures*, 18(9):095047, 2009.

[36] G. Li, E. Furman, and G. H. Haertling. Fabrication and properties of PSZT antiferroelectric RAINBOW actuators. *Ferroelectrics*, 188:223–236, 1996.

[37] M. W. Hooker. Properties and performance of RAINBOW piezoelectric actuator stacks. *Proceedings of the SPIE Smart Structures and Materials Symposium*, 3044:413–421, 1997.

[38] G. Li, E. Furman, and G. H. Haertling. Finite element analysis of RAINBOW ceramics. *Journal of Intelligent Material Systems and Structures*, 8(5):434–443, May 1997.

[39] M. W. Hyer and A. B. Jilani. Deformation characteristics of circular RAINBOW actuators. *Smart Materials and Structures*, 11(2):175–195, April 2002.

[40] R. F. Hellbaum, R. G. Bryant, and R. L. Fox. Thin layer composite unimorph ferroelectric driver and sensor, U.S. Patent 5,632,841, May 27, 1997.

[41] K. M. Mossi and R. P. Bishop. Characterization of different types of high-performance THUNDER actuators. *Proceedings of the SPIE Smart Structures and Materials Symposium*, 3675:43–52, 1999.

[42] J. P. Marouze and L. Cheng. A feasibility study of active vibration isolation using THUNDER actuators. *Smart Materials and Structures*, 11(6):854–862, 2002.

[43] Y. Kim, L. Cai, T. Usher, and Q. Jiang. Fabrication and characterization of THUNDER actuators – Pre-stress–induced nonlinearity in the actuation response. *Smart Materials and Structures*, 18(9):095033, 2009.

[44] K. J. Yoon, S. Shin, H. C. Park, and N. S. Goo. Design and manufacture of a lightweight piezo-composite curved actuator. *Smart Materials and Structures*, 11(1):163–168, February 2002.

[45] M. Syaifuddin, H. C. Park, and N. S. Goo. Design and evaluation of a LIPCA-actuated flapping device. *Smart Materials and Structures*, 15(5):1225–1230, 2006.

[46] S. M. Lim, S. Lee, H. C. Park, K. J. Yoon, and N. S. Goo. Design and demonstration of a biomimetic wing section using a lightweight piezo-composite actuator (LIPCA). *Smart Materials and Structures*, 14(4):496–503, 2005.

[47] W. Y. Shih, W. H. Shih, and I. A. Aksay. Scaling analysis for the axial displacement and pressure of flextensional transducers. *Journal of the American Ceramic Society*, 80(5):1073–1078, 1997.

[48] S. Aimmanee and M. W. Hyer. Analysis of the manufactured shape of rectangular THUNDER-type actuators. *Smart Materials and Structures*, 13(6):1389–1406, 2004.

[49] B. Xu, Q. M. Zhang, V. D. Kugel, Q. Wang, and L. E. Cross. Optimization of bimorph based double amplifier actuator under quasistatic situation. In *Proceedings of the Tenth IEEE International Symposium on Applications of Ferroelectrics, 1996. ISAF '96*, pages 217–220, 1996.

[50] A. Moskalik and D. Brei. Quasi-static behavior of individual C-block piezoelectric actuators. *Journal of Intelligent Material Systems and Structures*, 8(7):571–587, July 1997.

[51] A. Moskalik and D. Brei. Parametric investigation of the deflection performance of serial piezoelectric C-block actuators. *Journal of Intelligent Material Systems and Structures*, 9(3):223–231, March 1998.

[52] A. J. Moskalik and D. Brei. Force-deflection behavior of piezoelectric C-block actuator arrays. *Smart Materials and Structures*, 8(5):531–543, October 1999.

[53] J. S. Paine and Z. Chaudhry. The impact of amplification on efficiency and energy density of induced strain actuators. *Proceedings of the ASME Aerospace Division*, AD-52:511–516, 1996.

[54] F. K. Straub. A feasibility study of using smart materials for rotor control. *Smart Materials and Structures*, 5(1):1–10, February 1996.

[55] J. Garcia-Bonito, M. J. Brennan, S. J. Elliott, A. David, and R. J. Pinnington. A novel high-displacement piezoelectric actuator for active vibration control. *Smart Materials and Structures*, 7(1):31–42, 1998.

[56] P. Tang, A. Palazzolo, A. Kascak, G. Montague, and W. Li. Combined piezoelectric-hydraulic actuator based active vibration control for a rotordynamic system. *Journal of Vibration and Acoustics*, 117:285–293, July 1995.

[57] V. Giurgiutiu, Z. A. Chaudhry, and C. A. Rogers. Stiffness issues in the design of ISA displacement amplification devices: Case study of a hydraulic displacement amplifier. *Proceedings of the SPIE Smart Structures and Materials Symposium*, 2443:105–119, 1995.

[58] D. K. Samak and I. Chopra. Design of high force, high displacement actuators for helicopter rotors. *Smart Materials and Structures*, 5:58–67, February 1996.

[59] R. C. Fenn, J. R. Downer, D. A. Bushko, and N. D. Ham. Terfenol-D driven flaps for helicopter vibration reduction. *Smart Materials and Structures*, 5(1):49–57, February 1996.

[60] T. Lee and I. Chopra. Design issues of a high-stroke, on-blade piezostack actuator for helicopter rotor with trailing-edge flaps. *Journal of Intelligent Material Systems and Structures*, 11(5):328–342, May 2000.

[61] W. Xu and T. King. Flexural hinges for piezoelectric displacement amplifiers: Flexibility, accuracy, and stress concentration. *Precision Engineering*, 19(1):4–10, July 1996.

[62] M. Frecker and S. Canfield. Optimal design and experimental validation of compliant mechanical amplifiers for piezoceramic stack actuators. *Journal of Intelligent Material Systems and Structures*, 11(5):360–369, May 2000.

[63] M. Frecker, G. K. Ananthsuresh, S. Nishikawi, N. Kikuchi, and S. Kota. Topological synthesis of compliant mechanisms using multi-criteria optimization. *Journal of Mechanical Design, Transactions of the ASME*, 119(2):238–245, June 1997.

[64] B. Edinger, M. Frecker, and J. Gardner. Dynamic modeling of an innovative piezoelectric actuator for minimally invasive surgery. *Journal of Intelligent Material Systems and Structures*, 11(10):765–770, October 2000.

[65] A. E. Glazounov, Q. M. Zhang, and C. Kim. New torsional actuator based on shear piezoelectric response. *Proceedings of the SPIE Smart Structures and Materials Symposium*, 3324:82–93, 1998.

[66] A. E. Glazounov, Q. M. Zhang, and C. Kim. Piezoelectric actuator generating torsional displacement from piezoelectric d_{15} shear response. *Applied Physics Letters*, 72:2526, 1998.

[67] A. E. Glazounov, Q. M. Zhang, and C. Kim. Torsional actuator and stepper motor based on piezoelectric d_{15} shear response. *Journal of Intelligent Material Systems and Structures*, 11(6):456–468, 2000.

[68] S. Li, W. W. Cao, and L. Cross. The extrinsic nature of nonlinear behavior observed in lead zirconate titanate ferroelectric ceramic. *Journal of Applied Physics*, 69(10): 7219–24, 1991.

[69] V. Mueller and Q. M. Zhang. Shear response of lead zirconate titanate piezoceramics. *Journal of Applied Physics*, 83(7):3754–3761, 1998.

[70] D. Thakkar and R. Ganguli. Helicopter vibration reduction in forward flight with induced-shear based piezoceramic actuation. *Smart Materials and Structures*, 13(3): 599–608, 2004.

[71] L. R. Centolanza and E. C. Smith. Design and experimental testing of an induced-shear piezoelectric actuator for rotor blade trailing edge flaps. *Paper # AIAA-2000-1713, Proceedings of the 41st AIAA/ASME/ASCE/AHS/ASC Structures, Structural Dynamics, and Materials Conference*, Atlanta, GA, 2000.

[72] L. R. Centolanza, E. C. Smith, and B. Munsky. Induced-shear piezoelectric actuators for rotor-blade trailing-edge flaps. *Smart Materials and Structures*, 11(1):24–35, February 2002.

[73] L. R. Centolanza, E. C. Smith, and A. Morris. Induced shear piezoelectric actuators for rotor blade trailing edge flaps and active tips. *Paper # AIAA-2001-1559, Proceedings of the 42nd AIAA/ASME/ASCE/AHS/ASC Structures, Structural Dynamics, and Materials Conference*, Seattle, WA, April 2001.

[74] C. M. Bothwell, R. Chandra, and I. Chopra. Torsional actuation with extension-torsional composite coupling and magnetostrictive actuators. *AIAA Journal*, 33(4): 723–729, April 1995.

[75] A. P. F. Bernhard and I. Chopra. Hover testing of active rotor blade-tips using a piezo-induced bending-torsion coupled beam. *Journal of Intelligent Material Systems and Structures*, 9(12):963–974, December 1998.

[76] P. C. Chen and I. Chopra. Wind tunnel test of a smart rotor model with individual blade twist control. *Journal of Intelligent Material Systems and Structures*, 8(5):414–425, May 1997.

[77] G. A. Lesieutre, J. Loverich, G. H. Koopmann, and E. M. Mockensturm. Increasing the mechanical work output of an active material using a nonlinear motion transmission mechanism. *Journal of Intelligent Material Systems and Structures*, 15(1):49–58, January 2004.

[78] M. J. Brennan, J. Garcia-Bonito, S. J. Elliott, A. David, and R. J. Pinnington. Experimental investigation of different actuator technologies for active vibration control. *Smart Materials and Structures*, 8(1):145–153, February 1999.

[79] U. Schaaf. Pushy motors. *IEE Review*, 41(3): 105–108, May 1995.

[80] J. Wallaschek. Piezoelectric ultrasonic motors. *Journal of Intelligent Material Systems and Structures*, 6(1):71–83, 1995.

[81] T. P. Galante, J. E. Frank, J. Bernard, W. Chen, G. A. Lesieutre, and G. H. Koopmann. Design, modeling, and performance of a high force piezoelectric inchworm motor. *Journal of Intelligent Material Systems and Structures*, 10(12):962–972, December 1999.

[82] J. Park, G. P. Carman, and H. T. Hahn. Design and testing of a mesoscale piezoelectric inchworm actuator with microridges. *Journal of Intelligent Material Systems and Structures*, 11(9):671–684, September 2000.

[83] K. Duong and E. Garcia. Design and performance of a rotary motor driven by piezoelectric stack actuators. *Japanese Journal of Applied Physics, Part 1*, 35(12A):6334–6341, December 1996.

[84] J. E. Frank, G. H. Koopmann, W. Chen, E. M. Mockensturm, and G. A. Lesieutre. Design and performance of a resonant roller wedge actuator. *Proceedings of the SPIE Smart Structures and Materials Symposium*, 3985:198–206, 2000.

[85] E. M. Mockensturm, J. E. Frank, G. H. Koopmann, and G. A. Lesieutre. Modeling and simulation of a resonant bimorph actuator drive. *Proceedings of the SPIE Smart Structures and Materials Symposium*, 4327:472–480, 2001.

[86] J. E. Frank, E. M. Mockensturm, G. H. Koopmann, G. A. Lesieutre, W. Chen, and J. Y. Loverich. Modeling and design optimization of a bimorph-driven rotary motor. *Journal of Intelligent Material Systems and Structures*, 14(4-5):217–227, April/May 2003.

[87] S. Ueha and Y. Tomikawa. *Ultrasonic Motors: Theory and Application*. Clarendon Press, Oxford, 1993.

[88] T. Sashida and T. Kenjo. *An Introduction to Ultrasonic Motors*. Clarendon Press, Oxford, 1993.

[89] K. Uchino. *Piezoelectric Actuators and Ultrasonic Motors*. Kluwer Academic Publishers, Boston, 1997.

[90] J. N. Kudva, B. P. Sanders, J. L. Pinkerton-Florance, and E. Garcia. Overview of the DARPA/AFRL/NASA Smart Wing Phase II program. *Proceedings of the SPIE Smart Structures and Materials Symposium*, 4332:383–389, 2001.

[91] L. D. Mauck and C. S. Lynch. Piezoelectric hydraulic pump development. *Journal of Intelligent Material Systems and Structures*, 11(10):758–764, October 2000.

[92] D. Shin, D. Lee, K. P. Mohanchandra, and G. P. Carman. Development of a SMA-based actuator for compact kinetic energy missile. *Proceedings of the SPIE Smart Structures and Materials Symposium*, 4701:237–243, 2003.

[93] J. Sirohi and I. Chopra. Design and development of a high pumping frequency piezoelectric-hydraulic hybrid actuator. *Journal of Intelligent Material Systems and Structures*, 14(3):135–148, March 2003.

[94] K. Konishi, H. Ukida, and K. Sawada. Hydraulic pumps driven by multilayered piezoelectric elements – Mathematical model and application to brake device. *Proceedings of the 13th Korean Automatic Control Conference*, 1998.

[95] K. Konishi, T. Yoshimura, K. Hashimoto, and N. Yamamoto. Hydraulic actuators driven by piezoelectric elements (1st report, trial piezoelectric pump and its maximum power). *Journal of Japanese Society of Mechanical Engineering (C)*, 59(564):213–220, 1993.

[96] K. Konishi, T. Yoshimura, K. Hashimoto, T. Hamada, and T. Tamura. Hydraulic actuators driven by piezoelectric elements (2nd report, enlargement of piezoelectric pumps output power using hydraulic resonance). *Journal of Japanese Society of Mechanical Engineering (C)*, 60(571):228–235, 1994.

[97] K. Konishi, K. Hashimoto, T. Miyamoto, and T. Tamura. Hydraulic actuators driven by piezoelectric elements (3rd report, position control using piezoelectric pump and hydraulic cylinder). *Journal of Japanese Society of Mechanical Engineering (C)*, 61 (591):134–141, 1995.

[98] K. Konishi, H. Ukida, and T. Kotani. Hydraulic actuators driven by piezoelectric elements (4th report, construction of mathematical models for simulation). *Journal of Japanese Society of Mechanical Engineering (C)*, 63(605):158–165, 1997.

[99] M. J. Gerver, J. H. Goldie, J. R. Swenbeck, R. Shea, P. Jones, R. T. Ilmonen, D. M. Dozor, S. Armstrong, R. Roderick, F. E. Nimblett, and R. Iovanni. Magnetostrictive water pump. *Proceedings of the SPIE Smart Structures and Materials Symposium*, 3329:694–705, 1998.

[100] K. Nasser, D. J. Leo, and H. H. Cudney. Compact piezohydraulic actuation system. *Proceedings of the SPIE Smart Structures and Materials Symposium*, 3991:312–322, 2000.

[101] W. S. Oates and C. S. Lynch. Piezoelectric hydraulic pump system dynamic model. *Journal of Intelligent Material Systems and Structures*, 12(11):737–744, November 2001.

[102] E. H. Anderson, J. E. Lindler, and M. E. Regelbrugge. Smart material actuator with long stroke and high power output. *Papaer # AIAA-2002-1354, Proceedings of the*

43rd AIAA/ASME/ASCE/AHS/ASC Structures, Structural Dynamics, and Materials Conference, Denver, CO, April 2002.

[103] S. John, J. Sirohi, G. Wang, and N. M. Wereley. Comparison of piezoelectric, magnetostrictive, and electrostrictive hybrid hydraulic actuators. *Journal of Intelligent Material Systems and Structures*, 18(10):1035–1048, October 2007.

[104] K. Bridger, J. M. Sewell, A. V. Cooke, J. L. Lutian, D. Kohlhafer, G. E. Small, and P. M. Kuhn. High-pressure magnetostrictive pump development: A comparison of prototype and modeled performance. *Proceedings of the SPIE Smart Structures and Materials Symposium*, 5388:246–257, 2004.

[105] G. McKnight, L. Momoda, D. Croft, D. G. Lee, D. Shin, and G. P. Carman. Miniature thin film NiTi hydraulic actuator with MEMS microvalves. *Proceedings of the SPIE Smart Structures and Materials Symposium*, 5762:187–195, 2005.

[106] D. G. Lee, S. W. Or, and G. P. Carman. Design of a piezoelectric-hydraulic pump with active valves. *Journal of Intelligent Material Systems and Structures*, 15(2):107–115, February 2004.

[107] R. M. Tieck, G. P. Carman, Y. Lin, and C. O'Neill. Characterization of a piezohydraulic actuator. *Proceedings of the SPIE Smart Structures and Materials Symposium*, 5764:671–679, 2005.

[108] S. Shoji and M. Esashi. Microflow devices and systems. *Journal of Micromechanics and Microengineering*, 4:157–171, 1994.

[109] D. Accoto, O. T. Nedelcu, M. C. Carrozza, and P. Dario. Theoretical analysis and experimental testing of a miniature piezoelectric pump. *IEEE International Symposium on Micromechatronics and Human Science*, pages 261–268, 1998.

[110] N. W. Hagood, D. C. Roberts, L. Saggere, K. S. Breuer, K.-S. Chen, J. A. Carretero, H. Li, R. Mlcak, S. W. Pulitzer, M. A. Schmidt, M. S. Spearing, and Y.-H. Su. Microhydraulic transducer technology for actuation and power generation. *Proceedings of the SPIE Smart Structures and Materials Symposium*, 3985:680–688, 2000.

[111] A. Ullmann. The piezoelectric valve-less pump – performance enhancement analysis. *Sensors and Actuators*, A(69):97–105, 1998.

[112] A. Ullmann, I. Fono, and Y. Taitel. A piezoelectric valve-less pump-dynamic model. *Transactions of the ASME*, 123:92–98, March 2001.

[113] T. Gerlach, M. Schuenemann, and H. Wurmus. A new micropump principle of the reciprocating type using pyramidic micro flow channels as passive valves. *Journal of Micromechanics and Microengineering*, 6:199–201, 1995.

[114] J. Smits. Piezoelectric micropump with three valves working peristaltically. *Sensors and Actuators*, A(21-23):203–206, 1990.

[115] J.-Ho Park, K. Yoshida, and S. Yokota. Resonantly driven piezoelectric micropump-fabrication of a micropump having high power density. *Mechatronics: Mechanics, Electronics, Control*, 9(7):687–702, October 1999.

[116] J. Watton. *Fluid Power Systems*. Prentice-Hall, 1989.

[117] D. McCloy and H. R. Martin. *Control of Fluid Power: Analysis and Design*. Ellis Horwood Limited, Chichester, England, second (revised) edition, 1980.

[118] *Products for Micropositioning, US edition*. Physik Instrumente (PI), 1997.

[119] J. Sirohi and I. Chopra. Design and testing of a high pumping frequency piezoelectric-hydraulic hybrid actuator. *Proceedings of the 13th International Conference on Adaptive Structures and Technologies*, Potsdam, Germany, October 7–9, 2002.

[120] K. Nasser and D. J. Leo. Efficiency of frequency-rectified piezohydraulic and piezopneumatic actuation. *Journal of Intelligent Material Systems and Structures*, 11(10): 798–810, October 2000.

[121] C. Cadou and B. Zhang. Performance modeling of a piezo-hydraulic actuator. *Journal of Intelligent Material Systems and Structures*, 14(3):149–160, March 2003.

[122] H. Tan, W. Hurst, and D. Leo. Performance modeling of a piezohydraulic actuation system with active valves. *Smart Materials and Structures*, 14:91–110, 2005.

[123] S. John, C. Cadou, J. H. Yoo, and N. M. Wereley. Application of CFD in the design and analysis of a piezoelectric hydraulic pump. *Journal of Intelligent Material Systems and Structures*, 17(11):967–979, 2006.

[124] J. Sirohi, C. Cadou, and I. Chopra. Investigation of the dynamic characteristics of a piezohydraulic actuator. *Journal of Intelligent Material Systems and Structures*, 16(6): 481–492, 2005.

[125] C. Dorny. *Understanding Dynamic Systems: Approaches to Modeling, Analysis and Design*. Prentice-Hall, NJ, 1993.

[126] E. Doebelin. *System Dynamics Modeling and Response*. Charles E. Merrill Publishing Company, Columbus, OH, 1972.

[127] E. Doebelin. *System Modeling and Response, Theoretical and Experimental Approaches*. John Wiley & Sons, New York, 1980.

[128] T. Bourouina and J.-P. Grandchamp. Modeling micropumps with electrical equivalent networks. *Journal of Micromechanics and Microengineering*, 6(4):398–404, 1996.

[129] J. L. Shearer, A. J. Murphy, and H. H. Richardson. *Introduction to System Dynamics*. Addison Wesley, 1967.

[130] E. B. Wylie and V. L. Streeter. *Fluid Transients*. McGraw-Hill International Book Company, 1978.

[131] R. E. Goodson and R. G. Leonard. A survey of modeling techniques for fluid line transients. *Journal of Basic Engineering, Transactions of the ASME, Series D*, 94:474–482, June 1972.

[132] A. S. Iberall. Attenuation of oscillatory pressures in instrument lines. *Journal of Research, National Bureau of Standards*, 45:2115, July 1950.

[133] C. P. Rohmann and E. C. Grogan. On the dynamics of pneumatic transmission lines. *Transactions of ASME*, 79:853, 1957.

[134] N. B. Nichols. The linear properties of pneumatic transmission lines. *ISA Transactions*, 1(1):5–14, January 1962.

[135] F. T. Brown. The transient response of fluid lines. *Journal of Basic Engineering, Transactions of the ASME, Series D*, 84(4):547–553, December 1962.

[136] R. L. Woods. A first-order square root approximation for fluid transmission lines. In *Fluid Transmission Line Dynamics*, ASME Special Publication for the ASME Winter Annual Meeting, pages 37–50, Washington, DC, November 15–20, 1983.

[137] K. Suzuki. A new hydraulic pressure intensifier using oil hammer. In *Fluid Transients in Fluid-Structure Interaction*, ASME Special Publication for the ASME Winter Annual Meeting, pages 43–50, Boston, MA, December 13–18, 1987.

[138] R. W. Prouty. *Helicopter Performance, Stability and Control*. Robert E. Krieger Publishing Company, Inc., Malabar, FL, 1990.

[139] J. R. Olson. Reducing helicopter costs. *Vertiflite*, 39(1):10–16, January/February 1993.

[140] E. A. Fradenburgh. The first 50 years were fine... but what should we do for an encore? *The 1994 Alexander A. Nikolsky Lecture. Journal of the American Helicopter Society*, 40(1):3–19, January 1995.

[141] R. Ormiston. Aeroelastic considerations for rotorcraft primary control with on-blade elevons. *Presented at the American Helicopter Society 57th Annual Forum*, Washington, DC, May 9–11, 2001.

[142] P. C. Chen and I. Chopra. Hover test of a smart rotor with induced strain actuation of blade twist. *AIAA Journal*, 35(1):6–16, January 1997.

[143] N. A. Koratkar and I. Chopra. Analysis and testing of a Froude scaled rotor with piezoelectric bender actuated trailing-edge flaps. *Journal of Intelligent Material Systems and Structures*, 8(7):555–70, July 1997.

[144] M. V. Fulton and R. A. Ormiston. Hover testing of a small-scale rotor with on-blade elevons. In *53rd American Helicopter Society Forum*, pages 249–273, Virginia Beach, VA, May 1997.

[145] M. V. Fulton and R. A. Ormiston. Small-scale rotor experiments with on-blade elevons to reduce blade vibratory loads in forward flight. In *54th American Helicopter Society Forum*, Washington, DC, May 1998.

[146] A. P. F. Bernhard and I. Chopra. Hover testing of active rotor blade-tips using a piezo-induced bending-torsion coupled beam. *Journal of Intelligent Material Systems and Structures*, 9(12):963–974, December 1998.

[147] F. K. Straub, D. K. Kennedy, A. D. Stemple, V. R. Anand, and T. S. Birchette. Development and whirl tower test of the SMART active flap rotor. *Proceedings of the SPIE Smart Structures and Materials Symposium*, 5388:202–212, 2004.

[148] F. K. Straub, V. R. Anand, T. S. Birchette, and B. H. Lau. Wind tunnel test of the SMART active flap rotor. *Proceedings of the 65th Annual AHS Forum, Grapevine, TX*, 2009.

[149] D. Schimke, P. Jänker, A. Blaas, R. Kube, G. Schewe, and C. Keßler. Individual blade control by servo-flap and blade root control, a collaborative research and development programme. In *23rd European Rotorcraft Forum*, pages 46.1–46.16, Dresden, Germany, September 1997.

[150] D. Schimke, P. Jänker, V. Wendt, and B. Junker. Wind tunnel evaluation of a full-scale piezoelectric flap control unit. In *24th European Rotorcraft Forum*, Marseilles, France, September 1998. Paper TE-02.

[151] P. Jänker, V. Klöppel, F. Hermle, T. Lorkowski, S. Storm, M. Christmann, and M. Wettemann. Development and evaluation of a hybrid piezoelectric actuator for advanced flap control technology. In *25th European Rotorcraft Forum*, Rome, Italy, September 1999. Paper G-21.

[152] B. G. van der Wall, R. Kube, A. Büter, U. Ehlert, W. Geissler, M. Raffel, and G. Schewe. A multi concept approach for development of adaptive rotor systems. In *8th Army Research Office (ARO) Workshop on the Aeroelasticity of Rotorcraft Systems*, State College, PA, October 1999.

[153] T. Lorkowski, P. Jänker, F. Hermle, S. Storm, M. Christmann, and M. Wettemann. Development of a piezoelectrically actuated leading edge flap for dynamic stall delay. In *25th European Rotorcraft Forum*, Rome, Italy, September 1999. Paper G-20.

[154] F. Toulmay, V. Kloppel, F. Lorin, B. Enenkl, and J. Gaffiero. Active blade flaps-the needs and current capabilites. *Presented at the American Helicopter Society 57th Annual Forum*, Washington, DC, May 9–11, 2001.

[155] M. Tarascio, M. Gervais, T. Gowen, J. Ma, K. Singh, G. Gopalan, K. Kleinhesselink, Y. Zhao, and I. Chopra. Raven SAR Rotorcaft. Technical report, Alfred Gessow Rotorcraft Center, University of Maryland, College Park, MD, 2001.

[156] I. Chopra and J. L. McCloud. A numerical simulation study of open-loop, closed-loop and adaptive multicyclic control systems. *Journal of the American Helicopter Society*, 28(1):63–77, January 1983.

[157] W. Johnson. Self-tuning regulators for multicyclic control of helicopter vibration. NASA Technical Report TP 1996, March 1982.

[158] J. Epps and I. Chopra. Shape memory alloy actuators for in-flight tracking of helicopter rotor blades. *Smart Materials and Structures*, 10(1):104–111, 2001.

[159] K. Singh, J. Sirohi, and I. Chopra. An improved shape memory alloy actuator for rotor blade tracking. *Journal of Intelligent Material Systems and Structures*, 14(12):767–786, 2003.

[160] C. Liang, F. Davidson, L. M. Schetky, and F. K. Straub. Applications of torsional SMA actuators for active rotor blade control – opportunities and limitations. *Proceedings of the SPIE Smart Structures and Materials Symposium*, 2717:91–100, 1996.

[161] D. K. Kennedy, F. K. Straub, L. M. Schetky, Z. Chaudhry, and R. Roznoy. Development of a SMA actuator for in-flight rotor blade tracking. *Proceedings of the SPIE Smart Structures and Materials Symposium*, 3985:62–75, 2000.

[162] O. K. Rediniotis, D. C. Lagoudas, L. J. Garner, and L. N. Wilson. Development of a spined underwater biomimetic vehicle with SMA actuators. *Proceedings of the SPIE Smart Structures and Materials Symposium*, 3668:642–655, 1999.

[163] V. Giurgiutiu, C. A. Rogers, and J. Zuidervaart. Incrementally adjustable rotor-blade tracking tab using SMA composites. *Paper # AIAA-1997-1387, Proceedings of the 38th AIAA/ASME/ASCE/AHS/ASC*, Structures, Structural Dynamics and Materials Conference, Kissimmee, FL, April 1997.

[164] K. Ogata. *Modern Control Engineering*, 3rd edition. Prentice-Hall, Englewood Cliffs, NJ, 1997.

[165] L. C. Brinson. One-dimensional constitutive behavior of shape memory alloys: Thermomechanical derivation with non-constant material functions and redefined martensite internal variable. *Journal of Intelligent Material Systems and Structures*, 4(2):229–242, 1993.

[166] I. H. Abbott and A. E. von Doenhoff. *Theory of Wing Sections*. Dover Publications, Inc., New York, 1959.

[167] H. Prahlad and I. Chopra. Comparative evaluation of shape memory alloy constitutive models with experimental data. *Journal of Intelligent Material Systems and Structures*, 12(6):383–395, December 2001.

[168] C. Liang, C. A. Rogers, and E. Malafeew. Investigation of shape memory polymers and their hybrid composites. *Journal of Intelligent Material Systems and Structures*, 8(4):380–386, 1997.

[169] J. Epps and R. Chandra. Shape memory alloy actuation for active tuning of composite beams. *Smart Materials and Structures*, 6(3):251–264, 1997.

[170] C. A. Rogers and D. K. Barker. Experimental studies of active strain energy tuning of adaptive composites. *Paper # AIAA-1990-1086, Proceedings of the 31st AIAA/ASME/ASCE/AHS/ASC, Structural Dynamics, and Materials Conference*, Long Beach, CA, April 1990.

[171] A. Baz, K. Imam, and J. McCoy. Active vibration control of flexible beams using shape memory actuators. *Journal of Sound and Vibration*, 140(3):437–456, 1990.

[172] T. C. Kiesling, Z. Chaudhry, J. Paine, and C. Rogers. Impact failure modes of thin graphite epoxy composites embedded with superelastic nitinol. *Paper # AIAA-1996-1475, Proceedings of the 37th AIAA, ASME, ASCE, AHS, and ASC Structures, Structural Dynamics and Materials Conference*, Salt Lake City, UT, April 1996.

[173] D. C. Lagoudas and I. G. Tadjbakhsh. Deformations of active flexible rods with embedded line actuators. *Smart Materials and Structures*, 2(2):71, 1993.

[174] C. A. Rogers. Active vibration and structural acoustic control of shape memory alloy hybrid composites: Experimental results. *Journal of Acoustic Society of America*, 88: 2803–2807, 1990.

[175] H. Jia, F. Lalande, R. L. Ellis, and C. A. Rogers. Impact energy absorption of shape memory alloy hybrid composite beams. *Paper # AIAA-1997-1045, Proceedings of the 38th AIAA/ASME/AHS/ASC Structures, Structural Dynamics and Materials Conference*, Kissimmee, FL, April 1997.

[176] A. Baz, J. Ro, M. Mutua, and J. Gilheany. Active buckling control of Nitinol-reinforced composite beams. *Conference on Active Material and Adaptive Structures*, Alexandria, VA, pages 167–176, November 1991.

[177] W. Zhang, J. Kim, and N. Koratkar. Energy-absorbent composites featuring embedded shape memory alloys. *Smart Materials and Structures*, 12(4):642–646, 2003.

[178] T. Turner. A new thermoelastic model for analysis of shape memory alloy hybrid composites. *Journal of Intelligent Material Systems and Structures*, 11(5):382–394, 2000.

[179] B. Gabry, F. Thiebaud, and C. Lexcellent. Topographic study of shape memory alloy wires used as actuators in smart materials. *Journal of Intelligent Material Systems and Structures*, 11(8):592–603, 2000.

[180] Y. Xu, K. Otsuka, N. Toyama, H. Yoshida, H. Nagai, and T. Kishi. A novel technique for fabricating SMA/CFRP adaptive composites using ultrathin TiNi wires. *Smart Materials and Structures*, 13(1):196–202, 2004.

[181] R. L. Forward. Electronic damping of vibrations in optical structures. *Applied Optics*, 18(5):690–697, 1979.

[182] N. W. Hagood and A. von Flotow. Damping of structural vibrations with piezoelectric materials and passive electrical networks. *Journal of Sound and Vibration*, 146(2): 243–268, 1991.

[183] G. A. Lesieutre. Vibration damping and control using shunted piezoelectric materials. *The Shock and Vibration Digest*, 30(3):187–195, May 1998.

[184] J. Tang, Y. Liu, and K. W. Wang. Semiactive and active-passive hybrid structural damping treatments via piezoelectric materials. *The Shock and Vibration Digest*, 32(3): 189–200, May 2000.

[185] M. Ahmadian and A. P. DeGuilio. Recent advances in the use of piezoceramics for vibration suppression. *The Shock and Vibration Digest*, 33(1):15–22, January 2001.

[186] J. W. Dally, W. F. Riley, and K. G. McConnell. *Instrumentation for Engineering Measurements*. 2nd edition. John Wiley and Sons, New York, 1993.

[187] C. L. Davis and G. A. Lesieutre. An actively tuned solid-state vibration absorber using capacitive shunting of piezoelectric stiffness. *Journal of Sound and Vibration*, 232(3): 601–617, 2000.

[188] L. Meirovitch. *Elements of Vibration Analysis*. McGraw-Hill, New York, 1986.

[189] A. J. Prescott. Loss compensated active gyrator using differential-input operational amplifiers. *Electronic Letters*, 2:283–284, 1966.

[190] D. F. Berndt and S. C. DuttaRoy. Inductor simulation using a single unity gain amplifier. *IEEE Journal of Solid State Circuits*, SC-4:161–162, 1969.

[191] S. C. DuttaRoy and V. Nagarajan. On inductor simulation using a unity gain amplifier. *IEEE Journal of Solid State Circuits*, SC-5(95–98), 1970.

[192] S. Kim, S. Pakzad, D. Culler, J. Demmel, G. Fenves, S. Glaser, and M. Turon. Health monitoring of civil infrastructures using wireless sensor networks. *Proceedings of the 6th International Conference on Information Processing in Sensor Networks, April 25–27*, Cambridge, MA, pages 254–263, 2007.

[193] Y. Wang, K. J. Loh, J. P. Lynch, M. Fraser, K. Law, and A. Elgamal. Vibration monitoring of the Voigt Bridge using wired and wireless monitoring systems. In *The Proceedings of the 4th China-Japan-US Symposium on Structural Control and Monitoring*, October 16–17, 2006.

[194] N. E. duToit, B. L. Wardle, and S-G. Kim. Design considerations for MEMS-scale piezoelectric mechanical vibration energy harvesters. *Integrated Ferroelectrics*, 71:121–160, 2005.

[195] H. A. Sodano, G. Park, and Inman D. J. Estimation of electric charge output for piezoelectric energy harvesting. *Strain*, 40(2):49–58, 2004.

[196] J. L. Kauffman and G. A. Lesieutre. A low-order model for the design of piezoelectric energy harvesting devices. *Journal of Intelligent Material Systems and Structures*, 20(5): 495–504, 2009.

[197] G. K. Ottman, H. F. Hofmann, and G. A. Lesieutre. Optimized piezoelectric energy harvesting circuit using step-down converter in discontinuous conduction mode. *IEEE Transactions on Power Electronics*, 18(2):696–703, 2003.

[198] A. M. Wickenheiser and E. Garcia. Power optimization of vibration energy harvesters utilizing passive and active circuits. *Journal of Intelligent Material Systems and Structures*, 21(13):1343–1361, September 2010.

[199] Y. K. Tan and S. K. Panda. A novel piezoelectric based wind energy harvester for low-power autonomous wind speed sensor. *Proceedings of the 33th Annual IEEE Conference of Industrial Electronics Society (IECON07)*, Taipei, Taiwan, 2007.

[200] D. A. Wang and H. H. Ko. Piezoelectric energy harvesting from flow-induced vibration. *Journal of Micromechanics and Microengineering*, 20(2), 2010.

[201] W. P. Robbins, D. Morris, I. Marusic, and T. O. Novak. Wind-generated electrical energy using flexible piezoelectric materials. *IMECE2006-14050, ASME Publications-AD*, 71:581–590, 2006.

[202] M. Bryant and E. Garcia. Modeling and testing of a novel aeroelastic flutter energy harvester. *Journal of Vibration and Acoustics*, 133(1):011010.1-011010.11, 2011.

[203] J. Sirohi and R. R. Mahadik. Wind energy harvesting using a galloping piezoelectric beam. *Journal of Vibration and Acoustics*, 134(1): 011009, 2011.

[204] A. Erturk and D. J. Inman. An experimentally validated bimorph cantilever model for piezoelectric energy harvesting from base excitations. *Smart Materials and Structures*, 18(2):025009, 2009.

[205] N. G. Elvin, A. A. Elvin, and M. Spector. A self-powered mechanical strain energy sensor. *Smart Materials and Structures*, 10(2):293–299, 2001.

[206] W. P. Mason. *Piezoelectric Crystals and Their Application to Ultrasonics*. D. Van Nostrand Company, Inc., 1950.

[207] S. H. Crandall, D. C. Karnopp, E. F. Kurtz Jr., and D. C. Pridmore-Brown. *Dynamics of Mechanical and Electromechanical Systems*. Krieger Publishing Company, Inc., 1982.

[208] N. W. Hagood, W. H. Chung, and A. von Flotow. Modelling of piezoelectric actuator dynamics for active structural control. *Journal of Intelligent Material Systems and Structures*, 1(3):327–354, 1990.

[209] J. M. Dietl and E. Garcia. Beam shape optimization for power harvesting. *Journal of Intelligent Material Systems and Structures*, 21(6):633–646, 2010.

[210] J. Sirohi and I. Chopra. Fundamental behavior of piezoceramic sheet actuators. *Journal of Intelligent Material Systems and Structures*, 11(1):47–61, 2000.

[211] R. A. DiTaranto. Theory of vibratory bending for elastic and viscoelastic layered finite-length beams. *ASME Journal of Applied Mechanics*, 32:881–886, 1965.

[212] D. J. Mead and S. Markus. The forced vibration of a three-layer, damped sandwich beam with arbitrary boundary conditions. *Journal of Sound and Vibration*, 10(2):163–175, 1969.

[213] C. D. Johnson. Design of passive damping systems. *Journal of Mechanical Design*, 117 (B):171–176, 1995.

[214] C. T. Sun and Y. P. Lu. *Vibration Damping of Structural Elements*. Prentice Hall, Englewood Cliffs, NJ, 1995.

[215] J. M. Plump and J. E. Hubbard. Modeling of an active constrained layer damper. *Proceedings of the 12th International Congress on Acoustics*, Toronto, pages 24–31, 1986.

[216] A. Baz and J. Ro. Partial treatment of flexible beams with active constrained layer damping. *Recent Developments in Stability, Vibration and Control of Structural Systems, ASME*, New York, 167:61–80, 1993.

[217] W. C. Van Nostrand, G. J. Knowles, and D. J. Inman. Finite element model for active constrained layer damping. *Proceedings of the SPIE Smart Structures and Materials Symposium*, 2193:126–137, 1994.

[218] I. Y. Shen. Bending-vibration control of composite and isotropic plates through intelligent constrained layer treatments. *Smart Materials and Structures*, 3(1):59–70, March 1994.

[219] W. Liao and K. Wang. A new active constrained layer configuration with enhanced boundary actions. *Smart Materials and Structures*, 5(5):638–648, October 1996.

[220] A. Baz and J. Ro. Vibration control of plates with active constrained layer damping. *Smart Materials and Structures*, 5(3):272–280, June 1996.

[221] S. C. Huang, D. J. Inman, and E. M. Austin. Some design considerations for active and passive constrained layer damping treatments. *Smart Materials and Structures*, 5(3): 301–313, 1996.

[222] A. Baz. Optimization of energy dissipation characteristics of active constrained layer damping. *Smart Materials and Structures*, 6(3):360–368, June 1997.

[223] W. Shields, J. Ro, and A. Baz. Control of sound radiation from a plate into an acoustic cavity using active piezoelectric-damping composites. *Smart Materials and Structures*, 7(1):1–11, February 1998.

[224] C. Chantalakhana and R. Stanway. Active constrained layer damping of plate vibrations: A numerical and experimental study of modal controllers. *Smart Materials and Structures*, 9(6):940–952, December 2000.

[225] M. C. Ray, J. Oh, and A. Baz. Active constrained layer damping of thin cylindrical shells. *Journal of Sound and Vibration*, 240(5):921–935, 2001.

[226] J. A. Rongong and G. R. Tomlinson. Passive and active constrained layer damping of ring type structures. *Proceedings of the SPIE Smart Structures and Materials Symposium*, 3045:282–292, 1997.

[227] H. C. Park, S. H. Ko, C. H. Park, and W. Hwang. Vibration control of an arc type shell using active constrained layer damping. *Smart Materials and Structures*, 13(2):350–354, April 2004.

[228] A. Sampath and B. Balachandran. Studies on performance functions for interior noise control. *Smart Materials and Structures*, 6(3):315–332, 1997.

Index